CUCKOOS
OF THE WORLD

HELM IDENTIFICATION GUIDES

CUCKOOS
OF THE WORLD

Johannes Erritzøe, Clive F. Mann, Frederik P. Brammer
and Richard A. Fuller

Illustrated by Richard Allen, Jan Wilczur,
Martin Woodcock and Tim Worfolk

CHRISTOPHER HELM
LONDON

Published 2012 by Christopher Helm, an imprint of Bloomsbury Publishing Plc,
50 Bedford Square, London WC1B 3DP

Copyright © 2012 text by Johannes Erritzøe, Clive F. Mann, Frederik P. Brammer and Richard A. Fuller
Copyright © 2012 artwork by Richard Allen, Jan Wilczur, Martin Woodcock and Tim Worfolk
Copyright © 2012 photographs by photographers as named on the page

The right of Johannes Erritzøe, Clive F. Mann, Frederik P. Brammer and Richard A. Fuller to be identified as the authors of this work has been asserted by them in accordance with the Copyright, Designs and Patents Act 1988.

ISBN (print) 978-0-7136-6034-0
ISBN (e-pub) 978-1-4081-4268-4
ISBN (e-pdf) 978-1-4081-4267-7

A CIP catalogue record for this book is available from the British Library

All rights reserved. No part of this publication may be reproduced or used in any form or by any means – photographic, electronic or mechanical, including photocopying, recording, taping or information storage or retrieval systems – without permission of the publishers.

This book is produced using paper that is made from wood grown in managed sustainable forests. It is natural, renewable and recyclable. The logging and manufacturing processes conform to the environmental regulations of the country of origin.

Commissioning Editor: Nigel Redman
Project Editor: Jim Martin

Design by Julie Dando, Fluke Art

Printed in China by C&C Offset Printing Co., Ltd.

10 9 8 7 6 5 4 3 2 1

Cover art by Jan Wilczur
Great Spotted Cuckoo *Clamator glandarius* does not currently parasitise Iberian Azure-winged Magpie *Cyanopica cooki*, but this was not always the case. As recently as 100 years ago, Great Spotted Cuckoo brood parasitism of these corvids was fairly common. However, it seems that for the moment at least, the magpies have trumped the cuckoos; brood-manipulation studies show that the magpies remain very good at discriminating Great Spotted Cuckoo eggs (Avilés, 2004). Evolutionary change in host/model interactions can be rapid; it may be that in future the balance swings back to favour the cuckoos. The cover art for this book recreates the Spanish landscape of a century ago; a female Great Spotted Cuckoo sizes up the nest of a pair of Iberian Azure-winged Magpies, who have been distracted by the showy flight of the male.

CONTENTS

	Plate	Page
ACKNOWLEDGEMENTS		9
INTRODUCTION		11
LAYOUT OF THE BOOK		22
GLOSSARY		24
ABBREVIATIONS		26
PLATES		28
SPECIES ACCOUNTS		100
Genus *Guira*		101
Guira Cuckoo *Guira guira*	1	101
Genus *Crotophaga*		105
Greater Ani *Crotophaga major*	1	105
Smooth-billed Ani *Crotophaga ani*	1	108
Groove-billed Ani *Crotophaga sulcirostris*	1	112
Genus *Tapera*		115
American Striped Cuckoo *Tapera naevia*	1	115
Genus *Dromococcyx*		118
Pheasant-cuckoo *Dromococcyx phasianellus*	2	118
Pavonine Cuckoo *Dromococcyx pavoninus*	2	121
Genus *Morococcyx*		124
Lesser Ground Cuckoo *Morococcyx erythropygus*	2	124
Genus *Geococcyx*		126
Greater Roadrunner *Geococcyx californianus*	2	126
Lesser Roadrunner *Geococcyx velox*	2	131
Genus *Neomorphus*		133
Banded Ground Cuckoo *Neomorphus radiolosus*	3	133
Rufous-winged Ground Cuckoo *Neomorphus rufipennis*	3	135
Red-billed Ground Cuckoo *Neomorphus pucheranii*	3	137
Rufous-vented Ground Cuckoo *Neomorphus geoffroyi*	3	138
Scaled Ground Cuckoo *Neomorphus squamiger*	3	142
Genus *Centropus*		143
Buff-headed Coucal *Centropus milo*	4	143
Pied Coucal *Centropus ateralbus*	5	145
Greater Black Coucal *Centropus menbeki*	4	147
Biak Coucal *Centropus chalybeus*	6	149
Rufous Coucal *Centropus unirufus*	7	150
Green-billed Coucal *Centropus chlororhynchos*	8	152
Black-faced Coucal *Centropus melanops*	7	154
Steere's Coucal *Centropus steerii*	8	156
Short-toed Coucal *Centropus rectunguis*	6	158
Bay Coucal *Centropus celebensis*	7	160
Gabon Coucal *Centropus anselli*	9	162
Black-throated Coucal *Centropus leucogaster*	9	163
Senegal Coucal *Centropus senegalensis*	10	165
Blue-headed Coucal *Centropus monachus*	10	170
Coppery-tailed Coucal *Centropus cupreicaudus*	9	172

White-browed Coucal *Centropus superciliosus*	10	174
Javan Coucal *Centropus nigrorufus*	7	178
Greater Coucal *Centropus sinensis*	6	180
Goliath Coucal *Centropus goliath*	4	184
Madagascar Coucal *Centropus toulou*	8	186
African Black Coucal *Centropus grillii*	9	189
Philippine Coucal *Centropus viridis*	8	192
Lesser Coucal *Centropus bengalensis*	6	195
Violaceous Coucal *Centropus violaceus*	5	199
Lesser Black Coucal *Centropus bernsteini*	4	200
Pheasant-coucal *Centropus phasianinus*	5	202
Genus *Carpococcyx*		206
Sumatran Ground Cuckoo *Carpococcyx viridis*	11	206
Bornean Ground Cuckoo *Carpococcyx radiceus*	11	208
Coral-billed Ground Cuckoo *Carpococcyx renauldi*	11	210
Genus *Coua*		212
Crested Coua *Coua cristata*	11	212
Verreaux's Coua *Coua verreauxi*	11	214
Blue Coua *Coua caerulea*	7	215
Red-capped Coua *Coua ruficeps*	12	217
Red-fronted Coua *Coua reynaudii*	7	219
Coquerel's Coua *Coua coquereli*	12	221
Running Coua *Coua cursor*	12	223
Giant Coua *Coua gigas*	12	225
Snail-eating Coua *Coua delalandei*	13	227
Red-breasted Coua *Coua serriana*	12	228
Genus *Rhinortha*		230
Raffles's Malkoha *Rhinortha chlorophaea*	14	230
Genus *Ceuthmochares*		233
Whistling Yellowbill *Ceuthmochares australis*	15	233
Chattering Yellowbill *Ceuthmochares aereus*	15	235
Genus *Taccocua*		237
Sirkeer Malkoha *Taccocua leschenaultii*	16	237
Genus *Zanclostomus*		240
Red-billed Malkoha *Zanclostomus javanicus*	16	240
Genus *Phaenicophaeus*		242
Red-faced Malkoha *Phaenicophaeus pyrrhocephalus*	17	242
Genus *Rhopodytes*		244
Chestnut-bellied Malkoha *Rhopodytes sumatranus*	15	244
Blue-faced Malkoha *Rhopodytes viridirostris*	15	246
Black-bellied Malkoha *Rhopodytes diardi*	15	247
Long-tailed Malkoha *Rhopodytes tristis*	15	249
Genus *Rhamphococcyx*		251
Chestnut-breasted Malkoha *Rhamphococcyx curvirostris*	16	251
Yellow-billed Malkoha *Rhamphococcyx calyorhynchus*	16	254
Genus *Dasylophus*		256
Red-crested Malkoha *Dasylophus superciliosus*	17	256
Genus *Lepidogrammus*		258
Scale-feathered Malkoha *Lepidogrammus cumingi*	17	258
Genus *Clamator*		260
Chestnut-winged Cuckoo *Clamator coromandus*	18	260

Great Spotted Cuckoo *Clamator glandarius*	18	263
Levaillant's Cuckoo *Clamator levaillantii*	18	267
Jacobin Cuckoo *Clamator jacobinus*	18	270
Genus *Piaya*		274
Little Cuckoo *Piaya minuta*	21	274
Black-bellied Cuckoo *Piaya melanogaster*	13	277
Squirrel Cuckoo *Piaya cayana*	13	279
Genus *Coccyzus*		283
Dwarf Cuckoo *Coccyzus pumilus*	17	283
Ash-coloured Cuckoo *Coccyzus cinereus*	17	285
Dark-billed Cuckoo *Coccyzus melacoryphus*	21	287
Yellow-billed Cuckoo *Coccyzus americanus*	20	290
Pearly-breasted Cuckoo *Coccyzus euleri*	20	295
Mangrove Cuckoo *Coccyzus minor*	20	298
Cocos Cuckoo *Coccyzus ferrugineus*	20	301
Black-billed Cuckoo *Coccyzus erythropthalmus*	20	302
Grey-capped Cuckoo *Coccyzus lansbergi*	21	306
Genus *Hyetornis*		308
Chestnut-bellied Cuckoo *Hyetornis pluvialis*	21	308
Rufous-breasted Cuckoo *Hyetornis rufigularis*	21	310
Genus *Saurothera*		312
Jamaican Lizard Cuckoo *Saurothera vetula*	19	312
Puerto Rican Lizard Cuckoo *Saurothera vieilloti*	19	314
Great Lizard Cuckoo *Saurothera merlini*	19	316
Hispaniolan Lizard Cuckoo *Saurothera longirostris*	19	318
Genus *Pachycoccyx*		320
Thick-billed Cuckoo *Pachycoccyx audeberti*	22	320
Genus *Microdynamis*		323
Dwarf Koel *Microdynamis parva*	23	323
Genus *Eudynamys*		325
Common Koel *Eudynamys scolopaceus*	24	325
Genus *Urodynamis*		331
Long-tailed Koel *Urodynamis taitensis*	14	331
Genus *Scythrops*		334
Channel-billed Cuckoo *Scythrops novaehollandiae*	14	334
Genus *Chrysococcyx*		337
Asian Emerald Cuckoo *Chrysococcyx maculatus*	25	337
Violet Cuckoo *Chrysococcyx xanthorhynchus*	25	340
Dideric Cuckoo *Chrysococcyx caprius*	26	343
Klaas's Cuckoo *Chrysococcyx klaas*	26	348
Yellow-throated Cuckoo *Chrysococcyx flavigularis*	26	352
African Emerald Cuckoo *Chrysococcyx cupreus*	26	354
Horsfield's Bronze Cuckoo *Chrysococcyx basalis*	27	357
Rufous-throated Bronze Cuckoo *Chrysococcyx ruficollis*	25	361
Shining Bronze Cuckoo *Chrysococcyx lucidus*	27	363
White-eared Bronze Cuckoo *Chrysococcyx meyerii*	25	367
Little Bronze Cuckoo *Chrysococcyx minutillus*	28	369
Gould's Bronze Cuckoo *Chrysococcyx poecilurus*	28	373
Pied Bronze Cuckoo *Chrysococcyx crassirostris*	28	377
Genus *Misocalius*		379
Black-eared Cuckoo *Misocalius osculans*	27	379

Genus *Rhamphomantis*		382
Long-billed Cuckoo *Rhamphomantis megarhynchus*	14	382
Genus *Cacomantis*		384
Pallid Cuckoo *Cacomantis pallidus*	29	384
Chestnut-breasted Cuckoo *Cacomantis castaneiventris*	29	388
Fan-tailed Cuckoo *Cacomantis flabelliformis*	29	390
Banded Bay Cuckoo *Cacomantis sonneratii*	30	394
Plaintive Cuckoo *Cacomantis merulinus*	30	397
Grey-bellied Cuckoo *Cacomantis passerinus*	30	401
Brush Cuckoo *Cacomantis variolosus*	31	404
Rusty-breasted Cuckoo *Cacomantis sepulcralis*	31	409
Genus *Caliechthrus*		412
White-crowned Cuckoo *Caliechthrus leucolophus*	30	412
Genus *Cercococcyx*		414
Dusky Long-tailed Cuckoo *Cercococcyx mechowi*	32	414
Olive Long-tailed Cuckoo *Cercococcyx olivinus*	32	417
Mountain Long-tailed Cuckoo *Cercococcyx montanus*	32	419
Genus *Surniculus*		421
Fork-tailed Drongo-cuckoo *Surniculus dicruroides*	23	421
Philippine Drongo-cuckoo *Surniculus velutinus*	23	423
Square-tailed Drongo-cuckoo *Surniculus lugubris*	23	425
Moluccan Drongo-cuckoo *Surniculus musschenbroeki*	23	428
Genus *Hierococcyx*		429
Moustached Hawk-cuckoo *Hierococcyx vagans*	22	429
Dark Hawk-cuckoo *Hierococcyx bocki*	22	431
Large Hawk-cuckoo *Hierococcyx sparverioides*	22	433
Common Hawk-cuckoo *Hierococcyx varius*	22	436
Rufous Hawk-cuckoo *Hierococcyx hyperythrus*	33	439
Philippine Hawk-cuckoo *Hierococcyx pectoralis*	33	441
Javan Hawk-cuckoo *Hierococcyx fugax*	33	443
Whistling Hawk-cuckoo *Hierococcyx nisicolor*	33	445
Genus *Cuculus*		447
Black Cuckoo *Cuculus clamosus*	34	447
Red-chested Cuckoo *Cuculus solitarius*	34	451
Lesser Cuckoo *Cuculus poliocephalus*	35	455
Sulawesi Cuckoo *Cuculus crassirostris*	35	458
Indian Cuckoo *Cuculus micropterus*	35	459
Madagascar Cuckoo *Cuculus rochii*	35	463
African Cuckoo *Cuculus gularis*	34	465
Oriental Cuckoo *Cuculus saturatus*	36	468
Sunda Cuckoo *Cuculus lepidus*	36	473
Common Cuckoo *Cuculus canorus*	32	475
APPENDIX – English and scientific names of birds other than cuckoos mentioned in the text		482
BIBLIOGRAPHY		491
INDEX		537

ACKNOWLEDGEMENTS

We would like to make it obvious that we have depended on the help of many people to achieve as comprehensive, accurate and up-to-date a work as possible, and indeed we have met with great enthusiasm from many ornithologists who have supported us in a host of ways. Therefore we wish to express our thanks and debt to the following persons who have contributed to this monograph.

Oscar van Rootselaar, one of the original team, had to leave the project at an early stage due to private circumstances.

JE is grateful for assistance from the following: Des Allen, Dan Brooks, Michel Ciselet, David and John F. Cooper, Richard Craik, Siegfried Eck, Brian Gee, Steven Gregory, Michel Gosselin, Bill Harvey, Nicolas K. Haass, Paul Herroelen, Lothar Herzig, Ejlif Holle Jørgensen, Howard Martin King, Henning Kunze, Pete Leonard, Jochen Martens, Rainer Mönke, Sadegh Sadeghi-Zaqdegan, Sunjoy Monga, Danny Rogers, Morten Strange, Jim C. Wardill, Yeow Chin Wee and Jan Zwaaneveld.

Jevgeni Shergalin provided extracts of Russian works; Jírí Mlikovský provided a list of Czech and Slovak cuckoo references. Neil and Liz Baker supplied over 3,000 cuckoo records, along with some maps, from their Tanzania atlas project. Thomas S. Schulenberg supplied 1308 records from Peru. Anders Pape Møller kindly read and commented on the Common Cuckoo account, and contributed stimulating input. Niels Linneberg sorted out all computer problems for JE. Jon Fjeldså and Jan Bolding Kristensen (Zoological Museum, Natural History Museum of Denmark) supplied information and photos of specimens.

CFM warmly thanks Robert Prŷs-Jones, Mark Adams, Hein van Grouw, Douglas Russell, Alison Harding and 'Effie' Warr (BMNH) for considerable assistance with specimens, references and knowledgeable advice. Help is also acknowledged from P. Eckhoff and S. Frahnert (ZMB), S. van der Mije (Naturalis, Netherlands), L. Garetano (Bishop Museum, Honolulu), K. Garrett (Natural History Museum of Los Angeles County), S. Haber, P. Sweet and T. Trombone (AMNH), D. Willard and M. Hennen (FMNH), N. Rice (ANSP), J. Woods (DMNH) and K. Zyskowski (Yale Peabody Museum) with specimens housed at their museums. Librarians at the Natural History Museum, London were very helpful during many visits.

David Wells, Edward Dickinson and Robert Cheke were always forthcoming with expert advice. Thanks are also due to N. and E. Baker, M. Blair, Nick Brickle, A. Chia, J. Cracraft, Earl of Cranbrook, N. David, R. J. Dowsett, J. Eriksen, M. C. Jennings, R. S. Kennedy, N. Mita, N. Moores, T. Nuka, R. B. Payne, V. Rauste, N. Rice, R. Safford, V. Schollaert, D. Stanton, F. Steinheimer, L. Svensson, K. Ueda, K. Woods, Bas van Balen and Yong Ding Li who provided assistance in various ways or input to discussions.

FB would like to acknowledge the cooperation of: R. H. F. Macedo for unpublished information on *Guira*; R. P. Clay, H. de Castillo, A. de Lucca, L. Molinas and D. Unterkofler (Aves Argentinas/AOP) for assistance with literature; D. J. Agro; G. A. Bencke (Museu de Ciências Naturais, Fundação Zoobotânica do Rio Grande do Sul, Porto Alegre, Rio Grande do Sul, Brazil), D. M. Brooks (Houston Museum of Natural Science, Dept. of Vertebrate Zoology), C. J. Carlos (Universidade Federal de Pernambuco, Recife), A. de Almeida, M. R. de la Peña, J. N. M. Flanagan; C. S. Fontana and M. L. Gonçalves (for specimen data from Museu de Ciências e Tecnologia, PUCRS, Porto Alegre, Rio Grande do Sul, Brazil); M. Güntert (Naturhistorisches Museum der Burgergemeinde Bern, Switzerland) for data on 495 specimens; S. K. Herzog (Asociación Armonía – BirdLife International, Bolivia) supplied 1,713 records from Bolivia from the Armonía database); J. P. Kjeldsen; N. Krabbe sent data from c. 100 specimen records from Museo Ecuatoriano de Ciencias Naturales; B. D. Marks sent 302 specimen records from TCWC, Texas A & M University; A. Navarro-Sigüenza and A. T. Peterson provided 2,433 specimen records from Mexico; A. Pirie (MCZ) provided 2,886 specimen records; L. F. Silveira (MZUSP) for access to the bird collection, and for c. 1,000 specimen records. The librarians at MZUSP were most helpful during many visits, and with finding and sending large amounts of literature. H. Espersen kindly provided access to the library facilities at ZMUC. J. Ingels and T. S. Romdal sent much photocopied material. T. Rotenberry and The Peregrine Fund (Idaho) are thanked for forwarding instantly many hard-to-obtain papers.

Help and advice was received also from the following: J. I. Areta, M. J. Bechard, A. Bodrati, S. H. Borges, A. J. Bosso, R. Brito-Aguilar, A. Bruun Kristensen, G. Buitrón Jurado, D. Calderón Franco, P. Capllonch, K. Cockle, C. T. Collins, J. Cuello, A. M. Cuervo, K. Delhey, A. G. Di Giacomo, T. Dornas, J. Eberhard, J. C. Eitniear, R. M. Fraga, J. García-Moreno, A. R. Giraudo, F. I. de Godoy, G. Gorton, P. Grilli, K. E. Gustafsson, A. Hagerman, F. E. Hayes, J. B. Irusta, M. L. Isler, A. Jaramillo, J. Karubian, L. F. Kiff, G. M. Kirwan, S. Kreft,

C. Levy, M. W. Lockwood, S. McNeil, C. I. Miño, A. D. Mitchell, S. Mlodinow, S. Moore, Y. Oniki, J. F. Pacheco, M. Patrikeev, E. Paul, D. L. Pearson, M. A. Plenge, F. A. Pratolongo, M. A. Raposo, J. V. Remsen Jnr., S. Rick, C. Riehl, F. Schmitt, R. Scovell, S. G. Sealy, S. Seipke, C. J. Sharpe, C. Smith, P. Smith, S. M. Smith, P. van Gasse, G. E. Wallace, W. H. Weber, R. S. R. Williams, E. O. Willis and T. J. Zenzal.

The following museums also supplied specimen data to FB through A. Navarro-Sigüenza and A. T. Peterson: ANSP; Carnegie Museum of Natural History; Cornell University Museum of Vertebrates; University of Kansas Natural History Museum; Liverpool Museum; Museum of Natural Science, Louisiana State University; Muséum National d'Histoire Naturelle Zoologie, Paris; Museo de Historia Natural, Universidad Nacional Mayor de San Marcos, Lima; Museum of Vertebrate Zoology, University of California, Berkeley; Museum and Institute of Zoology, Warsaw; Naturhistorisches Museum Basel, Switzerland; Royal Ontario Museum; Museum of Zoology, University of Michigan; National Museum of Natural History, Smithsonian Institution; Burke Museum, University of Washington; Peabody Museum of Natural History, Yale University; and Zoologisches Forschungsinstitut und Museum Alexander Koenig, Bonn.

In addition to the institutions mentioned above RF received data for mapping from: Arizona State University Bird Collection; Australian Museum; Biologiezentrum Linz Oberoesterreich; Bombay Natural History Society; Borror Lab of Bioacoustics; Burke Museum of Natural History and Culture; Californian Academy of Sciences; Canadian Museum of Nature Bird Collection; Charles R. Conner Museum; Chicago Academy of Sciences; Colección de Aves y Mamíferos del Valle de Cuatrociénegas; Cornell Laboratory of Ornithology; Denver Museum of Nature and Science; Ditsong Museum of South Africa; Division of Genomic Resources, UNM; European Distributed Institute of Taxonomy; Ewha Womans University Natural History Museum; Facultad de Ciencias UNAM; Finnish Museum of Natural History; Fort Hays State University Museum of the High Plains; Gothenburg Natural History Museum; Hancock Museum; Ibaraki Nature Museum; Institut de recherche pour le Développement; Instituto de Biologia, Universidad Nacional Autonoma de Mexico; Jyväskylä University Museum; Kumamoto City Museum; Kyung Hee University Natural History Museum; Louisiana State University Museum of Zoology; Lund Museum of Zoology; Michigan State University Museum; Midwest Museum of Natural History; Miyazaki Prefecture Museum of Nature and History; Musée Zoologique de la Ville de Strasbourg; Museo Argentino de Ciencias Naturales; Museo de Zoología 'Alfonso L . Herrera'; Museo Nacional de Ciencias Naturales, Madrid; Museo Regionale di Scienze Naturali; Museo Universitario (University of Antioquia); Museu de Ciancies Naturals de Barcelona; Museum and Institute of Zoology, Polish Academy of Sciences; Museum Heineanum Halberstadt; Museum of Evolution, Uppsala; Museum of Nature and Human Activities; Museum of Southwestern Biology; Museum of Vertebrate Zoology; Museum of Zoology, University of Navarra; Museum Victoria; National Chemical Laboratory; National Museum of Ireland; National Museum of Namibia; National Museum of Natural History; National Museums and Monuments; Natural History Museum Rotterdam; Natural History Museum, University of Oslo; New Brunswick Museum; New Mexico Biodiversity Collections Consortium; Oklahoma Museum of Natural History; Oxford University Museum of Natural History; Provincial Museum of Alberta; Queen Victoria Museum Art Gallery; Queensland Museum, Brisbane; Rocky Mountain Bird Observatory; Royal British Columbia Museum; Royal Museum of Central Africa, Tervuren; Sam Noble Oklahoma Museum of Natural History; San Diego Natural History Museum; Santa Barbara Museum of Natural History; Slater Museum of Natural History; South African Museum; State Museum of Natural History, Stuttgart; Swedish Museum of Natural History; Tasmanian Museum and Art Gallery; Texas Cooperative Wildlife Collection; Tullie House Museum; United States National Museum; University Museum of Zoology Cambridge; University of Alaska Museum; University of Alberta Museums; University of Arizona Museum of Natural History; University of California, Davis; University of California, Los Angeles; University of Colorado Museum of Natural History; University of Connecticut Bird Collection; University of Michigan Museum of Zoology; University of Nebraska State Museum; University of Tennessee; Utah Museum of Natural History; Western Australia Museum; Western Foundation of Vertebrate Zoology; Western New Mexico University; Yokosuka City Museum; Zoological Museum Amsterdam; Zoologische Sammlung der Universität Rostock; Zoologische Staatssammlung München.

A glance at the bibliography shows that this monograph relies heavily on published material. We will here only call attention to Robert B. Payne's monograph *The Cuckoos* (2005) which has been a most valuable source of information, and we are very grateful for his painstaking scientific work.

Finally, we want to thank the project editor of this monograph, Jim Martin, at Christopher Helm/Bloomsbury for getting the book off the ground, Julie Dando at Fluke Art for design and layout, and our families who have been very tolerant and supportive.

INTRODUCTION

Systematics

The cuckoo family has a worldwide distribution, with the exception of the polar regions. The higher level taxonomic position of the cuckoos has long been subject to uncertainty and instability. Currently, the Family Cuculidae is typically considered the sole extant family in the Order Cuculiformes. The families Musophagidae (African turacos) and Opisthocomidae (South American Hoatzin) have historically been placed in this order (Peters 1940, Mayr & Amadon 1951, Wetmore 1960), but today's general consensus is that they belong elsewhere (e.g. Berger 1960, Sibley & Ahlquist 1990, Johansson *et al.* 2001, Mayr *et al.* 2003, Sorenson *et al.* 2003, Fain & Houde 2004, Ericson *et al.* 2006). Cuckoos differ from turacos in 12 anatomical characters, including nude vs. tufted oil gland, presence vs. absence of eyelashes and typical furcula, two vs. one bony canal in hypotarsus and perforated vs. notched atlas (Berger 1960). Anatomical, parasitological, behavioural and nuclear DNA characters have suggested that cuckoos and mesites (Mesitornithidae) may be sister taxa (Mayr & Ericson 2004).

However, recent anatomical studies (Livezey & Zusi 2007) suggest that Cuculiformes form a clade with Musophagiformes and Opisthocomiformes, which is sister to a clade consisting of Psittaciformes, Columbiformes, Passeriformes and the 'near-passerine' orders. Contrastingly, nuclear DNA sequence data of 32 kilobases from 19 loci suggest that cuckoos are a sister group to Gruiformes, the two forming a clade with the bustards (Otididae), and Hoatzin is basal (Hackett *et al.* 2008). Other DNA sequence studies have yielded different results (e.g. Fain & Houde 2004, Pacheco *et al.* 2011), such as grebes being closest to cuckoos (van Tuinen *et al.* 2000), although this may be related to limited taxon sampling or short DNA sequences (see Livezey & Zusi 2007), or failure to obtain sufficient resolution about the placement of the Cuculidae (e.g. Cracraft *et al.* 2004). Thus, the cuckoos are probably an ancient, monophyletic lineage with no close living relatives, or in other words, their placement in the phylogeny of birds is uncertain (e.g. Sibley & Ahlquist 1990, Sorenson & Payne 2005).

The oldest-known undoubted cuckoo fossils are *Cursoricoccyx geraldinae* (Martin & Mengel 1984) from the early Miocene of Colorado and *Thomasococcyx philohippus* (Steadman 2008), of the same age, from Florida; other genera are from the Pleistocene-Holocene of North America. *Thomasococcyx* is perhaps closest to New World terrestrial cuckoos such as *Morococcyx* and *Geococcyx*. More recent fossils are known from Australia and Madagascar. Earlier fossils have been claimed as cuculid, but supporting evidence is lacking.

Molecular studies, in some cases contradictory, have contributed to our understanding of relationships within the family (Sibley & Ahlquist 1990, Aragón *et al.* 1999, Johnson *et al.* 2000, Sorenson & Payne 2005), and future work is expected to shed further light on this area, and doubtless lead to some adjustments of species boundaries and ordering. Comprehensive sequences of nuclear DNA may be particularly suitable. Among other questions, resolving well the interrelationships among the cuckoos will help to answer where and how brood parasitism evolved in cuckoos, and whether it happened more than once (*cf.* Hughes 2000, Payne 2005, Sorenson & Payne 2005). It is thus intriguing that the two New World parasitic cuckoo genera, *Dromococcyx* and *Tapera*, are of quite different appearance from, and indeed only distantly related to, those of the Old World (Sorenson & Payne 2005).

This work recognises 144 species of cuckoo in 38 genera. Five subfamilies (following Dickinson 2003 and Payne 2005) are accepted, Crotophaginae and Neomorphinae in the Americas, Centropodinae (sometimes treated as a distinct family) and Couinae in the Old World, and Cuculinae occurring in both. The last is subdivided into three tribes, the Rhinorthini, Phaenicophaeini and Cuculini. As well as species with the English epithet 'cuckoo', the family contains koels, couas, coucals, malkohas and anis. English names are mostly those to be found in Dickinson (2003).

For the most part we have accepted the genera and species limits of Payne (2005), with some adjustments that we have deemed necessary. Species order follows Payne (2005), based on the mtDNA studies of Sorenson & Payne (2005). Numbers in brackets are the number of species for each genus accepted in this work.

Subfamily Crotophaginae Guira Cuckoo *Guira* (1) and the anis *Crotophaga* (3), confined to the New World; all indulge in group nesting, none is brood-parasitic.

Subfamily Neomorphinae New World: American Striped Cuckoo *Tapera* (1), Pheasant-cuckoo and Pavonine Cuckoo *Dromococcyx* (2); all three are brood-parasitic. Lesser Ground Cuckoo *Morococcyx* (1), roadrunners

Geococcyx (2), Banded, Rufous-winged, Red-billed, Rufous-vented and Scaled Ground Cuckoos *Neomorphus* (5); non brood-parasitic.

Subfamily Centropodinae Old World: coucals *Centropus* (26); non brood-parasitic.

Subfamily Couinae Sumatran, Bornean and Coral-billed Ground Cuckoos *Carpococcyx* (3), south-east Asia; couas *Coua* (10, one extinct), Madagascar; all non brood-parasitic.

Subfamily Cuculinae

Tribe Rhinorthini Raffles's Malkoha *Rhinortha* (1); aberrant, south-east Asian 'malkoha'; non brood-parasitic.

Tribe Phaenicophaeini Yellowbills *Ceuthmochares* (2), Afrotropical, non brood-parasitic; Sirkeer Malkoha *Taccocua* (1), Red-billed Malkoha *Zanclostomus* (1), Red-faced Malkoha *Phaenicophaeus* (1), Chestnut-bellied, Blue-faced, Black-bellied and Long-tailed Malkohas *Rhopodytes* (4), Chestnut-breasted and Yellow-billed Malkohas *Rhamphococcyx* (2), Red-crested Malkoha *Dasylophus* (1) and Scale-feathered Malkoha *Lepidogrammus* (1), all Oriental and non brood-parasitic; 'crested' cuckoos *Clamator* (4), Old World and brood-parasitic; Little, Black-bellied and Squirrel Cuckoos *Piaya* (3), Dwarf, Ash-coloured, Dark-billed, Yellow-billed, Pearly-breasted, Mangrove, Cocos, Black-billed, and Grey-capped Cuckoos *Coccyzus* (9), New World, two species facultative brood-parasites; Chestnut-bellied and Rufous-breasted Cuckoos *Hyetornis* (2) and lizard cuckoos *Saurothera* (4), Caribbean and non brood-parasitic.

Tribe Cuculini All brood-parasitic: Thick-billed Cuckoo *Pachycoccyx* (1), Afrotropical; Dwarf Koel *Microdynamis* (1), New Guinea; Common Koel *Eudynamys* (1), Oriental, Papuan-Australian; Long-tailed Koel *Urodynamis* (1), New Zealand and Pacific; Channel-billed Cuckoo *Scythrops* (1), Wallacea, New Guinea area, Australia; glossy cuckoos *Chrysococcyx* (13), Old World; Black-eared Cuckoo *Misocalius* (1), Australia, New Guinea; Long-billed Cuckoo *Rhamphomantis* (1), New Guinea; Pallid, Chestnut, Fan-tailed, Banded Bay, Plaintive, Grey-bellied, Rusty-breasted and Brush Cuckoos *Cacomantis* (8), Oriental and Australasian; White-crowned Cuckoo *Caliechthrus* (1), New Guinea; long-tailed cuckoos *Cercococcyx* (3), Afrotropical; drongo-cuckoos *Surniculus* (4), Oriental; hawk-cuckoos *Hierococcyx* (8), Oriental; Black, Red-chested, Lesser, Sulawesi, Indian, Madagascar, African, Oriental, Sunda and Common Cuckoos *Cuculus* (10), Old World.

Morphology

Cuckoos are highly distinct morphologically from other avian orders. The monophyly of cuckoos has been accepted for many decades, and an analysis of skeletal characters by Hughes (2000) identified fourteen skeletal characters found in all cuckoos, yet unique to the order. Several of the key characters relate to the details of the structure of the tarsometatarsus. However, Payne (2005) found that apart from the latter, and the shape of the maxillary bone and the humerus, these characters are variable amongst cuckoos and also occur in other avian orders.

The external morphology of cuckoos does not vary enormously across the order. Cuckoos range in size from about 15cm to 70cm in total length, and from 14.5g to 790g in mass. Cuckoos are typically slender, long-tailed birds, though the slender shape is most pronounced in the arboreal genera such as *Cuculus* and *Coccyzus*, while terrestrial genera such as *Carpococcyx* and *Geococcyx* are typically much more heavily built with thick and long tarsi. There is a zygodactyl arrangement of the toes (digits 1 and 4 pointing backwards), and the feet and legs are short to long. Some (including most coucals) have particularly long hallux claws. An interesting fact is that brood-parasitic cuckoos, in all three clades where they occur, have smaller brains than related nesting species.

In common with many avian orders, wing shape varies to a certain extent with migratory behaviour in the cuckoos. Wings are pointed in long distance migrants such as the migratory *Cuculus* species, and rounded in the sedentary coucals. Within the hawk-cuckoos, the migratory northern species have more pointed wings than the sedentary tropical species, showing that substantial variation in wing shape can arise relatively quickly. While many of the migrants are brood parasitic, there are many examples of brood parasitism in the non-migratory species with rounded wings, indicating that wing shape is more likely to be related to migratory behaviour than breeding biology (Payne 2005).

Plumage is rather lax, and varies from very bright and colourful to rather dull, almost cryptic. A number of species have obvious colourful bare patches of skin around the eyes, and the eye-lashes are often long. Bills are usually medium to longish, and fairly slender to robust, often with a markedly arched upper mandible.

Colourful facial skin occurs in many cuckoos and is most pronounced in the malkohas, roadrunners, couas, and ground cuckoos. Brightly coloured head and bill presumably relate to social signaling, but there has been little research on this. Sexual dimorphism is mainly manifested in size differences, with females usually being larger than males in species with parental care (e.g. coucals), and males being larger than females in brood-parasitic species. This suggests a role for co-evolution selecting for smaller females in parasitic species, perhaps so they can access small nests, lay eggs more closely matching hosts in size, or increase concealment when approaching host nests (Krüger *et al.* 2007). Striking plumage sexual dimorphism occurs in Common Koel *Eudynamys scolopaceus*, Raffles's Malkoha *Rhinortha chlorophaea* and several of the *Chrysococcyx* species.

Fledgling/juvenile plumage can be quite variable, as in the Thick-billed Cuckoo *Pachycoccyx audeberti*, and Levaillant's Cuckoo *Clamator levaillantii* (e.g. Chapin 1939). There is no agreement over whether this is genetic or environmental. Rufous versus green plumage in juvenile Dideric Cuckoos *Chrysococcyx caprius* may be due to bishop (*Euplectes*) foster parents feeding them on seeds, whereas weaver (*Ploceus*) fosters provide more insect food. Equally, it may be genetic, tied to the *gens* of the parents (Payne 2005).

In many nesting cuckoos, e.g. malkohas, and some brood-parasites, e.g. Jacobin Cuckoo *Clamator jacobinus*, juvenile plumage may be similar to adult. There may be an extra stage, the subadult, between juvenile and adult, as in the hawk-cuckoos *Hierococcyx*, and Sulawesi Cuckoo *Cuculus crassirostris*. Retained juvenile feathers are sometimes found in breeding birds indicating that they mature in their first year, and female Philippine Hawk-cuckoos *H. pectoralis* in juvenile/subadult plumage have been recorded laying eggs (Ripley & Rabor 1958). Some coucals, e.g. African Black Coucal *Centropus grillii* and Pheasant-coucal *C. phasianinus*, have distinct non-breeding plumages. Distinct colour morphs are found, usually in females, in some *Cacomantis* and *Cuculus* species, whereby individuals may have greyish or rufous/hepatic plumages. In the Pallid Cuckoo *Cacomantis pallidus* both sexes occur either as dark or pale morphs, but there is so much individual variation that it may be best seen as continuous. The significance of these colour morphs has not been fully elucidated, but may have some connection with brood-parasitism utilising different hosts and/or habitats.

Moult

Although moult is such an important part of a bird's life cycle, little is known about it in many species of cuckoo. Where it has been reported, often the information is confined to the dating of the process. However, sufficient is known so that some generalisations can be drawn. Moult and breeding are generally thought to be two discrete processes, both requiring considerable expenditure of energy and therefore occurring at different times. The anis *Crotophaga* and some coucals are examples where the two overlap, and moult may be year-round (e.g. Dickey & van Rossen 1938, Sibley 1951, Snow & Snow 1964). Post-juvenile and post-nuptial moults are usually complete, although in a number of species some feathers are retained into the next plumage; post-juvenile moult is recorded as partial in the Greater Roadrunner *Geococcyx californianus*, and post-nuptial complete, but remiges replaced slowly (Pyle 1995, Hughes 1996). Apparently neither juvenile nor adult moult is complete in the Common Hawk-cuckoo *Hierococcyx varius* (Ali & Ripley 1983). Length of complete moult varies from c. 80 days in some *Chrysococcyx* species to c. 100 days in the Common Cuckoo *Cuculus canorus* (Rowan 1983, Seel 1984b). The Great Spotted Cuckoo *Clamator glandarius* acquires complete adult plumage in six months, whereas in the African Black Coucal it takes two years (Stresemann & Stresemann 1961, 1969; Irwin 1988). The sequence of moulting the remiges is not recorded as regularly ascending or descending in any species. Across the family there are many examples reported of asymmetric, transilient and semi-transilient patterns, with primary moult at two, three or even four loci simultaneously (e.g. Stresemann & Stresemann 1961, 1969; Rowan 1983, Pyle *et al.* 1997, Higgins 1999, Wells 1999).

Vocalisations

Many species are known as well, if not better, from their vocalisations. The name 'cuckoo' in English, and similar names in other languages, as well as scientific nomenclatural derivatives such as Cuculiformes, *Cuculus* etc. of course come from the familiar sound of what is arguably the best known species – the Common Cuckoo. Many have a variety of, and more complex, vocalisations, often the first sign of their presence. Other vernacular names deriving from their calls include 'It will rain' or 'Piet mijn vrouw' (Afrikaans) for the Red-chested Cuckoo *C. solitarius*, 'Hototogisu' (Japanese) for the Lesser Cuckoo *C. poliocephalus*, and 'Ko-el' for *Eudynamys scolopaceus*.

Vocal communication is probably particularly important for forest species. A few species have breeding call-flights, but most call from perches; some call at night. Becking (1975), concerning the Lesser Cuckoo, wrote 'Voice, which is conservative in these cuckoos, probably constitutes a more important genetic isolation mechanism than any morphological character'. This observation could be expanded to include a number of other pairs/groups of species that are partially sympatric. It is a well-established and accepted fact that vocalisations are an extremely important pre-mating isolating mechanism (Remsen 2005). Suboscines, like cuckoos, have innate and not learned vocalisations. Isler *et al.* (2005) show that "vocal characters can vary clinally in a suboscine passerine and that this variation parallels a step-cline in plumage characters. Vocalisations can vary within a subspecies between sample points in a manner consistent with points along a cline". Closely related, sympatric bird species generally have different voices, but age, sex, individual, and ecological differences in vocalisations are also known, in addition to regional dialects. However, in cuckoos, it seems that some species may range over a very large geographical area, but show little or no regional dialects. Also bioacoustic properties vary with habitat. However, in some suboscines, it has been demonstrated that convergence in vocalisations occurs between two closely-related sympatric species, although there remains some element that is unique to each (J. Tobias *in litt.*). Such a situation may occur between two *Cacomantis* cuckoos, Plaintive *C. merulinus* and Banded Bay *C. sonneratii*, in peninsular Malaysia (Khoo Siew Yoong *in litt.*).

The question remains – how much difference is required to act as a reproductive barrier? Playback trials are often cited as evidence of conspecificity or otherwise. In some cases even crude imitations of vocalisations can generate a positive response, whereas in other cases a perfect recording can be ignored. Such experiments must be carefully designed and carried out over repeated trials with multiple individuals at various seasons of the year (Remsen 2005). In other words, broad geographical sampling is required to assess questions of species limits and it is not acceptable to compare vocalisations from distant localities and then assume that they represent a taxon as a whole, much less the intervening localities.

Behaviour and Feeding

Generally, cuckoos are not found in large groups except during or prior to migration. Some species, e.g. anis, are always found in groups, but the majority of species are found singly, in pairs, or in small family parties. Some species seem to be poor fliers (e.g. malkohas, coucals, roadrunners, and *Neomorphus* and *Carpococcyx* ground cuckoos), preferring to run to escape danger. However, most are proficient on the wing. Although the majority of species are sedentary, or undergo local or dispersal movements only, a number carry out long intercontinental migrations.

Anis become partially torpid at night, lowering their body temperature by as much as 8.2°C (from c. 40.8 to 32.6°C) at night or after fasting, but are still able to fly. The Greater Roadrunner is adapted to desert conditions in a number of ways such as lowering of body temperature at night, and use of the Sun's radiation to warm up melanin-pigmented dorsal skin in the morning; nasal salt glands resulting in less water being required for excretion; ability to re-absorb water through the cloaca; evaporation of body water through the skin (and gular-fluttering in nestlings) to lose heat; heat-exchange mechanism in the circulatory system to cool the brain; not requiring liquid water; regurgitation of watery liquid for young; and consumption of chick's watery faecal sacs (Hughes 1996). Sunning is found in many other species, including Guira Cuckoo, anis, coucals, couas, *Coccyzus* and *Chrysococcyx*, after rain as the plumage is not efficient at repelling water, or early in the morning (Payne 1997).

The majority of species feed on invertebrates and small vertebrates; larger species may take larger animals. Many species take substantial numbers of noxious caterpillars and grasshoppers avoided by other birds. Some include a certain amount of plant material in their diets, a few relying for a large part on fruit (e.g Common Koel). Despite the old name of the Asian ground cuckoos being 'fruit cuckoos' and the generic name *Carpococcyx* meaning the same, only one is known to eat any fruit. Some coucals may be omnivorous, being opportunistic feeders on fish or carrion, or taking birds from mist-nets. The Cuculinae generally glean insects from foliage at various heights and some employ 'flycatching', but Crotophaginae, Neomorphinae, Couinae and Centropodinae mostly ground-forage. Caterpillars are often de-gutted to remove indigestible and toxic leaf products before swallowing, the hairs forming a felted lining in the gut before being regurgitated as pellets. Some species feed on highly gregarious caterpillars (e.g. tent caterpillars in northern America) that in some cases are pests of commercially important trees. In the Neotropics ground cuckoos follow army ant swarms, peccaries and small primates, preying on arthropods flushed by them (Haffer 1977, Willis 1982, Siegel *et al.* 1989). Certain coucals may opportunistically feed on the edges of bush-fires for the same reason (C. F. Mann). Anis may take ticks from cattle (Rand 1953, Wetmore 1968).

Wyllie (1981) describes how a White-browed Coucal carried its young bird in flight, holding it between its legs. Such behavior is otherwise only reported from Red-tailed Hawk, Common Moorhen, Sungrebe, Eurasian and American Woodcocks, Common Sandpiper and Sunda Frogmouth (Campbell & Lack 1985, Mann 1991).

Many cuckoos demonstrate courtship feeding. 'If the biological significance of courtship feeding in parasitic cuckoos exists at all, it must be much less than found in the species in which the formation and maintenance of the sexual pair is advantageous to the rearing of the progeny. In many species of birds, food-begging of adult females is generally considered as part of infantile behaviour used in appeasement and courtship, which may act as a stimulus for parental behaviour or dominance on the part of the male. In cuckoos such a response obviously does not exist, so that the feeding by males may be interpreted as part of the ritualised courtship behaviour, the stimulus situation of which no longer involves parental drive. If so it is only the sexually-motivated males that respond to food begging of females and the feeding behaviour does not have to have an autochthonous functional' (Kikkawa 1968). Payne (2005) suggests that females may obtain a significant fraction of the energy needed to form eggs through courtship feeding.

Nestling Grey Warblers may require brooding for less of their nestling period than their brood-parasite Shining Bronze Cuckoo *Chrysococcyx lucidus* because they develop homeothermy earlier (Gill 1983a).

Breeding

The full gamut of breeding systems known in birds is found within the family. Normal nesting and rearing of their own young occurs by monogamous pairing (e.g. malkohas, couas), or is a polyandrous (African Black Coucal) or a polygynous (Guira Cuckoo *Guira guira* within a co-operative breeding context) situation. Co-operative breeding is also found in anis. Brood-parasitism, for which the family is notorious to the general public, is generally obligate (e.g. *Cuculus, Cacomantis*), but in a few species (e.g. Yellow-billed Cuckoo *Coccyzus americanus*) facultative, whereby the birds normally raise their own young and only occasionally deposit an egg (s) in the nest of another individual of the same or different species. In the classification adopted in this work, brood-parasitism, a phenomenon rare in birds, has evolved three times within the family, in

Yellow-billed Cuckoo *Coccyzus americanus* is among the species that usually build their own nest and raise their own young. Texas, USA, June (*Michael Patrikeev*).

Neomorphinae, Phaenicophaeni and Cuculini. It is highly unlikely that brood-parasitism was the ancestral condition in Cuculidae, and this cannot be shown one way or the other because of the unresolved phylogeny. Obligate brood-parasitism, known in less than 1% of bird species, is otherwise known only in the New World in cowbirds (Icteridae) and the Black-headed Duck, and in the Old World in honeyguides (Indicatoridae), and indigobirds, whydahs and Cuckoo Finch (Viduidae) (Davies 2000).

In a few species different genetic lineages – *gentes* (singular *gens*) – have evolved whereby females of different *gentes* lay differently-coloured eggs and utilise different hosts. It is thought that nestling cuckoos imprint on their foster parents and when they have grown and are ready for nesting themselves, they will utilise the same host as their parents, thus maintaining the line. It was previously believed that the different sympatric *gentes* were maintained by a genetic mechanism, probably by genes responsible for egg-coloration being carried only through the female line on the W chromosome, although the exact mechanism had not been elucidated. However, recent studies by Fossøy *et al.* (2011) demonstrate that the males also play a part in maintaining *gentes*, the genes involved presumably being autosomal. In the Common Cuckoo these *gentes* are thought to have arisen c. 80,000 years ago, but they are imperfectly maintained (Gibbs *et al.* 2000, Fossøy *et al.* 2011). How much of this breakdown is comparatively recent caused by human disruption of the environment, particularly by the spread and intensification of agriculture, is unknown.

The nests, eggs and nestlings of cuckoos are incompletely known. Of the 83 nesting cuckoo species, the nest is unknown for 17 (20 %). Eggs are unknown for 78 Cuculidae species (54 %), and the nestling is unknown for 64 cuckoo species (44 %). Eggs of parasitic cuckoos are relatively small, presumably to be a better match for their hosts. Those of nest-builders are of expected size for the female, whilst those of Crotophaginae are comparatively large.

Many cuckoos remove an egg of the host's clutch when they lay one of their own, presumably so that there is a greater chance that it will be adequately incubated, and not because the host will detect an extra egg. Removal of a host's egg when the adult cuckoo lays is reported in several species of *Cuculus* and *Chrysococcyx* (Friedmann 1968). But Hamilton & Orians 1965 claim "it may encourage acceptance by the host of the altered clutch, the size of which is perhaps sensed via the brood-patch". Interestingly, Black-eared Cuckoos *Misocalius osculans* ringed in September have a bare vascular area on the belly that looks identical to a brood-patch, though no report exists that this species ever incubates (Rogers *et al.* 1986 in Higgins 1999). No explanation for this has been suggested, and presumably the brood patch is vestigial.

Sometimes the eggs of more than one species of cuckoo are found in a single nest. Brooker and Brooker (1989) sometimes found nests in Australia parasitised by Shining Bronze Cuckoo where other cuckoo species also had laid an egg in the nest: 22 nests also with an egg of Horsfield's Bronze Cuckoo *Chrysococcyx basalis*, 14 with Fan-tailed Cuckoos *Cacomantis flabelliformis*, and one with Brush Cuckoo *C. variolosus*. Similarly in Africa, an egg of Red-chested Cuckoo has been found in the same nest as one of Dideric Cuckoo (Jackson 1938), and of African Emerald Cuckoo *Chrysococcyx cupreus* (Oatley 1980).

Where a brood may contain host offspring as well as a bood-parasite it has been claimed that the host parents preferentially feed the cuckoo chick rather than their own offspring. Larger nestlings by and large had faster growth rates than smaller nestlings, and parents preferred to feed larger nestlings than smaller ones (Skagen 1988).

One of the most fascinating aspects of brood parasitism is the 'arms race' – a co-evolutionary process between cuckoo parasite and host – whereby there is perpetual competition between the two. This has been a fruitful study over many years (von Haartman 1981, Davies 1989, Davies & Brooke 1989, 1991; Pehani 1993, Lotem *et al.* 1995, and many more recent references). This can result in better egg-mimicry by the parasite and enhanced detection by the host; thicker egg shell in the parasite to prevent puncture by the host (also less likely to break if laid from a height); host teaching of the unhatched young begging calls closely matching the host's incubation call which differ from the begging call of Horsfield's Bronze Cuckoo which cannot learn calls, so that she can recognise her own young (as in the Superb Fairy-wren – one of very few hosts of ejector cuckoos that detects and deserts cuckoo nestlings) (Colombelli-Négrel *et al.* 2010); and provisioning the brood-parasite young with lower quality food etc.

Bluer-greener eggs are thought to demonstrate greater fitness in a host (Soler *et al.* in press), and that blue-green coloration is more attractive to birds. Studying the Common Cuckoo, they found that host species parasitised with bluer-greener cuckoo eggs were more prone to accept non-mimetic model eggs. Furthermore, the variation in cuckoo egg coloration was more pronounced when parasitising hosts that lay blue-green eggs.

In many cases the hosts eggs and/or young are ejected by the parasitic nestling (e.g. in Common Cuckoo), but not in some other species (e.g. in Great Spotted Cuckoo). The nestling American Striped Cuckoo *Tapera naevia* kills host young with bites to the head and back using sharp hooks on the bill (Morton & Farabaugh 1979). 'Mafia' behaviour has evolved in at least one species, Great Spotted Cuckoo, whereby the brood-parasite, on later inspection of the nest, punishes the ejector host by damaging the host's eggs (Soler *et al.* 1995d). Tolerance, which varies geographically, by Common Magpies to Great Spotted Cuckoo parasitism may have evolved because of such behaviour; levels of tolerance to brood parasitism by different magpie populations were positively related to brood-parasite prevalence, but not to levels of resistance (Soler *et al.* 2011).

Although Jacobin Cuckoos do not evict host young, the latter rarely survive the nestling period. However, where the Southern Pied Babbler is its host in South Africa their young survive the nestling period and maintain a similar body mass to young babblers in unparasitised broods. But host young were less likely to survive to independence than young raised in unparasitised nests (Ridley & Thompson 2012). A recent study in China showed that over four years 6.9% (11 out of 159) of Chinese Babax nests were parasitised by Large Hawk-cuckoos *Hierococcyx sparverioides* laying immaculate white eggs (a previously unknown colour for this species). The babax eggs were blue. Most cuckoo eggs were accepted by the host, suggesting that this host–parasite system may be evolutionarily recent (Yang *et al.* 2012).

Hosts evolve anti-parasite defences like egg recognition and rejection. The basis for recognition of eggs by the host may be discordance and/or template-based mechanisms. In the former hosts recognise eggs that constitute the minority in a clutch as alien. In the latter hosts recognise eggs as alien when they do not match a template that can be innate or learnt. However, if a host species is polymorphic as regards egg colour (e.g. parrotbills *Paradoxornis*, hosts of Common Cuckoo) a host species male may learn one colour on one breeding attempt. If he subsequently mates with females laying differently coloured eggs, this could lead to him rejecting his own eggs (Liang *et al.* in press).

Many adult parasitic cuckoos feed on noxious caterpillars, which are perhaps inedible to their young. Finding substitute parents that would provide a more acceptable diet for their young could be a driving force in the evolution of brood-parasitism (e.g. Brooke & Birkhead 1991). Some adult *Chrysococcyx* cuckoos that sometimes utilise granivores as hosts have been observed feeding fledglings of their own species, e.g. Klaas's Cuckoo *Chrysococcyx klaas* (Nickalls 1983). Whether this is normal behaviour, compensating a previously inadequate diet, or it has some other explanation is unclear as the phenomenon is little-known and not well-studied. This behaviour is not confined to this genus but is reported elsewhere, e.g. in *Cacomantis*.

Female Brown Shrikes parasitised by the Indian Cuckoo *Cuculus micropterus* "fed their cuckoo chicks less willingly than did the males, and sometimes not at all. The young cuckoos soon learned to recognise the males and responded only to their calls" (Johnsgard 1997). This indicates a greater discriminatory ability in the female shrike than in her partner.

In Africa Levaillant's Cuckoos chiefly parasitise babblers, while the closely-related Jacobin Cuckoo primarily use bulbuls. In India, where Levaillant's Cuckoo is absent, Jacobin Cuckoos chiefly utilise babblers (Lack 1971). Jacobin Cuckoos have not been known to parasitise bulbuls in India, although they are very common there, and bulbuls are frequently parasitised in Africa (Becking 1981). This could imply that Indian bulbuls were used in the past, and have evolved strong host defence mechanisms, or that they are suboptimal hosts, and only used in Africa because of competition with Levaillant's Cuckoo. Experimental egg-recognition work comparing Indian and African bulbul species would help resolve this question. A particular advantage for Jacobin Cuckoos on the Indian subcontinent of parasitising *Turdoides* babblers is their co-operative breeding system, which involves assistance by nest-helpers in feeding the young. Cuckoos parasitising these species will gain additional benefit owing to their rapid rate of development compared to that of the host (Roberts 1991). A Fan-tailed Cuckoo in a nest of White-browed Scrubwrens was fed by four scrubwrens, and an adult female cuckoo was observed in close proximity to the nest. After fledging the cuckoo was seen being fed five times by the adult cuckoo and 14 times by the scrubwrens (Ambrose 1987).

The Indian, Fan-tailed and *Clamator* cuckoos, amongst others, are known to operate in pairs when laying, the male distracting the potential host so the female can approach the nest (Linton 1930, Neufeldt 1966). Whether parasitic cuckoos always lay directly into a host nest or not is highly controversial. Two observations have been made that suggest the second scenario may sometimes occur. The first concerns the placing of an egg into a domed nest. "The Fan-tailed Cuckoo deposits its egg in many different species' nests, all domed. One observer had seen how the male played the decoy and the female came to the nest with an egg in one of her feet, and after placing her egg in the nest flew away with another egg in her bill" (Linton 1930). The second concerns the placing of an egg in a nest with a side entrance. Netschajew (1977) describes how the Lesser Cuckoo often lays its egg on the ground and then takes it in its bill and places it in the nest of the most common host, the Japanese Bush Warbler. He also records nests and eggs of this host being destroyed and then deserted, presumably as a result of the cuckoo attempting to lay its eggs.

It is particularly difficult to discover the eggs of brood parasites – females collected with an egg ready to lay in the oviduct is the best proof but rarely obtained nowadays, and it is also difficult to make direct observations of female cuckoos laying (so that one can look into the nest immediately afterwards and see the cuckoo egg, but having also checked the nest beforehand to see if there were any host eggs, as the two may be very similar). Of course finding nests of nesting cuckoos is also often difficult because they are hidden, or high up in trees, etc.

Those species that construct their own nests may produce nests that are rather flimsy platforms through to quite substantial, in some cases globular, constructions, either on or close to the ground, or at varying heights in bushes or trees.

All known cuckoo nestlings are nidicolous – helpless, often with closed eyes at hatching, dependent on parents or fosterers for food, and mostly they remain in the nest while they grow. Natal down is completely or mostly absent in some species, and where it occurs it is generally not of the soft form of many birds, but as stiff, hair-like 'trichoptiles' that are particularly prominent in the coucals, but also found in some New World genera such as *Guira, Geococcyx* and *Coccyzus*, and in the Shining Bronze Cuckoo of Australia. Shelford (1900) shows that these thread-like structures are not down, but "abnormally elongated apexes of horny sheaths enveloping growing feathers". Nestlings of most brood-parasitic cuckoos are completely unfeathered (some *Chrysococcyx*, e.g. *lucidus, minutillus* and *russatus*, have small amounts feathering) perhaps improving tactile detection of host eggs or nestlings than need to be ejected from the nest (Miller 1924). The nestlings of Crotophaginae, Centropodinae, Couinae, malkohas, and *Coccyzus* have species-specific raised, contrasting, papillae and bars on the inside of the mouth, which are perhaps important to the parents for recognition.

Andersson (1995) suggests that coucals have a problem with egg-laying, with intervals of three or more days between eggs being reported. Coucals lay large eggs compared with their relatives, and these demand more resources. Further, he suggests that this has resulted in males being responsible for provisioning offspring and a sex reversal in size dimorphism. This may have led to polyandry in at least one species (African Black Coucal). Perhaps this habit is the precursor for parasitic laying.

Habitat

Cuckoos occur in wide a range of habitats from open savanna to dense tropical forests, and from sea level to 4,500m. The most highly specialised open country species are the roadrunners *Geococcyx*, year-round residents of lowland deserts, brushy woodlands and thickets in the USA and Mexico. The Greater Roadrunner is adapted to desert conditions in a number of ways (see under Behaviour). Other open country species include several of the Australian and southern African *Chryscococcyx, Clamator* species, some couas, and many other species in some parts of their geographic range, such as several of the *Cuculus* species in high latitudes or high altitudes, where savanna, alpine, farmland and moorland habitats are used, often relating to habitat choices of potential host species. Being freed from the necessity of building their own nests, many parasitic cuckoos have rather broad habitat tolerances. Food availability and breeding opportunities in high latitudes and in lower productivity open habitats tend to be seasonal, and thus many open country species are partial or complete migrants, and the degree to which seasonally-driven movements occur can vary from region to region. For example, forest-dwelling populations of African Black Cuckoos across equatorial Africa are largely sedentary, while those occupying more open habitats (savanna, *Acacia* thornveld, suburban gardens) in southern Africa migrate north during the non-breeding season.

Several cuckoo species are well adapted to urban habitats. Channel-billed Cuckoos *Scythrops novaehollandiae* and Common Koels are common summer visitors to suburban eastern Australia, where common urban-adapted species are parsitised (crows in the case of Channel-billed Cuckoos, and usually friarbirds or orioles in the case of Common Koels in Australia). Suburban gardens in southern Africa are commonly used by, among others, Klaas's, Dideric, Black *Cuculus clamosus* and Red-chested Cuckoos.

Despite this rather wide range of habitats used by cuckoos, forests and particularly tropical forests, house by far the greatest variety of species. The forest-dwelling nature and arboreal habits of most cuckoos makes them difficult to observe. Although most species are highly vocal, even to the point of monotony, they often sit still high in trees and obtaining good views can be infuriatingly difficult. Despite this, individual forest

Banded Ground Cuckoo *Neomorphus radiolosus* inhabits dense tropical forests, making them difficult to observe. Bilsa Reserve, Ecuador, October (*Murray Cooper*).

patches in south-east Asia can support 20 or more resident cuckoo species, often a combination of parasitic and non-parasitic species and with a range of dietary and microhabitat requirements. The extent to which species can tolerate deforestation leads to most of the conservation issues faced by cuckoos. Globally, about half of forest has been cleared (Boakes *et al.* 2010), and cuckoo populations, like those of most other forest-dwelling birds, must have dwindled substantially as a result of this loss.

Movements

Many species are more or less sedentary, a few perhaps are nomadic, some make local, usually altitudinal movements, whereas some are long-distance migrants. The Long-tailed Koel *Urodynamis taitensis* undertakes what is surely one of the most incredible, and is arguably the longest over-water migrations of any land bird, leaving breeding grounds in New Zealand and spreading out across hundreds of islands across the western Pacific as far east as Tuamotu, regularly making over-water flights of up to 3,000km.

Most long-distance migrants move at night using innate knowledge of stellar patterns for navigation. The Channel-billed Cuckoo is an exception, migrating by day. Within Africa some species undergo long migrations within the continent, moving from one hemisphere to the other. Elucidating details of such movements is not always straightforward, as some move into areas inhabited by resident individuals of the same species, and not all migratory populations are entirely so, with some individuals apparently sedentary. Movements in response to rain are found in a number of African species, whereby they may appear in an area around the onset of seasonal rains to breed. This has given rise to the vernacular names 'rain-bird' and 'storm-bird', applied to several African species, but also to several other cuckoos around the world (e.g. Common Koels that migrate to eastern Australia in the rainy summer season)

Long-distance post-breeding dispersal occurs in some species, with individuals appearing in localities far outside their normal range. Incredibly, a vagrant Common Cuckoo turned up in Barbados in November 1958. Yellow-billed and Black-billed Cuckoos *Coccyzus erythropthalmus* are highly sought-after transatlantic autumn vagrants in the other direction to Europe. Mangrove Cuckoos *C. minor* and Smooth-billed Anis *Crotophaga ani* have been recorded as far north as Quebec. Vagrant Long-tailed Koels have reached Palau, requiring a flight of nearly 6,000km over open ocean.

Conservation

Two cuckoos are known to have become extinct in historical times, both occurring on islands that have undergone complete deforestation. The Snail-eating Coua *Coua delalandei* is known with certainty only from 13 specimens collected on Île de Sainte-Marie, off the north-east coast of Madagascar. Early reports that it also occurred on the mainland of Madagascar may be erroneous, and given the total clearance of forest on Île de Sainte-Marie and no definite records since 1834, the species is now considered extinct (Collar *et al.* 1994). The other extinct species is the St Helena Cuckoo *Nannococcyx psix*, a very small forest cuckoo described by Olson (1975) from a fragment of a humerus collected in 1970 at Prosperous Bay Valley on St Helena by A. Loveridge. Complete deforestation of the island after it was discovered by the Iberian navigator João da Nova in 1502 presumably led to the demise of the St Helena Cuckoo.

At present, nine (6.25%) of the 144 cuckoo species are listed as globally threatened by BirdLife International (2011), substantially lower than the 12% of all bird species currently threatened with extinction (BirdLife International 2011). Many species have rather large distributions and there are relatively few extreme specialists or single-island endemics. Most of the globally threatened cuckoos are forest-dwelling, with their conservation status reflecting rapid deforestation rates across the tropics. Two species are threatened with imminent extinction, being listed as Critically Endangered in the IUCN Red List of threatened species. Endemic to Mindoro island in the Philippines, Steere's Coucal *Centropus steerii* was considered fairly common in the first half of the 20th century but had disappeared from many formerly occupied sites by the latter part of the century and currently occupies only about four sites with fewer than 250 birds estimated to remain. Continuing deforestation and harvesting of non-timber forest products in Mindoro gravely threatens the future of this species. Discovered in 1878, there were no records of the Sumatran Ground Cuckoo *Carpococcyx viridis* after 1916 until its sensational rediscovery in 1997 in Bukit Barisan Selatan NP (Zetra *et al.* 2002). It remains extremely rare, with few known localities and continues to be seriously threatened by deforestation and potentially as bycatch in snares set for Galliformes (BirdLife International 2011). Though not on the brink of extinction, two species are listed as Endangered. The Banded Ground Cuckoo *Neomorphus radiolosus* is a lowland forest specialist known only from seven areas in Colombia and Ecuador. The Rufous-breasted Cuckoo *Hyetornis rufigularis*, once widespread across Hispaniola now occurs only in deciduous forest below 900m at two or three sites in the Dominican Republic, under pressure from deforestation and forest degradation. Five cuckoo species, again all forest specialists, are listed as Vulnerable (Red-faced Malkoha *Phaenicophaeus pyrrhocephalus*, Cocos Cuckoo *Coccyzus ferrugineus*, Short-toed Coucal *Centropus rectunguis*, Javan Coucal *C. nigrorufus*, and Green-billed Coucal *C. chlororhynchos*), and a further nine species are Near Threatened (Moustached Hawk-cuckoo *Hierococcyx vagans*, Black-bellied Malkoha *Rhopodytes diardi*, Chestnut-bellied Malkoha *R. sumatranus*, Bornean Ground Cuckoo *Carpococcyx radiceus*, Verreaux's Coua *Coua verreauxi*, Violaceous Coucal *Centropus violaceus*, Biak Coucal *C. chalybeus*, Rufous Coucal *C. unirufus*, Scaled Ground Cuckoo *Neomorphus squamiger*). The subspecies *heinrichi* of Brush Cuckoo, sometimes split as Moluccan Cuckoo, has rarely been observed and is presently listed as Near Threatened.

The Common Cuckoo has declined in the United Kingdom by 65% since the 1980s, and the species has recently joined the Red List of Birds of Conservation Concern in that country (Eaton *et al.* 2009). Presumably as a result of climate change the Common Cuckoo has advanced its arrival date in Europe by 0.13 days per year, thus becoming increasingly miss-timed with its short-distance migratory hosts which are advancing on average by 0.38 days per year and long distance migrants by 0.16 days per year. These findings suggest that the *gentes* that rely on short-distance migrants such as Meadow Pipits, Winter Wrens and European Robins may decrease radically in abundance and ultimately become extinct (Møller 2011). However, a formal analysis suggests the primary driver of decline is probably not associated with changes in timing of host breeding (Douglas *et al.* 2010). Other suggested causes include a reduction in availability of hairy caterpillars in the UK, or deteriorating conditions on the non-breeding grounds in Africa. Declines are widespread across Europe (BirdLife International 2004), and although the species is not presently considered at elevated risk of extinction globally, this case highlights the rapid declines that can beset common species.

The effects of cuckoo parasitism on the conservation of their hosts should also be considered. There seems to be little evidence that such brood-parasitism detrimentally affects any species. The Helmeted Honeyeater is said to be in danger of extinction because of parasitism by the Pallid Cuckoo (Cooper 1967), but in general cuckoos seem to exert relatively little downward pressure on host populations. For example, despite 42–63% of all New Zealand Grey Gerygones being parasitised it seems not to depress their productivity because only late nests are affected, and few late young survive, even in the absence of parasitism (Gill 1983b). It is not clear why early nests are not parasitised.

African Emerald Cuckoo *Chrysococcyx cupreus* is more often heard than seen, but is one of the most spectacular of all cuckoos. Kongelai, Kenya, July (*Adam Kennedy*).

LAYOUT OF THE BOOK

All members of the family Cuculidae are included, each with a species account, map, and a variable number of illustrations to show different plumages, including markedly different subspecies. Photographs of almost all species are also included.

The Introduction comprises sections on Systematics, Morphology, Moult, Vocalisations, Behaviour and Feeding, Breeding, Habitat, Movements and Conservation. A Glossary is also included, concentrating on less well-known terms. The bulk of the book comprises the Species Accounts. A list of the scientific names of birds other than cuckoos mentioned in the text is also given. The Bibliography contains full references for all text citations.

In the species accounts section, there is an overview of each genus, followed by accounts of each species in that genus. The species accounts are subdivided as follows:

English name, scientific name, author, alternative English names.

Taxonomy: an overview of the taxonomic status and nomenclatural changes is presented. Includes synonyms, and often a discussion of contentious issues.

Field Identification: essential information on the species is given, beginning with non-morphological features such as flight and posture, followed by a brief description of the adult (including male, female, morphs and subadult where appropriate), juvenile and similar species.

Voice: vocalisations are important characters for identifying many cuckoos, which are often highly vocal. We have used transcriptions from different sources, and because not all humans interpret sounds in the same way, some of these interpretations may not be directly comparable.

Description: includes detailed information on the plumage of the adult (male, female, morphs, and subadults where appropriate), juvenile, nestling and bare parts.

Biometrics: comprises lengths of wing, bill, tail and tarsus, and mass, with number measured, mean and standard deviation, with sexes and ages differentiated where appropriate.

Moult: details of the timing of moult in adult and juvenile birds are given where known. The moult sequence of the primaries and tail feathers is descendent, i.e. numbering from innermost to outermost.

Geographical Variation: in polytypic species, differences from the form described above in 'Description' and 'Biometrics' are given, together with the distribution of these forms.

Distribution: breeding and non-breeding ranges are described in detail, usually working from north and west to south and east. Routes and timing of migration are given where relevant. The mapping of each species is based on a comprehensive collection of about 450,000 field records of cuckoos collated from museum collections, literature, and individual observers. Maps showing the distribution of individual records upon which these maps are based are available at http://www.fullerlab.org/cuckoos, together with contact details for obtaining access to the underlying data.

Key to the distribution:

Species present year-round, though not necessarily all individuals are resident. In some extreme cases, one population might leave an area seasonally, and be entirely replaced by another. Such cases are explained fully in the text.

Species present only during the breeding season.

Species present only during the non-breeding season.

Species present only on migration.

Scarce in any of the above categories.

Presence depends on local conditions, used primarily for African species where distributions are often rainfall-dependent.

X   Historical records only, not believed to be still present.

?   Areas of uncertain occurrence.

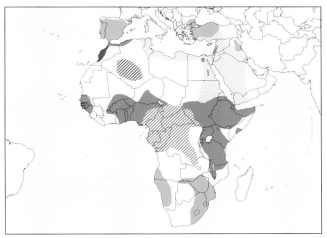
Sample distribution map.

Habitat: details of preferred habitat and altitudinal range are given. For migrants on passage, the habitats they use at stop-over sites are often not referred to because they are often atypical.

Behaviour: includes sociality, territoriality, courtship, parasitic and feeding behaviour.

Breeding: includes parasitism (hosts and localities); season (dates and localities); description of nests for non-parasitic species; eggs (number, description, measurements); incubation; chicks (development, behaviour); survival (including breeding success).

Food: includes information from unpublished sources and literature, and stomach contents from labels of museum study skins.

Status and Conservation: details the status or abundance of the species throughout its range; older information is given to indicate changes over time. Conservation projects, national parks and other protected areas important for the species are mentioned.

  BirdLife International global threat status is also given, as follows:

    Critically Endangered: species considered to be facing an extremely high risk of extinction in the wild.
    Endangered: species considered to be facing a very high risk of extinction in the wild.
    Vulnerable: species considered to be facing a high risk of extinction in the wild.

  Species considered to be at less risk are designated as follows:

    Near Threatened: species close to qualifying, or likely to qualify for a threatened category in
      the near future.
    Data Deficient: species for which there is inadequate information to make an assessment
      of its risk of extinction at the present time. This is not a category of threat but an acknowledgement
      that future research may show that threatened classification is appropriate.

  For further information, see the BirdLife website: www.birdlife.org

References are given at the end of each section or subsection. With important or contentious facts the reference is given within the text. In 'Status and Conservation', the references follow each geographical area where appropriate.

GLOSSARY

allopreening: preening of one individual by another.

allopatric: occurring in mutually exclusive geographical areas.

allospecies: member of a superspecies, having geographically separate breeding distributions but thought to be reproductively isolated; may have been treated previously as subspecies.

alluvial forest: forest associated with rivers; floods at times.

brachystegia: type of woodland in tropical Africa, also called miombo; named after a dominant tree genus in the family Caesalpinioideae.

cholla cactus: *Cylindropuntia* spp.; found in arid areas of the Americas.

clade: group of taxa that includes all descendants (i.e. a monophyletic group).

creosote bush: *Larrea tridentata* (family Zygophyllaceae); a bush in arid areas of the Americas.

dichromatism: where there are two genetically-determined basic colour or plumage patterns within an interbreeding population; usually associated with sex differences.

dimorphism: where there are two genetically-determined different forms within an interbreeding population; usually associated with sex differences.

dipterocarp: members of a family of trees (Dipterocarpaceae) dominant in many forests of south-east Asia.

ENSO: El Niño-Southern Oscillation; a worldwide periodic climatic event.

faecal sac: container of faeces and excretory matter usually produced by nestlings.

gens (**plural** *gentes*): an apparently genetic lineage that lays eggs of a particular coloration.

hallux: first, or inner, toe (normally points backwards).

holotype: a single specimen designated as representative of a species or subspecies.

huisache: *Vachellia farnesiana* (Fabaceae); a thorny shrub or small tree native to the New World.

kerangas forest: type of well-drained sandy/heath forest in south-east Asia, with dominant trees up to 25m.

mallee woodland: form of woodland in Australia typified by *Eucalyptus* species that have multiple stems.

mesa: flat, table-like hill.

mesic habitat: one neither overly wet or overly dry.

mesquite: leguminous deciduous trees (*Prosopis* spp.) found in dry regions of the New World, particularly Mexico and USA.

miombo: see brachystegia.

monotypic: genus (or higher rank) which only contains one species, or a species not divided into subspecies.

mopane: form of tropical African woodland characterised by *Colophospermum mopane*.

nominate (nominotypical): taxon that contains the type of a higher taxon, e.g. a subspecies that has the subspecific and specific names identical, such as *Cuculus canorus canorus*.

paloverde-saguaro: vegetation dominated by shrubs and small trees of the genus *Parkinsonia* (Fabaceae) and *Carnegiea gigantea* cacti.

polyandrous: breeding system where one female mates with more than one male.

polygynous: breeding system where one male mates with more than one female.

polymorphism: where there are two or more distinct genetically-determined forms, differing in colour, size or form, in an interbreeding population.

polytypic: taxon consisting of two or more immediately subordinate taxa, e.g. a species divided into subspecies, or a genus containing more than one species.

remiges (singular remex): flight feathers (primaries counted from innermost P1, P2, etc. and secondaries counted from outermost S1, S2 etc.).

Glossary

rectrices (singular rectrix): tail feathers (counted from middle outwards T1, T2, etc.).

riparian forest: forest along a natural watercourse.

river-bluff: cliffs along a river.

sclerophyll forest: form of vegetation with hard or leathery evergreen leaves; found particularly in Australia and other areas with a 'Mediterranean' climate.

sensu lato: in the broad sense.

sensu stricto: in the narrow sense.

superspecies: two or more closely related allopatric species.

sympatric: occurring in same geographical area.

synonym: in taxonomy, a scientific name no longer valid.

terra firme: upland forest (Neotropics) which never floods.

transilient moult: where moult occurs by jumps back and forth between the primaries or secondaries rather than ascendent (proceeding from outermost to innermost), or descendent (innermost to outermost).

trichoptile: form of down which is stiff and hair-like.

várzea: seasonally flooded forest (Neotropics).

yungas: region in east Andean slopes of Peru and Bolivia with humid evergreen transitional forest between the highlands and the lowland forest.

Greater Coucal *Centropus sinensis*. Adult, Gujerat, India, February (*Ram Mallya*).

ABBREVIATIONS

AMNH = American Museum of Natural History, New York
ANSP = Academy of Natural Sciences of Philadelphia
AOU = American Ornithologists' Union
BCSTB = Bird Conservation Society of Thailand Bulletin
BMNH = Natural History Museum, Tring, U.K.
CAR = Central African Republic
DRC = Democratic Republic of Congo (Kinshasha)
DMNH = Delaware Museum of Natural History
FMNH = Field Museum of Natural History, Chicago
FR = Forest Reserve
GR = Game Reserve
I., Is = Island, Islands
IC = Ivory Coast
IUCN = International Union for Conservation of Nature
LACMNH = Los Angeles County Museum of Natural History
MCZ = Museum of Comparative Zoology, Harvard
Mt., Mts = Mountain(s)
MVZ = Museum of Vertebrate Zoology, Berkeley
MZUSP = Museu de Zoologia da Universidade de São Paulo
NG = New Guinea
NMMZ = National Museums and Monuments of Zimbabwe
NP = National Park
NT = Northern Territory (Australia)
NZ = New Zealand
NSW = New South Wales
PNG = Papua New Guinea
R. = River
RC = Republic of Congo (Brazzaville)
ROM = Royal Ontario Museum
RSA = Republic of South Africa
SA = South Australia
SINGAV = Singapore Avifauna
SL = Sierra Leone
SUARENG = Suara Enggang
UAE = United Arab Emirates
UMMZ = Museum of Zoology, University of Michigan
USNM = United States National Museum
WA = Western Australia
WP = West Papua/Irian Jaya
WS = Wildlife sanctuary
ZMB = Zoologisches Museum, Berlin
ZMUA = Zoological Museum, Amsterdam

Yellow-billed Malkoha *Rhamphococcyx calyorhynchus*. Sulawesi, Indonesia (*Ingo Waschkies*).

PLATE 1: ANIS, AND STRIPED AND GUIRA CUCKOOS

Greater Ani *Crotophaga major*   Map and text page 105
Panama and South America. Wetlands, swamps, reed beds, mangroves, gallery forest, flooded forest, margins of oxbow lakes; to 800m, locally to 2,600m.
42–48cm. Larger than other anis, long-tailed, all black.
Adult Pale eyes, hump on bill, violet gloss.
Juvenile Dark eyes, no hump on bill, unglossed.

Groove-billed Ani *Crotophaga sulcirostris*   Map and text page 112
Central and south Texas, Middle America, west of Andes south to north Chile, Colombia, Venezuela. Open habitats such as savanna, forest edge, pastures, thickets. Mostly lowlands, in some regions mountains.
28–32cm. Black; dark eyes; smallest ani; similar to Smooth-billed; voice differs.
Adult Black; head and nape with glossy bluish scaling; curved grooves on bill sometimes difficult to see.
Juvenile Unglossed, no bill grooves.

Smooth-billed Ani *Crotophaga ani*   Map and text page 108
Florida, West Indies, Costa Rica, Panama, South America to central Argentina. Non-forest habitats, usually more humid than Groove-billed in areas of overlap in north-west South America; lowlands, locally to 2,800m.
33–35cm. Black; slightly larger than Groove-billed Ani; dark eyes; deep, laterally compressed bill. Voice distinguishes from Groove-billed.
Adult Hump on upper mandible, highest in males, notch between bill and forehead. Faint head and body scaling.
Juvenile Bill less deep.

American Striped Cuckoo *Tapera naevia*   Map and text page 115
South Mexico to central Argentina. Undergrowth, scrub, thickets, clearings, mangroves, etc.; mostly lowlands to *c.* 1,500m, sometimes higher.
27–30cm. Streaked above; ragged crest; pale eyebrow; often shy but vocal; large black alulae sometimes shown; tail narrow.
Adult Crest rufous streaked dark; upperparts boldly streaked black and buff.
Juvenile Upperparts spotted cinnamon.

Guira Cuckoo *Guira guira*   Map and text page 101
South and east of Amazonia south to central Argentina. Usually dry and open habitats such as savanna; also orchards, gardens, urban parks, etc. Mostly below 1,200m.
36–42cm. Shaggy reddish crest; dark wings; pale rump; pale tail base and tips; pale underparts; often vocal.
Adult Bill and iris yellow to orange; facial skin yellow; throat and chest streaked.
Juvenile Bill greyish or black-tipped; iris grey.

PLATE 2: MISCELLANEOUS TERRESTRIAL CUCKOOS

Pavonine Cuckoo *Dromococcyx pavoninus*   Map and text page 121

South America east of Andes, southward to north-east Argentina. Forest, sometimes with *Araucaria* or bamboo, often in dense understorey or vines; to 1,600m, but usually below 900m.

27–30.5cm. Crested; dark above; tail graduated and broad; uppertail-coverts long; lower underparts white. Also told from Pheasant-cuckoo by voice.

Adult Crown chestnut; narrow pale scaling on upperparts; sides of neck, throat and chest uniformly rich buff; small white tail tips.

Juvenile Crown and upperparts dark brown; throat and breast dull; no white tail tips.

Pheasant-cuckoo *Dromococcyx phasianellus*   Map and text page 118

South Mexico to north-east Argentina. A variety of habitats such as thickets, second growth forest, gallery forest, forest borders, seasonally flooded Amazonian forest; to 1,300m, but usually considerably lower.

33–41cm. Larger than Pavonine. Small head, crested, thin neck, dark above, uppertail-coverts long with white terminal spot (like Pavonine); long pale postocular stripe; cheeks dark brown; tail long, graduated and very broad; underparts mostly (buffy) white.

Adult Crown rusty-brown; pale edging on wing-coverts; spotted throat and breast.

Juvenile Sooty-brown crown with small buff spots; upperparts darker than adult, spotted buff; throat and chest unspotted.

Lesser Ground Cuckoo *Morococcyx erythropygus*   Map and text page 124

South-west Mexico to Costa Rica. Arid lowland scrub; edge of tropical deciduous forest and mangroves; to 1,500m.

25–28cm. Mostly terrestrial; difficult to see; often heard. Brown above, cinnamon to rufous below.

Adult Bare colourful skin around eye, surrounded by black; rufous below; white tail tips.

Juvenile Duller; bare orbital skin grey; outer rectrices indistinctly tipped buff.

Greater Roadrunner *Geococcyx californianus*   Map and text page 126

South-west USA and north Mexico. Scrub desert, mesquite, chaparral, etc.; to 2,500m.

52–58cm. Long legs, long, heavy, hooked bill, erectile dark crest.

Adult Streaked brownish and whitish with uniform whitish lower underparts; wings and uppertail glossy; orange, white and blue bare skin behind eye; white spots across outer primaries seen in flight.

Juvenile Duller, no gloss; rectrices more narrow and pointed, with wedge-shaped white tips.

Lesser Roadrunner *Geococcyx velox*   Map and text page 131

Mexico from south Sonora southward to central Nicaragua; north-west Yucatán Peninsula. Pine-oak forest, thorn scrub and other open dry habitats; to 2,800m but not down to sea level where forest occurs.

40–50cm. Smaller and more slender than Greater Roadrunner, and bill smaller.

Adult Throat and underparts buff, undertail-coverts tan; bare skin behind eyes red.

Juvenile Less sharply marked than adult, streaks and edgings of upperparts mostly dull buffy, not white.

PLATE 3: NEOTROPICAL GROUND CUCKOOS

Rufous-vented Ground Cuckoo *Neomorphus geoffroyi* Map and text page 138
Patchily from Nicaragua to Bolivia and south-east Brazil. Forest. Lowlands and foothills, in some regions to 1,600m.
45–51cm. Six subspecies (*aequatorialis* not illustrated; most similar to *salvini*). Black crest; black pectoral band; rufous belly; orbital skin bare, blue in most subspecies; runs swiftly on ground; often forages at army ant swarms, or with peccaries or primates; snaps bill.
Adult (*geoffroyi*; south Amazonian Brazil and presumably east Santa Cruz, Bolivia) Forehead and crown brown barred black; back and wings mostly glossy green; tail above glossed mauve; throat and breast heavily scaled; lower breast to belly greyish or whitish-buff.
Juvenile (*geoffroyi*) Mostly black; upperparts glossy; bill dark.
Adult (*australis*; south-east Peru, possibly including Huánuco and Pasco?, north-west Bolivia) Median portion of throat and breast down to breast-band unscaled; pectoral band narrower than in *geoffroyi*.
Adult (*salvini*; Nicaragua to north-west Colombia) Forehead and crown uniform cinnamon-brown (not barred); orbital skin slaty, becoming blue behind eye; upperparts bronzy brown; breast scaling faint.
Adult (*dulcis*; Espírito Santo and east Minas Gerais, south-east Brazil) Upperparts dark blue, breast scaled. Vent darker than *maximiliani*.
Adult (*maximiliani*; east Bahia, Brazil) Upperparts bronze-green, breast scaled.

Scaled Ground Cuckoo *Neomorphus squamiger* Map and text page 142
Small area south of Amazon River, Brazil. Lowland rainforest.
43cm. Similar to *N. g. geoffroyi*.
Adult Indistinct or lacking pectoral band; forehead and crown feathers brownish with buffy margins.
Juvenile Poorly known. Brown plumage; probably blue patch behind eye.

Banded Ground Cuckoo *Neomorphus radiolosus* Map and text page 133
South-west Colombia and north-west Ecuador. Wet foothill forest; to 500m.
46–51cm. Shape like other *Neomorphus* species.
Adult Black with bold pale scaling; wings and lower back chestnut; crown, crest and hindneck glossy black, forecrown scaled; orbital skin blue; tail black glossed dark purplish.
Juvenile Body scaling ochraceous. Forecrown not scaled.

Rufous-winged Ground Cuckoo *Neomorphus rufipennis* Map and text page 135
South Venezuela, west Guyana, north Brazil. Foothill rainforest; 100–1,100m.
43–48cm.
Adult Dark upperparts, head and breast; rufous wings; ashy-brown belly; bare skin around eyes pinkish-red; bill black tipped pale.
Juvenile Crown black; back and wing-coverts slaty-brown; rump black; underparts slaty; eye-ring dull red.

Red-billed Ground Cuckoo *Neomorphus pucheranii* Map and text page 210
Upper Amazonia in parts of Colombia, Ecuador, Peru and Brazil. Upland rainforest; 200–700m.
43–51cm. Shape like other *Neomorphus* species.
Adult Bill and bare skin around eye bright red; foreneck ash-grey; crown and crest glossy blue-black.
Juvenile Head and breast blackish; upperparts and belly brown; bill and feet black.

PLATE 4: COUCALS I

Buff-headed Coucal *Centropus milo* Map and text page 143
Solomons. Lowland and hill primary, and secondary, forest.
62–68cm. Large coucal; long, broad, graduated tail.
Adult (*milo*, south Solomons) Whole foreparts pale buffish, remaining plumage black.
Juvenile (*milo*) Buff and brown mottled and barred black.
Adult (*albidiventris*; west-central Solomons) Whole underparts, rump and lower back pale buffish.

Goliath Coucal *Centropus goliath* Map and text page 184
North Moluccas. Forest and forest edge, scrub, agricultural land; to 800m.
64–70cm. Largest cuckoo; long-tailed.
Adult (black morph; whole range) Black with conspicuous white patch in wings and stout black bill; some have few white patches on underparts.
Adult (whitish morph; Halmahera) Much more white on underparts.
Juvenile Variable. Dark brown, with variable whitish streaking and spotting on head, mantle and chin to upper breast; wings black with white streaking on coverts; breast to belly blackish barred whitish, or on Halmahera, irregularly blotched whitish, grey and blackish. Tail black glossed bluish.

Greater Black Coucal *Centropus menbeki* Map and text page 147
New Guinea. Forest and scrub; freshwater swamps; to 1,300m, but mostly below 800m.
60–69cm. Huge clumsy coucal with coarse contour feathering and 'hackles' on foreparts; pale stout bill.
Adult Black glossed blue-green above, and chin to breast; rest of underparts dull brown-black.
Juvenile Dull black, tinged blue-violet on upperparts, mantle indistinctly tipped brown; rectrices black banded brown.

Lesser Black Coucal *Centropus bernsteini* Map and text page 200
New Guinea. Grass around forests; vegetation near rivers and lakes; to 460m, rarely 900m.
45–52cm. Medium-sized coucal with broad tail about same length as body.
Adult Black with dark eyes.
Juvenile Black barred buff and rufous; throat white.

PLATE 5:   COUCALS II

Violaceous Coucal *Centropus violaceus* Map and text page 199
New Britain and New Ireland. Forest to 1,370m.
62–70cm. Huge coucal with rather long, decurved bill.
Adult Entirely black with strong violet sheen, becoming blue-black when worn. Naked skin around eyes whitish to red.
Juvenile Dull black; wings and tail glossed purple; underparts sooty-grey.

Pied Coucal *Centropus ateralbus* Map and text page 145
Bismarck Archipelago. Forests and edge, gardens, plantations; to 1,300m.
42–46cm. Medium-sized coucal; plumage variable.
Adult Varies from mostly black with purplish-blue gloss on head, wings and tail, and white area on wings and throat, to entirely white or pearly-grey.
Subadult Black with white feathers on neck and wings.
Juvenile Deep brown, wings and tail glossed purple, foreparts above striped buffish or rufous.

Pheasant-coucal *Centropus phasianinus* Map and text page 202
North and east Australia; New Guinea; Kai Is; Timor-Leste. Usually avoids forest; dense vegetation, swamps, savanna, gardens; up to 1,800m.
54–68cm. Great variation in size. Sexes similar but female larger.
Adult (*phasianinus*; eastern Australia; breeding) Head, neck and upper mantle black, lower mantle rufous streaked pale, wings red-brown barred and streaked buff and white, uppertail-coverts blackish-brown barred white and black; tail blackish barred and mottled brown, central rectrices tipped white.
Adult (*phasianinus*; non-breeding) Head, neck and underparts buffish; rufous upperparts streaked pale yellow.
Juvenile (*phasianinus*) As non-breeding adult but paler brown.
Adult (*spilopterus*; Kai Is) Chiefly blackish; rarely primaries barred pale; tail unbarred. No non-breeding plumage.
Adult (*mui*; Timor-Leste) Underparts and foreparts white; dark brown crown.

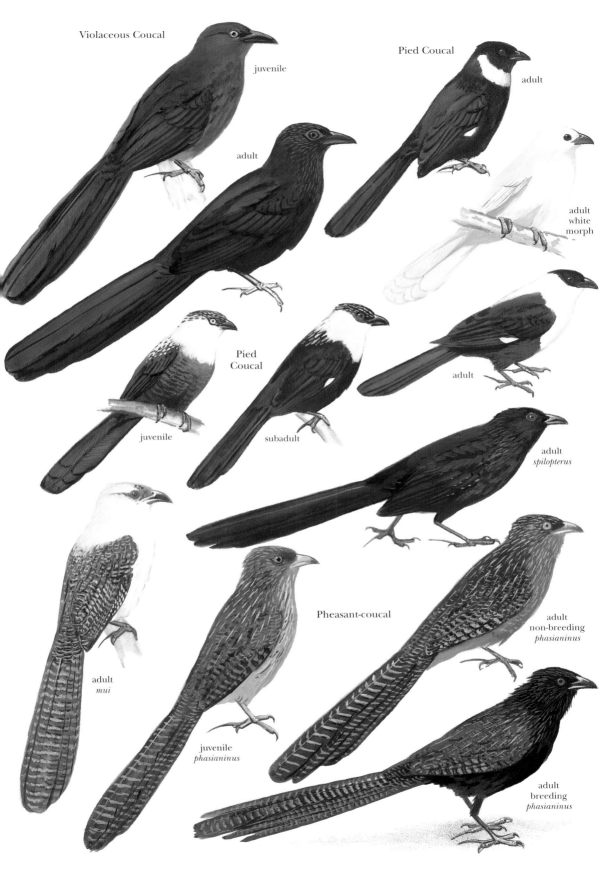

PLATE 6: COUCALS III

Short-toed Coucal *Centropus rectunguis*   Map and text page 158
Thai-Malay Peninsula; Sumatra; Borneo. Forest; to 600m, rarely 1,700m.
37–42cm. Shorter tail than other coucals.
Adult Black; rufous-chestnut mantle, back and wings.
Juvenile Pale chestnut-brown barred black above; dark brown barred whitish below; tail black barred white.

Biak Coucal *Centropus chalybeus*   Map and text page 149
Biak (Misori) and Supiori, Geelvink Bay, West Papua. Trees.
42–46cm. Medium-sized coucal with tail about length of body; yellow eyes.
Adult Black glossed purple above.
Juvenile Unbarred black washed rufous.

Greater Coucal *Centropus sinensis*   Map and text page 180
Indian subcontinent; south-east Asia; south China; Greater Sundas; Philippines. Areas with dense cover; to 2,200m.
48–53cm. Large coucal with stout decurved bill.
Adult (*sinensis*; northern Indian subcontinent to southern China) Black except for chestnut wings; wing-linings dark.
Juvenile (*sinensis*) Dark brown head and upperparts spotted and barred chestnut and white; wing-coverts chestnut-brown, remiges barred dark brown; underparts blackish barred off-white; graduated tail black narrowly barred whitish.
Adult (*parroti*; peninsular India and Sri Lanka) Forehead, face and throat dull brown. Mantle and back black, head and underparts black glossed blue to blue-green; juvenile as adult but much darker chestnut upperparts and wings unbarred.
Adult (*kangeanensis*, pale morph; Kangean Is) Head, neck and underparts pale buffish; tail greyish; wings rufous.
Adult (*kangeanensis*, dark morph) Head, back, tail and throat brownish-grey, breast mottled grey.
Adult (*andamanensis*; Andaman, Coco and Little Table Is.) Variable. Head, neck, mantle and underparts pale buffish-grey, wings brown, tail white above, rufous below; or brown with dusky abdomen and bronzy purple tail; or greyish head, mantle and underparts, dark maroon-brown wings, tail bronzy-purple; or dark brownish-grey head, mantle and underparts, wings dark chestnut, tail blackish-brown.

Lesser Coucal *Centropus bengalensis*   Map and text page 195
Indian subcontinent; Burma and Indochina; Thai-Malay Peninsula; China; Greater Sundas; Philippines; Wallacea. Scrub, forest edge, long grass, gardens, plantations, paddifields, marshes; to 2,000m.
31–35cm. Small coucal with short, stout, decurved bill. Sexes similar, but female larger.
Adult (breeding) Glossy blue-black with pale shaft streaks; mantle dull chestnut; wings dull rufous-brown with conspicuous buffish shaft streaks on shoulder, tips darker; tail black glossed green.
Adult (non-breeding) Dark brown head and mantle with prominent buff shaft streaks; rump dark brown barred rufous; underparts pale buffish, throat and breast streaked pale, sides of breast and flanks barred darker; undertail-coverts barred dark brown and dull buffish.
Juvenile As non-breeding adult but dark and light streaking on head and upperparts; wings rufous barred dusky; underparts off-white, flanks and thighs barred pale brown; tail rufous barred black.

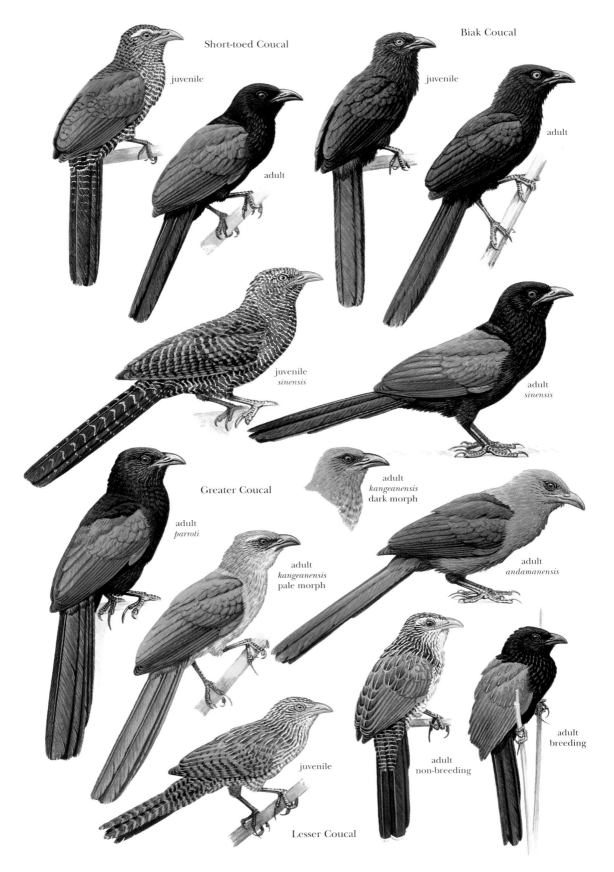

PLATE 7: COUCALS IV AND COUAS

Blue Coua *Coua caerulea* Map and text page 215
North and east Madagascar. Forests (including mangroves), secondary growth and dense clove plantations; to 1,800m.
48–50cm. Large blue coua; mostly arboreal.
Adult Blue with short crest; naked blue skin around eye.
Juvenile Mostly dark brown, black and dull blue; reduced bare area around eye.

Red-fronted Coua *Coua reynaudii* Map and text page 219
North and east Madagascar. Forest and edge; thick secondary growth, brushy areas and clearings; generally in dense vegetation; to 2,500m.
38–40cm. Medium-sized coua; long-tailed, short-legged, chiefly terrestrial.
Adult Above glossed olive-green; rufous crown, blackish face with bare skin around eye blue; green tail glossed blue and tipped black; dark grey below.
Juvenile Much more rufous than adult with coppery-green gloss; wing feathers with rufous tipping; no bare orbital skin.

Rufous Coucal *Centropus unirufus* Map and text page 150
Philippines (Catanduanes, Luzon and Polillo). Forest and bamboo thickets; to 1,200m.
38–42cm. Medium-sized coucal with tail longer than body, and stout decurved greenish bill tipped yellow.
Adult Plain rufous-chestnut, brighter below, and paler on chin.
Juvenile As adult but often with white trichoptiles on head.

Bay Coucal *Centropus celebensis* Map and text page 160
Sulawesi and offshore islands. Forest and edges, wooded savanna, cultivation, scrub, more rarely in mangroves; to 1,100m.
42–51cm. Medium-large coucal.
Adult (*celebensis*; north Sulawesi and Togian Is.) Rufous and pale grey.
Juvenile (*celebensis*) Paler than adult, especially on throat and breast.
Adult (*rufescens*; central and south Sulawesi; Butung, Labuan Blanda, and Muna Is.) More rufous than nominate.

Javan Coucal *Centropus nigrorufus* Map and text page 178
Java. Mangroves, coastal marshes, swamps, tall grass, flooded forest edges and sugarcane plantations.
46cm. Rather long-tailed coucal; short, strong bill.
Adult Glossy black, back strongly glossed purple, wings rufous tipped black on coverts; in flight wings rufous and black; juvenile similar.

Black-faced Coucal *Centropus melanops* Map and text page 154
Philippines (Biliran, Bohol, Leyte, Samar, Basilan, Dinagat, Mindanao, Nipa and Siargao). Forest and edge; to 1,200m.
42–48cm. Medium-sized coucal with stout decurved bill and red eyes.
Adult Black on sides of face, foreparts otherwise buffish-yellow, rest of upperparts black except for chestnut shoulders and wings; juvenile similar.

PLATE 8: COUCALS V

Steere's Coucal *Centropus steerii* Map and text page 156
Philippines (Mindoro). Forest and bamboo thickets; to 750m.
43–50cm. Medium-sized coucal, tail approximately as long as body, and strong, hooked bill.
Adult Head and throat black glossed green; rest of upperparts and wings blackish-brown glossed bronze-green; rest of underparts dark brown; tail blackish-green.
Juvenile Brown; chin and throat brownish-grey; tail dull black.

Philippine Coucal *Centropus viridis* Map and text page 192
Philippines. Grassland, forest, bamboo groves, secondary growth, scrub, cultivation; not in tree-tops; to 2,000m (much lower on some islands).
42–44cm. Medium-sized blackish coucal; rather short, decurved black bill; floppy flight. Sexes similar.
Adult Black glossed blue to green, wings chestnut tipped dark brown (obvious in flight). All black on Batan, Ivojos and Sabtang.
Juvenile Above dark brown with whitish shaft streaks from head to mantle; rump and uppertail-coverts dark brown tipped pale brown; wings chestnut barred dark brown; below blackish-brown marked buffish; tail brownish-black glossed greenish with some pale brown bars.

Green-billed Coucal *Centropus chlororhynchos* Map and text page 152
South-west Sri Lanka. Wet forest, bamboo, village gardens, abandoned swidden agriculture; to 1,450m.
43–46cm. Medium-sized, short-winged coucal with pale bill.
Adult Dull black glossed purplish on neck and tail; dark chestnut wings; and long pale greenish or ivory bill.
Juvenile As adult but barred blackish on chestnut wing-coverts; eyes grey.

Madagascar Coucal *Centropus toulou* Map and text page 186
Madagascar; Aldabra and formerly Assumption I.; Cosmoledo Atoll. Dense vegetation in forest, at edge, *Eucalyptus* woodlands, secondary growth, palm groves, mangroves, marshes, including *Phragmites* reedbeds, long grass, paddifields and gardens; to 1,800m
40–50cm. Medium-sized coucal. Sexes similar, but male smaller.
Adult (breeding) Black glossed blue; back and wings chestnut; tail black slightly glossed greeny-blue.
Adult (non-breeding) Head to upper back and breast dark brown finely streaked cream.
Juvenile (pale morph) As non-breeding adult, but more barred and speckled above with short shaft streaks; wings barred; belly slightly barred; tail barred. Intermediates occur.
Juvenile (dark morph) Darker above, mantle often more speckled and less barred cream and cinnamon-rufous; finer speckling on crown; wings and wing-coverts cinnamon-rufous to russet, barred brown; underparts darker than pale morph.

PLATE 9: COUCALS VI

African Black Coucal *Centropus grillii* Map and text page 189

Much of sub-Saharan Africa; migrant in some areas. Grassland, marshes, savannas, bracken briar, dry paddifields, forest clearings; occasionally reeds and papyrus; below 1,525m; occasionally to 2,000m.

Male 30cm, female 34cm. Rather small, skulking coucal that sometimes perches prominently. Females markedly larger, otherwise sexes similar.

Adult (breeding) Black with rufous-chestnut wings.

Adult (non-breeding) Above blackish-brown barred and streaked rufous, tawny, cream and rusty-buff; wings rufous; tail brownish-black; underparts tawny-buff to white streaked, spotted and barred cream and brown; undertail-coverts blackish barred buff.

Subadult Dark brown crown to mantle, streaked and spotted whitish and buff; back and rump dark brown streaked buff; uppertail-coverts dark brown barred buffish. Wings chestnut, inner secondaries buff barred brown. Mostly buff below, more cinnamon on breast, which is streaked blackish. Tail dark brown barred buff.

Juvenile Above mottled, barred and streaked buff, brown and cream; rufous remiges barred blackish-brown; brown rump and tail barred tawny-buff; underparts pale buff with cream streaks, belly and flanks barred and spotted.

Gabon Coucal *Centropus anselli* Map and text page 162

Cameroon, CAR and Gabon to RC and central DRC; north-west Angola. Forest and edge, secondary growth, old cultivation, grassy swamps; lowlands.

46–58cm. Large, but exceedingly difficult to see size.

Adult Black head, back and tail; wings chiefly dark chestnut; underparts tawny.

Juvenile Dark brown above with pale shaft streaks on head and upper back; wings rufous barred black; below rufous-buff, throat to upper breast barred dark brown.

Black-throated Coucal *Centropus leucogaster* Map and text page 163

West Africa, east to northern DRC. Dense undergrowth in forest, savanna, swamp edge, montane ravines; to 1,000m.

46–58cm. Largest African coucal.

Adult Black-glossed head to mantle and upper breast; rufous-chestnut wings; lower back to uppertail-coverts black barred buffy-white; lower breast to under tail-coverts white; tail black barred buff basally.

Juvenile Head and breast unglossed; streaked and spotted buff-white above; wings rufous barred blackish-brown; rump, upper tail-coverts and tail barred buff; throat and upper breast black spotted buff; rest of underparts unbarred fulvous to rufous-buff.

Coppery-tailed Coucal *Centropus cupreicaudus* Map and text page 172

Central and east Angola, north-east Namibia, north Botswana, Zambia, north and central Malawi, south-west Tanzania. Dense swamp and riparian vegetation, reeds, papyrus, long grass and bush; occasionally woodland; to 1,250m

42–50cm. Large coucal; deep bill.

Adult Black forehead to mantle glossed violet; chestnut wings; white below; blackish tail with coppery sheen.

Juvenile Less gloss on head and neck, head streaked buff, rest of upperparts mostly dark brown; some buff on breast; tail barred.

PLATE 10: COUCALS VII

White-browed Coucal *Centropus superciliosus*   Map and text page 174

Eastern Africa from Eritrea south to RSA; south-west Arabia; Sokotra. Long grass, reeds, thickets, scrub, forest edge, swamps, old cultivated fields, irrigation in arid country, bracken-briar; to 2,800m.

36–42cm. Medium-sized coucal.

Adult (*superciliosus*; east Sudan, Eritrea, Ethiopia and Somalia through Kenya and north-east Uganda to north-east Tanzania and north-east DRC) Blackish head streaked white, supercilium white; rufous-brown back; chestnut wings; creamy-white underparts; black tail tipped white.

Juvenile (*superciliosus*) Blackish crown and ear coverts streaked rufous or tawny-buff; buff-white supercilium; above black barred rufous and buff; remiges rufous-chestnut barred dark; greenish-black tail and uppertail-coverts barred buff; underparts buff-white streaked yellow, cream and black; flanks, legs and undertail-coverts buff-grey barred grey.

Adult (*loandae*; central and south Uganda, south-west Kenya and Tanzania through southern DRC, north Malawi and Zambia to Angola, north Botswana and north Zimbabwe) From *superciliosus* by unstreaked head with white supercilium.

Adult (*burchellii*; east Botswana, south Zimbabwe and south Malawi, south to RSA) Black head lacks streaking and supercilium. Intergrades occur.

Juvenile (*burchellii*) Darker head than juvenile *superciliosus*, and lacks supercilium.

Senegal Coucal *Centropus senegalensis*   Map and text page 165

Sub-Saharan Africa south to Namibia, Botswana and Zimbabwe; Egypt. Long grass and thickets in savanna, swamp, forest edges and clearings, sugarcane plantations, palm groves, rank vegetation along canals, edges of reedbeds and papyrus, dense riverine bush, gardens; mostly lowlands.

36–42cm. Medium-sized coucal.

Adult (*senegalensis*; white-bellied; west and east Africa south to north Angola and south-central DRC) Glossy black crown to nape, back and wings chestnut, underparts white.

Adult (*senegalensis*; rufous-bellied; west Africa) From white-bellied by deep rufous belly and flanks.

Juvenile (*senegalensis*) Crown brown or grey; rest of upperparts, including wings, barred chestnut and blackish; rump blackish barred buff; black tail faintly barred buff at tip; underparts buff/cinnamon streaked and barred dark.

Adult (*aegyptius*; Egypt) Larger than nominate; sootier crown and hindneck merges into rufous mantle; mantle, wing-coverts and inner secondaries darker, duller brown with almost no rufous.

Blue-headed Coucal *Centropus monachus*   Map and text page 170

West and east Africa from Eritrea south to DRC and north Angola. Dense cover in marshes, papyrus, river banks, long grass, forest edges, secondary growth, tea plantations and other cultivation; to 3,000m.

45–52cm. Medium-sized coucal.

Adult Black head to mantle glossed blue; rump black; wings mostly dark chestnut; whitish to pale buff below; tail black glossed bronze or greenish.

Juvenile Head and hindneck dull blackish streaked rufous-buff; dark chestnut wings barred blackish-brown; rump, uppertail-coverts and tail black barred buff; deeper buff below than adult, paler on belly; throat and breast spotted blackish.

PLATE 11: ASIAN GROUND CUCKOOS AND COUAS

Bornean Ground Cuckoo *Carpococcyx radiceus*   Map and text page 208
Borneo. Dense forest, usually in damp areas or near rivers, mangroves; extreme lowlands and low hills.
60cm. Large terrestrial forest cuckoo with strong feet, short rounded wings, broad rounded tail and slightly downcurved bill.
Adult Head and neck black glossed purple; green-blue naked skin around eyes; mantle, upper back and wings green washed purple; lower back rufous barred dark green; breast and sides of neck pale grey, rest of underparts off-white barred dusky; tail above glossy purplish-blue.
Juvenile From adult by mostly pale brown underparts.

Sumatran Ground Cuckoo *Carpococcyx viridis*   Map and text page 206
Sumatra. Hill forest with dense undergrowth; 300–1,400m.
55cm. Medium-large terrestrial cuckoo with rounded wings and tail; long strong legs and bill.
Adult Head, neck, upperparts, wings and tail glossy green; breast pale green and rest of underparts cinnamon-buff barred dusky.
Juvenile Brown head and upperparts; wings and tail green-black fringed red-brown.

Coral-billed Ground Cuckoo *Carpococcyx renauldi*   Map and text page 210
Thailand and Indochina. Broadleaved evergreen and semi-evergreen primary and secondary forest with dense ground cover; to 900m (occasionally 1,500m).
65–70cm. Pheasant-like cuckoo with long legs and stout red bill.
Adult Head, neck and upper breast glossy black; violet skin around eyes; upperparts and wings grey, primaries blackish; rounded blackish tail glossed violet; underparts whitish finely vermiculated.
Subadult From adult by russet forehead, crown nape and hindneck black glossed purple, lower back dusky mottled brown, rump white and orange-cinnamon barred dusky; skin around eye mauvish.
Juvenile Head dark brown, forehead rufous, grey skin around eyes; upperparts dark brown barred rufous on back; dark wings tinged purple/greenish and tipped dull rufous; underparts dull rufous-chestnut and whitish grey; lower flanks to undertail-coverts barred dark brown; tail black.

Crested Coua *Coua cristata*   Map and text page 212
Madagascar. Forest (prefers degraded), savanna, brush, palm groves, mangroves, sub-arid thorn scrub; to 900m.
40–44cm. Large, arboreal crested coua.
Adult (*cristata*; north and east Madagascar) Grey-green; long purple-blue tail tipped white; underparts grey, purplish to rufous and white.
Juvenile (*cristata*) Grey crown; back, uppertail- and wing-coverts edged rufous.
Adult (*pyropyga*; south-west Madagascar) Larger than *cristata*, and paler with broader white tail tip to tail and bright rufous-chestnut undertail and vent.

Verreaux's Coua *Coua verreauxi*   Map and text page 214
South and south-west Madagascar. Sub-arid thorn scrub; to 200m.
34–38cm. Small, crested, thin-billed arboreal coua.
Adult Green-grey above; white-tipped glossy blue tail; underparts grey and white; bare orbital skin blue with no black outline.
Juvenile Shorter crest than adult, and less white on tail.

PLATE 12: COUAS

Giant Coua *Coua gigas*   Map and text page 225

South and west Madagascar. Coastal lowlands, forest (not normally degraded or secondary), sub-arid thorn scrub, sub-desert; to 800m.

58–62cm. Largest coua; long-tailed, terrestrial.

Adult Above olive-grey to bronze-green; face black, blue and pink; below whitish, tan and rufous; black tail tipped white.

Juvenile Duller than adult; wings spotted; bare facial skin dull blue.

Red-breasted Coua *Coua serriana*   Map and text page 228

North-east Madagascar. Undisturbed rainforest; in more open areas than Red-fronted Coua; to 1,250m

42cm. Large, stocky, long-tailed, terrestrial coua.

Adult Above dark greenish-brown; black throat and dark rufous chest; rest of underparts brown; tail blue-black. Bare skin around eye sky-blue and ultramarine-blue.

Juvenile Head dark brown; rest of upperparts dark bronze spotted pale buff; wings edged buff; brownish-olive throat, breast dull chestnut, belly olive; dull blue around eye

Coquerel's Coua *Coua coquereli*   Map and text page 221

West Madagascar. Forest with sparse ground cover (not usually in degraded forest), edge of sub-desert, secondary growth; to 800m.

42cm. Long-tailed, slender terrestrial coua, also occurring in trees.

Adult Above olive-green; black, blue and pink face; white tipped black tail; below tan and rufous with blackish undertail coverts.

Juvenile Duller than adult; no black on face; below barred.

Running Coua *Coua cursor*   Map and text page 223

Southern Madagascar. Sub-arid thorn scrub, spiny desert, dry woodlands without ground cover, subdesert brush; secondary growth; to 200m.

34–40cm. Medium-sized coua; long-tailed, long-legged, terrestrial.

Adult Above grey-green above; facial blue and pink outlined by black feathers; purplish breast.

Juvenile Duller than adult; no black on face; bare skin around eye dull blue.

Red-capped Coua *Coua ruficeps*   Map and text page 217

West and south-west Madagascar. Dry deciduous forest and gallery forest (including degraded forest), sub-arid thorn scrub, secondary growth, woodland savanna, generally in more open areas of forests than other couas; to 850m.

42cm. Large, long-tailed and long-legged terrestrial coua; outermost rectrices much shorter than others.

Adult (*ruficeps*; west Madagascar) Rufous forehead and crown; indigo-blue bare facial skin rimmed with black feathering forming black band on nape; rest of upperparts greenish-brown; below white, tawny, purplish and rufous; dark tail edged whitish.

Subadult (*ruficeps*) Blackish crown; nape barred pale.

Juvenile (*ruficeps*) Paler; crown light brown; dark naked skin behind eye; upperparts and wings barred fawn.

Adult (*olivaceiceps*; south-west Madagascar) Forehead and crown greenish-brown; more contrast between breast and belly than in adult *ruficeps*.

Juvenile (*olivaceiceps*) More barring on upperparts, duller on crown and paler below than juvenile *ruficeps*.

PLATE 13: *PIAYA* CUCKOOS AND SNAIL-EATING COUA

Squirrel Cuckoo *Piaya cayana*   Map and text page 279

Mexico to Argentina. Various types of forest and woodland, humid to semi-arid and deciduous, second growth, plantations, etc.; to 2,500m.

43–46cm. Fourteen subspecies (six illustrated). Above rufous to chestnut; breast buff; belly grey; undertail-coverts grey to black; very long tail with broad white tips; bill yellow; eye-ring greenish-yellow west of Andes, red east of Andes; eyes red; runs and jumps along large branches; vocal.

Adult (*cayana*; east and south Venezuela, Guianas, north Brazil) Tail black below; under tail-coverts dark grey.

Juvenile (*cayana*) Eyes brown; eye-ring grey; bill greyish.

Adult (*circe*; south of Lake Maracaibo, Venezuela) Above slightly more rufous than *mehleri* and paler than *cayana*.

Adult (*macroura*; south-east Brazil, Paraguay, Uruguay, north-east Argentina) Upperparts chestnut; belly and undertail-coverts blackish; tail long.

Adult (*mesura*; Colombia east of Andes, east Ecuador) Belly and undertail-coverts blackish.

Adult (*obscura*; Amazonian Brazil south of Rio Solimões, from Rio Juruá to Rio Tapajós, and Peru from Junín south-east to north Bolivia) Darker, less rufous, above, than *cayana*, undertail-coverts black.

Adult (*mexicana*; West Mexico) Tail mostly brown below.

Black-bellied Cuckoo *Piaya melanogaster*   Map and text page 277

Amazonia and Guianas. Mostly canopy and subcanopy of primary upland forest; to 800m.

35–40.5cm.

Adult Rufous; grey cap; red bill; tail tipped white; blue around eye; yellow loral spot.

Juvenile Like adult; tail feathers narrower, with smaller tips.

Snail-eating Coua *Coua delalandei*   Map and text page 227

Previously Nosy Boraha, Île Sainte-Marie, Madagascar. Primary rainforest at sea level. Believed extinct.

56–57cm. Juvenile unknown.

Adult Black above with violet sheen, long violet-blue tail with white tips; chin to breast white, belly rufous.

See also Little Cuckoo, Plate 21

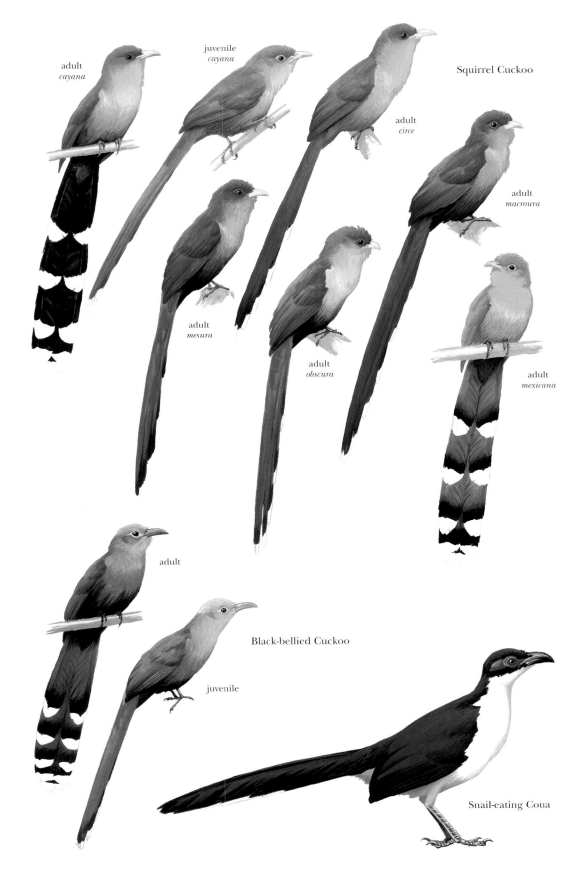

PLATE 14: MISCELLANEOUS ASIAN AND AUSTRALASIAN CUCKOOS

Raffles's Malkoha *Rhinortha chlorophaea* Map and text page 230
Thai-Malay Peninsula, Sumatra and Borneo. Forest (including mangrove and seccondary), wooded grassland, plantations, wooded gardens; lowlands and hills to 1,100m.

30–33cm. Smallest malkoha; pale green-blue bill and naked orbital skin; tail rather short.

Adult male Rufous, head and underparts paler; blackish tail broadly tipped white.

Adult female Head, neck, mantle, throat and breast pale grey; rectrices dark rufous.

Juvenile male Duller than adult; dark ends to remiges; smaller white tips to tail. (Juvenile female has less white on tail than adult.)

Long-billed Cuckoo *Rhamphomantis megarhynchus* Map and text page 382
New Guinea. Forest, including secondary and edges; scrub; lowlands.

18cm. Small, variable, dull, square-tailed, cuckoo; long, thin bill with decurved tip; perches upright; straight non-undulating flight.

Adult male Black head; red iris and eye-ring; dark brown upperparts, wings and tail; below greyish and dusky-buff or grey-brown.

Adult female More grey-brown on head and more cinnamon or red-brown above than male; dark brown iris, grey eye-ring.

Juvenile Variable. Broad circular area around eye and sometimes nape greyish-white, which may extend to streak behind eye; crown and ear-coverts darker; rest of upperparts brown to rufous-brown; tail sometimes barred. Below brownish or greyish and rufous; sometimes faint dusky vermiculations on abdomen and bars on undertail-coverts; underside of tail barred dark and tipped white.

Long-tailed Koel *Urodynamis taitensis* Map and text page 331
New Zealand and Pacific. Forest, scrub, coastal vegetation, plantations and other cultivation, gardens; lowlands.

40–42cm. Rather large, slender, falcon-like cuckoo with long tail; direct flight with unbroken wing-beats; often glides.

Adult Above brown heavily spotted, streaked and barred white, buff and rufous; white superciliary and malar stripes, broad dark brown eye-stripe; below white with long bold dark streaks, chevrons on sides of vent and undertail-coverts; tail dark brown tipped white and barred rufous.

Juvenile As adult but white-spotted upperparts; underparts rich buff with less dark streaking; tail shorter.

Channel-billed Cuckoo *Scythrops novaehollandiae* Map and text page 334
North and east Australia; New Guinea area; Wallacea; southern populations migrate northwards. Forest (including mangrove), woodland, savanna, scrub, gardens; to 1,100m.

60–70cm. Huge cuckoo with hornbill-like bill; in flight long pointed wings and long narrow tail; normally flies high, straight and hawk-like with slow wing beats.

Adult Head, neck and underparts pale grey; lores and skin around eyes red; upperparts, wings and tail pale grey-brown barred blackish; graduated tail with broad black subterminal band and tipped white. (Female has more barring on underparts.)

Juvenile Head and neck pale buffish; upperparts and wings grey barred and spotted buff; underparts buff finely barred on flanks and thighs; tail as adult.

PLATE 15: MALKOHAS I

Chattering Yellowbill *Ceuthmochares aereus*   Map and text page 235
West Africa to west Kenya; north-west Tanzania and extreme south Sudan. Mainly secondary and gallery forest; thick bush, dense secondary growth, arid thorn scrub; in vegetation about 8–30m above ground; to 2,000m.

33cm. Large; dark; long, graduated tail; yellow bill; red eye with yellow orbital skin. Skulks, and may be difficult see; flies with clattering wingbeats and long wavering glides.

Adult (*aereus*; east of River Niger) Usually appears blackish, although slaty-blue with bluish sheen above; paler below.
Juvenile (*aereus*) Duller than adult; sooty above; brownish buff throat; breast dark grey; bill brown; eye pale horn.
Adult (*flavirostris*; west of River Niger) Upperparts glossed violet; darker grey below than *aereus*.

Whistling Yellowbill *Ceuthmochares australis*   Map and text page 233
South-west Ethiopia and southern Somalia south to RSA. Secondary and gallery forest, particularly at edge; dense secondary growth, *Acacia* scrub, arid thorn bush, coastal scrub; to 2,000m.

33cm. Large; dark; long, graduated tail; yellow bill; red eye with blue orbital skin. Behaviour as Chattering Yellowbill.

Adult Plumage usually appears blackish in field, although dark slaty-blue with greenish and bluish sheen above; underparts paler. Crown and nape dark grey glossed green; lores greenish-grey. Back, wings and uppertail-coverts dark slate glossed greenish.
Juvenile Duller than adult; sooty above, with brownish-buff throat and breast dark grey; bill brown; eye pale horn.

Black-bellied Malkoha *Rhopodytes diardi*   Map and text page 247
Thai-Malay Peninsula; Sumatra; Borneo; ?Indochina. Forest (including mangrove and secondary) and edge; bamboo, plantations; mostly below 200m, but unusually to 2,000m.

36–38cm. Long tail approximately half total length; dark; large red area around eyes; dark grey belly; prominent yellow-green bill.

Adult (*diardi*; range of species except Borneo) Grey head, neck and upperparts; wings glossy blue-green; underparts grey, belly blackish-grey; tail tipped white below.
Juvenile (*diardi*) Duller than adult; bill darker and more slender.
Adult (*borneensis*; Borneo) Smaller than *diardi*; breast to belly rich buffy-grey.

Chestnut-bellied Malkoha *Rhopodytes sumatranus*   Map and text page 244
Thai-Malay Peninsula; Sumatra; Borneo. Forests (including mangrove and secondary) and edge, plantations; mostly lowlands, but to 1,300+m.

40–41cm. Grey and green malkoha with chestnut belly; long white-tipped tail.

Adult Head, neck, breast and flanks grey; green bill; red facial skin; wings glossy greenish-blue; belly and undertail-coverts chestnut; graduated tail glossed green and tipped white.
Juvenile Rectrices narrower than in adult, and white tips smaller.

Long-tailed Malkoha *Rhopodytes tristis*   Map and text page 249
North Indian subcontinent; south-east Asia; China; Sumatra. Forest (including secondary and edge of mangrove), scrub, plantations, orchards, edges of paddifields; to 2,000m.

51–60cm. Mostly green and grey; extremely long tail held straight when running lengthwise along branches giving squirrel-like appearance.

Adult (*tristis*; north Indian subcontinent) Head and neck grey, rest of upperparts glossy green; red facial skin; below grey washed ochraceous on throat and breast and finely streaked black; tail broadly tipped white.
Juvenile (*tristis*) Duller than adult, with rectrices shorter and white tips smaller.
Adult (*elongatus*; Sumatra) Darker than *tristis*; longer white tail spots; lacks streaking and ochraceous below.

Blue-faced Malkoha *Rhopodytes viridirostris*   Map and text page 246
Peninsular India and Sri Lanka. Lightly wooded areas, scrub, secondary growth; to 1,150m.

39cm. Mostly grey and green; blue skin around eyes; rarely flies; long-tailed.

Adult Blue skin around eyes; head and upperparts dark grey; wings and tail metallic greenish, latter broadly tipped white; underparts grey washed rufous on belly, streaked white on throat and breast.
Juvenile As adult but duller.

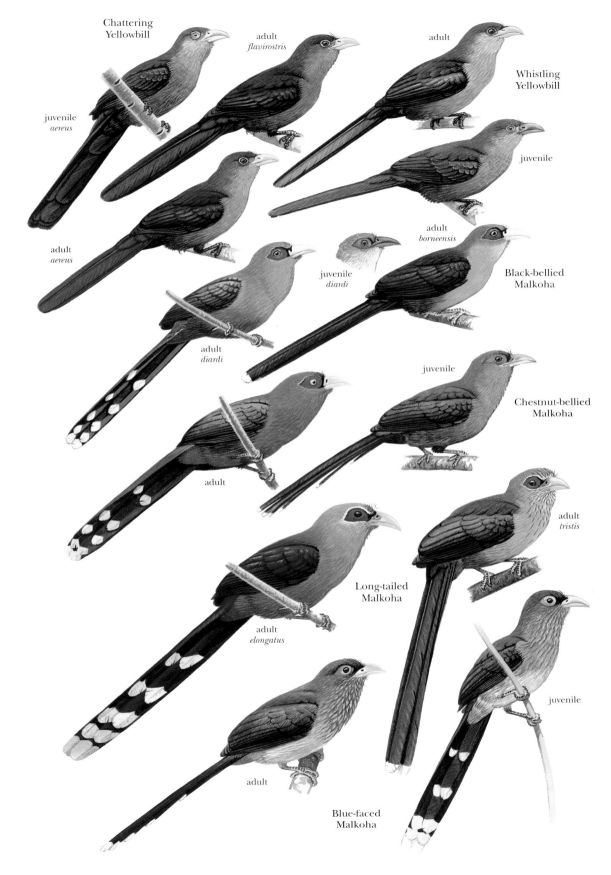

PLATE 16: MALKOHAS II

Sirkeer Malkoha *Taccocua leschenaultii* Map and text page 237
Indian subcontinent. Secondary forest, thinly-wooded areas, scrub, bamboo; bushes in stony areas or semi-desert; grass forest; to 2,100m.
42cm. Bushy-crested; tail about same length as body; terrestrial, flapping from one tree to another.
Adult (*leschenaultii*; south India and Sri Lanka) Sandy grey-brown above with fine blackish streaking; sandy grey-buff below becoming rufous on lower breast to belly (less so on female); undertail-coverts and tail dark brown, latter broadly tipped white, except central rectrices.
Juvenile (*leschenaultii*) As adult but edged pale rufous above; head duller with blackish-centred feathers, and chin, throat, breast, flanks and upper belly boldly streaked black; tail tipped buffish-white.
Adult (*sirkee*; Pakistan and north-west India) Paler on underparts and smaller-billed than *leschenaultii*.

Red-billed Malkoha *Zanclostomus javanicus* Map and text page 240
Thai-Malay Peninsula and Greater Sundas. Primary and secondary forest, and edge; scrub, grass swamp, plantations; to at least 1,750m.
42–44cm. Medium-sized malkoha; mostly grey and rufous; bill red.
Adult Crown, nape and hindneck grey; upperparts and wings dark grey glossed green; chin, throat and breast orange; flanks and upper belly pale grey; rest of underparts orange-chestnut; dark tail tipped white.
Juvenile As adult but primary-coverts and secondaries edged rufous; bill tipped black.

Yellow-billed Malkoha *Rhamphococcyx calyorhynchus* Map and text page 254
Sulawesi area. Primary and secondary forest, and edges; wooded savanna; palm plantations and other cultivation; to 1,650m.
51–53cm. Rather large malkoha; heavy arched yellow, red and black bill; very long rounded tail lacks white tipping giving squirrel-like appearance when creeping in thick vegetation; not shy; easy to observe.
Adult Top of head grey; iris red; facial skin red in male (small black area in female); sides of neck and upperparts chestnut; wings dark glossy purple; throat and breast chestnut; rest of underparts dark grey; tail blue-black.
Juvenile Duller than adult with brown, not red, iris; and bill black, later becoming yellow and red.

Chestnut-breasted Malkoha *Rhamphococcyx curvirostris* Map and text page 251
Thai-Malay Peninsula; Greater Sundas; Philippines (Palawan group). Forest (including peat-swamp, mangrove and secondary); coastal vegetation, grassland, plantations, gardens; to 1,500m.
42–50cm. Medium-sized malkoha; heavy yellowish and red bill; red facial skin; rather long rounded tail; creeps squirrel-like in thick vegetation swinging tail; flight usually brief.
Adult male (*curvirostris*; west and central Java) Forehead, crown and nape grey; red facial skin bordered grey; upperparts and wings glossy green; underparts including chin rufous except glossy green flanks, blackish abdomen and undertail-coverts; graduated tail glossy green, outer third mahogany-rufous without white tips.
Adult female (*curvirostris*) Grey stripe under red facial skin broader; chin grey.
Juvenile (*curvirostris*) Smaller area of red facial skin than adult; throat and breast rufous tinged purplish; abdomen and undertail-coverts blackish; rectrices narrower, central pair entirely green-grey or with few rufous subterminal spots.
Adult (*harringtoni*; Philippines) Head grey tinged olive-brown; broad grey stripe below eye; underparts often paler than in *curvirostris*. Sexes alike.
Adult (*oeneicaudus*; Mentawai Is.) Darkest subspecies; dark grey stripe below eye; chin to upper breast vinous-chestnut; lower breast, lower flanks and thighs glossy dark green; upper flanks, belly and vent brownish-black; no rufous in tail.
Adult (*borneensis*; Borneo, Natunas and Bangka) Chin pale rufous or greyish; crown and nape dark green-grey; rest of upperparts glossy bronze-green; underparts dark red-brown; tail shorter with T1 (sometimes also T2) occasionally entirely bronze-green.

See also Plates 7, 11 and 13

PLATE 17: MALKOHAS III AND NEOTROPICAL DWARF CUCKOOS

Red-faced Malkoha *Phaenicophaeus pyrrhocephalus*   Map and text page 242

Sri Lanka. Tall primary forest with dense undergrowth in wet zone; usually in canopy but also on ground; to 1,700m.

40–46cm. Rather large-billed and boldly-patterned malkoha; wings produce musical hum as it flies slowly and directly between trees.

Adult male Conspicuous large red facial area underlined by broad white stripe flecked black, and heavy green bill, upperparts metallic green-blue, throat and breast black, rest of underparts white, long black tail broadly tipped white.

Adult female From male by less white, and more black, on sides of head.

Juvenile Duller, with less red on head; shorter tail.

Red-crested Malkoha *Dasylophus superciliosus*   Map and text page 256

Philippines (Luzon, Catanduanes, Marinduque and Polillo). Forest (including secondary) and edges; secondary growth, grassland with bushes, river banks with trees; to 1,000m.

41cm. Blackish; red-crested; long black tail tipped white.

Adult (*superciliosus*; range of species except where *cagayanensis* occurs) Head, neck, upperparts and wings black glossed green-blue. Red superciliary crest of long hair-like feathers; underparts blackish; tail black glossed blue-green and broadly tipped white.

Juvenile (*superciliosus*) Browner and lacks red crest.

Adult (*cagayanensis*; Cagayan Province, north-east Luzon) Smaller than *superciliosus* with shorter red crest; underparts paler, tinged olive.

Scale-feathered Malkoha *Lepidogrammus cumingi*   Map and text page 258

Philippines (Luzon, Marinduque and Catanduanes). Forest; scrub; chiefly montane, but recorded 0–3,500m.

42cm. Tail just over half total length.

Adult Head grey; glossy blackish scales on crown, chin and throat; upperparts, wings and tail dark blue-green, tail tipped white; sides of chin and throat white; breast dark rufous becoming blackish-brown on rest of underparts.

Juvenile Mostly reddish-brown to rufous; wings blackish, coverts tipped cinnamon; tail shorter.

Ash-coloured Cuckoo *Coccyzus cinereus*   Map and text page 285

Peru and central Brazil to central Argentina. Chaco, wooded areas in pampas, dry woodlands, savanna, gallery forest; to 900m.

21–24cm. Small; tail rounded and rather short; bill black.

Adult Pale ashy-brown above; throat and breast light grey-brown; whiter belly; eye and eye-ring red; tail with black subterminal areas and white tips.

Juvenile Browner; no white tail tips; eyes brown.

Dwarf Cuckoo *Coccyzus pumilus*   Map and text page 283

Locally in Colombia and Venezuela. Savanna, gallery forest, parks and gardens in towns, tropical deciduous forest, secondary forest, orchards, plantations; to 2,000m Colombia but below 400m Venezuela.

20–22cm. Size and shape as Ash-coloured.

Adult Grey with chestnut throat and breast and whitish belly; rounded tail grey with black-and-white tips; bill black; eye-ring bright red; iris reddish-brown.

Juvenile Similar to juvenile Ash-coloured (no range overlap); eye-ring dull yellow.

PLATE 18: CRESTED CUCKOOS

Jacobin Cuckoo *Clamator jacobinus*   Map and text page 270
Sub-Saharan Africa and Indian subcontinent; some populations migratory. Secondary forest, woodland, scrub, semi-desert, cultivation; 0–3,800m.
34cm. Medium-sized slim cuckoo; straight, swift flight; erect posture with hanging tail when perched, horizontal with head and neck slightly raised when foraging; when alarmed, crest erected and tail wagged up and down.
Adult (*jacobinus*; south India, some migrating to southern Africa) Glossy black above; white below; prominent black crest; white wing patches; tail tipped white.
Juvenile (*jacobinus*) Brown to dull black above; shorter crest; white below, grey or washed buffish on throat and breast.
Adult (*pica*; Iran, Afghanistan, northern Indian subcontinent and Burma migrating to Africa, and sub-Saharan Africa south to northern Zambia and Malawi, northern populations migrating south) Washed pale yellow-buff below; some have fine dark streaks.
Adult (*serratus*; pale morph; southern Africa, migrating north) Greyish-white to white below; dark markings along throat and breast feather shafts; flanks dark grey.
Adult (*serratus*; melanistic morph) Glossy black; white wing patches.

Levaillant's Cuckoo *Clamator levaillantii*   Map and text page 267
Sub-Saharan Africa; some populations migratory, but movements not fully understood. Savanna woodland, scrub, woody growth along streams, *Acacia* savanna, agricultural areas and gardens, forest canopy; to 2,100m.
39cm. Long-tailed; crested; black and white.
Adult (pale morph) Mainly black above with white wing-bar and tail tips; conspicuous crest; white below with heavy black streaking on throat and breast.
Juvenile Differs from adult in having much brown above; brownish wash on chin to breast.
Adult (melanistic morph; east African coast; rare) All black with white wing-bar.

Chestnut-winged Cuckoo *Clamator coromandus*   Map and text page 260
Northern Indian subcontinent; northern south-east Asia; China; migrates south to south India, Sri Lanka, southern south-east Asia, Greater Sundas and Philippines. Wide variety of habitats with cover; to 2,000m.
38–46cm. Long-tailed, slender cuckoo with chestnut wings, white collar and long black crest; flight swift and direct with rapid wing-beats and with crest lowered.
Adult Top of head and upperparts to tail glossy black; white collar; throat to upper breast pale orange-brown, breast and upper belly white, rest of belly and undertail-coverts blackish; tail tipped white.
Juvenile Dark brown on head and upperparts, feathers edged paler; buffish collar; crest reduced; wing-coverts edged rufous-buff; underparts white; tail edged and tipped buffish.

Great Spotted Cuckoo *Clamator glandarius*   Map and text page 263
Portugal to Iraq; north Africa; sub-Saharan Africa; northern populations and those of southern Africa migrate to tropical Africa. Semi-arid areas, heathlands, dry *Acacia* savanna, open woodland; to 3,000m.
40cm. Rather large and long-winged; spotted dorsally; long, narrow tail drooping in flight.
Adult Dark grey-brown above spotted white; short grey crest, often flattened; below white tinged golden-buffish on breast; graduated tail brown broadly tipped white.
Juvenile Crown and face black; primaries rufous; smaller white spots than adult.

PLATE 19: LIZARD CUCKOOS

Puerto Rican Lizard Cuckoo *Saurothera vieilloti* Map and text page 314
Puerto Rico. Dry coastal forest, lowland swampy forest, brush in limestone hills, thick montane forest, coffee plantations; to 800m.
40–48cm. Large; long bill and tail.
Adult Throat and breast grey; lower underparts tawny-ochraceous; upperparts brown; red eye-ring.
Juvenile Above with rufous edges; belly and tail paler than adult; yellow eye-ring.

Hispaniolan Lizard Cuckoo *Saurothera longirostris* Map and text page 318
Hispaniola and surrounding islands. Highland pine forest, dense undergrowth and thickets, scant stands of bushes on barren mountainsides, coffee plantations; to 2,000m.
41–46cm.
Adult (*longirostris*; Hispaniola, Tortue and Saona) Above grey; chestnut wing patch; dull orange throat; breast pale grey; belly ochraceous; red eye-ring.
Juvenile As adult, but above brownish-grey, tail brown, rectrices narrower, with white ends tipped buff; eye-ring presumably yellow.
Adult (*petersi*; Gonâve I.) Throat whitish.

Jamaican Lizard Cuckoo *Saurothera vetula* Map and text page 312
Jamaica. Humid montane forest, dense limestone scrub, second growth; to 1,200m.
38–40cm.
Adult Crown brown, rest of upperparts grey; rufous wing patch; throat and chest white, belly ochraceous; tail tips very broad; red eye-ring.
Juvenile Rectrices narrow; secondaries tipped buff; eye-ring presumably yellow.

Great Lizard Cuckoo *Saurothera merlini* Map and text page 316
Cuba, Isla de la Juventud; Bahamas. Pine forest; woodland and shrub, mainly with many vines; abandoned coffee plantations, overgrown pastures; to 1,200m, sometimes higher.
44–55cm. Four subspecies (two illustrated). Breast grey, belly buff or ochraceous; rufous patch in primaries (not in Bahamas); bill long, rather straight; tail long, graduated, grey, below subterminally black tipped white.
Adult (*merlini*; Cuba) Above olive-brown; bare ocular area orange to red.
Adult (*decolor*; Isla de la Juventud) Greyish-brown above; bill shorter than *merlini*.
Juvenile Rectrices narrow, more pointed, without prominent subterminal bar, tips smaller and duller; bare ocular area yellow.

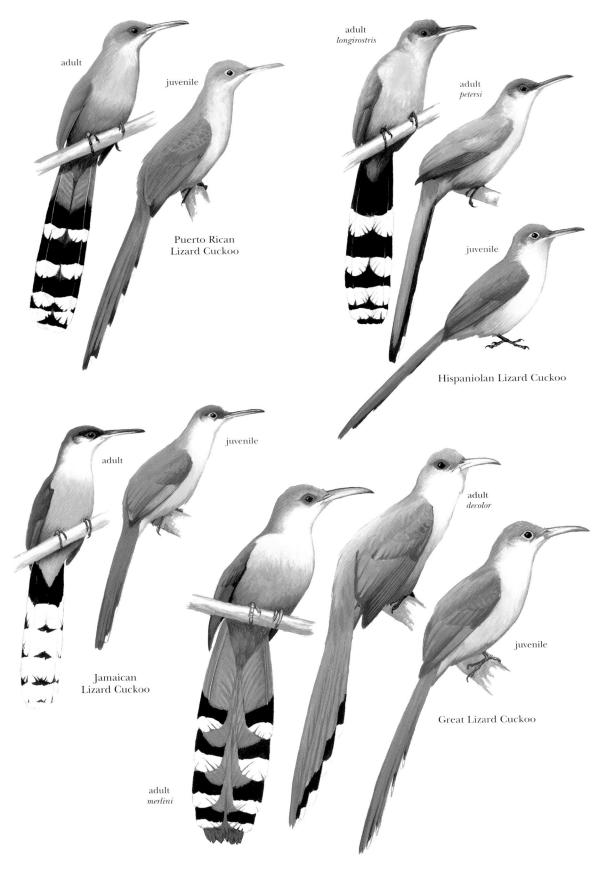

PLATE 20: *COCCYZUS* CUCKOOS

Black-billed Cuckoo *Coccyzus erythropthalmus*   Map and text page 302
North America, winters mostly Peru and Bolivia. Breeds in deciduous and coniferous forest and open woodland; lowlands, but higher (to 3,950m) in winter quarters or on passage.
27–31cm. Above brown, below greyish-white.
Adult Bill black; eye-ring red; tail below grey tipped white, subterminal band black.
Juvenile Eye-ring yellow; upperparts warmer rufous-toned, with some pale edging; breast buffy; rectrices narrower, with faint and narrow whitish tips (no black band); bill often paler, greyish or bluish.

Yellow-billed Cuckoo *Coccyzus americanus*   Map and text page 290
North America, winters South America south to central Argentina, apparently mainly south of Amazonia. Breeds in open woodland, often near water; to 1,500m, to 2,500m on passage.
28–32cm. Brown above; white below; much yellow on bill.
Adult Rufous in primaries (seen mostly in flight); tail graduated, dark below with broad white tips.
Juvenile Pale tail tips and outer edges; outer web of primaries and wing-coverts rufous.

Pearly-breasted Cuckoo *Coccyzus euleri*   Map and text page 295
Eastern South America. Breeds in humid forest and gallery forest; winters in Amazonian upland and river-bluff forest, etc.; mostly below 1,000m.
24–28cm. Very similar to Yellow-billed, but smaller; no rufous in primaries.
Adult Above greyish-brown; throat and chest (usually?) silvery-grey, remaining underparts white; bill like Yellow-billed.
Juvenile Pale edges on wings; rectrices narrow, spots indistinct.

Cocos Cuckoo *Coccyzus ferrugineus*   Map and text page 301
Cocos I., Costa Rica. Second growth forest, thickets, streamside vine tangles; to 570m.
33cm. Similar to Mangrove Cuckoo. Grey crown; dusky mask; greyish-brown upperparts; bright rufous wings; rich buff underparts.
Adult Eye-ring yellow or grey.
Juvenile Similar to adult, but tail pattern indistinct.

Mangrove Cuckoo *Coccyzus minor*   Map and text page 298
Mexico and Florida through Central America and Caribbean to coast of north Brazil. Scrubby woodland, forest (including mangroves) etc., to 1,200m.
28–34cm. Dark mask; crown grey; upperparts brown; bill and tail like Yellow-billed.
Adult Black mask; underparts cinnamon-buff to rufous.
Juvenile Like adult but mask and underparts paler; tail spots more diffuse.

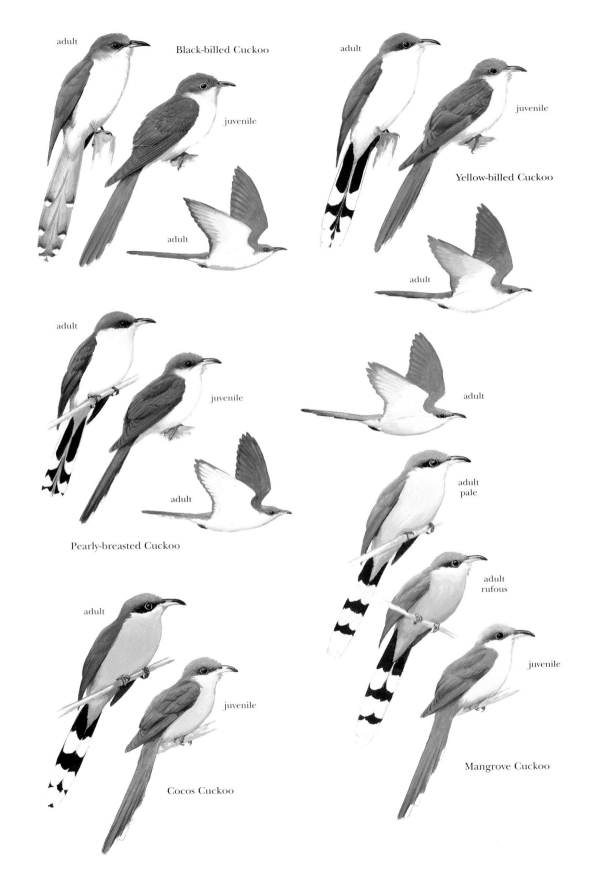

PLATE 21: MISCELLANEOUS NEOTROPICAL CUCKOOS

Dark-billed Cuckoo *Coccyzus melacoryphus*   Map and text page 287
South America south to central Argentina; Galapagos Is. Lowland humid forest edge and river-edge forest; mangroves; fence-rows, dry farmland, shrubby pastures, coffee plantations; to 1,200m, sometimes much higher (on passage?).
25–28cm. Dusky mask; buff underparts; pale grey band on side of neck; back and wings greyish-brown; bill black.
Adult Cap grey; prominent white tail spots.
Juvenile Brown crown and nape; sometimes some rufous in wings, coverts with dull buff tips; narrow pointed rectrices indistinctly tipped grey.

Grey-capped Cuckoo *Coccyzus lansbergi*   Map and text page 306
Locally in Venezuela to north-west Peru. Thickets and dense shrubbery in wet savanna; undergrowth of semi-deciduous and secondary woodland; to 900m, sometimes to 1,400m.
25–28cm.
Adult Rufous with slate-grey cap; white tail tips.
Juvenile Brown crown and less distinct pale tips on rectrices.

Rufous-breasted Cuckoo *Hyetornis rufigularis*   Map and text page 310
Hispaniola and Gonâve I. Mostly dry deciduous forest, perhaps mainly in narrow transition zone to broadleaf humid montane forest; to 900m, sometimes higher.
43–52cm.
Adult Stout, curved bill; throat and breast dark chestnut, upperparts grey, belly buff, chestnut wing patch, tail graduated, black tipped white.
Juvenile Rectrices narrow, T1 dark grey.

Chestnut-bellied Cuckoo *Hyetornis pluvialis*   Map and text page 308
Jamaica. Thickets, forest edge, wooded pastures, citrus plantations; 300–1,800m.
48–56cm. Crown dark grey; back, rump and wings dark olivaceous; throat white; lower underparts rufous-chestnut; tail long, graduated, rectrices broad, broadly tipped white.
Adult Tail faintly purple glossed black.
Juvenile Dark brown tail without sheen.

Little Cuckoo *Piaya minuta*   Map and text page 274
East Panama to Bolivia and Brazil. Dense shrubby vegetation near water; around oxbow lakes; humid forest edges, mangroves, savanna woodland, gallery forest, semi-deciduous forest; to 1,600m, but mostly in lowlands.
24–28cm. Like Squirrel Cuckoo, plumage rufous, tail graduated and white-tipped, and bill yellow, but much smaller size and tail relatively shorter.
Adult (*minuta*; east of Andes) Rufous-chestnut above; throat and chest tawny to rufous (paler than upperparts); belly buffy-grey; undertail-coverts blackish; orbital skin red.
Juvenile Dark brown; basal half of primaries rufous; tail blackish; bill dark.
Adult (*gracilis*; Panama to west Ecuador) Paler than nominate; rufous above; belly grey.

PLATE 22: THICK-BILLED CUCKOO AND HAWK-CUCKOOS I

Thick-billed Cuckoo *Pachycoccyx audeberti* Map and text page 320
Madagascar (rare); patchily over much of sub-Saharan Africa. Forest edge; gallery forest; moist open savanna woodland, including miombo and mopane; cocoa plantations; to 1,600m.
36cm. Large, heavy-billed, *Accipiter*-like cuckoo; perches upright; restless; often noisy; conspicuous when displaying.
Adult Dark slate-grey above; white below; tail dark-barred and spotted white on sides.
Juvenile Broad white edges to dark brown upperparts giving scaly appearance; underparts creamier than adult.

Large Hawk-cuckoo *Hierococcyx sparverioides* Map and text page 433
Himalayas to south-east Asia and China; northerly populations migrate south to Thai-Malay Peninsula, Greater Sundas and Philippines. Forest and woodlands, but more catholic on migration; to 3,500m.
38–40cm. Largest hawk-cuckoo; hawk-like; broad, rounded wings and tail; skulking; flight low and buoyant, fast wing beats followed by glide at the end to rise abruptly and land in tree.
Adult (dark morph; throughout range) Head grey, with yellow eye-ring, and entire upperparts and tail grey-brown; tail strongly barred black, tipped white; remiges strongly barred; below white streaked grey on throat, upper breast rusty striped blackish; broad black barring on lower breast, flanks and belly.
Adult (pale morph; Burma and Thailand; rare) Pale grey above with paler grey and rufous edging; upper breast barred rufous and white; lower breast to belly and flanks creamy-white with rufous and grey bars and chevrons.
Subadult Crown grey-bronze; upperparts and wings dull dark brown barred rufous-brown; white to pale buffish underparts boldly streaked black with chevrons on flanks.
Juvenile Brown upperparts edged paler; whitish underparts with drop-like brown streaks and bars.

Dark Hawk-cuckoo *Hierococcyx bocki* Map and text page 431
Peninsular Malaysia, Sumatra and Borneo. Montane forest; 800–2,000m.
30–32cm. Medium-sized cuckoo.
Adult Face dark; upperparts blackish grey; wings brown; chin and throat grey; breast orange-rufous; rest of underparts white barred black; tail dark brown broadly barred buffy-grey to greyish-white, tipped white.
Subadult Brown or bronze crown; sides of face dark grey; back dark brown barred rufous; wings brown with rufous and white notches; chin grey; rest of underparts white to buff-white, broadly barred and spotted or streaked blackish brown; belly and undertail-coverts white. Tail as adult but with some rufous in grey.
Juvenile Crown slaty-brown; few white feathers on nape; back, wing-coverts and rump brownish-black barred rufous; chin to upper breast slaty-brown; lower breast to undertail-coverts white barred dark; tail as adult.

Common Hawk-cuckoo *Hierococcyx varius* Map and text page 436
Indian subcontinent, including Sri Lanka. Woodland, mangroves, riverine forest, gardens, plantations, orchards and other cultivation; mainly below 600m, but as high as 2,200m.
34cm. Size as Common Cuckoo; hawk-like appearance and flight, but sometimes perches horizontally unlike hawk, with drooped wings and raised tail; tail less graduated than in other *Hierococcyx* species.
Adult Above grey-brown; ashy-grey head; pale red-brown breast; rest of underparts white with pale brown barring on belly and flanks; wings uniform; tail grey-brown with 3–5 white and black bars, and pale rufous tip.
Subadult Above as adult with rufous tips to wing-coverts; remiges blackish barred rufous. Less rufous below.
Juvenile Brown above barred rufous; underparts white variously tinged buffy and streaked or spotted blackish-brown; tail broadly tipped bright rufous.

Moustached Hawk-cuckoo *Hierococcyx vagans* Map and text page 429
Thai-Malay Peninsula; Borneo; once Sumatra, but may be more widespread there; west Java. Forest and edge, bamboo, secondary growth; to 915m.
26cm. Rather small hawk-cuckoo, slimly built, with long tail and distinctive facial pattern.
Adult Grey crown and nape; white ear-coverts; blackish down-curved moustache; upperparts grey-brown; wings, show white patch in flight; underparts whitish streaked blackish; tail grey with broadly barred black and tipped white.
Juvenile Crown, back and wings blackish; back and wings barred rufous; face, chin and throat blackish; broad black line in front of eye continuous with throat; breast streaked black; undertail-coverts unmarked white; tail grey barred black and tipped white.

PLATE 23: DRONGO-CUCKOOS AND DWARF KOEL

Fork-tailed Drongo-cuckoo *Surniculus dicruroides*   Map and text page 421
Northern Indian subcontinent (including Sri Lanka), Burma, Thailand, Indochina, south China; winters south to Thai-Malay Peninsula, south India, Sri Lanka?, Sumatra and west Java. Forest, scrub jungle, bamboo; to 2,600m.
25cm. Smallish black cuckoo; forked tail. White on thighs, and barring on underside of tail in all drongo-cuckoos.
Adult Plumage black glossed steel-blue and long tail obviously forked.
Juvenile Black spotted white.

Square-tailed Drongo-cuckoo *Surniculus lugubris*   Map and text page 425
North-east Indian subcontinent; Burma; Indochina; Thai-Malay Peninsula; Greater Sundas; Philippines (Palawan, Balabac, Calauit). Some northern populations migrate south. Forest (including secondary) and edge, scrub, orchards and plantations; to 2,250m.
24–25cm. Smallish cuckoo; tail more or less square. Sexes similar, but female slightly duller.
Adult Glossy blue-black, more green on wing-coverts; slightly forked or square tail.
Juvenile Duller than adult, and spotted white.

Philippine Drongo-cuckoo *Surniculus velutinus*   Map and text page 423
Philippines (Basilan, Biliran, Bohol, Bongao, Gigantes, Jolo, Leyte, Luzon, Malamaui, Mindanao, Mindoro, Negros, Panay, Samal, Samar, Tawi Tawi). Primary and secondary forest, forest edge, woodland, bamboo; to 1,000m.
24cm. Smallish velvety, purplish-black cuckoo, not glossed; slightly forked tail.
Adult Purplish-black, unglossed.
Juvenile Brown to rufous-brown.

Moluccan Drongo-cuckoo *Surniculus musschenbroeki*   Map and text page 428
Sulawesi, Butung, Halmahera, Bacan and Obi. Forest and wooded savanna; to 1,200m.
23cm. Smallish purplish-black cuckoo, with slightly forked tail.
Adult Velvety purplish-black; glossed green on wings.
Juvenile Duller than adult, and spotted white.

Dwarf Koel *Microdynamis parva*   Map and text page 323
New Guinea. Forests (including secondary and gallery) and edge; tall garden shade trees; to 1,400+m.
20–22cm. Small cuckoo; short, stout decurved bill, and medium length rounded tail.
Adult male Top of head, nape and malar stripe glossy blue, white line below eyes; upperparts, wings and tail brown; cheeks and chin orange-rufous; breast greyish-brown; abdomen and undertail-coverts orange-buff; rest of underparts brown.
Adult female More grey-brown than male, lacking blue and orange-rufous.
Juvenile Brown above barred and edged rufous and grey; lores and ear-coverts white; chin greyish-white; rest of underparts greyish-brown tinged rufous; all underparts indistinctly barred black.

PLATE 24: COMMON KOEL

Common Koel *Eudynamys scolopaceus* Map and text page 325

Indian subcontinent; south-east Asia; China; Greater Sundas; Wallacea; New Guinea; Australia. Some populations migratory. Forests (including coastal swamp, mangrove and secondary), woodland, savanna, gardens, plantations, orchards and other cultivation; to 1,800m.

39–46cm. Largish, robust cuckoo with long rounded tail; flight fast, direct and strong with rapid wing-beats; somewhat hawk-like, especially females.

Adult male (*scolopaceus*; Indian subcontinent) Glossy black with strong, decurved, pale green, ivory or horn-coloured bill; red eyes.

Adult female (*scolopaceus*) Dark brown crown streaked white and rufous; upperparts dark brown heavily spotted white or buff; dark brown wings barred and spotted white; chin to upper breast dark brown heavily spotted white or buffy-white; rest of underparts white to buffy-white, barred blackish-brown; tail dark brown barred and spotted white to rufous-white, bill as male.

Juvenile male (*scolopaceus*) Dull blackish, wing-coverts and tertials tipped white or buffish; underparts variably barred white and black; tail black, in some with rufous barring; eyes brown; bill blackish to greyish-buff.

Juvenile female (*scolopaceus*) As adult female, but more barred and less spotted above; bill as juvenile male.

Adult female (*chinensis*; Indochina and China, migrating south to Borneo) Larger than *scolopaceus*. No rufous on crown, sometimes head mostly blackish with long white malar stripe; spotted upperparts dark brown glossed greenish.

Adult female (*orientalis*; South and central Moluccas) Fewer, more rufous spots and bars above; rich cinnamon below with little or no barring; throat black or spotted black.

Adult female (*mindanensis*; Philippines, Talaud Is, Sangihe, Siau and Ruang) Browner above; more rufous on pale areas; more narrow ventral barring.

Adult male (*melanorhynchus*; Sulawesi, Bangka, Muna, Manterawu, Peleng, Talisei, Lembeh, Togian and Sula) Blue-violet iridescence; more rounded wings; bill black (on Sula may have some white on head).

Adult female (*melanorhynchus*) Polymorphic: (1) as male but blue-green gloss; (2) head black glossed green, black above glossed bronze, pale malar stripe, chin and throat black or grey barred black, orange-rufous on rest of underparts; (3) crown and face black, crown sometimes streaked rusty, upperparts barred rufous and black, whitish malar stripe which may extend to sides of neck, underparts barred buffish and black.

Adult female (*rufiventer*; New Guinea and islands off coast) Head largely rufous; whitish malar stripe; dark brown upperparts heavily spotted and streaked rufous; underparts ochraceous spotted from chin to upper breast, rest finely barred dark.

Adult female (*subcyanocephalus*; North and north-west Australia, wintering New Guinea) Polymorphic: (1) glossy black head with buff malar stripe broadening at neck; rest of upperparts blackish-brown spotted white; throat buff tipped blackish; rest of underparts buff barred blackish; (2) brown tipped buffish and whitish; breast pale buffish; rest of underparts white with brown fringes; brown streaking on throat; (3) rufous head and throat streaked dark; upperparts brown spotted white; breast pale buffish; rest of underparts white with prominent dark edges to feathers.

PLATE 25: BRONZE CUCKOOS I

Rufous-throated Bronze Cuckoo *Chrysococcyx ruficollis* Map and text page 361

New Guinea (Central Ranges, Vogelkop and Wandammen Mts.). Forest (including secondary) and edge; subalpine thickets; mostly 1,600–2,600m (occasionally 1,130–3,350m).

16cm. Very small cuckoo; rufous from chin to breast.

Adult male Forehead, face, sides of neck, chin, throat and upper breast rufous; top of head, nape and upperparts dark green with iridescent bronze or purple; crown sometimes greyer and duller; remiges dark brown; lower breast to undertail-coverts white barred glossy greenish-brown; closed tail bronze-green with broad dark subterminal band; when spread shows row of white, black and rufous bands. (Female lacks iridescence above.)

Juvenile Greenish upperparts; below grey barred on flanks.

White-eared Bronze Cuckoo *Chrysococcyx meyerii* Map and text page 367

New Guinea; Batanta I. Forest, also edge and clearings; gardens; mostly 800–1,600m (occasionally to 2,000m).

15cm. Very small bright, iridescent cuckoo with white ear patch.

Adult male Iridescent green on head, upperparts and wing-coverts, often metallic bronze on back; ear-coverts with white patch; remiges bright rufous basal half; underwing conspicuously rufous; underparts white broadly barred bronze-green; tail dark bronze-green tipped pale with wide white barring on outer rectrices.

Adult female From male by rufous-chestnut forehead and crown.

Juvenile Top of head green-grey; hint of white on sides of neck; upperparts olive grey-brown; wings grey-brown edged pale red-brown; underparts buffy white; tail red-brown, outer rectrices with two black bars on inner web.

Asian Emerald Cuckoo *Chrysococcyx maculatus* Map and text page 337

North-eastern Indian subcontinent; Burma; Thailand; Indochina; China. Some winter south to Andamans, Nicobars, Thai-Malay Peninsula and Sumatra. Forest (including swamp and secondary); plantations and gardens on migration; to 3,300m.

18cm. Small, brilliantly-coloured cuckoo with metallic green plumage. Flight swift and powerful, reminiscent of lorikeet, and showing broad white band on underwing at base of remiges.

Adult male Whole upperparts, head, wings and tail, and underparts from chin to upper breast glossy green with golden-bronze reflections; primaries blackish; lower breast and belly white barred glossy bronze-green.

Adult female Top of head to neck light rufous indistinctly barred dark; upperparts coppery-green; underparts white, washed rufous and barred coppery-brown.

Juvenile As adult female but variable. Unbarred or barred rufous top of head, nape and hindneck, or green with forecrown white barred black; less glossy green on upperparts and wings; rufous-orange barring on mantle, shoulders and wing coverts.

Violet Cuckoo *Chrysococcyx xanthorhynchus* Map and text page 340

North-east Indian subcontinent; south-east Asia; Yunnan; Greater Sundas; Philippines. Forest (including mangrove, secondary and logged) and edge, grassland, seashore, plantations, gardens, parkland; lowlands, exceptionally to 2,300m.

16cm. Small brightly-coloured cuckoo; undulating flight.

Adult male Purple-violet head, upperparts, throat, breast and tail (may appear blackish under certain light conditions); outer rectrices barred black and white; abdomen and flanks white broadly barred dark; bright red eye-ring and bill.

Adult female Top of head and mantle dark grey-brown (some with rufous wash); rest of upperparts more greenish-bronze; underparts white narrowly barred dark greenish-bronze.

Juvenile As female except more rufous top of head and upperparts.

PLATE 26: BRONZE CUCKOOS II

Klaas's Cuckoo *Chrysococcyx klaas*   Map and text page 348

Most of sub-Saharan Africa; much north-south migration, and movement in response to rainfall. Open forest and edge, woodland, savanna, thickets, long grass; to 3,000m.

18cm. Small cuckoo of well-wooded areas.

Adult male Glossy green above, with green shoulder patch; below white; often white mark behind eye; lores green. In flight underwing-coverts pure white.

Adult female Bronzy-brown barred greenish above; more closely barred bronzy-brown and pale buff below; throat centre usually white. (Some northern and western birds very similar or identical to male.)

Juvenile Bronzy upperparts barred greenish; brownish barring below. Usually shows whitish post-ocular patch, but often indistinct in field.

Yellow-throated Cuckoo *Chrysococcyx flavigularis*   Map and text page 352

West Africa east to west Uganda and south-west Sudan. Forest (including secondary) and edge; well-wooded savanna; to 1,200m.

18–19cm. Small dull, dark cuckoo; much darker than other African *Chrysococcyx*; rather difficult to see.

Adult male Head, upperparts and sides of breast bronzy with dull metallic green (sometimes appearing brown); vertical yellow stripe from chin to upper breast; underparts finely barred dark greenish-brown; outer rectrices white; eyes pale yellow.

Adult female Duller than male; finely barred below and on head sides.

Juvenile Like adult female, but feathers of back and wings narrowly fringed pale tawny and ground colour of underparts and head paler.

Dideric Cuckoo *Chrysococcyx caprius*   Map and text page 343

Much of sub-Saharan Africa; many populations migrate north-south, or undergo rains movements. Forest edge and clearings, open woodland, wooded and thorn savanna, steppes and semi-desert, edge of marshy and swampy areas, reed beds and papyrus, gardens; to 2,000m.

19cm. Small bright cuckoo; flies with irregular wing beats, straight or undulating, alternating flapping and gliding. Generally in more open habitats than other African *Chrysococcyx*.

Adult male Glossy green above; white below with green flank barring; white patches in front of and behind eye; thin green malar stripe extending onto white throat.

Adult female Rather like brown version of male; head patches and throat russet; thicker bars on flanks.

Juvenile Very variable (sometimes two morphs recognized, but many intermediates); thickly streaked on throat, merging into barring on belly, flanks and undertail; closely barred above; often coral-red bill; some have rufous head.

African Emerald Cuckoo *Chrysococcyx cupreus*   Map and text page 354

Much of sub-Saharan Africa; some movements occur. Forest (including gallery), thickly wooded riparian vegetation, coastal jungle, densely-wooded savanna, occasionally dense scrub and miombo; to 2,900m.

20–23cm. Small bright cuckoo; noisy, but often difficult to see.

Adult male Brilliant emerald-green upperparts, and throat to upper breast; rest of underparts golden-yellow.

Adult female Brown crown and nape; upperparts green barred rufous; remiges brown barred rufous; underparts white barred green.

Juvenile From adult female by green barring on white crown; forehead flecked green, brown and white.

PLATE 27: BRONZE CUCKOOS III AND BLACK-EARED CUCKOO

Black-eared Cuckoo *Misocalius osculans*　　　　　　　　　　　Map and text page 379

Australia; rare New Guinea, Moluccas and Lesser Sundas; some populations migratory or nomadic. Open woodland, savanna, saltflats, riverside thickets, mangroves; lowlands.

19–20cm. Small cuckoo with conspicuous broad white supercilium, broader blackish eye-stripe and faint glossy bronze tinge on upperparts. Tail rather long and slightly rounded. Flight low, swift and silent.

Adult (fresh) Top of head, nape, upperparts and wings brown-grey with fine bronze iridescence on back and wings; rump paler; whole underparts unmarked buffish to cream; tail black-brown tipped white. Feathers of back and wings edged pale grey giving scaly appearance.

Adult (worn) Pale feather edgings above much less obvious.

Juvenile As adult but facial stripe paler; supercilium greyer.

Horsfield's Bronze Cuckoo *Chrysococcyx basalis*　　　　　　Map and text page 357

Australia; southern populations migrate to New Guinea, Wallacea, Greater Sundas and Singapore. Forest (including mangroves) and edges; woodlands, coastal salt marsh, scrub, spinifex, cultivation and gardens; lowlands.

16cm. Very small cuckoo with swift and slightly undulating flight.

Adult Slightly glossy olive to bronze crown, nape, upperparts and uppertail; broad white supercilium; bold dark eye-stripe behind eye decurved to behind ear-coverts; wings glossy bronze-green with pale buff to whitish edges to some wing-coverts and remiges; white to pale rufous-brown bar on primaries of open wing, underwing with whitish stripe; underparts white with blackish bands on flanks; sides of closed tail rusty basally; undertail grey with red-brown base to outer feathers (often difficult to see) and prominently spotted black and white on inner webs, and tipped white.

Juvenile From adult by duller plumage; top of head and nape grey-brown; supercilium absent or only suggestion; underparts unbarred white with grey-buff tinge on breast and flanks; streaked undertail-coverts.

Shining Bronze Cuckoo *Chrysococcyx lucidus*　　　　　　　　Map and text page 363

New Zealand, Australia, south Solomons, New Caledonia, Loyalty Is, Vanuatu, Banks and Santa Cruz Is; southerly populations winter Solomons and south-west Pacific, Bismarck Is, New Guinea and Lesser Sundas. Forest (including mangrove and secondary), woodland, savanna, scrub, plantations, gardens; lowlands to 1,900m.

15–17cm. Very small green cuckoo with swift, slightly undulating and graceful flight.

Adult male (*lucidus*; New Zealand, Norfolk and Chatham Is; winters Solomons and south-west Pacific; on passage Tasmania and coastal east Australia, more rarely New Guinea) White face sparse dark mottling or barring; crown and upperparts shining green with bronze sheen; underparts white barred dark; little or no rufous in tail.

Adult female (*lucidus*) Much variation. From adult male by maroon-bronze crown, nape and mantle; few have speckling on white central forehead; may have more rufous in tail.

Juvenile (*lucidus*) As adult but duller; crown and nape browner; often less barring on underparts.

Adult male (*plagosus*; East and south-west Australia, and Tasmania; winters New Guinea, Bismarck Is, Lesser Sundas and Solomons) Much variation. Little or no green on grey crown and mantle, latter maroon-bronze contrasting with rest of iridescent green upperparts; often less mottling and more white on chin.

Adult female (*plagosus*) From adult male by heavier ventral barring (in some trace of pale supercilium).

Adult male (*layardi*; New Caledonia, Loyalty Is, Vanuatu, Banks and Santa Cruz Is). Top of head to mantle dull copper-bronze; less glossy green on upperparts.

PLATE 28: BRONZE CUCKOOS IV

Little Bronze Cuckoo *Chrysococcyx minutillus* Map and text page 369

Australia; New Guinea; Wallacea; Greater Sundas; Thai-Malay Peninsula; Cochinchina; south Cambodia. Open forest (including mangrove) and edge, woodland, thickets; to 1,400m.

15–17cm. Very small cuckoo. Very similar to Gould's Bronze Cuckoo, with many intermediates or hybrids in Australia. Swift direct flight, slightly undulating.

Adult (*minutillus*; North Australia, some possibly wintering in New Guinea and Wallacea) Top of head and nape glossy dark green; some white on forehead; white supercilium mottled dark; dark ear-coverts; upperparts slightly iridescent olive-green; wing-coverts fringed buff or rufous; remiges black; underparts white barred dark brown; some rufous on underside of tail; iris and broad eye-ring red.

Adult female (*minutillus*) Resembles male, but top of head and nape more concolorous with upperparts; chin and upper throat often unbarred; white barring on T5 tends to be broader; iris brown to cream, eye-ring whitish or grey.

Juvenile (*minutillus*) Duller than adult; crown and nape dull olive as rest of upperparts; sides of head grey-brown; underparts whitish with or without pale brown chevrons on flanks and sides of breast; tail as adult.

Adult male (*rufomerus*; Lesser Sundas) Much variation (or hybridisation?). Broad dark green spot on cheek; upperparts dark bronze-green; underparts strongly barred glossy greenish-brown.

Adult male (*albifrons*; Sumatra and north-west Java) White on forehead heavier and extensive; supercilium and face whiter; less barring below; centre of abdomen often white.

Adult male (*cleis*; Borneo) Dark bottle-green crown; white on forehead prominent; upperparts dark green washed purple; underparts heavily barred.

Gould's Bronze Cuckoo *Chrysococcyx poecilurus* Map and text page 373

Queensland (some perhaps migrate northwards); New Guinea; Sulawesi, Madu, Flores, Timor; Borneo; Philippines. Forest (including mangrove and secondary) and edge, woodland, bamboo, thickets, savanna, grassland, plantations and other cultivation; to 1,150m.

15–16cm. Very similar to Little Bronze Cuckoo. Many hybrids with this species from northern Queensland, and some from Port Moresby area, Papua New Guinea.

Adult male (*russatus*; North-east Queensland wintering New Guinea) Crown and nape glossy olive-green, with bronze iridescence; buffish to white supercilium; ear-coverts brown; upperparts iridescent olive-green with metallic bronze; uppertail-coverts similar but edged rufous; remiges dark brown; wing-coverts as upperparts but fringed rufous; underparts white barred dark brown, sometimes incomplete on belly and vent; strong rufous-brown wash on foreneck and breast; tail rufous above; iris and eye-ring red.

Adult female (*russatus*) As male, but duller, with more rufous on neck; iris brown, eye-ring yellowish or green.

Juvenile (*russatus*; little known) Faint barring on flanks; perhaps slightly more rufous and more obvious flank barring than juvenile Little Bronze Cuckoo.

Adult male (*jungei*; Sulawesi, Madu, Flores, Alor) More purplish above with less iridescence, and few or no white feathers on forecrown; underparts barred fainter; tail rarely with rufous trace.

Pied Bronze Cuckoo *Chrysococcyx crassirostris* Map and text page 377

Lesser Sundas; perhaps Moluccas and extreme western New Guinea. Woodland, scrub and forest edge; lowlands.

15–16cm. Very small cuckoo; white below with little barring.

Adult male Deep blue-green above, lacking bronze cast; very little or no barring below; large white wing patch; eye-ring red. (Males on Babar, *crassirostris*, more greenish above, with some dark barring below.)

Adult female Brown, slightly bronze above, less often dull green; some have white wing patch; white below with indistinct barring on flanks; eye-ring red.

Juvenile Rufous above; white below, some with dark barring; outer rectrices barred black and white.

PLATE 29: *CACOMANTIS* CUCKOOS I

Pallid Cuckoo *Cacomantis pallidus* Map and text page 384

Australia; some southern populations migrate to northern Australia, New Guinea, Lesser Sundas and ?Moluccas. Open woodland, mangroves, scrub, savanna, coastal dune-scrub, farmland, roadsides with remnants of trees, plantations, vineyards, orchards, gardens, parks; lowlands.

31cm. Medium-sized cuckoo with dark eye-stripe; lacks dark barring on underparts. Undulating woodpecker-like flight.

Adult male (pale morph) Head, nape and upper mantle pale grey; dark grey eye-stripe continuing down sides of neck; rest of upperparts darker grey to grey-brown, rump spotted white; remiges and coverts finely edged white; underwing-coverts creamy with fine grey barring, underside of remiges barred white on inner vanes; underparts pale grey; tail dark grey-brown, all rectrices except T1 spotted white. (Adult female pale morph intermediate between pale male and dark female.)

Adult female (dark morph) Top of head and nape blackish-brown streaked rufous-brown with white spot on nape; broad white and black eye-stripe; mantle dark brown fringed rufous; rest of upperparts brown to grey spotted white, buff and rufous; underparts pale grey unmarked or mottled rufous-brown on breast with pale grey barring on flanks and undertail-coverts, tail as pale male but all rectrices spotted or barred white or buffish. (Adult male dark morph has darker and browner upperparts and more white spotting on wings.)

Juvenile Head darker grey than pale morph; prominent white supercilium, and blackish cheeks and eye-stripe running down sides of neck to sides of breast; upperparts and wings grey heavily edged and spotted white; underparts white, chin, throat and breast streaked dark grey; tail as adult pale morph.

Chestnut-breasted Cuckoo *Cacomantis castaneiventris* Map and text page 388

North-east Australia, New Guinea and islands. Forest and edge, savanna, scrub; to 1,800m.

22–24cm. Small cuckoo.

Adult male (*castaneiventris*; North-east Australia and Aru Is.) Head dark grey with conspicuous pale yellow eye-ring; entire upperparts and wings dark blue-grey; underparts dark brown-rufous; tail above dark blue-grey, below tipped white with fine white spots along outer edges.

Juvenile (*castaneiventris*) Head, upperparts and wings brown to buffish brown with rufous edging; underparts off-white to brownish on foreneck and breast; tail spotted rufous.

Adult (*weiskei*; east Papua New Guinea) Largest subspecies. Generally much darker than adult *castaneiventris*; upperparts blackish glossed blue-green.

Fan-tailed Cuckoo *Cacomantis flabelliformis* Map and text page 390

Australia, New Guinea and south-west Pacific. Forest (including mangrove and secondary) and edge, woodland, parks, farmland, plantations, orchards; lowlands to 1,300m (exceptionally to 3,700m).

24–28cm. Medium-sized cuckoo with conspicuously long, rounded tail. Flight strong, with fast wingbeats and short glides; perches upright.

Adult male (*flabelliformis*; East, south and south-west Australia, and Tasmania; partly migratory to northern Australia and New Guinea) Head and neck ash-grey; eye-ring yellow; upperparts and wings uniform blue-grey; chin and upper throat ash-grey; rest of underparts including undertail-coverts pale buffish to pale rufous except for white vent; long fan-shaped tail blue-grey tipped white with many conspicuous broad white bands on underside.

Adult female (*flabelliformis*) Paler than male. White area below more extensive with pale rufous barring.

Juvenile (*flabelliformis*) Head, hindneck, upperparts and wings dark brown edged dark rufous; underparts off-white, tinged brown from chin to breast; brown spots from throat to belly; undertail-coverts edged brown; tail dark grey tipped white and banded buffish below.

Adult (*excitus*; New Guinea highlands) Upperparts darker than adult *flabelliformis* with green or blue tinge; chin dark grey; rest of underparts dark rufous-brown vermiculated grey; vent white.

Adult (*simus*; Fiji) Smallest subspecies; strong greenish wash on upperparts; underparts paler than *excitus*.

Adult (*simus*; melanistic morph) Mostly blackish to dark brown; white on underwing-coverts and bar on underside of remiges; tail lacks white.

PLATE 30: *CACOMANTIS* CUCKOOS II AND WHITE-CROWNED CUCKOO

White-crowned Cuckoo *Caliechthrus leucolophus*   Map and text page 412
New Guinea. Forest (including secondary) and edges; to 1,800m.
30–35cm. Medium-sized black cuckoo with undulating flight.
Adult Blackish with bluish gloss on upperparts and white-tipped tail; white stripe on top of head from bill to nape. (Female browner.)
Subadult Similar to adult but white stripe less prominent, more patchy; washed brown on mantle; barred whitish below. (Juvenile mostly white, with black face and throat to upper breast; grey remiges and rectrices.)

Banded Bay Cuckoo *Cacomantis sonneratii*   Map and text page 394
Northern Indian subcontinent; south-east Asia; south-west China; Greater Sundas; Palawan; (more northerly populations perhaps migratory). Forest, scrub, secondary growth, plantations and other agricultural land, gardens; to 2,440m.
22–24cm. Smallish cuckoo with rather long curved bill; often perches in upright position.
Adult Barred red-brown and blackish above; broad whitish supercilium, dark eye-stripe and whitish cheeks; dark red-brown crown finely cross-barred blackish; below whitish finely barred blackish; tail tipped white.
Juvenile As adult but more white on head and upper mantle; underparts less barred.

Grey-bellied Cuckoo *Cacomantis passerinus*   Map and text page 401
Indian subcontinent. Open woodland, secondary forest, scrub, grassy plains, swamps; gardens, tea plantations and other cultivation; to 2,300m.
22cm. Small cuckoo.
Adult male Mostly grey, slightly glossed above, with variable amount of white on underparts. (Adult blackish morph; rare; blackish and dark brown.)
Adult female (rufous morph; more common) Bright rufous head, upperparts and chin to breast, mostly barred dark brown; rest of underparts pale rufous barred black. (Adult female, grey morph, like male but abdomen barred whitish and rectrices more barred.)
Juvenile (grey morph) As adult but face darker, upperparts unglossed; underparts grey barred buffish-grey on flanks and belly; tail dark, notched white and rufous. (Variable; intermediates occur.)
Juvenile (rufous morph) Variable. From rufous adult female by heavier dark streaking and barring, and less rufous below.

Plaintive Cuckoo *Cacomantis merulinus*   Map and text page 397
Indian subcontinent; south-east Asia; China; Greater Sundas; Philippines; Sulawesi; northerly populations may migrate south. Open woodland, mangroves, secondary growth, scrub, savanna grassland; farmland and plantations; gardens; to 3,000m.
20–23cm. Small, pale cuckoo.
Adult male (*merulinus*; Philippines and Sulawesi) Pale grey head; rest of upperparts and wings dark grey slightly glossed, with some rufous on wing-coverts; chin to upper breast grey; rest of underparts pale rufous-buff; tail black tipped white.
Adult female (*merulinus*; rufous morph) Dark rufous heavily barred dark brown above; throat to upper breast rufous, rest of underparts whitish, barred black. (Adult female grey morph as male but abdomen barred whitish and rectrices more barred.)
Juvenile (*merulinus*) Variable from dark grey only with very small white bars on abdomen to rufous with dark brown barring on upperparts; underparts white barred black; tail black barred white, some edged rufous.
Adult male (*querulus*; Indian subcontinent, China and south-east Asia south to north Thai-Malay Peninsula) More strongly coloured than *merulinus*.

PLATE 31: *CACOMANTIS* CUCKOOS III

Brush Cuckoo *Cacomantis variolosus* Map and text page 404

Wallacea and Australasia. Forest (including mangrove and secondary) and edge, woodland, scrub, farmland, gardens; to 1,800m.

20–23cm. Small cuckoo with upright posture; great intraspecific variation. Flight fast and slightly undulating.

Adult male (*variolosus*; north and east Australia; southern populations migrate north to New Guinea, Moluccas and Lesser Sundas) Upperparts and wings grey-brown; head, neck and breast grey; dull buff belly and undertail-coverts; tail blackish-brown tipped and edged white and buff.

Adult female (*variolosus*; unbarred morph) As male but less buff below, breast finely barred grey.

Adult female (*variolosus*; barred morph; rare) Differs in having upperparts more or less barred and streaked pale buff, chin to undertail-coverts whitish barred dark.

Juvenile (*variolosus*) Above dark brown spotted and barred golden-rufous; whitish below streaked dark brown and washed buff.

Adult (*addendus*; Solomons) Upperparts blackish; chin and throat grey; rest of underparts variably dark rufous and grey.

Adult (*infaustus*; Moluccas; New Guinea and islands) Larger and darker than *variolosus*.

Adult ('*heinrichi*' = *infaustus*; Halmahera and Bacan) Darker and smaller than normal *infaustus*.

Rusty-breasted Cuckoo *Cacomantis sepulcralis* Map and text page 409

Thai-Malay Peninsula, Greater Sundas, Philippines and Wallacea. Forest (including mangrove and disturbed) and edge, woodland, beach scrub, plantations; to 2,500m.

22cm. Small cuckoo.

Adult male (*sepulcralis*; Thai-Malay Peninsula; Greater and Lesser Sundas; Philippines) Grey head, grey-brown upperparts, wings more bronze, rump grey; in flight underwing-coverts pink-rufous with white area; below clear, uniform pink-rufous with grey tinge on chin and throat gradually becoming pink-rufous on breast. Tail blackish-brown tipped white and edged with white spots, except T1; T4–5 spots larger becoming incomplete bars; underside paler strongly tipped white.

Adult female (*sepulcralis*; grey morph) As male but underparts paler and finely barred grey. (Adult female rufous morph dark brown above heavily barred and spotted rufous; below whitish heavily barred dark brown.)

Juvenile (*sepulcralis*) Head and upperparts buffish-rufous barred black; underparts white barred black, throat and breast washed buffish; tail dark brown notched dull rufous.

Adult (*virescens*; Sulawesi and related islands) Darker than *sepulcralis*; above more bluish-grey; below uniform dark chestnut, throat usually darker.

PLATE 32: COMMON CUCKOO AND LONG-TAILED CUCKOO

Common Cuckoo *Cuculus canorus* Map and text page 475

Europe through Russia to China, Korea and Japan; Indian subcontinent; most populations migrate south after breeding. Almost all habitats except arctic tundra and desert; to 5,250m.

32–34cm. Medium-sized cuckoo. Common posture rather horizontal with drooping wings, but when calling body more upright. Fast flight with constant wing-beats below horizontal; head, body and tail straight and bill pointing forward.

Adult male Grey upperparts, wings, throat and upper breast; rest of underparts white barred black, tail blackish tipped white.
Adult female (grey morph) As male but with some buffish wash on upper breast.
Adult female (hepatic morph) Red-brown barred black on upperparts and wings; barring on rump and uppertail coverts slight or absent.
Juvenile (dark morph) Dark brown head with white patch(es), upperparts and wings edged white; wings spotted rufous; underparts white barred dark brown; tail dark brown tipped and spotted white and rufous.
Juvenile (hepatic morph) Rufous-brown above fringed white and barred black; rump rufous slightly or extensively barred black with greyish-white tipping; uppertail-coverts rufous barred dark brown and tipped white; white patch(es) on head; wings barred blackish-brown and red-brown; below buffish and whitish barred blackish; tail barred black and red-brown, and broadly tipped white.

Mountain Long-tailed Cuckoo *Cercococcyx montanus* Map and text page 419

DRC and western Uganda to Zambia, Mozambique and Zimbabwe. Montane forest (*montanus*) to 2,800m; forest, miombo, coastal thicket and savanna to 2,100m (*patulus*).

33cm. All *Cercococcyx* cuckoos are small-bodied with long, full tail.

Adult (*montanus*; south-west Uganda, east DRC, Rwanda, Burundi and north-west Tanzania) Upperparts dark olive-brown with greenish sheen; buff-brown feather tips giving barred appearance above; wings barred buff; below whitish barred black brown and washed buff on upper breast and vent.
Adult (*patulus*; Kenya, east and south Tanzania, south Zambia, Malawi, Zimbabwe and Mozambique) Larger, paler and more heavily barred above, and less heavily barred below, than *montanus*.
Juvenile (*patulus*) Medium to dark brown above, flecked brownish-white on crown, otherwise barred dull rufous; remiges dark brown spotted rufous; tail dark brown barred rufous, barring whiter distally; underparts white washed tawny-buff with widely spaced dark brown streaks or bars.

Olive Long-tailed Cuckoo *Cercococcyx olivinus* Map and text page 417

West Africa (scattered localities) east and south to west Uganda, DRC, north-west Angola and north-west Zambia. Forest, secondary growth, thick bush; to 1,800m.

33cm.

Adult Upperparts plain olive-brown, sometimes glossed bronze; below whitish washed buff on breast and undertail-coverts barred dark; tail spotted and barred rufous, grey and white.
Juvenile Dark rufous crown; nape and wings to uppertail-coverts blackish-brown tipped rufous; underparts white washed tawny-buff with well-spaced dark brown bars; tail dark brown barred rufous.

Dusky Long-tailed Cuckoo *Cercococcyx mechowi* Map and text page 414

West Africa to south and west Uganda and extreme north-west Tanzania. Forest with dense undergrowth; tall secondary growth; to 1,800m.

33cm.

Adult Upperparts dark sooty grey-brown, with slight purple-blue iridescence; below white barred dark, breast and flanks variably washed buff; vent and undertail-coverts unmarked bright buff; white-tipped tail variably spotted and barred white and tawny-white.
Juvenile Similar to adult, but feathers of upperparts tipped rufous, giving barred appearance; rufous spots on wings. Blackish throat in some; rest of underparts barred white and blackish-brown; tail dark brown spotted and barred rufous.

PLATE 33: HAWK-CUCKOOS II

Javan Hawk-cuckoo *Hierococcyx fugax*   Map and text page 443
Thai-Malay Peninsula, Sumatra, Borneo and west Java. Forest (including secondary), bamboo, scrub, plantations; to 1,700m.
28–30cm. Smallish hawk-cuckoo; shy.
Adult Dark grey-brown above; white and pinkish-rusty on underparts with dark streaking; black spot on chin.
Subadult Black crown tipped buff; dark grey barred rufous on upperparts; underparts white with dark brown spots or streaks edged rufous.
Juvenile Grey above with buff edgings; white below broadly spotted dark.

Whistling Hawk-cuckoo *Hierococcyx nisicolor*   Map and text page 445
North-east Indian subcontinent; northern south-east Asia; southern China; northerly populations mostly migrate to southern south-east Asia and Greater Sundas. Forests, woodland, bamboo, plantations; also gardens on passage or winter; to 2,900m.
28–30cm. Smallish hawk-cuckoo.
Adult Blackish and dark grey above. Dark grey chin, face and sides of neck; undertail-coverts white; rest of underparts white streaked dark grey or brown strongly suffused rufous; tail grey banded black and tipped rufous.
Subadult Back and wings barred rufous; some rufous on tail; much less rufous below than adult.
Juvenile Mostly brown above with whitish tipping, more blackish on crown and nape with buff edging; white spot(s) on nape or sides of neck in some; remiges dark brown barred rufous and white; tail as adult with broader rufous tip, extreme tip buffy-white. Throat to upper breast dark brown with buff-white fringes, often with some white spotting; lower breast to belly creamy-white spotted brown, streaked on lower flanks; undertail-coverts white.

Rufous Hawk-cuckoo *Hierococcyx hyperythrus*   Map and text page 439
Ussuriland; Sakhalin; north-east and south China (some resident?); Korea and Japan; migrate south to south-east Asia and Borneo. Forest (including secondary), bamboo, plantations; to 2,800m.
28–30cm. Smallish hawk-cuckoo; shy.
Adult Slaty-grey above; variable amount of white and rusty pink on underparts; undertail-coverts white; cheeks and chin slaty-grey; narrow whitish vertical stripe from base of bill (in some extending in front of eyes) to chin and forming half-collar.
Subadult Grey head tipped buff, dark grey-brown barred rufous on upperparts, and underparts white with dark brown chevrons, spots or streaks and rusty suffusion.
Juvenile Dark grey-brown above with buff edgings; blackish throat; rest of underparts white streaked and spotted blackish from throat to belly.

Philippine Hawk-cuckoo *Hierococcyx pectoralis*   Map and text page 441
Philippines. Forest, often near water; to 2,300m.
28cm. Smallish hawk-cuckoo; shy and inconspicuous, more often heard.
Adult Head and upperparts slate-grey, paler on head; lores, throat and sides of neck white; breast and upper belly pale rufous; rest of underparts white; tail grey with black bands, tip rufous.
Subadult Head, neck, chin and sides of throat slate-grey washed brown; sometimes white hindneck band; upperparts and wings blackish-brown narrowly barred dark rufous; throat whitish finely streaked grey; rest of underparts buffy-white with dark brown chevrons, streaks or spots edged rufous; undertail-coverts buffish-white with dark edging; tail brownish-grey occasionally tinged rufous, tipped rufous with broad black subterminal band, and 2–3 narrow black bands.
Juvenile Black head barred rufous; upperparts grey barred rufous; underparts buffish-white heavily streaked and spotted dark brown with variable rufous tinge; tail as adult or washed rufous.

PLATE 34: *CUCULUS* CUCKOOS I

Red-chested Cuckoo *Cuculus solitarius* Map and text page 451
Much of sub-Saharan Africa; intra-African migrant. Forest and edge, woodland, savanna, avoiding extremely arid areas; to 3,000m.
31cm. Medium-sized cuckoo. Call loud and obvious.
Adult male Dark grey above, sides of head and throat pale grey; white barring and spotting on tail. Breast cinnamon, barred or unbarred; rest of underparts pale buff to creamy-white with narrow blackish-brown bars.
Adult female Similar to male but with less cinnamon on breast, and greater extent of barring.
Juvenile Blackish above with whitish edgings; white on hindcrown; throat to upper breast black, flecked or lightly barred white; rest of underparts barred black and buffish-white.

Black Cuckoo *Cuculus clamosus* Map and text page 447
Much of sub-Saharan Africa; *clamosus* mostly migratory. Forest, miombo, woodland in drier savannas, *Acacia* thornveld, thickets. suburban gardens; below 2,000m.
31cm. Medium-sized, fast-flying cuckoo; blackish, but underparts highly variable.
Adult male (*clamosus*; Eritrea south through eastern Africa to RSA, migrating to equatorial west Africa; black breasted) Mainly black with white-tipped tail that may have white spotting or barring; remiges barred dull white; underparts variable, may show white or buff or reddish barring, with variable amount of chestnut from chin to breast.
Juvenile Black, and lacks white tips to tail.
Adult male (*clamosus*; bar-breasted) From black-breasted by paler brown or whitish barring from breast to vent.
Adult female (*clamosus*) From male bar-breasted by heavier whitish barring from breast to vent.
Adult male (*gabonensis*; probably non-migratory in west Africa east to south-west Kenya, south to Angola) Chin (sometime dark grey) to upper breast rufous or chestnut (very little or absent in some); sometimes upper breast has dark barring; rest of underparts whitish or tawny with light or heavy dark barring. (Occasionaly whole underside dark with faint pale barring.)
Adult female (*gabonensis*) Duller than female *clamosus*. Throat mostly rufous (almost absent in some) with dark barring; rest of underparts whitish barred dark, in some giving a scaly appearance.
Juvenile (*gabonensis*) Mostly blackish; some paler markings on wings and underparts; no white in tail.

African Cuckoo *Cuculus gularis* Map and text page 465
Much of sub-Saharan Africa; intra-African migrant. Woodland and savanna; to 3,000m.
32cm. Long-tailed cuckoo with appearance of small raptor. Dashing, hawk-like flight.
Adult male Upperparts, including wings dark grey, paler grey on throat and breast, rest of underparts white barred dark. Yellow bill tipped dark, orange at base.
Adult female Less grey on chest than male and faintly barred; breast sometimes washed buff or tawny.
Juvenile (grey morph) From adult by white barring and tipping on upperparts, including wings, and patch on hind crown; pale tawny suffusion on head to upper back; uppertail-coverts brown broadly barred white; underparts creamy-white barred grey-brown, much more closely on chin to breast. Broad white spots and notches on tail, but no distinct barring except T5. (Juvenile hepatic morph brownish instead of grey, tawny or buff instead of white.)

See also Common Cuckoo, Plate 32

PLATE 35: *CUCULUS* CUCKOOS II

Indian Cuckoo *Cuculus micropterus*   Map and text page 459

From Pakistan, Himalayas, India to Siberia, Indochina, China, Korea, Greater Sundas and Philippines; northern populations migratory. Forest (including secondary) and edge, woodland, parkland, scrub; to 3,800m.
32–33cm. Medium-sized cuckoo with hawk-like jizz.
Adult male Head grey; upperparts and wings grey-brown; throat to upper breast grey, rest of underparts white with broad, widely spaced black barring. Tail grey with broad subterminal black band tipped white; white barring on T5 and shafts of T1.
Adult female Browner; throat to breast pale rufous barred dark.
Subadult Head and nape slaty-grey, sides of neck and breast rufous barred dark, back dark brown edged blackish, rump and uppertail-coverts grey-brown edged blackish and rufous; rest of underparts white barred black.
Juvenile Dark brown above; head to nape heavily spotted white to tawny-white; feathers of rest of upperparts broadly fringed white to rufous; blackish mask below eyes; below creamy-white variably barred brown.

Sulawesi Cuckoo *Cuculus crassirostris*   Map and text page 458

Sulawesi; Buton I. Forest (including secondary) and edge, often near water; 500 (exceptionally 200)–1,400m.
34–38cm. Medium-sized cuckoo. Considerable variation in plumage.
Adult Entire head ashy-grey; yellow eye-ring; upperparts, wings and tail dark rufous-brown; underparts white broadly barred black; tail barred black and white.
Subadult Head black and white; upperparts and wings rufous spotted dark brown, some remiges tipped white; below pale buff with sparse dark brown barring; nearly unbroken band across breast; undertail-coverts buffy-white. Tail black barred rufous and white, tipped white.
Juvenile Head white with some black; upperparts bright rufous; underparts buffy-white.

Lesser Cuckoo *Cuculus poliocephalus*   Map and text page 455

Afghanistan east to northern south-east Asia, China, Ussuriland, Sakhalin, Korea and Japan, migrating south in Asia and to eastern Africa from Kenya to Zimbabwe. Forest (including secondary), woodland, savanna; open scrub, plantations; to 3,360m.
25cm. Small, slim cuckoo with long narrow wings, often with upright stance. Flight direct and fast, with flat wing beats interrupted by short glides.
Adult male Slaty blue-grey above, paler on head, neck, throat and upper breast, more blackish on uppertail-coverts; lower breast and rest of underparts white with bold black barring; undertail-coverts pink-buff barred black; tail dark with white tips and white spots on all rectrices.
Adult female (grey morph) As male but washed rufous on upper breast.
Adult female (hepatic morph) Rufous above with blackish barring except on rump, and little on nape; underparts white barred black. Tail barred black and rufous, spotted and tipped white.
Juvenile (grey morph) Head blackish barred white; upperparts slaty-grey variably edged white to tawny; wings and coverts spotted and notched rufous and white; underparts barred black and white; tail blackish barred, notched and spotted white. (Juvenile hepatic morph barred tawny and black above, and brown and buff below.)

Madagascar Cuckoo *Cuculus rochii*   Map and text page 463

Madagascar; most migrate to eastern and southern Africa. Forest and edge, dense cover in savanna, scrub woodland, spiny subdesert woodland, marshes, plantations and other human-influenced habitats; to 1,800m
26–28cm. Smaller than other *Cuculus* in range except Lesser Cuckoo.
Adult male Dark grey above; throat and upper breast paler grey; undertail coverts pink-buff normally unbarred; rest of underparts white, barred heavily and broadly black. Tail dark grey, spotted and barred white.
Adult female As male, but with some tawny on breast and sides of neck; undertail-coverts with broken blackish bars. (Existence of female hepatic morph not proven.)
Juvenile Much darker grey than adult; tawny nape and sides of neck, edges to crown feathers and barring on wings, breast, throat and tail; wing feathers tipped white; undertail-coverts white barred black.

PLATE 36: *CUCULUS* CUCKOOS III

Oriental Cuckoo *Cuculus saturatus* Map and text page 468

Much of Palearctic and Oriental Regions, including Wallacea; New Guinea; south-west Pacific; Australia. Forest (including mangrove and secondary) and edge, woodland, savanna, secondary growth, plantations, gardens; to 4,500m.

32–33cm. Medium-sized cuckoo with hawk-like jizz; rapid undulating flight more graceful than Common Cuckoo.

Adult male (*saturatus*; southern Himalayas to northern south-east Asia and south China, wintering further south in Asia) Head, neck and upper breast plain ashy-grey; rest of upperparts and wings dark grey; underparts from hindbreast to undertail-coverts white with bold and regular black barring, and latter with buff wash; tail dark grey tipped white.

Adult female (*saturatus*; grey morph) Like male but neck and upper breast washed rufous.

Adult female (*saturatus*; hepatic morph) Rufous with dark barring on head, upperparts, wings and tail; white with fine black barring on sides of head, chin, throat and upper breast; rest of underparts white with black barring; tail tipped white.

Juvenile (*saturatus*; grey morph) Grey-brown above, most feathers finely edged buff-white, on wing and rump also dull rufous edging; underparts white to buffish-white barred black.

Juvenile female (*saturatus*; hepatic morph) Like adult female hepatic morph but with white feather fringes on upperparts and wings.

Adult male (*optatus*; European Russia east to Kamchatka, north China, Korea and Japan, wintering further south in Asia, Wallacea, New Guinea, south-west Pacific and Australia) Larger than *saturatus*, with slightly broader black barring on chest to belly; juvenile has white barring to black feathers of crown, nape, throat and breast, whereas nominate has buff-white edging.

Adult female (*optatus*; grey morph) Similar to *saturatus*, but usually less rufous below.

Adult female (*optatus*; hepatic morph) Similar to *saturatus*, but spacing of ventral barring greater.

Juvenile (*optatus*; grey morph) Similar to *saturatus*, but feather edging above whiter. (Juvenile hepatic morph as *saturatus* but larger.)

Sunda Cuckoo *Cuculus lepidus* Map and text page 473

Peninsular Malaysia; Greater and Lesser Sundas. Forest (including secondary; chiefly montane); 500–2,750m. 29–30cm. Smallish cuckoo.

Adult male Mostly dark grey above; tail barred white; grey chin to breast, rest of underparts white barred blackish; buff undertail-coverts mostly unbarred.

Adult female (rufous morph; Sumatra and Java only) Mostly barred chestnut and blackish above; rufous throat barred dark; rest of underparts whitish barred dark; undertail-coverts buffish. (Adult female grey morph similar to male but ochraceous on breast, and narrower ventral barring.)

Juvenile Dark grey to blackish above; tail barred white; white nape patch; below barred black and white, more buffy on undertail-coverts.

FAMILY CUCULIDAE
SUBFAMILY CROTOPHAGINAE

Genus *Guira*

Lesson, 1830 *Traité d'Orn.* livr. 2: 149. Type, by monotypy and tautonymy, *Cuculus Guira* Gmelin 1788. One species.

South America; non brood-parasitic; co-operative breeder. Medium-sized; sexes alike; eight rectrices; crested; streaked brownish plumage. Closest to *Crotophaga* (Posso & Donatelli 2001, Sorenson & Payne 2005).

GUIRA CUCKOO
Guira guira   Plate 1

Cuculus Guira J.F. Gmelin, 1788 (Brazil)

Alternative name: White Ani.

TAXONOMY In past in *Octopteryx* Kaup 1836 (e.g. von Pelzeln 1870). Monotypic. Synonyms: *Crotophaga piririgua* Buffon 1789, *Ptiloleptus cristatus* Swainson 1837, *pirigua* Strickland 1841.

FIELD IDENTIFICATION 36–42cm, including 20cm tail. Usually in flocks. **Adult** Reddish-ochre head with shaggy crest, yellow bare facial skin, yellow to orange bill and iris, dark brown back finely streaked pale, white to buff rump; underparts pale, streaked dark on throat and breast, tail long, basally yellowish-ochre, otherwise blackish, broadly tipped white. **Juvenile** Similar; remiges with small white tips, bill greyish or black-and-white, iris grey. **Similar species** None.

VOICE Sequences of mournful whistles *piiioo, piiioo*; alarm slow rough or gargled high-pitched trill or rattle, varying in pitch, tempo and amplitude, *keerrrrrrrrr*, lasting 0.5–2.5sec, given with head thrown back, crest raised and body shaking; occasional *creeep* of varying pitch; cacophonous series of warbling notes, gradually falling in pitch and becoming scratchy *kee-ay kee-ay kee-eh kee-orr keeoh cure cure* given by group members one after another, overlapping or simultaneously; quiet groans *yew yew yew yew…* in flight. Loud *kooeet* raptor warning. Fifteen calls recorded (Friedmann 1927, Eisentraut 1935, Davis 1940a, Belton 1984, Fjeldså & Krabbe 1990, Macedo 1992, Sick 1993, Fandiño 1986 in Sick 1993, de la Peña & Rumboll 1998, Restall *et al.* 2006).

DESCRIPTION Adult Head sandy-buff, paler around eyes; irregular, untidy, crest from forecrown to nape, each feather streaked black medially; upper back dark brown streaked white, wing-coverts black edged white, primaries mostly cinnamon, darker distally, secondaries dark brown, narrowly edged pale, lower back and rump buffy-white, uppertail-coverts orange-buff; underparts buffy-white, throat and breast with fine black shaft streaks; tail slightly graduated, central rectrices dark brown except at base, rest blackish broadly tipped white, and all rectrices with pale buffy bases. **Juvenile** Similar to adult, but remiges tipped with small white spot; terminal white tail band narrower. Mouth lining pink, palate white, tongue pink with subterminal black band and white tip. **Nestling** Eyes open upon hatching. Skin dark, greenish around eyes; creamy-white trichoptiles; iris dark brown; bill orangish with black line along culmen, and dark mandibular rami; palate pink, with large white ring, open laterally, on each side of palate, and white bars, spots and papillae; tongue pink with black band and white tip; feet dark grey. **Bare parts** Iris usually yellow or yellowish-white, in some birds orange; grey in juvenile. Bare facial skin pale yellow to pale pea-green. Bill normally yellow, in some orange or salmon, culmen sometimes brown; juvenile black and pinkish-white. Feet bluish to greenish-grey. Claws grey to black.

BIOMETRICS Wing male 170–187mm (n = 10), mean 178.5 ± 6.1, female 161–180mm (n = 8), mean 174.3 ± 5.9. Bill male 27–32mm, mean 29.2 ± 1.4, female 26–30mm, mean 27.8 ± 1.5. Tail male 214–235mm, mean 223.2 ± 6.6, female 198–243mm, mean 227.5 ± 13.2 (Payne 2005). Tarsus female 41mm (Pinto 1964), unsexed 38.1, 39.3, 39.3mm (Bugoni *et al.* 2002). Mass male 128.6–168.2g (n = 19), mean 138.7g, female 113.0–168.6g (n = 8), mean 144.8g (Payne 2005).

MOULT Body moult Mar, north Argentina. Espírito Santo, Brazil: wing and tail moult Jan, Apr, May (n = 7) (Willis & Oniki 2002, Di Giacomo 2005).

DISTRIBUTION South America. Resident. Brazil from south-east Amapá and islands in Amazon estuary southwards outside Amazonia; west to south Mato Grosso and Bolivia (Beni, Cochabamba, Santa Cruz, Chuquisaca and Tarija), Paraguay, Uruguay and Argentina south to Mendoza, east Neuquén and Chubut. Formerly occurred only in dry savannas, e.g. interior Brazil, but has expanded into deforested regions. Thus colonised Belém area, Pará, after 1982. Likewise, vanishes when open area is forested (von Pelzeln 1870, Allen 1893, Gazari 1967, Novaes 1974, Short 1975, Pinto 1978, Nores & Yzurieta 1981, da Silva & Oren 1990, Sick 1993, Arribas *et al.* 1995, da Silva *et al.* 1997, Henriques & Oren 1997, de la Peña 1999, Azpiroz 2001, Willis & Oniki 2002, Claramunt & Cuello 2004, Guyra Paraguay 2005, Veiga *et al.* 2005, Lees *et al.* 2008). Extralimital Curaçao (Voous 1955, 1983) almost certainly escape (Restall *et al.* 2006). Erroneously recorded Chile (Hellmayr 1932).

HABITAT Savanna, forest edge, fields, gardens, coastal dunes, pampas, cerrado and chaco woodlands, usually dry habitats, urban and suburban areas and parks, orchards, roadsides. To 1,200m or more; to 1,800m along Andes, reaching transition to temperate zone in Tucumán, Argentina (Reichholf 1974, Belton 1984, Fjeldså & Krabbe 1990, Canevari *et al.* 1991, Arribas *et al.* 1995, Stotz *et al.* 1996, de la Peña & Rumboll 1998).

BEHAVIOUR Conspicuous, often perching in open on bushes, wire fences and ground. Tame around farm buildings and cattle stations. Sometimes with Smooth-billed Anis. In early morning, towards evening and after rains, flock members sunbathe together, with backs turned towards sun, feathers fluffed and wings lowered. Communal night roost in favoured, dense tree or bamboo thicket, often in a closely packed circle, all birds facing outwards, and tails converging distally, or huddle side by side on branch. Some birds die during cold winter nights. Flight clumsy and slow, alternating series of wing beats with short glides. Group members allopreen. Mostly forages on ground, normally in small flocks, individuals c. 1m apart, running, walking and jumping after prey, or flies, glides or jumps to ground from perch, after first localising prey. Prey is swallowed whole. Peak feeding activity 10–13hrs. Catches flying termites, perching on top of tree, in middle of swarm. Strong body odour can attract carnivores and blood-sucking bats. When perched, cocks tail and throws it onto back. Regurgitates pellets. Scares hawks away. Attacked by other birds, e.g. Tropical Kingbird. Courtship: male dances with open wings around female on ground. Copulation on ground. Male stepped vigorously on tail of female, pinned her to ground, climbed onto her back and immediately jumped off to her side, and copulated while perched on ground, beating his wings and covering her with one wing, all the time pecking her head.

During breeding season groups of 2–18 (typically 6–8) keep close together, defend a territory and share single nest with communal clutch. Flocks up to 26 outside breeding season; some but not all of group related. Up to seven females lay in same nest, some in group lay more eggs than others. Fewer eggs often incubated than laid because eggs buried by adding leaves to nest lining or evicted throughout laying and incubation periods. Adults may carry eggs from nest in bill and drop them 10m away or more, or directly over rim of nest. Egg tossing more prevalent at onset of laying, but may happen all through incubation, including pipping eggs; unhatched eggs often removed after others hatch. Half of eggs laid evicted on average. A female that has not begun to lay often removes eggs of other females. Individual females lay eggs of variable size, shape, colour and speckling pattern, and are unable to tell their own eggs from those of other females and so will avoid risk of evicting their own eggs.

A few eggs buried inside nest may be uncovered and hatch. After brood hatches, one adult stays near nest while other adults forage away from nest, and with loud alarm calls summons other group members if threat approaches nest, and group then flees or confronts predator.

Nestlings frequently removed and/or killed by adults in group, mostly within first five days after hatching, sometimes leading to abandonment of nest. Partial or total loss of brood occurred in 69% of nests that produced nestlings. Large eggs result in heavier hatchlings less victimised by infanticide.

Most or all adults in breeding group feed nestlings, although effort often differs. Some nestlings in large broods receive more food than others, and 18% of 33 adults (at four of seven nests studied) fed certain chicks more than others. In nests with two or more breeding pairs, nestlings often have different parents, those that are related are usually half-siblings, and adults are polyandrous and polygynous.

Breeding group may nest up to five times in same territory in one season. Renesting occurs sooner (mean 35 days interval between first laying dates) after failed nests than after successful nests (mean 66 days) (Sclater & Hudson 1889, Friedmann 1927, Davis 1940a, Gazari 1967, Salvador 1981, Reichholf 1974, Belton 1984, Gallardo 1984, Storer 1989, Canevari et al. 1991, Cavalcanti et al. 1991, Macedo 1992, 1994, Sick 1993, Quinn et al. 1994, Macedo & Bianchi 1997, Macedo & Melo 1999, Macedo et al. 2001, 2004a,b, Martins & Donatelli 2001, Cariello et al. 2002, 2004, 2006, Willis & Oniki 2003, Di Giacomo 2005, Restall et al. 2006).

BREEDING Nonparasitic, but may egg-dump, e.g. in nests of Smooth-billed Ani, Southern Caracara, Chimango Caracara, White-tipped Plantcutter and Southern Lapwing (Serié 1923a, b, Eisentraut 1935, Sick 1993, Jenny 1997). Two eggs of parasitic Screaming Cowbird found with one cuckoo egg in cuckoo nest (de la Peña 2005). **Season** May Piauí, north-east Brazil. During rains, Jul–Mar, peak Sep–Oct, central Brazil. Testes enlarged Nov, ovaries enlarged Feb, nestlings late Apr, São Paulo. Aug–Feb Rio de Janeiro. Enlarged testes, nestlings being fed or in nest, juvenile with short tail, Nov–Feb Rio Grande do Sul. Nov–Dec (longer?) south Bolivia. Oct–Dec Uruguay. Oct–Mar Formosa, Santa Fe and Buenos Aires provinces, Argentina. Nov Tucumán and Jan Córdoba. Two broods sometimes raised per season (Euler 1867, Gibson 1885, Sclater & Hudson 1889, Reiser 1926a, Friedmann 1927, Smyth 1928, Eisentraut 1935, Salvador 1981, Belton 1984, Gallardo 1984, Cavalcanti et al. 1991, Macedo 1992, 1994, Melo & Macedo 1997, de Magalhães 1999, de la Peña 2005, Di Giacomo 2005). **Nest** Large open cup of sticks, roots and grass stalks, lined with green leaves, straws and flowers, outer diameter 15–40cm, inner 14–20cm, height 14–20cm, depth 10–15cm. Placed in isolated tree or on large cactus, in central Brazil usually in introduced thorny *Araucaria* trees, as high as possible, at 3.0–12.2m, mean 7.5 ± 2.2 (n = 79); 1.0–8.0m up (n = 38), Santa Fe, Argentina. Exposed on branches of trees, sometimes covered by vines, in thickets, growths of rushes and irises, on sheds, mills and other constructions. Old nests often renovated and reused, 98% of 86 breeding attempts in study area in Brazil; or group nests in another tree during same season. Also sometimes reuses or takes over nests of other birds, e.g. Chalk-browed Mockingbird, adding green leaves (Goodall 1923, Daguerre 1924, Friedmann 1927, Canevari *et al.* 1991, Cavalcanti *et al.* 1991, Macedo 1992, de la Peña 2005). **Eggs** Large in relation to body mass: 17–25%. Elliptical, sea-green, covered with latticework of calcareous white in high relief that may be rubbed off soon after egg laid, or dries hard; 37.6–48.8x27.8–37.6mm, mean (n = 148) $42.5 \pm 1.9 \times 31.8 \pm 1.4$. Communal clutch size variable, e.g. 4–20 eggs, mean (n = 33) 10.0 ± 3.7; up to 30 eggs recorded per nest. Egg number positively correlated with number of reproductive females in group. During one season 17 groups laid 3–28 eggs, mean 13.3 eggs, in 1–3 nesting bouts, 2–10 laying females per group. Mass, fresh, 19.7–22.4g, mean (n = 6) 20.8. Laid every second day, sometimes on two consecutive days. Two-four eggs commonly laid on same day, by different females. Each female can lay up to 5–7 eggs during one nesting bout. Hatching usually synchronous, e.g. within 24 hours in 75% of 28 clutches; longest span four days (Friedmann 1927, Salvador 1981, Macedo 1992, Sick 1993, Walters 1995, Macedo & Bianchi 1997, Macedo *et al.* 2001, Cariello *et al.* 2002, 2004, de la Peña 2005). **Incubation** At least sometimes by male; 9–16 days (Reiser 1926a, Azategui 1975, Salvador 1981, Cavalcanti *et al.* 1991, Sick 1993, Macedo & Bianchi 1997, Di Giacomo 2005, de la Peña 2005). **Chicks** Eye-like mouth markings probably frighten predators. Fed mostly insects, especially grasshoppers, also frogs, toads, lizards, snakes, birds, nestlings and mammals, most smaller than 6cm, but some larger than 12cm, and once 15cm worm lizard. Items swallowed whole. Feeding of nestlings all day, without peak. Nestling period 12–18 days. At 2 days, eyes brown, body covered with white trichoptiles. At 3–4 days begin to scramble about nest tree when adult arrives with food, and beg loudly, flapping wings. At 5 days, webs appear, tail with long pinfeathers, rest of body with white-tipped dark pinfeathers; nest smells strongly from excrement. At 7 days, wing and tail feathers dark, webs on sides of chest yellowish-white, tail 5cm; leave nest if disturbed. At 10 days, plumage similar to adult; may stay outside nest or hide in it on approach of observer; clamber about branches with support of beak and wings. If fall out of nest, can climb back up *Araucaria* tree, grasping trunk with wings, in spite of trunks and branches being covered by prickly, stiff leaves. More than half grown when leave nest. Fed by adults at least 3 weeks after fledging. Attain adult size at one month (Salvador 1981, Macedo 1992, 1994, Sick 1993, Macedo & Bianchi 1997, Melo & Macedo 1997, de la Peña 2005, Payne 2005, R. Macedo in Payne 2005). **Survival** 26% of eggs and 55% of chicks survived until fledging (Macedo 1992).

FOOD Hemiptera (including cicadas), beetles, mantids, ants, flies, swarming termites, locusts, butterflies and caterpillars; spiders; woodlice; small amphibians; lizards, snakes; eggs and nestlings of other birds, e.g. Screaming Cowbird, Pied Water Tyrant, Fork-tailed Flycatcher and House Sparrow; mice; kitchen rubbish (von Pelzeln 1868, Sclater & Hudson 1889, Daguerre 1922, Wetmore 1926, Friedmann 1927, Aravena 1928, Moojen *et al.* 1941, Hempel 1949, Schubart *et al.* 1965, Belton 1984, Gallardo 1984, Mason 1985, Lima & Lima 2004, de la Peña 2005).

STATUS AND CONSERVATION Common (Stotz *et al.* 1996). Abundant Rio Grande do Sul, Brazil (Belton 1984), and Buenos Aires province, Argentina (Narosky & Di Giacomo 1993), common San Luis (Nellar 1993) and Santiago del Estero (Nores *et al.* 1991). Scarce and local Neuquén (Veiga *et al.* 2005). Common to abundant Paraguay (Hayes 1995). Population probably increased due to deforestation in south-east and south Brazil, although does not tolerate short grass or intensive agriculture (do Rosário 1996, Willis & Oniki 2003). Not globally threatened (BirdLife International 2011).

Guira Cuckoo *Guira guira*. Adults. **Fig. 1.** Sunbathing. Ibera Marshes, Argentina, July (*Roy de Haas*). **Fig. 2.** Belém, Pará, Brazil, June (*Stefan Hohnwald*). **Fig. 3.** Foraging. Belém, Pará, Brazil, September (*Stefan Hohnwald*). **Fig. 4.** Belém, Pará, Brazil, June (*Stefan Hohnwald*). **Fig. 5.** Belém, Pará, Brazil, October (*Stefan Hohnwald*).

Genus *Crotophaga*

Linnaeus, 1758 *Syst. Nat.* (ed. 10) 1: 105. Type, by monotypy, *Crotophaga ani* Linnaeus 1758.
Three species.

New World; non brood-parasitic co-operative breeders. Medium-sized to large; sexes similar; eight rectrices; crestless; black plumage; deep, laterally compressed bill; naked lores. Sister of *Guira* (Posso & Donatelli 2001, Sorenson & Payne 2005).

GREATER ANI
Crotophaga major   Plate 1

Crotophaga major J.F. Gmelin, 1788 (in Cayenna = Cayenne, French Guiana)

Alternative name: Great Ani.

TAXONOMY Sister taxon to clade consisting of *C. sulcirostris* and *ani* (Posso & Donatelli 2001, Hughes 2003, Sorenson & Payne 2005). Monotypic. Isolated population in west Ecuador seemingly not morphologically distinct (Ridgely & Greenfield 2001). Synonym: *ivahensis* Sztolcman 1926.

FIELD IDENTIFICATION 42–48cm. Large, long-tailed. **Adult** All black glossed violet and bronze, conspicuous pale eyes and massive bill with high ridge on basal half giving a broken-nose effect. **Juvenile** Duller, with less metallic gloss. Bill without ridge, eyes dark. **Similar species** Unmistakable, even compared with other two anis, that both are smaller, dark-eyed, duller, and usually occur in different habitats. Juvenile also larger, and usually found with adults. Combination of size, glossy plumage, pale eyes and bill shape should rule out all other birds. Male Great-tailed Grackle is almost same size, with pale eyes, but has slender bill and different, creased tail. Carib Grackle much smaller, has proportionately shorter tail, and thin bill.

VOICE Often noisy; gives many different sounds. Most frequent low-pitched, guttural gobbling *kro-koro*, or *toodle-doodle-doodle*, repeated rapidly for up to 30+sec, often in loud chorus, with others in group giving *shhhrrrrrrrr*, all members often sitting on same branch, stopping abruptly; thought to serve to maintain group and mark territory. In flight repeated loud hoarse croak *kqua*; alarm call harsh rasping note repeated several times; danger note for flying predators three sharp croaks; throaty *kuk*. Low *coró-coró* while foraging on ground. Also croaks, hisses, whirs, grates and low-pitched, drawn-out growl. Grunted *scaub*, hollow squawk *skok*, hissing *scauhhhhh*, series of *tick* notes, loud *skort* notes followed by repeated snoring groans *cra-a-a-ah-hhhhh*, of which several are probably the same as those above (Davis 1941, Willis 1983a, Novaes & Lima 1998, de la Peña & Rumboll 1998, Ridgely & Greenfield 2001, Hilty 2003).

DESCRIPTION Adult All glossy blue-black or blackish steel-blue (becoming more nearly black with wear). Feathers on head, sides of face and throat narrow and pointed, feathers of forehead and crown and sides of head edged dull bronzy, feathers of neck and throat more broadly edged bright greenish-bronze, mantle, scapulars and lesser wing-coverts somewhat scaled or edged glossy bronze-green, proximal portion of primaries with greenish hue, underside of remiges glossy greenish-blue; tail long, loose-jointed, rounded, with purplish gloss above, glossy dark violet below. **Juvenile** Body and head dull black.

Nestling Hatches naked, with black skin and yellow gape. Growing feathers black, retaining tips of sheaths that fall off at fledging. **Bare parts** Iris light greyish-white, ivory-white, yellowish-white, greenish-white, light greenish-yellow, light green, or perhaps pale blue; dark brown in juvenile. Lores naked, black. Bill black, laterally compressed, culmen with elevated thin and arched ridge, forming hump on basal half to two thirds, highest anteriorly, sides of maxilla with several more or less distinct longitudinal ridges and grooves; smaller in juvenile. Feet black.

BIOMETRICS Wing male 184–217mm (n = 10), mean 202.6 ± 8.7, female 192–211mm (n = 8), mean 201.0 ± 6.7. Bill male 46–48mm, mean 47.2 ± 0.9, female 43–46mm, mean 45.1 ± 1.5. Bill depth at nostrils male 22.1–24.7mm, mean 23.3 ± 0.8, female 21.0–24.4mm, mean 22.1 ± 1.1. Tail male 252–282mm, mean 271.8 ± 9.1, female 248–276mm, mean 262.6 ± 10.2. Tarsus male 42–44mm, mean 42.6 ± 1.0, female 38–41mm, mean 39.4 ± 0.9 (Payne 2005). male tarsus up to 49.3mm, and female wing down to 178mm, tail down to 237mm and tarsus up to 46.8mm (Wetmore 1968). Mass male 139.5–259g, female 139.5–183g (Strauch 1977, Haverschmidt & Mees 1994, Payne 2005).

MOULT Tail moult continuous (Davis 1941). No further information.

DISTRIBUTION Neotropical. East Panama (from west Colón on Caribbean side and Canal area on Pacific side eastwards), and from north-west, north and east Colombia, Venezuela, Trinidad and Guianas southward, east of Andes to east Bolivia (Pando, Beni, La Paz, Cochabamba, Santa Cruz), Paraguay, west Uruguay and north Argentina (Jujuy, Salta, Tucumán, Formosa, Chaco, Misiones, Corrientes and north-east Santa Fe to La Pampa, Buenos Aires and Córdoba, regularly perhaps only to Corrientes and Río Uruguay valley, Entre Ríos/Uruguay; single record Santiago del Estero, very rare San Luis, casual La Pampa and La Rioja; not recorded Catamarca?). Also Ecuador west of Andes (Giacomelli 1923, Ménégaux 1925, Pereyra 1941, Meyer de Schauensee 1966, Höy 1969, Short 1975, Ochoa de Masramón 1983, Hilty & Brown 1986, Canevari *et al.* 1991, ffrench 1991, Nores *et al.* 1991, Nellar 1993, de la Peña 1994, 1997, Arribas *et al.* 1995, AOU 1998, de la Peña & Rumboll 1998, Azpiroz 2001, Ridgely & Greenfield 2001, Hilty 2003, Claramunt & Cuello 2004, Guyra Paraguay 2005, Restall *et al.* 2006). Two 1960 specimens south-east Tamaulipas, north-east Mexico (Olson 1978) vagrants or now-extirpated population as not encountered in region 1910 nor 1980s (Howell & Webb 1995). Sight record Aruba (Mlodinow 2006). Few sight records Tortuguero, Costa Rica, 2003–2004 (Garrigues & Dean 2007).

Resident in most of range, but movements not well understood. Seasonal occurrence or wandering reported from many regions, e.g. appears when riverine habitat floods

with onset of rains. In llanos savanna, Venezuela, occurs rainy season (Apr–Nov), with small numbers lingering until Jan, and few through dry season. Migratory Paraguay, with records Sep–May, once Jul. Probably also migratory Uruguay. Recorded mostly summer (Aug, Oct–Nov, Jan–Mar), Misiones, north-east Argentina, breeding not observed Santa Fe and Entre Ríos, migratory San Luis, and considered migratory in whole Argentinean range. Suspected to be only summer resident Rio Grande do Sul, south Brazil, with records mid Nov–early Mar, and likewise only recorded summer São Paulo (Pinto 1966, Thomas 1979, Ochoa de Masramón 1983, Belton 1984, Terborgh *et al.* 1984, Moskovits *et al.* 1985e, Chébez 1993, Davis 1993, Narosky & Di Giacomo 1993, Olmos 1993, Hayes *et al.* 1994, Stotz *et al.* 1996, de la Peña 1997, Robinson 1997, Barnett & Pearman 2001, Bencke 2001, Hilty 2003, Willis & Oniki 2003, Di Giacomo 2005, Guyra Paraguay 2005, Schulenberg *et al.* 2007, Azpiroz & Menéndez 2008).

HABITAT River edges, wetlands, swamps, reed beds, humid forest, mangroves, gallery forest, flooded Amazonian forest (igapó), margins of oxbow lakes. Generally below 800m. Mostly below 500m Colombia, but at 2,600m east Andes. To 200m Venezuela. Below 50m west Ecuador; to *c.* 1,000m east of Andes, most numerous below 400m. Maximum 900m Peru. Below 500m Bolivia, once 2,550m (Borrero 1946, Parker *et al.* 1982, Hilty & Brown 1986, Fjeldså & Krabbe 1990, ffrench 1991, Chébez 1993, Arribas *et al.* 1995, Stotz *et al.* 1996, 1997, Robinson 1997, Borges *et al.* 2001, Ridgely & Greenfield 2001, Hilty 2003, Restall *et al.* 2006, Schulenberg *et al.* 2007).

BEHAVIOUR More wary than other anis. Usually occurs independently of other birds, in groups of up to 20 or more Venezuela, more than 150 upper Amazonia and even hundreds Guyana. When breeding, 4–10/group. Flies better than the small anis, alternating flapping and sailing. When at rest, repeatedly swings long tail up over body, then lets tail fall. Sunbathes with wings spread on early mornings. Raises tail when alarmed and droops wings. Group sleeps together in dense tree or shrub, individuals adjacent to one another but without touching each other. At all times birds of group stay within calling distance of one another. Stays mostly quite low down and sneaks around dense shrubbery at margin of lagoon or creek, or sits in sun at edge. If frightened, they fly out one by one and dive into vegetation on opposite shore. Searches foliage for prey, hopping rather heavily and clumsily, and lunges awkwardly at prey. Also takes prey from ground. Flocks may follow Amazonian troops of capuchin and squirrel monkeys (*Cebus, Saimiri*) that flush insect and lizard prey in flooded forest and at oxbow lakes. Occasionally follows swarms of army ants, where behaves aggressively towards antbirds, sallying to ground or foliage, or hop-running along ground to capture prey, ending with tail up and spread. Scratches head under wing (W. Beebe in Chubb 1916, Belcher & Smooker 1936, Davis 1941, Wetmore 1968, Terborgh 1983, Willis 1983a, Ayres 1985, Hilty & Brown 1986, Haverschmidt & Mees 1994, Robinson 1997, Ridgely & Greenfield 2001, Hilty 2003, Riehl 2010).

Breeding communal, each group constructing single nest, behaving aggressively towards other groups, defending a territory, although wandering birds may be tolerated within territory without fighting. Eggs laid by more than one female in same nest. First-laid eggs usually ejected; four groups with ≥4 females ejected 8–19 eggs, then abandoned nests without initiating incubation. Groups drive away Snail Kites and Black-collared Hawks from nest with nestlings (Davis 1941, 1942, Short 1975, Sick 1993, Lau *et al.* 1998, Riehl & Jara 2009).

Male stood with chest high next to female on log with 16cm dead juvenile green iguana in bill, brought its head near female's bill, latter adopted typical precopulatory posture (slightly spread wings, tail and bill upwards) and took head of iguana into her mouth, male held iguana near tail, flapped wings and hopped onto female's back, lowered his tail and copulated, passing iguana into female's throat, released his grip, dismounted, perched beside female, extended wings, lowered head, erected body feathers and shook vigorously; after short while both birds flew up into tree. Another display begins with one individual, often male, giving series of loud, high-pitched cackling notes, other group members immediately fly in and perch beside caller, group forming circle with beaks pointing towards centre, bodies held horizontally, tails pointing outwards and all giving low gurgling vocalisation resembling idling motor for 10sec–2min. When group has nestlings, perform display with drooped wings and spread tails, vocalising loudly (Robinson 1997, Di Giacomo 2005, Logue 2007, Riehl & Jara 2009).

BREEDING Season Nest building from early May, laying Jun–Sep, mostly from mid-Jul, Panama. Eggs Aug–Nov, Trinidad. Two females in breeding condition Dec and Jan, north Colombia. May–Nov llanos savanna, north-central Venezuela. Apr–Sep Suriname; May–Dec Guyana; nest Apr French Guiana. Appears Aug/Sep Cocha Cashu, south-east Peru, breeding commences Dec. Fledglings Pará, Brazil, May; Argentina, Oct–Feb (Stone 1928, Young 1929, Belcher & Smooker 1936, Hellebrekers 1942, Thomas 1979, Hilty & Brown 1986, Tostain *et al.* 1992, Robinson 1997, Lau *et al.* 1998, de la Peña 2005, Di Giacomo 2005, Riehl & Jara 2009). **Nest** Bulky, open, flat cup of twigs broken off trees, sometimes also herbaceous stalks, bits of vine, Spanish moss *Tillandsia* sp., lichens, or fibres and strips of palm fronds, lined with leaves, with more fresh leaves being added during laying and incubation, in shrubs or trees in flooded margins of creek or lake, frequently above water and often protected from primate and rodent nest predators in clumps more than 1m from adjacent vegetation. 2–5m up in tree (Trinidad) or 2.5–8.5m from ground (Misiones and Formosa, Argentina). Diameter 27–30cm, cup 14–15cm, height 16–20cm, depth of cup 6–7cm. Up to eight birds may participate in construction which takes 3–7 days. Re-nesting within season after failure, not when brood successfully fledged. Reuse of nest following year (Young 1925, Davis 1941, Wetmore 1968, Willis & Eisenmann 1979, ffrench 1991, Haverschmidt & Mees 1994, Lau *et al.* 1998, Hilty 2003, de la Peña 2005, Di Giacomo 2005, Riehl & Jara 2009). **Eggs** Greenish-blue, covered with chalky coating which becomes partly worn off during incubation; broadly oval to almost spherical; 39.1–47.2x31.3–38.9mm, mean 43.5x36.0mm (n = 90). Fresh mass, Panama, 19.3–37.8g, mean 29.7 ± 2.9, (n = 343), Argentina 20.5–31.6g, mean 27.8g (n = 21). Sometimes become stained by nest material. 3–7 eggs/female/nesting attempt, mean 4.3 ± 0.9; communal clutches 6–17, mean 10.4 ± 2.3. Laying interval, individual females, *c.* two days. Clutch 4–10 eggs/nest (11 nests) Formosa, Argentina, 3–5 adults at each nest. Laying clutch of six occurred on successive days, with four adults near nest (Young 1925, Belcher & Smooker 1936, Schönwetter 1967, Di Giacomo 2005, Riehl & Jara 2009). **Incubation** 11–12 days for individual eggs. Both sexes have brood patch. Begins after third- or second-to-last egg laid, 86% of clutches, or when all eggs laid, 14%. Hatching interval 0–5 days, mean 2.5 ± 0.8 (Riehl & Jara 2009). **Chicks** Eyes open day after hatching, pinfeathers emerge at age 1–2 days. At age 5 days may leap from nest into water below if frightened, swim on surface up to several metres to shore, hide in vegetation for some min, and climb back into nest. When handled, nestling excretes dense fluid with repulsive odour from cloaca (probably secreted from very large and structurally unique anal glands), possible predator deterrent. At 6 days, feathers begin to unsheathe. Nestling period 12–13 days, although may leave nest and climb around nearby branches from day 6. Fledglings stay near nest until they can fly two weeks after leaving nest, entire group then often leaves area; juveniles fed for up to six weeks. Young birds often remain with group for at least six months before dispersing; young males sometimes stay with group and help at nest next season (Davis 1941, Quay 1967, Lau *et al.* 1998, Di Giacomo 2005, Riehl & Jara 2009). **Survival** In nests in clumps of vegetation surrounded by water, 4.8 (mean) young fledged per nest, versus 1.3 from nests in continuous riparian vegetation, due to more eggs hatching in isolated nests (Lau *et al.* 1998). Hatching success 84% of 196 eggs; survival of 165 nestlings 76% (Riehl & Jara 2009).

FOOD Earthworms; cockroaches, grasshoppers, katydids, mantids, termite alates, cicadas and other bugs, beetles, ants, wasps, caterpillars, fly larvae; spiders; fish; frogs; small snake brought to nestlings; lizards; heron eggs and nestlings; berries, euphorbia seeds, lauraceous and other fruits. Occasionally sallies for wasps and other flying insects (von Ihering 1930, Brodkorb 1937, Schubart *et al.* 1965, Wetmore 1968, ffrench 1991, Haverschmidt & Mees 1994, Robinson 1997, Di Giacomo 2005). In study, arthropod prey fed to nestlings during 40 events in 5h at two nests 55% Orthoptera, 17% caterpillars, 8% spiders, the rest other insects (Lau *et al.* 1998).

STATUS AND CONSERVATION Fairly common (Stotz *et al.* 1996). Locally common Panama (Wetmore 1968, Ridgely & Gwynne 1989). Locally common Colombia, especially numerous Amazonia (Hilty & Brown 1986). Rather local most of Venezuela except most arid regions in north-west and forested Bolívar (Hilty 2003). Common Guyana (Restall *et al.* 2006), Suriname (Haverschmidt & Mees 1994) and coastal French Guiana (Tostain *et al.* 1992). Fairly common Trinidad (ffrench 1991). Fairly common east of Andes, Ecuador and local in west (Ridgely & Greenfield 2001). Locally common Peru (Schulenberg *et al.* 2007). Rare or local arid biomes chaco (Arribas *et al.* 1995, Guyra Paraguay 2005) and caatinga (Olmos 1993). Few records Santa Catarina, south Brazil (do Rosário 1996, Piacentini *et al.* 2006), probably due to lack of study (Piacentini 2004). Few records Buenos Aires province, Argentina (Narosky & Di Giacomo 1993). Rare Uruguay (Azpiroz & Menéndez 2008) and most of Paraguay (Hayes 1995) but of least conservation concern (Guyra Paraguay 2005). Density of *c.* 18 groups/km river margin Venezuela (Lau *et al.* 1998). Eggs taken by snakes and monkeys (Riehl & Jara 2009). Not globally threatened (BirdLife International 2011).

Greater Ani *Crotophaga major.* **Fig. 1.** Adult, Yasuní NP, Orellana, Ecuador, November (*Dušan Brinkhuizen*). **Fig. 2.** Juvenile, San Isidro Lodge, Napo, Ecuador, November (*Roger Ahlman*).

SMOOTH-BILLED ANI
Crotophaga ani — Plate 1

Crotophaga Ani Linnaeus, 1758 (America, Africa = Jamaica)

TAXONOMY Monotypic. Synonyms: *ambulatoria* Linnaeus 1766, *minor* Lesson 1831, *rugirostra* and *laevirostra*, both Swainson 1837.

FIELD IDENTIFICATION 33–35cm. Flight weak and flapping with interspersed glides; on landing tail often thrown over back. Noisy and conspicuous, easy to observe. **Adult** Whole plumage black; head and body faintly glossed with bronze purple scaling, tail rounded, bill laterally compressed and high-ridged with curved hump on upper mandible higher than forecrown, highest in males, forming distinct notch between bill and forehead, bill usually smooth, occasionally with shallow grooves on basal half. **Juvenile** Plumage black with little or no gloss; bill usually lacks hump. **Similar species** From Groove-billed Ani by voice, adults of which have unhumped grooved bill (often difficult to see); but some, mainly juveniles, lack grooves. Adult Smooth-billed has more angled gonys, mandible typically narrows from angle to base, and strongly arched distal portion of cutting edge (must be seen when bill open); Groove-billed has even-width basal portion, and distal cutting edge rather straight.

VOICE Most commonly given is upslurred squeal *ah-nee* or *oooeeEEENK?*, ending in brief, slight fall in frequency, typically 0.5–0.8s; in flight and when alarmed, often repeated at 1–1.5s intervals, unlike any call of Groove-billed; variations include faster and higher-pitched when threatened, or rapidly repeated early morning and late evening in sleeping tree, or given by adults to call fledglings from nest. Repeated *chuck* or *conk* given by intruder being chased, intervals 0.5s; 'chuckle' when copulating; guttural note in confusion; low, soft 'whine' in courtship; *quack* warning of hawk; repeated *whew* notes from tree top when entering foreign territory. Screaming note lasting 1sec, intervals 2.5sec, up to 22 kHz; intense mobbing scream, *c.* 0.2sec, intervals 0.2–0.7sec (typically given when predator at nest); 'warble', 0.7sec, 1–9 kHz; 'squeak', *c.* 0.5sec, intervals 0.1sec. Song – series *glew-glew-glew...* . Thin, decending *teeew*. Series of musical, quiet warbling notes delivered by begging juvenile (Davis 1940c, Sick 1993, Quinn & Startek-Foote 2000, Sibley 2000, Hilty 2003, Willis & Oniki 2003, Schulenberg *et al.* 2007, Minns *et al.* 2010).

DESCRIPTION Adult Whole plumage faint glossy black, feathers on head, neck, upperparts, wing-coverts, throat and breast edged glossy purple and green, remiges with glossed violet; long rectrices black, truncate; bill high-ridged, laterally compressed and culmen sharply edged. Albinos known (Insfran 1931, Minns *et al.* 2010, MZUSP). **Juvenile** Uniform sooty-black, remiges and rectrices with slight violet gloss. Bill smaller than adult, without hump. Rectrices tapered. **Nestling** Naked, skin black, bill small, with egg tooth; eyes closed when hatched; mouth pink with white raised markings on palate and behind and under black-tipped tongue. **Bare parts** Iris dark brown. Bill, bare skin on lores and around eyes and feet black.

BIOMETRICS Wing male 143–159mm (n = 12), female 136–156mm (n = 12), mean 145.9 ± 6.0. Bill male 30–33mm, mean 31.8 ± 1.1, female 28–31mm, mean 29.6 ± 1.00. Tail male 172–191mm, mean 180.8 ± 5.8, female 160–188mm, mean 173.1 ± 8.7. Tarsus male 32–38mm, mean 34.2 ± 1.8, female 32–35mm, mean 33.5 ± 0.9. Mass male 81.9–133.1g (n = 11), mean 110.5, female 77.8–115.6g (n = 11), mean 92.6 (Payne 2005).

MOULT Year-round; moult remiges and rectrices also in breeding season, body feathers shed mostly in spring, Cuba. Adults moult flight-feathers Mar, Jun–Aug and Nov Trinidad, sequence irregular. Complete postbreeding moult Jun–Apr Florida. Juvenile's moult into adult plumage may be complete in Dec, or protracted over winter. Few remiges may be retained until spring (adults) or second postbreeding moult. Timing perhaps different in tropics (Davis 1940c, Snow & Snow 1964, Pyle *et al.* 1997).

DISTRIBUTION Nearctic and Neotropics. Resident. Breeds central and south Florida. Few records Florida until colonising after 1926 hurricane; first flock seen 1937; first breeding 1938; increased early 1960s, decreased 1978–1998, now scarce and local (Semple 1937, Sprunt 1939, Merritt 1951). Casual Ohio, Philadelphia area, North and South Carolinas and Georgia (Mlodinow & Karlson 1999); Louisiana and Indiana records questionable (Lowery 1974, AOU 1998). Two possible sightings south-east Texas (Oberholser 1974). Breeds irregularly Isla Cozumel and adjacent mainland of Quintana Roo, Mexico; few records Belize Cays including Ambergris; Honduras Bay Is.; Swan Is. (Honduras); Corn Is. (Nicaragua). Main islands, Bahamas; Caicos Is. and Grand Turk. Greater Antilles, Virgin and Cayman Is., Isla Providencia, West Indies; common Dominica, St Vincent, Grenada; uncommon Martinique and Guadeloupe; rare San Andrés; absent Barbados and islands north and north-west of Montserrat, but recorded St. Eustatius. Common breeder Tobago and Trinidad. First recorded south-east Costa Rica 1931 near Panama frontier, spread north-west and largely replaced Groove-billed Ani; now north-west to near Quepos. Common most of Panama

including Coiba I. and Pearl Is.; absent Bocas del Toro and Caribbean slope Veraguas; spread during 1900s. Tropical and subtropical lowlands of South America east of Andes, including Venezuela, Margarita and Patos Is., Guianas, Bolivia, Brazil, Paraguay and Uruguay, south to pampa region, central Argentina; west of Andes, Colombia and Ecuador; Andean valleys Peru to 2,800m. Coastal region and other open country Suriname. Argentina south to San Juan, San Luis, Córdoba, Santa Fe and northern Buenos Aires. Isla Gorgona, Colombia. Galapagos since 1960s, probably introduced. Cocos I., Costa Rica, 2006 (Peters 1929, Paynter 1956, Snyder 1966, Wetmore 1968, Brudenell-Bruce 1975, Bond 1979, Buden 1987, Ridgely & Gwynne 1989, Fjeldså & Krabbe 1990, Rosenberg *et al.* 1990, ffrench 1991, 1996, Stiles & Skutch 1991, Tostain *et al.* 1992, Grant & de Vries 1993, Haverschmidt & Mees 1994, Howell & Webb 1995, Raffaele *et al.* 1998, de la Peña 1999, Hilty 2003, Easley & Montoya 2006, Mata *et al.* 2006, Wiedenfeld 2006, Garrigues & Dean 2007).

HABITAT Marsh edges and brush country, Florida. Open lowland with scattered bushes and trees, Caribbean. Second growth and river island scrub. Semi-open grassy or agricultural areas and near rural human settlements, in undisturbed areas near water. Fencerows, roadside strips. To 1,200m Costa Rica. Lowlands Panama, locally to 1,500m. To 1,400m Ecuador, smaller numbers to 2,000m, to 2,400m in south. Mostly below 2,100m, but to 3,200m, Peru. Rarely to 2,700m Colombia (Belton 1984, Ridgely & Gwynne 1989, Fjeldså & Krabbe 1990, Stotz *et al.* 1996, Raffaele *et al.* 1998, Kaufman 2000, Ridgely & Greenfield 2001, Hilty 2003, Garrigues & Dean 2007, Schulenberg *et al.* 2007).

BEHAVIOUR Nearly always in loose, noisy groups of 10–12, occasionally 17, in winter 15–35, once 65 Florida. Often perches in open; one sentry when foraging while running on ground. Perch close together allopreening. After bathing lies on ground with outspread wings; sunbathing when air is damp and cool, e.g. mornings, with spread wings and tail and fluffed feathers on perch or open ground, often close together. Scratches head under wing. Perches on floating vegetation in streams to drink. Takes insects flushed by cattle, horses, etc., also follows lawnmowers, ploughs and sometimes armadillos; occasionally forages at army ant swarms emerging from forest. Prods soft earth and cattle dung with bill to get insects (Gosse 1847, Wetmore 1927, Merritt 1951, Haverschmidt 1968, Wetmore 1968, Smith 1971, Brudenell-Bruce 1975, Loflin 1983, Willis 1983a, Belton 1984, de Visscher & Moratorio 1984, Hilty 2003, Quinn *et al.* 2010).

Male offers or feeds a fruit, frog, lizard, spider, caterpillar, large harvestman or orthopteran to female in courtship, holding on to gift until copulation ends. Female may accept gift without copulating. Group defends territory, not tolerating members of other flocks to come close to nest; hierarchy within group. Nest shared by 2–5 pairs, sometimes unmated members included; some pairs have individual nests. All group members participate in nest building to varying degree. Up to five females/group, typically unrelated. Extra-pair fertilisations occur. Before laying, females may cover eggs of other female with leaves, with only top layer developing, or throw eggs out of nest. In buried eggs, embryo may develop to advanced stage before dying. When nest reused, often contains several layers of dead eggs separated from fresh eggs by decaying leaves; up to 8 layers with 22 unhatched eggs. More than 50% of eggs may be lost by tossing and burial in nest. Early-laying females attempt to initiate incubation and unbury eggs but are often driven off nest and eggs buried by later-laying females. Delayed onset of effective incubation also suggested by hatching dates closer together than laying dates, but incubation takes place before last egg laid. In single-pair nests, incubation begins soon after first egg. Male killed by conspecific in neighbouring territory when his mate attempted to lay in their still incomplete nest, after several intrusions for three days previously, including being chased and attacked by territory/nest owners, but succeeded in laying egg; nest abandoned and another initiated nearby. Intraspecific brood parasitism occurs when group's own nest predated or destroyed. Parasitic eggs may be buried within nest, or nest abandoned, if last female in group has not begun laying. Infanticide by adults joining group after completion of clutch, possibly in order to increase probability of renesting. Eggs not tossed or buried found with large hole believed to be pecked, whole clutches being destroyed leading to renesting (Köster 1971, Loflin 1982, 1983, Sick 1993, Quinn & Startek-Foote 2000, Payne 2005, Schmaltz *et al.* 2008, Quinn *et al.* 2010).

BREEDING Season First egg Apr–Sep Florida. Mar–Sep Costa Rica. All months but mostly Apr–Jun, Jamaica. Apr–Oct Cuba. Year–round Hispaniola, and apparently rest of West Indies. All months Trinidad, but 69% nests May–Oct. Colombia Aug–Dec. Year-round Suriname. Jan–Feb, south Brazil (Gundlach 1874, Belcher & Smooker 1936, Friedmann & Smith 1950, Snow & Snow 1964, Thomas 1979, Loflin 1983, Belton 1984, Hilty & Brown 1986, Stiles & Skutch 1989, ffrench 1991, Haverschmidt & Mees 1994, Raffaele *et al.* 1998, Latta *et al.* 2006, Haynes-Sutton *et al.* 2009). **Nest** Bulky cup of dry sticks, plant stems and vines, 0.6–15m up in dense thorny palm, medium-sized tree or bush. Lining of green leaves added just before laying. Outside diameter 17–30cm, depth 15cm, inside diameter 13–19cm, depth 5–13cm (Davis 1940c, Köster 1971, Quinn & Startek-Foote 2000). **Eggs** Subelliptical, blue-green with white chalky covering more or less worn off during incubation; 29.2–40.4x23.4–30.0mm (n = 63), mean 35.0x26.3. Laid at 2-day intervals. Each female lays 3–5 eggs, each *c.* 14% of female's body mass. Clutch completion 12–16 days (Bent 1940, Davis 1940c, Köster 1971, Quinn & Startek-Foote 2000). **Incubation** 12–15 days; last egg in communal clutch 16 days; mean for first egg 17.6 ± 2.1 days, 61 nests. Both sexes have brood patch. All group members incubate, but individual contributions vary greatly. Male incubates at night. Incubating females often fed by mates on nest. Hatching takes *c.* 4h, shells left in nest (Köster 1971, Harrison 1978, Loflin 1983, Sick 1993, Pyle *et al.* 1997, Quinn & Startek-Foote 2000). **Chicks** Beg with head and neck stretched and flapping wings. Fed by most or all adults in group, and by juveniles from previous nestings: grasshoppers, katydids, caterpillars, spiders and lizards. Eyes open after 2–3 days, sheaths break day 5, leave nest 10–11 days, fully volant 13–17 days. Fledging mass 36g, *c.* 40% of adult, growth completed after leaving nest. Usually disperse from territory after 8–12 months (Skutch 1966, Harrison 1978, Oniki & Ricklefs 1981, Loflin 1983, Quinn & Startek-Foote 2000, Payne 2005). **Survival** 24% of eggs produced surviving independent young (Davis 1940c). Up to 15 hatched/nest. Mean fledging success (number surviving to 20 days) greatest, 3.75/nest, in clutches with four laying females; mean number per female highest, 1.5, in clutches with two laying females (Loflin 1983).

FOOD Primarily grasshoppers; katydids, moths, caterpillars (including Noctuidae, Geometridae), adult butterflies (including Pieridae and purplewings, Nymphalidae), Hemiptera, beetles (including weevils and Carabidae), mole crickets, cockroaches, stick insects, cicadas, ants, mantids, Mallophaga, dragonflies, bees (including Andrenidae and Euglossini); spiders, harvestmen, ticks (from cattle, more important before pesticides used, and capybaras); rarely snails or frogs; small snakes, lizards; rarely birds' eggs, nestlings (Black-faced Grassquit, Vermilion Flycatcher), passerines in mistnets; mice; berries, fruits (including hot peppers), flower buds (during dry-season food shortage Venezuela) (Gosse 1847, Wetmore 1916, 1927, 1968, Bent 1940, Davis 1940c, Schubart *et al.* 1965, Gill & Stokes 1971, Olivares & Munves 1973, Loflin 1983, de Visscher & Moratorio 1984, Rosenberg *et al.* 1990, Dubs 1992, Haverschmidt & Mees 1994, Eitniear & Tapia 2000, Burger & Gochfeld 2001, Payne 2005, Sazima 2008).

STATUS AND CONSERVATION Common (Stotz *et al.* 1996), e.g. common to abundant Hispaniola; increased since 1930s as forest reduced (Latta *et al.* 2006). Readily colonises newly cleared areas (e.g. Wetmore 1968, Payne 2005). Sometimes disappears from small islands, then recolonises after some time (Raffaele *et al.* 1998). Ten of 23 Puerto Rico broods (43.5%) lost to predation by brown rats, Red-tailed Hawks or red fire ants *Solenopsis invicta* (Quinn *et al.* 2010). Not globally threatened (BirdLife International 2011).

Smooth-billed Ani *Crotophaga ani*. Adults. **Figs. 1–2.** Igarapé-Açu, Pará, Brazil, February (*Stefan Hohnwald*). **Fig. 3.** Osa Peninsula, Costa Rica, March (*Boris P. Nikolov*). **Fig. 4.** Chone, Manabí, Ecuador, March (*Roger Ahlman*). **Fig. 5.** Igarapé-Açu, Pará, Brazil, December (*Stefan Hohnwald*).

GROOVE-BILLED ANI
Crotophaga sulcirostris Plate 1

Crotophaga sulcirostris Swainson 1827 (Temiscaltipec = Temascaltepec de González, Mexico)

TAXONOMY Monotypic. Synonyms: *Casasii* Lesson 1828, *semisulcata* Swainson 1837, *rugirostris* Sclater 1858, *s. pallidula* Bangs & Penard 1921.

FIELD IDENTIFICATION 28–32cm. Flight slower than Smooth-billed Ani, on landing tail often thrown over back like Smooth-billed. **Adult** All black, tail long, rounded, loose-jointed, bill deep, culmen and crown forming unbroken arc, upper mandible curved, with 4–5 deep grooves (viewed closely). **Juvenile** Like adult, but bill smaller, with fewer, shorter and less distinct grooves, plumage unglossed brownish-black. **Similar species** See Smooth-billed Ani.

VOICE Most common call, often given in flight, high-pitched, whistled, sneezed, repeated disyllable *pee-ho pee-ho...*, second note lower-pitched, sometimes preceded by few throaty clucks, audible only at close range, *tuc tuc tuc pihuy pihuy pihuy*, very different from *ah-nee* call of Smooth-billed Ani. Also given a range of alert and alarm contexts, being varied according to state of alertness, e.g. with higher-pitched second note, and shorter interval between two notes, accelerated into *wicka-wicka-wicka*; apparently also varies geographically. Also short, repeated raspy notes; low-pitched, longer (c. 0.15sec) grating notes; low-amplitude, quickly repeated rattle, c. 16 notes/sec given when chased, low-pitched (1–2kHz) *conk* notes in threat; cat-like screaming notes (predator-deterrent?); soft, warbling 'whimper' given by individuals in body contact, e.g. on night roost and when brooding at nest; soft, deep 'chirps' repeated many times, c. 7 notes/sec, e.g. when trying to join new group. Nestling begs with variable, loud, high-pitched warbling or sizzling notes (Friedmann & Smith 1950, Skutch 1959, Vehrencamp *et al.* 1986, Ridgely & Gwynne 1989, Bowen 2002). Rarely heard song: fast, mellow clucks *whiuk-whiuk...* lasting several seconds (Howell & Webb 1995).

DESCRIPTION Adult Plumage black, head and nape with glossy bluish feather edges, scapulars and smaller wing-coverts with broad U-shaped submarginal bronze-green mark, chest with less distinct markings, underside of remiges and rectrices glossy greenish or bluish. Rectrices graduated. Bill deep, laterally compressed. Upper mandible with four or more curved furrows, parallel with culmen, reaching and forming serrations on cutting edge. Some males have deeper bills than females (some depth overlap). Long black eyelashes. Albinism reported (Carriker 1910, Huber 1932). **Juvenile** Brownish black with little or no bluish gloss. Rectrices tapered. Bill smaller, at first without furrows, during first year less than four shallow furrows on upper mandible not reaching cutting edge. **Nestling** Hatchling naked and black, eyes closed, bill with egg tooth. **Bare parts** Iris dark brown. Bare facial and loral skin dull black. Bill and feet black.

BIOMETRICS Wing male 129–138mm (n = 12), mean 134.7 ± 3.5, female 127–135mm (n = 10), mean 131.2 ± 2.7. Bill male 26–30mm, mean 29.0 ± 1.3, female 26–29mm, mean 27.5 ± 0.8. Tail male 168–184mm, mean 177.0 ± 6.9, female 166–177mm, mean 171.9 ± 4.0. Tarsus male 31–35mm, mean 33.2 ± 1.4, female 31–34mm, mean 33.1 ± 1.1. Mass male 61.1–94.5g (n = 14), mean 79.6, female 55.5–79.8g (n = 22), mean 66.6 (Payne 2005).

MOULT Slow and fairly continuous during most of year, sometimes overlaps with breeding Costa Rica. Postbreeding moult mid Jul–early Oct, one-year-old birds first. Complete post-juvenile body and tail moult and partial wing moult. Primary moult irregular, and some juvenile primaries retained until spring. Partial moult Mar–Apr; broad adult rectrices attained only in second autumn moult (Stone 1932, Dickey & van Rossem 1938, Foster 1975).

DISTRIBUTION Mostly resident. Breeds central and south Texas; Mexico on both slopes from south Sonora, north Coahuila and central Mexico, and Isla Cozumel, Isla Holbox and (perhaps irregularly) Ambergris Cay; Belize, Guatemala and El Salvador to west, north and central Costa Rica. Panama on Pacific slope from Chiriquí to east Panama province, and west Bocas del Toro. Locally north, central and east Colombia; west Ecuador and around Zumba; west slope of Andes and Marañón valley, Peru. North Chile, mainly in Lluta valley, Arica, to 800m; specimen, 1,200m, Tarapacá; sight 2,300m, Antofagasta. Aruba, Curaçao, Bonaire. Venezuela south to north-west Amazonas and Río Orinoco, and east to north Bolívar and Sucre. Has bred south Louisiana, and breeds casually west, north-central and south-east Texas. Migratory at northern edge, evacuating Texan breeding areas in autumn, few moving north to coastal Louisiana, Mississippi and Alabama (24 records), and even fewer reaching Florida Panhandle (>60 records), Oct–Apr. Rare rest of Texas, New Mexico and Arizona, vagrant north and east to south California, Nevada, Colorado, South Dakota, Minnesota, Ontario, Wisconsin, Michigan, Ohio, Maryland (Hellmayr 1932, Peña 1961, Russell 1964, Monroe 1968, Hilty & Brown 1986, Ridgely & Gwynne 1989, Stiles & Skutch 1989, Howell & Webb 1995, AOU 1998, Komar 1998, Russell & Monson 1998, Mlodinow & Karlson 1999, Ridgely & Greenfield 2001, Hilty 2003, Jaramillo 2003, Eisermann & Avendaño 2006, Piaskowski *et al.* 2006, Garrigues & Dean 2007, Schulenberg *et al.* 2007, Prins *et al.* 2009). Twice Salta, Argentina (Zotta 1937, Barnett & Pearman 2001). Once Manitoba (Kent 1988). Mostly early Jun–late Sep Sonora (van Rossem 1938a,

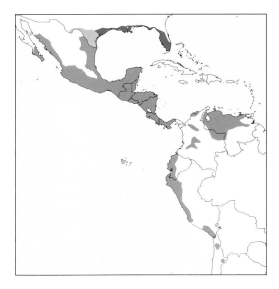

Russell & Monson 1998). Irregularly south Baja California (Howell & Webb 1992). Twice Socorro, Revillagigedo Is., 1990 (Wehtje et al. 1993).

HABITAT Dense thornbrush, Texas. Second growth, roadsides and pastures in humid areas Mexico. Savanna, plantations, lawns, open woods etc. Costa Rica. South America: forest edge, fields, scrub, waste areas, thickets and overgrown or neglected pastures in arid areas. Occurs with Smooth-billed Ani in transitional dry to moist areas and is replaced by it in humid zones. To 650m Sonora, to 1,800m Mexico–Nicaragua, to 1,500m Costa Rica, uncommonly to 2,200m. Lowlands Panama. To 500m Colombia, to 750m Venezuela, to 1,200m Ecuador, higher in El Oro and Loja, exceptionally to 2,750m, also Peru (Skutch 1959, Oberholser 1974, Hilty & Brown 1986, Stiles & Skutch 1989, Howell & Webb 1995, Ridgely & Gwynne 1989, Russell & Monson 1998, Ridgely & Greenfield 2001, Hilty 2003, Garrigues & Dean 2007, Schulenberg et al. 2007). Accidental 3,100m, Cuzco (Fjeldså & Krabbe 1990).

BEHAVIOUR Scratches head under wing. Perch close together, allopreening. Often forages around cattle and horses, capturing insects flushed by them. Picks insects from leaves and catches many grasshoppers on ground. Digs in cow dung for grubs and dung-beetles. Flycatches winged termites. Sunbathes early morning on top of bush. Forages at swarms of army ants. Breeds in pairs or in groups of 2–5 pairs that defend a group territory, area 1–11ha, build single nest and lay a large joint clutch. Males defend their mates vigorously from other males in group attempting extra-pair matings. Several nesting attempts may be made in a season. Young birds stay with group and serve as nonbreeding helpers later in season in which they have hatched; most breed following year. Female rolls eggs of other females out of nest before laying her own; last-laying female loses no eggs. Pair bond usually maintained from year to year. Courtship, on fence post or in tree: male approaches female with breast held up and out, holding insect or green leaf in bill, female often crouches, male mounts while female spreads wings, male flutters wings and wags tail, and after 5–20sec, lowers tail, cloacal contact <1sec, female often takes offering, male dismounts, often both preen. Reverse mounting, often without offering, is common, mostly before breeding season. Parents, especially males, defend brood against human intruder by striking from behind. Threat: lowers head and presents bill in profile, dropping near wing and raising far wing to enlarge body profile. In most intensive form with feathers erect and repeated 'conking', typically in territorial dispute between two males of neighbouring groups, males circle around each other on ground and vocalise. When threat mild, submissive response perhaps to sleek feathers, lower head, solicit allopreening, or turn or move away from aggressor. Territory invaders and rival males chased in flight. Group members often supplanted from nest by other members, especially during laying period, and by dominant pair (Miller 1947, Skutch 1959, Rand 1953, Wetmore 1968, Oberholser 1974, Bolen 1974, Vehrencamp 1977, 1978, Voous 1983, Willis 1983a, Vehrencamp et al. 1986, Bowen et al. 1989, 1991, Ridgely & Gwynne 1989).

BREEDING Season Eggs May–Jul, nestlings to Sep, Texas. Jun–Sep Sonora. May–Jul Oaxaca. Pair formation from Apr or May, beginning of rains, nesting in wet season, May–Nov, peak Jun–Oct, Guanacaste, Costa Rica. Eggs late Apr central Costa Rica where dry season less extreme; generally breeds from late May or Jun to Sep Guatemala, Costa Rica and north-west Panama. Jan, Oct Colombia. Jul–Nov Guárico, Venezuela; Mar–Aug Aragua. Feb–Apr, during rains, west Ecuador and north Peru. Presumably Oct–Mar coastal Peru. Nest with eggs Nov, north Chile (Dorst 1957, Marchant 1959, Skutch 1959, Johnson 1967, Oberholser 1974, Thomas 1979, Rowley 1984, Hilty & Brown 1986, Binford 1989, Russell & Monson 1998, Bowen 2002, Verea et al. 2009, M.A. Plenge in litt.). **Nest** Large bowl of coarse twigs, pieces of vines, weed stalks, grass tufts, roots, etc., lined with green leaves, outer diameter 25–40cm, inside 10–17cm, outside depth 18cm, inside 5–10cm. Well hidden in dense foliage, <1–13m up, mostly 2–4m, in large, dense bush or small tree, often with spines, commonly in *Citrus*, on branch, or in cactus or bamboo clump; low nests in sugar cane, marsh grass or *Mimosa* bush. Built by all pairs in group, male typically supplying materials to mate who places them in nest, takes 3 days to 3 weeks. May reuse old nest of Long-tailed Mockingbird or rarely of Boat-tailed Grackle, adding fresh leaf lining. Second nesting may be initiated a month after hatching of first brood (Huber 1932, Marchant 1959, Skutch 1959, Vehrencamp 1978, Bowen 2002). **Eggs** Elliptical, thick-shelled, turquoise blue with chalky covering, 22.2–25.4x28.6–35.7mm, mean 24.2x32.1 (n = 56), mass 11.0g ± 1.1. Group clutch, 1–5 laying females, 3–20; largest number known to hatch 13. Each female lays every 2–3 days, morning or around noon, 3–8 eggs per nesting bout, depending on group size, usually 4; because of burial in lining (mainly in large clutches) and egg tossing, c. 4 eggs incubated/female. Three eggs laid in 1.5h (Huber 1932, Skutch 1959, Schönwetter 1967, Vehrencamp 1978). **Incubation** 13–14 days. Begins gradually towards end of laying period. By both sexes during day, each adult taking short turns (few minutes) in beginning, later more than an hour; dominant male does all nocturnal incubation and more diurnal than others in group. When arriving to take over, incubator brings green leaf and adds it to nest lining (Skutch 1959, Vehrencamp 1977, 1978, 1982b). **Chicks** At hatching, cut egg in half with egg tooth, rotating inside egg, occasionally giving weak call. Often hatch over 1–2 days; up to 5 days between first and last. Feeding begins c. 8h; brood of eight fed 66 times in 2h, 4.1 times/nestling/h, by 3–4 adults; mainly grasshoppers, and cockroaches and other insects, spiders, some small lizards, and a few berries; by both sexes and by helpers from previous brood; two adults may cooperate to tear large prey apart before each one giving to nestlings. Brooded almost continuously first few days, by both sexes, first 7–8 nights by male. Adults ingest droppings of very small nestlings, older nestlings eject faeces over nest rim which becomes soiled. In large broods, strong competition for food; late-hatched sometimes die after few days, apparently from starvation. Growth rapid. Eyes open and pinfeathers begin to emerge after 2 days, pins long by day 5–6, rupture quickly day 6, down black; nestling climbs out of nest if disturbed, returning later to be brooded, until day 7–8, stays two nights in nest tree, leaves it day 9, hides in low vegetation until day 17, flies day 17–20, fed by adults until six weeks old (Skutch 1959, Vehrencamp 1977, 1978). **Survival** 48 of 196 juveniles stayed in natal territory until early in next breeding season, subsequently 25 dispersed and bred in nearby territories, while 15 bred within natal groups; males stay and breed more than females. Four-eyed Opossum *Metachirus nudicaudatus* probably took nestlings and brooding adult; other nest predators unknown (Bowen et al. 1989, Bowen 2002, Payne 2005).

FOOD Insects, mainly grasshoppers, also cockroaches, beetles, leafhoppers, harvestmen, spiders, termites, wasps, cicadas, caterpillars, flies, large bugs, and a few ants, berries and small lizards; ticks or other arthropods attached to cattle (Bendire 1895, Dickey & van Rossem 1938, Rand 1953, Dorst 1957, Skutch 1959, Wetmore 1968, Bolen 1974, Oberholser 1974, Voous 1983, Hernández *et al.* 1981 in Bowen 2002). Egg of Red-billed Pigeon (Eitniear & Tapia 2000), and of other bird, possibly a ground-sparrow (*Melozone* sp.) (Sandoval *et al.* 2008).

STATUS AND CONSERVATION Common (Stotz *et al.* 1996). Common Mexico–Nicaragua (Howell & Webb 1995), Costa Rica (Garrigues & Dean 2007), Panama (Ridgely & Gwynne 1989), Venezuela (Hilty 2003), Ecuador (Ridgely & Greenfield 2001). Locally common Colombia (Hilty & Brown 1986). Fairly common Peru (Schulenberg *et al.* 2007). Very rare Argentina (Barnett & Pearman 2001). Adults preyed upon by bat *Vampyrum spectrum*; one taken by Roadside Hawk; nest predation high, nocturnal incubator disappearing (Bowen 2002). Not globally threatened (BirdLife International 2011).

Groove-billed Ani *Crotophaga sulcirostris*. **Fig. 1.** Cojimies, Manabí, Ecuador, March (*Roger Ahlman*). **Figs. 2–3.** San Blas, Mexico, March (*Gary Thoburn*).

SUBFAMILY NEOMORPHINAE

Genus *Tapera*

Thunberg, 1819 *Göteborgs Kongl Wett. och Witt. Samh. Nya Handl.* 3: 1.
Type by monotypy *Tapera brasiliensis* Thunberg 1819 = *Cuculus naevius* Linnaeus 1766 (Lönnberg 1903, Peters 1940).
One species.

Neotropical; brood-parasitic. Medium-sized; sexes alike; streaked brown; crested; long, broad alula; long, narrow tail. Formerly *Diplopterus* Boie, 1826. Sister of *Dromococcyx* (Sorenson & Payne 2005).

AMERICAN STRIPED CUCKOO
Tapera naevia   **Plate 1**

Cuculus nævius Linnaeus, 1766 (in Cayania = Cayenne, French Guiana)

Alternative names: Striped/Four-winged/Brown Cuckoo.

TAXONOMY Monotypic. Payne (1997) recognised *excellens* for populations from Mexico to east Panama, but later (2005) considered species monotypic. *Chochi* often recognised (e.g. Peters 1940, Fjeldså & Krabbe 1990, Dubs 1992) for populations of south Brazil, north Argentina, Paraguay and Uruguay, based on larger size and colour of upperparts, but considerable overlap with populations in Venezuela and north Brazil (Wetmore 1926, Hellmayr 1929, Traylor 1958, Payne 1997). Synonyms: *Cuculus punctulatus* Gmelin 1788, *C. galeritus* Illiger 1816, *Coccyzus chochi*, *chiriri* and *ruficapillus*, all Vieillot 1817, *brasiliensis* Thunberg 1819, *septorum* Vieillot 1823, *Piaya brasiliana* Lesson 1839, *Diplopterus lessoni* Bonaparte 1850, *excellens* P.L. Sclater 1858, *T. naevia major* Brodkorb 1940.

FIELD IDENTIFICATION 27–30cm. Heard much more than seen. Often calls from exposed perch with raised crest and spread alulae, but can be difficult to locate, even when close. **Adult** Likened to overgrown, long-tailed sparrow. Crest ragged, rufous streaked dark; whitish eyebrow, broadening behind eyes. Upperparts, including long uppertail-coverts, pale brown boldly streaked black and buff, cheeks dusky, underparts whitish, narrow dark malar streak. **Juvenile** Upperparts spotted cinnamon, chest faintly barred dusky. **Similar species** Pavonine Cuckoo and Pheasant-cuckoo have much broader tail, longer bill, and dark, unstreaked upperparts; latter has spotted chest.

VOICE Most common vocalisation: loud far-carrying phrase of two separate, short, plaintive, whistled, ventriloquial notes, second higher than first, *sa-see… sa-see… sa-see…* , repeated at *c.* 5sec intervals, sometimes all day and all night, probably territorial, delivered swinging tail sideways and raising and lowering crest. Calling bird collected female. Series of usually 4–7 (1–16) short whistled *wee* notes, increasing slightly in amplitude and frequency with each note, sometimes with last note shorter, lower and weaker than penultimate, *wee-wee-wee-WEE-wi*, usually repeated with shorter intervals than *sa-see* call, commonly before dawn and at dusk, sometimes in duet. Several other short, soft and weak calls resembling abbreviated, condensed, or fragmented versions of these two types of vocalisations are also given, perhaps at short distance from conspecific during territorial disputes. Fledgling gives short whistles, 4.0–4.6kHz (Friedmann 1927, Davis 1940b, Sick 1953a, Morton & Farabaugh 1979, Belton 1984, Hardy *et al.* 1990, Haverschmidt & Mees 1994, Novaes & Lima 1998, Smith & Smith 2000, Hilty 2003).

DESCRIPTION Female slightly smaller than male. **Adult** Forehead, crown and short, somewhat ragged crest, rufous to cinnamon with black streaks (central stripe on each feather), eyebrow whitish, narrow back to over eyes, behind eyes broad. Upperparts, including uppertail-coverts (more than half length of tail, cinnamon to buff (more greyish with wear), back and scapulars with bold blackish stripes, rump and uppertail-coverts with narrow black shaft streaks, cheeks chestnut streaked dark; primaries and tertials greyish tipped pale, broad whitish stripe across base of primaries in flight, alulae long, prominent, black, independently movable, wing-coverts with irregular black marking in centre; underparts whitish, throat and breast sandy to greyish or tinged buff, undertail-coverts buff, narrow malar stripe black, tail long, somewhat graduated, tapering to two points, above brown, below greyish broadly edged cinnamon, narrowly tipped whitish. **Juvenile** Crest shorter, blackish-grey, feathers with large, pale buff terminal spots, above like adult but wing and tail-coverts, and tertials and primaries, all tipped cinnamon, throat and chest faintly barred dusky. **Nestling** Naked, pinkish, blackish around eyes, changing within three days to violet, bill yellowish. Both mandibles with sharp hooks, cutting edges serpentine-shaped and bill distally with open, crocodile-like gape. Mouth lining and cutting edges orange yellow, matching that of wren and spinetail hosts, becoming red with white cutting edges at 30–32 days. **Bare parts** Iris reddish-brown to yellowish-grey (Hilty 2003) or greenish, grey-brown in juvenile; eye-ring yellowish. Bill short, decurved, compressed, brown above, black along ridge of culmen, below horn to yellowish. Feet grey, tarsus scutes edged dull buff, claws fuscous to dark grey.

BIOMETRICS (Colombia, Venezuela, Guianas, north Peru, Ecuador and Brazil north of Amazon). Wing male 103–125mm (n = 13), mean 110.3 ± 6.4, female 101–109mm (n = 9), mean 105.1 ± 3.2. Bill male 16.4–19.7mm, mean 18.1 ± 0.9, female 15.3–18.7mm, mean 17.0 ± 1.1. Tail male 143–169mm, mean 154.8 ± 9.5, female 136–152mm, mean 148.8 ± 7.8. Tarsus male 25.5–34.4mm, mean 30.7 ± 2.3, female 26.9–32.0mm, mean 29.5 ± 1.9 (Payne 2005). Mass male 40.0–58.4g, female 40.4–59g (Haverschmidt & Mees 1994, Schmitt *et al.* 1997, de Magalhães 1999, Payne 2005).

MOULT Postjuvenile moult of wing and tail feathers, renewal of one or two outer rectrices Apr; adult annual moult mid Aug–late Sep, El Salvador; adult male worn, and wing moult Jun, Venezuela (Dickey & van Rossem 1938, Friedmann 1948a).

DISTRIBUTION Neotropics. Mexico from central Veracruz, north and east Oaxaca, west and north Chiapas, Tabasco, north Guatemala, east-central and south-east Quintana Roo, Belize and south on Pacific slope of Middle America in south Chiapas, south Guatemala and El Salvador to Panama, and on Atlantic slope in north Honduras, Nicaragua, Costa Rica and Panama (except Bocas del Toro province), and from Colombia, Venezuela (including Isla Margarita), Trinidad (not recorded Tobago) and Guianas southward, west of Andes to Lambayeque, north-west Peru, and east of Andes to east Bolivia, Paraguay, Uruguay and Argentina (to La Rioja, Mendoza, La Pampa and Buenos Aires). Resident in most of range; seasonal Uruguay and Argentina, e.g. Tucumán, Santiago del Estero, Santa Fe, Entre Ríos, Córdoba and Buenos Aires provinces (Mogensen 1927, Cuello & Gerzenstein 1962, Meyer de Schauensee 1966, Schulenberg & Parker 1981, Salvador 1982, Belton 1984, Ridgely & Gwynne 1989, Nores *et al.* 1991, Narosky & Di Giacomo 1993, Arribas *et al.* 1995, Howell & Webb 1995, Sibley 1996, de la Peña 1997, 1999, Hilty 2003, Guyra Paraguay 2005, Restall *et al.* 2006).

HABITAT Forest undergrowth, secondary scrub, edges, thickets and open country with bushes and bogs, early successional growth on river islands, locally in overgrown pastures, clearings and partially opened areas in humid forested regions, savanna, mangroves, restinga (dune scrub) in south Brazil. To 1,250m Mexico and north Central America; to 1,350m Guatemala. To 1,300m south Venezuela, mostly below 1,200m north Venezuela, occasionally to 2,500m. Mostly below 800m Ecuador, some to 1,500m, and locally (Loja) to 2,300m. Mostly below 1,000m Peru, but to 2,450m Marañón valley. At least to 1,400m Brazil. Possibly only to 900m Bolivia (Belcher & Smooker 1936, Land 1970, Meyer de Schauensee & Phelps 1978, Belton 1984, Sick 1993, Arribas *et al.* 1995, Howell & Webb 1995, Stotz *et al.* 1996, de la Peña & Rumboll 1998, Naka & Rodrigues 2000, Ridgely & Greenfield 2001, Hilty 2003, Restall *et al.* 2006, Schulenberg *et al.* 2007).

BEHAVIOUR Solitary and furtive; near ground in dense cover. Runs quickly on ground in short vegetation, often stopping to sway from side to side, raise crest, spread one or both wings and perform bouts of forward-directed semaphore-like flexes with alulae (to flush prey?), alternating with sudden scurries forward to catch grasshoppers etc. Also forages inside foliage; pounces from bush to catch grasshopper on ground. Usually calls from half-exposed perch on fence post, in top of bush or tree or telephone wires, often raising crest and flexing alulae. In social interactions and when frightened, also flashes dark alulae, very conspicuous against pale breast. When approaching playback of *sa-see* call, raises and lowers crest up to once/sec, often rhythmically, sometimes for several minutes, body usually hunched and wings held slightly out and down with alulae extended, back feathers sometimes ruffled, tail usually fanned, but rarely widely, body swaying from side to side. Rarely flies: steady wing beats on even keel. Dust-bathes shuffling around in sand. Scratches head under wing. Perhaps monogamous with exclusive individual territories (Chapman 1894, Naumburg 1930, Friedmann 1933, Sick 1953a, 1993, Haverschmidt 1955, 1968, Slud 1964, Stiles & Skutch 1989, Haverschmidt & Mees 1994, Smith & Smith 2000, Hilty 2003).

BREEDING Parasitism Known hosts: mostly furnariids (ovenbirds), with closed nests, often large, sometimes of thorny sticks, with narrow entrance and long entrance tube; Short-billed Canastero and Tufted Tit-spinetail (both Argentina); Spinetails: Azara's (north-west Argentina), Pale-breasted (Costa Rica, Trinidad, Suriname, Argentina), Spix's (Argentina and Brazil), Slaty (Colombia), Sooty-fronted (Argentina), Rufous-breasted (Guatemala), Plain-crowned (Suriname, Brazil), Stripe-breasted (Trinidad), Chotoy (Argentina, Brazil) and Yellow-chinned (Trinidad, Suriname, Brazil, Argentina); Thornbirds: Freckle-breasted, Greater and Russet-fronted (all Argentina), and Red-eyed (Brazil); Rufous-and-white Wren, Plain Wren and Black-striped Sparrow (all Costa Rica, Panama); White-headed Marsh Tyrant (Suriname), perhaps doubtful; flycatchers *Myiozetetes* spp.; Slate-headed Tody-tyrant. Some may not actually rear cuckoo, but Yellow-chinned, Spix's, Pale-breasted, Plain-crowned, Chotoy and Sooty-fronted Spinetails do (Hartert & Venturi 1909, Penard & Penard 1910, von Ihering 1914, Friedmann 1927, 1933, Mogensen 1927, 1930, Naumburg 1930, Snethlage 1935a, b, Belcher & Smooker 1936, Davis 1940b, Hellebrekers 1942, Loetscher 1952, Sick 1953a, 1993, Haverschmidt 1955, Nicéforo & Olivares 1966, Wetmore 1968, Land 1970, Kiff & Williams 1978, Morton & Farabaugh 1979, Salvador 1982, Ridgely & Gwynne 1989, de la Peña 1993, 2005, Haverschmidt & Mees 1994, Skutch 1999, Di Giacomo 2005). Buff-fronted Foliage-gleaner reported Paraguay (Fiebrig 1921), but author's nest description incorrect for that species.

Often lays just after dawn, when hosts away foraging, tears hole in one side of spinetail stick nest and enters chamber to lay, hosts possibly repair nest afterwards; or enters and leaves nest through narrow entrance tunnel (Rufous-breasted Spinetail; possibly Yellow-chinned Spinetail). Female ready to lay roams neighbourhood of host nest for up to a week, and may even sleep in nearby shrubbery, waiting for spinetail hosts to finish nest building. One cuckoo egg, Costa Rica, laid during host's laying period, while in Argentina some cuckoo eggs laid 1–3 days after completion of host clutches. Not known to remove host eggs. Usually one egg laid in each nest, but two eggs found in nests of Spix's Spinetail, Yellow-chinned Spinetail and Plain-crowned Spinetail, and two young in Yellow-chinned Spinetail nest Argentina (doubtful?). Three Argentina nests each contained two cuckoo eggs that were of different shape, perhaps laid by different females. Four eggs in same nest reported. Twelve of 45 Yellow-chinned Spinetail nests parasitised, 8 of 34 Sooty-fronted Spinetail, 4 of 11 Rufous-and-white Wren (Hartert & Venturi 1909, Penard & Penard 1910, da Fonseca 1922, Friedmann 1927, Snethlage 1935a, b, Belcher & Smooker 1936, Haverschmidt 1961, Schönwetter 1967, Wetmore 1968, Morton & Farabaugh 1979, Salvador 1982, Stiles & Skutch 1989, de la Peña 1993, Sick 1993, Di Giacomo 2005). **Season** Unshelled oviduct egg Jun, Oaxaca, Mexico. Vocal activity Jan–Jun, breeding condition female Jun, fledglings Aug, Panama. Probably year-round Trinidad. Sep (other months?) Guárico, Venezuela. Eggs all year Suriname and French Guiana, where vocally most active early dry season. Brazil: nestling Oct Rio de Janeiro, vocal Jul–Jan Minas Gerais. Argentina: Jul–Nov Misiones, Oct–Jan Tucumán, Santa Fe and Entre Ríos, eggs Nov–Feb Córdoba, and Formosa, where not vocal autumn and winter, and vocal from late Aug or early Sep (Snethlage 1928, Friedmann 1933, Loetscher 1952, Haverschmidt 1955, Wetmore 1968, Strauch 1977, Thomas 1979, Salvador 1982, Binford 1989, ffrench 1991, Tostain et al. 1992, Chébez 1993, Sick 1993, de la Peña 1997, Di Giacomo 2005). **Eggs** Pale greenish-blue, bluish-white or white, unglossed; many host eggs of same colours, suggesting mimicry; elliptical, rarely spherical, size and shape very variable; 18.7–23.5x14.1–17.3mm, mean 21.2x16.3 (n = 50) (Suriname); 19.8–23.5x15.4–17.2mm, mean 21.3x16.4 (n = 121) (northern South America); mass 2.6–3.5g, mean 3.22, heavier than spinetail hosts' 1.5–2.6g; 2.3–3.2g, mean 2.6 (n = 16) (Argentina) (Friedmann 1927, Snethlage 1935b, Belcher & Smooker 1936, Hellebrekers 1942, Haverschmidt 1961, Schönwetter 1967, de la Peña 1993, Haverschmidt & Mees 1994, Davies 2000, Di Giacomo 2005). **Incubation** 15 days Suriname (one host 18); 15–16 days Argentina (Haverschmidt 1961, Salvador 1982, Di Giacomo 2005). **Chicks** Mass at hatching 3.5g. Usually hatch before host young and grow quickly – one 3 days old and twice size of host young on hatching. Aggressive; at 3 days and mass 8g kills host young in less than 3 hours with bites. Foster parents remove dead young. Sprouting feathers from 4–5 days, eyes open at 8–9 days, fully feathered at 12–14 days, filling nest; at 16 days utters loud begging call constantly, fed by both foster parents 4–5 times/hour, each time single large insect; leaves nest at 16–18 days, not yet able to fly well but runs very quickly; excretes foul-smelling brown liquid around time of fledging; when leaving spinetail nest, body mass 43–47g, unable to pass through narrow entrance tunnel, and breaks open nest at top; flightless first week after leaving nest, remaining with foster parents; begs with wings tightly closed to body, vibrating large and contrasting black alulae, or flashing them in and out once/sec, or holds one or both extended for several sec (Friedmann 1933, Belcher & Smooker 1936, Haverschmidt 1961, 1970, Morton & Farabaugh 1979, Salvador 1982, Payne 2005). **Survival** Of eight eggs, seven hatched and six fledged (Salvador 1982).

FOOD Adults Grasshoppers, caterpillars, beetles, dragonflies, cockroaches, Hemiptera (including cicadas), mantids; arachnids; occasionally fruit. **Nestlings** Snails; caterpillars, Hemiptera, cockroaches, crickets; centipedes (von Pelzeln 1871, Layard 1873, Hallinan 1924, Friedmann 1927, Dickey & van Rossem 1938, Moojen et al. 1941, Loetscher 1952, Schubart et al. 1965, Haverschmidt 1968, Belton 1984, ffrench 1991, Lopes et al. 2005a).

STATUS AND CONSERVATION Generally common (Stotz et al. 1996). Fairly common to common Mexico and north Central America (Howell & Webb 1995), but locally very uncommon (Binford 1989). Common Costa Rica, smaller numbers in moist regions (Stiles & Skutch 1989). Fairly common Panama (Ridgely & Gwynne 1989). Frequent but quite local Colombia, common Guianas (Restall et al. 2006). Fairly common north Venezuela, local in forested south (Hilty 2003). Fairly common Trinidad (ffrench 1991). Ecuador: uncommon to fairly common west of Andes, east of Andes only in south-east, where recorded recent years at few sites (Ridgely & Greenfield 2001). Peru: uncommon arid tropical zone west of Andes and Marañón valley, fairly common east of Andes (Clements & Shany 2001, Schulenberg et al. 2007). Rare Pantanal, Mato Grosso, Brazil (Cintra & Yamashita 1990). Unrecorded from certain regions, e.g. parts of Rio Grande do Sul, Brazil, into which may have spread after mid-1900s (Belton 1984). Probably has expanded range and become more common in deforested regions (Sick 1953a, Ridgely & Gwynne 1989, Stiles & Skutch 1989, Ridgely & Greenfield 2001, Willis & Oniki 2003). Uncommon Formosa, Argentina (Di Giacomo 2005), rare Buenos Aires province (Narosky & Di Giacomo 1993), fairly common San Luis (Nellar 1993). Not globally threatened (BirdLife International 2011).

American Striped Cuckoo *Tapera naevia*. **Fig. 1.** Adult, Silanche, Pichincha, Ecuador, February (*Roger Ahlman*). **Fig. 2.** Juvenile, Brasso Seco, Trinidad, Trinidad and Tobago, March (*Michael Patrikeev*).

Genus *Dromococcyx*

Wied, 1832 *Beitr. Naturg. Bras.* 4: 351. Type, by monotypy, *Macropus phasianellus* von Spix 1824.
Two species.

Neotropical; brood-parasitic. Medium-sized; sexes alike; brown plumage; crested; slender bill; broad tail; upper tail-coverts reach end of tail. Includes *Geophilus* (Bertoni 1901). Sister of *Tapera* (Sorenson & Payne 2005).

PHEASANT-CUCKOO
Dromococcyx phasianellus   Plate 2

Macropus phasianellus von Spix, 1824 (forest of Rio Tonantins, Amazon Valley, Brazil)

TAXONOMY Monotypic. Populations in Mexico, Central America and north-west Venezuela often recognised as *rufigularis* Lawrence 1867 (e.g. Peters 1940, Wetmore 1968, Hilty 2003), although the type specimen juvenile (Ridgway 1916). Neither adults nor juveniles vary in plumage or in size across range (Payne 2005). Other synonyms: *Cuculus macrourus* Verreaux & Des Murs 1849, *mexicana* Bonaparte 1856, *Geophilus jasijatere* Bertoni 1901.

FIELD IDENTIFICATION 33–41cm. Small head, crested, thin neck, broad tail. Uppertail-coverts often raised, giving bird humped shape. **Adult** Throat and breast greyish-buff with blackish spotting, postocular stripe white, back and tail dark brown, scaled white. Wing-coverts narrowly edged whitish, usually forming narrow wing-bar. Rectrices broad, tipped white below. **Juvenile** Sooty-brown crown with small buff spots, upperparts darker than adult, spotted buff: cinnamon-buff postocular stripe, throat and chest unspotted. **Similar species** Pavonine Cuckoo smaller, unspotted breast, buff postocular stripe and less broad tail; voice usually different. American Striped Cuckoo smaller, much paler and streaked above, unspotted breast and even narrower tail; found in more open habitat.

VOICE Three far-carrying, melancholic whistled notes *se, see, werrrrrrr*, last note quavering, lasting *c*. 2sec, sometimes 4–6 notes in total, also with last note tremolo, repeated with short intervals, highly ventriloquial at close range, apparently used to mark territory; four notes of same quality, rising in pitch *sa, seh, si-see* when more excited (resembling Pavonine Cuckoo) or varied accelerating series *whee-whee whee whee-whee-whee-whee* etc; 4–5 rattling, clucking notes repeated every few minutes; quiet growling or snarling hiss *grrr* by female. High-pitched call from fledgling (Reiser 1926a, Sick 1953b, 1993, Rowley 1966, Wetmore 1968, Davis 1972, Willis & Eisenmann 1979, Stiles & Skutch 1989, Wilson 1992, Chébez 1993, Howell & Webb 1995, de la Peña & Rumboll 1998).

DESCRIPTION Adult Forehead, crown and short expressive crest sooty-brown tinged rusty, especially on longer feathers, forehead and crown with fairly distinct blackish-brown central streaks, crest sometimes uniform rufescent-brown; upperparts dark sooty-brown with faint purplish-bronze sheen, rump and uppertail-coverts dark greyish-olive with faint bronze sheen, and broadly edged light grey, uppertail-coverts very long, with small white terminal spots; scapulars, interscapulars and smaller lesser wing-coverts edged pale greyish-buff, these edges broader and often dull whitish on larger lesser wing-coverts, middle and greater coverts tipped with dull white or dull buffy-white, secondaries dusky-brown tipped whitish to pale brownish, outer primaries dusky-brown edged whitish to pale buff along middle portion, all primaries except one or two outermost with small terminal triangular pale spot, alula large and movable; axillaries and underwing-coverts white, underside of flight feathers greyish-brown to brownish-slate with white bases, outermost primary with inner half of inner web white, next with inner half mostly white, third-fifth with broad white bar across middle portion, sixth with pale grey transverse spot, innermost primaries and secondaries very pale grey on middle portion, fading into darker grey towards tips and into brownish-grey towards with base; white postocular stripe, lores, below eye and malar white, latter two narrowly and sparsely streaked blackish, auriculars blackish-brown, lower part streaked whitish, throat white with narrow blackish streaks, chest buffy with blackish spots, belly to undertail-coverts white; tail graduated and fan-like, feathers very broad, basally light brown, grading into greyish-brown distally, faintly glossed bronze, and broadly tipped white, underside of rectrices light grey with broad blackish subterminal area. **Juvenile** Forehead, crown and crest wholly dark sooty-brown with small buff spots, postocular stripe, below eye, malar, chin, throat and chest cinnamon-buff (latter without dark spotting), outlined dark brown. Further differs from adult in absence of white tips and blackish subterminal areas in narrower rectrices and pale edges on scapulars, interscapulars and wing-coverts, the latter instead with small terminal brownish-buff spots, also no white spots on uppertail-coverts. **Nestling** Unknown. **Bare parts** Iris white, yellow or brown; dark grey to greyish-brown in juvenile. Bare orbital area and lores yellowish-green, bluish-green or greenish-grey; dusky green in juvenile. Bill above blackish or dusky horn colour, paler along cutting edges, below pale plumbeous; juvenile lead grey. Female's gape dull brownish-orange. Feet pale grey to brownish-slate; juvenile grey.

BIOMETRICS Wing male 154–176mm (n = 22), mean 166.2 ± 5.9, female 153–173mm (n = 18), mean 162.8 ± 5.6. Bill male 17.7–25.2mm, mean 20.4 ± 2.0, female 17.5–23.6mm, mean 20.0 ± 1.6. Tail male 173.2–242mm, mean 209.4 ± 17.0, female 177–248mm, mean 203.1 ± 19.0. Tarsus male 27.0–35.7mm, mean 31.5 ± 2.8, female 27.3–36.9mm, mean 30.9 ± 3.2. Mass male 78–98g (n = 4), mean 85.0; laying female 98.1g (Payne 2005); female 86.8, 92.7g (Smithe & Paynter 1963); male 100g (Belton 1984).

MOULT In postjuvenile moult, P5, P7–10 and six outer secondaries retained (Dec, Apr). Limited body moult Apr, El Salvador (Dickey & van Rossem 1938).

DISTRIBUTION Resident south Mexico in east Puebla, south Veracruz, Guerrero, Oaxaca, Chiapas, Yucatán Peninsula, Belize, Guatemala, Honduras, east El Salvador, Nicaragua, Costa Rica (only Pacific slope), Panama (Pacific

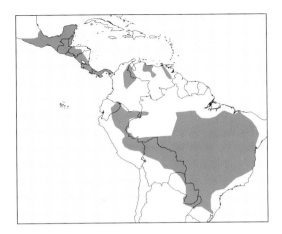

slope, and Canal area to west San Blas on Caribbean slope), and South America locally east of Andes in north and east Colombia, northern half of Venezuela, east Ecuador, east Peru and Brazil to central Bolivia (Pando, Beni, La Paz, Cochabamba, Santa Cruz), Paraguay (mostly in east, including Concepción, Paraguarí, Ñeembucú, Misiones, Itapúa and Cordillera, and north-east Alto Paraguay) and north-east Argentina (Chubb 1910, Warner & Beer 1957, Meyer de Schauensee 1966, Land 1970, Ridgely & Gwynne 1989, Stiles & Skutch 1989, Chébez 1992, Contreras 1993, Sick 1993, Arribas *et al.* 1995, Hayes 1995, Howell & Webb 1995, Sibley 1996, Lowen *et al.* 1997, AOU 1998, Hilty 2003, Guyra Paraguay 2005, Garrigues & Dean 2007, Schulenberg *et al.* 2007). Apparently under-recorded.

HABITAT Thickets, second growth forest, low, open forest, but not open areas, dense vegetation along dry to moist forest borders, and second-growth woodland, in Venezuela in drier regions and lower elevations than Pavonine Cuckoo. Semideciduous, evergreen, swamp, and lower reaches of cloud forest, Oaxaca, Mexico. Also seasonally flooded forest, Amazonia, and gallery forest, and arboreal, xeric, deciduous *caatinga* woodland. To 1,600m Mexico and northern Central America; to 1,200m Guatemala; to 1,000m Honduras; to 1,200m south Costa Rica and Panama, once 2,300m (Bangs 1902; unreliable?); to 1,300m Colombia; to 400m Venezuela; to 700m Ecuador; locally to 1,000m Peru; possibly not higher than 800m Bolivia (Sick 1953a, b, Smithe & Paynter 1963, Monroe 1968, Wetmore 1968, Land 1970, Hilty & Brown 1986, Binford 1989, Stiles & Skutch 1989, Willis & Oniki 1991, Olmos 1993, Arribas *et al.* 1995, Howell & Webb 1995, Stotz *et al.* 1996, 1997, Ridgely & Greenfield 2001, Hilty 2003, Garrigues & Dean 2007, Schulenberg *et al.* 2007).

BEHAVIOUR Furtive, hard to see. Perches on low branch with large tail hanging straight down. When vocalising or excited, may perch in top of up to 20m high tree. Of 44 encounters in Panama forest plot, 40% were on ground and 60% perched or flying in sub-canopy. Flight slow and undulating, with tail broadly spread, but quick when passing from one thicket to another; sometimes crosses open spaces with high lifts of wings and spread tail. Forages on ground, stands and holds head level with back, neck contracted and bobs body up and down, producing pulses of rattling noises (apparently vibrating wing and tail feathers, and clapping bill) on downward movements of body, and fanning tail downward and brushing surface of leaf litter. Movements and sounds possibly flush prey from litter. Then bird pauses for one second, sounds and movements stop, then runs straight forward for 4–12 short, quick steps with head outstretched and low. Primaries, partially extended out and downward from body, show small white spots (otherwise hidden) and flick forward intermittently, with alulae extended, and bill is clapped very rapidly, and fluttery rattle noise increasing in intensity. After 0.5–2sec all noise and movements suddenly cease, and bird begins to peck in the litter, capture prey, or just look at the ground while walking, for 2–15sec, sometimes longer if prey require much handling. Becomes very cryptic when immobile. When disturbed, flees running with much wing-flapping. Calls at dawn, dusk and during moonlit nights. Approaches upon whistled imitation of three-note call. In aggressive territorial display on ground, two birds walk 0.5–1m apart parallel and slowly, with frequent pauses to preen or stand briefly on logs or buttresses, with raised uppertail-coverts, heads and crests, partially extended primaries and alulae, and puffed-out spotted breast feathers. Courtship display on ground: two birds approach each other with wings partially outstretched and heads fully so, producing soft rattle, then circle each other once, heads to tails, curving long uppertail-coverts over their backs (Naumburg 1930, Sick 1953a, b, 1993, Wetmore 1968, Sieving 1990, Howell & Webb 1995, Giraudo 1996, Hilty 2003).

BREEDING Parasitism Known hosts: Eye-ringed Flatbill with pendulous nest, Costa Rica; eggs in domed nests of flycatcher *Myiozetetes* sp. and Pied Water Tyrant, and open cup nest of Barred Antshrike, Honduras. Juvenile in and near hanging closed nest with bottom entrance tunnel, and fed by pair of Yellow-olive Flycatchers, Mexico (Carriker 1910, von Ihering 1914, Schönwetter 1967, Snethlage 1913 in Schönwetter 1967, Wilson 1992). **Season** Oaxaca, Mexico: enlarged ovum Apr, oviduct eggs May, Jun. Yucatán, Mexico, breeding pair, female with nearly fully developed eggs, May. Quintana Roo, Mexico, juvenile Aug. Slightly enlarged ovaries May, north Guatemala. Egg, Apr, Honduras. Vocalises Mar–Aug Mexico and northern Central America. Breeding condition male, Colombia, Apr. In Mato Grosso, Brazil, vocalises mid–Jun to Oct, culminating in Jul with nocturnal singing, and female ready to lay Dec; egg Nov, Bahia, Brazil; male with slightly enlarged testes Rio Grande do Sul, Sep, reacted strongly to vocalisation playback (von Ihering 1914, Naumburg 1930, Sick 1953a, b, Paynter 1955, Smithe & Paynter 1963, Rowley 1966, Schönwetter 1967, Wetmore 1968, Belton 1984, Hilty & Brown 1986, Binford 1989, Wilson 1992, Howell & Webb 1995). **Eggs** Long, very narrow, subelliptical, whitish or faintly buff, slight or no gloss, with irregular bands of liver-brown dots around each end, band around smaller end incomplete. 25.2x14.3mm, 23.3x16.0mm, 25.6x16.9mm (Brazil, Honduras) (Naumburg 1930, Schönwetter 1967, Wetmore 1968). No further information.

FOOD Large grasshoppers, cockroaches, Hemiptera (including cicadas), beetles; spiders; small lizards, possibly snakes; bird nestlings (von Berlepsch & von Ihering 1885, Sick 1953a, b, Schubart *et al.* 1965, Wetmore 1968, Stiles & Skutch 1989, Sieving 1990).

STATUS AND CONSERVATION Uncommon, but of low conservation priority (Stotz *et al.* 1996). Fairly common Mexico and northern Central America (Howell & Webb 1995); uncommon Oaxaca (Binford 1989) and Honduras (Monroe 1968), threatened El Salvador (Komar 1998).

Widespread but apparently very rare Costa Rica, but perhaps more numerous than indicated by few records (Stiles & Skutch 1989). Uncommon Panama, but probably overlooked (Ridgely & Gwynne 1989). Local, few records, Colombia (Hilty & Brown 1986). Rare or uncommon and very local Venezuela, and less numerous than Pavonine Cuckoo, but perhaps locally fairly common (Hilty 2003). Rare to locally uncommon Ecuador, all records recent (Ridgely & Greenfield 2001). Rare Peru (Schulenberg *et al.* 2007) and Argentina (Chébez 1994). Possibly extirpated São Paulo state, Brazil (Willis & Oniki 2003). Not globally threatened (BirdLife International 2011).

Pheasant-cuckoo *Dromococcyx phasianellus*. **Figs. 1–2.** Adult, Ituberá, Bahia, Brazil (*Ciro Albano*).

PAVONINE CUCKOO
Dromococcyx pavoninus Plate 2

Dromococcyx pavoninus von Pelzeln, 1870 (Araguay [Goiás], Engenho do Gama [Mato Grosso] and Arimani [Roraima], Brazil)

Alternative name: Peacock Cuckoo.

TAXONOMY Monotypic. Synonyms: *gracilis* Ridgway 1885, *pavonicus* Dabbene 1910 (*lapsus*), *pavoninus perijanus* Aveledo & Ginés 1950.

FIELD IDENTIFICATION 27–30.5cm. Similar to Pheasant-cuckoo, with small head, thin neck and broad tail, with uppertail-coverts almost same length as tail. **Adult** Sides of neck, throat and breast uniformly rich buff, eye-brow and eye-ring buff, above blackish, with greyish feather edges giving scaly look. Wing-coverts narrowly edged whitish, usually forming narrow wing-bar. **Juvenile** Crown and upperparts dark brown without pale feather edges, tail long without white tips. **Similar species** Compare voice with Pheasant-cuckoo, which has spotted breast and broader tail. Similarly sized American Striped Cuckoo is paler and streaked above, including crest, and has narrow tail.

VOICE Series of five far-carrying, clear, monotonic notes with timbre of American Striped Cuckoo *hoo… hee… ho-hee-he*, first and third notes lower pitched than the others, first two notes slow, total duration *c.* 1sec, repeated at *c.* 15sec intervals, sometimes only four notes *hoo… hee… hee-he*, or two-note response *hoo… hee* very similar to call of American Striped Cuckoo. Higher-pitched than Pheasant Cuckoo and does not end in trill (Giai 1951, Remsen & Traylor 1983, Chébez 1993, Sick 1993, Schulenberg *et al.* 2007).

DESCRIPTION Adult Forehead, crown and short, bushy nuchal crest chestnut, bordered by long buff postocular stripe and eye-ring, face and ear patch dark rufous; upperparts dusky with pale edges giving scaled impression and usually single narrow, pale grey wing-bar, wings otherwise dark greyish-brown, alula large, uppertail-coverts almost as long as tail, blackish with small white spot at tip; cheeks and diffuse malar streak dusky-chestnut, throat whitish, sides of neck, throat and breast uniformly rich buff, rest of underparts white; tail graduated, long, above brown tipped white, below grey becoming dusky distally, tipped white. **Juvenile** Crown and upperparts dark brown without pale feather edges, cheeks greyish-brown, wings as adult, uppertail-coverts long, chin white, throat and breast grey, belly whitish; tail long without white tips. **Nestling** Undescribed. **Bare parts** Orbital skin greenish-grey, with yellow and black inner ring. Iris brown, including in juvenile. Bill almost straight, black above, grey below; in juvenile, upper mandible dark brown becoming paler towards gape, lower yellowish. Feet grey to greyish-brown, pale olive in juvenile; claws brownish.

BIOMETRICS Wing male 129–145mm (n = 13), mean 134.4 ± 4.7, female 126–142mm (n = 5), mean 133.8 ± 3.4. Bill male 18.3–22.2mm, mean 19.6 ± 1.4, female 18.4–19.7mm, mean 19.1. Tail male 132–174mm, mean 150.4 ± 13.0, female 135–170mm, mean 152.8. Tarsus male 26.6–33.2mm, mean 28.2 ± 1.2, female 26.2–26.9mm, mean 26.5 (Payne 2005). Mass male 50g, female 45.2g, unsexed 40.5–54g (n = 5) (Oniki 1972, Sick 1993, de Magalhães 1999, Hilty 2003, Payne 2005).

MOULT Unknown.

DISTRIBUTION Resident east of Andes. Colombia (few records); north-west, north-central and south Venezuela, Guyana, French Guiana, east Ecuador, east Peru. Locally Brazil (Amapá, Acre, Amazonas, Pará, Maranhão, Mato Grosso, Tocantins, Goiás, Bahia, São Paulo, Rio de Janeiro, Paraná, Santa Catarina, Rio Grande do Sul), north and east Bolivia (Pando, Beni, La Paz, Santa Cruz, Cochabamba), Paraguay (mostly east) and north-east Argentina (Misiones, Corrientes) (von Pelzeln 1870, Ihering & Ihering 1907, Ridgway 1916, Mogensen 1930, Giai 1949, 1951, Aveledo & Ginés 1950, Sick 1953a, b, 1997, Pinto 1964, Schubart *et al.* 1965, Snyder 1966, Hidasi 1973, Darrieu & Martínez 1984, Stotz & Bierregaard 1989, Willis & Oniki 1990b, Oren 1991, Chébez 1993, 1996, Arribas *et al.* 1995, Scherer-Neto & Straube 1995, de Albuquerque 1996, Giraudo 1996, Aleixo & Galetti 1997, Zimmer *et al.* 1997, Novaes & Lima 1998, de Magalhães 1999, Buzzetti 2000a, Rodner *et al.* 2000, Bencke 2001, Ridgely & Greenfield 2001, Salaman *et al.* 2001, Hilty 2003, Lima *et al.* 2003, Willis & Oniki 2003, Donatelli *et al.* 2004, Krauczuk & Baldo 2004, Azevedo & Ghizoni 2005, Guyra Paraguay 2005, Pacheco & Olmos 2005, 2006, Payne 2005, Schulenberg *et al.* 2007, Guilherme & Santos 2009, Planque & Vellinga 2010). Espírito Santo, Brazil mentioned e.g. by Ruschi (1979) but no specific records. Mentioned for Minas Gerais, Brazil (Ihering & Ihering 1907, Pinto 1938, Payne 2005), without detail, but not mentioned by Pinto (1952), and apparently no record.

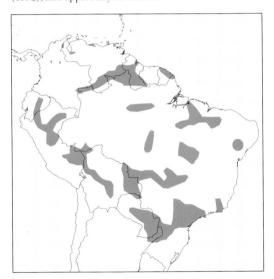

HABITAT Generally tall forest, but also secondary forest understorey. Humid foothill forest at east base of Bolivian Andes. Moist to humid forest-edge thickets, viny areas and dense patches of second growth, often perching under or in dense mats of vines, Venezuela. Humid and xerophytic forest, Paraguay. Recorded in *Araucaria* forest with much bamboo and dense understorey, south Brazil, also bamboo, Amazonia, and understorey of seasonally flooded transitional forest along rivers, with abundant *Heliconia* and figs, south-east Peru. To 1,600m, 350–900m south of Río Orinoco, 400–1,950m north of it, Venezuela. At least to 1,500m Ecuador. Locally to 900m Peru. Possibly not higher than 500m Bolivia (Remsen & Traylor 1983, Terborgh *et al.* 1984, Chébez 1993, Arribas *et al.* 1995, Stotz *et al.* 1996, Schulenberg & Awbrey 1997, Sick 1997, Hilty 2003, Azevedo & Ghizoni 2005, Guyra Paraguay 2005, Schulenberg *et al.* 2007).

BEHAVIOUR Elusive, skulking, solitary. Walks on ground opening wings and alulae. Mostly vocal before dawn and at dusk, but also at night; all day during breeding season. Calls with extended large twitching alulae, lowered tail and puffed long uppertail-coverts, perched in dense cover 3–9m up. Sometimes flies with slow mechanical wing beats, lifting wings above back and holding them there momentarily, with tail spread slightly. When nervous, raises and fans tail, and raises crest, then sneaks away into undergrowth almost without flapping wings (Sztolcman 1926, Giai 1951, Neunteufel 1951, Chébez 1993, Boesman 1998, Ridgely & Greenfield 2001, Hilty 2003, Willis & Oniki 2003).

BREEDING Parasitism Known hosts small suboscines with bag-shaped or closed nests: Plain Antvireo, Ochre-faced Tody-flycatcher, Eared Pygmy Tyrant, Drab-breasted Pygmy Tyrant, all considerably smaller than cuckoo. Unknown how egg is introduced into closed nests. Diameter of nest entrance of Ochre-faced Tody-Flycatcher is merely 3cm (Giai 1949, Neunteufel 1951, Chébez 1993). **Season** Vocalises irregularly late Jan–end May/Jun, occasionally Nov–Dec, Venezuela. Ovum 7mm, Jul, Guyana. Female ready to lay Nov, south-east Brazil. In Paraguay and north Argentina, vocalises during breeding season Sep–Dec, in beginning of season only giving first two notes of five-note call (Neunteufel 1951, de Magalhães 1999, Hilty 2003, Payne 2005). **Eggs** Four eggs from west Ecuador, elongated oval, pale yellow, sparsely but evenly marked with small, rounded, purplish-chestnut dots, 21.2–22.0x14.4–15.2mm, mean 21.5x14.8, each in clutch of two unidentified eggs (apparently two species of thamnophilids and/or tyrannids), were thought from size to be of this species (Schönwetter 1967), but this cuckoo is unknown there (Ridgely & Greenfield 2001). **Incubation** Unknown. **Chicks** Nestlings kill host young (Payne 2005). **Survival** Unknown.

FOOD Grasshoppers, earwigs (Forficulidae), cockroaches, cicadas, beetles; millipedes; spiders, scorpions (Sztolcman 1926, Sick 1953a, b, Schubart *et al.* 1965).

STATUS AND CONSERVATION Uncommon to locally fairly common Venezuela (Hilty 2003). Local Guyana (Snyder 1966). Rare Ecuador and Peru (Ridgely & Greenfield 2001, Schulenberg *et al.* 2007). Probably extirpated Rio de Janeiro state, Brazil (Alves *et al.* 2000, Gagliardi 2005) and possibly rare São Paulo (Willis & Oniki 2003), but not uncommon Paraná (e.g. Scherer-Neto & Straube 1995, Straube *et al.* 2005). Status Santa Catarina and Rio Grande do Sul uncertain, as only recorded in recent years at three sites (de Albuquerque 1996, Bencke 2001, Azevedo & Ghizoni 2005). Rare Paraguay (Hayes 1995). Sensitive to habitat disturbance (Stotz *et al.* 1996) – as densities are naturally low, forest fragmentation may increase risk of local extirpation (Azevedo & Ghizoni 2005). Not globally threatened (BirdLife International 2011).

Pavonine Cuckoo *Dromococcyx pavoninus*. **Figs. 1–4.** Adult, Las Piedras River, Madre de Dios, Peru, August (*Matthias Dehling*).

Genus *Morococcyx*

P.L. Sclater, 1862 *Cat. Amer. Birds* p. 322. Type, by monotypy, *Coccyzus erythropyga* Lesson 1842.
One species.

Neotropical; non brood-parasitic. Sexes alike; crestless; long legs; stout, curved bill; colourful bare facial skin; terrestrial. Basal to clade consisting of *Geococcyx* and *Neomorphus* (Sorenson & Payne 2005).

LESSER GROUND CUCKOO
Morococcyx erythropygus   Plate 2

Coccyzus erythropyga Lesson, 1842 (San-Carlos, Centre Amérique = La Unión, El Salvador)

Alternative names: Lesson/Rufous-rumped Ground-Cuckoo, Rufous-rumped Cuckoo.

TAXONOMY In error *erythropygius* or *erythropygia*. Polytypic. Synonyms: *e. dilutus* and *e. simulans*, both van Rossem 1938, in *mexicanus*; *e. macrourus* Griscom 1930 in nominate.

FIELD IDENTIFICATION 25–28cm. Chiefly terrestrial, strong-legged. Often heard; difficult to see. **Adult** Underparts cinnamon to rufous, eye-ring pale yellow and blue, surrounded by black, remiges with olive sheen, noticeable when flushed; tail tipped white. **Juvenile** Duller, outer rectrices indistinctly tipped buff. Similar species Mangrove Cuckoo arboreal (as other *Coccyzus*), lacks colourful bare skin around eyes (only narrow eye-ring), paler on underparts, grey on upperparts, with more white on undertail. Yellow-billed and Black-billed Cuckoos are less similar (even paler underparts, etc.).

VOICE Loud, rich, trilled whistles, initially slowly delivered single notes accelerating into fast series, then becoming slower *prree, prree, prree, prree- prreeprree… prree, prree, pree… prree, prree*. Sometimes only 2–3 introductory notes. Given from rock or small branch low over forest floor in disrupted shade, with head thrown back and tail jerking at each tremolo. Also burry or soft whistle, *whirrr* or *teeeee* on even pitch, repeated every 5–7sec, with raised head and hardly opening bill; rough, growling *ghaaaooow* and soft gargle (same as preceding?) given with raised head and open bill. At nest, on ground may deliver clicking sound resembling snapping of small dry sticks by clacking mandibles rapidly together (female?). Unsexed bird clacked mandibles while delivering high, trilled whistle (Slud 1964, Skutch 1966, Davis 1972, Rowley 1984, Stiles & Skutch 1989, Howell & Webb 1995).

DESCRIPTION Nominate. **Adult** Crown, nape, back, scapulars, wings and tail greyish-brown to olive-brown, crown feathers in part faintly edged with dull buff and with faint darker shaft-streaks, especially on forehead; bare area around eye surrounded by two black lines that diverge from base of bill, pass over and below bare area and meet near ear, usually with narrow pale buffy or greyish-white supercilium above black line; wings glossed olive, lower back and rump sooty-blackish, feathers tipped buff to olive-brown; underparts from chin and sides of neck to vent cinnamon-ochraceous to tawny-rufous, remiges plain brownish-grey below, outer primaries more dusky distally. Tail graduated, glossed purplish-bronze to bronze-green, outer rectrices tipped buffy-white with dusky subterminal area; underside light greyish-brown, lateral feathers with subterminal dull black area, largest on outermost feather, extending nearly to base. **Juvenile** Generally duller, feathers softer, above scaled greyish-buff, outer rectrices without black subterminal area or pale tip. **Nestling** Unknown. **Bare parts** Iris hazel or brown. Eye-ring yellow, bare triangular area in front yellow, bare triangle behind eye bright blue; bare orbital skin grey in juvenile. Bill orange-yellow, strongly decurved, with blackish or dark brown line along culmen, mouth-lining black; in juvenile, brown above, paler below. Feet bright orange-tawny, claws fuscous.

BIOMETRICS Nominate (Guatemala to Costa Rica). Wing male 93–105mm (n = 22), mean 98.7 ± 3.6, female 90–102mm (n = 15), mean 96.9 ± 3.8. Bill male 19.4–25.2mm, mean 22.2 ± 1.6, female 20–23.5mm, mean 21.8 ± 1.0. Tail male 126–140mm, mean 130.2 ± 6.7, female 120–142mm, mean 128.7 ± 7.4. Tarsus male 30.2–36.4mm, mean 33.6 ± 1.8, female 31.4–36.0mm, mean 33.5 ± 1.6. Mass *mexicanus* male 58–70.5g (n = 5), mean 63.1, female 62.9, 86.8g; nominate male 53.1–66.2g (n = 10), mean 61.1, female 56–76g (n =7), mean 65.3 (Payne 2005), female 55.2g (Tashian 1953).

MOULT Juvenile rectrices retained until Mar–Apr. Limited body moult, old birds and second-year birds, spring. Annual moult, adults, late Jul–mid Oct, varying individually, El Salvador; at least Jul–Aug Costa Rica; Aug Oaxaca, south Mexico (remiges, tail, body), in latter areas timing also apparently varies individually, and sometimes moult occurs in breeding condition. Primary and tail moult recorded in Jan, Guatemala (Dickey & van Rossem 1938, Tashian 1953, Foster 1975).

GEOGRAPHICAL VARIATION
M. e. erythropygus (Lesson, 1842). South Mexico (Oaxaca and Chiapas) to north-west Costa Rica.
M. e. mexicanus Ridgway, 1915. West Mexico south to Isthmus of Tehuantepec (Oaxaca). Larger and paler than nominate, upperparts more greyish-olive on average. Much individual variation in tone of underparts in both subspecies. Wing male 92–108mm, mean 100.5 (n = 11), female 95–107.5mm, mean 99 (n = 8) (Ridgway 1916).

DISTRIBUTION North and Central America. Resident Pacific slope, south-west Mexico, in south Sinaloa, Nayarit, west Jalisco, Colima, south Michoacán (also Río Balsas drainage), south Guerrero, extreme south-west Puebla, south Oaxaca, south and central Chiapas; south Guatemala, El Salvador, west Honduras, Nicaragua, north-west Costa Rica south to Nicoya Peninsula, Río Tárcoles and east to Villa Colón and Grecia in west Valle Central. Also Caribbean slope arid interior valleys, Guatemala and Honduras (Friedmann *et al.* 1950, Skutch 1966, Monroe 1968, Stiles & Skutch 1989, Sibley & Monroe 1990, Howell & Webb 1995, AOU 1998).

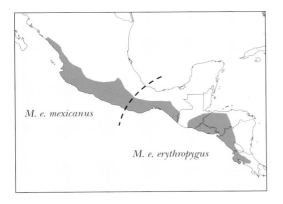

M. e. mexicanus

M. e. erythropygus

HABITAT Arid lowland scrub, edge of tropical deciduous forest, second-growth scrub, thorny thickets, sometimes with cacti, edge of mangrove. To 1,500m, locally to 1,850m, Mexico–Guatemala. Mainly below 1,000m El Salvador; to 1,200m Costa Rica (Skutch 1966, Land 1970, Stiles & Skutch 1989, Howell & Webb 1995, Stotz et al. 1996, Komar 1998).

BEHAVIOUR Furtive and skulking, but often inquisitive, not shy; in singles or pairs. Hops or walks on ground with jerky movements of head, slowly and deliberately; if alarmed, may freeze, and then becomes difficult to see; flushes with low flight up to 75m to cover. Walks or runs along branches dove-like, sometimes with tail cocked. Lands with short run. May expose itself, e.g. perching on lower strands of fence. Can run swiftly with tail and head held low. Picks up insects from ground, or leaps up to snatch them from foliage. Collects fine fallen twigs for nest. Distraction display near nest by incubating parent: walks away with mincing steps, stops, puffs out body feathers, droops wings, lowers and fans tail, moves back and forth several times with short, slow steps, beats ground with wings. On another occasion, crouched on ground, vibrated drooped wings, beating them against body, and perhaps scratched ground with feet. Incubating bird panted with open bill when reached by sun rays. Parent on nest exchanged calls with distant other parent before being relieved, and relieving parent brought fine nesting material and added it to lining of nest. Until end of incubation, brings an occasional stick to nest (Skutch 1959, 1966, Berger 1960, Slud 1964, Rowley 1984, Stiles & Skutch 1989, Howell & Webb 1995).

BREEDING Season During rainy season, May–Nov Oaxaca; Feb–May Costa Rica, apparently extending to Jul–Aug; incubation Jul, Guatemala (Skutch 1966, Foster 1975, Rowley 1984, Binford 1989, Stiles & Skutch 1989). **Nest** Shallow cup of twigs and dead leaves, diameter 130–150mm, inside of cup 8cm, depth 4.5cm; on ground, sometimes under cover (Skutch 1966, Rowley 1984, Howell & Webb 1995). **Eggs** 1–3; white or whitish, smooth, some chalky, elliptical, 27.8–30.3x20.4–22.6mm, mean 29.07x21.91mm (n = 7), Oaxaca; 27.0–27.8x20.6–21.0mm (n = 2), Guatemala (Skutch 1966, Rowley 1984, Howell & Webb 1995). **Incubation** One sex, possibly male, incubated most of time, including all night, while other sex incubated 3h or more in middle of day (Skutch 1966). No further information.

FOOD Beetles, grasshoppers and other insects (Dickey & van Rossem 1938, Tashian 1953, Stiles & Skutch 1989).

STATUS AND CONSERVATION Fairly common Mexico (Howell & Webb 1995), Guatemala (Land 1970) and Honduran lowlands (Monroe 1968); common Oaxaca (Binford 1989), Colima (Schaldach 1963) and El Salvador (Dickey & van Rossem 1938). Most common in lowlands, Mexico and Honduras (Davis 1972). Not globally threatened (BirdLife International 2011).

Lesser Ground Cuckoo *Morococcyx erythropygus*. Adults. **Fig. 1.** Guanacaste, Costa Rica, January (*Mike Danzenbaker*). **Fig. 2.** Costa Rica, November (*Kevin Easley*).

Genus *Geococcyx*

Wagler, 1831 *Isis von Oken* 24, col. 524. Type, by monotypy, *Geococcyx variegata*
Wagler 1831 = *Saurothera californiana* Lesson 1829.
Two species.

North and Central America; non brood-parasitic. Large, crested; sexes similar; mostly terrestrial. Sister of *Neomorphus* (Sorenson & Payne 2005).

GREATER ROADRUNNER
Geococcyx californianus   Plate 2

Saurothera californiana Lesson, 1829 (California)

Alternative names: Roadrunner, California Roadrunner.

TAXONOMY Larger, but otherwise similar extinct *G. conklingi* Howard 1931, from late Pleistocene–Holocene of Mexico, New Mexico and Texas, possibly subspecies of *californianus* (Miller 1943 and Harris & Crews 1983 in Hughes 1996). Larger in west, but variation clinal. Monotypic. Synonyms: *Cuculus viaticus* Lichtenstein 1830, *variegata* Wagler 1831, *Saurothera bottæ* Lesson 1831, *marginata* Kaup 1832, *Leptostoma longicauda* Swainson 1837, *mexicanus* Strickland 1842, *G. c. dromicus* Oberholser 1974 (see Browning 1978, 1990 and Hughes 1996).

FIELD IDENTIFICATION 52–58cm. Large, slender, streaked, long-tailed, ground-dwelling, fast-running bird with long, strong legs, long, heavy bill with hooked tip and prominent, bushy, erectile, blue-black crest. Occasionally flies short distances on rounded wings. Female slightly smaller than male. **Adult** Body plumage heavily streaked brownish and whitish with uniform whitish belly to undertail-coverts. Bold white spots across outer primaries in flight. Tail blackish edged white, outer rectrices boldly tipped white below. Blue or white area behind eye, becoming orange (usually hidden) on nape. **Juvenile** Duller; unglossed. **Similar species** Lesser Roadrunner smaller, bill shorter and more slender, throat and chest unstreaked buff, postorbital skin crimson.

VOICE Most frequently heard call male's 3–8 (usually 5) low-frequency (*c.* 6kHz), down-slurred *coo* notes, *co-coo-coo-coo-cooooooo*, accompanied by complex movements of head: from normal position, head is lowered with fully erected crest and exposed bare postocular skin, bill then moved back and forth from body at each note. Typically given from elevated perch, which may change during bout, at intervals of several min, from sunrise and continuing 30min to several hours. Also given by unattended juvenile. From ground, male (rarely) delivers similar *cooo-cooooo*. During 'flick-bow' display, both sexes give soft *coo*. Female and unattended juvenile give rapid series (2–22 notes) of sharp, short, low-frequency 'barks' reminiscent of coyote yelps, sometimes with raised crest and usually with exposed bare postorbital skin. Most frequent call during egg-laying and incubation, usually given by female at nest site, often in response to growl from nearby male. Both sexes and juveniles have 3–4 low-frequency growling notes, with bulging throat, fluffed feathers, fully erected crest and partially hidden bare skin behind eyes, given when birds are together, or to nestlings or fledglings. During precopulatory display, male emits soft, low-frequency, mechanical putting sounds alternating with slow whirs, *putt-putt-putt-putt-whirrrrrr-putt-putt-putt-putt-whirrrrrr*. Also long, deep whine, from both sexes together often repeated for up to 5min, usually from sitting or crouching position, moving head downward and shaking head horizontally 3–6 times, with sleeked crest and feathers, and obscured bare orbital skin. Nonvocal sounds: sharp *clack* produced by snapping mandibles together, in series of 5–6, accompanied by whine. Female clacks of higher frequency, slightly longer and at shorter intervals than male's. Also given by young at 16 days, when disturbed at nest. 4–8 loud pops by male by bringing wings rapidly inward to body during precopulatory display. Rasping buzz and hisses from third day from nestling in response to visual stimuli, louder as nestling grows. Begging call of older nestling whine similar to female (Muller 1971, Whitson 1971, 1975, Bahr *et al.* 1992, Hughes 1996).

DESCRIPTION Adult Crown blackish-brown, spotted cinnamon. Upperparts olive to dark grey-brown glossed bronze and coarsely striped pale buff. Wings and uppertail with bold green sheen. Remiges dark brownish-black glossed olive. Tertials rufous-bronze edged white, rest of wing-coverts like back, with broad russet edges. Large white spots over outer primaries form large crescent. Wing lining dark brownish-black. Face, throat and chest whitish; throat and chest striped cinnamon and blackish. Rest of underparts uniform greyish or pale buff. Tail graduated; T1 bronzy-olive narrowly edged white; other rectrices metallic blue-black on outer, bronze on inner webs. Undertail blackish, outer 3 rectrices with large oblong white tips (often tattered). **Juvenile** As adult, but with duller markings, without bronze gloss, more pointed feather tips, white U-shaped patches on tips of primary-coverts, secondaries slightly narrower and browner, rectrices more narrow and pointed, with wedge-shaped white tips. **Nestling** Mass at hatching *c.* 14g. Skin greasy dull black; egg tooth light-coloured. Scattered trichoptiles white. Eyes open day 4. Gape flange pink or dull flesh. Mandibles and tongue tip black. Mouth bright red with white hard palate and white markings in two soft dorsal areas at base of tongue, and sides of palate. **Bare parts** Iris pale brownish or glaucous, with straw yellow ring adjacent to pupil. In young nestlings, eyes black; in older nestlings, iris dark greyish-brown, pupil blue-black. Eye-ring cobalt to indigo blue, bare skin extending backwards as white (male) or light blue (female) band edged dark blue above and below, becoming orange towards nape, although sex differences not certain. Juvenile postorbital skin female; some juvenile males intermediate, white area patched blue. Orbital skin always exposed, but normally covered postorbitally. Orange area present at day 14. Bill slightly hooked, greyish, base of lower mandible pale bluish; black in nestling. Gape pink until day 50–55, when it becomes speckled black, at day 80–85 all black. Feet light dull blue or greyish, larger scales buff or cream, claws black; in nestling, feet black, later blue-grey.

BIOMETRICS Wing male 163–190mm, mean 178.3 ± 8.3 (n = 16), female 158–186mm, mean 170.7 ± 9.0 (n = 12). Bill male 43–53mm, mean 48.1 ± 3.2, female 41–51mm, mean 45.5 ± 3.2. Tail male 256–302mm, mean 283.6 ± 14.3, female 260–308mm, mean 275.5 ± 14.4. Tarsus male 56–66mm, mean 61.4 ± 2.7, female 56–64mm, mean 59.5 ± 2.8. Mass (California) male 275–430g, mean 344.0 (n = 8), female 264–352g, mean 308.5 (n = 6); (Texas) male 187–333g, mean 270.0 (n = 6) (Payne 2005); male 280–380g, mean 319 (n = 8), female 278–297g, mean 290 (n = 4) (Dunning 1993 and S. Vehrencamp in Hughes 1996). female mean 190g (n = 2) (Johnson 1968a), female (Coahuila, Mexico) 250, 274g (Miller 1955). Fat, winter, up to 588g (Geluso 1969).

MOULT Partial moult of remiges and rectrices Sep–Nov of hatching year. In adults, complete annual postbreeding moult May–Nov, peak Jun–Sep. Remiges replacement occurs throughout non-breeding season, slow and asymmetric, sequence irregular, each growing remex may be adjacent to fully grown feathers, and some juvenile remiges sometimes retained for several years (Pyle 1995, Hughes 1996).

DISTRIBUTION South-west USA and north Mexico. Resident. North-central (Sacramento Valley), east (Owens Valley) and south (widespread) California, west and south Nevada, extreme south-west Utah, Arizona (except north-east), central and south-east Colorado (east Fremont and south-west El Paso counties, south-east to south-west Baca County), New Mexico (except north-west), extreme south Kansas, central and east Oklahoma, Texas, south-west Missouri, central and west Arkansas, north and west Louisiana. Mexico: Baja California, west and central Sonora, north Sinaloa, Chihuahua (except western mountains), Coahuila, Nuevo León, north and central Tamaulipas, Durango, Zacatecas, San Luis Potosí, central and north-east Jalisco, Guanajuato, Querétaro, Hidalgo, north and east Michoacán, north México, Morelos, central Puebla, Veracruz. Reports from Distrito Federal, Mexico, probably erroneous. USA range has extended into Kansas, Missouri and Arkansas during 20th century, and contracted in parts of California (Ridgway 1916, Colvin 1935, Baerg 1950, Friedmann *et al.* 1950, Brown 1963, AOU 1983, 1998, Howell & Webb 1995, Hughes 1996, Rojas-Soto *et al.* 2001).

HABITAT Semiarid and arid open country with scattered 2–3m high scrub, sometimes open farmland and sparsely populated suburban developments. Piñon-juniper woodland and cholla cactus grasslands, Colorado foothills and mesas. Bottomland tamarix thickets, blackbrush and creosote scrub, and riparian areas, Utah and Nevada. Mesquite and juniper savanna, lowland and mesa riparian woodland, juniper-oak woodland, chaparral and huisache, Texas and south-west Oklahoma. Shortleaf-loblolly pine and hardwood uplands, Louisiana. Cedar glade vegetation on rocky terrain strewn with boulders, dominated by red juniper, woody shrubs and tall-grass prairie plants, south-west Missouri, west Arkansas and east Oklahoma. Grassland with edges of chaparral, California. Cultivated areas, saltbush and mesquite; also desert scrub, grassland, creosote bush and mountain shrub, Arizona. Open and semiopen arid areas and brushy woodland, Mexico. Between –76m (Death Valley, California) and 2,500m, occasionally to 3,000m. Prolonged cover of deep snow determines northern range limit (Sutton 1940, Behle 1943, Grinnell & Miller 1944, Hamilton 1962, Brown 1963, Goertz & Mowbray 1970, Folse & Arnold 1978, Norris & Elder 1982, Parker & Campbell 1985, Meinzer 1993, Howell & Webb 1995, Hughes 1996, Stotz *et al.* 1996).

BEHAVIOUR Solitary outside breeding season, in pairs when breeding. Holds head and tail parallel with ground, using latter as rudder to change direction, when running at high speed, reaching 29km/h. Often runs along roads or dry stream beds rather than through vegetation. Glides from high perch or nest; rarely flies 4–5m between treetops etc; may escape pursuit flying into brush. Bird flushed from nest to ground crouched on breast and fluttered wings. Scratches head under wing. Dustbathes squatting on breast, shuffling feet and fluttering wings. Sunbathes up to 2–3h early morning, or 30+min sessions throughout day in winter. In winter, sometimes roosts in old nest. Early in breeding cycle, male and female often roost in same or adjacent bushes. Reduces activity, resting in shade during hot midday (Sutton 1915, 1922, 1940, Bryant 1916, Sheldon 1922, Rand 1941, Calder 1968b, Kavanau & Ramos 1970, Harrison 1971c, Ohmart & Lasiewski 1971, Rylander 1972, Oberholser 1974, Whitson 1975, Ohmart 1989, Meinzer 1993, Hughes 1996).

Forages mostly on ground in areas with low or no vegetation, including mowed grass, but also sometimes in bushes or small trees, where gleans arthropods from leaves. Locates prey by sight, alternately moving forwards and pausing to scan, or flushes and captures prey while moving, often leaping up to 3m to catch flying insect. In winter, an individual walked slowly through 2.5cm deep snow, scrutinising dead 60cm high plants, peered under leaves and pecked at base of plants. Stalks small mammals or passerines. One followed pair of Gambel's Quail with 12 newly hatched chicks that would be easy prey, taking grasshoppers flushed by the quail. Scorpions attacked by tail. Partially plucks birds. Larger prey (rodents, snakes, some lizards, large grasshoppers) beaten repeatedly against hard substrate before swallowing. Texas Horned Lizards swallowed head first with ossified spiny dorsal head horns aimed down so as not to threaten bird's vital organs (captivity). Kills rodents and snakes by blow of bill to skull. Pair may cooperate in attacking snake. Knocks low-flying swifts to ground by lying in wait in dry creek bed, suddenly jumping into air. Takes birds from ringing traps, bird feeders, nest boxes and mistnets. Overturns large plates of caked dry mud to find crickets. Knocks cactus fruit to ground and tumbles it to remove spines (Anthony 1896, Sutton 1913, 1922, 1940, Bryant 1916, Gorsuch 1932, Jaeger 1947, Woods 1960, Calder 1967, Geluso 1970b, Zimmerman 1970, Wright 1973, Bleich 1975, Spofford 1976, 1978, Barclay 1977, Beal & Beal 1978, Beal & Gillam 1979, Beal 1981, Miller 1988, Ohmart 1989, Sherbrooke 1990, Meinzer 1993, Green 1994, Hughes 1996).

Foraging and breeding territory maintained by mated pair year-round. Territory divided between pair during nestling phase and winter, but otherwise forage together. Territorial male approaches intruding male with lowered head, bare skin on sides of head maximally exposed and tail held vertically and slowly swinging from side to side. Trespasser then turns and presents bare skin of head. Birds move from side to side, few metres apart, repeatedly presenting bare skin. A territorial male first gave *coo* call to trespasser, then approached in short bursts, popping wings every few steps. After retreating few metres beyond territory border, initiated mutual *coo* calling with neighbouring male. Resident female joined mate in defence of territory by going with male to boundary giving mandible snaps. Territories normally stable, pairs staying from year to year, but pairs may move 1.6km to new territory, or single birds may move without mate, change of place usually after nest failure. Early in breeding season, pair bond is formed or renewed through preliminary courtship displays: vigorous ground chase lasting up to several hours, running interspersed by low gliding flights and stops to rest, and accompanied by bill clacks and male's *coo* call from top of shrub or tree, and female's 'bark'. Occasionally, pursuing individual attacks the other with wings and tail raised and fully fanned. During intervals between chases, both sexes give one or more other displays (tail-wag, stick-offer, growl, whine, flick-bow). Stick-offer: bird of either sex approaches mate with stick or blade of grass in bill, often shaking it, then drops it in front of mate or transfers it to mate's bill. Precopulatory displays may last several minutes. In 'prance' display, male runs from mate with lifted wings and tail exposing black-and-white undertail, lowers wings and brings them toward sides of body to produce loud pop; repeated 4–5 times; postorbital skin maximally exposed, crest slightly to fully erected, contour feathers sleeked. 'Tail-wag' display: male faces female with fully exposed postorbital skin and erect crest, often with prey item or plant material in bill; tail wagged horizontally and head intermittently bowed and slowly lifted. When head down, tail is fanned. Gives putts and whirs, and usually approaches female, from *c.* 0.3–3m. Female flicks tail rapidly vertically with exposed postorbital skin and erect crest. Tail-wag usually followed by male mounting from rear with jump of 0.5–1m into air, rapidly flapping wings and fanning tail during descent, and giving whirring call during copulation, female takes prey item from male, and male dismounts. Both birds have raised tail and slightly erect crest. Copulation lasts 2–3min. Postcopulatory flick-bow display: male circles female rapidly, drops slightly fanned tail and points head upward. At intervals in circling, he stops, presents side view to female, bows head, lowers wings, demonstrating white wing spots, gives *coo*, then rapidly flicks tail and head upward, fully erecting crest and exposing postorbital skin. Female also flicks tail, simultaneously or alternately with male. Then walk away in opposite directions. Female eats presentation food, or feeds it to young if hatched. Sometimes both adults hold food during copulation, and both feed young afterwards.

Presentation food items: lizards (most frequent), scorpions, insects, birds, snakes, plants, small rodents. Copulation shortly after sunrise to just before sunset, mainly afternoon, usually <20m from nest, during all phases of breeding, but frequency less from onset of laying. Stick-offer display often followed by nest-building behaviour, the pair moving into shrubs or trees, male usually leading; female usually builds with sticks and twigs brought by male. Many sham nests initiated before final nest site selected. Nest-building until first egg laid lasts 3–6d, but continues until after hatching. Height of nest walls may be increased as chicks grow. Approaches nest by hopping from limb to limb; leaves in glide to ground. Pairs may begin to build new nest while still caring for fledglings (Bryant 1916, Rand 1941, Woods 1960, Calder 1967, Whitson 1975, Folse & Arnold 1978, Miller 1988, Meinzer 1993, Hughes 1996).

BREEDING Parasitism Brood parasitism may occur infrequently, but requires confirmation; eggs found in nest of Common Raven (Pemberton 1925) and Northern Mockingbird (Oberholser 1974). Intraspecific brood parasitism probably infrequently, suggested by large clutches, unusual laying intervals and observation of a third adult in a territory, once seen standing on nest (Hughes 1996). **Season** Varies geographically and/or with local climatic conditions; poorly known in east and north part of range and in Mexico. Mar–May, eggs sometimes Feb–Jul, south California. Four-egg clutch 2 Feb, probably laid late Jan, California. Eggs early Apr–mid Jul Kansas. Bimodal south Arizona: mid Apr–mid Jun, and late Jul–mid Sep; pattern caused by extreme heat and aridity late Jun–early Jul followed by rains late Jul. Pair-formation Feb or early Mar, eggs 5 Mar–10 Oct, Texas. Apr–Aug, Oklahoma. Peak possibly Jul Arkansas, recorded 23 Aug. Eggs 16 Apr–16 May Mexico, breeding May–Jul, Sonora. Commonly double-brooded, occasionally triple (Kelsey 1903, Sharp 1907, Willett 1912, Sutton 1940, 1967, Woods 1960, Johnston 1964, Ohmart 1973, Oberholser 1974, Whitson 1975, Cornett 1983, Hughes 1996, Russell & Monson 1998). **Nest** Shallow platform of thorny sticks loosely laid together, lined with leaves, grass, feathers, mesquite pods, snake skin, roots or dry manure; outside diameter 30–45cm, depth 15–20cm, inside diameter 15cm, depth 5–10cm; in thorny bush, small tree or cactus, rarely on ground, 0.4–6m up; usually in tree fork or on horizontal branch, near centre of bush, often well hidden; in isolated thicket, not in extensive tract of woody vegetation, and near area with vegetation lower than 5cm; also in abandoned oil derricks, neglected farm machine and cliff crevice. Few nests are reused next year or same breeding season (Kelsey 1903, Bryant 1916, Sutton 1922, 1940, 1967, Colvin 1935, Woods 1960, Ohmart 1973, Folse & Arnold 1978, Meinzer 1993, Pridgeon 1995, Hughes 1996). **Eggs** Short oval to short elliptical, smooth but not glossy, white overlain with slightly yellowish chalky film, sometimes stained brown or grey; 35.36–45.64x28.05–32.78mm. Clutch 2–7, depending on food availability; mean larger (5.7) Arizona during summer rains than spring dry season (4.6). Mean, south Texas, 4.0 eggs/clutch; Kansas 4.5. Clutches of up to 12 eggs probably due to more than one female laying in same nest. Laying interval 2 (1–4) days, reportedly up to 9 days (probably 'egg-dumping'). Clutch replacement common if eggs predated. First egg laid in new nest nine days after nest failure. Two captive females laid fertile eggs at 9 months (Sutton 1940, Johnston 1964, Ohmart 1973, Folse & Arnold 1978, Smith 1981, Vehrencamp 1982a, Meinzer 1993, Hughes 1996). **Incubation** Begins when first egg laid;

17–18 days; male incubates at night, both sexes during day, but female may take long bouts morning and afternoon. Captive female took half eggshell away from nest, broke and ate it (Pemberton 1925, Sutton 1940, Calder 1967, Ohmart 1973, Whitson 1975, Smith 1981, Vehrencamp 1982a). **Chicks** Hatching asynchronous, age difference up to seven days. Development of captive nestling: strong and active upon hatching; when dry, responds to touch by gaping. Mass at 24h *c.* 20g, at 3 days 34g, sheathed tips of primaries and rectrices emerging. At 4 days 46g, eyes open, black. At 6 days 69g, rises on tarsometatarsus and has no more head jerks to back/sides. At 8 days 94g, moves around; legs and feet growing quickly. At 11 days 132g, can walk; feet large, changing from black to grey, with soles flesh-colour; shows first signs of fear. At 14 days 154g, has plumage pattern and bare postocular skin colour of adult. At 16 days 174g, bobs tail, raises crest, runs rapidly, makes preening motions, picks up and swallows food items, but still begs for food. Until first-hatched young is 12–13 days old, at least one parent always at nest, brooding or shading. During remaining nestling period, parents absent increasingly longer, hunting. When shading, spreads and droops wings slightly, lowers breast and holds tail into breeze to funnel air down over young. Nestlings able to lower body temperature by gular flutter and staying in bands of shade over nest. Excretes salts through glands near eyes. Male broods at night, usually relieved by female shortly after dawn; parents take turns during day. Young fed mostly insects first few days, later mostly lizards, few snakes, nestlings, etc. Vertebrates are beaten into a flexible form and grasshoppers have their hind legs removed, but otherwise food items are delivered intact to nestlings. Until nestling age *c.* 14 days, parent inserts bill holding food item into nestling's mouth and passes a clear fluid of unknown nature to nestling before giving food. Parents ingest faecal sacs. Nestling sometimes eaten by parent after pecking at or shaking it, presumably to check lethargy. Leave nest at 17–19 days; 12–14 days if nest disturbed; fledging usually at 19–21 days, half adult mass, central rectrices 114mm; some pins beginning to rupture or ruptured. In broods of 4–6, first 4–5 may leave nest rather rapidly, others take up to week more. One or more eggs may be abandoned or fail because of erratic incubation after hatching of first eggs and large food demand by older chicks. Parents feed young until 30–40 days after leaving nest while foraging together. May initiate new nest while still caring for fledglings, perhaps when food plentiful (Finley & Finley 1915, Woods 1960, Calder 1967, 1968a, Muller 1971, Ohmart 1973, 1989, Whitson 1975, Folse & Arnold 1978, Vehrencamp 1982a, Meinzer 1993, Pridgeon 1995, Hughes 1996). **Survival** Proportion of eggs resulting in successful fledglings variable: 12% south Texas to 72% New Mexico, Oklahoma and west Texas. Ringed bird lived ≥7 years; captive male lived >9 years (Folse & Arnold 1978, Smith 1981, Hughes 1996, Payne 1997).

FOOD Opportunistic carnivore. Vertebrates taken mostly during breeding season, when most active, and fed to young. In winter, mostly insects and other arthropods. In California, 90% animal matter and 10% fruit and seeds (mostly of sumac *Rhus* spp.), food volume. Insects include beetles (18%), grasshoppers and crickets (37%), caterpillars (7%), cicadas and other Hemiptera (5%), ants, bees and wasps (4%); scorpions (4%); three species of lizards (4%); birds (1.5%); young cottontail rabbit, two mice of different species (84 gizzards). In Arizona, breeding season, grasshoppers constituted 62% of food volume, insects in total 86%, birds and reptiles (mostly lizards) each 6%, and vegetable matter 2% (*c.* 100 gizzards). In north-central Texas, Apr–Jun, insects constituted 77% of food items, mostly grasshoppers, lizards and snakes 15%, four amphibians and harvest mouse (16 gizzards) (Bryant 1916, Gorsuch 1932, Geluso 1969, 1970a, Sutton 1972, Parmley 1982). Food includes: snails; mantids, beetles, flies, Lepidoptera; tarantulas and other spiders; centipedes, millipedes; woodlice (*Porcellio* spp.); 7 species of snake, 12 of lizard; frogs, toads; at least 20 species of bird, nestlings and adults; 8 species of rodent, young bats; seeds and fruits of *Opuntia engelmannii* cactus in winter, New Mexico, infrequently fruit of tasajillo *O. leptocaulis* and prickly pear cacti (*O. phaeacantha* and *polyacantha*). Carrion when available. May drink, but able to maintain body mass without drinking if food has high water content (rodents, reptiles) (Anthony 1896, Fisher 1904, Sutton 1915, 1922, 1940, Bryant 1916, Gorsuch 1932, van Tyne & Sutton 1937, Monson 1946, Johnson *et al.* 1948, Herreid 1960, Woods 1960, Wilks & Laughlin 1961, Brown 1963, Geluso 1970a, Zimmerman 1970, Binford 1971, Mayhew 1971, Wright 1973, Bleich 1975, Dunson *et al.* 1976, Spofford 1976, 1978, Barclay 1977, Folse & Arnold 1978, Parmley 1982, Miller 1988, Garland 1989, Sherbrooke 1990, Meinzer 1993, Green 1994, Pridgeon 1995, Hughes 1996).

STATUS AND CONSERVATION Adults usually sufficiently swift to avoid terrestrial predators like Coyotes *Canis latrans* (Sutton 1940), although Bobcats *Felis rufus* and Ring-tailed Cats *Bassariscus astutus* are predators. Occasionally taken by Red-tailed Hawk, Cooper's Hawk and Harris's Hawk (van Tyne & Sutton 1937, Beal & Beal 1978, Mader 1979, Garland 1989). Raptors, feral cats and raccoons rarely predate incubating adults (Meinzer 1993, Hughes 1996). Snakes probably primary nest predators (Folse & Arnold 1978). Average territory size south Texas 0.5km^2; density 1.5–3.1 birds/km^2; minimum annual survival of adults 60% (Folse & Arnold 1978). Territory size south Arizona 1km^2 (Calder 1968b). Four birds/km^2 south California (Bryant 1916); 1.2 (Emlen 1974) and 10 (Tweit & Tweit 1986) birds/km^2 south Arizona; 0.65 birds/ km^2 on rocky slopes south California; lower in other desert habitats (Payne 2005). USA range roughly coincides with region with >140 clear days from sunrise to sunset/year (Root 1988). Areas of greatest density in USA are south-east California, Sonoran Desert in south Arizona, Chihuahuan Desert west of Pecos River, Texas, and south of Edwards Plateau, Texas (Root 1988, Price *et al.* 1995). No significant change in population numbers over most of range 1966–1993 (Price *et al.* 1995). Disappears from urbanised areas with disturbance from traffic and dogs and predation by cats diminishing breeding success and foraging activities (Emlen 1974), but can exist in reduced numbers in suburban areas with minimal exotic vegetation and remaining areas of native paloverde-saguaro vegetation (Arizona; Tweit & Tweit 1986). Many still shot illegally (Meinzer 1993 in Hughes 1996), but effect on populations not known. Fairly common (Stotz *et al.* 1996). Total estimated population 1.1 million (Harrison 2005). Not globally threatened (BirdLife International 2011).

Greater Roadrunner *Geococcyx californianus*. **Figs. 1–2.** Adult, Bosque del Apache, New Mexico, USA, April (*Håkan Sivencrona*).

LESSER ROADRUNNER
Geococcyx velox   Plate 2

Cuculus velox A. Wagner, 1836 (Mexico = outskirts of Mexico City)

Alternative names: Rusty (*affinis*)/Pale (*melanchima*)/Black-and-white (*longisignum*) Roadrunner.

TAXONOMY Putative subspecific characters not constant, and much intergradation between populations. Monotypic. Synonyms: *v. affinis* Hartlaub 1844, *v. melanchima* and *longisignum*, both Moore 1934, *v. pallidus* Carriker & Meyer de Schauensee 1935.

FIELD IDENTIFICATION 40–50cm. Crest often raised. Smaller version of Greater Roadrunner: slender, streaked, long-tailed, ground-dwelling, fast-running with long legs. **Adult** Crown and crest blackish-brown spotted whitish. Upperparts dark rufous-brown coarsely striped pale buff, wings and tail with strong metallic green sheen. Bold white spots on outer primaries in flight. Throat and underparts buff, undertail-coverts tan, sides of neck and chest streaked blackish. Tail blackish below, outer feathers boldly tipped white. **Juvenile** Similar to adult; markings of upperparts much less sharply defined. **Similar species** Greater Roadrunner larger, bill longer and thicker, throat and chest streaked, undertail-coverts whitish, not tan, bare skin behind eyes orange, not red.

VOICE Slowing, descending series of 3–7 low-pitched (*c.* 0.5kHz), moaning, somewhat dove-like coo notes, *oooah...* or *owoah...* lasting *c.* 0.5sec, often break into a pitch at 0.3kHz, successively earlier in each note, total length 5–6sec, *c.* 1 note/sec (Davis 1972, Hardy *et al.* 1990, Howell & Webb 1995, Payne 2005).

DESCRIPTION Adult Crown black slightly glossed bronze-green, spotted pale buff. Auricular region faintly streaked dusky and pale buffy. Lores buffy whitish with projecting black bristle-like feather-tips. Hindneck brownish-black feather edged broadly pale buffy. Scapulars and back brown sharply streaked pale buff. Rump brown, feathers indistinctly tipped pale. Primaries blackish, with oblique broad white band near middle of outer webs, and broadly tipped white. Wing-coverts black with broad white edges. Underparts buff in fresh autumn plumage becoming white in spring, sides of neck and chest broadly streaked black, sometimes with tawny-brown margins to streaks. Undertail-coverts dusky, sometimes tipped tawny or rusty-brownish. Uppertail-coverts and T1 brown glossed purplish-bronze and sharply edged white. T2 dusky-olive with small white terminal spot and narrow white edge to outer web. Rest of rectrices blackish tipped white, width of white tip increasing to outermost, where 30–45mm. Inner webs of undertail light grey, followed by black band of varying width and white tips. **Juvenile** Less sharply marked than adult, streaks and edgings of upperparts mostly dull buffy, not white. **Nestling** Undescribed. **Bare parts** Iris yellow to brown, or brown with dark buff, yellow or silvery-white ring around pupil. Orbital ring pale lavender to bright blue, extending backwards as blue or lilac band which becomes crimson on nape (latter usually hidden). Bill greyish above, pale blue-green below. Feet greyish.

BIOMETRICS Wing male 133–154mm (n = 11), mean 145.9 ± 6.1, female 132–156mm (n = 9), mean 142.1 ± 8.5. Bill male 34–40mm, mean 37.5 ± 1.9, female 33–39mm, mean 36.9 ± 2.3. Tail male 253–288mm, mean 271.8 ± 11.6, female 247–280mm, mean 262.2 ± 13.9. Tarsus male 46–52mm, mean 48.7 ± 1.8, female 43–50mm, mean 46.9 ± 2.1. Mass male 174–203g (n = 5), mean 186.0g, female 162.8–192g (n = 4), mean 174.0g (Payne 2005); male 207.6g (Davis 1944).

MOULT Male heavily moulting Aug, Sonora, presumably postbreeding, another in early stage Jalisco Jul, with tarsal scutes being replaced. Annual full moult Aug–Sep, El Salvador; partial body moult Feb–Mar, with replacement of variable number of remiges and rectrices in irregular sequence. Uncertain whether this tail moult is confined to second-year birds or includes all ages. Juvenile P6 retained at least until following spring (Dickey & van Rossem 1938, Selander & Giller 1959, Short 1974).

DISTRIBUTION North and Central America. Resident. Pacific slope and Río Balsas drainage of west and south Mexico in south Sonora, south-west Chihuahua, Sinaloa, Nayarit, west and central Jalisco, Michoacán, south México, Morelos, Guerrero, south Puebla, west-central Veracruz, central Oaxaca and central Chiapas, south-east in interior highlands of Guatemala, El Salvador, Honduras and north-central Nicaragua. Disjunctly Yucatán and north Campeche, north-west Yucatán Peninsula (Friedmann *et al.* 1950, Sibley & Monroe 1990, Howell & Webb 1995, AOU 1998).

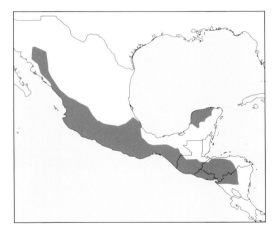

HABITAT Arid to semiarid brushy woodland, semiopen areas, pine-oak forest, open deciduous forest and savanna, thorn scrub, dense thorn forest, steep rocky slopes, farmland. In El Salvador, open habitat generalist, mainly occurs in highlands, e.g. grassy hillsides with scattered low trees above treeline. Lowlands to 2,800m, although restricted to high elevations in some areas where forest occurs at low elevations, such as in parts of El Salvador, and has spread downslope where forest has been cleared for cultivation. Down to sea level north-west Mexico and Yucatán Peninsula (Miller 1932, Moore 1934, Dickey & van Rossem 1938, Schaldach 1963, Land 1970, Short 1974, AOU 1983, Howell & Webb 1995, Stotz *et al.* 1996, Komar 1998, Russell & Monson 1998).

BEHAVIOUR Runs swiftly, cocks tail, droops wings. Mainly terrestrial but often perches in bushes or low in trees or on stone fences etc. When startled, jumps quickly to ground, runs into thick undergrowth to hide. Not shy but inquisitive. Perhaps somewhat arboreal, foraging in trees. Incubating bird did not flush from the nest until observer touched tip of bill (Salvin & Godman 1879–1904, Miller 1932, Zimmerman & Harry 1951, Short 1974, Howell & Webb 1995).

BREEDING Season Eggs Jun, pair displaying and copulating at nest Jul, Sonora, Mexico. Fresh nest 3 May Oaxaca; heavily incubated eggs 19 May; nest with young 15 Jul. Breeding condition female 2 Apr Chiapas. Slightly enlarged ovaries 24 and 26 Mar, Guatemala; eggs 3 Apr. Laying female 16 Mar El Salvador; eggs 3–12 Aug (Owen 1861, Miller 1932, Dickey & van Rossem 1938, Edwards & Lea 1955, Baepler 1962, Land 1962, Rowley 1984, Binford 1989, Russell & Monson 1998). **Nest** Cup of twigs and leaves, of finer material and more compactly built than Greater Roadrunner, lined with stalks of grass. Diameter 30cm, cup rather deep and 15cm wide. One 2.5m above ground in centre of large cholla cactus thicket, one in thorn bush, four others 1.5m above ground in small, thinly-leafed trees, visible from distance, one 5m up in fork of hecho cactus; generally in tree fork c. 3.5m up at locality where species very common (Owen 1861, Miller 1932, Dickey & van Rossem 1938, Rowley 1984, Howell & Webb 1995, Russell & Monson 1998). **Eggs** 2–3 (4); whitish or pure white, smooth; two-egg clutch, Oaxaca: 35.8x26.7, 34.9x25.8mm; Yucatán and Guatemala, 37x27mm; '*affinis*' 33.3–36.8x25.4–27.4mm, mean 35.2x26.1 (n = 5); 2 '*pallidus*' 31.5–35.8x24.8–25.6mm, (Owen 1861, Nehrkorn 1881, Dickey & van Rossem 1938, Schönwetter 1967, Short 1974, Rowley 1984, Howell & Webb 1995). **Incubation** Both sexes have brood patch (Miller 1932). No further information.

FOOD Grasshoppers, caterpillars, ants; small lizards (Miller 1932, Dickey & van Rossem 1938, Zimmerman & Harry 1951, Baepler 1962, Short 1974).

STATUS AND CONSERVATION Generally fairly common (Stotz *et al.* 1996), e.g. Río Tehuantepec basin and lowlands of Isthmus of Tehuantepec, Oaxaca (Binford 1989). Abundant interior Guatemala 1920s (Griscom 1932a), and fairly common there 1960s (Land 1970). 'Apparently uncommon' Colima and south Jalisco (Schaldach 1963). Threatened but not in danger of extirpation El Salvador (Komar 1998). Not globally threatened (BirdLife International 2011).

Lesser Roadrunner *Geococcyx velox*. **Fig. 1.** La Noria Road, Mexico, May (*Roy de Haas*). **Fig. 2.** La Mata, Municipality of Juchitán, Oaxaca, Mexico, May (*Jorge Montejo*).

Genus *Neomorphus*

Gloger, 1827 Froriep's *Notizen* 16, col. 278, note. Type, by original designation and monotypy, *Coccyzus geoffroyi* Temminck 1820.
Five species.

Neotropical; non brood-parasitic. Large, crested, sharply ridged bill; sexes alike. Mostly terrestrial; usually solitary; sometimes in pairs or small family parties. Often follow army ants, peccaries or troops of monkeys. Sister of *Geococcyx* (Sorenson & Payne 2005). Includes *Cultrides* Pucheran 1845.

BANDED GROUND CUCKOO
Neomorphus radiolosus   Plate 3

Neomorphus radiolosus P.L. Sclater & O. Salvin, 1878 ("Intaj" – Intac, Ecuador)

TAXONOMY Monotypic.

FIELD IDENTIFICATION 46–51cm. **Adult** Most of body black with bold pale scaling. Wings and lower back chestnut. Crown black. **Similar species** Only *Neomorphus* with such extensive scaling, reaching belly and undertail-coverts. See *N. geoffroyi* for separation in west Colombia.

VOICE Bill-snapping. Deep cow-like *moo*, 300–380Hz, lasting *c.* 2sec, repeated every *c.* 5sec, for *c.* 4min. Two, perhaps more, individuals separated by *c.* 80m, alternated these vocalisations, increasing intensity slowly, then decreasing to near-silence (López-Lanús *et al.* 1999).

DESCRIPTION Adult Crown, crest and hindneck glossy bluish-black. Forecrown, upper back and underparts black, feathers conspicuously edged buffy whitish, giving bold scaly effect. Lower back, rump and wing-coverts maroon-chestnut, outer primaries black, inner primaries and secondaries purplish-red, inner webs blackish, uppertail-coverts green. Tail black with dark purplish sheen, T1 with faint greenish to purplish tinge. **Juvenile** Back and underparts barred ochraceous not white, except parts of breast; head lacks bluish sheen (López-Lanús *et al.* 1999). **Nestling** Body covered with white down at hatching, head featherless until day 5 when pin feathers appear, and dark feathers appear under body down. By day 10, has prominent crest on head, body equally covered with dark feathers and light-coloured down. By day 15, almost covered in dark feathers, and bare skin behind eyes turning blue. At fledging on day 20, plumage all dark, similar to adult, including crest and blue skin around eyes, but without iridescence; tail less than 25% adult length; bill smaller (Karubian *et al.* 2007). **Bare parts** Iris dark brown. Bare skin around eye blue; juvenile greyish behind eye. This area can be expanded and contracted (signal function?; López-Lanús *et al.* 1999). Bill blackish, tip bluish-grey. Feet bluish-grey.

BIOMETRICS Wing male 165, 167mm, female 162–172mm (n = 5), mean 167.0 ± 4.1. Bill male 45, 46mm, female 42–46mm, mean 44.8 ± 1.6. Tail male 236, 242mm, female 236–254mm, mean 245.4 ± 7.0. Tarsus male 71, 74mm, female 68–74mm, mean 69.8 ± 3.4. Unsexed holotype, wing 177, tail 256 (Payne 2005). Male wing 170mm, tarsus 70mm (Hartert 1898a). Male Paramba, Ecuador, tail 230, 232mm (Haffer 1977). Female Río Michengue, Colombia, bill 48mm (Bond & Meyer de Schauensee 1940).

MOULT Body and remiges slight moult Dec, heavy May with active brood patch (Karubian & Carrasco 2008).

DISTRIBUTION North-western South America. Resident, Pacific slope of western Andes, south-west Colombia (Valle, Cauca, Nariño; Collar *et al.* 1994) and north-west Ecuador (Esmeraldas, Imbabura, Pichincha; Ridgely & Greenfield 2001, Solano Ugalde & Arcos 2007).

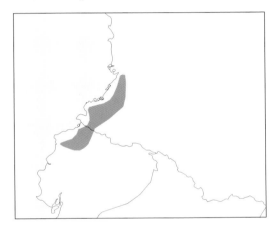

HABITAT On or near ground in wet foothill forest Ecuador, to 500m in recent years, in past to 1,200m, perhaps 1,525m. Very broken terrain Munchique area, Colombia. Requires large areas of primary forest, but uses adjacent secondary forest when following army ants. Sole nests found were on hillside in primary forest with open understorey, closed canopy 25–30m high and visibility of *c.* 20m. Common trees included *Otoba gordonifolia* (Myristicaceae) and *Gustavia dodsonii* (Lecythidaceae) (Lönnberg & Rendahl 1922, Chapman 1926, von Sneidern 1954, Haffer 1977, Collar *et al.* 1992, López-Lanús *et al.* 1999, BirdLife International 2000, Ridgely & Greenfield 2001, Karubian *et al.* 2007).

BEHAVIOUR Often in pairs, even when not nesting, maintaining vocal but not visual contact, and often foraging more than 20m apart. Infrequent forager at army ant swarms, sometimes in mixed-species flocks (in Ecuador, mostly Ocellated Antbird; also Plain-brown Woodcreeper, Immaculate and Bicoloured Antbirds; in Colombia, Esmeraldas, Bicoloured and Immaculate Antbirds, Plain-brown Woodcreeper and Checker-throated Antwren). Forages by scouring live leaves and stems of understorey, examining tree trunk bases from ground, or standing still with crest rising and falling rhythmically, running quickly to capture prey. When consecutively catching prey items, immediately performed excited zigzag run kicking up dead leaves. Has favourite viewpoints on fallen logs. Grasshoppers caught by flushing and pursuing. Observed following band of Collared Peccaries *Tayassu tajacu*. Sometimes nervous or alarmed at human presence. Nest probably attended by pair only. Access either by running up and down 60° inclined trunk, or hopping up lower branches of tree with vertical trunk, and gliding down (Negret 1991, López-Lanús *et al.* 1999, Karubian *et al.* 2007, Karubian & Carrasco 2008).

BREEDING Season Mar–May (Karubian *et al.* 2007). **Nest** Large, open cup 4–5m up in fork close to trunk in understorey tree *Miconia* sp. Two nests 37–38x24–25x13–15cm, walls *c.* 6cm thick. Constructed from leaves, mostly of fern *Diplazium* sp., with few added daily until chick fledges (Karubian *et al.* 2007). **Eggs** Round; cream; *c.* 4.5x4.0cm. As incubation progresses, gradually acquire small brown spots. Clutch probably one. One hatched during night or morning (Karubian *et al.* 2007). **Incubation** At least 13 days. Parents contribute equally, and either parent may incubate at night. Parents travel >400m from nest between incubation bouts. Incubating parent alert but not nervous, and rounds back and extends tail and wings, covering cup, during rain. At changeover, parents do not vocalise or interact. Egg moved, probably turned, once/120min (Karubian *et al.* 2007). **Chicks** Brooding sessions 2.1–9.5h, mean 4.7 ± 2.2h, n = 17, parents contributing equally to brooding and provisioning of nestling. Nestling stage 20 days. Fed frogs, anoles, coral snake, insects, spiders, and earthworms; grasshoppers 32% of 71 recorded items, 28% unidentified. Before fledging begins to catch small insects within nest, and one caught flying dipteran. Not known to vocalise in nest but snaps bill from day 15, when parents away, often answered from parents out of sight. Parents continue feeding fledgling at least four days, and fledgling apparently also dependent on parents for protection during this time (Karubian *et al.* 2007). **Survival** Unknown.

FOOD Arthropods (López-Lanús *et al.* 1999).

STATUS AND CONSERVATION Individual with radio transmitter probably predated by mammal. Home range 50ha (95% kernel analysis), core area 3.4ha (50% kernel analysis) (Karubian & Carrasco 2008). Recorded in only three areas in Colombia and four Ecuador, 1988–2000 (BirdLife International 2000). Very rare Ecuador, recently from only three localities: Bilsa and north-west of Alto Tambo, both Esmeraldas, and Mangaloma, Pichincha (Hornbuckle 1997a, López-Lanús *et al.* 1999, Ridgely & Greenfield 2001, Karubian *et al.* 2007, Solano Ugalde & Arcos 2007). Threats: forest clearance, agricultural and hydroelectric projects in Munchique National Park (Cauca, Colombia), gold-mining Anchicayá, Colombia, deforestation along rivers and railways as human population increases and agriculture expands in Ecuador (Collar *et al.* 1992). Relatively low-density species; seen every few months by hunters in Junín area, Nariño, Colombia in 1990s (BirdLife International 2000). Presence in forests at Bilsa somewhat reassuring (Ridgely & Greenfield 2001) although may depend on large tracts of undisturbed forest for long-term survival, and low density, presumed poor dispersal ability, and small clutch size may contribute to vulnerability (Karubian *et al.* 2007, Karubian & Carrasco 2008). Current rapid and uncontrolled conversion of lowland wet forest to agriculture in the Chocó presents an imminent threat, and conservation hampered by inadequate and ineffective governmental protective measures of national parks (López-Lanús *et al.* 1999). Suggested future conservation measures include surveys of foothill forest in species' range, studies of its ecology, and development of network of effectively protected reserves in its foothill range (López-Lanús *et al.* 1999, BirdLife International 2000). Globally threatened. IUCN Red List category: Endangered (BirdLife International 2011).

Banded Ground Cuckoo *Neomorphus radiolosus*. **Fig. 1.** Bilsa Reserve, Ecuador, October (*Murray Cooper*). **Fig. 2.** Adult, Bilsa, Esmeraldas, Ecuador, October (*Dušan Brinkhuizen*).

RUFOUS-WINGED GROUND CUCKOO
Neomorphus rufipennis Plate 3

Cultrides rufipennis G.R. Gray, 1849 ("Supposed to be a native of Mexico", corrected to lower Orinoco River, Venezuela; Chapman 1928).

Alternative names: Chapman's Ground Cuckoo (*nigrogularis*).

TAXONOMY May form superspecies with *pucheranii* (Haffer 1977). Monotypic. Synonym: *nigrogularis* Chapman 1914.

FIELD IDENTIFICATION 43–48cm. **Adult** Dark, almost black, upperparts, head and breast, ashy-brown belly, pink naked area around eyes obvious in field. **Similar species** As other *Neomorphus*, may be mistaken for chachalaca or small guan, although those are not usually terrestrial.

VOICE Single or double loud bill snapping. Single *whóu*, reminiscent of hoot of pigeon or owl, also similar to call of Blue Ground Dove, *c.* 0.2sec duration, given repeatedly at 3–6sec intervals for up to several min walking on ground or perched 0.5–3m up on branch or log. Low guttural *gr'r'r* and loud bill-snaps at army ant swarms and in agonistic encounters (Haffer 1977, Hardy *et al.* 1990, Zimmer & Hilty 1997, Hilty 2003).

DESCRIPTION Adult Head, crest, neck and upper breast deep purplish-blue, back metallic purplish-olive, secondaries purplish-maroon, primaries bluish-black; throat ashy to dark grey, lower throat feathers edged dark, giving scaly appearance, lower breast and belly ashy-brown, undertail-coverts dusky. Central rectrices metallic purple, outer greenish-black. **Juvenile** Crown black, back and wing-coverts slaty-brown, flight feathers as adult, rump black; underparts slaty; rectrices narrow, black glossed purple and green. **Nestling** Hatchling unknown. Fledgling has black skin below and some dark brown trichoptiles attached to tips of crown feathers. **Bare parts** Iris brown in adult and juvenile. Bare skin around eyes pinkish-red; dull red in juvenile, with cerulean blue skin behind ear, covered with black feathers. Bill blackish except grey-blue tip of upper mandible; juvenile, black and less deep than adult. Feet olivaceous.

BIOMETRICS Wing male 164–176mm (n = 6), mean 169.5 ± 4.2, female 152–174mm (n = 6), mean 164.3 ± 8.3. Bill male 39–42mm, mean 40.7 ± 1.4, female 38–44mm, mean 41.3 ± 2.0. Tail male 262–290mm, mean 275.7 ± 12.9, female 266–300mm, mean 288.5 ± 14.5. Tarsus male 63–71mm, mean 67.4 ± 3.4, female 68–71mm, mean 69.8 ± 1.6. Mass male 350, 520g, female 315, 340g (Payne 2005).

MOULT Unknown.

DISTRIBUTION Northern South America. Resident. South Venezuela in Bolívar and north and central Amazonas; west Guyana; Roraima, north Brazil (von Pelzeln 1870, Pinto 1966, Snyder 1966, Moskovits *et al.* 1985, Borges 1994, Hilty 2003, Ridgely *et al.* 2005). Unconfirmed sight record Surinam (Hilty & Brown 1986). Also sight records upper Rio Negro, Amazonas, Brazil (Sick 1997), and Sierra de Chiribiquete, Caquetá, Colombia (Stiles *et al.* 1995).

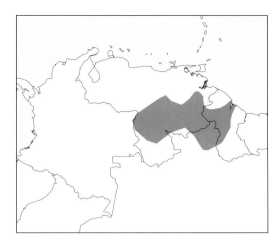

HABITAT Rainforest. Foothill zone of mountains, 100–1,100m. Upland (terra firme) and seasonally flooded (várzea) forest Junglaven, Venezuela. Mostly undisturbed forest Iwokrama Forest Reserve, Guyana, and rarely disturbed forest (Haffer 1977, Meyer de Schauensee & Phelps 1978, Zimmer & Hilty 1997, Ridgely *et al.* 2005).

BEHAVIOUR Terrestrial forager, fast runner when startled. Solitary, wary and restless; sometimes allows close approach, but generally difficult to find and observe; sometimes perches low in trees, in pairs or (more infrequently) in small family groups. Follows herds of peccaries *Tayassu* sp. and army ant swarms, taking arthropods flushed from forest floor, but at least as often forages away from either. Sometimes also follows below groups of small monkeys such as *Saguinus*, taking fallen fruit and flushed insects. Forages alone, but lives in pairs with large home ranges. Erects and lowers crest, raises and lowers tail (in excitement?), dust-bathes, responds to play-back of voice, sometimes with bill-snaps, or running silently towards source of playback on zigzagging course, pausing occasionally on logs or low branches to peer above. Parents accompany and care for fledgling, as an emaciated, newly fledged young was collected on same anthill where an adult male and female had been taken few days previously (Meyer de Schauensee & Phelps 1978, Zimmer & Hilty 1997, Hilty 2003, Payne 2005, Ridgely *et al.* 2005).

BREEDING Probably nonparasitic. **Season** Testes enlarged Mar, south Venezuela (Chapman 1914); female with large ova Sep, recently fledged Mar, Guyana (Payne 2005). **Nest** Unknown. **Eggs** Yellowish white with small bulges on surface, 37.1–41.5x29.2–32.0mm (n = 8), mean 40.1x30.8 (Schönwetter 1967). No further information.

FOOD Grasshoppers, crickets, spiders (von Pelzeln 1870).

STATUS AND CONSERVATION Local (Phelps & Phelps 1958, Snyder 1966). Poorly known and in need of research (Stotz *et al.* 1996). Rare and normally at low densities (Hilty 2003). Uncommon in Iwokrama Forest Reserve, Guyana, but still may be more numerous there than elsewhere (Ridgely *et al.* 2005). Not globally threatened (BirdLife International 2011).

Rufous-winged Ground Cuckoo *Neomorphus rufipennis*. **Figs. 1–2.** Adult, Rio Grande, Venezuela, February (*Pete Morris*).

Red-billed Ground Cuckoo *Neomorphus pucheranii*. **Fig. 1.** Adult, taken with camera trap, Reserva de Desenvolvimento Sustentável do Uacari, Carauari, Amazonas, Brazil (*Daniel P. Munari*).

RED-BILLED GROUND CUCKOO
Neomorphus pucheranii   Plate 3

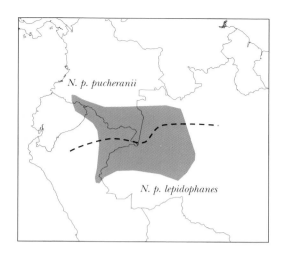

Cultrides Pucheranii Deville, 1851 ("L'Ucayale et l'Amazone" = Río Yaguas, north-east Peru [Peters 1940]).

TAXONOMY Red-billed Ground Cuckoo previously used for Asian *Carpococcyx renauldi* (Coral-billed Ground Cuckoo). May form superspecies with *rufipennis* (Haffer 1977). Polytypic. Synonym: *napensis* Chapman 1928 in nominate.

FIELD IDENTIFICATION 43–51cm. **Adult** Bill and bare skin around eye bright red. Foreneck ash-grey; crown and crest glossy blue-black. **Juvenile** All dark, head and breast blackish, belly brown, bill and feet black. **Similar species** From congeners by red bill combined with black crown contrasting with face and foreneck. No other confusion species.

VOICE Loud bill-snapping likened to sticks breaking. Roaring, almost guttural noise similar to curassow *Crax* sp. heard near army ant swarm accompanied by antbirds immediately before two *N. p. lepidophanes* appeared from direction of voice. Called *c.* 03:00 hrs to sunrise and again *c.* 10:00 hrs (Todd 1925, Gyldenstolpe 1945a, Siegel *et al.* 1989).

DESCRIPTION Nominate. **Adult** Crown including crest glossy blue-black or with purple gloss, forehead sometimes brown, rest of upperparts bronzy-green, rump bronzy-purple; wings rufous, outer primaries black with violet gloss; face, foreneck and breast ashy-grey scaled black, bordered below by black breast-band, belly buffy to light grey or clay colour; tail dark chestnut, central rectrices olive with green sheen, tail black below. **Juvenile** Above unglossed brown, crest blackish, throat and breast blackish, belly brown, bill and feet black. **Nestling** New head feathers with buffy filamentous natal down attached to tips, plumage uniformly dark brown, bill black. **Bare parts** Iris brown. Bill bright red tipped green or yellowish, red of bill continuous with bright red bare lores and orbital skin, which becomes blue behind eye. Orbital skin bare in juvenile, colour unrecorded. Feet dark grey.

BIOMETRICS Nominate. Wing male 166–178mm (n = 6), mean 171.7 ± 4.8, female 160–168mm (n = 6), mean 165.8 ± 3.0. Bill male 48–56mm, mean 51.9 ± 3.2, female 44–55mm, mean 50.7 ± 3.6. Tail male 266–285mm, mean 273.8 ± 7.3, female 262–273mm, mean 267.0 ± 3.6. Tarsus male 65.0–68.5mm, mean 66.9 ± 1.4, female 60.5–67.2mm, mean 65.9 ± 1.1. Mass male 330g (Payne 2005).

MOULT Jul specimen moulting from juvenile to adult has blackish throat and chest, sprinkled with a few new grey feathers, and some dark brown juvenile feathers on whitish belly and on greenish nape and scapulars. Male moulting from juvenile to adult has brown forehead, upper Río Negro, undated (Haffer 1977, Willis 1982).

GEOGRAPHICAL VARIATION Two subspecies.
N. p. pucheranii (Deville, 1851) Amazonian Peru and west Brazil north of Amazon; possibly this subspecies east Ecuador and south-east Colombia. Forehead light brown. Fairly uniform breast. Lower breast and belly, below breast-band, buffy to light grey. Described above.
N. p. lepidophanes Todd, 1925. Amazonian Peru and west Brazil south of Amazon. Forehead black as crown. Breast conspicuously scaled black. Lower breast and belly, below breast-band, clay colour. Wing male 162–182mm (n = 6), mean 170.2 ± 7.4, female 162–183mm (n = 5), mean 169.8 ± 8.0 (Payne 2005).

DISTRIBUTION West Amazonia. Resident. Sight records Colombia (Willis 1982, Cuadros 1991) and east Ecuador (López-Lanús 1999, Ridgely & Greenfield 2001). North-east Amazonian Peru in Loreto and north Ucayali. Upper Amazonian Brazil, north to upper Rio Negro and south to Rio Purús, Amazonas state (Haffer 1977, 1997). Sight records of unidentified *Neomorphus*, Jaú National Park, west of lower Rio Negro, Amazonas (Borges *et al.* 2001).

HABITAT Terra firme forest and forest on moderately hilly, well-drained soils, canopy 20–25m high with some emergent trees over 30m (north-east Peru); *c.* 200–700m (Siegel *et al.* 1989, Stotz *et al.* 1996, Ridgely & Greenfield 2001).

BEHAVIOUR Terrestrial; wary and swift runner. Associates with army ants and probably with peccaries. Sometimes follows mixed-species troops of tamarins *Saguinus fuscicollis* and *S. mystax*, walking and running on ground below or behind tamarins and taking partially-eaten fruits dropped on ground and arthropods such as large katydids flushed from trees to ground by tamarins; rested and sunned in coordination with tamarins' long resting periods, much of the time partially hidden in thick bushes, or perched on horizontal tree trunk 1.5m up, close to tamarins, preening rectrices (Todd 1925, Siegel *et al.* 1989, Cuadros 1991, Peres & Whittaker 1991, Peres 1992, Sick 1997, López-Lanús 1999).

BREEDING Presumably nonparasitic. Pair incubated (probably information from Indian; Castelnau 1855 in Sick 1993). **Season** Nestling Feb, Peru (Payne 1997). **Nest** Unknown. **Eggs** Clutch possibly two (Castelnau 1855 in Sick 1993). No further information.

FOOD Insects. Fruit dropped by tamarins (see Behaviour) including *Inga punctata*, *Salacia juruana*, *Abutta* sp., *Rheedia* sp., Sapotaceae, Flacourtaceae, Leguminosae, Hyppocrataceae, Myrtaceae and Menispermaceae; faeces of tamarins (Siegel *et al.* 1989, Payne 1997, 2005).

STATUS AND CONSERVATION Status little known. Rare and local Ecuador, known from only two recent sightings (Ridgely & Greenfield 2001). Rare Peru (Parker *et al.* 1982, Clements & Shany 2001). Not globally threatened (BirdLife International 2011).

RUFOUS-VENTED GROUND CUCKOO
Neomorphus geoffroyi Plate 3

Coccyzus geoffroyi Temminck, 1820 (No locality; vicinity of the city of Pará (=Belém), Brazil, designated by Peters 1940)

Alternative name: Rufous-vented Cuckoo.

TAXONOMY Two or more species may be involved; more study needed. *Squamiger* sometimes treated as subspecies of *N. geoffroyi* (e.g. Haffer 1977, Payne 2005). *Salvini* occasionally considered separate species (Carriker 1910, Ridgway 1916, Davis 1972). Polytypic. Synonym: *g. amazonicus* Pinto 1964 in nominate.

FIELD IDENTIFICATION 45–51cm. Bill curved, greenish-yellow or horn. In juvenile, bill blackish-brown, shorter than in adult. Bare orbital skin dull bluish to greyish. Triangular bare area behind eye covered with feathers when bird at ease. **Adult** Brown head (no black) and brownish underparts. Thin black pectoral band (irregular, broken into line of spots, or lacking in some subspecies), rufous belly. **Juvenile** Short tail, blackish plumage and blue mark on side of head. **Similar species** Seen poorly, may be mistaken for chachalaca or guan. Resembles other *Neomorphus* in size, shape and behaviour. Plumage most similar to Scaled Ground Cuckoo (small area south of lower Amazon, Brazil), which is mainly distinguished by indistinct breast-band. In west Colombia, separated from Banded Ground Cuckoo (no known range overlap) by general brown colours, rufous vent (not black head, scaled underparts and upper back), breast band and yellow bill (not dark). From Red-billed Ground Cuckoo by bill colour of and bare orbital skin (not red). Rufous-winged Ground Cuckoo very dark on head and breast, with red orbital skin and dark bill.

VOICE Dry cracking produced by snapping mandibles. When excited, repeats snaps in series, forming loud rattle. Bill snaps may serve as communication between conspecifics. "Song", presumably of *dulcis*, low descending note, *OOOOoo*, with timbre of Grey-fronted Dove or moan of curassow *Crax* sp., delivered every 3–4sec for several minutes. Low muffled *woof* when feeding at ant swarm. In Amazonian Brazil, soft dovelike moaning *oooooo-oóp* may also be the same sound. Loud, explosive *kchak* to frighten other bird from prey. Adult gave low *hooooooooooh* groans from 3m up when observer approached young (Sick 1949, 1953, 1993, Slud 1964, Willis 1982, Hilty & Brown 1986, Stiles & Skutch 1989).

DESCRIPTION *N. g. salvini*. **Adult** Forehead and crown cinnamon-brown, occipital crest tipped glossy blue-black, protruding 2–3cm behind nape, feathers broad and round-tipped. Auricular region brown streaked blackish, upper edge all blackish. Hindneck, upper back and wing-coverts bronzy olive-brown, lower back and scapulars bronze-brown to purplish, rump and upper tail-coverts deep purplish-brown, glossed bronzy-purple. Secondaries olive-green, primaries darker and more bluish, outer primaries glossed purplish. Underside of remiges blackish-brown slightly glossed, under wing-coverts brown, some with fawn tips. Malar area and below eyes light brown, like forehead or paler, chin and throat pale buff, breast buff to cinnamon, or greyish-brown, feathers with dusky semicircular bands giving scaly effect. Black band across breast, broken in some. Below latter, greyish-buff grading into rufous or chestnut on flanks, lower belly and vent. Tail black glossed purple, or purple and green. Tail below blackish. **Juvenile** Forehead blackish, crown and crest black glossed green and dark blue, back and wing-coverts black glossed coppery-green. Primaries dusky with green gloss, secondaries mostly metallic purple, underparts blackish, tail metallic purple and greenish in irregular pattern. **Nestling** Unknown. **Bare parts** Iris yellow, red, yellowish-red, light red, or outer ring red and inner ring yellow (Haffer 1977); brown in *salvini* (Stiles & Skutch 1989). Bill curved, greenish-yellow to horn-colour. Bare orbital skin slaty, becoming bright blue behind eye in *salvini* (Stiles & Skutch 1989), bright blue in other sspp. Blue in juvenile (MZUSP). Feet bluish-grey. In fledgling, mouth lining rose pink with small white rugosities on palate; iris dark brown; feet light blue-grey or plumbeous; bill black; bare orbital skin deep charcoal grey with bright blue spot behind eye.

BIOMETRICS *N. g. salvini*. Wing male 167–184mm (n = 11), mean 172.4 ± 5.1, female 161–194mm (n = 15), mean 170.1 ± 7.9. Bill male 40.0–52.3mm, mean 46.5 ± 4.5, female 41.0–51.5mm, mean 43.3 ± 3.0. Tail male 250–281mm, mean 263.3 ± 10.7, female 247–278mm, mean 264.3 ± 10.0. Tarsus male 61–71mm, mean 67.6 ± 3.2, female 63–71mm, mean 67.0 ± 2.9. Mass *salvini* male 350g, female 375, 400g; *aequatorialis* male 340g; *australis* male 370g (Payne 2005); unsexed 340g (Terborgh *et al.* 1990); *dulcis* male 339–355g (n = 3), female 302, 349g (Sick 1949, Willis 1982).

MOULT Dec–Jan, after breeding, Espírito Santo, southeast Brazil (Sick 1953a). Juvenile *geoffroyi* (MZUSP), west Rondônia, Apr, with few new breast feathers (whitish with two dark bars, similar to adult feathers), had moulted all feathers on lores and above eyes, and some on central forehead.

GEOGRAPHICAL VARIATION Six subspecies.

N. g. geoffroyi (Temminck, 1820). Central and south Amazonian Brazil, south of Amazon (Rondônia, Mato Grosso, Pará, north Maranhão); old specimen from unknown locality at Rio Araguaia, Goiás or (more likely) Tocantins (Deville 1851 in Haffer 1977); also uncertain record Peixe, upper Rio Tocantins (Pinto 1964). Specimens from Balta, Ucayali, Peru, may belong here or may be intermediate between *geoffroyi* and *australis* (Haffer 1977). Sight records east Santa Cruz, Bolivia (Bates *et al.* 1989), are presumably this subspecies. Forehead and crown brown barred black, hindneck, mantle and scapulars glossed green; rest of upperparts olive-brown with some green gloss, uppertail-coverts glossed greenish, tail above glossed mauve; throat and breast much more heavily scaled with unbroken band. Lower breast to belly greyish or whitish-buff, indistinctly barred on flanks.

N. g. salvini P.L. Sclater, 1866. Nicaragua, Caribbean Costa Rica and Panama to Chocó, Antióquia and Córdoba, Colombia. Described above.

N. g. aequatorialis Chapman, 1923. West Caquetá (and possibly west Putumayo), Colombia, Sucumbíos, Napo, Pastaza and Morona–Santiago, east Ecuador, and Huánuco, Pasco, and possibly north-west Loreto, east Peru. Forehead unbarred brown. Breast band broad and complete. Wing male 162–188mm (n = 8), mean 170.4 ± 8.7, female 162, 170mm (Payne 2005).

N. g. australis Carriker, 1935. Cuzco, Puno and Madre de Dios, possibly also Ucayali (Balta), south Peru, and Pando, Beni and La Paz, Bolivia (Remsen & Traylor 1989, Arribas *et al.* 1995). Almost plain light grey throat and central breast (Haffer 1977); forehead unbarred.

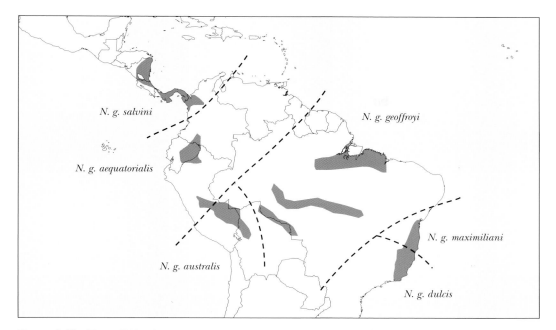

N. g. maximiliani Pinto, 1962. Bahia, east Brazil. Upperparts bronze-green, breast scaled.

N. g. dulcis Snethlage, 1927. Espírito Santo, Minas Gerais (Rio Doce valley), south-east Brazil. Old record of pair from Cantagalo, east Rio de Janeiro (Cabanis 1874), not always accepted (e.g. Pinto 1938, 1962, 1978). Upperparts dark blue, breast scaled. Abdomen darker than *maximiliani*. Iris yellow (de Melo 1998). Wing male 172, 173mm, female 163mm, unsexed 170, 177mm (Payne 2005).

DISTRIBUTION Central and South America. Resident. Nicaragua, Caribbean slope of Costa Rica (locally approaching Pacific slope), Panama, north-west Colombia south to Baudó Mountains and east along north base of Andes (Córdoba); south-east Colombia in Caquetá and possibly in Putumayo, east Ecuador, east Peru, north Bolivia, south-central and south-east Amazonia south of Amazon river in Brazil; east Bahia, east Minas Gerais, Espírito Santo and east Rio de Janeiro (only Cantagalo), Brazil (Wetmore 1968, Hilty & Brown 1986, Willis 1988, Ridgely & Gwynne 1989, Stiles & Skutch 1989, Clements & Shany 2001, Ridgely & Greenfield 2001).

HABITAT On or near ground inside tall, virgin lowland terra firme forest in Amazonia and humid Atlantic Forest, east Brazil. Dry forest, La Paz, Bolivia. Open river-bluff forest dominated by *Cecropia* sp., Rondônia, Brazil. High ground forest (canopy 40–50m) with clear dark understorey, and seasonally inundated forest with abundant *Heliconia* and *Ficus*, south-east Peru. Riverine or slope forests, usually in areas with many treefalls and vines, mature forest perhaps unsuitable in some regions. Requires extensive expanses of undisturbed forest, and seems to occur at very low densities. Lowlands and foothills, locally to 900m, Costa Rica. Lowlands and foothills Panama, to 1,450m. To 1,000m Colombia; record from Caquetá, east slope of Andes, at 1,100m. To *c.* 400m Ecuador. To 1,600m Peru. To 1,300m Bolivia, locally to 1,650m (Wetmore 1968, Willis 1974, 1988, Terborgh *et al.* 1984, Hilty & Brown 1986, Ridgely & Gwynne 1989, Stiles & Skutch 1989, Sick 1993, Arribas *et al.* 1995, Stotz *et al.* 1996, 1997, Perry *et al.* 1997, Ridgely & Greenfield 2001, Garrigues & Dean 2007, Schulenberg *et al.* 2007).

BEHAVIOUR Mostly terrestrial, but often perches on branch to preen, or rest, supporting wings on branch; sleeps lying on branch. May perch high in tree if flushed. Follows army ants *Eciton* and peccaries. At swarms of army ants, catches arthropods flushed from litter by ants, usually by running forward rapidly and pecking at prey on ground. May also capture prey atop logs, by tossing leaf litter, from twigs or trunks up to 2m above ground. May run through dense vine tangles. Being largest bird at ant swarms, sometimes supplants smaller birds or scatters them when running for prey. Investigates heaps of dry branches, armadillo burrows and empty terrestrial termitaria. Routinely associates with primates such as *Saimiri* and *Cebus*, and catches insects flushed by them. Associated with Purplish Jay and Crested Oropendola at army ants in dry forests, La Paz, Bolivia; with flock of antshrikes (*Thamnomanes*, *Thamnophilus*), antwrens (*Myrmotherula*) and Rose-breasted Chat, Rondônia. Eats some fruit and may thus be partly independent of mammals and army ants. Scratches head under wing; preens and yawns. Raises and lowers crest when excited. One crouched on ground, quivering wings and opening bill (Goeldi 1903, Carriker 1910, Snethlage 1927, Sick 1953a, 1993, Slud 1964, Willis & Eisenmann 1979, Willis 1982, Terborgh 1983, Stiles & Skutch 1989, Perry *et al.* 1997, Whittaker 2004).

BREEDING Pairs accompanying, caring for and carrying food to young, south-east Brazil, Panama and east Ecuador (Sick 1949, 1953, 1962, 1993, Willis 1982, P. Donahue in Ridgely & Gwynne 1989). **Season** Northern summer or wet season south to north Colombia; fledged young Jul and Dec, Nicaragua and Costa Rica; downy young out of nest Aug, Panama (*salvini*). Egg Sep, Mato Grosso (nominate). Young out of nest Dec and Jan south-east Brazil indicating breeding rainy season/summer (*dulcis*) (Sick 1953a, Roth 1981, Willis 1982, Robbins *et al.* 1985, Stiles & Skutch 1989). **Nest** Loose assemblage of heavy sticks with flat cup, lined with green leaves (fresh leaves added periodically during incubation), external diameter 25cm, flat cup 12cm wide, well hidden in dense foliage in fork in 5m high *Miconia* shrub 2.5m above ground in dense swampy second growth, Mato Grosso (Roth 1981). **Eggs** Clutch 1. Unspotted yellowish white, 40x32mm (nominate). Oviduct egg white, slightly rough, 43x32mm (*salvini*) (Wetmore 1968, Roth 1981). No further information.

FOOD Grasshoppers, cockroaches, Hemiptera, beetles, ants; spiders, harvestmen, scorpions; centipedes; small frogs; lizards; small birds; occasionally fruit. Juvenile *salvini* had eaten hairy caterpillar and two 5mm hard seeds (Howell 1957, Schubart *et al.* 1965, Terborgh 1983, Stiles & Skutch 1989, de Melo 1998, Payne 2005).

STATUS AND CONSERVATION Rare Costa Rica (Stiles & Skutch 1989). Rare and local Panama (Ridgely & Gwynne 1989) and Colombia (Hilty & Brown 1986). Only six localities Ecuador, 1970s – *c.* 2001 (Ridgely & Greenfield 2001). Rare Peru (Clements & Shany 2001, Schulenberg *et al.* 2007); 0.25 pairs/100 ha in floodplain forest (Terborgh *et al.* 1990). Among first birds to disappear from forests disturbed by man or dogs (Sick 1969); very sensitive to slightest modifications of habitat (Sick & Teixeira 1979). Significant habitat destruction occurred throughout range, particularly Bahia and Espírito Santo, south-east Brazil (Ruschi 1975, Whittaker 2004). Not recorded Bahia (*maximiliani*) since 1980s (Silveira 2008). *N. g. dulcis* possibly extinct (Knox & Walters 1994); most recent Minas Gerais record (sight) early 1980s (de Melo 1998), except for camera trap record, Parque Estadual do Rio Doce, *c.* 2007 (L. Scoss pers. comm.); last recorded Sooretama reserve, Espírito Santo, 1977 (Sick & Teixeira 1979, Scott & Brooke 1985). However, may also survive at 1–2 sites in Espírito Santo (Knox & Walters 1994), but intensive searches should be undertaken, especially in Linhares/Sooretama reserves and Rio Doce park, to evaluate status of any remaining populations, and if located, basic habitat requirements and other aspects of natural history should be studied (Silveira 2008). *N. g. dulcis* considered threatened Minas Gerais (de Melo 1998), critically endangered Espírito Santo (Simon *et al.* 2007), extinct Rio de Janeiro (Alves *et al.* 2000), and threatened Brazil (Silveira 2008). Estimated range 612,000 km²; not globally threatened (BirdLife International 2011).

Rufous-vented Ground Cuckoo *Neomorphus geoffroyi*. **Figs. 1–3.** Adult, Manu, Madre de Dios, Peru, July (*Matthias Dehling*).

SCALED GROUND CUCKOO
Neomorphus squamiger   Plate 3

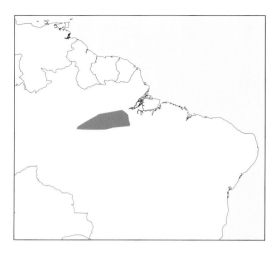

Neomorphus squamiger Todd, 1925 (Colonia do Mojuy, Santarém, Brazil).

TAXONOMY Closely related to, and probably forming superspecies with, *N. geoffroyi* (Haffer 1977, Sibley & Monroe 1990). Has been considered subspecies of *N. geoffroyi* (e.g. Peters 1940, Haffer 1977, Payne 2005), but distribution of *squamiger* and *geoffroyi* south of Amazon poorly known. Here accorded specific status while awaiting future clarification. *Squamiger* said to occur without intergradation close to *geoffroyi* along Amazon near Rio Tapajós (Sibley and Monroe 1990, Sick 1997), but contrastingly Payne (2005) claims intergradation between *N. g. geoffroyi*, *N. g. australis* and *squamiger*. However, full diagnosability of *squamiger* in plumage characters claimed (without stating which), concluding that it must be considered separate species (Raposo *et al.* 2009). Sighting of bird without prominent breast-band north Mato Grosso (Lees 2003) complicates situation. More material from upper Tapajós valley, and Xingu and lower Madeira rivers needed to elucidate nature of geographical variation (Haffer 1977). Monotypic. Synonym: *s. iungens* Griscom & Greenway 1941.

FIELD IDENTIFICATION 43cm. **Adult** Similar to *N. geoffroyi*, especially nominate. **Juvenile** Known from single specimen (see Description). **Similar species** From Rufous-vented chiefly by indistinct, or lack of, breast-band.

VOICE Unknown, but probably similar to *N. geoffroyi*.

DESCRIPTION Adult Very similar to *N. g. geoffroyi*. Indistinct black pectoral band, lacking in some from near Amazon. Forehead and crown feathers brownish with buffy margins, crest purplish-blue. **Juvenile** Unscaled dark brown forehead, rest of upperparts like adult, tail shorter, breast uniform dark brown (new scaled feathers appearing), same colour extending down over flanks and belly except centre which is pale brown as in adult (MZUSP). **Nestling** Unknown. **Bare parts** Iris cherry red, bare skin behind eye cobalt blue, bill greenish with yellowish tip and bluish-dusky base, feet slaty. Juvenile colours unrecorded.

BIOMETRICS Wing male 163, 164, 171mm, female 160, 170mm (Payne 2005). Wing, unsexed, 160–161mm, tail 270mm, culmen 42–45mm, tarsus 65–66mm (Todd 1925, Pinto 1964). Mass male 340g (Graves & Zusi 1990).

MOULT Juv moulting underparts Mar (MZUSP).

DISTRIBUTION Small area of central Amazonian Brazil. Resident. Pará, south of Amazon River on both sides of lower Rio Tapajós, on south bank of Amazon River west and east of mouth of Rio Tapajós (Pinto 1964, Haffer 1977) and near lower Rio Xingu (Graves & Zusi 1990).

HABITAT Tropical Amazonian rainforest, well-drained to hilly (Willis 1982).

BEHAVIOUR Follows army ants (Willis 1982).

BREEDING Presumably nonparasitic. **Season** Juveniles Mar, May (Willis 1982). Nothing else known.

FOOD Unknown.

STATUS AND CONSERVATION Rare (Stotz *et al.* 1996). Highly sensitive to human disturbance. Threats include logging, deforestation owing to road construction, ranching, smallholder agriculture, mining and hydroelectric development. Considered globally near-threatened because of its small range and destruction of its habitat (BirdLife International 2011).

SUBFAMILY CENTROPODINAE

Genus *Centropus*

Illiger, 1811 *Prodromus* p.205. Type, by subsequent designation, *Cuculus aegyptius* Gmelin
(G.R. Gray *List Gen. Birds*: 56).
Twenty-six species.

Old World; non brood-parasitic. Large to very large cuckoos, mostly terrestrial; stout bill; short, rounded wings, long, broad tails; well-developed eyelashes; usually only one testis develops; most with long, straight hallux claw; weak flyers. Includes *Polophilus* (Leach 1814), *Corydonyx* (Vieillot 1816), *Nesocentor*, *Centrococcyx*, *Pyrocentor* (last three Cabanis & Heine 1863), *Megacentropus* and *Grillia* (both Roberts 1922).

BUFF-HEADED COUCAL
Centropus milo   Plate 4

Centropus Milo Gould, 1856 (Guadalcanal, Solomon Islands)

TAXONOMY Mayr and Diamond (2001) treated *milo* and *ateralbus* as two megasubspecies. Mason *et al.* (1984) suggested it belongs to species group together with *goliath* and *violaceus*. MtDNA analysis places this species as sister to clade containing *ateralbus*, *menbeki* and *chalybeus* (Sorenson & Payne 2005). Polytypic.

FIELD IDENTIFICATION 62–68cm. Large coucal with long, broad, graduated tail hanging down when perched, and flapping flight. Sexes similar. **Adult** Whole foreparts pale buffish, remaining plumage black; whole underparts, rump and lower back also pale buffish in *albidiventris*. **Juvenile** Buff and brown mottled and barred black, wings and tail barred black and rufous. **Similar species** No other coucals in range. Female Common Koel smaller with whitish streak from below eye to whole side of neck, and conspicuously broad, black malar stripe.

VOICE Load raucous *na-ow*, likened to lion roaring. Its call when together with partner *urrrh-uh* or guttural *kkkow-kkk-kk-kk*; also high and throaty barks *erch-erch*, or deeper gobbling *argh-a-argh* heard when birds presumably foraging on ground; young birds have loud stuttering series of short calls *koko-kokokkokoko*. Voice of subspecies *albidiventris* rendered as *craw-aw-aw-aw* given in a slow series; perched bird uttered series of low-pitched mellow and ventriloquial grunts. Bird mobbed by flycatcher gave low mammal-like *growl* (Sibley 1951, Cain & Galbraith 1956, Doughty *et al.* 1999, Diamond 2002).

DESCRIPTION Nominate. **Adult** Whole head, neck, mantle, chin, throat and breast buffish-white; lower back, rump, wings and tail black with slight purplish gloss; abdomen, undertail-coverts and flanks blackish-brown. **Juvenile** Head buff-brown or crown to nape and throat to upper breast blackish-brown; rest of body darker rufous- or buff-brown with variable sooty-black mottling; back and rump with buff-white bases to feathers; wings and tail rufous-brown barred blue-black. **Nestling** No information. **Bare parts** Iris red to red-brown, juvenile dull greyish to brown. Bill black; juvenile brown, lower mandible paler. Feet bluish-grey.

BIOMETRICS Nominate. Wing male (n = 9) 260–283mm, mean 270.6 ± 7.0, female (n = 9) 245–290mm, mean 266.8 ± 16.5. Bill male 52–63mm, mean 58.4 ± 2.5, female 56–73mm, mean 63.9 ± 5.8. Tail male 325–410mm, mean 358.0 ± 25.0, female 312–372mm, mean 346.2 ± 21.5. Tarsus male 60–77mm, mean 65.6 ± 6.4, female 67–74mm, mean 70.4 ± 3.0. Mass male (n = 5) 742–790g, mean 769.4g (Payne 2005).

MOULT Adult with large testis in heavy moult on head, throat, wings and tail, 23 Oct (Sibley 1951).

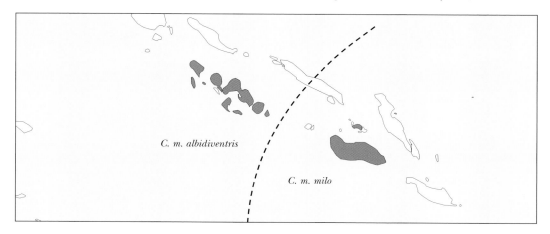

GEOGRAPHICAL VARIATION Two subspecies.

C. m. milo Gould, 1856. Florida and Guadalcanal Is. (south Solomons). Described above.

C. m. albidiventris Rothschild, 1904. Vella Lavella, Bagga, Ganonga, Gizo, Simbo, New Georgia, Kulambangra (Kolombangara), Kohinggo, Vangunu, Gatukai, Rendova, Tetipari Is. (west-central Solomons). Whole underparts, rump and lower back buffish. Juvenile like nominate but bars on back and tail broader.

DISTRIBUTION Solomons. Being weak flier may explain absence from some smaller islands close to larger islands populated by it, e.g. occurs on Kohinggo but not nearby Vonavona (Wana Wana) despite only 2km separating them, or in Florida group where it occurs on Big Nggela but not on ecologically-similar Small Nggela only 150m distant (Mayr & Diamond 2001, Diamond 2002).

HABITAT Lowland and hill primary, and secondary forest (Cain & Galbraith 1956, Doughty *et al.* 1999).

BEHAVIOUR Easily found by its loud calls; any loud noise, not only imitated voice, stimulates calling. Clumsy, gliding flight, mostly only from tree to tree. When foraging on ground walks slowly with tail drooping (Sibley 1951, Cain & Galbraith 1956, Diamond 2002).

BREEDING Season Breeding condition males New Georgia, Feb, May, 23 Oct; Guadalcanal, Jun; Villa Lavella, Feb and Jun (Sibley 1951, Payne 2005, BMNH). No other information.

FOOD Insects such as large stick insects, grasshoppers and other large Orthoptera, beetles and lepidopteran pupae found in stomach; probably large poisonous centipedes. Feeds mainly on ground. Mobbing Steel-blue Flycatcher suggests nest predation (Sibley 1951, Cain & Galbraith 1956).

STATUS AND CONSERVATION Fairly common to uncommon throughout central Solomons (Doughty *et al.* 1999). Calls were 'constant sound' Feb 1945, New Georgia (Sibley 1951). Absence on Buka and Bougainville and north Solomons may be due to introduced domestic animals (Diamond 2002). Not globally threatened (BirdLife International 2011).

Buff-headed Coucal *Centropus milo*. **Figs. 1–2.** Adults, *C. m. milo*, near Honiara, Guadalcanal, Solomon Islands, July (*Kevin Vang*).

PIED COUCAL
Centropus ateralbus   Plate 5

Centropus ateralbus Lesson, 1826 (New Ireland)

Alternative names: White-necked/New Britain Coucal.

TAXONOMY Mayr and Diamond (2001) treated it and *milo* as two megasubspecies, and Mason *et al.* (1984) thought it related to *phasianinus*, *spilopterus* and possibly *bernsteini*. MtDNA analysis places it as sister clade to *menbeki* and *chalybeus* (Sorenson & Payne 2005). Monotypic.

FIELD IDENTIFICATION 42–46cm. Medium-sized coucal. Sexes similar. **Adult** Varies from mostly black with purplish-blue gloss on head, wings and tail, and white area on wings and throat, to entirely white or pearly grey. **Subadult** Black with white feathers on neck and wings. **Juvenile** Deep brown with purplish gloss on wings and tail, foreparts above striped buffish or rufous. **Similar species** Only black-and-white coucal in its range. Violaceous Coucal much larger with no trace of white.

VOICE Most frequent call resembles distant pounding on two hollow metal drums *soo-hoo*, one after another, given by two dueting birds, lasting *c.* 17sec, rising in pitch and becoming faster and faster; one bird may start and second bird joins after several calls, and infrequently they get out of phase. Alarm call short, sharp *chit-chick-krek* or *chunk*. Nasal tinny *k-k-k-naaah*. Calls day and night (Diamond 1972a, Beehler 1978, Coates 1985, Payne 1997).

DESCRIPTION **Adult** Great variation. Crown black, sides of head, nape, sides of neck and upper mantle white; upperparts, wings and tail black with purplish-blue gloss, white patch on primary wing-coverts; chin, throat and breast white, belly and graduated tail purplish-black. Also from entirely black except white wing patch to greyish-white to grey with white as in normal individuals, or black with scattered white feathers. Grey and white morphs have dark around eyes, and crown mostly black. **Subadult** Blackish, with some white feathers on crown, neck and wings; some with large white area on nape, neck and chin to breast. **Juvenile** Deep brown glossed purplish on wings and tail, buff-brown shaft streaks on foreparts of body above, and dark sooty-brown to slate below tipped paler on throat and breast; some with large white area as subadult. **Nestling** No information. **Bare parts** Iris dark red, juvenile grey to brown. Bill black, juvenile lower mandible and tip of upper brown or whitish. Feet blue-black to slaty-blue, juvenile blue-grey.

BIOMETRICS Wing male (n = 7) 192–217mm, mean 201.6 ± 8.8, female (n = 7) 210–227mm, mean 214.1 ± 10.4. Bill male 39.5–44.0mm, mean 42.5 ± 1.6, female 38.0–43.5mm, mean 41.2 ± 1.9. Tail male 252–270mm, mean 263.0 ± 8.8, female 195–227mm. Tarsus male 44.5–52.0mm, mean 49.4 ± 2.9, female 45–54mm, mean 48.8 ± 3.2 (Payne 2005). Mass male 330g, immature female 342g (Gilliard & LeCroy 1967).

MOULT No information.

DISTRIBUTION Resident throughout Bismarck Archipelago, including New Ireland, Dyaul, Lolobau, New Britain and Umboi (Rook Island) (Gilliard & LeCroy 1967, Coates 1985, Mayr & Diamond 2001).

HABITAT Primary and secondary forests, forest edges, gardens; coconut, oil palm and ulatawa plantations. Lowlands to 1,300m, but uncommon throughout uplands (Gilliard & LeCroy 1967, Coates 1985, Eastwood & Gregory 1995, Fletcher 2000).

BEHAVIOUR Mostly in pairs but also groups of up to four birds. Prefers shrubbery but also forest floor and canopy of high trees. Rather active bird easy to observe when it walks and hops around on branches or 'planes' down to lower part of next tree; also described as skulking. When duetting both birds may perch in different trees about 10m apart or sit on same branch (Diamond 1972a, Coates 1985, Gregory 2007).

BREEDING Season Nestlings recently fledged Nov-Feb, recently hatched Feb, feathered nestlings May, and Umboi, Jul or Aug (Dahl 1899, Rothschild & Hartert 1914a, Payne 2005). **Nest** Chamber or hollow of dry stems and grass leaves lined with green vegetation, sometimes with two holes on opposite sides (Meyer 1936). **Eggs** 2 or 3; white; 40.8x33.0mm (Meyer 1936, Schönwetter 1967). No further information.

FOOD Large stick insects (swallowed whole); longicorn beetles; lizards; probably other small animals (Dahl 1899, Coates 1985).

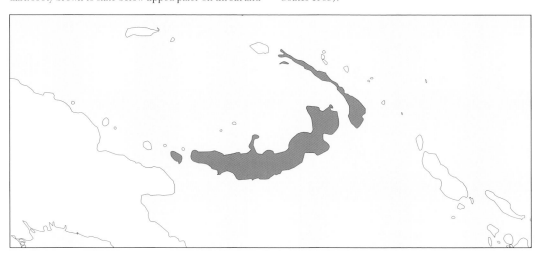

STATUS AND CONSERVATION Common in lowlands and hills but rare in lower mountains (Coates 1985). Found fairly common Kimbe and Walindi, New Britain, 1993–94 (Eastwood & Gregory 1995), and widespread and common throughout lowlands and foothills of Nakanai Mts., west New Britain, Dec 1979 (Bishop & Jones 2001). Restricted range species. Not globally threatened (BirdLife International 2011).

Pied Coucal *Centropus ateralbus*. **Fig. 1.** Adult, Lelet Plateau, New Ireland, June (*Heidi Doman*). **Fig. 2**. Subadult, Kilu Ridge, Walindi Plantation, New Britain, PNG, June (*Nik Borrow*). **Fig. 3**. Adult. Walindi Plantation, New Britain, PNG, August (*Jon Irvine*). Both adults and young show great variation in plumage.

GREATER BLACK COUCAL
Centropus menbeki Plate 4

Centropus Menbeki Lesson & Garnot, 1828 (New Guinea = Manokwari)

Alternative names: Black Jungle/Menbek's Coucal.

TAXONOMY Previously thought to form superspecies with *goliath*, but mtDNA analysis places it closest to *chalybeus*, in clade with *milo* and *ateralbus* (Sorenson & Payne 2005). Name variously spelled *menebiki/Menebiki/menebikii*. Polytypic. Synonym: *m. jobiensis* Stresemann & Paludan 1932 in nominate.

FIELD IDENTIFICATION 60–69cm. Huge clumsy coucal with coarse contour feathering and 'hackles' on foreparts; stout pale bill; noisy when moving around. Sexes similar. **Adult** Entirely black with blue-green gloss above, and chin to breast; rest of underparts dull brown-black. **Juvenile** Dull black, tinged blue-violet on upperparts, mantle indistinctly tipped brown, rectrices black banded brown. **Similar species** Violaceous Coucal of Bismarck Archipelago larger, has strong violet wash and longer bill. Lesser Black Coucal nearly one third smaller with dark bill but otherwise similar; where size difficult to judge different voices may help. Pheasant-coucal has darker bill and variable brown mottling and is rarely found in forest.

VOICE Call rather similar to Pheasant-coucal's but louder and deeper, booming, slow *HOOO-HOOO* either both at same level or last higher or lower pitched, and often repeated many times, each interval 5–10sec. Duet of two or three notes *HOO-HOO* or *HOO-HOO-HU* where last note is lower; often answered with *HOOO-HOOOO-HU-HU-HU-HU-HU-HU-hu*, higher pitched and more rapid than first bird, or *WHOOD-HOOO-hoodi-hoodi-hoodi-hoodi-hoodi*. Cough-like call becoming rattling chatter. Call far-carrying and reminiscent of Papuan Harpy Eagle. Grunt like bullfrog followed by rattled, fast series of rising notes. Captive fledgling's call cooing, followed by incessant chattering; adults' call starts with 2–3 short *coos* followed by reverberating, resonant, descending and far-carrying *coos*. Calls by day, at dawn and after dark (Gilliard & LeCroy 1966, Diamond 1972b, Coates 1985, Beehler *et al.* 1986).

DESCRIPTION Nominate. **Adult** Above, and chin to breast, black glossed blue to blue-green all over wearing to dull brown-black glossed oil-green; bristle-like tips to feathers, or 'hackles', on forepart of body. Remiges and rectrices greenish when new. Belly to vent dull brown-black. Tail broad and graduated. **Juvenile** Dull black tinged blue-violet on upperparts, feathers on mantle indistinctly tipped brown, rectrices black with narrow dark red-brown banding, some only basally. **Nestling** Fledglings entirely black. **Bare parts** Iris red, juvenile white or light grey (male), orange (female) to bright brown. Bill ivory or horn to olive-grey with dark base, particularly upper mandible. Feet dark grey to black, with curved hind claws.

BIOMETRICS Nominate. Wing male (n = 13) 210–240mm, mean 224.5 ± 9.1, female (n = 10) 209–240mm, mean 220.8 ± 8.3. Bill male 46–54mm, mean 49.1 ± 3.5, female 44–55mm, mean 50.3 ± 3.8. Tail male 344–411mm, mean 367.9 ± 23.0, female 326–391mm, mean 355.2 ± 17.5. Tarsus male 55–64mm, mean 59.0 ± 3.2, female 57–61mm, mean 59.6 ± 1.4 (Payne 2005). Mass male Oct 575g (Mees 1982), female 505g (Ripley 1964).

MOULT Moulting Jan (male), Feb, Jun, Oct (both sexes); juvenile to adult moult west NG, Sep (Junge 1937a, Mayr & Rand 1937, Mees 1982, BMNH).

GEOGRAPHICAL VARIATION Two subspecies; poorly differentiated.

C. m. menbeki Lesson & Garnot, 1828. NG; WP islands; Numfor and Yapen Is. Described above. Those of Yapen (Jobi) ('*jobiensis*') more blue, less green than some from NG but similar blue plumage occurs in NG and Misool, and bird from last locality has straighter, less heavy bill (Rothschild & Hartert 1907, Rothschild *et al.* 1932, Mayr & Meyer de Schauensee 1940c, Payne 2005). Described above.

C. m. aruensis (Salvadori, 1878). Aru Is. Larger; bill more curved. Upperparts and tail dark purplish-blue.

DISTRIBUTION Resident. NG; Misool, Salawati and Batanta (WP islands); Yapen and Numfor Is, Geelvink Bay; Aru Is. (Stresemann & Paludan 1932, Mayr & Rand 1937, Ripley 1964, Mees 1982).

HABITAT Primary and secondary forest, forest edge, monsoon scrub and small areas with monsoon forest in savanna; lowlands to 1,300m, but most common below 800m. Freshwater swamps at Bintuni Bay, WP (Ripley 1964, Coates 1985, Beehler *et al.* 1986, Erftemeijer *et al.* 1991).

BEHAVIOUR Usually alone but often heard duetting suggesting some sort of pair bonding and/or territoriality. Moves around in dense foliage and tangles of vines in clumsy

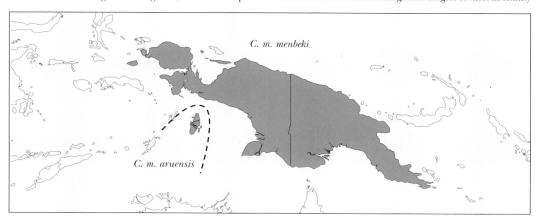

and noisy manner, but also frequents forest floor, switching tail from side to side when hopping up to trunk. Flies infrequently. Sometimes shy and retiring (Rand & Gilliard 1967, Diamond 1972b, Coates 1985, Beehler *et al.* 1986).

BREEDING Season Oviduct egg north WP, Apr; fledgling Fly R., Oct; four females in breeding condition Snow Mountains, Apr; fledgling middle Sepik, Jan (Rand 1942a, b, Gilliard & LeCroy 1966). **Nest** Large mass of leaves (Payne 2005). **Eggs** Oviduct egg white; 37x30.3mm; 36.9x26.7mm (Rand 1942b, Schönwetter 1967). No other information.

FOOD Large insects including caterpillars, grasshoppers and cicadas; small birds, including net-captures; small snakes; frogs. Of eight examined stomachs six contained insects, one a frog, one a small bird and one a 30cm long snake; only insects in five birds (Rand 1942a, b, Diamond 1972b, Coates 1985).

STATUS AND CONSERVATION Population near Brown R. estimated one/10ha (Bell 1982), which Coates (1985) believed fairly typical for all NG, but difficult to observe. Not globally threatened (BirdLife International 2011).

Greater Black Coucal *Centropus menbeki*. **Fig. 1.** Adult, *C. m. menbeki*, Variata, PNG, June (*Ashley Banwell*).

BIAK COUCAL
Centropus chalybeus   Plate 6

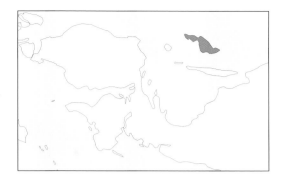

Nesocentor chalybeus Salvadori, 1876 (Biak)

Alternative names: Biak Island/Biak Bronze Coucal

TAXONOMY Affinities unclear; may be related to *violaceus* (Mayr & Diamond 2001), but mtDNA analysis places it close to *menbeki* (Sorenson & Payne 2005). Monotypic.

FIELD IDENTIFICATION 42–46cm. Medium-sized coucal with tail about length of body; yellow eyes. Sexes similar. **Adult** Black glossed purple above. **Juvenile** Unbarred black washed rufous. **Similar species** Only coucal on Biak. Greater Black Coucal on mainland much larger with green gloss, red eyes and heavier, whitish bill, and juvenile barred. Lesser Black Coucal, also on mainland, has dark brown eyes.

VOICE Call far-carrying. Harsh and broken rasp; falling series of accelerating and upslurred hoots; repeated *bup* (Mayr & Meyer de Schauensee 1939, Beehler *et al.* 1986).

DESCRIPTION Adult Whole plumage black with purple gloss on upperparts and wings, duller on lower back and uppertail-coverts, foreparts with spiny feathers, and faint pale feather shafts on head and neck, belly often tinged brownish; graduated tail black glossed purple. **Juvenile** Plain dull black-tinged rufous. **Nestling** Undescribed. **Bare parts** Iris yellow, orbital skin dark. Bill black. Feet dark.

BIOMETRICS Wing male (n = 5) 189–214mm, mean 203.0, female (n = 6) 189–220mm, mean 203.0 ± 9.7. Bill male 44–50mm, mean 47.2, female 40.0–49.3mm, mean 45.0 ± 3.1. Tail male 260–285mm, mean 274.2, female 253–315mm, mean 280.3 ± 23.0. Tarsus male 47–54mm, mean 50.4, female 50–57mm, mean 53.6 ± 2.4 (Payne 2005).

MOULT No information.

DISTRIBUTION Resident Biak (Misori) and Supiori, Geelvink Bay, WP (Mayr & Meyer de Schauensee 1939, Rand & Gilliard 1967, Beehler *et al.* 1986).

HABITAT Trees, including thick secondary growth (Rand & Gilliard 1967, Beehler *et al.* 1986).

BEHAVIOUR Noisy. Frequents both ground and trees. Not shy; makes awkward and heavy short flights among bushes (Mayr & Meyer de Schauensee 1939, Beehler *et al.* 1986).

BREEDING No information.

FOOD Unknown.

STATUS AND CONSERVATION Restricted range species. Common on Biak (Beehler *et al.* 1986); Eastwood (1996) saw only one bird but heard it often, and other authors describe it as probably not uncommon and probably more common on Supiori, where more forest still exists (Bishop 1982, Gibbs 1993 in BirdLife International 2001). On Biak heavy forest logging has changed large areas to farmland, but its rather tolerant habitat preference and its occurrence in 110km^2 Biak-Utara Protected Area give some hope for future of this species. Near threatened; population estimate 10,000–19,999 (BirdLife International 2011).

Biak Coucal *Centropus chalybeus*. **Fig. 1.** Adults, after heavy rainfall, South-West Biak Island, January (*Mehd Halaouate*).

RUFOUS COUCAL
Centropus unirufus
Plate 7

Pyrrhocentor unirufus Cabanis and Heine, 1863 (Luzon)

TAXONOMY Placed in superspecies with *celebensis* (White & Bruce 1986), but vocalisations different, and relationship is not supported by mtDNA analysis (Sorenson & Payne 2005). Monotypic. Synonym: *polillensis* Hachisuka 1930.

FIELD IDENTIFICATION 38–42cm. Medium-sized coucal with tail longer than body, and stout decurved greenish bill tipped yellow. Sexes similar. **Adult** Plain rufous-chestnut, somewhat brighter below, and paler on chin. **Juvenile** As adult but often with some white trichoptiles persisting on head. **Similar species** None in range.

VOICE Snapping staccato *squip-whip* or *squip whip-whip-whip-whip* somewhat reminiscent of child's high-pitched toy. Also some squeaky and short metallic notes not unlike seabird chicks. Contact note shrill, plaintive *kaow* (Payne 1997, Kennedy *et al.* 2000).

DESCRIPTION Adult Entire plumage deep rufous-chestnut including underside of wings but shade paler on underparts; feathers on forehead, crown, and nape to mantle with glossy shafts. **Juvenile** As adult but with white trichoptiles on head; rectrices narrower. **Nestling** Yellowish white trichoptiles on head. **Bare parts** Iris light brown, orbital skin yellowish. Bill green tipped yellow, juvenile black. Feet slate-grey to olive-brown, hind claw short, 16–20mm.

BIOMETRICS Wing male (n = 11) 145–162mm, mean 155.7 ± 5.6, female (n = 8) 155–165mm, mean 161.5 ± 3.3. Bill male 32–36mm, mean 33.8 ± 1.1, female 33–38mm, mean 35.3 ± 1.6. Tail male 213–242mm, mean 228.2 ± 1.9, female 220–262mm, mean 237.6 ±12.8. Tarsus male 36–45mm, mean 40.2 ± 3.3, female 39–45mm, mean 41.8 ± 2.5. Mass male (n = 6) 145.7–187.7g, mean 168.3, female (n = 8) 146–227g, mean 201.4 (Payne 2005). Mass female Jun 162g (J. Erritzøe).

MOULT Very worn female, Jun (J. Erritzøe).

DISTRIBUTION Resident Catanduanes, Luzon and Polillo, Philippines (Kennedy *et al.* 2000).

HABITAT Hill forest with dense undergrowth and vines; bamboo thickets. Primary or selectively-logged forest at 550 and 760m Mt. Isarog NP. To 1,200m, but presumably mainly lowland (Goodman & Gonzales 1990, Poulsen 1995, Kennedy *et al.* 2000, BirdLife International 2001).

BEHAVIOUR Singly, in pairs or noisy groups of up to 10–12 birds moving around in dense undergrowth or on forest floor; habits more like those of malkohas, and follows mixed-species flocks (Poulsen 1995, Kennedy *et al.* 2000).

BREEDING Season Male with enlarged gonads, early Apr, and fledgling late Nov (Goodman & Gonzales 1990, Payne 2005). No further information.

FOOD Passerine and insect in stomach (J. Erritzøe).

STATUS AND CONSERVATION Already Ogilvie-Grant (1894) declared it "not very common and difficult to obtain". Rare (Hachisuka 1934), only collected by three or four collectors; rare (Delacour and Mayr 1946); uncommon and local (Dickinson *et al.* 1991). Up to three Quezon NP and Angat Watershed, Jan–Feb 1994 (Hornbuckle 1994). Not particular in choice of habitat, even living close to humans (Poulsen 1995), but because of rare status and loss of forests considered near threatened (BirdLife International 2011).

Rufous Coucal *Centropus unirufus*. **Fig. 1.** Adult, Zambales, Luzon, Philippines, August (*Tina Sarmiento Mallari*). **Fig. 2.** Adults, sunning, Subic Forest, Olongapo, Philippines, July (*William Lim*). Coucals often spread their wings and tail to dry after rain.

GREEN-BILLED COUCAL
Centropus chlororhynchos   Plate 8

Centropus chlororhynchos Blyth, 1849 (Sri Lanka).

Alternative names: Sri Lanka/Ceylon Coucal.

TAXONOMY Original spelling erroneously amended to *chlororhynchus* (e.g. Blyth 1867). "This species seems to have no near relatives; in spite of its very different coloration it appears to be structurally closer to *C. [sinensis] andamanensis* than to any other representatives of the genus" (Peters 1940). Monotypic.

FIELD IDENTIFICATION 43–46cm. Medium-sized, short-winged coucal. Sexes similar. **Adult** Dull black with purplish gloss on neck and tail, dark chestnut wings, and long pale greenish or ivory bill. **Juvenile** As adult but barred blackish on chestnut wing-coverts, and eyes grey. **Similar species** Sympatric Greater Coucal larger, has paler chestnut on wings with more contrast between wings and body, has black bill, not green. Voices of both species rather similar but this species has fewer syllables, and lacks water-bottle phrase.

VOICE Two to three syllables; far carrying, sonorous and mournful *hooo-poop*, *hooo-poo-poop*, with *poop* being lower pitched than *hooo*, but taped call consisted of short double syllable *hu, hu* (*hu* as 'oo' in 'book'), both notes identical at same pitch. Deep *whoop*, *whoop*, *whoop*, *whoop* and at times *chowk, chowk, chowk* or *chouk, chouk* repeated *c.* 20 times. Deep coughing sound and sleepy *hmmm, hmmm*. Series of paired pipe-organ-like coos *fWOO-HOop-fWOO-Hoop*. On alighting on nest female uttered soft *whoop, whoop*. Alarm call loud *dhjoonk*. Pairing-note *chewkk*. Most vocal at dawn and sunset. Voice easily mistaken for that of Greater Coucal but lacks water-bottle expression, and also that of Brown Fish Owl, which has shorter intervals between calls (Legge 1880, Henry 1955, Hoffmann 1989, de Silva Wijeyeratne *et al.* 1997, Wijesinghe 1999, Payne 2005, Rasmussen and Anderton 2005).

DESCRIPTION Adult Head and body dull brown-black with purple gloss on head, neck and broad graduated tail, upper back, throat and breast with coppery-bronze tinge, wings dark maroon-chestnut, underwing-coverts black; tail black glossed greenish-blue. **Juvenile** Like adult but chestnut wing-coverts barred and spotted black. **Nestling** Black skin with white trichoptiles. **Bare parts** Iris red to brown; female pure white; juvenile grey. Bill pale green or ivory (during breeding: da Silva Wijeyeratne 1999), blackish at base and nostrils; juvenile dark greenish, base and culmen dark grey; nestling has ivory stripe from base of upper mandible to tip, lower mandible brownish-black. Feet black, claws dusky, hind claw rather short; juvenile feet flesh.

BIOMETRICS Wing male 158–173mm, (n = 7) mean 167.3 ± 4.9, female 168–184mm, (n = 7) mean 174.6 ± 7.7. Bill male 36–41mm, mean 38.9 ± 1.7, female 38.9–47.0mm, mean 42.0 ± 2.9. Tail male 222–252mm, mean 234.3 ± 8.8, female 225–254mm, mean 239.3 ±10.7. Tarsus male 40–46mm, mean 43.3 ± 2.1, female 42–47mm, mean 42.9 ± 2.6 (Payne 2005).

MOULT Male moulting tail, Dec (Legge 1873).

DISTRIBUTION Resident south-west Sri Lanka; restricted to wet zone. Distribution formerly extended north and east of present known range at Sinharaja, Bodhinagala and occasionally Kitulgala. Common from Deduro Oya in north

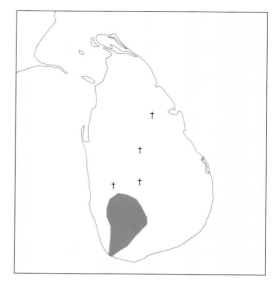

to hills of Galle and coffee districts of Morawak Korale; also numerous Ratnapura District and up into Peak Wilderness Forest (Legge 1880).

HABITAT Lowlands to 1,100m, recently 1,450m. Primary wet forests with dense undergrowth of dwarf bamboo and other luxuriant vegetation, such as cane brakes. 'Coucal' bamboo *Ochlandra stridula* in swamps. Village gardens with coffee, coconut and areca, without bamboo but not far from forest; abandoned swidden agriculture in spite of site being surrounded by primary forest; some pairs maintained territories along forest outskirts bordering village and along old logging trails, Sinharaja FR (Henry 1955, Ali & Ripley 1983, Hoffmann 1989, Kotagama & Fernando 1994, Jones *et al.* 1998, Wijesinghe 1999, de Silva Vijeyeratne *et al.* 2000, Rasmussen & Anderton 2005, Gunawardena *et al.* in prep.).

BEHAVIOUR Shy; never far from cover. Poor flyer, flapping from tree to tree. Occurs early mornings and after heavy rains in open areas in forest or roadsides, hopping about searching for food. Feeds up to 15m in trees. Often suns itself in mornings with outstretched wings in open. When calling dips head after each syllable (Henry 1955, Hoffmann 1989, Wijesinghe 1999, Rasmussen & Anderton 2005).

BREEDING Season Jan–Jul, but fledgling 20 Sep, and nest-building 8 Oct, later abandoned and another finished, 5 Nov, suggesting season possibly year-round (Wijesinghe 1999). **Nest** Domed, of sticks, roots and grass and lined with green leaves, in thorny bush *c.*1.5m high in deep, evergreen forest. Another shaped like rugby ball, primarily of bamboo leaves, 60cm tall, 45cm wide with side entrance and placed 3.65m above ground about 40m from stream in Wana-idala tree *Wendlandia bicuspidata* (Henry 1955, Wijesinghe 1999). **Eggs** 2–3; chalky white; mean 34.7x27.0mm (n = 9) (Baker 1934, Henry 1955). **Incubation** *c.*17 days (Wijesinghe 1999). **Chicks** Fed by both parents; fledge after *c.*15 days (Wijesinghe 1999). **Survival** No information.

FOOD Skinks and other lizards, snakes; frogs; termites, moths; worms; snails; fruit (Henry 1955, Wijesinghe 1999).

STATUS AND CONSERVATION Sri Lanka's rarest endemic bird, only found in fragmented wet forests, coexisting peacefully with Greater Coucal, but perhaps 'only a few hundred pairs' remain (Hoffmann 1989). In high densities Bodhinagala Forest, lower densities Sinharaja (de Silva Wijeyeratne *et al.* 2000). Even 150 years ago considered very rare (Layard 1854 in Legge 1873). Greatest threat extensive logging and degradation of the few forest remains, then converted to agriculture or plantations; also burning, wood-cutting, settlements, gem mining and perhaps competition with Greater Coucal contribute to its decline (BirdLife International 2000, 2001), although last point contradicts Hoffmann (1989). Occupies different niches in disturbed areas; stronghold Sinharaja National Heritage Wilderness Area, a World Heritage Site, but apparently unable to move between forest fragments resulting in genetic isolation (BirdLife International 2001). "Fairly common wherever the wet-zone forest was spared by the axe" (Henry 1955); however since then most forest has disappeared from its limited range, so Fleming (1977) considered it "one of the rarest birds in Sri Lanka". Survey of over 200 forest sites during 1991–1996 found it only in 12 forests. Very optimistic estimate "between 2,500 and 10,000, but most likely number is only a few thousand individuals" (BirdLife International 2000, 2001). Legally protected and occurs in several national parks and forest reserves which is no guarantee of its survival. Protection of few remaining forests against illegal logging, education of villagers to instill responsibility for biological resources, and effective management, are most important targets. Its flexibility in habitat choice, however, gives some hope for survival. Vulnerable (BirdLife International 2011).

Green-billed Coucal *Centropus chlororhynchos*. **Fig. 1.** Adult, Sinharaja, Sri Lanka, April (*Uditha Hettige*). **Figs. 2–3.** Adult, Sinharaja, Sri Lanka, July (*Gehan de Silva Wijeyeratne*).

BLACK-FACED COUCAL
Centropus melanops　　　　　　　Plate 7

Centropus melanops Lesson, 1830 (Java, error = Mindanao)

TAXONOMY MtDNA analysis places it as sister to clade consisting of *steeri* and *rectunguis* (Sorenson & Payne 2005). Birds from south Philipines ('*banken*') said to be larger, and have more black on face, but large sample shows no consistent differences. Monotypic. Synonyms: *nigrifrons* Peale 1848, *m. banken* Hachisuka 1934.

FIELD IDENTIFICATION 42–48cm. Medium-sized coucal with stout decurved bill and red eyes. Sexes similar. **Adult** Black on sides of face, foreparts otherwise buffish-yellow, rest of upperparts black except chestnut shoulders and wings. **Juvenile** As adult. **Similar species** Buffish-yellow colour on foreparts distinguishes it from smaller Philippine Coucal which is all black with blackish-brown wings.

VOICE Call booming *wooooop-wooop-wooop-wooop* repeated 4–5 times, first note louder and longer, following by descending softer notes, lasting *c.* 5sec, quite different from Philippine Coucal, and easily located. Another call reminiscent of Philippine Coucal *wooop-boop-boop-boop-boop*, first louder, then descending (duPont & Rabor 1973b, Kennedy *et al.* 2000).

DESCRIPTION Adult Forehead, crown, nape, neck, mantle buffish-yellow; lores, eyebrows, cheeks, and behind eye to sides of chin black; wings and scapulars bright chestnut, tips of primaries brown, back, rump and uppertail-coverts black glossed bluish-green; underwing-coverts black; chin, throat and breast yellow-buff, more buffish (feather wear) on breast and mantle, belly blackish-brown, undertail-coverts and graduated tail black glossed bluish-green. **Juvenile** Like adult but rectrices narrower (Payne 2005). **Nestling** Undescribed. **Bare parts** Iris red. Bill black. Feet black; hind claw short, 15–18mm.

BIOMETRICS Wing male (n = 8) 152–167mm, mean 162.0 ± 4.1, female (n = 5) 158–172mm, mean 162.0 ± 5.4. Bill male 30–37mm, mean 34.6 ± 2.9, female 33–36mm, mean 34.17 ± 1.2. Tail male 222–252mm, mean 237.7 ± 10.6, female 215–270mm, mean 243.8 ± 20.2. Tarsus male 34–47mm, mean 42.0 ± 5.5, female 39–44mm, mean 41.3 ± 1.7. Mass male (n = 10) 197.5–237.3g, mean 211.1, female (n = 12) 214–265.5g, mean 237.9 (Payne 2005).

MOULT Primary moult Nov (BMNH).

DISTRIBUTION Philippines. Resident on Biliran, Bohol, Leyte, Samar, Basilan, Dinagat, Mindanao, Nipa and Siargao (Dickinson *et al.* 1991).

HABITAT Lowland primary and secondary forest and forest edge, preferring dense vines and tangles in middle and upperstories; below 1,000–1,200m (duPont & Rabor 1973b, Dickinson *et al.* 1991, Kennedy *et al.* 2000).

BEHAVIOUR Shy and difficult to observe. Usually singly or in pairs but on Bohol often in small groups. Favours treetops (duPont & Rabor 1973b, Brooks *et al.* 1995a, Kennedy *et al.* 2000).

BREEDING Season Oviduct egg, Apr; birds in breeding condition, May–Jun (Rand & Rabor 1960, Kennedy *et al.* 2000). **Eggs** White; smooth unglossed, broad oval, one end very slightly more pointed; 37.4x28.8mm (n = 2) (Rand & Rabor 1960, Schönwetter 1967). No further information.

FOOD Unknown.

STATUS AND CONSERVATION Fairly common (Dickinson *et al.* 1991, Kennedy *et al.* 2000). 58 records Rajah Sikatuna NP, Bohol, 23–27 Jul 1994 (Brooks *et al.* 1995a). Forest degradation taking place on large scale all over Philippines may, however, soon be severe threat. Not globally threatened (BirdLife International 2011).

Black-faced Coucal *Centropus melanops*. **Fig. 1.** Adult, Rajah Sikatuna NP, Bohol Island, Philippines, August (*Kevin Vang*). **Figs. 2–3.** Adults, near Bislig, Mindanao, Philippines, May (*Paul Noakes*).

STEERE'S COUCAL
Centropus steerii Plate 8

Centropus steerii Bourns and Worcester, 1894 (Mindoro, Philippines)

Alternative name: Black-hooded Coucal.

TAXONOMY MtDNA analysis places this in clade with *rectunguis* and *melanops* (Sorenson & Payne 2005). Monotypic.

FIELD IDENTIFICATION 43–50cm. Medium-sized coucal, tail approximately as long as body, and strong, hooked bill. Sexes similar. **Adult** Head and throat black glossed green, neck, upperparts, wings blackish-brown glossed bronze-green, rest of underparts dark brown, tail blackish-green. **Juvenile** Brown; chin and throat brownish-grey; tail dull black (Payne 2005). **Similar species** Philippine Coucal on Mindoro smaller, with smaller and less decurved bill, and body colour black glossed blue-green.

VOICE Series of 5–9+ *hooot-hoot-hoot* notes, first louder and longer, and descending at end, each series lasting 4–5sec, and continued after pause. Also explosive *tchoc-cou* reported, but there is some doubt about identity of bird giving this call (Dutson *et al.* 1992, Kennedy *et al.* 2000).

DESCRIPTION **Adult** Whole head and nape black glossed green with shiny shaft streaks, hindneck and sides of neck and mantle blackish-brown glossed bronze-green, back and rump slaty-black, each feather tipped greenish; wings blackish-brown glossed bronze-green except outer four primaries which have no gloss, underwing-coverts and axillaries blackish-brown glossed green and flight feathers blackish-brown; chin, throat and upper breast greenish-black, lower breast and belly dark brown, flanks, thighs and undertail-coverts slaty-black; tail and uppertail-coverts dull metallic green with jet black shafts, underside of tail black with faint metallic blue gloss. **Juvenile** Mainly brown without shiny shaft streaks; chin and throat brownish-grey; tail dull black (Payne 2005). **Nestling** Undescribed. **Bare parts** Iris brown to red. Bill black, juvenile paler. Feet black, hind claw stout and not exceeding hind toe in length.

BIOMETRICS Wing male (n = 8) 152–171mm, mean 159.4 ± 5.7, female (n = 9) 157–170mm, mean 164.4 ± 5.0. Bill male 35–40mm, mean 36.9 ± 1.9, female 35–41mm, mean 37.1 ± 1.9. Tail male 217–253mm, mean 238.6 ± 11.5, female 232–277mm, mean 251.3 ± 16.6. Tarsus male 35–45mm, mean 39.6 ± 3.4, female 38–45mm, mean 43.2 ± 3.7 (Payne 2005). Mass male (n = 4) 175–200g, mean 183.7, female (n = 4) 190–238g, mean 214g (Ripley & Rabor 1958).

MOULT Primary moult Nov (BMNH).

DISTRIBUTION Resident Mindoro, Philippines (Dickinson *et al.* 1991).

HABITAT Primary and advanced secondary forest from lowlands to 750m, also vine-covered scrub, along riversides, and bamboo thickets within primary forest. In dipterocarp

forest, most commonly 30–60m (Ripley & Rabor 1958, Collar *et al.* 1999, Kennedy *et al.* 2000).

BEHAVIOUR Shy and secretive; not easy to find. Occurs in dense vines and foliage both in understorey and canopy. Runs along branches. Mostly solitary and reluctant to fly (Collar *et al.* 1999, Kennedy *et al.* 2000).

BREEDING No information.

FOOD Larger insects noted on skin label (Collar *et al.* 1999, BMNH).

STATUS AND CONSERVATION Early collectors such as Ogilvie-Grant and Whitehead recorded it as relatively common in low-altitude forest (Dutson *et al.* 1992). As late as 1954 considered "fairly common" (Ripley & Rabor 1958) but since then population has decreased rapidly and today within its altitudinal range only found in small and fragmented forests except in forest at Malpalon where ground is so rough that it evades any cultivation attempt. During four week's research in 1991 on Mindoro only two birds heard calling and one seen at Malpalon on 1 Oct, and two singles seen at Siburan on 22 Dec (Dutson *et al.* 1992). Since 1980 only recorded from four localities, Puerto Galara, 1992 (not confirmed, Collar *et al.* 1999), MUFRC Experimental Forest, Siburan, Sablayan Penal Colony (closed Apr 2009 due to apparent insurgent activity: A. Lewis in Collar & Sykes 2009) and Malpalon. In Mt. Iglit Baco NP not observed since 1979, perhaps due to shifting cultivation, and at other sites rattan collection and illegal tree-cutting are serious disturbances for this shy bird. Replaced by Philippine Coucal where forests are degraded. Restricted species seriously threatened by forest destruction. Critically endangered; population estimate 50–249, decreasing over its 700km^2 range (BirdLife International 2011).

Steere's Coucal *Centropus steerii*. **Fig. 1.** Adult, Suburan, Mindoro, Philippines, March (*Markus Lagerqvist*). **Fig. 2.** Adult, Suburan, Mindoro, Philippines, March (*Paul Noakes*).

SHORT-TOED COUCAL
Centropus rectunguis Plate 6

Centropus rectunguis Strickland, 1847 (Malacca)

TAXONOMY MtDNA analysis places this species in clade with *steerii* and *melanops* (Sorenson & Payne 2005). Monotypic.

FIELD IDENTIFICATION 37–42cm. Typical coucal except for shorter tail. Sexes similar. **Adult** Black with rufous-chestnut on mantle, back and wings. **Juvenile** Above pale chestnut-brown barred black; underparts dark brown barred whitish, tail black with fine white barring. **Similar species** Resembles Greater but is much smaller, lacks dark tips to primaries, and tail shorter; juvenile Greater has black on foreparts and underparts barred and spotted whitish, and black barring on chestnut mantle and wings more prominent. Lesser more similar in size, but more glossy black, mantle and wings brighter chestnut, under wing-coverts chestnut, and best diagnosed by longer tail; juvenile *rectunguis* buffish with some black barring on whole underparts.

VOICE Territorial call series of 4–5 hoarse ascending *boop-boop-boop-boop* notes descending towards end, also rendered as *whu-huup-huup-huup-huup*, slower and deeper than Greater. At dusk more rapid *boop* series on rising scale. Other notes such as squirrel-like *hut-hut-hut* or explosive *jeézaw* are also ascribed to this species but not confirmed. Most vocal Nov–Mar, but calling also heard May and Aug (Jeyarajasingam & Pearson 1999, Wells 1999, Robson 2000).

DESCRIPTION Adult Black glossed blue on head, neck, upper mantle, whole underparts and tail, more purplish-blue on nape, breast and tail; lower mantle, back and wings bright chestnut, underwing-coverts black; less gloss on lower breast to vent; some with chestnut on flanks. **Juvenile** Head, neck, chin, throat and upper breast pale rufous-brown with 2–3 black-brown bars on each feather, pale buffish supercilium; back and wings plain chestnut, tertials and wing-coverts chestnut edged black-brown, underwing-coverts barred black and white, rest of underparts variable blackish to brown or buffish barred whitish, barred rufous on flanks; tail black with fine white barring. **Nestling** No information. **Bare parts** Iris red, juvenile grey. Bill black, juvenile brown, lower mandible paler. Feet black, almost straight claw on first digit short, male 8–12mm, female 9–19mm.

BIOMETRICS Wing male (n = 4) 156–170mm, mean 165.5, female 166, 186mm. Bill male 35–37mm, mean 36, female 34, 42mm. Tail male 192–204mm, mean 198, female 194, 238mm. Tarsus male 42–46mm, mean 44.3, female 45, 52 (Payne 2005). Mass adult 237.5g (Wells 1999).

MOULT Two adults in active wing moult at up to four loci, Aug; juvenile to adult moult Malay Peninsula, Aug; interrupted wing moult Jul (Wells 1999, BMNH).

DISTRIBUTION Resident Thai-Malay Peninsula from 5°20'N southwards, and on many off-shore islands on west coast; Batam I. in Singapore Strait and Tioman, Ko Samui, and Ko Tao I. on east coast. Extreme south Thailand at Hala-Bala Wildlife Sanctuary at 5°58'N in 1998, and Khlong Hala in 1999. On Sumatra Gunung Singgalang, Padang Highlands, Pasir Ganting, and south Barisan Range, all old records; also Sungai Ketalo 1976 and Sungai Tembesi 1976. On Borneo only four skins known until 1960, all Sarawak and Brunei, but now 10 localities Sarawak, 4 Brunei, 6–7 Sabah, 3–4 East Kalimantan, 2 Central Kalimantan and 1–2 West Kalimantan, but unrecorded South Kalimantan (Smythies & Davison 1999, Wells 1999, Robson 2000, BirdLife International 2001, Mann 2008).

HABITAT Dense primary forest and occasionally scrub and grass in small clearings in forest, but rarely at forest edge like Greater Coucal. To 600m Thai-Malay Peninsula; at Taman Negara often near small streams. On Sumatra also selectively logged primary forest and at forest edges to *c.*1,700m Mt. Singgalang which is an extraordinary record. Extreme lowland (to 600m) closed-canopy forest, Borneo, in lowland dipterocarp, peat swamp and alluvial forest, also kerangas, logged and secondary forests, and perhaps

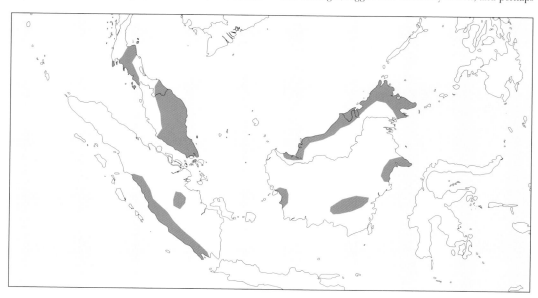

coastal scrub-forest, beach vegetation and long grass; scrub jungle and shrubby riverside areas but close to dense forest also reported (van Marle & Voous 1988, Smythies & Davison 1999, Wells 1999, J.E. Duckett and S. Harrap in BirdLife International 2001, Mann 2008). On paddifields, Sabah (Gore 1968), but this and other unusual habitats doubted (e.g. Smythies & Davison 1999) due to similarity to Greater Coucal.

BEHAVIOUR Skulking, solitary species, difficult to observe because occurs in densest parts of forest mainly foraging on forest floor and understratum. Reported at forest edge in company with Greater Coucal (Jeyarajasingam & Pearson 1999, Wells 1999, Robson 2000).

BREEDING Season Fledglings attended by adults 17 and 21 May, and nest with adults Sep, Thai-Malay Peninsula. Chick Sumatra, Mar (van Marle & Voous 1988, Wells 1999). **Nest** Globular; untidy construction of leaves, some of them with twigs attached, lined with palm-frond pith and placed 2m high in palm (Wells 1999). **Eggs** White; 36.5x29.5, 37.3x30.3mm (Schönwetter 1967). **Chicks** Both parents care for offspring (Wells 1999). No further information.

FOOD Presumably insects as other coucals. One killed and ate large frog. Kills all birds caught in mistnets (S. Sreedharan in Smythies & Davison 1999, Wells 1999).

STATUS AND CONSERVATION Difficult to monitor Thai-Malay Peninsula due to uncertainty as to whether only territory-holders are vocal; uncommon to more or less common (Wells 1999). Scarce to uncommon (Robson 2000). Rare Borneo (Mann 2008). It may be presumed that some records of this species, especially those from higher elevations, are misidentified Greater Coucals. Perhaps overlooked on Sumatra (van Marle & Voous 1988). Rather limited habitat preference for lowland forest, which is being lost at alarming rate, puts it at great risk, and now considered vulnerable; population estimate 10,000–19,999 (BirdLife International 2011).

Short-toed Coucal *Centropus rectunguis.* **Figs. 1–2.** Adults, Danum, Sabah, Borneo, May (*James Eaton*).

BAY COUCAL
Centropus celebensis   Plate 7

Centropus celebensis Quoy and Gaimard, 1830 (Manado, Sulawesi).

Alternative names: Brown/Celebes/Sulawesi Coucal.

TAXONOMY Claimed to form superspecies with *unirufus* (White & Bruce 1986) but vocalisations differ. MtDNA analysis shows no close relative (Sorenson & Payne 2005). Delacour (1947) included *kangeanensis* (Kangean I.), but later authors placed this in *sinensis* (Peters 1940, White & Bruce 1986, Payne 2005). Polytypic. Synonyms: *Pyrrhocentor bicolor* Cabanis & Heine 1863 in nominate; *Aethostoma celebense rufofuscum* and *c. trigeminus*, both Stresemann 1931, in *rufescens*.

FIELD IDENTIFICATION 42–51cm. Medium-large coucal. Sexes similar. **Adult** Overall colour rufous and pale grey. **Juvenile** Paler than adult, especially on throat and breast. **Similar species** On Sulawesi cannot be confused with other species except perhaps juvenile Lesser Coucal which is barred and spotted with black on rump and belly and heavily streaked pale buffish on head and mantle.

VOICE 2–10 *woop-woop* notes slightly higher pitched at end, five phrases/2sec; about eight deep *hoo-hoo-hoo* notes which accelerate to its highest pitch, and then slows down and ends slightly higher, duration of whole call 2.3–3.6sec, often started by one bird and joined by others and rapidly speeds up; high pitched *duli-duli-duli* repeated about 12 times/5sec. On Togian Is. *coung-coung* repeated 3–25 times increasing in tempo and pitch. *Wheeze*, often by solitary birds. Often duets (Stresemann 1940–41, Watling 1983, Coates & Bishop 1997, Indrawan *et al.* 2006).

DESCRIPTION Nominate. **Adult** Crown greyish, rest of head and mantle uniform buff-grey with buff shafts to each feather, rest of upperparts and wings rufous with fine violet gloss, underparts pale grey-buff becoming pale rufous on belly and undertail-coverts; graduated tail plain rufous with violet gloss. **Juvenile** Like adult but paler, crown less grey, and rectrices narrower. **Nestling** No information. **Bare parts** Iris red, juvenile grey. Bill black tipped horn brown to yellowish white; lower mandible and tip of upper pale in juvenile. Feet black.

BIOMETRICS Nominate. Wing male (n = 10) 164–184mm, mean 173.2 ± 7.1, female (n = 8) 172–187mm, mean 181.6 ± 5.9. Bill male 31–38mm, mean 34.6 ± 1.9, female 33–41mm, mean 38.1 ± 3.0. Tail male 250–302mm, mean 267.1 ± 14.3, female 282–333, mean 296.0 ± 22.7. Tarsus male 36–44mm, mean 40.6 ± 2.2, female 38–50mm, mean 41.6 ± 3.9. Hallux claw short, 16–20mm (Payne 2005).

MOULT Nov–Mar; primary moult Oct–Nov (Riley 1924, BMNH).

GEOGRAPHICAL VARIATION Two subspecies.
C. c. celebensis Quoy & Gaimard, 1830. North Sulawesi and Togian Is. Described above. One seen on Togian had black lores indicating that it may be *rufescens* (Indrawan *et al.* 2006).
C. c. rufescens (A.B. Meyer & Wiglesworth, 1896). Central and south Sulawesi; Butung, Labuan Blanda, and Muna Is. Generally more rufous than nominate; lores often black.

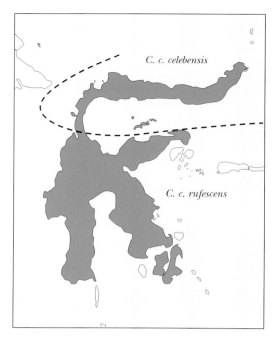

DISTRIBUTION Resident Sulawesi and offshore islands (White & Bruce 1986, Coates & Bishop 1997).

HABITAT Primary and secondary lowland and hill forest, but also at forest edges, in scrub and more rarely in mangroves; to 1,100m. Also selectively logged or disturbed primary forest, and palm vegetation, or wooded savanna, forest on ultra-basic rock, and cultivated areas with settlements (Coates & Bishop 1997, Wardill *et al.* 1999, Bororing *et al.* 2000, Riley *et al.* 2003).

BEHAVIOUR Presumably strongly bonded as usually found in pairs, but also joins foraging Yellow-billed Malkohas or troops of Sulawesi Macaques *Macaca nigra*; prefers dense vegetation and therefore difficult to observe. Climbs through lianas and thick foliage like malkoha. Can move very fast on forest floor (Stresemann 1940–41, Watling 1983, Coates & Bishop 1997).

BREEDING Season Female with enlarged ovary, Feb (Payne 2005). **Nest** Flat and open like pigeon's, of brushwood, built in trees (Meyer & Wiglesworth 1896). No other information.

FOOD Orthoptera, beetles, caterpillars; spiders; fruit, e.g. nutmeg (Meyer & Wiglesworth 1896, Stresemann 1940–41).

STATUS AND CONSERVATION Sparse in most parts of Sulawesi and neighbouring islands, but locally moderately common (Watling 1983, Coates & Bishop 1997). Locally common Dumoga-Bone NP (Rozendaal & Dekker 1989). Seems to have great habitat tolerance, even utilising cultivated areas (Wardill *et al.* 1999) and therefore not in any immediate danger, but rattan collection serious threat (Payne 2005). Not globally threatened (BirdLife International 2011).

Bay Coucal *Centropus celebensis*. **Fig. 1.** Adult, *C. c. celebensis*, Tangkoko NP, Sulawesi, August (*Marc Thibault*).

GABON COUCAL
Centropus anselli   Plate 9

Centropus ansellii Sharpe, 1874 (Danger River, Gabon).

TAXONOMY Considered allospecies of *leucogaster*, supported by mtDNA analysis (Payne 2005). *Leucogaster neumanni* may belong to this species, or may be specifically distinct; possibly some introgression between *anselli* and *neumanni* in DRC (Louette 1986). Monotypic.

FIELD IDENTIFICATION 46–58 cm. Exceedingly difficult to see despite large size. Sexes similar. **Adult** Black crown, sides of neck and head, and back; wings chiefly dark chestnut; underparts tawny; broad tail black. **Juvenile** Dark brown above with pale shaft streaks to head and upper back; wings rufous barred black; below rufous-buff, throat to upper breast narrowly barred dark brown. **Similar species** From similar-sized Black-throated Coucal by entirely fulvous underparts, and subadult from same stage of that species by lack of blackish throat, and generally darker below. Rather larger than Blue-headed Coucal which is all whitish to pale buff below. Juveniles similar to *leucogaster neumanni*, but head browner, chin grey barred buffy-white, throat to breast rufous-buff barred dark brown, as opposed to throat to upper breast blackish finely streaked buffy-white.

VOICE Call 9–12 loud, descending deep notes; melancholy bass *ouh ouh ouh ouh…* like Black-throated Coucal. Female has deeper voice than male. Trill more rapid and modulated than in other large African coucals, starting at 7–8 notes/sec slowing to five near end. In duet one bird trills while other gives longer notes; very similar to *leucogaster. Ukuk…… ukuk.* Often higher pitched and more varied than congeners, including more drawn-out syllables and series of rapid yelping notes. Calls at dawn and dusk (Chapin 1939, Heinrich 1958, Chappuis 1974, 2000, Dowsett & Dowsett-Lemaire 1993, Christy & Clarke 1994, Borrow & Demey 2001).

DESCRIPTION Adult Crown, lores, ear coverts, face and sides of neck, and mantle black glossed purple; rufous-buff back barred blackish; lower back and rump buffy-white barred black; long, disintegrated uppertail-coverts black barred buff; dark brown inner secondaries, remaining remiges dark chestnut with large blackish tips and washed bronze; underwing-coverts tawny. Tawny or rufous-buff below, including undertail-coverts which are broadly barred black; paler on centre of belly. Tail blackish-bronze glossed green; central rectrices (sometimes others) narrowly barred buff particularly proximally. One from Cameroon had few black feathers with white bases on throat (Louette 1986). **Subadult** (Non-breeding birds in first year only). Forecrown orange-buff streaked black with buff shaft streaks; rest of head to upper back dark brown with faint pale shaft streaks; face dark brown streaked buff; throat to belly rufous-buff narrowly barred dark brown particularly on flanks and undertail-coverts; wings and wing-coverts rufous barred black; inner secondaries dark brown, rest of remiges dark chestnut; tail as juvenile. **Juvenile** Differs from adult by black areas more brownish unglossed; forehead buff in some; yellowish buff shaft streaks on head and breast; chin grey barred or streaked buffy-white, throat to breast rufous-buff barred or streaked dark brown; rest underparts rufous-buff variably barred or streaked black, centre of belly unbarred buff; undertail-coverts, lower back and rump buffy-white barred black; wings and wing-coverts rufous barred black; uppertail-coverts and tail black barred buff-white. **Nestling** No information. **Bare parts** Iris red in male; female reddish-brown, some with whitish outer ring; juvenile greyish-brown. Bill black; juvenile dark grey above, greenish-black below. Feet black, juvenile slate to bluish-grey.

BIOMETRICS Wing male 178–197mm (n = 11), mean 186.0 ± 7.0, female 180–209mm (n = 14), mean 194.8, ± 11.8. Bill male 30–37mm, mean 33.1 ± 1.9, female 31–38, mean 34.8 ± 2.2. Tail male 242–302mm, mean 266.2 ± 24.5, female 280–380mm, mean 302.2 ± 22.7. Tarsus male 40–48mm, mean 44.2 ± 2.2, female 45–52mm, mean 48.9 ± 2.3. Mass male 210g (Payne 2005). Wing up to 200mm male, 212mm female (Irwin 1988).

MOULT No information.

DISTRIBUTION Afrotropical resident. South Cameroon, south-west CAR and Gabon to RC and central DRC (east along Congo R. to Isangi, south to Lusambo); north-west Angola (Chapin 1939, Good 1952, Louette 1981a, 1986, Irwin 1988, Dowsett 1989, Green & Carrroll 1991, Payne 1997, 2005, Dean 2000).

HABITAT Primary forest, undergrowth in swampy forest, forest edge along rivers, secondary growth, old cultivation and grassy swamps (Chapin 1939, Irwin 1988, Payne 2005).

BEHAVIOUR Terrestrial; territories of few hectares. Very skulking; calls mostly morning and evening. When giving long call, head lowered and throat puffed out. Feeds chiefly on or near ground. Scavenges around camps and human habitation (Chapin 1933, Mackworth-Praed & Grant 1970, Irwin 1988, Payne 1997).

BREEDING Season Begins with rains, or in short dry season in wettest forest regions. Lays Dec–Feb, north-east Gabon; young Nov, Uelle, DRC; ready to lay Mar, Angola (Chapin 1939, Heinrich 1958, Brosset & Érard 1986, Payne 1997). No further information.

FOOD Grasshoppers, katydids, beetles; molluscs; frogs; lizards, small snakes; small birds, eggs and nestlings (Bates 1930, Bannerman 1933, Mackworth-Praed & Grant 1970, Brosset & Érard 1986, Irwin 1988, Payne 1997).

STATUS AND CONSERVATION Common south Cameroon, south-west CAR, Equatorial Guinea, Gabon and RC (Borrow & Demey 2001). Population density *c.* 3–5 pairs/100 ha Gabon, where territories mutually exclusive to those of Blue-headed Coucal (Brosset & Érard 1986). Not globally threatened (BirdLife International 2011).

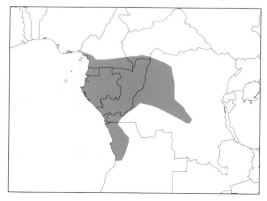

BLACK-THROATED COUCAL
Centropus leucogaster **Plate 9**

Polophilus leucogaster Leach, 1814 (New Holland; in error for Ghana)

Alternative name: Great Coucal.

TAXONOMY Could be considered allospecies with *anselli*, and mtDNA analysis places these as sister taxa, forming sister clade to *senegalensis, monachus, cupreicaudus* and *superciliosus* (Sorenson & Payne 2005). *Neumanni* Alexander 1908 has been considered closer to *anselli*, and Louette (1986) suggests it is a separate species based chiefly on size, but there is overlap, and some confusion due to specimens not being sexed. Some introgression between *neumanni* and *anselli* in DRC (Louette 1986). Polytypic. Synonym: *leucogaster var chalybeiceps* Reichenow 1902 in nominate.

FIELD IDENTIFICATION 46–58cm. Largest African coucal. Sexes similar. **Adult** Black glossed head to mantle and upper breast; rufous-chestnut wings; lower back to uppertail-coverts black barred buffy-white; lower breast to under tail-coverts white; tail black barred buff basally. **Juvenile** Lacks gloss on head and breast; buff-white streaking and spotting above; wings, wing-coverts and scapulars rufous barred blackish-brown; rump, upper tail-coverts and tail barred buff; throat and upper breast black spotted buff; rest of underparts unbarred fulvous to rufous-buff. **Similar species** Only coucal with black-and-white underparts (black and fulvous in juvenile), distinguishing it from similar sized Gabon Coucal. Juveniles of both species extremely similar (Louette 1986).

VOICE 'Dove' call deep, slow, ventriloquial and resonating bubbling of 8–20 *boo* notes 4/sec, starting fast, slowing and falling in pitch, then rising in pitch and speeding up towards end in 'water-bottle' call. Deep and booming; deeper than Senegal Coucal. Voice very similar to that of Gabon Coucal. Young birds give soft *coo* every 10–15sec. Sometimes calls at night. Series of 10–20 deep, resonant *hoots*, similar to Senegal Coucal, but lower and slower, not accelerating at end. Pairs duet at similar pitch but one almost twice as fast (Chapin 1939, Irwin 1988, Dowsett & Dowsett-Lemaire 1993, Payne 1997, Borrow & Demey 2001).

DESCRIPTION Nominate. **Adult** Head, neck and mantle black, feathers broadly tipped deep glossy violet-blue particularly on nape and mantle; mantle feathers broad and scallop-shaped, not lanceolate as other African coucals. Back, rump and uppertail-coverts black with buff barring. Deep chestnut wings; coverts and scapulars darker; dark rufous inner secondaries and tips to all remiges. Chin to upper breast black, feather tips deep glossy green. Lower breast and belly white, feather tips buff; flanks, thighs and undertail-coverts pale rufous. Tail black glossed greenish-blue, often barred buff, particularly basally. Variant has bright buff breast becoming paler on belly and vent (Rossouw & Lindsell 2001). **Juvenile** As adult but lacks gloss on head and breast; buff-white shafts to feathers, and some on mantle have terminal shaft spots; remiges, wing-coverts and scapulars rufous barred dark brown; throat and upper breast black with small buff spots; rest of underparts unbarred fulvous to rufous-buff, not white; rump, upper tail-coverts and tail dark brown narrowly barred buff. **Nestling** Dense white trichoptiles over upper surface; tongue has black 'U'-shaped mark (Chapin 1939); when well-feathered resembles juvenile, but head, neck and throat feathers tipped with 18–20mm horn-white trichoptiles. **Bare parts** Iris red; juvenile grey or reddish-brown circled with bluish slate. Bill black, lower mandible horn in juvenile. Feet black to blue-grey.

BIOMETRICS Nominate. Wing male 182–207mm (n = 8), mean 190.3 ± 7.4; female 187–213mm (n = 8), mean 198.4 ± 8.6. Bill male 31–46mm, mean 35.1 ± 5.0; female 34–41mm, mean 37.1 ± 2.6. Tail male 244–320mm, mean 276.5 ± 23.0; female 262–310mm, mean 306.0 ± 20.0. Tarsus male 39–46mm, mean 45.1 ± 3.6; female 46–54mm, mean 51.2. Also wing female (n = 6) 197–223mm (Payne 2005). Mass (*efulenensis*) male 293g, female 327, 346g; male 270g (Eisentraut 1963, BMNH).

MOULT Juvenile to adult moult Cameroon, Dec, primary moult Feb; juvenile to adult moult SL, Feb; primary moult IC, Nov (BMNH).

GEOGRAPHICAL VARIATION Three subspecies.
C. l. leucogaster (Leach, 1814). South Senegal and Guinea-Bissau to south Nigeria; south Mali (Lamarche 1980) but questioned (R.J. Dowsett in Lachenaud 2003); reported Niger (Debout *et al.* 2000) but considered doubtful (Lachenaud 2003). Described above.
C. l. efulenensis Sharpe, 1904. South-west Cameroon, intergrading with nominate at Yenagoa and Umuagwu, south-east Nigeria (Marchant 1953, Serle 1957). Reported Gabon (Bannerman 1933) but rejected by Dowsett & Dowsett-Lemaire (1993). Head, neck and mantle glossed green; inner secondaries olive-brown; wings of juvenile barred blackish-brown. Wing male (n = 8) 177–200mm, mean 189; female (n = 6) 195–210mm, mean 198 (Irwin 1988). Louette (1986) suggests not separable from nominate.
C. l. neumanni Alexander, 1908. North-east DRC north of equator on Mbomou R., Uele R. and right bank of Congo R. (Payne 2005), once Semliki NP, Uganda (Rossouw & Lindsell 2001). Smaller than *efulenensis*, and separated from it by 1,000km (Irwin 1988). Some have few pale feathers on throat suggesting introgression with Gabon Coucal (Louette 1986). Wing male 174, 180, 182mm; female 176mm (Payne 2005); also wing unsexed 165–198.5mm (n = 36), mean 180 (Louette 1986).

DISTRIBUTION Afrotropical resident. South Senegal, south Mali, Guinea-Bissau, Guinea, SL, Liberia, IC, Togo, Nigeria, west Cameroon, Gabon (Ogwe R.), northern DRC (Oriental Province from equator to Uelle R., and 21–30°E) (Irwin 1988, Demey 1995, Cheke & Walsh 1996, Demey & Rainey 2004, Payne 2005).

HABITAT Dense undergrowth in rainforest or savanna. Prefers secondary growth, gallery forest, forest edge, partially cleared forest, secondary growth and rank grass, generally avoiding forest interior; often along streams and at swamp edge; lowlands to 1,000m; montane ravines in Liberia (Chapin 1939, Eisentraut 1973, Irwin 1988, Gatter 1997).

BEHAVIOUR Solitary or paired. Generally shy and skulking. Puffs out feathers when calling, with head bent forward and downward to touch breast. Feeds on or near ground (Irwin 1988, Payne 1997, 2005).

BREEDING Season SL, Dec; Liberia, Nov–Feb, young attended by parents Mar and independent Apr, young just left nest Dec; Ghana, Aug (laying) and Oct; Cameroon, Jun–Nov (enlarged ovary Nov); Nigeria, Jun and Aug in

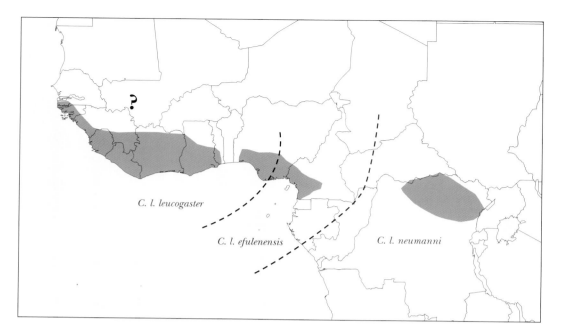

south; north DRC (Uelle and Ituri), Mar–Dec, probably breeding almost all year; sings throughout year Yapo Forest, IC (Chapin 1939, Serle 1957,1965, Mackworth-Praed & Grant 1970, Grimes 1987, Irwin 1988, Demey & Fishpool 1994, Gatter 1997, BMNH). **Nest** Ball of dry leaves and grass lined with green leaves, c.30cm diameter, up to 30cm up in bush or long grass (Chapin 1939). **Eggs** 2; dirty white; elliptical; 38x28mm, but size doubted; female in Cameroon had one yolked egg and four recently ruptured follicles (Chapin 1939, Serle 1965, Mackworth-Praed & Grant 1970, M.P. Walters in Payne 2005). **Incubation** By male (and female?) (Chapin 1939). No further information.

FOOD Caterpillars, Hemiptera, mantids, beetles, grasshoppers and other insects; spiders; snails; frogs (Bannerman 1933, Chapin 1939, Irwin 1988, Gatter 1997).

STATUS AND CONSERVATION Not uncommon in much of west Africa (Borrow & Demey 2001). Uncommon Ghana (Grimes 1987); rare Togo (Cheke & Walsh 1996); common Ouari Maro FR, Benin (Demey 2009). Not globally threatened (BirdLife International 2011).

Black-throated Coucal *Centropus leucogaster*. **Fig. 1.** Adult, *C. l. leucogaster*, Aboabo, Ghana, December (*Ian Merrill*).

SENEGAL COUCAL
Centropus senegalensis   Plate 10

Cuculus senegalensis Linnaeus, 1766 (Senegal)

Alternative name: Rufous-bellied Coucal ('*epomidis*').

TAXONOMY MtDNA analysis places this species as sister to *monachus*, *cupreicaudus* and *superciliosus* (Sorenson & Payne 2005). Dark rufous morph '*epomidis*', from Ghana and coastal Nigeria, interbreeds freely with white-bellied morph. Polytypic. Synonyms: *epomidis* Bonaparte 1850, *tschadensis* Reichenow 1915 and *incertus* Granvik 1923 in nominate.

FIELD IDENTIFICATION 36–42cm. Medium-sized coucal. Sexes similar. **Adult** Glossy black crown to nape, back and wings chestnut, underparts white. **Juvenile** Crown brown or grey, rest of upperparts, including wings, barred chestnut and blackish; rump blackish barred buff; black tail faintly barred buff at tip; underparts buff to cinnamon streaked and barred dark. **Similar species** From Coppery-tailed and Blue-headed Coucals by smaller size and bill, redder on back, and blacker, less shiny crown. From 'Burchell's' Coucal (*superciliosus burchelli/fasciipygialis*) by plain, unbarred uppertail-coverts. Dark rufous morph in west Africa from Black Coucal by rufous belly.

VOICE Calls mostly in breeding season, although duetting occurs throughout year, and often at night. Deep 'dove' call of up to 25 spaced booming notes/6sec, falling in pitch then rising at end. Pitch varies, generally higher than larger coucals, but can be deep and close to call of Blue-headed or White-browed Coucal. Often duets, with deeper-voiced bird, perhaps male, initiating; faster tempo than White-browed Coucal. 'Water-bottle' call consists of 15–20 fast tooting or stuttering notes *but-ut-ut…* over 8–10sec, on even pitch, becoming slower towards end. Series of deep *coo* notes; short hoot; sharp *guk, guk* if disturbed at nest, or in alarm or excitement; harsh call. *Hoo-hoo-hu-hu-hu-hu-hu-hu-hu-hu* and *tock-tock-toc-toc-toc-toc-toc-toc-toc-toc*; pairs closely duet. Series of deep *coo* notes, first two often shorter than others, 3–4/sec, sometimes dropping in pitch towards end. Nestlings give nasal wheezing sound (Chapin 1939, Maclaren 1950, Chappuis 1974, 2000, Irwin 1988, Stjernstedt 1993, Christy & Clarke 1994, Payne 2005).

DESCRIPTION Nominate. **Adult** Forehead to nape glossy blue-black; black ear-coverts; back dark rusty-brown to chestnut, rump and uppertail-coverts dark olive-brown; wings chestnut or rusty, primaries tipped brown. Below creamy-white, with stiff yellowish feather shafts on breast, flanks barred blackish, underwing-coverts light rufous, undertail-coverts buff. Tail blackish with dull greenish gloss, although some lack gloss on head and tail. **Juvenile** From adult by brown or grey crown, rest of upperparts including wing-coverts narrowly barred chestnut and blackish; rump blackish with buff barring; primaries narrowly barred with black on distal third, secondaries progressively more barred, inner ones like back. Buff below, more cinnamon on breast with darker shafts, barred brown; centre of belly pale with little or no barring. Tail black barred whitish to buff. **Nestling** Leathery black above, dark pink below; covered with long, stiff, trichoptiles. Pink inside mouth, greyer towards base; dark red tongue with glossy black line across tip and small forward-projecting white papillae. **Bare parts** Iris red, female orange-red (sometimes yellow in west Africa); juvenile olive-brown to yellow-brown; nestling grey.

Bill black (pale pink with prominent egg-tooth in nestling). Feet blackish-grey to slate-grey, grey in nestling.

BIOMETRICS Nominate. Wing male 150–176.5mm (n = 9), mean 159.7 ± 7.5; female 155–185mm (n = 10), mean 164.3 ± 10.0. Bill male 28.6–31.8mm (n = 9), mean 29.6 ± 1.7; female 27–33.4mm (n = 10), mean 30.5 ± 2.0. Tail male 179–195mm (n = 9), mean 188.9 ± 11.1; female 172–224mm (n = 10), mean 195.7 ± 16. Tarsus male 36.1–40mm (n = 9), mean 37.7 ± 1.7; female 36.5–42mm (n = 10), mean 38.3 ± 1.6 (Payne 2005). Mass 152g (Irwin 1988).

MOULT Juvenile moults progressively into adult plumage, barred tail changing first, then crown and nape becoming plain olivaceous brown; wings and wing-coverts do not lose barring until crown has become black. Post-breeding moult complete; in Egypt Jun–Nov, occasionally to Jan (Cramp 1985, Irwin 1988, BMNH).

GEOGRAPHICAL VARIATION Three subspecies.

C. s. senegalensis (Linnaeus, 1766). Mauritania, Senegambia and Guinea east through Nigeria and Chad to Sudan, Uganda, west Kenya, Ethiopia and north-west Somalia, and southwards to north Angola and south-central DRC (Kasai northwards). Head, nape and upper mantle glossed green; uppertail-coverts and tail purplish to olive-brown glossed green. Smaller than *flecki*. In higher rainfall areas in west Africa morph '*epomidis*' occurs, constituting about half of coastal Nigerian population, decreasing inland and not known more than from 200km from coast supporting Gloger's Rule, and to 500km inland in IC; forested zone d'Adiopodoumé south of Korhogo, IC; Liberia, Ghana; possibly Niamey, Niger. Chin, throat, crown and breast dark brown to black with green gloss, belly, flanks and undertail-coverts uniform dark rufous. Mixed pair nested Ibadan, Nigeria (Elgood 1955, 1973, Brunel & Thiollay 1969, Debout *et al.* 2000, Lachenaud 2003). Some birds near Lake Chad ('*tschadensis*') paler with greyish-brown crown and pale rufous-brown back (Louette 1986, Irwin 1988, Payne 1997, Demey & Rainey 2004). Described above.

C. s. aegyptius (J.F. Gmelin, 1788). Lower Egypt from Faiyum and Nile south to El Minya (28°N). From smaller nominate by sootier crown and hindneck merging into rufous mantle; mantle, wing-coverts and inner secondaries darker, duller brown with almost no rufous. Wing male 170–183mm (n = 16), mean 175; female 177–187mm (n = 13), mean 183 (Irwin 1988).

C. s. flecki (Reichenow, 1893). South-east Angola, northern Namibia and north Botswana, Upemba and Shaba (DRC), Zambia, Malawi, Zimbabwe, west Mozambique and parts of Tanzania. From nominate by lack of blue gloss on head and neck. Wing male (n = 32) 163–176mm, mean 169; female (n = 32) 170–186mm, mean 176) (Irwin 1988).

DISTRIBUTION Chiefly Afrotropical. Non-migratory, or locally migratory (as in arid regions of Mali and Kordofan, Sudan). Senegambia, Mauritania (Sahel, rarely north of 17°N), Burkina Faso and Guinea-Bissau, including Bijagos Archipelago, throughout IC from Basse Côte to extreme north, to north-west Angola, CAR, RC, central DRC, west Kenya; Sudan south of 15°N; Ethiopia and north-west Somalia; Tanzania, Malawi, Zambia, south Mozambique and Zimbabwe to north Botswana and north-east Namibia; Egypt north of 29°N (Faiyum, Nile and Delta), recorded three times south to 24°N; in wet season southern edges of Sahara in north Mali and Western Kordofan, Sudan, north to Tazza Well, Niger, and Aïr, central Sahara (Lynes 1925, Chapin 1939, Bannerman 1951, Brunel & Thiollay

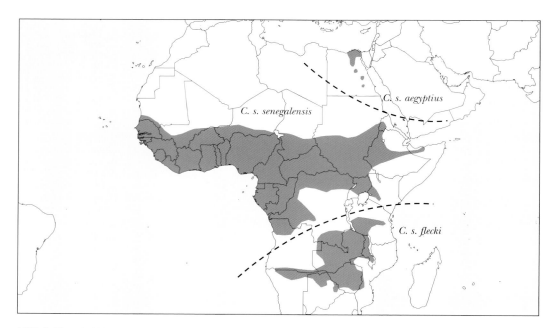

1969, de Naurois 1969, Benson & Benson 1977, Snow 1978, Lamarche 1980, 1988, Nikolaus 1987, Irwin 1988, Dowsett 1989, Dowsett-Lemaire *et al.* 1993, Germain & Cornet 1994, Rodwell 1996, Isenmann *et al.* 2010).

HABITAT Long grass and thickets in savanna, swamp, forest edges and clearings, elephant grass, sugarcane plantations (Zimbabwe); palm groves and rank vegetation along canals and Nile (Egypt); edges of reedbeds, papyrus, dense riverine bush, scrub and gardens. Borders of dry forest (southwest Tanzania). Less associated with wet areas than other African coucals, but near water in Northern Territories, Ghana. Competes with White-browed Coucal which occurs in more humid areas, but in south savannas latter found in drier habitats (Lowe 1937, Chapin 1939, Ripley & Heinrich 1966b, Rowan 1983, Irwin 1988, Lewis & Pomeroy 1989, Maclean 1993, Christy & Clarke 1994, Cheke & Walsh 1996, Zimmerman *et al.* 1996, Payne 1997, 2005).

BEHAVIOUR Shy; creeps about in dense cover; rather cumbersome when walking and hopping; sometimes runs after prey. Takes cover in trees, sometimes in crowns, whereas White-browed Coucal generally skulks in long grass. Poor flier, with rapidly beating wings, often launching itself from tall vegetation and gliding. May crash into vegetation on alighting. Often perches in tree and swoops on prey. Can swim, and will wade in water to catch frogs. Suns itself by fluffing up feathers and spreading and drooping wings and spreading tail. Calls with downward pointing tail, bill touching breast, vibrating wings and body, and sides of throat inflating and deflating. Most active early morning, and late afternoon onwards. Sometimes 'ants'. Rarely drinks; gular flutters; scratches head under wing. Will kill all birds in mistnets before feeding. Harasses snakes, e.g. Boomslang *Dispholidus typus*. Takes insects from animal dung and edges of grass fires (Bannerman 1933, Hamling 1937, Rowan 1983, Irwin 1988, Payne 1997, 2005, C.F. Mann).

BREEDING Monogamous and territorial; territory *c.* 6ha (Payne 2005). **Nest** Domed, of grass, roots, twigs and leaves, lined with fresh leaves, and with substantial base *c.* 45x25–40cm, with high side entrance; from ground level to 10m in dense thorny bush or tree; in Gambia used crown of citrus tree as natural roof (Granvik 1923, Lynes 1925, Bannerman 1933, 1951, Chapin 1939, Steyn 1972, Tarboton 2001). **Season** Generally in rains. Egypt, Mar–Aug, eggs Apr; Senegambia, Feb, May–Nov; north Senegal, May–Oct and perhaps other months; Gambia, Jul–Nov; SL, Sep–May, mainly before and after main rains; Liberia, Mar–Apr; nest south Guinea, Feb; carrying nest material north Burkina Faso, Aug; nestlings Togo, May, fledglings Jul and Oct, adult carrying food Jul; Mali, Jul–Oct; south Ghana, Mar–Jun; north Ghana Jul–Aug; Nigeria, mainly Mar–Aug, young in nest north Nigeria 1 Jul; Sudan, Mar, Aug–Oct; DRC, Oct–Feb in south-east, Sep–Nov in Uelle, Aug–Nov in north; fledgling Sudan, Jul; fledglings Somalia, Oct; Ethiopia, Apr–Aug; north and east Uganda, west Kenya, Mar–May; Malawi, Nov–May; Zambia, Oct–May; Zimbabwe and Botswana, Oct–Mar (Reichenow 1902, Lynes 1925, Chapin 1939, Serle 1943, 1957, Bannerman 1951, Etchécopar & Hüe 1967, Mackworth-Praed & Grant 1970, Urban & Brown 1971, Benson & Benson 1977, Greig-Smith 1977a, Brown & Britton 1980, Irwin 1981, 1988, Louette 1981b, Anon. 1985, Grimes 1987, Nikolaus 1987, Goodman *et al.* 1989, Gore 1990, Morel & Morel 1990, Elgood *et al.* 1994, Halleux 1994, Cheke & Walsh 1996, Skinner 1996, Balança & de Visscher 1997, Gatter 1997, Payne 1997, Vernon *et al.* 1997, Ash & Miskell 1998, BMNH). **Eggs** 2–5 (usually 3 laid over four days, often commencing before nest completed); thick-shelled, chalky-white becoming stained, slightly glossed; 30.0–38.0x24.0–26.7mm (n = 300) mean 33.1x25.4. Probably multi-brooded (Granvik 1923, Bannerman 1933, Irwin 1988). **Incubation** By male, starting with first egg; 18–19 days with hatching asynchronous (Steyn 1972). **Chicks** Cared for by both sexes. Quills appear 4th day; eyes open 6th day; down above and below 9th day; feather-sheaths burst by 12th day; fledges 15–20th days, well before they can fly, and may leave nest earlier; smallest chick may be ejected by parent; from 6th day excrete nauseating liquid in defence, as well as normal encapsulated faeces; *flecki* mass 145g on 18th day (Steyn 1972, Payne 1997). **Survival** No information.

FOOD Grasshoppers, distasteful bush locust *Phymantous viridipes* avoided by other animals, ants, *Bellicositermes* termites, beetles, bugs, caterpillars; centipedes; crabs; snails; frogs; small snakes, and lizard *Agama agama* swallowed whole head first, after hammering on ground; small birds up to size of doves, including Red-billed Quelea, nestlings, eggs; rodents. Scavenges dead fish. Takes insects from dung of large herbivorous mammals (Hamling 1937, Chapin 1939, Dekeyser 1956, Serle 1957, Brotherton 1965, Steyn 1970, Rowan 1983, Ewbank 1985, Irwin 1988, Payne 1997, Goodwin 2001).

STATUS AND CONSERVATION Rare east of 35°E; generally common, although less so where in competition with White-browed Coucal (Irwin 1988). Abundant SL, except in tall forest. Expanded range during 20th century in Egypt, and now quite common in Cairo, Rosetta Nile, Nile Delta and lower valley (Goodman *et al.* 1989). Abundant Macenta Prefecture, Guinea (Halleux 1994). Rare on outer edge of forest, Yapo Forest, IC (Demey & Fishpool 1994). Very common Togo (Cheke & Walsh 1996), southeast Nigeria (Marchant 1953) and Ghana (Grimes 1987). Fairly common Sudan (Cave & MacDonald 1955), later uncommon (Nikolaus 1987). Locally common Ethiopia (Ash & Atkins 2009). Locally common Malawi (Dowsett-Lemaire & Dowsett 2006). Not globally threatened (Birdlife International 2011).

Senegal Coucal *Centropus senegalensis*. **Figs. 1–2.** Adult, *C. s. aegyptius*, Al Abassa, Egypt, April (*Daniele Occhiato*). **Fig. 3.** Adult, *C. s. aegyptius*, Al Abassa, Egypt, April (*Daniele Occhiato*). **Fig. 4.** Adult, *C. s. senegalensis*, Shonga, Nigeria, January (*John Sawyer*). **Fig. 5.** Juvenile, *C. s. senegalensis*, Shonga, Nigeria, September (*John Sawyer*). This individual has already acquired partial adult plumage. **Fig. 6.** Juvenile, *C. s. senegalensis*, Lake Bunyonyi, Uganda, March (*Tadeusz Rosinski*). **Fig. 7.** Juvenile moulting in to adult plumage, *C. s. flecki*, Chobe, Botswana, July (*Lyn Francey*).

BLUE-HEADED COUCAL
Centropus monachus **Plate 10**

Centropus monachus Rüppell, 1837 (Kulla, northern Ethiopia)

TAXONOMY Similar vocalisations to *cupreicaudus*, and has been considered conspecific, and there are possible intermediates, but overlap in distribution in Shaba Province, DRC (Louette 1986). MtDNA analysis places it as sister to *cupreicaudus* and *superciliosus* (Sorenson & Payne 2005). Polytypic. Synonyms: *heuglini* Neumann 1910, from Sudan, has been separated because of blue not violet head, but thought to be immature *fischeri* (Friedmann 1930), but is smaller, wing male (n = 10) 159–171mm, female (n = 5) 167–177mm (Payne 2005); smaller *verheyeni* Louette 1986, from Shaba, south-east DRC, overlaps with *fischeri*, and not accepted by Payne (2005), and var *nigrodorsalis* Reichenow 1902, all in *fischeri*; *m. angolensis* Neumann 1908 in *occidentalis*.

FIELD IDENTIFICATION 45–52cm. Medium-sized coucal. Sexes alike. **Adult** Black head to mantle glossed blue; rump black; wings mostly dark chestnut; whitish to pale buff below; tail black glossed bronze or greenish. **Juvenile** Head and hindneck dull blackish streaked rufous-buff; dark chestnut wings barred blackish-brown; rump and uppertail-coverts faintly barred buff; tail black barred buff; deeper buff below than adult, paler on belly; throat and breast feathers spotted blackish. **Similar species** From larger, thicker-billed Coppery-tailed Coucal by usually unbarred uppertail-coverts. From White-browed by unstreaked head. Larger and heavier-billed than Senegal with which it is often confused, and darker, less chestnut, above. Juvenile from juvenile Senegal by broader barring on wings extending to whole distal half of primaries, and fine buffy shaft streaks or spots on crown; breast rich buff.

VOICE Slow, deep, resonating and ventriloquial 'dove' and 'water-bottle' calls, up to 25 notes, beginning with two *coo* notes, lasting 6sec, deeper than other coucals; descending series of *toot* notes, rising at end very similar to Senegal. Gruff monkey-like note; also bark and raucous chuckle. Often duets (pair having different pitch but same rhythm). Series of 'toots' rising at end, very similar to Senegal. Slow, deep, resonant series of notes *hoo, hoo, hoo-wu-wu-wu-wu-wu hoo hoo, hu,* first two short and separated by pause from rest of notes which are slightly lower and given at rate of 4/sec (Bates 1930, Chapin 1939, Chappuis 1974, 2000, Irwin 1988, Christy & Clarke 1994, Zimmerman *et al.* 1996, Borrow & Demey 2001).

DESCRIPTION Nominate. **Adult** Forehead to mantle black with blue gloss; lores and ear-coverts dull back; back dark chestnut-brown; rump and uppertail-coverts dull black, latter sometimes barred; wings dark chestnut, with inner secondaries and tips to primaries dark olive-brown; underwing-coverts buff; whitish to pale buff below, darker on flanks; tail black glossed greenish or bronze, with narrow white tips when fresh. **Juvenile** Pale rufous-buff shaft streaks to dull blackish crown and hindneck; broad black or very dark brown barring on wings; inner secondaries pale brown; rump and uppertail coverts faintly barred buff; deeper buff below than adult, but paler on belly, some throat feathers tipped dark, breast with small black spots; tail black, all except central rectrices barred, although these may be barred at tip. **Nestling** Black skin partially covered with long, white trichoptiles. **Bare parts** Iris dark red. Bill black. Feet dark grey to black.

BIOMETRICS Nominate. Wing male (n = 3) 188–220mm, mean 201.7, female (n = 14) 182–226mm, mean 198.6 ± 12.3. Tail male 218–237mm, mean 225.7, female 208–253mm, mean 229.3 ± 13.3. Bill (*occidentalis*) male (n = 8) 30–34mm, mean 31.8 ± 1.2, female 32–35mm, mean 33.0 ± 1.1. Tarsus (*occidentalis*) male (n = 8) 43–47mm, mean 45.3 ± 2.0, female (n = 6) 43–48mm, mean 44.5 ± 1.9 (Payne 2005). Mass male (n = 5) 163–177g, mean 171; female 206, 222, 283g (Irwin 1988).

MOULT Primary moult Nigeria, Jun and Dec, Cameroon, Jan and Nov, Kenya and Sudan, Feb. Juvenile to adult moult Cameroon, Jan, Oct and Dec, Ubangi R., Nov, Sudan, Jan and Mar, Ethiopia, Sep, Oct and Dec (BMNH).

GEOGRAPHICAL VARIATION Three subspecies.
C. m. monachus Rüppell, 1837. Highlands of Eritrea and Ethiopia to west and central Kenya. Described above.
C. m. fischeri Reichenow, 1887. Upper Nile basin in south Sudan (south of 14°N), Uganda, west Kenya in Lake Victoria basin, north-west Tanzania, Rwanda, Burundi, north-eastern and south-eastern DRC. Smaller, with back and wings less bright and clear chestnut, with more olive-brown on coverts than nominate, juvenile darker below. Wing male (n = 12) 171–182mm, mean 177, female (n = 10) 180–202mm, mean 189. '*Heuglini*' (south Sudan) wing male (n = 10) 159–171mm, mean 166.5 ± 4.7; female (n = 5) 167–177mm, mean 170.8 ± 4.1. Tail male 181–223mm, mean 197.8 ± 11.5; female 188–230mm, mean 190.8 ± 15.6 (Traylor 1960, Dowsett & Prigogine 1974, Pinto 1983, Irwin 1988, Germain & Cornet 1994, Payne 1997, 2005).
C. m. occidentalis Neumann, 1908. Mali, Guinea, Liberia, IC, Togo and Ghana through south Nigeria, Cameroon, CAR, Gabon, north Angola (Cuanza Norte, north Malanje) and extreme western RC and DRC. Back and wings rufous-brown; larger than *fischeri*; wing male (n = 8) 180–191mm, female (n = 6) 176–196mm, mean 185.5 ± 7.1 (Payne 2005).

DISTRIBUTION Afrotropical resident. Mali, Guinea to Nigeria, Cameroon, CAR, Gabon, RC, DRC and north Angola, south Sudan north to 14°N, Ethiopian and Kenyan highlands, Lake Victoria basin, Uganda and north-west Tanzania (Lynes 1925, Schouteden 1971, Britton 1980, Nikolaus 1987, Irwin 1988, Dowsett-Lemaire 1990, Germain & Cornet 1994, Payne 1997).

HABITAT Dense cover in swamps, marshes, papyrus, river banks, long grass, forest edges, mesic savannas near forest or water, secondary growth, tea plantations, old cassava fields, edges of villages, farmland with or without trees. West Cameroon to 1,200m; 1,000–2,000m in wetter areas with 1,000+mm annual rainfall in Kenya and at higher altitudes than Senegal Coucal; east Africa generally 700–2,000m; to 2,400m Kivu volcanoes, and to 2,500–2,600m in *Hagenia-Hypericum* in Rwanda. 1,500–2,420m Eritrea; 510–2,730m Ethiopia. Bush and scrub to 3,050m, south Sudan (Bannerman 1933, Chapin 1939, Cave & MacDonald 1955, Smith 1957, Eisentraut 1963, 1973, Britton 1980, Irwin 1988, Lewis & Pomeroy 1989, Dowsett 1990, Bowden 2001, Ash & Atkins 2009).

BEHAVIOUR Usually on ground. Stays in deep cover much of time, and although often flounders off on being disturbed, flight can be strong and direct. Kills small birds in mistnets. Puffs out feathers when calling, and pushes

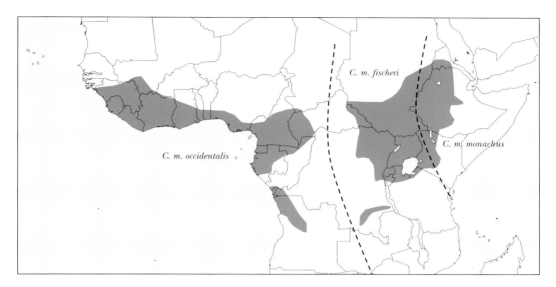

head forwards and downwards with bill touching breast (Irwin 1988, C.F. Mann).

BREEDING Probably monogamous and territorial; male attends nest when eggs present (Payne 2005). **Nest** Oval mass of dry grass and sedges, or sticks and dry leaves, bent over, lined with green leaves, with side entrance, well concealed in sedges, bushes or elephant grass, often in marshy areas, 0.2–3m above ground; or loose mass of dry sticks and leaves in fork of small tree (Bates 1909, 1930, Bannerman 1933, Chapin 1939, Brosset & Érard 1986, Kizungu 2000). **Season** Togo, 4 young Anécho, 5 Jul; Nigeria, Mar–Aug, eggs south-west, 26 Jun; Cameroon, Jan, Apr–Sep; north-east Gabon, Aug–Mar; DRC, Apr–May, in north (Uelle), early May–Nov, in Irangi, Dec, in south probably Nov–Mar; south Sudan, Jun, Aug; Ethiopia, possibly May, Dec; Uganda generally Mar, Jun; south and west Uganda, west Kenya, Lake Victoria basin, Feb–Jun, Sep; north and east Kenya, north-east Tanzania, Jun (Bates 1909, van Someren 1916, Millet-Horsin 1921, Chapin 1939, Serle 1950, 1965, Mackworth-Praed & Grant 1970, Brown & Britton 1980, Nikolaus 1987, Irwin 1988, Christy & Clarke 1994, Elgood *et al.* 1994, Cheke & Walsh 1996, Kizungu 2000, BMNH). **Eggs** 3; long oval with truncated pointed end; white becoming stained, lightly glossed with smooth soapy surface. 29.1–37.5x23.5–29.5mm (n = 13), mean 34.5x26.5; also 37.8–38.9x25.0–25.6mm (Chapin 1939, Serle 1965, Irwin 1988). **Incubation** Probably by male. No other information.

FOOD Grasshoppers, locusts, beetles and other insects; slugs, snails (including *Limicolaria*); tree-frogs; chameleons and other small lizards, small snakes, including venomous ones; birds' eggs, nestlings, small birds (*Bradypterus* and *Acrocephalus* warblers swallowed whole, including those shot by collectors and not retrieved immediately); mice, rats; 1.2m cobra killed by attacking head; slugs in captivity (Bates 1909, 1930, Lynes 1925, Chapin 1939, Verheyen 1953, Eisentraut 1963, Royston 1981, Irwin 1988).

STATUS AND CONSERVATION Generally common to uncommon western Africa (Borrow & Demey 2001). Very common Lake Victoria and Lake Kyoga basins, Uganda, and west of country, but less numerous Kenya (Irwin 1988, Payne 1997, Rossouw & Sacchi 1998). Rare central mountains Eritrea (Smith 1957). Common to very common Ethiopia (Ash & Atkins 2009). Common resident Sudan south of 14°N (Nikolaus 1987). Common around villages north-east Gabon (Christy & Clarke 1994); common grasslands and forest clearings, Cameroon (Serle 1950); rare south Togo (Cheke & Walsh 1996); common secondary forest in Irangi Forest in east (Kizungu 2000) and very common savannas in north and east DRC south to Marungu (Chapin 1939). Not globally threatened (BirdLife 2011).

Blue-headed Coucal *Centropus monachus*. **Fig.1.** Adult, *C. m. occidentalis*, Shonga, Nigeria, June (*John Sawyer*).

COPPERY-TAILED COUCAL
Centropus cupreicaudus   Plate 9

Centropus cupreicaudus Reichenow, 1896 (Okavangoland and south Angola)

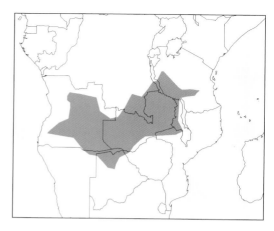

TAXONOMY Once considered conspecific with *monachus*, but overlap in Shaba Province, DRC (Louette 1986). MtDNA analysis indicates clade with *senegalensis*, *monachus* and *superciliosus* (Sorenson & Payne 2005). Smaller, darker, large-billed *songweensis* Benson, 1948, R. Songwe (north Malaŵi and Tanzania) overlaps with other populations in these characters. Monotypic.

FIELD IDENTIFICATION 42–50cm. Large coucal; deep bill. Sexes similar. **Adult** Black forehead to mantle glossed violet, chestnut wings, entirely white below, and coppery sheen to blackish tail. **Juvenile** Less gloss on head and neck, head streaked buff, rest of upperparts mostly dark brown; tail barred; some buff on breast. **Similar species** Larger than other coucals in area. Barred uppertail-coverts also separate it from most Blue-headed Coucals. Heavier bill, barred uppertail-coverts, dark brown trailing edge to wings noticeable in flight and darker upper back also distinguish it from Senegal Coucal. Lacks streaked head of White-browed Coucal.

VOICE Has 'water bottle' and 'dove' calls which are deep, resonant and loud, former 10–20 low pitched descending notes. Other calls include series of hoots on descending scale, and series of notes that begin by accelerating and then slowing down, which is perhaps 'dove' call. Pairs duet. Contact call, answered by mate, fast low *cou…cou… cou* repeated. Long series of *coo* notes on one pitch or descending. Most vocal early morning and evening; also calls at night (Vincent 1946, Chappuis 1974, Irwin 1988, Stjernstedt 1993, Payne 1997).

DESCRIPTION Adult Forehead to mantle black glossed violet; ear-coverts black; back reddish-brown; uppertail-coverts blackish-brown variably barred dull rufous; wings reddish-chestnut, with inner secondaries and tips to all remiges olive-brown. Underparts creamy-white, occasionally with buff on breast; undertail-coverts sometimes barred dark (retained juvenile feathers?). Tail blackish-brown with coppery sheen and indistinct barring towards base. **Juvenile** Differs from adult in having less violet gloss on head and neck, head with buff shaft streaks, and back, rump, scapulars, secondaries and secondary-coverts dark olive-brown; primaries tipped brown; inner remiges barred or flecked olive-brown; some buff on breast; tail barred. **Nestling** Blackish skin with whitish trichoptiles. Feathered nestling has unglossed dull bluish-black crown, remiges slightly barred dark brown towards tips, and some barring on chestnut of upperparts and dark brown mantle, otherwise immaculate; tail blackish-brown (Benson & Pitman 1964). **Bare parts** Iris scarlet. Bill and feet black.

BIOMETRICS Wing male 200–219mm (n = 25), mean 211; female 217–233mm (n = 23), mean 223; '*songweensis*' unsexed 195–203mm. Bill male 33–39mm (n = 25), mean 36; female 35–42mm (n = 24), mean 38. Tail male 223–244mm (n = 19), mean 233; female 227–246mm (n = 23), mean 235. Tarsus male 46–52mm (n = 25), mean 49; female 46–54mm (n = 24), mean 50. Mass male 250–293g (n = 6), mean 272; female 245–342g (n = 5), mean 299 (Irwin 1988). Also wing male 202–224mm; female 216–236mm (Payne 2005).

MOULT Jul, Oct and Nov, Malaŵi; remiges moulting Botswana, Dec, primary moult, May (Benson 1948, BMNH).

DISTRIBUTION Afrotropical resident, but local movements occur in response to rainfall, burning etc. Central and east Angola, extreme north-east Namibia, north Botswana, once north Zimbabwe (Hornbuckle 1997b), Zambia, once north-west Mozambique (USNM), north and central Malaŵi, south-west Tanzania (Rukwa north to Karema and Uhehe), south-east DRC (Schouteden 1964, Irwin 1988, Payne 1997, Vernon *et al.* 1997).

HABITAT Swampy savannas and flood plains in dense swamp and riparian vegetation, reeds, papyrus, long grass and bush; occasionally woodland. Lowlands to 1,250m (Irwin 1988, Hustler 1997, Payne 1997).

BEHAVIOUR Territorial throughout year. Perches on tops of reeds and papyrus. Most active early morning and evening. Puffs out feathers, bends head forward and down so bill touches breast when calling. Active predator, walking on ground, and carrying prey away in flight. Forages up to 1,000m from nest. Scavenges for dead fish and other foods in open. Takes weaver nestlings by breaking open nests (Irwin 1988, Hustler 1997).

BREEDING Probably monogamous. Both parents carry out nesting duties, but perhaps more by male (Payne 2005). **Season** DRC, Dec–Feb, eggs Lubumbashi, south DRC, 15 Feb; Angola, Feb, feathered nestlings 24 Mar; Zambia, Sep–Feb usually during rains, egg 3 Sep; Zimbabwe in rains, Jan–Apr; Botswana, Nov (Vincent 1946, Benson & Pitman 1964, Irwin 1988, Hustler *et al.* 1996, Hustler 1997). **Nest** Up to 35cm diameter; coarse dome of dried grass, reeds and twigs lined with leaves, with side entrance; low down in reed bed, often with runway. One nest in DRC 3m up in small thorn tree supported all around by twigs and foliage, attached by loops of grass, constructed of coarse grass with some *Mimosa* foliage at base; also sphere of fresh grass, coarse straws or twigs, lined with leaves, with side-entrance, sometimes with runway of flattened grass to entrance within 0.5m of water level (Vincent 1946, Payne 2005). **Eggs** 2–4, laid at intervals, sometimes before nest is complete; white, but become stained, oval, smooth and lightly glossed; 36.0–41.0x24.0–28.6mm (n = 12), mean 38.0x26.0; 38.0x26.7 (2), 38.3x26.9 (Vincent 1946, Benson & Pitman 1964, Keith & Vernon 1969, Irwin 1988, Payne 1997). **Incubation** Begins with first egg, sometimes before nest complete; chicks hatch at intervals. **Chicks** May leave nest when only half-grown, or at least when fully-feathered but unable to fly; fed on frogs and locusts. Wing-flutter and give continuous *chuck* calls when begging. Predated by monitors *Varanus spp.* and mongoose *Atilax paludinosus* (Irwin 1988, Maclean 1993, Hustler 1997, Payne 2005). **Survival** No information.

FOOD Grasshoppers and other large insects; crabs; snails; fish; frogs; snakes; lizards; *Ploceus* weaver nestlings, adult African Blue Quail swallowed whole; some vegetation. Crop contained only green grass and water weed, DRC (Vincent 1946, Irwin 1988, Payne 1997).

STATUS AND CONSERVATION Sparsely distributed but locally common, e.g. upper Zambezi R. (Irwin 1988, Payne 1997). Not globally threatened (BirdLife International 2011).

Coppery-tailed Coucal *Centropus cupreicaudus.* **Fig. 1.** Adult, Okavango, Botswana, June (*Robert van Zalinge*).

WHITE-BROWED COUCAL
Centropus superciliosus   Plate 10

Centropus superciliosus Hemprich & Ehrenberg, 1828 (Arabia and Ethiopia)

Alternative names: Lark-heeled Cuckoo, Burchell's Coucal (*burchellii/fasciipygialis*).

TAXONOMY On mtDNA evidence forms clade with *cupreicaudus*, *monachus* and *senegalensis* (Sorenson & Payne 2005). *Burchellii*, with black head and neck, and lacking supercilium, has been considered specifically distinct, but intermediates (hybrids?) between birds with supercilium and brown crown and neck, and those without supercilium and with black crown and neck, known from north Malawi, Zambia, Mozambique and Zimbabwe. Apparently more northerly forms retain into adulthood more juvenile type plumage (neotenous), and southern form *burchelli* has additional plumage that includes glossy black head and neck. Southern birds may breed in streaked plumage, sometimes only one of pair streaked (Lawson 1962, Payne 2005). *Fasciipygialis* from east Zimbabwe and Mozambique to east Tanzania has been synonymised with *burchelli*, but moult sequence different. Clancey (1989) separates *burchellii* (with *fasciipygialis*) at species level from *superciliosus*. Polytypic. Synonyms: *meridionalis* Madarász 1914, *intermedius* van Someren 1921 (said to be darker and smaller, wing 140–155mm, than *sokotrae*), *furvus* Friedmann 1926 (replaces pre-occupied *intermedius*) and *niloticus* Stolzmann 1924 in nominate; *pymi* Roberts 1914 in *burchelli*.

FIELD IDENTIFICATION 36–42cm. Medium-sized coucal. Sexes similar. **Adult** Rufous-brown back, chestnut wings, creamy-white underparts and black tail with white tips. Forms *loandae*, *superciliosus*, *sokotrae* and subadult *burchellii* have streaked heads and white supercilium; *fasciipygialis* and adult *burchellii* have unstreaked black crown to mantle and lack supercilium. **Juvenile** Blackish crown and ear coverts streaked rufous or tawny-buff; indistinct buff-white supercilium; above black barred rufous and buff; remiges rufous-chestnut with dark barring; narrow buff barring on greenish-black tail and uppertail-coverts barred buff; underparts buff-white streaked yellow, cream and black; flanks, legs and undertail-coverts buff-grey barred grey. **Similar species** White supercilium of *loandae*, *superciliosus*, *sokotrae* and subadult *burchellii* distinguish them from all other coucals; *fasciipygialis* and adult *burchellii* distinguished from slightly smaller Senegal Coucal by finely barred, not black, rump.

VOICE 'Water-bottle' call begins rather quavering, developing into "fast gulping or glugging sound falling in pitch", about 8–20 bubbling *coo* notes/sec, lasting 1–3sec, and increasing in tempo. 'Dove' call, slow series of *coo* notes, accelerating and falling in pitch like 'water-bottle' call, then decelerates and rises to resemble original notes, about 30 notes in all lasting 4sec. Hissing alarm; explosive *cluck* repeated few times; repeated hoarse call similar to that of francolin; buzzing note used by nestling when begging. Mostly calls in breeding season; pairs may duet and sometimes several birds call together. Sometimes one begins as another is finishing, often with some overlap. Voice not as low as Gabon Coucal. 'Tooting' call softer and lower than Senegal or Blue-headed, notes running together more rapidly. 'Water-bottle' call rapid series of bubbling *coo* notes, about 8/sec over 2–3sec, first note longer, and series often descends in pitch and accelerates; another slower, more deliberate bubbling series of notes that fall and rise, and accelerate and then slow down; single bubble call. Descending series more hurried than other coucals. Rapid chatter of 6–7 notes/sec when excited. Young in nest make hissing wheezing noise when disturbed (van Someren 1916, Chapin 1939, Gibbon 1951, North 1958, Stannard 1966, Chappuis 1974, 2000, Gillard 1987, Irwin 1988, Stjernstedt 1993, Christy & Clarke 1994, Skead 1995, Stevenson & Fanshawe 2002, Payne 2005).

DESCRIPTION *C. s. loandae*. **Adult** Forehead, crown and ear coverts blackish-brown, sometimes dusky olive-brown, with broad white supercilium; nape, sides of neck and mantle blackish streaked whitish or cream; back rufous-brown; rump and uppertail-coverts blackish narrowly barred buff. Primaries chestnut with olive-brown tips; outer secondaries chestnut, inner secondaries and tertials olive-brown to chestnut-brown; lesser wing-coverts red-brown streaked cream; rest of coverts chestnut. Below creamy-white narrowly streaked cream, whiter on centre of belly; sides of breast have fine blackish flecks; flanks have variable fine buff and brown barring; buff undertail-coverts with narrow blackish bars; underwing-coverts rufous-buff to chestnut. Tail black glossed green narrowly tipped white. **Subadult** (*burchellii*) Pale buffish supercilium, bars on remiges and rectrices gradually replaced, and hindneck, upper back and scapulars with straw-coloured streaks, referred to as '*loandae*' stage, apparently not found in other forms. **Juvenile** From adult by rufous or tawny-buff streaking on blackish crown and ear coverts; indistinct buff-white supercilium; back barred rufous and black; rump and uppertail-coverts blackish barred buff. Rufous-chestnut remiges barred dark brown to black distally, heavier on secondaries; wing-coverts rufous-chestnut barred brown. Below buff-white streaked yellow on throat and breast, paler on centre of belly; sides of neck and breast have fine cream and black serrated streaks; flanks, leg feathering and undertail-coverts buff-grey irregularly barred grey; greenish-black tail and uppertail-coverts narrowly barred buff. **Nestling** Hatchling black, with head and back covered with long, stiff cream trichoptiles. **Bare parts** Iris red, juvenile brown to red. Bill black. Feet blue-grey to black.

BIOMETRICS Nominate. Wing male (n = 6) 135–152mm, mean 145.8 ± 7.5; female (n = 6) 148–158mm, mean 154.2 ± 4.2. Bill male 27.7–29.3mm, mean 28.8 ± 0.6; female 28.6–33.3mm, mean 31.3 ± 1.8. Tail male 176–200mm, mean 189.3 ± 8.5; female 185–207mm, mean 197.3 ± 7.5. Tarsus male 28.9–38.0mm, mean 33.3 ± 3.7; female 33.7–38.1mm, mean 36.0 ± 1.5 (Payne 2005). Mass nominate male (n = 6) 110–140g, mean 124; female (n = 7) 125–170g, mean 136; *loandae* male (n =14) 142–213g, mean 180, female (n = 10) 153–212g, mean 180 (Irwin 1988); *burchellii* female 159g (Payne 2005); *sokotrae* (4 male, 3 female) mean 100g (Ripley & Bond 1966).

MOULT Series of moults to reach adult plumage not fully understood. Nape and mantle become streaked, with bars on tail being lost, and later those on remiges. Adult plumage may not be completely attained, particularly in black-headed forms. Subadult plumage of *burchellii*, with pale buffish supercilium, bars on remiges and rectrices gradually replaced, and hindneck, upper back and scapulars with straw-coloured streaks, referred to as '*loandae* stage'; *fasciipygialis* lacks '*loandae* stage' between juvenile and adult (Irwin 1988). Primary moult RSA, Apr; juvenile to adult moult south-east DRC, Jan (egg laying), Malawi, Apr (with enlarged testes) and Oct, Kenya, Feb, Uganda, Jan, Sokotra, Mar, Tihama, Jan, Zambia, May (BMNH); moult Malawi mostly Jan–Jun (Hanmer 1995).

GEOGRAPHICAL VARIATION Five subspecies.

C. s. superciliosus Hemprich & Ehrenberg, 1828. East Sudan along Nile north to 19°N, Eritrea, Ethiopia and Somalia through Kenya and north-east Uganda to north-east Tanzania and north-east DRC (Rutshuru, Kivu Province) (Chapin 1939, Louette 1986, Payne 2005). From larger *loandae* by paler, less blackish crown; paler and duller, less reddish-chestnut upperparts; only centre of belly white; juvenile paler, crown streaked rufous-buff.

C. s. sokotrae C.H.B. Grant, 1915. Sokotra; east Tihamah region, Yemen (south of 19°N) and Jebal Bura eastwards to west Hadramaut (48°E); eastern Somalia? Not very distinct from nominate; lighter brown, and greyer on crown and nape, paler secondaries, underparts whiter, tail with greener and less bluish gloss (Meinertzhagen 1954, Ripley & Bond 1966, Brooks *et al.* 1987, Irwin 1988). Wing male (n = 6) 149–160mm, mean 155.2 ± 4.0; female 165, 168mm (Payne 2005).

C. s. loandae C.H.B. Grant, 1915. Uganda (except north), south-west Kenya and Tanzania (including Zanzibar and Pemba) through southern DRC, north Malawi and Zambia to Angola, north Botswana and north Zimbabwe (Irwin 1988); once Djéno, RC (Chapin 1939, Louette 1986, Dowsett-Lemaire *et al.* 1993). Population from north-east DRC apparently intergrade between this form and nominate and not separable (Louette 1986). Wing male (n = 6) 157–168mm, mean 161.5 ± 3.7; female (n = 10) 160–172mm, mean 167.4 ± 8.2 (Payne 2005); also wing male (n = 31) 155–176mm, mean 163; female (n = 28) 161–180mm, mean 167 (Irwin 1988). Described above.

C. s. burchellii Swainson, 1838. East Botswana, south Zimbabwe and south Malawi, south to RSA. South-east Zambian birds close to this form (Aspinwall 1975). Intergrades with *loandae* in Viphya Mts. and Karonga, Malawi (Benson & Benson 1977). Distinctive. Crown and ear coverts to mantle unstreaked glossy blue-black, and lacks supercilium (or trace in some); more buff below. Juvenile has blackish crown with dark rufous spots and feather shafts, darker rufous barring on back than *loandae*, black cheeks and no supercilium (or trace), pale rufous below, centre of belly whitish, and flanks indistinctly barred dark. In intermediate plumage has pale buff supercilium, brown crown, and blackish neck with straw-coloured streaks (Irwin 1988). Wing male (n = 5) 164–181mm, mean 170; female (n = 6) 163–184mm, mean 172 (Lawson 1962); also wing male from 158mm (Payne 2005).

C. s. fasciipygialis Reichenow, 1898. East Zimbabwe, Mozambique south to Beira, south-east Tanzania and Mafia I. Whiter below than larger similar *burchellii* lacking cream or buff wash, and having plain flanks and undertail-coverts. No intermediate '*loandae*' stage (Irwin 1988). Some variation in plumage in east Tanzania. Wing male (n = 6) 149–165mm, mean 156.4 ± 6.05; female (n = 4) 156–166mm, mean 161.5 ± 4.4 (Payne 2005).

DISTRIBUTION Afrotropical. Non-migratory, but may move so as to appear with rains in some more arid areas. From 19°N in Nile Valley through east Africa west to Western Rift Valley; Socotra; south-west Arabia (Yemen); RC (Djéno, 21 Jul 1991: Dowsett-Lemaire *et al.* 1993); south-east DRC; south of 5°S, south and east through east Africa (including Zanzibar, Pemba and Mafia Is.; absent from much of dry east Kenya), to Malawi, Mozambique, Zambia, north Namibia, north and east Botswana, Zimbabwe (Zambezi Valley and east), Transvaal, Swaziland, Lesotho and coastal RSA to Cape Town (Vincent 1934, Meinertzhagen 1954, Ripley & Bond 1966, Cornwallis & Porter 1982, Irwin 1988, Lewis & Pomeroy 1989, Porter *et al.* 1996, Payne 2005, Jennings 2010).

HABITAT Moister savannas in long grass, reeds and thickets, old cultivated fields, around irrigation in arid country, bracken-briar, and arid country in north-east of range, also swamps, forest edge, *Acacia* scrub and mopane.

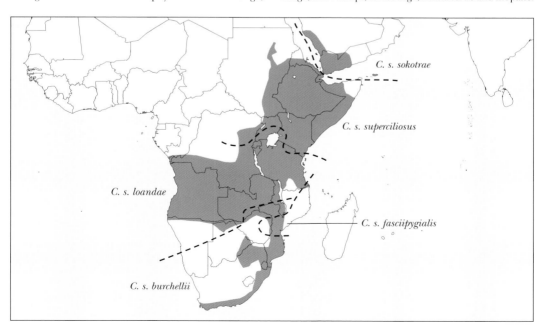

To 1,000m Yemen. Confined to lake edge where range overlaps with Senegal Coucal, which is found on tributary streams, and generally in more marshy areas than that species. Mostly lowland, but to 2,800m east Africa, to 1,870m Zambia, 2,000m Malawi, and 1,300m Mozambique. *Salvador, Ziziphus, Hyphaene* scrub, rank grass and flooded crops to 1,830m, Eritrea and Ethiopia. Absent from extremely arid areas of Kenya (van Someren 1916, 1956,Vincent 1934, Smith 1957, Benson & Benson 1977, Stead 1979, Marshall 1985, Brooks *et al.* 1987, Irwin 1988, Lewis & Pomeroy 1989, Payne 1997, 2005, Ash & Atkins 2009).

BEHAVIOUR Not shy, but skulks in deep cover moving with agility. Flight weak, horizontal and usually short with dangling feet, flopping ungainly into cover. Heavy, ungainly walk, but runs fast. May forage within 1–2m of grass fires; can swim but lacks buoyancy; suns itself with spread wings and tail atop bush to dry out when waterlogged by heavy dew or rain. Calls from cover near nest or from prominent perch. When calling vibrates upright body, puffs out neck, and bends head down until bill touches breast. Becomes habituated to mistnets, killing all trapped birds before beginning to eat. Most food items swallowed whole, but snail shells may be removed by beating. Although food normally carried in bill, larger items may be carried in feet. Male feeds female large insect at copulation (Irwin 1988, Brook *et al.* 1990, Payne 1997, 2005, C.F. Mann).

BREEDING Monogamous; territorial; may breed while still in subadult/juvenile plumage (Lawson 1962). **Season** Generally during rains. In DRC, nest with one remaining young, Apr, Bukama; in east Dec–Feb, in south-east Dec–May, lower Congo R., Apr at least; laying before full adult plumage acquired, Jan. Sudan south of 15°N, May–Jun. Eggs Ethiopia and Eritrea, Mar–Jun, Sep, mainly Apr–May. South Somalia, fledglings Oct; west coast May, Jul, Sep, Oct. Uganda generally, Jan–Oct, varying locally. Kenya all months, Ngong mainly in wet, Apr–Aug; west Jul; south-west Mar–Jun, Aug–Oct; south, central and east Jan–Jul, Sep–Dec; coastal May, Jul, Sep, Oct. Tanzania (except north-east) Jan–Apr, Jun, Dec; north-east Jan–Jul, Sep–Dec; Zanzibar and Pemba Jun, nests Apr–Jul, copulation 16 Oct and 11 Dec, enlarged ovary 4 Nov, building 26 Dec, young in nest Feb. Angola, oviduct egg Nov. Zambia Dec–Feb, young Mar–Apr. Malawi Oct–Mar, Jun, mostly Dec–Feb. Zimbabwe Sep–Feb. RSA Sep–Feb, peak Nov, south-west Cape, Aug–Jan, nestlings Transvaal, 7 Jan (Fischer 1880, van Someren 1916, 1956, Schuster 1926, Pitman 1928, Friedmann 1930, Chapin 1939, Pakenham 1943, Johnson 1968b, Mackworth-Praed & Grant 1970, Benson *et al.* 1971, Greenberg 1975, Benson & Benson 1977, Brown & Britton 1980, Urban & Brown 1980, Rowan 1983, Nikolaus 1987, Maclean 1993, Skead 1995, Skinner 1996, Zimmerman *et al.* 1996, Vernon *et al.* 1997, Ash & Miskell 1998, Dowsett-Lemaire & Dowsett 2006, Ash & Atkins 2009, BMNH). **Nest** Large, untidy, domed; 20–30cm diameter, 15–20cm deep, of grass and twigs lined with leaves with side entrance. In reeds, dense tangled vegetation or thick bush, from ground level to 10m, once 22m up in crown of palm. May be built into bower of living twigs in bush, hedge or thorn tree (Pitman 1928, Chapin 1939, Johnson 1968b, Irwin 1988, Tarboton 2001). **Eggs** Normally 4 (3–5, rarely 2 or 6); oval, white; first may be laid before nest completed; double- or triple-brooded. 30.8–39.0x25.5–28.0mm (n = 40), mean 34.5x26.1; mass 10.2g; also 32.9x24.7mm, 30.1–34.9x22.5–26.6mm, Uganda; laying intervals 24–48hr (Schuster 1926, Pitman 1928, Chapin 1939, Dean 1971, Irwin 1988, Skead 1995, Payne 1997). **Incubation** Probably mainly by male, beginning after first egg laid; 14–16 days, complete clutch up to 5 days to hatch. **Chicks** Fed by male, rarely by female, mostly early morning and evening. Young fed at 10–35min intervals. Void encapsulated faeces and viscous foul smelling liquid if alarmed, and hiss. Both parents clean nest. Young leave nest from 14 days onwards (usually 18–20 days if not disturbed.) before completely feathered; adults thought to carry young between their thighs in flight to escape danger (Bannerman 1933, Johnson 1968b, Wyllie 1981, Irwin 1988, Payne 1997, 2005). **Survival** Chick killed by Lizard Buzzard (Irwin 1988).

FOOD Grasshoppers, locusts, *Enyaliopsis* crickets, beetles (mainly Curculionidae), ants, other hymenoptera, cockroaches, shield bugs, cut-worms; spiders (including 70mm long mygalomorph), once 50mm scorpion; crabs; snails (including large *Achatina*); frogs; lizards, snakes (one 600mm swallowed whole); birds caught in mistnets, eggs of Grosbeak Weaver, nestlings up to size of small dove, 21 day old guineafowl in captivity; mole or mole-rat, striped field-mouse, other unspecified rodents; fruits, including loquats and Buffalo Thorn *Ziziphus mucronata,* and other plant material. Larger prey broken up or pulped; snails beaten against stone (Pakenham 1943, Johnson 1968b, Lack & Quicke 1978, Rowan 1983, Freere 1984, Irwin 1988, Payne 2005, C.F. Mann).

STATUS AND CONSERVATION Generally common; one pair/0.5km on Nylsvlei, Transvaal, RSA. Commonest coucal east Africa. Generally avoided as food by humans, but are sometimes eaten (C.F. Mann); once eaten by Chimpanzee *Pan troglodytes* (Irwin 1988). Common below 1,830m Eritrea (Smith 1957). Common to rare DRC (Chapin 1939). Common Sudan from Kiteiyab (Berber) to Rejaf (Equatoria) (Cave & MacDonald 1955, Nikolaus 1987). Locally common to very common Eritrea and Ethiopia (Ash & Atkins 2009). Rare Chingola District, Zambia (Hall 1970). Common throughout Malawi (Dowsett-Lemaire & Dowsett 2006). Uncommon Tihama, Saudi Arabia (Jennings 1981). Not globally threatened (BirdLife International 2011).

1

White-browed Coucal *Centropus superciliosus*. **Fig. 1.** Adult, *C. s. superciliosus*, Tsavo NP, Kenya, March (*Artur Bujanowicz*). **Fig. 2.** Juvenile, *C. s. superciliosus*, Murchinson Falls NP, Uganda, August (*Jacques Erard*). **Fig. 3.** Adult, *C. s. burchellii*, Kruger NP, RSA, February (*Gordon Holtshausen*). **Fig. 4.** Subadult, *C. s. burchellii*, Kruger NP, RSA, April (*Gordon Holtshausen*). This stage, often referred to as '*loandae*' stage, is not found in other subspecies. **Fig. 5.** Adult, *C. s. burchelli*. Kruger NP, RSA, November (*Rick van der Weijde*).

JAVAN COUCAL
Centropus nigrorufus   Plate 7

Cuculus nigrorufus Cuvier, 1816 (Java).

Alternative name: Sunda Coucal.

TAXONOMY MtDNA analysis places it close to *sinensis* (Sorenson & Payne 2005). Monotypic. Synonym: *purpureus* Shelley 1891.

FIELD IDENTIFICATION 46cm. Rather long-tailed coucal; short, strong bill. Sexes similar. **Adult** Glossy black, back strongly glossed purple, wings rufous tipped black on coverts; in flight wings rufous and black; in some areas birds much blacker. **Juvenile** Similar to adult. **Similar species** Greater Coucal larger with rufous mantle, no black on wing-coverts and long, broad tail black glossed green. Lesser Coucal in breeding plumage much smaller, lower back and rump brown; non-breeding and juveniles streaked.

VOICE Similar to Greater Coucal (BirdLife International 2001).

DESCRIPTION Adult Head and neck black with broad glossy purplish edges to all feathers, and stiff pointed neck feathers, upperparts glossy black with purplish gloss to feather edges on back, rump and uppertail-coverts; wings rufous on outer webs, black on inner webs and tips, inner secondaries black, wing-coverts rufous tipped black on inner webs to rufous-buff on lesser wing-coverts; underparts black glossed purplish; tail glossy black, underside glossed purplish. In some areas birds are much blacker, perhaps due to inbreeding (N. Brickle pers. comm.). **Juvenile** Similar to adult, some with pale barring. **Nestling** Undescribed. **Bare parts** Iris red. Bill black. Feet black; hind claw long, 25–42mm.

BIOMETRICS Wing male (n = 9) 195–218mm, mean 207.4 ± 6.4, female (n = 4) 210–227mm, mean 220.8. Bill male 37–41mm, mean 39.0 ± 1.5, female 40–41mm, mean 40.7 ± 0.6. Tail male 182–242mm, mean 221.1 ± 20.7, female 242–262mm, mean 250.3 ± 10.4. Tarsus male 57–60mm, mean 58.0 ± 1.7, female 57–70mm, mean 62.3 ± 6.8 (Payne 2005).

MOULT No information.

DISTRIBUTION Resident Java. Occurrence on Sumatra based on single trade skin of doubtful provenance, and more recent unconfirmed sightings. Most records from north-west coast of Java, only three from south coast, another concentration around Surabaya, but none from extreme east (Andrew 1990, BirdLife International 2001).

HABITAT Mangroves, coastal marshes and tall grass *Imperata* near mangroves, but last is probably secondary habitat of less importance. *Nypa* palm swamps, and swamp vegetation along rivers and estuaries, but formerly found far inland. In wet season grass fields, flooded forest edges and sugarcane plantations and in dry season meadows with high grass, and dry marshland (Bartels 1915–1931, MacKinnon & Phillipps 1993, BirdLife International 2001, Ir Darjono in Payne 2005).

BEHAVIOUR Similar to other coucals. Forages on foot near puddles, and on dry marshland and tall grass meadows (Bartels 1915–1931, MacKinnon & Phillipps 1993).

BREEDING Season Eggs Jan–Mar and Jun (Hellebrekers & Hoogerwerf 1967, Arifin 1997). **Nest** Untidy construction of ferns and grass leaves, *c.* 30cm in diameter in clearing along river, or 3–5m high in mangroves (Bartels 1915–1931, Arifin 1997). **Eggs** 1–5; white; 37.0–41.5x29.7–31.9mm (n = 6), mean 39.0x30.9 (Hellebrekers & Hoogerwerf 1967, Arifin 1997). No further information.

FOOD Probably omnivorous. Grasshoppers, caterpillars, beetles, large moths, pupae, dragonflies, cicadas and other large Hemiptera, insect eggs; snails, slugs; frogs; geckos, tree and water snakes; small mammals such as rats; 40% amphibians and reptiles, 34% insects, 2% mammals (Bartels 1915–1931, Sody 1989, Arifin 1997).

STATUS AND CONSERVATION Wetlands of Java have been almost totally converted to rice fields, shrimp ponds and industrial sites (Jepson 1997) and most of this species' habitats have therefore disappeared. After 1980 only recorded from: Muara Angke Reserve, near Jakarta (only 15ha), most recently two 12 Aug 2009 (Wedderburn 2009), but site now nearly completely surrounded by houses; Muara Gembong, near Jakarta; Cangkring, Indramayu county, 1988 and 1991; Muara Cimanuk, near Indramayu, 1988; Ujung Pangkah, Brantas delta, Surabaya county, 1990 and 1999; Sidoarjo south of Surabaya, 1992; Muara Bungin (numerous references in BirdLife International 2001); Ujong Kulon NP; Segara Anakan NR; Tanjung Sedari; possibly Muara Bobos (BirdLife International 2011); Ujang–Cilacap 1996 (Condole 1997); Cirebon and Indramayu 1988 (Robson 1989); Pacinan near Situbondo (van Balen *et al.* 2011). High numbers in zoos and markets (often exceeding Greater Coucal). Considered vulnerable due to habitat destruction, heavy fragmentation of few existing mangroves, trapping, and species' rapid decline; population estimate 2,500–2,999 (BirdLife International 2011).

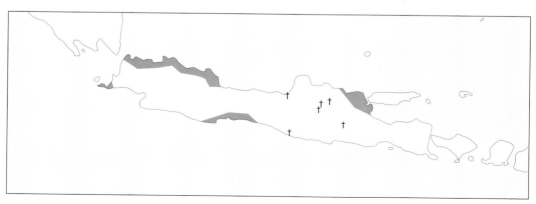

Javan Coucal

Javan Coucal *Centropus nigrorufus*. **Figs. 1–3.** Adults, Muara Angke, near Jakarta, Java, October (*Fabio Olmos*). **Fig. 4.** Adult, captivity (*Mehd Halouate*). In some areas birds show much more black than in others, perhaps due to inbreeding.

GREATER COUCAL
Centropus sinensis Plate 6

Polophilus sinensis Stephens, 1815 (Ning Po = Ningbo, Zhejiang, China)

Alternative names: Common Crow-pheasant; Lark-heeled Cuckoo; Brown or Andaman Coucal (both *andamanensis*); Southern Coucal (*bubutus*).

TAXONOMY Results of mtDNA analysis place it as sister to *nigrorufus* (Sorenson & Payne 2005). *Andamanensis* sometimes treated as separate species due to paler coloration (Ripley 1982, Ripley & Beehler 1989, Dickinson 2003, Rasmussen & Anderton 2005, but not Payne 2005); *kangeanensis* from southern part of range similar, and *bubutus* on Bawean also paler. *Parroti* has been treated as separate species (Rasmussen & Anderton 2005). *Anonymus* erroneously written *anonymous* (Peters 1940) and *kangeanensis* as *kangeangensis* (Ostende et al. 1997). Polytypic. Synonym: *eurycercus* Blyth 1845 in *bubutus*.

FIELD IDENTIFICATION 48–53cm. Large coucal with stout decurved bill, often seeking escape by running rather than flight. Flight laboured with legs partly dangling. Sexes similar. **Adult** Black except for chestnut wings; wings with dark linings, long tail black. Subspecies from Andamans and Kangean Is. differ substantially (see Geographical Variation). **Juvenile** Dark brown head and upperparts spotted chestnut and white on top of head and nape, mantle barred chestnut, wing-coverts chestnut-brown, flight feathers barred dark brown; underparts blackish barred off-white; graduated tail black narrowly barred whitish. **Similar species** Large size distinguishes it from three quarters-sized Lesser which has shorter and more strongly curved bill, and duller mantle and wings, often with some pale shaft streaks, and chestnut underwing-coverts; non-breeding and juvenile plumages have prominent pale shaft streaks. Lesser often found at higher elevations than Greater. Green-billed from Sri Lanka has green bill and black mantle. Short-toed from Malay Peninsula, Sumatra and Borneo similar but much smaller and shorter-tailed. Philippine Coucal on Basilan and Jolo much smaller and glossed green.

VOICE Series of far-carrying primate-like *hoop-hoop-hoop-hoop-hoop* notes, first descending then rising towards end; or *hoop-oop-oop-oop* answered by mate at higher pitch. Other calls include popping *tok-tok-tok* or rasping *shaaah* or explosive *kwisss* when mobbing other animals, and in courtship *djoonk* like stone dropped into water. In south-east Asia series of notes *bi-buup-buup-buup*, where *bi* is higher pitched. On Andamans deep *boom-boom-boom*. Pairs often duetting, accurately at first, but getting more and more out of step. Bornean birds call *bu-bu-bu*, giving name to subspecies *bubutus*. Fledgling's begging call *kren-kren*. Often heard at night (Aagaard 1930, Nichols 1937, Henry 1955, Khajuria 1984, Tikader 1984, Lekagul & Round 1991, Grimmett et al. 1998, Smythies & Davison 1999, Wells 1999, Robson 2000).

DESCRIPTION Nominate. **Adult male** Head, nape, mantle, throat and breast black glossed purple-blue, shining black shaft streaks on head, neck, mantle and breast, wings and scapulars chestnut, remiges tipped dull brown, underwing-coverts black, rarely with narrow white bars, below black glossed purple-blue on throat and breast, greenish-blue on rest of underparts; broad graduated tail black glossed greenish. **Adult female** Often more sooty or dusky on wing-coverts. **Subadult** Head and body duller than adult. **Juvenile** Head black spotted whitish, crown edged rufous, nape and hindneck black, upperparts and wings chestnut barred black; uppertail-coverts blackish barred buffish, remiges tipped blackish; underparts black narrowly barred whitish, denser on throat, with yellow shaft streaks; tail black narrowly barred whitish. **Nestling** Skin black at hatching with long white trichoptiles, forming fringe over eyes; centre of belly pinkish. **Bare parts** Iris red, juvenile greyish-white; nestling brown. Bill black; juvenile horn, lower mandible partly paler; nestling black edged pink, gape yellow. Feet black, rather straight claw on hind toe elongated (22–30mm); nestling brownish-grey.

BIOMETRICS Nominate. Wing male (n = 9) 197–208mm, mean 203.0 ± 3.9, female (n = 10) 202–228mm, mean 214.6 ± 9.1. Bill male 31–34mm, mean 32.8 ± 1.1, female 34–38mm, mean 36.1 ± 1.6. Tail male 230–256mm, mean 245.8 ± 9.1, female 241–277mm, mean 262.9 ± 11.3. Tarsus male 53–58mm, mean 54.9 ± 2.2, female 56–64mm, mean 60.1 ± 2.7. Mass (different subspecies and sexes) 208–380g (Payne 2005); female 416.6g (Wells 1999); male *bubutus* 300g (G. Nikolaus in Wells 2007).

MOULT Sep–early Feb, primaries replaced at up to four loci simultaneously, often asymmetrical. No wing moult Apr–Aug, Thai-Malay Peninsula (Wells 1999, YPM in Wells 2007).

GEOGRAPHICAL VARIATION Seven subspecies.

C. s. sinensis (Stephens, 1815). Pakistan (Sind to Punjab and Indus plain), Kashmir, Himalayas, Nepal, Gangetic Plain, Bengal, Sikkim, north Assam, Bhutan, and east to southern China (Zhejiang, Fujian, Guangdong, Guanxi Zhuang, Guizhou; Hekou in Yunnan). Described above.

C. s. parroti Stresemann, 1913. Peninsular India and Sri Lanka. Forehead, face and throat dull brown. Mantle and back black, head and underparts black glossed blue to blue-green; juvenile resembles adult but much darker chestnut upperparts and wings unbarred. Wing male 173–195mm, female 178–210mm (Ali & Ripley 1983).

C. s. intermedius (Hume, 1873). Bangladesh, south Assam, Burma, Thailand, north Thai-Malay Peninsula, Indochina to west and south-west Yunnan and Hainan. Smaller, richer rufous, and tail shorter but broader. Wing male 185–210mm, female 205–216mm (Wells 1999). Leucistic birds (BMNH).

C. s. bubutus Horsfield, 1821. Thai-Malay Peninsula except north, Sumatra, Java, Bali, Nias I., Mentawai Is, Borneo and west Philippines (Balabac, Bugsuk, Cagayan Sulu, Palawan). Larger, bill more massive, and wings paler. Fresh birds glossed bluish on head, worn birds purplish. On Bawean (north of Java) paler. Wing male 202–224mm, female 226–236mm (Wells 1999).

C. s. anonymus Stresemann, 1913. South-west Philippines (Basilan, Jolo, Sanga Sanga, and Tawi-Tawi). Wings shorter, and dark brown. Wing 182–199mm (Hachisuka 1934).

C. s. kangeanensis Vorderman, 1893. Kangean Is. (Java Sea). Pale morph head, neck and underparts pale buffish, tail greyish, wings rufous. Dark morph has head, back, tail and throat brownish-grey, breast mottled grey; larger than *andamanensis*. Wing male 192–205mm (Payne 2005).

C. s. andamanensis Beavan, 1867. Andaman Islands; Great and Little Coco Is, and Little Table I. off south Burma. Variable. Head, neck, mantle and underparts pale buffish-grey, wings brown, tail white above, rufous below; or brown

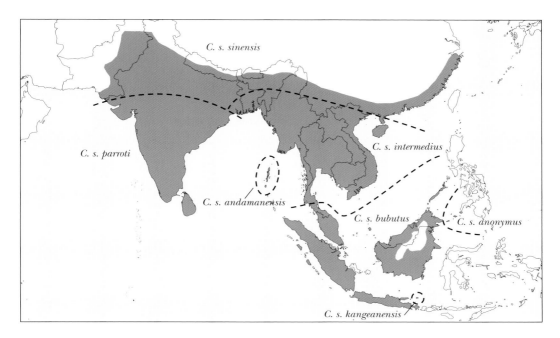

with dusky abdomen and bronzy purple tail; or greyish head, mantle and underparts, dark maroon-brown wings, tail bronzy-purple; or dark brownish-grey head, mantle and underparts, wings dark chestnut, tail blackish-brown. Wing 173–195mm (Ali & Ripley 1983, Robson 2000).

DISTRIBUTION Oriental. Resident except perhaps northwest India. Indian subcontinent. South-east Asia. South China (Hebei?). Greater Sundas. Philippines (Balabac, Basilan, Bugsuk, Cagayan Sulu, Jolo, Palawan, Sanga Sanga, Tawi-Tawi) (Cheng 1987, Roberts 1991, MacKinnon & Phillipps 1993, Wells 1999, Kennedy et al. 2000). Both *sinensis* and *parroti* occur around Delhi (Abdulali 1956). Probable South Korea, May (Birds Korea 2009).

HABITAT Tall grassland, secondary forest, groves, thickets, bamboo, gardens and orchards. In Nepal to 365m, in summer 900m; 2,200m India; 400–1,600m Bhutan; to highest areas, Sri Lanka. Prefers vicinity of dark mango groves Karnataka, India. Foothills at edges of swamps bordering wooded areas, Nepal. Stronghold around edges of seasonal swamps, lakes and canals, Pakistan. On Malay Peninsula plantations, riverine vegetation, roads, grazing-land, reservoirs and edges of mangroves to 700m. Common in large gardens Bangkok, Thailand; brush land around ruins of old temples in north. Railway clearings Singapore and Penang (peninsular Malaysia). On Greater Sundas gardens, reedy riverbanks, mangroves, open country, secondary and dipterocarp forest, and forest edge, kerangas forest, beach vegetation, *Pinus/Acacia*, cocoa, and *Albizia* plantations, to 750m, but generally avoids primary forest interior. In Philippines also grassland, scrub and forest edge (Aagaard 1930, Deignan 1945, Fleming & Traylor 1961, Ghorpade 1973, Bucknill & Chasen 1990, Roberts 1991, MacKinnon & Phillipps 1993, Grimmett et al. 1998, Jeyarajasingam & Pearson 1999, Wells 1999, Kennedy et al. 2000, Mann 2008).

BEHAVIOUR Flight weak, walking or running pheasant-like with tail horizontal. Forages mostly on ground where it can be fast when hunting prey, but clambers also with ease in trees by hopping from branch to branch. Skulking; rarely comes in to open places and therefore more often heard than seen. Usually hunts solitarily, but mates in neighbourhood can be heard by their contact notes. When flying takes off from top of tree with rapid fluttering and glides to bottom of next tree where on alighting spreads and erects broad tail vertically as brake. Pair bond long-lasting. In courtship both sexes raise breast and flank feathers, droop wings with tail spread and raised. Both contribute to building nest and incubation, one nest built in three days. Basks on tops of bushes in mornings and after rain. Fledgling begs with drooping and quivering wings (Aagaard 1930, Henry 1955, Khajuria 1984, Roberts 1991, Grimmett et al. 1998, Smythies & Davison 1999, Wells 1999, C.F. Mann). Only takes to wing when hard pressed and with reluctance, but established on Krakatau in Sunda Strait, involving over-water flight of ten miles (Robinson 1927).

BREEDING Season Almost all year but varying locally peaking Jul–Sep, India; may be double-brooded north India. Feb–Sep and Oct–Dec Sri Lanka, peaking Mar–Apr; Apr–Aug Pakistan; Feb–Jul Andamans; Dec–Aug Thai-Malay Peninsula, eggs Dec, chicks Apr. Once Java, Feb. Nest with three eggs Hainan, 1 Jun; eggs south-east China, Jun. In Borneo nest building Sarawak, 19 Dec; eggs Sabah, Mar, Jul–Sep, chicks Feb, Mar, Jul, Nov and Dec. Eggs China, Jun, young in nest, Aug (Vaughn & Jones 1913, Law 1928, La Touche 1931, Kuroda 1933–36, Gibson-Hill 1950, Henry 1955, Anon. 1958, Ali & Ripley 1983, Tikader 1984, Roberts 1991, Grimmett et al. 1998, Wells 1999, Sheldon et al.2001, Madoc bequest in Wells 2007, Mann 2008). **Nest** Untidy dome of twigs and elephant grass or bamboo leaves with side entrance placed low in bamboo clump, dense or thorny bushes, dense hanging creepers or rarely crown of palm or in rice-field; or built only of grass or coconut leaves, its form rarely rough saucer and nearly always extremely difficult to find; like rugby football, *c.* 35x45cm, built by both sexes; up to 4m above ground in screw-palms and cane-grass clumps. Nest, built in three days, internally 28x25cm with entrance hole 15x19cm (Büttikofer 1900, Law 1928, Henry 1955, Singh-Dhindsa & Toor 1981, Ali & Ripley 1983, Smythies 1986, Roberts 1991, Grimmett *et al.* 1998, Smythies & Davison 1999, Madoc bequest in Wells 2007, Mann 2008). **Eggs** 3–4 (5), oval and chalky or slightly glossy white, but most eggs soiled; west Java 35.3–42.7x26.7–31.1mm, (n = 13) mean 38.3x29.6; India (n = 50) mean 35.9x28.0mm (Hellebrekers & Hoogerwerf 1967, S. Baker in Ali & Ripley 1983). **Incubation** 17–18 days by both parents. In one nest five eggs laid at one, two and three day intervals, last hatching within 12 hours of first. In another nest first two eggs hatched after 18 days and last 16 days (Singh-Dhindsa & Toor 1981). **Chicks** Hatched blind with black skin and long white trichoptiles on dorsal surface, ventral surface naked and centre of belly pinkish; bill black with cutting edges of upper mandible pinkish and egg-tooth *c.* 2mm from tip, gape yellow, iris brown and legs greyish. Eyes open 4th day; remiges break through 5th day; all feathers in pin 9th day. Mass newly hatched, mean 16.08g (n = 3), 11th day 175g (Singh-Dhindsa & Toor 1981). **Survival** Nesting success in nest with five eggs poor, three chicks probably predated. In study in India 77% of eggs hatched and 67% fledged; much nest-robbing by Jungle Crows. Highly infested with endoparasites (Singh-Dhindsa & Toor 1981, Natarajan 1997).

FOOD Mice and bats; eggs and young birds; lizards, including geckos, and small snakes (one 450mm); frogs; fish; insects; spiders; centipedes; crustaceans; molluscs including snails (such as large *Achatina* which are first smashed); carrion; ripe berries. Often springs in air after lizard or grasshopper sitting on overhanging twig. Garden Snail *Helix vittata* main food around villages in Tamil Nadu (Ali 1941, Henry 1955, Ali & Ripley 1983, Roberts 1991, Natarajan 1993a, b, Wells 1999, SUARENG 1999 in Wells 2007). Commercially important pest of oil palm fruit (Rasmussen & Anderton 2005).

STATUS AND CONSERVATION Common and widespread species in most of its huge range, except Bhutan where uncommon (Grimmett *et al.* 1998), and on Philippines, except Jolo where common (Kennedy *et al.* 2000), and subspecies *bubutus* rare Palawan (Hachisuka 1934). Common Pakistan (Roberts 1991). Common Manipur, north-east India (Choudhury 2009). Very common Thailand (Lekagul & Round 1991). Common Burma (Smythies 1986). Common Borneo where not hunted by natives because of foul taste, although chicks are used as cure-all medicine (Smythies & Davison 1999), especially in bone-healing (Duckett 1990). In India, however, relished as delicacy and also used against pulmonary diseases (Ali & Ripley 1983). Accounts for 3.9% of all road-kills Rajasthan, India (Chhangani 2004). Fairly common south China (Cheng 1987), including Hainan (Shing 2005). Common in many localities Laos and Cambodia (Duckworth *et al.* 1998, Thomas & Poole 2003). On Sumatra off, "seen or heard in all locations visited during May 1990 and clearly widespread" Nias I. (Dymond 1994), scarce in Way Kambas NP (Parrott & Andrew 1996), common Siberut, Mentawai Is. (Kemp 2000); 74 recorded in bird trade, Medan, 1997–2001 (Shepherd 2006). Not globally threatened (BirdLife International 2011).

Greater Coucal *Centropus sinensis*. **Fig. 1.** Adult, *C. s. sinensis*, Rajasthan, India, January (*Harri Taavetti*). **Fig. 2.** Adult, *C. s. parroti*, Mysore, Karnataka, India, July (*Subharghya Das*). **Fig. 3.** Adult, *C. s. andamanensis*, Wandoor, Andamans, India, November (*Niranjan Sant*). **Fig. 4.** Adult, *C. s. intermedius*, Kolkata, West Bengal, India, January (*Abhishek Das*); this individual shows brown on head more typical of *C. s. parroti* and may be an intergrade. **Fig. 5.** Adult, *C. s. andamanensis*, Port Blair, Andamans, October (*Subhasis Roy*). **Fig. 6.** Adult carrying snake (probably Banded Krait) for young, *C. s. bubutus*, Kuching, Sarawak, Borneo, March (*Amar-Singh HSS*).

GOLIATH COUCAL
Centropus goliath   Plate 4

Centropus goliath Bonaparte, 1850 (Halmahera)

Alternative names: Giant/Large Coucal.

TAXONOMY Considered to form superspecies with *menbeki* (White and Bruce 1986), but mtDNA analysis places it as sister to *toulou* (Sorenson & Payne 2005). Monotypic.

FIELD IDENTIFICATION 64–70cm. Largest cuckoo, although female Pheasant-coucal can be similar size. Clumsy, long-tailed, with weak flight. Sexes similar. **Adult** Black with conspicuous white patch in wings and stout black bill; some have few white patches on underparts. Whitish morph on Halmahera. **Juvenile** Variable. Dark brown; head, mantle and chin to upper breast streaked and spotted whitish, in some throat chiefly whitish; wings black, large white patch on coverts, secondaries sometimes streaked white; breast to belly blackish barred whitish or (on Halmahera) irregularly blotched whitish, grey and blackish. Tail black. **Similar species** Common Koel much smaller and all black in adult male, female superficially like young Goliath but has barred tail. Lesser Coucal much smaller with brown wings, juvenile barred and streaked brown, wing and tail barred black and off-white; immature yellowish with dark spots on underparts and tail black barred rufous.

VOICE Far-carrying call deep *ooom-ooom* repeated about twice/sec, continuing for 4–12sec. Alarm call harsh and guttural *kcau* or *kcau-kuc* repeated every 1–4sec (Coates & Bishop 1997).

DESCRIPTION **Adult, black morph** Entire plumage black except for white patch on greater wing-coverts, wings with bluish gloss, broad graduated tail black. **Adult, whitish morph** (Halmahera). Whitish, with buffish head; others pied or black with some irregular white spots. **Juvenile** Variable. Dark brown, with whitish shaft streaks and diamond-shaped spots of varying size on head, mantle and chin to upper breast, lower breast to belly barred whitish; in some whitish spotting on chin and throat very extensive; wings black, large white patch or white streaking on coverts, secondaries sometimes streaked white; breast to belly blackish, or on Halmahera, irregularly blotched whitish, grey and blackish. Tail black glossed bluish. **Nestling** White trichoptiles on head. **Bare parts** Iris deep brown, grey in nestling. Bill black, pale grey below in nestling. Feet black, pale in nestling; hind claw long, 22.5–28mm.

BIOMETRICS Wing male (n = 10) 244–290mm, mean 261.6 ± 14.6, female (n = 8) 255–298mm, mean 274.9 ± 13.8. Bill male 48.5–58.0mm, mean 50.6 ± 4.0, female 48.0–60.4mm, mean 50.0 ± 5.5. Tail male 390–434mm, mean 421.6 ± 22.1, female 360–425mm, mean 394.5 ± 26.6. Tarsus male 46.5–55.5mm, mean 52.4 ± 3.2, female 48–57mm, mean 55.1 ± 4.0. Mass male (n = 7) 340–453g, mean 400.7, female (n = 2) 605, 638g (Payne 2005).

MOULT No information.

DISTRIBUTION Resident north Moluccas (Morotai, Halmahera, Tidore, Bacan, Obi) (White & Bruce 1986); Kasiruta? Status Ternate uncertain (Peters 1940, van Bemmel 1948), probably extirpated (White & Bruce 1986).

HABITAT Lowland forest; dense primary forest, selectively-logged or regenerated forest, and forest edge, as well as scrub and edge of agricultural land, or near rivers and clearings. To 250m Bacan. Not in mangroves, but in swamp forest and to 800m, Halmahera (Lambert & Yong 1989, Coates & Bishop 1997, Poulsen & Lambert 2000, Strange 2001).

BEHAVIOUR Social; usually in pairs or 3–4 birds together, foraging both in trees mainly up to midstorey, more rarely higher up, and on forest floor. Slow clumsy bird, easily overlooked outside breeding season. Poor flyer, and usually restricts its flights to short distances between trees or across roads (Lambert & Yong 1989, Coates & Bishop 1997).

BREEDING No information.

FOOD Grasshoppers, large crickets, cicadas and phasmids (Rand & Gilliard 1967, Coates & Bishop 1997).

STATUS AND CONSERVATION Common parts of Halmahera, uncommon others; uncommon Bacan, rare or local Obi (Coates & Bishop 1997). Seen 112 occasions Halmahera on two visits 1994–95 (Poulsen & Lambert 2000). Common Domato, south Halmahera, Jul 1987 (Lambert & Yong 1989). Probably common as almost every collector has collected it (White & Bruce 1986). Due to its varied habitat preference not considered globally threatened (BirdLife International 2011).

Goliath Coucal *Centropus goliath*. **Fig. 1.** Adult, Bale Village, Halmahera, Moluccas, Indonesia, July (*Hanom Bashari*).

MADAGASCAR COUCAL
Centropus toulou Plate 8

Cuculus Toulou P.L.S. Müller, 1776 (Madagascar)

TAXONOMY Once considered conspecific with *grillii*, and occasionally *bengalensis*, but voices differ (Roché 1971, King *et al.* 1975, Dowsett & Dowsett-Lemaire 1993). MtDNA analysis places it closest to *goliath* (Sorenson & Payne 2005). Polytypic. Synonyms: extinct *assumptionis* Nicoll 1906, in *insularis*; *Polophilus tolu* (Stephens 1815) in nominate.

FIELD IDENTIFICATION 40–50cm. Medium-sized coucal. Sexes similar, but male smaller. **Adult breeding** Black glossed blue, back and wings chestnut; tail black slightly glossed greeny-blue. **Adult non-breeding** Head to upper back and breast dark brown finely streaked cream. **Juvenile** Similar to non-breeding adult, but more barred. **Similar species** Only coucal in range; no other species on Madagascar shows this plumage combination.

VOICE On Madagascar long bubbling call in falling cadence, longer than White-browed Coucal. 'Waterbottle' call (resembles sound of water being poured from narrow-necked bottle) when duetting described as series of *toulou, toulou...* notes, about 6/sec, by one bird, and second bird joining with higher pitched series of notes, dropping in pitch, starting at 14/sec and slowing to 10/sec; sudden guttural *coogoo*, muffled *toogoo toogoo toogoo...* decreasing in volume; 5–10 hoots decreasing in volume. On Aldabra male call, lower-pitched than female, falling series of 12–20 stacato *hoo* notes; lower pitched call usually starts duet, and continues until higher notes finish; male may call alone; hissing and loud chattering like very slow alarm call of Eurasian Blackbird. Also 8 *cook* notes, first very faint; 6 *pop* notes; chattering *chuka-k-k-k-k* from juvenile; low, almost inaudible *i-k-k-k* may precede water bottle call; *coo-cook-cook* also preceding water bottle call; single hoarse *tweet*. On Aldabra descending *tou-lou-lou*, male higher pitched; greetings at nest described as variety of low rattling, chuckling and other harsh noises. Nestling hisses like snake, as does adult when alarmed. Most frequently calls early morning and late afternoon, and sometimes at night; 2 or 3 birds may counter-call (Rand 1936, van Someren 1947, Gaymer 1967, Benson & Penny 1971, Roché 1971, Milon *et al.* 1973, Frith 1975, Woodell 1976a, Langrand 1990, Randrianary *et al.* 1997, Morris & Hawkins 1998).

DESCRIPTION Nominate. **Adult breeding** Mostly black except for unbarred rufous-chestnut middle and lower back, and wings, last tipped dark brown; greater underwing-coverts rufous-chestnut, rest black and grey; tail glossy black. **Adult non-breeding** Head to mantle, and chin to breast brown to dark brown with creamy shaft streaks giving streaked appearance; wings unbarred rufous-chestnut; uppertail-coverts black barred tawny-white; belly and undertail-coverts blackish-brown. Tail blackish-brown faintly barred pale. Perhaps dimorphic on Aldabra as some individuals breed in non-breeding plumage (Frith 1975). **Juvenile** Two morphs. As non-breeding adult, but more barred and speckled above with short shaft streaks; wings barred; belly slightly barred and has shaft streaks; tail barred. All show barring on upper back, less in speckled birds, least in Aldabra specimen. Tail barring varies considerably but no consistent differences between morphs. Intermediates occur. Possibly connected to camouflage (Woodell 1976b). **Pale (barred) morph** Dusky brown with some black above, heavily barred buff-yellow on mantle, merging into dense, more cinnamon-rufous, speckling on crown; wings and upperwing-coverts barred cinnamon-rufous and brown; tail and coverts brown barred buff-yellow, and washed green; underparts heavily barred buff-yellow and brown, darker on abdomen. **Dark (speckled) morph** Darker above, mantle often more speckled and less barred cream and cinnamon-rufous; finer speckling on crown; wings and wing-coverts cinnamon-rufous to russet, barred brown; underparts darker than pale morph, but Aldabra specimen has throat and chest more striated, less barred. **Nestling** Black on hatching with dense trichoptiles on dorsal surface. Older, feathered nestling has short, pale shaft streaks. **Bare parts** Iris red; *insularis* with navy blue rim (Benson 1967). Bill black, rose-brown or pale to dark horn in non-breeding plumage; nestling black with remains of white egg-tooth; mouth and palate pink, whitish tongue with black 'U' mark near tip (Payne 2005). Feet grey to black.

BIOMETRICS Nominate. Wing male (n = 25) 135–173mm, mean 145.9; female (n = 15) 160–174mm, mean 166.3. Bill male 25.3–35.2mm, mean 28.9; female 25.8–34.6mm, mean 30.8. Tail male (n = 24) 161–269mm, mean 231.2; female 215–265mm, mean 248.3; unsexed to 273mm. Tarsus male (n = 23) 34.0–42.4mm, mean 38.0; female 37.5–46.2mm, mean 41.2 (BMNH). Also wing female 156–176mm, tail to 270mm (Payne 2005). Mass male 139g, female 189g (BMNH); male (n = 4) 125–150g, mean 135, female 185, 210, 220g (Goodman & Benstead 2003).

MOULT Tail moult Assumption I., Mar; wing and tail moult Aldabra, Feb; tail moult Madagascar, Jan. Male moulting into breeding plumage almost complete Assumption I., 18 Sep (Benson 1967, BMNH).

GEOGRAPHICAL VARIATION Two subspecies.
C. t. toulou (P.L.S. Müller, 1776). Madagascar. Juvenile plumage tends to pale morph in humid eastern regions, dark morph in drier western areas (Woodell 1976b). Described above.
C. t. insularis Ridgway, 1894. Aldabra. Smaller than nominate, and paler in breeding plumage; female differs from nominate by having black on thighs and undertail-coverts only, rather than whole abdomen, and rest buffy-white (Benson 1967); juvenile dark morph (Woodell 1976b). Wing male 149, 150, 154mm, female 157, 165, 174mm (BMNH).

DISTRIBUTION Resident Madagascar, Aldabra and formerly Assumption I.; Cosmoledo Atoll. May have reached Aldabra via Comoros, where it no longer occurs, or via Gloriosa, Astove and Cosmoledo, or possibly colonised western Indian Ocean islands from Asia (Moreau 1966, Benson 1967, Benson & Penny 1971, Payne 1997).

HABITAT Dense vegetation or undergrowth in forest, occurring in primary forest, including spiny and gallery forests, and at edge, in clearings, *Eucalyptus* woodlands, secondary growth, dense champignon thickets, littoral forest, palm groves, mangroves, marshes, including *Phragmites* reedbeds, long grass, paddifields and gardens; to 1,800m (Rand 1936, Benson & Penny 1971, Langrand 1990, Goodman *et al.* 2000).

BEHAVIOUR Keeps low, chasing and pouncing on prey in thick vegetation on ground or in bushes. Rarely in open, but in early mornings or after rain may sun itself on bush, with drooping wings, spread tail and ruffled feathers. Pulls bark off trees to get lizards. Often feed in pairs, one defending

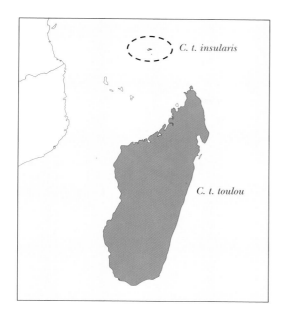

prey against Crested Drongo on Aldabra. Infrequently fly by vigorous flapping and gliding in to vegetation. Before copulation pair wag tails side to side occasionally calling, female lowers and extends wings, male then mounts, and feeds female with insect almost immediately after; copulation takes 45sec (Benson & Penny 1971, Frith 1975, Prŷs-Jones & Diamond 1984, Langrand 1990).

BREEDING Nesting duties mostly, but not entirely, performed by male, but claimed that both parents incubate. Both, or neither, of pair breeding in non-breeding plumage on Aldabra (Frith 1975, Langrand 1990). **Season** Sep–Mar; eggs Aldabra, Dec–early Apr; eggs Assumption I., Mar; enlarged testes Tulear, south-west Madagascar 30 Jan, and eggs 2 Feb; north Madagascar 15 Oct; Tabiky, east-central Madagascar, 6 Nov; active ovary Vondrozo, south-east Madagascar 14 Jun; Manonbo 3 Oct; ready to lay Soalala, north-west Madagascar, 27 Feb; laying Marotony, north Madagascar, 4 Jan; breeding Befandrana, south-west Madagascar, 24 Nov; collecting nest material Sahavondronina, 11 Dec (Ridgway 1895, Rand 1936, van Someren 1947, Benson 1967, Milon et al. 1973, Woodell 1976a, Appert 1980, Langrand 1990, Goodman et al. 1997, BMNH). **Nest** Built by both sexes on Aldabra, where first egg laid before nest completed; male did most of building, female helped by weaving and shaping. Large, domed, spherical, woven from dry grass, bark, coconut and other leaves, and *Passiflora* stems and tendrils, with entrance at one end; 1–4m above ground in dense bush; open platform also recorded; lined with fresh green leaves; placed near tree trunk in dense vegetation (Abbott in Ridgway 1895, Rand 1936, Frith 1975, Woodell 1976a, Langrand 1990). **Eggs** 2–3 (sometimes 4–5) Madagascar, 2–4 Aldabra, 2 Assumption I.; smooth and chalky white; 33x26, 30.4x21.1, 27.7x23.5mm; mean mass 8.3g; also 28.0x23.0, 27.5x23.0, 25.0x20.0mm, mass 6.9, 7.7, 8.2g, but highly variable in size and shape (Ridgway 1895, Rand 1936, Frith 1975, Woodell 1976a, Langrand 1990). **Incubation** On Aldabra mostly by male. Clutch of two hatched 7–8 days apart, total time 14 days; on Madagascar eggs laid at intervals of 3–9 days, total incubation 14–16 days, hatching 3–9 days apart (Woodell 1976a). **Chicks** On Aldabra both chicks fledged together at about 19 days, having reached 90% of fledging weight in 15 days. Hiss like snake and excrete sticky foul-smelling liquid; burst through rear of nest to escape (Frith 1975). **Survival** No information.

FOOD Large insects such as mantids, cockroaches, beetles (Elateridae, Scarabaeidae, Tenebrionidae), cicadas and other Hemiptera, Hymenoptera, grasshoppers, crickets including Tettigoniidae, and caterpillars; centipedes; spiders; molluscs; chameleons, geckos and skinks; chicks and eggs; rats; nestlings on Aldabra fed on geckos, skinks, moths, mantids and other large insects. Food captured on ground and swallowed whole. Raids nests of Madagascar Paradise-flycatcher, Red Fody and Madagascar Munia (Gaymer 1967, Benson & Penny 1971, Milon et al. 1973, Frith 1975, Langrand 1990, Goodman et al. 1997).

STATUS AND CONSERVATION Widespread and fairly common throughout Madagascar, except denuded central plateau (Delacour 1932a, b, Langrand 1990); common in highlands (van Someren 1947). On Aldabra where it is tame, population estimated 400–800 pairs (Rocamora & Skerrett 2001); widely but sparsely distributed (Betts 2002). Previously common and tame on Assumption I. (Nicoll 1906a, b), last recorded 1937 (Vesey-Fitzgerald 1940), extirpated 1937–1958 due to bird guano collecting and resulting destruction of soil and vegetation (Stoddart et al. 1970). Not globally threatened (BirdLife International 2011).

Madagascar Coucal *Centropus toulou*. **Fig. 1.** Adult, *C. t. toulou*, breeding plumage, Antananarivo, Madagascar, September (*Bruno Boedts*). **Fig. 2.** Adult, *C. t. toulou* non-breeding plumage, Salary, Madagascar, August (*Bruno Boedts*). **Fig. 3.** Adult breeding, *C. t. toulou*, Masoala, Madagascar, December (*Henry Cook*). **Fig. 4.** Adult non-breeding *C. t. toulou* attacking a chameleon. Ifaty spiny forest, Madagascar, October (*Mark Rentz*).

AFRICAN BLACK COUCAL
Centropus grillii   Plate 9

Centropus Grillii Hartlaub, 1861 (Gabon)

Alternative names: Black/Black-chested/Black-bellied Coucal

TAXONOMY Once considered conspecific with *toulou* and *bengalensis*, but voices differ (Roché 1971, King *et al.* 1975, Dowsett & Dowsett-Lemaire 1993), and ranges of *grillii* and *bengalensis* very disjunct. MtDNA analysis puts this species closest to *bengalensis* and *viridis*, this being closest to clade consisting of *violaceus*, *bernsteini*, *nigrorufus* and *phasianinus*, with *toulou* more distant, clustering with *goliath* (Sorenson & Payne 2005). Monotypic. Synonyms: *thierryi* Reichenow 1899, *toulou caeruleiceps* Neumann 1904, and *toulou wahlbergi* C. Grant 1915.

FIELD IDENTIFICATION Male 30cm, female 34cm. Rather small, skulking coucal that sometimes perches prominently. Females markedly larger; otherwise sexes similar. **Adult breeding** Black with rufous-chestnut wings. **Adult non-breeding** Above blackish-brown barred and streaked rufous, tawny, cream and rusty-buff; wings rufous; tail brownish-black; underparts tawny-buff to white streaked, spotted and barred cream and brown; undertail-coverts blackish barred buff. **Juvenile** Above mottled, barred and streaked buff, brown and cream; rufous remiges barred blackish-brown; brown rump and tail barred tawny-buff; underparts pale buff streaked cream (feather shafts), belly and flanks barred and spotted. **Similar species** Black underparts in breeding plumage distinguishes it from other African coucals; non-breeding plumage and immature birds from White-browed Coucal by shorter tail, and lack of supercilium. '*Epomidis*' morph of Senegal Coucal appears very dark, but back, breast and belly chestnut (not black). Juvenile separated from other juvenile coucals by pale overall appearance, and very streaky head and mantle, rest of upperparts heavily barred.

VOICE Female calls only in breeding season, and male much less. Advertisement call fast double note repeated at 2sec intervals up to 50 times, *kuk kuk*, sometimes up to 105min; *dodder… dodder…dodder*; *kuk… kutuk… kutuk*; up to ten low hoots, 1/sec, *hoo-hoo*, rather ventriloquial and may be interrupted by advertisement call. 'Water-bottle' call, faster, descending, higher pitched than that of Senegal and White-browed Coucals, consists of rapidly repeated hooting *kok…kok…kok…kok* or *huk…ku ku ku* and lacks final rising notes; in flight and when perched. Male's response to female's advertising call is rapid *too loo too*. Male's alarm on nest slow clucking *tuck-tuck*. Laughing series of hollow *cow* notes, repeated 8–10 times, recalling Eastern Grey Plantain-eater; also *ku* and *kutuk*; *huk ku-ku-ku-ku-ku-ku-ku*, slightly falling, less rich or vigorous than Senegal; female gives thin, weak *pik*. Double *kop-kop* at 2sec intervals; series of rising and falling *kop* notes, 6 notes/sec for 2–3sec or more on one pitch; excitement or aggressive call series of hash *shrehhh* notes; alarm slow clucking *tuck-tuck* (Chapin 1939, Holman 1947, Chappuis 1974, 2000, Irwin 1988, Stjernstedt 1993, Payne 1997, 2005).

DESCRIPTION Adult breeding Head and mantle glossy blue-black, feather shafts stiff and shiny; dark olive-brown back; lower back barred buff; uppertail-coverts black glossed greenish; wings rufous-chestnut, remiges tipped olive-brown; inner secondaries dark brown, scapulars dark olive-brown; underwing-coverts chestnut; underparts glossy blue-black; black tail glossed bluish (some retain juvenile buff-barred rectrices). **Adult non-breeding** Above blackish-brown, back to tail barred rufous; head to mantle with creamy shaft streaks; remiges dull chestnut or rufous tipped dark brown with few spots on outer vanes; wing-coverts dull chestnut or rufous-barred dark brown with creamy shaft streaks; scapulars and tertials barred dull chestnut with creamy shaft streaks. Below tawny-buff with cream feather shafts; sides of neck and breast spotted brown; buff flanks barred brown; belly narrowly barred brown, centre of belly whitish; undertail-coverts blackish narrowly barred buff; underwing-coverts dark rufous. Tail brownish-black. **Subadult** Dark brown crown to mantle, streaked and spotted whitish and buff; back and rump dark brown with variable buff streaking; uppertail-coverts dark brown barred buffish. Wings chestnut, inner secondaries buff barred brown. Mostly buff below, more cinnamon on breast, which is streaked blackish. Tail dark brown barred buff. **Juvenile** Similar to non-breeding adult. Above light rufous and dark brown, pale buff triangular tips to feathers give head streaked and mottled appearance; nape and upper back streaked white, rufous and dark brown; back barred light rufous and dark brown; wing-coverts and remiges barred rufous and tipped dark brown, rump and uppertail-coverts dark brown narrowly barred buff, tail dark brown finely barred rufous. Underwing-coverts pale chestnut finely barred black, cheeks whitish, chin and throat whitish finely streaked black; breast whitish buff with paler shaft streaks, barred black on flanks; belly whitish, lower belly grey, undertail-coverts blackish barred buff (Payne 2005). **Nestling** Black with sparse long white trichoptiles on upperparts forming fringe over forehead. **Bare parts** Iris dark brown, juvenile pale grey. Bill black in breeding plumage; non-breeding brown above, lower mandible horn to cream; juvenile pale yellowish to horn; nestling black with white egg-tooth. Feet black, juvenile blue-grey.

BIOMETRICS Wing male (n = 14) 146–158mm, mean 152.3 ± 3.9, female (n = 14) 164–173mm, mean 169.1 ± 3.50. Tail male 122–170mm, mean 149.2 ± 14.0, female 132–170mm, mean 159.0 ± 10.1. Bill male 19–25mm, mean 22.29 ± 1.6, female 35–42mm, mean 38.9 ± 2.5. Tarsus male 29–40mm, mean 34.6 ± 3.8, female 35–42mm, mean 38.9 ± 2.5 (Payne 2005). Mass male (n = 6) 94–108g, mean 100, female 151g (Irwin 1988).

MOULT May take two years to attain full adult plumage, retaining barred remiges and rectrices in to first breeding and second non-breeding plumages. Juvenile to adult moult Cameroon, Oct, Nov; Sudan, Jun; Kenya, Nov; Tanzania, Jan; Angola, Sep; DRC, Mar; Malawi, May (enlarged ovary), Nov, Dec; Mozambique, Jan; Zimbabwe, Dec (Irwin 1988, BMNH).

DISTRIBUTION Sedentary and intra-African migrant; resident in areas with permanent fresh water. West Africa, from Senegambia, south of 15°N east to south Sudan (but apparently absent from much of eastern Nigeria and Cameroon), south-west Ethiopia, west, central and south Kenya north to Tana R., south to Malawi, Angola, north and Botswana, north Zimbabwe, Mozambique, Namibia (Caprivi); irregular summer visitor RSA in north-east Natal, Kruger NP and Roodevaal, Transvaal. Mainly summer visitor to eastern RSA, Zimbabwe, Zambia and Malawi, Dec–Mar, sometimes Apr, rarely overwintering; south Ghana May–Aug,

north during rains Jul–Sep; locally resident and partially migratory Uganda, north-west May; Nyanza, Kenya, Jan–Mar and Jun–Jul, occasional highlands May–Aug. Breeding visitor Tanzania (Serengeti Plains Dec–Mar, Mikindani Feb–Mar; north-west and scattered records elsewhere). In west Africa resident at low latitudes in and around rainforest, but rains migrant further north to 13°N in Guinean and Soudanian savannas, Apr–Oct. Gambia, Aug–Oct with rains, some until Dec. Resident south Nigeria; north Nigeria, Apr–Oct in rains, previously on islands of Lake Chad, Nov–Feb. Apparently resident SL; in wet and dry seasons in inundation zone of R. Niger, Mali; Gabon, Jan–Feb in dry season; Mekrou, Niger, Nov; south Togo, Nov–Jul in dry season, north Togo, Jul–Aug in wet season when in breeding plumage; south Benin, Jan–Jun; resident Burkina Faso; presumed resident Sudan south of 9°N; resident Ethiopia; disappears from seasonally wet areas in Kenya during dry periods; Zambia, Dec–May on Kafue floodplain where breeds, Oct–Apr; Okavango, perhaps resident; appears east Botswana with heavy rains; Zimbabwe Dec–Apr (Bannerman 1951, Rand et al. 1959, Duhart & Descamps 1963, Dowsett 1969, Urban & Brown 1971, Aspinwall 1972, Elgood et al. 1973, Osborne 1973, Berry 1974, Greig-Smith 1976, Britton 1978, 1980, EABR 1980, 1982, 1983, Irwin 1981, 1988, Brosset & Érard 1986, Grimes 1987, Nikolaus 1987, Giraudoux et al. 1988, Lewis & Pomeroy 1989, Dowsett & Dowsett-Lemaire 1993, Cheke & Walsh 1996, Skinner 1996, Barlow et al. 1997, Vernon et al. 1997, Anciaux 2000, Brewster 2000, Barlow 2003, Carswell et al. 2005, C.F. Mann). Vagrant Zanzibar during mainland dry season (Pakenham 1979).

HABITAT Moist and partially flooded grassland and marshes; woody vegetation; occasionally reeds and papyrus; grassland, palm and bushy savannas; bracken briar on Nyika Plateau, Malawi; in west Africa in high rank grass, dry paddifields, forest clearings, airstrips. Usually below 1,525m, but to at least 2,000m Kenya. Previously in Sahel at Lake Chad. Generally areas of 500mm annual rainfall, Kenya, but also some arid/semi-arid areas (Chapin 1939, Cave & MacDonald 1955, Elgood et al. 1973, Dowsett et al. 1974, Stead 1979, Thiollay 1985, Irwin 1988, Lewis & Pomeroy 1989, Gatter 1997, Barlow 2003, C.F. Mann).

BEHAVIOUR Skulking, but perches prominently to call or preen and dry wet plumage. Creeps through dense grass and difficult to flush. Makes short, heavy flights, rapid wing-beats alternating with gliding. May fly straight to nest, or perch nearby. When flushed from nest flies short way and disappears into grass; may jump to catch insects. One female dominates several males in polyandrous relationships, and initiates courtship with both sexes silently quivering lowered wings prior to copulation, when male feeds female (Winterbottom 1938, Vincent 1946, Vernon 1971a, Rowan 1983, Irwin 1988).

BREEDING Usually monogamous, occasionally polyandrous; in latter case male carries out all duties, including nest building, and after laying female plays no further role; in former case both sexes tend young; female territorial, or pair defends territory; female observed with three males, each with nest. Territory of pair c. 5ha, of polyandrous female up to 10ha. May breed co-operatively. Female may initiate new cycle as soon as young independent. 14.2% of chicks result from extra-pair paternity. May breed in juvenile plumage (Winterbottom 1938, Vernon 1971a, Rowan 1983, Irwin 1988, Barlow 2003, Muck et al. 2009). **Season** Gambia, nesting activity Aug–Nov when grass high, laying late Oct, fledging Dec. Breeding plumage Liberia, Feb–Aug. Togo, territorial behaviour Jul. Coastal Ghana, Apr–Jul, north Ghana, Jul–Sep. Nigeria, Jun–Aug in north, also young just fledged south-east Nigeria, 20 Nov, breeding condition female north Nigeria, 2 Sep. DRC, Jan–May, enlarged testes Dec, south-west DRC. Tanzania (except north-east) Feb–Mar; Kenya and north-east Tanzania, Apr, Jun; north Tanzania (Serengeti and Mkomazi), Dec–Feb, Katavi Plain GR, Feb. Zambia, Nov–Apr. Zimbabwe, Dec–Apr. Malawi, Jan–May, once Sep. Botswana Jan–Feb. RSA, Natal Oct (Serle 1943, Vincent 1946, Holman 1947, Marchant 1953, Dekeyser 1956, Cawkell 1965, Mackworth-Praed & Grant 1970, Vernon 1971a, Benson & Benson 1977, Greig-Smith 1977, Britton 1980, Brown & Britton 1980, Irwin 1981, 1988, Medland 1993, 1995, Elgood et al. 1994, Cheke & Walsh 1996, Skinner 1996, Zimmerman et al. 1996, Gatter 1997, Vernon et al. 1997, Aspinwall & Beel 1998, Barlow 2003, Finch 2010, BMNH). **Nest** Built by male. Oval ball 17–25cm diameter; 15–45cm above ground well concealed in dense tangled grasses; living sedges or grasses form outer frame concealing main nest of dry grass, thickest at base and lined with leaves, with side entrance. Lower nests better concealed because of thicker grass; nests have strong smell; may have second escape hole. Territory of pair c. 5ha; up to 10ha. for polyandrous female. Female dominates several males and initiates non-elaborate courtship. In polyandrous situation, each male tends one nest (Chapin 1939, Vincent 1946, Vernon 1971a, Rowan 1983). **Eggs** 4 (3–6) laid over period up to 9 days; multiple brooded; lost clutches may be replaced. One female laid 24 eggs in six clutches for three males. Oval, pure white, but may be stained, slightly glossed. 27.2–35.5x23.0–25.0mm (n = 15), mean 30.6x 24.0; 26x22, 25x22mm; 31.6x23.9–29.0x24.3 (n = 8), mean 29.2x23.8 (Vincent 1946, Holman 1947). **Incubation** Male incubates. Starts with first egg; c. 14–16 days, eggs hatching at intervals. **Chicks** Male tends nestlings. Black; upperparts sparsely covered with white trichoptiles most prominently on head; bill black, egg-tooth white; eyes closed. By 7th day feather quills appear, by 11th day eyes well open, and body covered with feather sheaths, well-camouflaged and may leave nest temporarily. Make snake-like hissing if threatened, defaecating foul-smelling liquid as well as normal encapsulated faeces. Finally leave nest at 18–20 days, or earlier if disturbed, and male continues to feed for another 9–12 days when they can fly fully; fledging probably 18–20 days (Vernon 1971a, Irwin 1988, Payne 1997, Barlow 2003). **Survival** Group of four adults reared c.19 young in six breeding attempts (Vernon 1971a).

FOOD Green grasshoppers, crickets, locusts, mantids, beetles, caterpillars, ants, Hemiptera; spiders; small reptiles; seeds (Chapin 1939, Vincent 1946).

STATUS AND CONSERVATION Scarce over much of range. Uncommon, or locally common throughout west Africa except arid north (Borrow & Demey 2001). Not uncommon in suitable habitat throughout SL (Irwin 1988, Payne 1997). Common in inundation zone, Mali (Duhart & Descamps 1963). Uncommon migrant Togo (Cheke & Walsh 1996). Numerous Keta, Ghana in rains (Holman 1947). Fairly common south-east Nigeria (Marchant 1953). In Sudan uncommon north to 9°N (Nikolaus 1987). Rare south-west Ethiopia (Ash & Atkins 2009). Generally uncommon east Africa (Britton 1980). Fairly common Malawi (Stead 1979). Destruction of wetlands habitat by conversion to sugarcane or grazing has local effects on status (Payne 2005). Not globally threatened (BirdLife International 2011).

African Black Coucal *Centropus grillii*. **Fig. 1.** Adult, breeding plumage, Maasai Mara NP, Kenya, June (*Adam Scott Kennedy*). **Fig. 2.** Subadult moulting into adult plumage, Shonga, Nigeria, August (*John Sawyer*).

PHILIPPINE COUCAL
Centropus viridis   Plate 8

Cuculus viridis Scopoli, 1786 (Antigua, Panay)

Alternative name: Green Coucal.

TAXONOMY MtDNA analysis places this species in clade with *bengalensis* and *grillii* (Sorenson & Payne 2005). Subspecies *mindorensis* misspelt *mindoroensis* (Ripley & Rabor 1958). Polytypic. Synonyms: *Cuculus Philippensis* Cuvier 1817, *Corydonix pyrropterus* Vieillot 1819, *Centropus molkenboeri* Bonaparte 1850, all in nominate.

FIELD IDENTIFICATION 42–44cm. Medium-sized blackish coucal with rather short, decurved black bill and floppy flight. Sexes similar. **Adult** Black glossed blue to green, wings chestnut tipped dark brown (obvious in flight). All black on Batan, Ivojos and Sabtang. **Subadult** Above dark brown, streaked pale from head to mantle; wing-coverts chestnut barred dark with pale streaks and spots, remiges barred rufous and dark brown; throat to breast whitish barred brown; rest of underparts dark brown with pale shaft streaks; tail blackish with few pale rufous bars distally. **Juvenile** Above dark brown with whitish shaft streaks from head to mantle, rump and uppertail-coverts dark brown tipped pale brown, wings chestnut barred dark brown; below blackish-brown marked buffish, tail brownish-black glossed greenish with some pale brown barring distally. **Similar species** Lesser occurs on many islands with this species and adult birds difficult to separate. Former smaller, wings paler, has shiny shaft streaks on upperparts, rectrices (often worn) brownish-black narrowly tipped buffish; iris in adult chestnut not red; juvenile has conspicuous buff shaft streaks to dark brown upperparts, and tail dark brown barred rufous on outer third. Steere's Coucal on Mindoro larger, has stronger and more decurved bill, black body colour glossed bronze-green, and claw on hind toe never longer than toe as in *viridis*.

VOICE Fast, loud, staccato *coo-coo-coo* or *boop-boop-boop*, repeated at *c.* 8sec intervals, but sometimes delivered more slowly. Also *chi-gook-gook* or *chi-go-go-gook* or *chi gook*, giving it many of its native names. Mournful *coo*, repeated several times at same pitch, often at night (duPont & Rabor 1973b, Kennedy *et al.* 2000).

DESCRIPTION Nominate. **Adult** Black heavily glossed oil-green, and hackle-like feathers on head, back and breast with glossy black shafts, lower back and wings chestnut, inner secondaries darker than outer secondaries and primaries, all remiges tipped brown; underwing-coverts black glossed green edged white, below black with buffish shaft streaks; tail black. Presumed leucistic bird Luzon (BMNH) white with dark feather bases, yellow bill and dark feet. **Subadult** Above dark brown with pale shaft streaks on head, neck, including sides, and mantle; back, rump and uppertail-coverts dark brown, some pale streaks anteriorly. Wing-coverts deep chestnut with some dark barring and pale shaft streaks and spots; remiges barred rufous and dark brown. Throat to breast whitish with yellowish shaft streaks and brown barring; rest of underparts dark brown with pale shaft streaks. Tail blackish with greenish-bronze sheen, and few pale rufous bars distally. **Juvenile** Head and neck dark brown with buff shaft streaks broadening to spots on mantle; back barred dark brown and dark rufous, rump and uppertail-coverts brown narrowly fringed pale rufous. Wings dark rufous broadly barred dark brown. Throat mottled whitish and brown; rest of underparts brown barred pale fawn, less so on belly, flanks and undertail-coverts. Tail dark brown glossed greenish-bronze, narrowly barred pale rufous distally. **Nestling** Naked at hatching, with dark skin (Rabor 1977). **Bare parts** Iris red, orbital skin grey, juvenile iris brown. Bill black, juvenile lower mandible mottled grey and black. Feet black, juvenile grey. Hind claw 16.6–28.2mm.

BIOMETRICS Nominate. Wing male (n = 8) 142.5–155.5mm, mean 148.4 ± 4.7, female (n = 8) 151–178mm, mean 159.9 ± 8.5. Bill male 26.9–29.2mm, mean 28.1 ± 1.0, female 27.6–31.5, mean 29.8 ± 2.2. Tail male 223–234mm, mean 228.2 ± 5.6, female 235–265mm, mean 247.6 ± 9.8. Tarsus male 35.5–38.5mm, mean 37.0 ± 0.9, female 37.3–42.0mm, mean 40.0 ± 1.8. Mass male 100.2–126.7g, mean 112.1, female 108.4–152.1, mean 133.5 (Payne 2005). Mass *major* male 155, 169g, female 212, 219, 223g (Ross & Ramos 1992); *mindorensis* male (n = 6) 104–140g, mean 123, female 150, 165g (Ripley & Rabor 1958); *carpenteri* male 153, 170g, female (n = 6) 179–253g, mean 209 (Payne 2005).

MOULT Male moulting primaries and rectrices, and female rectrices, both Apr, Bohol; wing moult Aug and Sep (*viridis*), Nov (*mindoroensis*) (Rand & Rabor 1960, BMNH).

GEOGRAPHICAL VARIATION Four subspecies.
C. v. viridis (Scopoli, 1786). Apo, Balut, Banga, Bantayan, Banton, Basilan, Biliran, Bohol, Boracay, Cagayancillo, Calagna-an, Calicoan, Camiguin Sur, Carabao, Catanduanes, Cebu, Cuyo, Dinagat, Gigantes, Guimaras, Jolo, Leyte, Lubang, Luzon, Malamaui, Marinduque, Masbate, Mindanao, Negros, Olango, Panay, Pan de Azucar, Polillo, Romblon, Samar, Siargao, Siasi, Sibuyan, Siquijor, Tablas, Talicud and Ticao (Kennedy *et al.* 2000). Described above.
C. v. major Parkes & Niles, 1988. Calayan, Dalupiri and Fuga (and probably Camiguin Norte: Kennedy *et al.* 2000). As *viridis* but larger. Wing male (n = 5) 163–182mm, mean 171.0, female (n = 5) 175–190.5mm, mean 182.6 (Parkes & Niles 1988, Ross & Ramos 1992). Chestnut-winged birds on Babuyan and Pamoctan Claro presumed this form (Allen *et al.* 2006).
C. v. mindorensis (Steere, 1890). Caluya, Mindoro, Semirara and Sibay (Kennedy *et al.* 2000). Smaller than *carpenteri* and wings also blackish (few have faint rufous markings on wing-coverts). Wing male (n = 9) 143–158mm, mean 153.7, female (n = 5) 163–175mm, mean 171.8 (Ripley & Rabor 1958).
C. v. carpenteri Mearns, 1907. Batan, Ivojos and Sabtang (Kennedy *et al.* 2000). Larger; generally black glossed bluish, purplish and greenish without chestnut on wings. Wing male 166mm (Mearns 1907); male 173mm, female (n = 4) mean 189.5mm (Parkes & Niles 1988).

DISTRIBUTION Resident most Philippine islands (Kennedy *et al.* 2000).

HABITAT Great variety of habitats: grassland, forest, bamboo groves, disturbed secondary growth, scrub and mixed cultivation to 2,000m, but unlike Steere's Coucal never found in treetops. In Mt. Isarog NP 300–700m. Preferred habitat on Siargao and Dinagat dense tall grass, especially Kogon *Imperata* spp. and Talahib *Saccharum spontaneum* var. *indicum*, but also bamboo thickets; not above 300m (duPont & Rabor 1973b, Gonzales 1983, Goodman & Gonzales 1990, Evans *et al.* 1993, Kennedy *et al.* 2000).

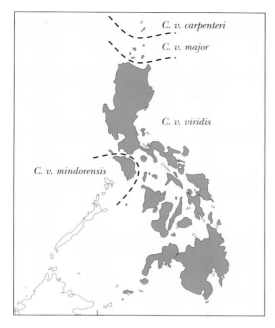

or other dense vegetation; singles or pairs. When disturbed flies with fast wing strokes alternating with gliding on outstretched wings low over ground, normally not more than 10–15m before landing and hiding among high grass (duPont & Rabor 1973b, Kennedy *et al.* 2000).

BREEDING Season May Negros; said to be Jun, Luzon, but male with enlarged gonads south Luzon, 28 Mar, and two nestlings, Apr. Five females in breeding condition Bohol, Apr. Nest with broken egg Catanduanes, May, and birds in breeding condition, Jun (Wolfe 1938, Rand & Rabor 1960, Rabor 1977, Gonzales 1983, Goodman & Gonzales 1990). **Nest** Dome-shaped with side-entrance, constructed of grass and leaves, well-concealed among tall grass up to 2m from ground (Kennedy *et al.* 2000). **Eggs** 2–3; white, chalky; 32.0–34.1x24.1–26.6mm, mean 32.7x25.4mm (Schönwetter 1967). **Incubation** *c.* 2 weeks (Rabor 1977). **Chicks** No information on fledging period. **Survival** Highly infested with endoparasites (Gonzales 1983).

FOOD Insects, e.g. caterpillars, ants and beetles; spiders; small lizards; occasionally carrion (Gonzales 1983).

STATUS AND CONSERVATION Common throughout Philippines. Often in cultivation and therefore future status seems secure (Dickinson *et al.* 1991, Kennedy *et al.* 2000). On Dinagat and Siargao found commonly in 1972 (duPont & Rabor 1973b). 38 recorded Aug 1991, Siquijor (Evans *et al.* 1993). Common Rajah Sikatuna NP, Bohol (Brooks *et al.* 1995a). Not globally threatened (BirdLife International 2011).

BEHAVIOUR Mostly on or near ground. Shy species difficult to observe because normally hidden in tall grass

Philippine Coucal *Centropus viridis*. **Fig. 1.** Adult, *C. s. viridis*, Subic, Zambales, Philippines, August (*Tina Sarmiento Mallari*). **Fig. 2.** Adult, *C. s. viridis*, Morong, Bataan, Luzon, Philippines (*Tina Sarmiento Mallari*). **Fig. 3.** Adult, *C. v. viridis*, leucistic individual, Laurel, Batangas, Luzon, Philippines, April (*Tina Sarmiento Mallari*). **Figs. 4–5.** Adult, *C. v. mindorensis*, sunning, Siburan, Mindoro, Philippines, May (*Paul Noakes*).

LESSER COUCAL
Centropus bengalensis   Plate 6

Cuculus bengalensis J.F. Gmelin, 1788 (Bengal)

Alternative names: Small Coucal, Lark-heeled Cuckoo.

TAXONOMY Once considered conspecific with *grillii* and *toulou*, but due to widely separated ranges and very different vocalisations here treated as three species (Roché 1971, King *et al.* 1975, Dowsett & Dowsett-Lemaire 1993). MtDNA analysis places *bengalensis* and *viridis* as sister taxa, and together forming sister clade to *grillii*, whereas *toulou* is closest to *goliath* (Sorenson & Payne 2005). Population in Western Ghats, India, darker, possibly undescribed subspecies. Polytypic. Synonyms: *bengalensis molkenboeri* Bonaparte 1850, error, is synonym of *C. viridis* (Wolters 1976, Dickinson *et al.* 1991); *b. takatsukasai* Momiyama 1932 in *lignator*; *aegyptius var g[amma]* Latham 1790 and *toulou chamnongi* Deignan 1955 in nominate; *Centrococcyx lepidus* Salvadori 1879 in *javanensis*.

FIELD IDENTIFICATION 31–35cm. Small coucal with short, stout decurved bill. Sexes similar; female larger. **Adult breeding** Glossy blue-black, often with pale shaft streaks, mantle feathers dull chestnut and wings dull rufous-brown with conspicuous buffish shaft streaks on shoulder, wings tipped darker; tail black glossed green. **Adult non-breeding** Dark brown head and mantle with prominent buff shaft streaks, rump dark brown barred rufous, below pale buffish streaked pale on throat and breast and barred darker on sides of breast and flanks; undertail-coverts barred dark brown and dull buffish. **Juvenile** As non-breeding adult but dark and light streaking on head and upperparts, wings rufous barred dusky, underparts off-white barred pale brown on flanks and thighs, tail rufous with narrow black barring, broadening near tip. **Similar species** Greater Coucal much larger with black, not chestnut, underwing-coverts and brighter chestnut wings, mantle also chestnut, and never any buff streaking; no non-breeding plumage. Philippine Coucal, sympatric on many islands, larger, head and underparts black glossed green, wings darker brown, no shiny shaft streaks on upperparts, tail black glossed blue-green, and iris in adult red; birds from Mindoro uniformly black, and juvenile lacks buff shaft streaks to dark brown and black barred upperparts, tail black glossed green and often barred chestnut at tip. Short-toed Coucal on Thai-Malay Peninsula, Sumatra and Borneo has bright chestnut wings and black underwing-coverts. Rare Javan Coucal from Java has black in chestnut wings.

VOICE In India series of *pwoop-pwoop*, *pwoop* notes first ascending and then descending in faster and more interrogative way than Greater Coucal; also 5–6 *whoot* notes, at first slow then faster. Rarer call is *kurook* or *kulook* or *kirook*. In Thai-Malay Peninsula series of sharp staccato *hup-hup-hup…tokato-tokato-tokato*, also rendered as *hup-hup-hup-t-t-tok, t-t-tok, t-t-tok*, and another accelerating call from same area *haup-haup-haup-haup… hau-hau-hau… ho-ho-ho-ho*. From south-east Asia in general voice described as series of metallic cluckings *thicthicthicthicyhic-thuc-thuc-thuc-thuc-thucthucthucthucthucucucucuc*, faster at beginning and end, or *wup-wup-wup/K-DA-KDA/droop-KDA/droop*, last like wooden knocking reminiscent of Blue-throated Barbet. Call in China and Greater Sundas three *hup* notes followed by series *logokok-logokok-logokok*. In Borneo 3–4 *whoop* notes followed by staccato *boob-boob-boob, kok-kok-oo, kok-kok-oo, kok-kok-oo*, all high pitched. In Philippines single *hoop* followed by loud series of *tu-dut* or *tu-dut-dut* repeated seven or more times in sharp stuttering manner. Sometimes calls at night (MacKinnon & Phillipps 1993, 2000, Grimmett *et al.* 1998, Jeyarajasingam & Pearson 1999, Smythies & Davison 1999, Wells 1999, Kennedy *et al.* 2000, Robson 2000, Payne 2005, Rasmussen & Anderton 2005).

DESCRIPTION Nominate. **Adult breeding** Head, neck, upper mantle and entire underparts black, black feather shaft streaks on head, and forepart of body conspicuously glossy, wings, wing-linings and wing-coverts dull rufous-brown, coverts with pale shaft streaks, remiges tipped dark brown, lower mantle and upper back chestnut, underwing-coverts rufous, long uppertail-coverts and tail black with bronze-green gloss, sometimes tipped white. **Adult non-breeding** Dark rufous-brown head and upperparts with conspicuously pale to whitish shaft streaks, wings dull rufous; underparts pale buffish with fine barring on breast and flanks, long uppertail-coverts rufous finely barred black; tail blackish tipped whitish or fawn. **Subadult** Like non-breeding adult but head and upperparts brown finely streaked white with black barring in rufous wings; tail barred as juvenile. **Juvenile** Similar to non-breeding adult but paler with less obvious streaking; tail dark brown barred rufous-brown. **Nestling** Hatches naked, skin black, long grey-white or buffish trichoptiles on head and upperparts; iris grey, dark red-pink mouth lining (Wells 1999, Payne 2005). **Bare parts** Iris red, juvenile brown with orbital area yellowish. Bill black, juvenile fleshy yellow-brown, culmen darker and black at base, lower mandible fleshy. Feet black, juvenile slaty; hind claw longer than toe, 21.2–25.6mm.

BIOMETRICS Nominate. Wing male (n = 7) 138–150mm, mean 145.0 ± 6.0, female (n = 12) 158–174mm, mean 165.9 ± 4.9. Bill male 21.2–24.4mm, mean 22.8 ± 1.2, female 22.6–26.0mm, mean 24.4 ± 1.2. Tail male 161–188mm, mean 172.3 ± 10.3, female 180–215mm, mean 194.0 ± 11.9. Tarsus male 34–38mm, mean 36.1 ± 1.9, female 40–45mm, mean 42.7 ± 2.2 (Payne 2005). Mass male (n = 3) 79.3–92.0g (Wells 1999).

MOULT Irregular transilient moult with up to three loci for primaries. Adult wing moult Aug–Nov, once Feb (Stresemann & Stresemann 1961, Wells 1999). Central rectrices last to be moulted into adult plumage 24 Jun, central Thailand (Riley 1938a).

GEOGRAPHICAL VARIATION Six subspecies, mostly differing in size.
C. b. bengalensis (J.F. Gmelin, 1788) India, Nepal, Bangladesh, Burma, Thailand and Indochina. Described above.
C. b. lignator Swinhoe, 1861. China; Taiwan. Larger than nominate. Wing male 148–159mm, (n = 5) mean 153.2, female165–174mm, (n = 11) mean 169.5 (Stresemann 1912).
C. b. javanensis (Dumont, 1818). Thai-Malay Peninsula, Sumatra, Riau and Lingga archipelagos, Banka, Belitung, Java, Bali, Borneo, Natuna Is. to south-west Philippines (Balabac, Bongao, Calamianes, Calauit, Jolo, Palawan, Sanga Sanga, Tawi-Tawi). Smaller than nominate; black glossed blue, mantle dark brown. Underwing-coverts brown, or brown and black. Wing male 125–147mm (n = 29) mean 133.6, female 150–166mm, (n = 25) mean 158.9 (Stresemann 1912).
C. b. philippinensis Mees, 1971. Philippines except south-west (Bantayan, Basilan, Batan, Batbatan, Biliran, Bohol, Calagna-an, Cebu, Fuga, Gigantes, Leyte, Luzon, Manuk,

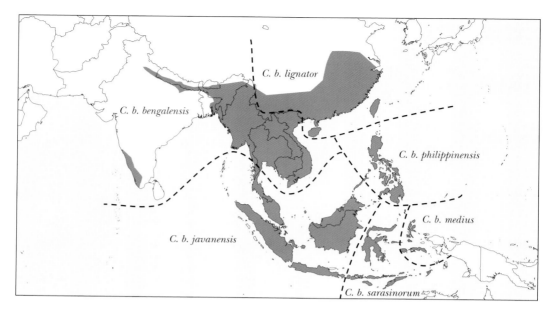

Manka, Mindanao, Mindoro, Negros, Panay, Sabtang, Samar, Semirara, Siasi, Sibay, Siquijor). Mantle dark brown with buffish shaft streaks, tail brownish-black tipped buff. Presumably this form Camiguin Norte and Caucauayan (Babuyans) (Allen *et al.* 2006). Payne (1997, 2005) lumps with *javanensis*, considering differences due to plumage wear or age.

C. b. sarasinorum Stresemann, 1912. Sulawesi, Banggai Is, Sula Is. and Lesser Sundas. Larger and darker than nominate. Wing male 146–159mm, (n = 5) mean 153.4, female 175–188mm, (n = 5) mean 178.2 (Stresemann 1912).

C. b. medius Bonaparte, 1850. Moluccas except south-east and possibly Tanimbar Is. Larger than *sarasinorum*. Wing male 160–177mm, (n = 9) mean 171.7, female 190–205mm, (n = 12) mean 199.0 (Stresemann 1912).

DISTRIBUTION Oriental. Resident over whole range, possibly with exception of Himalayas, where mostly summer, and Hebei, east China. India in south-west, and north-east from north Uttar Pradesh to Arunachal Pradesh, Bhutan and Bangladesh. Burma and Indochina. All divisions of Thai-Malay Peninsula. Riau Archipelago on Bintan, Bukit and Karimun Besar; Natuna Islands. China in Hebei, south Henan, Yunnan, south Guizhou, Hubei, Hunan, Anhui, Jiangxi, Jiangsu, Zhejiang, Fujian, Guangdong, Hong Kong, Guangxi Zhuang, Hainan, and Taiwan. Greater Sundas. Sulawesi group on Bangka, Butung, Kalao, Lembeh, Manterawu, Muna, Peling, Ruang, Sangihe, Siau, Sulawesi, Talaud, Tukangbesi (Kaledupa) and Tanahjampea; Togian. Lesser Sundas on Adonara, Atauro, Besar, Flores, Kisar, Lembata (Lomblen), Padar, Paloe, Pantar, Poeloe Endeh, Rinca and Timor; Alor and Roti. Moluccas on Amboina, Batjan, Buru, Ceram, Halmahera, Morotai, Obi, Ternate and Tidore, Taliabu and Tanimbar. Philippines (Inskipp & Inskipp 1985, Smythies 1986, White & Bruce 1986, Cheng 1987, Stones *et al.*1997, Grimmett *et al.* 1998, Wells 1999, Kennedy *et al.* 2000, King *et al.* 2001, Payne 2005, Trainor 2005a, b, Zheng *et al.* 2005, Indrawan *et al.* 2006, J. Hornskov *in litt.*, C.F. Mann). Havelock I., Andamans, 5–6 April 2004 (Chandi 2007); possibly Great Nicobar I. (Sivakumar 2000 in Chandi 2007). Twice South Korea, May and Jun (Birds Korea 2009). Once (*sarasinorum*) Ashmore Reef, Australia, 28–30 Oct 2005 (Carter *et al.* 2010). Old trade skin from Sri Lanka of doubtful provenance (Rasmussen & Anderton 2005).

HABITAT In India at forest edges with dense scrub, tall grass, reedbeds to 900m in south-west India; in Nepal also in *Salina* grass mixed with bushes, trees and bamboos; to 1,800m Himalayas. In Thailand favours marshy areas; to 1,800m. Preferred habitat Thai-Malay Peninsula open grassland with low bushes, both in dry and marshy areas, also paddifields and reedbeds; to 1,500m Cameron Highlands. Occasionally scrub and cultivation central Laos. Agriculture, Taliabu, Moluccas. Also frequents hillside scrub, China. In Borneo open country, *Pandanus* fringe and young *Albizia* plantations, also cultivation including paddifields, *Acacia* plantations, secondary scrub and long grass. On Java three breeding pairs reported in large garden. On Greater Sundas to 1,000m, rarely 1,500m, but 2,000m Borneo. Grassland and thick scrub, Bali. In Philippines never forest or forest edge, only grassland. Grassland and disturbed secondary growth to 1,000m Taliabu (Stresemann 1913a, Spennemann 1928, Lekagul & Round 1991, MacKinnon & Phillipps 1993, 2000, Thewlis *et al.* 1996, Stones *et al.* 1997, Grimmett *et al.* 1998, Smythies & Davison 1999, Wells 1999, Kennedy *et al.* 2000, Mann 2008, Rheindt 2010).

BEHAVIOUR Usually solitary. Forages mostly on ground like pheasant, but perches inactive for long periods often on top of grass. Unlike Greater Coucal never perches or clambers in trees, and has slow flapping flight low above grass. Three pairs in garden in Java had territories in all covering *c.* 4ha; male often seen with drooping wings and tail raised and spread. Supposed courtship feeding with leaf and grasshopper. Grasshoppers fed three times to female, first after *c.* 30sec copulation where female drooped wings and turned her straightened head alternately left and right. Both sexes said to take part in nest-building, incubation and feeding offspring, but apparently male performs most if not all parental duties which makes it possible for female to build up reserves for laying more eggs. Faecal sac carried 60m away from nest before dropped (Spennemann 1928, Aagaard 1930, Ali & Ripley 1983, Grimmett *et al.* 1998, Smythies & Davison 1999, Wells 1999, Kennedy *et al.* 2000, C.F. Mann).

BREEDING Season Mar–Oct India, varying locally. Nests with eggs Thailand, May–Aug. Thai-Malay Peninsula nearly year-round, Dec–Jul, fledgling early Oct. Dec–Oct south-east Asia generally. Eggs Hainan, May–Jun and south China, Jun. Nests Sabah, Borneo, Dec–May, one feeding nestling, 12 Sep, and oviduct eggs, 9 Feb; nesting Kelabit uplands, Sarawak, Jan–Feb; breeding activity Brunei, May–Jul; nestling East Kalimantan, 11 Sep. Breeding Flores, Feb–May. Breeding condition Negros, Philippines, May (Hartert 1910, La Touche 1931, Riley 1938b, Smythies 1957, Pfeffer 1960, Ottow & Verheijen 1969, Rabor 1977, Mann 1991a, Grimmett *et al.* 1998, Smythies & Davison 1999, Wells 1999, Robson 2000, Sheldon *et al.* 2001). **Nest** Untidy dome 25x18cm, of twigs, grass-blades and leaves with large entrance on side and occasionally sparse green leaves in egg-chamber, in open grassland, dense bush, rhododendron, various grasses or bamboo clump 0.2–1.5m high (Spittle 1950, Grimmett *et al.* 1998, Wells 1999). **Eggs** 2–3; chalky or faintly glossy white eggs, often heavily soiled, normal to longish elliptical; 28.7–31.8x22.8–26.7mm (n = 6); India, mean (n = 50) 28.2x23.8mm; mean 29.0x24.5mm; west Java 26.1–34.8x21.5–27.1mm, (n = 29) mean 30.1x23.9 (Hellebrekers & Hoogerwerf 1967, Schönwetter 1967, S. Baker in Ali & Ripley 1983, Wells 1999). **Incubation** Both parents incubate (Riley 1938b); period unknown. **Chicks** (See Description). **Survival** Two birds re-trapped after 15 months (Wells 1999).

FOOD Principal food grasshoppers. One stomach contained beetles and grasshoppers, another 13 hemiptera, two hairy caterpillars, two locusts, lizard's bone and bird or mammal flesh; nestling fed with gecko; mantis carried to nest; stick insects, other insects, small birds, reptiles (La Touche 1931, Deignan 1945, Grimmett *et al.* 1998, Smythies & Davison 1999, Wells 1999).

STATUS AND CONSERVATION Locally common India and Nepal (Grimmett *et al.* 1998). Common resident Thailand (Lekagul & Round 1991). Common and regular Thai-Malay Peninsula (Wells 1999). Fairly common to common resident throughout south-east Asia; ?uncommon winter visitor north-east Thailand, uncommon passage east Tonkin (Robson 2000). Common Myingyan District, Burma (MacDonald 1906), but local and rare (Smythies 1986), and rare north Burma 1998–99 (King *et al.* 2001). Rare north Cambodia, Laos and Vietnam. Common Greater Sundas (Nash & Nash 1985b, MacKinnon & Phillipps 1993, Parrott & Andrew 1996, Smythies & Davison 1999, Mann 2008). Fairly common China (Cheng 1987) and Philippines (Dickinson *et al.* 1991). Regular Buru, Moluccas (Jepson 1993). Common Damar (Trainor 2007) and Taliabu (Rheindt 2010). Vagrant South Korea (Birds Korea 2009) and Australia (Carter *et al.* 2010). Not globally threatened (BirdLife International 2011).

Lesser Coucal *Centropus bengalensis*. **Fig. 1.** Adult, breeding plumage (with traces on non-breeding plumage), *C. b. bengalensis*, Dhikala, Corbett NP, India, May (*Rohan Kamath*). **Fig. 2.** Adult, non-breeding plumage, *C. b. javanensis*, Ipoh, Perak, Malaysia, November (*Amar-Singh HSS*). **Fig. 3.** Adult, breeding plumage, *C. b. lignator*, Hong Kong, May (*Martin Hale*). **Fig. 4.** Adult, breeding plumage (with traces of non-breeding plumage) carrying prey for young, *C. b. javanensis*, Ipoh, Perak, Malaysia, June (*Amar-Singh HSS*).

VIOLACEOUS COUCAL
Centropus violaceus   Plate 5

Centropus violaceus Quoy & Gaimard, 1830 (Carteret Harbour, New Ireland)

Alternative names: Violet/Giant Forest/Bare-eyed Coucal.

TAXONOMY Affinities unclear; perhaps close to *chalybeus*. Results of mtDNA analysis place it in clade with *bernsteini* and *phasianinus* (Sorenson & Payne 2005). Populations on both islands seem to be phenotypically stable in spite of long residence there, and seem to avoid crossing water gaps (Mayr & Diamond 2001). Monotypic.

FIELD IDENTIFICATION 62–70cm. Huge coucal with rather long, decurved bill. Sexes similar. **Adult** Entirely black with strong violet sheen. In worn plumage colour changes to blue-black. Naked skin around eyes whitish to red; feet pale. **Juvenile** Dull black, wings and tail glossed purple, underparts sooty-grey. **Similar species** None in range.

VOICE Deep hollow far-carrying booming repeated many times. Duetting sounds like giant hiccough. Loud, deep plopping *woo-woo-woop*, first note softer (Diamond 1972a, Orenstein 1976, Mayr & Diamond 2001).

DESCRIPTION Adult Head, upperparts, wings, underparts and broad, graduated tail black glossed violet. **Juvenile** As adult, but upperparts dull black, below sooty or dark grey-brown; wings and tail glossed purple; feathers more loose. **Nestling** Skin black with long white trichoptiles on head, upperparts and wings; underparts naked. **Bare parts** Iris red, juvenile grey, orbital skin white to grey or reddish, eye-ring black. Bill black, nestling's black tipped white. Feet pale slaty-horn, light brown or whitish to yellow; nestling bluish-grey. Hallux claw 20–25mm.

BIOMETRICS Wing male (n = 6) 258–292mm, mean 271.2 ± 12.5, female (n = 5) 245–288mm, mean 261.2 ± 16.0. Bill male 61.5–69.0mm, mean 65.3 ± 2.9, female 62–77mm, mean 69.1 ± 5.9. Tail male 340–405mm, mean 379.6 ± 18.4, female 310–400mm, mean 356 ± 34.9. Tarsus male 61.5–68.0mm, mean 65.3 ± 2.9, female 62.5–70.0mm, mean 66.3 ± 3.2 (Payne 2005). Mass female >500g (Gilliard & LeCroy 1967).

MOULT Adult male in tail and body moult, 17 Nov; most of eight birds in both wing and tail moult Jan–Apr (Hartert 1925a, 1926).

DISTRIBUTION Resident New Britain and New Ireland, but absent from nearby small islands (Mayr & Diamond 2001).

HABITAT Lowland and hilly primary forest with dense vines or twining ferns. To *c.*1,370m Nakanai Mountains, New Britain; *c.* 900m Whiteman and Talawe Mts. but more typically lowland forest to *c.* 760m (Hartert 1926, Coates 1985, Bishop & Jones 2001).

BEHAVIOUR Solitary or in pairs and difficult to find in dense vegetation. Works its way up in trees and then glides down to next tree. When duetting perch on branch face to face; when giving booming calls neck pulled down and tail lowered. Three birds sunning while pair of Pied Coucals foraged in same tree. Courtship feeding with large insect in bill (Gilliard & LeCroy 1967, Diamond 1972a, Fletcher 2000, Coates 1985, Payne 2005).

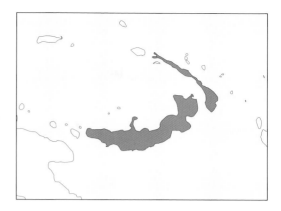

BREEDING Season Three nestlings early Jan, two nestlings and fledgling late Nov (Gilliard & LeCroy 1967). **Nest** Crow-like loose structure of dry twigs and green leaves on tops of tall trees (Meyer 1931, Gilliard & LeCroy 1967, Payne 2005). **Eggs** White with dull gloss; 41.6x33.9, 43.5x34.5mm (Meyer 1931, Schönwetter 1967). **Incubation** No information. **Chicks** Two nestlings, 37 and 42g, still with egg-tooth; another 151g, with feathers in pin and remiges emerging from sheaths (Gilliard & LeCroy 1967). **Survival** No information.

FOOD Stomach contents large insects (up to 18+mm), frogs and tiny snails (Gilliard & LeCroy 1967).

STATUS AND CONSERVATION Widely but thinly distributed on both islands, most common on New Britain (Coates 1985). Widespread Nakanai Mts, west New Britain to 760m, but uncommon in lowland forest (Bishop & Jones 2001). Low population density with less than 1,000 pairs in 10,000km^2 (Mayr and Diamond 2001). Restricted range species. Considered near threatened by conversion of forests into oil palm plantations; population estimate 10,000–19,999 decreasing (BirdLife International 2011).

Violaceous Coucal *Centropus violaceus*. **Fig. 1.** Adults, Pokili Forest, New Britain, PNG, June (*Nik Borrow*).

LESSER BLACK COUCAL
Centropus bernsteini   Plate 4

Centropus Bernsteini Schlegel, 1866 (Salawati, New Guinea)

Alternative names: Bernstein's/Scrub/Black Scrub Coucal.

TAXONOMY MtDNA analysis indicates it forms clade with *violaceus*, *nigrorufus* and *phasianinus* (Sorenson & Payne 2005). Birds from Manam I. said to be larger but great overlap in size between those and birds of NG (Payne 2005). Monotypic. Synonyms: *spilopterus* Sharpe 1876 and *bernsteini manam* Mayr 1937.

FIELD IDENTIFICATION 45–52cm. Medium-sized coucal with broad tail about same length as body. Sexes similar. **Adult** Black with dark eyes. **Juvenile** Black barred buff and rufous; throat white. Long hallux claw straight. **Similar species** Biak Coucal all black but with strong purple wash above and yellow iris, only found on Biak I., Geelvink Bay. Sympatric Greater Black Coucal about one third larger, with pale bill and red eyes, but otherwise determined by voice. Juvenile could be confused with much larger Pheasant-coucal, but former has dark eye.

VOICE Deep *hou* repeated up to ten times and descending towards end, or fast double note. Three singing together: single bird gave repeated *hui* and pair answered with similar but deeper and slower *whui*. Slow descending *woop-woop-woop* reminiscent of Brown Cuckoo-dove. Calls mornings and before sunset (Coates 1985, Beehler *et al.* 1986).

DESCRIPTION Adult Black with glossy green wash on upperparts and shining black shaft streaks on head, neck, mantle, throat and upper breast. **Juvenile** Blackish with rich rufous barring on upperparts, chestnut-brown on neck; wings and tail narrowly barred paler chestnut or buffish; underparts black-brown with finer pale barring, except white throat, and grey centre of abdomen barred dusky; feathers on sides of neck and back with brown shaft streaks or tipped brown; small brown bars on uppertail-coverts, flanks and thighs (Junge 1937a). **Nestling** Face feathers with small white tips; no white trichoptiles as in most other coucals (Payne 2005). **Bare parts** Iris dark brown, juvenile grey. Bill black, juvenile brown. Feet black, juvenile grey.

BIOMETRICS Wing male (n = 12) 164–193mm, mean 173.8 ± 8.5, female (n = 8) 176–196mm, mean 180.0 ± 8.0. Bill male 27–33mm, mean 31.3 ± 6.0, female 28–35mm, mean 31.8 ± 6.2. Tail male 220–265mm, mean 247.4 ± 13.6, female 230–265mm, mean 247.8 ± 10.9. Tarsus male 39–44mm, mean 41.2 ± 1.9, female 38–44mm, mean 41.4 ± 2.3. Mass male (n = 5) 130–160g (n = 5), mean 146, female 160, 200g. Hallux claw male 21–26mm, mean 23.9 ± 1.4, female 24–26mm, mean 25.1 ± 0.6 (Payne 2005).

MOULT Juvenile Jan (Junge 1937a).

DISTRIBUTION Resident north and west NG east to Fly R. in south and Huon Peninsula in north; Manam (= Vulcan) I. (Coates 1985).

HABITAT Areas with tall grass and cane grass surrounding primary and secondary forests, roadside vegetation, near rivers and lakes; lowlands to 460m, rarely 900m (Coates 1985, Beehler *et al.* 1986).

BEHAVIOUR Shy and difficult to observe. Singly or in pairs. Sometimes perches in open areas like (Gilliard & LeCroy 1966, Coates 1985).

BREEDING Season Presumably year-round. Eggs Manam I., Dec; fledgling Idenburg R., Dec; females in breeding condition, Mar–Apr; laying female Kaku, Nov (Rothschild & Hartert 1915, Rand 1942b, Payne 2005). **Nest** Oval with untidy side entrance, 210x380mm, inside 145x250mm, constructed only of flat dried grass leaves placed among grass stems (Rand 1942b). **Eggs** Broadly oval; white, slightly glossed; 31–33x25–26mm; 31.2–34.0x25.6–28.0mm, mean 33.3x27.0mm; eggs Manam I. much larger, 33–38x27–29mm (Rothschild & Hartert 1915, Rand 1942b, Schönwetter 1967). **Incubation** Nest also attended by male (Payne 2005). No further information.

FOOD Small lizards and snakes; butterflies and other insects in grass (D'Albertis & Salvadori 1879, Gyldenstolpe 1955, Payne 2005).

STATUS AND CONSERVATION Locally not uncommon, but secretive (Gilliard & LeCroy 1966, Coates 1985). Very little known and monitoring therefore desirable. Not globally threatened (BirdLife International 2011).

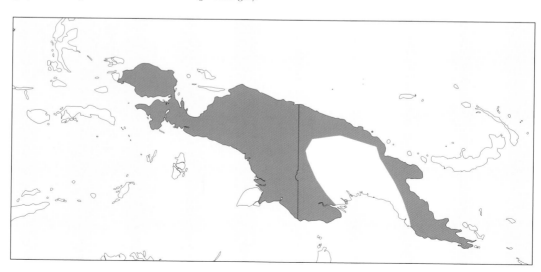

Lesser Black Coucal *Centropus bernsteini.* **Fig. 1.** Adult, Kiunga, Western Region, PNG, September (*Philip Barden*).

PHEASANT-COUCAL
Centropus phasianinus   Plate 5

Cuculus phasianinus Latham, 1801 (New Holland = New South Wales)

Alternative names: Swamp Pheasant; Spur-footed Cuckoo, Timor Coucal (*mui*), Kai Coucal (*spilopterus*).

TAXONOMY MtDNA analysis places it in clade with *bernsteini* and *violaceus* (Sorenson & Payne 2005). Subspecies *mui* only known from one female from Los Palos swamp, Timor-Leste, 14 May 1974 (Mason & McKean in Mason *et al.* 1984), and therefore taxonomic status uncertain; may represent allospecies of *phasianinus* (White & Bruce 1986), or variant within nominate (Payne 2005). *Melanurus* subsumed in nominate by Payne (2005) due to overlap both in size and tail patterns, but Higgins (1999) acknowledges it as subspecies with narrow zone of secondary hybridisation south of Ayr, Queensland, and is accepted here. *Spilopterus* G.R. Gray, 1858, a melanistic form, has been specifically separated as Kai Coucal (Peters 1940, Payne 1997). In east Australia abrupt change in wing and tail lengths in Burdekin Valley indicating past geographical isolation and secondary contact (Mason *et al.* 1984). Polytypic. Synonyms: *Poliophilus phasianus (lapsus)* Leach 1814. *Polophilus Gigas* Stephens 1815, *Corydonix giganteus* Vieillot 1819 and *Polophilus p. yorki* Mathews 1916, all in nominate; *macrourus* and *melanurus*, both Gould 1847, *Polophilus phasianinus keatsi* Ashby 1915, *p. melvillensis* Mathews 1919 and *p. highami* Mathews 1922, all in *melanurus*; *p. obscuratus* Mayr 1937 in *nigricans*.

FIELD IDENTIFICATION 54–68cm. Great variation in size, and largest female near size of Goliath Coucal, with long tail and pheasant-like plumage, but often rather untidy. Difficult to flush and flies only short distances low over ground on broad, rounded wings. Sexes similar but female larger. **Adult breeding** Head, neck and upper mantle black, lower mantle rufous streaked pale, wings red-brown barred and streaked buff and white, uppertail-coverts blackish-brown barred white and black; tail blackish barred and mottled brown, central rectrices tipped white; in flight broad dark trailing edges to remiges. **Adult non-breeding** Head, neck and underparts buffish; rufous upperparts streaked pale yellow. **Juvenile** As non-breeding adults but paler brown. **Similar species** Unmistakable due to its huge size and brown barred wings and tail, only confusion species in NG is young Lesser Black Coucal, but it is much smaller with dark brown iris. Greater Black Coucal has pale bill and prefers forests. Only coucal in Australia.

VOICE Up to 25 hollow notes *hooo-hoo-hoo-hoo-hu-hu-h-h-h…h-h-h-hu-hu-hu-hu-hu*, first slow then faster and slower again at end, lasting 4+sec; also rendered as *oop-oop-oop*. Slow call of 20 notes *whu-whu-whu-whu…* lasting *c.* 6sec. Also *schiaw-shiaw-shiaw* 1.3–2.4 notes/sec. Alarm call sharp *tschew*, or scolding *chew*, or harsh *pthuck*; grunt or soft hoots often mumbled when foraging or when close to nest. Double alarm call by parents causes young to leave nest; *chuff-chuff* when chasing juvenile away from territory. Female most vocal during dawn chorus; male's call higher pitched than female and pairs sometimes duet. Audible up to 2km. 'Kai' Coucal has very similar calls. Series of far-carrying *bob* notes, changing between ascending and descending, starting slowly but becoming gradually faster dying out at end, lasting 6sec. Another call harsh, medium-pitched *chuk-chk-chk* of 8–10sec, often in answer to first-mentioned call. Vocalises also by night but rarely heard outside breeding period. Calls NG and Australia similar (Rand 1942a, Hindwood 1957, Diamond 1972b, Mackness 1979, Coates 1985, Beehler *et al.* 1986, Taplin & Beurteaux 1992, Coates & Bishop 1997, Higgins 1999).

DESCRIPTION Nominate. **Adult male breeding** Head, neck and upper mantle black to black-brown with glossy shaft streaks, more pronounced on neck and mantle; lower part of mantle, scapulars and tertials rufous-brown barred and vermiculated black; rufous-brown and off-white with cream-white spiny streaks on mantle and shoulder; back and rump dark grey-brown to black-brown vermiculated buff to whitish; long uppertail-coverts dark brown mottled pale brown to rufous-brown and barred whitish; remiges buffish on outer webs grading to rufous-brown on inner webs with black half-bars not reaching shafts, and on outer web, with bold blackish subterminal bars and rufous-brown

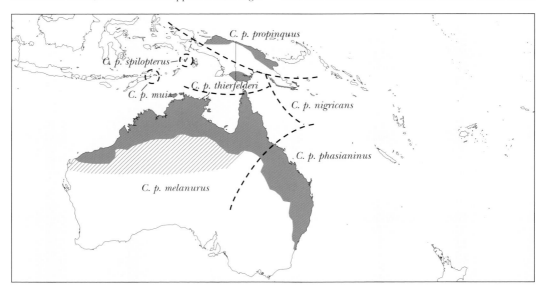

tips; wing-coverts and alula pale buff mottled rufous-brown on inner webs and broad dark brown on outer webs, and creamy shaft streaks; underwing light rufous barred dark brown, coverts vermiculated buff and dark brown with creamy shaft streaks, greater coverts barred dark grey-brown and rufous-brown; underparts black-brown rarely with some light brown mottling on centre of breast, thighs and vent; tail black-brown, in fresh plumage tipped white and narrowly barred pale brown. **Adult female breeding** Like male but larger. Black bars on inner webs of primaries shorter, becoming rufous-brown from P3–8, and pale and dark tail barring broader. **Adult male non-breeding** Head and neck dark rufous-brown with creamy shaft-streaks, top of head with black-brown vermiculations and creamy shaft-streaks bordered black; mantle rufous-brown with creamy shaft-streaks bordered dark brown, and feathers with fine dark brown vermiculations at tip, rest of upperparts and wings like male in breeding; underparts buffish to dark cinnamon with creamy shaft-streaks; tail as in breeding but pale barring often broader. **Adult female non-breeding** Like male but wings and tail pattern as adult female breeding. **Juvenile** Head and neck buffish to red-brown mottled dark brown, top of head darker, mantle and shoulder feathers dark brown, creamy-buffish spot in middle of each feather, spots vary in size or rarely missing; rest of upperparts dark brown narrowly barred rufous-brown and tipped off-white; wings as adult but narrower markings; chin and throat buff edged dark brown, midline of chin and throat paler, rest of underparts brownish variably spotted, streaked and barred buff, broadest on midline; rectrices narrower than adult with more pointed tips, pale barring slightly broader. **Nestling** Naked grey-black skin on most of upper- and underparts with creamy long, wiry trichoptiles on top and sides of head, neck, mantle, sides of back and rump, shorter on breast, flanks, wings and tail. **Bare parts** Male iris bright red, outside breeding red to pale brown, chestnut or yellowish spotted brown; female orange-red to bright red, outside breeding whitish or creamy with brown centre to pinkish straw-yellow to grey-brown, juvenile brown, orbital ring dark grey; nestling dark brown. Bill grey-black, outside breeding pale horn, juvenile paler lower mandible, nestling black; mouth whitish to dark flesh, nestling red. Feet bluish-grey or greenish-grey to black, nestling leaden-grey; soles dirty grey, claws grey-black; hallux claw 20–27mm.

BIOMETRICS Nominate. Wing male (n = 23) 205–248mm, mean 225.6 ± 10.29, female (n = 31) 235–267mm, mean 249.3 ± 8.26. Bill male (n = 21) 33.1–37.5mm, mean 35.5 ± 1.27, female 35.2–42.6mm, mean 38.6 ± 1.75. Tail male breeding (n = 13) 282–342mm, mean 310.2 ± 18.67, non-breeding (n = 6) 323–356mm, mean 338.3 ± 12.26, female breeding (n = 12) 305–361mm, mean 337.1 ± 21.22, non-breeding (n = 9) 333–393mm, mean 360.2 ± 22.1. Tarsus male (n = 23) 46.3–53.8mm, mean 50.5 ± 1.63, female (n = 32) 49.5–56.7mm, mean 53.2 ± 1.82. Mass male Australia (n = 16) 200–364g, mean 302.1 ± 36.56, female (n = 22) 242–520g, mean 444.9 ± 63.21 (Higgins 1999). Mass *thierfelderi* male Aug 250, 281g, Mar 300g, female Apr 375g (Mees 1982).

MOULT 3 Mar and 1 Sep (*nigricans*). Male and female full moult WA, 28 Jun and 2 Jul; male body moult WA, 19 Jun; two NT, Nov. Type of *mui* in partial moult Timor-Leste, 14 May. T1 replaced twice per year, others once (Hall 1974, Mason & McKean 1984, Higgins 1999).

GEOGRAPHICAL VARIATION Seven subspecies.
C. p. phasianinus (Latham, 1801). Eastern Australia. Some intergrades with *melanurus*. Described above.

C. p. propinquus Mayr, 1937. North NG east to Astrolabe Bay. Like *nigricans* but smaller and paler, white markings on wing. Wing male (n = 5) 194–219mm, mean 202.4, female (n = 5) 207–237mm, mean 228.8 (Payne 2005).
C. p. nigricans (Salvadori, 1876). South-east NG from Hall Sound eastwards; Yule I. and D'Entrecasteaux Is. Variable; lacks more pronounced wing-barring of *thierfelderi* and yellow underwing bars narrower than black bars and less bright plumage. No non-breeding plumage. Wing male (n = 10) 200–229mm, mean 216.8, female (n = 10) 230–248mm, mean 236.3 (Payne 2005).
C. p. spilopterus G.R. Gray, 1858. Kai Is. Chiefly blackish; primaries rarely with narrow pale bars; tail unbarred. No non-breeding plumage. Wing male (n = 9) 212–241mm, mean 228.9, female (n = 4) 249–268mm, mean 256.0 (Payne 2005).
C. p. thierfelderi Stresemann, 1927. South NG from Merauke, WP to Fly R. and north-west Torres Strait islands. Wing shafts reddish-brown, underwing bars rufous and same width or wider than black bars, underparts blackish. No non-breeding plumage (Mees 1982). Wing male (n = 10) 205–222mm, mean 214.3, female (n = 8) 222–251mm, mean 238.3 (Payne 2005).
C. p. melanurus Gould, 1847. North and north-west Australia. Larger than nominate with broader black bars on rectrices (third dark bar from tip of T1 male, 10–16mm, female 11–18mm). Wings male (n = 36) 226–249mm, female (n = 21) 247–280mm (Mason *et al.* 1984). Some intergrades with nominate.
C. p. mui Mason & McKean, 1984. Timor-Leste. Much white on foreparts with some dark brown on crown, and white below; bill proportionately larger. Only known from type. Wing female 250mm (Mason & McKean 1984).

DISTRIBUTION Widespread resident in whole north and east Australia from Exmouth Gulf in north-west to around Sydney in south-east. Presumably some movements in non-breeding season but may be confounded by marked seasonality in conspicuousness. Rarely extends range beyond coastal watershed, i.e. 320km from coast. In last atlas project only one record from central Australia, probably due to its weak flying ability and absence of dense cover in this arid area. Resident in east NG except highlands and most of Gulf Province; Yule I., D'Entrecasteaux Is.; Kai Is.; Timor-Leste. Entered Australia from NG, and then from Australia to Timor. Unrecorded from some islands in Torres Strait suggesting no dispersal between NG and Australia (Hindwood 1957, Mason *et al.* 1984, Higgins 1999, Barrett *et al.* 2003, Payne 2005). *Nigricans* collected Fergusson Island, 1894 (Rothschild & Hartert 1907), and type of *obscuratus*, Nov 1928 (Mayr 1937) but status now uncertain.

HABITAT Generally absent from forests. Common in pastures and *Lantana* thickets Atherton Shire, Queensland, and *Pandanus* swamp in sandstone gorge near Joint Hill, WA. In Darwin area, NT, in open forest with undergrowth of long grass, and in savanna. Plentiful along railways, Townsville, Queensland. Dense riverine vegetation, cane fields and dense grass along roadsides. In NG prefers savanna and grassland with scrub, forest edges or road edges through forest, more rarely reed beds and monsoon forest; gardens and regrowth; most common in lowlands and hills but to 1,800m Morobe Province. Wooded cultivation, forest edges, scrub and small forest patches from lowlands to 200m Kai I. (Hopkins 1948, Bravery 1970, Crawford 1972, Hall 1974, Mason *et al.* 1984, Coates 1985, Beehler *et al.* 1986, Coates & Bishop 1997).

BEHAVIOUR Solitary or in pairs, supposedly forming long-term pairs. Pair has 6ha territory; calling males may have territorial quarrels with neighbouring males; *c.* 10month old bird showed territorial behaviour. Mostly terrestrial with very weak flight. Runs at high speed with wings lowered, tail vertical and neck stretched out. In open places most often seen morning, late afternoon and after rain, and much less shy than other coucals. In long grass often impossible to flush. May plane down from high in tree to another tree or ground, crash-landing clumsily. Forages in thickets, occasionally upside-down. Flushes food by walking through undergrowth. Sunbathes, and spreads wet wings and tail to dry in sun; dusts in cold ashes from man-made fire. When calling nape feathers erected. Courting male behind female moved his head from side to side and up and down, then female ran ahead with body close to ground, male following her; then male drew himself up, spread and lowered wings, female stopped running, raised her body and male mounted her, and copulation took place. Courtship feeding observed. Parent observed carrying young in claws but losing it before flying 3m; egg-carrying from disturbed nest reported but not confirmed. Not determined if both sexes incubate, but male plays major role in feeding young; during three warmest hours male perched near nest, gular-fluttering and ignoring begging of nestlings; in captivity only male incubates. Observed falling from trees on several occasions (Hindwood 1957, Chisholm 1962, Rand & Gilliard 1967, Bell 1970, Hall 1974, Mackness 1979, Coates 1985, Taplin & Beurteaux 1992, Frith & Frith 1995, Coates & Bishop 1997, Higgins 1999, Pratt 1971 in Higgins 1999, Payne 2005).

BREEDING Season Australia generally Sep to May; Queensland Oct–Feb; subcoastal NT Feb–Apr; WA Nov–Feb; eggs NSW Oct–Dec. Breeding condition Mar–Dec, NG; May central NG; adult carrying half-grown young Dec, Amazon Bay, NG; nest with young mid Jun; tailless fledgling Sep; adult carrying nest materials May; these records suggest breeding in dry season in NG. Nestling, Timor-Leste, Mar (Rand 1942a, Mayr & Gilliard 1954, Frith & Davies 1961, Bell 1970, Bravery 1970, Kenneally *et al.* 1979, Coates 1985, Higgins 1999, Gilbert 1923 in Higgins 1999, C.R. Trainor *in litt.*). **Nest** In herbaceous growth or tall grasses such as sword-grass, spike-rush, thick tussocks, bushy shrub, leafy palm or *Lantana* scrub. Dome-shaped *c.* 20cm in diameter inside and *c.* 46cm external, constructed of fresh twigs broken off living trees, rarely with *c.* 20cm long tunnel or ramp of grass leading to small chamber lined with fibres and large leaves; may be constructed of cane-trash and lined with gum leaves. Many nests have two side openings, one for head and another for tail; if only one opening tail pushed over head. Nest 0.3–1.6m up, mean 0.5m (Edwards 1925, Hindwood 1957, Coates 1985, Taplin & Beurteaux 1992). **Eggs** 2; white, NG; in Australia 3–4 (7) dull white, faintly glossed and round-oval, 35.3–42.4x26.7–31.2mm, (n = 10) mean 38.7x29.3, up to four clutches per season; nominate 37.6x29.4mm (Schönwetter 1967, Coates 1985, Taplin & Beurteaux 1992, Higgins 1999). **Incubation** 15+ days (Taplin & Beurteaux 1992). **Chicks** Skin leaden-black and wrinkled with long creamy trichoptiles on head and back for first two weeks; hatching asynchronous, often spread out over long period, and fledge at 10–15 days (mean 12.6 for nine young), at about 40% of adult mass and still unable to fly, but may leave nest earlier, presumably adaptation to predation. When disturbed hiss snake-like, and release foul-smelling excreta, and burst through rear of nest. Parents remove faecal sacs. Mass at hatching 18.6g, and 123.5g at fledgling, mean increase 9.33g/day (n =3). Only male fed fledglings; one seen in parents' territory nearly three months after fledgling, but normally fledglings driven away from territory at 65–70 days (Hindwood 1957, Mackness 1979, Taplin & Beurteaux 1992). **Survival** 78.6% fledgling success of four nests with 14 nestlings (Taplin & Beurteaux 1992).

FOOD Insects, e.g. grasshoppers and butterflies; scorpions; snails; frogs; small snakes and lizards; eggs and chicks, and birds caught in mistnets; mice and other small mammals; seeds. Four stomachs contained only insects, another four had grasshoppers, both imagines and nymphs, stick insects and caterpillars. Hunts 71% on ground, and 29% foliage-gleans (Mayr & Rand 1937, Rand 1942a, Coates 1985, Hicks & Restall 1992, Taplin & Beurteaux 1992, Higgins 1999).

STATUS AND CONSERVATION Density 0.02–0.4 bird/ha NT, but decrease in numbers Australia in recent decades (Higgins 1999). In NG locally common west from Astrolabe Bay (Coates 1985). Common to very common Kai Kecil and Kai Besar (Coates & Bishop 1997). Locally common Timor-Leste (Trainor *et al.* 2008b). Often road-killed (Bravery 1970, Gill 1970) and presumably often bothered with ticks, e.g. *Haemaphysalis bancrofti* (Hill 1911, Hoogstraal 1982, Loye & Zuk 1991). Birds infested with endoparasite *Porrorchis hylae* also described, probably from eating frogs (Mackness 1979). Foxes and cats serious predators of eggs and young (Lord 1956, Warham 1957). Not globally threatened (BirdLife International 2011).

Pheasant-coucal *Centropus phasianinus*. **Fig. 1.** Adult, breeding plumage, *C. p. melanurus*, Broome, Western Australia, January (*Rohan Clarke*). **Fig. 2.** Adult, breeding plumage, *C. p. melanurus*. Darwin, Northern Territory, Australia, November (*Colin Trainor*). **Fig. 3.** Adult, non-breeding plumage, *C. p. melanurus*, Weipa, Queensland, Australia, October (*Rohan Clarke*). **Fig. 4.** Adult, *C. p. spilopterus*, Kai Island, Moluccas, Indonesia, July (*Jon Hornbuckle*). **Fig. 5.** Nestling, *C. p. mui*, Parlemento, Timor, Lesser Sundas, Indonesia, March (*Colin Trainor*).

SUBFAMILY COUINAE

Genus *Carpococcyx*

G.R. Gray, 1840 *List Gen. Birds* p. 56 Type, by monotypy, *Calobates radiceus* Temminck 1832. Three species.

South-east Asia and Sundaland; non brood-parasitic. Large; bare skin on face; sexes similar. Formerly included in Neotropical *Neomorphus* but now recognised as convergence due to same terrestrial behaviour.

SUMATRAN GROUND CUCKOO
Carpococcyx viridis   Plate 11

Carpococcyx viridis Salvadori, 1879 (Mt. Singgalang, east Sumatra)

Alternative names: Green-billed/Sunda/Malay Ground Cuckoo (also for Borneo Ground Cuckoo).

TAXONOMY Until recently treated as subspecies of *radiceus*, but due to considerable plumage and size differences now considered separate species (Collar & Long 1996). Monotypic.

FIELD IDENTIFICATION 55cm. Medium-large ground dwelling cuckoo with rounded wings and tail and long strong legs and bill. **Adult** Head, neck, upperparts, wings and tail glossy green, breast pale green and rest of underparts cinnamon-buff with dusky barring. **Juvenile** Brown on head and upperparts, wing and tail feathers green-black with red-brown fringes. **Similar species** None in range.

VOICE Harsh, shrieking *Waaa-aaaa, waa-aa-aa* (transcribed from recording from BirdLife International 2009). Squealing *waaa-aaaa-waaa* or *waa-aaa*, first (and third) note rising, second falling (transcribed from recordings from N. Brickle).

DESCRIPTION Adult Forecrown blackish gradually becoming blackish-green on crown centre and bottle-green on hind crown, nape and neck; mantle and upper back dull green, lower back and rump dull chestnut broadly barred greenish-brown. Bare orbital skin verditer green above and in front of eye, pale lilac behind, pale indigo blue on cheek (Robinson & Kloss 1924). Wings glossy green-black on primaries with cobalt wash in some lights, secondaries and wing-coverts glossy bottle-green. Chin dull black extending below to behind orbital skin; throat and upper breast dull pale grey, rest of underparts cinnamon-buff on flanks, rufous and denser barred brownish-green on breast, and with brown barring on flanks and abdomen; graduated tail glossy dull oil-green, in some lights grey-black. **Juvenile** Head brownish. Bright rufous-chestnut barred brown above; abdomen and vent rufous-buff with brown barring, area on chin and upper breast more mottled than barred; remiges plain rufous tinged green and edged chestnut, some wing-coverts washed dull oil-green; tail barred dark brown proximally, unbarred distally, faintly tinged green or dull glossy green tinged chestnut. **Nestling** Undescribed. **Bare parts** Iris blood-red to reddish-brown, juvenile dark grey; orbital skin green above eyes and lores, pale blue at cheek and pale vinous-red behind eyes, bare skin reduced in juvenile. Bill pale green, base bluish, lower mandible pale green, juvenile blackish green-brown, lower mandible whiter. Feet greenish to grey-green, juvenile dark grey.

BIOMETRICS Wing adult unsexed 181–225mm (n = 4), mean 203.7 ± 14.85; juvenile 191, 194mm. Bill adult male 38.3–42.8mm, mean 40.72 ± 2.63; juvenile 31.6, 39.5mm. Tail adult 239–285mm, mean 260.5 ± 16.59; juvenile 210, 250mm. Tarsus adult 64.8–69.6mm, mean 69.26 ± 2.74; juvenile 65.5, 68.9mm (Collar & Long 1996).

MOULT No information.

DISTRIBUTION Resident south-west Sumatra at Gunung Singgalan, Air Njuruk (Pasemah), Muara Sako, Padang Highlands, Rimbo Pengadang (Bengkulu), Way Titias, Bukit Barisan Selatan NP (Lampung), Kerinci Seblat NP, and perhaps Bukit Rimbang Baling (Riau) and above Tapan (Collar & Long 1996, Zetra *et al.* 2002, Brickle 2007, Dinata *et al.* 2008, M. Catsis pers. comm.).

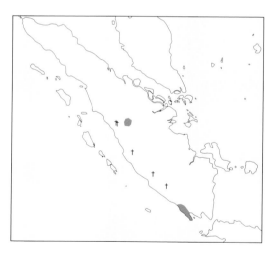

HABITAT Confined to hilly forest. 300–1,400m; 1997 record in primary evergreen rainforest with sloping hills and understorey of ferns, palms, *Pandanus* and rattan at *c.* 500m (Collar & Long 1996, Zetra *et al.* 2002).

BEHAVIOUR Ground-living. No further information.

BREEDING Season Juvenile Sep (Payne 2005). No further information.

FOOD Insects (Finsch 1898, Robinson & Kloss 1923).

STATUS AND CONSERVATION Discovered in 1878, and only eight skins known in museums (Collar & Long

1996, Holmes 1996). No records after 1916 and for long thought extinct, but in 1997 live-trapped, photographed and released in Bukit Barisan Selatan NP in Lampung Province. Since then also seen in Bukit Rimbang Baling WS, Riau in 2000, but last sighting unconfirmed (Zetra *et al.* 2002), like several others in Kerinci Seblat NP and above Tapan (BirdLife International 2001); photographed, taped and seen in Kerinci Seblat and Bukit Barisan Selatan NPs, 2006–7 (Brickle 2007, Dinata *et al.* 2008). Pair Way Titias, Aug 2010 (M. Catsis pers. comm.). Deforestation of lowland dry forest takes place at alarming rate in Sumatra. There is illegal timber cutting and encroachment in area; one caught in mammal trap demonstrates vulnerability to trapping and perhaps snaring (Zetra *et al.* 2002). Hopefully it is just as inedible as its Bornean relative so indigenous people will release it when trapped. Critically endangered, with population estimated at 50–249 (BirdLife International 2011).

Sumatran Ground Cuckoo *Carpococcyx viridis*. **Figs 1–2**. Adult, Bukit Barisan Selatan NP, Lampung Province, Sumatra, Indonesia, January (*Nick Brickle/WCS Indonesia*). **Figs. 3–4**. Adult. Bukit Barisan Selatan NP, Sumatra, February (*Filip Verbelen*).

BORNEAN GROUND CUCKOO
Carpococcyx radiceus   Plate 11

Calobates radiceus Temminck, 1832 (Pontianak district, west Borneo)

Alternative names: Green-billed/Sunda/Malay Ground Cuckoo (includes Sumatran Ground Cuckoo).

TAXONOMY *Calobates radiceus* was given opposite the plate with first description, but in index *radiatus* appears; the former has been in use since 1899 by substantial majority of authors and therefore following ICZN Article 23.9 has priority, and *radiatus* is a junior synonym. However, more recently the latter has been used (e.g. Collar and Long 1996, Payne 2005). Until the issue is resolved, and following advice from N. David (pers. comm.), this work uses *radiceus*. Previously Sumatran population treated as subspecies of *radiceus* but now generally treated as separate species *C. viridis* (Collar & Long 1996). Monotypic.

FIELD IDENTIFICATION 60cm. Large as medium-sized pheasant. Terrestrial forest cuckoo with strong feet, short rounded wings, broad rounded tail and slightly downcurved bill. **Adult** Head and neck black glossed purple with prominent green-blue naked skin around eyes and bill; mantle, upper back and wings green with purplish wash, lower back rufous barred dark green, breast and sides of neck pale grey, rest of underparts off white with dusky barring; tail glossy purplish-blue above, glossy grey-black below. **Juvenile** As adult but underparts mostly pale brown. **Similar species** None on Borneo.

VOICE 'Main self-advertisement call' far-carrying *thook-toor*, first note rising, last falling and not unlike a hornbill's call lasting *c.* 1sec, with *c.* 4sec between calls, also rendered as *boot-boooooo, boot-booooooo* or *tok-terr*. A variation is *aaw-oo* where both notes falling, or a deep *pooppoo* rising then falling; 'roll call' is a strong *torrmmm*, rolling towards end. Alarm call snarling *ark, herk* or *hark*, also transcribed as harsh *khaaa*. When released, e.g. after ringing, *heh-heh-heh*. Three different calls presumably related to breeding: one harsh and persistent, falling dove-like cooing, and lamb-like bleating. Regularly repeated guttural call (?male), and less-frequent soft clucking (?female) (Long and Collar 2002, Hill & Hill 2010).

DESCRIPTION Adult Whole head, nape, chin and upper throat black, crown and nape glossed purplish, mantle and upper back dull green with a purplish wash and coppery-red reflections, lower back and rump dull rufous barred dark green; wing-coverts dull green tinged purplish, remiges dull purplish-blue with coppery wash, underwing-coverts dark rufous; upper breast and lower part of neck sides pale grey-purple forming collar, rest of underparts off-white to pale buffish, tinged rufous on visible part of flanks, and upper flanks (hidden by closed wings) nearly unbarred cinnamon-chestnut, otherwise barred greenish-grey to dull brown, barring broadest on lower flanks; tail glossy purplish-blue above and glossy grey-black below. **Juvenile** As adult but greyish throat, crown dark greenish-brown, lower back, rump and greater wing-coverts edged rusty-brown, underparts pale buffy or rusty-brown, some birds with a little barring on breast and flanks. **Nestling** Undescribed. **Bare parts** Iris bright brown with grey outer ring, orbital skin bluish-green. Bill grey-blue to green; feet pale brown.

BIOMETRICS Wing male 240–272mm (n = 15), mean 258.9 ± 10.4, female 242–272mm (n = 8), mean 252.5 ± 8.9. Bill male 44–56mm, mean 47.5 ± 3.0, female 43–53mm, mean 47.6 ± 3.5. Tail male 290–326mm, mean 303.7 ± 9.8, female 286–326mm, mean 299.3 ± 12.0. Tarsus male 78–94mm, mean 85.5 ± 4.3, female 82–95mm, mean 87.9 ± 5.2 (Payne 2005). Mass adult male Aug, 455g; adult female Sep, 540g (Long & Collar 2002).

MOULT Adult moulting P3, P10, S4 and S7, 14 Apr; captive bird had almost completed primary moult Oct. Juvenile moulting to adult 7 Nov (Thompson 1966, Smythies & Davison 1999).

DISTRIBUTION Borneo. Patchily distributed resident over whole island at 50+ localities (10–11 in Sabah, 15+ Sarawak, 5 Brunei, 13 in East, 4 Central, and 5–7 in West Kalimantan); unrecorded South Kalimantan (numerous references in Long & Collar 2002, Mann 2008).

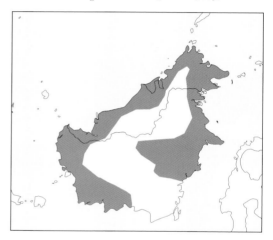

HABITAT Extreme lowland species occurring in dense forest over dry ground lowland dipterocarp forest on alluvial terraces near river. Undulating lowland and low hilly forest probably with strong preference for areas near rivers. Limestone soils mentioned but not confirmed. Nipah and mangroves with some large trees, and bordering kerangas forest 7km from river mouth, and low swampy riverside; prefers swampy patches within drier forest and alluvial forest (Duckworth *et al.* 1997, Laman *et al.* 1997, Smythies & Davison 1999, Long and Collar 2002, K. Phillipps and J.R. Howes in Long & Collar 2002, Fredriksson & Nijman 2004, Mann 2008).

BEHAVIOUR Shy and little known, preferring to escape danger by running, but also flies through understorey or hops from branch to branch with wings half spread. In Gunung Palung NP, West Kalimantan only sighted six times in seven years research, and in 44 months study in Sungai Wain Protection Forest, East Kalimantan only encountered 32 times, demonstrating difficulty of observation. Raises tail and jerks wings down with each call. Display reported where it stands on forest floor with both wings lowered, tail half spread horizontally and head lowered showing black nape; when alarmed posture adopts same posture, but gapes with head raised. Sunbathes in sun-flecks on ground. Perches low in trees from where it often calls. Probable male raised and threw back head, with tail moving up and down, while making guttural call; probable female gave clucking call while making semicircular moves around partner; this performance lasted 25 minutes (Laman *et al.* 1997, Smythies & Davison 1999, Fredriksson & Nijman 2004, Hill & Hill 2010).

BREEDING Season Judging from juvenile bird skins Feb–Jul is tentative estimate. Calls regularly Jan–Jul and Nov (Holmes 1997, Long & Collar 2002). **Nest** Unknown. **Eggs** Elliptical, smooth and white unglossed; 44.4–49.1x33.2–36.1mm (n = 16), mean 46.7x35.2 (Schönwetter 1967); however all these eggs are from one apparently unmated captive female, so this information tentative (Long & Collar 2002). **Incubation** No information. **Chicks** No information. **Survival** Nearly 18 years in captivity (Beddard 1901).

FOOD Two followed Bearded Pigs *Sus barbatus* picking up arthropods ploughed up from soil, and pigs ignored birds although less than 1m away; this feeding behaviour may be rather common because three of five local names from East Kalimantan are translated 'pig bird'. Beetles, large ants and small seeds found in stomach; other insects; worms; small snakes; frogs; some fruit (Büttikofer 1900, Laman *et al.* 1997, Payne 1997, Smythies & Davison 1999, Long & Collar 2002).

STATUS AND CONSERVATION Found in most parts of Borneo but rare (Fogden 1976, Smythies & Davison 1999, Mann 2008), as at Gunung Palung NP, West Kalimantan (Laman *et al.* 1997). Widespread and common upper Mahakam region East Kalimantan (Holmes 1997), and recent pheasant survey found it common at many localities, also finding it in secondary habitats and lower hill country. Released from snares by villagers due to its bad taste (BirdLife International 2001), and many museum skins from snares (Long & Collar 2002). However, with more and more fragmented lowland forests due to legal and illegal logging, uncontrolled fires which in recent years have devastated huge areas of forest, replacement of lowland forest with oil palm plantations, and lack of proper management of protected areas, its survival in near future is precarious and therefore considered near threatened (BirdLife International 2011).

Bornean Ground Cuckoo *Carpococcyx radiceus*. **Fig. 1.** Adult, River Kinabatangan, Sabah, Borneo, May (*James Eaton*).

CORAL-BILLED GROUND CUCKOO
Carpococcyx renauldi Plate 11

Carpococcyx Renauldi Oustalet, 1896 (Quang-tri Province, Annam)

Alternative names: Renauld's/Annam/Red-billed Ground Cuckoo (last also for Neotropical *Neomorphus pucheranii*).

TAXONOMY Monotypic.

FIELD IDENTIFICATION 65–70cm. Pheasant-like cuckoo with long legs and stout red bill. **Adult** Head, neck and upper breast glossy black; violet skin around eyes; upperparts and wings grey, primaries blackish, and rounded tail blackish glossed violet; underparts whitish finely vermiculated. **Juvenile** Head dark brown, grey skin around eyes, and forehead rufous; upperparts dark brown barred rufous on back; dark wings tinged purple and greenish with dull rufous tipping; underparts dull rufous-chestnut and whitish grey; lower flanks, thighs and undertail-coverts barred dark brown; tail black. **Similar species** Adults unmistakable; juvenile Lesser Coucal much smaller and tail dark brown finely barred buffish.

VOICE Deep rolling whistle. Male territorial call loud, mellow, moaning *woaaaab-wooaa* or *wohaaau* repeated every 5–10sec, shorter *pohh-poaaah* or loud and rolling *wb-ohh-whaaaaohu*; quite different deep and grumbling *grrrro-grrrro* or *whrrro-whrrro* and *grrroah-grrroah*. Three different duets: (1) three-note *whoop-cuhh-uhh* often abbreviated to *whoop-cuh*, (2) sliding whistle *whooooooooo*, rising and falling, and (3) fluttering *blirrrrrrrr*, last only given by female. Quiet duets recorded when close to one another. Participate in both simultaneous and antiphonal duets. Duetting, *whup—whoo-up* answered with rolling gargle (Lekagul & Round 1991, Hughes 1997a, Robson 2000, Strange 2000).

DESCRIPTION Adult Head, neck and upper breast glossy black with large area of violet naked skin around eyes upperparts grey, rump dark grey often vermiculated dark rufous, primaries black, secondaries and wing-coverts grey tinged violet, strongest posteriorly; underwing-coverts rufous-chestnut, lower breast, flanks, belly and undertail-coverts whitish with fine dusky barring on flanks, more strongly on vent and undertail-coverts; tail black with strong violet gloss. **Subadult** (Male) Forehead russet, crown nape and hindneck black glossed purple, back and scapulars olive-grey, lower back dusky mottled brown, rump white and orange-cinnamon with irregular narrow dusky bars, uppertail-coverts dusky-green; primaries violet tipped cinnamon, base dusky-greenish, secondaries dark olive-grey glossed green and purplish, wing-coverts olive-grey, larger feathers tipped cinnamon; underparts white with fine dusky barring, flanks with some orange-cinnamon feather tips; tail dusky-violet, below dusky tipped brown, longer feathers dusky-green with cinnamon tips (Riley 1938a). **Juvenile** Head and face brown, naked grey skin around eyes; forehead dull rufous; upperparts dark grey washed greenish on mantle, barred rufous on back, rump brown; primaries and primary-coverts blackish former glossed purplish, secondaries and upperwing-coverts dark greyish-brown barred rufous and washed greenish, inner primaries, shoulder, all secondaries and coverts tipped dull rufous. Unbarred dull rufous-chestnut to grey on chin, throat, breast and upper flanks, paler on rest of underparts, centre of lower breast to vent, flanks and thighs whitish grey; lower flanks and thighs whitish barred dark brown, undertail-coverts rufous finely barred black. Tail unbarred black. **Nestling** Naked skin and down brownish, palate red with two broad white raised marks on either side and raised U-shaped shield behind that and arc of small white papillae anterior to marks; tongue red with black central mark; colours persist until independent (Payne 2005). Feather sheaths at 3–4 days dark grey. **Bare parts** Iris yellow to dull orange, bare orbital skin violet and red, juvenile greyish to dark brown. Bill coral-red, base of lower mandible yellow-orange, juvenile brown-black. Feet red, juvenile dark brown.

BIOMETRICS Wing male 260–290mm (n = 12), mean 278.4 ± 9.5, female 266–294mm (n = 7), mean 278.2 ± 9.7. Bill male 44–49mm, mean 46.3 ± 1.7, female 43–47mm, mean 44.8 ± 1.7. Tail male 294–348mm, mean 325.1 ± 16.2, female 310–362mm, mean 330.0 ± 17.0. Tarsus male 87–95mm, mean 90.4 ± 2.6, female 82–95mm, mean 89.0 ± 5.2 (Payne 2005). Mass 44 day-old captive 400g (Robiller *et al.* 1992).

MOULT Captive moulted Jul; juvenile moulting remiges and rectrices, Aug. Complete adult plumage attained at *c.* 90 days (Delacour 1927, Riley 1938a, Payne 2005).

DISTRIBUTION Resident Thailand and Indochina. North and east Thailand. Throughout Cambodia; north, central, and south Laos; Vietnam (?Tonkin) (Lekagul & Round 1991, Robson 2000).

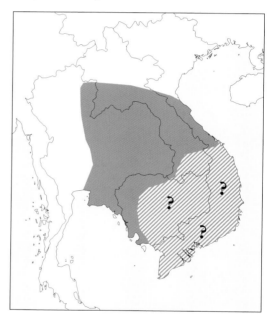

HABITAT Broadleaved evergreen and semi-evergreen primary and secondary forest with dense ground cover. To 900m Thailand, exceptionally 1,500m Indochina (Round 1988, Lekagul & Round 1991, Robson 2000).

BEHAVIOUR Walks on forest floor like pheasant and perches low in trees, and very difficult to observe because it hides at least sound in dense vegetation and rarely flies, but capable of strong fast flight. Life history little known and mostly from captivity. Roosts in trees. Lived in captivity peacefully with small passerines (Wildash 1968, Anon. 1980, Robiller *et al.* 1992, Payne 1997, Strange 2000).

BREEDING Season May–Aug; oviduct egg 12 Jun; eggs laid Jun, Khao Yai NP, Thailand (Walters 1996, Robson 2000, Pobprasert & Pierce 2010). **Nest** In captivity cup-like structure on ground, or 3–4m high in trees; only known wild nest of large sticks 4.85m above ground in dense vegetation (Robson 2000, Pobprasert & Pierce 2010). **Eggs** 2–4 (6), white; 46.0–49.0x38.5–40.0mm (n = 7), mean 48.1x39.3, laid at 2–3 days intervals in captivity; one egg 44.4x34.0mm (Schönwetter 1967, Robiller *et al.* 1992). **Incubation** 18–19 days in incubator at 37.6°C and 60–70% humidity. Both parents incubate and brood young (Atkinson 1982, Robiller *et al.* 1992, Pobprasert & Pierce 2010). **Chicks** Hatched naked, dark brown skin; after 3–4 days first pins in wings grow out, on 5th day eyes open, make faecal sacs; fed by both parents; fledge when 17–19 days old, starts feeding itself from 28 days and after 50–60 days independent, and colour of palate becomes pale. All information from captivity (Atkinson 1982, Robiller *et al.* 1992). **Survival** Wild nest predated by Pig-tailed Macaque *Macaca nemistrina* (Pobprasert & Pierce 2010).

FOOD Insects and other small animals. In captivity nestlings were fed with mixture of cow hearts, fish, newly hatched chickens and dog food, adults given mixture of meat and insects, but mice, earthworms and corn were not taken; six in captivity fed also on earthworms, bread and boiled maize. Young in wild fed nestlings; lizards, snake; frogs; earthworms; other invertebrates (Delacour & Ezra 1927, Wildash 1968, Robiller *et al.* 1992, Pobprasert & Pierce 2010).

STATUS AND CONSERVATION Uncommon Thailand (Lekagul & Round 1991), threatened by hunting and capture for live bird trade (Round 1988). Scarce to locally common (Robson 2000). At least six recorded 27 Jan–8 Feb 1990 in Bach Ma NP, Vietnam and also common Cat Bin area (Robson 1990), but otherwise considered rare Vietnam where little lowland forest exists today. In Laos local to common on Nakay Plateau (Evans & Timmins 1998) but otherwise few records. Only recent record from Cambodia from Cardamom Mts. (Eames *et al.* 2002). Bred successfully in captivity (Robiller *et al.* 1992). Not globally threatened (BirdLife International 2011).

Coral-billed Ground Cuckoo *Carpococcyx renauldi*. **Fig. 1.** Adult, Khao Yai NP, Thailand, February (*Peter Ericsson*).

Genus *Coua*

Schinz, 1821 *Das Thierreich* 1: 661. Type, by monotypy,
Cuculus madagascariensis Gmelin = *Cuculus gigas* Boddaert 1783 (Peters 1940).
Ten species, one of which extinct.

Madagascar; non brood-parasitic. Large; terrestrial and arboreal; sexes alike; soft silky plumage of mostly pastel hues. Includes *Cochlothraustes* (Cabanis & Heine 1863) and *Coccycus* (Temminck 1827).

CRESTED COUA
Coua cristata   Plate 11

Cuculus cristatus Linnaeus, 1766 (Madagascar)
Alternative name: Crested Madagascar Coucal.

TAXONOMY MtDNA evidence suggests it forms clade with *verreauxi* and *caerulea* (Johnson *et al.* 2000, Sorenson & Payne 2005). Recent speciation may have separated *cristata* and *verreauxi* which exclude each other. Polytypic. Synonym: *typica* Milne-Edwards 1879 in nominate.

FIELD IDENTIFICATION 40–44cm. Large, arboreal coua. **Adult** Crested grey-green bird with long purple-blue tail tipped white, underparts grey, purplish to rufous and white. **Juvenile** Grey crown, rufous edging on back, uppertail- and wing-coverts. **Similar species** Larger than Verreaux's Coua with grey chin and throat, and orange-fawn breast in adult.

VOICE Clear, loud *coy coy coy...* , notes well separated and decreasing in volume, and others counter-call, choruses just before sunset and occasionally at night; muted grunts. Loud, descending *koa-koa-koa*, or *guay-guay-guay-guay-gwuck* mainly before sunset; loud *guilp*; chicken-like *wuk-wuk-wuk*; low cooing like Madagascar Hoopoe counter-calling creates 'a pleasantly melodic twilight chorus' (Langrand 1990, Morris & Hawkins 1998).

DESCRIPTION Nominate. **Adult** Crested. Head and neck grey; rest of upperparts, including wings, green-grey, darker on rump; long blue tail glossed purplish-blue, rectrices tipped white except central. Chin and throat grey, upper breast purplish-maroon, lower rufous or orange-fawn; belly to undertail-coverts white. **Juvenile** Crown grey, back and wing-coverts edged rufous; feathers of upperparts and wings have russet apical spot; chin to breast grey, breast unbarred; skin around eye almost completely feathered. **Nestling** Naked on hatching with purplish-black skin, darker on dorsum, becoming well-feathered by 6–8 days; bill reddish. Palate bright red, with white ring around white central spot. Black tongue with raised outer shield white posteriorly becoming light blue in middle with central ring attached to shield by spoke on each side (Bluntschli 1938, Appert 1980, Marcodes & Rinke 2000). **Bare parts** Skin around eye violet, becoming light blue behind eye, outlined with black; iris dark brown; bill black, pale in juvenile; feet black (Langrand 1990, Payne 1997, BMNH).

BIOMETRICS Nominate. Wing male (n = 9) 126–141mm, mean 134.8; female (n = 7) 128–140mm, mean 135.7; unsexed (n = 6) 133–155mm, mean 142.5. Bill male 20.8–27.0mm, mean 22.3, female 21.7–27.4mm, mean 25.4. Tail male 175–199mm, mean 192.2, female 183–212mm, mean 199.3, unsexed 185–223mm, mean 198.5. Tarsus male 37.4–46.6mm, mean 40.9, female 40.0–45.5mm, mean 41.9, unsexed 35.8–43.5mm, mean 40.4. Mass unsexed 136g (BMNH); unsexed (n = 6),135–152g, mean 144.4 (Goodman *et al.* 1997).

MOULT Tail moult, Mar, no locality; Tulear, Oct; tail and wing moult Ihosy, Aug and Tabiky, Nov (BMNH).

GEOGRAPHICAL VARIATION Four subspecies.
C. c. cristata (Linnaeus, 1766). North and east Madagascar south-west to Mahajanga. Described above.
C. c. dumonti Delacour, 1931. West Madagascar from Mahajanga to Morondava. Averages larger than nominate, but there is considerable overlap; generally paler; crest longer; broader white tips to tail; faint rufous undertail-coverts. Juvenile has very little rusty on breast; mostly grey; primaries, scapulars, wing-coverts, back to uppertail-coverts tipped rufous; greyish throat to breast; rest of underparts whitish; iris grey-brown, bill dark horn, slate or pale fleshy-grey. Wing male (n = 5) 132–150mm, mean 142.6; female 135, 144, 145mm (BMNH).
C. c. pyropyga A. Grandidier, 1867. South-west Madagascar between Morondava and Toliara, south to Amboasary. Larger, paler with broad white tip to tail and bright rufous-chestnut undertail and vent. Wing male (n = 7) 153–166mm, mean 158.3; female (n = 4) 150–171mm, mean 163.5 (BMNH). Intergrades with *dumonti* at Morondava (Appert 1968).
C. c. maxima Milon, 1950. South-east Madagascar near Tolagnaro. Known from unique specimen, possibly hybrid

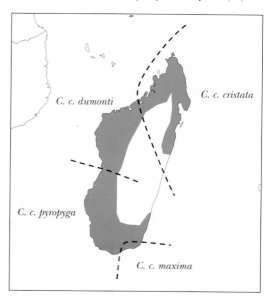

(Goodman *et al.* 1997). Larger and darker than other forms. Wing unsexed 175mm, tail 232.5mm (Benson *et al.* 1976–77).

DISTRIBUTION Resident throughout much of Madagascar at appropriate altitude (Bangs 1918, Langrand 1990).

HABITAT Primary and secondary forest, savanna, brush, vegetation on calcareous hills (sw. coast), palm groves, mangroves; sub-arid thorn scrub and adjacent areas. Prefers degraded forest, but not on coast where forest totally cleared; to 900m. In canopy of littoral forest; in damper habitat than Verreaux's Coua (Appert 1968, Payne 1997, 2005, Sinclair & Langrand 2003).

BEHAVIOUR Arboreal, walking, running or hopping along branches, but sometimes hops on ground; alone, in pairs or family groups; often in canopy where it glides from tree to tree. Where it occurs with Coquerel's and Red-capped Couas uses higher levels in vegetation (5+m) exclusively. Gleans prey from trunks and branches of trees; glides down from tree-tops. When excited stretches neck, raises crest and slowly wags tail; hides in foliage if disturbed, climbs upwards, then glides short distance to next tree; sunbathes in early morning at top of bush or tree, with feathers ruffled and wings drooping. Calling bird raises crest, and bobs tail and wings (Hartlaub 1877, Langrand 1990, Urano *et al.* 1994, Goodman *et al.* 1997).

BREEDING Both sexes build nest and involved in other duties (Milon 1952). **Season** Half-grown chick Nov, Isalo NP; nearly ready to lay Vohemar, Sep – *cristata*; male building nest Bezona, Nov – *cristata*; laying Anaborano, Nov – *cristata*; fledgling near Mamopikony, 9 Dec – *dumonti*; enlarged ovary Tsiandro, Jul – *dumonti*, Nov, no locality – *pyropyga*; nest with 2 eggs, Nov – *pyropyga*; fledglings late Oct–early Feb; active gonads Oct–Nov (Milon 1952, Goodman *et al.* 1997, Payne 1997, BMNH). **Nest** Bulky but shallow bowl of twigs and rootlets in tree 4–15m above ground (Langrand 1990). **Eggs** 2, dull white; 34.7x26.5mm; *cristata* 31.5–37.0x26.0–28.0mm (n = 18), mean 33.2x26.8; mass 10.2–12.52g (n = 16), mean 11.79; *pyropyga* 38x28, 37x27mm, mass 13.82, 14.45g (Milon 1952, Langrand 1990). No further information.

FOOD Large insects, including caterpillars, grasshoppers and crickets, mantids, phasmids, beetles (Curcilionidae), cicadas and other Hemiptera; snails; chameleons, geckos and other lizards; fruits including *Terminalia* and *Commiphora*, berries, seeds and *Albizia* gum (van Someren 1947, Charles-Dominique 1976, Langrand 1990, Goodman *et al.* 1997, BMNH).

STATUS AND CONSERVATION Widespread and locally common; can exist in degraded forests (Payne 1997). Uncommon in east, common in north and west to Mahajanga (nominate); common west from Mahajanga to Morondava (*dumonti*); common west and south from Morondava to Toliara, and uncommon in south to Amboasary (*pyropyga*); common around Ampijoroa and Berenty (Lavauden 1937, Milon 1950,1952, Milon *et al.* 1973, Langrand 1990, Morris & Hawkins 1998, Sinclair & Langrand 2003); *maxima* apparently extinct (Goodman & Wilmé 2003). Not globally threatened (BirdLife International 2011).

Crested Coua *Coua cristata*. **Fig. 1.** Adult, *C. c. pyropyga*, Berenty Estates, Toliara, Madagascar, October (*Mike Danzenbaker*).

VERREAUX'S COUA
Coua verreauxi   Plate 11

Coua Verreauxi A. Grandidier, 1867 (Cape Sainte-Marie, Madagascar)

Alternative names: Southern Crested Coua, Southern Crested Madagascar Coucal.

TAXONOMY Perhaps recently separated specifically from larger *cristata*, and the two may competitively exclude each other. MtDNA analysis indicates that it forms clade with *caerulea* and *cristata* (Sorenson & Payne 2005). Monotypic.

FIELD IDENTIFICATION 34–38cm. Small, crested, thin-billed arboreal coua. **Adult** Crested; green-grey above, with white-tipped glossy blue tail, and grey and white underparts; bare orbital skin blue with no black outline. **Juvenile** Shorter crest than adult, and less white on tail. **Similar species** Smaller than Crested Coua, pale grey to white below. Blue Coua wholly bluish and much larger.

VOICE Fairly vocal; call series of squawking notes often at dusk *crick-crick-crick-corick-corick* higher pitched, briefer and more rasping than *coy coy* call of Crested Coua; loud growl *quark quark* followed by soft *coo coo* descending in pitch; loud *trew-ee trew-ee trew-ee* followed by less audible *crow crow crow* often resulting in counter-calling (Rand 1936, Langrand 1990, Morris & Hawkins 1998).

DESCRIPTION Adult Crest light grey tipped dark. Green-grey above; remiges glossy greenish-blue; tail glossy dark blue with indistinct dark barring, and all except central rectrices broadly tipped white. Underparts grey becoming whitish on belly and flanks, with small pinky brown area on flanks. **Juvenile** From adult by shorter crest and lack of blue facial skin; narrower, more pointed rectrices with smaller white tips. **Nestling** Unknown. **Bare parts** Bare skin around eye dark blue in front, pale blue behind, not outlined in black; iris yellow to brown or grey-brown, or red. Bill black, occasionally grey at base; pale in juvenile. Feet black.

BIOMETRICS Wing male (n = 9) 129–137mm, mean 133.2 ± 3.1; female (n = 6) 128–132mm, mean 130.3 ± 1.6. Bill male 15.2–18.4mm, mean 16.5 ± 1.0, female 16.4–17.3mm, mean 16.9 ± 0.4. Tail male 173–195mm, mean 187.9 ± 3.7; female 182–198mm, mean 189.4 ± 4.8. Tarsus male 38.3–41.1mm, female 31.7–42.7mm, mean 37.8 ± 4.6 (Payne 2005).

MOULT Primary moult Lac Tsimanampetsotsa, Feb (3) (BMNH).

DISTRIBUTION Madagascar. Resident in south and south-west from just south of Toliara to just east of Cap Ste. Marie (Morris & Hawkins 1998).

HABITAT Sub-arid thorn scrub, especially in *Euphorbia* and *Alluaudia* brush, on sandy and calcareous soils; to 200m. Occupies drier habitats than Crested Coua. Coral rag scrub near Toliar (Langrand 1990, Morris & Hawkins 1998).

BEHAVIOUR Active arboreal forager; singly, in pairs or family groups; sometimes with Crested Coua; generally feeds in bushes, but occasionally descends to ground to capture prey; often sits motionless in upper part of bush or tree; flies with swift wing beats alternating with glides. Calls from tree tops. Puffs itself into ball when sunbathing (Langrand 1990, Payne 1997).

BREEDING Season Males call Nov. No further information (Payne 1997).

FOOD Insects; gekkos and small chameleons; *Cassia* fruit (Rand 1936, Benson *et al.* 1976, Langrand 1990, Goodman & Benstead 2003).

STATUS AND CONSERVATION Fairly common between Onilahy and Menarandra Rs; recently east of Menarandra R. at Cap Sainte-Marie, Beloha, Tsiombe and Berenty (Langrand 1990). Near threatened due to degradation of habitat (BirdLife International 2011).

Verreaux's Coua *Coua verreauxi*. **Fig. 1.** Adult, La Table, Madagascar, October (*Jacques Erard*).

BLUE COUA
Coua caerulea
Plate 7

Cuculus cæruleus Linnaeus, 1766 (Madagascar)

TAXONOMY Evidence from mtDNA analysis suggests it forms clade with *verreauxi* and *cristata* (Johnson *et al.* 2000, Sorenson & Payne 2005). Monotypic.

FIELD IDENTIFICATION 48–50cm. Large coua; mostly arboreal. **Adult** Blue with short crest, and naked blue skin around eye. **Juvenile** Mostly dark brown, black and dull blue; skin around eye feathered with much reduced bare area. **Similar species** None in range.

VOICE Varied, but unmelodious; brief, trilled *brrreee-ee*, increasing in volume; loud, low pitched *coy coy coy coy*, decreasing in intensity, similar to Crested and Red-breasted Couas but louder and lower; grunted *kroo kroo* or *krong krong* like Brown Lemur *Eulemur fulvus*. Loud penetrating *karr-ow*, accented on first syllable, short and rasping, often repeated (van Someren 1947, Langrand 1990).

DESCRIPTION Adult Short crest. Dark blue above and below; wings and tail have violet sheen. **Juvenile** Blackish brown on back, rump and lower belly; wings dull blue; tail lacks violet sheen; skin around eye feathered. **Nestling** Undescribed. **Bare parts** Naked skin around eye blue, mostly feathered in juvenile; iris brown; bill black or pale to dark horn; feet black.

BIOMETRICS Wing male (n = 12) 176–206mm; mean 194.2; female (n = 10) 191–206mm, mean 198.6; unsexed (n = 20) 181–223mm, mean 198.3. Bill male 23.4–30.3mm, mean 27.0; female 24.8–33.2mm, mean 28.7. Tail male 220–255mm, mean 241.0; female 237–260mm, mean 248.9; unsexed 213–265mm, mean 241.9. Tarsus male 48.8–62.0mm, mean 55.9; female 51.3–58.5mm, mean 54.9 (BMNH). Mass male 225, 235.5, 257g; female 240, 268g (Payne 2005); unsexed (n = 6) 225–257g, mean 235.3 ± 11.7 (Goodman *et al.* 1997).

MOULT Remiges moulting May (2), Jun (BMNH).

DISTRIBUTION Resident Madagascar. Mainly north and east (Sambava to Tolagnaro; Tsaratanana and Sambirano forests; Ankarana and Analamera forests; locally around Antsohihy – Bora forest) (Langrand 1990).

HABITAT Primary rainforest, secondary growth, dense clove plantations, deciduous forest and mangroves; to 1,800m (Langrand 1990, Payne 2005).

BEHAVIOUR Arboreal, from undergrowth to tree tops, mainly midstorey. Also runs on ground. Courtship-feeding; may follow troops of Brown Lemurs. Not shy; singly, in pairs or family groups; hops or runs on branches with tail raised; glides between trees, and crosses larger gaps by heavy gliding flight (Langrand 1990, Goodman *et al.* 1997, Payne 1997).

BREEDING Season Mainly in rains, Jul–Dec. Active ovaries Jun and Jul Vondrazo, Aug Vohibe; laying Nov, Tsarakibany and Anaborano; large testes Oct (Rand 1936, Langrand 1990, Goodman *et al.* 1997, BMNH). **Nest** Domed, of interlaced dry plant material, in dense foliage 3.5–10m above ground; also open bowl-shaped nest (Langrand 1990, Payne 2005). **Eggs** 1, white; 37x28.5mm (Schönwetter 1967). No further information.

FOOD Carefully inspects foliage for food. Cicadas and other bugs, phasmids, locusts, crickets, bees, beetles, flies, caterpillars; millipedes; centipedes; spiders; crabs; frogs; small reptiles such as chameleons; fruits such as *Cussonia*; *Sloanea rhodantha* resin rich in polysaccharides (Charles-Dominique 1976, Langrand 1990, Goodman *et al.* 1997).

STATUS AND CONSERVATION Hunted for food. Survives in man-influenced habitats (Langrand 1990). Common in west and south, uncommon in east (Sinclair & Langrand 2003). Not globally threatened (BirdLife International 2011).

1

Blue Coua *Coua caerulea*. **Figs 1–3**. Adult. Masoala, Madagascar, October (*Jon Irvine*).

RED-CAPPED COUA
Coua ruficeps   Plate 12

Coua ruficeps G.R. Gray, 1846 (Madagascar)

Alternative name: Red-capped Madagascar Coucal.

TAXONOMY *Olivaceiceps* previously placed in *Sericosomus* (Sharpe 1873). Usually treated as two subspecies, but both have been collected in breeding season (Dec) at Mampikohy, and considerable plumage differences suggest treating as distinct species (Payne 1997); songs very different supporting this (Huguet & Chappuis 2003, Dowsett-Lemaire 2008). However, Payne (2005) does not separate. Polytypic. Synonym: *typicus* Milne-Edwards & Grandidier 1879 in nominate.

FIELD IDENTIFICATION 42cm. Large, long-tailed and long-legged terrestrial coua, with outermost rectrices much shorter than others. **Adult** Rufous forehead and crown (greenish-brown in *olivaceiceps*), indigo-blue bare facial skin rimmed with black feathering forming black band on nape, dark tail edged whitish; rest of upperparts greenish-brown; white, tawny, purplish and rufous below. **Subadult** Blackish crown and nape barred pale. **Juvenile** Paler, crown light brown; dark naked skin behind eye; upperparts and wings barred fawn. **Similar species** Rufous crown (greenish-brown in *olivaceiceps*), purplish breast, and blue face without pink patch distinguishes this species from congeners. Slimmer, longer-necked and smaller-headed than Coquerel's; larger than Running Coua with solid blue skin around eyes, and much heavier black rim.

VOICE Loud *hug yew yew yew kuh kuh* (*kuh*), last two (three) notes lower, and may descend in scale; song more rapid than other couas in area loud *coy coy coy coy* lower than Crested Coua and usually at dusk; grunts. Loud *koa-koa-koa* mainly at dusk. Rapid, ringing *quer-quer-quer-quee-quee* (Langrand 1990, Morris & Hawkins 1998, Sinclair & Langrand 2003).

DESCRIPTION Nominate. **Adult** Forehead and crown rufous; black line around bare ultramarine skin on face, much broader below and behind eye forming band on nape; tail dark brown to purplish broadly tipped white which may be lacking on longest rectrices; rest of upperparts light greenish-brown. Upper throat white, lower throat tawny; breast light purplish; belly pale rufous to whitish. **Subadult** (*olivaceiceps*; perhaps transitional) Blackish-brown crown barred buff-white; nape barred blackish and greyish-white; remiges glossed greenish-brown with dark subterminal bar and buff white edges and tips; wing-coverts greenish-brown with dark subterminal bar and rufous-buff edging; uppertail-coverts brown with dark subterminal bar and tipped rufous-buff; tail purplish-brown slightly glossed with narrow blackish subterminal band and narrow buff-white edging and tipping to T1–2, outer rectrices broadly tipped white with few brown spots. Buff-white below, with dark brown barring from lower throat to breast; belly whitish. Ultramarine bare skin behind eye. **Juvenile** (*olivaceiceps*) Duller, with light brown crown barred dark brown; dark naked skin behind eye only; nape to upper back light brown barred dark brown and whitish, forming indistinct pale collar; rest of back, mantle and upperwing-coverts light brown with slight bronze-green sheen barred dark brown and buff-white; remiges brown with bronze-green sheen, with dark subterminal band and off-white tips and edges (more buff-white in nominate); uppertail-coverts dark brown barred tawny and tipped tawny-white. Underparts whitish, throat and breast more grey barred brown; remiges below as above. Tail (above and below) has T1–3 dark brown with purple-blue sheen, edged rufous (T1) and indistinctly barred dark; T4–5 dark brown with large white tips. **Nestling** Hatches naked, with purplish-black skin, reddish bill and closed eyes. Palate bright pinkish-red with raised white ring around raised white spot. Black tongue has white ring attached to white central boss by four radial spokes; this regresses with age (Appert 1967, 1970, 1980). When feathered has bare pigmented area around eye. **Bare parts** Bare skin around eye blue; iris dark brown, subadult brown; bill black, base of lower mandible pale, sometimes bluish, juvenile flesh coloured, culmen sepia; feet black.

BIOMETRICS Nominate. Wing male (n = 5) 161–172mm, mean 166; female 153, 160, 162mm. Bill male 27.1–30.3mm, mean 29.0; female 27.3, 27.7, 28.1mm; unsexed 26.7mm. Tail male 233–250mm, mean 238.8; female 213, 231, 231mm; unsexed 232mm. Tarsus male 58.6–62.6mm, mean 59.2; female 57.6, 59.9, 61.1mm; unsexed 57.4mm (BMNH). Also wing female 154, 159, 167mm, mean 167.9 ± 4.9 (Payne 2005). Mass unsexed 190g (Payne 1997); unsexed *olivaeiceps* 182g (Goodman & Benstead 2003); male 202g (BMNH).

MOULT Wing moult Tsimanampetsoa, Feb – *olivaceiceps*; tail moult Tulear, Oct; primary moult Lac Lotry, Nov – *olivaceiceps*; Nov; primary moult Tsimanampetsoa, Feb – *olivaceiceps*; tail and wing moult Soalala, Feb – *ruficeps*; tail moult Anaboratabe, Mar – *ruficeps*; tail and wing moult Mampikony, Dec – *ruficeps*; subadult to adult moult Soalala, Feb, Mampikony and Tulear, Dec – *ruficeps* (BMNH).

GEOGRAPHICAL VARIATION Two subspecies.
C. r. ruficeps G.R. Gray, 1846. West Madagascar from Mahanjanga west and south to near Morondava (Langrand 1990). Described above.
C. r. olivaceiceps (Sharpe, 1873). South-west Madagascar from Morondava south to Lac Anony, including Tulear region and Lake Tsimanampetsotsa (Bangs 1918, Langrand 1990). Paler above and below than nominate, with greenish-brown crown. Juvenile crown grey slightly barred, back barred, wing-coverts tipped buff, breast white barred grey (Payne 1997). Iris red. Wing **male** (n = 8) 146–175mm, mean

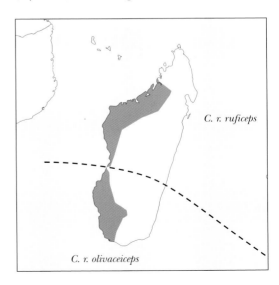

165; female (n = 9) 156–171mm, mean 165.7 (BMNH). Also wing male (n = 6) 164–172mm, mean 167.5 ± 2.8; female (n = 7) 161–173mm, mean 167.9 ± 4.9 (Payne 2005).

DISTRIBUTION Resident mainly west and south Madagascar (Bangs 1918, Langrand 1990).

HABITAT Dry deciduous forest, sub-arid thorn scrub, secondary growth, woodland savanna with grass cover not too dense, gallery forest along rivers, forested bottomlands; only coua to inhabit degraded forests; generally in more open areas of forests than other couas, and may occur on tracks; to 850m. May benefit from burning of habitat as prey is then more easily detected (Appert 1968, Langrand 1990, Chouteau 1997, Payne 1997).

BEHAVIOUR Where sympatric with Coquerel's and Crested Couas forages mainly on ground like former, but does so more frequently and efficiently on trails and more open areas. Also occurs with Running Coua. Exposes area of black skin on rump to early morning sun after cold night or suns itself atop bush at daybreak with feathers ruffled and wings drooping. Singly, in pairs or family parties; walks deliberately with body and tail horizontal; when disturbed "draws itself up, stretches neck, raises tail and walks away"; if alarmed "runs away in a succession of swift bounds, or takes wing for heavy glide of a few meters"; may find refuge in low tree branches, bush or behind termite mound (Appert 1968, Langrand 1990, Urano *et al.* 1994, Garbutt *et al.* 1996).

BREEDING Season Nov–Dec; enlarged testes Tabiky, Nov – *olivaceiceps*; enlarged ovaries Soalala, Feb, Ambararatabe, Apr, Tabiky, Nov, Lac Lotry, Dec (4 females); nests with eggs, Jan – *olivaceiceps*; juveniles Tabiky and Befandriana, Nov, and Lac Lotry, Dec – *olivaceiceps*; nests with young Manja, south-west Madagascar, Jan and Dec; nests Nov–Jan, southwest (Rand 1936, Milon 1952, Appert 1970, 1971, Langrand 1990, BMNH). **Nest** Shallow bowl of thin branches, bark and creepers, lined with finer material, in tree, 15x25cm, 5–12cm thick, 2–10m up; 19 out of 37 nests in canopy, others in tree fork or bush; *olivaceiceps* 1.5–6m above ground (Milon 1952, Appert 1970, Langrand 1990, Payne 2005). **Eggs** 1–3, usually 2; white with dull blue tinge; 34.8x27.8mm; *olivaceiceps* 33.5x29.0, 33.5x29.0, 34.0x29.0, 34.5x29.0mm, 13.82, 13.85, 14.07, 14.29g (Langrand 1990). **Incubation** No information. **Chicks** *Olivaceiceps* 10–12 days in nest (Milon 1952, Payne 2005). **Survival** No information.

FOOD Orthoptera, beetles; fruits and berries; seeds, including rice (Milne-Edwards & Grandidier 1879, Rand 1936, Langrand 1990, Goodman *et al.* 1997, Payne 2005, BMNH).

STATUS AND CONSERVATION Common throughout range, and only common terrestrial coua in degraded wooded habitats; commonly hunted (Langrand 1990, Morris & Hawkins 1998). Nominate very common Ankarafantsika Nature Reserve and other protected areas (Payne 1997). Where sympatric with Coquerel's Coua seems better adapted to degraded forest and open areas such as woodland savanna (Urano *et al.* 1994), and unlike that species may benefit from burning (Chouteau 1997). Much nest predation (Payne 2005). Not globally threatened (BirdLife International 2011).

Red-capped Coua *Coua ruficeps*. **Fig. 1**. Adult, *C. r. olivaceiceps*, Berenty Estates, Toliara, Madagascar, October (*Mike Danzenbaker*). **Fig. 2**. Adult, *C. r. ruficeps*, Ampijoroa, Mahajanga, Madagascar, October (*Mike Danzenbaker*).

RED-FRONTED COUA
Coua reynaudii   Plate 7

Coua Reynaudii Pucheran, 1845 (Madagascar)

Alternative names: Reynaud's Coua, Red-fronted Madagascar Coucal.

TAXONOMY Monotypic.

FIELD IDENTIFICATION 38–40cm. Medium-sized coua; long-tailed, short-legged, chiefly terrestrial. **Adult** Above glossed olive-green, rufous crown and blackish face with bare skin around eye blue; long green tail glossed blue and tipped black; dark grey below. **Juvenile** Much more rufous than adult with coppery-green gloss; wing feathers with rufous tipping; no bare orbital skin. **Similar species** Adult only coua all grey below with bright rufous crown. Smaller than Red-breasted and Coquerel's Couas, with shorter legs, and from latter by green upperparts and lack of pink facial spot. Much rufous above and below, and yellow bill, distinguishes juvenile from congeners.

VOICE Brief, raucous, plaintive *koo-ah* decreasing in volume and repeated several times (Payne 1997).

DESCRIPTION Adult Glossed dark olive-green upperparts, with rufous forehead and crown; nape and neck dark olive-green; long dark olive-green tail glossed green and blue; blackish stripe around bare facial skin; underparts grey, darker on belly and undertail-coverts. **Juvenile** Head and neck dull rufous-brown; rest of upperparts, including wings and tail, glossed coppery-green, T1 more bronzy; primaries more bluish; remiges and wing coverts tipped rufous; underparts dull rufous; skin around eye dull and mostly feathered. **Nestling** Undescribed. **Bare parts** Bare orbital skin ultramarine blue around and in front of eye, sky blue behind; eye-ring blue; iris brown; bill black, yellow in juvenile (Payne 2005); feet dark grey to black.

BIOMETRICS Wing male (n = 24) 128–150mm, mean 143.3; female (n = 17) 127–155mm, mean 137.9; unsexed (n = 7) 138–161mm, mean 145.4. Bill male 20.0–28.1mm, mean 23.5; female 22.1–26.4mm, mean 22.9. Tail (mostly very worn) male (n = 22) 188–225mm, mean 213.8; female (n = 16) 182–230mm, mean 200.8. Tarsus male (n = 24) 42.9–51.7mm, mean 48.0; female (n = 17) 45.5–50.2mm, mean 47.4 (BMNH). Mass male 128, 151g; female 163g (Payne 2005); unsexed (n = 5) 128–175g, mean 153.2 (Goodman & Benstead 2003).

MOULT Juveniles to adult moult, adults body, or wing and/or tail moult, Aukafana, Mar; juvenile tail and body moult Anaborano, Nov; adults in heavy moult May in north-east, and Nov in north (BMNH).

DISTRIBUTION Resident mainly north and east Madagascar. East from Tolagnaro to Sambava, and in north-west in Tsaratanana and Sambirano forests (Langrand 1990, Sinclair & Langrand 2003).

HABITAT Undisturbed rainforest, forest edge, thick secondary growth, brushy areas and clearings; also in dry forest in west of range; generally in dense vegetation; to 2,500m (Langrand 1990, Payne 1997).

BEHAVIOUR Terrestrial; shy and secretive. Usually in pairs. Walks slowly on forest floor, usually in dense undergrowth; also in low herbs, branches, and along sloping trunks to height of 8m. When alarmed runs with head, body and tail horizontal, into dense vegetation, or may glide for short distance on short wings before running off. May roost 2–3m above ground (Hartlaub 1877, Langrand 1990).

BREEDING Season Aug–Jan, peaking Oct–Dec, once Jun. Two laying females Anaborano, Nov; active ovary Vondrozo, Jun; 3 females laying Bezona, Nov; male with enlarged testis and female with large follicle south-east Madagascar, Aug; nest building mid Nov (Rand 1936, Langrand 1990, Goodman *et al.* 1997, BMNH). **Nest** Bowl, *c.* 19cm diameter, 9cm high, 5cm thick wall, inside of bowl 10x6cm, of dry stalks, palm fibres and large leaves at 2–7m above ground on *Pandanus* or arborescent fern (Langrand 1990, Payne 2005). **Eggs** 2, dull chalky white elongated oval, with creamy overlay at broad end; 38x28 (in oviduct), 36.1x27.5, 36.0x27.7, 36.2x29.2mm (2) (Rand 1936, Benson *et al.* 1976, Hawkins *et al.* 1998). No other information.

FOOD Beetles (Cerambycidae, Curculionidae, Elateridae, Scarabaeidae), grasshoppers and crickets, phasmids, Lepidoptera (including caterpillars); spiders; sometimes lizards; fruits and seeds (Langrand 1990, Goodman *et al.* 1997, Payne 1997).

STATUS AND CONSERVATION Fairly common in its restricted range; commoner above 800m; hunted for food (Langrand 1990, Morris & Hawkins 1998). Reported from more than 30 localities (Payne 2005). Not globally threatened (BirdLife International 2011).

Red-fronted Coua *Coua reynaudii*. **Fig. 1.** Adult sitting tight on the nest. Perinet, Madagascar, October (*Jon Irvine*). **Fig. 2.** Adult, Perinet, Madagascar, May (*Roy de Haas*). **Fig. 3.** Adult, Andasibe, Madagascar, December (*Henry Cook*).

COQUEREL'S COUA
Coua coquereli **Plate 12**

Coua coquereli A. Grandidier, 1867 (Morondava, Madagascar)

Alternative name: Coquerel's Madagascar Coucal.

TAXONOMY MtDNA analysis suggests closest to *cursor* (Sorenson & Payne 2005). Monotypic. Synonym: *hartlaubi* Grandidier 1869.

FIELD IDENTIFICATION 42cm. Long-tailed, slender terrestrial coua, that also occurs in trees. **Adult** Olive-green above, with black, blue and pink face; white tipped black tail; mostly tan and rufous below with blackish undertail coverts. **Juvenile** Duller than adult, lacks black on face and barred below. **Similar species** Considerably smaller than similarly plumaged Giant Coua. Running Coua grey-green above with tawny chin to throat and purplish breast, and lacks blackish undertail-coverts. Red-capped has chestnut (nominate) or rufous-olive (*olivaceiceps*) crown, greenish upper surface and no bare pink skin on sides of face.

VOICE Loud, clear *kew-kew-kew* or *kewkiwkewkewkew*, which may be answered by others; *ayoo-ew* when breeding, higher pitched than Giant Coua; muted grunts (Langrand 1990, Sinclair & Langrand 2003).

DESCRIPTION Adult Olive-green above; bare facial skin dark blue in front of eye, pale blue and pink-lilac behind, all outlined with black; light edging to primaries, rectrices white-tipped except T1–2. Chin to throat whitish, becoming tan on upper breast, rest of breast rufous, belly and undertail-coverts dull brown to black. **Juvenile** Duller than adult, lacking black on face; brown wing-coverts edged buff; fawn apical spot on wing feathers and scapulars, below grey-brown barred whitish; skin around eye slightly feathered. **Nestling** Naked on hatching, skin purplish-black, bill light reddish. Palate bright pinkish-red, with fold on each side supporting raised white rosette forming raised thin white ring around small white central ring. Tongue bright red with black dorsal surface supporting bright blue raised shield outlining edge and central blue wing joined to outer shield at base, and by radial spoke on either side (Appert 1980). **Bare parts** Bare skin around eye blue with pink patch behind eye outlined in black, all bare skin dull blue in juvenile; iris brown to red-brown, dark brown in juvenile; bill black, juvenile flesh with sepia culmen; feet black, grey in juvenile (Payne 1997).

BIOMETRICS Wing male (n = 12) 129–158mm, mean 145.4; female (n = 6) 140–152mm, mean 144.5. Bill male 22.8–29.0mm, mean 26.2; female 23.9–27.0mm, mean 25.6. Tail male 190–238mm, mean 215.7; female 205–245mm, mean 222.8. Tarsus male 47.0–52.4mm, mean 50.3; female 46.8–51.4mm, mean 49.3; unsexed 53.3mm. Mass juvenile 135g (BMNH).

MOULT No information.

DISTRIBUTION Madagascar. Resident mainly west southwards to Morombe and Sakaraha, and north to Antsiranana (Ankarana and Analamera forests); Sambirano; Berevo and upper Tsiribihina R. (Bangs 1918, Langrand 1990).

HABITAT Dry and humid forest with sparse ground cover; occasionally secondary growth; to 800m. Edge of south-west sub-desert; replaced south of range by Running Coua. Not generally found in degraded forest. Occurs with Giant, Crested and Red-capped Couas. Where occurs with Crested and Red-capped Couas feeds mainly on ground, like latter and unlike former, but called, rested and preened more frequently than latter 1–5m up in vegetation. Foraged less frequently and less efficiently on trails and in open areas than Red-capped. Burning unfavourable and birds have to enlarge feeding territories (Appert 1968, Langrand 1990, Urano *et al.*1994, Chouteau 1997, Payne 1997).

BEHAVIOUR Terrestrial; secretive; singly or in pairs; moves slowly on ground with tail held line of back, but runs swiftly when alarmed. Feeds on forest floor and trails, and middle levels to 5m. Exposes area of black skin on rump to early morning sun after cold night (Langrand 1990, Garbutt *et al.* 1996, Payne 1997).

BREEDING Season Nov–Mar. Ready to lay Maromandia, 27 Jan; laying Namoroka, 13 Mar. **Nest** Bowl of twigs, small branches, petioles and bark, lined with leaf petioles, in dense bush *c.* 2m above ground. **Eggs** 2; dull white; 33.5x25.2mm (Langrand 1990). **Chicks** Leave nest unable to fly at 9 days, and fed on ground by both parents – perhaps strategy to reduce nestling predation (Chouteau & Pedrono 2009). No further information.

FOOD Insects including Orthoptera, Lepidoptera, Hemiptera; seeds, berries, other fruits; fruits 20% of diet (Langrand 1990, Payne 1997).

STATUS AND CONSERVATION Common north, north-west and west, less so south; hunted (Langrand 1990, Sinclair & Langrand 2003). Particularly common where protected (Payne 1997). Less well-adapted to degraded forest and open areas such as savanna woodland where sympatric with Crested Coua (Urano *et al.* 1994). Adversely affected by burning, unlike Red-capped Coua, which may benefit (Chouteau 1997). Not globally threatened (BirdLife International 2011).

Coquerel's Coua *Coua coquereli*. **Fig. 1**. Adult, Baobaby, Madagascar, October (Zdeněk Hašek) **Fig. 2.** Adult, Ankarafantsika, Madagascar, October (*Jacques Erard*).

RUNNING COUA
Coua cursor Plate 12

Coua cursor A. Grandidier, 1867 (Cape Sainte-Marie and Machikora, Madagascar)

Alternative name: Running Coucal.

TAXONOMY MtDNA evidence places it closest to *serriana* (Johnson *et al.* 2000) or *coquereli* (Sorenson & Payne 2005). Monotypic.

FIELD IDENTIFICATION 34–40cm. Medium-sized coua; long-tailed, long-legged, terrestrial. **Adult** Grey-green above; face blue and pink outlined in black; naked skin deep ultramarine blue around and in front of eye, bright pink behind, all outlined by black feathers; purplish breast. **Juvenile** Duller than adult, lacks black on face, and bare skin around eye dull blue. **Similar species** From Giant Coua by much smaller size, paler above and lacking rufous below. From Coquerel's by purplish breast, grey flanks and undertail-coverts. From Red-capped by blue and pink naked skin around eye, and lacking red spot near eye, grey tail and lack of red cap, and tan sides of neck, although *olivaceiceps* also lacks red cap.

VOICE Fairly vocal. Loud, clear *kewkewkewkookoor*; calls from ground *ayreeyoo*; muted grunts (Langrand 1990).

DESCRIPTION Adult Pale grey-green above, darker on uppertail-coverts, with black line around face; tail grey. Throat tawny-buff, centre whitish, breast light purplish-maroon, belly whitish, flanks and undertail-coverts grey. Bare skin around eye deep ultramarine-blue around and in front of eye, bright pink behind, whole area outlined with black feathers; primaries and outer secondaries olive brown; T1 greenish-brown with very indistinct darker bars; other rectrices dark olive-brown with broad white tips. **Juvenile** Duller, with no black on face; skin around eye mostly feathered; back, rump, wing coverts and tips of remiges edged buff; fawn apical spot on feathers of wings and upperparts. **Nestling** Hatchling naked. Palate bright pink with raised white ring and central spot. Tongue black with raised shield with white proximal corners and blue outer circle and inner circle linked to it by two lateral radial spokes (Appert 1980). **Bare parts** Bare orbital skin ultramarine blue below, sky blue above and behind, with pink patch at rear, all surrounded by black line, thicker below. Iris brown; bill black, pale in juvenile; feet black.

BIOMETRICS Wing male (n = 5) 126–146mm, mean 133; female 130, 131, 139mm. Bill male (n = 5) 22.5–27.1mm, mean 25.1; female 24.0, 25.7, 26.2mm. Tail male 182–200mm, mean 191.4; female 195, 197, 200mm. Tarsus male 40.8–47.5mm, mean 43.4; female 43.5, 44.7, 44.7mm (BMNH). Also wing male (n = 7) 125–135mm, mean 129.8 ± 3.8; female (n = 6) 130–142mm, mean 134.7 ± 5.4. Bill male 19.0–21.7mm, mean 21.0 ± 1.5, female 20.2–20.5mm, mean 20.4. Tail male 180–200mm, mean 190.5 ± 8.3, female 182–223mm, mean 200.5 ± 15.2. Tarsus male 40–45mm,

mean 41.7 ± 2.0; female 40–43mm, mean 42 (Payne 2005). Mass female 118g (Goodman *et al.* 1997).

MOULT Body moult Ampotaka, 24 Mar (BMNH).

DISTRIBUTION Madagascar. Generally resident south-west, south northwards to Morombe, east to western slopes of Anosyenne Mts.; Mahamavo, Ifaty, and south and east of Toliara and Berenty, but absent from parts of south-west Jan, Feb, May, Aug; Tulear region and Lake Tsimanampetsotsa (Bangs 1918, Appert 1968, Langrand 1990, Payne 1997, 2005).

HABITAT Sub-arid thorn scrub, spiny desert, dry woodlands without ground cover, subdesert brush, low forest bush on calcareous plateau, secondary growth; not in pure *Euphorbia* bush; to 200m. Occurs with Giant Coua in scrub-desert of south (Appert 1968, Langrand 1990, Payne 1997).

BEHAVIOUR Terrestrial; secretive. Singly or in pairs. Walks and hops on ground, running quickly to escape danger; flight heavy. Exposes area of black skin on rump to early morning sun after cold night (Langrand 1990, Garbutt *et al.* 1996, Payne 2005).

BREEDING Season Breeds in rains. Lays Oct; young juveniles Feb–Mar; nest Morombe, Dec; enlarged ovaries Oct (Appert 1968, Goodman *et al.* 1997, Payne 1997). **Nest** Bowl of twigs and bark lined with leaf stalks placed 2m up in bush. **Eggs** 2; whitish; 34.4x22.6mm (Langrand 1990). No further information.

FOOD Spiders; beetles (Curculionidae), cicadas, ants; plant material (Goodman *et al.* 1997).

STATUS AND CONSERVATION Replaces Coquerel's Coua in south-west Madagascar. Thinly distributed (Langrand 1990). Uncommon, but locally common, as at Ifaty (Payne 1997, Sinclair & Langrand 2003). Least numerous of four *Coua* species near Mananara R., Andohahela NP (Goodman *et al.* 1997, Morris & Hawkins 1998). Not globally threatened (BirdLife International 2011).

Running Coua *Coua cursor*. **Fig. 1.** Adult, Ifaty spiny forest, Toliara, Madagascar, October (*Mike Danzenbaker*). **Figs. 2–3.** Adult. Tulear, Madagascar, October (*Jon Irvine*).

GIANT COUA
Coua gigas   Plate 12

Cuculus gigas Boddaert, 1783 (Madagascar)

Alternative name: Giant Madagascar Coucal.

TAXONOMY MtDNA analysis places it in clade with *delalandei* and *serriana* (Sorenson & Payne 2005). Monotypic. Synonym: *Cuculus madagascariensis* Cuvier 1816.

FIELD IDENTIFICATION 58–62cm. Largest coua; long-tailed, terrestrial. **Adult** Olive-grey to bronze-green above, with black, blue and pink face; mostly whitish, tan and rufous below; white tipped black tail. **Juvenile** Duller than adult with spotting on wings; bare skin on face dull blue. **Similar species** Considerably larger than similarly plumaged Coquerel's Coua. Much smaller Running Coua is grey-green above with tawny chin to throat and purplish breast. Much smaller Red-capped has rufous (nominate) or olive green-brown (*olivaceiceps*) crown, greenish upper surface and no bare pink skin on sides of face.

VOICE Vocal, generally calling from ground or low branch. Deep muted *wok wok wok*... , audible over short distance often precedes sonorous, guttural *ayoo-ew;* also clear resonant *kookoogogo* or *kookookookoogogo*, last two notes lower; short muted grunts; 'mewing growl' (Langrand 1990, Dowsett-Lemaire 2008).

DESCRIPTION Adult Above dull olive-grey to bronze-green, with more bronze on nape and upper mantle, darkest on uppertail-coverts, black face with blue and pink naked skin; pale edging to primaries and coverts. Chin and throat creamy-white, becoming greyish-bronze on upper breast which merges with broad buff band; mid to lower breast and flanks deep tan to orange-rufous; belly and undertail-coverts blackish-brown. Tail glossy black with purplish-blue tinge; all rectrices except T1 tipped white. **Juvenile** Duller than adult; crown brownish, with fawn spots and tips on olive-grey remiges, wing-coverts and scapulars; belly and flanks barred buff and dark brown. **Nestling** Undescribed. **Bare parts** Bare skin ultramarine-blue below and behind eye, light blue above with pink spot behind eye, whole area outlined in black, thicker below; in juvenile bright greenish-blue above, pink to purplish below and behind, and greyish-blue in front of eye (Payne 2005). Eye-ring blue; iris dark brown to red-brown; feet black; bill black, juvenile flesh.

BIOMETRICS Wing male (n = 9) 200–221mm, mean 210.4; female (n = 6) 211–216mm, mean 213.3; unsexed 223mm. Bill male 27.7–38.8mm, mean 35.7; female 34.7–39.8mm, mean 38.0. Tail male 280–315mm, mean 294.7; female 280–302mm, mean 291.8. Tarsus male 62.0–73.3mm, mean 67.3; female 64.9–75.5mm, mean 68.0 (BMNH). Mass unsexed 410, 415g (Goodman *et al*. 1997).

MOULT Primary moult Tulear (2), Dec; remiges moult just completed Namoroka, Mar (BMNH).

DISTRIBUTION Resident Madagascar. Coastal lowlands of south and west, including Upper Tsiribihina R., and north to Betsiboka R., Antisingy and Befandriana (Delacour 1930, 1932a, b, c, Bangs 1918, Langrand 1990, Payne 1997).

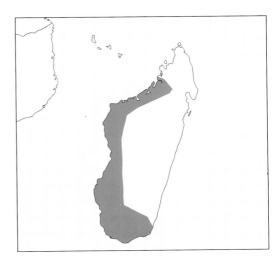

HABITAT Deciduous forest, gallery forest, forest on sand, subarid thorn scrub on calcium-rich soil, large trees with little understorey or grass, brush on sand, sub-desert; not normally in degraded or secondary forest; to 800m. Occurs with Red-capped and Coquerel's Couas, and in scrub-desert in south with Running Coua (Appert 1968, Langrand 1990).

BEHAVIOUR Secretive, but not shy; chiefly terrestrial, walking on ground with head raised and tail horizontal; singly, in pairs or small family groups; forages unhurriedly among dead leaves on forest floor; runs swiftly if flushed, and seldom flies, and only for short distances; may rest on low stump or branch; suns itself on ground with ruffled feathers and drooping wings. Calls from ground or in low tree. Runs quickly after prey, and leaps in air to catch insects (Langrand 1990, Goodman *et al*. 1997, Payne 1997).

BREEDING Season Nest building late Oct–late Dec; enlarged testes late Dec; chicks Jan. **Nest** In trees 3–10m up; bowl of twigs, bark and large leaves, lined with leaf petioles. **Eggs** 3; dull white; 43.5x32.3mm. **Survival** One caught and eaten by large boa *Acrantophis dumerilii* (Appert 1970, Milon *et al*. 1973, Langrand 1990, Goodman *et al*. 1997, Morris & Hawkins 1998). No further information.

FOOD Reptiles; centipedes, millipedes; grasshoppers, beetles (Carabidae, Curculionidae, Scarabaeidae, Tenebrionidae), ants, flies (Asilidae), Lepidoptera including caterpillars; occasionally seeds (Benson *et al*. 1976, Langrand 1990, Goodman *et al*. 1997).

STATUS AND CONSERVATION Uncommon in south, more common in west; may be very common and tame where protected, but hunted and trapped. Common where good forest still exists, but this is rapidly disappearing in many areas. Densest populations in closed-canopy gallery forest with thick leaf litter (Appert 1970, Milon *et al*. 1973, Langrand 1990, Goodman *et al*. 1997, Payne 1997, Morris & Hawkins 1998). Not globally threatened (BirdLife International 2011).

Giant Coua *Coua gigas*. **Fig. 1.** Adult. Berenty, Madagascar, October (*Jon Irvine*). **Fig. 2.** Adult, Tsingy of Bemahara, Bekopaka, Madagascar, July (*Bruno Boedts*).

SNAIL-EATING COUA
Coua delalandei   Plate 13

Coccycus Delalandei Temminck, 1827 (Madagascar)

Alternative name: Delalande's Coua.

TAXONOMY Previously placed in monotypic *Cochlothraustes* (Cabanis & Heine 1863). MtDNA analysis places it in clade with *serriana* and *gigas* (Sorenson & Payne 2005). Monotypic. Synonym: *lalandii* J.E. Gray 1829.

FIELD IDENTIFICATION 56–57cm. Believed extinct. **Adult** Black above with violet sheen, long violet-blue tail with white tips; chin to breast white, belly rufous. **Juvenile** Unknown. **Similar species** None.

VOICE Unknown.

DESCRIPTION Adult Black above with violet-blue sheen; remiges black glossed purple-violet and blue. Chin to breast white; belly and flanks to undertail-coverts rufous, brighter on undertail-coverts, one skin has rufous confined to flanks and sides of belly; underwing-coverts bright red-brown; long tail violet-blue, T3–5 (one specimen), or T4–5 (one specimen) with large white tips (BMNH). **Juvenile** Unknown. **Nestling** Unknown. **Bare parts** Bill black; iris red or dark brown or yellow (Temminck 1827, Ackerman 1841); bare facial skin blue extending broadly behind eyes, bordered with black feathers; feet black, blue-grey or sooty (Langrand 1990, Payne 2005).

BIOMETRICS Wing unsexed 217–226mm (n = 4), mean 221; bill to skull unsexed 35–42mm, mean 38.1, depth 17mm, width at base of mandible 15mm; tarsus 68–80mm, mean 73.0 (Payne 2005). Tail unsexed 265, 268mm; bill to skull unsexed 48.9, 45.6mm (BMNH).

MOULT No information.

DISTRIBUTION Previously Madagascar, but believed extinct. Occurred Nosy Boraha, Île Sainte-Marie. That it occurred also at Pointe-à-Larrée on opposite mainland of east Madagascar, and probably from head of Antongil Bay south to Tamatave (Peters 1940), or between Fito and Maroantsetra, have no factual basis (Langrand 1990).

HABITAT Primary rainforest near sea level (Langrand 1990).

BEHAVIOUR Terrestrial; shy. Moves tail up and down when jumping from perch to perch (Ackermann 1841).

BREEDING No information.

FOOD Large forest snails broken on stone anvils (Ackermann 1841).

STATUS AND CONSERVATION Believed extinct (BirdLife International 2011). Last sighting Nosy Boraha 1834; 13 museum specimens, none collected after 1834 (Greenway 1967, Benson & Schüz 1971). Deforestation, and hunting for feathers and meat, and possibly introduction of Black Rats *Rattus rattus* that preyed on molluscs, are believed to be causes (Langrand 1990). However in 1930s "a very reliable native who knew exactly what bird was being referred to" reported that it still survived on mainland but was very rare and shy (Lavauden 1932), but area has not been surveyed since (Collar & Stuart 1985).

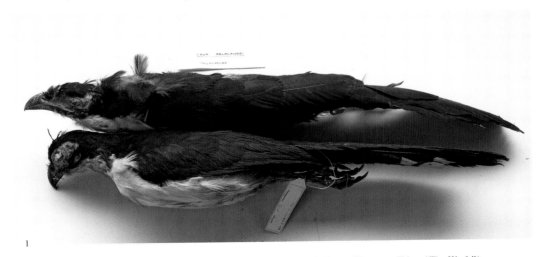

Snail-eating Coua *Coua delalandei*. **Fig. 1.** Adult skins from The Natural History Museum, Tring (*Tim Worfolk*).

RED-BREASTED COUA
Coua serriana   Plate 12

Coua serriana Pucheran, 1845 (Madagascar)

Alternative name: Rufous-breasted Madagascar Coucal.

TAXONOMY MtDNA evidence places it closest to *cursor* (Johnson *et al.* 2000), or *delalandei* and *gigas* (Sorenson & Payne 2005). Monotypic. Synonym: *serrisiana* Bonaparte 1854.

FIELD IDENTIFICATION 42cm. Large, stocky, long-tailed, terrestrial coua. **Adult** Dark greenish-brown above, remiges blue-black; black throat, rest of underparts brown with dark rufous chest; rectrices blue-black. Bare skin around eye sky-blue above, ultramarine-blue in front, behind and below eye (Payne 2005). **Juvenile** Dark brown head, rest of upperparts dark bronze spotted pale buff; wing-coverts edged buff, remiges with blackish subterminal bar and edged buff; brownish-olive throat, breast dull chestnut, belly olive; rectrices dull black. Area around eye dull blue. **Similar species** Only coua with dark rufous breast.

VOICE Noisy. Loud, deep *hoor ha ha* repeated; high-pitched melodic *tee oooo*; disyllabic *chee-guall*, second syllable growl-like; alarm harsh resonant growl *eeowll*. Individuals respond to each other's calls (Rand 1936, Langrand 1990, Morris & Hawkins 1998, Sinclair & Langrand 2003).

DESCRIPTION Adult Crown to nape dark greenish-brown; rest of upperparts greenish-brown, paler than crown, with coppery cast to mantle and upper back, remiges blue-black; black chin and throat, dark rufous chest, rest of underparts dark olive-brown to blackish-brown. Tail glossy blue-black. **Juvenile** Upperparts dark bronze spotted pale buff; wing-coverts edged buff, remiges with blackish subterminal bar and edged buff; brownish=olive throat, breast dull chestnut, belly olive; rectrices dull black. **Nestling** Unknown. **Bare parts** Bare skin around eye sky-blue above, ultramarine-blue in front, behind and below, juvenile dull blue; eye-ring blue; iris brown (deeper than other couas except Great), or red; bill black, pale tipped dark in juvenile; feet dark grey.

BIOMETRICS Wing male (n = 11) 156–176mm, mean 166; female (n = 7) 158–175mm, mean 166. Bill male 24.8–30.4mm, mean 25.6; female 25.0–30.5mm, mean 27.9. Tail male 200–232mm, mean 218.7; female 210–250mm, mean 225.3. Tarsus male 56.5–65.5mm, mean 60.1; female 56.3–61.1mm, mean 59.1 (BMNH). Mass unsexed 298g (Payne 1997).

MOULT Remiges moulting Maroantsetra, May (6)–Jun (5) (BMNH).

DISTRIBUTION Madagascar. Resident; chiefly north-east from Zahamena north to Sambava; Sambirano (Langrand 1990).

HABITAT Undisturbed rainforest; in more open areas and at lower altitudes than Red-fronted Coua; to 1,250m (Rand 1936, Langrand 1990, Payne 1997, Goodman *et al.* 2000).

BEHAVIOUR Mainly terrestrial, perching immobile on ground, or on stump or branch at least 80cm up; secretive and shy; singly, in pairs or family groups; forages by walking or running on ground, or on horizontal branches; if alarmed runs with body horizontal into cover; seldom flies; warms or dries itself in sunny spots with feathers ruffled, wings drooping and tail fanned widely. Forages beneath other frugivorous birds, collecting fallen fruit (Langrand 1990, Payne 1997).

BREEDING Season Chicks in nest, Oct; Nov. **Nest** Bowl of intertwined branches 2–4m up in epiphytic fern or *Pandanus*. **Eggs** 2; white (Langrand 1990, J. Irvine *in litt.*). No further information.

FOOD Berries and fruits; beetles and Diptera (Milon *et al.* 1973).

STATUS AND CONSERVATION Locally fairly common. Common north and east-central (Sinclair & Langrand 2003). Common in limited area eastwards from Sihanaka Forest (Delacour 1932a) and north to just west of Sambava (Benson *et al.* 1976). Very common around Maroansetra and in Sihanaka Forest (Milon *et al.* 1973), but rare to south of Sihanaka (Dee 1986). Occurs in several protected areas e.g. Mantadia NP and Masoala Peninsula NP (Payne 1997). Not globally threatened (BirdLife International 2011).

Red-breasted Coua *Coua serriana*. **Fig. 1.** Adult on the nest. Masoala, Madagascar, October (*Jon Irvine*). **Fig. 2.** Adult. The bird is holding in an insect in its bill. Masoala, Madagascar, October (*Jon Irvine*). **Fig. 3.** Adult. Masoala, Madagascar, October (*Jon Irvine*). **Fig. 4.** Adult bringing food to its hungry chicks; the nest is very well hidden. Masoala, Madagascar, October (*Jon Irvine*).

SUBFAMILY CUCULINAE

Genus *Rhinortha*

Vigors, 1830 *Mem. Raffles.* p. 671. Type, by monotypy, *Cuculus chlorophaeus* Raffles 1822.
One species.

Oriental; non brood-parasitic. Sometimes placed in *Phaenicophaeus*, but very different to that genus and only malkoha with marked sexual dichromatism. Sorenson & Payne (2005), using mtDNA analysis, place it in its own tribe, Rhinorthini, as sister to Phaenicophaeini and Cuculini. Includes *Anabaenus* (Swainson 1838).

RAFFLES'S MALKOHA
Rhinortha chlorophaea Plate 14

Cuculus chlorophæus Raffles, 1822 (Sumatra)

Alternative name: Little Malkoha.

TAXONOMY Subspecies formerly described may be due to incomplete post-juvenile moult, and size differences not maintained over larger series (Payne 2005). Monotypic. Synonyms: *Anabaenus rufescens* Swainson 1838, *Anadænus ruficauda* Peale 1848, *c. fuscigularis* Baker 1919, *facta* Ripley 1942, *bangkanus* Meyer de Schauensee 1958; '*mayri*' Delacour 1947 (smaller form in south Borneo not formally described).

FIELD IDENTIFICATION 30–33cm. Smallest malkoha; pale green-blue bill and naked orbital skin. Weak flight on soft, rounded wings; tail rather short. **Adult male** Rufous, palest on head and underparts, with blackish tail broadly tipped white. **Adult female** Whole head, neck, mantle, throat and breast pale grey; rectrices plain dark rufous except for black subterminal band and white tip both of about equal width. **Juvenile** Similar to adult. **Similar species** Cannot be confused with other cuckoos. Long-tailed Malkoha has similar bill, but much larger, has red facial skin, upperparts dark grey with green gloss on wings and extremely long tail. Chestnut-bellied and Black-bellied Malkohas also have pale green bill, but facial skin red, ground colour dark grey with green wash on wings and long tails, and former has chestnut on abdomen appearing black in most light conditions.

VOICE Cat-like descending *kiaow-kiaow-kiaow-kiaow*, *kiau-kiau-kiau* (Lekagul & Round 1991), or *mew*, quiet chirp (Hume & Davison 1878, J. Motley in Smythies & Davison 1999, Wells 1999).

DESCRIPTION Adult male Bright rufous on head, neck, upperparts and wings, palest on head, neck with pale orange-buff bases to feathers showing through, uppertail-coverts blackish-rufous, remiges tipped dusky, underwing nearest body rich rufous, coverts paler; underparts rufous to orange, becoming brownish-black on belly, thighs greyish and undertail-coverts dark grey tinged rufous, lower back, rump and uppertail-coverts finely barred dull bronze. Graduated tail blackish with variable bronze barring, darkest distally and broadly tipped white. **Adult female** Whole head, neck and mantle pale grey, rest of upperparts and wings deep rufous, remiges tipped dusky, rump blackish; underparts pale grey becoming buffish on belly, with rufous flanks, thighs grey mixed with rufous, black undertail-coverts. Rectrices plain dark rufous except for black subterminal band and broad white tip. **Juvenile** As adult but rectrices indistinctly dusky barred in male, and head and throat in female buffish to rufous; rectrices narrower and white tips much smaller. **Nestling** Undescribed. **Bare parts** Iris dark brown male, hazel to blackish-brown female; orbital skin pale green to light green-blue. Bill light green, upper mandible turquoise-blue at base. Feet green-grey to slate-blue.

BIOMETRICS Wing male (n = 19) 112–122mm, mean 114.6 ± 2.5, female (n = 17) 113–118.5mm, mean 115.9 ± 1.7. Bill male 23–31mm, mean 27.1 ± 2.3, female 24–29.5mm, mean 27.8 ± 1.9. Tail male 154–183mm, mean 167.2 ± 10.1, female 155–187mm, mean 176.0 ± 7.0. Tarsus male 21–27.7mm, mean 25.9 ± 1.6, female 23–29.8mm, mean 25.7 ± 2.1. Mass male (n = 14) 45–62g, mean 53.2, female (n = 7) 41–62g, mean 49.3 (Payne 2005).

MOULT 46 adults in wing moult, Jul–Nov, peak Aug, and 13 late May (Wells 1999, 2007).

DISTRIBUTION Resident Tenasserim and south-west Thailand from about 15°12'N, south through Thai-Malay Peninsula to Sumatra and Borneo. In Thai-Malay Peninsula found in all divisions except Phuket and Pattani; also Penang I. In Sumatra throughout lowlands, and Bangka and Tanahmasa, Batu Is. In Borneo throughout lowlands, including north Bornean and Natuna Is. (Ripley 1942, Pfeffer 1960, van Marle & Voous 1988, MacKinnon & Phillipps 1993, Wells 1999, Mann 2008).

HABITAT Evergreen and semi-deciduous forest in lowland and hill-slopes to *c.* 1,000m in Thai-Malay Peninsula, but mostly found in dense regrowth, also peatswamp forest and rubber plantations. On Sumatra frequents lowland primary and secondary forest with dense growth of creepers, vines and epiphytes, also mangroves and forest remains to 700m. In Borneo lowland dipterocarp, riverine, peat-swamp, old-logged and secondary forests, forest edge, old rubber, *Albizia* and cocoa plantations, grass swamp and wooded gardens to 1,100m (Thompson 1966, Gore 1968, Smythies 1981, van Marle & Voous 1988, Smythies & Davison 1999, Wells 1999, Sheldon *et al.* 2001).

BEHAVIOUR Singly or in pairs, occasionally small parties. Moves from tree to tree in loose flocks, lowering its wings and spreading tail. Forages in midstorey foliage-gleaning for insects, often frequenting dense masses of creepers covering trees, climbing squirrel-like. Male involved in nest-building (Robinson & Kloss 1911, Smythies & Davison 1999, H.S.S. Amar-Singh *in litt.*).

BREEDING Season Nest building Mar and Aug, eggs Jan, Apr–May, Thai-Malay Peninsula; dependent young May–Aug. Eggs 20 Jan, Deli, 16 May, Langkat, Sumatra.

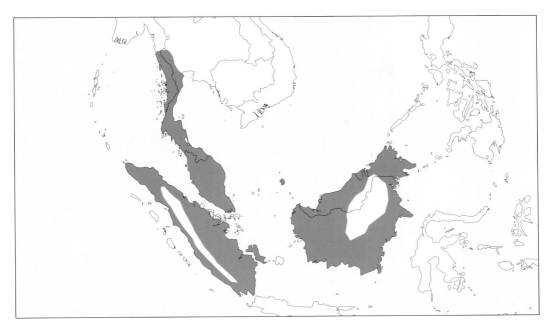

Oviduct egg Borneo, 26 Jan (Smythies 1981), egg 7 Feb, Sabah, carrying nesting material, 3 Feb, Brunei, nest Brunei, Apr, and male carrying food 23 Jun, Central Kalimantan (Robinson 1927, van Marle & Voous 1988, Nash & Nash 1988, Vowles & Vowles 1997, Wells 1999, Sheldon *et al.* 2001, various references in Wells 2007, Mann 2008). **Nest** Shallow saucer of twigs lined with green leaves, 3m high (Wells 1999). **Eggs** 2–3; glossy pure white to dirty white and chalky; 25.8x20.0mm (Robinson 1927, Wells 1999). No further information.

FOOD Caterpillars, cicadas, crickets, beetles and locustids in stomach; phasmids; large spiders (Smythies & Davison 1999, Wells 1999, H.Y. Cheng *in litt.*).

STATUS AND CONSERVATION Regular and common Thai-Malay Peninsula, including Tenasserim (Wells 1999, Robson 2000), but last recorded Singapore 1895 (Jeyarajasingam & Pearson 1999). Common throughout Borneo to 1,100m (Mann 2008). Common Sumatra (van Marle & Voous 1988). Preference for regrowth forest after logging or storm-damage suggests that habitat loss is not yet threatening this species. Not globally threatened (BirdLife International 2011).

Raffles's Malkoha *Rhinortha chlorophaea*. **Fig. 1.** Adult male, Gunung Mulu NP, Sarawak, Borneo, September (*Rohan Clarke*). **Fig. 2.** Adult female, Panti Forest, Kota Tinggi, Johor, Malaysia, March (*H. Y. Cheng*). **Fig. 3.** Adult male, Ulu Kinta FR, Perak, Malaysia, July (*Amar-Singh HSS*).

Genus *Ceuthmochares*

Cabanis & Heine, 1863 *Mus. Hein.* Th. 4, Heft 1: 60. Type, by monotypy, *Cuculus aereus* Vieillot, 1817.
Two species.

Afrotropical; non brood-parasitic. Sexes similar. Short, rounded wings; long, broad, graduated tail; short, rounded wings; brightly coloured bare skin around eye; bill short, broad and swollen. MtDNA analysis places this genus as sister to clade consisting of all other Phaenicophaeini excluding *Rhinortha* (Sorenson & Payne 2005). Sole African genus of Phaenicophaeini, other members Asian.

WHISTLING YELLOWBILL
Ceuthmochares australis   Plate 15

Ceuthmochares australis Sharpe, 1873 (Natal, South Africa)

Alternative names: Green Malkoha, Green/South African Coucal.

TAXONOMY Previously considered conspecific with *aereus* (e.g. Peters 1940), but separated by Payne (2005). Monotypic. Synonym: *dendrobates* Clancey 1962.

FIELD IDENTIFICATION 33cm. Large, dark, clumsy cuckoo with long, graduated tail, striking yellow bill and red eye with blue orbital skin. Often skulks through thick vegetation and given its size it can be surprisingly difficult to obtain good views, although it takes flight with clattering wingbeats followed by long wavering glide and unsteady landing. **Adult** Plumage usually appears blackish in field, although dark glossed greenish or bluish, paler underneath especially towards throat. **Juvenile** Duller than adult, sooty above, with brownish-buff throat and breast dark grey, tipped buff; bill brown and eye pale horn. **Similar species** None within range.

VOICE In coastal east Africa series of low-pitched nasal whistles, beginning with *si* notes at 6/sec, followed by low-pitched whistles becoming shorter and more rapid at end of song; long *si-si…tsik tsik…tik tik…teeew…teew…tew… wip wip…wipwipwip*, lasting *c.* 6sec on one pitch except for initial *si* notes; songs in RSA broadly similar, but in few may be longer and whistles without overtones (Payne 2005).

DESCRIPTION Adult Crown and nape dark grey glossed green; lores greenish-grey. Back, wings and uppertail-coverts dark slate glossed greenish. Chin and throat mid grey, breast grey with olive-buff wash; belly grey becoming blackish on undertail-coverts. Long, graduated tail bright glossy green above and below. **Juvenile** As adult, but upperparts duller and less glossed; dull sooty above, breast dark grey tipped buffy, remiges and rectrices less glossy than adult, and latter narrower. Throat and breast brownish, and wing-coverts with buff tips above and below. **Nestling** Much darker than adult; grey of head and neck washed olive-green, wings and tail dark bluish-green. **Bare parts** Iris crimson or chestnut-brown, brown in sub-adults; orbital skin in front of eye yellow to greenish-yellow, behind greenish to bluish. Bill bright lemon-yellow, blackening toward base of culmen; horn-brown in nestling. Feet black, soles white or yellowish-grey.

BIOMETRICS (Malawi and Mozambique). Wing male 120–131mm (n = 6), mean 125.8 ± 3.9; female 126, 127, 138mm. Tail male 210–222mm, mean 214.2 ± 6.3; female 212, 215, 218mm (Payne 2005). Bill (Tanzania) male (n = 6) 23–26mm, mean 24.6 ± 1.0, female (n = 7) 23–28mm, mean 25.2 ± 1.7. Tarsus (Tanzania) male 27–29mm, mean 27.7 ± 0.9, female 27–29mm, mean 27.7 ± 0.9 (Payne 2005). Mass (RSA) male 63.1–79.2g (n = 4), mean 69.5; female 60.4, 63.7, 72.4g (Maclean 1985); male 61–74g (n = 9), mean 67.6, female 52–75g (n = 8), mean 64.2 (Irwin 1988).

MOULT Secondaries moulting Ethiopia, Jan; primaries moulting RSA, Apr (BMNH).

DISTRIBUTION Afrotropical; mainly resident. South-west Ethiopia, north-east Uganda (once Nyakwai Hills), southern Somalia, east Kenya south through Tanzania (including Pemba, Zanzibar and Mafia islands), north and east Zambia, Mozambique, Malawi and south Zimbabwe to eastern RSA. Birds may move away from coast in Kenya and Tanzania, Oct–Apr, and bird killed at lighthouse on Pemba I., suggesting migratory tendency and further evidence of movement in Tanzania (Benson *et al.* 1971, Johnstone-Stewart 1984, Irwin 1988, N. & E. Baker 2009 *in litt.*).

HABITAT Mainly patches of secondary and gallery forest, particularly near forest edge; dense secondary growth, often at edges of clearings. Also in riparian *Acacia* scrub and arid thorn bush in Ethiopia; mainly coastal scrub RSA. Feeds in vegetation about 8–30m above ground with dense lianas and tangled creepers, either within canopy or on forest edge. To 2,000m east Africa; breeds to 1,300m south Malawi (Johnstone-Stewart 1984, Irwin 1988, Payne 1997).

233

BEHAVIOUR Creeps unobtrusively with short hops through thick vegetation, solitary or in pairs or small groups, using tail for balance. Frequently calls while foraging, but generally difficult to observe. Occasionally joins mixed feeding flocks. Probably territorial and monogamous. Pair observed in ritual behaviour of unknown purpose facing each other, slowly moving their tails from side to side through 90° then one bird spread its tail. Male approaches female with food item, mounts her and passes food during copulation. No calling during copulation (Irwin 1956, 1988, Rowan 1983, Payne 1997).

BREEDING Season Tanzania, May, Jul–Sep, Nov, Dec, eggs Nov; southern Africa usually Oct–Dec; Malawi Dec; Zimbabwe, Nov; Angola Jul–Nov, oviduct egg May; RSA, Nov–Dec (Dean 1971, Benson & Benson 1977, Rowan 1983, Brown & Britton 1980, Maclean 1993, Vernon et al. 1997, Baker & Baker in prep.). **Nest** Few sticks with leaves still attached forming untidily assembled platform, 2–5m from ground in fork in dense vegetation, occasionally domed RSA. Usually unlined with shallow central depression (Stark & Sclater 1903, Holman 1947, Morrison 1947, Tarboton 2001). **Eggs** 2 (1–4); ovate-elliptical, smooth, occasionally some chalky areas, pure white or creamy, often stained; 27x22.0–22.6mm (Holman 1947, Dean 1971). **Chicks** Fed by both parents (Rowan 1983). No further information.

FOOD Most stomachs contained Orthoptera and caterpillars (hairy and hairless); also termite alates, beetles, bees, cockroaches, Hemiptera; spiders; slugs; tree-frogs; kills and eats small birds in mistnets; fruits, seeds and leaves (Sclater & Moreau 1932, Mackworth-Praed & Grant 1970, Zimmerman 1972, Rowan 1983, Irwin 1988, Payne 1997).

STATUS AND CONSERVATION Very rare south-west Ethiopia (Ash & Atkins 2009). Fourteen records Somalia (Ash & Miskell 1998). Locally common east Africa (Stevenson & Fanshawe 2002). Common in 'thickets of charcoaled forest' near Cape coast (Payne 2005). Not globally threatened (BirdLife International 2011).

Whistling Yellowbill *Ceuthmochares australis*. **Fig. 1.** Adult, north of Shimoni, Kenya, October (*Bernd de Bruijn*). **Fig. 2.** Adult, Ubombo, South Africa, November (*Jim Scarff*).

CHATTERING YELLOWBILL
Ceuthmochares aereus **Plate 15**

Cuculus æreus Vieillot, 1817 (Malimbe, Angola)

Alternative names: Blue Coucal/Malkoha, Congo (*flavirostris*)/Western/Senegal Yellowbill.

TAXONOMY *Flavirostris* placed in *Zanclostomus* (Swainson 1837). Following Payne (2005) *australis* is considered separate species. Polytypic. Synonyms: *intermedius* Sharpe 1884, *aeneus* Shelley 1891 and *extensicaudus* Clancey 1962 in nominate.

FIELD IDENTIFICATION 33cm. Large, dark, clumsy cuckoo with long, graduated tail, striking yellow bill and red eye with yellow orbital skin. Often skulks through thick vegetation and given its size it can be surprisingly difficult to obtain good views, although takes flight with clattering wingbeats followed by long wavering glide and unsteady landing. **Adult** Usually appears blackish in field, although slaty-blue, slightly paler underneath especially towards throat, with bluish sheen above. **Juvenile** Duller than adult, sooty above, with brownish buff throat and breast dark grey, tipped buff; bill brown and eye pale horn. **Similar species** None within range.

VOICE Very distinctive series of notes beginning with slow squeaky clicks, followed by long accelerating series of popping notes that descend in pitch and terminating in long trilling rattle reminiscent of whinnying Little Grebe *kuk, kuk, kuk kuk kukkukkukkukukukkkkkkkrrrrrr*. Main call hoarse mournful *courlee*, recalling Eurasian Stone-curlew, rising, then dropping at end. Scolding squirrel-like *tsik-tsik* given at rate of two notes/sec; clicking notes used as contact call; various nasal sounds e.g. muffled *ououaaaaaaa*. West African birds, including those of Rwanda, lack introductory nasal cries of east African birds and referred to as Guineo-Congolian dialect. "One call mid-way between a whistle and a whine, another being a prolonged clicking chatter, first notes separate, but later running together more rapidly" – *flavirostris*. High-pitched notes becoming trill or rattle in Cameroon and south-east Nigeria, and apparently similar calls, introduced by short, high-pitched *tsik* notes, in north-east DRC, Rwanda and west Kenya. Notes on one pitch in Guinea-Bissau and IC. Call from SL described as "notes in a descending chromatic scale commencing slowly and gradually increasing in rapidity". Vocalisation in Ghana of nominate similar to *flavirostris* in central Africa (Kelsall 1914, Bates 1930, Chapin 1939, Dowsett 1990, Dowsett-Lemaire 1990, Christy & Clarke 1994, Borrow & Demey 2001, Chappuis 2005, Payne 2005, C.R. Barlow in Payne 2005).

DESCRIPTION Nominate. **Adult** Crown and nape dark grey. Lores greenish-grey. Back, wings, uppertail-coverts and long and graduated tail dark slate glossed bronzy, greenish-blue or violet. Chin and throat mid grey, darkening to slate on belly and undertail-coverts; undertail glossed purple. **Juvenile** As adult, but upperparts duller and with less glossy sheen; dull sooty above, breast dark grey tipped buffy, remiges and rectrices less glossy than adult. Throat and breast brownish, and wing-coverts with buff tips above and below. **Nestling** Much darker than adult; grey of head and neck washed olive-green, wings and tail dark bluish-green (van Someren 1916). **Bare parts** Iris crimson or chestnut-brown, brown in sub-adults; orbital skin yellow, tinged greenish particularly in front of eye. Bill bright lemon-yellow, blackening toward base of culmen; horn-brown in nestling. Feet black, soles white or yellowish-green.

BIOMETRICS Nominate. Wing male (n = 8) 111–120mm, mean 114.5 ± 3.3, female (n = 9) 114–128mm mean 119.9 ± 4.4; tail male 176–206mm, mean 190.1 ± 10.9, female 172–225mm, mean 197.7 ± 14.7; bill male 23.5–26.0mm, mean 24.8 ± 0.8, female 22.9–26.1mm, mean 24.0 ± 1.0. Also wing male 107–132mm, female 110–130mm (Verheyen 1953, Irwin 1988, Payne 2005). Mass (east Africa) male 55–72g (n = 17), mean 61.9, female 52–72g (n = 8), mean 62.1; (DRC) male 57–80g (n = 10), mean 68.5 (Irwin 1988).

MOULT Adults moulting Dec–Apr, Liberia. Three centres of primary moult: P10–9–8–7, sometimes P10–8–9–7; P6–5–4–3; P2–1; usually two complete moults annually in south DRC in May–Jun, Dec–Jan or Feb–Mar, but primary and rectrices moult interrupted by breeding (Verheyen 1953, Colston & Curry-Lindahl 1986, Gatter 1997).

GEOGRAPHICAL VARIATION Two subspecies.
C. a. aereus (Vieillot, 1817). From Niger R. in Nigeria east through Cameroon, CAR, Gabon, Equatorial Guinea, north Angola and throughout DRC, Ruanda, Burundi to Uganda, west Kenya and north-west Tanzania and extreme south Sudan; Bioko; south-west Niger? Increases in size from west to south-east (Irwin 1988). Described above.
C. a. flavirostris (Swainson, 1837). Niger R. in Nigeria west along coastal west Africa to Gambia, including Bijagos Archipelago and islands of SL. Upperparts glossed violet; darker grey below. Wing male (n = 8) 112–127mm, mean 116.3 ± 4.7; female (n = 5) 117–123mm, mean 120.4 ± 2.2. (Payne 2005).

DISTRIBUTION Afrotropical. Resident from coastal Senegambia and Guinea, along narrow band across coastal west Africa south of 10°N through CAR and whole of DRC to Uganda and extreme south Sudan; west Kenya and north-west Tanzania; north Angola, Gabon and Equatorial Guinea. Resident Bioko. Present Nov–Jun in forest islands in Gambia. Some northward movements associated with rains in Nigeria (Elgood *et al.* 1994, Barlow *et al.* 1997, Demey & Rainey 2004).

HABITAT Mainly patches of secondary and gallery forest, particularly near forest edge; dense secondary growth, often at edges of clearings; riparian *Acacia* scrub and arid thorn bush Ethiopia; *Musanga*-dominated secondary growth thickets at 700–900m, Mt. Nimba, Liberia. Feeds in vegetation about 8–30m above ground with dense lianas and tangled creepers, in canopy or on forest edge. To 1,600m west Cameroon; to 2,000m east Africa. Thick bush in savanna, south-east Nigeria (Chapin 1939, Marchant 1953, Colston & Curry-Lindahl 1986, Irwin 1988, Payne 1997).

BEHAVIOUR Creeps unobtrusively with short hops through thick vegetation, solitary or in pairs or small groups, using tail for balance. Frequently calls while foraging. Occasionally joins mixed feeding flocks. Pair, in ritual behaviour of unknown purpose facing each other, slowly moving their tails from side to side through 90° then one bird spread its tail. Male approaches female with food item, mounts her and passes food during copulation. No calling during copulation (Chapin 1939, Walker 1939, Irwin 1988, Dowsett-Lemaire 1990, Payne 1997).

BREEDING Probably territorial and monogamous. **Season** In tropical west and central Africa, dates scattered throughout year with no clear pattern. SL, Feb, Oct–Dec; fledglings Liberia, Jan and Oct; laying female IC, Dec;

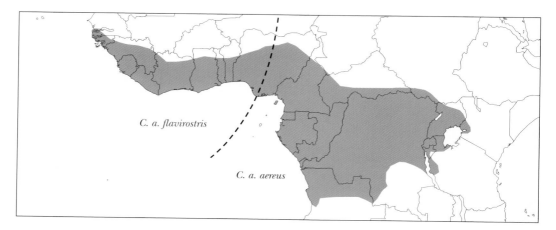

Guinea-Bissau Sep; Ghana, Mar and May, Jul; Nigeria, Jun, Aug; west Cameroon, Nov; fledgling Gabon, Jan; Uelle, north DRC, laying Jul; south DRC, Feb, Jun, Dec, granular ovaries Dec, ovaries developing Jun, enlarged testes Dec–Jan, Jun; nestling Uganda, Nov; central and south Uganda and south-west Kenya Aug, Oct; eggs north-west Tanzania, Nov; Jul–Aug Angola, nest Nov, oviduct egg May (van Someren 1916, Bannerman 1933, Chapin 1939, Holman 1947, Morrison 1947, Serle 1950, 1957, Verheyen 1953, 1957, Heinrich 1958, Traylor 1960, 1963, Mackworth-Praed & Grant 1970, Pinto 1983, Colston & Curry-Lindahl 1986, Grimes 1987, Christy & Clarke 1994, Elgood et al. 1994, Gatter 1997, Dean 2000, Baker & Baker in prep.).
Nest Few sticks with leaves still attached assembled platform, 2–5m from ground in fork in dense vegetation, occasionally domed in Uganda. Usually unlined with shallow central depression (Holman 1947, Morrison 1947, van Someren & van Someren 1949). **Eggs** 2 (1–4); ovate-elliptical, smooth, occasionally with some chalky areas, pure white or creamy and often stained; 27.8–31.9x21.0–25.9mm (n = 9), mean 30.3x23.0mm; also 27x21.5, 27x22.6, 27x22.0mm, Ghana (Chapin 1939, Holman 1947, James 1970, Irwin 1988). No other information.

FOOD Most stomachs contained Orthoptera and caterpillars (hairy and hairless), also termite alates, beetles, bees, cockroaches, Hemiptera; spiders; slugs; tree-frogs; kills and eats small birds in mistnets; fruits, seeds and leaves (Kelsall 1914, Bates 1930, Bannerman 1933, Chapin 1939, Eisentraut 1963, 1973, Mackworth-Praed & Grant 1970, Irwin 1988).

STATUS AND CONSERVATION Often common in secondary forests, even if heavily degraded (Borrow & Demey 2001). Apparently declined Gambia where once scarce, and now only vagrant. Population density c. 15–20 pairs/km^2 in Gabon (Brosset & Érard 1986). Common south-west Senegal (Morel & Morel 1990), Bijagos Archipelago, Guinea-Bissau (C.R. Barlow in Payne 2005), Liberia (Gatter 1997), in forest and gallery woodlands, IC (Thiollay 1985), Macenta Prefecture, Guinea (Halleux 1994), Togo (Cheke & Walsh 1996), Bioko (Pérez del Val 1996), Gabon (Brosset & Érard 1986), and very common Mayombé zaurois, DRC (Dowsett 1989). Previously common in parts of Ghana, such as Kumasi (Holman 1947). Quite common south-east Nigeria (Marchant 1953). Not uncommon in extreme south Sudan, west of R. Nile (Cave & MacDonald 1955), although later considered uncommon (Nikolaus 1987). Once south-west Niger (Borrow & Demey 2001). Not globally threatened (BirdLife International 2011).

Chattering Yellowbill *Ceuthmochares aereus*. **Fig. 1.** Adult, *C. a. flavirostris*, Kakum NP, Ghana, January (*Yvonne Stevens*). **Fig. 2.** Adult, *C. a. aereus*, Nyungwe Forest, Rwanda, March (*Ron Hoff*).

Genus *Taccocua*

Lesson, 1830 *Traité d'Orn.* livr. 2: 143. Type, by monotypy, *Taccocua leschenaultii* Lesson 1830.
One species.

Indian subcontinent; non brood-parasitic. Sexes similar. Sometimes subsumed in *Phaenicophaeus*, but quite different coloration without iridescence, and feathered face, justify this monotypic genus. MtDNA analysis places this as sister to *Zanclostomus* and *Phaenicophaeus* (*sensu lato*) (Sorenson & Payne 2005).

SIRKEER MALKOHA
Taccocua leschenaultii   Plate 16

Taccocua Leschenaultii Lesson, 1830 (Madras)

Alternative names: Sirkeer (Cuckoo), Southern Sirkeer.

TAXONOMY Once placed in *Centropus* (Gray 1831). Polytypic. Synonyms: *affinis* Blyth 1846 in *infuscata*, *vantynei* Koelz 1954 in *sirkee*.

FIELD IDENTIFICATION 42cm. Bill powerful and somewhat decurved; bushy-crested; tail about same length as body; terrestrial. Flight laboured and flapping and only from one tree to another. **Adult** Head, hindneck, upperparts and wings uniform sandy grey-brown, with fine glistening black feather shafts on top of head, neck and upperparts; underparts paler sandy grey-brown with black shaft streaks, belly rufous, undertail-coverts dark brown; tail blackish-brown broadly tipped white except T1. **Juvenile** As adult but pale rufous edging on back, scapulars, tertials and wing-coverts; head duller and chin to flanks and upper belly boldly streaked black, tail tipped buffish-white. **Similar species** Greater Coucal larger and more coarse with shorter tail lacking white tips; black on head, neck, underparts and tail; female and juvenile barred on wings and tail which also lacks white tips. In bad light could be confused with Blue-faced Malkoha but latter distinguished by pale green bill, more slaty-grey and bluish ground colour and arboreal behaviour. Long-tailed Malkoha has much longer tail with large white tips to T1, and pale green bill.

VOICE Often silent. Short harsh alarm call and low chuckling note when foraging which can only be heard few metres distant. Sharp *kek-kek-kek-kerek-kerek-kerek* or *kik-kik-kik* reminiscent of Ring-necked Parakeet, presumably in courtship. *CHLIK-CHLIK-CHLIK-CHLIK-CHLIK,CHLIK-CHLIK-CHLIK-CHIKCHIK\jzeeuw*, starting very slowly, then accelerating near end and ending with strange hollow buzzing. Soft *kokh-kokh-kokh* like Greater Coucal; high-pitched *kwit* after dark at roost (Baker 1927, Henry 1955, Ali and Ripley 1983, Roberts 1991, Rasmussen and Anderton 2005).

DESCRIPTION Nominate. **Adult** Forecrown, crown, nape, hindneck and upperparts plain sandy grey-brown with shining black shaft streaks on head and nape, faint trace of white supercilium, small feathers on eyelid white; short rounded wings sandy grey-brown, underwing-coverts light brown; chin, throat and long upper breast feathers pale sandy grey-buff streaked pale, lower breast becoming rufous (female less so), darkest on belly, undertail-coverts dark brown; graduated tail blackish-brown broadly tipped white, central rectrices same colour as upperparts and lacking white tips; tail, back and wings faintly tinged greenish. **Juvenile** As adult but feathers on back, scapulars, tertials and wing-coverts edged pale rufous, head duller with blackish-centred feathers, and chin, throat, breast, flanks and upper belly boldly streaked black, tail tipped buffish-white. **Nestling** Undescribed. **Bare parts** Iris crimson, reddish-brown to brown, narrow orbital skin blue. Bill red tipped yellow, cutting edges brown and base darker. Feet slaty-brown to dark grey.

BIOMETRICS *T. l. infuscata*. Wing male (n = 15) 147–169mm, mean 157.6 ± 5.6, female (n = 10) 148–163mm, mean 153.5 ± 4.4. Bill male 24.9–33.5, mean 29.4 ± 3.1, female 27.0–30.5mm, mean 29.1 ±1.5. Tail male 216–262mm, mean 239.8 ±14.8, female 213–254mm, mean 235.2 ±12.7. Tarsus male 36.5–43.5mm, mean 39.7 ±2.6, female 36.3–41.2, mean 39.3 ± 1.8. Mass male (n = 4) 174–249g, mean 218.5, female 132,148g (Payne 2005). Wing unsexed 153–186mm (BMNH).

MOULT Primary moult Aug–Dec, secondary moult Apr, India (BMNH).

GEOGRAPHICAL VARIATION Three subspecies which all intergrade.
T. l. leschenaultii Lesson, 1830. Sri Lanka and south India north to *c.* 19°N. Wing unsexed (n = 27) 148–167mm (Ali & Ripley 1969). Described above.
T. l. sirkee (J.E. Gray, 1831). Pakistan and north-west India south to *c.* 22°20'N. Palest on underparts and more rufous with smaller bill. Wing unsexed (n = 24) 153–168mm (Ali & Ripley 1969).

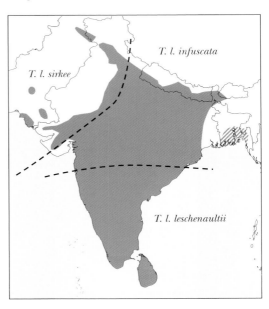

T. l. infuscata Blyth, 1845. North India in sub-Himalayas from Simla and Kumaon and south to *c.* 19°N; Nepal to Bhutan, west Assam, west Bengal; southern Bangladesh. Darkest and largest.

DISTRIBUTION Indian subcontinent; sedentary. In two isolated areas in Pakistan, north-east Punjab along frontier with India, with recent records around Islamabad and Sialkot, and small precarious population in south-east in Hab Valley on border of Sind. In Nepal occasionally Sukla Planta and Bardia in south-west but otherwise rarely recorded. Throughout most of India except much of north-west, north-east and east coast. Sri Lanka in dry zone from Uva sub-montane area up to Maduru Oya. North-west and south-east Bangladesh (Harvey 1990, Inskipp & Inskipp 1991, Roberts 1991, Grimmett *et al.* 1998, Kotagama & Fernando 1994).

HABITAT In India secondary forest, scrub, and mixed bamboo and scrub; *Acacia* bushes in stony areas or semi-desert; grass forest. In Sri Lanka in thinly-wooded areas with *Lantana*, manna-grass and illuk undergrowth. In Pakistan only in uncultivated tracts with dry, deciduous thorny scrub. To 915m, once 1,370m, Nepal. Lowlands and foothills to 1,000m peninsular India, but to 2,100m in Himalayas (Baker 1927, Henry 1955, Ali & Ripley 1983, Roberts 1991, Grimmett *et al.* 1998).

BEHAVIOUR Poor flier, preferring to move through thickets, often in thorny bushes and trees when disturbed. Terrestrial when feeding. When running to cover keeps head down and body and tail horizontal, confusable with mongoose. Does not flush until nearly trodden on. Shy; mostly singly or in pairs. In courtship display one bird circles the other with repeated bowing, its plumage fluffed and tail raised and fanned to display conspicuous white tips; female alone, in other cases both birds, bob with erected tails, open bills and head thrust high into air after each bow (Menesse 1939, Ali & Ripley 1983, (Grimmett *et al.* 1998, Rasmussen & Anderton 2005).

BREEDING Season Eggs Mar–Sep, India and Pakistan, varying locally but chiefly Jun–Aug (Baker 1927, Ali & Ripley 1983, Roberts 1991, Grimmett *et al.* 1998). **Nest** Untidy, rather flat, cup of twigs lined with green leaves in tree fork, or *Euphorbia*, or scrub 2–6m from ground (Henry 1955, Grimmett *et al.* 1998). **Eggs** 2–3; chalky white; 33.0–35.0x24.8–27.2mm, (n = 10) mean 33.9x26.1 (Baker 1927). **Incubation** Both sexes incubate (Ali & Ripley 1983). No further information.

FOOD Insects such as grasshoppers, beetles, locusts; molluscs dug from soil; lizards; fledgling birds; mice; seeds; berries and other fruits (Baker 1927, Henry 1955, Roberts 1991).

STATUS AND CONSERVATION Local north-east and south-east Pakistan, rare elsewhere (Roberts 1991). Frequent far west but uncommon elsewhere, Nepal; generally uncommon India (Grimmett *et al.* 1998). Rare Bangladesh (Harvey 1990). Rare Sri Lanka except in some parts of Uda Walawe NP (Payne 1997) and range has become much restricted in dry lowlands (Henry 1955, Grimmett *et al.* 1998). Does not extend into tea and rubber plantations (Payne 2005). Expanded range in Sindh from 1930s along perennial canal system into desert areas (Menesse 1939). Although generally rare or uncommon, seemingly not in danger because of large range and preference for dry sterile areas. Not globally threatened (BirdLife International 2011).

Sirkeer Malkoha *Taccocua leschenaultii*. **Fig. 1.** Adult, *T. l. leschenaultii*, Tanamalwila, Sri Lanka, March (*Amila Salgado*). **Fig. 2.** Adult, *T. l. leschenaultii*, Hampi, Karnataka, India, January (*Niranjan Sant*). **Fig. 3.** Adult, *T. l. leshenaultii*, with lizard prey, Hidkal, Belgaum, Karnataka, India, March (*Niranjan Sant*).

Genus *Zanclostomus*

Swainson, 1837 *Class. Birds* 2: 323. Type, by monotypy, *Phoenicophaeus javanicus* Horsfield 1821.
One species.

South-east Asia and Sundaland; non brood-parasitic. Sexes similar. Sometimes subsumed in *Phaenicophaeus* but its plumage coloration, sickle-shaped bill, and only small unfeathered area on face set it apart. MtDNA evidence suggests it is sister to *Phaenicophaeus* (*sensu lato*) (Sorenson & Payne 2005).

RED-BILLED MALKOHA
Zanclostomus javanicus — Plate 16

Phænicophaus Javanicus Horsfield, 1821 (Java)

TAXONOMY Polytypic. Synonym: *natunensis* Chasen 1934 in *pallidus*.

FIELD IDENTIFICATION 42–44cm. Medium-sized malkoha; bill red. **Adult** Top of head, nape and hindneck grey, upperparts and wings dark grey glossed oily green; lower part of head, chin, throat and breast orange, flanks and upper belly pale grey, rest of underparts orange-chestnut. Tail dark oily green becoming steely blue towards end, and tipped white. **Juvenile** As adult but primary-coverts and secondaries tinged and edged rufous, bill tipped black. **Similar species** Red bill and orange on underparts distinguish it from all other cuckoos. Raffles's Malkoha smaller, with pale green bill and facial skin, warm brown on upperparts, wings and tail, white tail tip broader. Brush Cuckoo much smaller, has thin dusky bill, upperparts, wings and tail brown, not grey, and whole underparts rufous.

VOICE Single hoarse *turk-urk* or repeated *turk-urk—turk-urk*, or *kuk* repeated many times. Soft *taup* or *k'taup* close by. On Borneo clear whistle *who-oo* repeated every 10sec; also clicks and rattles (Smythies & Davison 1999, Wells 1999).

DESCRIPTION Nominate. **Adult** Forecrown, crown, nape and hindneck plain dark grey, palest on head, gradually becoming glossy blue-green on back and wings, inner webs of primaries dusky, underwing-coverts pearl grey with small fulvous area near carpal joint, rump grey; lores, and chin to upper breast rufous-orange, flanks, lower breast and upper belly pale grey, lower belly and undertail-coverts deeper rufous-orange than on throat and breast; graduated tail blackish with oily green gloss becoming steely-blue towards end and tipped white. **Juvenile** As adult, but paler, greater primary coverts and secondaries tinged and edged dull rufous, and rectrices narrower and white tips to central rectrices smaller, 8mm versus 18mm in adult. **Nestling** Undescribed. **Bare parts** Iris brown to red or whitish, skin around eye light purplish-blue to grey, eyelid blue. Bill coral-red, ridge of culmen sometimes black, juvenile paler red tipped black. Feet bluish-slate to greenish.

BIOMETRICS *Z. j. pallidus* Wing male (n = 16) 136–160mm, mean 144.3 ± 6.2, female (n = 15) 140–154mm, mean 147.3 ± 4.9. Bill male 31–37mm, mean 33.8 ± 2.1, female 31.5–37.0mm, mean 34.1 ± 1.6. Tail male 225–275mm, mean 249.4 ± 13.8, female 233–288mm, mean 265.6 ± 17.1. Tarsus male 26.5–31.0mm, mean 30.8 ± 2.7, female 29–34mm, mean 30.6 ± 1.9. Mass 97–98g (Payne 2005); male (n = 4) 100.3–122.1g, female 133.5g (Medway 1972).

MOULT Primaries replaced at one or two loci; wing moult Thai-Malay Peninsula early May–Sep and Nov; earliest completion 20 Aug (Wells 1999, 2007).

GEOGRAPHICAL VARIATION Two subspecies, not well marked.

Z. j. javanicus (Horsfield, 1821). Java. Lower breast dark grey. Wing male (n = 6) 135–147mm, female (n = 3) 135–143mm, mean 138.3 (n = 3) (Payne 2005). Described above.

Z. j. pallidus Robinson & Kloss, 1921. Thai-Malay Peninsula, including Tenasserim, Sumatra, Borneo. Paler below, rufous area less intense, and grey paler and more washed with buff.

DISTRIBUTION Sundaland. Resident Tenasserim from *c.* 14°N, and southern Thailand from *c.* 13°N, through peninsular Malaysia south to Johor. Sumatra, Borneo, North Natuna Is. and Java (Horsfield 1921, van Marle & Voous 1988, Wells 1999, Mann 2008).

HABITAT Evergreen and semi-evergreen primary and secondary forest, 0–1,200m Thai-Malay Peninsula; dense elfin forest and scrub, or bamboo jungle. Primary and old secondary forests with dense growth of creepers and vines to *c.* 1,000m, Sumatra, once *c.* 1,550m. Lowland and hill dipterocarp, peat-swamp, logged and secondary forests, grass swamp, ladang, forest edge, cocoa, *Eucalyptus* and mature *Albizia* plantations to 1,750+m, Borneo (Robinson 1928, van Marle & Voous 1988, Holmes 1996, Smythies & Davison 1999, Wells 1999, Mann 2008).

BEHAVIOUR Monogamous, living in pairs; also singly (Wells 1999, C.F. Mann).

BREEDING Season Egg Thai-Malay Peninsula, 1 Jun. Borneo, breeding East Kalimantan, Feb; pair in breeding condition, Sabah, 22 Mar; nest building, Central Kalimantan, Apr (Holmes & Burton 1987, Holmes 1997, Wells 1999, Sheldon *et al.* 2001). **Nest** Of twigs and grass blades placed in thick bush (Robinson 1928). **Eggs** Long oval; white; 31.5x23.0mm, 34.1x23.0mm; from Borneo (n = 2) mean 29.1x23.2mm; from Java (n = 3) mean 35.0x27.1mm, but some doubt about last mentioned because of large size (Schönwetter 1967, Smythies & Davison 1999). No further information.

FOOD Caterpillars (including Sphingidae) up to 78mm long and 5g, pupae (Pieridae), beetles (including Scarabaeidae), grasshoppers (Acrididae, Tettigonidae), termite alates, cockroaches, phasmids, Hemiptera (including cicadas); spiders; crustaceans (Sody 1989, Smythies & Davison 1999, Wells 1999, Sheldon *et al.* 2001, Payne 2005).

STATUS AND CONSERVATION Very rare Tenasserim (Hume and Davison 1878). Common to uncommon all divisions of Thai-Malay Peninsula; not yet threatened (Wells 1999); last recorded Singapore 1895 (Jeyarajasingam & Pearson 1999). Uncommon Borneo (Mann 2008) and considered rarest malkoha there (Smythies & Davison

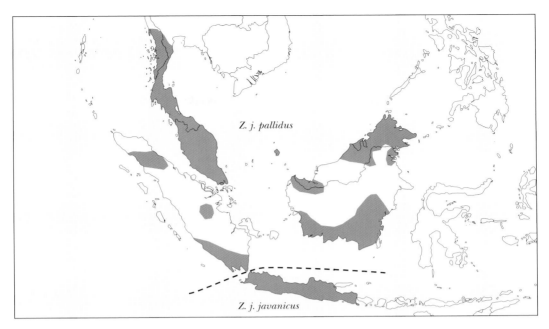

1999), but prefers primary canopy and therefore may be overlooked (Gore 1968); uncommon but regular Natai Lengkuas area, Tanjung Puting NP, Central Kalimantan (Nash & Nash 1988). Scarce Way Kambas NP (Parrott & Andrew 1996), and recorded Gunung Leuser NP, Sumatra (Buij *et al.* 2006). Occasional Alas Purwo NP, east Java (Grantham 2000). Not globally threatened (BirdLife International 2011).

Red-billed Malkoha *Zanclostomus javanicus*. **Fig. 1.** Adult, *Z. j. pallidus*, Panti Forest, Kota Tinggi, Johor, Malaysia, March (*H. Y. Cheng*).

Genus *Phaenicophaeus*

Stephens 1815 Shaw's *Gen. Zool.* 9, part 1: 58. Type, by subsequent designation,
Cuculus pyrrhocephalus Pennant 1769 (Peters 1940: 56).
One species.

Sri Lanka; non brood-parasitic. Sexes alike; large, slender, long-tailed, crestless malkoha; brightly coloured bare face. Payne (2005) includes *Rhopodytes* and *Rhamphococcyx* in this genus.

RED-FACED MALKOHA
Phaenicophaeus pyrrhocephalus   Plate 17

Cuculus pyrrhocephalus Pennant, 1769 (Sri Lanka)

Alternative name: Crimson-faced Malkoha.

TAXONOMY Monotypic.

FIELD IDENTIFICATION 40–46cm. Rather large-billed and boldly-patterned malkoha with slow, direct flight between trees on short, rounded wings which produce musical hum. **Adult** Conspicuous large red facial area underlined by broad white stripe flecked black, and heavy green bill, upperparts metallic green-blue, throat and breast black, rest of underparts white, long black tail broadly tipped white. **Juvenile** Duller, with less red on head; shorter tail. **Similar species** In range can only be confused with Blue-faced Malkoha, which is slightly smaller, has blue around eyes, greenish-grey upperparts, belly grey (not white), long, graduated tail black with wider white terminal band.

VOICE Rather silent, only uttering soft grunt-like calls. Low *kaa* in flight; on alighting, low petulant *kra*, soft low *krrr* like purring of cat, *kree-kree-kree*, *kok*, and short yelping whistles. Main call *grrr-GRRRRR-GRRRRR-(gt't'i')* (Ali & Ripley 1983, Fuller & Erritzøe 1997, Rasmussen & Anderton 2005, Salgado 2006).

DESCRIPTION Adult Forehead, crown and nape glossy black with variable white stripes, large area on face with small crimson bristly hair-like feathers which on many birds protrude into middle above crown; nape and sides of neck black with variable white flecking; upperparts and wings black glossed green and blue; chin and broad line beneath red face white flecked black, throat and upper breast black, lower breast, flanks, belly and undertail-coverts white, long graduated tail black with very broad white terminal bands, undertail nearly white. Female has less white, and more black, on sides of head. **Juvenile** Less brick-red on face and only around eyes, ground colour of crown, nape and breast brownish and stripes on crown, nape and chin off-white; less glossy on upperparts, remiges and rectrices browner and more pointed, and tail shorter with narrower white terminal band. **Nestling** Undescribed. **Bare parts** Iris brown, female white. Bill apple-green, lower mandible slightly paler with dusky base and around nostrils, juvenile brown. Feet bluish, bluish-green or slate-blue.

BIOMETRICS Wing male (n = 6) 148–169mm, mean 154.7 ± 6.7, female (n = 6) 151–169mm, mean 157.0 ± 8.1. Bill male 33.0–35.0mm, mean 34.1 ± 0.8, female 32.0–35.4mm, mean 33.9 ± 1.4. Tail male 279–318mm, mean 290.2 ± 14.5, female 272–300mm. Tarsus male 29.2–34.2, mean 31.8 ± 1.7, female 31.0–34.8mm, mean 33.1 ± 1.4 (Payne 2005).

MOULT Tail moult, Dec; primary moult Oct (Legge 1873, BMNH).

DISTRIBUTION Resident Sri Lanka, perhaps with some altitudinal movements. Rarer in dry zone of north and east than in wet zone of south-west. Doubtfully recorded south Kerala and west Tamil Nadu, south India (Ali & Whistler 1936, Biddulph 1956, Hoffmann 1984, 1996, Nanda 1996).

HABITAT Tall primary forest with dense, tangled undergrowth in wet zone, rarely seen near cultivation. Prefers canopy but also forages on ground. To 1,700m, perhaps seasonal altitudinal migrant; records over 920m rare and old, suggesting it may have changed habitat to lower altitudes perhaps due to forest fragmentation (Wait 1931, Fuller & Erritzøe 1997, D. Warakagoda in Fuller & Erritzøe 1997).

BEHAVIOUR Feeds alone or 2–4, but up to 6, in mixed foraging groups led by Orange-billed Babblers, and other mainly medium-sized birds, mostly in wet season. In 48% of mixed flocks Sinharaja. Leaf-gleans. Rather shy; most active morning and afternoon; restless. Forages mostly in forest canopy, or near ground where understorey thick. Flight is often "no more than a 'hop' from one thicket to another". Both parents participate in nest building (Henry 1955, Fleming 1977, Ali & Ripley 1983, Fuller & Erritzøe 1997, S. Perera in Fuller & Erritzøe 1997, Payne 1997, Kotagama & Goodale 2004).

BREEDING Season Ali and Ripley (1983) suggest whole year, but dated records and bird-skin specimen data Jan–May, possibly also Aug–Sep (Henry 1955). **Nest** One *c.* 10m high in fork of kunumella tree on thick layer of twigs each 15–20cm long; deep cup lined with leaf midribs, and

outside covered with leaves from this tree; shallow cup lined with fresh leaves; also grass and roots used (Baker 1934, Ali & Ripley 1983, S. Perera in Fuller & Erritzøe 1997). **Eggs** 2–3; rounded to oval, practically equal at both ends; white and chalky; 34.0–39.4x25.1–39.3mm (Schönwetter 1967). **Incubation** No information. **Chicks** Both parents feed nestlings (Payne 2005). **Survival** No information.

FOOD Fruit and berries from trees but probably largely insectivorous. Caterpillars and other insects in stomach (Henry 1955). Female ate 25cm-long giant stick insect *Palophus sp.* after removing legs and thorax; mantids, grasshoppers, moths and cicadas (Salgado 2006).

STATUS AND CONSERVATION Since 1873 considered extremely rare to quite common (Fuller & Erritzøe 1997). "Most of the forests and heavily-clad jungles of the low country" (Legge 1880). "Not uncommon in the wilder stretches of forests" (Wait 1931). Scarce (Henry 1955); "one of the rarest Indian birds" (Ali and Ripley 1983). "Fairly good numbers" in Uva and Ratnapura provinces and especially in Nilgala, Henebedda, Timbolketiya, Embilipitiya, Demaliyagalge and Kalawewa; Wanadivi, 1972 (Fernando of Dehiwela in Tirimanna 1981). Population has seriously declined mainly due to deforestation and selective logging where largest trees are targeted and thus closed canopy (preferred habitat) is destroyed (Kotagama 1994 in BirdLife International 2001). Sinharaja World Heritage Reserve of 11,187ha foremost stronghold today (Salgado 2006). Fragmentation of forests may also soon become serious problem causing inbreeding. Deforestation is occurring at alarming rate, and only 9% of wet zone in south-west Sri Lanka where it primarily occurs still covered with forest, but still recorded locally from riverine forests of dry zone. Wilpattu and Yala East NP and many sanctuaries are no guarantee for its future because of lack of effective protection where profit has become all-important (Hoffmann 1996, Fuller & Erritzøe 1997, BirdLife International 2001). Because of its shyness, lowland preference and avoidance of cultivated areas its future depends on preservation of few remaining undisturbed forests in Sri Lanka. Vulnerable; population estimate 2,500–9,999 (BirdLife International 2011).

Red-faced Malkoha *Phaenicophaeus pyrrhocephalus*. **Fig. 1.** Adult female, feeding on the endemic 'mega-stick' phasmid, *Phobaeticus hyparphax*, Sinharaja, Sri Lanka, December (*Amina Salgado*). **Fig. 2.** Adult male. Morapitya, Sri Lanka, January (*Gehan de Silva Wijeyeratne*). **Fig. 3.** Adult female. Sinharaja, Sri Lanka, July (*Gehan de Silva Wijeyeratne*).

Genus *Rhopodytes*

Cabanis & Heine, 1863 *Museum Heineanum* Th. 4, Heft 1: 61. Type, by subsequent designation, *R. diardi* = *Melias diardi* Lesson 1830 (Sharpe 1873, *Proc. Zool. Soc. London* p. 604). Four species.

Oriental; non brood-parasitic. Medium to large malkohas. Sexes similar. Sometimes placed in *Phaenicophaeus* (e.g. Sorenson & Payne 2005, on mtDNA grounds), but *diardi*, *sumatranus*, *tristis* and *viridirostris* show distinct plumage similarities that set them apart from *Phaenicophaeus pyrrhocephalus*.

CHESTNUT-BELLIED MALKOHA
Rhopodytes sumatranus   Plate 15

Cuculus Sumatranus Raffles, 1822 (Sumatra)

Alternative name: Rufous-bellied Malkoha.

TAXONOMY Bornean population smaller but overlaps with other populations; *rodolphi* of Batu I., Sumatra, said to be larger with stouter bill. Monotypic. Synonyms: *minor* Riley 1938 and *rodolphi* Ripley 1942.

FIELD IDENTIFICATION 40–41cm. Grey and green malkoha with chestnut belly and long tail tipped white; weak, seldom flies actively, more often gliding from treetop to another tree. **Adult** Head, neck, breast and flanks grey with prominent green bill and red facial skin, wings glossy greenish-blue; belly and undertail-coverts diagnostically chestnut but often difficult to see in poor light where they appear black. Dark graduated tail glossed green and tipped white. **Juvenile** As adult but rectrices narrower and white tips smaller. **Similar species** Black-bellied has black belly but is otherwise similar, except for larger white tail tips, facial skin red rather than orange, and trifle smaller in size, and bill less prominent. Both species have similar range and habitat preference. Green-billed is larger, has tail proportionally much longer, 150% length of body, and tail with much larger white tip. Chestnut-breasted has sides of head and entire underparts chestnut, and lacks white tail tips, terminal half being chestnut.

VOICE Rather silent. Call high-pitched mewing notes; *tok-tok* or *chi-chi*; call like Long-tailed Malkoha. Knocking *kokokokokoko*, c. 12 notes/sec; *chak*; thin, high-pitched descending *mew* (Hume & Davison 1878, Hails & Jarvis 1987, MacKinnon & Phillipps 1993, Jeyarajasingam & Pearson 1999, Robson 2000, Supari 2003, Payne 2005).

DESCRIPTION Adult Head grey with broad black lores, bare facial skin bordered with narrow black line along upper margin in most specimens, sometimes some traces also visible on lower margin; upperparts glossy green, wings glossy green-blue; chin, throat and breast grey, belly, vent and undertail-coverts deep chestnut. Tail bluish-grey with green sheen, and tipped white. **Juvenile** As adult but rectrices narrower with smaller white tips. **Nestling** Black naked skin with pale yellow hair-like down. **Bare parts** Iris pale blue or whitish, juvenile brown; facial skin orange to orange-red, shading to blood-red on posterior portion. Bill pale yellow-green. Feet greenish, slate-blue or greyish.

BIOMETRICS (Malay Peninsula). Wing male (n = 5) 134–145mm, mean 139.4 ± 4.3, female (n = 5) 135–142mm, mean 138.6 ± 2.7. Bill male 33–37mm, mean 34.7 ± 1.5, female 29–35mm, mean 33.4 ± 2.5. Tail male 220–244mm, mean 232.4 ± 11.1, female 219–252mm, mean 233.8 ± 18.6. Tarsus male 28–31mm, mean 29.9 ± 1.2, female 29–34mm,

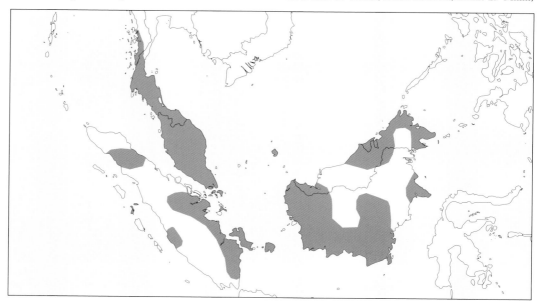

mean 31.1 ± 2.0. Mass male (n = 5) 82–95g, mean 86.0, female 90, 102, 105g (Payne 2005).

MOULT Active moult late May–late Aug; primaries moulting at one or two loci simultaneously (Wells 1999).

DISTRIBUTION Thai-Malay Peninsula, including south Tenasserim; Sumatra, and Riau and Lingga Archipelagos, Pini, Bangka, Belitung, Batu and Mendanau Is; Borneo and North Natunas. Only malkoha now in Singapore (Ripley 1942, Smythies 1986, Hails & Jarvis 1987, van Marle & Voous 1988, Round 1988, Wells 1999, Robson 2000, Mann 2008).

HABITAT Forests, including mangroves, forest edge and plantations, Malay Peninsula. Common in secondary scrub Tenasserim. On Borneo lowland and hill dipterocarp, and peat-swamp forests, kerangas, mangrove, logged and secondary forests, cocoa and *Albizia* plantations, and fire-padang, to 1,300+m. In Sumatra primary and secondary forests, overgrown rubber plantations and mangroves to 500m. To 330m North Natuna Is. 'Plains-level species' Thai-Malay Peninsula (Hume and Davison 1878, Oberholser 1932, Smythies 1957, Pfeffer 1960, van Marle & Voous 1988, Jeyarajasingam & Pearson 1999, Smythies & Davison 1999, Wells 1999, Sheldon *et al.* 2001, Mann 2008).

BEHAVIOUR Skulks singly or in pairs in dense foliage of smaller trees, occasionally in small parties. Moves slowly and can be quite motionless for prolonged periods (Hails & Jarvis 1987, MacKinnon & Phillipps 1993, Wells 1999).

BREEDING Season Copulation Menggala, Sumatra, 29 Aug. On Borneo females in breeding condition Sarawak, Oct–Dec, and Sabah, Mar, and dependent young Central Kalimantan, Jul. Nest building Thai-Malay Peninsula, mid Jan, Jun and mid Jul suggesting two clutches per year. Fledgling Aug, Singapore (Smythies 1957, Nash & Nash 1988, Holmes 1996, Wells 1999, Sheldon *et al.* 2001, Low 2008c). **Nest** Pigeon-like platform of sticks lined with leaves built 1.5–6m up in young trees (Payne 1997, Wells 1999). **Eggs** 2; chalky white; 27.5–28.2x23.0–24.0mm, (n = 3) mean 28.0x23.5; Borneo 27.1x23.7mm (Schönwetter 1967, Payne 1997, Wells 1999). **Incubation** Both sexes participate (Payne 2005). **Chicks** Nestling period 14 days (Payne 2005). **Survival** No information.

FOOD Foliage-gleans taking stick insects, katydids, crickets, grasshoppers, mantids, cicadas and other insects on leaves; small fruits and seeds; caterpillar and agamid lizard found in stomachs; flying lizard *Draco* fed to chick (Hails & Jarvis 1987, Wells 1999, Smythies & Davison 1999, Low 2008c).

STATUS AND CONSERVATION Uncommon Borneo (Mann 2008). Least frequently seen malkoha in Thai-Malay Peninsula, except Ayer Keroh Forest, Melaka, where very common; absence from hill-slopes and dependence on natural forest places it at greater risk than other malkohas (Wells 1999). Very common around Mergui, Burma (Smythies 1986). Not immediately at risk Thailand, where found in four wildlife sanctuaries, but because of its preference for lowland forests, which are being destroyed rapidly, its status may soon be critical (Round 1988, Robson 2000). Considered near threatened due to loss of habitat, and apparently decreasing population (BirdLife International 2011).

Chestnut-bellied Malkoha *Rhopodytes sumatranus*. **Fig. 1.** Adult, Kuala Selangor, Malaysia, August (*Marc Thibault*). **Fig. 2.** Adult feeding nestling with large orthopteran, Singapore, August (*Lee Tiah Khee*).

BLUE-FACED MALKOHA
Rhopodytes viridirostris Plate 15

Zanclostomus viridirostris Jerdon, 1840 (Bottom of Coonoor Pass, south India)

Alternative names: Lesser/Small Green-billed Malkoha.

TAXONOMY Sri Lankan birds have somewhat heavier bill, white tips to T4–5 larger (Ali & Ripley 1983), but not formally named. Monotypic.

FIELD IDENTIFICATION 39cm. Rarely flies but flight, slow, direct and laboured on short, rounded wings; long-tailed. **Adult** Blue naked skin around eyes diagnostic. Head and upperparts dark grey, wings and tail more metallic greenish, tail broadly tipped white, underparts grey darkest on breast, washed rufous on belly, feathers on throat and breast streaked white. **Juvenile** As adult but duller. **Similar species** Long-tailed Malkoha, confusion species only in north and east India, much larger with extremely long tail, and red eye patch, not blue.

VOICE Alarm call low *kraa*, but usually silent (Henry 1955, Rasmussen & Anderton 2005).

DESCRIPTION Adult Head dark grey; upperparts grey glossed olive-green, wings glossy blue-green; underparts dark grey streaked white on chin, throat and breast where feathers are narrow and bifurcated at tips, belly washed rufous; broad graduated tail black glossed green and broadly tipped white. **Juvenile** Duller, and less glossed; outermost primary broader and more rounded as compared with sickle-shape of adult, rectrices narrower narrowly tipped buffy-white, most obvious on T1. **Nestling** Black skin and long, white trichoptiles (Payne 2005). **Bare parts** Iris deep brown to red with white outer ring, orbital skin warty and pale blue, palest around eye. Bill pale apple green. Feet plumbeous.

BIOMETRICS Wing male (n = 13) 132–141mm, mean 135.0 ± 3.7 female (n = 4) 127–138mm, mean 131.0. Bill male 28.0–31.7mm, mean 30.2 ± 1.2, female 29–31mm, mean 30.3. Tail male 220–247mm, mean 233.6 ± 10.3, female 222–229mm, mean 224.3. Tarsus male 24–34mm, mean 30.6 ± 2.7, female 31.0–33.6mm, mean 32.2. Mass male 77.9g (Payne 2005).

MOULT Post-nuptial and post-juvenile moults complete (Ali & Whistler 1936).

DISTRIBUTION Resident peninsular India and Sri Lanka. In India from Baroda (Gujarat) 22°20'N on western side, and Cuttack (Orissa) 22°25'N on eastern side, south throughout peninsula (Ali & Ripley 1983, Grimmett *et al.* 1998).

HABITAT In India lightly wooded and deciduous scrub country, open secondary jungle and broken foothills to 1,150m; to 330m Sri Lanka, preferring thorny scrub or areas with *Euphorbia* in dry lowlands (Ali & Whistler 1936, Henry 1955, Ali & Ripley 1983, Kotagama & Fernando 1994, Grimmett *et al.* 1998).

BEHAVIOUR Clambers in dense tangled herbage scrutinising leaves for food, but also on ground; seldom in open areas. Solitary or in pairs (Baker 1927, Henry 1955, Payne 2005).

BREEDING Season Breeds year round India, peak Mar–May, and Sri Lanka (Butler 1881, Baker 1927, Ali & Whistler 1936, Henry 1955, Ali & Ripley 1983). **Nest** Saucer-shaped,

of thorny green twigs about 1m up in cactus; or of twigs and lined with few green leaves few metres high in small tree, or thorny bush or bamboo clump (Ali & Whistler 1936, Henry 1955, Ali & Ripley 1983). **Eggs** 2, occasionally 3; chalky white; 27x22, 27x21mm, 31.0x24.5mm, mean (n = 6) 29.4x24.7mm (Baker 1927, Ali & Whistler 1936, Henry 1955). No other information.

FOOD Mainly fruit, but also crickets and mantids; lizards and other small vertebrates (Henry 1955, Ali & Ripley 1983).

STATUS AND CONSERVATION Locally fairly common India; common Sri Lanka (Grimmett *et al.* 1998). Wide habitat tolerance makes it less vulnerable to human disturbance. Not globally threatened (BirdLife International 2011).

Blue-faced Malkoha *Rhopodytes viridirostris*. **Fig. 1.** Adult, Bondala Goa, India, March (*Niranjan Sant*).

BLACK-BELLIED MALKOHA
Rhopodytes diardi   Plate 15

Melias Diardi Lesson, 1830 (Java, error = Sumatra)

Alternative name: Lesser Green-billed Malkoha.

TAXONOMY Polytypic. Synonym: *Phænicophæus nigriventris* Peale 1848 in nominate.

FIELD IDENTIFICATION 36–38cm. Long tail approximately half total length; dark appearance with large red area around eyes, dark grey belly and prominent yellow-green bill. **Adult** Grey head, neck and upperparts, wings glossy blue-green, underparts grey, belly blackish-grey, tail as wings but tipped white below. **Juvenile** Duller; bill slender and darker. **Similar species** Chestnut-bellied slightly larger with chestnut belly and undertail-coverts, bill stronger and skin around eyes more pinkish-orange. Long-tailed larger with extremely long tail about 150% length of rest of bird, and broadly tipped white; underparts streaked dark; red facial skin bordered white.

VOICE Far-carrying *taup* or *pauk*, and frog-like *gwaup*, or faster *gwagaup*. First mentioned difficult to distinguish from Banded Pitta. On Borneo sharp *pwew-pwew* or single soft *taup* (MacKinnon & Phillipps 1993, Smythies & Davison 1999, Wells 1999, Robson 2000).

DESCRIPTION Nominate. **Adult** Head and neck plain slate-grey, feathers mostly faintly darker shafted with only hint of white around crimson facial patch, narrow black lores; upperparts slate-grey, upper back slightly glossed greenish, wings glossy blue-green; underparts grey, chin, throat and breast paler than head or belly; flanks, belly and undertail-coverts blackish-grey, fine shaft streaks on throat and breast; graduated tail glossy blue-green broadly tipped white. **Juvenile** As adult but duller wings, throat more whitish and tail narrower with shorter white tips. **Nestling** Undescribed. **Bare parts** Iris pale blue to dark brown, juvenile brown, facial skin rich crimson. Bill yellow-green, juvenile dark grey. Feet dull lead-grey often washed green.

BIOMETRICS Nominate. male wing (n = 10) 127–138mm, mean 131.9 ± 2.8, female (n = 11) 127–137mm, mean 130.9 ± 3.0. Bill male 25.4–32.0mm, mean 29.8 ± 1.6, female 28–31mm, mean 30.2 ± 1.0. Tail male 215–240mm, mean 227.3 ± 8.4, female 207–241mm, mean 222.7 ± 9.3. Tarsus male 28.5–30.5mm, mean 29.2 ± 0.9, female 28–32mm, mean 30.0 ± 1.5 (Payne 2005). Mass male Jul 55.8, Sep 58.2g (Thompson 1966); male 72.2g (Medway 1972); male 66g, female 75g; *borneensis* male 62g, female 65g (Payne 2005).

MOULT Moulting all months except Dec–Jan (no Apr birds examined); wing moult May–Nov, primaries moulted at up to three loci simultaneously (Wells 1999).

GEOGRAPHICAL VARIATION Two subspecies.
R. d. diardi (Lesson, 1830). Tenasserim, Thai-Malay Peninsula, Sumatra; once Vietnam; Laos and Cambodia? Described above.
R. d. borneensis Salvadori, 1874. Borneo. Smaller; forehead slightly paler grey; throat pale grey, breast to belly rich buffy-grey, flanks sooty. Wing male (n = 7) 121–131m, mean 126.1 ± 3.5, female (n = 5) 121–136mm, mean 126.0 (Payne 2005).

DISTRIBUTION Thai-Malay Peninsula, Sumatra and Borneo; resident. Tenasserim, north to Tavoy. Khao Sok NP, Thailand, and wildlife sanctuaries: Khao Phanom Bencha, Thaleban, Khlong Nakha, Khlong Saeng and Khao Banthad. In Malay Peninsula south to Johor, but last recorded Singapore 1950. Rare resident south Laos? and north Cambodia? Mainland Sumatra to 900m. Mainland Borneo to 915m, once 1,700m (Smythies 1986, van Marle & Voous 1988, Round 1988, Sibley & Monroe 1990, Payne 1995, Jeyarajasingam & Pearson 1999, Mann 2008). Once Vietnam (Quý & Cu 1995).

HABITAT In Thailand lowlands and lower hills with

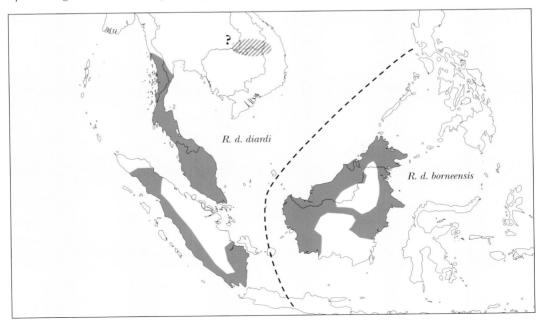

evergreen forest below 200m. Similarly in Thai-Malay Peninsula but also peatswamp forest, overgrown rubber plantations and disturbed forest, mostly to *c.* 200m. On Sumatra primary and secondary forest, mangroves, riverine swamp forest and edges, *Melaleuca* forest edges, peatswamp forest edges and patches of bamboo, to 900m but 2,000m Gunung Leuser NP. In Borneo lowland dipterocarp, peatswamp, mangrove and secondary forests, cocoa and *Albizia* plantations to 915m; once 1,700m Gunung Kinabalu due to lowland fires (Nash & Nash 1985, van Marle & Voous 1988, Round 1988, Lekagul & Round 1991, Smythies & Davison 1999, Wells 1999, Buij *et al.* 2006, Mann 2008).

BEHAVIOUR Seldom flies; keeps to thick foliage of trees and bushes, often preferring canopy and forest edges, climbing up trunks in spiral fashion. Often single but sometimes up to five together. Clumsy hover-snatching of prey from leaf-tip (Smythies & Davison 1999, G.W.H. Davison in Wells 2007).

BREEDING Season Jan–May mainland south-east Asia. Eggs Thai-Malay Peninsula, 26 Mar, and pair building nest early Apr; feeding fledgling Kedah, 7 Jul; nestlings Khao Luang NP, 5 Apr. Fledgling begging food south Sumatra, 19 Sep. Active gonads Sabah, Borneo, May (Nash & Nash 1985, Wells 1999, 2007, Robson 2000, 2008, Mann 2008). **Nest** Untidy shallow cup nest, 3–4cm deep, of dead twigs lined with small green leaves, very similar to that of pigeon; both sexes gather nest materials (Chasen 1939, Wells 1999, Robson 2000). **Eggs** 2; chalky-white; nominate 27.5–33.6x22.5–26.5mm, mean 30.2x24.5, *borneensis* 28.3–29.9x22.9–24.6mm, mean 29.3x24.0 (Schönwetter 1967). No further information.

FOOD Beetles, caterpillars, ants, grasshoppers, large cockroach (Smythies & Davison 1999, Wells 2007).

STATUS AND CONSERVATION Common resident extreme south peninsular Thailand (Lekagul & Round 1991). Uncommon and local throughout lowlands of Borneo (Mann 2008). Silent, secretive behaviour results in rather few records, but wide habitat preference, even regrowth, ensures no immediate risk in Thai-Malay Peninsula (Wells 1999). However, forest fragmentation great threat throughout range, and preference for lowland forests which are much threatened may result in status becoming critical (Round 1988, Lambert & Collar 2002). Near threatened due to habitat loss and decreasing population (BirdLife International 2011).

Black-bellied Malkoha *Rhopodytes diardi*. **Fig. 1.** Adult, *R. d. diardi*, Singapore, March (*H. Y. Cheng*).

LONG-TAILED MALKOHA
Rhopodytes tristis **Plate 15**

Melias tristis Lesson, 1830 (Sumatra; error = Bengal)

Alternative names: (Greater/Large) Green-billed Malkoha.

TAXONOMY Payne (2005) placed *hainanus* Hartert 1910, *saliens* Mayr 1938, *Phoenicophaeus longicaudatus* (Blyth 1843) and *P. elongatus* (S. Müller 1835) in nominate, claiming no consistent size differences. Polytypic. Synonym: *nigristriatus* Baker 1926 in nominate.

FIELD IDENTIFICATION 51–60cm. Extremely long tail held straight when running lengthwise along branches giving squirrel-like appearance. Slow clumsy flight. **Adult** Head and neck ashy-grey with red facial skin, above glossy green; below grey becoming darker on breast and dark grey on rest of underparts; tail black with strong green gloss and broadly tipped white. **Juvenile** Duller than adult, with rectrices shorter and white tips smaller. **Similar species** Blue-faced Malkoha smaller, with shorter tail and conspicuous blue orbital skin.

VOICE Low croaking *ko-ko-ko* and cat-like chuckle when disturbed, also rendered as *kuk-kuk* and sometimes followed by deep *grah-grah-grah-grah-grah* or *gurh-gurh-gurh*. Territorial call well-spaced and slightly nasal *ob-ob-ob*, *ko-ko-ko* occasionally at end. Hoarse *kaup* and sharper *krup* or faster *kluk-rup-rup*. Soft, wheezing *whiow*. Croaking, frog-like *ko, ko, ko, ko* sometimes ending with gruff flurry *co-co-co-co* (Hume & Davison 1878, Smythies 1940, 1986, Ali & Ripley 1969, MacKinnon & Phillipps 1993, Wells 1999, Robson 2000, Rasmussen and Anderton 2005, Loke Wan Tho in Wells 2007).

DESCRIPTION Nominate. **Adult** Head and neck deep grey with tiny black shaft streaks, white above black lores often continued over or completely around crimson facial skin; black bristles around gape; upperparts dark ashy-grey glossed green, wings metallic oil-green; underparts grey becoming darker on belly with ochraceous wash on throat and breast; fine black shaft streaking on chin, throat and breast; undertail-coverts sooty-black; long graduated tail glossy dark oil-green both above and below and very obvious broad white tips below, but soon becomes abraded and white tail tips can be entirely lost in summer. **Juvenile** As adult but greener and duller, and wear changes gloss on wings and tail from greenish to bluish; rectrices shorter and narrower and more pointed, white tips smaller and less clearly defined. **Nestling** Undescribed. **Bare parts** Iris brown to red-brown or crimson, orbital skin bright crimson in male, duller in female, juvenile reduced and dark brown. Bill apple-green, washed red at base and around nostrils, gape red, juvenile bill shorter and slate-horn on upper mandible, lower greenish-horn. Feet dark greenish-plumbeous, soles white.

BIOMETRICS Nominate. Wing male (n = 12) 160–168mm, mean 163.3 ± 2.2, female (n = 11) 151–165mm, mean 161.2 ± 4.5. Bill male 30–34mm, mean 31.8 ± 1.3, female 31–32mm, mean 31.4 ± 0.5. Tail male 340–408mm, mean 374.4 ± 26.8, female 320–430mm, mean 369.8 ± 29.9. Tarsus male 33–37mm, mean 35.2 ± 1.3, female 33–36mm, mean 34.3 ± 1.0. Mass male 116, female 114g; *kangeangensis* male 120, 128g, female 100g (Payne 2005).

MOULT Adult male in fresh plumage Nepal, 13 Nov; two young birds just beginning complete moult, Oct. Primaries moulted at up to three loci simultaneously May–Oct (Diesselhorst 1968, Wells 1999).

GEOGRAPHICAL VARIATION Six subspecies; some overlap in measurements and intergradation in first five.

R. t. tristis (Lesson, 1830). North India patchily from Himalayas to Vijayawada in south-east India, Nepal, Sikkim, Bhutan, Assam and Bangladesh. Birds from Assam have slightly more bluish gloss on wings and tail. Described above.

R. t. hainanus E. Hartert, 1910. Hainan, south-west Guangdong. Said to be smaller than nominate, tail 40–70mm shorter, and forehead darker; similar to *saliens* but wings more greenish, breast duller with more brownish wash.

R. t. saliens Mayr, 1938. Bhamo and upper Chindwin district, Burma, south-west China, Tonkin south to mountains of north Thailand, north Laos, north Annam; Mishmi Hills, Arunachal Pradesh and Nagaland where it intergrades with nominate. Smaller than nominate. Similar to *hainanus*, but larger with much longer tail, and more bluish. Very little ochraceous below; crown, back and underparts rather dark; white on forehead and supercilium not extensive; bluish gloss on wings; white tips to rectrices narrow.

R. t. longicaudatus (Blyth, 1843). South Indochina, Burma, Thailand, Thai-Malay Peninsula. Small, but tail long. White tips on T4 very broad; very light above, and particularly below, with almost no ochraceous wash; black shaft streaks on throat narrower and less conspicuous than in nominate; usually extensive white on forehead and supercilium; bluish gloss on wings and tail.

R. t. elongatus (S. Müller, 1835). Sumatra. Darker green-grey, with long white tail spots; lacks ochraceous streaking below.

R. t. kangeangensis Vorderman, 1893. Kangean I. Dark grey with olive gloss on upperparts, white spots on tail nearly twice as long as nominate. Wing male 160–170mm, (n = 6) mean 167.2 ± 3.4, female 160–163mm, mean 161.7 (Payne 2005).

DISTRIBUTION Oriental; resident throughout most of range. In Nepal all year Bardia, Chitwan and near Dharan, but only Apr–Sep Kathmandu Valley. Disjunct distribution in India; in Himalayas from north Uttar Pradesh to Arunachal Pradesh in east, around Calcutta, and isolated population near Jamshedpur in north to near Vijayawada in south; Manipur; Bangladesh. Burma; Thailand (local movements?); throughout Thai-Malay Peninsula south to Terengganu Valley and Genting Highlands, also Koh Samui and Penang; Sumatra. China in Yunnan, Guanxi Zhuang, Guangdong (Guangzhou Bay) and Hainan (Cheng 1987, Inskipp & Inskipp 1991, Grimmett *et al.* 1998, Wells 1999, Choy Wai Mun *in litt.* to Wells 2007, Choudhury 2009).

HABITAT In India dense vegetation, shrubbery in evergreen forests and moist deciduous thickets to 1,800m. Usually below 700m, but to 2,000m, Nepal. Thick bamboo jungles and secondary forests favoured in Thailand. In wooded patches near villages and in hill forest, Bangladesh. Often fruit-gardens, Hong Kong. In Thai-Malay Peninsula orchards, mountains with bamboos and rotan palms, and edges of paddifields, edges of mangrove forest, paperbark stands and grazing grounds. On Sumatra pristine and disturbed hill and lower montane forests, scrub in secondary forests, and rubber plantations, 700–1,500m, rarely down to 200m. Below 700m, Nepal, occasionally to 2,000m. To 1,600m, Burma (Gyldenstolpe 1913, Robinson 1928,

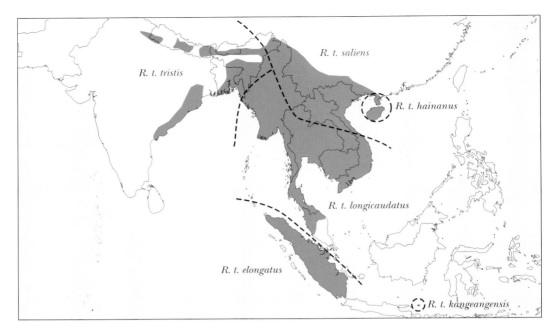

Aagaard 1930, Ali & Ripley 1983, Smythies 1986, van Marle & Voous 1988, Grimmett et al. 1998, Wells 1999).

BEHAVIOUR Often follows flocks of laughing-thrushes in Thailand, concealing itself in thick foliage and moving silently among branches; shy but curious. Mostly only seen in short, weak and laboured flight from one thicket to another, but more graceful when gliding from branch to branch, sometimes flicking long tail. Occasionally on ground (Gyldenstolpe 1913, Baker 1927, Ali & Ripley 1983).

BREEDING Season Apr–Aug India. Eggs Thailand 3 Apr–13 Aug (peaking May), fully grown young 10 Jun, young in nest 12 Apr. Two 'well-feathered' chicks Malay Peninsula, 10 Mar. Nestlings Tenasserim, 1 Mar; Dec–Jun, south-east Asia (Robinson 1928, Berwick 1952, Ali & Ripley 1983, Smythies 1986, Robson 2000, Round 2008). **Nest** Few sticks, lined with fresh green leaves, replaced at intervals, diameter 18cm, at 2.5–6m in small trees; flimsy structure like that of pigeon and with no special effort at concealment; 1–3m up (Robinson 1928, Aagaard 1930, La Touche 1931, Ali & Ripley 1983, Wells 1999). **Eggs** 2–3 (4); cylindrical-oval; white, chalky; mean (n = 50) 33.8x25.8mm (Robinson 1928, Aagaard 1930, Ali & Ripley 1983). **Incubation** Both sexes incubate (S. Baker in Ali & Ripley 1983). No further information.

FOOD Caterpillars, mantids, crickets, grasshoppers and other insects; lizards and perhaps other small vertebrates; fruit, seeds (Baker 1927, Ali & Ripley 1983).

STATUS AND CONSERVATION Common Nepal below 700m, uncommon to 2,000m; locally fairly common India, foothills to 1,800m, and Bangladesh (Grimmett et al. 1998); common Manipur (Choudhury 2009). Uncommon 200–1,400, occasionally to 1,900m, Bhutan (Spierenburg 2005). Very common resident Thailand (Lekagul & Round 1991). Uncommon to common Thai-Malay Peninsula depending on habitat (Wells 1999). Locally common throughout Burma (Smythies 1986). Frequent along Barisan chain, Sumatra, and common Kangean I. (MacKinnon & Phillipps 1993). Rare China (Cheng 1987). Not at immediate risk but continued clearance of forest could soon become problematic. Not globally threatened (BirdLife International 2011).

Long-tailed Malkoha *Rhopodytes tristis.* **Fig. 1.** Adult, *R. t. longicaudatus*, Air Hitam Dalam, Penang, Malaysia, May (*Choy Wai Mun*).

Genus *Rhamphococcyx*

Cabanis & Heine, 1863 *Mus. Hein.* Th. 4, Heft 1: 65. Type, by original description,
Phaenicophaeus calyorhynchus Temminck 1825.
Two species.

Sundaland and Sulawesi; non brood-parasitic. Large, long-tailed malkohas, typified by large bill, and separated from other malkohas (except *Dasylophus* and the here separated *Lepidogrammus*) on mtDNA analysis (Sorenson & Payne 2005). Sexes similar. Sometimes placed in *Phaenicophaeus* or *Zanclostomus* (e.g. Dickinson 2003), but due to distinctive bill, red facial skin and different plumage colour here used for *calyorhynchus* and *curvirostris*, although on mtDNA evidence Sorenson & Payne (2005) place the latter in *Phaenicophaeus*. Includes *Rhinococcyx* (Sharpe 1873), *Dryococcyx* (Sharpe 1877) and *Urococcyx* (Shelley 1871).

CHESTNUT-BREASTED MALKOHA
Rhamphococcyx curvirostris   Plate 16

Cuculus curvirostris Shaw, 1810 (western Java)

Alternative name: Palawan Malkoha (*harringtoni*).

TAXONOMY Polytypic. Deignan (1952) found *Phoenicophaeus erythrognathus* Hartlaub 1844 unavailable and replaced it with *singularis* Parrot 1907. Other synonyms: *Malcoha rouverdin* Levaillant 1806 in nominate; *microrhinus* Berlepsch 1895 in *borneensis*.

FIELD IDENTIFICATION 42–50cm. Medium-sized malkoha with heavy yellowish and red bill, large red patch of naked facial skin, and rather long rounded tail giving squirrel-like appearance when it creeps around in thick vegetation and swinging tail. Flight usually brief and low on short, rounded wings. **Adult** Top of head and nape grey, sides of neck rufous, upperparts and wings glossy green, underparts rufous, flanks glossy green and lower belly and undertail-coverts dark rufous to blackish, tail glossy green broadly tipped rufous. **Juvenile** Less red facial skin, lower mandible mostly black and rectrices narrower. **Similar species** None in range. Only malkoha without white tail tips in Thai-Malay Peninsula.

VOICE Chicken-like *tok-tok-tok*, faster in flight, or *kok-kok-kok*. When breeding single, gentle *konk*; two foraging birds uttered harsh *miaou* like cat, and nestlings have bubbling hiss. Mellow whistle *wee-oo, wee-oo* (MacKinnon & Phillipps 1993, Smythies & Davison 1999, Wells 1999, Robson 2000, SUARENG 2003 in Wells 2009).

DESCRIPTION Nominate. **Adult male** Forehead, crown and nape medium grey, fine grey stripe bordering dorsal crescent of red facial skin ending at bill; upperparts and wings deep glossy green which becomes blue in worn plumage, underwing-coverts dark rufous; underparts including chin rufous except glossy green flanks, blackish abdomen and undertail-coverts, graduated tail glossy green, outer third mahogany-rufous without white tips, tail rufous-chestnut below. **Adult female** Grey stripe under red facial skin broader, and chin also grey. **Juvenile** Red facial skin reduced, throat and breast rufous tinged purplish, abdomen and undertail-coverts blackish, rectrices narrower and central pair entirely green-grey or with few rufous subterminal spots. **Nestling** Skin black with hair-like down, and gape red. **Bare parts** Iris bluish, sometimes whitish or grey, with paler outer ring (male), yellow, sometimes cream or red (female), juvenile dark brown, red-brown changing to white (Hume & Davison 1878 in Payne 2005) facial skin dark scarlet. Upper mandible mostly yellow to yellow-green, darker at base, lower mandible brownish horn to dark yellowish; entire bill black to slaty-blue in juvenile, not arched and swollen. Feet dark grey.

BIOMETRICS *R. c. singularis*. Wing male (n = 9) 160–180mm, mean 169.1 ± 7.9, female (n = 11) 166–182mm, mean 171.5 ± 5.5. Bill male 37.0–42.5mm, mean 40.4 ± 1.9, female 37.0–42.5mm, mean 39.9 ± 1.5. Tail male 240–274mm, mean 256.7 ± 11.4, female 245–290mm, mean 264.3 ± 12.6. Tarsus male 36.5–44.5mm, mean 39.1 ± 2.3, female 37–41mm, mean 38.6 ± 1.9 (Payne 2005). Mass adult 188, 190g (Wells 1999); female 167.6g (Medway 1972); nominate male 145, 151, 160g, female 163, 165g; *microrhinus* male (n = 4) 134–155g, mean 142.3, female (n = 7) mean 141.9g; *harringtoni* male (n = 6) 111–177.1g, mean 152.3, female 137.5, 146.5, 148.6g (Payne 2005).

MOULT Late Apr–early Nov peaking Jul–Aug, occasionally outside this period, Thai-Malay Peninsula; primaries usually replaced at two loci, occasionally three (Wells 1999).

GEOGRAPHICAL VARIATION Six subspecies.
R. c. curvirostris (Shaw, 1810). West and central Java. Wing male (n = 6) 166–177mm, mean 171.2 ± 4.5, female (n = 6) 170–174mm, mean 171.7 ± 1.9 (Payne 2005). Described above.
R. c. singularis Parrot, 1907. Thai-Malay Peninsula, Pulau Siantan (Anambas) and Sumatra. Cheeks and chin normally grey but much variation, red facial skin flanked by fine black-and-white feathers, abdomen black, bill green, red below round to oval nostril and on lower mandible. T1 entirely dark bronze-green in some from Burma and Sumatra (BMNH).
R. c. oeneicaudus J. & E. Verreaux, 1855. Mentawai Is. Darkest subspecies, with dark grey stripe below eye, chin to upper breast vinous-chestnut, lower breast, lower flanks and thighs glossy dark green, upper flanks, belly and vent brownish-black; no rufous in tail; T1 entirely dark glossy green with some purple, other rectrices glossed dark green, blue and purple. Iris light blue (male), dark brown (female). Bill as *singularis*. Wing male (n = 6) 159–174mm, mean 165.4, female (n = 6) 163–175mm, mean 168.7 ± 4.2 (Payne 2005).
R. c. deningeri (Stresemann, 1913). Bali and east Java. Face pale grey with paler rufous throat, and paler all over. Bill as nominate.
R. c. borneensis Blasius & Nehrkorn, 1881. Borneo, Natunas and Bangka. Some have whole head dark grey, some with chin pale rufous, others greyish; crown and nape dark green-grey,

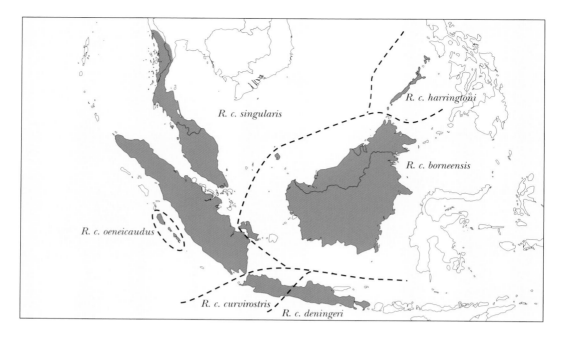

rest of upperparts glossy bronze-green, underparts dark red-brown, tail shorter and less rounded, bill green with red lower mandible, ungrooved, oval, sometimes whitish or grey, nostril wider than deep. T1 (sometimes also T2) may be entirely bronze-green (BMNH). Wing male 150–172mm (n = 8), mean 164.1, wing female 157–164 (n = 8), mean 160.5 (Kloss 1930); wing male (n = 4) 154–165mm, mean 159.0, female (n = 7) 156–169mm, mean 162.4 ± 4.7 (Payne 2005).

R. c. harringtoni (Sharpe, 1877). South-west Philippines on Balabac, Busuanga, Calauit, Culion, Dumaran and Palawan. Head grey tinged olive-brown with broad grey stripe below eye, underparts often paler, chestnut central rectrices vary considerably in length; bill ungrooved yellow or green, lower mandible red or blackish; no sexual dichromatism. Wing male (n = 8) 171–181mm, mean 175.5 ± 2.8, female (n = 6) 167.0–180.5mm, mean 175.4 ± 4.2 (Payne 2005).

DISTRIBUTION Sundaland. Resident. Thai-Malay Peninsula in all divisions, including Tenasserim, except Singapore; also Langkawi, Penang and Tioman islands. Sumatra, including Bangka, and Pagai, and Simabugai, Siberut and Sipura (Mentawai Is). Borneo, including North Natunas and Anambas. Java, Bali and Philippines (Palawan group) (van Marle & Voous 1988, Dickinson et al. 1991, MacKinnon & Phillipps 1993, Wells 1999, Mann 2008).

HABITAT Lowland forest. Variety of habitats in Thai-Malay Peninsula from evergreen to semi-deciduous lowland forests, and peat swamps; also mangroves Langkawi. Prefers dense foliage like other malkohas but also frequents selectively-logged patches. On Borneo to 1,220m in lowland and hill dipterocarp, peat-swamp, alluvial, logged and secondary forests, *Imperata* grassland, mangroves, coastal vegetation, gardens, *Albizia*, citrus, cocoa and rubber plantations. Also coffee plantations on Bali. On Sumatra lowlands, and to 1,500m Berastagi, Batak Highlands; mangroves. Below 1,000m Philippines. To 1,172m, Java (Stresemann 1913a, van Marle & Voous 1988, MacKinnon & Phillipps 1993, Smythies & Davison 1999, Wells 1999, Kennedy et al. 2000, Sheldon et al. 2001, Mann 2008).

BEHAVIOUR Singly or in small family groups, often in thickets in forest where it may sit motionless for prolonged time at tops of trees. Sometimes joins mixed-species foraging groups of other malkohas, drongos, Black Magpies and Large Woodshrikes. Forages in low, dense scrub (MacKinnon & Phillipps 1993, Jeyarajasingam & Pearson 1999, Smythies & Davison 1999, Wells 1999, Mann 2008).

BREEDING Season In Thai-Malay Peninsula eggs Jan–Mar and Jun–Jul, nestlings Jun–Jul and begging fledgling 26 Sep; fledgling being fed Pulau Pinang, 15 Oct. In Sarawak, Borneo, nesting Nov, active ovary Feb, juvenile Jun, and seen carrying twigs 30 Aug; in Brunei incubating Apr and fledglings Sep; in Sabah eggs Feb, active ovaries May and Jun, eggs Feb and active testes Mar. Eggs Java Feb (1), Mar (4), Apr (10), May (6), Jun (1) (Smythies 1957, Hellebrekers & Hoogerwerf 1967, Duckworth & Kelsh 1988, Vowles & Vowles 1997, Wells 1999, Sheldon et al. 2001, C. Robson in Mann 2008, Mann 2008, Robson 2008). **Nest** In fork of sapling 2.5–10m up; built on loose platform of sticks still with their leaves attached, and constructed with finer twigs, 35cm across, 13cm deep, nest cup diameter 11cm. Male brings new nest materials until incubation commences (Edgar 1933, Smythies & Davison 1999, Wells 1999). **Eggs** 2–3; broad oval; plain chalky white, little or no gloss, often stained with dirt; Borneo 34x27.5–29mm; Java mean 37.6x30.0mm (n = 48) (Hellebrekers & Hoogerwerf 1967, Smythies & Davison 1999, Wells 1999). **Incubation** 13 days or longer; both parents contribute to breeding duties (Wells 1999, Payne 2006). No further information.

FOOD Hairy and non-hairy caterpillars, ants, grasshoppers, cockroaches, beetles, orthopterans, cicadas, butterflies, 15cm phasmid; crabs; 28cm green lizard, skink and 10cm snake; bird nestlings; whole small banyan figs (Smythies & Davison 1999, Wells 1999, R.R. & V.M. Kersley in Wells 2007).

STATUS AND CONSERVATION Regular and common Thai-Malay Peninsula (Wells 1999). Common Borneo (Mann 2008), Sumatra (van Marle & Voous 1988), Java and Bali (MacKinnon & Phillipps 1993). Common Philippines (Dickinson *et al.* 1991). Although preferred habitat lowlands it seems not to be at risk because of its extensive use of marginal habitats (Johns 1986), but Lambert & Collar (2002) do not necessarily concur. Not globally threatened (BirdLife International 2011).

Chestnut-breasted Malkoha *Rhopodytes curvirostris*. **Fig. 1.** Adult female, *R. c. singularis*, Taman Negara, Malaysia, May (*H. Y. Cheng*). **Fig. 2.** Adult female, *R. c. singularis*, Ipoh, Perak, Malaysia, August (*Amar-Singh HSS*). **Fig. 3.** Adult male, *R. c. singularis*, Ipoh, Perak, Malaysia, January (*Amar-Singh HSS*).

YELLOW-BILLED MALKOHA
Rhamphococcyx calyorhynchus Plate 16

Phænicophæus calyorhynchus Temminck, 1825 (Sulawesi)

Alternative names: Fiery-billed/Celebes Malkoha.

TAXONOMY Sometimes placed in *Zanclostomus* (Dickinson 2003) but due to distinctive bill, red facial skin and different plumage colour here placed in separate genus with *curvirostris*, with which it has been considered to form a superspecies (White & Bruce 1986). Birds from central Sulawesi intermediate between *meriodionalis* and nominate. Polytypic. Synonym: *centralis* Riley 1918 in *meridionalis*.

FIELD IDENTIFICATION 51–53cm. Rather large malkoha with heavy arched bill and very long rounded tail without any white tipping giving squirrel-like appearance when it creeps around in thick vegetation. Not especially shy and easy to observe. Short, rounded wings. **Adult** Top of head grey, large red area of facial skin in male, small black area in female, sides of neck and upperparts chestnut, wings dark glossy purple, throat and breast chestnut, rest of underparts dark grey, tail blue-black. **Juvenile** Duller than adult with brown, not red, iris and (initially) black, not yellow and red bill. **Similar species** None on Sulawesi.

VOICE Common voice mammal-like rattle lasting *c*. 1.5sec, first accelerating then dying away. Other mammal-like calls and note resembling creaking branch (Watling 1983, Coates & Bishop 1997).

DESCRIPTION Nominate. **Adult** Sexes similar except for orbital skin. Forecrown, lores, crown, nape and hindneck grey, ear-coverts, cheeks, sides of neck and upperparts chestnut to rufous-maroon, lower back, rump, wings and greater wing-coverts dark purple-blue, rest of wing-coverts blackish-rufous, underwing-coverts black; underparts rufous except dark grey lower breast, belly and undertail-coverts, tail blackish-blue tinged purple. **Juvenile** Duller than adult and rectrices narrower. Crown rufous-grey, lower belly brownish-grey. **Nestling** Undescribed. **Bare parts** Iris red, juvenile brown; bare orbital skin red in male, female black (Riley 1924). Bill yellow, distally black tipped white, base, and below horizontal nostril and lower mandible red; juvenile smaller bill wholly blackish. Feet black.

BIOMETRICS Nominate. Wing male (n = 10) 164–178mm, mean 175.4 ± 5.3, female (n = 8) 172–189mm, mean 177.8 ± 4.9. Bill male 40.5–45.0mm, mean 42.8 ± 1.3, female 41–45mm, mean 42.8 ± 1.5. Tail male 302–344mm, mean 320.9 ± 17.8, female 324–344mm, mean 331.3 ± 5.5. Tarsus male 34–38mm, mean 37.0 ± 1.4, female 34.5–39.0mm, mean 37.1 ± 1.9 (Payne 2005).

MOULT Tail moult Sep (BMNH).

GEOGRAPHICAL VARIATION Three subspecies.
R. c. calyorhynchus (Temminck, 1825). North, east and south-east Sulawesi; Togian I. Described above.
R. c. meridionalis (A.B. Meyer & Wiglesworth, 1896). West and south-west Sulawesi. Paler on crown and underparts. Wing male (n = 8) 172–193mm, mean 180.4 ± 7.3, female (n = 6) 168–187mm, mean 179.8 ± 8.1 (Payne 2005).
R. c. rufiloris (E. Hartert, 1903). Butung I. Paler than nominate. Supraloral feathers rufous rather than grey. Averages smaller: wing male 165, 174mm; female 162, 174mm (Payne 2005).

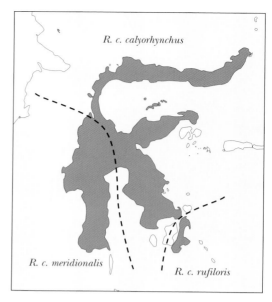

DISTRIBUTION Sulawesi area. Widespread and sedentary Sulawesi, Togian and Butung Is. (Coates & Bishop 1997).

HABITAT Secondary forest and forest edge, open woodland and riverine vegetation near villages, to 1,300m, where it forages in thick foliage; in Manembonembo NR also in primary forest and palm vegetation, 425–625m; wooded savanna, forest on ultra-basic rock and cultivated areas with settlements; secondary growth of coconut plantations. 1,650m Lore Lindu NP. Primary, secondary and selectively-logged forest, 0–570m, Panua NR, north Sulawesi (Coomans de Ruiter 1951, Watling 1983, White & Bruce 1986, Coates & Bishop 1997, Bororing *et al.* 2000, Grantham 2000, Riley *et al.* 2003).

BEHAVIOUR Usually solitary, but sometimes in groups of up to ten, foraging near Crested Macaque *Macaca nigra*, Hair-crested Drongo and Bay Coucal. Creeps around among thick vine tangles like squirrel, often with tail or head downwards, gleaning foliage for insects (Coates & Bishop 1997, Bororing *et al.* 2000, Strange 2001, Cochrane & Cubitt 2004).

BREEDING Presumably nonparasitic, but juvenile repeatedly fed by Eastern Crimson Sunbird (Rozendaal & Dekker 1989). **Season** Female with large eggs Pinedapa, 11 Jan (Riley 1924); dependent juvenile 8 Jan (Rozendaal & Dekker 1989). **Nest** Pigeon-like platform of twigs (Meyer 1879). **Eggs** Pure white; 36x31mm (Schönwetter 1967). No further information.

FOOD Insects including hairy caterpillars, beetles, locusts, mantids and large phasmids (Meyer 1879, Stresemann 1940–41, Coates & Bishop 1997).

STATUS AND CONSERVATION Widespread and moderately common throughout Sulawesi (Coates & Bishop 1997). Wide habitat preference suggests it is relatively secure. Occurs Lore Lindu NP, central Sulawesi (Cochrane & Cubitt 2004) and Panua NR, north Sulawesi (Riley *et al.* 2003); moderately common Manembonembo NR (Bororing *et al.* 2000); common Gunung Ambang NR (Riley & Mole 2001). Not globally threatened (BirdLife International 2011).

Yellow-billed Malkoha *Rhamphococcyx calyorhynchus*. **Fig. 1.** Adult female, race uncertain. Sulawesi, Indonesia (*Ingo Waschkies*). **Fig. 2.** Adult female, *R. c. calyorhynchus*, Tangkoko, Sulawesi, September (*Nigel Voaden*).

Genus *Dasylophus*

Swainson, 1837 *Class. Birds* 2: 324. Type, by monotypy, *D. superciliosus*
Swainson = *Phænicophaus* (sic) *superciliosus* Dumont, 1823.
One species.

Philippines; non brood-parasitic. Large slender malkoha; sexes similar. Sometimes included in *Phaenicophaeus* but here given monotypic genus due to its unique crest contrasting with rest of plumage. Payne (2005) includes *Lepidogrammus* because of similarity of nestlings, but states that nestling of *Dasylophus* is undescribed. MtDNA analysis shows its nearest neighbour to be *Lepidogrammus*, with *Rhamphococcyx calyorhynchus* more distant, and none close to *Phaenicophaeus* (Sorenson & Payne 2005).

RED-CRESTED MALKOHA
Dasylophus superciliosus   Plate 17

Phænicophaus superciliosus Dumont, 1823 (Philippines)

Alternative name: Rough-crested Malkoha.

TAXONOMY Once placed in *Malcoha* (Guérin-Méneville 1838). Polytypic.

FIELD IDENTIFICATION 41cm. Blackish, red-crested and short-winged, with long black-and-white tail. Short floppy flight, but like other malkohas more often seen climbing around in dense foliage. **Adult** Head, neck, upperparts and wings black glossed green-blue. Diagnostic red superciliary crest of long hair-like feathers; underparts blackish; tail black glossed blue-green and broadly tipped white. **Juvenile** Browner and lacks red crest. **Similar species** Scale-feathered Malkoha has grey head without red crest but with shining black scale-feathers on top of head and in triangular patch from chin to upper throat. Philippine Coucal has brown wings, black bill and tail plain black without white tips.

VOICE Soft guttural *cheuk* with slightly metallic tone, repeated once or twice (Kennedy *et al.* 2000).

DESCRIPTION Nominate. **Adult** Head black with faint green-blue gloss, elongated, loosely-webbed shining red superciliary crest from eyes running backwards, between and behind eyes; black rictal bristles form small tuft directed forwards, crown black, some (old birds?) with few red feathers on mid-crown; nape, upperparts and wings black with metallic green-blue, underwing blackish-brown slightly glossed blue, entire underparts dull black faintly tinged green-blue; graduated tail black glossed green-blue and broadly tipped white. **Juvenile** Head and body browner, and lacks red crest; tail shorter. **Nestling** Undescribed. **Bare parts** Iris yellow, juvenile dark brown; bare skin around eye orange-red in male, female deep yellow (Ogilvie-Grant 1894), juvenile yellow. Bill pea green, skin at base orange, juvenile dark brown. Feet green anteriorly, yellow posteriorly, juvenile brown.

BIOMETRICS Nominate. Wing male (n = 12) 143–160mm, mean 150.8 ± 6.0, female (n = 9) 148–166mm, mean 156.5 ± 7.0. Bill male 33–39mm, mean 36.3 ± 1.9, female 33.0–39.5mm, mean 36.7 ± 2.0. Tail male 215–244mm, mean 229.0 ± 8.1, female 213–235mm, mean 224.0 ± 7.8. Tarsus male 30.0–39.2mm, mean 33.3 ± 2.7, female 32.0–38.3, mean 36.8 ± 2.0 (Payne 2005). Mass male (n = 5) 118–130g, mean 121.6, female (n = 8) 105–138g, mean 123.4 (Goodman & Gonzales 1990); male 100, 112g (J. Erritzøe).

MOULT Juvenile female to adult moult, 24 Mar (Goodman & Gonzales 1990).

GEOGRAPHICAL VARIATION Two subspecies.
D. s. superciliosus (Dumont, 1823). Luzon, west and south of *cagayanensis*; Catanduanes, Marinduque and Polillo. Described above.
D. s. cagayanensis (Rand & Rabor, 1967). Cagayan Province, north-east Luzon. Smaller with short red superciliary crest, underparts paler, tinged olive. Wing male (n = 4) 147–155mm, mean 150.5 ± 4.1, female (n = 4) 140–151mm, mean 147.0 ± 5.0 (Payne 2005).

DISTRIBUTION Resident Philippines on Luzon, Catanduanes, Marinduque and Polillo (Kennedy *et al.* 2002).

HABITAT Lowland forest and forest edges, second growth, grassland with bushes to 1,000m; primary and secondary forest, creeks, lakes and especially river banks with trees. Mt. Isarog 300–760m (Gilliard 1950, Gonzales 1983, Goodman & Gonzales 1990, Kennedy *et al.* 2000).

BEHAVIOUR Singly or in small parties, but up to 30+ recorded, creeping and hopping among tangles of vines and often on ground catching insects disturbed by movements above. Frequents all forest strata, although usually near ground; when flushed flies short distance then dives to

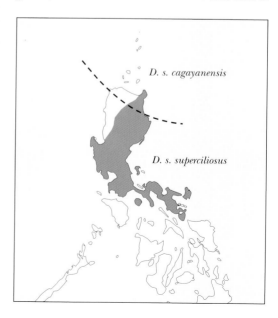

ground and runs. Often in mixed flocks of Scale-feathered Malkoha, Sooty Woodpecker, Greater Flame-backed Woodpecker, Blackish Cuckoo-shrike, Black-and-white Triller, Philippine Bulbul, Balicassiao, Philippine Oriole and Slender-billed Crow (Delacour & Mayr 1946, Gilliard 1950, Poulsen 1995, C.F. Mann).

BREEDING Season Breeding condition May–Jun; nestlings first half Apr; fledglings first half Apr–May and Aug; male with large brood patch, 15 Aug (Goodman & Gonzales 1990, Dickinson *et al.* 1991, Kennedy *et al.* 2000, J. Erritzøe). **Chicks** Fledge when *c.* 60g (Payne 1997). No further information.

FOOD Beetles, katydids, other large insects, and ants; lizards; worms; some vegetation; perhaps snails (Gonzales 1983, Payne 1997).

STATUS AND CONSERVATION Common Mt. Isarog NP 1961, but uncommon 1988 (Goodman & Gonzales 1990). Common Bataan 1947 (Gilliard 1950). Common Catanduanes 1968–1971, especially at Dugui-Too and Mabini Viga (Gonzales 1983). 30+ in mixed feeding party Dec 1982, Quezon NP, Luzon (C.F. Mann) and 10–11 there daily Jan–Feb 1994 (Hornbuckle 1994). Heavy deforestation even in national parks suggests this species requires special monitoring. Not globally threatened (BirdLife International 2011).

Red-crested Malkoha *Dasylophus superciliosus*. **Fig. 1.** Adult, *D. s. superciliosus*, Los Baños, Laguna, Luzon, Philippines, May (*Tina Sarmiento Mallari*).

Genus *Lepidogrammus*

Reichenbach, 1849 *Av. Syst. Nat.* pl. xlvii. Type, by subsequent designation, *Phoenicophaus* (sic) *cumingi* Fraser (Bonaparte 1854, *Consp. Vol. Zygod.*: 5). One species.

Philippines; non brood-parasitic. Large slender malkoha; sexes similar. Payne (2005) subsumed this in *Dasylophus* due to similarity of crest contrasting with rest of plumage, and nestling plumage (although states that nestling of *Dasylophus* is unknown), to *superciliosus*, but because of its unique scale-feathers is here placed in monotypic genus. MtDNA analysis suggests proximity to *Dasylophus*, and *Rhamphococcyx calyorhynchus* forms sister group (Sorenson & Payne 2005).

SCALE-FEATHERED MALKOHA
Lepidogrammus cumingi   Plate 17

Phoenicophaus Cumingi Fraser, 1839 (Luzon)

Alternative names: Scaly/Scale-feathered Cuckoo.

TAXONOMY Monotypic. Synonym: *Phoenicophaus* (sic) *Barrotii* Eydoux & Souleyet 1841.

FIELD IDENTIFICATION 42cm, tail forming just over half total length. **Adult** Head grey with glossy blackish scales on top, and on chin and throat. Upperparts, wings and tail dark blue-green, tail tipped white; sides of chin and throat white, throat with glossy blue-black feathers, breast dark rufous becoming blackish-brown on rest of underparts. **Juvenile** Mostly reddish-brown to rufous; blackish wings with cinnamon tips to coverts; and tail shorter. **Similar species** Smaller, glossy blackish Red-crested Malkoha has untidy bushy red superciliary crest and yellow feet.

VOICE Long metallic whistles, usually higher pitched than those of Red-crested Malkoha; explosive *quizzzz-kid* or *whizzzz-kid*. Strange hissing, wheezing and clicking sounds. In captivity, snoring sounds rendered as nasal *blaeaeae* or hoarse *raeaeaeb*, also faint *ptr* or deeper *krr* followed by higher *krii-aeae-aeaeae* (Gonzales 1983, Gonzales & Rees 1988, Raethel 1992, Kennedy *et al.* 2000, M. Widmer *in litt.*).

DESCRIPTION Adult Sides of head ash grey, grading into elongated feathers with white subterminal areas on forehead, crown and nape; tips of these feathers, together with those of white chin and central throat feathers, form glossy black, flat paper-like spangles with blue or greenish iridescence, and in consistency not unlike red secondary tips in waxwings; eyelashes strong and black, hindneck and sides cinnamon-brown to deep rufous-brown grading into dark glossy blue-green on rest of upperparts, wings purplish-blue-black, underwing-coverts cinnamon-brown; sides of chin and throat white, breast cinnamon-brown gradually becoming dark brown to blackish on abdomen, undertail-coverts and thighs; graduated tail blue-black with purplish wash and broadly tipped white, undertail black faintly tinged bluish with broad white tips. **Juvenile** Few or no scale-feathers, crown to upper back, throat and line to breast, belly, flanks and vent dark reddish-brown, lower back to rump and wings blackish slightly glossed bluish-green, wing-coverts tipped cinnamon; sides of head cinnamon becoming darker and more reddish on sides of breast; much shorter tail grey-black with large white tips. **Nestling** Sparse chestnut bristles above bill on forehead extending onto dark brownish crown, most body feathers tipped warm chestnut, remiges show purple and green iridescence; dark chestnut collar, chin rufous-grey; rectrices basally brownish-black tipped white. **Bare parts** Iris and orbital skin red, latter covered with small warts; juvenile dark brown iris, eye-ring pinkish. Bill pale yellowish-brown, nostrils partly concealed by loral feathers, juvenile basally horn, distal third deep chestnut-brown. Feet greyish.

BIOMETRICS Wing male (n = 9) 150–169mm, mean 156.9 ± 6.9, female (n = 12) 145–168mm, mean 159.0 ± 6.3. Bill male 30.5–41.0mm, mean 38.1 ± 2.1, female 30.5–41.0mm, mean 36.5 ± 3.0. Tail male 220–247mm, mean 241.4 ± 15.2, female 220–245mm, mean 233.7 ± 8.7. Tarsus male 30.5–42.0mm, mean 34.9 ± 3.1, female 34.0–39.5mm, mean 37.1 ±1.6. Mass male (n = 18) 148.5–204.3g, mean 174.1 ± 16.9, female (n = 24) 121.1–194.8g, mean 167.2 ± 19.8 (Payne 2005).

MOULT Primary moult May (BMNH).

DISTRIBUTION Philippines on Luzon, Marinduque and Catanduanes (Kennedy *et al.* 2002).

HABITAT Primary and secondary montane forest, sometimes to 40m high in trees in humid forests with vines, or in crowns of pine forests, but also undergrowth and nearby shrub-grown watercourses. 450–900m Mount Isarog; all types of forest Sierra Madre, 0–1,650m. Presumed resident with some dispersal movements; fairly common to 2,000m;

Dalton Pass at 3,500m (Gilliard 1950, Gonzales 1983, Goodman & Gonzales 1990, Dickinson *et al.* 1991, Alonzo-Pasicolan 1992, Poulsen 1995, M. Widmer *in litt.*).

BEHAVIOUR Poor flier; usually in tangle of forest understorey, sometimes bounding along large boughs like squirrel; also in canopy or high vines; caught on ground in mammal trap. Singly or in small groups, often in mixed-species flocks. Quite conspicuous although skulking (Gonzales & Rees 1988, Goodman & Gonzales 1990, Poulsen 1995, Kennedy *et al.* 2000, B. Gee *in litt.*, C.F. Mann).

BREEDING Season Enlarged gonads Mar–May (Goodman & Gonzales 1990, Kennedy *et al.* 2000), and nestling Apr or Nov (BMNH), but no sign of breeding in birds collected May–Jun (Gonzales 1983). **Nest** Flimsy, cup-shaped; in understorey tangle or high in trees (local observations). **Eggs** 2–3 (Raethel 1992). **Incubation** 22 days for captive bird, but eggs did not hatch (Raethel 1992). No further information.

FOOD Predominantly carnivorous; insects, particularly caterpillars up to 110mm; centipedes; scorpions; snails; worms; small snakes and lizards; small passerines caught in mistnets; vegetation supplementary (Delacour & Mayr 1946, Gonzales 1983, Gonzales & Rees 1988, Goodman & Gonzales 1990, Raethel 1992, BMNH).

STATUS AND CONSERVATION Fairly common in most forest types (Gilliard 1950, Gonzales & Rees 1988), but uncommon in ultrabasic forest (Danielsen *et al.* 1994). Not in any immediate danger. Fairly common in Patapat-Calbario NP, Pasaleng Bay, Ilocos Norte Province. Common to very common in Sierra Madre (Poulsen 1995). Common 300–760m (Rabor 1966). Continued large-scale deforestation in Philippines may become serious threat in near future and hunting for food and trade is perhaps already a threat (Alonzo-Pasicolan 1992, M. Widmer *in litt.*, J. Erritzøe). Populations not yet monitored and no estimates known. Further collection of basic biological data urgently required to aid understanding of importance of threats. Not globally threatened (BirdLife International 2011).

Scale-feathered Malkoha *Lepidogrammus cumingi*. **Fig. 1.** Mt. Makiling, Luzon, Philippines, March (*Stijn De Win*). **Fig. 2.** Los Baños, Laguna, Luzon, Philippines, May (*Tina Sarmiento Mallari*).

Genus *Clamator*

Kaup, 1829 *Skizz. Entw.-Gesch. Eur. Thierw.* p.53. Type, by original designation and monotypy *Cuculus glandarius* Linnaeus 1758.
Four species.

Old World; brood-parasitic; crested; sexes similar. MtDNA analysis places *Clamator* as clade within Phaenicophaeini, closest to *Coccyzus* and *Coccycua* and not within Cuculini (Sorenson & Payne 2005). Includes *Oxylophus* (Swainson 1837), *Coccystes* (Gloger 1842) and *Melanolophus* (Roberts 1922).

CHESTNUT-WINGED CUCKOO
Clamator coromandus   Plate 18

Cuculus coromandus Linnaeus, 1766 (Coromandel Coast, south-east India).

Alternative name: Red-winged (Crested) Cuckoo.

TAXONOMY Monotypic. Synonyms: *Cuculus collaris* Vieillot 1817, *C. coromandelicus* Gray 1870, *Oxylophus coromandus var. rubramus* Gray 1846.

FIELD IDENTIFICATION 38–46cm. Slightly larger than congeners. Flight swift and direct with rapid wing-beats like Common Koel, and with crest lowered. Long-tailed, slender cuckoo with chestnut wings, white collar and long black crest. **Adult** Top of head and upperparts to tail glossy black, throat to upper breast pale orange-brown, breast and upper belly white, rest of belly and undertail-coverts blackish, tail tipped white. **Juvenile** Dark brown on head and upperparts, feathers edged paler, crest reduced, underparts white; wing-coverts edged rufous-buff and tail edged and tipped buffish. **Similar species** Jacobin Cuckoo is black and white with shorter crest. Philippine, Green-billed, Short-toed, Lesser and Greater Coucals, sympatric in various areas, have chestnut wings, but are crestless and with black underparts.

VOICE Territorial call run of metallic notes *thu-thu, thu-thu, thu-thu*, resembling Moustached Hawk-cuckoo, but with shorter pause between each double note, also rendered as *TEE-TSEE–TEE-TSEE*; guttural, descending crackling rattle *ghee-ghe-ghuh-ghuh-ghuh-ghuh*. Fast, grating, woodpecker-like *crititititit*; *klinck-klinck* in flight and when disturbed. Monotonous *bee bee* repeated every few seconds in Philippines, often at night. Apparently silent in non-breeding range in India and Thai-Malay Peninsula, although noisy during breeding (Ali & Ripley 1983, Chattopadhyay 1987, Wells 1999, Kennedy *et al.* 2000, Robson 2000, Rasmussen and Anderton 2005).

DESCRIPTION Adult Top and sides of head metallic blue-black. Bifurcated crest long and erectile, black with blue tinge, conspicuous white collar on hindneck; mantle, scapulars, inner upperwing-coverts and tertials black, strongly tinged green, uppertail-coverts black washed bluish, remiges and coverts chestnut, flight feathers tipped dark brown, underwing paler, coverts and axillaries buffy to pale brown; chin and throat pale rust, breast and belly white; lower belly, flanks and thighs ashy-grey, undertail-coverts black glossed purplish and tipped rusty, tail purplish-black, outermost feathers tipped white on outer webs. **Juvenile** Crest begins to develop at about five weeks and adult-like plumage attained by three months, upperparts including all coverts dark olive-brown with slight gloss and feathers tipped fulvous-white tips, collar rufescent-white; underparts white, smoky on flanks, tertials olive-brown, secondaries brown washed chestnut on outer webs; terminal half of primaries brown, basal half chestnut, all tipped pale, tail blackish, slightly glossy and with broad fulvous-white tips. **Nestling** Undescribed. **Bare parts** Iris pale red-brown, juvenile hazel. Bill black with greenish wash, base of lower mandible and gape yellow-orange in juvenile, gape and mouth salmon in adult. Feet grey.

BIOMETRICS Wing male (n = 11) 154–167mm, mean 161.3 ± 4.0, female (n = 10) 155–165mm, mean 160.1 ± 3.4. Bill male 24.0–28.8, mean 25.8 ± 1.5, female 22.0–26.6mm, mean 24.7 ± 1.8. Tail male 216–247mm, mean 232.7 ± 8.5, female 208–248mm, mean 230.6 ± 14.5. Tarsus male 22.5–26.6mm, mean 24.9 ± 1.6, female 22.0–26.6mm, mean 23.8 ± 1.7 (Payne 2005). Mass unsexed adults (n = 44) 66.4–83.8g (Wells 1999); autumn daytime up to 86g (P.R. Kennerley in Wells 2007).

MOULT Primaries moulted in sequence P6 and P9, pause, P7 and P10, pause, P8 and P5, or P5 then P8; irregular transilient moulting of secondaries; rectrices replaced one by one, each feather dropping after previous has fully grown; adult plumage fully attained at about six months; complete moult lasts about 100 days. Adults arrive Thai-Malay Peninsula Sep having completed moult, while juveniles attain adult plumage Jan–Feb (Stresemann & Stresemann 1961, 1969, Wells 1999).

DISTRIBUTION Oriental. Mostly summer in north; winters in south and east Asia. Himalayan foothills in north and north-east India (resident Manipur), Nepal, Bhutan, Bangladesh, Burma, north and east-central Thailand, Indochina and China; accidental Taiwan. China (summer; resident south?): Jiangsu, Fujian, Guangdong, Hainan, east Sichuan, south Yunnan north to Gansu and south Shaanxi. Winters south India, Sri Lanka (Oct/Nov–Apr/May); Thai-Malay Peninsula, Nov–Feb, but 14 Sep–31 Mar, Singapore; Sumatra including islands of Aruah, Lingga, Bangka and Siberut, Oct–2 Apr; rare Java and Bali; scarce passage and winter migrant Borneo to Sarawak, Brunei, Sabah, once Kalimantan, 22 Sep–8 Mar; Sep–Mar Philippines on Luzon, Mindanao, Palawan, Sanga Sanga, Siquijor and Tawi-Tawi (Makatsch 1971, Cheng 1987, van Marle & Voous 1988, MacKinnon & Phillipps 1993, Condole 1997, Wells 1999, Kennedy *et al.* 2000, Rasmussen & Anderton 2005, Low 2008a, d, Mann 2008, Choudhury 2009). Three records Sulawesi (White & Bruce 1986). Once Spratly Is. 26 Sep (Davison 1999). Four–five records, once suspected breeding, South Korea (Park & Kim 1995, Birds Korea 2009). Maldives Dec (Robson 2007). *c.* 5 records Japan (Honshu, Iriomote I., Tokara Is, Okinawa) (Brazi! 1991, CCJB 2000). Palau, west Micronesia (Pratt *et al.* 1987).

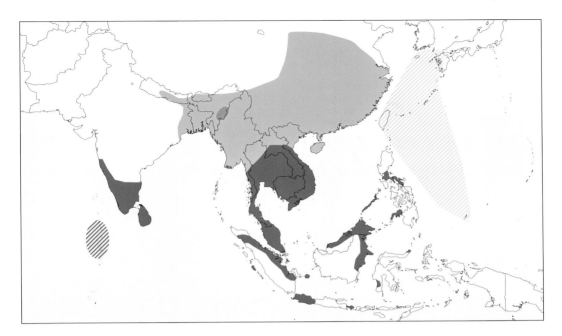

HABITAT Lowlands to 2,450m. Wooded areas from secondary scrub to open forest with low trees, bushes along rivers, farmland and gardens in small villages. Swamp forest and riparian woodland Kaziranga NP, Assam, India. In mangroves on passage and in non-breeding range; orchards, plantations, tall reedbeds and scrub near cultivation Thai-Malay Peninsula and not above 400m. In Himalayas to 1,500m, although in Nepal mainly 250–365m; Bhutan 400–1,600m; Sri Lanka to 2,000m. Foothills to 1,900m Burma. Lowland dipterocarp and secondary forests, open areas, oil palm plantations, hill rice and gardens in lowlands, Borneo. To 1,500m in scrub, forest clearings, bushes along rivers, cultivation and gardens, Sumatra. In south-east Asia to 1,525m. Breeding occasionally to 1,500m east, central, south-west and south-east China, Hainan; rarely Taiwan (Smythies 1953, Ali & Ripley 1983, Inskipp & Inskipp 1985, Cheng 1987, van Marle & Voous 1988, Tymstra *et al.* 1997, Barua & Sharma 1999, Wells 1999, Robson 2000, Rasmussen & Anderton 2005, Mann 2008).

BEHAVIOUR Shy; singly or small, loose groups in non-breeding season. Scrambles about in low vegetation and flies quickly from one small tree to another. Nocturnal migrant. Captive bird constantly jerked wings and tail in quick, nervous manner. Often follows mixed flocks through forest (Medway & Wells 1976, Rasmussen & Anderton 2005).

BREEDING Parasitism Eviction of eggs or chicks never observed, but nests containing only parasite chick are frequent. More than one parasite egg frequent, and up to five recorded. Mainly laughing-thrushes, similar in size to parasite (Johnsgard 1997). Most frequent hosts India Greater Necklaced and Lesser Necklaced, but also Striated, Rufous-vented, Blue-winged and Grey-sided Laughing-thrushes, in decreasing frequency of parasitism (Becking 1981); nest of first two sometimes contain single eggs of Large Hawk-cuckoo also (Osmaston 1916, MacKenzie 1918). Lesser Necklaced Laughing-thrush may account for up to 45% of all host records, but following species also recorded: Rufous-necked, Chestnut-crowned, White-crested and Rufous-chinned Laughing-thrushes, Red-faced Liochichla and Rusty-fronted Barwing. Usual hosts in south-east Asia Lesser and Greater Necklaced Laughing-thrushes. Suspected host South Korea introduced Common Magpie. In China probably Melodious and Masked Laughing-thrushes. Also Fork-tailed Drongo, Oriental Magpie-robin, Slaty-backed Forktail, Blue Whistling, Orange-headed and Black-breasted Thrushes, Moustached and Spot-breasted Laughing-thrushes, White-headed Babbler, Rusty-cheeked Scimitar Babbler and Long-tailed Shrike (La Touche 1931, Makatsch 1971, Johnsgard 1997, Robson 2000, Birds Korea 2009, P.E. Lowther *in litt.*). **Season** In India breeds in rainy season (Apr–Aug), mostly May/Jun. Breeding in Burma governed by host and mainly Mar–May with second broods extending into Aug (Ali & Ripley 1969, Smythies 1953). Suspected breeding South Korea, late Apr/May (Birds Korea 2009). **Eggs** Similar to hosts' but mostly unglossed, slightly different shade of blue and rounder and thicker-shelled, and less tapered than is usual in laughing-thrush eggs; 26.9x22.8mm (n = 50); 25.4–29.9x20.3–24.4. Eggs in White-headed Babbler nest deep turquoise blue, closely matching host's (Baker 1934, Schönwetter 1967, Makatsch 1971, Becking 1981). **Chicks** No information. **Survival** Nest 17 May, Maymyo, Burma, with two healthy cuckoo chicks and three young host chicks, one dead, one dying and last apparently starving; four days later all host chicks gone. Survival of two cuckoo chicks to fledging age recorded (Osmaston 1916, Becking 1981, Johnsgard 1997).

FOOD Mostly large insects such as caterpillars, beetles, ants, mantids and large orthopterans; spiders; small fruits. When given caterpillar, tame but free flying young cuckoo cut off both ends and then flicked it violently from side to side against branch expelling stomach contents; caterpillar was gulped down after repeating this operation several times; also fed on cockroaches and beetle larvae (Osmaston 1916, Ali & Ripley 1983, SUARENG 2004 in Wells 2007).

STATUS AND CONSERVATION Although uncommon, has large global range. Uncommon summer Bhutan (Spierenburg 2005). Rare India, local Bangladesh, rare winter Sri Lanka (Grimmett *et al.* 1988). Uncommon breeding visitor Burma (Robson 2000). Uncommon Thailand (Lekagul & Round 1991), but apparently formerly common migrant throughout Malay Peninsula, and Langkawi and Aroa Is. (Robinson & Kloss 1910–11, 1923), but now uncommon Sep–May, Thai-Malay Peninsula (Wells 1999). Uncommon breeding visitor Laos, east Tonkin and north-central Annam, Vietnam, and scarce to uncommon winter visitor south Thailand, peninsular Malaysia and Singapore. Status Tenasserim, Cambodia, south Annam and Cochinchina uncertain (Robson 2000). Fairly common China; accidental Taiwan (Cheng 1987). Scarce Borneo (Mann 2008) and Sumatra (van Marle & Voous 1988); very rare Java (MacKinnon & Phillipps 1993). South Korea, 4–5 records, possibly breeding (Birds Korea 2009). Vagrant Sulawesi (White & Bruce 1986), Maldives (Robson 2007), Spratly Is. (Davison 1999), Japan (Brazil 1991) and Palau, west Micronesia (Pratt *et al.* 1987). Not globally threatened (BirdLife International 2011).

Chestnut-winged Cuckoo *Clamator coromandus.* **Figs. 1–3.** Adult, Ipoh, Perak, Malaysia, November (*Amar-Singh HSS*).

GREAT SPOTTED CUCKOO
Clamator glandarius   Plate 18

Cuculus glandarius Linnaeus, 1758 (Northern Africa and southern Europe = Gibraltar)

TAXONOMY Southern African population '*choragium*' averages smaller, has paler throat and shaft streaks less prominent, but great overlap in size with northern populations (Payne 2005). Monotypic. Synonym: *choragium* (Clancey 1951).

FIELD IDENTIFICATION 40cm. Rather large and long-winged, conspicuously spotted dorsally; long, narrow tail drooping in flight, and rather long slightly decurved bill. **Adult** Dark grey-brown above spotted white, with short grey crest, often flattened; below white tinged golden-buffish on breast, graduated tail brown broadly tipped white. **Juvenile** Crown and face black, primaries rufous, with smaller white spots on wing-coverts and upperparts. **Similar species** No other cuckoo in range is spotted white on wings and back except juvenile Thick-billed Cuckoo, but latter's short, thick bill and white underparts distinguish it. Both Levaillant's and Jacobin Cuckoos have glossy black upperparts and white wing-bars.

VOICE Large repertoire. Male advertising call rasping *keeow-keeow*, or accelerating *keyar-keyar-keyar-keyar*, or raptor-like *kleeok-kleeok-kleeok*, or rapid *kow-kow-kow-kow* or *kioc-kioc*. Last also when male is greeting female. Male in courtship makes grinding clockwork sound followed by descending *kianm-kianm-kian-kiacq-kkiau*. When chasing other cuckoos rasping *kzek* and bubbling *ge-ge-ge-ge-gerrr*, followed by *keera-keera-keera-keera-keera*, or *gak-gak-gak* or *gah-gah-gah*. Alarm call crow-like *krak*. Female gives resonant *gi-gi-gi* or *ku-lu-ku* like Green Woodpecker, often mixing *gi* and *ku* notes; bubbling call rather reminiscent of something between voice of female Common Cuckoo and a tern; pre-copulatory call *chet-chet-chet-chet* and pre-laying call *woig-woig-woig* or *woing-woing-woing*. Male calls both perched and in flight. Nestling food-begging call resembling that of young Common Magpie, Pied Crow or Pale-winged Starling. Silent during migration (Mundy 1973, Glutz von Blotzheim & Bauer 1980, Wyllie 1981, Cramp 1985, Beaman & Madge 1998, Stevenson & Fanshawe 2002, Payne 2005).

DESCRIPTION Adult Crown and ear-coverts silvery-grey, feathers white at base, with fine black shaft streaks and elongated short crest, sides of face from lores, below eyes to nape darker grey, dark brown on middle of nape; upperparts dark grey-brown edged with triangular white spots in fresh plumage especially prominent on scapulars, long uppertail-coverts with black shaft streaks; wings and coverts dark grey-brown tipped and spotted white, markings on tips of remiges crescent-shaped, primaries dark olive-brown, in fresh plumage with light metallic sheen, underside of wing grey-brown, coverts golden-buffish; underside creamy-white with faint buffish tinge, cheek, throat, sides of neck and upper breast golden-buffish to straw-yellow; graduated tail in fresh plumage dark grey, dark olive-brown when worn, all except central pair broadly tipped white (2–4cm), broadest on outermost pair, underside grey-brown broadly tipped white. **Juvenile** Crown and side of face black, crest shorter than adults and tipped white in fresh plumage, white spots on brown upperparts and wings smaller, primaries except outermost rufous, P10 and all secondaries brownish-black, all remiges tipped white, primary-coverts broadly tipped white; chin, throat, sides of neck and breast yellow, rarely orange-yellow, rest of underparts white, flanks and vent often washed olive-brown; uppertail-coverts and tail brownish-black, white area on tail feather tips about half length in adult. **Nestling** Naked with pinkish or yellow skin with pinkish mouth lining and white spots on palate. **Bare parts** Iris dark brown, eye-ring grey or dull, red to orange-red in juvenile. Bill blackish, underside of lower mandible pale yellow. Feet grey to black. Albino, RSA (Roberts 1928).

BIOMETRICS Wing male (n = 11) 194–215mm, mean 210.2 ± 6.5, female (n = 7) 191–218mm, mean 202.7 ± 9.9. Bill male 23–28mm, mean 25.1 ± 1.4, female 22–28mm, mean 25.3 ± 2.9. Tail male 192–230mm, mean 218.7 ± 11.1, female 192–211mm, mean 203.4 ± 5.9. Tarsus male 28–34mm, mean 31.0 ± 2.0, female 30–34mm, mean 31.5 ± 1.5. Mass (RSA) male (n = 10) 128–158g, mean 139.3g, female (n = 9) 118–148g, mean 128.7g (Payne 2005); (Spain) male (n = 6) 153–192g, mean 169 ± 14.5; female 138g; lean male winter 139g; (Nigeria) breeding female (n = 10) mean 130g (Peréz Chiscano 1971, Valverde 1971, Mundy & Cook 1977 all in Cramp 1985).

MOULT Jul–Feb with peak Sep–early Dec (Cramp 1985). Primary moult sequence P6 and P9, pause, P7 and P10, pause, P8 and P5, or P5 then P8; irregular transilient moulting of secondaries; rectrices dropped one by one, each feather dropping out after previous has fully grown; adult plumage fully attained at about six months; complete moult *c.* 100 days (Stresemann & Stresemann 1961, 1969).

DISTRIBUTION Palearctic and Afrotropical. Spain holds *c.* 98% of whole European population Apr–early Aug, adults already departing mid Jun. Annual migrant Balearics Jan–Aug, but no breeding records. Also populations in Portugal, south France and west Italy. Before 1964 only 26 records from Italy, and first bred that year in Sardinia, but now breeding along coast of Tuscany, north Lazio, Puglia and Sicily; 15–25 pairs. Breeds Bulgaria (probably more than ten pairs), and once Montenegro; attempted/suspected breeding former Yugoslavia, eight years between 1889 and 1974. Mostly late Mar–early May, Greece; first bred 1978 (twice) and 1981; several territorial and calling Kos Apr–May 1965 and 1968; few autumn, mostly Crete and Dodecanese. In Turkey widespread interior high plateau and in west, but none at Black Sea; stronghold west and south. Many on spring migration Cyprus, Apr, where breeds, but no autumn records. In north and central Israel rare breeding but common passage migrant mainly Mar–Apr and Jul–Aug, and bred twice Eilat. Breeds north Iraq and adjacent extreme west Iran; "abundant" south Iran 1880s, and in spring (earliest 25 Apr) around Tehran. Twice bred Lebanon, and uncommon passage migrant mid Feb–late Apr and mid Aug–late Sep; twice Syria, Feb and Sep, breeding suspected. Feb–early May, Jordan, once Aug; breeding suspected. In Arabia scarce passage migrant in west and vagrant in east, has wintered in north, and scarce on passage Kuwait. European population may winter no farther south than 10°N in northern tropics of west Africa, e.g. Senegambia Nov–mid Feb. In Mediterranean Africa breeds Egypt (now rare), Algeria (formerly) and Tunisia (formerly bred and not uncommon; now infrequently recorded). Occasional in winter (less frequent autumn passage) Morocco and bred eight times in 1990s. Once Ahaggar, south Algeria, Dec. Passage early Jul–mid Sep, very few winter and spring, Mauritania. Sedentary breeding population in much of tropical Africa from Senegambia, SL, Guinea, Burkina Faso, Ghana, Mali, Nigeria, Sudan, and in eastern Africa south through Eritrea, Ethiopia, Somalia, Kenya, and Tanzania to north Mozambique. Mainly Oct–Mar with peak passage north Tanzania Jan–Mar. Mostly May–Sep south-east Sudan. "Considerable numbers on Eritrean coastal plains in Dec–Mar suggest overwintering". Common breeding resident Somalia, mostly in north. Summer visitor southern Africa Oct–Mar (few overwinter) in all eastern parts of RSA from Port Elizabeth in south, north to Mmabatho at border with Botswana, and from there north-west to Kaukau Veld in Namibia, south-west to Windhoek, and again north-west to eastern Damaraland. Uncommon but widespread Kruger NP, RSA, Oct–Mar (Norman 1888, Whitaker 1905, Bannerman & Bannerman 1958, 1983, Passburg 1959, Vaughan 1960, Kumerloeve 1961, Heim de Balsac & Mayaud 1962, Smith 1965, Etchécopar & Hüe 1967, Bauer et al. 1969a, b, Bundy 1976, Romè & Tomei 1977, Glutz von Blotzheim & Bauer 1980, Newman 1980, Rowan 1983, Cramp 1985, Gallagher 1986, Štumberger 1987, Hollom et al. 1988, Irwin 1988, Maumary & Duperrex 1991, Kasparek 1992, Miltschew 1992, Dowsett & Dowsett-Lemaire 1993, Andrews 1995, Porter et al. 1996, Shirihai 1996, Hagemeijer & Blair 1997, Handrinos & Akriotis 1997, Ash & Miskell 1998, Isenmann & Moali 2000, Stevenson & Fanshawe 2002, Yosef 2002, Thévenot et al. 2003, Isenmann et al. 2005, 2010, Payne 2005, Ruggieri & Festari 2005, Serra et al. 2005, Brichetti & Fracasso 2006, Nikolov et al. 2007, Murdoch & Betton 2008, Ramadan-Jaradi et al. 2008, Ash & Atkins 2009, de Bont 2009). Three old records Lesotho (Bonde 1993). Birds breeding in RSA winter in tropical Africa where usually indistinguishable from sedentary breeding population (Irwin 1988). Many vagrants reported, especially dispersing juveniles, e.g. Britain and Ireland 50+, Finland 2, Norway 4, Sweden 5, Denmark 7, Belgium 7, Netherlands 14+, Germany c. 20, Switzerland 9, Austria 5+, Poland 2, Slovakia 1, Hungary 1, Slovenia 1, Croatia 7, Serbia 1, Macedonia 3, Albania 1, Ukraine/Moldova 2, Russia, Armenia, Turkmenistan 1, Libya, Canary Is. 5+, Madeira 2, Cape Verde Is. 1, Bahrain 2, Qatar 3, Seychelles 1 (Dement'ev & Gladkov 1966, Bannerman 1963, Bannerman & Bannerman 1965, Jacobson & Wallin 1969, Glutz von Blotzheim & Bauer 1980, Flint et al. 1984, Cramp 1985, Babó 1987, Hollom et al. 1988, Stipcevic 1991, 1992, Olsen 1992, Anon. 1995, Perrins 1998, Nankinov 1999, van den Berg & Bosman 1999, Skerrett & Seychelles Bird Record Committee 2001, Payne 2005, Clarke 2006, Dudley et al. 2006, Maumary et al. 2007, Slack & Wallace 2009, Balmer & Murdoch 2010, Anon. 2011b).

HABITAT In Europe in lowlands rarely 500+m in semi-arid regions like heathlands often dominated by cork oaks and stone pines; olive groves. Low-lying areas, Israel. Only breeds in mountains, Lebanon, presumably due to hunting pressure. Dry *Acacia* savanna, avoids thick woodland, Africa. Open habitat with scattered trees and low bushy undergrowth, often near water. In west and central Africa rare or absent from heavy forest. In Somalia to 1,500m particularly open thorn bush; to 3,000m east Africa. Open woodlands, especially thornveld, RSA (Chapin 1939, Benson 1970, Payne 1973, Rowan 1983, Cramp 1985, Irwin 1988, Barlow et al. 1997, Shirihai 1996, Ash & Miskell 1998, Davies 2000, Mullarney et al. 2000, Stevenson & Fanshawe 2002).

BEHAVIOUR Singly or in pairs, occasionally in larger groups foraging in foliage or more rarely hops and runs on ground. Shy and unobtrusive. Territorial behaviour not definitely solved. In Spain territories 1–3.7km² encompassing up to 40 pairs of Common Magpies; in Nigeria one territory c. 36ha, but many reports of several females sharing same host nests. Mating system flexible but due to too few detailed observational studies incompletely understood, but monogamy predominates. Courtship feeding followed by mating lasting up to 2min; male only releases food when copulation successfully completed. Male often assists female at laying by diverting attention of hosts, and after hosts leave nest female lays egg in 2–3sec. Two captive females laid at two-day intervals totals of 15 and 16 eggs. Does not remove host eggs when laying but often damages eggs which then are removed by host; presumably not incidental consequence of rapid laying but adaptation. Rejection rate by Common Magpie varies between populations depending on duration of sympatry, but presence of cuckoo near magpie nest does not increase probability of egg rejection. Hatching, fledging and breeding success of parasitised Common Magpie low compared with non-parasitised, on average only 0.6 fledged, due to egg damaging and food competition between fast-growing cuckoos and smaller host chicks, but in nests where cuckoo eggs were laid late compared with incubation of hosts' eggs, number of fledged host chicks higher. Fledging success lower in nests of Carrion Crows than in nests of magpies, due to crow nestlings being larger. Cuckoo chicks mimic young of host both in appearance before feathers develop and in begging calls, but different appearance in feathered stage compared with host chicks is perhaps compensated for by exaggerated begging. However, magpie hosts preferentially feed parasitic chicks with beetles, which are lower-quality prey because of higher percentage of chitin, maybe countermove because it punishes hosts (mostly Common Magpies) if they reject its egg by host clutch destruction, e.g. 86% of Common Magpie nests with rejection predated, but only 12% of 117 nests with acceptance (mafia behaviour). Common Magpie's defence against parasitism is breeding at high

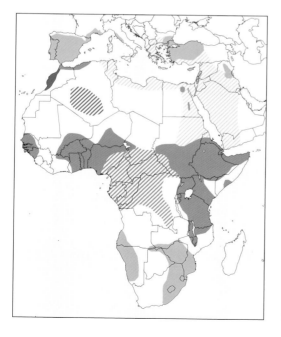

density, laying synchrony and perhaps group defence. Azure-winged Magpie has not in recent years been found as host for cuckoo chicks in Spain, and is found to have high level (74%) of discrimination and rejection of models of Great Spotted Cuckoo eggs. Fledged cuckoos often gather in flocks and are fed by all Common Magpies in group, not only foster parents. Adult cuckoos visit oft-parasitised nests in late phase of nestling period, and after chicks have fledged, to imprint chicks on their own species; defence of parasitised nests reported. Fledglings receive parental care from their foster parents for more than one month. Adult fed juvenile two caterpillars, north Nigeria, 16 Sep. European adults migrate on average two months before juveniles (Bannerman 1933, von Frisch 1969, Mundy & Cook 1977, Arias de Reyna et al. 1982, 1987, Rowan 1983, Irwin 1988, Soler 1990, Soler & Møller 1990, Soler & Soler 1991, Soler et al. 1994, 1995a, b, c, d, 1999, 2000b, 2002, Martinez et al. 1996, 1998, Davies 2000, Soler & Martínez 2000, Avilés 2004, Soler 2008).

BREEDING Parasitism Mainly parasitises Common Magpie southern Europe and Morocco, but some populations also parasitise European Roller, Common Hoopoe, Azure-winged Magpie, Red-billed and Alpine Choughs, Carrion/Hooded Crows and Eurasian Jackdaw. Whereas most parasitised Common Magpie chicks die, most host chicks in Carrion Crow and Azure-winged Magpie nests survive (Redondo & Arias de Reyna 1989), and maybe therefore rejection mainly occurs in Common Magpies (Soler 1990). Most common host in Israel Hooded Crow, and rarely Brown-necked and Fan-tailed Ravens, and Eurasian Jay, and recently House Crow and Eurasian Jackdaw. Hooded Crow, Egypt. Brown-necked Raven, Mauritania. Pied Crow, Senegal and Nigeria. Pied and Cape Crows most common hosts RSA, but also Cape Glossy, Red-winged, and African Pied Starlings, all hole-nesters. Pale-winged, Burchell's, Long-tailed, Greater Blue-eared, Rüppell's Glossy, Hildebrandt's, Superb, White-crowned, and Splendid Glossy Starlings, Common Myna, Common, Fan-tailed and Brown-necked Ravens also in different parts of Africa (Ivy 1901, Mundy & Cook 1977, Conradty 1979, Limbrunner 1979, Rowan 1983, Cramp 1985, Irwin 1988, Lamarche 1988, Goodman et al. 1989, Morel & Morel 1990, Shirihai 1996, Handrinos & Akriotis 1997, Davies 2000, Yosef 2002, Thévenot et al. 2003, Charter et al. 2005, D.A. Turner pers. comm.). **Season** Laying period for two marked females in Spain 1.5 months, in Mediterranean late Apr–early Jun. Juveniles Italy, Jun, Montenegro, Jul, Bulgaria, Jun. Nestlings Greece, May. Eggs/young Turkey, May–Aug. Young Morocco Apr/May. Eggs Algeria/Tunisia late May–early Jun. Nestlings south Tunisia, Jul. Nestlings Israel, Jun–Jul. Eggs Jan–Jun Egypt, Apr–May and Aug Senegambia, Apr–Jun Mali, Apr–Jul Nigeria, Jan, Apr–Jun north Somalia, Jun in south. Oct–Jan Cape Province, Natal, Transvaal, and Zimbabwe; Dec–Mar, May Namibia (Ivy 1901, Heim de Balsac & Mayaud, 1962, Lévêque 1968, Conradty 1979, Limbrunner 1979, Rowan 1983, Cramp 1985, Štumberger 1987, Miltschew 1992, Handrinos & Akriotis 1997, Ash & Miskell 1998, Martinez et al. 1998, Maumary & Duperrex 1991, Yosef 2002, Thévenot et al. 2003, Klepfer & Di Fraia 2006, Kirwan et al. 2008, Ash & Atkins 2009, B. Cornelissen in litt., F.P. Brammer). **Eggs** Lays perhaps up to 23 eggs per season (Payne 1973); in captivity up to 16, much smaller than crows' eggs and slightly larger than starlings' (Irwin 1988). Nest with two or three eggs often found and fledge more parasite chicks than those with only one egg (Soler 1990). Up to 13 eggs in one nest, RSA (Rowan 1983). Lays egg every second day. In Europe all eggs resemble Common Magpie and Carrion Crow eggs, elliptical greenish-blue with brown spots (Davies 2000), and with ultraviolet reflectance (Avilés et al. 2006). Eggs with pink ground colour (erythrism) also known. In RSA egg colour also greenish-blue but less heavily spotted red-brown and with underlying lilac; 29.0–37.0x21.0–27.0mm, (n = 43) mean 33.4x24.5 (Rowan 1983), but not resembling host species in size or colour. Eggs thick-shelled. No *gentes* (Friedmann 1948b, Etchécopar & Hüe 1967, Soler 1990, Soler & Soler 1991, Martinez et al. 1998, Soler et al. 2003). **Incubation** Short period, 12.8 days (n = 5) compared with those of host, e.g. 18 days for Common Magpie; in Nigeria 14–15 days; in captivity 13.5 days (von Frisch 1969, Mundy & Cook 1977, Soler 1990). **Chicks** Naked with pinkish skin and orange mouth with two conspicuous white papillae on palate, and gape edged white. Does not evict host eggs or chicks. Growth average 9–10g per day. Eyes open at 7 days, well-feathered by 16 days but still some pins on crown, reaching maximum mass 120–130g after 20 days, fledge after 20–26 days. Post-fledging dependence period unusually long, 25–59 days, mean 33.2 days (n = 25). Perhaps chicks learn begging calls from foster chicks (Mundy & Cook 1977, Irwin 1988, Soler et al. 1994, 1995a, b, c, d, Madden & Davies 2006). **Survival** Of nine eggs six failed and one disappeared. In Israel mean of 1.8 cuckoos fledged per parasitised Eurasian Jackdaw nest which suggests Jackdaw suitable host in Israel, in contrast to Spain where only one of nine nests fledged cuckoos. Fledging success lower in nests of Carrion Crow than in nests of Common Magpie as nestlings of former larger (Mundy & Cook 1977, Soler 2002, Soler et al. 2002, Charter et al. 2005).

FOOD Most frequent food caterpillars, hairy or hairless, including those of Gypsy Moth *Lymantria dispar;* grasshoppers, locusts, termites, moths and beetles; small lizards (Limbrunner 1979, Rowan 1983, Irwin 1988).

STATUS AND CONSERVATION European population estimated 55,000–65,000 pairs and Turkish population 1,000–5,000, range increasing and expanding (Cramp 1985, Hagemeijer & Blair 1997). Relatively common, early 20th century Egypt, now much rarer breeder (Goodman et al. 1989). Rare breeder Italy, probably <10 pairs, and rare migrant (Ruggieri & Festari 2005). Formerly not uncommon breeder north Tunisia (Whitaker 1905). In Africa local, frequent or uncommon (Irwin 1988). Common at times Eritrea and Ethiopia (Ash & Atkins 2009). Uncommon but widespread Kenya and north Tanzania (Zimmerman et al. 1996). Common RSA, summer (Roberts 1928). Density 3–9 birds/km^2 in optimal habitat, south Spain (Díaz et al. 1996 in Payne 2005). In southern Europe hunting and traffic accidents main causes of mortality (Glutz von Blotzheim & Bauer 1980). Preyed upon by Eleonora's Falcon north-east Crete (Walter 1979). Not globally threatened (BirdLife International 2011).

Great Spotted Cuckoo *Clamator glandarius*. **Fig. 1.** Adult, Segura, Portugal, May (*David Monticelli*). **Fig. 2.** Adult, Tejo International NP, Segura, Portugal, May (*David Monticelli*). **Fig. 3.** Juvenile being fed by Burchell's Starling, Waterberg, Namibia, May (*Bram Cornelissen*). **Fig. 4.** Juvenile, Kgomo Kgomo, North-west Province, RSA, February (*Niall Perrins*).

LEVAILLANT'S CUCKOO
Clamator levaillantii   **Plate 18**

Coccyzus levaillantii Swainson, 1829 (Senegal)

Alternative names: (African) Striped/Striped Crested Cuckoo.

TAXONOMY Sometimes placed with *jacobinus* in *Oxylophus* (Swainson 1837) based on wing shape and plumage characters. Monotypic. Synonyms: *Cuculus cafer* Lichtenstein 1793 (*Cuculus clamosus* had been known as *Cuculus cafer*, but this name officially suppressed to avoid confusion: ICZN 1956); *Coccystes albonotatus* Shelley 1881 once considered valid subspecies (e.g. Sclater 1924) or morph of *Oxylophus jacobinus serratus* (Reichenow 1902); *C. caroli* Norman 1888.

FIELD IDENTIFICATION 39cm. Long-tailed, crested, pied. **Adult** Mainly black above with white wing-bar and tail tips; conspicuous crest; white below with heavy black streaking on throat and breast. Melanistic morph ('*albonotatus*') all black with white wing-bar. **Juvenile** Differs from adult in having much brown above, and on chin to breast. **Similar species** Jacobin Cuckoo is smaller and lacks heavy ventral streaking; immature of this species has buff, not rusty, tail tip. Black Cuckoo differs from melanistic morph in lack of wing-bar, voice and smaller size.

VOICE Loud, clear, fluty *piu* or *peee-u*, 21 times/20sec, and alternates with harsh chattering of 7–10 notes for *c.* 1sec. Hollow *kur, kur, kur...* repeated about 12 times, followed after brief pause by rapid series of about 20 *kwi* notes dying away at end. Juvenile's begging call harsh chattering *ker, ker, ker* similar to host's call. Series of clear, loud *KEEow* notes, ending in 'staccato chatter (like machine gun burst)' (Chapin 1939, Irwin 1988, Borrow & Demey 2001).

DESCRIPTION Adult Above, including ear-coverts, black with greenish or bluish sheen; wings with prominent white bar across base of outer primaries which extends along inner webs for more than half their length; crest on crown extends to nape. Dull white or cream below, with throat and sides of neck heavily streaked with black, breast more lightly so, amount of streaking variable; light streaking on flanks and thighs, and creamy undertail-coverts with fine shaft streaks. Dull white underwing-coverts with black streaking, axillaries sooty-black; large white patch on underside of primaries. All except central rectrices prominently tipped white, underside of tail being greyer with dull white tips. Melanistic morph (rare; east African coast) entirely glossy black, except for white wing patch and much reduced tail tips; may have broad white tips to undertail-coverts. **Juvenile** Dull or rusty-brown, or cinnamon, with dark brown remiges and rectrices, becoming glossy black 5–6 weeks after fledging, with short crest; primaries blackish, slightly glossed, lacking white patch; tail dull grey-brown, sometimes tipped rusty-buff, and with small amount of white on outer edges of rectrices; chin to breast dull brown, sometimes washed rusty, with indistinct black streaking; rest of underparts dull greyish-white or buffy; large white patch on underside of primaries; lacks white tips to rectrices. Fledgling in nest of Capuchin Babbler in DRC very rufous above (Chapin 1939), and others from east DRC uniformly rufous above and below and lacking white wing patch (Payne 2005). **Nestling** Naked; dark pink skin matches that of host nestling, becoming blackish by 5th day; spiky trichoptiles, black on dorsal surface, white on ventral, grow by 7th day; upper mandible black; gape brighter orange-red than babbler; thick yellow eyelids. **Bare parts** Iris brown, bluish or bluish-grey in juvenile. Bill black; gape bright orange in juvenile. Feet grey or blue-grey, with yellowish-grey soles.

BIOMETRICS Wing male (n = 30) 170–187mm, mean 180; female (n = 30) 171–189mm, mean 178. Bill unsexed (n = 20) 24–28mm, mean 25.8. Tail male 215–238mm, mean 226; female 210–242mm, mean 223. Tarsus unsexed 25–28mm, mean 27.0. Mass male (n = 8) 106–140g, mean 123.0; female (n = 8) 102–141g, mean 122.0; fledgling 45–50g five days after hatching (Irwin 1988).

MOULT Primaries moult after all secondaries and tail; in post-juvenile moult rectrices replaced last. Primaries moulting SL, Feb and Aug, Liberia, Nov, Kenya, Jan, Tanzania, Mar; tail moult south Ethiopia, Jun; Malawi, Feb; wing and tail moult Nigeria, Dec, Cameroon, Feb, Lake Victoria, Tanzania, Nov, south Uganda, Jul; body, wing and tail moult west Tanzania, Nov; body and wing moult south-west Uganda, May; interrupted wing moult west Uganda, Oct, Namibia, Mar. Post-juvenile moult SL, Feb, Apr, Liberia, Jan, Mar, Nov, Dec, Ghana, Oct, south Nigeria, Dec, Cameroon, Jul, DRC, May, Sudan, Aug, Uganda, Feb, Jun, Jul and Sep, Kenya, Sep, Zambia, Nov, Malawi, Apr, Mozambique, Mar, Cape Province, RSA, Oct. Post-nuptial moult north Tanzania, Feb–Apr (Friedmann 1948b, Stresemann & Stresemann 1961, BMNH).

DISTRIBUTION Intra-African migrant. Mauritania and Senegambia to Eritrea, Ethiopia and Somalia, south to north Namibia, Transvaal, Natal and stragglers to Eastern Cape. Visits southern African savannas to breed Oct–Mar/Apr; movement north continuing to first half May, but recorded Jul, Transvaal, Zimbabwe, Zambia and Angola (some may not migrate), and Jul–Aug, Malawi. South-east Tanzania Nov–Apr; north-west Tanzania Oct–Jun; west and south Uganda Dec–Jun; west Kenya May–Sep (breeding), east Kenya Nov–Apr (rains), mostly absent Aug-Feb, except fledglings, Central Rift, Kenya; black morph only in coastal zone of south Somalia (once), Kenya and Tanzania from Malindi to Usambaras, Soga and Utete mainly Mar–Sep, but perhaps resident. Resident and partial migrant savannas, DRC; visitor Ethiopia, but some perhaps resident lowlands; Eritrea, Nov; Somalia (4 records) Mar, Jun. Presumed migrant breeder Sahelian Mauritania along Senegal valley, rarely further north. Common breeding visitor Senegambia to Cameroon, wet season Jun–Nov; in rains Mali, May–Oct, rare Apr, Nov; Togo, few perhaps resident, late Dec–Apr in south, Feb–Nov in north; Nigeria, May–Nov, most Jun and Aug in north-west, but most southern records in dry season Nov–Apr, occasionally Aug–Sep, passing through south-east Dec–Mar, perhaps returning Aug. North Guinean and adjacent parts of Soudanian savanna zones, but much less common in Chad and south Sudan, in latter uncommon non-breeding visitor Mar–Oct, moving north with rains where it probably breeds; migrates to south Guinean zone, dry season Dec–May, to borders of rainforest; some resident for whole or most of year as in IC and Liberia; Ghana Jun–Sep; south Ghana and south Nigeria, but many in last case move to north in dry season; south Guinea Apr–Jun; rare north Burkina Faso, Sep; Niger; RC, Dec; common migrant throughout Togo, late Dec–Apr, few perhaps resident (Marchant 1949, Bannerman 1951, Cave & MacDonald 1955, Cawkell & Moreau 1963, Brunel & Thiollay 1969, Elgood *et al.* 1973, Benson & Benson 1977, Britton 1980,

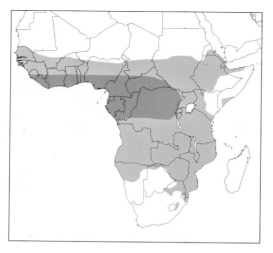

Lamarche 1980, 1988, Nikolaus 1987, Giraudoux et al. 1988, Irwin 1988, Dowsett & Dowsett-Lemaire 1989, Medland 1992, Halleux 1994, Cheke & Walsh 1996, Balança & de Visscher 1997, Ash & Miskell 1998, Zinner 2001, de Bont 2002, Hockey et al. 2005, D.A. Turner pers. comm.).

HABITAT Dense, closed, humid savanna woodland; scrub, woody growth along streams, *Acacia* savanna, open woodland, gardens, forest canopy, old farms. All altitudes Eritrea. East Africa generally in areas of 500+mm annual rainfall, occasionally to 2,100m. Does not reach Sahel in Nigeria. Malawi to 1,370m (Smith 1957, Elgood et al. 1973, Benson & Benson 1977, Irwin 1988, Lewis & Pomeroy 1989, Stevenson & Fanshawe 2004).

BEHAVIOUR Generally solitary, but in pairs or up to 4 together in breeding season; generally unobtrusive, but can be noisy. Males territorial, females less so or not at all. Perches horizontally with tail above level of primaries. Feeds in trees and bushes, but also hopping on ground. Two calling birds may face each other crests raised, bobbing heads, with half-open wings and tail fanned. while mounted male feeds a caterpillar to female. When attempting to lay male provokes vigourous attack by babblers. Female usually less vigourously attacked, and sneaks in to deposit egg in nest, sometimes puncturing or ejecting host egg. This process may take several days (Lorber 1984, Irwin 1988, Payne 1997, Leonard 1998).

BREEDING Parasitism More than one female may lay in same nest. Host eggs may be punctured or ejected. Hosts almost all babblers, principally *Turdoides jardineii* superspecies. RSA north to south DRC almost all Arrow-marked, rarely Bare-cheeked Babblers. Arrow-marked Babbler, RSA, east Africa including Kenya and Tanzania, Zambia, Zimbabwe, Malawi, Mozambique, south DRC. Hartlaub's Babbler, Zambia. Southern Pied Babbler, Namibia. Brown Babbler, west Africa including Mauritania, Nigeria, and Mali. Blackcap Babbler, west Africa. Chestnut-bellied Starling, Mauritania, Mali and Gambia. Common Bulbul, RSA and Uganda. Unconfirmed hosts: Black-faced Babbler; Cape Robin-chat, RSA; Green Wood-hoopoe, RSA; Greater Blue-eared Starling, Mali; probably once Capuchin Babbler, DRC (van Someren 1916, Belcher 1930, Bradfield 1931, Priest 1934, Vincent 1934, 1946, 1947, Chapin 1939, Roberts 1939, Serle 1939, Winterbottom 1942, Bell Marley and A. Loveridge in Friedmann 1948b, Friedmann 1949, Milstein 1954, Benson et al. 1971, Benson & Benson 1977, Colebrook-Robjent 1977, Lamarche 1980, Moore 1984, Jones 1985, Irwin 1988, Allan 2002, P. Roberts *in litt.*, D.A. Turner pers. comm.). 7.5% Arrow-marked Babblers parasitised, Zimbabwe; one group raised four cuckoos in succession in five months (Irwin 1988). In small area of north Nigeria co-operatively breeding Brown Babbler lays aberrant eggs (bright or pale pink, mauve, grey or blue-grey, often varying within clutch) as opposed to immaculate blue eggs throughout rest of range, perhaps adaptation to brood parasitism by this cuckoo which may have developed undetectable blue eggs (Serle 1977). Arrow-marked Babblers' nests in RSA and south DRC each had two cuckoo eggs and three host eggs, all well incubated (Vincent 1946). **Season** Mainly in rains. Gambia, Jun–Nov, copulation Aug; Senegambia, Aug–Dec; SL, Sep; copulating Togo, 31 Jan; young west Liberia, Jan and Mar; Mali, Jun–Oct. Nearly finished laying Ghana, 15 Feb, young with host Jan, Jun, Jul and Nov; north Nigeria eggs May–Jun, active testis Jul, Apr–Sep, probably Oct–Nov; enlarged testes Jan, Mar; developing eggs Cameroon, Mar, enlarged ovary Apr. Oviduct egg Eritrea, Jul; Ethiopia, May–Sep; may breed Sudan in rains. DRC, Feb–Oct; east DRC nestlings May and Oct; fledgling Beni Oct; eggs Apr in south; enlarged testes Katanga, Mar and Oct during rains, egg-laying female Feb, fledglings Apr, May, Jul; fledgling South Kivu May. Kenya, Mar–May, Sep-Oct; central and south Uganda and south-west Kenya, Apr–May; Tanzania young being fed, 23 Feb, Mar; enlarged testes north Tanzania, Dec; active ovaries and laying and/or enlarged testes south Tanzania, Jan, Mar (3 females) and Apr. Malawi, Dec–Apr, occasional young to Jun, also Oct, also enlarged testes Feb and May; Zambia, Nov–Apr, in south peak Feb; breeding Mozambique, Mar, enlarged ovaries, Apr and Nov, eggs Jul, Oct, fully grown fledgling Apr. Enlarged ovary Angola, Oct; Zimbabwe, enlarged testes Nov, eggs Dec and Feb, Oct–Jun extending into dry season; RSA (Transvaal), Nov–Jun; enlarged testes Namibia, Mar, juvenile Mar; north Namibia, Nov; laying Botswana, Jun, fledglings Mar, Jun (Blanford 1870, Alexander 1902, Neumann 1905, Townsend 1921, Loveridge 1922, Belcher 1930, Bowen 1932, Priest 1934, Vincent 1934, 1946, Lowe 1937, Chapin 1939, Serle 1939, Friedmann 1948b, J. Büttikofer in Friedmann 1948b, Verheyen 1953, Smith 1957, M. Gore in Cawkell & Moreau 1963, Benson & Benson 1971, Benson et al. 1971, Elgood et al. 1973, 1984, Colebrook-Robjent 1977, Brown & Britton 1980, Grimes 1987, Nikolaus 1987, Irwin 1988, Cheke & Walsh 1996, Payne 1997, P. Roberts *in litt.*, D.A. Turner pers. comm., BMNH). **Eggs** Clutch up to 4; sometimes 2 eggs in one nest (Colebrook-Robjent 1977), but may be from more than one female (Jones 1985). Oval, slightly rounder and broader than host eggs, slightly glossy and finely pitted; bluish-white to turquoise-blue like host, or (north Nigeria) pink with darker speckles (Payne 1997) matching Brown Babbler's eggs which may also be pink, or mauve, grey or blue-grey (Serle 1977); uniform greenish-blue (Chapin 1939); white in Common Bulbul's nest (Milstein 1954). In RSA blue eggs match those of host Arrow-marked Babbler, but slightly larger and pitted (Vincent 1946); also blue-green RSA (Jensen & Jensen 1969). 23.8–27.6x19.6–22.1mm (n = 13), mean 25.1x20.6; shell mass 0.45–0.56g (Irwin 1988). **Incubation** Embryo develops for one day before laying. Probably at least 11 days; once 12 days (Jones 1985, Payne 1997). **Chicks** Do not evict host young. Nestling period 9–10 days, rapidly outgrowing foster siblings. Voids foul-smelling liquid from 5th day, and from this time solicits food more forcibly than babblers, but using same call. After leaving nest dependent on foster parents for further 3–5 weeks (up to 36 days), and consorts with young babblers. Fledgling imitates call of Arrow-marked Babbler. Older young still fed when sitting, but mobbed when flying (Mundy 1973, Irwin 1988, Payne 1997). **Survival** No information.

FOOD Insects; chiefly hairy and spiny caterpillars, but also termite alates, ant alates, beetle larvae, locusts and grasshoppers. Stomach contents of one bird included berries and young tree shoots. Matted hair or fur in one stomach, RSA (Roberts 1928, Chapin 1939, Brooke 1965, Irwin 1988, Payne 1997, BMNH).

STATUS AND CONSERVATION Not uncommon over much of range; common Gambia Jun–Sep (Cawkell & Moreau 1963) and north Mali in rains, Jun–Oct, rare Apr, Nov (Lamarche 1980); common in savanna, DRC (Chapin 1939); was considered scarce Darfur, Sudan (Lynes 1925); uncommon Eritrea (Smith 1957) and later considered rare (Zinner 2001); uncommon Ethiopia (Ash & Atkins 2009); rare north Tanzania (Payne 1997); rare Yapo Forest, IC (Demey & Fishpool 1994); common migrant throughout Togo (Cheke & Walsh 1996); irregular but not uncommon Ghana (Grimes 1987); common south-east Tanzania Nov–Apr, north-west Tanzania Oct–Jun, west and south Uganda mainly Dec–Jun, west Kenya mainly May–Sep (Britton 1980); fairly common to common Malawi (Stead 1979); quite common Zambia (Colebrook-Robjent 1977). Uncommon RSA, rare Eastern Cape (Hockey *et al.* 2005). Not globally threatened (BirdLife International 2011).

Levaillant's Cuckoo *Clamator levaillantii*. **Figs. 1–2.** Adult, Masai Mara, Kenya, June (*Adam Scott Kennedy*). **Fig. 3.** Juvenile, Gola Forest, Sierra Leone, January (*David Monticelli*).

JACOBIN CUCKOO
Clamator jacobinus **Plate 18**

Cuculus Jacobinus Boddaert, 1783 (Coromandel Coast, south-east India)

Alternative names: Pied/Black-and-white/Pied Crested/Black Crested Cuckoo.

TAXONOMY Sometimes placed with *levaillantii* in *Oxylophus* (Swainson 1837), based on differences in wing shape and plumage characters; also in *Melanolophus* (Roberts 1922). Melanistic morph of *serratus* sometimes considered separate species *Coccystes hypopinarus* (Cabanis & Heine 1863, Harrison 1971a) but now widely accepted as a morph as interbreeding occurs and there are no structural differences (Snow 1978, Rowan 1983). *Coccystes caroli* Norman 1888 from Gabon now considered synonym of *levaillantii*, as is *Coccystes albonotatus* Shelley 1881 (Friedmann 1930). Somewhat smaller than *jacobinus*, birds from north-west Sri Lanka have been described as *taprobanus* Hartert 1915, but here considered to belong to nominate. Polytypic. Other synonyms: *Cuculus melanoleucus* J.F. Gmelin 1788 and *Oxylophus serratoides* Hodgson 1844, both in nominate.

FIELD IDENTIFICATION 34cm. Medium-sized slimly-built cuckoo with straight, swift flight. Erect posture with hanging tail when perched, although horizontal posture with head and neck slightly raised when foraging in foliage. When alarmed, crest erected and tail wagged up and down. **Adult** Glossy black above, white (*jacobinus*) or pale yellow-buff (*pica*) below with prominent black crest. White wing patches and tail tipped white conspicuous both when perched and in flight. In Africa *serratus* similar except underparts greyish-white and throat more or less streaked, and melanistic morph completely black except for white wing patches. **Juvenile** Brown to dull black above with short crest, and white below, grey or washed buffish on throat and breast. **Similar species** Unmistakable in Oriental Region. In Africa confusion possible with sympatric Levaillant's Cuckoo; pale morph Jacobin distinguished by smaller size, shorter tail and white throat and breast (heavily streaked black in Levaillant's). Dark morph birds easily confused with rare dark morph Levaillant's, mainly occurring in coastal east Africa, but latter distinguished by voice, larger size and stronger build. Juvenile Levaillant's has blacker bill than juvenile Jacobin, bushier crest, grey-brown chin and throat to breast often tinged rusty and indistinctly streaked black. Black Cuckoo lacks white wing patch, has shorter tail and is more stoutly built.

VOICE More varied repertoire with less monotonous repetition than most cuckoos; however *serratus* considered rather quiet. Both sexes give loud metallic, ringing and descending *kleeuw, kleeuw, kleeuw* repeated 5–12 times, male often continues with faster ascending series *kwik-kwik-kwik*. Call single *kweek*. In courtship *kruu-kru-kru-kleeuuu*. Female gives quiet or soft *chuka* note before egg-laying to stimulate male to perform distracting manoeuvre. Male post-laying calls *chit-it-chee* and *clueoo* answered by female with loud, rapid series of gurgling notes, often persisting for ten minutes. Alarm call low rattling *chuka-chuka-chuka-chuka*, followed by *kleeeuu* if bird takes flight. Contact call harsh and rattling. Other interpretations: *quer-qui-qui-quik, tyo-whi-tyo, tyo-whi-tyo, ker-wi-wi* or *kikikiki*, shrill metallic *pliu* or *pio–piu–pee-pee-piu* and metallic yelping *kyaOW-pi, kyaOW-pi, kyaOW-pi*, also rendered as *kyit-it-it–(it)*. Call also described as 'somewhat unmusical whistle, almost shriek *pie-ou*, becoming more metallic as breeding season draws to a close'. Often calls at night, but mostly at dawn and dusk; calls commonly in flight. Silent by late Nov, Pakistan. Very noisy during breeding by day and night in south Asia, but quiet otherwise. Calls of all subspecies similar (Swynnerton 1908, Benson 1945, Bates & Lowther 1952, Gaston 1976, Gallagher & Woodcock 1980, Roberts 1991, Zimmerman *et al.* 1996, Payne 1997, 2005, Irwin 1998, Robson 2000, Rasmussen & Anderton 2005).

DESCRIPTION Nominate. **Adult** Head from bill, just below eyes to nape, upperparts and tail black glossed green or blue-green with black, pointed, erectile crest, wings black washed slightly green but fading to blackish-brown with wear, conspicuous white wing-bar across base of P2–9; entire underparts including neck sides white, black tail graduated, with white tips broadening towards outer feathers. **Juvenile** Bronze-brown to dull black above, darkest on crown and mantle with shorter, more rounded crest than adult, wing-coverts and scapulars narrowly tipped white, wing patch smaller and off-white; whitish below with grey on throat and red-brown on breast; rectrices with narrower buffish tips. **Nestling** Naked at hatching, skin pinkish to orange-pink; after 48h purplish-brown with white egg tooth and black point at tongue tip. Feather tracts appear 2nd day and wing quills erupt 11th day; eyes open 6th day. **Bare parts** Iris dark brown, juvenile pale yellow; nestling has yellow eyelids. Down-curved bill black, slightly paler basally, in juvenile browner, base of lower mandible yellow and gape bright red; nestling mouth scarlet and gape yellow. Feet grey.

BIOMETRICS *C. j. pica*. Wing male (n = 17) 143–154mm, mean 147.2 ± 2.8, female (n = 26) 140–156mm, mean 148.5 ± 3.9. Bill male 19–23mm, mean 21.3 ± 1.4, female 19–22mm, mean 20.1 ± 1.1. Tail male 160–174mm, mean 167.4 ± 6.1, female 155–173mm, mean 161.2 ± 5.7. Tarsus male 23–25mm, mean 24.0 ± 0.8, female 22–26mm, mean 23.7 ± 1.1. Mass *serratus* male (n = 11) 69.2–87.6g, mean 78.6 ± 5.0, female (n = 14) 80.9–104.0g, mean 90.1 ± 6.5 (Payne 2005).

MOULT Primary moult sequence P6 and P9, pause, P7 and P10, pause, P8 and P5, or P5 then P8; irregular transilient moult of secondaries; rectrices replaced one by one, each feather dropping after previous has fully grown. Adult plumage fully attained at about six months; complete moult lasting about 100 days. *Pica* moulting RSA, Oct–Apr. Post-juvenile moult begins shortly after fledging, or soon after arrival in non-breeding quarters. Moulting south India 2 Oct (Stresemann & Stresemann 1969, Cramp 1985, Irwin 1998, N. Srinivasamurthy *in litt.*).

GEOGRAPHICAL VARIATION Three subspecies; two differ mainly in size.

C. j. jacobinus (Boddaert, 1783). India south of 15°N (Mysore, Kerala, Tamil Nadu) and Sri Lanka; migrants recorded eastern Zimbabwe, Mozambique and Natal (Irwin 1998), but most sedentary (Ali & Ripley 1983, Kotagama & Fernando 1994, Grimmett *et al.* 1998) or local migrant (Gaston 1976). Smallest subspecies; underparts white. Wing male (n = 5) 140–148mm, mean 145.5 ± 2.2, female (n = 6) 137–160mm, mean 143.5 ± 8.8 (Payne 2005). Described above.

C. j. pica (Hemprich & Ehrenberg, 1833). India and sub-Saharan Africa south to northern Zambia and Malawi. North of 20°N in north-west India to Nepal, south-east Tibet, Bangladesh to south-central Burma; migrant breeder Manipur, north-east India (Choudhury 2009); vagrant

Thailand, but up to three calling May–Jun 2008 may indicate colonisation (Robson 2008); another population in southeast Iran, Afghanistan and eastern Pakistan, most migrating to northern tropical Africa south to central Tanzania; some pass through southern Arabia (e.g. Balmer & Murdoch 2009a). African population from south Mauritania (presumed migrant breeder Sahel during rains, Jun–Nov), south Mali, Burkina Faso, Senegambia to Ethiopia and Somalia (perhaps in more arid habitats than *serratus*), south to Tanzania, Zambia and Malawi; may winter in southern savannas in north Namibia, north Botswana, Zimbabwe and Transvaal. Asian population indistinguishable from African. Medium-sized subspecies with underparts pale yellow-buff, greyer in worn plumage. Some individuals have pale, hair-like streaks on throat and breast.

C. j. serratus (Sparrman, 1786). Southern Africa south of Zambesi R.; Cape to east Botswana, Namibia, Zimbabwe and south Zambia; winters in northern savannas of Kenya, Eritrea, Ethiopia and Sudan, west possibly to Chad (Irwin 1998), but this or *pica* reported Western Cape, RSA, 4 Mar (Demey 2010). Largest subspecies. Two colour morphs: one glossy black above and greyish-white to white below with dark markings along throat and breast feather shafts, and flanks dark grey; other entirely glossy black except white wing patch (50% of Natal birds) (Snow 1978), only 1%, but varying from year to year, in Zimbabwe (Irwin 1981). This subspecies also characterised by host choice (Snow 1978) and egg colour (Harrison 1971a). Wing male (n = 40) 148–165mm, mean 156, female (n = 20) 148–160mm, mean 153 (Irwin 1998). Also wing male (n = 12) 146–167mm, mean 153.9 (Payne 2005).

DISTRIBUTION Afrotropical and Oriental. Sub-Saharan Africa, and Iran to Sri Lanka and Burma; occasional Tibet. Resident Uganda, west Kenya, Rwanda, Burundi and north Tanzania. North African subspecies (*pica*) winters in tropics south of Equator and southern subspecies (*serratus*) winters north of their breeding grounds. Another population (nominate) from eastern Pakistan, north-west India south to Sri Lanka and east to Tibet and Burma. Asian migrants (*pica*) on passage in UAE, Oman and Saudi Arabia (>20 records), northern Yemen, Djibouti, Zanzibar. Most appear to winter in Africa south to Tanzania Dec–Apr. Presumed breeding Mauritanian Sahel during rains, Jun–Nov, and wintering Senegambia, Nov–Dec; Mali May–Oct; north Togo Apr–Sep (once Dec), passage south Feb, Mar, Jun, Jul, Dec; north Nigeria Apr–Sep, in south Nigeria in dry season and in north Tanzania migrating Feb–Mar. Black morph of *serratus* winters north to Chad, Sudan, Eritrea, Ethiopia and Kenya, although also recorded west Africa, and birds from west Africa moving south crossing equator in dry season. Nominate apparently common in Zimbabwe and Mozambique Sep–Apr, perhaps having flown directly across Indian Ocean, although no ringing recoveries; common Ruvuma Delta, north Mozambique, May (Becking 1981, Cheng 1987, Dowsett & Dowsett-Lemaire 1993, Lamarche 1988, Cheke & Walsh 1996, Irwin 1998, Payne 1997, Borrow & Demey 2001, Borghesio *et al.* 2009, Isenmann *et al.* 2010, C.F. Mann). Nine records Lesotho (Bonde 1993). Once Phuket I., Thailand (Round 1995). Pulau Pinang, Dec 2009 (Choo 2009), and two further unconfirmed peninsular Malaysia (A. Jeyarajasingam in Bakewell 2010). UAE, Jan (Balmer & Murdoch 2009b). Oman, Sep–Nov (Balmer & Murdoch 2009a). Ten records Seychelles (Skerrett *et al.* 2006, Demey 2009). Iriomotejima, Ryukyu Is, Japan, 1 Jun 1997 (A. Kamata in Allen *et al.* 2006). Dalupiri, Baluyan Is, Philippines, 21 May 2004 (Allen *et al.* 2006).

HABITAT Dry, open woodland and scrub, cultivation, including gardens and forest plantations, but entirely absent from forests. In west Africa common in semi-arid parts of Sahel and thick swampy bush, also savanna with *Mimosa* and *Acacia*. In east, central and south Africa thorny bush, scrub, woodland and cultivation. In Middle East in thorny scrub, woodland and cultivated areas usually below 1,000m. Breeds in India in open woodlands, groves, gardens, near villages and towns, both in dry semi-desert and moist deciduous areas. In Nameri NP, India, in secondary forest and disturbed areas like cultivation, settlements and edges of parks. Lowlands to 2,600m (3,800m), India; to 4,270m on migration in Himalayas. In Africa lowlands to 1,500m (3,000m). Nepal mainly below 365m, but to

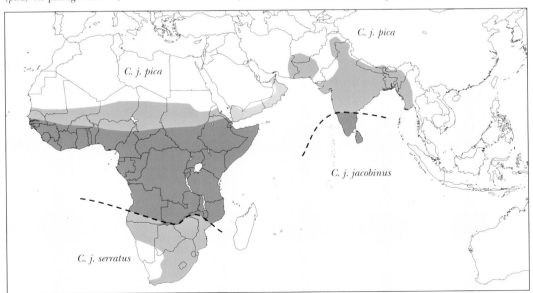

3,660m; plains to 2,600m (3,800m) India; Sri Lanka to *c.* 1,300m (Bannerman 1933, Cramp 1985, Porter *et al.* 1996, Sibley 1996, Payne 1997, Grimmett *et al.* 1998, Barua & Sharma 2005).

BEHAVIOUR Solitary during breeding season except during courtship, and on arrival where often seen in small flocks in Pakistan indulging in much excited calling and aerial pursuit, although many records in RSA were of pairs. Serious aggressive encounters when three or more birds seen together. Male probably promiscuous except in areas of low density. Arboreal and not shy, often foraging in open, in low bushes or hopping awkwardly on ground, or calling from telephone wires. Gut contents of caterpillars usually removed before ingestion and hairy caterpillars 'de-haired' before being swallowed. Flutters into air after flying insects. Flies freely, keeping near ground. Female quietly surveys host nests sites from conspicuous position. Every second day, when eggs ready for laying, female calls male to assist in distracting hosts. Courtship feeding and mating follow egg-laying, which can be accomplished within ten seconds. Most eggs laid 07:00–09:00hrs. Male in courtship has 'butterfly flight' with slow, irregular wing strokes below level of back, and tail spread with head up and calling; female calls in response, followed by courtship feeding and copulation. No records of female removing host's egg when egg-laying, but female collected with nearly completed egg in oviduct and part of egg shell and yolk, probably from bulbul *Pycnonotus sp.* in crop. Anti-predator behaviour of fledglings, even when capable of flying well, is to freeze in cover. Young in Pakistan apparently migrate later than adults. Diurnal migrant east Africa (Godfrey 1939, Liversidge 1961, Payne 1968, 1997, Liversidge 1969, Johnsingh & Paramanandham 1982, Irwin 1988, Roberts 1991, Grimmett *et al.* 1998).

BREEDING Parasitism In north sub-Saharan Africa Fulvous Babbler, Rufous Chatterer and Common Bulbul. In RSA Laughing Dove, Red-faced Mousebird, one cuckoo egg with three host's, Speckled Mousebirds' eggs in nest, Common Fiscal, Southern Boubou (1), Bokmakierie (3), Fork-tailed Drongo, Common (also Zimbabwe), Cape and Red-eyed Bulbuls, Sombre and Terrestrial Greenbuls, Black Cuckoo-shrike (1), Southern Pied Babbler, Chestnut-vented Warbler (1), Fiscal-flycatcher (1), Cape Wagtail, Yellow-throated Petronia. Elsewhere in Africa African Black Coucal, and Brown-hooded Kingfisher, one egg with two hosts'. In India (*pica*) mainly *Turdoides* such as Common, Jungle and Large Grey Babblers; also Streaked Laughing-thrush. White-headed Babbler, Sri Lanka. Most common hosts in Burma Lesser Necklaced Laughing-thrush, Striated and White-throated Babblers. Striated Grassbird, Thailand. Numerous unconfirmed or accidental hosts, e.g. Brown and Hinde's Babblers; probably White-rumped Babbler and Rufous Chatterer, south Ethiopia, and Cape White-eye, but this cuckoo is known for 'egg-dumping', and cuckoo nestlings can easily be misidentified (Ivy 1901, von Erlanger 1905, Butler 1908, Wood 1911, Schönwetter 1928, Bradfield 1931, Sparrow 1936, Roberts 1939, Benson 1945, Vincent 1946, Friedmann 1948b, Gaston 1976, Becking 1981, Payne 1997, Irwin 1998, Robson 2000, 2011, Ash & Atkins 2009; Ridley & Thompson 2012). Rate of parasitism often high, e.g. 71% of Jungle Babbler nests, 39 % of Common Babbler nests (Gaston 1976) and 36% of Cape Bulbul nests (Liversidge 1969). Two Jacobin and one Great-spotted Cuckoos' eggs in nest of Sombre Greenbul, RSA, Nov (Ivy 1901). In garden in Lilongwe, Malawi, several Arrow-marked Babblers observed apparently co-operatively defending nest by mobbing two adult Jacobin Cuckoos, Dec, even though their eggs were near to hatching, and thus could not be parasitised by cuckoos (Clark & Clarke 1985). No accounts of rejection behaviour, but Cape Bulbul often deserts parasitised nest (Liversidge 1969). **Season** Breeding coincides with rains in Africa and Asia. Eggs Ethiopia Mar–Jul, Oct. Somalia May. Senegambia and Mali Mar–Aug. Mauritanian Sahel presumed Jun–Nov. North Nigeria May–Jul. RSA, late Oct–Mar. Zimbabwe Nov–Feb. Fledgling south Ethiopia, Jun. Pakistan Jun–Aug. Generally all months India, varying locally; north-east May–Sep, Delhi area Oct–Nov. Central Burma May–Nov. North Thailand Jun (Benson 1945, Bannerman 1953, Irwin 1981, Rowan 1983, Roberts 1991, Payne 1997, Ash & Miskell 1998, Grimmett *et al.* 1998, Ash & Atkins 2009, Isenmann *et al.* 2010, Robson 2011, S. Prakash pers. comm.). **Eggs** Laid every second day, laying period may extend to ten weeks (Irwin 1998). Eggs in India and north Africa most like those of babbler hosts, but larger, rounder, paler blue (blue-green in *pica*) and unglossed. Quite different from bulbul eggs (Irwin 1998), suggesting this family has not yet evolved rejection mechanisms sufficient to force parasite to lay mimetic eggs. In RSA 11 of 50 eggs were laid on or before day of first host egg; *pica* Africa: 22.1–26.0x18.1–22.0mm, (n = 125) mean 24.1x19.8mm; India: 23.9x18.6mm; RSA (*serratus*), and few Mali and Kenya, white and larger, 23.0–28.5x19.0–24.0mm, (n = 78). Commonly more than one cuckoo egg in a nest. In RSA estimated total of 19–25 eggs laid by each female (Bates & Lowther 1952, Liversidge 1969, Payne 1973, Becking 1981, Irwin 1988). **Incubation** 11–12 days, *c.* 2 days shorter than African bulbuls (Liversidge 1969, Irwin 1998), although Friedmann (1948b) states 14–15 days. **Chicks** Ejection of host's egg not recorded. Host eggs and chicks often trampled, or chicks starve because young cuckoo lays 'spread-eagled' over them, although there are records of host young being successfully reared alongside cuckoo. Chicks reared by Common Fiscals produce foul-smelling excretion in alarm. Begging call similar to that of host chicks. Feather tracts appear 2nd day; eyes open 6th day; wing quills erupt 11th day; well-feathered by 15 days and usually fledge at *c.* 17 days, although will leave at 11 days if disturbed. Fed by host parents for further 15–25 days (Liversidge 1969, Gaston 1976, Irwin 1988, Maclean 1993). **Survival** Near Port Elizabeth, RSA, 20% of 50 eggs reached fledging stage, and hosts' (Cape Bulbul) reproductive output was reduced by *c.* 33%. Current longevity record from sparse ringing data 12 months (Liversidge 1969, Rowan 1983, McClure 1998).

FOOD Mainly hairless, hairy or spiny caterpillars, other large insect larvae, termites, including alates, beetles, ants, including alates, mantids, tree-crickets, grasshoppers; snails; passerine eggs; berries e.g. blackberries and raspberries; occasionally green leaves. (Inglis 1909, Vincent 1934, Henry 1955, Mackworth-Praed & Grant 1957, Ali & Ripley 1983, Roberts 1928, 1991, Irwin 1998).

STATUS AND CONSERVATION Common and widely distributed throughout range. Locally common in dry woodland, Africa (Irwin 1998), but in RSA numbers fluctuate over years (Rowan 1983). Rare to uncommon Senegambia (Barlow *et al.* 1997); seasonally common Kenya and north Tanzania (Zimmerman *et al.* 1996); common and widespread Somalia (Ash & Miskell 1998); common Ethiopia (Ash & Atkins 2009). Common breeding visitor Pakistan, plentiful around Islamabad and lower Margalla hills (Roberts 1991); uncommon summer visitor Nepal; locally common India in summer; local breeding visitor Bangladesh (Grimmett *et al.* 1998). Fairly common breeding visitor Burma (Smythies 1986, Robson 2000). Common lowland breeding resident Sri Lanka (Kotagama & Fernando 1994). Much vagrancy. Tolerant of habitat degradation in semi-arid areas, therefore no major threats currently, and not globally threatened (BirdLife International 2011).

Jacobin Cuckoo *Clamator jacobinus*. **Fig. 1.** Adult, *C. j. pica*, Kolkata, India, September (*Abhishek Das*). **Fig. 2.** Adult, *C. j. jacobinus*, Yala, Sri Lanka, January (*Gehan de Silva Wijeyeratne*). **Fig. 3.** Juvenile, *C. j. serratus*, being fed by Dark-capped Bulbul (out of photo), Kruger NP, RSA, March (*Ian Nason*).

Genus *Piaya*

Lesson, 1830 *Traité d'Orn.* livr. 2: 139. Type, by original designation, *Cuculus cayanus* Linnaeus 1766. Three species.

Neotropics; non brood-parasitic. Small to large, wings short and rounded; crestless; sexes alike; arboreal (*minuta* less so). Includes *Pyrrhococcyx* Cabanis 1862 and *Coccyzusa* Heine 1863. *P. minuta* placed in *Coccycua* Lesson 1830, along with *Coccyzus pumilus* and *cinereus*, based on mtDNA data (Sorenson & Payne 2005).

LITTLE CUCKOO
Piaya minuta   Plate 21

Coccyzus minutus Vieillot, 1817 (No locality = Cayenne, *ex* Latham and Brisson)

TAXONOMY Sometimes placed in monotypic genus or subgenus *Coccycua* Lesson 1830 on basis of smaller size, and relatively shorter tail and outermost primary (Ridgway 1916), and of mtDNA data (Hughes 2006). Based on other mtDNA genes, placed in enlarged genus *Coccycua* with *cinereus* and *pumilus* (Sorenson & Payne 2005), followed by Banks *et al.* (2006). Polytypic. Synonyms: *Cuculus cayanensis* Var. A. Latham, *Turdus macrourus* A. Lichtenstein 1793 (Meise 1950), *Cuculus rutilus* Illiger 1811 (invalid: Zimmer 1930), *Macropus caixana* von Spix 1824, *Coccycua monachus* Lesson 1831, *m. orinocensis* Cherrie 1916, and *m. barinensis* Aveledo & Pérez 1994, all in nominate; *m. panamensis* Todd 1912 in *gracilis*; *m. chaparensis* Cherrie 1916 in nominate or *gracilis*.

FIELD IDENTIFICATION 24–28cm. **Adult** Uniformly rufous-chestnut above, throat and chest tawny to rufous (paler than upperparts), belly buffy-grey, undertail-coverts blackish, graduated tail below rufous-black tipped white. Bill greenish-yellow. Bare orbital skin red. **Juvenile** Darker; primaries dusky, no white tail tips, dark bill. **Similar species** Similar to much larger Squirrel Cuckoo. Also, proportionately shorter tail, and no pure grey on breast, as in many Squirrel Cuckoo populations. Black-bellied Cuckoo also much larger, with red bill, yellow loral spot and grey cap. Dwarf Cuckoo also has rufous chest, but even smaller and shorter-tailed, has pale belly, dark bill, grey upperparts.

VOICE Often quiet. Contact call, low clucking *czek!* or *tchek* like crackle of static electricity, is most common. Sharp *chik! wreeanh* or soft *geep-were* like muted Squirrel Cuckoo. Nasal chatter *kyaah hoo-hoo-hoo-hoo* lasting *c.* 1.2sec or *nyaa-nyaa-nyaa*, first note clearly separate and drawn-out, rest fast descending phrase, sometimes ascending (Boesman 1999, J. C. Minns recording). Slow series of descending *kewl* notes (Hilty & Brown 1986, Ridgely & Gwynne 1989, Hardy *et al.* 1990, ffrench 1991, Ridgely & Greenfield 2001, Hilty 2003, Schulenberg *et al.* 2007).

DESCRIPTION Nominate. **Adult** Crown cinnamon-brown, hind crown feathers rather long and erectile, forming bushy crest, nape to uppertail-coverts uniform rufous-chestnut, wings bright chestnut, underwing-coverts rufous to brownish-buff, flight feathers below cinnamon tipped brownish-slate, throat and breast tawny, belly brownish-grey, undertail-coverts and tibia greyish-black; tail graduated for nearly half length, above rufous-chestnut as upperparts, below black or dark brown with broad white tips. Eyelashes distinct. **Juvenile** Head, upperparts and wings dark liver-brown, very faintly tipped with cinnamon, rump barred with cinnamon-rufous and fuscous, basal half of primaries rufous, underparts dull, dark brown, belly and undertail-coverts blackish, tail narrower and darker (blackish) than adult, without white tips, in some indistinct, very narrow reddish-brown tips. **Nestling** Unknown. **Bare parts** Iris reddish-brown to red, brown in juvenile; eyelid carmine. Eye-ring red, greyish in juvenile. A male had skin below eye greenish-yellow. Bill rather small, strongly decurved, greenish-yellow with dusky tip; dark brown to blackish in juvenile. Inside of mouth and top of tongue deep black, underside of tongue bluish-grey. Feet greyish-black or bluish-grey, black in juvenile, claws black.

BIOMETRICS Nominate (Guyana, Suriname, French Guiana and north-east Brazil to Pará and Mato Grosso). Wing male 101–107mm (n = 13), mean 105.2 ± 2.4, female 104–108mm (n = 5), mean 105.0 ± 2.0. Bill male 13.8–23.5mm, mean 20.1 ± 2.1, female 16.8–22.0mm, mean 18.8 ± 3.2. Tail male 138–158mm, mean 147.9 ± 6.4, female 146–163mm, mean 156.2 ± 6.9. Tarsus male 20.5–27.9mm, mean 23.8 ± 2.2, female 22.0–26.4mm, mean 24.1 ± 1.7 (Payne 2005). Mass male 26–54g, female 31–50g (Haverschmidt 1948, 1968, Payne 2005).

MOULT Female moulting Dec, Trinidad. Irregular transilient primary moult (Junge & Mees 1958, Stresemann & Stresemann 1961).

GEOGRAPHICAL VARIATION Two subspecies.
P. m. minuta (Vieillot, 1817). East of eastern Andes, Colombia (Meyer de Schauensee 1949), Venezuela, Guyana, Surinam, French Guiana, central and Amazonian Brazil, east Ecuador and Peru; Trinidad. Possibly north Bolivia. Described above.
P. m. gracilis (Heine, 1863). East Panama, Colombia and west Ecuador. Possibly north Bolivia. Bolivian birds intermediate in breast colour between *minuta* and *gracilis* (Payne 2005). Paler than nominate, rufous above (Chapman 1917, Ridgely & Greenfield 2001). Wing male 102–109mm (n = 5), mean 104.6 ± 3.6, female 97–108mm (n = 7), mean 103.4 ± 4.3 (Payne 2005).

DISTRIBUTION Central and tropical South America. Resident. East Panama lowlands (east Colón, Canal area, east Panama Province, Darién), north and extreme north-west Colombia, Magdalena and Cauca valleys, and eastern lowlands, Venezuela (except arid north-west and north-east), Trinidad and Guianas. West of Andes in west Ecuador (Esmeraldas south to El Oro and Loja), east of Andes in east Ecuador, east Peru, north and east Bolivia (Pando, Beni, La Paz, Cochabamba, Santa Cruz) and Amazonian and central Brazil south and east to Amapá, Maranhão, Tocantins, Goiás and Mato Grosso do Sul (Chapman 1917, Pinto 1938, 1964, Meyer de Schauensee 1949, Snyder 1966, Haffer 1975, Hilty & Brown 1986, Ridgely & Gwynne

1989, ffrench 1991, Tostain *et al.* 1992, Forrester 1993, Haverschmidt & Mees 1994, Arribas *et al.* 1995, Sick 1997, Rodner *et al.* 2000, Ridgely & Greenfield 2001, Hilty 2003, Schulenberg *et al.* 2007). Migration suggested west Ecuador (Freile & Chaves 2004).

HABITAT Normally near water in dense shrubby vegetation such as second growth mixed with small openings, around oxbow lakes and along humid forest edges. Mainly mangroves Trinidad. Savanna woodland (stunted, scrubby forest growing on sandy soil; south Venezuela), and gallery forest and semi-deciduous forest Pantanal and Amapá, Brazil. Lowlands to 750m Panama; to 1,600m Colombia. To 950m Venezuela, but only to 700m north of Río Orinoco. In Ecuador, mostly below 600m, but following river valleys in east to 800–900m, locally to 1,500m in west, once 1,900m. Lowlands of east Peru, locally to 1,050m. To 500m Bolivia (von Pelzeln 1870, Wetmore 1968, Remsen & Parker 1983, Hilty & Brown 1986, Cintra & Yamashita 1990, ffrench 1991, Arribas *et al.* 1995, Robinson 1997, Silva *et al.* 1997, Stotz *et al.* 1997, Zimmer & Hilty 1997, Ridgely & Greenfield 2001, Hilty 2003, Freile & Chaves 2004, Schulenberg *et al.* 2007).

BEHAVIOUR Shy and rather skulking. Singly, in pairs or small groups foraging independently and often not very close, usually in cover <5m up, sometimes on ground. Peers about carefully, then hops short distances or lunges after prey. Rarely follows army ants. Raises and lowers tail gently. Scratches head under wing (von Pelzeln 1870, Willis 1983b, ffrench 1991, Sick 1993, Haverschmidt & Mees 1994, Novaes & Lima 1998, Hilty 2003, Schulenberg *et al.* 2007).

BREEDING Season Jun–Sep, Surinam. Recently fledged Mar, French Guiana or Surinam; eggs in nest Jul, Trinidad; three birds in breeding condition Jan–May, Colombia; young foraging together with adult Apr, Belém, Brazil; large ova Jul and Nov, Bolivia (Belcher & Smooker 1936, Haverschmidt 1955, 1968, Willis 1983b, Hilty & Brown 1986, Tostain *et al.* 1992, Haverschmidt & Mees 1994, Payne 2005). **Nest** Open cup of twigs in dense vegetation, e.g. bamboo, 25cm deep, with very shallow depression on top, 2–5m up (Stone 1928, Herklots 1961, Tostain *et al.* 1992). **Eggs** 1–2; dull white, elliptical; 22.9–26.7x18.2–20.5mm (Belcher & Smooker 1936, Hellebrekers 1942, Schönwetter 1967). No further information.

FOOD Wasps, caterpillars, beetles, cicadas, flies; spiders; centipedes (Layard 1873, Salvin & Godman 1879–1904, Haverschmidt 1968, Wetmore 1968, ffrench 1991).

STATUS AND CONSERVATION Fairly common to locally very common, but rare to absent in arid regions, Venezuela (Hilty 2003). Locally abundant e.g. Amapá, Brazil (Sick 1993). Uncommon Peru (Terborgh *et al.* 1984, Clements & Shany 2001, Schulenberg *et al.* 2007) and Trinidad (ffrench 1991). Possibly becomes less common southward: rare Pantanal, Mato Grosso, Brazil (Cintra & Yamashita 1990). Probably overlooked because of quiet, skulking habits (e.g. Wetmore 1968). 1.5 individuals captured/100m mistnets/100h, French Guiana (Reynaud 1998). Generally fairly common and of low conservation and research priorities (Stotz *et al.* 1996). Not globally threatened (BirdLife International 2011).

Little Cuckoo *Piaya minuta*. Adults. **Fig. 1.** *P. m. minuta*, Guatemala road, near Kourou, French Guiana, June (*Michel Giraud-Audine*). **Fig. 2.** *P. m. minuta*, Pantanal, Brazil, August (*Laurens Steijn*). **Fig. 3.** *P. m. minuta*, Pariacabo, French Guiana, June (*Alexandre Vinot*). **Figs. 4–5.** *P. m. gracilis*, Silanche, Pichincha, Ecuador, December (*Roger Ahlman*).

BLACK-BELLIED CUCKOO
Piaya melanogaster   **Plate 13**

Cuculus melanogaster Vieillot, 1817 (Java, error; Cayenne, von Berlepsch & Hartert 1902)

TAXONOMY Monotypic. Synonyms: *brachyptera* Lesson 1831 and *ochracea* Cory 1915.

FIELD IDENTIFICATION 35–40.5cm. Rarely glides across light gaps as Squirrel Cuckoo, and much less vocal. Size, proportions and mainly rufous plumage similar to Squirrel Cuckoo. **Adult** Bill red, prominent bare orbital area blue, loral spot yellow, crown grey, lower belly to undertail-coverts black; underside of long graduated tail with tipped broadly white blackish subterminal band; rest of plumage rufous to chestnut. **Juvenile** Similar to adult. **Similar species** If head not seen, from below distinguished from range-overlapping subspecies of Squirrel Cuckoo by lack of broad grey band on belly (between rufous breast and sometimes black undertail-coverts, but latter grey in e.g. nominate subspecies of Squirrel Cuckoo). Slightly smaller and shorter tailed than Squirrel Cuckoo. Superficially resembles Grey-capped Cuckoo (limited or no range overlap), but that species much smaller, has black bill and no black on undertail-coverts.

VOICE Loud *jjit, jjit-jjit-jjit* and scratchy, descending *yaaaaa* followed by dry rattle. Similar to Squirrel Cuckoo but harsher and heard far less often. Steadily repeated *jerreé-jew, jerreé-jew, jerreé-jew...*, sometimes lasting minute or more, but difficult to find motionless vocalising bird in dense cover. Other growling notes; strident *pi* or *peeee*; rattle and chitter *i-wik-i-yer* in dispute; alarm *stit-stit*; repeated hoarse scream *swearrrr*; loud *huweep-hew* or *eep, were*, repeated once/sec for up to a minute; second note sometimes drawn out, falling, and sometimes more hoarse (Haverschmidt 1968, Willis 1983b, Hilty & Brown 1986, Hardy *et al.* 1990, Sick 1993, Ridgely & Greenfield 2001, Payne 2005).

DESCRIPTION Adult Crown and lores grey, contrasting with otherwise rufous upperparts, outer webs of P7–10 brownish-black, primaries with blackish tips, throat and breast cinnamon-rufous becoming black on belly and vent, tail long, graduated, tinged maroon above, black below with large white tips (12–14mm). **Juvenile** Similar to adult, but rectrices narrower with smaller white tips (8–12mm). **Nestling** Unknown. **Bare parts** Iris dark brown to dark red. Bare orbital area pale blue, large loral spot yellow, rear corner of eye greenish. Bill bright deep-red, sometimes black at very tip, sometimes violet at base (Surinam, Venezuela). Feet lead-grey to black.

BIOMETRICS Wing male 131–142mm (n = 12), mean 135.8 ± 5.2, female 129–141mm (n = 10), mean 135.4 ± 3.7. Bill male 30.0–35.5mm, mean 32.9 ± 1.7, female 27.7–35.5mm, mean 33.6 ± 2.3. Tail male 193–239mm, mean 221.1 ± 12.6, female 193–242mm, mean 218.5 ± 17.1. Tarsus male 30.0–37.4mm, mean 33.6 ± 2.3, female 29.0–37.0mm, mean 32.2 ± 2.2 (Payne 2005). Mass male 73–111.5g, female 79–118g (Haverschmidt 1968, Novaes & Lima 1991, Haverschmidt & Mees 1994, Payne 2005).

MOULT Irregular transilient primary moult (Stresemann & Stresemann 1961).

DISTRIBUTION Tropical South America. Resident. East of Andes from south-east Colombia (from south Meta, north-

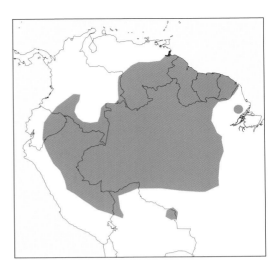

east Guainía and Vaupés southwards), south Venezuela (Amazonas, Bolívar, once Delta Amacuro; Hilty 2003) and Guianas south through east Ecuador and east Peru to north Bolivia (Pando, La Paz, Santa Cruz) and west Amazonian Brazil (Roraima, Amazonas, west Pará, Acre, Rondônia, north Mato Grosso) (Snyder 1966, Hilty & Brown 1986, Parker & Remsen 1987, Bates *et al.* 1992, Tostain *et al.* 1992, Haverschmidt & Mees 1994, Sick 1997, Ridgely & Greenfield 2001, Schulenberg *et al.* 2007).

HABITAT Canopy and subcanopy of terra firme forest and campinarana (low forest on sandy soil). Occasionally in forest borders and scrubby savanna woodland, but found less often in borders and openings, and more restricted to primary forest, than often-sympatric Squirrel Cuckoo. Lowlands to 800m, perhaps most common below 500m, once 900m, Ecuador (Sibley & Monroe 1990, Servat & Pearson 1991, Stotz *et al.* 1996, 1997, Borges *et al.* 2001, Ridgely & Greenfield 2001, Henriques *et al.* 2003, Hilty 2003, Borges 2004).

BEHAVIOUR Similar to Squirrel Cuckoo, but more in canopy, staying in foliage cover, not often gliding across open spaces. Accordingly, difficult to see, and thus often overlooked. Solitary (92% of observations) or in pairs, foraging in live foliage. Joins mixed-species bird flocks in canopy. Forages with a few hops or scrambling runs along branches, alternating with periods of immobility and careful peering. Staying mostly in canopy, does not forage near army ants, except when ants climb a trunk (Pearson 1971, Willis 1983b, Stotz & Bierregaard 1989, Servat & Pearson 1991, Thiollay & Jullien 1998, Henriques *et al.* 2003, Hilty 2003).

BREEDING Season Enlarged gonads Apr, Venezuela. Feeding young in nest Jul, French Guiana (Friedmann 1948a, Tostain *et al.* 1992). **Nest** Undescribed. **Eggs** Unglossed white, 30.0–33.0x23.8–25.3mm, mean 31.1x24.5 (n = 3) (Schönwetter 1967). No further information.

FOOD Grasshoppers, katydids, Hemiptera, beetles, ants, flies, caterpillars; spiders; centipedes (Schubart *et al.* 1965, Haverschmidt 1968, Payne 2005).

STATUS AND CONSERVATION Considered common (recorded more than once every other day) Cachoeira

Nazaré, Rondônia (Stotz *et al.* 1997). Generally uncommon Venezuela (Stotz *et al.* 1996, Hilty 2003), and Alta Floresta, Mato Grosso (Zimmer *et al.* 1997), Tapajós National Forest (Henriques *et al.* 2003) and Jaú NP (Borges *et al.* 2001). Uncommon Peru (Clements & Shany 2001). Much less common than Squirrel Cuckoo, Suriname (Haverschmidt & Mees 1994). Rare to uncommon and perhaps local Ecuador (Ridgely & Greenfield 2001). Sensitive to habitat disturbance (Stotz *et al.* 1996). Not globally threatened (BirdLife International 2011).

Black-bellied Cuckoo *Piaya melanogaster*. **Figs. 1–3.** Adult, INPA tower, 60km north of Manaus, Amazonas, Brazil, March (*Andrew Whittaker*). **Fig. 4.** Adult, Parque Nacional Yasuní, Ecuador, January (*Ian Davies*).

SQUIRREL CUCKOO
Piaya cayana   **Plate 13**

Cuculus cayanus Linnaeus, 1766 (Cayana = Cayenne, French Guiana)

TAXONOMY Mexican and Central American subspecies *mexicana* and *thermophila* separated specifically (Navarro-Sigüenza & Peterson 2004). Detailed study of geographic variation including nature of contact zone in Oaxaca needed (Binford 1989). Polytypic. Synonyms: *P. c. stirtoni* van Rossem 1930 in *P. c. thermophila* (but recognised by Wetmore 1968); *c. incincta* Griscom 1932 in *thermophila* (Peters 1940, Wetmore 1968); *c. extima* van Rossem 1930 in *mexicana* (Wetmore 1968); *c. venezuelensis* Cory 1913 in *circe* (Junge 1937c); *c. inexpectata* Todd 1947 and *Pyrrhococcyx columbianus* Cabanis 1862 in *mehleri* (Junge 1937c, Phelps & Phelps 1958); *c. boliviana* Stone 1909 in *obscura* (Peters 1940, Payne 2005), not in *nigricrissa* as in Dickinson (2003); *caucae* Stone 1909 in *nigricrissa* (Cory 1919); *c. cearae* Cory 1915 in *hellmayri* (Dickinson 2003) or *pallescens* (Pacheco 2000); *cayana* var. *guarania* von Ihering 1904 in *macroura* (Peters 1940); *Pyrrhococcyx guianensis* Cabanis & Heine 1863 in nominate (Cory 1919). *Cuculus ridibundus* J.F. Gmelin 1788 apparently synonym of *P. c. mexicana* and *Coccyzus macrocercus* Vieillot 1817 of nominate (Allen 1893).

FIELD IDENTIFICATION 43–46cm. Leaps or runs squirrel-like along branches or up through vine tangles; loose-jointed. **Adult** Large, slender, long-tailed, arboreal. Bill and bare orbital ring pale greenish-yellow; orbital ring red in subspecies east of Andes. Eyes red. Above rufous to chestnut, throat and upper breast pale buff, belly light grey; undertail-coverts grey to black (subspecies vary). Very long graduated tail dark chestnut, the feathers with broad round white tips below. Tail below blackish in some subspecies, rufous to chestnut with dark subterminal band and white tips. **Juvenile** Eyes brown, eye-ring grey, bill greyish, feet grey. **Similar species** Plumage and bill colour similar to much smaller Little Cuckoo. Size, proportions and plumage similar to Black-bellied Cuckoo, which has grey cap, red bill, different coloured bare skin of face and lacks broad grey band on belly.

VOICE Loud whistled series *whit whit whit whit whit*... , each note 0.2sec, 2 notes/sec, total 5–8 notes, or almost without stop for hours, proclaiming territory. In Mexico and Central America, given mostly Apr–Aug. Musical *tek-wong* or *ik-wank*, first note rising, second note falling, total duration 0.3sec, some seconds between calls. Prolonged quiet dry rattle *chrrrrrrrr* given in interactions. Low dry nasal rolling *hic-a-roo* lasting 0.8sec, last note often prolonged, given especially during breeding season, e.g. at changeover of incubation. Loud, explosive metallic alarm *kip!* or *stit-it* often in flight. Accented *wík-i-y'wer* in disputes. Fledglings call sharp *eee-ka*. Young bird gaped, lowered head, gave loud squeaks *scraahh* and flapped wings when fed by adult (Slud 1964, Skutch 1966, Snyder 1966, Willis & Eisenmann 1979, Willis 1983b, Hilty & Brown 1986, Stiles & Skutch 1989, Sick 1993, Howell & Webb 1995, de la Peña & Rumboll 1998, Oniki & Willis 1999, Payne 2005).

DESCRIPTION *P. c. mexicana.* **Adult** Upperparts light cinnamon-rufous, becoming gradually paler on crown and deeper on tail, underwing-coverts pale grey becoming whitish distally; remiges tipped light greyish-brown, inner webs becoming dark brown distally; chin, throat and upper chest salmon-buff, deepening on sides of head and neck into nearly colour of crown and hindneck, rest of underparts plain pale bluish-grey, gradually becoming light slate-grey on undertail-coverts; tail long and graduated, rectrices below cinnamon-rufous with contrasting brownish-black subterminal band preceding white tip, on outer feathers black-and-white areas of approximately equal width, but the white tips gradually decreasing in width to middle pair, where white very narrow or almost lacking. **Juvenile** Similar to adult, plumage more lax and fluffy, rectrices narrower and more pointed, middle pair of tail below without white tip, and subterminal band brown. Tail short in recently fledged. **Nestling** Skin black with sparse trichoptiles; when feathered, mouth interior bright red. **Bare parts** Iris red; (reddish) brown in juvenile. Eyelids apple-green. Bare orbital region yellow-green (red in subspecies east of Andes); grey in juvenile. Bill apple-green, greyish horn-colour in juvenile, inside of mouth blue-black. Feet grey, also in juvenile, claws black.

BIOMETRICS *P. c. thermophila.* Wing male 141–157mm (n = 12), mean 146.1 ± 4.7, female 138–152mm (n = 12), mean 145.7 ± 4.2. Bill male 23.5–28.4mm, mean 26.6 ± 1.4, female 23.4–27.8mm, mean 25.4 ± 1.4. Tail male 246–284mm, mean 270 ± 12.8, female 245–272mm, mean 263.2 ± 12.1. Tarsus male 33.8–38.8mm, mean 35.4 ± 1.7, female 33.9–39.1mm, mean 36.0 ± 1.6 (Payne 2005). Mass male 73.0–137.0g (n = 33), mean 104.0, female 76.0–129.4g (n = 28), mean 100.3 (Payne 2005). Male with unossified skull and undeveloped testes, Venezuela, 69g, (Pérez-Emán *et al.* 2003). Female *macroura*, Corrientes, Argentina, and female Bolivia, 134.0g (Contreras 1979, Bates *et al.* 1989).

MOULT Irregular transilient primary moult. Juvenile remiges and rectrices normally shed second autumn, during first annual moult (Jul–Sep), and limited spring body moult in Feb–Mar, with some juvenile rectrices replaced, El Salvador. In south Venezuela: male (apparently young) in wing, tail and body moult in Mar (Dickey & van Rossem 1938, Stresemann & Stresemann 1961, Pérez-Emán *et al.* 2003).

GEOGRAPHICAL VARIATION Fourteen subspecies. Ranges follow Payne (2005) unless otherwise stated.

P. c. cayana (Linnaeus, 1766). East and south Venezuela in south-east Monagas (but see *insulana*), Delta Amacuro, south Apure, Amazonas and Bolívar (Hilty 2003), Guianas, north Brazil north of Rio Amazonas. Above like *thermophila*. Belly ashy-grey, under tail-coverts darker grey, tail black below.

P. c. mexicana (Swainson, 1827). West Mexico (Sonora to Oaxaca). Intermediates between *mexicana* and *thermophila* with rusty tint on underside of rectrices Isthmus of Tehuantepec (Salvin & Godman 1879–1904, Binford 1989), but otherwise intergradation between two forms rare. Described above.

P. c. thermophila P.L. Sclater, 1860. East Mexico (east San Luis Potosí and south Tamaulipas) to Panama and north-west Colombia (west of Gulf of Urabá; Haffer 1975). Dark rufous-chestnut above, tail shorter than in *mexicana*, below blackish preceding white tip (not cinnamon-rufous with dark subterminal band).

P. c. nigricrissa (Cabanis, 1862). West Colombia, west Ecuador, north-west Peru south to Piura. Eye-ring greenish-yellow (Ridgely & Greenfield 2001, Schulenberg *et al.* 2007). Darker than *thermophila*, rich rufous-chestnut above, belly and undertail-coverts blackish.

P. c. mehleri Bonaparte, 1850. North-east Colombia east of Golfo de Urabá to Magdalena valley and along west slope of

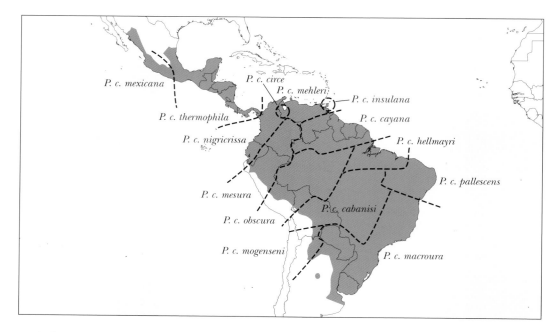

east Andes, east in Venezuela east of Andes and north of Río Orinoco to Península de Paria (Haffer 1975, Hilty 2003). Eye-ring greenish-yellow (Hilty & Brown 1986, Restall *et al.* 2006). Plumage more rufous than *mexicana*, lighter throat and breast grade into light grey belly, undertail-coverts blackish, tail below rufous.

P. c. mesura (Cabanis & Heine, 1863). Colombia east of Andes, east Ecuador. Eye-ring red. Similar to *nigricrissa*, but black on lower belly and undertail-coverts less extensive.

P. c. circe Bonaparte, 1850. West Venezuela south of Lake Maracaibo (Hilty 2003). Above slightly more rufous than *mehleri* and paler than *cayana*.

P. c. insulana Hellmayr, 1906. Trinidad. Also reported Monagas, Venezuela (Cherrie 1916a). Like *cayana*, but undertail-coverts black.

P. c. obscura Snethlage, 1908. Amazonian Brazil south of Rio Solimões, from Rio Juruá to Rio Tapajós (Amazonas, Acre and Rondônia; Peters 1940, Novaes 1957, Pinto 1978) and Peru from Junín south-east to north Bolivia. Darker, less rufous, above, than *cayana*, undertail-coverts black.

P. c. hellmayri Pinto, 1938. Right bank of lower Rio Amazonas from Rio Tapajós eastward to Ilha de Marajó (Pará) and Maranhão, Brazil. Upperparts less rufous, more tan, than *cayana*.

P. c. pallescens (Cabanis & Heine, 1863). North-east Brazil from Piauí and Ceará to Rio Grande do Norte, Paraíba, Pernambuco, Alagoas and north Bahia and north Goiás, and possibly Tocantins (Hellmayr 1929, Naumburg 1930, Peters 1940, Lamm 1948, Pinto & Camargo 1961, Pinto 1964, 1978, Olmos 2003). Feet grey, soles yellowish (Forbes 1881). Upperparts pale rufous.

P. c. cabanisi J.A. Allen, 1893. Central-west Brazil (north and central Goiás, Mato Grosso and Mato Grosso do Sul) (Naumburg 1930, Pinto 1938, 1964, Peters 1940, Stager 1961), south-east Bolivia (Santa Cruz) and north-west Paraguay (Short 1975). Upperparts more rufous than in *pallescens*, throat rusty, belly and undertail-coverts light grey.

P. c. macroura Gambel, 1849. South-east Brazil (from south Bahia and south Goiás to São Paulo, Paraná, Santa Catarina and Rio Grande do Sul), Paraguay, Uruguay, north-east Argentina (east Formosa, east Chaco, Misiones, Corrientes, north Santa Fe, north and east Entre Ríos, north Buenos Aires, single specimen from east Córdoba) (Anon. 1937, Peters 1940, Rosillo 1943, Pinto 1964, 1978, Nores & Yzurieta 1979, Nores *et al.* 1983, Contreras *et al.* 1990, Sibley & Monroe 1990, Narosky & Di Giacomo 1993, do Rosário 1996, de la Peña 1997, 1999). Upperparts chestnut (Short 1975), belly and undertail-coverts blackish, tail long.

P. c. mogenseni J.L. Peters, 1926. South Bolivia (Santa Cruz, Tarija) and north-west Argentina (Jujuy, Salta, Tucumán, north-west Santiago del Estero, Catamarca, north Córdoba) (MacDonagh 1934, Anon. 1937, Nores *et al.* 1991, de la Peña 1994, 1999). Upperparts light rufous, throat rusty, belly and undertail-coverts grey.

DISTRIBUTION Neotropics; resident. West Mexico in south-east Sonora, Sinaloa, south-west Chihuahua, west Durango, west Zacatecas, Nayarit, west Jalisco, Colima, south Michoacán east to México, Guerrero, south Oaxaca, Isthmus of Tehuantepec, south Chiapas and south Guatemala to El Salvador and on Pacific slope to Panama, east Mexico in south Tamaulipas, east San Luis Potosí, east Puebla and Veracruz to Isthmus of Tehuantepec, Tabasco, north Oaxaca, north Chiapas, Yucatán Peninsula, Belize, north and central Guatemala and north Honduras south on Atlantic slope to Panama, and Colombia, Venezuela, Trinidad (not Tobago) and Guianas southward, west of Andes to north-west Peru and east of Andes to east Ecuador, east Peru, Brazil, Bolivia, Paraguay, north Argentina and Uruguay (Sibley 1996).

HABITAT Humid to semiarid and deciduous forest, including flooded lowland forest, thorn forest, pine-oak woodland in Mexico, riparian and gallery woodland, savanna, second growth, clearings and open and cultivated country with scattered trees, including palm and banana plantations, occasionally mangroves. To 2,400m Colima, Mexico; to 2,200m but most abundant below 1,500m

Costa Rica; to 2,500m Venezuelan Andes (usually below 1,200m), to 1,800m Venezuela south of Orinoco, and sometimes to 2,700m on Andean slopes Colombia, east Ecuador and perhaps west Ecuador and north-west Peru, and to *c.* 2,500m east Peru; to 2,350m Bolivian yungas, and lower from Cochabamba, Bolivia, to Catamarca, Argentina, but to 2,000m Tucumán, Argentina (Haverschmidt 1955, Schaldach 1963, Skutch 1966, Wetmore 1968, Hubbard & Crossin 1974, Olrog 1979, Skutch 1983, Hilty & Brown 1986, Fjeldså & Krabbe 1990, Howell & Webb 1995, Stotz *et al.* 1996, Ridgely & Greenfield 2001, Hilty 2003).

BEHAVIOUR In pairs all year. When foraging, mates usually straggle along one behind the other, keeping in contact by voice, often several trees apart. Combines deliberate motion and careful scrutiny with sudden darts and leaps up to a metre, and usually well concealed inside foliage; turns head slowly from side to side, scanning leaf surfaces. May beat prey against perch before ingestion. Rarely undertakes long flights. Works its way upwards in trees, jumping from branch to branch and hopping or running along horizontal or ascending branches. When reaching top of tree or hillside, makes short downward glide, uttering sharp metallic *kip* notes. Male presents insect or caterpillar to female in courtship, females grasps it, male mounts female while still holding insect, then male leans sideways, crossing his tail beneath hers, and copulation occurs. Pair may tug at insect if female does not let male mount. Nest building mostly late morning. Not known to collect nesting material on ground. A bird pulls off leaves or dead twigs with beak, jerking head quickly sideways, flies to low in nest tree, hops up through crown holding long twigs near one end. Presumed male brings material to female that, sitting on site, then inserts it beneath her. Female also collects a little nesting material. Male returns with material every 1–7min, catching and eating some insects in between. Leaves are added after incubation begins. Sometimes lays in nest of other female.

May join mixed bird flocks in forest canopy. Has been seen catching flying ants in sallies from treetop. A frog was beaten repeatedly on ground and log. Sometimes follows army ants, taking arthropods flushed by ants, mostly Nov–Dec at end of wet season, Panama; in forest interior, but liana tangles and dense foliage above old tree falls and second-growth canopy favoured. One taken by White-tailed Kite that plummeted from high up. Scratches head over wing. Sunbathes on cool mornings with back to sun, wings and tail spread. Night rest in canopy, sheltered among branches and vine tangles. Uses claws to preen auricular and superciliary regions (Sutton 1951, Skutch 1966, 1983, Wetmore 1968, Willis & Eisenmann 1979, Slud 1980, Willis 1983b, Cintra & Sanaiotti 1990, Sick 1993, Haverschmidt & Mees 1994, de Vasconcelos 1998, Jullien & Thiollay 1998, Oniki & Willis 1999).

BREEDING Season Birds in breeding condition collected early May–early Jun at north end of range in Sonora, Mexico. Laying female Jalisco, Jul. Eggs May, nest with young Jul, Oaxaca. Jul, Yucatán. Laying female El Salvador, Jul. Mainly Apr–Jun Costa Rica, also incubation Jan, and fledgling Oct, thus in both dry and wet seasons; fledgling Sep, Honduras. May, Jun, Panama. Jan (copulation), Apr (laying female), May (nestling, fledgling), Jul (eggs), Colombia. Apr–Sep Venezuela. Two nests with eggs Trinidad, Jul (Belcher & Smooker 1936), also breeding Jan, May and Oct; Mar–Jul and Sep Surinam; Apr and Jul–Sep Guyana; Mar and Jul–Aug French Guiana. Brazil: eggs and young Nov–Dec (south-east?), enlarged testes Feb and ovaries undeveloped Jul, São Paulo, eggs Jan, Mar, May and Jul, Pará; fledglings early Aug Amapá; season Sep–Nov, Mato Grosso; adult feeding nestlings Sep, Rio Grande do Sul. Egg in nest Sep, Bolivia (Allen 1893, Cherrie 1916a, W. Beebe in Snethlage 1928, Guimarães 1929, Dickey & van Rossem 1938, Hellebrekers 1942, Bond & Meyer de Schauensee 1943, Pinto 1953, Schäfer & Phelps 1954, Friedmann & Smith 1955, Selander & Giller 1959, Skutch 1966, Novaes 1974, Willis & Eisenmann 1979, ffrench 1980, Belton 1984, Rowley 1984, Hilty & Brown 1986, Binford 1989, Tostain *et al.* 1992, Russell & Monson 1998, Magalhães 1999, Payne 2005). **Nest** 3–20m above ground, two 75–100cm up in tangle of bracken and straggling composite, or dense vine inside crown of isolated tree, in hedge, thick shrub or clump of tall bamboo, usually well concealed by foliage; loose shallow platform of twigs up to at least 25cm, supporting thick mass of green and yellowing leaves, or rootlets; 15–60cm in diameter and 6–30cm high, with shallow depression in top *c.* 9cm wide (Cherrie 1892, von Ihering 1902, Stone 1928, Belcher & Smooker 1936, Skutch 1966, 1983, Rowley 1984, Oniki & Willis 1999). **Eggs** 2–3; opaque white, elliptical, 31.2–36.5x22.4–27.3mm. Two eggs laid with less than 23 hour interval. Stained brown by leaves in nest (Cherrie 1892, Todd & Carriker 1922, Skutch 1966, 1983, Schönwetter 1967, Howell & Webb 1995). **Incubation** C.19 days. Apparently begins when first egg laid, but somewhat irregular first two days; male apparently at night, beginning well before dark and ending well after sunrise. Later, incubation continuous, with only two changeovers each day, one parent, probably female, staying 7–8 hours during day. Tail may be held upward at sharp angle (Skutch 1966, Oniki & Willis 1999). **Chicks** Large mangled insects or larvae, including caterpillars with stinging spines, are fed about once/hour by both parents. Parents carry droppings away from nest, which nevertheless becomes somewhat soiled. After a week, nestling is feathered and may stand on nest rim and flap wings, and take food on or beside nest. Parents brood nestlings all night, early morning, late afternoon, and when it rains, until they leave nest. Fledging gradual, with some days spent moving much around, and resting in, vicinity of nest, often returning (Skutch 1966). **Survival** Unknown.

FOOD Mainly caterpillars (smooth green and hairy stinging), and winged insects, chiefly grasshoppers and stick insects (up to at least 10cm long), also mantids, Hemiptera (including Pentatomidae and treehoppers), beetles, moths, wasps, winged ants, bees (Apoidea and large stingless Meliponidae), dragonfly (Libellulidae), cicadas; millepedes; spiders; tree-frog killed but abandoned; geckos and other lizards; Indigo Bunting in mistnet partially eaten; small fruits (Todd & Carriker 1922, Dickey & van Rossem 1938, Moojen *et al.* 1941, Blake & Hanson 1942, Friedmann & Smith 1950, Schubart *et al.* 1965, Skutch 1966, 1983, Cintra & Sanaiotti 1990, Haverschmidt & Mees 1994, de Vasconcelos 1998, Piratelli & Pereira 2002, Komar & Thurber 2003).

STATUS AND CONSERVATION Generally common (Stotz *et al.* 1996). Uncommon Trinidad (Herklots 1961). Rare Santa Fe and Entre Ríos provinces (de la Peña 1997), and Buenos Aires (Narosky & Di Giacomo 1993), Argentina. C. 4 pairs/100 ha in floodplain forest in south-west Amazonia (Terborgh *et al.* 1990). Not globally threatened (BirdLife International 2011).

Squirrel Cuckoo *Piaya cayana*. Adults. **Fig. 1.** *P. cayana cabanisi*, Pantanal, Brazil, August (*Laurens Steijn*). **Fig. 2.** *P. c. thermophila*, Canal Zone, Panama, November (*Mike Danzenbaker*). **Fig. 3.** *P. c. nigricrissa* Pichincha, Ecuador, November (*Mike Danzenbaker*). **Fig. 4.** *P. c. mesura*, Wildsumaco Lodge, Napo, Ecuador, December (*Roger Ahlman*). **Fig. 5.** *P. c. mesura*, San Isidro Lodge, Napo, Ecuador, November (*Roger Ahlman*).

Genus *Coccyzus*

Vieillot, 1816 *Analyse d'une nouvelle Ornithologie élémentaire* p. 28.
Type by monotypy, 'Coucou de la caroline' Buffon = *Cuculus americanus* Linnaeus 1758.
Nine species.

New World; mostly non brood-parasitic. Small to medium-sized; crestless; sexes alike. Based on mtDNA data, *pumilus* and *cinereus* placed in *Coccycua* Lesson 1830, and *Coccyzus* considered to include *Hyetornis* and *Saurothera* (Sorenson & Payne 2005), latter followed in AOU Checklist (Banks *et al.* 2006). Includes *Micrococcyx* (Ridgway 1912) and *Coccygus* (Cabanis & Heine 1863).

DWARF CUCKOO
Coccyzus pumilus   Plate 17

Coccyzus pumilus Strickland, 1852 (Trinidad, error = Ciudad Bolívar, Venezuela (Phelps & Phelps 1958)

TAXONOMY Placed in *Coccycua* Lesson 1830 (Sorenson & Payne 2005), or *Micrococcyx* Ridgway 1912 (e.g. Cherrie 1916a, Hughes 2006). Forms superspecies with *C. cinereus* (Short 1975, Sorenson & Payne 2005). Monotypic.

FIELD IDENTIFICATION 20–22cm. **Adult** Cap grey, face, throat and breast chestnut, belly to undertail whitish to buff, back greyish-brown; rounded tail grey with black distal half tipped white; bill black, iris reddish-brown. **Juvenile** Pale grey throat, obscure or lacking tail tipping, brown eyes and yellow eye-ring. **Similar species** No other *Coccyzus* cuckoo has chestnut throat and breast, and tail is short for genus.

VOICE Grating call *trrr trrr trrr...*, normally 4–25 notes, *c.* 2/sec, given perched and in flight, both when inside foliage and in top of tree, when foraging, when moving to or from nest and on nest. Dawn call *kööa kööa*. Disturbance calls low *cluck*, repeated at *c.* 11sec intervals and accompanied by shake of body; and complaining, long *mew*. Begging creak of nestlings and fledglings low, grating *eeeeee* (Ralph 1975, Thomas 1978, Hardy *et al.* 1990).

DESCRIPTION Adult Crown grey, blending into smooth brownish-grey mantle and lower back, wings and tail browner, primaries blackish, especially distally; face, chin, throat and upper breast chestnut or dull rufous, sides of body and undertail-coverts buff, rest of underparts white, tail seemingly rounded, not obviously graduated as in most *Coccyzus*, grey with black distal half and narrow, pale tips. **Juvenile** Upperparts and remiges with tannish cast, throat pearly-grey. **Nestling** Eyes closed at hatching. Skin dark grey, tinted yellowish-green, dorsal tracts with scattered tan trichoptiles, otherwise naked. Mouth lining bright pink with fine black lines and pattern of white bumps and spots with serrations directed backwards. **Bare parts** Iris bright red; olive-brown in juvenile. Eye-ring bright red; dull yellow in juvenile; eye-lids carmine. Bill black, rather thin, slightly curved; mouth lining black. Feet slate-grey.

BIOMETRICS Wing male 92–112mm (n = 17), mean 102.2 ± 5.5, female 97–110mm (n = 10), mean 102.3 ± 4.2. Bill male 16.2–23.0mm, mean 18.9 ± 1.8, female 16.9–21.0mm, mean 19.1 ± 1.7. Tail male 95–111mm, mean 105.2 ± 7.9, female 94–122mm, mean 106.0 ± 9.2. Tarsus male 17.8–24.5mm, mean 20.6 ± 2.1, female 17.5–23.3mm, mean 20.3 ± 2.2. Mass male 45.7g, female 39.1g (Payne 2005), unsexed 36g (Ralph 1975), 33.2g (Thomas 1990).

MOULT Adults sometimes moult body feathers during nest-building and incubation, and tail and remiges during all stages of nesting. Juvenile 41–49 days old had scattered rusty feathers on throat, but still had juvenile brown iris. Irregular transilient primary moult (Stresemann & Stresemann 1961, Ralph 1975).

DISTRIBUTION Colombia and Venezuela in semi-arid regions. Breeding and non-breeding range, as well as migratory movements, poorly known. Has been recorded Valle (including Isla Punta Arenas on west coast), Risaralda, Atlántico, north Magdalena, La Guajira, central Magdalena valley in south César, Cundinamarca, Tolima, east of Andes in Norte de Santander, north Arauca, Meta and west Caquetá, Colombia; Venezuela: Zulia, Lara, Táchira, Apure, Cojedes, Guárico, south Anzoátegui, Monagas, Delta Amacuro, west Amazonas, north-west Bolívar and Isla Margarita. Sight Roraima, north Brazil, and Panama. Only present during rainy season, Apr–Nov, in central Venezuelan savanna, but resident Valle, Colombia (von Berlepsch & Hartert 1902, Cherrie 1916a, Meyer de Schauensee 1949, Phelps & Phelps 1958, Nicéforo & Olivares 1966, Thomas 1978, 1979, 1990, Naranjo 1982, Hilty & Brown 1986, Braun & Wolf 1987, Whittaker 1995, Hilty 2003). Spread south into Magdalena and Cauca valleys, Colombia, during 20th century, with forest clearance (Ralph 1975), recorded on western edge of Amazon forest in Caquetá (Nicéforo & Olivares 1966) and breeding on wet Pacific coast near Buenaventura, Valle (Ralph & Chaplin 1973).

HABITAT Bushes and trees of wet savanna flooded by rains May–Oct, gallery forest, parks and gardens in towns, tropical deciduous forest, secondary forest, secondarily in tree-lined pastures, orchards, cleared forests, citrus groves, coffee

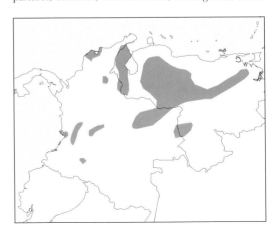

plantations. To 2,000m, Colombia; once 2,600m (Oct) eastern Andes. To 400m north of Río Orinoco, Venezuela, to 200m south of it (Borrero 1946, Ralph 1975, Thomas 1978, 1979, Naranjo 1982, Hilty & Brown 1986, Stotz et al. 1996, Hilty 2003).

BEHAVIOUR Sluggish, inconspicuous, usually solitary. When foraging, peers slowly at both outer and inner foliage of small trees and bushes, descends to clear ground under trees and enters dense grass. Usually monogamous, but female may mate with two males simultaneously, and both males help rearing young. In courtship display, male brings caterpillar or furry moth to female, both may give stuttering *trr* calls as they approach each other, and may perch side by side on horizontal branch and tug at insect for up to 2 minutes. If female receptive, crouches slightly, holds quivering wings out from body. Male then hops onto her back, letting her grasp insect so that both hold it, and after few to 25sec, male pumps and rotates his tail and slides down to side until achieving cloacal contact. This position held for a few seconds, male dismounts and usually lets female eat insect, sometimes followed by some more stuttering *trr* calls and short chase. Courtship feeding occurs in all stages of nesting and in between. Home range in pasture with rows and clumps of trees *c.* 10ha, broadly overlapping with neighbours, and pairs sometimes nest down to 40m from each other. Area within 10–20m from nest defended, intruders supplanted or chased, but incubating bird ignores alien conspecifics in nest tree. Smooth-billed Ani attacked and chased away; pair of Scaled Doves and pair of nesting Pied Water Tyrants were tolerated. Only two aggressive encounters were observed with pair of Great Kiskadee nesting less than 60cm away, during 23 days of observation. Reaction to Smooth-billed Ani: fluffs feathers, hunches back, drops wings and fans tail, while giving choked-up, gagging *trr*, facing away from or toward ani. If mate nearby when aggressive encounter begins, participates in confrontation. Nest building continues after eggs are laid, especially during early incubation, tapers off and ends in first days after eggs hatch. Twigs usually brought by bird arriving to relieve mate. If incubating or brooding bird does not go off nest, arriving bird, often male, gives twig to mate who adds it to nest. In intense afternoon nest-building bout, male brought eight twigs in 18 minutes and five twigs in 12 minutes. Nest and ground and foliage below splattered with droppings of incubating birds and nestlings. Male incubates and broods at night. If eggs lost, usually renests. After successful brood, sometimes begins another immediately; nesting cycle down to 55 days, but sometimes with non-breeding intervals of up to *c.* 3 months, not synchronised among pairs, although breeding may be (more) synchronised, during rainy season, central Venezuela (Ralph 1975, Thomas 1978, Naranjo 1982, Hilty & Brown 1986).

BREEDING Season At least Dec–Sep, probably year-round, Colombia. Jun–Aug Venezuela (Friedmann 1948a, Ralph 1975, Thomas 1978, 1979, Naranjo 1982). **Nest** Flimsy platform of tendrils and twigs (length 6–13cm, diameter 1–3mm), sometimes lined with 1–2 dry leaves, or without lining but cup reinforced by finer twigs, diameter 11–18cm, height 25mm, 1–6.5m up (n = 29), mean 2.6m, usually in bush or crown of small tree, sometimes on mistletoe or small bromeliads, or terminal fork of horizontal branch, often in thorny tree and/or near wasp's nest; nest tree usually isolated or facing open area, with view of area around tree from nest (Ralph 1975, Thomas 1978, Naranjo 1982). **Eggs** 2–3, mean 2.2 (n = 24). Unmarked glossy white with dull white layer that wears off, oval; mean 24.3x19.2mm (n = 15). First egg laid 2–4 days after onset of nest construction, when nest is still rather scanty. Eggs laid with 1–3 day intervals; female laid two eggs in each of two nests, in six days. Sometimes lays in nest of other female (Ralph 1975, Naranjo 1982). **Incubation** Both parents incubate, begins when first egg laid, lasts *c.* 13 days, parents keep nest covered almost continuously, sessions 30–156min, mean 77min (n = 8) (Ralph 1975). **Chicks** Asynchronous development. Shading, brooding and feeding nestlings during day equally by both parents. Brooding sessions 3–75min, mean 23min (n = 41). Each adult visits nest with food for nestlings 0–3 times/h, 1–6 total feedings/h, mean 3.0, each nestling fed 0.4–4.5 times/h, mean 1.8. When begging, young nestling quivers wings, and begins to creak when 1–4 days old. Pin feathers on body, wings and tail, and can hold on to twigs with feet by day 4. Eyes almost open on day 5. Can stand and defecate over rim of nest, and some feathers erupt from sheaths, on day 6. First preening, scratching, panting and gular fluttering on day 7. May sit on side of nest, unsteadily, on day 8–9, and gape conspicuously yellowish. Well feathered on day 10, and begins to leave nest, e.g. when disturbed, and stays 4–11more days climbing around nest tree, until fully volant. Still has contrasting mouth marks day 11. On day 13, attains 65% of adult body mass (Ralph 1975, Naranjo 1982). **Survival** Nest success in Colombia 56% (n = 25); predation caused 57% of egg and nestling mortality (Ralph 1975).

FOOD Adults mostly caterpillars (13–51mm); *c.* 1/3 prey items treehoppers, mostly nymphs, but latter have average body mass only half of caterpillars; few distasteful lightning bugs (lampyrids) and other beetles. Nestlings are fed many, probably almost exclusively, caterpillars (Ralph 1975).

STATUS AND CONSERVATION Fairly common Colombia (Hilty & Brown 1986) and Venezuela, but scarce or locally absent during dry season (Hilty 2003). Not globally threatened (BirdLife International 2011).

Dwarf Cuckoo *Coccyzus pumilus*. **Fig. 1.** Adult, Hato El Cedral, Los Llanos, Venezuela, January (*João Quental*).

ASH-COLOURED CUCKOO
Coccyzus cinereus Plate 17

Coccyzus cinereus Vieillot, 1817 (Paraguay)

TAXONOMY Forms superspecies with *pumilus* (Short 1975). Sometimes placed in *Micrococcyx* Ridgway 1912 (e.g. Cory 1919, Pinto 1938, Hughes 2006), or *Coccycua*, based on mtDNA sequences (Sorenson & Payne 2005). Monotypic.

FIELD IDENTIFICATION 21–24cm. **Adult** Pale ashy-brown above, throat and breast light grey-brown, contrasting with whiter belly, tail not strongly graduated, with small black subterminal area and narrow white tips. Bill rather short, curved, black, eye and eye-ring red. **Juvenile** Browner above, greyer on throat and breast, white tail tips lacking, eyes brown. **Similar species** Tail shorter, and without large white spots, than in Dark-billed, Black-billed, Pearly-breasted and Yellow-billed Cuckoos. Black-billed Cuckoo has white (not grey) throat and breast. Similar to smaller juvenile Dwarf Cuckoo (no range overlap), but latter has yellow eye-ring.

VOICE Mournful series of up to 15–18 notes *cow-w cow-w cow-w...* , each note *c.* 0.3sec long, falling from 1.0 to 0.8kHz, 2 notes/sec for 9–12sec; also given by female. Also loud harsh call (Hudson 1920, Wetmore 1926, Hardy *et al.* 1990, Payne 2005).

DESCRIPTION Adult Crown and sides of head grey, back and mantle greyish-brown, remiges brown; throat and breast ashy white or light grey-brown, rest of underparts white with slight greyish-buff wash on undertail-coverts and flanks, axillaries, underwing-coverts and basal section of inner webs of primaries buff, rest of primaries ashy-brown; tail only slightly graduated, above ash-brown shading into blackish towards end, below dull brownish-grey with black subterminal area and narrow white tip on each feather except T1. **Juvenile** Crown and back brown tinged rufous, greyer on throat and breast than adult; tail brown, tips pointed and lacks white. **Nestling** Hatchling has eyes half open, skin dark greenish, down long, pale grey, mouth rosy, with white papillae, two pairs around pharynx, one or two on tongue and five or six on palate (Wetmore 1926, de la Peña 2005). **Bare parts** Bill black with silver base of mandible. Mouth interior and tongue black. Eye-ring and iris red, iris brown in juvenile. Feet silvery-grey to black. Claws black (female Argentina; Wetmore 1926).

BIOMETRICS Wing male 97–111mm (n = 11), mean 106.8 ± 3.8, female 104–116mm (n = 9), mean 109.6 ± 4.3. Bill male 17.6–23.0mm, mean 20.4 ± 2.2, female 17.5–24.0mm, mean 21.7 ± 2.2. Tail male 98–114mm, mean 106.0 ± 5.4, female 105–114mm, mean 109.4 ± 4.2. Tarsus male 17.5–23.3mm, mean 20.9 ± 1.9, female 19.0–23.0mm, mean 21.0 ± 1.5. Mass male 34.8, 37.3, 39.0g, female 43.0–57.0g (n = 5), mean 45.4g (Contreras 1979, Belton 1984, Payne 2005).

MOULT Wing moult Jun, Paraguay. Irregular transilient primary moult. Remiges moulting Apr, Argentina (Wetmore 1926, Stresemann & Stresemann 1961, Short 1976).

DISTRIBUTION South America. North and east Bolivia (Pando, Beni, Cochabamba, Santa Cruz, Tarija), Paraguay (Alto Paraguay, Presidente Hayes, Boquerón, Misiones, Ñeembucú and Alto Paraná); very locally Brazil, Uruguay, north Argentina (Formosa, Chaco, Misiones, Santiago del Estero, Salta, Tucumán, Catamarca, La Rioja, Córdoba, Santa Fe, Corrientes, Entre Ríos, Buenos Aires, San Luis, La Pampa, Río Negro). Visitor, austral winter, lowlands, south-east Peru, very rarely further north to Amazon river (sight records, mostly Jul–Sep). Sight record extreme south-east Colombia, Jul. Resident Paraguay, or possibly migratory. Migratory southern part of Argentinian breeding range. In north possibly resident Chaco province (Mar, May, Jul, Oct), but migratory at study sites in Formosa and Salta, and in provinces Santa Fe, Entre Ríos, Santiago del Estero, San Luis and Buenos Aires; mostly Oct–Apr Córdoba. Probably migratory Uruguay. Presumed resident Rio Grande do Sul, Brazil, although not confirmed; eight of nine records Nov–May, including copulation and oviduct egg Nov, and recently possibly transients (Mar and Nov). No nests or eggs Brazil; recorded Dec Santa Catarina; not recorded Paraná; Jan–Mar São Paulo; Feb Mato Grosso do Sul; Sep and Oct Distrito Federal; Jun Rio de Janeiro; May Mato Grosso and Goiás; Apr Bahia; Aug Paraíba; wet season Ceará; Jul Amazonas; Jun and Jul Rondônia. Bolivia: Beni, May and other dates?; Pando, Jul; Cochabamba, sight Feb; Santa Cruz, Jan, Mar, Apr, Nov and other dates?; Tarija, Feb (Salvadori 1897, Reiser 1926b, Anon. 1937, Laubmann 1939, Bond & Meyer de Schauensee 1943, Rosillo 1943, Gyldenstolpe 1945a, Steullet & Deautier 1945, Cuello & Gerzenstein 1962, Meyer de Schauensee 1966, Short 1975, 1976, Gore & Gepp 1978, Pinto 1978, Parker 1982, Nores *et al.* 1983, Belton 1984, Willis & Oniki 1985, 2003, Hilty & Brown 1986, Narosky & Yzurieta 1987, Capurro & Bucher 1988, Contreras *et al.* 1990, Nacinovic *et al.* 1990, Nores *et al.* 1991, Forrester 1993, Narosky & Di Giacomo 1993, Nellar 1993, Pacheco 1993, Sick 1993, Parker *et al.* 1994, Whitney *et al.* 1994, Arribas *et al.* 1995, Hayes 1995, Scherer-Neto & Straube 1995, Straube & Bornschein 1995, do Rosário 1996, de la Peña 1997, 1999, Stotz *et al.* 1997, Maurício & Dias 2000, Azpiroz 2001, Bencke 2001, Carvalhães 2001, Clements & Shany 2001, Parker & Hoke 2002, Claramunt & Cuello 2004, Di Giacomo 2005, Guyra Paraguay 2005, Lopes *et al.* 2005b, Payne 2005, Schulenberg *et al.* 2007,

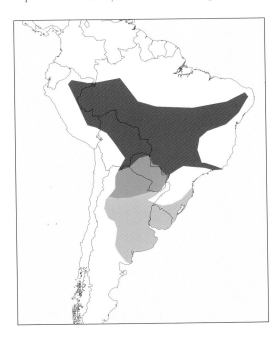

Azpiroz & Menéndez 2008, Medcraft 2009, de Godoy 2010, FMNH, MZUSP).

Reportedly breeds Bolivia (Hilty & Brown 1986), but in spite of records from supposed austral breeding season, documentation is lacking (e.g. Payne 2005). Thought to have extended range southwards and become much more common Argentina (Sclater & Hudson 1889, Hudson 1920).

HABITAT Chaco, wooded areas in pampas, chaco scrub, dry woodlands, savannas, gallery forest. To 900m; only to 500m Bolivia (Short 1975, Arribas *et al.* 1995, Hayes 1995, Stotz *et al.* 1996, Guyra Paraguay 2005).

BEHAVIOUR Shy, difficult to see, solitary, sometimes in pairs, stays hidden in foliage at 2–5m in bushes and thickets, possibly higher up in trees in austral winter in Amazonia, sometimes joins mixed-species bird flocks in autumn. May sometimes perch in sun in top of tree. Flight usually low over ground, and direct. Sunbathes. Calls all day long during breeding season (Sclater & Hudson 1889, Goodall 1923, Wetmore 1926, Parker 1982, Willis & Oniki 2003, Di Giacomo 2005, de Godoy 2010).

BREEDING Season Enlarged gonads Dec and Jan, west Paraguay. Enlarged testes Jan, Entre Ríos. Nests, eggs and nestlings Oct–Dec (once Mar) Santa Fe, Argentina, nests Nov–Dec Formosa, in summer Entre Ríos. Nests with eggs Dec, La Pampa. Copulation, and female with shelled oviduct egg Nov, Rio Grande do Sul (Pereyra 1937, Short 1976, Contreras 1979, Belton 1984, de la Peña 1997, 2005, Di Giacomo 2005). **Nest** Built in 6–8 days. Simple platform of sticks, sometimes thorny, and herbaceous and vine stalks, lined with grass straws, leaf ribs, leaves and/or petioles, 0.75–5m up in bush or tree, in fork, exposed or semi-hidden, sometimes covered by vine, in open bushy areas. Outer diameter 10–15cm, internal diameter 8cm, height 3.0–8.5cm, internal depth 0.5–3.5cm (n = 19) (Goodall 1923, Rosillo 1943, de la Peña 2005, Di Giacomo 2005). **Eggs** 3–4 (2) Argentina; laid on successive days; opaque/unglossed white, elliptical; 24.4–26.4x18.3–21.5mm (n = 13), mean 25.4x19.3mm; mass 4.8, 5.4, 5.7g; 4.0–5.4g (n = 8), mean 4.7g. Shells of hatched eggs not removed from nest (Hartert & Venturi 1909, Goodall 1923, Pereyra 1937, Schönwetter 1967, Short 1975, de la Peña 2005, Di Giacomo 2005). **Incubation** 12–13 days. **Chicks** Mass at hatching 3.6g. Day 3, 11.7g, eyes open, brown, feet dark, bill dark grey, gape paler, skin dark, scarce pale grey down, grey primary sheaths; calls. Day 6, 16.0g; white feathers on underparts, flanks cinnamon, brown head, back, wing-coverts and tail. Day 7, black spots on tongue; throws head back until it touches back. Day 8, mass 18.7g, fully feathered, bill black, upperparts brownish-grey, throat and breast pale grey, abdomen ochraceous whitish, wings brown, tail brownish. Day 9, in full juvenile plumage, nest very dirty with faeces. Fledge at 13–15 days. Parents feed nestlings caterpillars, other insects and spiders (de la Peña 2005). **Survival** Unknown.

FOOD Caterpillars, beetles and other insects (Aravena 1928, Payne 2005).

STATUS AND CONSERVATION Generally uncommon. Medium sensitivity to habitat disturbance (Stotz *et al.* 1996). Rare south-east Peru (Schulenberg *et al.* 2007). Rare and rather local Paraguay (Hayes 1995, Guyra Paraguay 2005). Thought to be scarce Uruguay (Gore & Gepp 1978), but many recent records (Azpiroz & Menéndez 2008). Scarce Formosa (Di Giacomo 2005), Santa Fe and Entre Ríos (de la Peña 1997). Fairly common Santiago del Estero (Nores *et al.* 1991). Apparently rare, but possibly under-recorded, Chaco province (Contreras *et al.* 1990). Rather scarce Córdoba (Nores *et al.* 1983). Less than 0.15 individuals/ha at study site in Salta, Argentina (Capurro & Bucher 1988). Not globally threatened (BirdLife International 2011).

Ash-coloured Cuckoo *Coccyzus cinereus.* **Fig. 1.** Adult, Barão de Melgaço, Mato Grosso, Brazil, March (*João Quental*).

DARK-BILLED CUCKOO
Coccyzus melacoryphus   Plate 21

Coccyzus melacoryphus Vieillot, 1817 (Paraguay)

Alternative name: Azara's Cuckoo.

TAXONOMY In error *melanocoryphus*. Monotypic. Synonyms: *Piaya melanorhynchus* Lesson 1830, *Coccygus seniculus* Wied 1832, *Coccyzus minor* Hartlaub 1847, *surniculus* Burmeister 1855.

FIELD IDENTIFICATION 25–28cm. Relatively small. **Adult** Dusky mask, buff underparts, prominent white tail spots, narrow pale grey band below cheeks, running down sides of neck and breast; back and wings greyish-brown, cap grey, bill black. Eyes may appear black in the field. **Juvenile** Brown crown and nape, sometimes some rufous in wings, coverts with dull buff tips; narrow pointed rectrices with indistinctly tipped grey. **Similar species** Grey-capped Cuckoo slightly larger with grey contrasting more with rufous, not brown, upperparts; richer rufous, not buff, below, and lacks dusky mask and grey band on side, giving impression of less contrasting uniformly rufous body. Larger Mangrove Cuckoo (limited range overlap) has yellow on lower mandible and warmer brown upperparts.

VOICE Infrequently given *cu-cu-cu-cu-cu-kolp, kolp, kulop*, sometimes without last three notes, not loud, 6–9 notes total, rising then falling slightly in pitch, each note lasting 0.2–0.3 sec, 3 notes/sec, with intervals up to several minutes, resembling Yellow-billed Cuckoo, but slower, shorter and weaker. Low rattle *dddddrr* or purring *grrrr*, and a *graa-ak*. Also harsh rapid *charr-charr-chao-chur-chur* and *cherrrroouuuu*. Nestling gives shrill *peep* note (Friedmann 1927, Hilty & Brown 1986, Hardy *et al.* 1990, de la Peña & Rumboll 1998, Ridgely & Greenfield 2001, Hilty 2003, Payne 2005, Schulenberg *et al.* 2007).

DESCRIPTION Adult Crown and nape slate-grey, sometimes tinged brown, rest of upperparts greyish-brown, underwing-coverts light buff, inner webs of remiges becoming light buff on edges, except distally; lores and auricular region slaty-black, pale grey band from lower cheeks down sides of neck and breast, underparts buff, usually deepest on thighs; tail graduated, T1 like back, T2 black narrowly tipped white, other rectrices black, broadly tipped white. **Juvenile** Crown and nape brown, sometimes some rufous in remiges and wing-coverts, mostly on edges and tips; throat and breast paler than in adult, rectrices narrow and rather pointed, with indistinct grey tips with little contrast to bases. **Nestling** At hatching, skin orange, skin over eye dusky greenish-grey, white trichoptiles, gape white and swollen (Friedmann 1927), bill and feet grey, interior of mouth rosy with three white spots on palate and one on tongue (de la Peña 2005). **Bare parts** Iris dark brown. Eye-ring greyish or yellow. Bill and mouth interior black. Feet lead grey to black, bluish-grey or pale blue, claws black.

BIOMETRICS (Paraguay). Wing male 113–123mm (n = 9), mean 117.9 ± 3.2, female 115–123mm (n = 11), mean 119.2 ± 2.6. Bill male 21.2–24.0mm, mean 22.3 ± 0.8, female 20.2–24.5mm, mean 22.6 ± 1.5. Tail male 126–148mm, mean 135.6 ± 7.5, female 134–144mm, mean 137.2 ± 5.0. Tarsus male 22.8–26.1mm, mean 24.3 ± 0.9, female 23.2–25.2mm, mean 24.0 ± 0.6. Mass male 42–54g (n = 14), mean 45.6g, female 50–66.5g (n = 7), mean 54.2g (Payne 2005).

MOULT Juveniles may moult rectrices before next breeding season, Uruguay. Adults moulting head, neck and breast Apr, Tucumán, Argentina, along with asymmetrical primary moult. Body moult Jun, Venezuela. Female in worn plumage Aug, east Colombia. Irregular transilient primary moult (Wetmore 1926, Friedmann 1948a, Stresemann & Stresemann 1961, Olivares & Hernández 1962).

DISTRIBUTION South America and Galapagos Is. From Colombia (except west of Andes), Venezuela (patchily; including Islas Patos and Margarita), Trinidad (not Tobago) and Guianas southward; west of Andes to west Ecuador and west Peru and east of Andes including all of Brazil to Uruguay and to La Rioja, San Juan, Córdoba, San Luis, La Pampa, Buenos Aires, Neuquén and Río Negro, Argentina; possibly Mendoza. Resident Isabela, Fernandina, Santa Cruz, San Cristóbal, Floreana, Pinzón (uncommon), Santiago and Santa Fe (rare), Galapagos; also Plaza Sur (sight). Records associated with severe weather and unusual wind directions from Genovesa, Española, Champion and Daphne Major. Very rare visitor Trinidad (has bred); formerly resident. Apparently partly resident, partly austral migrant Colombia, most specimens Apr–Oct. Breeding resident Venezuela, and probably also austral migrant, May–Oct. Apparently breeds south-west Ecuador, where specimens in breeding condition Feb, but other records may represent austral migrants. Status western Peru uncertain, but probably rare and/or local resident, south to Arequipa. Austral migrant east Ecuador and Peru Mar–Oct, and east Bolivia (recorded Pando, Beni, La Paz, Cochabamba, Santa Cruz and Tarija). Jun–Aug Guyana. Probably only migrant Suriname and French Guiana, May–Sep. Recorded Apr–Sep Belém area, Pará, Brazil, with no indication of breeding, and considered austral migrant central Amazonia, with records May to Sep or Oct, but resident Rondônia. Sep–Apr, probably breeds, São Paulo, and most records Oct–Feb Rio de Janeiro and Rio Grande do Sul. Recorded on passage Bolivian chaco Apr and Oct–Nov, May and Sep north Bolivia. Breeding populations Argentina and Paraguay migratory (summer visitors); few birds remain Paraguay during austral winter (May–Aug). Arrives Argentine pampas Sep. Breeds Uruguay,

and only recorded austral summer (Vieillot 1817, Hudson 1920, Wetmore 1926, Belcher & Smooker 1936, Pereyra 1937, Gyldenstolpe 1945a, Junge & Mees 1958, Herklots 1961, Cuello & Gerzenstein 1962, Snyder 1966, Harris 1973, Short 1976, Gore & Gepp 1978, Meyer de Schauensee 1982, Millington & Price 1982, Nores et al. 1983, 1991, Belton 1984, Hilty & Brown 1986, Capurro & Bucher 1988, Curry & Stoleson 1988, ffrench 1991, Tostain et al. 1992, Narosky & Di Giacomo 1993, de la Peña 1994, 1999, Haverschmidt & Mees 1994, Hayes et al. 1994, Arribas et al. 1995, Cohn-Haft et al. 1997, de la Peña 1997, Sick 1997, J.F. Pacheco in Sick 1997, Stotz et al. 1997, de la Peña & Rumboll 1998, Novaes & Lima 1998, Azpiroz 2001, Ridgely & Greenfield 2001, Salaman et al. 2001, Jahn et al. 2002, Hilty 2003, Willis & Oniki 2003, Di Giacomo 2005, Guyra Paraguay 2005, Payne 2005, Veiga et al. 2005, Restall et al. 2006, Wiedenfeld 2006, Schulenberg et al. 2007, Vogt 2007, Kenefick et al. 2008, Quiñonez & Tello 2011).

Specimen Arica, north Chile, 1960, and photo, window strike, Santiago Jan 2011 (Philippi 1968, Johnson 1972, Barros & Schmitt 2011). Specimen Clipperton Island (Stager 1964). Accidental Grenada (Schwartz & Klinikowski 1965), Panama (Ridgely & Gwynne 1989), Falkland Is. (Bennett 1937), Tristan da Cunha (Brooke 1979). Specimen Texas, provisionally mentioned (AOU 1998), of uncertain provenance (Robbins et al. 2003).

HABITAT Lowland humid forest edge and river-edge forest (terra firme, várzea and gallery forest), scattered trees along fencerows, dry farmland, shrubby pastures, woodlots. Mangroves and shade trees in coffee plantations. To 400m, in mangroves, arid *Opuntia* zone and humid *Scalesia* zone, Galapagos. Generally to 1,200m or more. To 2,100m Colombia, with sightings to 2,400m. To 500m north of, and to 950m south of Río Orinoco, Venezuela. Mostly below 1,000m, occasionally to 1,900m or even higher, perhaps when migrating, Ecuador; reportedly to 2,800m. Once 3,600m, Peru, Oct. To 500m, rarely 2,500m, Bolivia (Ménégaux 1911, Chapman 1926, Lévêque 1964, Fjeldså & Krabbe 1986, Hilty & Brown 1986, Haverschmidt & Mees 1994, Arribas et al. 1995, Stotz et al. 1996, Ridgely & Greenfield 2001, Hilty 2003, Restall et al. 2006, Schulenberg et al. 2007, Vogt 2007).

BEHAVIOUR Graceful, calm; moves quietly, sometimes furtively. Solitary, sometimes in pairs. One remained immobile for long. May sit still openly on low branch or fence wires. Runs swiftly on ground. Flies low with rapid wing-beats. Calls from thicket. Also vocalises at night. Forages 3–6m up among leafy branches, or in dense bushes below 1.5m. Occasionally follows mixed-species bird flocks. Male feeds insect to female in courtship. Sometimes lays in nest of other female. Shells of hatched eggs not removed from nest (Gifford 1919, Hudson 1920, Wetmore 1926, Friedmann 1927, Eisentraut 1935, Novaes 1957, Lévêque 1964, Ralph 1975, Harris 1982, Jackson 1985, Hilty & Brown 1986, Sick 1993, Hilty 2003, Maldonado-Coelho & Marini 2003, Di Giacomo 2005, Restall et al. 2006).

BREEDING Season Jan–May, rainy season, Galapagos, some of both sexes with small gonads Apr, May, and no calling May–Jun (perhaps in year with shorter rainy season). Eggs Oct Trinidad. Both sexes breeding condition Oct, Colombia. Aug, central Venezuela, large gonads May (to Jul?), Venezuela. Enlarged testes Oct, nest and eggs Nov, Rio Grande do Sul, Brazil. Nest and enlarged testes Dec, south Bolivia. Argentina: active nests Oct–Jan Formosa, Dec Tucumán and Jan Entre Ríos, Oct–Feb Santa Fe, Dec–Apr Córdoba. Fully grown juveniles and parent feeding young Feb, Uruguay (Wetmore 1926, Gifford 1919, Friedmann 1927, Eisentraut 1935, Belcher & Smooker 1936, Freiberg 1943, Lévêque 1964, Thomas 1979, Harris 1982, Nores et al. 1983, Belton 1984, Hilty & Brown 1986, Ervin 1989, de la Peña 2005, Di Giacomo 2005, Payne 2005). **Nest** Flimsy, rather translucent platform in fork or on slender horizontal branch, more or less hidden in tree, spiny bush or dense thicket, often covered by vine, 1–4m up, built of spiny twigs, greenish beard lichens (*Usnea sulcata*) and herbaceous and vine stalks, lined with grass stems, moss, lichens, tendrils and/or petioles, outer diameter 8–15cm, height 4–8cm, inner depth 0.5–1cm (n = 22); cup diameter 5–6cm; construction 6–8 days. Twigs broken from trees, not taken on ground; nest so small and flat that eggs often fall out. Rim stained with parents' excrement (Gifford 1919, Hudson 1920, Goodall 1923, Friedmann 1927, Eisentraut 1935, Rosillo 1943, Ervin 1989, de la Peña 2005, Di Giacomo 2005). **Eggs** 3–4 (2–6, 5 unusual, 6 recorded twice – Argentina and Galapagos – in latter, sixth egg hatched six days after fifth, whereas four or five eggs hatched in just two days); pale greenish-blue, bluish-white, pale greyish-green or green, with chalky layer, oval; 28.0–32.0x21.5–24.0mm (n = 25), mean 30.2x22.8mm; 24.5–30.4x21.5–23.2mm (n = 13), mean 28.6x22.2mm, Argentina. Mass 6.6–8.4g, mean 7.3g (n = 13), masses of 5–egg clutch, Galapagos, 8.0–10.5g, mean 9.1g. Laid on successive days (Hartert & Venturi 1909, Gifford 1919, Hudson 1920, Goodall 1923, Friedmann 1927, Snethlage 1928, Eisentraut 1935, Belcher & Smooker 1936, Schönwetter 1967, de la Peña 1983, 1988, 1994, 2005, Belton 1984, Ervin 1989, Di Giacomo 2005). **Incubation** 7–12 days. Nest material also brought during incubation. Hatching over several days (Ervin 1989) or synchronous? (Hudson 1874). **Chicks** Mass at hatching 8.5–9.0g; eyes half open. Large size difference between nestlings. Eyes open 3rd day, mass 16.5–23.5g (n = 4), mean 20.4g. Three of those chicks weighed 27.5–32.5g, mean 29.7g, 6th day. When handled, defecate sticky brown liquid consisting of digested hawkmoth caterpillars. Fourth day shows feather sheaths, tail 1.5cm; 5th day skin dark grey; still has white spots in mouth. Fledge at 9–10 days, but may leave nest and climb around nearby branches from 6th day. Bittern-like pose with bill pointing up and neck stretched, near nest (Ervin 1989, de la Peña 2005, Di Giacomo 2005). **Survival** Unknown.

FOOD Cicadas, grasshoppers, stick insects, small beetles, moths, caterpillars (including hawkmoth larvae), ants; fruit (von Pelzeln 1870, Gifford 1919, Hudson 1920, Friedmann 1927, Belcher & Smooker 1936, Harris 1982, Klimaitis & Moschione 1987, Ervin 1989, Haverschmidt & Mees 1994, Piratelli & Pereira 2002).

STATUS AND CONSERVATION Fairly common across range (Stotz et al. 1996). Common on most larger islands and found on some smaller ones, Galapagos (Wiedenfeld 2006), and became much more common during 1982–83 ENSO event (Jackson 1985) of unusually wet weather with strong winds (Ervin 1989), and also bred on Genovesa, where does not normally occur, 1984 (Curry & Stoleson 1988). Fairly common Colombia (Hilty & Brown 1986) and Venezuela (Hilty 2003). Uncommon Guyana (Restall et al. 2006), Surinam (Haverschmidt & Mees 1994) and French Guiana (Tostain et al. 1992). Very rare Trinidad (Kenefick et al. 2008). Uncommon to locally fairly common austral migrant east of Andes, Ecuador; west of Andes believed to breed, perhaps locally, and occur as austral migrant; few records from central valley in north, also presumed austral migrants (Ridgely & Greenfield 2001). Uncommon east of Andes, and rare western Peru (Koepcke 1970, Clements & Shany 2001, Schulenberg *et al.* 2007). Relatively common Brazil (Sick 1997). Uncommon Paraguay (Hayes *et al.* 1994, Hayes 1995). Very common Uruguay (Gore & Gepp 1978). In Argentina, uncommon in Chaco province, but can survive in man-altered areas (Contreras *et al.* 1990); fairly common Santiago del Estero (Nores *et al.* 1991); common Buenos Aires province (Narosky & Di Giacomo 1993) and San Luis (Nellar 1993), generally scarce Córdoba, but in some years more frequent (Nores *et al.* 1983), once Neuquén (Veiga *et al.* 2005). Remains found at nest of Stygian Owl, central Brazil (Lopes *et al.* 2004). Not globally threatened (BirdLife International 2011).

Dark-billed Cuckoo *Coccyzus melacoryphus*. **Figs. 1–2.** Juvenile, Buenos Aires, Argentina, February (*Marc Guyt*). **Fig. 3.** Adult, Guatemala road, near Kourou, French Guiana, June (*Michel Giraud-Audine*).

YELLOW-BILLED CUCKOO
Coccyzus americanus   Plate 20

Cuculus americanus Linnaeus, 1758 ("In Carolina" = South Carolina)

TAXONOMY Western population sometimes separated as *occidentalis*, mainly based on slightly larger average size and slightly greyer crown and upperparts (e.g. Peters 1940, Hughes 1999b). However, biometric analyses (Banks 1988b, 1990, Franzreb & Laymon 1993) and mtDNA sequences (Pruett *et al.* 2001, Fleischer 2001, Farrell 2006) render diverging results. Payne (2005) did not recognise *occidentalis* because of considerable overlap, wintering birds cannot be traced to eastern or western breeding population, and lack of ringing recoveries from non-breeding range. More research needed. Most closely related to Pearly-breasted Cuckoo (Hughes 1999b, Sorenson & Payne 2005). Monotypic. Synonyms: *Cuculus dominicus* Linnaeus 1766, *carolinensis* Wilson 1811, *pyropterus* Vieillot 1817, *Cuculus cinerosus* Werner 1827, *flavirostris* Gloger 1854, *bairdi* Sclater 1864, *C. a. occidentalis* Ridgway 1887.

FIELD IDENTIFICATION 28–32cm. Medium-sized cuckoo, slender, long-tailed; like Black-billed Cuckoo hides in dense vegetation, flies low, with deep, loose wing-beats between cover, often gliding. **Adult** Plain brown above, rufous in primaries obvious in flight, underparts white, six broad white spots on underside of graduated dark tail, lower mandible mostly yellow, rufous wash on underwings. **Juvenile** Similar to adult, pale tail tips and outer edges, outer web of primaries and wing-coverts rufous, bill without yellow first two months. **Similar species** Most similar to smaller Pearly-breasted Cuckoo which has plain brown wings. Black-billed Cuckoo lacks rufous in wings, tail spots duller and much smaller, bill black, and adult has red eye-ring. Juvenile Black-billed has only faint rufous on inner webs of primaries, i.e. less rufous also than juvenile Yellow-billed (best mark when seen only in flight, and undertail and bill not seen). Mangrove Cuckoo has black mask and buff underparts. Dark-billed Cuckoo lacks yellow on bill, is buff below and less white under tail.

VOICE Most common call 8–12 note staccato, guttural *kuk-kuk-kuk* that slows and descends, ending *kakakowlp-kowlp*, usually at intervals of >10 min, most frequently early morning. Rapid, harsh rattle *kow-kow-kow-kow-kow-kow…*, up to *c.* 20 notes, 5–7 notes/sec, often given several times and by both members of pair when close to each other, including from nest. Series of 5–11 soft *coo* notes which resembles call of Greater Roadrunner, but quieter and higher-pitched, duration 3–7sec, given 4–6 times/minute with 7–10sec pauses in between, often for more than an hour, given with closed bill, and gular region inflating to size of golf ball with each note, by unmated males in pre-breeding season, calling bird exposing white breast, perched in open, high on dead snag. Softer *coo* notes from female during courtship. Rapid *cuk-cuk-cuk* given to newly hatched chick and when arriving at nest with food. Whining given with distraction display, alternating with rattle. Nestling from first day utters quick series of clicks *qua-a-a-a-a*, like faint, insect-like hiss when given by more than one nestling, becoming louder in 2–3 days. Begging call: quiet *cuk-cuk-cuk-cur-r-r-rrr*; after feeding, short *curr* or *cuk-currrr*. Older nestling gives croaking resembling call of grey tree-frog *Hyla versicolor*. Mostly vocal Texas late Apr–Jul. Not very vocal Arizona before Jul. Sometimes vocalises on spring migration and vocal Feb–Dec Puerto Rico, but not known to call in Costa Rica, Panama, Colombia, Ecuador or Peru (Bent 1940, Preble 1957, Hamilton & Hamilton 1965, Oberholser 1974, Kepler & Kepler 1978, Potter 1980, Scott *et al.* 1983, Hilty & Brown 1986, Bosque & Lentino 1987, Ridgely & Gwynne 1989, Stiles & Skutch 1989, Hughes 1999b, Wilson 2000, Ridgely & Greenfield 2001, Schulenberg *et al.* 2007). 'Singing' birds in South America (e.g. Sick & Pabst 1968) may be *C. euleri*.

DESCRIPTION Adult Head to below eyes, upper side of neck and entire upperparts, wings and middle pair of rectrices plain brown, base of primaries and secondaries on inner web rufous, wing underside plain brown with large rufous area and buffy-white underwing-coverts; underparts from throat to undertail-coverts white, lower flanks pale grey; graduated tail except middle pair of rectrices blackish-brown with large (2cm) white tips, outermost also with white outer web. **Juvenile** Like adult but duller, throat and breast tinged greyish-buff, outer web of primaries rufous, wing-coverts tipped cinnamon brown, tail below dark grey, rectrices narrower and tapered distally, outer rectrices dusky above, olive brown below, tips below dull white and not so sharply contrasting; bill smaller. Suspected hybrid with Black-billed Cuckoo (Parkes 1984) is juvenile male Yellow-billed (Payne 2005). **Nestling** Skin and hair-like sheathed down feathers on dorsal surface and thighs grey, unlike Black-billed Cuckoo which has black skin and white sheathed feathers. Frontal apterium light grey, paler than surrounding skin and bill. Bill light slate-blue, upper mandible tipped white, lower tipped black. Cutting edges white, gape including unswollen corners red with white disc on tongue which has black tip, palate with white pads that grow and spines that increase in number as nestling grows. Feet light slate-blue. Fledgling's bill rather stubby, upper mandible dark grey, lower light grey becoming darker towards tip. Yellow on bill appears at *c.* 2 months. **Bare parts** Iris brown, also in juvenile, fledgling grey. Bill black, yellow to orange yellow on base of cutting edges of upper mandible and on lower mandible, latter tipped black. Eye-ring grey or yellow in breeding birds, grey at fledging, yellow in juvenile Sep–May, then grey. Eye-ring of female changed from grey to yellow during Jun–Jul nesting (S. McNeil *in litt.*). Feet grey. Claws brownish-black.

BIOMETRICS Wing male 126–142mm (n = 11), mean 138.8 ± 5.8, female 139–150mm (n = 12), mean 144.2 ± 3.6. Bill male 22.5–28.7mm, mean 25.6 ± 1.7, female 23.2–27.5mm, mean 25.5 ± 1.1. Tail male 138–156mm, mean 145.5 ± 5.9, female 132–157mm, mean 147.3 ± 7.5. Tarsus male 22.3–25.4mm, mean 23.9 ± 0.8, female 23.4–25.7mm, mean 24.4 ± 0.9. Mass, breeding season, male 42.8–70.9g (n = 11), mean 58.3, female 53.6–78.0g (n = 10), mean 65.3. Up to 110g when fat before migrating, and down to 31g after long overwater flight (Payne 2005).

MOULT Adults moult Sep–Mar, beginning in breeding area with head and body, often retaining some secondaries and rectrices. Juvenile body moult Aug–Oct. Remiges and rectrices mostly moulted in winter quarters. Some individuals retain some juvenile flight and tail feathers until following summer or autumn. Several alternate primaries, never neighbouring ones, grow simultaneously, sequence irregular (Partridge 1961, Stresemann & Stresemann 1961, Potter & Hauser 1974, Cramp 1985, Hughes 1999b).

DISTRIBUTION Breeds North America, Mexico and

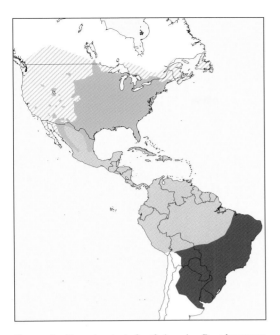

Migrates through southern USA, Central America, Bermuda and West Indies. Regular autumn Nova Scotia. Passage east Mexico Apr–mid Jun and Aug–early Dec; uncommon transient interior and west Mexico. Guatemala Apr–May and Sep–Dec. Common Apr–Jun and Sep–Nov Honduras. Costa Rica Sep–mid Nov and Apr–late May, Panama Sep–Dec, peak Oct, and Apr, rarely to mid–May. Common transient Oct south Bahamas. Caribbean breeders migrate south Sep–Oct. Common spring and autumn Hispaniola, mid Mar–2nd week May (peak last week Apr and first two weeks May), early Sep onwards, peak Oct, latest 24 Nov. Sep–Nov Cuba. Aug–Nov and late Apr–Jun, once Jan, Cayman Is. Northward passage Greater Antilles Mar–Apr, sometimes in large numbers. Uncommon Jamaica Apr–May, rare Virgin Islands and Lesser Antilles. Early Oct–early Nov, perhaps also later, Barbados. Apr–early Jun and mainly mid Oct–early Nov, and few Aug, Dec–Feb, Netherlands Antilles, where sometimes hundreds in autumn, often arriving in poor condition, weighing as little as 29g. Apr–May and Oct–Dec Trinidad & Tobago. Colombia: passage Sep–Dec in north, Oct–Nov and Apr–May Cauca, Apr passage eastern Andes, few linger to Jun central and north Colombia. Oct–May north-west Amazonia; Feb record Nariño. Venezuela: late Aug–early May except south-east, most transients Sep–early Nov and late Mar–mid May, few Jun. Many cross high Andean passes Oct, Dec and Apr–May; rare in north in winter. Only transient Guyana? Oct and Feb–Apr French Guiana. Oct–Nov and Apr–Jun Suriname. Mar–Apr and Sep–Dec Ecuador (Lönnberg & Rendahl 1922, Todd & Carriker 1922, Chapman 1926, Wallace 1958, Snyder 1966, Monroe 1968, Land 1970, Lack 1976, Russell 1980, Voous 1983, Hilty & Brown 1986, Bosque & Lentino 1987, Buden 1987, Thurber et al. 1987, Ridgely & Gwynne 1989, ffrench 1991, 1996, Tostain et al. 1992, Haverschmidt & Mees 1994, Howell & Webb 1995, AOU 1998, Raffaele et al. 1998, Bradley 2000, Ridgely & Greenfield 2001, McNair et al. 2002, Hilty 2003, Keith et al. 2003, Kenefick 2004, Latta et al. 2006, Garrigues & Dean 2007, Ruiz-Guerra et al. 2007, Kirkconnell et al. in prep.).

Greater Antilles; migrates to South America. Breeds western Washington (formerly), south-west and south-east Idaho, south-east Montana, east, south-west and north-central Wyoming, south Manitoba, Canada (has bred), south-east North Dakota, South Dakota, Minnesota (except northeast), Wisconsin, Michigan, south Ontario and south-west Quebec (Canada), New York, Vermont, New Hampshire, south Maine, south New Brunswick? (Canada), extending southward through interior and south-east California (rare and local), south Nevada, north, east and south Utah, locally in central and south Arizona and New Mexico, west and east Colorado, Nebraska, Kansas, Oklahoma, Texas, and east to Gulf coast and Florida; in Mexico, extreme northeast Baja California Norte (perhaps only formerly); south Baja California Sur; east Sonora and west Chihuahua to west Zacatecas, east Nayarit and central Jalisco; north-east Coahuila and Tamaulipas to east San Luis Potosí and north Veracruz; north Yucatán Peninsula; possibly (irregularly?) south Veracruz, Guatemala and El Salvador; Hispaniola, Gonâve Island (Haiti), Cuba with surrounding cays uncommonly Apr–Jul, Puerto Rico, rarely Jamaica and St. Croix, possibly Bahamas; recorded North Caicos, Middle Caicos and Grand Turk. Probably transient, Isle of Pines (Cuba), Beata, Navassa, Swan and Mona Is, St. Thomas and Lesser Antilles (Newton & Newton 1859, Scott 1892, Field 1894, Cherrie 1896, Bond 1956, Edwards 1957, Andrle 1967, Klaas 1968, Brudenell-Bruce 1975, Kepler & Kepler 1978, Wendelken & Martin 1986, Buden 1987, Thurber et al. 1987, Howell & Webb 1995, Sibley 1996, AOU 1998, Hughes 1999b, Keith et al. 2003, Kirkconnell et al. in prep).

Spring arrival on breeding grounds in east of range generally late Apr–early May, 2–3 weeks earlier Florida, later at higher elevations and in north. In western part of breeding range from mid to late May, peak early to mid Jun. Autumn departure in east generally late Sep–early Oct, varying from late Aug–Nov, rarely Dec. In west, beginning late Aug, most gone by mid Sep, very rarely after early Oct (Franzreb & Laymon 1993, Hughes 1999b).

Winters east of Andes from northern South America south to northern Argentina, casually north to southern USA, Aug–early May. Few Panama. Recorded most of Brazil, status and distribution poorly known, but probably winters mainly south of Amazonia, where recorded May–Jun. Sometimes several birds at single localities São Paulo. East Peru Oct–May, Bolivia Nov–Apr. Extreme dates Paraguay 12 Aug–1 May. Dec–Apr Uruguay. Argentina north to La Rioja, Santa Fe, north-east Buenos Aires and Córdoba, few records; Nov–Mar at Formosa site (Lönnberg 1903a, Laubmann 1930, Gyldenstolpe 1945b, Pinto 1964, 1978, Belton 1973, 1984, Short 1976, Willis & Oniki 1985, 2003, Remsen et al. 1986, Ridgely & Gwynne 1989, Fjeldså & Krabbe 1990, Nores et al. 1991, Stotz et al. 1992, Narosky & Di Giacomo 1993, Sick 1993, Arribas et al. 1995, Sibley 1996, de la Peña 1997, 1999, AOU 1998, Sferco & Nores 2003, Brooks et al. 2005, Di Giacomo 2005, Guyra Paraguay 2005, Garrigues & Dean 2007, Schulenberg et al. 2007, Azpiroz & Menéndez 2008, Renaudier & CHG 2010).

Casual north to south-east Alaska (Gibson & Kessel 1992, 1997, Tobish & Isleib 1992), south-west British Columbia, central Alberta, south Saskatchewan, south Manitoba, Labrador, east Newfoundland (Godfrey 1986, AOU 1998) and Clipperton Island (Ehrhardt 1971); once west Greenland (Boertmann 1994), Iceland (three dead, latest Oct 1987; Hjartarson 1995); Britain (59) and Ireland

(9) to end 2007; France (2), Belgium (2), Denmark (1), Norway (2), Italy (9), Azores (>20), most mid Sep–mid Nov (Lewington *et al.* 1991, Slack & Wallace 2009); Morocco (1) (Thévenot *et al.* 2003); Mallorca, Spain (1) (de Juana *et al.* 1996). Half of records from British Isles of dead or dying birds (Cramp 1985). Sudden invasion of thousands 9 Oct 1849 Bermuda (Bent 1940).

HABITAT Breeds in open woodland, often near watercourses, overgrown orchards, thickets along streams, willow-dogwood wetlands, dense stands of hawthorn, hammocks in south-east USA, willow-cottonwood riparian woodlands in south-west USA, abandoned farmland and mesquite scrub along rivers, Texas. Breeds mostly in riparian woodland Mexico, to 1,500m; to 2,500m on passage. On migration mostly in second growth, coastal scrub and forest borders, in winter gallery forest and mangroves, Costa Rica. Deciduous, gallery and secondary forest, scrub, bushy savanna and woodland, South America. Also mangroves Suriname. Humid forest and forest edge Peru. Dry thorny woodland on migration Venezuela. To 700m Hispaniola. Transients commonly at 2,500–3,000m east Andes, Colombia; two found dead 4,200m Venezuela, but mostly below 1,400m; below 2,600m Colombia. To 2,800m Ecuador; to 1,200m Peru; to 500m Bolivia (Wetmore & Swales 1931, van Tyne & Sutton 1937, Gaines 1974, Meyer de Schauensee & Phelps 1978, Hilty & Brown 1986, Bosque & Lentino 1987, Fjeldså & Krabbe 1990, Stiles & Skutch 1991, Haverschmidt & Mees 1994, Arribas *et al.* 1995, Howell & Webb 1995, Stotz *et al.* 1996, Hughes 1999b, Ridgely & Greenfield 2001, Hilty 2003, Schulenberg *et al.* 2007).

BEHAVIOUR Forages skulkingly in dense canopy and subcanopy by waiting for movement of prey. May move along branch until prey can be taken by stretching; sometimes hops to branch below cluster of caterpillars at tip of other branch, then flutters up awkwardly to take caterpillar. Works spiny caterpillars through bill before swallowing. Caterpillars often wiped on branch and shaken to remove guts. Occasionally sallies for flying insects or hops on ground pursuing grasshoppers. Joins subcanopy mixed-species foraging birds flocks in high-ground Amazonian forest. Adults and nestlings sunbathe in warm weather. Migrates at night, orienting by stars (Bender 1961, Cracraft 1964, Hamilton & Hamilton 1965, Stoddart & Norris 1967, Potter & Hauser 1974, Potter 1980, Crawford & Stevenson 1984, Terborgh *et al.* 1984, Sick 1993, Robinson *et al.* 1995, Hughes 1999b, Payne 2005, Marler & Hamilton 1966 in Payne 2005).

Flies into cover in tree when disturbed, turns cryptically coloured upperparts towards intruder, white underparts shaded, and remains immobile for protracted length of time. Adult flushed from nest drops slowly through nest tree, with tail spread and fluttering wings high over back, flashing rufous of flight feathers, and giving whining mew call and *kow-kow-kow* rattles. If nest more in open, flushed bird flies away waveringly and silently low over ground, exhibiting white of tail and rufous of wings (Hamilton & Hamilton 1965).

Male brings twig or caterpillar to female in courtship. Female crouches on branch in flight intension posture with horizontal tarsi and head and tail lifted, emits repeated, faint *coo* calls and/or lowers tail straight down, then lifts it straight up, repeating this pumping movement 6–10 times, immediately followed by male flying in and copulating. Copulation lasts 3–5sec. During copulation, both birds hold twig, or male grasps female's bill. After copulation, female drops twig, places it below her, or eats insect. A male afterwards perched nearby with head down and tail up. Display and copulation may be repeated. Female sometimes holds insect in bill for more than one copulation. Both members of pair build nest. Twigs broken off vegetation. Construction continues for some days after laying begins. Breeding pair sometimes with male helper that may supply up to 40% of food to young, which allows dominant pair to initiate second brood. Female occasionally abandons nestlings and initiates new nest with another male who may help to feed young of first nest, being tolerated by first male. Infanticide involving removal of youngest nestling sometimes occurs, and probably committed by unrelated helper male (Clay 1929, Bent 1940, Hamilton & Hamilton 1965, Hendricks 1975, Eaton 1979, Potter 1980, Wilson 2000, Laymon 2001, Pistorius 1985 in Payne 2005, Halterman 2009).

BREEDING Parasitism Facultative brood parasite. Sometimes has helper male at nest which may not be related. Normally builds nest and rears own young, but sometimes lays in nests of conspecifics, which is apparently linked to abundant food, e.g. periodical cicada eruptions, or lays in nests of other bird species: Mourning Dove, Black-billed Cuckoo, Dickcissel, American Robin, Wood Thrush, Cedar Waxwing, Northern Cardinal, Eastern Towhee, Black-throated Sparrow?, Red-winged Blackbird, Grey Catbird. Almost fully feathered nestling of Yellow-billed or Black-billed Cuckoo in nest of Blue Jay (Widman 1882, Bent 1940, Wiens 1965, Fleischer *et al.* 1985, Lasley & Sexton 1985, Darwin 1990, Hughes 1997c, 1999, Yasukawa 2010). Cuckoos hatched in nest of Mourning Dove (Wolfe 1994), reared until fledging by Red-winged Blackbird (Nickell 1954) and Black-billed Cuckoo (Hughes 1999b).
Season Late May–early Aug Canada and central and northeast USA, from Apr Florida, peak Midwest USA Jun–Aug. Laying mid May–late Aug Indiana. Sometimes into Sep, e.g. Ohio, Kansas, South Dakota, Missouri, Kentucky, Illinois. Eggs Texas 22 Mar–5 Sep. Western populations Jun–Aug, peak mid Jul–early Aug. Eggs 15 May–20 Aug California. Active nests late Jul–early Sep Arizona. May–Aug Mexico; Jun–Aug Sonora. Apr–Jun West Indies, May–Aug Hispaniola, May–Jul Puerto Rico, Cuba and Jamaica (Gundlach 1874, Scott 1892, Field 1894, Hess 1910, Adams 1933, Bent 1940, Trautman 1940, Wood 1951, Mengel 1965, Oberholser 1974, Nolan & Thompson 1975, Kepler & Kepler 1978, Raffaele 1989, Howell & Webb 1995, Raffaele *et al.* 1998, Russell & Monson 1998, Hughes 1999b, Keith *et al.* 2003, Latta *et al.* 2006, Halterman 2009). **Nest** Platform of sticks and twigs, sometimes oblong, without lining, or lined with dry and green leaves, pine and spruce needles, moss, dry flowers, catkins, bark strips, etc, often loose and can be seen through from below or more sturdy, components not intertwined, outside diameter 10–35cm, inside 10–18cm, outside depth 4–10cm, inside 2–4cm. On horizontal branch, or in fork in tree, e.g. pine, elm, various other broadleaf trees or in bush or hedge, in western USA, e.g. in cottonwood, mesquite, alder, soapberry or willow. At 1–6m above ground, sometimes to 27m. Lining and twigs may be added after incubation begins. Male brings twigs, female positions them. Regularly double-brooded. Second clutch initiated as soon as young leave first nest, tended by male, sometimes before a week old. Up to three broods can be raised in years of unusual food abundance (Bendire 1895, Schneider 1900, Adams 1933, Hanna 1937, Bent 1940, Hamilton & Hamilton

1965, Potter 1980, Hughes 1999b, Laymon 2001, Halterman 2009). **Eggs** 2–5 (mean 2.5 in west, 3.75 in east). Pale blue to pale greenish-blue, smooth, unglossed, elliptical to subelliptical; 27.4–35.5x20.8–25.4mm. Laid on successive days within 24 hours of nest initiation; intervals up to 5 days probably result of laying by two females in same nest (Bendire 1895, Bent 1940, Hamilton & Hamilton 1965, Schönwetter 1967, Nolan & Thompson 1975, Potter 1980, Raffaele *et al.* 1998, Hughes 1999b, Garrido & Kirkconnell 2000, Keith *et al.* 2003, Halterman 2009). **Incubation** 9–11 days. Begins when first egg laid, sometimes from deposition of last egg? By male at night, mostly by female during day. Parents call frequently from nest and nearby perches day and night during incubation. Eggshells left in nest (Bendire 1895, Hamilton & Hamilton 1965, Potter 1980, Payne 2005, Halterman 2009). **Chicks** Hatching mass 8–9g. First day fed regurgitated food from within an hour after hatching. Subsequently fed large crickets when numerous, other insects, spiders, tree-frogs and large lizards; 6–10 items/nestling/day, mean 7.9 (3 nests). Female ceases feeding nestlings after 4–7 days. Male does most provisioning to nestlings. Rapid growth, fledging mass reached in 4 days. From start of incubation to fledging 17–21 days. Can raise head over nest rim 1.5h after hatching. Stand and flutter wings, stretch neck up vertically and open mouth fully when begging from first day, begin begging when adult steps on nest, covered with greyish 4–5cm feather sheaths, eyes open and hunch down silently and motionlessly when threatened on day 3, sheaths burst and are preened off day 6–7, fully feathered within 2h, tail still short, no longer brooded during day, all sheaths broken open except those over bill, removed by rubbing head under wing, by day 8, fledge day 8–9 at *c.* 38g, 1–3 days later when weather is cool. Parents ingest or remove faecal sacs; young may defecate over nest rim from day 6. Fledglings fed up to *c.* 23 days, mostly by male (Bendire 1895, Preble 1957, Potter 1980, Hughes 1999b, Payne 2005, Halterman 2009). **Survival** Nest success >70% in west, <35% in east. Fledgling stayed within 150m of nest for eight days after fledging (Best & Stauffer 1980, Halterman 2009).

FOOD Mostly caterpillars, including poisonous, spiny and hairy species (hairs sometimes coat inside of stomach), often in large quantities; large katydids; when abundant, annual and periodic cicadas, gypsy moths, tent caterpillars and fall webworms. Also important: crickets, beetles, mayflies, dragonflies and sawfly larvae. Also: phalangid harvestmen; bugs, ants, wasps, grasshoppers and other insects; few frogs and lizards. Few raspberries, elderberries and mulberries summer and autumn; seeds and fruits winter. Nestlings fed on butterflies, katydids, beetles, grasshoppers, large crickets, caterpillars, spiders, tree-frogs and large lizards (Bent 1940, Hamilton & Hamilton 1965, Oberholser 1974, Haverschmidt & Mees 1994, Payne 2005, Rappole *et al.* 1983 in Payne 2005).

STATUS AND CONSERVATION Fairly common (Stotz *et al.* 1996). Highest breeding population densities in parts of east Kansas, Oklahoma and Texas, and in Louisiana (Price *et al.* 1995). Density in breeding range fluctuates, apparently in response to outbreaks of cicadas and caterpillars. Numbers in eastern North America increased 1965–1979, declined 1984–1993 throughout North America (Payne 2005). Populations in western part of range declined substantially during 20th century, and are now restricted to river valleys in New Mexico, Arizona and California. Extirpated British Columbia 1920s, Washington by 1934, Oregon by 1945; accidental since (Gaines & Laymon 1984, Laymon & Halterman 1987, Johnson & Tweit 1989, Siddle 1992, Hughes 1999b). Common Hispaniola (Keith *et al.* 2003), but uncommon breeder (Latta *et al.* 2006). Locally common in remainder of West Indian range, but rare Puerto Rico (Bond 1956); uncommon summer resident Cuba, more frequent on autumn passage (Kirkconnell *et al.* in prep.). Fairly common breeder Mexico; uncommon as transient in interior and west (Howell & Webb 1995). Uncommon Costa Rica (Garrigues & Dean 2007). Scarce Trinidad & Tobago (White *et al.* 2007). Locally fairly common Colombia (Hilty & Brown 1986). Uncommon Guyana (Restall *et al.* 2006). Rare Suriname (Ottema *et al.* 2009), French Guiana (Tostain *et al.* 1992), Ecuador (Ridgely & Greenfield 2001) and Peru (Schulenberg *et al.* 2007). Uncommon to rare Paraguay (Hayes 1995). Rare Buenos Aires province (Narosky & Di Giacomo 1993). Very rare Santa Fe province, Argentina (de la Peña 1997) and Uruguay (Azpiroz & Menéndez 2008). Remnants found in stomach of Tiger Shark Mexican Gulf (Saunders & Clark 1962). Adults eaten by Aplomado Falcon (Hector 1985). Eggs and nestlings possibly taken by mammals, snakes and Blue Jays (Nolan 1963, Potter 1980). Not globally threatened (BirdLife International 2011).

Yellow-billed Cuckoo *Coccyzus americanus*. Adults. **Fig. 1.** Bentsen State Park, Hidalgo County, Texas, USA, June (*Michael Patrikeev*). **Fig. 2.** Big Lake NWR, Arkansas, USA, June (*Tadao Shimba*). **Fig. 3–5.** Point Pelee National Park, Ontario, Canada, September (*Tadao Shimba*).

PEARLY-BREASTED CUCKOO
Coccyzus euleri   Plate 20

Coccygus Euleri Cabanis, 1873 (Cantagalo, Rio de Janeiro, Brazil)

TAXONOMY Name *Coccyzus julieni* Lawrence 1864, long and often used, sometimes as subspecies of *americanus*, has priority, but officially suppressed by ICZN (Willis & Oniki 1990a, Tubbs 1992); see also Banks (1988). Willis & Oniki (2003) suggested that *euleri* could be subspecies of *americanus*, based on similarities in vocalisations. Monotypic. Synonyms: *julieni* Lawrence 1864 and *Coccygus lindeni* Allen 1876.

FIELD IDENTIFICATION 24–28cm. Resembles Yellow-billed Cuckoo, but without rufous in primaries, and with greyish-brown upperparts. **Adult** Throat and chest silvery-grey, with slight contrast to white remaining underparts; not necessarily diagnostic. Bill brown to black above, mostly orange or yellow below, similar to Yellow-billed. **Juvenile** Pale edges on wings, rectrices narrow, spots indistinct. **Similar species** Very similar to Yellow-billed Cuckoo; smaller; darker and more grey upperparts. Reported difference in colours of breast at best doubtful if not useless for identification – more study needed. Lack of rufous in primaries most useful in flight. At rest, rufous in Yellow-billed sometimes difficult to see; care needed, but probably safest distinguishing feature. Inner webs of primaries partly white (brown in Yellow-billed), but only of use when bird in hand. Underwing-coverts white (buffy-white in Yellow-billed). Yellow-billed Cuckoo normally silent in South America (Hilty & Brown 1986, Hilty 2003). From Black-billed differs mainly by broad white tail tips and yellow mandible, rather than all-black bill. Dark-billed lacks yellow on bill, and has buff underparts.

VOICE Slow series of 5–20 ventriloquial disyllabic *kyoa* or *kuoup* notes, 1.0–1.4sec between notes, slower towards end, each note lasting 0.2–0.3sec, dropping from 1 to 0.5kHz. Less often (Venezuela) short ascending and guttural series followed by 4–9 accented notes similar to Yellow-billed Cuckoo *tuctuctuctuctuctuc, tówlp, tówlp, tówlp, tówlp*. 5–6 accented rattles falling in pitch and amplitude through series, each rattle consisting of four notes and lasting *c.* 0.15sec (Hilty & Brown 1986, Hardy *et al.* 1990, Sick 1993, Ridgely & Greenfield 2001, Hilty 2003).

DESCRIPTION Adult Front part of crown, and lores, cinereous, ear-coverts pale brown, darkening anteriorly, upperparts uniform earth-brown slightly glossed bronze; wing-coverts and outer webs of remiges almost same colour, inner webs partly white; underparts white, throat and chest more or less shaded grey, extension of this apparently varies, e.g. may be restricted to sides of breast; axillaries and underwing-coverts whitish; tail long, graduated, T1 brownish-olivaceous tipped darker, rest blackish with broad V-shaped white tips, and outermost rectrices with white outer edge. **Juvenile** Ear-coverts not darker than crown. Wing brown edged pale, rectrices narrower and more pointed than adult, tail spots smaller, whitish, grading into brown. **Nestling** Unknown. **Bare parts** Iris brown to black. Eye-ring grey to black (adult female) or at least sometimes yellow or red. Bill brown to black above, edge at base of upper mandible, and lower mandible, orange or yellow, lower mandible with dark or black tip (Cherrie 1916a, Hellmayr 1929, Pinto 1964, Hilty & Brown 1986, Narosky & Yzurieta 1987, de la Peña 1988, 1994, Canevari *et al.* 1991, Haverschmidt & Mees 1994, de la Peña & Rumboll 1998, de Magalhães 1999, Ridgely & Greenfield 2001, Restall *et al.* 2006). Mouth interior black. Feet bluish-grey with yellowish sole, claws black. Juvenile iris, bill and feet as in adult.

BIOMETRICS Wing male 119–135mm (n = 24), mean 128.2 ± 4.3, female 130–138mm (n = 9), mean 133.0 ± 3.0. Bill male 22.1–26.4mm, mean 24.3 ± 1.2, female 23.5–29.3mm, mean 25.9 ± 1.8. Tail male 116–140mm, mean 125.7 ± 7.5, female 131–138mm, mean 133.4 ± 2.4. Tarsus male 18.2–25.4mm, mean 21.5 ± 1.7, female 23.4–25.7mm, mean 24.4 ± 1.8 (Payne 2005). Mass male 45, 53.5g, female 54, 61g (Haverschmidt & Mees 1994, Payne 2005), laying female 61.4g (de Magalhães 1999).

MOULT Irregular transilient primary moult (Stresemann & Stresemann 1961).

DISTRIBUTION South America. Breeding range and seasonal distribution poorly known. Confusion in the field with *americanus*, especially before recognition as separate species, probably obscures knowledge which is therefore based mostly on specimens. North and extreme east Colombia (Meyer de Schauensee 1949, Carriker 1955), Ecuador (Ridgely & Greenfield 2001), Venezuela (Isla Margarita, Mérida, Distrito Federal, south-east Apure, north and south-west Bolívar, west Amazonas; Hilty 2003), Guyana (Chubb 1916, Snyder 1966, Ridgely *et al.* 2005), few records Suriname (Hellmayr 1913, Haverschmidt 1955, 1968, Haverschmidt & Mees 1994), and French Guiana (Claessens *et al.* 2011).

In Brazil, Roraima (Apr, May, Oct), Amazonas (May, Jul), Pará (Apr, Jun, Sep), Rondônia, Amapá, Maranhão (May; Dec sight), Ceará (Dec), Rio Grande do Norte (Feb), Paraíba (Jun), Pernambuco (May), Alagoas (sight), Bahia (Apr, Dec), Mato Grosso (Oct, Nov), Mato Grosso do Sul, Minas Gerais (Nov, Dec); sight record Espírito Santo (Dec or Jan); São Paulo (Mar, Nov, Dec); Rio de Janeiro; Paraná (Jan, Feb). Perhaps Parque Estadual do Turvo, Rio Grande

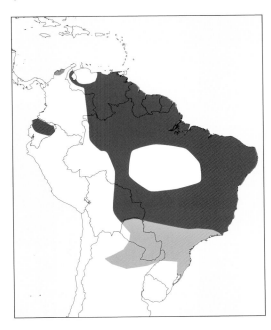

do Sul (von Pelzeln 1870, Cabanis 1873, Allen 1876, Riker 1891, Snethlage 1914, 1928, Naumburg 1930, Pinto 1935, 1938, 1964, 1966, Pinto & de Camargo 1955, 1961, Novaes 1970, Hidasi 1973, Coelho 1979, 1987, Scott & Brooke 1985, Straube & Bornschein 1989, 1995, de Mattos *et al.* 1993, Forrester 1993, Pacheco 1993, Cohn-Haft *et al.* 1997, Parker & Goerck 1997, Stotz *et al.* 1997, Novaes & Lima 1998, de Magalhães 1999, Parrini *et al.* 1999, Buzzetti 2000b, Bencke 2001, Silva & Albano 2002, de Almeida *et al.* 2003, Olmos 2003, Willis & Oniki 2003, Antunes 2005, Payne 2005, Straube *et al.* 2005, Telino *et al.* 2005, de Vasconcelos *et al.* 2006, Albano & Girão 2008, FMNH, LACMNH, MZUSP, L.F. Silveira pers. comm.).

Migratory Argentina (Barnett & Pearman 2001), where recorded Misiones (e.g. Partridge 1961, Saibene *et al.* 1996, Bodrati 2005, Bodrati *et al.* 2007, 2010), and with single sight records Chaco (Bodrati 2005), Formosa (Di Giacomo 2005) and Santiago del Estero provinces (Contreras *et al.* 1990; considered doubtful by Chébez 1994). Few records Paraguay, Oct–Jan (Capper *et al.* 2001, Guyra Paraguay 2005, Payne 2005, Cockle *et al.* 2005), Colombia Jan, sight May (Meyer de Schauensee 1949, Carriker 1955, Hilty & Brown 1986, Strewe & Navarro 2004), Ecuador, Mar, Apr, Sep, including two juvenile specimens (Ridgely & Greenfield 2001, Claessens *et al.* 2011) and Bolivia (May; Remsen *et al.* 1986; sight Oct, Santa Cruz; Vidoz *et al.* 2010); several sight records east Bolivia (Sibley 1996). Status uncertain Venezuela; may be mostly migrant from southern South America, but indications of breeding Jun and Aug (Cherrie 1916a, Hilty 2003, Payne 2005); only other available date Apr (Cherrie 1916a). Austral migrant Amazonia (Pacheco 1993, Cohn-Haft *et al.* 1997) and probably Surinam (Haverschmidt & Mees 1994). Once Sombrero I., Lesser Antilles, Oct 1863, type of *julieni*, presumably accidental (Greenway 1978, Banks 1988a).

HABITAT On breeding grounds in Atlantic (humid evergreen) forest, gallery forest and secondary forest. Primary terra firme forest and river-bluff forest dominated by *Cecropia* on Amazonian wintering grounds. Gallery forest, scrub and sandy-belt woodland, forest edge (Venezuela and Colombia). Generally to 900m; once 1,300m Ecuador. In Venezuela, only to 700m, and to 350m south of Río Orinoco. To 1,050m east Brazil (Partridge 1961, Meyer de Schauensee & Phelps 1978, Hilty & Brown 1986, Canevari *et al.* 1991, Stotz *et al.* 1996, Cohn-Haft *et al.* 1997, Ridgely & Greenfield 2001, Hilty 2003, Guyra Paraguay 2005, Straube *et al.* 2005).

BEHAVIOUR Sluggish; usually inside cover and easily overlooked. Forages in midstorey, canopy and forest edge; alone or with mixed-species canopy bird flocks in Amazonia. Individual approached another with small twig then copulated (Stotz & Bierregaard 1989, Stotz *et al.* 1996, 1997, Cohn-Haft *et al.* 1997, Capper *et al.* 2001).

BREEDING Season Eggs ready to lay and fully active testes (2 females, 4 males), Misiones, Argentina, Oct–Dec. Eggs Dec Tucumán and Buenos Aires, Argentina. Laying female Nov, São Paulo, south-east Brazil. Enlarged ovary Bolívar, Venezuela; male taken there sang vigorously and continually, Jun. Nearly-grown juvenile Aug, Amazonas, Venezuela, and Jan Cartagena, Colombia (Cherrie 1916a, Smyth 1928, Pereyra 1933, Partridge 1961, de Magalhães 1999, Payne 2005). **Nest** Loosely built of sticks, in tree fork, over water, Argentina, as *C. americanus* (Pereyra 1933). **Eggs** 4, Argentina, said to be *C. americanus*, most likely *euleri*: pale bluish-green, elliptic, 28–32.5x21–24mm (Smyth 1928, Pereyra 1933). Specimen label indicates clutch of two eggs (Payne 2005). No further information.

FOOD Caterpillars (Haverschmidt & Mees 1994).

STATUS AND CONSERVATION Nowhere common (Sick 1997), and distribution poorly known, but can occur in secondary forest, is not very sensitive to habitat disturbance, and considered of lowest conservation priority (Stotz *et al.* 1996). Erratic and local Venezuela, sometimes absent for year or more, then present one year in low numbers, but easily overlooked (Hilty 2003). Has become rare for unknown reasons São Paulo state, Brazil (Willis & Oniki 2003). Rare central Amazonia (Cohn-Haft *et al.* 1997). Not globally threatened (BirdLife International 2011a).

Pearly-breasted Cuckoo *Coccyzus euleri*. **Fig. 1.** Adult, Reserva Ecológica de Guapiaçu, Brazil, December (*Sandro Henrique*). **Figs. 2–5.** Adults, Serra de Baturité, Ceará, Brazil (*Ciro Albano*).

MANGROVE CUCKOO
Coccyzus minor **Plate 20**

Cuculus minor J.F. Gmelin, 1788 ("in Cayenna" = Cayenne, French Guiana)

TAXONOMY Formerly divided into 9–14 subspecies (e.g. Peters 1940) based on minor size and colour differences and on small samples, but analysis of larger series documented great individual variation within populations and broadly overlapping wing lengths (Banks & Hole 1991). Further investigation may reveal few definable subspecies (Hughes 1997b). Island endemic *C. ferrugineus* (see separate account) included in *minor* (Peters 1940), but later returned to species status (Slud 1967, AOU 1983, 1998). Sibley & Monroe (1990) treated them as superspecies, but perhaps conspecific; they are each others' closest relatives (Sorenson & Payne 2005). Monotypic. Synonyms: *Cuculus seniculus* Latham 1790, *Coccygus helviventris* Cabanis 1848, *Coccygus nesiotes* Cabanis & Heine 1863, *maynardi* Ridgway 1887, *dominicae* Shelley 1891, *abbotti* Stone 1899, *m. vincentis* A.H. Clark 1905, *m. grenadensis* Bangs 1907, *m. palloris* and *m. rileyi* Ridgway 1915, *m. caymanensis* Cory 1919, *m. teres* Peters 1927, *m. continentalis* and *m. cozumelae* van Rossem 1934.

FIELD IDENTIFICATION 28–34cm. Slender, streamlined; flight direct with several strong flaps and short glides. **Adult** Crown grey, upperparts and wings plain brown, broad black mask through eyes to ear-coverts; underparts buff, long graduated tail black below with bold white tips, T1 uniform brown. **Juvenile** Like adult but mask and underparts paler, tail spots more diffuse. **Similar species** Combination of black mask and buff underparts diagnostic, only shared with allopatric Cocos Cuckoo. Black- and Yellow-billed Cuckoos both have white underparts and lack black mask. Black-billed has narrow white tail tips. Yellow-billed has similar tail spots, but rufous on inner webs of primaries visible in flight. Dark-billed Cuckoo smaller, mask paler, grey neck sides, no yellow on bill. Grey-capped Cuckoo more rufous on upperparts and wings, cap grey, no mask, black bill.

VOICE Series of 8–25 or more nasal to rasping notes *ahrr-ahrr...*, ending *ah-ahr*, or *ga-ga-ga-ga-ga-ga-gau-gau-gau-go*, like Yellow-billed Cuckoo but slower and more nasal, not slowing suddenly. Sharp, barking *whip*, guttural, fast *coo-coo-coo-coo-coo*, and clucking and squirrel-like notes. Calls most frequently Apr–Aug Florida, mostly at dawn and before rain (Bond 1979, Stiles & Skutch 1991, Howell & Webb 1995, Hughes 1997b, Raffaele *et al.* 1998).

DESCRIPTION Adult Forehead and crown grey; nape, hindneck, upperparts and wings grey-brown, wings less grey and outer webs of outer secondaries brighter brown, underwing-coverts whitish to dark buffy tinged rufous; broad black line from before and below eyes and broadening at ear-coverts, chin and side of neck white, throat and breast whitish buff grading to cinnamon-buff on rest of underparts; in some, underparts more richly coloured, having belly to undertail-coverts rufous; long graduated tail dark olive-brown and black below broadly tipped with oval white spots and with some white markings on T3–T5, T1 plain olive-brown. **Juvenile** Paler and indistinct mask, crown brown, wing-coverts, secondaries and tertials edged pale buff; underparts whitish buff; rectrices narrower and more pointed, more grey or brownish, tips whitish, smaller and more diffuse. **Nestling** Unknown. **Bare parts** Iris dark brown to reddish-brown, orbital ring yellow. Curved bill black, lower mandible yellow to orange tipped blackish; in juvenile black above and paler below. Feet dark bluish-grey, claws brown.

BIOMETRICS Wing male 125–147mm (n = 8), mean 137.3 ± 7.7, female 125–144mm (n = 9), 131.8 ± 6.9. Bill male 25.8–31.1mm, mean 29.0 ± 1.6, female 27.0–31.7mm, mean 28.0 ± 4.3. Tail male 154–176mm, mean 163.3 ± 7.8, female 157–169mm, mean 163.1 ± 4.6. Tarsus male 23.1–30.5mm, mean 27.1 ± 2.4, female 25–29mm, mean 27.4 ± 1.0 (Payne 2005). Mass male 51.5–76.6g (n = 9), mean 64.3, female 62.5–75.0g (n = 11), mean 67.0 (Hughes 1997b).

MOULT Female in heavy body moult Mar, Panama. Some flight feathers may be retained until following spring or summer. Body plumage probably moulted autumn. Sequence of remiges' replacement sometimes irregular (Bent 1940, Wetmore 1968, Pyle 1995, Hughes 1997b).

DISTRIBUTION Mainly Central America and Caribbean. Resident. Coastal south Florida and Florida Keys. From Sonora, Mexico, south on Pacific slope and locally in interior, to Panama, and from south Tamaulipas on Atlantic slope to east Nicaragua, including Holbox, Mujeres and Cozumel Is (Mexico), and Bay Is off Honduras. Uncommon Costa Rica (breeding?) in north Pacific lowlands and foothills Dec–Jun, and extreme north-west Caribbean slope; possibly small breeding population Guanacaste. Female in breeding condition May Veraguas strongest indication of

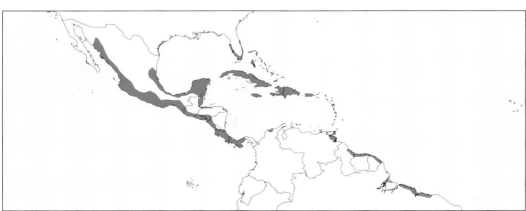

breeding Panama, where recorded Pacific lowlands east to Panama City, islands off south coast, Canal area on Caribbean slope and Bocas del Toro. Most of West Indies, including Jamaica, Virgin Is, Puerto Rico, Hispaniola, Isla Beata, Tortue, Ile à Vache, Saona, Mona, Vieques, Barbuda, Antigua, Guadeloupe, Désirade, Dominica, Martinique, Santa Lucia, St. Vincent, Grenada, Grenadines, Montserrat, Bahamas, Caicos Is, Grand Turk, Cayman Is, Swan I., Isla Providencia, Corn Is. Mainland Cuba mostly eastern coastal areas and offshore cays; recorded locally from Matanzas to Guantánamo provinces. Coastal Guyana, Suriname and French Guiana (Chubb 1916, Ridgway 1916, Bond 1939, 1950, 1956, Snyder 1966, Monroe 1968, Wetmore 1968, Brudenell-Bruce 1975, AOU 1983, 1998, Buden 1987, 1990, Raffaele 1989, Ridgely & Gwynne 1989, Stiles & Skutch 1991, Tostain *et al.* 1992, Olson 1993, Haverschmidt & Mees 1994, Howell & Webb 1995, Raffaele *et al.* 1998, Russell & Monson 1998, Bradley 2000, Latta *et al.* 2006, Garrigues & Dean 2007, Haynes-Sutton *et al.* 2009, Kirkconnell *et al.* in prep.).

Unknown Barbados (Buckley *et al.* 2009). May not occur any longer on Isla San Andrés, west Caribbean (Colombia), where recorded 1887 (Stone 1899) and 1948 (Bond 1950), but not later (Paulson *et al.* 1969, Russell *et al.* 1979, Tye & Tye 1991). Unknown Isle of Pines (Bond 1979). Non-breeding visitor Curaçao and Bonaire; once Apr Aruba, Netherlands Antilles (Prins *et al.* 2009). Scarce breeder Trinidad (White *et al.* 2007), no records Tobago 1934–1981 (ffrench 1984), but several records since (e.g. ffrench & White 1999, ffrench & Kenefick 2003). Only 'Bogotá' skins Colombia (Hilty & Brown 1986), also recently north Colombia (Strewe & Navarro 2004). Twice Delta Amacuro, north-east Venezuela, and (sight?) Islas Los Roques (Hilty 2003). Once confirmed Brazil – two collected 1835 north-east of Belém, Pará (von Pelzeln 1870); sight, Iguaiba, north Maranhão (Forrester 1993) and photo, coastal Piauí (Mota 2010). Vagrant Quebec and possibly Nova Scotia, Canada (Aubry *et al.* 1992). Accidental Texas and north, peninsular and interior Florida (Hughes 1997b).

HABITAT Mangroves and tropical hardwood hammocks, Florida. Dense jungly coppice and rather open woodland Bahamas. Scrubby woodland, forest and mangroves, to 1,200m, Mexico. Gallery forest, forest edges, woodland with scrubby clearings, more often found near water inland than in coastal swamps, Costa Rica and Panama. Mostly dry scrub, also shade coffee plantations and mangroves, Caribbean. Dense thickets overgrown with vines along small streams Puerto Rico, but most common on small islands with dry brush, also cactus scrub Hispaniola. Mostly dry forest and shrubland, usually not mangroves, Cayman Is. Mangroves, Suriname. Locally to 1,100m Costa Rica, once 1,370m El Salvador (Wetmore 1916, 1968, Wetmore & Swales 1931, Dickey & van Rossem 1938, Haverschmidt 1968, Brudenell-Bruce 1975, Ridgely & Gwynne 1989, Stiles & Skutch 1991, Howell & Webb 1995, Raffaele *et al.* 1998, Hughes 1999a, Bradley 2000).

BEHAVIOUR Inactive; fearless; skulking and secretive; silent autumn and winter. In morning often perches quietly in sunshine on dead limbs, sometimes calling. Moves slowly, pausing for long to look around. Gleans insects from twigs and leaves. May beat large prey or crush in bill before swallowing. Scratches head under wing. Male fed spider to female in courtship. Female on branch lifted bill, head and tail, pumped tail rhythmically up and down through 180°, *c.* 20/min, calling softly, 1–2min; male landed on her back, grasped her bill while copulating for 6sec, without courtship feeding; female had eaten large orthopteran previously, captured on ground (Wetmote 1916, 1968, Bradley 1985, Langridge 1990, McNair 1991, Hughes 1997b).

BREEDING Season Eggs May–Jul, Florida. May Bahamas (once); Apr–Jul Cuba; Mar–Jun Jamaica; Feb–Jul, but copulation early Dec, Hispaniola; Mar–Oct Puerto Rico; Mar–Jul Lesser Antilles; Jul–Sep (and earlier?) Trinidad. Breeding female late Aug, Sonora, May–Jul Oaxaca. Enlarged testes Apr, Belize. May, Panama; Jun Suriname (Wetmore 1916, Bent 1940, Hellebrekers 1942, Russell 1964, Wetmore 1968, Buden 1987, Binford 1989, ffrench 1991, Russell & Monson 1998, Payne 2005, Latta *et al.* 2006, Haynes-Sutton *et al.* 2009, Kirkconnell *et al.* in prep.). **Nest** Loose platform of coarse twigs, lined with few leaves, 15–25x5–8cm, inside 8–12x2–4cm, in vertical fork or on horizontal branch, 2–3m up in mango, mangrove or acacia tree, built by both members of pair (Belcher & Smooker 1936, Bond 1941, Rowley 1984, DeGraaf & Rappole 1995, Hughes 1997b). **Eggs** 2–3 (1–4); broadly elliptical, bright blue-green, smooth, without gloss, 27.19–33.81x20.65–26.12mm (n = 26) mean 30.53 ± 1.78x22.96 ± 1.28 (Belcher & Smooker 1936, Hellebrekers 1942, Rowley 1984, Stockton de Dod 1987, Hughes 1997b). **Incubation** Unknown. **Chicks** Fed insects by both parents (Bent 1940, Hughes 1997b). **Survival** Unknown.

FOOD Slow-moving, large insects, including orthoptera such as large green locusts, caterpillars, cockroaches, sugar cane borers; large spiders; tree-frog *Hyla*; *Anolis* lizards. Fifteen stomachs, Puerto Rico, contained 50% grasshoppers, 30% caterpillars, mole cricket, phasmids, cricket, mantids, earwigs, cicadas, stink bugs, moth, weevils and few other beetles; large spiders; snail; vegetation. Two stomachs, Florida, contained hairy caterpillars, mantids, grasshoppers, grasshopper eggs and spiders (Wetmore 1916, Bent 1940, Robertson 1978, Wunderle 1981,Voous 1983, Stiles & Skutch 1991, Payne 2005).

STATUS AND CONSERVATION Uncommon Florida coast and keys, threatened by destruction of habitat, but persists where large areas of protected mangroves, e.g. Key Largo Hammocks State Botanical Preserve and Everglades and Biscayne NPs. 372ha of mangroves and deciduous forest acquired 1992–1999 to create reserve network in Florida Keys (Hughes 1999a). Fairly common Mexico (Howell & Webb 1995); rare Guatemala (Eisermann & Avendaño 2006) and El Salvador (Komar 1998). Common Isla Providencia (Bond 1950). Very rare Cuba; regular coastal areas, eastern main island, rather common on cays (Kirkconnell *et al.* in prep.). Fairly common Jamaica (Haynes-Sutton *et al.* 2009) and Hispaniola, more so since 1970s (Latta *et al.* 2006), and West Indies generally (Raffaele *et al.* 1998). Scarce Trinidad & Tobago (White *et al.* 2007). Fairly common Suriname (Haverschmidt & Mees 1994), uncommon Guyana (Restall *et al.* 2006). Much more common on small Mona I. than in similar habitat on nearby Puerto Rico, ecological release? (Faaborg 1980). Not globally threatened (BirdLife International 2011).

Mangrove Cuckoo *Coccyzus minor*. **Fig. 1.** Adult, Puerto Rico, December (*Alfredo D. Colón Archilla*). **Fig. 2.** Adult, Puerto Rico, January (*Alfredo D. Colón Archilla*). **Fig. 3.** Adult, Puerto Rico, March (*Alfredo D. Colón Archilla*).

COCOS CUCKOO
Coccyzus ferrugineus Plate 20

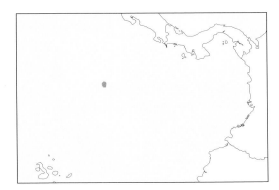

Coccyzus ferrugineus Gould, 1843 (Cocos Island, east Pacific Ocean)

Alternative names: Cocos Island Cuckoo.

TAXONOMY Considered subspecies of *minor* by Peters (1940), Morony *et al.* (1975) and Wolters (1975–1982), but most authors have considered it separate (e.g. Stotz *et al.* 1996, Payne 1997, 2005, AOU 1998, Dickinson 2003). Sibley & Monroe (1990) treated *minor* and *ferrugineus* as superspecies. Most closely related to *minor*, based on mtDNA analysis (Sorenson & Payne 2005). Monotypic.

FIELD IDENTIFICATION 33cm. Only cuckoo on Cocos I. **Adult** Grey crown, dusky mask, greyish-brown upperparts with bright rufous wings, rich buff underparts, graduated black tail with large white terminal spots underneath, yellow eye-ring, black bill, with yellow lower mandible tipped black. **Juvenile** Similar to adult, but less distinct tail pattern. **Similar species** Adult Mangrove Cuckoo has brown mantle and greyish wings, Black-billed has white underparts and all-black bill, Yellow-billed is white below, Dark-billed Cuckoo has dark brown wings, grey neck sides and all-black bill; none recorded Cocos I. (Montoya 2007).

VOICE 5–8 deep, dry, coughing *kcha* notes, sometimes preceded by rolling or rattling notes. Also guttural *k'k'k'k'ru'hoo* (Slud 1967, Stiles & Skutch 1989).

DESCRIPTION Adult Forehead and crown greyish-black, gradually becoming greyish-brown on back, scapulars, proximal wing-coverts, secondaries, rump and uppertail-coverts; wings above mostly deep rufous, primaries greyish-brown distally, underwing-coverts warm buff, remiges below similarly buff, deepening through more rufescent medially into greyish-brown distally; narrow blackish mask from lores to auriculars, contrasting with crown; below mask, including malar region and chin, buffy-white, becoming rich buff on rest of underparts, especially on flanks, thighs and undertail-coverts; T1 greyish-brown, faintly glossed greenish-bronze, and becoming darker subterminally, with narrow tip pale brownish, rest of rectrices black, broadly tipped white, white tip 35mm in extent on outermost and 10mm on T2, black portion edged on both webs pale buff, T2 with approximately proximal third greyish-brown. **Juvenile** Similar to adult, but tail pattern indistinct. **Nestling** Unknown. **Bare parts** Iris dark brown. Eye-ring yellow. Upper mandible black, lower bright yellow with black tip. Feet slate.

BIOMETRICS Wing male 126–137mm (n = 13), mean 131.5 ± 3.6, female 131–137mm (n = 7), mean 134.2 ± 1.7. Bill male 30.0–32.3mm, mean 31.0 ± 0.7, female 26.4–31.0mm, mean 28.5 ± 2.0. Tail male 152–177mm, mean 163.1 ± 5.8, female 158–164mm, mean 161.2 ± 3.0. Tarsus male 22.1–28.4mm, mean 26.6 ± 2.3, female 27.5–28.6mm, mean 28.1 (Payne 2005). Mass 70g (Stiles & Skutch 1989).

MOULT No information.

DISTRIBUTION Resident Cocos I. (Isla del Coco), Costa Rica, eastern Pacific Ocean, *c.* 500km from Costa Rican coast.

HABITAT Second growth forest, *Hibiscus* thickets and streamside vine tangles, banana trees; to 570m (Gifford 1919, Stiles & Skutch 1989, BirdLife International 2000).

BEHAVIOUR Furtive, skulking, preferring to run and hop through dense vegetation rather than fly. Flies with few quick flaps, and long glides. Forages like other *Coccyzus* cuckoos, alternately looking around deliberately and quietly, and suddenly hopping quickly up or along branch or among branches, then stopping again (Slud 1967, Stiles & Skutch 1989).

BREEDING Season Approximately Jan–Feb or more extended (F. López pers. comm.). Otherwise, nothing known.

FOOD Sphingid larvae, cicadas, possibly crickets; *Anolis* lizards (Gifford 1919, Slud 1967, Stiles & Skutch 1989).

STATUS AND CONSERVATION Widespread and probably under-recorded in suitable habitat, but extent of such habitat only *c.* 25km². Least common of four endemic land bird species of island (Stiles & Skutch 1989, BirdLife International 2000). Threats: potentially predated by cats and rats; habitat grazed by feral goats, deer and pigs; understorey degraded and regeneration of forest inhibited by pigs, but may tolerate some modification of habitat; increasing tourism may cause disturbance. So far no evidence of decline in population or range. Although Cocos I. designated national park, measures needed to control populations of introduced rats, cats, deer, pigs and goats. Better estimate of population size, study of impact of introduced mammals, and, where feasible, initiation of eradication of mammals, required (BirdLife International 2000). Total population probably much less than 1,000, perhaps only 250. Considered globally threatened. IUCN Red List category: Vulnerable (BirdLife International 2011).

Cocos Cuckoo *Coccyzus ferrugineus*. **Fig. 1.** Juvenile, Cocos I., Costa Rica (*Kevin Easley*).

BLACK-BILLED CUCKOO
Coccyzus erythropthalmus Plate 20

Cuculus erythrophthalma (sic) A. Wilson, 1811 (Locality not stated; probably near Philadelphia, Pennsylvania).

TAXONOMY In error *erythrophthalmus* (e.g. Herrick 1910, Cramp 1985). Monotypic. Synonym: *dominicus* Nuttall, 1832.

FIELD IDENTIFICATION 27–31cm. Medium-sized, slender, long-tailed; hides in dense vegetation, flies low, with deep, loose wing-beats between cover, often gliding. Secretive. **Adult** Bill black (at most weak blue-grey base on lower mandible). Eye-ring red. Above brown, below greyish-white. Tail below grey tipped white (sometimes worn off), subterminal band black. **Juvenile** Similar to adult, but eye-ring yellow, upperparts warmer rufous-toned, with some pale edging, and slightly more rufous in primaries, breast buffy, rectrices narrower, with faint and narrow whitish tips (no black band), bill often paler, greyish or bluish. **Similar species** Adult Yellow-billed Cuckoo has yellow on bill, often yellow eye-ring (like juvenile Black-billed) and much more and brighter rufous in wings (Black-billed is only faintly rufous on inner webs of primaries), tail below darker and with larger white spots, and throat and breast paler. Juvenile Yellow-billed Cuckoo has wider pale tail tips, and outer edges, yellow on bill (after *c.* 2 months), more rufous in primaries than both adult and juvenile Black-billed. Bill smaller than in Mangrove and Yellow-billed Cuckoos.

VOICE Phrases of 2–4 evenly pitched, monotonous notes *cu-cu-cu-cu*, repeated at *c.* 1sec intervals, sequence of more than ten phrases, but sometimes more than 200, without cessation. Sometimes preceded by *c.* five short, descending, croaking notes *krak-ki-ka-kruk-kruk* that may also be given alone. Also slower version, *kow-kow-kow* etc. Low dove-like *coo-oo-oo*, 2–5 notes, with very short intervals, given with wide-open bill in presence of predator during incubation. Loud mewing given by female during courtship. Alarm *cuck-a-ruck* or *koor-uck-uck-úk*; guttural *erratt-o-too*; soft *pruh-pruh-pruh* and single *chuck* from nest. Nestling begs with insect-like buzz or wheezing, larger nestling gives *ker-ut-ut-út* and explosive squeal *kar-achít*; utters *cuck* when not hungry. Calls day and night. In midsummer performs nocturnal calling in flight, often high up. Not known to vocalise Colombia, Ecuador or Peru (Thayer 1903, Herrick 1910, Forbush 1927, Mousley 1931, Bent 1940, Spencer 1943, Scott *et al.* 1983, Hilty & Brown 1986, Hardy *et al.* 1990, Raffaele *et al.* 1998, Hughes 2001, Ridgely & Greenfield 2001, Schulenberg *et al.* 2007).

DESCRIPTION Adult Forehead, crown, ear-coverts, side and back of neck, whole upperparts, wings and tail plain brown, washed grey on forehead, wing-coverts glossed bronze-olive, underwing brown with pale buffish area, underwing-coverts white; underparts white tinged pale grey on chin, throat and upper breast; graduated tail brown above, below brown-grey, with dusky subterminal band and tipped off-white. Leucism and perhaps albinism recorded (Hughes 2001). **Juvenile** Similar to adult, but upperparts slightly more rufous and faintly scaly-looking, wing-coverts narrowly fringed pale greyish-buff to whitish; remiges and greater upperwing-coverts strongly tinged rufous basally and medially, greater primary-coverts, secondaries and inner primaries narrowly fringed buff on tips; underparts white tinged buff on chin, throat and chest; rectrices narrower, with inconspicuous whitish to buffy-white tips, *c.* 2mm wide, and without subterminal bar. Buff-edged wing-coverts sometimes retained into first breeding season. Suspected hybrid with Yellow-billed Cuckoo (Parkes 1984) is juvenile male Yellow-billed Cuckoo (Payne 2005). **Nestling** Hatchling 7.5g. Skin blackish, contrasting with white hair-like sheathed down feathers (3–10mm) on dorsal surface and thighs, bill and frontal apterium dark, eyes closed, cutting edges white, corners of gape red and unswollen, mouth interior red with white markings, similar to Yellow-billed Cuckoo, tongue red with black tip, feet

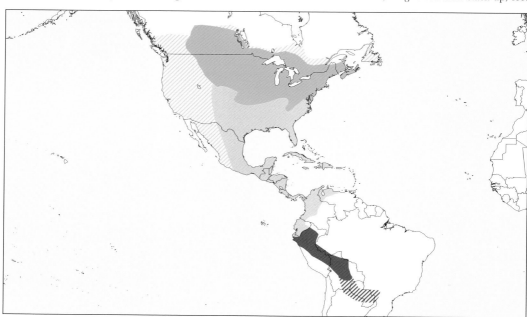

blue-grey. **Bare parts** Iris dark brown, eye-ring red, yellow in juvenile, and normally yellowish in non-breeding season. Slightly curved bill black, base of lower mandible paler, latter often blue-grey in juvenile. Feet grey, dark cinnamon in juvenile; claws slate.

BIOMETRICS Wing male 133–142mm (n = 12), mean 135.2 ± 2.5, female 133–145mm (n = 12), 139.2 ± 4.1. Bill male 21.1–22.6mm, mean 22.0 ± 0.4, female 20.7–25.3mm, mean 22.6 ± 1.2. Tail male 142–161mm, mean 150.9 ± 5.3, female 144–160mm, mean 154.4 ± 5.0. Tarsus male 20.2–24.8mm, mean 22.9 ± 1.6, female 21.6–24.9mm, mean 23.4 ± 1.1. Mass male 41.6–50.0g (n = 21), mean 46.6, female 46.2–62.8g (n = 19), mean 54.1 (Payne 2005).

MOULT Remiges and rectrices replaced during non-breeding season. Irregular sequence of flight feather replacement. Few secondaries and rectrices sometimes retained. Juvenile begins moult on summer grounds and resumes on wintering grounds (Pyle *et al.* 1997).

DISTRIBUTION Breeds North America: Canada from east-central and south Alberta, south Saskatchewan, south Manitoba, central Ontario, south-west Quebec, New Brunswick, Prince Edward Island and Nova Scotia, southwards in USA to Montana east of Rocky Mts. (rare in west), south-east Idaho (formerly), central and south-east Wyoming, extreme north-east Colorado, most of Nebraska, central and east Kansas, north-central Oklahoma, north-west Arkansas (rare), Missouri except south-east, Illinois, Indiana, Ohio, east Kentucky, east Tennessee, north Alabama (formerly), north Georgia (formerly), west and south-east Virginia, west North Carolina, extreme west South Carolina, south Maryland. Bred Louisiana (at least once), west Tennessee, south Texas (once) and possibly north Utah. Summer records west Colorado, south-west Idaho, British Columbia and Washington. Autumn records California. Spring arrival mostly May, e.g. 7–30 May Kansas (n = 11), sometimes from early Apr (e.g. Alabama, Minnesota, Pennsylvania, Massachusetts) or early Jun (e.g. New England, Canada). Autumn departure eastern Canada late Aug–late Sep, sometimes until mid Oct, extreme date Nova Scotia 22 Nov. Often departs USA Oct, sometimes Nov. Recorded 15 Dec Pennsylvania (Behle & Selander 1952, Johnston 1964, Hayward *et al.* 1976, Fischer 1979, Sibley 1996, AOU 1998, Hughes 2001, Graves *et al.* 2002).

Migrates regularly through south-east USA, Mexico and Central America, e.g. passage Texas Apr–May and Sep–Oct, but rare Florida Apr–Jun; uncommon Mexico (absent north-west) to Nicaragua Apr–early Jun and Aug–Nov. Mostly Guanajuato and Guerrero southwards in Mexico; also Cozumel. Casual Arizona, New Mexico, Sinaloa and Newfoundland. Fairly common transient Apr–early May and Sep–early Nov Honduras (Peterson 1960, Monroe 1968, Tuck 1968, Vickery 1977, Stevenson & Anderson 1994, Howell & Webb 1995, AOU 1998). Rare passage migrant Costa Rica late Sep–mid Nov and Apr–mid May (Garrigues & Dean 2007). Transient Panama Sep–Nov; two Apr records (Ridgely & Gwynne 1989). Rare passage Cuba, mostly west and central, mid Sep–mid Nov and late Mar–early May (Kirkconnell *et al.* in prep.). Regular but rare transient Apr–May and Sep–Dec Hispaniola (Latta *et al.* 2006), very rare New Providence, Andros and Grand Bahama, vagrant Jamaica, Puerto Rico, Antigua, Dominica, St. Lucia, Barbados (Raffaele *et al.* 1998). Nov specimen Barbuda (Seaman 1957). Once Grand Cayman (Bradley 2000). Rare transient (e.g. Rodner *et al.* 2000, Restall *et al.* 2006) and winter visitor Colombia early Nov–early Apr (Hilty & Brown 1986). Rare transient and possible winter visitor late Oct–Nov and Feb, very few records, north-west Venezuela (Hilty 2003). Mainly Mar–Apr Ecuador, few Nov and Feb (Ridgely & Greenfield 2001).

Rare north-west Peru southward to La Libertad, and lower east Andean slopes and adjacent lowlands and Marañón valley, Sep–May but mostly Feb–Apr. Most Dec–Feb records from east Peru (e.g. Amazonas, San Martín, Ucayali, Madre de Dios, Puno) and east Bolivia (also Mar, Nov; La Paz, Cochabamba, Santa Cruz) (Remsen & Ridgely 1980, Remsen & Traylor 1983, Pearman 1993, Arribas *et al.* 1995, Payne 2005, Schulenberg *et al.* 2007). Once Trinidad Sep (ffrench 1991). Twice Argentina: Feb specimen Misiones (Partridge 1961) and ringed Jujuy (Olrog 1979). Twice Paraguay (Short 1972, Guyra Paraguay 2005). Singles Acre, Brazil, Feb (Whittaker & Oren 1999), Paraná, south Brazil (no details; Scherer-Neto *et al.* 2001) and Española, Galapagos, May (Harris 1973), and twice, Apr and Sep, Santa Cruz I. (MVZ).

Vagrant Bermuda late Jul–mid Nov, and three late Apr–early Jun (Bent 1940, Amos 1991). Rare autumn vagrant west Greenland (3 records; Boertmann 1994), west Europe (Iceland, British Isles, Denmark, Germany, Italy) late Sep–early Nov, peak Oct, France Jul; three records Azores (Hartert & Ogilvie-Grant 1905, Bannerman & Bannerman 1966, Cramp 1985, Lewington *et al.* 1991). Thirteen records Britain, one northern Ireland, until end 2007. One at sea 1,650km east of New York 30 Aug 1965 (Slack & Wallace 2009).

HABITAT Deciduous and coniferous forests and open woodland, both dry and humid, North America. Forest and edge Mexico to Nicaragua. Lowland moist and dry forests, scrubland and mangroves, West Indies. Nearly everywhere Panama, preferring open woodlands. Mostly lowlands Costa Rica. To 900m when wintering, higher only on passage. To 1,600m (or higher) Colombia, 350–1,100m Venezuela; to 1,000m, rarely 3,950m, dry scrub to borders and canopy of moist forest, Ecuador. To 2,400m Peru, forest borders, dry and humid forest. Early successional riverine vegetation of *Tessaria* composite trees and *Gynerium* cane, lowland Peru. Grasslands, dry Andean valleys and Amazonian forest, Bolivia (AOU 1983, Hilty & Brown 1986, Ridgely & Gwynne 1989, Stiles & Skutch 1989, Arribas *et al.* 1995, Howell & Webb 1995, Robinson *et al.* 1995, Stotz *et al.* 1996, Raffaele *et al.* 1998, Ridgely & Greenfield 2001, Hilty 2003, Schulenberg *et al.* 2007).

BEHAVIOUR Perhaps even more elusive than Yellow-billed Cuckoo. Alone or in pairs. May associate with other species on migration. Migrates at night. Forages in trees, sometimes on ground, skulking quietly in foliage, looking for movement of prey, and flies to trunks and branches to snatch insects. Caterpillars sometimes hammered against branch, shaken or crushed in bill, removing gut contents. Forages with mixed-species flocks of resident birds in Amazonian forest, Peru. Regurgitates pellet of caterpillar hairs accumulated in stomach (Bender 1961, Terborgh *et al.* 1984, Munn 1985, Robinson *et al.* 1995, Kaufman 2000, Hughes 2001, Forbush & May 1939 in Hughes 2001, Payne 2005). Courtship: male calls with caterpillar in bill, female spreads and pumps tail up and down 4–5 times and gives mewing call, and male mounts, copulation lasts a few seconds, takes place on branch near nest, sometimes during incubation period (Spencer 1943, Potter 1980). Male may feed caterpillar to female (Nero 1988).

BREEDING Parasitism Facultative brood parasite. Occasionally lays in nests of other Black-billed Cuckoos or of other bird species. Most frequent interspecific hosts: Yellow-billed Cuckoo, American Robin, Grey Catbird, Chipping Sparrow and Wood Thrush. Others: Eastern Wood Pewee, Cedar Waxwing, Veery, Yellow Warbler, Northern Cardinal?, Yellow-breasted Chat. Cuckoo nestlings evicted nestlings of Chipping Sparrow and raised by foster parents, and crowded out Yellow Warbler nestlings that subsequently died of starvation. Parasitised nest of Yellow-breasted Chat abandoned before eggs hatched. Selects hosts with similar eggs in 75% of cases (Bendire 1895, Herrick 1910, Roberts 1936, Bent 1940, Thomas 1995, Hughes 1997c, 2001, Sealy 2003). Sometimes parasitised by Yellow-billed Cuckoo (Nolan & Thompson 1975). **Season** Between May and Sep in general, with regional and annual variation, e.g. laying 21 May–10 Aug, mode 5 Jun Kansas (n = 17). Southern populations arrive 2–4 weeks earlier on breeding grounds, and depart 1–2 weeks later, than north-eastern populations. Breeding timed to coincide with food availability, e.g. peak of annual cicadas, Michigan. Two weeks earlier in years with abundance of caterpillars, Manitoba. Late nesting (Aug–Sep) not proven to represent double-broodedness (Johnston 1964, Nolan & Thompson 1975, Sealy 1978, Hughes 2001, Payne 2005). **Nest** Rather frail platform built of twigs, sticks, weed stems, bark and rootlets, lined with green and dead leaves, pine needles and other plant material, hidden on horizontal branch of shrub or tree, or in clump of saplings, 0.5–6m up. Highest 13.5m. Rarely on ground. Outside diameter 14–18cm, inside 7–10cm, outside depth 3–10cm, inside 2–3cm. Probably built by both parents, lining material added during incubation (Bent 1940, Spencer 1943, Johnston 1964, Peck & James 1983, Hughes 2001, Payne 2005). **Eggs** 2–4; up to eight in single nest, but large clutches probably result of laying by more than one female; oval, dull greenish-blue, smooth, not glossy; 22.4–30.5x18.5–22.9mm (n = 90), mean 27.2x20.5mm ± 0.11. Laid in the morning, at least sometimes at two-day intervals, but intervals possibly 1–3 days (Bendire 1895, Bent 1940, Spencer 1943, Schönwetter 1967, Sealy 2003). **Incubation** Begins when first egg laid, by both sexes, time sitting during day 15–115min, mean 90min, period 10–11 days (Spencer 1943); nocturnal incubation unstudied. **Chicks** Hatching asynchronous; normally early morning. Dry at hatching, 6–7cm, mass 7.5g. Shells left in nest or partly eaten by parent. Mass one day old 7.5–9g, mean 8.5 (n = 9). Eyes begin to open day 2. Beg with neck stretched up and wings held straight out to side and flapped, giving wheezing call. Fed hairy and smooth caterpillars, grasshoppers and other insects, by both parents. Faecal sacs removed or ingested by parents. Mass 26g day 6. Fully feathered except tail day 7, when 5cm long sheaths burst through within 6–7 hours, and leaves nest; nine nestlings left nests at 7–9 days. Unable to fly for another 14 days; feet strong, enabling chick to pull itself upright again if it falls, also using bill and only one foot. At age 1–3 weeks attains erect posture with outstretched neck, bill pointing up and open eyes, probably in response to predator. Stub-tailed fledgling mass 17.9–33.3g, mean 23.8 (n = 4) (Herrick 1910, Mousley 1931, Spencer 1943, Sealy 1985, Payne 2005). **Survival** Of 18 eggs laid in six nests over two breeding seasons, 14 hatched (78%) and 10 chicks fledged successfully (71%); 55% of eggs resulted in fledged young (Spencer 1943).

FOOD Large, slow-moving katydids and caterpillars (including deforesting pests such as gypsy moths, fall webworms, army worms and tent caterpillars), grasshoppers, crickets, beetles, sawfly larvae, bugs, annual and periodic cicadas; Monarch butterflies when plentiful in autumn; spiders and phalangid harvestmen; occasionally small fish and tree-frogs, and eggs of other birds; rarely small molluscs and aquatic larvae; fruit and seeds commonly in winter (Bent 1940, Dawn 1955, Oberholser 1974, Stiles & Skutch 1989, Hayes *et al.* 1990, Agro 1994, Payne 2005).

STATUS AND CONSERVATION Common (Cramp 1985), becoming less common southward in breeding range (Hughes 2001). Rare Colombia (Hilty & Brown 1986), Venezuela (Hilty 2003), Ecuador (Ridgely & Greenfield 2001), Peru (Schulenberg *et al.* 2007) and Argentina (Chébez 1994). Very rare Trinidad (White *et al.* 2007). Decline across breeding range throughout 20th century, mostly 1980s and 1990s (Slack & Wallace 2009). Eggs of Brown-headed Cowbird occasionally found in cuckoo nests (Friedmann 1971, Friedmann *et al.* 1977, Peck & James 1983). Predators include *Accipiter* hawks (Storer 1966) and Aplomado Falcon (Hector 1985). Eggs and nestlings probably taken by snakes and chipmunks (Nolan 1963), nestlings possibly by Common Grackle (Nolan & Thompson 1975). Not globally threatened (BirdLife International 2011).

1

Black-billed Cuckoo *Coccyzus erythropthalmus*. **Fig. 1.** Adult, Jorupe, Loja, Ecuador, April (*Roger Ahlman*). **Figs. 2–3.** South Walsingham Forest, Norfolk County, Ontario, Canada, June (*Michael Patrikeev*).

GREY-CAPPED CUCKOO
Coccyzus lansbergi   Plate 21

Coccyzus lansbergi Bonaparte, 1850 (Santa Fé de Bogotá, Colombia)

TAXONOMY Monotypic. In error *landsbergi* (e.g. Hardy *et al*. 1990).

FIELD IDENTIFICATION 25–28cm. **Adult** Rufous overall, slightly paler below, with contrasting slate-grey cap, extending to below eyes, dark bill (sometimes small yellow spot on base of lower mandible). Tail graduated with large white terminal spots (mainly visible from below). **Juvenile** Brown crown and less distinct pale tips on rectrices, otherwise as adult. **Similar species** No other *Coccyzus* so richly rufous. From Dark-billed, which shares grey cap, by lack of dusky mask.

VOICE Rapidly repeated hollow *cucucucucucucu-cu*, total 6–8 notes, often with pause before last note, 7/sec, falling near end, series sometimes repeated several times in quick succession (Hilty & Brown 1986, Hilty 2003, Restall *et al*. 2006).

DESCRIPTION Adult Forehead, crown and sides of head down to upper edge of malar region slate-grey, hindneck, back, scapulars, tertials, rump and uppertail-coverts dark rufous-brown, wings more rufescent, primaries more greyish-brown on terminal portion of inner webs, underwing-coverts pale cinnamon-buff, flight feathers below dull russet, fading to pale cinnamon-buff toward edge of inner web proximally; malar region, chin, throat and chest deep rufous-buff, fading to cinnamon-buff on belly, undertail-coverts ochraceous-tawny; tail very steeply graduated and blackish-brown, rectrices blunt-ended, T1 narrowly, others broadly, tipped white. **Juvenile** As adult, but crown brown and less distinct pale tips to rectrices. **Nestling** Naked at hatching, skin dark (Marchant 1960). **Bare parts** Iris dark brown. Eye-ring grey, white, dull yellowish-grey or yellow. Bill black, sometimes with small grey or yellow spot on base of lower mandible. Feet grey.

BIOMETRICS Wing male 107–118mm (n = 8), mean 113.9 ± 4.2, female 110–117mm (n = 8), mean 113.2 ± 1.6. Bill male 22.6–27.0mm, mean 24.1 ± 1.7, female 21.4–27.0mm, mean 23.8 ± 2.4. Tail male 117–156mm, mean 130.0 ± 14.0, female 120–142mm, mean 137.2 ± 4.6. Tarsus male 22.4–28.5mm, mean 25.9 ± 1.6, female 24.0–26.0mm, mean 24.4 ± 0.8. Mass male 46–56g (n = 4), mean 50.3 (Payne 2005).

MOULT No information.

DISTRIBUTION North-western South America. North Colombia, north-west Venezuela, west Ecuador and north-west Peru (Hilty & Brown 1986, Ridgely & Greenfield 2001, Hilty 2003, Schulenberg *et al*. 2007). No definite breeding records Venezuela (Hilty 2003). Once Valle, west Colombia (Gochfeld *et al*. 1980). Mainly from Manabí and Los Ríos southward, in some years further north, Ecuador, mostly Jan–Jun; numbers higher in years of high rainfall, or less vocal in drier years. Numbers may vary with outbreaks of caterpillars. Regular non-breeding species Peru, but only very rarely on coast south to Arequipa, Aug–Sep (Stotz *et al*. 1996, Ridgely & Greenfield 2001, Hilty 2003, Schulenberg *et al*. 2007). Only Jun–Aug (breeding not recorded) at site in central Venezuela (Thomas 1979). Few sight records Panama Dec–Feb (Braun & Wolf 1987), specimen Bonaire, Netherlands Antilles, Oct (Voous 1982); vagrant, no details, Galapagos (Ridgely & Greenfield 2001). Irregular or seasonal occurrence in many areas and records of apparent vagrants, including at high elevations suggest some kind of movements – migration, dispersal, nomadism? Movements may be confined to within two areas of occurrence, not between them (Marchant 1958, Fjeldså & Krabbe 1990, Lyons & Perez 2000, Ridgely & Greenfield 2001, Buitrón & Freile 2006).

HABITAT Bushes of wet savanna in central Venezuela. Thickets and dense shrubbery near water, undergrowth of moist to dry semi-deciduous and secondary woodland, clearings and borders. To 600m Colombia; 1,400m Venezuela. Mostly below 900m, exceptionally to 1,800m, Ecuador. Below 800m north-west Peru and rarely further south along coast; once 4,000m Apurímac (Thomas 1979, Hilty & Brown 1986, Fjeldså & Krabbe 1990, Clements & Shany 2001, Ridgely & Greenfield 2001, Hilty 2003, Schulenberg *et al*. 2007).

BEHAVIOUR Solitary; hides low in vegetation (Thomas 1979, Hilty 2003).

BREEDING Season Feb–Mar (rainy season) south-west Ecuador, perhaps sometimes Jan–Jun. Male, breeding condition May, north Colombia (Marchant 1959, Hilty & Brown 1986, Ridgely & Greenfield 2001, Knowlton 2010). **Nest** Platform of thick twigs, lining of lichens, on horizontal branches in thicket or shrubbery. In nine tree species, tree height 3–7m, mean 4.7m, well concealed in all directions by foliage, 0.4–5m up (n = 22), mean 2.5m, external diameter 13–25cm, mean 17.9cm, inside diameter 9–19cm, mean 14.6cm, outside height 4–13cm, mean 7.4cm, inside 3–6cm, mean 4.3cm (Marchant 1960, Knowlton 2010). **Eggs** 2–5, mean 3.7; uniform greenish-white, bluish-green, or white, chalky and rough, both ends blunt and rounded; 25.2–28.0x20.5–21.5mm (n = 6), mean 26.3x21.1mm (Marchant 1960, Schönwetter 1967, Knowlton 2010). **Incubation** 9–12 days. Adult sits very closely (Marchant 1960, Knowlton 2010). **Chicks** Hatching asynchronous, and nestlings noticeably different in size until fledging. Quills grow rapidly, apparently burst together over short period. Development quick. Fledging at *c.* 8–13 days, long before able to fly; number of young in nest fell from three to none over 4–5 days. Nestling period, 22 nests, 9–12 days (Marchant 1960, Knowlton 2010). **Survival** No information.

FOOD Insects, mostly caterpillars, and some wasps (Payne 2005).

STATUS AND CONSERVATION Rare and local Venezuela (Hilty 2003). Few records Colombia (Hilty & Brown 1986). Rare, but occasionally fairly common, Ecuador (Ridgely & Greenfield 2001). Rare Peru (Schulenberg *et al.* 2007). Not globally threatened (BirdLife International 2011).

Grey-capped Cuckoo *Coccyzus lansbergi*. **Fig. 1.** Adult, Jobi, Chocó, Colombia, June (*Digiscoping Juan D. Ramírez Rpo*).

Genus *Hyetornis*

P.L. Sclater, 1862 *Cat. Amer. Birds* p. xiii, 321. New name for *Ptiloleptis* Bonaparte 1854.
Type, by monotypy, *Cuculus pluvialis* Gmelin 1788.
Two species.

Greater Antilles; non brood-parasitic. Large; long-tailed; short-winged; crestless; sexes alike; arboreal. Treated as subgenus in *Piaya* (Peters 1940), and returned to *Piaya* by Stotz *et al.* (1996), but genus usually recognised (e.g. Shelley 1891, Ridgway 1916, Cory 1919, Wolters 1975–1982, Sibley & Monroe 1990, Payne 1997, AOU 1998, Dickinson 2003). Placed in *Coccyzus* based on mtDNA data (Sorenson & Payne 2005), followed by Banks *et al.* (2006); see *Saurothera*.

CHESTNUT-BELLIED CUCKOO
Hyetornis pluvialis   Plate 21

Cuculus pluvialis J.F. Gmelin, 1788 (Jamaica)

Alternative names: Jamaican Hyetornis, Old Man Bird.

TAXONOMY Monotypic. Synonyms: *Piaya cinnamomeiventris* Lafresnaye 1846, *Coccyzus jamaicensis* Hartlaub 1846.

FIELD IDENTIFICATION 48–56cm. Heard more often than seen; large; wings round, tail long; stout, curved bill. **Adult** Upperparts brown, throat white, chest pale grey, contrasting with chestnut lower breast to undertail-coverts. **Juvenile** Like adult, but unglossed tail. **Similar species** Jamaican Lizard-Cuckoo smaller, has long, straight slender bill, and rufous in primaries.

VOICE Guttural *u-ak-u-ak-ak-ak-ak-ak-ak-ak-ak*, *c.* 4 notes/sec, accelerating towards end; soft *qua*; hoarse croak (Hardy *et al.* 1990, Bond 1993, Raffaele *et al.* 1998).

DESCRIPTION Adult Crown dark grey, back, rump and wings dark olivaceous, throat white, chest pale grey, lower breast, belly and undertail-coverts rufous-chestnut; underwing-coverts rufous; tail long, graduated, rectrices broad, faintly purple glossed black, broadly tipped white. **Juvenile** Similar to adult, but with dark brown tail without sheen. **Nestling** Unknown. **Bare parts** Iris red to brown. Eyelids blackish. Bill thick, curved, maxilla blackish, mandible pale grey. Mouth interior black. Bare skin around eye blackish. Feet dark grey to bluish-grey.

BIOMETRICS Wing male 165–189mm (n = 8), mean 177.4 ± 8.4, female 189.5–200mm (n = 7), mean 190.5 ± 4.8. Bill male 39.2–42.0mm, mean 40.2 ± 1.0, female 36.4–47.0mm, mean 42.0 ± 5.2. Tail male 279–309mm, mean 290.4 ± 12.4, female 273–318mm, mean 289 ± 16.4. Tarsus male 39.5–47.5mm, mean 41.8 ± 3.3, female 41–44mm, mean 42.0 ± 1.4. Mass male juv. 129, 130g, female adult 189g (Payne 2005), unsexed 163g (Faaborg 1985 in Payne 2005).

MOULT No information.

DISTRIBUTION Resident Jamaica (Raffaele *et al.* 1998).

HABITAT Thickets in semi-open hill and mountain districts; forest edge, wooded pastures, citrus plantations; 300–1,800m. Infrequent lowlands, mostly winter (Cruz 1973, Bond 1993, Raffaele *et al.* 1998, Stattersfield *et al.* 1998).

BEHAVIOUR Usually alone, sometimes pairs, foraging members stay separated up to several trees apart, keeping in vocal contact. Forages in midstorey and canopy. Gleans prey from branches, twigs and leaves, and (rarely?) catches insects in air. Hops from branch to branch; sometimes runs along limbs. Flight slow, alternating few flaps with gliding, for short distance, landing in cover of shrubbery or trees (Cruz 1973, Lack 1976, Stotz *et al.* 1996, Raffaele *et al.* 1998).

BREEDING Season Mar–Jun; incubating female Feb (Scott 1892, Levy 1996, Raffaele *et al.* 1998). **Nest** Shallow saucer of twigs; up to 10m up in bush or tree (Bond 1993, Levy 1996, Payne 1997). **Eggs** 2–4, yellowish white, 39.8–40.8x30.2–32.1mm, mean 40.3x31.7 (n = 6) (Schönwetter 1967, Bond 1993). No further information.

FOOD Mostly grasshoppers, beetles and caterpillars; 15–17cm phasmids, mantids and other insects; slugs, snails extracted from shells; tree-frogs; anoles; birds' eggs and nestlings; mice (Cruz 1973, Payne 1977, 2005, Raffaele *et al.* 1998).

STATUS AND CONSERVATION Common (Raffaele *et al.* 1998). Small range may render it vulnerable; survey of status and population desirable (Stotz *et al.* 1996, Payne 1997). Periodic hurricanes appear to represent less of a problem than to rarer Jamaican endemics (Varty 1991, Stattersfield *et al.* 1998). Population believed to be larger than 10,000 and stable, and range 11,000km². Not globally threatened (BirdLife International 2011).

Chestnut-bellied Cuckoo *Hyetornis pluvialis*. Adults. **Figs. 1–3.** Port Royal Mountains, Jamaica, November (*Yves-Jacques Rey-Millet*).

RUFOUS-BREASTED CUCKOO
Hyetornis rufigularis   Plate 21

Coccyzus rufigularis Hartlaub, 1852 (mountain forests of Spanish Santo Domingo = Dominican Republic)

Alternative names: Bay-breasted Cuckoo, Haitian Hyetornis.

TAXONOMY Monotypic. Synonyms: *Piaya Pauli Guilielmi* Hartlaub 1852, *fieldi* Cory 1895.

FIELD IDENTIFICATION 43–52cm. Stout, curved bill. **Adult** Throat and breast dark chestnut, upperparts grey, belly buff, chestnut wing patch, tail graduated, black tipped white. **Juvenile** Similar to adult. **Similar species** Sympatric Hispaniolan Lizard-Cuckoo has long, slender bill, grey breast, red eye-ring. Chestnut-bellied Cuckoo (no range overlap) has grey breast and chestnut lower underparts. Mangrove Cuckoo much smaller, has dark mask, yellow lower mandible.

VOICE Loud *cua* or *oowack*; tremolo bleating like large lamb; hoarse, accelerating *u-ak-u-ak-ak-ak-ak-ak-ak-ak-ak* (Bond 1928, Hardy *et al.* 1990, Raffaele *et al.* 1998).

DESCRIPTION Female slightly larger. **Adult** Upperparts, including wing-coverts, sides of head and neck olivaceous-grey; primaries mostly deep chestnut, distally similar to rest of upperparts, but darker, secondaries chestnut or bay, broadly edged greyish; underwing-coverts and axillaries tawny-ochraceous, remiges below cinnamon-rufous, distally abruptly becoming greyish-brown; chin, throat, central chest and breast dark chestnut, flanks, belly and thighs light ochraceous-tawny, fading into buffy-white on undertail-coverts; tail graduated for half of length, rectrices mostly blue-black (only about distal third of T1), tipped white, tips smallest on T1. **Juvenile** Similar to adult, but rectrices narrower, T1 dark grey. **Nestling** Unknown. **Bare parts** Iris brown, eyelids yellow to grey. Bill narrow and deep, curved, with very fine serration along cutting edges, upper mandible black, lower mandible paler, middle lower part yellowish. Feet horn.

BIOMETRICS Wing male 161–182mm (n = 15), mean 171.6 ± 8.8, female 167–188mm (n = 9), mean 180.8 ± 6.9. Bill male 34–41mm, mean 38.4 ± 3.0, female 46–50mm, mean 47.5 ± 1.5. Tail male 243–274mm, mean 260.4 ± 11.8, female 248–295mm, mean 272.2 ± 13.6. Tarsus male 35.5–42.0mm, mean 39.6 ± 2.0, female 40.0–45.0mm, mean 41.9. Mass male 128g (Payne 2005).

MOULT Nothing known.

DISTRIBUTION Resident Hispaniola and Gonâve I. (Raffaele *et al.* 1998).

HABITAT Mostly dry deciduous forest at low to moderate elevation, also locally in dense broadleaf humid montane forest, dense lowland arid scrub, patchy broadleaf woodland composed of exotic fruit trees, yards, gardens. Possibly prefers narrow transition zone between dry forest and moist broadleaf forest. To 900m, sometimes higher (Collar *et al.* 1992, Stotz *et al.* 1996, Raffaele *et al.* 1998, Keith *et al.* 2003, BirdLife International 2011).

BEHAVIOUR Retiring but active; prefers canopy. Moves quickly through forest, running along branches; flight

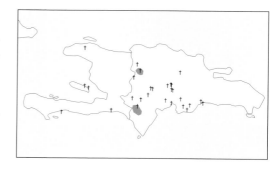

laboured (Cherrie 1896, Bond 1928, Collar *et al.* 1992, Raffaele *et al.* 1998).

BREEDING Season Feb–Jun, tied to onset of wet season with bloom of cicadas (Raffaele *et al.* 1998, BirdLife International 2000, L. Woolaver in BirdLife International 2010). **Nest** Loose stick nest 3–6m up in tree (jabilla or mango), hidden in clump of bromeliad epiphytes or by dense foliage (Bond 1984, A.S. Dod in Collar *et al.* 1992, BirdLife International 2011). **Eggs** 2 (once 3); dirty white with chalky coating; 37.2–39.0x24.7–26.1mm, (n = 2) (Schönwetter 1967, Bond 1984, Collar *et al.* 1992, Raffaele *et al.* 1998, L. Woolaver in BirdLife International 2011). **Incubation** Unknown. **Chicks** Fed almost exclusively cicadas (L. Woolaver in BirdLife International 2011). **Survival** Unknown.

FOOD Locusts, mantids, beetles, pentatomid bugs, grubs, caterpillars; frogs; lizards; birds' eggs and nestlings; mice. One stomach contained 92% lizard remains and 8% insects (Danforth 1929, Wetmore & Swales 1931, Raffaele *et al.* 1998, Payne 2005).

STATUS AND CONSERVATION Last recorded Gonâve I. 1928, when "not uncommon" (Bond 1928). Last recorded mainland Haiti 1983 (Collar *et al.* 1992), and now extremely rare, if not extirpated. Dramatic decline in range and numbers during 20th century due to habitat destruction for agriculture and degradation from grazing, and hunting for medicine and food. Breeding populations now only known to occur at two sites in Dominican Republic, each probably less than 50 pairs: near Río Limpio (village at base of Nalga de Maco NP on lower north slope of Cordillera Central), and near Puerto Escondido (north slope of Sierra de Bahoruco). Convincing local reports from El Tetero on lower south slope of Cordillera Central, on edge of José del Carmen Ramírez NP. Suggested conservation measures: targeted survey using play-back during breeding season throughout potential range; research into basic ecology, territory size, population estimate, impact of habitat modification, hunting and agrochemicals; effective protection of Sierra de Bahoruco and Nalga de Maco NPs; education and awareness programme in communities near remaining known populations (Raffaele *et al.* 1998, BirdLife International 2011). Creation of Loma Charco Azul Biological Reserve north-west of Sierra de Bahoruco NP in 2009 will help conserve this species (American Bird Conservancy press release 20 Oct 2009). Considered globally threatened. IUCN Red List category: Endangered. Population estimate 2,500–9,999, decreasing; range estimate 880km^2 (BirdLife International 2011).

Rufous-breasted Cuckoo *Hyetornis rufigularis.* **Figs. 1–2.** Adult, La Placa, Sierra de Bahoruco, Dominican Republic, June (*Eladio Fernández*).

Genus *Saurothera*

Vieillot, 1816 *Analyse d'une nouvelle Ornithologie élémentaire* p. 28.
Type, by monotypy, 'Coucou à long bec' Buffon = *Cuculus vetula* Linnaeus 1758.
Four species.

West Indies; non brood-parasitic. Large; crestless; sexes alike; short-winged; long-billed. Based on mtDNA data forms clade with *Coccyzus* and *Hyetornis*, in which *Saurothera* and *Hyetornis* are sisters, embedded in *Coccyzus*, and therefore merged into it (Sorenson & Payne 2005); this decision followed in AOU Checklist (Banks *et al.* 2006).

JAMAICAN LIZARD CUCKOO
Saurothera vetula   Plate 19

Cuculus Vetula Linnaeus, 1758 (Jamaica)

Alternative name: Old Woman Bird.

TAXONOMY Considered conspecific with *longirostris* and *vieilloti* (Peters 1940), but separate by most (e.g. AOU 1998, Payne 2005). Apart from plumage, the four species of *Saurothera* differ greatly in length of bill and wing (Lack 1976). Monotypic. Synonym: *jamaicensis* Lafresnaye 1847.

FIELD IDENTIFICATION 38–40cm. Large; long, straight bill. **Adult** Crown brown, rest of upperparts mostly grey, rufous patch in primaries, throat and chest white, belly ochraceous, all rectrices very widely tipped white, so tail looks nearly all white below. **Juvenile** Similar to adult, but rectrices narrow, and secondaries tipped buff. **Similar species** Large size, but smaller than Chestnut-bellied Cuckoo, which has curved bill, grey throat and breast, dark chestnut belly, no rufous in primaries, no red around eye. Mangrove and Yellow-billed Cuckoos smaller, former with dark ear-patch, latter all-white below.

VOICE Rapid, nasal *cak-cak-cak-ka-ka-ka-k-k*... lasting 5-6sec, accelerates slightly from 14 to 17 notes/sec and falls in pitch in second half; higher tempo and pitch, and longer, than voice of Chestnut-bellied Cuckoo. Rasping call; braying *gaah* (Curio 1970, Levy 1984, Hardy *et al.* 1990, Raffaele *et al.* 1998, Payne 2005).

DESCRIPTION Adult Crown, nape and upper hindneck brown, contrasting with brownish-grey mantle, back, uppertail-coverts and wings, except primaries that are bright cinnamon-rufous tipped greyish-olive, outer secondaries and primary coverts tinged rufous; cheeks, lower ear-coverts, chin and throat white, slightly tinted grey on sides of throat, rest of underparts rufous-buff, except pinkish or creamy-buff underwing-coverts, and pale buff undertail-coverts; tail long, graduated, T1 grey, other rectrices bluish-black, all very broadly tipped white with black subterminal area. **Juvenile** Similar to adult, but plumage fluffy, rectrices narrow, T1-2 without black subterminal area, secondaries tipped buff. **Nestling** At *c.* 7 days, has reddish-brown in wing, whitish throat. Otherwise unknown. **Bare parts** Iris brown. Small bare area around eye scarlet. Bill long, straight, above black, below grey. Feet grey.

BIOMETRICS Wing male 120–128mm (n = 10), mean 123.8 ± 3.0, female 117–136mm (n = 6), mean 128.8 ± 3.7. Bill male 43–51mm, mean 48.2 ± 2.6, female 38.7–49.0mm, mean 43.9 ± 4.5. Tail male 194–203mm, mean 199.0 ± 3.8, female 190–227mm, mean 202.0 ± 13.5. Tarsus male 29.0–32.4mm, mean 30.2 ± 1.4, female 30–32mm, mean 30.9. Mass, unsexed 86.0, 104.6g (Payne 2005).

MOULT No information.

DISTRIBUTION Resident Jamaica (Raffaele *et al.* 1998, Haynes-Sutton *et al.* 2009).

HABITAT Lowland limestone scrub to humid montane forest and wooded ravines, mostly at mid-elevations, usually in dense thickets and woodland, including second growth, rarely in more open country; to 1,200m. Nest in plantation including pine 100m from primary montane forest at 1,140m. (Cruz 1975, Levy 1984, Raffaele *et al.* 1998, Stattersfield *et al.* 1998).

BEHAVIOUR Usually singly; inconspicuous; more often heard than seen. Moves slowly through undergrowth, 3–4m up, peering around; mostly forages among inner branches of tall shrubs and trees, occasionally on ground; mostly gleans prey from ground or branches, sometimes probes into rotting logs. When detecting prey, moves forward rapidly to seize it. Pecks at head of prey or squashes it in bill before swallowing. Female holding green caterpillar in bill crouched on branch with raised belly feathers, holding tail horizontally and jerking it upwards rhythmically, male arrived, copulated for 2–3sec, then immediately disappeared; female kept caterpillar in bill during copulation, perhaps giving it to male afterwards. Approaches nest with young from below, hopping up branches (Curio 1970, Cruz 1975, Lack 1976, Levy 1984, Raffaele *et al.* 1998).

BREEDING Season Mar–Aug; copulation Oct (Curio 1970, Levy 1984, Raffaele *et al.* 1998, Haynes-Sutton *et al.* 2009). **Nest** Loose platform of sticks lined with leaves, low to medium height, diameter 17–21cm, depth 2.5cm, many loose twigs and small branches surrounding nest up to 5cm, well hidden in tangle (March 1863, Levy 1984, Raffaele *et al.* 1998). **Eggs** 3–4; dull white to greenish or pale blue with thin outer layer; 32.5x24.3mm (March 1863, Scott 1892, Nehrkorn 1910, Schönwetter 1967, Raffaele *et al.* 1998).

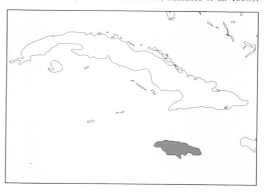

Incubation By both parents (Levy 1984). **Chicks** Stretch neck upwards and hiss when begging for food. Fed lizards by both parents (Levy 1984). **Survival** Unknown.

FOOD Mantids, cicadas, caterpillars, locusts, beetles; spiders; rarely tree-frogs; *Anolis* lizards; bird nestlings; mice (Gosse 1847, Danforth 1928, Cruz 1975, Payne 2005).

STATUS AND CONSERVATION Uncommon (Cruz 1975, Stotz *et al.* 1996) or fairly common (Haynes-Sutton *et al.* 2009), not as common as Chestnut-bellied Cuckoo, but widespread (Raffaele *et al.* 1998). Scott (1892) thought it had suffered from predation by mongoose. Not globally threatened (BirdLife International 2011).

Jamaican Lizard Cuckoo *Saurothera vetula*. **Figs. 1–2.** Adult, Ecclesdown, Jamaica, November (*Yves-Jacques Rey-Millet*). **Fig. 3.** Adult, Jamaica, November (*Wenfei Tong*).

PUERTO RICAN LIZARD CUCKOO
Saurothera vieilloti Plate 19

S[*aurothera*] *vieilloti* Bonaparte, 1850 ('Antilles' = Puerto Rico)

Alternative name: Vieillot's Lizard Cuckoo.

TAXONOMY Sometimes misspelled *vieillotii*. Forms superspecies with *vetula* and *longirostris*, and possibly also with *merlini* (AOU 1983, Sibley & Monroe 1990). Monotypic. Synonym: *S. vieilloti* var. *rufescens* Bryant 1866.

FIELD IDENTIFICATION 40–48cm. Large; tail very long, underparts two-toned. **Adult** Throat and breast grey, belly and undertail-coverts tawny-ochraceous, upperparts brown, no contrasting rufous in wings, bill long and straight. **Juvenile** Above with rufous edges, breast with cinnamon wash, belly and tail paler than adult. **Similar species** Smaller Mangrove Cuckoo has dark mask. Hispaniolan Lizard Cuckoo (no range overlap) has grey upperparts, rufous wing patch, sometimes ochraceous throat.

VOICE Forceful staccato series *ka-ka-ka-ka...*, 8–10 notes/sec lasting 6–9sec, sometimes increasing in amplitude or accelerating, falling in pitch towards end, or first few notes slightly longer and slower; two raucous notes, each lasting c. 0.7sec, total 1.7sec, repeated after 3–4 sec interval (Hardy *et al*. 1990).

DESCRIPTION Adult Males slightly paler. Above plain greyish-brown, secondaries and tail more olivaceous, outer webs of primaries brown to greyish-brown, proximal half of inner webs cinnamon-rufous; chin and throat whitish, grading into pale buffy-grey breast, belly to undertail-coverts and underwing-coverts deep tawny-ochraceous; rectrices grey with black subterminal area and white tips, both areas increasing from central to outermost pair. **Juvenile** Similar to adult, but upperparts edged rufous, wing-coverts and secondaries indistinctly edged dull tawny-brown; belly paler rufous, rectrices more pointed, T1 without white tip and black subterminal area, on rest of rectrices much less sharply defined and smaller, white tips tinged brown. **Nestling** Unknown. **Bare parts** Iris brown. Bare ocular skin red, yellow in juvenile. Bill long, straight, black above, pale below. Feet grey.

BIOMETRICS Wing male 122–138mm (n = 11), mean 128.5 ± 4.2, female 125–134mm (n = 10), mean 130.4 ± 3.6. Bill male 46.9–50.0mm, mean 47.6 ± 1.8, female 43.5–50.0mm. Tail male 224–251mm, mean 237.8 ± 9.7, female 230–267mm, mean 244.9 ± 11.4. Tarsus male 28.8–39.8mm, mean 32.1 ± 3.0, female 29–36mm, mean 34.2 ± 2.0. Mass male 74.5, 83.3, 83.4g, female 87.9, 96.9g, laying female 107.0, 110.5g, unsexed 68–104g (n = 23), mean 81.5 ± 10.6 (Payne 2005). Another male 73.6g (Olson & Angle 1977).

MOULT No information.

DISTRIBUTION Resident, Puerto Rico; formerly nearby

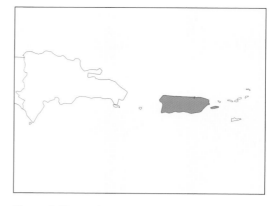

Vieques I. (Newton & Newton 1859, Sorrié 1975, AOU 1998, Raffaele *et al*. 1998). Old specimen St. Thomas, U.S. Virgin Is. (Shelley 1891) considered doubtful (Peters 1940) or stray (Raffaele *et al*. 1998).

HABITAT Woodland, brush in limestone hills, coffee plantations, thick montane forest, haystack hills, dry coastal forest, lowland swampy forest; to 800m, sometimes higher (Wetmore 1927, Bond 1993, Stotz *et al*. 1996, Raffaele *et al*. 1998).

BEHAVIOUR Rather inactive, staying quietly inside cover of foliage, not very alarmed by approach of observer; more often heard than seen. Forages in midstorey and canopy, occasionally on ground. Early morning may perch in open for sunning, with lowered wings and ruffled feathers (Wetmore 1916, Stotz *et al*. 1996, Raffaele *et al*. 1998).

BREEDING Season Recently fledged Sep, nest-building Feb; also Apr, May, Nov, Dec, thus perhaps breeds all year (Struthers 1923, Wetmore 1927, Raffaele *et al*. 1998, Payne 2005). **Nest** Loose structure of sticks and leaves (Raffaele *et al*. 1998). **Eggs** 2–3; white to pale blue with thin white outer layer, 32.0–35.1x23.5–26.5mm (n = 3), mean 33.8x25.1 (Nehrkorn 1910, Schönwetter 1967, Raffaele 1989, Raffaele *et al*. 1998, Gleffe *et al*. 2006). **Incubation** Unknown. **Chicks** Unknown. **Survival** 50% eggs fledged in secondary forest (four nests, mean 1.00 ± 1.15 fledged of 2.00 eggs/nest); nest in coffee plantation failed (Gleffe *et al*. 2006).

FOOD In contents of 11 stomachs, at least four species of anoles (small lizards) constituted 79%; rest large spiders, and cicadas, caterpillars, moths, beetles, earwigs and grasshoppers (Wetmore 1916). In another sample insects, mostly beetles, dominated (Bowdish 1902).

STATUS AND CONSERVATION Fairly common (Raffaele *et al*. 1998). Formerly more common Puerto Rico (Taylor 1864, Wetmore 1916, 1927, Bond 1947). Possibly eradicated by hunting on Vieques in 19th century (Sorrié 1975). Not globally threatened (BirdLife International 2011).

Puerto Rican Lizard Cuckoo *Saurothera vieilloti.* **Figs. 1–2.** Adults, Maricao State Forest, Puerto Rico, USA (*Lucas Limonta*).

GREAT LIZARD CUCKOO
Saurothera merlini   Plate 19

Saurothera merlini d'Orbigny, 1839 (Cuba)

Alternative name: Cuban Lizard Cuckoo.

TAXONOMY Polytypic. Synonym: *andria* G.S. Miller 1894 in *bahamensis*.

FIELD IDENTIFICATION 44–55cm. Largest lizard-cuckoo; only one in Cuba and Bahamas. Frequently glides. **Adult** Above olive-brown, greyish-brown or grey depending on subspecies, breast grey, belly buff or ochraceous. Rufous patch in primaries in Cuba. Bill long, rather straight. Tail long, graduated, grey, below subterminally black tipped white. **Juvenile** More indistinct tail pattern. **Similar species** Three *Coccyzus* cuckoos from Cuba and Bahamas much smaller, have white underparts, and shorter bills and tails.

VOICE Harsh, grating, repeated, loud *tac-o* or *tacóó*. Two-part *tacooo-tacooo, ka-ka-ka-ka-ka-ka...*, second part gradually increasing in volume, accelerating slightly from 11 to 12.5, then slowing to 11 notes/sec towards end, whole series lasting *c.* 9sec; louder, longer and faster than similar call of Mangrove Cuckoo. Croaking *cra-cra-cra*, loud *ack-ack-ack* (Gundlach 1874, Barbour 1923, Brudenell-Bruce 1975, Hardy *et al.* 1990, Raffaele *et al.* 1998, Garrido & Kirkconnell 2000).

DESCRIPTION Nominate. **Adult** Above rufous, more brown on crown, back and mantle, some olive gloss on secondaries and rectrices; primaries bright rufous with broad olive-brown tips, inner webs of secondaries mostly cinnamon-rufous, especially outer ones, underwing-coverts rufous-buff; chin and throat white to pale greyish-buff, breast grey, belly and undertail-coverts golden-chestnut. Rectrices except T1 broadly tipped white and with even broader subterminal bars, below pale brownish-grey before black bar. **Juvenile** Similar to adult, but rectrices narrower, more pointed, without prominent subterminal bar, tips smaller and duller. **Nestling** Unknown. **Bare parts** Iris grey to brown, brown in juvenile. Bare ocular area orange to red, yellow in juvenile. Bill very long, only slightly downcurved, blue-grey, darkest above. Feet grey to pink or horn-coloured.

BIOMETRICS (Cuba; presumably nominate). Wing male 163–182mm (n = 6), mean 174.0 ± 6.6, female 172–196mm (n = 6), mean 180.0 ± 10.3. Bill male 49–63mm, mean 54.2 ± 5.2, female 50.0–59.5mm, mean 54.8 ± 3.4. Tail male 257–352mm, mean 312.0 ± 31.5, female 280–362mm, mean 320.0 ± 31.0. Tarsus male 38.0–46.5mm, mean 41.8 ± 3.0, female 40–49mm, mean 45.8 ± 5.3. Mass male 124, 145g, female 127, 186g (Payne 2005).

MOULT Six outer primaries usually moulted in sequence 6–9–7–10–8–5 (or 5–8) (Piechocki 1971).

GEOGRAPHICAL VARIATION Four subspecies.
S. m. merlini d'Orbigny, 1839. Cuba. Described above.
S. m. bahamensis Bryant, 1864. Andros, New Providence and Eleuthera (with Harbour I.), Bahamas (Brudenell-Bruce 1975). Smaller, grey above, belly buff; little rufous in remiges, middle rectrices tipped black (Ridgway 1916, Bond 1993, Payne 2005).
S. m. decolor Bangs & Zappey, 1905. Isla de la Juventud (=Isle of Pines, Isle of Youth, Isla de Pinos), Cuba. Greyish-brown above (Bond 1993). Bill shorter than *merlini* (Bangs & Zappey 1905).
S. m. santamariae Garrido, 1971. Islands north of Camagüey and Ciego de Ávila provinces, off north-central Cuba (Garrido & Kirkconnell 2000, Dickinson 2003). Smaller, pale above, longer bill (Payne 1997, 2005).

DISTRIBUTION Resident. Cuba, Isla de la Juventud; some islands in Bahamas (Bond 1993).

HABITAT Forest, woodland, shrub, mainly with many vines, abandoned coffee plantations, overgrown pastures, limestone hills, pine forest. To 1,200m, sometimes higher (Barbour 1943, Stotz *et al.* 1996, Raffaele *et al.* 1998, Garrido & Kirkconnell 2000).

BEHAVIOUR Very tame, mostly solitary; poor flier, often gliding from treetops; runs and forages on ground, with tail straight but hanging down when in trees. In midstorey and canopy, hopping upwards from branch to branch, then

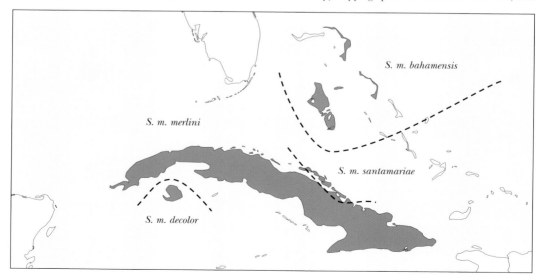

flying clumsily to nearby tree or ground. Two birds in sight of each other lowered heads, extended throat feathers, spread wings and tail fully, and gave loud chuckling notes. Very vocal, but silent during midday heat. Occasionally joins mixed-species bird flocks (Bangs & Zappey 1905, Barbour 1943, Vaurie 1957, Stotz *et al.* 1996, Raffaele *et al.* 1998, Garrido & Kirkconnell 2000, Hamel & Kirkconnell 2005).

BREEDING Season Mostly Apr–May, rarely Oct; large ova May (Gundlach 1874, Northrop 1891). **Nest** Flimsy cup-shaped stick nest lined with leaves, low to medium height, well hidden in tree or bush; small for size of bird (Raffaele *et al.* 1998, Garrido & Kirkconnell 2000). **Eggs** 2–3; white, slightly shiny, not very smooth, 38.8–42.3x29.0–31.5mm (n = 6), mean 40.3 x 30.0 (Gundlach 1874, Schönwetter 1967, Bond 1993, Garrido & Kirkconnell 2000). No further information.

FOOD Grasshoppers, beetles, caterpillars, large moths, bees, wasps; frogs; lizards, small snakes; birds' eggs and nestlings; sometimes mice, small fruits and berries (Gundlach 1874, Northrop 1891, Bangs & Zappey 1905, Garrido & Kirkconnell 2000, Payne 2005).

STATUS AND CONSERVATION All three Cuban subspecies common and widespread (Todd 1916, Bond 1956, Wallace *et al.* 1996, Garrido & Kirkconnell 2000); uncommon Bahamas (Raffaele *et al.* 1998), though fairly common New Providence and "not uncommon" on east coast, Andros (Bond 1956). Not globally threatened (BirdLife International 2011).

Great Lizard Cuckoo *Saurothera merlini*. **Fig. 1.** Adult, *S. m. merlini*, Zapata, Matanzas province, Cuba, March (*Jos Wanten*).

HISPANIOLAN LIZARD CUCKOO
Saurothera longirostris Plate 19

Cuculus longirostris J. Hermann, 1783 (Hispaniola)

Alternative name: Haitian Lizard Cuckoo.

TAXONOMY Polytypic. Synonyms: *longirostris* has priority over *dominicensis* Lafresnaye 1847 (Stresemann 1920); *l. saonae* Bond 1933 in nominate (Payne 1997, Keith *et al.* 2003).

FIELD IDENTIFICATION 41–46cm. **Adult** Above grey with chestnut wing patch, throat dull orange, whitish in Gonâve subspecies, breast pale grey, belly ochraceous, tail long, graduated, broadly tipped white; straight, slender bill. **Juvenile** See Description. **Similar species** From Rufous-breasted Cuckoo by grey on breast, paler throat, red around eyes and long bill. On mainland Hispaniola, dull orange throat different from other *Saurothera* (all allopatric). Yellow-billed, Black-billed and Mangrove Cuckoos smaller.

VOICE Rapid, long, guttural *ka-ka-ka-ka-ka-ka-ka-ka-ka-ka... kau-kau-ko-ko*, 11 notes/sec, total 7–10sec, falling in pitch and slowing a bit on last 4–5 notes. Nasal, repeated *check, checker*, interval 0.6sec, or quick *checkerer* lasting c. 0.2sec, interval c. 0.8sec (Hardy *et al.* 1990, Raffaele *et al.* 1998).

DESCRIPTION Nominate. **Adult** Above grey, crown and nape slightly more brownish and darker, primaries bright cinnamon-rufous, broadly tipped greyish-brown; sides of head and upper throat ochraceous, lower throat and upper chest ashy, lower chest and underwing-coverts to undertail-coverts ochraceous-buff; tail above olive-greyish, T1 broadly tipped black and narrowly margined grey or whitish at tips, remaining rectrices black below, broadly tipped white. **Juvenile** As adult, but above brownish-grey, tail brown, rectrices narrower, with white ends tipped buff. **Nestling** Unknown. **Bare parts** Iris brown to reddish-brown. Bare skin around eyes red, presumably yellow in juvenile as in congeners. Bill dull black above and on tip below, rest pale grey. Feet grey, toes yellowish below.

BIOMETRICS Nominate. Wing male 126–139mm (n = 11), mean 134.3 ± 4.4, female 132–142mm (n = 8), mean 137.8 ± 3.8. Bill male 43–53mm, mean 49.4 ± 3.4, female 40.0–51.5mm, mean 48.2 ± 2.7. Tail male 197–241mm, mean 217.7 ± 12.8, female 228–238mm, mean 232.3 ± 5.5. Tarsus male 31–35mm, mean 33.5 ± 1.3, female 34.5–39.0mm, mean 36.3 ± 2.1. Mass male 83.4–99.5g (n = 6), mean 92.3g, female 92.5–128.5g (n = 6), mean 105.4g (Payne 2005).

MOULT Unknown.

GEOGRAPHICAL VARIATION Two subspecies.
S. l. longirostris (Hermann, 1783). Hispaniola, Tortue and Saona (Bond 1928, Wetmore & Swales 1931, Payne 1997). Described above.
S. l. petersi Richmond & Swales, 1924. Gonâve I., off west coast of Haiti (Peters 1940). Throat whitish, upperparts, belly, undertail-coverts and underwing-coverts paler than in nominate.

DISTRIBUTION Resident, Hispaniola and surrounding

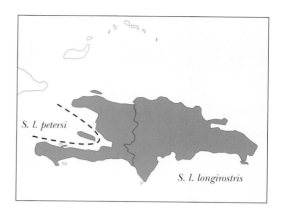

islands: Gonâve, Saona, Tortue (Bond 1993); Isla Beata and Isla Catalina (breeding not confirmed) (Keith *et al.* 2003).

HABITAT Forested and wooded areas, shade coffee plantations, highland pine forest, dense undergrowth and thickets, scant stands of bushes on barren mountainsides, sometimes gardens. To 2,000m, rarely to 2,200m (Wetmore & Swales 1931, AOU 1983, Stotz *et al.* 1996, Wunderle & Latta 1996, Raffaele *et al.* 1998, Keith *et al.* 2003, Latta *et al.* 2006).

BEHAVIOUR Very confident. Moves in leisurely manner, walks with long strides along branches, crouches and proceeds stealthily when hunting, creeps and crawls along branches near ground, often rests quietly for several minutes. Attempts to escape danger by climbing swiftly up tree, only taking flight as last resort, glides to low in nearby tree. Early morning, fluffs feathers in sun. Forages between understorey and canopy. In shade coffee forages only in canopy, dominated by *Inga vera*. Prey captured from branches, foliage and epiphytic lichens by gleaning and leaping. Catches lizards waiting motionless, then darts forward. Occasionally in mixed-species flocks in pine forest (Cory 1881, Kaempfer 1924, Wetmore & Swales 1931, Latta & Wunderle 1996, Raffaele *et al.* 1998, Wunderle & Latta 1998).

BREEDING Season Mar–Jun; also nestling Dec (Kaempfer 1924, Danforth 1929, Wetmore & Swales 1931, Raffaele *et al.* 1998). **Nest** Concealed, flat, coarse stick nest lined with leaves, on tree stump 50cm up (sometimes higher?) (Kaempfer 1924, Raffaele *et al.* 1998). **Eggs** 2–3; white or pale blue with thin white outer layer, 36.7x25.4mm (Schönwetter 1967, Raffaele *et al.* 1998). No further information.

FOOD Large grasshoppers, cockroaches, mantids, cicadas, caterpillars, damselflies; anoles, snakes (Christy 1897, Kaempfer 1924, Danforth 1929, Wetmore & Swales 1931).

STATUS AND CONSERVATION Common (Bond 1993, Raffaele *et al.* 1998, Latta 2008). Possibly less common Gonâve (Danforth 1929) and Tortue (Bond 1928). Declined 88% at Los Haïtises NP 1976–1996 (Keith *et al.* 2003). Not globally threatened (BirdLife International 2011).

Hispaniolan Lizard Cuckoo *Saurothera longirostris*. **Figs. 1–2.** Adults, *S. l. longirostris*, Parque Nacional del Este, Dominican Republic. January (*Nick Kontonicolas*). **Fig. 3.** Adult, *S. l. longirostris*, Santo Domingo, Dominican Republic, March (*Oda Gallart-Brea*).

Genus *Pachycoccyx*

Cabanis 1882 *J. Orn.* 30: 230. Type, by monotypy *Cuculus validus* Reichenow 1879.
One species.

Afrotropical; brood-parasitic. Broad bill, culmen strongly-curved; slit nostrils; sexes similar. MtDNA analysis places it in sister clade to *Microdynamis* and *Eudynamys* (Sorenson & Payne 2005).

THICK-BILLED CUCKOO
Pachycoccyx audeberti   Plate 22

Cuculus audeberti Schlegel, 1879 (Ambodikilo, Antongil Bay, Madagascar)

TAXONOMY Placed in *Coccystes* by Oustalet (1886). Polytypic. Synonyms: *validus canescens* Vincent 1933 in *validus*.

FIELD IDENTIFICATION 36cm. Large, rather *Accipiter*-like cuckoo; restless; often noisy; may be conspicuous as when display-flighting. **Adult** Dark slate-grey above and white below. Bill hook-tipped and heavier than in many cuckoos. Tail dark-barred and spotted with white on sides. Perches rather more upright than many cuckoos. Voice distinctive. **Juvenile** Broad white edges to dark brown upperparts, imparting a scaly appearance. Underparts creamier than adult, particularly on belly and flanks. **Similar species** Scaly pattern of juvenile recalls adult Great Spotted Cuckoo, and white underparts of adult and juvenile eliminate all but *Clamator* cuckoos. However, Thick-billed lacks crest, has entirely dark wings, and tail noticeably shorter and thicker than in *Clamator*.

VOICE Loud, far-carrying, rising series of paired notes *wirr-wi*, *whuee-di* or *hwee-wik*. Loud whistled *whee-whee-wheep*. Also loud intermittent chattering *chee-cher-cher-cher! Cher-cher!* and an undulating call starting low and slow, building to crescendo and then falling away again *kloo kloo kla kla kla klo kloo kloo*. In flight, sometimes gives a soft *weedidi weedidi weedidi*. Fledglings give a disyllabic squeak, like call of host Retz's Helmet-shrike. Calls embellished excitedly when two or more birds together (Haydock 1952, Asterley Maberley 1961, Benson & Irwin 1972, Irwin 1988, Borrow & Demey 2001, Stevenson & Fanshawe 2002).

DESCRIPTION *P. a. validus*. **Adult** Crown, nape and upperparts uniform blackish-brown above when fresh, fading to dark brown with time. Ear-coverts and lores more persistently slate-grey, although lores sometimes whitish. Remiges and wing-coverts blackish-brown, thinly margined and tipped white, although may be lost rapidly through abrasion (Rowan 1983). Underside of remiges pale grey, underwing-coverts white. Underparts entirely white with some variable greyish wash, undertail-coverts white with one or two widely spaced brown or black bars, giving overall effect of four to five bars, and broadly tipped white. Tail dark grey, with indistinct broad blackish-brown bars. Undertail with broad greyish-white bars, broad dark subterminal bar and white tip. **Juvenile** Forehead whitish, feathers of crown, mantle and wings with broad white tips, giving scaly appearance. Underparts whitish but washed buff on belly and flanks; extent and tone of buff varies, and in some white may be more grey; underwing coverts buff-white to deep buff. Broad white tips to rectrices. Some are more pied, with white head, and broad white tips to wing-coverts, remiges and uppertail-coverts. **Nestling** Hatchling has orange skin, unlike purple-skinned young of host; bill and feet yellow, gape orange, darkening within a few days to leave only two orange spots. Quills erupt eighth days, and bill darkens, becoming black by day 12, when other feathers erupt. Upperparts pale creamy-buff at 14–18 days, primaries dark brown basally. One in Zambia reddish-brown with black bill, yellow-orange inside (Benson & Irwin 1972). **Bare parts** Iris brown, narrow eye-ring yellow. Bill black, base of lower mandible yellowish or greenish, entirely black in juvenile. Feet yellow or orange-yellow, pale yellow in juvenile, claws black (Rowan 1983, Irwin 1988).

BIOMETRICS *P. a. validus*. Wing male (n = 33) 214–240mm, mean 225; female (n = 18) 218–236mm, mean 224. Bill male (n = 18) 26–32mm, mean 28.8; female (n = 10) 25–32mm, mean 28.7. Tail male (n = 30) 166–205mm, mean 183; female (n = 19) 163–198mm, mean 180. Tarsus unsexed (n = 10) 25–27mm, mean 26.0. Mass male 92g, female (n = 4) 100–120g, mean 115 (Irwin 1988); female (n = 5) 83–120g, mean 105 (Verheyen 1953); fledgling 84, 86g (Benson & Irwin 1972).

MOULT Interrupted primary moult Malawi, Dec (BMNH).

GEOGRAPHICAL VARIATION Three subspecies, poorly differentiated.

P. a. audeberti (Schlegel, 1879). Madagascar. Larger than other two subspecies, bill sometimes all dark, no greyish on head or neck sides (Friedmann 1948b). Dusky bars on rectrices virtually obsolete, much more obvious white fringing on remiges and coverts (Benson & Irwin 1972). Wing male 237mm, female 238mm (Payne 2005).

P. a. validus (Reichenow, 1879). East DRC to extreme south Sudan, north-west and south Uganda and south-east Kenya (few records east central and extreme north) and Tanzania, through Malawi, Zambia, Zimbabwe and Mozambique to east Transvaal, RSA, and isolated population in Angola; at least eight records Botswana; once Natal (Benson & Irwin 1972, Irwin 1988). Described above.

P. a. brazzae (Oustalet, 1886). Isolated populations in SL, Guinea, Liberia, IC, Mali, Ghana, Togo, Benin, Nigeria, Cameroon, CAR, Gabon, RC and extreme west DRC. Smaller than *validus*, blacker and more slaty above, and lores darker. Said to intergrade with *validus* in Kivu Province and north-east DRC and south-west Sudan (Benson & Irwin 1972), although large gap in distributions. Wing male 204, 218, 232mm, female 216, 233mm (Payne 2005).

DISTRIBUTION Afrotropical. Sedentary, but may undergo local seasonal movements. Discontinuous isolated populations in forest fragments in west Africa from SL, Mali, Liberia, IC, Togo north to 10°21'N, Benin (two localities), Guinea north to 09°30'N, Burkina Faso (once), Ghana, CAR to RC, DRC and Gabon, few records Uganda and Angola. Seasonal movements west Africa, but resident

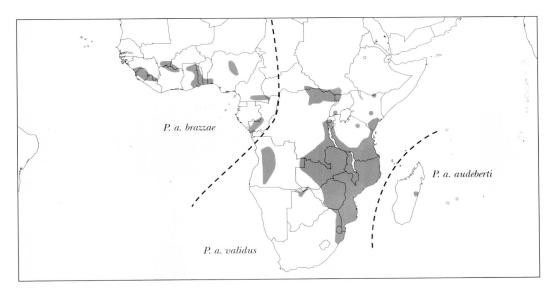

Nigeria. More continuous distribution in east Africa, from coastal south-east Kenya and south Sudan (where possibly only rare non-breeding visitor, although claimed resident in Equatoria) to north Botswana, north-east RSA and south Mozambique. No distinct migration apparent across most of range, although possibly altitudinal migrant RSA, where retreats from high altitudes during late austral summer, following breeding pattern of host. Absent Malawi May and Nov. Sep–May, Zambia, perhaps resident (Sclater & Mackworth-Praed 1919, Chapin 1939, Cave & MacDonald 1955, Pinto 1959, Benson et al. 1971, Benson & Irwin 1972, A.D. Forbes-Watson in Benson & Irwin 1972, Benson & Benson 1977, Rowan 1983, Thiollay 1985, Grimes 1987, Irwin 1988, Dowsett 1989, Thonnérieux et al. 1989, Cheke & Walsh 1996, Payne 1997, Borrow & Demey 2001, van den Akker 2003, Dowsett & Dowsett-Lemaire 2005, Demey 2005, 2007, S. Dark & J. Squire *in litt.*). Resident Madagascar which was presumably colonised from Africa (Benson & Irwin 1972), where now very rare with just two records since 1922, a bird that responded to playback at Maromiza forest 1992, and sighted Andringitra,1993 (Payne 1997, 2005); also previously recorded Maroanstra, Mananara, Sihanaka forest and near Rogez, Andekaleka (Langrand 1990).

HABITAT Differs across range. In west Africa favours forest edge and gallery forest, in east and south moist open savanna woodland, with stands of miombo, teak *Baikaiea plurijuga* and mopane *Colophospermum mopane*, but also in riparian forest on alluvial soils particularly at lower altitudes, and cocoa plantations on edge of forest. In east Africa, distribution coincides with area of 500+mm rainfall. To 1,400m, but 1,600m Zimbabwe. Undisturbed rainforest to 900m, Madagascar (Langrand 1990), but avoids evergreen forest in much of Africa except in DRC, although favours gallery forest over heavy forest (Chapin 1939, Benson & Irwin 1972, Irwin 1981, Rowan 1983, Allport & Fanshawe 1994, Payne 1997).

BEHAVIOUR Restless and vocal. Usually solitary, but sometimes in conspicuous groups of up to five birds. Usually seen during display flights high above canopy, visible at >1km distance, possibly associated with territorial advertisement, after which lands on exposed branch or plunges into canopy. Females make display flights over larger area, thought to announce presence. Courtship feeding observed at termination of display flight. Flight slow and buoyant, wings performing 'rowing' action. Despite *Accipiter*-like appearance, not frequently mobbed by passerines. May ruff out feathers forming frontal crest (Townley 1936, Payne 1968, Rowan 1983, Vernon 1984, Dyer et al. 1986, Irwin 1988).

BREEDING Parasitism Confirmed hosts: Retz's Helmet-shrike, Zimbabwe, nine of 24 nests parasitised; Red-billed Helmet-shrike. White Helmet-shrike DRC. In west Africa, Retz's Helmet-shrike does not occur, and Red-billed Helmet-shrike presumed host there, although White Helmet-shrike perhaps used. Chestnut-fronted Helmet-shrike Angola, but not White Helmet-shrike. Possibly Amethyst Starling, Zimbabwe. Chestnut-fronted Helmet-shrike possibly south-east Kenya. Retz's Helmet-shrike, Tanzania. Madagascar host unknown, possibly Chabert Vanga. 37–55% of host nests within study area in Zimbabwe parasitised. Perhaps help to feed own young (Townley 1936, Newby-Varty 1946, Pinto 1959, Vernon 1984, Irwin 1988, Louette 1989, Payne 1997, R. Peek in Payne 2005, R. Marais *in litt.*). **Season** Probably pre-rains/early rains breeder south Tanzania, Malawi, Zambia and Zimbabwe, although mainly late rains, Sep–Mar, particularly Oct–Nov, Zimbabwe. Oviduct eggs: Cameroon, Mar and Jul; Kenya, Sep–Nov; Tanzania, Oct. Nestlings: east DRC, Jan and Mar; Zambia, Sep–Oct. Fledgling south Tanzania, Jan. About to lay south DRC, Jun. Female with yolking eggs Zimbabwe, Jan, and egg-laying observed there twice Sep, eight times Oct, seven times Nov and once Dec, fledged young Jan and May, possibly Dec; RSA, Jan; active testes Tanzania, Oct, and fledged young with Retz's Helmet-shrike Jan (Townley 1936, Jackson 1938, Chapin 1939, Newby-Varty 1946, Friedmann 1948b, Verheyen 1953, Irwin 1957, 1988, Vernon 1971b, 1978, 1984, Benson & Irwin 1972, Louette 1989, R. Marais *in litt.*). **Eggs** Highly mimetic. Pale creamy-green or blue-green ground colour, spots and blotches of pale brown, grey, grey-brown or lilac, merging toward blunt end, similar in colour to Retz's Helmet-shrike, but slightly larger. Fewer spots than eggs of host. One confirmed egg 23.8x17.2mm, within normal size range of host eggs. Oviduct egg, east Africa, pale greenish-blue with scattered small brown and grey-brown speckles. Laying commences when host clutch complete, thus about four days into host incubation period. Cuckoo attacks incubating helmet-shrike, often flying directly at it, and removes host egg, presumably laying subsequently, although laying has not been directly observed (Fischer 1879, Chapin 1939, Jensen & Jensen 1969, Benson & Irwin 1972, Rowan 1983, Vernon 1984). **Incubation** 13 days or less, at least 4 days shorter than incubation period of host (Rowan 1983). **Chicks** Resemble foster parent, White Helmet-shrike, so closely as to be mistaken for it. Evicts remaining host eggs then young at 2–4 days. By 14–18 days, chick fills nest, gapes to show black and orange mouth pattern and lunges at intruders, raising head and tail feathers. Voids evil-smelling fluid if handled. Quills erupt 8th day, and bill darkens, becoming black by 12th day, when other feathers erupt. Fledge at 28–30 days and fed by host for another 50 days. As cuckoo nears independence, hosts may mob it or call at it when it flies, although feed it intermittently when perched (Rowan 1983, Irwin 1988, Louette 1989). **Survival** 18 of 29 eggs hatched and six young survived to fledging, thus 21% overall survival (Vernon 1984).

FOOD Orthoptera dominant in six stomachs, another stomach contained hairy caterpillars and mantid; caterpillars, Kenya; marula worms. Caterpillars, Orthoptera and mantis, DRC (Fischer 1879, Chapin 1939, Benson *et al.* 1971, Benson & Irwin 1972, Rowan 1983, Payne 1997).

STATUS AND CONSERVATION Uncommon to rare across much of range, particularly in west Africa where many populations isolated (Irwin 1988, Payne 1997). Not rare Lamto, IC (Brunnel & Thiollay 1969). Rare Togo (Cheke & Walsh 1996). Uncommon resident Zande district, Equatoria province, Sudan (Cave & MacDonald 1955), otherwise rare south Sudan (Sclater & Mackworth-Praed 1918–1920). Rare DRC (Chapin 1939). Few records Uganda (Demey 2005, S. Dark & J. Squire *in litt.*). Regular Kenya's coastal forests from Sokoke to Lower Tana R., otherwise scarce (Britton 1980, C.F. Mann). Local Tanzania, mostly in west, some east, south and central (Baker & Baker in prep.). Sparse Zambia (Benson *et al.* 1971); widespread but not common Malawi (Benson & Benson 1977); local Zimbabwe (Payne 1997). Fragmentation of wooded savanna and removal of trees for charcoal may threaten species in RSA (Parker 1999). Fragmentation of forest in west Africa is of concern because this species requires forest for breeding (Allport & Fanshawe 1994). Nearly extirpated from Madagascar, but recorded 1992 and 1993; habitat destruction presumably main cause as most evergreen forest in areas collected was destroyed long ago (Benson & Irwin 1972, Payne 2005). Not globally threatened (BirdLife International 2011).

Thick-billed Cuckoo *Pachycoccyx audeberti.* **Fig. 1.** Adult, *P. a. brazzae*, Gbadagba, Benin, February (*Bruno Portier*). **Fig. 2.** Fledgling, *P. a. validus* accompanying Retz's Helmet-shrikes, Selous GR, Tanzania, January (*Riaan Marais*).

Genus *Microdynamis*

Salvadori 1878 *Ann. Mus. Civ. Genova* 13: 461. Type, by monotypy, *Eudynamis parva* Salvadori 1876.
One species.

New Guinea; presumably brood-parasitic. Small cuckoo; nostrils slit-like; marked sexual and age dichromatism. Shows similarities in vocalisations to *Eudynamys*, but due to morphology generally placed in monotypic genus. MtDNA analysis places it as sister to *Eudynamys* (Sorenson & Payne 2005).

DWARF KOEL
Microdynamis parva   Plate 23

Eudynamis parva Salvadori, 1876 (Tidore [error]; north-west New Guinea)

Alternative names: Black-capped Koel/Cuckoo, Little Koel.

TAXONOMY Polytypic. Synonym: *Rhamphomantis rollesi* Ramsay 1883 in nominate.

FIELD IDENTIFICATION 20–22cm. Small cuckoo; short, stout decurved bill, and medium length rounded tail. **Adult male** Top of head, nape and malar stripe glossy blue, white line below eyes; upperparts, wings and tail rich brown, cheeks and chin orange-rufous, breast greyish-brown, orange-buff on abdomen and undertail-coverts, rest of underparts brown. **Adult female** Resembles male but lacks blue and orange-rufous, ground colour more grey-brown. **Juvenile** Rufous edging to brown head feathers; back feathers brown indistinctly barred rufous and edged grey; uppertail coverts brown edged rufous; remiges brown washed olive, outer edges margined rufous; secondaries have indistinct black bar; lores and ear-coverts white, chin greyish-white barred black, rest of underparts greyish-brown tinged rufous indistinctly barred black (Junge 1937a). **Similar species** Male unmistakable but more drab female confusable with Long-billed Cuckoo, but latter distinguished by much longer bill, lack of white stripe below eyes, more blackish head and paler underparts. Common Koel cannot be confused with this species, but calls are similar. Tawny-breasted Honeyeater has large area of bare yellow skin around eyes.

VOICE 10–30 ascending notes *touei-touei-touei* lasting 30sec or more given at uneven intervals; fast rising *tewhodohodohodohodo* lasting 2.5–5sec. Calls of Common Koel very similar; shorter series of notes, faster, and not so high pitched or loud; second rapid and ascending but changing to level steady pitch towards end (Coates 1985, Beehler *et al.* 1986).

DESCRIPTION Nominate. **Adult male** Forehead, crown, nape and side of head to below eyes metallic blue-black with white subocular stripe becoming orange-rufous behind eyes extending to shoulders; upperparts and wings brown, remiges and coverts narrow edged rufous, underwing-coverts light brown; prominent malar stripe metallic blue-black; chin and upper throat white, lower throat orange-rufous; rest of underparts brown, more light and ochraceous on belly and undertail-coverts; tail dark brown and narrowly tipped white. **Adult female** Generally dull grey-brown above, totally lacking glossy blue colour which is replaced by grey; malar stripe and subocular stripe off-white; throat whitish grey; underparts dull pinkish-brown with some faint brown streaks on breast and flanks. **Subadult male** Dark brown crown with few scattered blue feathers, subocular stripe grey, posterior ochre. Otherwise as adult. **Subadult female** Grey-brown crown to nape; mantle to tail brown, rufous wash on mantle; wings brown with rufous edging to remiges, coverts more rufous; chin white; throat grey with white flecking and some rufous feathers; rest of underparts rufous-brown with indistinct paler barring; underwing-coverts as chest but brighter; underside of remiges rufous-brown becoming brown distally; tail unmarked dark brown above and below; unsexed grey-brown crown with rufous and buff barring; nape to uppertail-coverts barred dark brown and rufous-brown, uppertail dark rufous-brown with small rufous notches on T1–2; upperwing-coverts rufous-brown; remiges dark brown with rufous edging; chin and throat barred grey-brown and pale buff; rest of underparts barred brown and dull buff, more rufous on breast; underwing and tail as previous. **Juvenile** As female but narrow buffish fringes on crown to back feathers, remiges indistinctly barred rufous on outer webs; underparts finely barred pale and dark brown; tail brown, T1 indistinctly notched rufous. **Nestling** Undescribed. **Bare parts** Iris bright red, female hazel. Bill blackish. Feet slate.

BIOMETRICS Nominate. Wing male (n = 12) 102–117mm, mean 105.4 ± 4.6, female (n = 8) 95–106mm, mean 101.5 ± 4.0. Bill male 16.1–19.8mm, mean 18.0 ± 1.0, female 16.7–19.2mm, mean 18.0 ± 0.9. Tail male 88–105mm, mean 95.1 ± 6.0, female 81–98mm, mean 87.4 ± 5.3. Tarsus male 18.6–22.7mm, mean 20.5 ± 1.2, female 18.6–20.8mm, mean 20.1 ± 0.7. Mass male (n = 4) 39.3–54.0g, mean 43.3, female 49, 54g (Payne 2005).

MOULT Adult secondary moult, Jan (BMNH).

GEOGRAPHICAL VARIATION Two subspecies.
M. p. parva (Salvadori, 1876). NG, except northern area between Humboldt Bay and Kumusi R.; D'Entrecasteaux Is. Described above.
M. p. grisescens Mayr & Rand, 1936. Northern NG between Humboldt Bay and Kumusi R. Brown-grey above; breast rufous-brown; belly and undertail-coverts grey.

DISTRIBUTION Presumed sedentary. Widespread NG, Fergusson, Goodenough and D'Entrecasteaux Is. (Mayr & Rand 1937, Rand & Gilliard 1967, Coates 1985, Beehler *et al.* 1986). Once Dauan I. (north Torres Strait), Australia (Carter 1999). Once Babar, Lesser Sundas (Bruce 1987).

HABITAT Rainforest and monsoon forest, forest edges, well-grown secondary and gallery forests, or in tall garden shade trees. To 1,400+m (Coates 1985, Beehler *et al.* 1986).

BEHAVIOUR Solitary or in pairs; rarely seen. Inhabits treetops. Moves between perches when feeding, but remains motionless for prolonged periods. Male feeds female with a fruit during courtship in May. Observed in

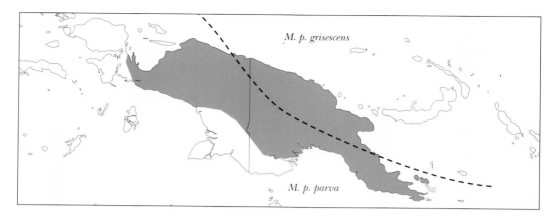

fruit trees with fruit doves, and may chase other foragers away (Coates 1985).

BREEDING Parasitism Presumably parasitic. **Season** Both sexes with enlarged gonads 24 Jun (Diamond 1972b). No further information.

FOOD Maybe entirely frugivorous, preferring figs. One stomach contained fruit. Fed on yellow fruit on vine, and red berries (Rand 1942a, Diamond 1972b, Beehler *et al.* 1986, Fletcher 2000).

STATUS AND CONSERVATION Often considered rare (Ogilvie-Grant 1915, Rand & Gilliard 1967, Diamond 1972b), but Coates (1985) considered it locally fairly common to common. Difficult to estimate population because so seldom seen and voice difficult to separate from Common Koel. Monitoring of status and investigation of its nearly unknown biology necessary. Not globally threatened (BirdLife International 2011).

Dwarf Koel *Microdynamis parva*. **Figs. 1–2.** Adult male, *M. p. parva*, Kiunga, Western Province, PNG, June (*Pete Morris*).

Genus *Eudynamys*

Vigors & Horsfield, 1826 *Trans. Linn. Soc. London* 15(1): 303.
Type, by subsequent designation, *Cuculus orientalis* Linnaeus 1766 (Gray 1840, *List Gen. Bds.*: 57).
One species.

Oriental and Australasian; brood-parasitic. Marked sexual dichromatism; nostrils slit-like. MtDNA analysis places this species close to *Microdynamis parva*, the two forming sister clade to *Pachycoccyx* (Sorenson & Payne 2005). Genus requires urgent revision.

COMMON KOEL
Eudynamys scolopaceus Plate 24

Cuculus scolopaceus Linnaeus, 1758 (Malabar)

Alternative names: Asian/Indian/Pacific Koel (*scolopaceus* group), Black-billed/Moluccan Koel (*melanorhynchus*); Australian Koel.

TAXONOMY Some authors (e.g. White & Bruce 1986, Sibley & Monroe 1990, Andrew 1992, Johnsgard 1997, Schodde & Mason 1997, Christidis & Boles 2008) divide this species into three: *scolopaceus* (nine subspecies) from southeast Asia to central Indonesia, monotypic *melanorhynchus* from Sulawesi, and *cyanocephalus* (seven subspecies) in Australia, NG and eastern Indonesia, due to differences in female and juvenile plumages, and in brood parasitism. Penhallurick (2010) accepts three species, monotypic *melanorhynchus*, with *picatus*, *everetti*, *rufiventer*, *minimus*, *salvadorii*, *alberti*, *cyanocephalus* and *subcyanocephalus* in *orientalis*, and all other forms in *scolopaceus*. There are great gaps in our knowledge of this complex, e.g. variation in plumage and breeding information of several subspecies, and whether many records refer to resident or non-breeding migrants. Therefore, following Higgins (1999), Dickinson (2003) and Payne (2005) we prefer to retain all forms in one species and await future investigation. Polytypic. Synonyms: *Eudynamis* (sic) *honorata* Linneus 1766 replaced by *scolopaceus*; *facialis* Wallace 1863 in *melanorhynchus*; *Eudynamis* (sic) *cyanocephala everetti* Hartert 1900 in *picatus*; *mindanensis* var. *sanghirensis* Blasius 1888, *s. corvina* Stresemann 1931, and *paraguena* and *onikakko* both Hachisuka 1934, all in *mindanensis*; *maculata* J.F. Gmelin 1788 and *s. enigmaticus* Rothschild 1926 in *chinensis*; *ransomi* Bonaparte in *orientalis*; *Cuculus crassirostris* Stephens 1816, *C. cinereus* Meuschen 1787 and *dolosus* Ripley 1946 in nominate. *Sinensis* in error for *chinensis*.

FIELD IDENTIFICATION 39–46cm. Flight fast, direct and strong with rapid wing-beats. Male often fans tail and droops wings when female nearby, simultaneously making short staccato calls. **Adult male** Large glossy black bird with strong, decurved, pale green, ivory or horn-coloured bill (black on Sulawesi), red eyes, strong legs and feet and long rounded tail. **Adult female** Hawk-like, with dark brown crown streaked white and rufous, upperparts dark brown heavily spotted white or buff, dark brown wings barred and spotted white; chin to upper breast dark brown heavily spotted white or buffy-white; rest of underparts white to buffy-white, barred blackish-brown; tail dark brown barred and spotted white to rufous-white, bill as male. **Juvenile** Male dull blackish, wing-coverts and tertials tipped white or buffish; underparts variably barred white and black, tail black, in some with rufous barring, eyes brown and bill blackish or greyish-buff; female as adult but more barred and less spotted above. **Similar species** Drongo-cuckoos much smaller, tail slightly forked or square, never rounded, decurved black bill slender and some white barring on undertail-coverts, and white on thighs. Blue-faced Malkoha from India and Sri Lanka has greenish bill, but broadly tipped white on long tail, naked blue area around eyes, upperparts and tail glossy dark greenish-grey and underparts dark grey. Greater Black Coucal from NG more than one third larger, more robust, and sexes similar. Philippine Coucal about same size as Koel, has stout black bill, normal morph with dark brown wings, but subspecies from Mindoro and Semirara Is. entirely black; sexes in coucal similar; but tail shorter in Koel.

VOICE Ascending, loud *ko-el-ko-el-ko-el-ko-el-ko-el-ko-el* beginning slowly and then accelerating, or faster and bubbling *koel-koel-koel-koel*; first call repeated 2–15+ times, one note lasting 1.5–2.5sec, last 4–6 calls also rendered *wuroo-wuroo-wuroo* or *wrep-wrep-wrep* with four notes in 1.25sec or less, rising at start and falling at end, sometimes finishing with drawn-out *wheeoo*. 4–8 short *duli-duli-duli* often directly followed by ascending 5–10 nasal whistles *kweel-kweel-kweel*, suggestive of bird being tortured, its different calls always given in hottest weather. Males during flight-chases and when excited *kwa-kwa-kwa-keow-keow*. Female has repeated hoarse trumpet-like call, and rapid repeated *kik-kik-kik-kik* given both in flight and when perched. Flying fledglings give loud and harsh *kaaa*, rather like young crow. On Lembata, Lesser Sundas, soft *hmm*, followed by raucous *waaa* repeated at 5sec intervals. Six different calls identified near Brisbane, Australia; most common *cooee* or *ko-el*, *whik* and *wurroo*, last usually given when conspecifics present; *whik* given in connection with *wurroo* calls either at start or end, and clucking *wuk-wuk-wuk* when courting female, only audible at close range; female gives fast series of *keek-keek-keek-keek* phrases; begging call of juvenile loud sharp trills *wheeet-oop-wheet-wheeet-wheeet-oop*. Mated male and female duet at beginning of breeding season, sometimes with other conspecifics. Begging call of fledgling series of sharp trills *wheet-oop-wheet-wheet-wheet-oop*. Vocalisations rather similar over its large range. Calls by day and often by night, but mostly at dawn and dusk; less vocal in winter. On Mindoro calling very loudly by day and night Apr–May, call far-carrying. Harsh *chuck, chuck, chuck...* from female involved in fighting another female (Henry 1955, Ripley & Rabor 1958, Ali & Ripley 1983, Coates 1985, Roberts 1991, Maller & Jones 2001, Payne 1997, 2005, Grimmett *et al.* 1998, Maller 1997, Kennedy *et al.* 2000, Trainor 2003, SUARENG 1997 in Wells 2007).

DESCRIPTION Nominate. Adult male Entirely black glossed blue-green, in some lights purplish; long rounded tail. **Adult female** Dark brown crown streaked white and rufous, rest of upperparts dark blackish-brown heavily spotted white, blackish-brown wings barred and spotted white; chin to upper breast blackish-brown heavily spotted white or buffy-white; rest of underparts white to buffy-white, barred blackish-brown; underwing-coverts white or buffy-white barred dark brown; underside of remiges dark brown barred and spotted white; tai blackish-brown barred and spotted white to rufous-white. **Subadult female** Smaller rufous spots dorsally, distinctly barred back; below dark bars slightly broader and rectrices narrower. **Subadult male** Unrecorded. **Juvenile** Male and female similar to their respective sexes but male duller, paler on rump with some whitish or chestnut-buff tipping on rump and wing-coverts and fine tips to remiges, underparts dull blackish-brown with narrow white bars on breast to undertail-coverts; underwing-coverts barred dark brown and white, with some off-white edging and barring to inner webs of remiges. Female darker, more sooty and greyish-black above, with more barred than spotted pattern; head and face black; underparts black barred white; tail black barred pale rufous and tipped white. **Nestling** Hatches with pinkish-red skin soon becoming black. First feathering black, tipped white to pale brown, both sexes resembling male, similar to young of host House Crow; female little browner on crown and mantle and barred below and on tail, male has some chestnut-buff spots on wing-coverts. **Bare parts** Iris red in male, female brown to orange-red, juvenile brown. Bill ivory-green, blackish at base and around nostrils, nestling dirty pink, fledgling blue-black, juvenile black to greyish-buff tipped white; adult and juvenile gape crimson. Feet blue-grey, nestling pale pink, sole whitish to grey-buff.

BIOMETRICS Nominate. Wing male (n = 10) 187–209mm, mean 194.8 ± 7.0, female (n = 17) 172–195mm, mean 188.6 ± 4.4. Bill male 26–32mm, mean 28.3 ± 2.1, female 26–30mm, mean 27.6 ± 1.2. Tail male 170–205mm, mean 188.6 ± 12.7, female 175–200mm, mean 185.2 ± 7.5. Tarsus male 31–36mm, mean 33.4 ± 1.8, female 30–35mm, mean 32.8 ± 1.6 (Payne 2005). Mass unsexed India (n = 10) 136–190g, mean 167 (Ali & Ripley 1983); *cyanocephalus* male (n = 16) 175–340g, mean 261.9, female (n = 23) 167–330g, mean 257.0 (Higgins 1999).

MOULT Two juvenile males moulting to adult Daru, NG, 6–7 Mar. Moulting remiges NG and Moluccas, Dec–Apr. Juvenile male moulting to adult Sri Lanka, Jul. Two moulting north Australia, Jan and Mar; three *subcyanocephalus* had suspended primary moult, Feb–Jun. Moulting Bali Feb/Mar. Juvenile moult may be interrupted. Post-juvenile head/body moult before flight feathers. P1, P2, P3, P4, then P9 and P7, then P8 and P6, then P10 and P5, but no obvious rule (Mayr & Rand 1937, Meise 1941, Stresemann & Stresemann 1961, 1966, Higgins 1999, Payne 2005, Wells 2007, BMNH).

GEOGRAPHICAL VARIATION Eighteen subspecies, differing mostly in female plumage.

E. s. scolopaceus (Linnaeus, 1758). Pakistan, Nepal, India, Sri Lanka, Lakshadweep and Maldive Is.; Andamans and Nicobars (not resident: Rasmussen & Anderton 2005). Scarce winter visitor Iran (Hollom *et al.* 1988). Populations from north India perhaps winter farther south. Described above.

E. s. malayanus Cabanis & Heine, 1863. Assam, Bangladesh, Burma, central and south Thailand, Thai-Malay Peninsula, Sumatra (including Nias, Tello and Tana Massa islands), Bangka, Borneo (winter visitor, perhaps resident), Lombok, Sumbawa and Flores. Slightly larger than nominate; female more rufous on pale areas. Wing male (n = 12) 199–238mm, mean 213.0 ± 11.7, female (n = 9) 195–213mm, mean 205.0 ± 5.8 (Ripley 1944, Payne 2005).

E. s. chinensis Cabanis & Heine, 1863. Indochina, and west, central and south China (Sichuan and Yunnan east to Fujian and north to Gansu, Shaanxi and Henan), migrating south to Borneo. Larger than nominate. Male with steel-blue gloss; female without rufous on crown, sometimes head mostly blackish with long white malar stripe, and spotted upperparts dark brown glossed greenish. Wing male (n = 14) 192–213mm, mean 203.0 ± 7.4, female (n = 14) 187–227mm, mean 201.9 ± 12.4 (Payne 2005).

E. s. harterti Ingram, 1912. Hainan. Bill longer than *chinensis*. Wing male (n = 13) 199–220mm, mean 206.9 ± 5.5, female (n = 12) 196–217mm, mean 204.4 ± 8.1; bill male 25–32mm, mean 29.0 ± 1.6, female 25–31mm, mean 28.1 ± 2.0 (Payne 2005).

E. s. simalurensis Junge, 1936. Simeulue, and Babi and Kokos islands, north-west Sumatra. Smaller; female has large rufous spots on upperparts, underparts strongly tinged red-brown. Wing male (n = 11) 179–200mm, mean 194, female (n = 6) 189–200mm, mean 195 (Junge 1936). Payne (2005) includes in *malayanus*.

E. s. mindanensis (Linnaeus, 1766). Philippines (except where *frater* occurs), and Talaud, Sangihe, Siau, Ruang and ?Manterawu. Female browner above, more rufous on pale areas and more narrowly barred on ventral side than *malayanus*; rufous on upperparts paler than *simalurensis*. Wing male Palawan (n = 16) 177–217mm, mean 198.4 ± 6.8, female Luzon (n = 6) 185–203mm, mean 195.7 ± 8.4 (White & Bruce 1986, Payne 2005).

E. s. frater McGregor, 1904. Cagayan and Fuga islands, north-east Philippines. Similar to *mindanensis* but larger. Wing male (n = 10) 196–273mm, mean 234.3 ± 21.5, female (n = 4) 204–226mm, mean 215.5 ± 9.0 (Payne 2005), who includes in *mindanensis*.

E. s. picatus S. Müller, 1843. Kai islands and Sumba to Timor and Roma. Female variable, black malar stripe bordered above by rufous-buff streak broadening distally, upperparts spotted whitish, wings and tail brown, black bars of varying thickness on underparts, or throat black or mottled black and ochre. Payne (2005) includes in *orientalis*.

E. s. orientalis (Linnaeus, 1766). South and central Moluccas on Buru, Manipa, Kelang, Seram, Ambon, Tujuh and Watubela. Larger; female has fewer, more rufous spots and bars above and rich cinnamon below with little or no barring, throat black or spotted black; juvenile pale rufous-buff. Wing male (n = 17) 201–232mm, mean 220.8 ± 7.7, female (n = 13) 209–229mm, mean 217.5 ± 6.7 (Payne 2005).

E. s. corvinus Stresemann, 1931. North Moluccas on Morotai, Halmahera, Ternate, Tidore, Moti and Bacan. Juvenile smoky black. Wing male (n = 7) 197–215mm, mean 208.5 ± 5.4, female 189, 214mm (Payne 2005) who includes in *mindanensis*.

E. s. melanorhynchus S. Müller, 1843. Sulawesi, Bangka, Muna, Manterawu, Peleng, Talisei, Lembeh, Togian and Sula. Male blue-violet iridescence, more rounded wings, bill black. Female polymorphic: (1) as male but blue-green gloss; or (2) head black glossed green, black above glossed bronze, pale malar stripe, chin and throat black or grey barred black, orange-rufous on rest of underparts; or (3) crown and face black, crown sometimes streaked rusty, upperparts barred rufous and black, whitish malar stripe

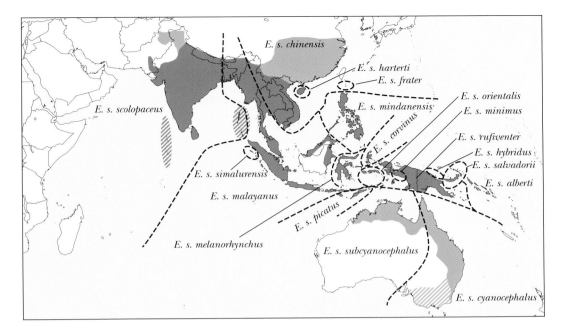

which may extend to sides of neck, underparts barred buffish and black. Male in Sula ('*facialis*') black often with some white on head. Juvenile sooty-black or like female (3), except head, neck and throat blue-black. Wing male (n = 19) 187–211mm, mean 203.9 ± 4.9, female (n = 11) 180–213mm, mean 190.5 ± 6.6 (Payne 2005).

E. s. rufiventer (Lesson, 1830). NG and islands off west coast. Smaller. Male glossy blue-black with pale brown or blue-grey bill. Female head largely rufous, whitish malar stripe, dark brown upperparts heavily spotted and streaked rufous, underparts ochraceous spotted from chin to upper breast, rest finely barred dark, bill pale. Wing male (n = 10) 184–200mm, mean 191.8 ± 4.7, female (n = 12) 181–198mm, mean 188.5 ± 5.0 (Payne 2005).

E. s. minimus van Oort, 1911. WP. Smaller. Male as *rufiventer*. Female crown and face black with pale rufous lines under eyes, small white and rufous spots on back, underparts plain rufous except black on sides of throat and darker rufous on sides of upper breast. Wing male 168mm, 178mm, female 170mm (Payne 2005).

E. s. hybridus Diamond, 2002. Long, Tolokiwa and Crown Is. (PNG). Similar size to *rufiventer*, and female belly plumage, intermediate between *rufiventer* and *salvadorii*; tail pattern like *salvadorii*. Wing male (n = 14) 184–201mm, mean 189.5 ± 4.4, female (n = 5) 180–196mm, mean 187.2 ± 6.7 (Payne 2005).

E. s. salvadorii Hartert, 1900. Bismarck Archipelago. Larger than *rufiventer*. Males less greenish, more bluish. Females similar to *rufiventer* except for whitish and buffish spots on wing-coverts, underparts buffy-white narrowly barred blackish-brown, with broader black tail bands; underwing-coverts dark buff finely barred black. Wing (n =17) 203–220mm, mean 210 (Diamond 2002).

E. s. alberti Rothschild & Hartert, 1907. Bougainville and Solomons. Much smaller; male more bluish-black on head and above, female rufous barred dark brown, no white spots on wing-coverts, underparts finely barred black and rufous. Wing male (n = 12) 167–201mm, mean 183.3 ± 8.5, female (n = 11) 165–184mm, mean 175.2 ± 5.9 (Payne 2005).

E. s. subcyanocephalus Mathews, 1912. North and north-west Australia, wintering NG and perhaps Aru, Kai and Banda Is. Smaller than *cyanocephalus*. Three female morphs (1) glossy black head with buff malar stripe broadening at neck, rest of upperparts blackish-brown spotted white, throat buff with few blackish tips, rest of underparts buff barred blackish; or (2) brown tipped buffish and whitish, breast pale buffish, rest of underparts white, throat streaked brown, rest of underparts with fine or broad brown fringes to feathers; or (3) rufous head and throat with dark streaks, upperparts brown spotted white, breast pale buffish, rest of underparts white with more prominent dark edges to feathers (BMNH). There is some confusion as to which subspecies (this or next) some specimens belong. Wing male (n = 22) 193–221mm, mean 212.3 ± 6.61, female (n = 9) 193–214mm, mean 204.9 ± 6.23 (Higgins 1999). Juvenile male paler, less rufous than *rufiventer* (Rand 1942a).

E. s. cyanocephalus (Latham, 1801). North-east Queensland to east NSW, sporadically Victoria and SA. Partly migratory, probably wintering Indonesia (recorded Buru, Seram Laut, Ambon, Banda, Watubela, Tayandu, Kai, Komodo, Sumba, Alor, Timor, Wetar, Moa, Tanimbar and ?Romang: White & Bruce 1986) and perhaps Philippines. Larger. Female often slightly more streaked on head than *subcyanocephalus*, crown black (see previous subspecies). Leucistic NSW (Gosper 1962). Wing male (n = 23) 204–230mm, mean 217.2 ± 6.87, female 200–225mm, mean 214.6 ± 6.7 (Higgins 1999).

DISTRIBUTION Oriental and Australasian; widespread. Two populations in Pakistan: one north-east and one south-east; southernmost from Hyderabad and south resident. Occurs Nepal year-round at Bardia and east Tarai and foothills, but summer visitor Kathmandu Valley, Mar–Oct, and Chitwan, Mar–Sep. Most of India and Sri Lanka; resident Manipur. Thai-Malay Peninsula and on Penang, Langkawi group, Singapore, Great Redang and islands of Bandon Bight. Summer visitor south and west China, and Dabie Shan, central China. Winter visitor (*malayanus*) throughout Sumatra and islands; *simalurensis* presumably breeds on Simulue. Common to local winter visitor, perhaps few resident, Borneo, including South Natunas (*chinensis*) and Spratly Is; commonest form *malayanus*. Uncommon (?resident) Java, Bali. Philippine subspecies *mindanensis* on Apo, Balut, Bantayan, Basilan, Bohol, Bongao, Boracay, Bucas, Burias, Cagayancillo, Cagayan Sulu, Cahayagan, Calagna-an, Calayan, Calicoan, Caluya, Camiguin Norte, Camiguin Sur, Carabao, Catanduanes, Cayoagan, Cebu, Cuyo, Dinagat, Fuga, Gigantes, Guimaras, Jolo, Leyte, Loran, Lumbucan, Luzon, Malamaui, Malanipa, Marinduque, Masbate, Mindanao, Mindoro, Negros, Omapoy, Palawan, Panay, Pan de Azucar, Polillo, Romblon, Samar, Sanga Sanga, Sarangani, Semirara, Siargao, Sibay, Sibutu, Sibuyan, Sicogon, Simunul, Sipangkot, Siquijor, Tablas, Talicud, Tawi-Tawi, Tumindao and Calayan. In Wallacea on Sulawesi, and Bangka, Lembeh, Manterawu, Muna, Peleng, Ruang, Siau, Sangihe, Sulawesi, Talaud, Talisei and Tongian; Moluccas on Ambon, Bacan, Banda, Buru, Halmahera, Kai, Kelang, Manipa, Morotai, Moti, Seram, Seram Laut, Sula, Tayandu, Ternate, Tidore, Tujuh and Watubela; Lesser Sundas on Alor, Flores, Komodo, Lembata, Lombok, Moa, Romang (?), Sumba, Sumbawa, Tanimbar, Timor, Wetar, Adonara, Satonda and Obi. In Australia *subcyanocephalus* from Kimberley in west over Top End in NT to Queensland; *cyanocephalus* east Australia south of 22°S. In NG *rufiventer* widely scattered on mainland, also Traitor's I., Geelvink Bay, Tarawai, Schouten, Manam, Kakar and Bagabag, off north coast, and Daru and Aru Is. south of western NG; *hybridus* on Long, Tolokiwa and Crown islands, where *salvadorii* is also recorded, and latter also on Lolobau, New Britain, Sakar, and Umboi in Bismarck Archipelago (Robinson 1927, La Touche 1931, Coates 1985, White & Bruce 1986, Bruce 1987, Cheng 1987, van Marle & Voous 1988, Inskipp & Inskipp 1991, Roberts 1991, MacKinnon & Phillipps 1993, Lambert 1994, Grimmett et al. 1998, Higgins 1999, Wells 1999, Kennedy et al. 2000, Diamond 2002, Trainor 2002, Allen et al. 2006, Liu et al. 2008, Mann 2008, Choudhury 2009). Oman, Jan–Apr and Oct–Nov (Gallagher & Woodcock 1980, Eriksen 1991 in Payne 2005); UAE, Feb–Mar (Richardson 1990 in Payne 2005); once Kuwait, Feb; Bahrain; once Sokotra (Balmer & Murdoch 2009a, b, Al Sirhan & Al Hajji 2010). Once Tobi, Palau, 1978 (Engbring 1983). Two or three South Korea, Apr–May, apparently *chinensis* (Birds Korea 2009, 2010). Once north-east Afghanistan (Payne 2005); accidental south-east Iran in 1970's (Porter et al. 2010). Once Taiwan (Cheng 1987). Five ringed Point Calimere, south India, Dec recovered up to 160km away within less than one month (McClure 1998); one ringed Beerburrum, Queensland, found 34 months later Bundaberg 240km distant; juvenile ringed Lane Cove, NSW recovered 15 months later at Old Iare, PNG, distance of 2,950km (Higgins 1999).

HABITAT In India open woodland, groves near villages and farms, orchards, gardens, parks, towns and cities. In Australia dense rainforest, moist eucalypt or monsoon forest, also along watercourses often with paperbarks, *Acacia* scrubland, wooded grassland and human settlements, even streets with fig and mango trees, rarely heathlands and borders of coastal mangrove forest. In NG forest, gallery woodland, also secondary forest, gardens and locally savanna. On Sumatra mangroves, coastal swamp forest, small coral islands, cultivation and gardens in lowlands; on Mentawai Is. occasionally in freshwater and peatswamp forest. In Philippines lowland and mountain forest, secondary forest, coconut plantations, cultivation with trees, and near villages. To 1,800m Himalayas and 1,000m peninsular India. On Borneo kerangas forest, mangroves, secondary forest, coastal scrub forest, cocoa and oil palm plantations, cultivation, gardens and forest edge; lowlands. On Sulawesi lowland forest, forest edge and secondary forest, also deep in forest especially along rivers to 1,400m. To 1,500m mainland NG (Gonzales 1983, Watling 1983, Blakers et al. 1984, Coates 1985, van Marle & Voous 1988, Grimmett et al. 1998, Higgins 1999, Kemp 2000, Kennedy et al. 2000, Mann 2008).

BEHAVIOUR Arboreal; mostly quiet when feeding; usually concealed and secretive in dense foliage but in early mornings may sun-bathe in treetops. More often heard than seen; often among dense foliage and tangles of vines at tops of tall trees. Calls from very tall tree-tops overlooking open spaces, particularly favouring trees with wide canopy and dense foliage, usually species of *Ficus*. Groups often seen together at trees with ripe fruit, but adults never seen on ground. Often mobbed by crows and male performs distraction behaviour when female laying. Probably pair-bond in breeding season as pairs duet, and perhaps male defends territory, but some promiscuity. Male displays to female by bowing low with lowered tail, female then also starts to bow, followed by male giving four loud and harsh calls, answered by female with similar call. Courtship feeding once recorded, but usually copulation without any preceding display. In Australia night-flying migrants travel singly, in pairs or few together. Confrontation and fighting between two females in presence of male (Dharmakumarsinji 1954, Henry 1955, Ripley & Rabor 1958, Lamba 1969, Diamond 1972b, Guthrie 1973, Denny & Dudman 1979, Crouther 1985, Roberts 1991, Grimmett et al. 1998, Higgins 1999, Maller & Jones 2001, SUARENG 1997 in Wells 2007).

BREEDING Parasitism Hosts: House and Jungle Crows, and Common Myna, India. In Pakistan also Black Drongo. Long-tailed Shrike, House Crow and Common Myna, Bangladesh. House Crow and Common Myna, Thai-Malay Peninsula. Black-collared Starling common host south-east China, including Hong Kong Island, and Red-billed Blue Magpie; latter also Burma. Slender-billed and Jungle Crows most common hosts, west Java. Orioles, Greater Sundas. Fledgling fed by Sulawesi Starling, Sulawesi. Hill Myna, Palawan. Large-billed and Flores Crows, Flores, Lesser Sundas. Mobbed by Yellow-faced Myna, NG. Blue-grey Robin reported host New Britain. Magpie-lark preferred host, NSW. Over whole of Australia mostly friarbirds and figbirds; following species recorded with either egg, young in nest or fledgling fed by foster parents: Magpie-lark (24), Black Butcherbird (1), Grey Butcherbird (1), Dusky Wood Swallow (1), White-breasted Wood Swallow (1), Black-faced Cuckoo-shrike (1), White-bellied Cuckoo-shrike (1), Blue-faced Honeyeater (11), Red Wattle Bird (8), Silver-crowned Friarbird (9), Little Friarbird (31), Helmeted Friarbird (9), Noisy Friarbird (32), Noisy Miner (2), Yellow-throated Miner (3) Olive-backed Oriole (13), Australasian Figbird (35), Victoria's Riflebird (1), Spangled Drongo (4). Hill Myna, Palawan. Two fledglings fed by one House Crow, India (Inglis 1908, Vaughan & Jones 1913, Osmaston 1916, Whitehead in Stresemann 1927, Gosper 1962, Hellebrekers & Hoogerwerf 1967, Ottow & Verheijen 1969, Harrison 1971b, Gonzales 1983, Coates 1985, Andrew & Holmes 1988, Roberts 1991, Corlett & Ping 1995, Lewthwaite 1995, Payne

1997, Grimmett *et al.* 1998, Brooker & Brooker 1999, various references in Wells 2007, Begum *et al.* 2011, C. Roy *in litt.*). **Season** Pakistan mostly Jun–Jul. India Mar–Oct, varying locally peaking May–Jul. Sri Lanka Apr–Aug. Burma May. Thai-Malay Peninsula Jan, Apr–Jul. Singapore Apr–Oct. Laying north Sulawesi, 8 Mar. In north Australia arrives with wet season, sometimes up to two months later than usual, Nov–Feb, unusually Oct and Mar, and in SA Oct–Jan, rarely Sep, once Mar. In east highlands of NG male with somewhat enlarged gonads, 11 Sep. On Luzon, Philippines, female with enlarged ovary 26 Apr (Osmaston 1916, Henry 1955, Diamond 1972b, Ali & Ripley 1983, White & Bruce 1986, Goodman & Gonzales 1990, Roberts 1991, Grimmett *et al.* 1998, Higgins 1999, various refs. in Wells 2007, Y.C. Wee *in litt.*). **Eggs** Normally egg laid in crow's nest after first host egg laid, rarely after two or three eggs laid. Up to 13 Koel eggs in one crow's nest. Female removes one foster parent's egg when laying. Normal to broad oval, smooth or slightly rough, little or no gloss; ground colour in crow's nests grey-blue, pale green-yellow to dirty yellow spotted reddish-brown and black; other eggs yellowish, pale reddish-salmon to dirty olive-green, spotted and speckled olive to olive-sepia, red-brown, dull purplish-red, or nearly black; *scolopaceus* 28.0–34.4x21.6–24.6mm, mean 30.9x23.2, *malayanus* 33.8x25.5mm, *chinensis* 32.5x24.2mm, *salvadori* 39x26mm; west Java (n = 6) mean 34.5x25.1mm. Nests closest to fruit trees more likely to be parasitised (Baker 1927, Hellebrekers & Hoogerwerf 1967, Lamba 1967, Schönwetter 1967, Crouther 1985, Begum *et al.* 2011). **Incubation** 13–17 days in House Crows' nests; 15 days in Figbird's nest (Lamba 1963, Johnsgard 1997, Higgins 1999). **Chicks** Hatch naked and blind, at five days feathered, no downy stage, eyes open at seven, fairly well-feathered at 10 days; fledge 19–28 days. Australian subspecies which parasitises smaller hosts often eject host eggs or chicks within first three days, however in Asia Koels parasitising larger crows, commonly one or two host siblings starve, but others survive. Hosts seen feeding fledgling for more than 14 days after leaving nest, young begging very noisily and generally fluttering wings; may be fed for 3–4 weeks (Roberts 1991, Higgins 1999, Payne 2005). **Survival** 11 of 22 eggs hatched, and eight fledged (Gosper 1962).

FOOD Fruit and berries, especially *Ficus*; Yellow Oleander, mulberry, sandalwood, nuts of Fishtail Palm, poisonous fruit of *Thevatia neriifolia*, and flower nectar; flowers of *Clitoria ternatea* and *Carica papaya*; fruit of latter, *Pithecellobium dulce* and guava; hairy caterpillars and other insects; snails; birds' eggs. In Thai-Malay Peninsula Yellow-vented Bulbuls' eggs. Sri Lanka fruits of various ornamental garden palms. In Pakistan flowers, nectar, and eggs of small passerines such as Red-vented Bulbuls. In Hong Kong *Livistona chinensis* most favoured fruit. 82.1% fruit, 17% invertebrates. In Australia Morton Bay Fig and mulberry preferred, also fruit trees in gardens frequented as are clumps of fruiting mistletoe; eggs of Silvereye. Takes fruits up to 41mm diameter in NG, but chicks fed with insects by foster parents. In Hong Kong swallows fruit up to 17.5mm whole and rapidly regurgitates seeds (Henry 1955, Gosper 1962, Rajasingh & Rajasingh 1970, Ali & Ripley 1983, Coates 1985, Roberts 1991, Corbett and Ping 1995, Santharam 1996, Grimmett *et al.* 1998, Higgins 1999, Walker 2007, SUARENG 2001 in Wells 2007, R.J. Phukan *in litt.*, A. Thakurta *in litt.*).

STATUS AND CONSERVATION Locally common Pakistan, India, Sri Lanka and Bangladesh; widespread Maldives, uncommon Bhutan (Grimmett *et al.* 1998). Common throughout Nepal (Inskipp & Inskipp 1991) and Andaman and Nicobar Is. (Tikader 1984). Common resident near human settlements Thailand (Lekagul & Round 1991). Increased greatly on Hong Kong Island, presumably due to spread of preferred host, Black-collared Starling (Corlett & Ping 1995). Common in summer south China (Cheng 1987). On Sumatra common winter visitor (*malayanus*) Oct–Dec (once late Jul, perhaps breeding); *simalurensis* very common Simeulue where presumably breeds but no evidence (van Marle & Voous 1988). Local to scarce winter visitor Borneo, where few may be resident (Mann 2008). Very common in all habitats Gunung Ambang NR, Sulawesi (Riley & Mole 2001). Presumably uncommon north Moluccas, but common Morotai and Flores (White & Bruce 1986). Abundant Lembata (Lesser Sundas) where up to five birds heard calling simultaneously (Trainor 2003). Locally common to fairly common NG (Coates 1985). Widespread Australia (Higgins 1999). Common Sulu islands (duPont & Rabor 1973a). Frequent road-kill (10 in 4.5yrs) Kumbhalgarh WS, India (Chhangani 2004). Not globally threatened (BirdLife International 2011).

Common Koel *Eudynamys scolopacea*. **Fig. 1.** Adult male, *E. s. malayanus*, Ipoh, Perak, Malaysia, Feburary (*Amar-Singh HSS*). **Fig. 2.** Adult female, *E. s. malayanus*, Bukit Maung, Penang, Malaysia, September (*Choy Wai Mun*). **Fig. 3.** Adult female, probably *E. s. subcyanocephalus*, Ashmore Reef, Australia, October (*Rohan Clarke*). **Fig. 4.** Juvenile, *E. s. malayanus*, Nakorn Nayok, Thailand, July (*Peter Ericsson*). **Fig. 5.** Juvenile, probably of race *E. s. cyanocephalus*, with Magpie-lark foster-parent, Eastern Queensland, Australia, December (*Bruce Thomson*).

Genus *Urodynamis*

Salvadori, 1880 *Orn. Pap. e delle Mol.* 1: 370. Type, by original designation and monotypy, *Cuculus taitensis* Sparrman 1787.
One species.

NZ and Pacific; brood-parasitic. Sexes alike; nostrils oval slits; long wedge-shaped tail. Sometimes subsumed in *Eudynamys* (e.g. Mayr 1944), but apparently not close. MtDNA analysis places it as sister to *Scythrops* (Sorenson & Payne 2005).

LONG-TAILED KOEL
Urodynamis taitensis   Plate 14

Cuculus taitensis Sparrman, 1787 (Tahiti)

Alternative names: Long-tailed (New Zealand/Pacific) Cuckoo.

TAXONOMY Monotypic. Synonyms: *Cuculus tahitius* Gmelin 1788, *C. perlatus* Vieillot 1817, *C. fasciatus* Forster 1844, *Eudynamis cuneicauda* Peale 1848, *pheletes* Wetmore 1917 and *belli* Mathews 1918.

FIELD IDENTIFICATION 40–42cm. Rather large, slender falcon-like cuckoo with conspicuously long tail (c. 24cm), giving appearance of flying cross; direct flight with unbroken wing-beats; often glides. **Adult** Top of head dark brown heavily streaked buff, white superciliary and malar stripes, broad dark brown eye-stripe; above brown heavily spotted and barred white and rufous, underwing-coverts creamy to buffish; below white with long bold dark streaks, chevrons on sides of vent and undertail-coverts; tail dark brown tipped white and barred rufous. **Juvenile** As adult but upperparts spotted white, underparts rich buff with fewer and finer dark streaks, and tail shorter. **Similar species** Female and juvenile Common Koel have fine barring on underparts, conspicuous white moustachial stripe, no white supercilium, bill of juvenile wholly grey-black and iris of adult red (yellow to red-brown in Long-tailed) and tail slightly shorter. Can easily be confused with small birds of prey. New Zealand Falcon has shorter tail. Juvenile New Caledonian Hawk heavier, shorter and with shorter, more curved bill, and underparts crossbarred. Fiji Goshawk juvenile has striated underparts, feet yellow and more rounded wings in flight.

VOICE Long, harsh screech *zzheesht* often repeated every few sec. Long, rapid, ringing *zip-zip-zip-zip* or noisy or ringing *rrp-pe-pe-pe-pe-pe*, or sharp *pe-pe-pe*, first also described as shrill upslurred *zhiz-z-z-z-z*, or *zzwheesht* between shriek and chatter, and last as *wheet-wheet-wheet*, or ringing *zip-zip-zip*, or loud *pe-pe-pe*. Chicks have quiet cricket-like trill *cheep-sheep*, and fledglings constantly trill when foster-parents nearby *whirr-r-r-r-r* resembling that of Whitehead's or Yellowhead's fledglings. Often heard chattering when flying; more vocal at night. Calling very loud and difficult to locate but responds to playback; both shriek and chatter notes in calling display. Mostly silent in winter quarters (Falla *et al.* 1981, McLean 1985, Robertson 1985, Bregulla 1992, Higgins 1999).

DESCRIPTION Adult Forehead, crown, nape and hindneck brown with narrow buffish shaft-streaks, hindneck also with buff-brown spots, supercilium off-white with slight brown streaking bordered by bold dark brown eye-stripe, dark moustachial stripe bordered white above and below, lores and ear-coverts brown with pale streaks; back, scapulars, upperwing-coverts, rump and uppertail-coverts dark brown with broad rufous bars, prominent white tips to wing-coverts and uppertail-coverts, small white tips to greater primary-coverts, remiges dark grey-brown with transverse white, buff-white or creamy bars, more rufous near base, and often finely tipped white, underwing-coverts and axillaries buffish; underparts whitish, pale buff or rufous with variable number of dark brown longitudinal shaft streaks, broader at feather tips and most prominent on breast and flanks, chin and upper throat mostly white, undertail-coverts white with blackish chevrons or broken bars; graduated tail with rufous and blackish-brown bars of about same width and tipped white. **Juvenile** As adult but upperparts from nape to lower back and wings heavily spotted off-white, with some dark rufous; rump and uppertail-coverts barred rufous and reddish-brown; shaft streaks on top of head cream, broad supercilium yellow-brown to buffish from upper mandible to side of neck, dark brown eye-stripe below supercilium and down side of neck; cheeks, rest of sides of neck, chin and throat yellow-brown to buffish with faint dark tips to many feathers, rest of underparts from breast to vent pale brownish-yellow with much narrower streaking than adult, flanks buffish with dark brown chevrons or spots, undertail-coverts creamy with few brown chevrons; slightly shorter dark brown tail with rufous bars and tipped off-white, undertail paler. **Nestling** Hatches naked with grey-black or dark brown skin (Wilkinson 1947). **Bare parts** Eye-ring greenish. Iris pale yellow to light brown or red-brown, juvenile brown; orbital ring greenish-yellow to olive-green, grey-brown in juvenile. Strongly curved bill greyish-brown, lower mandible paler cream to yellow or pinkish-grey; nestlings have orange to yellow gape. Feet olive-green to greyish-slate or brownish, juvenile normally has yellow edges to scales, soles yellowish, claws black or dark grey.

BIOMETRICS Wing male (n = 23) 183–197mm, mean 189.0 ± 3.49, female (n = 25) 176–196mm, mean 185.5 ± 5.1. Bill male 29.5–35.0mm, mean 33.4 ± 1.3, female 32.0–35.7mm, mean 33.5 ± 0.7. Tail male 211–236mm, mean 226.8 ± 6.0, female 204–237mm, mean 220.2 ± 7.6. Tarsus male 33.4–37.3mm, mean 34.9 ± 1.1, female 32.5–36.6mm, mean 34.7 ± 1.0. Mass male (n = 9) 74–153.9g, mean 110.4, female (n = 10) 93–148g, mean 124.2 (Higgins 1999).

MOULT Presumably adult post-breeding moult starts after Feb–Mar when migrating to winter areas. May begin May and end after Jul with all odd-numbered primaries moulted first followed by all even-numbered after first are fully grown, but P1–2 often not in this sequence. Post-juvenile primary moult Nov–Mar/Apr in following succession: P2, P3, P7 or P9, then P6, P8 or P10, P1, P4, and P5 without consistent sequence (Stresemann & Stresemann 1966).

DISTRIBUTION NZ and Pacific. Wide winter range in central and south tropical Pacific; perhaps most remarkable over-water migrant of any land bird. Breeding range restricted to NZ, Stewart I. and offshore islands where arrives early Oct and mostly departs Feb–Mar, but many records Apr or later. Unconfirmed breeding Norfolk I. Most birds migrate to spend austral winter west of NZ to New Caledonia, Norfolk I., Solomons, Caroline Is. (Kusaie, Lukunor, Ponape, Truk and Yap) and Palau I., or north to Fiji and Marshall Is (Auru, Bikini, Elmore, Jaluit and Wotze), and east to Tonga, Cook Is, Society Is, Samoa, Marquesas, Pitcairn and Tuamotu. Principal wintering area Fiji, Tuvalu, Tonga, Samoa, Tokelau, Cook Is, Society Is and east Tuamotu Archipelago. Also Tubai Is, Kiribati, Niue, Vanuatu, and Wallis and Futuna. Vagrant Caroline Atoll, Line Is. Uncommon resident Kermadec Is; vagrant Snares Is; irregular Lord Howe I.; irregular on passage Norfolk I.; also Misima I., Louisiade Archipelago, PNG, and Bismarck Archipelago. Throughout Fiji, Mar–Dec peaking Apr–Oct; Polynesia, May–Sep (Lack 1959, Baker 1951, Hadden 1981, Watling 1982, Sagar 1977, Schodde *et al.* 1983, Clunie & Morse 1984, Robertson 1985, Ellis *et al.* 1990, Hutton 1991, Higgins 1999, DMNH). Some remain all year on NZ, probably young birds, and stragglers Auckland and Chatham Is. (Bogert 1937), presumably non-breeding young birds (Robertson 1985). Some state that young birds only return to NZ in their second year (e.g. Stresemann & Stresemann 1966). Makes non-stop flights of up to 3,000km (Robertson 1985) but vagrants to Palau I. cover nearly 6,000km over open sea (Dorst 1962). Once Queensland (Barrett *et al.* 2003), an earlier record, and two NT, Australia, considered unconfirmed (Higgins 1999). 5–6 May 2008, Choshi City, Chiba prefecture, Japan (Haga & Nuka 2009).

HABITAT Lowland primary and secondary forest as well as scrub, cultivated land and gardens; on North I., NZ, also pine plantations, and in winter coconut plantations. Most often in primary montane forests with beech, other broadleaf trees or *Podocarpus*, or mixture, with or without undergrowth, sometimes sub-alpine scrub near bush-line or short coastal vegetation. Primarily wooded areas on Fiji, Tonga and Samoa. To 1,200m; only 5.4% of Whitehead's nests parasitised below 250m, but 37.5% above 250m on Little Barrier I.; parasitises at lower elevations, Stewart I. (Watling 1982, Robertson 1985, McLean 1988, Higgins 1999).

BEHAVIOUR Secretive species of forest canopy and not often seen; usually solitary but also in pairs or in small groups up to five which in breeding season call and display together. When calling often spreads tail and flaps wings slowly; this behaviour occurs during mating and after completing breeding. Often mobbed by other birds. When hunting lizards creeps snake-like among rocks. Commonly perches lengthways on branches with wings lowered. Promiscuous without pair-bonding, presumably with lek mating system. Fledglings beg while hidden in foliage and often fed by other species, but adult cuckoos often watching nearby. Migrates in groups or alone, usually at night. More males than females when first arriving on breeding grounds; rather quiet for first days, but in Nov become more active. When leaving assembles in groups up to *c.* 12. Maximum flight-speed 80km/hour or more (Bogert 1937, Sparrow 1984, Robertson 1985, Pratt *et al.* 1987, McLean 1988, Bregulla 1992, Higgins 1999).

BREEDING Parasitism On North I., NZ, Yellowhead most common host, on South I. Pipipi and Whitehead; also New Zealand Bellbird, Tui, New Zealand Fantail, New Zealand Robin, Silver-eye, Song Thrush, House Sparrow and European Greenfinch (Falla *et al.* 1981, Robertson 1985, Higgins 1999). Grey Warbler (St. Paul 1976) not confirmed. **Season** Nov–Dec, dependent fledglings early Jan–early Mar (Fulton 1904, McLean 1987, Higgins 1999). **Eggs** Little variation irrespective of host; broad oval, one end more pointed; white, slightly glossy with creamy wash, or very pale pink spotted purplish-brown with underlying grey spots,

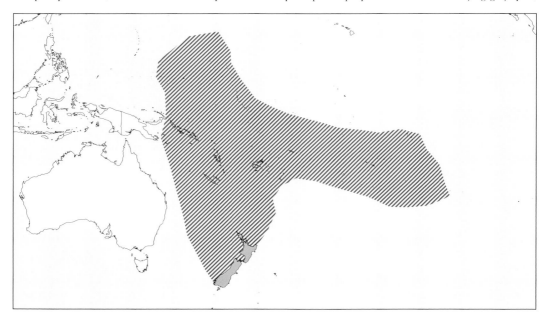

spotted more at broader end; 22.5–23.5x17.0–18.0mm (n = 4) (Stead 1936, Cunningham 1948). **Incubation** Probably *c.* 16 days; host Whitehead 18 days, Yellowhead *c.* 21 days and Pipipi 17–21 days (McLean 1988, Johnsgard 1997). **Chicks** Hatch naked. Most foster chicks do not survive, but whether evicted or starved not known, although two cases known where Yellowhead nestlings survived. Nestling period estimated 21 days, young Jan–Feb, but Whiteheads feed their young for many months (McLean 1982, 1987, McLean & Gill 1988, Elliot 1990, last two in Higgins 1999). **Survival** No information.

FOOD Principally caterpillars; other insects, including those flying above canopy; arachnids; crab and fish remains; lizards; eggs and chicks from bird nests, occasionally adult birds up to thrush size; fruit. One stomach contained Auckland Weta *Hemideina thoracica* nymph, adult katydid *Caedicia simplex*, four males and three females of Smooth Stick Insect *Clitarchus hookeri*, and female Prickly Stick Insect *Acanthoxyla prasina*; another contained six *Trifolium* seeds, some vegetable rootlets and some grit 1–2mm diameter (Bogert 1937, Gill 1980b, Reed 1980, Falla *et al.* 1981, Watling 1982, Robertson 1985, Higgins 1999).

STATUS AND CONSERVATION 262+ adults Little Barrier I. (McLean 1988). Large numbers winter Fiji, Tonga and Samoa (Watling 1982). Rare Vanuatu (Bregulla 1992). Widespread but uncommon throughout Solomons chiefly on smaller outer islands (Doughty *et al.* 1999). Some vagrancy. Difficult to monitor as it perches in canopy among dense foliage and often along branches presumably for camouflage, and ventriloquial calls difficult to locate, and often quiet outside breeding season. Widespread deforestation on NZ and many wintering areas may be partly responsible for decline in breeding area in NZ, but declining population of Yellowhead host perhaps also to blame. Killed by collisions with windows, wires or buildings, or taken by cats (Higgins 1999). Close monitoring of this species in near future important. Not globally threatened (BirdLife International 2011).

Long-tailed Koel *Urodynamis taitensis.* **Fig. 1.** Adult, Canterbury, New Zealand, December (*Peter Smith*).

Genus *Scythrops*

Latham, 1790 *Index Orn.* 1: 141. Type, by monotypy, *Scythrops novaehollandiae* Latham 1790.
One species.

Wallacean and Australasian; brood-parasitic. Very large, with huge bill. MtDNA analysis shows this genus to be sister to *Urodynamis* (Sorenson & Payne 2005).

CHANNEL-BILLED CUCKOO
Scythrops novaehollandiae — Plate 14

Scythrops novæ Hollandiæ Latham, 1790 (New South Wales)

Alternative names: Giant Cuckoo.

TAXONOMY Polytypic. Synonyms: *psittaceus* Kerr 1792, *australasiae* Shaw 1812, *australis* Swainson 1837, *Cuculus praesagus* Bonaparte 1850 and *novaehollandiae neglectus* Mathews 1912, all in nominate.

FIELD IDENTIFICATION 60–70cm. Huge cuckoo with hornbill-like bill, and in flight long pointed wings and long narrow tail, underwing grey, tipped and edged blackish. Normally flies high, straight and hawk-like with slow wing beats. **Adult** Pale grey on head, neck and underparts, lores and skin around eyes red, upperparts, wings and tail pale grey-brown barred blackish, broad black subterminal band on graduated tail tipped white. Female smaller with more barring on underparts. **Juvenile** Plain pale buffish on head and neck, upperparts and wings dark grey to grey with buff bars and spots, underparts pale buff with some fine barring on flanks and thighs, tail like adult. **Similar species** No confusion species in Australia. In NG Grey Crow superficially similar but has unbarred dark brown upperparts, much shorter tail without black barring, and shorter bill.

VOICE Loud, rising, raucous *oik* or *awk*, repeated, or far-carrying *crerk-crerk-crerk*, changing to *tawark-tawark-tawark* and on perching *ko-ko-ko-ko-ko*. In flight loud, raucous, ascending *quorrk* repeated 4–12 times; both sexes have low guttural call. Harsh croak when mobbing. In Wallacea loud, ascending and guttural *crrrRRRROW* or *crrRRRRAY*, often repeated several times; also hysterical descending, *CLOo-CLOo-Clu-Clu-clu*. Call of female inviting copulation quiet, thin trumpeting. Roosting birds resemble herd of pigs grunting, squealing and fighting. Fledgling has voice similar to young Pied Currawong, but more raucous and louder; one month after fledging captive bird had adopted adult's call. Calls day and night, flying or perching; very vocal in flight (Hindwood & McGill 1951, Lord 1956, LeCroy & Peckover 1983, Coates 1985, Coates & Bishop 1997, Pizzey & Knight 1997, Higgins 1999).

DESCRIPTION Nominate. **Adult male** Whole head, neck to upper mantle plain light grey; lower mantle, back, shoulders, rump and uppertail-coverts grey tinged brown, each feather boldly tipped black-brown giving scalloped appearance, rump more strongly washed brown and dark feather-tips narrower, lateral uppertail-coverts have smaller black-brown tips, two outermost often with white outer edges, remiges and coverts grey tinged brown broadly tipped black-brown, shortest lesser coverts lacking dark tips, underside of remiges light grey tipped black-brown, axillaries and underwing-coverts grey finely barred light grey; underparts pale grey to grey-white on belly; flanks, thighs and occasionally abdomen faintly barred grey, vent and undertail-coverts finely barred dark brown. Graduated white-tipped tail brownish-grey with *c.* 70mm wide dark brown subterminal band on T1 to *c.* 30mm on T5, and series of white half-bars on inner webs of T2–6, all with wide black-brown lower margins visible from underside. **Adult female** As male but bill shorter, paler on underparts, especially breast, flanks and belly, and barring slightly more conspicuous. **Juvenile** Bill shorter, crown to nape and below eye plain light rufous-yellow, each feather with concealed grey base and darker rufous subterminal band; upperparts as adult but feathers smaller and tipped rufous-yellow, on lower back and rump becoming more brownish-grey narrowly tipped buff, and uppertail-coverts tipped buff with dark subterminal bands; wings as adult but with pale rufous-brown tips, buffish to whitish in worn plumage; underparts pale rufous-brown on chin and throat, grading to greyish-white on breast and becoming whiter, sometimes with cream tinge, on belly and undertail-coverts, flanks and thighs barred dark brown, barring on belly grey-brown; tail as adult but narrower and may be slightly longer. **Nestling** Hatches naked, skin bronze-coloured. **Bare parts** Iris red to ruby-red, juvenile brown, orbital ring and facial skin to bill pinkish-red to dark red, skin greyish-brown in juvenile. Bill slightly grooved, pale grey, tip cream, female has larger part of bill cream distally; at first whole bill dirty pink in juvenile, later becoming dark pink at base of upper mandible and dark grey, mouth pinkish-white. Feet lavender to grey or dark grey, juvenile greyish-green, soles greenish.

BIOMETRICS Nominate. Wing male (n = 21) 337–367mm, mean 353.9 ± 7.8, female (n = 9) 336–368mm, mean 346.8 ± 9.1. Bill male 83.7–93.5mm, mean 86.4 ± 2.3, female 70.3–79.0mm, mean 74.5 ± 3.0. Tail male 253–278mm, mean 266.5 ± 7.4, female 253–276mm, mean 262.8 ± 7.2. Tarsus male 41.8–49.8mm, mean 45.8 ± 2.3, female 38.7–46.3mm, mean 44.0 ± 2.0 (Higgins 1999). Juveniles smaller but have longer tails than adults: male 267–286mm, female 269–287mm (Mason & Forrester 1996). Mass male (n = 12) 560–935g, mean 708.2 ± 113.5, female (n = 12) 560–800g, mean 660.0 ± 83.7 (Higgins 1999).

MOULT In Australia Feb–Apr, but some suspend moult of primaries until after northerly migration where it finishes May–Jun. Sequence of primary moult poorly understood, often different on left and right wings of same bird. Post-juvenile moult often begins soon after fledging, but on some delayed until Mar–Apr and some moult first primaries after migration. Some adults may retain few juvenile remiges. Primary moult semi-transilient ascending (Stresemann & Stresemann 1961, Higgins 1999, Payne 2005).

GEOGRAPHICAL VARIATION Three subspecies.
S. n. novaehollandiae Latham, 1790. Sub-coastal north and east Australia, and Babar, Flores and Sumba (Lesser

Sundas), wintering in north NG, Moluccas, Lesser Sundas and presumably Bismarck Archipelago. Described above.

S. n. fordi Mason, 1996. Sulawesi, Muna, Butung, Banggai, Tukangbesi Is and Sula Is; Buru? Larger, especially bill. Wing male (n = 3) 350–365mm, female (n = 4) 344–352mm; bill (to nostril) male 65–76mm, female 67–73mm; tail male 260–300mm, female 262–300mm (Payne 2005).

S. n. schoddei Mason, 1996. New Britain and Duke of York I. Populations on Admiralty Is, St. Matthias group and New Ireland undetermined. Large, especially bill and tail. Barring on upperparts stronger. Wing male mean (n = 4) 352.5mm, female mean (n = 3) 346.5mm; bill (to nostril) male mean 80.5mm, female mean 79.7mm; tail male mean 275.0mm, female mean 274.5mm (Payne 2005).

DISTRIBUTION Australasia and Wallacea. Breeding range sub-coastal north and east Australia, Bismarck Archipelago and Wallacea. Australian population winters north PNG to south Indonesia and probably rarely Bismarck Archipelago, Apr–Oct. Possibly small breeding population south PNG, as occurs all months, and preferred host in Australia, Australian Crow, also common there and has been seen mobbing cuckoo. Migrants Torres Strait, Aug–Sep and Mar. Average arrival date 27 Sep Queensland and departure Mar, but in dry periods when figs fruited sparsely breeding not recorded and departure as early as 21 Jan. Few records Australia 15 Jun–15 Aug. Wallacea: Alor, Ambon, Babar, Bacan, Banggai, Buru, Butung, Flores, Halmahera, Kai Is, Karakelong and Kaburuang (Talaud Is), Mangole, Muna, Obi, Paloe, Seram, Siau, Sula, Sumba, Tanimbar Is, Ternate, Tukangbesi and Wetar; once west Timor; once Timor-Leste; breeding Babar, Buru, Flores, Muna, Butung, Sulawesi, Banggai and Sumba (Lord 1943, Coates 1985, Draffan *et al.* 1983, Coates & Bishop 1997, Noske 1994, Barrett *et al.* 2003). Three records Solomons (Savo, Simbo and Rennell) (Diamond 2002). Two collected Kiriwina, Trobriand Is. (Rothschild & Hartert 1896). Twice Tasmania, Nov 1867 and Feb 1943 (Littler 1910, Sharland 1943). Twice NZ, Dec 1924 and Oct 1986–Jan 1987 (Higgins 1999). Once New Caledonia (Hindwood 1953).

HABITAT In Australia eucalypt and rainforest with rivers or streams with abundance of figs; broadleaf gallery forest and scrub, wet sclerophyll swamp woodland and farm woodland; gardens with fruiting figs. In NG edge of primary forest, clearings, woodland and savanna; often along roadsides. In Wallacea monsoon forest, forest edge, open areas with trees, and mangroves, from lowlands to 1,000m Sulawesi, 340m Flores. To 650m NG, once 1,200m, 25 Mar, apparently migrating bird (Hindwood & McGill 1951, Lord 1956, Bravery 1970, Blakers *et al.* 1984, Coates 1985, Mason & Forrester 1996, Coates & Bishop 1997, Pizzey & Knight 1997).

BEHAVIOUR Solitary or in pairs; outside breeding roosts communally in high trees. During migration in flocks, e.g. at Bensbach, where assembly of 60+ birds Nov, presumably before flight across Torres Strait although claimed to migrate singly or less than 15 together; night migration reported. Feeding flocks of 60 birds, which roosted together. Usually high in trees, feeding noisily, but may sit motionless or preening for prolonged period. If python nearby will fly furiously from branch to branch calling excitedly. In courtship feeding male first approached female with stick in bill, and after copulation fed female with insect; male fed female with masticated leaves as she lay flat on branch with lowered wings, male raising his tail before copulation. Male distracts hosts while female lays egg in nest. Nestling only gives begging call in horizontal posture with quivering wings when host nearby. Adult cuckoos seen near nest with their offspring, and adults also join young cuckoo before migration (Barnard 1926, Clarke 1982, Draffan *et al.* 1983, Coates 1985, Larkins 1994, Higgins 1999).

BREEDING Parasitism Brood parasite of crows, and in Australia, other species. Following list gives all recorded hosts with either egg, young in nest or fledgling fed by foster parents: Brown Falcon, Collared Sparrowhawk, Black-throated Butcherbird (1), Australian Magpie (14), Pied Currawong (66), Magpie-lark (2), White-winged Chough (2), Torresian Crow (25), Little Crow (4), Forest Raven (1), Australian Raven (17). Slender-billed Crow parasitised

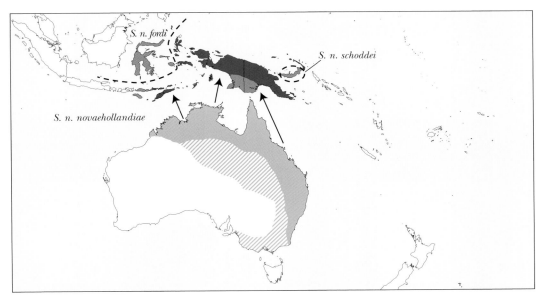

on Sulawesi, Jungle Crow on Flores and Sumba, and Flores Crow on Flores (Coates & Bishop 1997, Brooker & Brooker 1999, Higgins 1999). **Season** Australia generally Aug–Dec; in Alice Springs, central Australia, eggs laid Jan and young fledged late Feb. No information from other parts of breeding range (Mason & Forrester 1996, Higgins 1999). **Eggs** Mimic those of Pied Currawong in Australia, but quite unlike those of Magpie-lark and Torresian Crow; up to eight eggs in one nest, but mean 2.5 eggs present in parasitised nests of crow species, and 1.7 in Pied Currawong nests. Oval to elongated oval, warm buff and coarsely spotted with chestnut, brown or dark brown with pale purplish-brown to violet-grey underlying markings, and most spots at rounder end; 40.6–47.0x28.7–32.0m, (n = 6) mean 43.7x30.5; also dull pinkish-grey spotted violet-grey (Schönwetter 1967, Goddard & Marchant 1983, Higgins 1999). **Incubation** No information. **Chicks** Fully feathered at four weeks (Higgins 1999). Eviction of host young not recorded, but only few records of nests where host chicks survived, probably cuckoo simply outgrows host chicks which starve from food shortage, but recorded case where crow family was successful in getting two of their own offspring fledged. Fledge after 17–24 days (Salter 1978, Goddard & Marchant 1983, Payne 2005). **Survival** Host usually raises only one young (Lord 1956).

FOOD Mainly fruit, but also some insects such as stick insects and leaf insects. Figs preferred fruit, but also mulberries, mistletoe berries, and seeds. Captive cuckoos ate beetles larger than 2cm and grasshoppers, but preferred caterpillars and flying insects caught in air. Takes eggs and chicks of other birds (Lord 1956, Gilliard & LeCroy 1966, Mitsch 1979, Blakers *et al.* 1984, Coates 1985, Mason & Forrester 1996, Higgins 1999).

STATUS AND CONSERVATION Locally fairly common, but generally scarce NG (Coates 1985). Widespread but local Wallacea, but locally and seasonally common Flores and Sulawesi (Coates & Bishop 1997). Common Oct–Feb east Flores and north coast of central areas, rare along south coast and west Flores (Verhoeye & Holmes 1999). Uncommon Australia (Higgins 1999). Deforestation and clearing should be advantage for this species because of increased population of crow hosts (Dawson *et al.* 1991 in Higgins 1999). Not globally threatened (BirdLife International 2011).

Channel-billed Cuckoo *Scythrops novaehollandiae*. **Fig. 1.** Adult, *S. n. novaehollandiae*, Calliope, Queensland, Australia, November (*Bruce Thomson*). **Fig. 2.** Adult, *S. n. novaehollandiae*, Calliope, Queensland, Australia, November (*Bruce Thomson*). **Fig. 3.** Juvenile, *S. n. novaehollandiae*, Ingham, Queensland, Australia, May (*Rohan Clarke*).

Genus *Chrysococcyx*

Boie, 1826 *Isis von Oken* Band 2, col. 977. Type, by monotypy *Cuculus cupreus* Latham = *Cuculus cupreus* Shaw 1792. Thirteen species.

Old World; brood-parasitic; small. Plumage usually glossy bronze or green. Some members previously placed in *Chalcites*, but Berger (1955) concluded that all are congeneric and *Chrysococcyx* has priority. However, Christidis and Boles (2008) cite mtDNA evidence to retain *Chalcites* for Australo-Papuan species. MtDNA analysis places *Chrysococcyx* (*sensu lato*) as sister clade to that consisting of *Cacomantis*, *Cercococcyx*, *Surniculus*, *Hierococcyx* and *Cuculus* (Sorenson & Payne 2005). Also includes *Trogon* (J.F. Gmelin 1788), *Neochalcites* (Mathews 1912), *Lampromorpha* (Vigors 1831), *Lamprococcyx* (Cabanis & Heine 1863), *Chalcococcyx* (Hume & Davies 1873), *Heterococcyx* (Salvadori 1874), *Metallococcyx* (Reichenow 1896) and *Adamatornis* (Roberts 1922).

ASIAN EMERALD CUCKOO
Chrysococcyx maculatus   Plate 25

Trogon maculatus J.F. Gmelin, 1788 (Sri Lanka, corrected to Pegu, Burma; Robinson & Kloss 1923)

Alternative name: Oriental Emerald Cuckoo.

TAXONOMY MtDNA analysis places it closest to *xanthorhynchus* (Sorenson & Payne 2005). Monotypic. Synonyms: *Cuculus lucidus* (Blyth 1843), *Chalcites xanthorhynchus* (J.E. Gray 1844), *smaragdinus* Blyth 1846 (not *Chalcites smaragdineus* Swainson 1837, which is *Chrysococcyx cupreus*), *Cuculus malayanus* (G.R. Gray 1847), *Chrysococcyx plagosus* Bonaparte 1850 (not *plagosus* Latham 1801, a subspecies of *lucidus*), *hodgsoni* Horsfield & Moore 1858, *schombourki* Gould 1864.

FIELD IDENTIFICATION 18cm. Brilliantly-coloured small cuckoo with metallic green plumage, and in flight broad white band on underwing at base of remiges. Flight swift and powerful reminiscent of lorikeet. **Adult male** Whole upperparts, head, wings and tail, and underparts from chin to upper breast glossy green with golden-bronze reflections, primaries blackish; lower breast and belly white barred glossy bronze-green. **Adult female** Top of head to neck light rufous with indistinct dark barring, upperparts coppery-green, and whole underparts white, washed rufous and barred coppery-brown. T1 black with coppery-green iridescence; others dark rufous broadly barred dark brown, with white tips on T4, T5 white broadly barred dark brown, inner webs have white areas becoming rufous on inner two thirds. **Juvenile** As female but variable, with unbarred or barred rufous top of head, nape and hindneck; less glossy green on upperparts and wings, with rufous-orange barring on mantle, shoulders and wing-coverts. Some have forecrown finely barred black and white, and hindcrown green. **Similar species** Within range only female Violet Cuckoo confusable, which has more grey-brown to coppery head and neck and more dark brown above, with barring on rufous top of head, nape and hindneck, and more rufous underwing bar.

VOICE Three-note ascending whistle; descending *kee-kee-kee*; in flight sharp *chweek* or *chutweek*. From sonogram clear whistle *fwi-\FWI-\FWI-FWI-FWI*. Calls day and night (Inglis 1909, Inskipp *et al.* 1999, Robson 2000, Rasmussen & Anderton 2005).

DESCRIPTION Adult male Whole head, neck and upperparts metallic green with golden-bronze reflections, remiges blackish tipped bluish with whitish longitudinal band on inner web on underside of primaries in flight and white shafts to P9–10; chin, throat and upper breast glossy green like upperparts, rest of underparts clean-cut white evenly barred metallic bronze-green, undertail-coverts emerald-green spotted white; tail as upperparts, outermost feathers black with three white bars and outermost two tipped white. **Adult female** Top of head, nape and hindneck light rufous faintly barred bronze; upperparts and outer vanes of remiges bronze-green, entire underparts white barred bronze-green, barring more narrow on cheek, chin and throat, and with rufous wash on sides of neck; central rectrices blackish glossed green, rest barred rufous and black glossed green with rufous tips, and T4–5 tipped white, outermost also spotted white. **Juvenile** Like female but variable; above barred rufous and brown to greenish-brown or almost plain brown, some with elongated dark marks on rufous crown, others white barred or barred white and green, remiges dark brown washed green, outer edges and narrow tips chestnut, chin and throat dull rufous with black streaks, rest of underparts off-white barred brown, tail as female but outermost rectrices barred brown and rufous spotted white. **Nestling** Unknown. **Bare parts** Iris male bright red to red-brown, female brown, eye-ring coral to orange. Bill yellow to orange tipped black, juvenile black with flesh-coloured base, tongue yellow. Feet green to olive, juvenile grey, claws black.

BIOMETRICS Wing male (n = 21) 104–113mm, mean 108.8 ± 2.5, female (n = 15) 103–114mm, mean 109.1 ± 3.7. Bill male 14.0–15.9mm, mean 14.9 ± 0.5, female 14.3–16.4mm, mean 15.0 ± 0.8. Tail male 66–76mm, mean 70.8 ± 3.1, female 66–78mm, mean 70.8 ± 3.3. Tarsus male 13.8–16.2mm, mean 15.0 ± 0.9, female 14–16mm, mean 14.8 ± 0.6. Mass male (n = 3) 23–27g, mean 25.0, female 30g (Payne 2005).

MOULT Three juveniles Dec and early Jan had fresh remiges, another renewing P1–5 and P7–8 (Wells 1999).

DISTRIBUTION Oriental. Breeds in Himalayas from Garhwal to Arunachal Pradesh, north-east India, Nepal, Bhutan, Bangladesh, north and north-west Thailand, Burma, Cambodia, north Laos, Annam, and west and east Tonkin. In China summer visitor Sichuan, Hubei and Guizhou; resident west and south Yunnan, and Hainan; south-east Tibet? Winters south India south to Andaman and Nicobar Is, and Sumatra (Cheng 1987, MacKinnon & Phillipps 2000, Robson 2000). Rare winter visitor Thai-Malay Peninsula (Wells 1999, 2007), also female and juvenile Singapore, 31 May (Lim 2008c). Once Hong Kong (Viney *et al.* 1994).

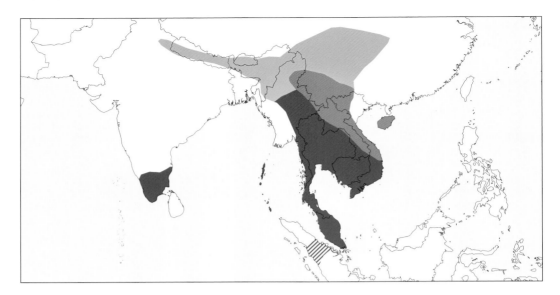

HABITAT Dense jungle, broadleaved evergreen forest, also secondary forest, swamp forest; when migrating and during winter in gardens and plantations. Dense evergreen forest in China. South-east Asia to 2,440m; may only breed above 600m. Nepal 250–1,800m; Bhutan 300–2,200m, once 2,745m. To at least 1,800m Thailand. To 3,300m Muli region, south Sichuan, China (Robinson & Kloss 1911, Meyer de Schauensee 1984, Cheng 1987, Lekagul & Round 1991, Grimmett et al. 1998, Robson 2000).

BEHAVIOUR Prefers canopy where its colour makes it extremely difficult to find. Shy and retiring. Normally very sluggish, keeping to canopy in dense evergreen forest where it sits motionless for long periods and perches along branch rather than across (Inglis 1909, Robinson & Kloss 1911, Smythies 1986, Grimmett et al. 1998).

BREEDING Parasitism Hosts: Eastern Crimson Sunbird, Mrs. Gould's Sunbird and Little Spiderhunter (Becking 1981). Claims of Zitting Cisticola, Grey-bellied Tesia, Brownish-flanked Bush Warbler, Brown Bush Warbler, Common Tailorbird, Blyth's Leaf Warbler, Chestnut-crowned Warbler (Sikkim), Rufous-capped and Grey-throated Babblers (Stevens 1926, Baker 1942, Friedmann 1968) perhaps erroneous. **Season** Breeding Duars, India, mid Apr–end Jul (Inglis 1909). **Eggs** Close mimics of Little Spiderhunter with fewer rusty spots, more olive-brown; elongated and slightly ovate, glossy pale blue or white with rufous to rufous-olive spots and blotches, often most at broader end; 17.4–20.5x11.9–13.8mm, (n = 33) mean 18.5x13.0; another eight from India, mean 17.6x12.3mm (Harrison 1970, Becking 1981). Some confusion between eggs of this species and those of Violet and Plaintive Cuckoos (Payne 2005). **Incubation** No information. **Chicks** Probably evict nest mates as do congeners, but not confirmed. **Survival** No information.

FOOD Caterpillars, cicadas, ants, grubs and other insects; spiders (Inglis 1909, Tikader 1984). Often takes insects in flight as flycatcher (Grimmett et al. 1998).

STATUS AND CONSERVATION Rather rare, both formerly and today; perhaps overlooked due to behaviour and coloration. Rare north Malay Peninsula (Robinson and Kloss 1911). Very rare throughout peninsular Thailand but commoner in north (Robinson and Kloss 1919). Wells (1999) only found one record for Thai-Malay Peninsula, but added eight records, involving up to 13 individuals Jan, Feb, Apr and Oct, 2000–2003 (various references in Wells 2007). Claimed as rare Thailand (Deignan 1945), but uncommon resident and winter visitor (Lekagul and Round 1991). Eight records Nepal, Apr–Sep; frequent summer visitor Bhutan (Inskipp et al. 1999). Rare India; local summer visitor Bangladesh (Grimmett et al. 1998). Sparingly distributed Burma (Smythies 1986). Uncommon China (Cheng 1987). Four specimens Sumatra (van Marle & Voous 1988). Not globally threatened (BirdLife International 2011).

Asian Emerald Cuckoo *Chrysococcyx maculatus*. **Fig. 1.** Adult male with traces of juvenile plumage on head, Khao Yai NP, Thailand, December (*Peter Ericsson*). **Fig. 2.** Adult female, Bangkok, Thailand, January (*Peter Ericsson*). **Fig. 3.** Adult male, Panwari Forest Range, Kaziranga, Assam, India (*Anand Arya*).

VIOLET CUCKOO
Chrysococcyx xanthorhynchus   Plate 25

Cuculus xanthorhynchus Horsfield, 1821 (Java)

Alternative name: Asian Violet Cuckoo.

TAXONOMY MtDNA analysis places it closest to *maculatus* (Sorenson & Payne 2005). Andaman population may be undescribed subspecies but few specimens exist (Rasmussen & Anderton 2005). Polytypic. Synonyms: *Cuculus Malayanus* Raffles 1822, *Chalcites* (*Chalcococcyx*) *xanthorhynchus bangueyensis* Chasen & Kloss 1929 and *limborgi* Tweeddale 1877, all in nominate.

FIELD IDENTIFICATION 16cm. Small size and undulating flight combined with *kie-vik* flight call at each dip make identification of male easy. **Male** Purple-violet on head, upperparts, throat, breast and tail, though under some field conditions this colour appears black, outer rectrices barred black and white and tipped white, abdomen and flanks white broadly barred dark, bright red eye-ring conspicuous. Males on some Philippine islands have blue upperparts. **Female** Top of head and mantle dark grey-brown, rest of upperparts more greenish-bronze; underparts white with narrower dark greenish-bronze barring. **Juvenile** As female except more rufous top of head and upperparts. **Similar species** Adult male's purple-violet upperparts unmistakable in good light. Female of Asian Emerald Cuckoo has top of head, nape and hindneck light rufous, not grey-brown, glossy green-bronze upperparts and more prominent white underwing bar, juvenile also more green on upperparts than juvenile Violet but identification can in some cases be difficult, especially in females. Little Bronze is more brown-bronze on upperparts, barring on chin and throat coarser, and bill black.

VOICE Loud whistle *pe-a-wit-pee-pee-pee-pee-pee-pee-pee-pe-a-wit-pe-a-wit-pe-a-wit-*(*pe-a-wit*) where middle section downward trill, whole lasting *c*. 4sec; flight call *kie-vik* repeated every sec audible over several hundred metres, or *tee-wit* and *seer-se-seer-seeseeseesee*. Rarely shrill *kree-cha* or twittering trill. Male advertises territory with long call-flights over canopy or calls from exposed perches and counter-calls to neighbours; call of female unknown (Wells 1999, Kennedy *et al.* 2000, Robson 2000).

DESCRIPTION Nominate. **Adult male** Whole head, neck, upperparts, exposed parts of wings, throat, breast and tail uniformly iridescent purple-violet, inner webs of remiges black, contrasting white belly to undertail-coverts broadly barred metallic violet or green; remiges below dark brown glossed violet, with large white area on secondaries and inner primaries; underwing-coverts white barred dark brown with violet gloss; rectrices black with violet gloss, T3–5 tipped white, outermost barred white more heavily on outer web. **Adult female** Crown grey-brown to coppery, nape, hindneck and entire upperparts and wings uniform dark bronze-green, inner webs of remiges below have broad chestnut patch on 2/3 of length, underwing-coverts white barred brown with bronzy sheen; sides of head and whole underparts white barred bronze-green, barring on head finer than rest; some females have rufous wash on throat and breast; tail glossy green to green-rufous, T1–2 with indistinct subterminal band and tipped buffish, T3–5 rufous barred green spotted white on outer web, T4–5 tipped white, underside of T1 brown, T2 brown with blackish subterminal bar and broad rufous tip, T3–4 broadly barred rufous and black and tipped white, T5 black with large rufous notches on inner web, rufous and white patches on outer web and tipped white. **Juvenile** More rufous than female, crown and nape rufous barred and spotted bronze; upperparts and wing-coverts rufous broadly barred brown with greenish sheen, remiges dark brown edged and tipped rufous; sides of head and underparts off-white with irregular dark brown barring, and no rufous on throat and breast, tail dark brown, T1–2 notched and tipped rufous, T3–5 barred rufous or rufous and white and tipped white. **Nestling** Like juvenile but whole head and neck streaked black instead of barred. **Bare parts** Iris and eye-ring dark red, female and juvenile iris dark brown, eye-ring pinkish-grey. Bill yellow with vermilion-red base, female and juvenile dark horn with yellow base. Feet olive-green, juvenile grey, claws yellowish.

BIOMETRICS Nominate. Wing male (n = 20) 93–104mm, mean 100.2 ± 2.8, female (n = 18) 93–106mm, mean 99.6 ± 4.2. Bill male 14.2–16.7mm, mean 15.6 ± 0.7, female 13.7–17.5mm, mean 15.4 ± 1.1. Tail male 63–73mm, mean 68.1 ± 3.9, female 58–76mm, mean 67.8 ± 4.8. Tarsus male 12.4–15.6mm, mean 13.6 ± 0.9, female 13.1–16.5mm, mean 14.4 ± 0.9 (Payne 2005). Migrants with longer wings 105–114mm (n = 4) at Fraser's Hill, peninsular Malaysia, may have been from unknown extralimital population (Medway & Wells 1976). Mass male (n = 10) 19.9–24.0g, female (n = 7) 20.5–25.6g (Wells 1999).

MOULT Adults wing-moulting late Apr–Oct, peak Aug–Sep, first finished Jul. Post juvenile moult Mar–May, though first-year migrants in late Aug still in moult or moult suspended, once suspended 13 Jun. Juvenile to adult moult Palawan, Jan. Primary moult starts P10, P8 and P6 (Stresemann & Stresemann 1969, Wells 1999, 2007, BMNH).

GEOGRAPHICAL VARIATION Two subspecies.
C. x. xanthorhynchus (Horsfield, 1821). North-east India, Bhutan, Bangladesh, Burma, Thailand, Thai-Malay Peninsula, Yunnan, Indochina, Greater Sundas and Palawan; south-east Tibet? Male metallic purple-violet on upperparts, female grey-brown crown (some with rufous on head) and greenish-bronze on upperparts. Described above.
C. x. amethystinus (Vigors, 1831). Philippines on Basilan, Catanduanes, Cebu, Leyte, Luzon, Mindanao, Mindoro and Samar (Kennedy *et al.* 2000). Said to be larger; male metallic blue with violet wash on upperparts. Female indistinguishable from nominate female. Wing male 101mm, tail 70mm, bill 19.6mm (BMNH).

DISTRIBUTION Oriental. India in Arunachal Pradesh, Assam, Manipur, Meghalaya, Mizoram, Nagaland and Tripura, where resident or summer visitor, but winter status unknown. Only five records Bangladesh to 1997. In Thailand on north and east plateaux, central plains, south-east, west and peninsular provinces. Summer visitor Bhutan. Presumably chiefly resident Thai-Malay Peninsula. Andaman and Nicobar Is. In Burma Arakan, Thayetmyo, Karenni, Pegu and Tenasserim; once Htinzin, Upper Chindwin. Cambodia, south Laos, Cochinchina and Annam. North-west Yunnan (and south-east Tibet?). Sumatra, including Sipora, Simeulue, Mentawai Is. and Lingga Archipelago. Resident throughout lowlands of Borneo, including Banggi I.; North Natuna Is. Java. Philippines (Ripley 1944, Deignan 1963, Cheng 1987, Ali & Ripley 1983, Smythies 1986, Cheng 1987, van Marle & Voous 1988, Davidar *et al.* 1997, Grimmett *et al.* 1998, Quý & Cu 1999, van Balen 1999, Wells 1999, Grantham 2000, Kennedy *et al.* 2000, Mackinnon & Phillipps 2000, Robson 2000, Thompson &

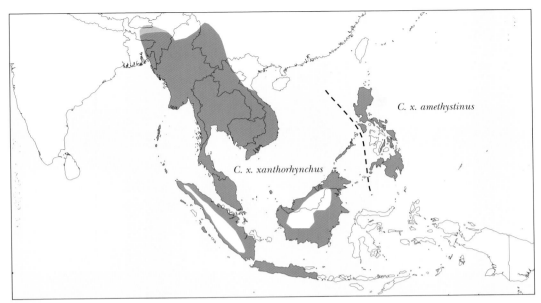

Johnson 2003, Spierenburg 2005, Mann 2008, B. van Balen pers. comm.). Caught at night Dalton Pass, Luzon Sep–Dec, and 36 larger migrants caught at Fraser's Hill, peninsular Malaysia, 7 Aug–31 Dec, and one Pisang I., Malacca Strait, and at light at 1,300m, Rinangisan, Sabah, Apr (Gonzales 1983, McClure 1998, Sheldon et al. 2001), evidence of migration and/or dispersal.

HABITAT Dry lowland semi-evergreen and evergreen forests, both primary and logged, forest edge, occasionally secondary growth, plantations, gardens and parklands to 500m, Thai-Malay Peninsula. Lone male in thick bamboo forest on east Java. Also swamp and peat swamp forests, kerangas forest, fire padang and grassland, edges of mangroves, on seashore, cocoa and *Albizia* plantations, to 1,300m, even heavily-logged peat swamp and riparian forests in Borneo. To 1,500m (exceptionally 2,300m) India. 200–1,900m, mostly below 600m Bhutan. Usually below 600m but to 1,100m Thailand. To 700m Sumatra. Usually below 1,000m Philippines. In China prefers forest edge, gardens and plantations rather than primary forest (van Marle & Voous 1988, Lekagul & Round 1991, van Balen & Prentice 1997, Grimmett *et al.* 1998, Smythies & Davison 1999, Wells 1999, Grantham 2000, Kennedy *et al.* 2000, MacKinnon & Phillipps 2000, Spierenburg 2005, Mann 2008).

BEHAVIOUR Often inconspicuous; singles or pairs, mostly on outer branches in canopy, but obvious when calling in flight. Runs sideways along horizontal branches snapping prey from overhead foliage; males counter calling. Long and circular calling flights over forest canopy (Wells 1999, Kennedy *et al.* 2000, Rasmussen & Anderton 2005, C.F. Mann).

BREEDING Parasitism Eastern Crimson Sunbird and Little Spiderhunter main hosts India (Becking 1981); possibly Grey-hooded Warbler, Common Tailorbird and Zitting Cisticola (Baker 1934, 1942, Harrison 1970), but contested (Becking 1981). Pair of Brown-throated Sunbirds feeding fledgling Singapore; Little Spiderhunter, peninsular Malaysia; Olive-backed, Purple-throated and Brown-throated Sunbirds, Singapore; also Abbott's Babbler, Eyebrowed Wren-babbler, Thick-billed Flowerpecker, Nepal Fulvetta (Friedmann 1968, Wells 1999, SUARENG 2004 in Wells 2007, Low 2008b, P.E. Lowther *in litt.*). Host unknown, Philippines (Kennedy *et al.* 2000). **Season** Fledglings Thai-Malay Peninsula late May–late Jun; calls all months Brunei; oviduct egg Sabah, Apr; oviduct egg Luzon, Philippines, 4 Apr (Goodman & Gonzales 1990, Wells 1999, Sheldon *et al.* 2001, SUARENG 1999 in Wells 2007, Mann 2008). **Eggs** Oviduct egg off-white washed violet with small black-and-white dots evenly distributed all over but colouring may be incomplete; laid eggs broad oval, moderately glossy, whitish-buff to pink, profusely speckled and blotched vinaceous-red or violet with underlying olive-brown; 15.9–16.8x10.8–12.9mm, (n = 8) mean 16.4x12.3 (Becking 1981, Sheldon *et al.* 2001). No further information.

FOOD Leaf-eating beetle larvae; flies, ants, beetles and other insects, and fruit. Often 'flycatches' insects (J. Motley and Lord Medway in Smythies & Davison 1999, Wells 1999, Rasmussen & Anderton 2005).

STATUS AND CONSERVATION Rare India and Bangladesh (Grimmett *et al.* 1998). Uncommon Bhutan (Spierenburg 2005). Very rare China (Cheng 1987). Common resident and winter visitor Thailand (Lekagul & Round 1991), but Robson (2000) found it scarce winter visitor. In Thai-Malay Peninsula locally fairly common and ability to inhabit man-used habitats suggest that it is not at risk in spite its lowland preference (Wells 1999); formerly considered rare (Robinson and Kloss 1911) suggesting population has increased. Scarce or overlooked Burma (Smythies 1986). Local but not uncommon Borneo (Mann 2008). Common Sumatra (van Marle & Voous 1988.). Vagrant Andaman and Nicobar Is. (Tikader 1984). Threatened by forest fragmentation Java (van Balen 1999). Uncommon Philippines (Kennedy *et al.* 2000); not found during four weeks fieldwork in forest on Mindoro, Philippines in 1992 (Dutson *et al.* 1992), and probably now extinct Cebu (Brooks *et al.* 1995b). Not globally threatened (BirdLife International 2011).

Violet Cuckoo *Chrysococcyx xanthorhynchus*. **Fig. 1.** Adult male, *C. x. xanthorhynchus*, Singapore, January (*H. Y. Cheng*). **Fig. 2.** Juvenile, *C. x. xanthorhynchus*, Khao Yai NP, Thailand, February (*Peter Ericsson*).

DIDERIC CUCKOO
Chrysococcyx caprius   Plate 26

Cuculus caprius Boddaert, 1783 (Cape of Good Hope, South Africa)

Alternative names: Diederik/Diederic/Didric Cuckoo.

TAXONOMY MtDNA analysis places this species in clade with *klaas*, *flavigularis* and *cupreus* (Sorenson & Payne 2005). '*Chrysochlorus*' (Cabanis & Heine 1863) from Senegal based on shorter wing measurement, but longer-winged birds also occur there, and shorter-winged birds occur RSA. Clancey (1990) believes this subspecies valid, and that non-breeding movements of longer-winged southern African breeding birds into areas where shorter-winged birds occur confuses issue. Variation in colour of iridescence clinal, but amount of white mottling on wing-coverts of both sexes markedly less in northern birds, and white frontal streak of males broader and white elliptical spotting on rectrices coarser and more spaced in southern birds. Formerly known as *Chalcites cupreus* Kirk 1864 and *Chrysococcyx cupreus* Reichenow 1887, so care must be taken in evaluating early material, when African Emerald Cuckoo was known as *C. smaragdineus* Swainson 1837. Monotypic. Other synonyms: *Chalcites auratus* Lesson 1830 and *Lampromorpha chalcopepla* Vigors 1831.

FIELD IDENTIFICATION 19cm. Small cuckoo; flies with irregular wing beats, with alternating periods of flapping and gliding, either straight or undulating. **Adult male** Glossy green above, largely white below with variable amount of green flank barring. White marks in front of, and behind eye, and thin green malar stripe extending onto otherwise white throat and upper breast. **Adult female** Rather like brown version of male, although throat russet and thicker bars on flanks. **Juvenile** Variable, but thickly streaked on throat, merging into barring on belly, flanks and undertail; closely barred above and often shows coral-red bill. **Similar species** Combination of green upperparts and whitish underparts of male separate it from all other African cuckoos except Klaas's. Unlike Klaas's, Dideric has much white on head, including strong white supercilium extending in front of eye; throat sides, and front part of ear-coverts also white. Dark green malar stripe. Iris and eye-ring red. Flanks strongly barred dark green. Underwing barred heavily and appears rather dark in flight, also comparatively little white in tail. Adult female has strong supercilium extending in front of eye, with russet throat and upper breast, and broad whitish areas on many wing-coverts further separating it from other *Chrysococcyx* cuckoos. Juvenile, always heavily streaked on throat (rather than barred as in congeners), shows white in front and behind eye, and frequently coral-red bill, particularly when recently fledged.

VOICE Male gives memorable *dee-dee-dee-dee-derik*, sometimes pure and fluty, but frequently accelerated and slurred, and first notes often rising in pitch; number of *dee* notes varies; such calls may be given in aerial chases between males; repeated at regular intervals with *c*. 2sec pause between phrases; insistent *dee dee dee dee dee* by male; wavering *weahweahweahweah* by male when offering caterpillar to female; female gives plaintive *deea deea deea* during courtship. Persistent high pitched cheeping given by nestlings and fledglings, which apparently differs depending on host species (McLachlan & Liversidge 1957, Reed 1968, Chappuis 1974, 2000, Irwin 1988, Stjernstedt 1993, Payne 1997, 2005, Borrow & Demey 2001).

DESCRIPTION Adult male Forehead, lores, crown and ear-coverts glossy bronze-green, narrow white lateral crown-stripe over head from near bill base, broad white supercilium in front and behind eye. Back and rump bronze-green, and uppertail-coverts green with broad white tips. Wings green, with secondary and lesser coverts broadly tipped and barred white. Remiges dark grey, tipped green, barred and spotted white on inner webs. Underwing and its coverts dark grey barred white. Dark green malar stripe, chin, throat, front part of ear-coverts, upper breast and belly white. Sides of breast, flanks and under tail-coverts variably barred bronze-green, undertail sometimes entirely white. Tail green, with outermost feathers barred white, sometimes almost entirely white with small dark spots, and remainder variably spotted white, T1 often entirely dark with very narrow white tips. **Adult female** Variable, upperparts normally more bronzy than male and generally duller. Variably spotted bronze and green on mantle and back. Wing patterning as male, but green areas duller, outer webs of remiges with small rusty spots, inner webs barred white with some rusty areas. White areas on head duller, malar stripe buff, and strong buffish suffusion to breast; belly usually white but with variable dark bronzy barring. Central rectrices green, others rusty-red, barred with green, and outermost feathers variably spotted or barred white. Females of different egg *gentes* not different in appearance or size (Payne 2005). **Juvenile** Two colour morphs sometimes distinguished, but in reality very variable both within and between populations. Rufous birds apparently occur in areas where young are raised by seed-eating bishops, whereas green morph young occur where common hosts are *Ploceus* weavers (Payne 2005). Most birds similar to adult female, but with rusty-red barring on forehead and crown extending variably onto rest of upperparts. Some birds much more rufous, with all upperparts, wings and tail rusty-red variably barred darker greenish-bronze. These birds also tend to have rufous wash on throat and breast. Underparts usually with dull white ground colour, although some have dull olive throat. Throat usually thickly streaked, with streaks merging into spots on breast and belly. Flanks and undertail-coverts barred dark brown. **Nestling** Hatchling naked, with pink skin that darkens to blackish within two days (Irwin 1988). Distinguished from congeners by coppery top of head, and back not greenish. May be quite rufous, especially head. Bill orange-red, gape vermilion edged yellow; iris grey. **Bare parts** Iris red edged whitish in adult male, hazel brown to grey in female, sometimes edged yellow, and dark brown in juvenile. Eye-ring orange or red in male, grey in female. Bill dark grey to black with greyish base, in juvenile bright coral-red or orange-red, darkening with age. Feet of adult grey with yellow soles, dark brown in juvenile.

BIOMETRICS Wing male 114–120mm (n = 6), mean 116.5 ± 2.2; female 117–128mm (n = 23), mean 121.3 ± 2.9; tail male 82–92mm, mean 85.3 ± 4.2; female 87–98mm, mean 90.7 ± 2.8; bill male 15.3–17.6mm, mean 16.4 ± 1.0; female 16.6–19.9mm, mean 17.7 ± 0.9; tarsus male 17.3–20.4mm, mean 18.3 ± 1.2; female 17.4–20.0mm, mean 18.3 ± 0.8. Also wing male 103–116mm, female 105–116mm (Payne 2005). Mass male 24–36g (n = 24), mean 29, female 29–44g (n =14), mean 35 (Irwin 1988).

MOULT Gradual, lasting several months. Juvenile primary moult, lasting 80 days, often interrupted and incomplete, begins at four months. Transilient primary moult from two loci. Active tail and body moult during breeding season.

However, most arrive RSA in fresh plumage, presumably having moulted prior to migration. Young birds may retain juvenile remiges into start of first breeding season. Female just completed moult from juvenile plumage, Darfur, Sudan, 3 Nov; immature to adult moult Kenya, Dec (Lynes 1925, Friedmann 1948b, Traylor 1963, Rowan 1983, Cheesman & Sclater 1935, Hanmer 1995, Payne 2005, C.F. Mann).

DISTRIBUTION Afrotropical. Resident across central Africa from south Senegal and SL to Ethiopia and east Africa. Common breeding visitor north to *c.* 15°N. Southern limit of residents from Equatorial Guinea to central Tanzania, although some possibly resident south DRC and south Tanzania. Rains visitor north Senegal. Migrant or resident breeder south Mauritania (Senegal valley; some north to 18°N during rains). Resident Bioko, Zanzibar and Pemba (brief visitor?); but migration into Zanzibar Oct–Nov to breed, leaving Mar; possibly only breeding visitor Mafia. Breeding visitor north and south of residents' range, frequently timed to coincide with rains. Breeds in every southern African country; birds reach Zambia by Aug, and extreme south of continent by mid Oct–Nov. Malawi mostly Oct–Mar, but some throughout year; Zambia Oct–Apr and during rains south Africa. Departure from RSA begins Jan, and elsewhere in southern Africa most birds departed by Mar, some remaining until May. Northward movements coincide with rains, typically Apr–Oct in Sudan, but resident extreme south, CAR and Chad (in last early Jul–mid Oct, but absent in less wet years), and Aug–Oct Sahel zone; Gambia, Jun–Nov. Non-breeder savannas and Sahel of Gabon, Jan–Feb, but resident around weaver colonies, and in forest clearings all year, but mostly Mar–Sep; mainly wet season visitor Ghana; R. Niger inundation zone, Jun–Nov. Resident south Togo and south Benin, Apr–Aug in north. North Burkina Faso in rains, Jun–Sep; Dosso, Niger, Apr–May. In wet season Apr–Oct, occasionally Feb–Mar, north Nigeria to 15°N, and more common in south in rains where present throughout year. Throughout Cameroon in rains, but absent north of 6°N Nov–Mar in dry season. Resident Gulf of Guinea islands; mostly dry season Sep–Mar Bioko, but some Jun–Aug. Visitor during rains in Sudan to 16°N, in Darfur, arriving Jun, departing Sep, small numbers breeding, others non-breeders; resident extreme south.

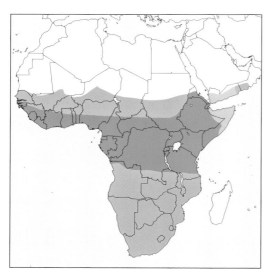

Resident Eritrea and Ethiopia, but also immigrants late May–late Oct, Eritrea, and late Mar–late Oct, Ethiopia. In much of east Africa resident in wet areas, and wet season visitor to dry areas. Resident and migrant Kenya, but rare arid north and east; occurs in dry areas of Kenya during rains (Lynes 1925, Jackson 1938, Pakenham 1943, Bannerman 1951, Smith 1957, Elgood *et al.* 1973, 1994, Greig-Smith 1976, Benson & Benson 1977, Nikolaus 1987, Lack *et al.* 1980, Snow 1978, Stevenson 1980, Louette 1981a, Brosset & Érard 1986, Grimes 1987, Giraudoux *et al.* 1988, Irwin 1988, Lamarche 1988, Lewis & Pomeroy 1989, Gore 1990, Morel & Morel 1990, Christy & Clarke 1994, Cheke & Walsh 1996, Zimmerman *et al.* 1996, Balança & de Visscher 1997, Barlow *et al.* 1997, Pérez del Val *et al.* 1997, Vernon *et al.* 1997, Quantrill & Quantrill 1998, Anciaux 2000, Anon. 2010, Ash & Atkins 2009). Absent Somalia 10 Nov–14 Mar, but once breeding 15 Jan; passes through north Somalia Mar–early Jun and Sep (Ash & Miskell 1998). Southern breeders probably overfly equatorial forest and east African savannas when moving to northerly non-breeding areas (Clancey 1990). Breeding visitor east Yemen and west Oman (Dhofar) May–Oct; few records, possibly passage, west Yemen (Gallagher & Woodcock 1980, Brooks *et al.* 1987, Porter *et al.* 1996). Vagrants Cyprus, Jun 1982 (Lobb 1983) and Israel, Mar 1994 (van den Berg 1994).

HABITAT Generally in more open habitats than other African *Chrysococcyx*. Wide range of habitats used including forest edge and clearings, edge of marshland and swampy areas, reed beds and papyrus, gardens, open woodland, wooded and thorn savanna, steppes and semi-desert. Bushed and wooded country to 2,000m Kenya, avoiding arid areas without permanent water. Often around weaver colonies. To 2,000m, although only common below 1,200m, and generally below 1,780m DRC (Chapin 1939, Friedmann 1948b, Benson *et al.* 1971, Lewis & Pomeroy 1989, Gore 1990, Elgood *et al.* 1994, Payne 1997).

BEHAVIOUR Relatively conspicuous for a cuckoo. Usually forages in canopy, but sometimes on ground. Prey typically beaten and gut contents removed prior to ingestion. Male territorial in breeding season, calling from prominent perch and chasing away other males. Male regularly advertises territory by calling from stereotyped positions, also performs aerial display by 'sweeping in wide arc', alternating flapping and gliding with wings held high. Interactions among territorial males frequent, and typically involves frequent calling and noisy pursuits, often with several birds (males and females) participating. Territory size appears to be determined by habitat type and song post availability, usually 4–5ha. Two males observed in leafless tree with tails fully depressed, occasionally raised over back, wings drooped but sometimes raised half open over back, in crouched position uttering *dee dee dee* call and hopping from branch to branch around each other. Courtship feeding frequently observed with male feeding female on hairy caterpillars, accompanied by intricate behavioural ritual by male prior to food presentation involving drooped wings, cocked tail and extended neck; up to eight caterpillars/15min fed to female. Female raises and fans tail before copulation, which may take place after courtship feeding. Generally fledgling ignores other species and only begs from foster; fledgling fed by Southern Grey-headed Sparrow, but later driven off by Southern Red Bishop which then fed it (Chapin 1939, Friedmann 1948b, Maclaren 1952, Reed 1968, Irwin 1988, Borrow & Demey 2001, Payne 2005). Adult observed feeding fledgling (Iversen & Hill 1983), and similarly fledgling fed by male Dideric Dec, Transkei, RSA (Maclaren 1952) and Ethiopia (Ash & Atkins 2009), may have been courtship feeding, male mistaking fledgling for female, or due to many fosters being granivorous, or nest parasitism is comparatively recent phenomenon (Friedmann 1948b, 1956).

BREEDING Parasitism Probably monogamous, although pair bond very brief and serial polygamy possible (Irwin 1988). "Females that parasitise different species of hosts may mate with same male" (Verheyen 1953) suggesting polygyny or serial monogyny. Female removes and eats one egg from hosts' nest (Friedmann 1948b, Calder 1951, Skead 1952, Payne 2005), however Reed (1953) claims that no host egg is eaten or removed. Nest of African Pied Wagtail in Uganda contained three hosts' eggs, one of Red-chested Cuckoo, and one of this species (Jackson 1938). Concentrates on weavers, so host range only overlaps with Klaas's and African Emerald slightly. Host choice reflects use of open country habitats. Hosts, with number of known cases, as follows. Common Bulbul, east Africa (2); Zambia; Natal. African Golden Oriole, Zambia (1). African Paradise-flycatcher, RSA; Cameroon; Uganda. Stout Cisticola, east Africa. Red-faced Cisticola, Uganda. Red-pate Cisticola, Uganda. Rattling Cisticola, Zimbabwe (1). Karoo Prinia, RSA (5). Black-chested Prinia, possibly RSA. Little Rush Warbler. Chestnut-vented Warbler, RSA (6); Namibia. Yellow-bellied Eremomela, possibly Mauritania. White-throated Robin-chat, RSA; Cape Robin-chat, possibly RSA; Karoo Scrub Robin, RSA (3). White-browed Scrub Robin, RSA (3). Kalahari Scrub Robin. Rufous Scrub Robin. Mountain Wheatear, RSA (4), Namibia. Familiar Chat, possibly RSA. Mariqua Flycatcher, Zimbabwe (2). African Grey Flycatcher, Kenya (1). Pale Flycatcher, possibly RSA. Olive-bellied Sunbird, Bioko. Possibly Western Olive Sunbird, Ethiopia. Bronzy Sunbird. Red-chested Sunbird. White-browed Sparrow-weaver, possibly RSA and Zimbabwe. Grey-headed Social Weaver, east Africa (3). Cape Sparrow, RSA (118); Namibia. Southern Rufous Sparrow, RSA (5); Namibia; east Africa. Northern Grey-headed Sparrow, east Africa. Southern Grey-headed Sparrow, RSA; Namibia; Zimbabwe. Chestnut Sparrow, east Africa (3). Yellow-spotted Petronia, east Africa (1). Thick-billed Weaver, possibly southern Africa. Spectacled Weaver, RSA (13); Zimbabwe; Zambia. African Masked Weaver, RSA (219); Namibia; Zimbabwe; Botswana; Zambia. Cape Weaver, RSA (40). Bocage's Weaver, Zambia (1). African Golden Weaver, main host Zanzibar and Tanzanian littoral; RSA (8). Village Weaver, southern and east Africa (25); RSA; south Tanzania; south Mozambique; Uganda?; DRC, frequent; west Africa (including Senegal), probably main host; Bioko; RC; Zambia; Malawi. Vieillot's Weaver, Uganda; common DRC; Cameroon (4). Black-necked Weaver, east Africa; common Nigeria. Holub's Golden Weaver, Zimbabwe (8); Zambia; Malawi; east Africa. Lesser Masked Weaver, RSA (9); south Mozambique; Zambia (5) and Malawi; east Africa. Baglafecht Weaver, east Africa (8). Chestnut Weaver, east Africa (4). Speke's Weaver, east Africa (4). Heuglin's Masked Weaver, common Mali. Golden-backed Weaver, several east Africa. Northern Masked Weaver, east Africa (2); DRC. Southern Brown-throated Weaver, Malawi (1). Rüppell's Weaver, Oman. Possibly Golden Palm Weaver. Black-headed Weaver, Mauritania. Crested Malimbe, east Africa (1). Red-headed Weaver, RSA (26); Zimbabwe; Zambia; Malawi; east Africa. Black-winged Bishop, Zimbabwe (1); Zambia (8); Malawi; not proven Upemba, south-east DRC and southern Africa. Southern Red Bishop, RSA (245); Zimbabwe; Zambia; Malawi; east Africa. Zanzibar Red Bishop, east Africa. Yellow Bishop, possibly RSA and Zimbabwe. Red-collared Widowbird, possibly RSA and Zimbabwe. White-winged Widowbird, several RSA; Zimbabwe; Zambia; Angola. African Pied Wagtail, frequent Uganda. Cape Wagtail, RSA (15); Zimbabwe. Possibly African Pipit, RSA. Yellow-crowned Canary, possibly southern Africa. Cabanis's Bunting, possibly southern Africa. Golden-breasted Bunting, RSA (5); Zambia. Cape Bunting, RSA (Sharpe 1875, Fischer 1880, Nicholson 1897, Petit 1899, Ivy 1901, Bates 1905, van Someren 1916, Roberts 1939, Priest 1934, Sparrow 1936, Jackson 1938, Chapin 1939, 1954b, Vincent 1947, Benson & Benson 1949, Friedmann 1948b, 1956, Haydock 1950, 1951, Miles 1951, Skead 1952, Benson 1953, Reed 1953, Verhayen 1953, Serle 1954, Lamm 1955, C. Belcher *in litt.* to Friedmann 1956, Fry 1961, Farkus 1962, 1966, Reed 1968, Jensen & Jensen 1969, MacLeod 1969, Benson *et al.* 1971, Aspinwall 1973, Benson & Benson 1977, Colebrook-Robjent 1977, Gallagher & Woodcock 1980, Irwin 1988, Lamarche 1988, Morel & Morel 1990, Hockey *et al.* 2005, Payne 2005, Isenmann *et al.* 2010, BMNH). Arrow-marked Babbler, RSA (Winterbottom 1951) and Zimbabwe (Swynnerton 1911, both doubtful (Friedmann 1956). 'Egg-dumping' may have successful outcome, e.g. fledgling once fed by Yellow-tufted Malachite Sunbird (Richardson 1983); similar cases involving Scarlet-chested Sunbird, Zambia (Benson *et al.* 1971), and Copper Sunbird, Zambia (Colebrook-Robjent 1977, 1984).

Female extremely secretive during egg laying period, but observed moving very rapidly through hosts' territories using direct hawk-like flight (Irwin 1988). Rate of parasitism of nests very variable; 31–33%, although ranging from 0–66% (Southern Red Bishop), 8.5% (Cape Sparrow), 33% (Heuglin's Masked Weaver); many other weavers average 10–20% (Irwin 1988). Two cuckoo eggs in one nest of Cape Weaver, but no host eggs, and one with one cuckoo egg and two host's, RSA (Vincent 1934). Intra-specific nest parasitism in Northern Masked Weaver of much greater importance than that of Dideric Cuckoo (23–35 % nests parasitised (n = 645), less than 1% by cuckoo) at Lake Baringo, Kenya, and therefore considered unimportant in selecting for egg-colour variability and egg-recognition by weaver host (Jackson 1992). In southern Africa Red Bishops did not reject model eggs on basis of size, increase in clutch size, or stage at which egg appeared in nest, but only if colour was significantly different (Lawes & Kirkman 1996). Pair of Spectacled Weavers lost 8.8% body mass in raising young cuckoo in RSA (Chalton 1976). Variation in egg morphs within one weaver colony, and observations on individually-recognisable females, suggest that more than one female may lay in one colony (Hunter 1961, Friedmann 1968, Collias & Collias 1970, Macdonald 1980). **Season** Year-round breeding in equatorial Africa, often contingent on rains. Distinctly seasonal further north and south; north Senegal, Aug–Oct; south Senegal and SL, Aug–Nov; Gambia, Jul–Nov; Mali, Aug–Oct; juvenile IC, early Feb; calls in lowland rice fields Liberia all year, peaks May–Jun, Sep–Oct; north Burkina Faso, Jun–Sep in rains; courtship feeding Togo, Sep; coastal Ghana, Mar–Jul; Nigeria, Jan–Oct, near coast Apr–Sep, north Nigeria Apr–Sep; fledgling Bioko, 2 Sep, Oct; enlarged testes north Nigeria, 12 May; Cameroon, Mar, Jun, Dec, fledglings Jan, Mar–Jul, Dec, mating Sep, calling throughout year, juveniles Mar and Aug; during rains DRC, in south-east enlarged testes Dec–Feb, active ovaries Dec; Darfur, Sudan, Aug, also Oct, Sudan; young just fledged Eritrea, Oct; Ethiopia, Jun–Sep, possibly Apr, May, Oct; enlarged ovary south Ethiopia, Mar; juvenile being fed Somalia, 15 Jan; nestling Mpumu, Uganda, Aug; egg Entebbe, Uganda, Mar central and south Uganda and south-west Kenya, Apr; Kenya (except south-west and coast) and north-east Tanzania, Mar–Aug, Dec; Tanzania (except north-east) Feb; Zanzibar and Pemba, Nov–Mar; Angola, Dec–Jun; Kenya, at semi-arid Lake Baringo, Apr–May (short rains) and Nov–Jan (long rains); Zambia, Aug–Apr, once May, once Dec; Malawi, Oct–Apr; Zimbabwe, Oct–Apr; Botswana, Oct, Jan, fledglings Oct–Apr; Zambia, Aug–Apr; Eastern Cape, RSA, late Oct–mid Jan, month or two later in north; Namibia, Dec–Feb (Lynes 1925, Bannerman 1933, Jackson 1938, Pakenham 1979, Chapin 1939, Serle 1943, 1950, 1954, 1965, Benson 1945, Vincent 1946, Friedmann 1948b, 1956, Benson & Benson 1949, 1977, Verheyen 1953, Benson & Serventy 1957, Smith 1957, Fry 1961, Reed 1968, Jensen & Jensen 1969, Benson et al. 1971, Aspinwall 1973, Eisentraut 1973, Colebrook-Robjent 1977, Bruggers & Bortoli 1979, Brown & Britton 1980, Macdonald 1980, Irwin 1981, 1988, Rowan 1983, Grimes 1987, Nikolaus 1987, Morel & Morel 1990, Maclean 1993, Demey & Fishpool 1994, Elgood et al. 1994, Hanmer 1995, Payne 1995, Cheke & Walsh 1996, Skinner 1996, Zimmerman et al. 1996, Balança & de Visscher 1997, Barlow et al. 1997, Gatter 1997, Vernon et al. 1997, Ash & Miskell 1998, Quantrill & Quantrill 1998, Dean 2000, BMNH). **Eggs** 16–21 eggs laid, usually on alternate days, but female can retain egg in oviduct for further 24 hours if necessary (Irwin 1988); usually pause of few days after each set of three eggs (Payne 2005). Elongated ovate, smooth, glossy, usually greenish, bluish or whitish usually with some spotting; or pinkish with reddish-brown or greyish spots; 19.6–25x13.7–16mm, also 24.9x16.4–16.7mm. About ten gentes identified, with colour and degree of host matching varying with host choice. Southern Red Bishop and Red-headed Weaver, eggs blue, unglossed, occasionally with greenish spotting, thus highly mimetic; African Masked Weaver, eggs highly variable and mimetic or non-mimetic, frequently rejected; Cape Weaver, eggs blue and highly mimetic; Bocage's Weaver, eggs blue, highly mimetic; Cape Sparrow, eggs mimetic pale bluish-white with fine sparse brown speckling, lighter than host with less speckling; Cape Wagtail, eggs pure white, partially mimicking host eggs, which are creamy and lightly speckled (van Someren 1916, Chapin 1939, Vincent 1946, Serle 1950, Hunter 1961, Markus 1964, Reed 1968, Irwin 1988, Lawes & Kirkman 1996). Egg thought to be of this species in nest of Black-chested Prinia resembled type of egg normally laid in Southern Red Bishop's nest, nest structure of bishop and prinia being similar (Reed 1968). Eggs rejected by Southern Red Bishops only if poor colour match, and not by size, clutch size or time of laying. Negative correlation between size of colony of Southern Red Bishop and rate of parasitism (Lawes & Kirkman 1996). In captivity, Village Weaver, which also lays variety of egg types not markedly different in size from cuckoo's, will reject foreign eggs on colour and spotting even if own eggs are not available for comparison; greater the difference, greater the rate of rejection (Victoria 1972); the greater the difference in brightness, colour and spot density, the greater the rejection rate, and rejection more likely if host clutch already complete, RSA (Lahti 2006). 19.6–25x13.7–16mm, also 24.9x16.4–16.7mm (Vincent 1946, Serle 1950, Irwin 1988). **Incubation** 12 days (11–13), 10–14 days in Cape Sparrow's nest, 11–12 days in African Masked Weaver's nest; 9–10.5 days, hatching two days before host in Spectacled Weaver's nest, Natal, RSA (Skead 1952, Reed 1968, Irwin 1988, Chalton 1991). **Chicks** Eject host young and eggs when 2–4 days old, host young may survive if hatched before cuckoo. Eyes open at 7–10 days, quills emerge at 8 days and erupt by 12 days. Initially silent, but calls persistently during last few days in nest; hisses and lunges with open bill if disturbed. Typically spends 21 (18–26) days in nest, and three weeks as dependent fledgling; may be attended by African Masked Weaver for 18–38 days. Cuckoo young seen fed by Cape Sparrow who also fed fledgling sparrows; 2 cuckoo chicks recorded in nest of Southern Red Bishop, but only one after few days. Begging calls may differ depending on host (Friedmann 1948b, Reed 1968). Adults recorded feeding cuckoo young just able to fly, Eritrea (Smith 1957). Fledgling fed by male Dideric Dec, Transkei, RSA (Maclaren 1952), may have been courtship feeding, male mistaking fledgling for female (Friedmann 1956). **Survival** Young frequently leave nest early, possibly resulting in source of mortality, particularly where host nest is small and unable to accommodate large nestling. Adult male killed by African Masked Weaver. Oldest 6+ years (Friedmann 1948b, Irwin 1988).

FOOD Mainly insects, occasionally seeds. Caterpillars, mainly hairless but some hairy, found in most stomachs examined; also grasshoppers, termites, pyrrhocorid bugs, adult butterflies, chrysalids, and beetles. Caterpillars and adults of distasteful *Acraea* butterflies included in diet. Also eggs removed from hosts' nests. Parasitises mainly granivorous hosts, so young may be fed mainly seeds by some host parents. Large caterpillars may be gripped at one end and twirled around for up to 30sec before being swallowed whole, and beak wiped; smaller caterpillars may be gripped in middle and brushed against branch before being swallowed. Caterpillars of *Acraea* and Lymantridae given in courtship feeding, up to three being given to female per min. Young fed by Southern Red Bishop chiefly on grass seeds; stomach of fledgling Zambia contained mostly grass seeds, with few parts of beetles (Chapin 1939, Maclaren 1952, Benson & Serventy 1957, Reed 1968, Irwin 1988).

STATUS AND CONSERVATION Common over much of its very large range, and no large declines reported. Tolerates human modification of habitats fairly well e.g. will use planted *Eucalyptus* and other exotics for song posts (Friedmann 1948b), and is quite common in suburban habitats (Rowan 1983). Was considered common in summer Darfur, Sudan (Lynes 1925); common in Sudan north to Kosti, being commoner in north Jun–Sep, moving south other months (Cave & MacDonald 1955). Widespread in DRC, although rare Katanga (Chapin 1939). Common resident southern half Togo (Cheke & Walsh 1996). Previously common Ghana (Holman 1947) and Eritrea (Smith 1957). Abundant Kumba and Bamenda Divisions, Cameroon, 1947 (Serle 1950). Widespread, locally abundant, but patchily distributed south-east Nigeria (Marchant 1953). Common breeding migrant Eritrea and Ethiopia (Ash & Atkins 2009). Common east Africa (Stevenson & Fanshawe 2004). Annual survival rate c. 59% (Hanmer 1995). Common near colony-nesting weavers, their main hosts, also frequently associated with human settlement. Not globally threatened (BirdLife International 2011).

Dideric Cuckoo *Chrysococcyx caprius*. **Fig. 1.** Adult male, Tala Ranch, near Camperdown, KwaZulu-Natal Province, RSA, October (*Adam Riley*). **Fig. 2.** Adult male, Kruger NP, RSA, November (*José Kemp*). **Fig. 3.** Adult male, Solio, Kenya, February (*Adam Scott Kennedy*). **Fig. 4.** Adult male, Kenema, Sierra Leone, August (*David Monticelli*). **Fig. 5.** Adult female, So-Ava, Benin, January (*Bruno Portier*). **Fig. 6.** Adult male, retaining red bill and dark eye of juvenile, Salalah, Oman, November (*Aurélien Audevard*).

KLAAS'S CUCKOO
Chrysococcyx klaas   **Plate 26**

Cuculus Klaas Stephens, 1815 (Cape Province, South Africa)

TAXONOMY MtDNA analysis places this in clade with *flavigularis*, *cupreus* and *caprius* (Sorenson & Payne 2005). In error, *Chalcites klaasii* Lesson 1830. Monotypic. Synonym: *klaas arabicus* Bates 1937 from Taif, Asir, Saudi Arabia, separated on differences in patterning of outer rectrices, but species poorly known in Arabia, and more information needed.

FIELD IDENTIFICATION 18cm. Small cuckoo of well-wooded areas, but not deep forest. **Adult male** Glossy green above, green forming shoulder patch; white below through to undertail. Often white mark behind eye, but lores entirely green. In flight, underwing-coverts pure white, and underside of flight feathers barred dusky and white. **Adult female** Bronzy-brown barred with greenish above, more closely barred bronzy-brown and pale buff below, throat centre usually white. Some northern and western birds can look very similar or even be identical to male (Serle 1965, Irwin 1987). **Juvenile** Bronzy upperparts barred greenish, and variable but usually fine brownish barring below. Usually shows whitish post-ocular patch, but often indistinct in field. **Similar species** Male quickly distinguished from all other African cuckoos apart from Diederic by combination of glossy green above and white below. Distinctions from Diederic include entirely green head apart from narrow white throat bordered with green, which extends to form wedge on shoulder (throat, shoulder and front part of ear-coverts white in Diederic), and usually no white in front of eye. Klaas's has much more white in outer tail than Diederic, and cleaner flanks with little barring. In flight, underwing appears much paler than Diederic. Some adult females, particularly southern birds, can be rather like other African *Chrysococcyx* females, so care needed in separation. Variable, but usually bronzy barred greenish above with underparts finely barred brown and buff. It is more finely barred below than African Emerald and Diederic, and usually has white post-ocular patch, although sometimes not obvious in field. Lack of streaking on throat distinguishes juvenile Klaas's from young Diederic, but very similar to juvenile African Emerald, and good views required for safe identification. Any bird showing distinct whitish post-ocular patch will be Klaas's, and bird with coral-red bill and throat streaking will be Diederic, but in absence of these features, critical examination of colour and width of underparts barring is necessary to separate juvenile Klaas's from African Emerald. Klaas's has thin bars relatively widely and variably spaced, unlike thicker regular barring on young African Emerald. Extreme care needed in identification of nestling, and should only be attempted on advanced or fledged birds: Klaas's is told from African Emerald by forehead barred brown rather than white and uppertail-coverts with whitish margins, and from Diederic by back greenish rather than coppery-brown (Friedmann 1948b).

VOICE Male gives pure whistled disyllabic *whee-hew* repeated monotonously, or *dee da*, often stretched to *huee-ti huee-ti*, or *meit-jie*, first note slurred up, second down. In south-east Nigeria disyllabic *chee-wee*. Also continuous *dew dew dew dew*, with 3 notes/sec. Phrase repeated about four times in 3–4sec, and whole call repeated at intervals of 15sec for up to 30min. Male adopts motionless upright posture when calling, often given throughout day. Female occasionally calls on her own and may respond to male with plaintive note recalling first syllable of its call. When courting, male gives soft *swii... hii*, while performing body and wing movements, gradually merging into *whit whit* and terminating with long *sweeeeeeeeee*. Agitated male makes harsh chatter *cheecheechee*. Female gives *quip quip* followed by trill; male makes whistle while feeding caterpillar to female in courtship. Juvenile makes querulous high-pitched chittering when begging food, or high thin ascending trill when begging from Long-billed Crombec (Chapin 1939, Marchant 1953, Serle 1965, Irwin 1988, Payne 1997, 2005, Stjernstedt 1993, Borrow & Demey 2001).

DESCRIPTION Adult male Crown, ear-coverts, upperparts including wings brilliant metallic green with variable copper wash. White patch immediately behind eye, and occasionally small white patch in front of eye. Outer webs of remiges metallic green, but inner webs dusky-grey and barred or blotched white, thus appearing barred from underneath. Underwing-coverts white, with some faint dusky barring. Chin, throat and central belly white, but metallic green half-collar extending onto breast sides from lower ear-coverts. Variable amount of indistinct dusky barring on belly side and flanks, and faint green streaking on sides of thighs. Two central pairs of rectrices metallic green, next pair white, barred dull green on inner webs and with metallic green subterminal spot on outer webs, other rectrices white except for bar or distal spot on outer web and variable dark barring on inner web. Undertail-coverts white, underside of tail white with blackish terminal spots to outer feathers. **Adult female** Varies across range. Larger than male. Southern birds consistently different from males, but many northern and western birds more similar (or even identical) to male. Typical southern birds have all upperparts bronzy-brown, pale buffy mark behind and sometimes in front of eye, lores and ear-coverts brown. Remiges and wing-coverts blackish-brown, outer webs barred thickly with chestnut and inner webs barred whitish grading to buff at base of feathers. Chin whitish, merging into buffy throat and upper breast, darkening toward sides with variable amount of barring. Dusky-brown half collar extends onto breast sides from ear-coverts, reminiscent of pattern in male. Belly, flanks and undertail-coverts thinly barred olive- or dusky-brown. Northern birds more like male, with variable amount of metallic green wash on upperparts shot with iridescent purple, and with restricted barring on whiter underparts. Some specimens from Cameroon, Nigeria, Uganda and Kenya resemble male (Type 1 female); some from Liberia, Cameroon, Ghana, Bioko, Kenya and Tanzania similar to male but less greenish with purple above, on sides of throat and half-collar (Type 2 female); others from Liberia, Ghana, Nigeria, Cameroon and Gabon are similar to Type 2 but have less iridescence on sides of neck and half collar, more marked barring on chest and flanks, and extensive chestnut barring on wings (Type 3 female). Some birds from SL, Cameroon, CAR, Sudan, Ethiopia, DRC, Uganda, Kenya, Tanzania, Malawi, Mozambique, Angola and Cape Province, RSA are more sexually dimorphic, with crown and back metallic purple, less glossy and more matt, sides of throat and half-collar non-metallic dark brown with hint of purplish suffusion, underparts extensively barred, washed buff, wings metallic green barred chestnut, with primaries extensively barred and spotted white in female (Type 4 female). Other birds from Malawi, Zambia and Cape Province, RSA are fully dimorphic, with non-metallic sides of neck and half-collar, finer barring on underparts with clearer vermiculations in female (Type 5 female). Types 1, 2 and 3 unknown south of Tanzania, whereas Type 5 unknown from east or equatorial Africa (Irwin 1987, 1988). **Nestling** Newly-hatched nestling has yellowish olive-brown skin, slightly darker above, and soon darkening to deep blackish-olive, but said to be almost black on hatching (van Someren 1916). Bill dark horn, eye grey-brown and gape bright orange (Payne 2005). **Bare parts** Iris dark brown in male, lighter grey in female and hazel-brown with deep-blue pupil in juvenile, grey-brown in nestling; eyelids and eye-ring green. Bill olive or pale green, darker in female often with black tip; nestling dark horn with orange gape. Feet olive-green, duskier in female; claws black.

BIOMETRICS Wing male (n = 42) 98–108mm, mean 103; female (n = 18) 96–106mm, mean 102. Bill 15–17.5mm, mean 16.4. Tail male (n = 40) 69–80mm, mean 73; female (n = 16) 66–75mm, mean 73. Tarsus 14–15mm, mean 14.4. Mass male (n = 15) 21–31g, mean 25.6; female (n = 10) 28–34g, mean 30.2 (Irwin 1988); male (n = 30) 22.7–32.8g, mean 26.2; female (n = 38) 24.7–35.1, mean 28.6 (Hanmer 1995). Also wing male 90mm, female 92mm (Payne 2005).

MOULT Transilient primary moult from two loci; timing unclear. Body moult RSA, Oct when in breeding condition. Juvenile primary moult begins at four months and lasts about 80 days, often interrupted (Rowan 1983, Hanmer 1995).

DISTRIBUTION Afrotropical. Resident across central Africa in broad belt from SL to Sudan, Ethiopia, Somalia (although mostly Feb–Nov) and south to central Angola, Zambia and Mozambique. Resident Bioko, occurring on no other islands, except once São Tomé. Breeding visitor north (Mauritanian Sahel, Senegal, Mali to Eritrea, northern Ethiopia and south-west Arabia) and south (south Angola, Namibia, Botswana, RSA) of this range usually coinciding with rains. Birds breed along coast of RSA Oct–Mar, absent from inland eastern RSA, south Namibia and south Botswana. In RSA, some early arrivals and late departures may relate to overwintering birds, particularly in mesic habitats where some birds are known to overwinter. In Namibia, arrival delayed to Jan if rains late. Northward movements also coincide with rains, with birds reaching about 16°N. Non-breeding visitor to Senegambia, Oct–Feb; rainy season visitor to north Senegal; southern Nigeria all months, in north Mar–Sep during rains, but does not reach Sahel; north Burkina Faso, rainy season Jun–Sep, but young in nest Nov; resident south Togo, rainy season visitor north Togo; Mali May–Oct; Ghana mainly wet season; in parts of Cameroon present and calling all year; Gulf of Guinea resident; in south DRC Jul–Apr; Sudan north to Er Roseires, and Kulme, Darfur in summer, moving to south Equatoria in winter, appearing with rains to breed in north, but resident Imatong Mts. in south; Malawi Jul–Apr; recorded all months Zambia; mainly migrant Zimbabwe, but some calling Jul; apparently resident, calling in winter, Cape Town area, RSA (Friedmann 1948b, Serle 1950, Amadon 1953, Verheyen 1953, Cave & MacDonald 1955, Eisentraut 1973, Elgood et al. 1973, Lamarche 1980, Grimes 1987, Nikolaus 1987, Ash & Miskell 1983, Eccles 1988, Irwin 1988, Gore 1990, Morel & Morel 1990, Medland 1995, Cheke & Walsh 1996, Hazevoet 1996, Rodwell et al. 1996, Balança & de Visscher 1997, Payne 1997, 2005, Pérez del Val et al. 1997, Aspinwall & Beel 1998, Quantrill & Quantrill 1998, Sheehan 1999, Borrow & Demey 2001, Bowden 2001, Ash & Atkins 2009, Isenmann et al. 2010); Yemen, late spring and summer (Brooks et al. 1987, Porter et al. 1996); rare breeder north to Tihamah, coastal Saudi Arabia (Jennings 1981).

HABITAT Riparian forest, forest edge, clearings, savanna woodland and gallery forest, thickets, *Acacia* savanna, and gardens in urban and suburban areas; long grass close to reeds; to 3,000m. Resident equatorial and west African rainforest zone and adjacent mesic savanna woodlands. In Kenya usually in denser and moister habitats than Dideric Cuckoo, but more open habitats than African Emerald Cuckoo; 94% of records within 250+mm, and 84% within 500mm annual rainfall regions. Usually below 1,630m in DRC; below 2,000m Rwanda; *Combretum* woodland below 1,220m Eritrea. Does not reach Sahel in Nigeria (Chapin 1939, Friedmann 1948b, Smith 1957, Elgood et al. 1973, Rowan 1983, Irwin 1988, Lewis & Pomeroy 1989, Dowsett-Lemaire 1990).

BEHAVIOUR Forages unobtrusively in vegetation, sometimes flycatching continually from perch. May join mixed feeding flocks. Often fairly high in tree. Courtship display accompanied by excited posturing, including twisting movements, wing-drooping and tail-cocking; head held low, and bill gaping. Often culminates in male finding food and offering it to female, and frequently repeated. Male strongly territorial when breeding; male holds loose territory of up to 30ha, while female has smaller territory. Territories possibly established serially during season. Probably monogamous. Pair bond only brief during breeding season, but courtship feeding observed outside breeding season (Brook et al. 1934, Rowan 1983, Irwin 1988, Halleux 1994). Friedmann (1948b) suggests *Chrysococcyx* cuckoos feed their own young after fledging, as many fosters are granivorous, and this recorded Malawi, Sep (Benson & Benson 1977).

BREEDING Parasitism Degree of host-specificity unknown. Host egg usually removed from nest before cuckoo lays; removed egg sometimes eaten (Friedmann 1948b). Two young being fed in nest of Long-billed Crombec, Transvaal, RSA (O.P.M. Prozesky *in litt.* to Friedmann 1956). Largely concentrates on warblers and sunbirds reflecting its use of well-vegetated but non-forest habitat. Known hosts with number of instances and localities if recorded: Common Bulbul, Mali (3), Kenya and Angola (1). Black-throated Wattle-eye, east Africa (1). Cape Batis, RSA (10). Pririt Batis, Namibia (5). Chinspot Batis, RSA (4), Zimbabwe, Malawi.

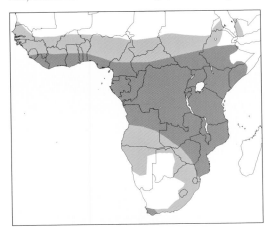

African Paradise-flycatcher, Cameroon (3). Uganda, commonly Kenya, Zambia (1) and Cape Province, RSA; possibly DRC. Piping Cisticola, RSA (1). Singing Cisticola, east Africa (1). Red-faced Cisticola, Uganda and Kenya. Stout Cisticola, Uganda and Kenya. Chattering Cisticola, Cameroon (1). Tawny-flanked Prinia, Ghana (1); 7% of nests of this species parasitised in Zambia; Kenya. Bar-throated Apalis, RSA (13). Grey-backed Camaroptera, Uganda and Kenya. Greater Swamp Warbler, south-east DRC (1). African Reed Warbler, east Africa (1). Yellow-bellied Eremomela, RSA (10), Namibia, Zimbabwe, Zambia, east Africa, perhaps Mauritania. Long-billed Crombec, RSA (13), Namibia, Zimbabwe. Green Crombec, Nigeria (1). Buff-bellied Warbler, Ethiopia. Rockrunner, Namibia. Arrow-marked Babbler. African Yellow White-eye, Ghana (2). African Stonechat, Malawi (1). Pale Flycatcher, Zambia (3). African Dusky Flycatcher, RSA (7), east Africa. Swamp Flycatcher, Lake Victoria, Tanzania (1). Grey Tit-flycatcher, Malawi (1). Sunbirds: Collared, RSA (4); Pygmy, Mali; Mouse-coloured, Kenya (1), Zululand, RSA; Green-headed, east Africa (2); Amethyst, RSA (5), Malawi; Scarlet-chested, Zimbabwe (2), Zambia, Malawi; Bronzy, east Africa (13); Yellow-tufted Malachite, RSA (2); Beautiful, Gambia (3), Mauritania, Mali, Burkina Faso; Mariqua, RSA (4), east Africa; Greater Double-collared, RSA (13), one of chief hosts in Cape Province; Southern Double-collared, RSA (1); White-breasted, RSA (2), Malawi; Variable, Ethiopia, east Africa (9); Dusky Sunbird, Namibia (7); Copper, Angola (1), Botswana (1); Red-chested regular host Entebbe, Uganda; Cameroon, Nigeria. Weavers: Vieillot's, Cameroon (1); Baglafecht, Uganda, Kenya, Tanzania; Black-headed, Kenya; Lesser Masked, Uganda, Zimbabwe; Village, frequent south-east Tanzania, south-east Kenya. Northern Grey-headed Sparrow, Ethiopia. Southern Red Bishop, Malawi. Yellow-fronted Canary, Malawi (1). African Pied Wagtail, Ethiopia (1), Somalia (1). Cabanis's Bunting, Zimbabwe (3), Mozambique, Tanzania, probably Zambia (Erlanger 1905, Haagner & Ivy 1906, van Someren 1916, 1922, 1932, 1939, Grote 1924, Skinner 1929, Belcher 1930, 1949, Sclater & Moreau 1932, Braun 1931, 1934, Vincent 1934, Sparrow 1936, Jackson 1938, Chapin 1939, Roberts 1939, Winterbottom 1939, 1951, Benson 1945, 1946, Joubert 1943, Meneghetti 1944, Williams 1946, Holman 1947, Friedmann 1948b, 1949b, 1956, G.L. Bates and R.F. Cheesman *in litt.* to Friedmann 1948b, Haydock 1950, MacLeod & Murray 1952, Skead 1952, 1954, Stoneham 1952, Benson 1953, Chapin 1953, C.R.S. Pitmann *in litt.* to Chapin 1954, Mackworth-Praed & Grant 1957, C.F. Belcher, J.P. Chapin, H.M. Miles, O.P.M. Prozesky and W. Stanford in Friedmann 1956, Pitman 1957, Serle 1965, Martin 1968, Jensen & Jensen 1969, Benson *et al.* 1971, Benson & Benson 1977, Lamarche 1980, Hanmer 1982, Irwin 1988, Balança & de Visscher 1997, Ash & Miskell 1998, Quantrill & Quantrill 1998, Ash & Atkins 2009, Isenmann *et al.* 2010, V. Schollaert *in litt.* 2010, I. Nason pers. comm., R.A. Cheke pers. comm., R. Sternstedt pers. comm.). Possibly Black-backed Puffback and Red-headed Weaver, Zimbabwe (Masterson 1953), and White-browed Sparrow-weaver, Zimbabwe, rejected by Jensen & Jensen (1969). Other unconfirmed hosts: Cape Wagtail, RSA; Long-tailed Glossy Starling; House Sparrow, RSA, Cape Sparrow, RSA and White Nile/Kenya Rufous Sparrow, east Africa (Ivy 1901, Bannerman 1933, Jerome 1943, Joubert 1943, Vincent 1946, Belcher 1949). Female cuckoo attacked savagely by pair of Blue Crested Flycatchers, north-east DRC (Chapin 1939). Rate of parasitism of nests 2.6% of Long-billed Crombec, 1.4% of Bar-throated Apalis, 2.4% of Amethyst Sunbird, 7% of Pririt Batis and 8% of Dusky Sunbird (Payne & Payne 1967, Jensen & Clinning 1974). 7% of nests of Tawny-flanked Prinia parasitised in Zambia (Pitman 1957). 1 pair Collared Sunbirds raised three cuckoos successively (Irwin 1988). Female Klaas's Cuckoo recorded feeding chicks of own species, Kenya, on more than one occasion (Nickalls 1983). **Season** Just after wet season in much of equatorial Africa, but distinctly seasonal to north and south. Mauritania Sept–Oct, Senegambia, Sep–Nov; yolking ovaries Liberia, Oct; south Mali, Jun–Nov; calls all year IC, juvenile end Feb, courtship feeding Feb; Burkina Faso, Oct–Nov; Togo, courtship feeding Apr, recently fledged juvenile Jun; Nigeria, Mar–Sep, but calls throughout year in south-east, laying south-east Jul; Cameroon, Feb, fledgling Mar, juvenile Apr, Nov, calls all year Kunda; Gabon calls Sep–Feb; DRC, probably all months, mainly in rains, Aug–Jan, well-feathered nestling in south-east late Feb, in south Aug–Jan; Angola, Nov–Apr, oviduct egg Sep, nestling Nov, breeding condition Oct–Feb; Darfur, Sudan Aug, other parts of Sudan, perhaps Oct–Jan; Ethiopia, Sep–Oct, ?Mar, fledgling 7 Jun; active testes Eritrea, Feb; Somalia, Apr–Jun; young Uganda and Kenya, Jan, Feb, Apr–Jul; also breeding north and east Uganda and west Kenya, Sep, central and south Uganda and south-west Kenya, Dec; Tanzania (except north-east), Jan, Apr, Sep–Oct, Dec; Kenya (except west and coast) and north-east Tanzania, Jan–Apr, Jun, Dec; coastal Kenya, Apr–Jun; Angola Sep–Dec, fledgling mid Apr; Zambia, Oct–Feb; Malawi, Sep, Nov–Feb; Zimbabwe, Sep–Apr; RSA, Aug–Feb; RSA, Cape, Jul–Jan, occasionally May, usually in rains Natal and Transvaal, Sep–Feb; Eastern Cape, Oct–Dec, occasionally late as May, Cape during winter rains Jul–Sep; Namibia, Dec–Apr (Erlanger 1905, van Someren 1916, 1922, Lynes 1925, 1938, Belcher 1930, Braun 1931, 1934, Sparrow 1936, Jackson 1938, Roberts 1939, Winterbottom 1939, Meneghetti 1944, Benson 1946, Williams 1946, Friedmann 1948b, Serle 1950, MacLeod & Murray 1952, Skead 1952, 1971, Marchant 1953, Masterson 1953, Verheyen 1953, J.P. Chapin and O.P.M. Prozesky *in litt.* to Friedmann 1956, Serle 1957, 1965, Smith 1957, Heinrich 1958, Martin 1968, Jensen & Jensen 1969, Benson *et al.* 1971, Dean 1971, 2000, Urban & Brown 1971, Elgood *et al.* 1973, Benson & Benson 1977, Brown & Britton 1980, Morel & Morel 1990, Irwin 1981, 1988, Pinto 1983, Rowan 1983, Nikolaus 1987, Lamarche 1988, Christy & Clark 1994, Demey & Fishpool 1994, Cheke & Walsh 1996, Balança & de Visscher 1997, Payne 1997, Ash & Miskell 1998, Quantrill & Quantrill 1998, Salewski 2000, V. Schollaert *in litt.*). **Eggs** Laid in sets of 3–4 eggs, with one day between eggs, and about 24 eggs per season (Payne 1973a, 1974); occasionally before host's eggs (Holman 1947), but usually after. Non-mimetic, variable white, pale greenish-blue or pale blue, and ranging from unspotted to densely marked with brick red, brown or purple-grey, sometimes concentrated near broad end; 20–23x13–14.3mm, also 16–19x12–13mm (Chapin 1939, Holman 1947, Friedmann 1948b, Jenson & Clinning 1974). **Incubation** 11–12 days (possibly 14), but always shorter than that of sunbird hosts (Irwin 1988). **Chicks** Eject host young and eggs usually within 4 days of hatching. Eyes open at 6 days, begins preening at 9 days. Hisses and lunges if disturbed, and will void oily blackish-brown fluid and encapsulated faeces if handled. Typically spends 19–21 days in nest, and 2–3 weeks as dependent fledgling. May continue begging for up to 25 days. Begs with husky *tsseek* repeated to become rapid stuttering while quivering wings at approach of foster parent (MacLeod & Hallack 1956, Payne 2005). **Survival** No information.

FOOD Insects; occasionally fruit. Caterpillars, mainly hairless but some hairy, found in all stomachs examined. Also adult *Acraea* butterflies, cotton-stainer bugs (*Dysdercus*), bug nymphs (Pyrrhocoridae), beetles and small orthopterans, termites. Juvenile observed deliberately drinking on hot day. Young in Angola fed on berries (Chapin 1939, Friedmann 1948b, Moreau 1949, Benson & Benson 1977, Rowan 1983, Irwin 1988, Payne 1997).

STATUS AND CONSERVATION In west Africa usually rarer than Dideric; apparently some losses in Gambia in past to plumage hunters (Friedmann 1948b, Cawkell & Moreau 1963). Can tolerate degraded forest, where it parasitises birds nesting in dense secondary regrowth (Friedmann 1948b), and urban areas with flowers attracting sunbirds (Payne 1997). Frequent Macenta Prefecture, Guinea (Halleux 1994). Common resident southern half of Togo (Cheke & Walsh 1996). Rare Eritrea (Smith 1957). Common resident and partial migrant Eritrea and Ethiopia (Ash & Atkins 2009). Fairly common Sudan (Cave & MacDonald 1955). Fairly common DRC (Chapin 1939). Common Angola (Dean 2000). Common in low country Natal, RSA, where replaces Dideric Cuckoo, but rare on higher ground (Vincent 1946). Not globally threatened (BirdLife International 2011).

Klaas's Cuckoo *Chrysococcyx klaas*. **Fig. 1.** Adult male, Kenema, Sierra Leone, August (*David Monticelli*). **Fig. 2.** Adult male, Lake Bunyoni, Uganda, March (*Gabi Bujanowicz*). **Fig. 3.** Juvenile, Koeberg NR, Western Cape, RSA, March (*Ian Nason*). **Fig. 4.** Adult female, Ngorongoro Conservation Area, Tanzania, February (*Martin Goodey*). **Fig. 5.** Juvenile, Ngorngoro Conservation Area, Tanzania, July (*Martin Hale*).

YELLOW-THROATED CUCKOO
Chrysococcyx flavigularis Plate 26

Chrysococcyx flavigularis Shelley, 1880 (Elmina, Ghana)

Alternative names: Yellow-throated Green/Yellow-throated Glossy Cuckoo.

TAXONOMY MtDNA analysis places this closest to *cupreus*, in clade with *klaas* and *caprius* (Sorenson & Payne 2005). Monotypic. Populations of Cameroon, DRC and Uganda separated as *parkesi* Dickerman, 1994, as female and immature male browner below, less creamy-white ventrally especially on undertail-coverts, but most specimens seem indistinguishable from populations further west (Payne 2005).

FIELD IDENTIFICATION 18–19cm. Small dull, dark cuckoo, rather difficult to see. **Adult male** Much darker than other African *Chrysococcyx*; bronzy with dull metallic green (sometimes appearing brown) head, upperparts and sides of breast. Most noticeable feature is vertical yellow stripe running from chin to upper breast, although not visible when viewed from side. Underparts finely barred dark greenish-brown. Eyes pale yellow. Striking white outer rectrices, particularly in flight. **Adult female** Similar but lacks yellow throat stripe; instead is entirely finely barred below and on head sides, unlike Dideric, Klaas's and African Emerald Cuckoos. **Juvenile** Like adult female, but feathers of back and wings with narrow pale tawny fringes. Ground colour of underparts and head paler than in adult. **Similar species** In all plumages noticeably darker than any other African *Chrysococcyx*. Yellow throat stripe of male diagnostic if seen well. In all plumages, fine barring of underparts, leading to dark appearance, characteristic.

VOICE Calls throughout year, often from high tree in early morning. Male's call far-carrying series of 9–12 pure thin whistles, first note longest, then gradually accelerating and decreasing in volume. Call first rises slightly, then descends. Series, *tui, tiu-uhuhuhuhuhuh*, lasts *c.* 3sec, with long intervals between 3–4 bouts. Series of short, sweet whistles, slightly descending; two loud, unmusical, notes not unlike call of Blue-throated Roller. Both sexes give two-note whistle, second note slightly lower, *dhuiit-tiu* or *hee-huu* or *chuwee-chwee* (Chapin 1939, Friedmann 1969, Irwin 1988, Christy & Clarke 1994, Payne 1997, Borrow & Demey 2001, Stevenson & Fanshawe 2002).

DESCRIPTION Adult male Head, neck sides, back, scapulars, upperwing-coverts and uppertail-coverts dark coppery-brown washed with dull metallic olive-green. Remiges purple-brown, with faint paler brown barring on secondaries. Underwing-coverts closely barred dark brown and buff, remiges have large buff-white bars on inner webs. Chin, and central areas of throat and upper breast bright yellow, and sides of throat and breast dark metallic green. Lower breast and belly buff closely barred with dark brown, bars more widely spaced on undertail-coverts. T1–2 coppery bronze, whereas T5 strikingly white with narrow subterminal dark bar. Sometimes bilateral asymmetry in yellow tinge to outer rectrices. **Adult female** Top of head dark bronzy green with fine whitish bars. Rest of upperparts dull bronzy green, wing-coverts with fine pale rufous bars. Ear-coverts, lores, chin, throat and whole of underparts pale buff finely barred dusky-brown. Underwing-coverts as breast; remiges brown with large pale buff spots on inner webs giving barred effect. T1–2 blackish-brown, T3 mostly white, but dark brown proximally and with broad subterminal band of same colour, T4–5 mostly white with broad dark brown subterminal bars. **Juvenile** Upperparts dark bronzy green with distinct narrow but broadly spaced bars of pale tawny. Underparts buff barred dusky-brown as in adult female. Eventually, head and back feathers lose pale tips, and yellowish tinge may develop on chin and throat. Dusky bars on underparts darken with age. **Nestling** No information. **Bare parts** Iris creamy yellow in male, creamy-white to yellow in female; bare skin around eye greenish-yellow in both sexes; eyelids light yellowish-green, dark brown in female (Friedmann 1969), in juvenile yellowish or greyish-white. Bill greenish-yellow in adult male, blackish at base, nostrils olive; in female black with greenish wash at base, extending further along lower mandible than upper mandible; in juvenile brown, lower mandible mostly yellow. Feet of adult dull yellow to yellowish-green, soles dull yellow, claws black; yellow in juvenile.

BIOMETRICS Wing male (n = 10) 91–98mm, mean 94.6 ± 2.8; female (n = 8) 91–99mm, mean 94.8 ± 2.4. Bill male 14.7–18.2mm, mean 16.1 ± 1.1; female 14.4–18.2, mean 17.3 ± 1.6. Tail male 56–72mm, mean 65.0 ± 6.0; female 56–69mm, mean 61.6 ± 4.4. Tarsus male 13.1–14.9mm; mean 13.8 ± 0.7; female 12.3–15.0mm, mean 14.1 ± 1.2 (Payne 2005). Mass male (n = 7) 27.5–31g, mean 29; female 30g, laying 30.5g (Friedmann 1969). Tail length increases from east to west (Irwin 1988).

MOULT Male attains yellow throat stripe during post-juvenile moult. Male moulting primaries Cameroon, Jan; female tail moult west DRC, Oct (Irwin 1988, BMNH).

DISTRIBUTION Afrotropical resident. In west Africa scattered records from lowland forests in SL, Liberia, IC, Ghana, Togo and Nigeria. Slightly more continuously distributed in broad band centred on DRC extending to west Uganda in east, and Cameroon, CAR, Equatorial Guinea, RC and Gabon in west. Once extreme south-west Sudan (Traylor & Archer 1982, Irwin 1988, BirdLife International 2003, Payne 2005, Anon. 2010).

HABITAT From sea level in west to 1,200m DRC and Uganda. Marked preference for canopy of thick lowland forest, in both primary and recovering secondary forests, but also forest edge, clearings and gallery forest; logged forest; occasionally well-wooded savanna (Chapin 1939, Irwin 1988, Stevenson & Fanshawe 2002, Demey 2009).

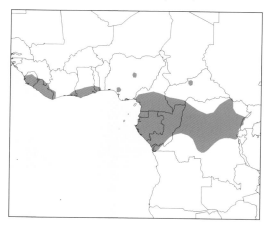

BEHAVIOUR Rarely-seen forest cuckoo, usually in canopy and frequently solitary. Usually detected by its call, given from high perch and often early in morning. Droops wings when resting. May feed acrobatically. Apparent display behaviour involves male on small horizontal branch of dead tree scurrying sideways backwards and forwards continuously jerking tail up and down; on reaching end of branch, bird hops around 180º and moves sideways back again; two males performed this behaviour together in Uganda in Nov, suggesting little territoriality. Males respond readily to imitation of feeding call, often from several hundred metres (Chapin 1939, Friedmann 1969, Irwin 1988, Stevenson & Fanshawe 2002).

BREEDING Parasitism Only known host Grey-throated Tit-flycatcher (Mills *et al.* 2007). **Season** Young fed out of nest Gabon, 13 Oct. Males call throughout year. Gonads enlarged Uganda, Jul, Sep and Dec. Oviduct egg Liberia, Mar (Friedmann 1948b, 1969, van Someren & van Someren 1949, Ridley *et al.* 1953, Irwin 1988, Mills *et al.* 2007). **Eggs** Oviduct egg pale green blotched blackish (Dickerman 1994). No further information.

FOOD Stomachs of eight Ugandan birds contained almost entirely caterpillars, with some beetle fragments and scale insects; birds feeding on caterpillar swarms in Nigeria and DRC; fruit (Chapin 1939, Friedmann 1948b, 1969, Ridley *et al.* 1953, Brosset & Érard 1986, Irwin 1988, Ash 1990).

STATUS AND CONSERVATION Although relatively frequent Cameroon, Gabon, RC and DRC, and regular central Nigeria, very rare SL to Togo (Chapin 1939, Irwin 1988, Cheke & Walsh 1996, Payne 1997); locally common Dadieso and Boin R. to Bobiri FR, Ghana (Demey 2009); uncommon south-west Uganda (Stevenson & Fanshawe 2004). One calling bird/12ha forest in Gabon (Brosset & Érard 1986). Appears to tolerate secondary forest and has been recorded in well-wooded savanna, although many records from primary forest. Despite being very poorly known has large global range and not of immediate conservation concern. Not globally threatened (BirdLife International 2011).

Yellow-throated Cuckoo *Chrysococcyx flavigularis*. **Fig. 1.** Adult male, Atwa Farmbush, south-east Ghana, November (*Kevin Easley*). **Fig. 2.** Adult male, Semliki, Uganda, July (*Niall Perrins*).

AFRICAN EMERALD CUCKOO
Chrysococcyx cupreus   Plate 26

Cuculus cupreus Shaw, 1792 (Gambia)

Alternative names: Golden/Green Cuckoo.

TAXONOMY MtDNA analysis places it as sister to *flavigularis*, and in same clade as *klaas* and *caprius* (Sorenson & Payne 2005). Sometimes split into several subspecies on tail length and undertail markings. Northward cline of increasing tail length from RSA (where short-tailed birds sometimes recognised as *sharpei* van Someren 1922) to Ethiopia (northern longer-tailed birds treated as nominate). Gulf of Guinea island forms *insularum* Moreau & Chapin, 1951 (Príncipe, São Tomé, Annobón, Pagalu Is) and *intermedius* Hartlaub 1857 (Cameroon, Bioko, possibly east to Kenya) less frequently recognised, separated on tail length and undertail-covert pattern; in Cameroon both undertail patterns are found, otherwise tail pattern resemble 'nominate', but considerable overlap in tail length and undertail patterns between populations (Serle 1965, Payne 1997, 2005). Dideric Cuckoo was formerly known as *Chrysococcyx cupreus* Reichenow, 1887, so care must be taken in evaluating early material, when African Emerald Cuckoo was known as *Chalcites smaragdineus* Swainson 1837. Correct spelling of Dideric's specific name *caprius* (Grant 1915), also for African Emerald Cuckoo, *cupreus* Shaw, predates *smaragdineus* Swainson 1837, resulting in confusing swap of specific names between the two forms (Bannerman 1922). Reichenow 1902 used *Metallococcyx smaragdineus*. Monotypic. Other synonyms: *auratus* Sclater & Mackworth-Praed 1919, *Cuculus splendidus* G.R. Gray 1847.

FIELD IDENTIFICATION 20–23cm. Small cuckoo; noisy, but often difficult to see. **Adult male** Stunning bird, only cuckoo showing combination of bright golden-yellow below and brilliant emerald-green above. **Adult female** Shorter-tailed, lacks white behind eye shown by female Klaas's and Dideric, and has underparts strongly barred green. **Juvenile** Distinctive green and white barred throat; forehead flecked green, brown and white. Very restricted narrow white edging to lateral uppertail-coverts. Transitional immature males show bright green and white barring below and strong rufous barring in wings. In flight shows dark head, yellow belly, barred underwing-coverts and plain dark undersides to primaries. **Similar species** Male unmistakable, female and juvenile similar to Klaas's Cuckoo, but African Emerald never shows white behind eye and has thicker greener barring on underparts. Female Klaas's and Dideric show white behind eye. Juvenile Klaas's has throat barred dark and light brown, Dideric has plain buff; forehead of Klaas's barred dark and light brown; Klaas's shows white belly and underwing-coverts, and Dideric has underwing entirely barred; outer webs of lateral uppertail-coverts white in Klaas's.

VOICE Loud ringing *chi-wu chu chi* lasting 1.5–2.0sec. Repeated every 2–3sec or more irregularly; memorably translated as 'hel-lo geor-gie' or 'hel-lo, ju-dy'. First and last notes one tone higher than middle two. First and second notes slurred together, while last two distinct from each other. Also rapid *jujujujujujuju*. Trisyllabic in south-east Nigeria, loud and ringing *téw-téw, ee*. Apparently call only when breeding, e.g. rarely heard Apr–Aug in southern Africa. Whistled three-note call *hew-tu-whee* and explosive and melodious series of notes in DRC (Chapin 1939, Marchant 1953, Irwin 1988).

DESCRIPTION Adult male Head, breast, back, scapulars, upperwing-coverts, uppertail-coverts brilliant emerald-green. Remiges above also brilliant emerald-green with basal two-thirds of inner webs white. Underwing primary-coverts iridescent green with dark centres to feathers, secondary-coverts yellowish white; large white areas on inner webs of remiges forming large patch on open wing below. Belly to vent rich golden-yellow. T5 wholly brilliant emerald-green, T3–5 broadly tipped white, outermost pair additionally broadly and variably banded white. Undertail-coverts sparsely (northern birds, 'nominate') to strongly banded green (southern birds, '*sharpei*'), although much overlap apparent if long series examined. **Adult female** Forehead to nape dark olive-brown more or less distinctly finely barred whitish, rest of upperparts metallic glossy green barred pale rufous. Primaries dull brown, thick rufous bands across outer webs, basal two-thirds of inner webs and all secondaries barred whitish. Inner secondaries barred rufous and metallic green, outer feathers patterned like primaries. Underwing-coverts barred bronze-green and white, underside of primaries dark brown. Shorter-tailed than male. Central rectrices indistinctly banded rufous and green, three outer pairs banded rufous and green on inner web, whitish on outer web or with variable green and rufous spotting. Underparts strongly barred whitish and dark greenish, bars narrowing and becoming tinged with coppery-brown toward throat. **Juvenile** Similar to adult female, but white tips or narrow bars on feathers of crown and nape. Uppertail-coverts narrowly edged white. Green brighter in young male than young or adult female. **Nestling** Upper mandible whitish, gape orange. Hatchling has pinkish-yellow skin, shading to mauve on dorsal surface and darkening with age. Identification of nestling should only be attempted on advanced or fledged birds: from young Dideric by greenish rather than coppery-brown back, from Klaas's by forehead barred white rather than brown and uppertail-coverts without whitish margins (Friedmann 1948b). **Bare parts** Iris dark brown in adult; orbital ring blue-green in male and bright blue in female. Bill, sometimes tipped black, bright blue-green above, lower mandible pale bluish or grey in adult male; dull greenish or blackish in adult female; gape blue. Feet bluish (males) to lead-grey (females); claws black.

BIOMETRICS Wing male (n = 8) 98–108mm, mean 103.6 ± 3.5; female (n = 8) 97–110mm, mean 101.9 ± 4.2; tail male 84–106mm, mean 95.1 ± 7.1; female 72–91mm, mean 82.4 ± 6.5; bill male 15.2–17.2mm, mean 16.5 ± 0.8; bill female 15.6–17.6mm, mean 16.7 ± 0.7; tarsus male 13.8–16.3mm, mean 15.2 ± 0.9; tarsus female 14.7–16.1mm, mean 15.5 ± 0.6. Also wing male (n = 21) 110–120mm; mean 114; female (n = 6) 105–116mm, mean 112. Mass male (n = 21) 33–46g, mean 38.3; female (n = 13) 30–46.3g, mean 37.4 (Irwin 1988, Payne 2005).

MOULT Pre-breeding moult begins southern Africa, Oct, soon after arrival (Friedmann 1948b).

DISTRIBUTION Afrotropical Region. Apparently resident in broad band across central Africa from Eritrea, Sudan (south of 12°N) and Mozambique to north Angola, CAR, and narrow band along Gulf coast to SL and Guinea. Mainly resident east Africa. Probably resident Bioko, Príncipe, São Tomé and Annobón Islands, although Keulemans (1907) claims only present Bioko and Príncipe Feb–Nov. In west Africa partial wet season migrant north into well-wooded savanna zone, typically extending inland c. 500km from coast (Apr–Oct); in Nigeria some migrate north as far as Kano at

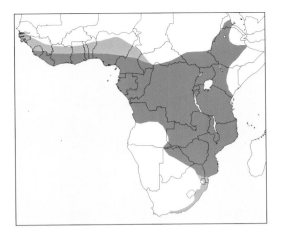

height of rains. In southern Africa partial wet season migrant (Sep–Mar) from DRC and Kenya (absent from eastern half and much of north: Lewis & Pomeroy 1989) into south Zambia, Malawi (Oct–May, Jul, Aug, but some resident?), Zimbabwe, south Mozambique and coastal RSA to Knysna in west; vagrant to Western Cape and north Botswana. Regular central Nigeria, although birds occasionally reach as far as Senegal, Gambia and north Nigeria, usually Jul. Birds may be year-round residents in forested habitats in some parts of southern range, and it is possible that Mar–Sep records represent migrants from another part of breeding range. Apparently no breeding movements made from core year-round range in east Africa (Elgood et al. 1973, Britton 1980, Irwin 1988, Gore 1990, Medland 1992, 1993, 1995, Halleux 1994, Harrison et al. 1997, Payne 1997).

HABITAT Primarily forest, including evergreen and gallery forests, but also thickly wooded riparian vegetation, coastal jungle and densely-wooded savanna. Occasionally dense scrub; less common in woodlands such as miombo in Zambia; females collected in dense undergrowth, Ghana; shady plantations on São Tomé and Príncipe. Occurs in drier habitats less frequently throughout range; open savanna woodland, including miombo, used particularly when breeding or on passage in south Africa. Also uses wooded suburban habitats, and will perch on alien trees; secondary growth; in Kenya 500–2,000m in forest islands in semi-arid areas, almost always in areas with 500+mm annual rainfall. From sea level west and southern Africa to 2,900m Ethiopian plateau, but only to 1,220m Eritrea. Below 2,000m Rwanda. Breeds to 1,500m south Malawi (Vincent 1934, Lowe 1937, Friedmann 1948b, Moreau & Chapin 1951, Benson & White 1957, Smith 1957, Benson et al. 1971, de Naurois 1983, Rowan 1983, Johnstone-Stewart 1984, Irwin 1988, Lewis & Pomeroy 1989, Dowsett-Lemaire 1990).

BEHAVIOUR Canopy species, moving quietly and inconspicuously through dense vegetation when not calling. Usually singly or in pairs, although sometimes in groups, particularly in response to food availability when not breeding. Observed in flock with five other cuckoo species east Zimbabwe, Nov. Joins mixed species flocks in IC. Females even more difficult to see than males. Males territorial when breeding, faithful to particular calling trees to extent of driving away intruders from favoured tree. Usual calling perch is thick branch rather than twig. Male observed feeding female with large hairy caterpillar while bobbing in front of her with his wings drooped and tail erect; after female accepted food male called loudly with head thrown back and unsuccessfully attempted copulation. Calling male raises and spreads tail and droop wings in presence of female, moving sideways along branch while singing; female droops and quivers wings as prelude to copulation, which is then followed by male flying off to collect insects, which he feeds to female on return. Sometimes mobbed by small passerines, including bulbuls, warblers and sunbirds (Haydock 1950, Friedmann 1948b, van Someren 1956, Rowan 1983, Irwin 1988, Demey & Fishpool 1994). Males twice observed feeding fledgling cuckoos (Worman 1930, Millar 1943).

BREEDING Parasitism Single host egg, sometimes all, usually removed by female before laying. Confusion of nestlings with those of Dideric Cuckoo means there are fewer than expected confirmed records of brood parasitism (Irwin 1988, Payne 1997). Female lays in groups of 3–4 eggs on successive or alternate days, about 20/season (Payne 1973, 1974). Egg of this species and one of Red-chested Cuckoo together in nest of White-starred Robin, RSA (Oatley 1980). Mainly uses insectivorous forest passerines. Confirmed hosts, with numbers and localities where recorded: Black-throated Wattle-eye, east Africa (2), Mozambique. Brown-throated Wattle-eye, Gabon (1). Black-backed Puffback, Zimbabwe (2) and Mozambique. African Paradise-flycatcher, Gabon (3), DRC, east Africa and Malawi. Príncipe White-eye and Príncipe Speirops, Príncipe. Common Bulbul, Ethiopia; east Africa' (12); Zambia; RSA. Yellow-whiskered Greenbul (1 nest in 160 parasitised), Gabon. Grey-backed Camaroptera, east Africa (4); RSA. Lesser Swamp Warbler, Zanzibar. Barratt's Scrub Warbler, RSA. Brown Illadopsis, Gabon (1). Ashy Flycatcher, RSA. Sunbirds: Western Olive, DRC; Eastern Olive, Zanzibar and Malawi; Newton's (3 out of many nests), São Tomé; Scarlet-chested, Malawi; Yellow-tufted Malachite; Red-chested (2); Olive-bellied; Amethyst; Bronzy. Cape Sparrow, RSA, usually doubted despite bird attaining yellow underparts before dying, because unlikely to parasitise open country species. São Tomé Weaver (20 out of 100+ nests), São Tomé. Baglafecht Weaver, east Africa (2); southern Africa (3). Crested Malimbe, Uganda. Zanzibar Red Bishop, Zanzibar (Keulemans 1866, 1907, Fischer 1880, Haagner & Ivy 1906, van Someren 1932, 1939, Vincent 1934, Jackson 1938, Chapin 1939, 1953, Benson & Benson 1947, van Someren & van Someren 1949, Guichard 1950, Winterbottom 1951, Benson 1953, Ryves 1959, Jensen & Jensen 1969, Stagg 1973, Brosset 1976, de Naurois 1979, Harrison et al. 1997, Cheke & Mann 2001, Mills 2010, R. Stjernstedt in litt.). Two cuckoo young in one nest of Príncipe Speirops fed by many small birds as well as foster parents, Príncipe (Keulemans 1866, 1907). Possible hosts: young in nest of Green Barbet, RSA?; Red-bellied Paradise-flycatcher and Annobón White-eye, Annobón; Yellow-throated Woodland Warbler, RSA; Wailing Cisticola; Tawny-flanked Prinia, RSA and Kenya; Dohrn's Thrush-babbler, Príncipe; Cardinal Woodpecker, RSA; Grey-backed Cisticola, RSA; Cape White-eye, RSA; Speckled Mousebird, RSA. Young fed by Chestnut Wattle-eye, Gabon (Keulemans 1866, Jackson 1938, Friedmann 1948b, 1949, Holliday & Tait 1953, Fry 1961, Oatley & Pinnell 1968, Jensen & Jensen 1969, Brosset 1976, Irwin 1988). Village Weaver frequently parasitised Kenya and Uganda (Jackson 1938), Spectacled Weaver, Kenya, and Grosbeak Weaver, RSA, considered unsubstantiated (Friedmann 1948b). **Season** Breeds mainly in rains. Juveniles IC, Dec–Jan, calls throughout year Yapo Forest; calls Liberia Mar–Oct, and oviduct egg Aug; breeding Príncipe, Jul; calls throughout year Cameroon, and breeding condition female, Jan, Mar and Nov, and juvenile with unossified skull, Feb; calling May–Sep, Gabon; calls all year, but less May–Jul, south-east Nigeria; Sudan Apr–Jun; east Africa, Feb–Oct; calls Zanzibar, Apr–Jun. Active gonads Nov–Dec Zambia. Across southern Africa calling commences synchronously, Oct. Laying Liberia, Aug–Sep; São Tomé, early Oct–late Dec; Gabon, Feb–Mar; nestlings May, north-east DRC, and breeding at least May–Sep; south Ethiopia, probably Mar–Jun, Dec; in long rains iserwhen hosts breed, Kenya; nestlings Apr–Jul central and south Uganda, and south-west Kenya; Kenya (except west and coast) and north-east Tanzania, Jan–May, Sep–Oct; Angola, oviduct egg Feb; Malawi, Sep–Nov; breeding condition Zambia, Dec; Mozambique, fledgling Mar and Feb; Zimbabwe, Nov; RSA, Sep–Feb, mainly Sep–Nov (Fischer 1880, Keulemans 1907, van Someren 1916, Pitman 1929, Vincent 1934, Jackson 1938, Moreau & Sclater 1938, Winterbottom 1939, Benson 1945, 1953, Benson & Benson 1947, 1977, Friedmann 1948b, G.L. Bates in Friedmann 1948b, Marchant 1953, Serle 1954, 1965, Chapin 1939, Jensen & Jensen 1969, Benson *et al.* 1971, Dean 1971, 2000, de Naurois 1979, Brown & Britton 1980, Brosset & Érard 1986, Colston & Curry-Lindahl 1986, Nikolaus 1987, Christy & Clarke 1994, Demey & Fishpool 1994, Rodewald *et al.* 1994, Gatter 1997, Harrison *et al.* 1997, Payne 1997, Mills 2010). **Eggs** Non-mimetic, variable plain white, pinkish, pale blue, greyish-green, rose-red or salmon ground colour somewhat freckled, often with ring of speckles at broad end, or streaked all over pale olive-brown, dull brown and pale brown; 21.8x15.0, 19.1x18.0, 18.7x11.8mm. No evidence of egg-mimicry or *gentes* (Benson & Benson 1947, Rowan 1983, Irwin 1988, Tarboton 2001). **Incubation** 13 days maximum Gabon, 16 days RSA (Brosset 1976, Hockey *et al.* 2005). **Chicks** Eject host nestlings, but not eggs. Fed on insects or fruit, e.g. berries of *Erythrococca rigidifolia* (Friedmann 1948b) depending on host diet; Yellow-whiskered Greenbul raised young cuckoo on diet of fruit; nestlings fed berries by Common Bulbul, Kenya (Moreau 1949). Eyes open and quills formed on all feather tracts by 7–8th day, sheaths breaking in next three-four days. Remains in nest 18–20 days, and dependent on foster parent for up to two weeks after fledging (Irwin 1988, Brosset 1976, Oatley 1980). **Survival** No information.

FOOD Mainly hairy caterpillars, including *Agathodes thomensis* on São Tomé, taken from middle and upper canopy, although smooth green caterpillar consumed while perched in mid-canopy of thick riparian forest in Uganda. Takes insects from lichens on branches. Also grasshoppers and other insects, including whole adult *Acraea* butterflies, distasteful to many other predators; hard-bodied insects; birds' eggs; fruit, including wild peach *Kiggelaria africana*, and tree seeds (Chapin 1939, Haydock 1950, Friedmann & Williams 1969, Friedmann 1970, Rowan 1983, Payne 1997, 2005).

STATUS AND CONSERVATION Fairly common to abundant across much of current range, although present at low densities in some southern breeding areas (Parker 1999). Deforestation implicated in decline in Senegambia, where now great rarity (Morel 1972, Gore 1990). Common RSA. Common eastern SL, but not elsewhere. Numbers in Togo reduced through hunting (Payne 1997), but considered not uncommon resident by Cheke & Walsh (1996), although rare migrant to northern savannas. Uncommon Macenta Prefecture, Guinea (Halleux 1994). Common central Nigeria in rains (Dyer *et al.* 1986), and Owerri, south-east Nigeria (Marchant 1953). Abundant São Tomé, less common on Príncipe (de Naurois 1983). Rare Eritrea (Smith 1957), where decreased rapidly, but locally common Ethiopia (Ash & Atkins 2009). Rare Sudan, becoming less so near Ethiopian border (Cave & MacDonald 1955). Very uncommon Malawi (Medland 1992, 1993, 1995). Common Angola (Dean 2000). Formerly hunted for plumage RSA (Taylor 1906), but otherwise main threat forest destruction and degradation (Harrison *et al.* 1997). Not globally threatened (BirdLife International 2011).

African Emerald Cuckoo *Chrysococcyx cupreus*. **Fig. 1.** Adult male, Karkloof, near Howick, KwaZulu-Natal Province, RSA, January (*Adam Riley*). **Fig. 2**. Adult female, showing more brownish tones on head than is commonly seen, Gola Forest, Sierra Leone, July (*David Monticelli*).

HORSFIELD'S BRONZE CUCKOO
Chrysococcyx basalis   Plate 27

Cuculus basalis Horsfield, 1821 (Java)

Alternative names: Narrow-billed/Rufous-tailed Bronze Cuckoo.

TAXONOMY Monotypic. Synonyms: *Cuculus neglectus* Schlegel 1864, *Lamprococcyx modesta* Diggles 1876; *Chrysococcyx b. mellori* and *b. wyndhami* both Mathews 1912.

FIELD IDENTIFICATION 16cm. Very small cuckoo with swift and slightly undulating flight. Sexes similar. **Adult** Slightly glossy olive to bronze top of head, nape, upperparts and uppertail, with diagnostic broad white supercilium and bold dark eye-stripe behind eye decurved to behind ear-coverts; wings glossy bronze-green with pale buff to whitish edges to some wing-coverts and remiges, white to pale rufous-brown bar on primaries of open wing, underwing dark grey with conspicuously whitish stripe; underparts white with bold blackish transverse bands on flanks; sides of closed tail rusty basally; undertail grey with red-brown base to outer feathers (often difficult to see) and prominently spotted black and white on inner webs, and tipped white on all except T1. **Juvenile** Like adult but duller, top of head and nape grey-brown, supercilium absent or only suggestion, and underparts normally unbarred white with grey-buff tinge on breast and flanks, and streaked undertail-coverts. **Similar species** Shining Bronze Cuckoo is slightly larger, lacks prominent white supercilium and blackish eye-stripe, instead side of head is white with fine dark mottling, crown and nape red-brown; upperparts more shining iridescent emerald-green without scaling and no visible red-brown at base of closed tail; juvenile Shining has dark barring on flanks and fine barring on throat and breast. Both adult and juvenile Black-eared Cuckoo larger, have unbarred underside, lack red-brown base to outer rectrices and have less green upperparts with paler rump. The many subspecies of Little Bronze Cuckoo have little or no dark eye-stripe, prominent red eyes, narrower barring on whole underparts, and, only visible with tail spread, all except T5 and T1 red-brown below, T5 broadly barred black and white on inner web; juvenile more glossy green on upperparts and no visible red-brown at base of closed tail. Gould's and Little Bronze Cuckoos also have red-brown in tails, but only on juvenile Little Bronze is colour visible on upperside of closed tail. Pied Bronze Cuckoo has dark sides to face, and males and some females have white wing patch.

VOICE Descending call, similar to Black-eared Cuckoo's, but shorter, louder and repeated more often, drawn-out *whe-o*, *tseeeeeuw* or *peeer*, wide carrying and plaintive, repeated 15–32 times, and each call nearly 1sec duration. In flight or perched sparrow-like *chirrup* and in winter chattering *cheer-r-r-r* when in small flocks. In courtship behaviour high-pitched *pee-eeee-eep*; before copulation female utters soft trill. Juveniles imitate begging calls of hosts. Calls most often during spring and summer, by day and night, often from exposed perch; outside breeding period generally quiet, but in Atherton region calls all year (Kloot 1969, Blakers *et al.* 1984, Debus 1989, Tarbuton 1993, Coates & Bishop 1997, Higgins 1999).

DESCRIPTION Adult male Forehead, crown, nape and hindneck brown, in bright light with bronze sheen, broad white supercilium from bill to sides of neck, lores white with narrow dark brown stripe to around eye becoming broader behind eye curveding to behind ear-coverts; upperparts glossy green, in fresh plumage narrowly edged buffish most conspicuously on shoulder, but olive-brown with olive gloss on upper mantle and here without buff fringes to feathers, rump with dusky subterminal bands, lateral uppertail-coverts tipped white, remiges dark brown narrowly edged off-white to buff and with glossy olive to outer webs of secondaries, when wings fully spread white to pale rufous-brown on basal half to inner webs of P2–8, primary coverts and alula dark brown, rest of wing-coverts iridescent green edged rufous-brown to buffish, but greater secondary-coverts and tertials only tipped buff; in worn plumage buff scalloping paler and narrower, underside of wing glossy grey-brown edged paler with white base to all remiges except outermost forming broad white band, underwing-coverts white with dark brown barring; chin and throat off-white with faint dark brown chevrons or shaft-streaks on chin and upper throat, rest of underparts white broadly barred dark brown, breast and sides of breast occasionally tinged buffish, midline on lower breast and abdomen often quite white, undertail-coverts white with bold dark brown shaft spots; tail glossy olive with faint dark brown subterminal band, diffuse on central feathers, and rufous-brown on basal half of T2–4 with row of bold white spots on blackish inner web and small white spots on outer web of T5; undertail like uppertail but ground colour light grey and T4 has diagnostic basal half rufous-brown and at distal end 2–3 white spots, T5 grey-black with white and black spots on inner web. **Adult female** Not always separable except for iris colour where there is much overlap. Fewer streaks on chin and brown on mantle more prominent, breast less heavily barred, more mottled, and barring paler brown; 40% have rufous-brown on base of outer rectrices, but rest lack any trace of rufous-brown there. **Juvenile** Paler and more greyish-brown on top of head, supercilium only suggestion, and area on ear-coverts pale brown, upperparts and wings less glossy than adults but with same fringes to feathers but narrower, in worn plumage absent; underwing-coverts have fainter to almost no brown barring; whole underparts lack barring, off-white to pale grey-buff on throat, upper breast and flanks, and white undertail-coverts have dark brown shaft streaks; some birds have faint brown barring on flanks, rarely almost as in adults; tail like adults but more often with rufous-brown on base of T5. **Nestling** Hatches blind and naked, skin flesh-coloured with grey on head and lower back, later when in pin throat grey, pins white and breast skin dark pink; when feathers emerged ear-coverts dark brown-grey, throat grey and wing-coverts brownish-grey with green tinge and edged coppery. **Bare parts** Male iris red to dark brown, rarely with whitish, yellow or greenish outer ring, orbital ring grey to dark grey-brown; female brown to pale brown or off-white with brown wash, juvenile cream to light grey or brown-grey with inner or outer darker ring, orbital ring pale blue-grey washed yellow or brown. Bill grey-black to black, rarely with paler base to lower mandible, female dark brown to black, often with lower mandible paler; gape and palate black; juvenile dark grey with pale yellowish base to lower mandible, gape pale yellow or pale yellow-brown, palate pink becoming black; nestling has white gape, becoming creamy-white later, palate pale yellow, later orange-yellow. Feet grey to dark grey, often with olive wash and with whitish border to scutes, claws dark grey and soles off-white to yellow-grey, nestling pink, becoming greyish later.

BIOMETRICS Wing male (n = 78) 97–108mm, mean 102.7 ± 2.4, female (n = 32) 95–108mm, mean 100.7 ± 2.8. Bill male 14.0–18.3mm, mean 16.7 ± 0.9, female 15.2–18.7mm, mean 17.2 ± 0.9. Tail male 64–75mm, mean 70.9 ± 2.2, female 63–76mm, mean 69.3 ± 2.8. Tarsus male 16.2–19.8mm, mean 18.0 ± 0.9, female 15.9–19.2mm, mean 17.9 ± 0.7. Mass male (n = 22) 17–30g, mean 22.0g, female (n = 16) 20–32g, mean 24.3g (Higgins 1999).

MOULT Complete post-breeding moult starting (Dec) Jan–Feb and ending Mar–Apr (Jun). Moulting sequence of remiges and rectrices erratic without any apparent rules. Adult female in active primary moult at three loci (P4, P6–7 and P9–10) and T1. Post-juvenile moult complete or nearly so (Nov) Dec–Jun, peak Mar–Apr; many adults retain some juvenile secondaries or rectrices (Higgins 1999, Wells 1999).

DISTRIBUTION Australasian and Oriental. Two main areas of distribution: one south-west, other south-east Australia, but breeding records from all parts of Australia and Tasmania with less reported from great parts of central Australia and Cape York. North Australian population resident but most birds from more southerly parts and Tasmania migrate in winter to north Australia. Sometimes considered more nomadic than migratory because of its irregular appearance related to rainfall and accessibility of food. 24 Jul 2005, Sitiawan, Perak, peninsular Malaysia; eleven records Singapore, 23 May–20 Aug. Very scarce Borneo, Sarawak (two–three localities), Brunei (four localities), South Kalimantan, North Natunas and Spratly Is, May–Sep, once Dec. Bali, 25–28 May and 15–19 Aug. Six Sulawesi, Jun, Palu Valley, and once Lompobattang Massif. Common Flores, Apr–May, and other Lesser Sundas (Lombok, Komodo, Kalaotoa, Sumbawa and Timor), and Kangean I., Mar–Oct. Aru Is Mar–Sep; Sulawesi; Halmahera (Sharpe 1878, Hartert 1896, Peters 1940, Brooker et al. 1979, Cameron 1983, Watling 1983, Ash 1984, White & Bruce 1986, Mann 1987, 1991a, 2008, Coates & Bishop 1997, Duckworth et al. 1997, Smythies & Davison 1999, Chan 2005, Lim 2008a, b, Chow & Lim 2009, R.A. Fuller). Believed to occur south NG Mar–Oct, however, due to confusion with Gould's Bronze Cuckoo and juveniles of other species, only one confirmed record, 13 Jun 1977 (Heron 1977, Coates 1985). Sumatra, twice Jun (Bengkulu and Belitung), once Jun, Belimbing (Holmes 1996), and once Palembang, undated (van Marle & Voous 1988). Once Christmas I. (Andrews 1900). Once Khao Yai NP, Thailand, Sep (per R.A Fuller).

HABITAT In Australia open *Eucalyptus*-dominated wooded habitats and forests, woodlands with grass, scrub, and open or closed heath; in wheat belt of WA mostly concentrated in large remnants of fragmented habitats; open bushy plains with few trees, or low bushes; rare dense wet sclerophyll and rainforest; *Acacia* and mulga woodlands, and river gums along creeks; edges and clearings within forests, coastal salt marsh and spinifex; saltbush, mangroves, roads and golf-courses; mallee woodlands; in Rutherglen District, Victoria, orchards, gardens, vineyards and towns. In Malay Peninsula also open, sandy ground. In Wallacea scrub, open and dry woodland, usually near coast and infrequently in mountains; lowlands to 1,830m Sulawesi; 400m Lombok (Kikkawa et al. 1965, McEvey 1965, Bravery 1970, Hall 1974, Matheson 1976, Johnstone 1983, Blakers et al. 1984, Brooker & Brooker 1989b, 2003, Sibley & Monroe 1990, Coates & Bishop 1997, Pizzey & Knight 1997, Wells 1999).

BEHAVIOUR Mostly found singly but groups of eight or more chattering birds seen chasing one another. Courtship behaviour and feeding by two pairs late afternoon 15 Oct described thus. Individuals of one pair flew at each other with high-pitched *pee-eeee-eep* calls, almost clashing in mid-air. After settling on separate limbs of same tree, spread wings and moved them up and down three or four times; one bird then moved in front of other with drooping wings, bobbing up and down, and raising and depressing tail. After 1–2min both moved quickly away searching for food among branches. This was repeated six times, but no copulation observed, nor did they feed one another. Another pair 30m away displayed in similar way, but one, presumably male, did feed female (Kloot 1969). Male feeds female in courtship, and occasionally fledgling cuckoos. In wheat belt of WA 1993–1998 no birds parasitised one year, suggesting

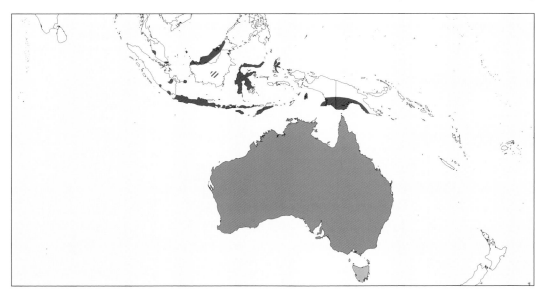

some nomadism. Mobbed by Ashy Tailorbird, unknown as host, in winter quarters, Singapore, Jul. On Flores up to ten birds together, Apr–May, also found in mixed-species flocks; sometimes in small parties in non-breeding quarters (up to five Brunei), where frequently feed on ground (Sedgwick 1951, Friedmann 1968, Bell 1986, Mann 1987a, Brooker & Brooker 1989b, 2003, Tarburton 1993, Coates & Bishop 1997, Seng 2008, C.F. Mann).

BREEDING Parasitism Parasitises mostly small dome-nesting passerines. Fairy-wrens (38.5%) most common hosts, followed by thornbills 22.4%, robins 8.7%, chats 4.6% and whitefaces 2.4% (Brooker & Brooker 1989b). Recorded hosts as follows. Diamond Dove (1). Fairy-wrens: Superb (230), Splendid (216), White-winged (81), Variegated (72), Red-backed (22), Blue-breasted (80), Red-winged (4), and Purple-crowned (1). Rufous-crowned (2) and Southern (9) Emu-wrens; Striated (8) and Thick-billed (4) Grasswrens. Honeyeaters: Brown (3), Scarlet (1), Black (7), Yellow-faced (6), White-eared (2), Yellow-plumed (2), Fuscous (2), White-plumed (1), Brown-headed (4), Black-headed (1), Tawny-crowned (8), Crescent (1), New Holland (14), White-cheeked (4), White-fronted (4), and Bar-breasted (1); Eastern Spinebill (2), Western Spinebill (1); Crimson (11), Orange (8), and White-fronted (52) Chats. Spotted Pardalote (1), Chestnut-rumped Heathwren (4), Shy Heathwren (8), Striated Fieldwren (9), Redthroat (2), Speckled Warbler (23), White-browed (39), Large-billed (3), and Yellow-throated (3) Scrubwrens, Weebill (17). Gerygones: Brown (4), Mangrove (4), White-tailed (4), Fairy (1), White-throated (9), and Large-billed (1). Thornbills: Brown (44), Inland (41), Tasmanian (3), Chestnut-rumped (44), Slaty-backed (9), Western (38), Buff-rumped (50), Slender-billed (8), Yellow-rumped (87), Yellow (21), and Striated (28). Southern Whiteface (28), Banded Whiteface (10),White-browed Babbler (1), White-browed Woodswallow (1), Varied Sitella (10), Rufous Whistler (1), Grey Fantail (18), Rufous Fantail (2), Willie-wagtail (4), Leaden (1), Satin (1), and Restless (3) Flycatchers. Robins: Scarlet (53), Red-capped (76), Flame (5), Eastern Yellow (9), Hooded (5), and Rose (2); Jacky Winter (8). Welcome Swallow (1), Golden-headed Cisticola (12), Little Grassbird (7), Spinifexbird (1), Australian Reed Warbler (2), Silver-eye (17), Mistletoebird (5), Olive-backed Sunbird (1), House Sparrow (1). Finches: Red-browed (7), Zebra (6), Double-barred (7), Black-throated (2), and Plum-headed (1); Diamond Firetail (1). European Greenfinch (1), European Goldfinch (5) (Brooker & Brooker 1999). Near Perth, WA, 21% of 562 nests of Banded Fairy-wren parasitised (Rowley & Russell 1989). One Red-browed Finch apparently also parasitised by Shining Bronze Cuckoo. Fairy-wrens, Rufous-crowned Emu-wren, scrubwrens, thornbills, Scarlet Robin and Silver-eye often bury cuckoos' eggs in nest lining if laid before their own eggs (Brooker & Brooker 1989a, 1999, Higgins 1999, Langmore et al. 2003, 2008). **Season** Eggs all months; in SA peak Aug–Nov, in central Australia two peaks, one spring and one autumn; in north Australia breeding birds found all months except Apr and Jul; over whole Australia eggs all months except May. In WA laying period 8–11 weeks (Brooker & Brooker 1989b, 2003). **Eggs** Elongated oval; ground colour white tinged pinkish, spotted and freckled all over with light brownish red, resembling many host species, Blue Fairy-wren and Western Thornbill in particular; 16.0–19.6x11.2–13.2mm (n = 108), mean 17.9x12.1; *c*.1g. Laying female enters nest 08:00–09:00hrs, head first with tail and wing tips visible outside entrance of nest; laying takes 3.3–5sec; emerges backwards from nest with host egg in bill carrying it 50m (Grant 1966, Ashton 1985, Brooker *et al.* 1988, Brooker & Brooker 1989b). **Incubation** From completion of host clutch 12–13.5 days (Brooker & Brooker 1989b). **Chicks** At hatching blind, naked and flesh-coloured; mass about 1.4g. Skin becomes darker with age, eyes open at four days, pin-feathered at about six days and completely feathered at 14 days. 24–30 hours after hatching evicts eggs or chicks of host; perhaps eviction takes place early morning when any host young are still lethargic. Nesting period 15–19 days (n = 10), mean 16.7. Fledgling began to forage alone on hairy caterpillars after three days, but was still fed by foster parents until two weeks old. Normally fledgling cuckoo fed by foster parents for 4–21 days by Splendid Fairy-wren and 10–20 days by Western Thornbill (Alley 1978, Brooker & Brooker 1989a,b, Payne & Payne 1998, Davies 2000). **Survival** Breeding success with Splendid Fairy-wren 43% (n = 95) but Western Thornbill 65% (n = 26); 80% Blue-breasted Fairy-wren fosters produced nestlings, and 76% resulted in fledglings between 1993 and 1998. 14 out of 20 Superb Fairy-wrens accepted cuckoo chicks (Brooker & Brooker 1989a, 2003, Langmore et al. 2003). Superb Fairy-wren is one of very few hosts of ejector cuckoos that detect and desert cuckoo nestlings. This is due to hosts calling to unhatched young who then develop begging calls that closely match parent's incubation call, and differ from begging calls of cuckoo which cannot learn call (Colombelli-Négrel *et al.* 2010).

FOOD Hairy caterpillars, and those of Coster Butterfly *Acraea violae* and Lemon Emigrant Butterfly *Catapsilla pomona*, Singapore; green grasshoppers, wasp, earwigs, Diptera; spiders; birds eggs; perhaps purplish berries. Forages most often in low bushes and trees, but in open woodland also on ground; sallies for flying insects or caterpillars hanging on sticky threads. 13 brown and red worms 25mm long from ground swallowed whole (Storr 1953, Kloot 1969, Hall 1974, Alley 1978, Blakers *et al.* 1984, Mann 1987a, Coates & Bishop1997, Higgins 1999, Seng 2008, Seow *et al.* 2010).

STATUS AND CONSERVATION Common over most parts of breeding range. Density Armidale, NSW, 0.04 birds/ha and Boola Boola, Victoria 0.1–0.2 birds/ha, both near or in eucalypt forest (Blakers *et al.* 1984). At many sites 1962–1970 in Kimberley and Karratha, both northern Australia, and common Mt. Shenton, WA, but otherwise only few birds seen or heard (Hall 1974). Declined in wheat belt of WA, presumably because of reduction of host population (Higgins 1999), and also because cuckoo prefers large remnants in this fragmented habitat with many of its host species breeding, and hosts have high reproduction rate in small remnants, but short life span due to predation (Brooker & Brooker 2003). Uncommon to rare in western south-east Asia in winter (van Marle & Voous 1988, Wells 1999, Mann 2008). Not globally threatened (BirdLife International 2011).

Horsfield's Bronze Cuckoo *Chrysococcyx basalis*. **Fig. 1.** Adult male, Melbourne, Victoria, Australia, July (*Rohan Clarke*). **Fig. 2.** Juvenile, Onslow, Pilbara, Western Australia, May (*Colin Trainor*).

RUFOUS-THROATED BRONZE CUCKOO
Chrysococcyx ruficollis   Plate 25

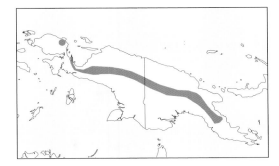

Lamprococcyx ruficollis Salvadori, 1876 (Hatam, Arfak Mts., New Guinea)

Alternative names: Reddish-throated Bronze/Mountain Bronze Cuckoo.

TAXONOMY Monotypic.

FIELD IDENTIFICATION 16cm. Very small cuckoo; male iridescent above; both sexes rufous from chin to breast. **Adult male** Forehead, face, sides of neck, chin, throat and upper breast diagnostically rufous, top of head, nape and upperparts dark green with iridescent bronze or purple, crown sometimes greyer and duller, remiges dark brown, underparts from lower breast to undertail-coverts white barred glossy greenish-brown, closed tail bronze-green with broad dark subterminal band, when spread row of white, black and rufous bands visible. **Adult female** As male but no purplish-bronze iridescence to upperparts. **Juvenile** Greenish upperparts; below grey barred on flanks. **Similar species** White-eared Bronze Cuckoo female has chestnut only on forehead and forecrown, none on sides of face or throat. Only other bronze cuckoo with which it can be confused is allopatric subspecies *harterti* of Shining Bronze on south Solomons, which has no rufous on forehead and upper breast.

VOICE Series of seven-nine slightly descending but high-pitched and drawn-out *seeuw-seeuw* notes, *c.* 2/sec, or *peer-peer-peer* like White-eared Bronze Cuckoo, but higher pitched; single *tseew* repeated at irregular intervals. Often calls from high exposed perch in forest clearing or at edge (Coates 1985, Beehler *et al.* 1986).

DESCRIPTION Adult male Forehead to above eyes, lores, ear-coverts, cheeks, chin, throat, sides of neck and upper breast cinnamon-rufous, eyebrow behind eyes, cheeks and ear-coverts washed dusky, upper breast with few diffuse black bars, crown, nape and upperparts dark green with glossy bronze and purple reflections, crown often greyer and duller, on some pale rufous edgings to outer webs of uppertail-coverts; remiges dark brown, coverts as upperparts, white to cinnamon wing-bar on inner webs of basal half of underwing; underparts from lower breast to undertail-coverts white boldly barred glossy greenish-brown, most heavily on flanks, and often broken in middle of belly. T1 bronze-green with blackish subterminal area, blackish nearest shaft grading to pale rufous at edges; T2 outer web dark bronze-green edged pale rufous, inner web dark brown-green with pale rufous on outer third with white spot and subterminal band on both webs; T3 dusky bronze-green with black subterminal band and outer web edged rufous towards base, inner web black with small rufous spots on outer third, and tipped white; T4 dark brown-green on outer web with black subterminal band and tipped grey, edged rufous towards base, inner web also black, outer third rufous with pointed white area; T5 white with greyish bands on outer web which do not reach edge of feathers, two outer bands nearly joined and washed rufous, inner web black, outer third rufous with large pointed white area. **Adult female** Like male but upperparts and greater wing-coverts bright green or blue-green without any purplish-bronze iridescence. **Juvenile** Greenish on upperparts, grey below with diffuse grey barring on flanks, belly white. **Nestling** No information. **Bare parts** Iris male red-brown, female brown, juvenile dark brown, orbital ring grey-green. Bill black. Feet dark olive-green.

BIOMETRICS Wing male (n = 9) 89–101mm, mean 94.8 ± 3.6, female (n = 9) 94–96mm, mean 95.5 ± 1.2. Bill male 13.2–15.0mm, mean 14.1 ± 0.7, female 13.2–15.8mm, mean 14.0 ± 0.8. Tail male 58–72mm, mean 65.0 ± 4.6, female 60–69mm, mean 63.8 ± 2.9. Tarsus male 15.0–16.4mm, mean 16.0 ± 0.5, female 14.4–16.7mm, mean 15.7 ± 0.8. Mass male 21.0g, female 23.5g (Payne 2005).

MOULT Adult has complete post-breeding moult. Sequence of primary moult: P9 and P7 simultaneous, P5–P8–P6, P1 and P4 simultaneous, P2–P3, and P10 when P9 fully grown (Stresemann & Stresemann 1969).

DISTRIBUTION Resident at higher elevations in mountains of Central Ranges, Vogelkop and Wandammen Mts., NG (Beehler *et al.* 1986).

HABITAT Reported from higher montane primary forest, at forest edge, in secondary forests and subalpine thickets; 1,130–3,350m, mostly 1,600–2,600m (Diamond 1972b, Coates 1985, Johnsgard 1997).

BEHAVIOUR Solitary; in canopy of highest trees as well as shrubbery close to ground. Occasionally with White-eared Bronze Cuckoo at caterpillar infestations; 'flycatches' flying insects. May sit quite motionless for short time and then catches prey by quick movements. Occasionally in mixed feeding parties (Mayr and Rand 1937, Rand & Gilliard 1967, Coates 1985, Beehler *et al.* 1986).

BREEDING Presumably brood-parasitic. **Season** Female with large ovary, Jul (Payne 2005). No further information.

FOOD Caterpillars and other insects (Coates 1985).

STATUS AND CONSERVATION Rare and local (Rothschild & Hartert 1907, Rand & Gilliard 1967, Diamond 1972b, Coates 1985). Very few juveniles known. Research on its biology and status urgently required. Not yet globally threatened (BirdLife International 2011).

Rufous-throated Bronze Cuckoo *Chrysococcyx ruficollis*. **Fig. 1.** Adult male, Tari, PNG, August (*Jon Hornbuckle*). **Fig. 2.** Adult with insect larva prey, Tari, PNG, August (*Jon Hornbuckle*).

SHINING BRONZE CUCKOO
Chrysococcyx lucidus **Plate 27**

Cuculus lucidus J.F. Gmelin, 1788 (Queen Charlotte Sound, New Zealand)

Alternative names: Golden/Golden Bronze/Shining/Broad-billed Cuckoo.

TAXONOMY MtDNA analysis places it as sister to *ruficollis* (Sorenson & Payne 2005). Polytypic. Synonyms: *Cuculus chalcites* Temminck 1838 and *Lamprococcyx l. australis* Mathews 1916 in nominate; *L. poliurus* Salvadori 1889, *Chrysococcyx plagosus tasmanicus* and *p. carteri* Mathews 1912 in *plagosus*; *Chalcites l. aeneus* Warner 1951 in *layardi*.

FIELD IDENTIFICATION 15–17cm. Very small green cuckoo with swift, slightly undulating and graceful flight. Young and moulting adults especially difficult for visual identification. **Adult male** In NZ has white face with sparse dark mottling or barring, most with some speckling on white central forehead; crown and upperparts shining green with bronze sheen; underparts white with bold transverse dark barring and tail with little or no rufous. Australian subspecies differs in having little or no green on crown and mantle but maroon-bronze contrasting with rest of iridescent green upperparts, less mottling and more white on chin, but much variation. **Adult female** Maroon-bronze on crown, nape and mantle, few with some speckling on white central forehead; underparts white with bold transverse dark barring and tail with little or no rufous. Most often resembles adult male of Australian subspecies though crown rarely as glossy green, side of head usually more mottled brown and chin often more white, but much variation and often difficult to separate from nominate. **Juvenile** As adult but duller and often with less barring on underparts, and Australian form generally even less barred on underparts. Nominate subspecies from NZ during migration also occurs in eastern coastal Australia. **Similar species** Sympatric Horsfield's Bronze Cuckoo has white supercilium and dark eye-stripe, upperparts less green, wings with rufous fringes; juvenile has unbarred underparts and grey not white throat. Little Bronze Cuckoo male has conspicuously red eyes and eye-ring, female brown iris with yellow to white eye-ring and both sexes narrower, finer barring on white underparts, plain off-white in juvenile, tail with much rufous in T2–4. Gould's Bronze Cuckoo is bronze-green on upperparts, male with red iris, rufous wash on breast and rufous on all rectrices, juvenile with some rufous in wings and barred flanks. Pied Bronze Cuckoo has dark sides to face, and males (and some females) have white wing patch.

VOICE Call like man whistling his dog *fwee-fwee-fwee* ascending for each note followed by descending staccato *pee-eeerr*, last like call of Horsfield's Bronze Cuckoo but faster, *kui-kui-kui-kui-kui-kui-kui-kui......tiu-tiu-tiu*. In display *wit-wit-here-er* and occasionally *pee-pee-pee*, or trilling and descending *piu-piu*. Call downslurred *tsee-ew*, or *tsiu* often heard when more gather or when flying at night. Many authors state that its voice is markedly ventriloquial and so high-pitched that it is inaudible to older people and difficult to locate, perhaps because bird moves its head when calling. When calling flicks out wings partly or fully. Begging call of nestling mimics that of young warblers *eee-eee-eee*. Calls from exposed branches or dense foliage. Quiet *cheep* when courtship feeding. Most often calls in spring and summer (Fulton 1910, Watson and Bull 1950, Sedgwick 1955, Turbott 1967, Falla *et al.* 1981, Gill 1982, Robertson 1985, Bregulla 1992, Strahan 1994, Pizzey & Knight 1997, Higgins 1999, St. Paul 1976).

DESCRIPTION Nominate. **Adult male** Forehead, crown and nape glossy green, sometimes faintly tinged maroon-bronze, with varying white-speckled central patch, sides of head white with some brown barring, strongest on ear-coverts and broad white supercilium; upperparts shining green, mantle faintly glossed bronze, white spots on lateral uppertail-coverts; remiges dark brown with increasingly glossy green on outer web of inner secondaries, wing-coverts and tertials glossy green, alula and primary-coverts dark grey-brown with little gloss on innermost coverts; underside of remiges grey with white base to all except two outermost forming broad white band, coverts white with dark grey-brown barring; underparts white with faint dark brown barring on chin and throat, rest of underparts broadly barred dark brown with strong green gloss, bars about half width of white bars but wider on belly; upperside of tail olive-brown glossed green with dark subterminal band, underside grey with blackish subterminal band, tipped white, grey on T1 and T4, and T5 black with large white spots, T4 often with some rufous. **Adult female** Very similar to male of subspecies *plagosus*; differs from nominate male in having little, or seldom no, white speckling on forecrown, occasionally as in some *plagosus* males; face not so white with no, or only hint of, white supercilium and ear-coverts darker, crown, nape and hindneck glossy maroon-bronze, upperparts like male but more maroon-bronze wash on mantle nearest neck, rarely whole mantle, and white on uppertail-coverts slightly narrower; barring on underparts less iridescent and more maroon; tail often with little more rufous on T4 and white spots on T5 tinged buffish. **Juvenile** Top of head shining green but lacks white on forehead; sides of head, chin and throat white mottled and barred brown; upperparts like adult but little less glossy; and lateral uppertail-coverts normally only edged white on inner web and when white on outer web too, this marking narrower than in adult; underparts white with diffuse, dense brown barring on breast, flanks and undertail-coverts, broadly barred dark brown with slight green gloss, centre of belly white; tail olive-brown with darker subterminal band, white tips normally smaller or wholly absent, undertail mainly grey but white spots on T5 extending to both inner and outer webs, and often lacks buff wash, T4 has less or no rufous. **Nestling** Hatchling naked, either pinkish-yellow, similar to chicks of Superb Fairy-wren, or black (Langmore *et al.* 2003) only sparsely covered with short white trichoptiles on head, nape and back, greenish-grey head and lower back and dark apricot shoulders becoming orange-brown on shoulders and dark purple-grey on back, and later with yellowish pins around area below; below skin of chin dark maroon-brown, feathers on ear-coverts grey, throat becoming speckled cream and grey, and with grey line between throat and breast, wing-coverts greyish-brown tinged green (Brooker & Brooker 1986). **Bare parts** Iris typically red-brown but varies from light brown to yellow or golden in male, in female pink, juvenile dark brown to pale grey, eye-ring pale green to dark grey, broader in juvenile. Bill dark brown to blackish with base of lower mandible often pale bluish-grey to brown or flesh; fledgling dark grey with pink base ventrally; gape flange white (Payne 2005); juvenile black with dirty yellow or dirty pink base to lower mandible but soon becomes as adult, gape and palate black, mouth also reported fleshy to orange in male. Feet grey to dark grey, in some blackish, occasionally washed olive or brown, claws dark grey, soles pale grey to yellow or buffish; nestling has orange feet when newly hatched, soon becoming dark greenish-grey to blue-grey.

BIOMETRICS Nominate. Wing male (n = 40) 102–112mm, mean 104.1 ± 2.09, female (n = 38) 99–109mm, mean 104.2 ± 1.99. Bill male 16.9–19.7mm, mean 18.2 ± 0.65, female 17.0–19.5mm, mean 18.6 ± 0.62. Tail male 63–75mm, mean 68.7 ± 2.33, female 64–73mm, mean 68.8 ± 2.19. Tarsus male 16.8–20.7mm, mean 18.6 ± 0.71, female 17.6–20.2mm, mean 18.8 ± 0.57. Mass male (n = 16) 16.7–32.0g, mean 25.6, female (n = 12) 18.5–29.0g, mean 24.4 (Higgins 1999); up to 50g before migration (Payne 1997); male 20g, immature female 20, 22.8, 24.1, 25.8g (Kratter *et al.* 2001a).

MOULT Complete adult post-breeding moult in winter range. Sequence of primary moult: P9 and P7, P5–P8–P6, P1 and P4, P2–P3, and P10 after P9 fully grown. Complete post-juvenile moult also takes place in wintering area and finished before migration to breeding area in Aug–Oct; latest record of full juvenile plumage Mar–Apr, Victoria and WA (Stresemann & Stresemann 1961, Higgins 1999).

GEOGRAPHICAL VARIATION Four subspecies.
C. l. lucidus (J.F. Gmelin, 1788). NZ, Norfolk and Chatham Is. Winters Solomons and other islands in south-west Pacific; on passage also Tasmania and coastal east Australia, more rarely NG. Bill broad unsexed 5.72–5.86mm (n = 73), mean 5.8 (Higgins 1999). Described above.
C. l. plagosus (Latham, 1801). East and south-west Australia, and Tasmania. In winter to NG, Bismarck Is, Lesser Sundas (rarely) and Solomons. Much variation. Male often as female nominate. Head browner than back; slightly less green on upperparts than nominate; throat may be whiter and less mottled; dark barring below narrower; bill narrower and shallower. Female lacks white spots on forehead and over eye, crown and upperparts bronze; sides of face less white without white scallops on forehead, and more brown or bronze, less green gloss on back and on broader ventral bars. Iris typically dark brown but orange, grey or white recorded; pale area on lower mandible smaller; fledgling's gape flange yellow (Payne 2005). Wing male (n = 26) 99–110mm, mean 106.3 ± 2.2, female (n = 11) 102–110mm, mean 105.6 ± 2.9; bill width unsexed (n = 45) 4.7–5.3mm, mean 4.94 (Higgins 1999). Not all birds separable from nominate.
C. l. layardi Mathews, 1912. New Caledonia, Loyalty Is, Vanuatu ('*aeneus*'), Banks and Santa Cruz Is.; apparently sedentary. Smaller. Top of head to mantle dull copper-bronze, less glossy green on upperparts; bill broad and longer. Wing male (n = 17) 94–101mm, mean 99.5, female (n = 7) 93–100mm, mean 97.1 (Mayr 1932, Payne 2005).
C. l. harterti (Mayr, 1932). Rennell and Bellona Is, south Solomons; probably sedentary. As *layardi* but smaller. Adult male has crown, neck and upperparts with strong coppery tinge, and underwing barred; adult female differs in having dusky-purple on crown; hindneck and mantle bronze with slight green gloss, throat and upper breast rufous and tail more rufous. Wing male (n = 3) 90–93mm, mean 92.7, female (n = 3) 90–95mm, mean 93.0 (Mayr 1932, Payne 2005).

DISTRIBUTION Australasia and Wallacea. Nominate breeds in much of NZ south to Stewart Is, occurring late Sep–late Mar, with few wintering. Probably adults migrate before juveniles as calling stops Jan. Supposedly some of population traverses some 3,200km of ocean from its breeding grounds in NZ to its winter range in Solomon Is, New Britain, Woodlark and Nissan, perhaps without feeding en route except perhaps on Norfolk and Lord Howe Is. However, it is only supported by single bird taken at sea *c.* 65km east of Lord Howe I. (Hindwood 1940). Bird taken before migrating was twice normal mass (Payne 1997). However, at least some birds take longer route over eastern Australia. Australian population divided into two, one south-west corner from west Great Australian Bight and north to Shark Bay, and eastern population from Tasmania in south over Eyre Peninsula to Cape York in north. Populations in temperate zone south of 35°S (Tasmania, Victoria and south highlands of NSW) are migratory presumably taking northerly direction, those from south-west Australia most likely migrate to Lesser Sundas, and those from south-east Australia to NG and nearby islands, remaining north of 35°S sedentary. Uncertainty due to lack of ringing recoveries.

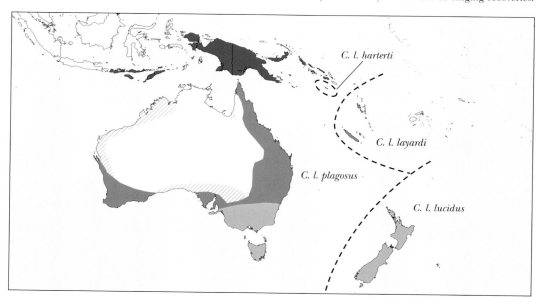

Widely distributed early Mar–late Oct in NG and nearby islands, on Solomons Mar–Apr departing Aug–Sep, and on Santa Isabel Apr–Sep. Lesser Sundas on Lombok, Sumbawa, Flores, Sumba, Dao, Roti, Wetar and Timor-Leste. Vagrants Kermadec, Aukland and Snares Is.; two seen Jul 1996 Kapingamarangi Atoll, Micronesia (Gill 1979, 1983c, Hadden 1981, Coates 1985, Robertson 1985, White & Bruce 1986, Webb 1992, Buden 1998, Higgins 1999, Trainor *et al.* 2008b). Lord Howe I. (McKean & Hindwood 1965). Presumably *lucidus* and *plagosus* are not totally allopatric, as Mees (1964) reported *lucidus* at Merauke, WP, 8 Apr 1959, and seven specimens from Guadalcanal, Solomons, included both subspecies (Galbraith & Galbraith 1962).

HABITAT In NZ pine and native forests, and often frequents willow trees along rivers; cultivation with scattered trees and gardens. WA in peppermint thickets and open bush land, in Queensland tropical savanna and forest with streams and rivers. NSW in open forest and mixed woodland. Rare in mangroves near Sydney. Dense eucalypt woodlands and forests receiving 380+mm/year of rain. In NG secondary forest, scrub, savanna, gardens and *Casuarina* growth, sometimes *Pinus* plantations and mangroves. Frequent Vanuatu in primary and secondary forest, and park-like country and gardens in both lowlands and mountains, but more common at lower altitudes. To 1,200m NZ. Lowlands, and to 1,920m PNG and 1,500m east WP, to 1,280m Karkar I. and to 1,000+m Bougainville. To 1,000m North Solomons (Hindwood 1935, Hall 1974, Schodde & Tidemann 1976, Falla *et al.* 1981, Hadden 1981, Coates 1985, Robertson 1985, Bregulla 1992).

BEHAVIOUR Territory in temperate rainforest of south North I., NZ, *c.* 21.9ha, and on south-east South I. in modified forest 4.9ha. Up to ten birds gathered at one site gave communal display by calling and flying from perch to perch, and also calling in such a 'parliament' sitting with very erect posture and flicking wings; one gathering in Mar or Apr consisted of 250–300 individuals, presumably pre-migration gathering. Caterpillars shaken vigorously and passed through bill from end to end before being swallowed, or empty skin is ejected. Feeds in foliage in canopy of forests and in scrub, more rarely on ground; catches flying insects; often hunts by sitting still on bare vantage branch under canopy waiting for prey. In mixed-species flocks of insectivorous birds. Courtship feeding reported (Smith 1930, Watson & Bull 1950, Serventy 1958, Fitzgerald 1960, Blackburn 1963, Kikkawa 1966, Schodde & Tidemann 1976, Falla *et al.* 1981, Robertson 1985, Bell 1986, Higgins 1999).

BREEDING Parasitism Mainly parasitises thornbills (64%) in Australia and, presumably because of very different egg colour compared with host's white eggs, most often in domed nests very dimly lit inside (Brooker & Brooker 1989a); also gerygones (9.5%), sericornis (5%) and fairy-wrens often reported from Australia. In NZ and Norfolk I., host-specific only to Grey Warbler of which 42–63% are parasitised in NZ, but only in their second clutch (Falla *et al.* 1981, Gill 1983a, b), and on Chatham I. Chatham Island Gerygone (Oliver 1955). Fan-tailed Gerygone favoured, south-west Pacific (Bregulla 1992). Hosts with either egg, young in nest or fledgling fed by foster parents with recorded instances as follows. Red-browed Treecreeper (1). Fairy-wrens: Superb (30), Red-winged (1), Variegated (3), White-winged (2), Red-backed (5) and Splendid (9); Southern Emu-wren (2). Honeyeaters: Black (1), Rufous-throated (2), Yellow-faced (2), White-eared (1), Yellow-tufted (1), Yellow-plumed (4), Brown (1), Brown-headed (1), Tawny-crowned (3), White-cheeked (1) and New Holland (4); Eastern Spinebill (2); Chats: Orange (1), White-fronted (6) and Crimson (1). Spotted Pardalote (1), Striated Pardalote (2), Scrub Tit (4), Striated Fieldwren (1), Speckled Warbler (17), Yellow-throated Scrubwren (9), White-browed Scrubwren (2), Weebill (28). Gerygones: White-tailed (9), Mangrove (9), Large-billed (15), Brown (17), White-throated (30), Fairy (3), Chatham Island and Fan-tailed. Grey Warbler, Inland (46). Thornbills: Yellow-rumped (264), Tasmanian (16), Western (60), Striated (49), Yellow (46), Brown (79), Buff-rumped (56) and Chestnut-rumped (4). Black-faced Woodswallow (1), Masked Woodswallow (1), White-winged Triller (1), Grey Shrike-thrush (1), Grey Fantail (9), Willie-wagtail (2), Rufous Fantail (3), Spectacled Monarch (1), Leaden Flycatcher (4), Satin Flycatcher (2). Robins: Scarlet (7), Flame (1), Pink (2), White-browed (1), Ashy (2), Red-capped (4), Hooded (1) and Dusky (1); Jacky Winter (3). Welcome Swallow (2), Tree Martin (1), Gold-headed Cisticola (4), Australian Reed Warbler (2), Silver-eye (8), Mistletoebird (1), Olive-backed Sunbird (3), House Sparrow (1), Crimson Finch (1), Red-browed Finch (5) (once apparently also parasitised by Horsfield's Bronze Cuckoo), Zebra Finch (2), European Goldfinch (2) (Brooker & Brooker 1999, H.A. Britton pers. comm.). New Zealand Robin, not a known host, frequently rejected artificial cuckoo eggs, suggesting that it may have been parasitised by cuckoos in past (Briskie 2003). **Season** In NZ eggs mid Oct–Jan. In WA 20 Aug–20 Nov. On Vanuatu eggs Sep–Jan (Robinson 1955, Gill 1982, Bregulla 1992). **Eggs** Egg broad oval to elliptical; 17.1x12.3mm (n = 4); plain olive-green with olivaceous tinge. Lays in mornings directly into nest, afterwards emerges backwards from nest with host egg in bill, whole process taking <20 sec, or forces itself out through opposite wall and host repairs damage afterwards (Gill 1982), this colour removed with moisture to reveal pale blue shell (Campbell 1900, Brooker *et al.* 1988, Brooker & Brooker 1989a, Campbell & White 1910 in Higgins 1999). **Incubation** Hatches in 13–14 days in nests of Yellow-rumped Thornbills, 13–17 days in nests of Grey Warbler in NZ, four days less than eggs of host. **Chicks** Hatch nearly naked and blind, and evict nest-mates at three-seven days. Nestling period 19–22 days. Fledgling fed by host parents up to 28 days (Gill 1979, 1982, 1983b, Brooker & Brooker 1989a), and often also by other bird species other than host, including Shining Bronze Cuckoos (Schodde & Tidemann 1976). **Survival** About half of cuckoo eggs become fledglings, among nestlings which failed to survive, three were predated, and one died because of bad weather. All nests (n = 4) with pale morph nestlings were abandoned by Superb Fairy-wrens (Gill 1983b, Langmore *et al.* 2003). See under Horsfield's Bronze Cuckoo.

FOOD Hairy caterpillars, flies, other insects; spiders; molluscs; woodlice. Caterpillars infesting citrus in orchards, and fig leaves in gardens. In NZ nearly half of all food items caterpillars, and beetles comprise nearly one third, some caterpillars and beetles being toxic (Smith 1930, Lord 1939, Michie 1948, Bravery 1970, Gill 1980a, Higgins 1999).

STATUS AND CONSERVATION Common and widespread NZ in spite enormous areas in last century being converted to pasture, as both cuckoo and host, Grey Warbler, have adapted to new habitat (Gill 1982, Robertson 1985). Often killed by cats when foraging on ground or flying into windows, particularly those under verandas (Robertson 1985). In Australia density at Armidale 0.02–0.42 birds/

ha, at Wollomombi 0.1, Canberra 0.02 and Boola Boola 0.2–0.9 (Blakers *et al.* 1984). Subspecies *layardi* formerly widespread Vanuatu but now only on Santo, Malakula, Efate and Ambrym where uncommon (Bregulla 1992), although Doughty *et al.* (1999) describe it as fairly common resident in Vanuatu, New Caledonia, Santa Cruz, Banks Is, and Rennell and Bellona Is. in Solomons. Nominate winters in good numbers throughout Solomons (Doughty *et al.* 1999). Common Vanuatu in 1920s and 1930s but unknown before that time, which Diamond and Marshall (1977) ascribed to immigrations and extinctions not influenced by man. Rare Timor-Leste (Trainor *et al.* 2008b). Not globally threatened (BirdLife International 2011).

Shining Bronze Cuckoo *Chrysococcyx lucidus*. **Fig. 1.** Adult female, probably *C. l. lucidus*, Brisbane, Queensland, Australia, July (*Rohan Clarke*). **Fig. 2.** Adult female, probably *C. l. lucidus*, Brisbane, Queensland, Australia, July (*Rohan Clarke*). **Fig. 3.** Adult male, probably *C. l. lucidus*, Ashmore, Queensland, Australia, April (*Rohan Clarke*). **Fig. 4.** Juvenile moulting to adult plumage, probably *C. l. plagosus*, Ashmore, Queensland, Australia, April (*Rohan Clarke*). **Fig. 5.** Juvenile, probably *C. l. plagosus*, Ashmore, Queensland, Australia, April (*Rohan Clarke*).

WHITE-EARED BRONZE CUCKOO
Chrysococcyx meyerii Plate 25

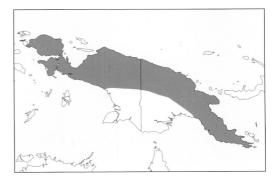

Chrysococcyx meyerii Salvadori, 1874 (Hatam, Arfak Mts., NG)

Alternative names: Meyer's Bronze/Mountain Bronze/White-eared Cuckoo.

TAXONOMY In error *meyeri* (e.g. Peters 1940). MtDNA analysis indicates it is closest to *minutillus* (Sorenson & Payne 2005). Monotypic. Synonym: *splendidus* A.B. Meyer 1874, replaced by *meyerii* (not *Cuculus splendidus* G.R. Gray 1847 which is *Chrysococcyx cupreus*).

FIELD IDENTIFICATION 15cm. Very small cuckoo. **Adult male** Iridescent green on head, upperparts and wing-coverts, often metallic bronze on back, ear-coverts with diagnostic white patch, flight feathers bright rufous on basal half, outer half dark brown, underwing conspicuously rufous; underparts white with broad bronze-green bars, tail dark bronze-green tipped pale with wide white barring on outer rectrices. **Adult female** Like male but forehead and crown rufous-chestnut. **Juvenile** Top of head green-grey, hint of white on sides of neck, upperparts olive grey-brown, wings grey-brown edged pale red-brown; underparts off-white with buffish wash, tail red-brown, outer feathers with two black bars on inner web. **Similar species** Adults cannot be confused with other bronze cuckoos due to white crescents. Young Rufous-throated Bronze is greenish above and grey below with diffuse grey barring on flanks. Juvenile Shining Bronze has top of head shining green, green-grey in White-eared. Juvenile Little Bronze Cuckoo is dull grey-brown above and juvenile Horsfield's Bronze is also dull grey-brown above but sometimes edged buffish. Juvenile Gould's Bronze has faint pale brown chevrons on flanks.

VOICE Common call series of 5–8 high-pitched, plaintive *peer-peer-peer* notes, dropping slightly in pitch towards, similar to Brush Cuckoo, but with less volume and more leisurely tempo, lower pitched than Rufous-throated Bronze; or *sieeu-sieeu-sieeu*, *c.* 4 notes/5sec. Another, reminiscent of cross between *Accipiter* and cuckoo, of four pairs of downslured notes, rising and falling in tone. Soft and loud *hyije*, easy to imitate and lure (Stein 1936, Diamond 1972b, Coates 1985, Beehler *et al.* 1986).

DESCRIPTION Adult male Forehead, crown, nape and hindneck metallic green, face green spotted white on ear-coverts; upperparts metallic bronze-green, wings dusky tipped grey-brown, basal half of remiges rufous on outer edge, wing-coverts metallic bronze-green, underwing-coverts green barred white; underparts white barred glossy green. T1 green, T2–T4 green tipped white, T5 black barred white on inner web and spotted white on outer web. **Adult female** As male but forehead and crown rufous-chestnut. **Juvenile** Head and upperparts yellowish-brown with ashy wash on head and hindneck, flight feathers dark brown, basal half of primaries edged chestnut, bases of all remiges tinged buffish with cinnamon-red, underwing-coverts cream; underparts greyish buff, rectrices brown with bold subterminal bar, outer web of T5 with two black and one white patch near base, and inner web blackish at base with three black-and-white bars on inner half. **Nestling** Not described. **Bare parts** Iris dark brown, female red-brown; narrow eye-ring orange to red, female blue-grey. Bill black, base of lower mandible paler in juvenile. Feet dark brown to dark grey.

BIOMETRICS Wing male (n = 8) 86–91mm, mean 88.6 ± 2.0, female (n = 12) 85–91mm, mean 89.1 ± 1.6. Bill male 14.0–15.4mm, mean 14.6 ± 0.5, female 14.7–17.0mm, mean 15.5 ± 0.6. Tail male 57–62mm, mean 59.8 ± 1.9, female 57–66mm, mean 62.7 ± 4.5. Tarsus male 13.6–16.4mm, mean 14.8 ± 0.9, female 13.6–18.4, mean 16.3 ± 1.2. Mass male (n = 6) 15.0–19.5g, mean 17.8, female (n = 6) 19.0–27.4g, mean 21.2 (Payne 2005).

MOULT Adult has complete post-breeding moult. Sequence of primary moult: P9 and P7 simultaneous, P5–P8–P6, P1 and P4 simultaneous, P2–P3, and P10 after P9 fully grown. Female primary moult, Jan, Dec (Mayr & Rand 1937, Stresemann & Stresemann 1969, BMNH).

DISTRIBUTION Resident NG, where it occurs widely; also Batanta I. (Beehler *et al.* 1986). Not reported from large flat lowlands such as Trans-Fly region (Coates 1985).

HABITAT Primary rainforest, monsoon and secondary forests, clearings, forest edges and gardens. Prefers lower altitudes than sympatric Rufous-throated Bronze, in forest to *c.* 2,000m, mostly 800–1,600m (Rand & Gilliard 1967, Diamond 1972b, Coates 1985).

BEHAVIOUR Solitary and sluggish, usually 5–13m in midstorey. Often found in mixed feeding passerine flocks consisting of Yellow-bellied Gerygone, Frilled Monarch, Wallace's Fairy Wren, Golden Monarch, Yellow-breasted Broadbill, Chestnut-bellied Fantail and Grey Whistler (Diamond 1972b, Coates 1985).

BREEDING Presumably brood-parasitic. **Season** Two females breeding condition Jan, Mar and Oct (Mayr 1931, Mayr & Rand 1937). No further information.

FOOD At caterpillar infestations, often with Rufous-throated Bronze Cuckoo. Two stomachs contained hairy caterpillars (Mayr & Rand 1937, Coates 1985).

STATUS AND CONSERVATION Uncommon, but widespread (Rand & Gilliard 1967, Diamond 1972b), but Coates (1985) describes it as not uncommon locally. May be overlooked due to silent and sluggish demeanour. Research of its biology and status required. Not globally threatened (BirdLife International 2011).

White-eared Bronze Cuckoo *Chrysococcyx meyerii*. **Fig. 1.** Adult male, Dablin Creek, Tabubil, PNG, May (*Nik Borrow*). **Fig. 2.** Adult male, Tabubil, PNG, August (*Jon Hornbuckle*).

LITTLE BRONZE CUCKOO
Chrysococcyx minutillus   **Plate 28**

Chrysococcyx minutillus Gould, 1859 (Port Essington, NT, Australia)

Alternative names: Red-eyed Bronze Cuckoo; Green-cheeked/Dark-backed/Lesser Sundan Bronze Cuckoo (*rufomerus*); Malay (Green) Cuckoo.

TAXONOMY MtDNA analysis shows that *minutillus* (*sensu lato*) is closest to *meyeri* (Sorenson & Payne 2005). Some authors, e.g. Payne (2005), consider it conspecific with *poecilurus* due to supposed hybridisation in north-east Queensland and Cape York, but in north Borneo both species overlap in breeding range without hybridisation. However, since both are rare in Borneo, and breeding is not actually recorded, this is probably irrelevant (Mann 2008). Supposed hybrids in north-east Australia may possibly be related to great individual variation, as both species have to some degree different habitat preferences and hosts. Moreover, both phenotypes are still maintained in Queensland (Nielsen 1996), which supposes some reproductive isolation (Helbig *et al.* 2002). Based on molecular studies Joseph *et al.* (2011) consider all Australian forms to be conspecific. In many works Gould's and Little Bronze Cuckoo are known as *Chrysococcyx 'malayanus'* group, but type of '*malayanus*' is female *xanthorhynchus* (Blyth 1865, Oates 1882, Parker 1981). Payne (2005) considers *cleis* to be a synonym of *aheneus*, stating there are no consistent differences in plumage or bill size. Dickinson (2003) split *crassirostris* as separate species, Pied Bronze Cuckoo, with *salvadorii* as subspecies, as here, and Parker (1981) and White and Bruce (1986) separated *rufomerus* as Green-cheeked Bronze Cuckoo, whereas Payne (2005) lumped *rufomerus*, *salvadorii* and *crassirostris* in *minutillus*; *rufomerus* may intergrade with *crassirostris*. Calls of all forms appear very similar, and there are no confirmed cases of sympatric breeding of two forms (Payne 2005). Birds from Timor discussed by Hartert & Stresemann (1925) and Parker (1981) may well be an undescribed form. Much work is still required on this complex group. Polytypic. Synonyms: *Chalcococcyx innominatus* Finsch 1900 in *rufomerus*; *Cuculus neglectus* Schlegel 1864, *Chalcococcyx Nieuwenhuisi* Vorderman 1898, *Chrysococcyx minutillus perplexus* Mathews 1912 and *C. m. melvillensis* Zietz 1914 in nominate.

FIELD IDENTIFICATION 15–17cm. Very small cuckoo. Very similar to Gould's Bronze Cuckoo, with many intermediates or hybrids found in Australia. Swift direct flight, slightly undulating. Sexes similar. **Adult** Top of head and nape glossy dark green contrasting in colour with paler upperparts, with some white on forehead, white and dark mottled supercilium and dark ear-coverts, iris and broad eye-ring red, upperparts slightly iridescent olive-green, wing-coverts as upperparts but greater-coverts with narrow buffish or rufous fringes in fresh plumage, flight feathers black, underwing-coverts white boldly barred dark, flight feathers dark grey with prominent white bar across outer secondaries and centre of primaries; underparts white with narrow dark brown barring, more narrow on chin, throat and abdomen; upperside of closed tail glossy olive with diffuse dark subterminal band and tipped white on all feathers except central pair, when spread row of white bars on inner webs of outer feathers visible, undertail as above but rufous banded black and white on two outer feathers. **Juvenile** Duller, crown and nape dull olive as rest of upperparts, sides of head grey-brown, underparts whitish with or without pale brown chevrons on flanks and sides of breast, tail as adult. **Similar species** Gould's and Little Bronze sympatric in north Queensland and north Borneo; intermediates or hybrids described in detail under Gould's Bronze Cuckoo. Normal individuals of Gould's have rufous on breast, Little has no rufous there, contrast between crown and upperparts weak, little or no white on forehead, ear-coverts pale, barring on underparts broader, tail more rufous without white barring on sides of outer rectrices, and black barring narrower and bill broader at base than in Little. Juvenile has less rufous in rectrices and mostly barred on flanks, unbarred in Gould's. Shining Bronze female has brownish crown and nape contrasting with glossy green upperparts, less pronounced in male, broad dark barring, no rufous in tail and pale to dark brown iris and grey eye-ring. Horsfield's Bronze in all plumages has dark decurved eye-streak, brown on top of head and nape, male has streaked, and female pale brown mottling on throat, iris brown to red, eye-ring grey. Rufous-throated Bronze has diagnostic rufous face, chin and throat. Pied Bronze Cuckoo has dark sides to face, and males and some females have white wing patch. Female Violet Cuckoo has rusty crown, bill yellow-brown to red, but eye-ring red as in Little and Gould's. Banded Bay and juvenile Brush Cuckoos much larger.

VOICE Slightly descending territorial call *chiwchiwchiw-chiw*, or *rhew-rhew-rhew-rhew* or *eug-eug-eug-eug*, repeated 3–5 times and sometimes interrupted by rising screeching *wireeg-reeg-reeg*. Long high-pitched trill slowly falling away. Calls mostly from high exposed perch. Voice undescribed in Wallacea (van Balen & Prentice 1997, Coates & Bishop 1997, Jeyarajasingam & Pearson 1999, Robson 2000).

DESCRIPTION Nominate. **Adult male** Forehead, crown and nape glossy dark green, usually contrasting with upperparts, forehead spotted white, hindneck duller olive-green with slight bronze iridescence, supercilium white with dark grey base and glossy green tips to each feather giving barred appearance, forehead mottled white to varying degree, sides of head white, each feather with dark barring, strongest on ear-coverts; upperparts slightly glossed olive-green, mostly with faint bronze gloss, uppertail-coverts glossy green with white to pale rufous tips to lateral feathers; upperwing dark brown on remiges, alula and primary-coverts, inner primaries and outer secondaries narrowly edged rufous-brown in fresh plumage, soon becoming off-white, wing-coverts glossy olive-green occasionally edged grey-brown, greater secondary-coverts duller and outer webs edged rufous-brown, changing to buffish, and absent in worn plumage; underwing glossy grey with broad white wing-bar formed by white bases to secondaries and inner webs of primaries, last washed pinkish buff, underwing-coverts white barred dark brown with green gloss; underparts white with 1–2mm wide dark brown bars glossed green or bronze in sunlight, bars narrowest at chin and throat and often broken on middle of abdomen, rarely with washed buffish on sides of breast; outer web of T5 on uppertail dark brown with white spots, inner web broadly barred black-brown and white, T4–2 have brownish-olive outer webs and rufous-brown inner webs, on T4 often barred black-brown, T1 glossy olive, on all subterminal black-brown bands with white spots on inner webs except central feathers where band is diffuse with no white spots, underside of tail with brownish-grey ground colour becoming light grey on outer feathers, pattern of rufous, black and white as uppertail. **Adult female** Resembles male, but top of head and nape more concolorous with upperparts, chin and upper throat often unbarred and white barring on T5 tends to be broader. **Juvenile** Forehead, crown, nape and hindneck olive with faint gloss, face pale brown-grey, upperparts olive and not as glossy as adults, white spots on lateral uppertail-coverts smaller than in adult; wing-coverts glossy olive-brown, tipped light brown on secondary-coverts, underwing brown, barring on white coverts narrower than in adult; underparts white, becoming grey-brown on breast, few diffuse chevrons on flanks; tail as adult but narrower. **Nestling** Hatches naked except pale yellowish plumes on crown and along back; skin dusky. **Bare parts** Male iris red-brown to red, broad orbital ring red to orange, female iris brown to dirty cream, orbital ring greenish-white to olive-grey, juvenile iris grey-brown to brown, eye-ring narrower than in adult, light grey to pale yellow or brown to dull red. Bill narrow and blackish, occasionally pale grey to blue-grey base to lower mandible, pinkish in juvenile, mouth black, juvenile whitish to pale yellow. Feet dark grey.

BIOMETRICS Nominate. Wing male (n = 17) 92–99mm, mean 95.1 ± 2.2, female (n = 4) 92–95mm, mean 93.0. Bill male 15.6–18.4mm, mean 17.1 ± 0.7, female 16.6–18.2mm, mean 17.7 ± 0.7. Bill width male 5.0–6.2mm, mean 5.7, female 5.4–6.5mm, mean 6.1. Tail male 57–66mm, mean 62.1 ± 3.4, female 57–61mm, mean 58.8 ± 1.6. Tarsus male 13.2–17.3mm, mean 15.7 ± 0.9, female 14.7–17.5mm, mean 15.9. Mass male (n = 5) 14.5–17.0g, mean 15.3 ± 1.12, female (n = 3) 15.4–17.0g, mean 16.5 (Higgins 1999).

MOULT Adult has complete post-breeding moult. Sequence of primary moult: P9 and P7 simultaneous, P5–P8–P6, P1 and P4 simultaneous, P2–P3, and P10 after P9 fully grown (Stresemann & Stresemann 1969).

GEOGRAPHICAL VARIATION Six subspecies.
C. m. minutillus Gould, 1859. North Australia, some possibly wintering in NG, Lesser Sundas, Moluccas, and Peleng (Sulawesi). Bill width 5.7–6.1mm (Parker 1981). Described above.
C. m. cleis Parker, 1981. Borneo. Dark bottle-green crown, white on forehead prominent, upperparts dark green with purplish wash, barring below heavy and eye-ring coral-red. Bill narrower (4.2–4.7mm) and sharply ridged. Wing male (n = 6), shorter, 88.7–92.8mm, mean 90.8 ± 1.7 (Parker 1981). Payne (2005) synonymises with *aheneus* (here subspecies of *C. poecilurus*).
C. m. rufomerus Hartert, 1900. Sedentary on Lesser Sundas (Damar, Kisar, Leti, Moa, Romang, Sermata). Broad dark green spot on cheek, upperparts dark bronze-green, crown slightly greener in birds from Leti, Moa and Romang, some white wing-covert fringes in fresh plumage, underparts strongly barred glossy greenish-brown, tail blackish green with white markings. Some show great individual variation (Parker 1981). Hybrids or intergrades (with *minutillus*?) on Moa, Leti, Romang and Sermata (Coates & Bishop 1997). Wing male (n = 21) 93.0–98.5mm, mean 95.9 ± 1.52 (Parker 1981).
C. m. albifrons (Junge, 1938). Resident Sumatra (few records, none breeding, Aceh, Lampung, Utara; van Marle & Voous 1988) and north-west Java. White on forehead heavier and extensive, superciliary and face also whiter; less barring below, which is narrower, with centre of abdomen often white. Female eye-ring green-yellow. Wing male (n = 10) 90–99mm, mean 94.1 ± 2.8. Bill narrower than *peninsularis*, mean 5.4mm (Junge 1938, Parker 1981).
C. m. peninsularis Parker, 1981. Resident Thai-Malay Peninsula; southernmost Vietnam and south Cambodia. Crown darker bottle-green contrasting with dorsum, face mottled white, female with green eye-ring. Wing male (n = 6) 90.7–97.0mm, mean 94.2 ± 2.0; bill longer, mean 18.2mm, width 5.9mm (Parker 1981).
C. m. barnardi Mathews, 1912. South-east Queensland from Yamala and Byfield and south to Clarence R. in north-east NSW, migrating to north Queensland and south NG. Upperparts darker than nominate, white fringes of

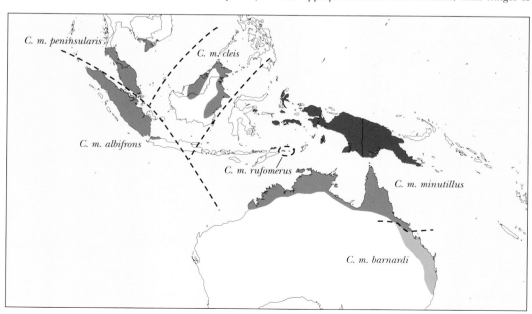

uppertail-coverts more prominent and female duller with deep yellow eye-ring. Longer winged, male (n = 4) 102.0–106.5mm, mean 104.3 ± 2.0 (Parker 1981).

DISTRIBUTION Australasian and Oriental. Sundaland, Wallacea, NG and Australia. Populations Kimberley, WA and around Darwin, NT sedentary. Presumably most other birds also resident, but great difficulty in identification of this species from Gould's and intermediates in north Queensland have complicated the case. However, those recorded on Booby and other islands in Torres Strait in Feb, Apr–Sep and Nov–Dec may be non-breeding migrants. Nominate also recorded Sulawesi (Peleng), Moluccas on Ambon, Buru, Halmahera, Kai (Kai Besar), Morotai?, Seram, Seram Laut (Gorong), and Lesser Sundas on Babar, Leti, Moa, Romang and Wetar. Resident Thai-Malay Peninsula from 09°10'N south to Singapore. On Sumatra in Aceh (Gunung Leuser NP and Gayo Highlands), Utara (Tanjung Kassau). Lampung (Bandar Lampung, Kotabumi, Menggala), Riau (Bangkinang, Lower Rokan); unconfirmed records Utara (Lake Toba) and Lampung (including Way Kambas Reserve). On Borneo in Brunei, Sabah and Kalimantan; others in Sarawak, Sabah, Kalimantan and Maratuas could be this species or Gould's (Parker 1981, White & Bruce 1986, van Marle & Voous 1988, Holmes 1996, Jeyarajasingam & Pearson 1999, Wells 1999, 2007, Robson 2000, Mann 2008). Cambodia and Cochinchina mentioned in error by many authors (Parker 1981), but now known from Cochinchina and Prey Nup, south Cambodia (Robson 2000, 2010).

HABITAT Thai-Malay Peninsula lowlands in mangroves, at forest edge, in scrub and gardens; swampy riverine forest Java. In NT commonly in mangroves and monsoon vine forests; in Australia generally in dense vegetation in tall mangroves, at edges of monsoon forest, in dense riparian vegetation, paperbark forest, vine thickets, open dry forest and woodland near rainforest. Gallery forests and forest edge preferred in south-east Queensland and north-east NSW; also sclerophyll forest, eucalypts, paperbarks or *Casuarina* woodlands. To 1,400m NG. To *c.* 500m Gunung Leuser, Sumatra; *c.* 1,000m Java (Crawford 1972, Ford 1981, Parker 1981, Coates 1985, Higgins 1999).

BEHAVIOUR Forages mostly in canopy where normally unobtrusive, but also on lower branches, rarely on ground. Courtship feeding in mangroves, Dec. Joins mixed bird flocks (Brooker *et al.* 1990, Crawford 1972, McKinnon & Phillipps 1993, Jeyarajasingam & Pearson 1999).

BREEDING Parasitism Like Shining Bronze Cuckoo only uses hosts with enclosed nests with very dimly lit interiors, presumably due to different egg colour (Brooker & Brooker 1989b). Following are recorded hosts from WA, NT and NSW (all regions where Gould's Bronze Cuckoo absent) with either egg, young in nest or fledgling fed by foster parents, with incidences if recorded. Bar-breasted Honeyeater (1), Rufous-throated Honeyeater (1); Gerygones: Large-billed (15), White-throated (8), Mangrove (8), Green-backed (5) and Dusky (2) (Brooker & Brooker 1989b). Lovely Fairy-wren, Red-backed Fairy-wren, Fairy Gerygone, Double-barred and Zebra Finches also claimed (P.E. Lowther *in litt.*), and possibly Rufous-sided Gerygone (Parker 1981). Higgins (1999) doubts that honeyeaters, Red-backed Fairy-wren or finches are hosts. Also, in Malay Peninsula Golden-bellied Gerygone (Batchelor 1958, Jeyarajasingam & Pearson 1999), 'heavily parasitising' this species in Singapore, where Olive-backed Sunbird also host (Low 2008e). Host unknown Wallacea (Coates & Bishop 1997). **Season** Oct–Mar (May) in north-east Australia. In Darwin region, five records Apr–Aug; breeding NT Feb–Apr and Jul–Nov; *barnardi* from Queensland breeds Sep–Jan. Fledglings Singapore, Dec and Oct. Java, Mar (Hartert & Stresemann 1925, Thompson 1982, Lavery 1986, Higgins 1999, Noske 2001, BIRDLINE 1996 in Wells 2007, Low 2008e). **Eggs** Elongated oval to elliptical, glossy, olive-bronze, greenish-bronze or brownish bronze, sometimes freckled blackish, mostly at larger end; 18.5x13.2, 19.3x14.2mm (Australia); 21x14mm (Java); 21.9x14.6mm (Flores); 20.5x14.7mm (NG) (Hellebrekers & Hoogewerf 1967, Ottow & Verheijen 1969, Mees 1982, A.J. Campbell in Higgins 1999). **Incubation** 13–15 days (Noske 2001). **Chicks** Nestling period 16–25 days and fed by hosts after fledgling for at least 30 days (Noske 2001). **Survival** Two cases of ejection dark cuckoo nestling by host Mangrove Gerygone, whose own nestlings are pinkish-grey (Tokue & Ueda 2010).

FOOD Insects and their larvae such as ants, bees, sawflies, wasps, butterflies, arboreal caterpillars, ladybirds and bugs (Higgins 1999, Wells 2007).

STATUS AND CONSERVATION Near Palmerston 0.01 birds/ha, 0.1 birds/ha South Alligator R. and 0.07–0.72 birds/ha Howard Peninsula, all NT (Higgins 1999). Cyclone said to have reduced population by 50% at Holmes Jungle, NT (J.L. McKean in Higgins 1999). In lowlands of NG fairly common, but scarce above 500m (Coates 1985). More or less common south Peninsular Malaysia, elsewhere regular (Wells 1999, Robson 2000). Possibly more common on Sumatra than van Marle and Voous (1988) suggest, as nine records since 1980, one of three birds (Holmes 1996), but absent Gunung Leuser NP Mar 1995–Dec 2000 (Buij *et al.* 2006). Now regular south Cambodia (Robson 2010). Rare Java (MacKinnon & Phillipps 1993); scarce Borneo (Mann 2008). Not globally threatened (BirdLife International 2011).

Chrysococcyx

Little Bronze Cuckoo *Chrysococcyx minutillus*. **Fig. 1.** Adult male, *C. m. minutillus*, Broome, Western Australia (*Adrian Boyle*). **Fig. 2.** Adult male, *C. m. peninsularis*, interior of Tambun, Perak, Malaysia, October (*Amar-Singh HSS*). **Fig. 3.** Adult female, *C. m. peninsularis*, interior of Tambun, Perak, Malaysia, October (*Amar-Singh HSS*). **Fig. 4.** Juvenile male, *C. m. peninsularis*, Ingham, Queensland, Australia, May (*Rohan Clarke*). **Fig. 5.** Juvenile, *C. m. peninsularis*, fed by Olive-backed Sunbird, Singapore, March (*Lee Tiah Khee*). **Fig. 6.** Adult male, *C. m. rufomerus*, Romang Island, Lesser Sundas, Indonesia, October (*Colin Trainor*).

GOULD'S BRONZE CUCKOO
Chrysococcyx poecilurus   Plate 28

Chrysococcyx poecilurus G.R. Gray, 1862 (Misool)

Alternative names: Rufous/Rufous-breasted Bronze Cuckoo.

TAXONOMY *Poecilurus* has priority over *russatus* Gould 1868. Many authors (e.g. Payne 2005) consider it conspecific with *minutillus* (see under that species). Parker (1981) separated *poecilurus* as monotypic species, and placed all other forms (here included in *poecilurus*) in *russatus*. Birds from Timor discussed by Hartert & Stresemann (1925) and Parker (1981) may well be an undescribed endemic form. Further work required. Polytypic. Synonyms: *Lamprococcyx poeciluroides* Salvadori, 1878 in nominate.

FIELD IDENTIFICATION 15–16cm. Very similar to Little Bronze Cuckoo. **Adult male** Forecrown grey-brown, crown, nape, sides of neck and upperparts glossy green-bronze, in sunlight profusely iridescent orange and purple, uppertail-coverts edged rufous; hint of white eyebrow before eye, little more pronounced behind eye, sides of head and underparts white, barred and mottled dark on cheeks, ear-coverts and chin, rest barred blackish, boldest on undertail-coverts, with diagnostic rufous tinge on throat and upper breast; uppertail mainly rufous with diffuse dark brown subterminal band, underside of tail rufous with black-and-white subterminal spots, diffuse on central rectrices. Iris red-brown to red. **Adult female** As male, but iris dark brown or yellowish-brown, with yellow to pale green eye-ring. **Juvenile** Poorly known – see Description. **Similar species** Little Bronze Cuckoo has more white flecking on forecrown and lores, white on superciliaries more pronounced and ear-coverts darker, less glossy green on top of head, nape and upperparts with only slight bronze iridescence in strong light, but contrast between crown and upperparts strong, and weak in Gould's; white spots on sides of rump, fringes of wing-coverts mostly pale buff to whitish, but strongly rufous in Gould's; normally lacks any rufous tinge on white underparts, banding below finer than Gould's and size of black bars on second outermost rectrices large, upperside of tail glossy green in Little, rufous in Gould's, and in flight Little has no trace of rufous on underwings. Bill narrower at base than Gould's. Juvenile has less rufous in rectrices and rarely barred on flanks. Shining Bronze Cuckoo female has brownish crown and nape contrasting with glossy green upperparts, less pronounced in male, broad dark barring, no rufous in tail and pale to dark brown iris. Horsfield's Bronze Cuckoo has in all plumages dark decurved eye-streak, brown on top of head and nape, male has streaked, and female pale brown mottling on throat and iris brown to red, eye-ring grey. Rufous-throated Bronze Cuckoo has diagnostic rufous face, chin and throat. Female Violet has pale rusty suffusion below, particularly on chin to breast where the barring is much closer, crown more bronze, with little or no white speckling, but eye-ring red as in both Little and Gould's. In north Queensland to Cape York, and Port Moresby region, PNG, intermediates between Little and Gould's occur, varying much and often impossible to determine and therefore best identified by combination of call and habitat; whistling call shorter in Little, 4–6 notes against 4–8 in Gould's, and Little prefers more open and drier habitat such as open forest, forest edge and woodland versus more dense vegetation in riverine forest, thickets, monsoon rainforest and mangroves in Gould's. Pied Bronze Cuckoo has dark sides to face, and males and some females have white wing patch. Banded Bay and juvenile Brush Cuckoos much larger.

VOICE In Australia plaintive grasshopper-like high-pitched trill, occasionally descending; 4–8 note plaintive whistle of *tew-tew-tew-teew...* (Higgins 1999). On Sulawesi high-pitched whistled trill lasting 2sec, or tinkling *tete-te-te-te*. On Timor *kiri-kiri-kiri-kiri* (Coates & Bishop 1997).

DESCRIPTION *C. p. russatus*. **Adult** Forecrown, crown and nape glossy olive-green, with bronze iridescence in sunlight, some have few white or buffish spots on forecrown and brown lores, suggestion of buffish to white superciliary, ear-coverts varying from brown to dark brown often mottled white; upperparts iridescent olive-green with metallic bronze and in strong sunlight tinge of orange and purple, uppertail-coverts same colour but edged rufous, occasionally with some white spots on sides; remiges, alula and primary-coverts dark brown, flight feathers edged narrow rufous in fresh plumage, secondaries keep this colour even when worn, wing-coverts same colour as upperparts but with rufous fringes; underwing glossy grey and broadly edged pale rufous on remiges with broad pale wing-bar at base of secondaries and inner primaries, in some wing-bar is pale rufous, in others whitish at base becoming pale rufous distally, underwing-coverts white with dark brown barring with green or bronze gloss, most of them tipped or edged rufous; underparts mainly white with dark brown 1.5–2.5mm wide barring with some purplish gloss in sunlight, broadest on undertail-coverts and narrower on chin, throat and in midline, in some broken on belly and vent, diagnostic strong rufous-brown wash on foreneck and breast, palest at breast centre; tail rufous above with diffuse subterminal dark brown band, T5 with 4–5 blackish-brown bars on inner webs distally with white patch on all except central feathers, on outer webs less regular and dark brown, more olive-brown distally, undertail as uppertail except olive-brown areas light grey. **Intermediates** In Cape York 50%, and in wet tropical Queensland, 20%, of population appear to varying degrees intermediate between Gould's and Little, as are few to north NSW in range of *C. minutillus barnardi*, and identification problematic; sometimes considered hybrids between the two species, or variation within Gould's. Most common is 'Gould's-like' intermediates with more white on face and particularly on supercilium, upperparts with strong bronze gloss, upperwing-coverts with narrower rufous fringes but rufous tips to underwing-coverts like Gould's, with less rufous but more bars below, less rufous on uppertail and only on feather edges, underside of T5 less rufous, and white areas same breadth as rufous bars or broader giving nearly black and white impression. 'Little-like' intermediates differ from Little in having small white spots on forecrown, rufous fringes to greater secondary-coverts broader, rufous or buffish tips to greater underwing-coverts and underwing bar washed rufous, and only suggestion of pale rufous on breast and sides of foreneck (Ford 1981, Parker 1981). **Juvenile** Not well known, only two skins available and these perhaps hybrids: glossy olive on top of head, rest of head pale greyish-brown, upperparts faintly glossed olive with some rufous-brown tips to feathers; underparts plain white with grey-brown on breast, flanks faintly barred with pale brown chevrons, tail as adult. **Nestling** Naked at hatching with black skin and four white tufts on nape. **Bare parts** Iris red-brown to red, female dark brown or yellow-brown to dirty cream, juvenile grey-brown, broad orbital ring scarlet to orange-red, female pale ochre-yellow to pale green, juvenile narrower and grey to pale yellow. Bill broader than *minutillus*, dull grey-black, base of lower mandible often blue-grey, juvenile dark grey, base of lower mandible buff-yellow, mouth off-white to pale yellow. Feet of adults and juvenile dark grey.

BIOMETRICS *C. p. russatus*. Wing male (n = 17) 92–102mm, mean 95.9 ± 3.0, female (n = 7) 94.0–99.7mm, mean 96.7 ± 1.7. Bill male (n =16) 15.9–19.3mm, mean 18.0 ± 1.0, female (n = 6) 18.5–20.3mm, mean 19.4 ± 0.7. Bill width male 5.0–6.8mm, mean 6.2, female 6.5–6.8mm, mean 6.6. Tail male 56–72mm, mean 63.9 ± 3.86, female 62–68mm, mean 64.9 ± 2.5. Tarsus male 14.3–16.8mm, mean 15.7 ± 0.6, female 15.7–16.9mm, mean 16.2 ± 0.5. Mass male (n = 7) 16.9–20.2g, mean 18.5 ± 1.0, female (n = 4) 17–21g, mean 18.8 ± 1.7 (Higgins 1999).

MOULT Just after wet season Sep–Apr, sometimes finishing May–early Jun. Female in breeding condition and body moult, 4 Jul. Sequence of primary moult: P7 and P9, then P5–P8–P6, and P1 and P4 simultaneously, P2–P3 then P10 but much variation. Post-juvenile moult started in fledgling still fed by hosts, early Dec (Stresemann & Stresemann 1961, Thompson 1966, Higgins 1999).

GEOGRAPHICAL VARIATION Five subspecies.

C. p. poecilurus G.R. Gray, 1862. West and south-west NG at Mimika R., Setakwa R. and ?Merauke, and Misool I; Wetar?; Timor? Male glossy dark green crown, rest of upperparts mid-green lacking rufescent or purplish iridescence; ventral barring as *misoriensis*, but centre of lower belly unbarred; flanks and head frosting similar to *misoriensis*; T1 outer web light green edged rufous, inner green rufescent distally, T2 outer web dull green with brownish subterminally, inner web rufous with black subterminal band and small white apical spot, T3 as T2 but two black bands in rufous area of inner web and apical spot larger, T4 as T3 but rufous areas on inner web have white spot or band, T5 has outer web banded grey and white proximally, last two or three grey bands form dusky rufous-tinged area reaching tip and edge, inner web banded black and white with white apical spot. Female has crown concolorous with back, pale rufous tinge to breast and head-frosting; eye-ring pale greenish-white (Parker 1981).

C. p. aheneus (Junge, 1938). North and east Borneo and southern Philippines. Crown green with no, or only slight, contrast to upperparts which are more purplish instead of rufous and more strongly metallic than other subspecies, some white on forecrown, supercilium more distinct, freckled white, below and underwing-coverts white barred blackish, no rufous tinge on breast and throat, tail fuscous-green washed rufous on outer web of T5 and inner web banded black and white, T4 purplish-fuscous on outer web, inner web rufous, rest fuscous glossed green or purplish. Female duller with no darker green on crown and pale to yellow-ochre eye-ring. Wing male (n = 8) 90.8–98.3mm, mean 93.8 ± 2.5, bill longest of all subspecies, mean 18.9mm, and wider (5.2–6.3mm) than sympatric *minutillus cleis* (Parker 1981). Strong bronze gloss separates most from *albifrons*; very little white on forehead and restricted to area before eyes; barring on breast and abdomen as in nominate. BMNH specimen from Mt. Kinabalu, Borneo, has head much darker bluish-green, rest of upperparts greener, much darker than nominate and perhaps represents undescribed subspecies. Specimen collected in daylight at 1,300m, Fraser's Hill, peninsular Malaysia, 17 Dec 1969, tentatively identified by S.A. Parker as this form (Wells 1999).

C. p. jungei (Stresemann, 1938). Sulawesi, Madu, Flores, Alor; Wetar?; Timor? As last; more purplish above but duller, less iridescence and fewer white feathers on grey-brown forecrown, underparts barred fainter than other subspecies, crown concolorous with upperparts, T5 broadly banded black and white, rarely with rufous trace. Wing male (n = 5) 87.6–91.4mm, mean 89.6 ± 1.5, tail shorter 55.7–60.0mm, mean 58.3 ± 1.6, against mean 62–67mm in

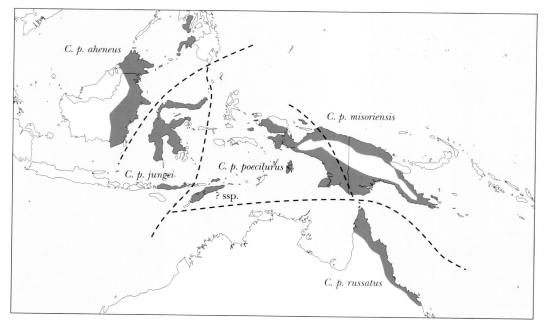

all other subspecies (Parker 1981, Trainor 2005b, Trainor *et al.* 2008a).

C. p. misoriensis (Salvadori, 1876). NG, and islands, except where nominate occurs. Forehead greyish, white mottling of lores, forehead and eyebrow reduced or missing; upperparts with more rufescent bronze-green gloss, wings lack white; most with faint rufous tinge on sides of breast, rest of underparts more heavily barred than nominate, tail less rufous. Female has duller crown, and barring on underparts less pronounced. Great variation as in nominate. Wing male (n = 15) 88–95mm, mean 92.2 ± 2.1 (Parker 1981); female 84mm (Payne 2005).

C. p. russatus Gould, 1868. Coastal and sub-coastal north-east Queensland from Cape York to Bowen, Townsville and Ayr; Hichinbrook I., Sir Charles Hardy Is, and Bushy I.?; some winter west and south-west coast NG from Merauke to Port Moresby district, Daru I. and Fergusson I. (Parker 1981). Most rufous subspecies. Hybrids or great individual variation commonly seen where plumage intermediate between Little and Gould's Bronze Cuckoos. Described above.

DISTRIBUTION Australasian and Oriental. Queensland from top of Cape York along east coast south to Pleystowe, south of Bowen; west border between Moonlight Creek and Mcarthur River in Gulf of Carpentaria; islands in Torres Strait and offshore islands of east coast. Sedentary through most of its Australian range, longer winged southern population maybe migratory to north Cape York and south NG (Parker 1981), but not proven and difficult to judge because bronze cuckoos are silent and secretive in winter (Ford 1981). In Atherton region population only 7.5% in winter compared with summer (Blakers *et al.* 1984). Intermediates between *poecilurus* and *minutillus* found from north of Watson R. on west side of Cape York to Bowen on east coast of Queensland and to north end of Cape York. Within this area there are three sites where Gould's strongly dominates: 1) Cairns to Cardwell, 2) Iron Range to McIlwraith Range, and 3) tip of Cape York Peninsula (Ford 1981). In NG nominate found from more widely separated sites and subspecies *misoriensis* inhabits lowlands, and islands of Aru, Batanta, Biak, Dampier, Daru, Fergusson, Goodenough, Karkar, Manam, Tarawai, Trobriand (?error, E. Mayr *in litt.* to Parker 1981), Vulcan, Waigeu; Misool off west NG; central and south-west Sulawesi, Madu, Flores, Timor. In Philippines on Basilan, Bongao, Mindanao, Negros, Sanga Sanga and Tawi-Tawi. On Borneo in Sarawak, Sabah and Kalimantan; others recorded in Sarawak, Sabah, Kalimantan and Maratuas could be this species or Little Bronze Cuckoo (Mees 1965, Parker 1981, Coates 1985, White & Bruce 1986, Buck 1988, Smythies & Davison 1999, Kennedy *et al.* 2000, Mann 2008).

HABITAT In Queensland prefers wet rainforest, but many other habitats reported: dense forest along riversides, mangroves, monsoon forest, thickets of paperbarks, *Melaleuca* gallery forest, open forest, woodland near forest and dune woodland. In Wallacea primary and well-grown secondary forests and edge, open woodland, bamboo thickets, grassy savannas, slightly wooded cultivation and along tree-lined streets in towns. In Sulawesi open woodland and grassy hills; in south-east Sulawesi in wooded savanna, forest on ultra-basic rock, and cultivated areas with settlements. On Borneo probably same as Little, i.e. mainly lowlands in dipterocarp, riverine, peat-swamp and secondary forest, open woodland and grassy hillsides, *Albizia* and cocoa plantations. In NG to 150m, once *c.* 700m. On Sulawesi from lowland to 800m, Flores 1,150m and Timor to 300+m (Stresemann 1940–41, Ford 1981, Parker 1981, White & Bruce 1986, Coates & Bishop 1997, Higgins 1999, Wardill *et al.* 1999, Mann 2008).

BEHAVIOUR Usually solitary and mostly recognised when calling from an exposed high perch, but can also be seen in company with other cuckoos at caterpillar infestations. Two cuckoos, supposedly paired, stayed near nest of Olive-backed Sunbird for more than one week and laid an egg before nest was finished, and sunbirds took no notice and continued building; female cuckoo seen twice coming to nest of sunbirds with egg in her bill; female reported clinging to nest of gerygone without entering nest which contained cuckoo egg; adult seen carrying egg of Fairy Gerygone (Barnard 1926, Seaton 1962, Coates 1985, Coates & Bishop 1999, Higgins 1999).

BREEDING Parasitism Nearly all records of known hosts do not separate Gould's and Little Bronze Cuckoos, therefore following list from Queensland are '*malayanus* cuckoos' hosts with either egg, young in nest or fledgling fed by foster parents, with incidences where recorded. Red-backed Fairy-wren (4); Honeyeaters: Bar-breasted (1), Brown-backed (1), Lewin's (1), Mangrove (1), Red-throated (1), Yellow (1) and Macleay's (1); Gerygones: Large-billed (80), Fairy (16), White-throated (4), Brown (3) and Mangrove (2); Northern Fantail (1), Spectacled Monarch (1), Ashy Robin (1), Olive-backed Sunbird (7), Double-barred Finch (1) and Zebra Finch (1) (Brooker & Brooker 1999). Rufous-sided Gerygone (P.E. Lowther *in litt.*). On Flores Golden-bellied Gerygone common host (Ottow & Verheijen 1969). However, Higgins (1999) does not mention fairy-wrens, honeyeaters or finches as hosts. **Season** 83% '*malayanus*' breed Oct–Feb, rest Jul–Mar north-east Australia (Brooker & Brooker 1989b). In north Borneo females in breeding condition Jul and Nov, and male Aug (Thompson 1966). **Eggs** Elongated to oval, pointed and glossy; uniform brown, or buffish to buffish olive freckled dark brown, most at pointed end; 20.0x12.5mm in nest of Large-billed Gerygone (Mack 1930, Beruldsen 1980); 19.2–21.2x13.3–14.3mm, (n = 5) mean 20.0x13.6 (Higgins 1999). **Incubation** No information. **Chicks** Hatch naked with four white tufts of down on crown as chicks of Large-billed Gerygone (Higgins 1999). **Survival** From four nests one hatched and fledged (Barnard 1926, Seaton 1962).

FOOD Caterpillars, beetles, bugs and sawfly larvae (Coates & Bishop 1997, Higgins 1999).

STATUS AND CONSERVATION Status in NG difficult to estimate because until recently considered conspecific with Little Bronze Cuckoo, but nominate probably only rare migrant in southern part of island (Beehler *et al.* 1986). *Misoriensis* widespread in NG and nearby islands, but otherwise status unknown. Nominate uncommon Kanganaman region, NG (Gilliard & LeCroy 1966). In Queensland no estimates due to misidentification (Higgins 1999). Status in Borneo unclear, presumably resident (Smythies & Davison 1999, Mann 2008) and said to be common in cocoa plantations at 190m, Sabah (Thompson 1966). Common and widespread Sulawesi and in hills of Flores; moderately common Timor (Coates & Bishop 1997). On Sulawesi 16 individuals heard over two hours, Jul (Andrew & Holmes 1988). In Philippines rare and poorly known (Kennedy *et al.* 2000). Not globally threatened (BirdLife International 2011).

Gould's Bronze Cuckoo *Chrysococcyx poecilurus*. **Fig. 1.** Adult female, *C. p. russatus*, Kewarra Beach, Queensland, Australia, July (*Greg Miles*). **Fig. 2.** Adult male, *C. p. jungei*, Tambun, Sulawesi, Indonesia, October (*David Beadle*). **Fig. 3.** Adult male, *C. p. russatus*, Torres Strait, Australia, July (*Rohan Clarke*). **Fig. 4.** Unsexed juvenile, *C. p. russatus*, Torres Strait, Australia, September (*Rohan Clarke*).

PIED BRONZE CUCKOO
Chrysococcyx crassirostris   Plate 28

Lamprococcyx crassirostris Salvadori, 1878 (Tual, Little Kei I.)

TAXONOMY MtDNA analysis places it close to *minutillus* and *meyeri*; included in former (Sorenson & Payne 2005). *C. malayanus salvadorii* (Hartert & Stresemann 1925) is intermediate, and may be hybrid between this species and *rufomerus* (Ford 1981, Parker 1981), but here tentatively accepted. Polytypic.

FIELD IDENTIFICATION 15–16cm. Very small cuckoo; white below with little barring. **Adult male** Deep blue-green above, without bronze cast; very little or no barring below; conspicuous white wing patch. **Adult female** Brown, slightly bronze above, less often dull green, some with white wing patch; white below with indistinct barring on flanks. **Juvenile** Rufous above, white with some barring below; outer rectrices barred black and white. **Similar species** From Little, Gould's and Horsfield's Bronze Cuckoos by dark face, and male, and some females, by white patch on wing. Juvenile unbarred below.

VOICE Rapid trill of high notes on one pitch, increasing in volume then fading away. On Tanimbar Is. descending series of 3–4 short, moderately high-pitched *pi* notes that tails off towards the end lasting 1.4sec, and repeated at intervals of *c*. 10sec to many minutes (Coates & Bishop 1997).

DESCRIPTION Nominate. **Adult male** Face, nape and upperparts blackish-blue glossed green, wings dark blue, white patch on greater and median secondary-coverts, underparts white, occasionally with few light brown bars on flanks, rarely sides of throat and breast; T1 blue-black, T2–4 blackish with white spot on tip of outer web, T5 black tipped and barred white. **Adult female** Crown brownish tinged green; rest of upperparts above dull oil-green, including wings, which may or may not have white patch; face brown or brownish-green, breast and flanks white indistinctly barred brown, centre of belly unbarred white; rectrices brown marked as male. **Juvenile** Face, head, back, wings and tail rufous; T1 with grey subterminal band, T5 barred black and white on inner webs; underparts unbarred white. **Nestling** Undescribed. **Bare parts** Eye-ring red; iris brown to red; bill black; feet grey to black.

BIOMETRICS Nominate. Wing male (n = 9) 83–97mm, mean 90.0 ± 3.9, female 88, 90, 91mm; tail male 57–60mm, mean 58.6 ± 1.2, female 56, 61, 61mm; bill male 14.2–15.4mm, mean 14.8 ± 0.4, female 13.4, 14.6, 14.9mm; tarsus male 14.2–15.4mm, mean 14.8 ± 0.8, female 13.4, 14.6, 14.9mm (Payne 2005).

MOULT No information.

GEOGRAPHICAL VARIATION Two subspecies.

C. c. crassirostris (Salvadori, 1878). Range of species except for Babar I. Described above.

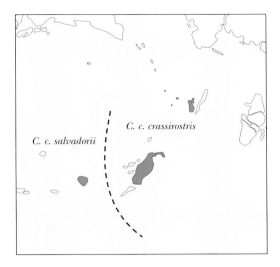

C. c. salvadorii (Hartert & Stresemann, 1925). Babar I., Lesser Sundas; perhaps migratory. As nominate, but male is greener above, underparts irregularly, but more extensively barred brownish or blackish, particularly on flanks than nominate. Juveniles very similar to immature nominate but former duller, more brownish-green on head, nape and upperparts, and uppertail-coverts dull oil-green. Wing male 89.0, 92, 96.6mm. Possibly hybrid (see Taxonomy).

DISTRIBUTION Wallacea. Apparently sedentary on Lesser Sundas on Tanimbar Is. (Larat, Yamdena) and Babar, and Moluccas on Tayandu Is. (Kur, Taam), Kai Is. (Tual, Rumadan); perhaps Seram Laut (Gorong), Halmahera and Ambon, and Sorong (extreme west NG). Records from Ambon, Gorong, Halmahera and Sorong questioned because putative host, Rufous-sided Gerygone, unknown from these islands, but records could be due to migration or dispersal (Parker 1981, White & Bruce 1986). Specimen from Ternate is misidentified *C. poecilurus* (Payne 2005).

HABITAT Woodland, scrub and forest edge on smallish islands within range of the Rufous-sided Gerygone (Coates & Bishop 1997).

BEHAVIOUR Singly, and in groups of 2–4 birds. Mainly frequents the treetops; often calls from exposed bare branches; sometimes feeds on caterpillars in low forest-edge scrub.

BREEDING No information. Rufous-sided Gerygone suggested host (Parker 1981).

FOOD Caterpillars (Coates & Bishop 1997).

STATUS AND CONSERVATION Common Yamdena (Coates & Bishop 1997). Not globally threatened (BirdLife International 2011).

Pied Bronze Cuckoo *Chrysococcyx crassirostris*. **Fig. 1.** Adult male, *C. c. crassirostris*, Kai Islands, Lesser Sundas, Indonesia, July (*Jon Hornbuckle*).
Fig. 2. Adult female, *C. c. crassirostris*, Kai Islands, Lesser Sundas, Indonesia, July (*Jon Hornbuckle*).

Genus *Misocalius*

Cabanis & Heine, 1863 *Mus. Hein.* Th. 4, Heft 1: 16. Type, by monotypy, *Cuculus palliolatu*s of authors, not of Latham = *Chalcites osculans* Gould 1847.
One species.

Australasian; brood-parasitic; sexes similar. MtDNA analysis shows *osculans* to be deeply embedded within *Chrysococcyx* (Sorenson & Payne 2005), in which genus those authors place it, but markedly different plumage, and eggs that have superficial pigments, support separate genus. Includes *Owenavis* (Mathews 1912).

BLACK-EARED CUCKOO
Misocalius osculans   Plate 27

Chalcites osculans Gould, 1847 (New South Wales)

TAXONOMY Monotypic. Synonym: *Owenavis o. rogersi* Mathews 1912.

FIELD IDENTIFICATION 19–20cm. Small cuckoo with conspicuously broad white supercilium, a still broader blackish eye-stripe and faint glossy bronze tinge on upperparts. Tail rather long and slightly rounded. Flight low, swift and silent. **Adult** Top of head, nape, upperparts and wings brown-grey with fine bronze iridescence on back and wings, rump paler; whole underparts unmarked buffish to cream, tail black-brown tipped white. **Juvenile** As adult but black facial stripe paler. **Similar species** In Australia Horsfield's Bronze Cuckoo has the head pattern of Black-eared, but adult birds easily identified by their more metallic green upperparts, heavy transverse dark barring on underparts and rufous on base of tail which is not tipped white; young birds more similar to Black-eared but more glossy green on upperparts without paler rump, and diagnostic rufous on tail.

VOICE Fine, far carrying descending and mournful call *peeeeeeer* or *feeeeeee*, repeated up to about eight times. Courtship call staccato *pee-o-wit, pee-o-weer.* Number of males together utter animated *pee-o-wit-pee-o-weer.* Soft *cheep-cheep* from bird perched in dead tea-tree. Young fed by fieldwrens uttered low twittering note like host's (Alexander 1925, Hall 1974, Johnsgard 1997, Pizzey & Knight 1997, Higgins 1999).

DESCRIPTION Adult (fresh plumage) Forehead, crown, nape and hindneck brown-grey bordered black on top of head; broad white supercilium running from base of upper mandible, over eye ending at side of neck, some birds with buffish wash on supercilium nearest bill and rest finely tipped brown-grey making supercilium less visible; lores with black patch in front of eye but behind eye large blackish decurved ear-patch, and below eye white moustachial stripe to below ear-patch; upperparts brownish-grey with faint bronze-green or olive-bronze gloss only visible in bright light, rump and uppertail-coverts paler unglossed, last with white fringes and tips on lateral feathers; wings dark brown on remiges, alula and primary-coverts with light rufous-brown band on middle of inner webs of primaries and base of secondaries, concealed on closed wing, remiges have narrow off-white fringes, buffish on outer edges of secondaries; wing underside warm buffish on lesser and median-coverts, greater coverts glossy grey tipped buffish on primaries and orange-buff on secondaries, remiges glossy grey with concealed broad white wing band at base; chin and throat orange-buff becoming buff to light rufous-brown on rest of underparts except white undertail-coverts with broad dark brown bars or chevron-shaped spots; tail dark brown to brown-grey at fringes and tip and black-brown shafts, rectrices broadly tipped white; T5 with broad white bars on inner webs, T4 similar but bars less obvious, undertail glossy grey with more obvious white barring. **Adult (worn plumage when breeding)** Supercilium always white, more black on lores and ear-patch, often narrow black stripe along underside of white moustachial stripe, underparts pale buffish to off-white, sometimes with dark brown barring on thighs and rear flanks otherwise like adult in fresh plumage. **Juvenile** Like adult in worn plumage but top of head, nape and hindneck slightly paler and more grey, ear spot smaller and dark brown with no connection to eye, lores off-white with blurred brown spot near eye; upperparts without iridescence, wings with light brown fringes on primaries and secondary-coverts in fresh plumage which fade into grey-buff or off-white and soon lost because of wear; underparts plain grey-buff, undertail-coverts unbarred pink-buff, barring on outer rectrices paler and denser, e.g. on underside of T5 normally 4–5 bars, in adults 2–4, colour in fresh plumage off-white washed orange-buff. **Nestling** Hatches naked with black skin, eyes open by nine days. **Bare parts** Iris black-brown, rich brown to yellow-brown, female dark brown to dull olive-brown, eye-ring grey to dark grey; juvenile iris grey-brown. Bill grey-black to black, juvenile with yellow gape. Feet dark or light grey (Higgins 1999, BMNH).

BIOMETRICS Wing male (n = 26) 114–123mm, mean 118.4 ± 1.9, female (n = 16) 113–122mm, mean 116.9 ± 2.3. Bill male 17.3–21.3mm, mean 19.5 ± 1.1, female 17.2–21.2mm, mean 19.5 ± 1.1. Tail male 83–92mm, mean 88.2 ± 2.4, female 83–92mm, mean 87.3 ± 2.3. Tarsus male 16.9–20.9mm, mean 19.2 ± 0.8, female 18.2–22.0mm, mean 19.4 ± 0.9. Mass male (n = 9) 25.5–34.4g, mean 29.4 ± 3.5, female (n = 5) 27.0–38.5g, mean 32.2 ± 5.0 (Higgins 1999).

MOULT Adult has complete post-breeding moult. Remiges and rectrices moulted Nov–Mar. Primary moult not fully understood, skins show many different sequences, but up to four primaries can grow simultaneously on each wing and P10 follows when P9 complete. Post-juvenile moult starts moulting Feb–Jun, often without replacing remiges and rectrices before second summer (Higgins 1999).

DISTRIBUTION Chiefly Australia; main range south-east and south-west, but scattered records from most parts of continent, although few from arid inland parts, and also uncommon in fertile Cape York Peninsula. Rare visitor south NG. Breeds only in Australia, mostly south of 26°S. Authors divided in opinions about its movements from migratory to

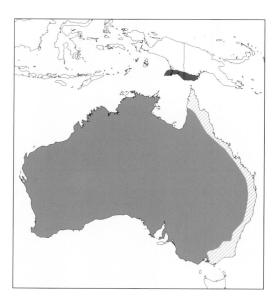

seasonally- or partly-nomadic to resident. Absent Jan–Aug in southernmost range, but year-round in north Australia. Twice north Tasmania. Most likely migratory in southern range, partly nomadic elsewhere. Two ringing recoveries (n = 79), both < 10km from ringing site (Ford & Stone 1957, Rounsevell et al. 1979, Draffan et al. 1983, Blakers et al. 1984, Frith 1984, Higgins 1999). Vagrant Aru, Bacan and Kai Is, Moluccas, and Romang, Luang and Babar, Lesser Sundas (White & Bruce 1986, Coates & Bishop 1997).

HABITAT Drier areas than *Chrysococcyx* species. More open woodlands and scrublands in interior of Australia, or drier coastal regions, never in dense forest. Drier woodlands dominated by *Eucalyptus*, mulga, mallee, broombush, *Lignum* or sheoak, saltflats with samphire or riverside thickets in north and central Australia, eucalypt woodlands with paperbarks and mangroves on coast, and tropical savanna woodland or open grassland; open *Acacia* woodlands with mulga, myall, boree and pindan (Friedmann 1968, Hall 1974, Pizzey 1981, Blakers et al. 1984, Higgins 1999).

BEHAVIOUR Solitary, occasionally two or three together or larger groups at beginning of breeding season when males chase each other noisily with spread tails (Hall 1974, Ryan 1980, Pizzey 1981). Mating described as follows: "The display appeared to be largely instigated by the hen calling continuously until male arrived. With his appearance the length of the hen's call was curtailed, and the tempo increased considerably. The cock joined in with similar calls and the hen commenced quivering her wings and crouching on the branch. The calls of both birds became excited and higher in tone than normal. Mating occurred on the branch upon which the hen had been perched" (Rix 1976). Mostly forages on ground, but also in trees and bushes (Higgins 1999).

BREEDING Parasitism 23 host species among 163 incidents of parasitism in Australia, most hosts with domed nests on or near ground, and dark brown eggs. Most common Speckled Warbler in eastern Australia (49.1%) and Redthroat in rest of breeding range in south, central and west (25.2%). However, figures may be biased as far more widespread Redthroat's nest is extremely difficult to find (Brooker & Brooker 1989). Other hosts: Variegated, Superb and White-winged Fairy-wrens, Yellow-rumped Thornbill, Shy Heathwren, Striated Fieldwren, Rufous Fieldwren, Yellow-throated Scrubwren, White-browed Scrubwren, Southern Whiteface, White-browed Babbler and Red-capped Robin. Superb and White-winged Fairy-wrens, Yellow-rumped Thornbill and White-browed Babbler have contrasting egg colours, and Red-capped Robin builds open nests (Higgins 1999). **Season** Jun–Oct WA, Aug–Dec east Australia; few arid inland breeding records Mar–Aug, as rain determines start of breeding by host species there (Brooker & Brooker 1989). Nest contained 3 eggs of Superb Fairy-wren, one of Shining Bronze Cuckoo and one of present species (Campbell 1906). **Eggs** Elongated oval; plain chocolate with reddish tinge, 19.6–22.4x14.2–17.5mm, (n = 5) mean 21.1x15.1. Egg-ejection behaviour not reported by any host species, presumably because egg greatly resembles eggs of Speckled Warbler but larger and more elongate (Brooker & Brooker 1989), though its pigment is superficial and easy to remove by rubbing with wet finger, which is not possible with eggs of Speckled Warbler (Gilbert & Keane 1913). Female removes host egg when laying (Friedmann 1968). **Incubation** Unknown but that of the hosts *c.* 12 days (Johnsgard 1997). **Chicks** Hatch blind and featherless and in this condition eject other eggs or nestlings (Chisholm 1973). All feathers in pin at 7 days, most feathers emerged at 12–14 days and fully fledged at 18 days (Higgins 1999). **Survival** No information.

FOOD Chiefly caterpillars and moths, also beetles and sandflies; yellow seeds (Hall 1974, Higgins 1999).

STATUS AND CONSERVATION Abundance poorly known. Brooker and Brooker (1989) found it uncommon. Harold Hall Australian expeditions 1962–70 collected only seven specimens (Hall 1974) and in most regional lists declared rare. Not globally threatened (BirdLife International 2011).

Black-eared Cuckoo *Misocalius osculans*. **Fig. 1.** Adult, Round Hill, NSW, Australia, September (*Simon Bennett*). **Fig. 2.** Juvenile being fed by its foster parent, a Speckled Warbler. Cocoparra, New South Wales, Australia, April (*David Cook*). **Fig. 3.** Juvenile, Broome, Western Australia, November (*Adrian Boyle*).

Genus *Rhamphomantis*

Salvadori, 1878 *Ann. Mus. Civ. Genova* 13: 459. Type by monotypy, *Cuculus megarhynchus* G.R. Gray 1858. One species.

Australasian; presumably brood-parasitic. Owing to bulging base of lower mandible *megarhynchus* has been considered most closely related to *Eudynamys* and *Microdynamys* (Ripley 1964); plumage, vocalisations and behaviour suggest a more close relationship to *Cacomantis* (Coates 1985); placed in *Chrysococcyx* on mtDNA evidence as sister to clade consisting of other Australo-Papuan species of that genus (Sorenson & Payne 2005). Bill shape among other characters support retention of this monotypic genus. Includes *Thelazomenus* (Reichenow 1915).

LONG-BILLED CUCKOO
Rhamphomantis megarhynchus   Plate 14

Cuculus megarhynchus G.R. Gray, 1858 (Aru Islands)

Alternative name: Little Koel.

TAXONOMY Gyldenstolpe (1955) and Rand and Gilliard (1967) claim that birds from Vogelkop Peninsula and Misool differ, but not supported by few specimens available (Payne 2005). Monotypic. Synonyms: *Thelazomenus poecilocercus* Reichenow 1915 and *m. sanfordi* Stresemann & Paludan 1932.

FIELD IDENTIFICATION 18cm. Small, rather dull cuckoo with long, thin bill with decurved tip and bulging base to lower mandible, square-ended tail, upright posture when perching, and straight non-undulating flight. Highly variable. **Adult male** Black head with conspicuous red iris and eye-ring, uniform dark brown upperparts, wings and tail; throat and upper breast greyish grading into dusky-buff or grey-brown. **Adult female** As male but more grey-brown on head and more cinnamon or red-brown above, with dark brown iris and grey eye-ring. **Juvenile** Variable. Broad circular area around eye and sometimes nape greyish-white, sometimes extending to streak behind eye; crown and ear-coverts darker; rest of upperparts brown to rufous-brown, tail sometimes barred. Upper breast brownish or greyish, lower breast pale rufous; rest of underparts pale rufous, in some faint dusky vermiculations on abdomen and bars on undertail-coverts; underside of tail barred dark and tipped white (Coates 1985). **Similar species** Brush Cuckoo has shorter and less decurved bill, head pale grey, and undulating flight. Female can be confused with Tawny-breasted Honeyeater which has naked pale yellow skin around eye, and in most subspecies, small yellow ear-stripe, slender bill and never upright posture.

VOICE Usually quiet. Loud and far-carrying trill in descending series lasting 4sec repeated at length every 5–7sec, similar to Little Bronze Cuckoo but louder; higher pitched and not so loud as Lesser Yellow-billed Kingfisher. Sweet, two-note upslurred whistle. Accented three-note whistle *hüo HÜTT tühü* (Stein 1936, Coates 1985, 2001, Beehler *et al.* 1986, J. Diamond in Payne 2005).

DESCRIPTION Adult male Head greyish-black, upperparts and wings uniform dark red-brown to dark sootybrown, remiges with buffish inner edges, underwing-coverts buff often washed pinkish; throat yellowish-grey to sooty grading to grey-brown on rest of underparts, variably tinged buff on belly and flanks; tail as upperparts but with narrow paler tips, faint subterminal band and strong barring on outer rectrices. **Adult female** As male but head dark grey-brown, cheeks to malars dark grey; upperparts, wings and tail dark cinnamon; throat to breast buff finely barred grey; lower, belly to undertail dull rufous, or grey finely barred rufous. **Juvenile** Head pale grey and brown with area around eyes and often nape greyish-white, often dark curved stripe from eye to upper ear-coverts; rest of upperparts rufous-brown to brown; throat greyish-white grading first to brown then to pale rufous on lower breast, belly creamy-brown, and undertail-coverts barred dusky; tail rufous-brown, some with faint barrings, rectrices below barred dark and tipped pale. **Nestling** Undescribed. **Bare parts** Iris and orbital ring bright red in male; female and juvenile dark to light brown iris, grey eye-ring in female, juvenile blackish. Bill black, juvenile brown with yellow gape. Feet greyish-blue, juvenile black to greyish-brown.

BIOMETRICS Wing male (n = 4) 96–100mm, mean 97.8, female (n = 6) 96–104mm, mean 99.1 ± 2.0. Bill male 22.4–24.8mm, mean 23.6, female 22.2–25.5mm, mean 22.6. Tail male 72–84mm, mean 79.1, female 66–82mm, mean 76.2 ± 4.8. Tarsus male 15.8–16.2mm, mean 16.0, female 14.8–18.6mm, mean 16.3. Mass female 31.5g (Payne 2005); juvenile 24g (Ripley 1964).

MOULT No information.

DISTRIBUTION Probably nomadic. NG, Aru, Waigeo and Misool Is. WP at Vogelkop, Mamberano R. and Humboldt Bay; PNG at Sepik R., Kumusi R. and Port Moresby, and Kiunga area, Western Province (Coates 1985, Gregory 1997).

HABITAT Lowlands in rainforest, forest edges, monsoon forest, secondary forest and scrub (Coates 1985).

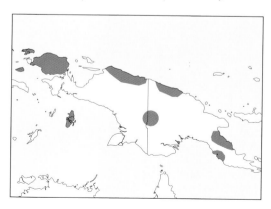

BEHAVIOUR Inconspicuous and usually solitary. Often with other cuckoo species and similar looking Tawny-breasted Honeyeater feeding at Crow Butterfly (*Euploea*) caterpillar-infested trees (Coates 1985).

BREEDING Season Female with oviduct egg, Sep (Payne 1997). No further information.

FOOD Caterpillars, including those of Crow Butterfly; winged insects (Coates 1985).

STATUS AND CONSERVATION Considered rare, occurring at few widely scattered sites, but probably overlooked due to its cryptic plumage and quiet habits (Coates 1985). Not globally threatened (BirdLife International 2011).

Long-billed Cuckoo *Rhamphomantis megarhynchus*. **Fig. 1.** Adult male, Boystown Road, near Kiunga, PNG, June (*Nik Borrow*). **Fig. 2.** Adult male. Kiunga, PNG, July (*Jerry R. Oldenettel*).

Genus *Cacomantis*

S. Müller, 1843 *Verh. Nat. Gesch. Nederl., Land- en Volkenk.*, part 6: 177, note. Type, by subsequent designation *Cuculus flavus* Gmelin = *Cuculus merulinus* Scopoli 1786 (Salvadori 1880, *Orn. Pap. delle Mol.* vol. 1: 331). Eight species.

Subsumed in *Cuculus* in past. Includes *Polyphasia* (Blyth 1843), *Penthoceryx* and *Heteroscenes* (both Cabanis & Heine 1863), *Ololygon* (Hume 1873) and *Vidgenia* (Mathews 1918). Oriental and Australasian; brood-parasitic. Small to medium-sized; long-tailed; round nostrils; plumage mostly browns and greys. MtDNA analysis places *Cacomantis* as sister-group to *Surniculus*, *Cercococcyx*, *Cuculus* and *Hierococcyx* (Sorenson & Payne 2005).

PALLID CUCKOO
Cacomantis pallidus   Plate 29

Columba pallida Latham, 1801 (New South Wales)

Alternative names: Unadorned Cuckoo, Brain-fever Bird (also used for *C. passerinus*, *C. merulinus*, *Hierococcyx varius* and *H. sparverioides*).

TAXONOMY Sometimes placed in *Cuculus* (White & Bruce 1986), or monotypic *Heteroscenes* between *Cuculus* and *Cacomantis* (e.g. Wolters 1976). Müller (1843) and Payne (2005) place it in *Cacomantis* due to vocalisations, unbarred underparts and mtDNA. Results of mtDNA studies show it to be sister to *leucolophus* (here placed in monotypic *Caliechthrus*), these two forming sister group to all other *Cacomantis* (Sorenson & Payne 2005). Monotypic. Synonyms: *Heteroscenes occidentalis* Cabanis & Heine 1863, *H. pallidus tasmanicus* Mathews 1916, *Cuculus variegatus* Vieillot 1817, *C. cinereus* Vieillot 1817, *inornatus* and *albostrigatus* both Vigors & Horsfield 1826, *Chalcites simplex* Lesson 1844 and *C. poliogaster* S. Müller 1845.

FIELD IDENTIFICATION 31cm. Medium-sized cuckoo. All ages have diagnostic dark eye-stripe and lack dark transverse barring on underparts. Often undulating woodpecker-like flight. **Adult male, pale morph** Head, nape and upper mantle pale grey with concealed white nape spot best seen in contrary wind, and contrasting dark grey eye-stripe which continues down sides of neck; rest of upperparts darker grey to grey-brown, rump spotted white; remiges and coverts like upperparts but finely edged white; underwing-coverts creamy with fine grey barring, underside of remiges barred white on inner vanes; underparts pale grey; tail dark brown, all rectrices except T1 spotted white. **Adult male, dark morph** Upperparts darker and browner, rufous area on nape and more white spotting on wings. **Adult female, dark morph** Top of head and nape blackish-brown streaked rufous-brown with white spot on nape, broad white and black eye-stripe; mantle dark brown fringed rufous, rest of upperparts paler brown mottled white, buff and rufous; wings dark brown spotted rufous-brown and pale buffish; underparts pale grey unmarked mottled rufous-brown on breast with pale grey barring on flanks and undertail-coverts, tail as male but all rectrices spotted or barred white or buffish. **Adult female, pale morph** Intermediate between pale male and dark female. **Juvenile** Head darker grey with prominent white supercilium, and blackish eye-stripe running down sides of neck to sides of breast; upperparts and wings grey heavily spotted and edged white; underparts white, chin, throat and breast streaked dark grey; tail as adult pale morph. **Similar species** Confusion only possible with much smaller Black-eared Cuckoo which is paler, ear spot darker and running only from lores to ears.

VOICE Monotonous 8–19 ascending whistled notes, or may first increase and then decrease or vice-versa, and often given continuously for up to one hour by male only. Pursuing males give rising staccato *pip-pip-pip-pip*; female a hoarse, brassy whistle. Female also gives harsh single *cheer*, or repeated higher pitched *wheeya*. Another call like horse, uttered by both sexes; during courtship feeding soft *keer-keer-keer*. In flight *kew-kew-kew-kew* when male flies towards female, or by flying female; also *toy-it-yer, toy-it-yer*. Often calls after rain or storm, hence vernacular weather names, and most often at nights with full moon (Higgins 1999). So-called 'scale call' described as "beads in water" (Lord 1956, Hall 1974, Keast 1993, Strahan 1994, Johnsgard 1997, Pizzey & Knight 1997, Higgins 1999, Campbell 1989 in Higgins 1999).

DESCRIPTION Variation between morphs perhaps continuous. **Adult male, pale morph** Forecrown, crown, nape and hindneck grey (when worn dark feather bases visible and forehead uneven pale brown), hidden white spot on nape, sides of head pale grey with striking grey-brown to blackish-brown eye-stripe running from upper mandible through eye to end of ear-coverts and down sides of neck; rest of upperparts plain grey-brown, feathers narrowly fringed paler; outer edges of longest uppertail-coverts spotted white; worn feathers browner, particularly near shafts, resulting in some striping on rump and uppertail-coverts; wing-coverts grey-brown edged whitish, broadest and more irregular on longest coverts and tertials, primaries blackish-brown in fresh plumage finely edged white around tips, secondaries grey-brown edged white when not worn, forewing with white patch. Underwing-coverts creamy with fine pale grey barring, primaries brown-grey with large white spots on inner webs on proximal end, merging together to a conspicuously broad white wing-bar at border to coverts, secondaries uniform grey-brown. Underparts pale grey becoming white on vent and undertail-coverts, latter narrowly barred grey-white. Tail dark brownish-grey, end white, T5 darkest broadly marked with triangular white to buffish spots, T2–4 incompletely barred white with small pale spots near shafts. **Adult male, dark morph** Slightly darker and more brown on head and upperparts with rufous spot below, white spot on nape; primaries narrowly edged or spotted white; pale spots, often with buffish wash, on secondaries, tertials and greater wing-coverts larger than in pale morph. **Adult female, dark morph** Top of head, nape and hindneck blackish-brown streaked pale rufous-brown in middle with small white spot on nape; broad white supercilium gradually becoming buffish near nape, dark brown to blackish eye-stripe running from middle of lores, through eyes and lower ear-coverts to sides of neck; some individuals have dark moustachial and malar stripes; rest of upperparts dark brown mixed with rufous-brown, paler on rump; wings as upperparts but heavily spotted rufous-brown; underparts whitish grey striped dark grey-brown stripes on chin, with gorget of dark rufous barring on lower throat and upper breast, most heavily on sides of neck; rest of underparts finely barred pale grey-brown. Tail as male but spotting more often buffish. In worn plumage all rufous-brown spotting become more buffish and ground colours greyish-brown. **Adult female, pale morph** Intermediate between dark morph of adult male and dark morph of adult female, but lacking gorget. **Juvenile** Differs from adult in more white spots on upperparts and wings. Compared with dark morph female: forehead and lores darker, nape and hindneck white with few small brown spots, blackish-brown eye-stripe bolder and continues to sides of breast; upperparts white with row of blackish-brown arrow-like spots on scapulars, rump white striped pale brown, wings dark brown spotted and fringed white, secondary coverts white with few dark brown spots; underparts white heavily streaked brown on chin, throat, breast and middle of belly. **Nestling** Undescribed. **Bare parts** Iris dark to medium brown, often with white or yellow outer ring, orbital skin yellow, iris of juvenile pale yellow to olive-green. Bill dull black to grey-black, base paler brown, green or buffish, juvenile grey, cutting edges and tip whitish, gape of adult yellow to orange, juvenile whitish washed pinkish. Feet olive to dull yellow; juvenile pale pink, grey, or olive-green to olive-brown.

BIOMETRICS Wing male (n = 69) 179–204mm, mean 193.0, female (n = 48) 182–204mm, mean 191.5. Bill male 23.8–29.3mm, mean 26.9, female 24.8–29.1mm, mean 26.8. Tail male 152–177mm, mean 163.4, female 152–178mm, mean 160.6. Tarsus male 19.5–25.3mm, mean 21.9, female 20.2–25.7mm, mean 22.2. Mass male (n = 22) 64–118g, mean 89.5, female (n = 18) 63–106g, mean 85.6 (Higgins 1999).

MOULT Post-breeding moult of wing and tail Jan–Apr, usually complete but one old secondary and 1–2 old primaries observed. Body feathers moulted earlier, normally Dec–Jan and completed Apr. Sequence of primary moult: outer group (P9,7)–P5–P8–P6, and inner group (P1,4)–P2–P3, both groups moulting at about same time (Stresemann & Stresemann 1961, Higgins 1999).

DISTRIBUTION Australasia and Wallacea. Main range within Australia, including Tasmania. Nearly ubiquitous Australia but rarer in north, most breeding records south of 20°S, chiefly Victoria, NSW and south-west WA, perhaps because these areas have most dense human population. Migration patterns poorly known in spite of wealth of sightings, but migration movements recorded from Murray-Darling district and Murphy's Creek, both along coast of east Australia. Most likely whole or part of northern population sedentary as breeding found in all months except Apr. Southern populations partly migrating to northern Australia, immatures later; few reach north-west and south NG, Ternate, Halmahera, Flores, Babar, Ashmore Reef and Timor (perhaps resident). Very rare winter, Tasmania (Rand & Gilliard 1967, Stronoch 1981, Mees 1982, Blakers *et al.* 1984, Coates 1985, White & Bruce 1986, Coates & Bishop 1987, Payne 1997, AMNH). Vagrants: Lord Howe I. 1877 (Hindwood 1940); Christmas I. 20 Oct 1977 (Stokes *et al.* 1987); Norfolk I. 23 May and Jun 1984 (Hermes *et al.* 1986); Macquarie I. 23 Sep 1990

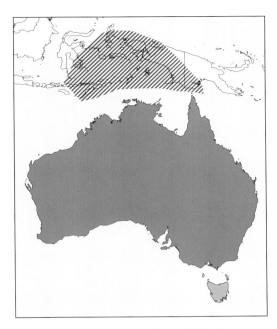

(Tasmanian Bird Report 20 in Higgins 1999); NZ at Craig Flat, Otago May–Oct 1939, 1940, 1941, Jan 1990, Okarito 1941, Greymouth 1942 and Wairarapa 1977 (Falla *et al.* 1981, Heather & Robertson 1996).

HABITAT Very catholic. Open woodland and scrub, often dominated by *Eucalyptus* and *Acacia*, farmland with grassy paddocks, along roadsides with remnants of trees and scrub and gardens, parks and suburbs, rail reserves, savanna woodland, coastal dune-scrub, dry creeks with gum trees, *Pinus* plantations, vineyards, orchards, mangroves, newly-logged forest and deserts with scattered shrubs. Near rivers, but more often in dry areas and least often in rainforest (Hall 1974, Blakers *et al.* 1984, Johnsgard 1997, Pizzey & Knight 1997, Higgins 1999).

BEHAVIOUR Usually solitary. Forages in foliage and on ground, in latter case takes small hops like Common Cuckoo. Post-breeding birds forage in mixed-species flocks. Flight either straight and rapid, or like woodpecker, undulating with periodic wing-closure. Courtship feeding during mating, both male and female seen feeding other sex, which has been misinterpreted as feeding fledgling. Female took food from bill of male and at same time quivered body and vibrated half-spread and lowered wings as begging young. After copulation both sexes rotate tails for 1–2min. Probably nocturnal migrant. Although common, no other information (Gilbert 1934, 1935, Robinson 1950, Kikkawa 1966, Crawford 1972, Hall 1974, Klapste 1981, Bell 1986, Higgins 1999).

BREEDING Parasitism Only 2% of parasited nests dome-shaped or cavity. 1048 records from 113 hosts; nearly 50% of hosts are honeyeaters with open cup nests. Following species recorded with either egg, young in nest or fledgling fed by foster parents. Honeyeaters: Bar-breasted (3), Black-chinned (11), Black-headed (39), Blue-faced (4), Brown (6), Brown-headed (31), Crescent (1), Fuscous (20), Grey-fronted (11), Grey-headed (2), Lewin's (9), Yellow-plumed (7), Mangrove (3), Painted (2), Pied (2), Purple-gaped (3), Red-throated (1), Regent (14), Singing (44), Spiny-cheeked (4), Striped (6), Strong-billed (16), Tawny-crowned (15), Varied (1), White-cheeked (2), White-eared (24), White-fronted (2), White-gaped (1), White-naped (41), White-plumed (51), White-throated (5), Yellow (2), Yellow-faced (57), Yellow-throated (15), Yellow-tufted (38) and New Holland (16); Little (3), Noisy (4) and Helmeted (2) Friarbirds; Eastern Spinebill (15), Western Spinebill (7); Little (17), Red (59) and Yellow (4) Wattlebirds; Yellow-throated Miner (major host WA), Crimson Chat (1), White-fronted Chat (6) – total *c.* 620 honeyeater records of 43 species. Rainbow Bee-eater (1) where chick was found in nest tunnel, White-throated Treecreeper (1), Southern Emu-wren (1), Striated Fieldwren (3); Large-billed (1), White-browed (2) and Tasmanian Scrubwrens (2); Brown Thornbill (2), White-browed Babbler (2), Chiming Wedgebill (2), Pied Butcherbird (1), Grey Butcherbird (1). Woodswallows: Black-faced (6), Dusky (17), Little (1), Masked (8), White-breasted (7) and White-browed (11). Black-faced Cuckoo-shrike (4), Cicadabird (1), White-winged Triller (13), Varied Sittella (2), Eastern Shrike-tit (4). Whistlers: Gilbert's (1), Golden (15), Mangrove Golden (1), Olive (1), Red-lored (1) and Rufous (29). Australasian Figbird (2), Olive-backed Oriole (12), Grey Shrike-thrush (15), Stripe-breasted Shrike-thrush (1), Crested Bellbird (4), Spangled Drongo (1), Willie-wagtail (35), New Zealand Fantail (4), Rufous Fantail (2), Magpie-lark (11); Leaden (8), Restless (9) and Satin (7) Flycatchers. Robins: Flame (3), Red-capped (4), Scarlet (6), White-breasted (1) and Eastern Yellow (8); Jacky Winter (13). Southern Scrub-robin (1), Red-whiskered Bulbul (4), Rufous Songlark (2), Australian Reed Warbler (5), Silver-eye (3), Eurasian Blackbird (4), House Sparrow (1), New Zealand Pipit (5), European Greenfinch (4), European Goldfinch (16). In WA parasitises large honeyeaters, mean mass 60g, but in eastern Australia small honeyeaters, mean 19g, as here larger Common Koel parasites the others (Cheney 1914, Brooker & Brooker 1989b, 1999, Johnsgard 1997, Higgins 1999, Johnstone & Storr 2004). **Season** WA Jul–Jan peaking Sep; eastern Australia mainly Aug–Jan peaking Oct–Nov; northern Australia mainly Aug–Jan, but recorded all months except Apr; Tasmania Oct–Dec, rarely Sep and Jan (Higgins 1999). **Eggs** Female removes one host egg at laying. Eggs elongated oval, glossy and flesh pink, some with few darker pink small spots at brooder end, closely similar to those of some honeyeater species; 20.8–26.2x15.0x19.6mm, (n = 126) mean 23.9x17.4 (Brooker & Brooker 1989b). **Incubation** 12–14 days (A.G. Campbell in Higgins 1999). **Chicks** Eject host eggs and nestlings or trample weak chicks to death as early as 48 hours after hatching. Many reports of fledglings being fed by species other than host parents or cuckoo fledgling flying to nest other than its host's and there being fed, the parents ignoring their own offspring. Also report of adult cuckoos six time bringing mistletoe berries to fledgling in spite of being mobbed by many passerines. Pair of Jacky Winters both attacked and fed fledgling, which even begged when attacked. Fledglings fed by hosts up to six weeks (Brendt 1905, Cole 1908, Cooper 1958, Kikkawa & Dwyer 1962, Chaffer 1973, Brooker & Brooker 1989b, Johnsgard 1997). **Survival** 32.4% eggs hatched and 13.5% fledged (n = 37). Most parasitised hosts have eggs with great similarity to Pallid Cuckoo's egg, but where there is little likeness rejection percentage high (Johnsgard 1997, Higgins 1999).

FOOD Mainly hairy caterpillars; also beetles, grasshoppers, butterflies, Bag Moth from Boree trees favoured on northbound migration in spring; spiders; earthworms; mistletoe berries. Plant foods given by hosts include berries of *Lantana* and nectar (Cooper 1958, Hobbs 1961, Hall 1974, Higgins 1999).

STATUS AND CONSERVATION Widespread and common Australia. Perhaps more sparse northern Australia as fewer records there, but more probably because lower human population. Density 0.25 birds/ha Inverell, NSW (Baldwin 1975), Boola Boola, Victoria, 0.22 birds/ha (Blakers *et al.* 1984). Not globally threatened (BirdLife International 2011).

Pallid Cuckoo *Cacomantis pallidus*. **Fig. 1.** Adult, Canberra, Australia, November (*Simon Bennett*). **Fig. 2.** Adult, Broome, Western Australia, December (*Adrian Boyle*). **Fig. 3.** Adult, King's Canyon, Northern Territory, Australia, August (*Rohan Clarke*). **Fig. 4.** Juvenile, Onslow, Pilbara, Western Australia, May (*Colin Trainor*). There may be continuous variation between pale and dark morphs, and perhaps also between the sexes, resulting in considerable difficulty in ascribing an individual in the field to a particular sex or morph.

CHESTNUT-BREASTED CUCKOO
Cacomantis castaneiventris   Plate 29

Cuculus castaneiventris Gould, 1867 (Cape York, Queensland)

Alternative names: (Chestnut-breasted) Brush-cuckoo.

TAXONOMY Polytypic. Synonyms: *bihagi* Mathews 1914 in *weiskei*; *Vidgenia yorki* Mathews 1922 in nominate.

FIELD IDENTIFICATION 22–24cm. Small cuckoo. Sexes similar. **Adult** Head dark grey with conspicuous pale yellow eye-ring, entire upperparts and wings dark blue-grey, underparts dark brown-rufous; tail above dark blue-grey, below tipped white with fine white spots along outer edges. **Juvenile** Head, upperparts and wings plain brown to buffish brown, underparts off-white to brownish on foreneck and breast. **Similar species** Fan-tailed Cuckoo larger, with only breast pale rufous, female duller, rest of underparts off-white to buffish, tail longer, with more and wider white bars below; juvenile striated and spotted brown on head, upperparts grey-brown, feathers edged rufous, below not barred but with arrow-like spots. Brush Cuckoo larger but with proportionally shorter tail, upperparts somewhat lighter, underparts grey tinged more or less rufous, undertail-coverts lighter rufous.

VOICE Mournful descending trill of <1sec, very similar to Fan-tailed Cuckoo and virtually indistinguishable from Lesser Yellow-billed Kingfisher, also rendered as grasshopper-like *chirrip*. Slow repeated *wee-oo-whew* or *seei-to saai*, second note lower and third ascending, resembling start of Brush Cuckoo's call, but much slower. Begging call of nestling high-pitched thin *siiiaar-swee-sweep*. Calls at various times of year (Coates 1985, Beruldsen 1990, Higgins 1999).

DESCRIPTION Nominate. **Adult** Head, nape, hindneck, upperparts and wings dark blue-grey, some males with faint brown edgings to some greater and median secondary-coverts, white area on carpal joint; remiges grey-brown, undersurface glossy dark grey with large white area at base of primaries (except outer two), continuing on secondaries to form large bar at base, underwing-coverts buffy-white; below uniform deep brown-rufous running from chin, foreneck to sides of neck in sharp contrast to grey head, chin in favourable light mottled grey, rarely totally grey; graduated tail slightly darker bluish-grey than wings, central rectrices narrowly tipped white, rest have slightly larger white tips with variable white barring, bars broadest near shaft on inner webs. **Juvenile** Forehead, crown, nape and hindneck brown, sides of head paler, more rufous-brown above, brown to bright rufous-brown on rump and uppertail-coverts, latter with slightly exposed shaft-streaks; remiges and wing-coverts dark brown finely edged and tipped rufous-brown, wing underside like adult but underwing-coverts dark-buff white, and cream and white wing-bar gradually becoming pale brown distally on primaries, last indistinctly barred rufous and dark brown; chin pale brown-rufous, throat off-white, or buffy-white, with diffuse pale brown tips to some feathers, breast plain dull brown, belly whitish, flanks and undertail-coverts light brown, rectrices tipped and edged rufous, T5 barred rufous. **Nestling** No information. **Bare parts** Iris dark brown, eye-ring pale yellow; juvenile iris black-brown or grey, eye-ring brown-grey. Bill blackish-grey with orange-pink base to lower mandible in some birds, mouth salmon to orange-red; juvenile bill black with brown wash around nostrils, base of lower mandible yellow-brown and gape buffish yellow. Feet yellow to orange-yellow, juvenile dull pink, claws black.

BIOMETRICS Nominate. Wing male 104.0–114.9mm, (n = 9), mean 110.2 ± 3.3, female 111, 114mm. Bill male 20.7–23.4mm, mean 21.6 ± 0.9, female 19.9, 21.7mm. Tail male 104.0–124.9mm, mean 115.2 ± 6.6, female 114mm. Tarsus male 17.2–19.8mm, mean 18.0 ± 0.8, female 15.1, 18.2mm. Mass male Jan 34g, Jul 31.9g, Nov 26, 28g, adult female Aug 32g (Higgins 1999); male 32–38g (n = 4), mean 35.8, female 35, 37, 38g (Diamond 1972b).

MOULT Full adult plumage attained at *c.* 1 year; in Australia active primary moult late Jul–Jan, moulting 2–4 alternating primaries simultaneously leaving at least one fully-grown primary between each moulting section (Higgins 1999). Interrupted wing moult, Australia, Jul; body moult, *arfakianus*, WP, Oct (BMNH).

GEOGRAPHICAL VARIATION Three subspecies.
C. c. castaneiventris (Gould, 1867). North-east Australia (Cape York and north-east Queensland) and Aru Is. Described above.
C. c. arfakianus Salvadori, 1889. West NG islands: Salawati, Misool, Jobi; WP from Vogelkop to Weyland and Snow Mts., and east to Fly R. in PNG. Underparts darker chestnut than nominate, upperparts glossed blue, sometimes indistinct dark barring on flanks; juvenile whiter below. Wing male 108–117mm (n =11), mean 112.5 ± 3.3, female 102–116mm (n = 9), mean 109.6 ± 6.0 (Payne 2005). Variable; some females tinged greenish on remiges (Junge 1937a).
C. c. weiskei Reichenow, 1900. PNG from Sepik region eastwards. Upperparts blackish markedly glossed blue-green, underparts often darker chestnut than nominate. Largest subspecies. Wing male 102–115mm (n = 15), mean 112.1 ± 3.9, female 102–118mm (n = 11), mean 112.3 ± 5.0 (Payne 2005).

DISTRIBUTION Australasia. Misool, Salawati, Aru I., NG and Yapen I.; Cape York Peninsula, south to Cooktown area, with few records further south at Mt. Lewis and Julatten. Probably some movements across Torres Strait as recorded Booby I. Dec, and two at sea west of Booby I. Oct. However, ringing records in NG and Australia suggest

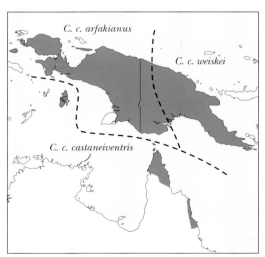

resident; of 62 ringed all 8 recoveries < 10km from ringing site (Higgins 1999).

HABITAT Dense moist monsoon forests, forest edges, *Eucalyptus* forest, scrub and vegetation along rivers, Australia. Bushy savanna, forest edges and forest, to 1,800m, NG; more often at lower elevations than sympatric Fan-tailed Cuckoo (Gilliard & LeCroy 1961, Coates 1985, Higgins 1999).

BEHAVIOUR Usually in canopy feeding among foliage, but also feeds on ground and takes insects by hovering. Solitary, more rarely seen in small loose feeding flocks and usually recognised by voice; sometimes mobbed by flycatchers (Coates 1985).

BREEDING Parasitism Following hosts recorded but must be taken with reservation because eggs of Brush and Fan-tailed Cuckoos very similar: Lovely Fairy-wren, Little Scrubwren, Large-billed Scrubwren and Crimson Finch (Higgins 1999). **Season** Particularly active Australia Jun–Aug, Sep–Oct and Dec–Jan; slightly enlarged gonads east NG, Jul–Aug (Diamond 1972b, Payne 1997, Higgins 1999, BMNH). **Eggs** Near oval, ground colour whitish with fine freckles of brown to pinkish; 20.8x14.7mm (Higgins 1999). No further information.

FOOD Beetles, grasshoppers, caterpillars; ticks (MacGillivray 1914, Hall 1974).

STATUS AND CONSERVATION Status little known. Fairly common to common NG (Coates 1985), but uncommon eastern highlands (Diamond 1972b). Uncommon Australia (Pizzey & Knight 1997); mainly Cape York region, rarely farther south (Higgins 1999). Not globally threatened (BirdLife International 2011).

Chestnut-breasted Cuckoo *Cacomantis castaneiventris*. **Fig. 1.** Adult. Tabubil, PNG, August (*Mark Harper*).

FAN-TAILED CUCKOO
Cacomantis flabelliformis Plate 29

Cuculus flabelliformis Latham, 1801 (Sydney area, NSW)

Alternative names: Ash-coloured Cuckoo, Fan-tailed Brush Cuckoo.

TAXONOMY Previously known as *C. pyrrhophanus* but *flabelliformis* has priority (Mason 1982). Incorrectly *pyrrophanus*. Breeding female from Kulombangra (Kolombangara) I., central Solomons, is "distinct in plumage and smaller than other cuckoos of this species, wing 116, tail 127" (Payne 2005). Polytypic. Synonyms: *Sylvia rubricata* Latham 1802, *Cuculus rufulus* Vieillot 1817, *prionurus* Lichtenstein 1823, and *cineraceus* and *incertus*, both Vigors & Horsfield 1826, *rubricatus athertoni* and *r. albani* both Mathews 1912, and *r. eyrie* Mathews 1918, all in nominate; *infuscatus* (black morph) Hartlaub 1866 in *simus*; *Cacomantis bronzinus* Gray 1859 and *meeki* Rothschild & Hartert 1902 in *pyrrhophanus*.

FIELD IDENTIFICATION 24–28cm. Medium-sized cuckoo with conspicuously long, rounded tail. Flight strong, with fast wingbeats and short glides; perches upright. **Adult** Head and neck ash-grey, eye-ring yellow; upperparts and wings uniform blue-grey, chin and upper throat ash-grey, rest of underparts pale buffish to pale rufous, breast strongly washed rufous; long fan-shaped tail blue-grey tipped white with many conspicuous broad white bands on underside. Black morph (*simus* only) mostly blackish to dark brown; white on underwing-coverts, and bar on underside of remiges; tail lacks white. **Juvenile** Head, hindneck, upperparts and wings dark brown, each feather indistinctly edged dark rufous; underparts off-white with brown tinge from chin to breast; and brown arrow-like spots from throat to belly, undertail-coverts edged brown; tail dark grey tipped white and banded buffish below. **Similar species** Chestnut-breasted Cuckoo smaller, has bright chestnut underparts and white bands on underside of tail much finer; juvenile more plain brown on upperparts and lacks spotting on plain brown breast, and rest of underparts off-white. Brush Cuckoo smaller with shorter, square-ended tail, and subspecies with same range as Fan-tailed Cuckoo has grey breast with only faint wash of buffish on belly and undertail-coverts, except for banded morph of which female has breast off-white finely edged pale brown.

VOICE Slow descending trill *peeeeer*, less than 1sec duration and usually given in clusters lasting 2–3min with maximum 12–14 trills/min; possibly territorial and very similar Chestnut-breasted Cuckoo but longer. Whistled *p-whee* followed by brief syllable, then repeated, but more tremulous *p-whee*, whole taking less than 2sec and repeated every 10–15sec. These calls probably only uttered by male. Female call high-pitched excited *chiree*. Alarm call (Fiji) brief scream *ki-ki-ki-ki*, and advertising call wavering and monotonous whistle *towtee*, occasionally in three syllables, also described as *foo-feeee*. Begging call of fledgling (Fiji) continuous cicada-like *zep-zep-zep*. Calls at night and when migrating (Gilbert 1935, Clunie 1973, Watling 1982, Clunie & Morse 1984, Coates 1985, Higgins 1999).

DESCRIPTION Nominate. **Adult male** Forecrown, crown, nape, hindneck, ear-coverts and upperparts plain blue-grey, except for lateral uppertail-coverts which are narrowly edged white and have small white half-bars on outer webs; wings dark brown-grey on remiges, primary-coverts and alula, secondary-coverts dark blue-grey, white area near carpal joint, underside of remiges grey, with white band from centres of primaries and bases of secondaries, front of wing light rufous-brown or buffish finely vermiculated grey; greater underwing-coverts grey, greater secondary-coverts tipped cream; chin and upper throat pale grey, rest of underparts pale rufous-brown, darkest on breast and with small amount of white on vent and thighs; flanks and belly often finely vermiculated grey; rufous fades occasionally to buffish; undertail-coverts pale rufous-brown, rarely narrowly and diffusely tipped white; tail grey-black tipped white with many white bands, largest on T5, and small white spots along shafts of T2–5. **Adult female** Like male but rufous on underparts paler and more extensively barred, rufous-brown not extending to lower breast, white area on vent usually larger and often vermiculated grey, and undertail-coverts often tipped white; however, owing to feather wear safe sexing only possible in fresh plumage, i.e. Apr–Oct (Higgins 1999). **Juvenile** Top of head, nape, hindneck and ear-coverts dark brown, in some mottled paler, or grey-brown narrowly fringed rufous-brown with blackish subterminal bands to feathers; rest of upperparts plain dark brown, in fresh plumage with narrow rufous-brown fringes, in few, rump and uppertail-coverts grey-brown, remiges dark grey-brown edged pale brown on inner primaries, secondaries and tertials, wing-coverts grey-brown fringed brown on secondary-coverts and white mixed with brown on primary-coverts, underwing like adult except buffish to cream leading edge, and greater secondary-coverts tipped dark grey with light brown subterminal spots; underparts off-white on chin and throat merging to buff on foreneck and rufous-brown on breast to more white on flanks, belly and vent, all irregularly barred and spotted grey-brown, undertail-coverts pale rufous-brown barred dark brown; tail black-brown tipped white and notched rufous along sides, underside dark brown barred buffish. **Nestling** Hatches blind and naked; skin pinkish for first few days, becoming dark brown. **Bare parts** Iris dark brown to reddish-brown or yellow-brown, sometimes with paler ring, eye-ring pale yellow to yellow-orange; iris of juvenile grey-brown to black-brown, before fledging creamy-grey in some, orbital ring pale yellow tinged green or grey. Bill black to grey-black, base of lower mandible often yellowish to olive-brown (more often in females?), gape pale buff-yellow, pink or orange, mouth red-orange to yellow or flesh-pink; bill of juvenile light brown, culmen dark brown with pinkish-brown base. Feet pale yellow to orange-yellow, scales often olive-brown in centres, claws dark grey; juvenile tinged pink on feet but soon attains adult colour.

BIOMETRICS Nominate. Wing male 138–151mm (n = 42), mean 144.4 ± 3.4, female 134–147mm (n = 18), mean 140.6 ± 3.8. Bill male 18.2–23.5mm, mean 21.5 ± 1.3, female 19.0–22.6mm, mean 21.4 ± 1.1. Tail male 130–149mm, mean 140.5 ± 4.3, female 123–151mm, mean 135.4 ± 6.4. Tarsus male 18.0–22.2mm, mean 19.7 ± 1.0, female 18.3–20.2mm, mean 19.6 ± 0.6. Mass male 42.6–65.0g (n = 32), mean 49.8 ± 4.4, female 45.0–53.5g (n = 7), mean 49.9 ± 3.1 (Higgins 1999).

MOULT Some flexibility in moulting process in response to environmental conditions. Moulting Dec–May, rarely as early as Sep, beginning with body feathers, tail often before remiges. Sequence of primary moult differs individually: normally P9 and P7 simultaneously, then P5 and P10, P8 and

P6, and simultaneously P4 and P1, then P2–3, or P4 and P2, P1 and P3. Post-juvenile moult Dec–Mar nearly complete but often retains some secondary coverts and secondaries, and rarely few primaries (Higgins 1999).

GEOGRAPHICAL VARIATION Five subspecies.

C. f. flabelliformis (Latham, 1801). East, south and south-west Australia, and Tasmania. Partly migratory to northern areas of Australia and NG lowlands. Described above.

C. f. excitus Rothschild & Hartert, 1907. Highlands of NG. Upperparts darker with green or blue tinge, chin dark grey, rest of underparts dark rufous-brown vermiculated grey. Wing male 138–145mm (n = 8), mean 141.4 ± 2.4, female 137–143mm (n = 8), mean 140.4 ± 2.1 (Payne 2005).

C. f. pyrrhophanus (Vieillot, 1817). New Caledonia, Solomons and Loyalty Is. Darker upperparts than nominate, underparts deep rufous with few white spots on tail; bill slender. Wing male 139–145mm (n = 12), mean 141.8 (Amadon 1942), female 139–146mm (n = 4), mean 142.5 (Payne 2005).

C. f. schistaceigularis Sharpe, 1900. Vanuatu and Banks Is. Like *pyrrhophanus* but grey colour on head continues to throat and upper breast, wings and tail shorter and bill heavier. Wing male 129–140mm (n = 17), mean 134 (Amadon 1942), female 127–137mm (n = 6), mean 130.8 ± 3.9 (Payne 2005).

C. f. simus (Peale, 1848). Fiji. Smallest subspecies with broader bill and strong greenish wash on upperparts, underparts paler. Black morph '*infuscatus*' has blackish-brown head, nape, mantle and rump; back dark brown with blackish-brown fringes; wings dark brown; underparts dark brown becoming paler on lower breast to undertail-coverts; underwing primary coverts whitish suffused brown; broad white band on secondaries and inner primaries below; tail uniform blackish-brown. Intermediates occur; juvenile has either sooty-brown and whitish barred underparts; tail black with small pale tips or as black morph adult. Wing male 128–145mm (n = 18), mean 134.8, female 128, 128, 130mm (Amadon 1942b, Payne 2005).

DISTRIBUTION Australasia and Pacific. NG, Australia (including Tasmania), and south-west Pacific in Fiji (Kandavu, Makongai, Mathuata, Mbengga, Navandra, Ovalau, Taveuni, Vanua Kula, Vanua Levu, Vatu-I-Ra, Viti Levu, Wakaya and Yasawa), Vanuatu (Ambae, Ambrym, Aniwa, Efate, Emae, Epi, Erromango, Futuna, Gaua, Lopevi, Malakula, Malo Paama, Nguna, Santo, Tanna, Ureparapara, Valua and Vanua Lava, where it may be at least partly migratory), Banks Is, New Caledonia, Solomons and Loyalty Is. In Australia movements complex with no clear pattern. Common throughout year at many sites, in others only in some years, which may be explained by influxes from other areas or different rainfall that influences availability of food. Long, pointed wings suggest it has adapted to special Australian circumstances where long unpredictable drought periods follow rain, just as Budgerigar, Cockatiel and Flock Pigeon, as all appear suddenly in large number in areas where they have not been seen for years. Ringing recoveries (65 of 1194) all within 14km of ringing site. Some move north from SA in winter. Vagrant: NZ, single 1960, three 1991; Lord Howe I. 1882 and 1911; Espiritu Santo, Vanuatu 20 and 29 Oct 1944 (Amadon 1942a, Scott 1946, Medway & Marshall 1975, Blakers et al. 1984, Bregulla 1992, Higgins 1999).

HABITAT No special preference. Tropical, subtropical and temperate rainforests with understorey, forest edge, wet *Eucalyptus* forest, riversides and gullies, dry sclerophyl forest or woodland with heath or grass ground cover, second growth, logged or burnt forest, mangroves, orchards, wooded farmland, pine plantations, parks and towns. Lowlands to 1,300–3,700m, NG; to 1,350m, Viti Levu, Fiji (Hindwood 1935, Kikkawa et al. 1965, McEvery 1965, Baldwin 1975, Gorman 1975, Coates 1985, Higgins 1999).

BEHAVIOUR Singly or in pairs, occasionally small groups. Reclusive and inconspicuous. Unlike other *Cacomantis* typically forages on ground or low in bushes and trees, and only occasionally higher in foliage. Follows mixed-species feeding flocks of thornbills, Blue Fairy-wren, Varied Sitella and White-throated Gerygone, all major hosts of cuckoos, changing feeding behaviour and foraging at same height as

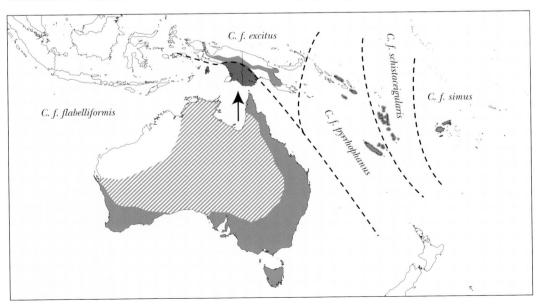

flocks. Some evidence of territoriality, i.e. one roosted for prolonged period in same tree, and adult female came back to same site in two subsequent years. However, aggression against conspecifics rarely seen and only in breeding season. On alighting often fans or gently bobs tail; fans tail while calling. Courtship feeding observed where female gave *chiree* call several times. Copulation seen between adult male and juvenile female, latter with wings lowered and tail slightly arched, copulation lasting *c.* 4sec; no courtship feeding observed. Male and female co-operate in egg-laying where male distracts host. Often mobbed by small birds, Fiji (Linton 1929, Lord 1956, Marchant & Höhn 1980, Watling 1982, Smedley 1983, Clunie & Morse 1984, Coates 1985, Bell 1986, Ambrose 1987, Payne 1997).

BREEDING Parasitism Most hosts have domed nests on or near ground. 503 of 662 records were of nest with egg or successfully fledged nestlings. 45.7% of these scrubwrens, 31.0% thornbills, 6.8% fairy-wrens, 7.4% Rock Warblers, 5.2% Striated Fieldwren and 4.0% heathwrens. Most common host species depends on area, e.g. around Sydney Brown Thornbill, Rock Warbler, Striated Fieldwren and White-browed Scrubwren; north of Sydney Large-billed and Yellow-throated Scrubwrens. Also Fairy-wrens: Red-backed, Red-winged, Superb, White-shouldered (probable), Splendid, Blue-breasted and Variegated; Southern Emu-wren, Striated Grasswren. Honeyeaters: Lewin's, Mangrove, Bar-breasted, Brown-headed, New Holland, White-eared, White-naped, White-plumed, Yellow-faced, Yellow-plumed, Yellow-throated, Yellow-tufted, Purple-gaped, Black-headed, Crescent and Striped; Noisy Miner, White-fronted Chat. Eastern Bristlebird, Rufous Bristlebird, Pilotbird, Scrubtit, Chestnut-rumped Heathwren, Shy Heathwren, Redthroat, White-tailed Gerygone, White-throated Gerygone; Buff-rumped, Inland, Yellow-rumped and Yellow Thornbills; White-winged Triller, Varied Sittella, Golden Whistler, Rufous Whistler, New Zealand Fantail, Rufous Fantail, Willie-wagtail. Robins: Dusky, Eastern Yellow, Pink, Red-capped, Rose, Hooded, Pale-yellow and Scarlet. Australasian Reed Warbler, Silver-eye, Mistletoebird, Rock Warbler, Speckled Warbler; Tasmanian, Chestnut-rumped and Striated Thornbills; Mangrove, White-tailed and Brown Gerygones; Eastern Shrike-tit, Black-faced Woodswallow, Dusky Woodswallow, Spectacled Monarch, Leaden Flycatcher, Restless Flycatcher, Little Grassbird, House Sparrow, Red-browed Finch, Beautiful Firetail and European Goldfinch. In WA White-browed Scrubwren and Inland Thornbill occasional hosts. In Tasmania White-browed Scrubwren, Tasmanian Thornbill and Tawny Grassbird. In NG Large Scrubwren, White-shouldered Fairy-wren and Short-bearded Honeyeater. In Fiji, Fiji Warbler and various fantails. In Vanuatu Scarlet Robin, Broad-billed Flycatcher and various fantails and trillers (North 1911, Dove 1916, Hindwood 1930, Favaloro 1933, Diamond 1972b, Watling 1982, Coates 1985, Brooker & Brooker 1989b, Bregulla 1992, Higgins 1999, Johnstone & Storr 2004, P.E. Lowther *in litt.*). A cuckoo fledgling raised by White-browed Scrubwrens was fed 14 times by scrubwrens and five times by the adult cuckoo (Ambrose 1987). **Season** Most Aug–Jan but also Jul, Feb and Mar in Australia, earliest in west; Nov–Feb Fiji; Aug–Jan Vanuatu (Clunie & Morse 1984, Brooker & Brooker 1989b, Bregulla 1992). **Nest** Domed nests used by most hosts, on or near ground (Brooker & Brooker 1989b). **Eggs** Rounded to oval, white with very small purplish-brown dots and underlying grey spots; 19.8–24.6x14.5–16.8mm (n = 61), mean 21.4x15.2mm (Brooker & Brooker 1989b).

Incubation <13 days 5 hours for egg in nest of thornbill (Brooker & Brooker 1989b). **Chicks** Evict host's eggs and young. Feathers in pin on body and wings 7th day, all feathers except on back emerged 11th day, feathers on back followed two days later and fully feathered 15th day, fledging 16–17th day and fed by hosts for 3–4 weeks after hatching (Brooker & Brooker 1989b, Payne 1997, Higgins 1999). **Survival** From 23 eggs 43.5% hatched and 21.7% fledged (BANRS in Higgins 1999).

FOOD Mainly caterpillars; many other insects such as beetles, flying ants, adult moths; spiders (Bell 1986, Payne 1997).

STATUS AND CONSERVATION Nearly all habitats except desert. In Australia reported common by many observers, e.g. Marchant & Höhn (1980) estimated six birds in <1ha of woodland at Moruya, NSW, and at Boola Boola State Forest, Victoria up to 53 territories per km^2 (Loyn 1980 in Higgins 1999). Rare eastern highlands NG (Diamond 1972b). Fairly common Fiji (Watling 1982). Uncommon New Caledonia and Vanuatu (Doughty *et al.* 1999). Not globally threatened (BirdLife International 2011).

Fan-tailed Cuckoo *Cacomantis flabelliformis*. **Fig. 1.** Adult male, *C. f. flabelliformis*, Dryandra State Forest, Wheatbelt, Western Australia, September (*Ron Hoff*). **Fig. 2.** Adult male, *C. f. flabelliformis*, Stirling Range, Western Australia, September (*Ron Hoff*). **Fig. 3.** Juvenile, *C. f. flabelliformis*, Billiatt, South Australia, October (*Rohan Clarke*). **Fig. 4.** Juvenile, *C. f. flabelliformis*, Billiatt, South Australia, October (*Rohan Clarke*).

BANDED BAY CUCKOO
Cacomantis sonneratii **Plate 30**

Cuculus Sonneratii Latham, 1790 (North Cachar Hills, India).

Alternative name: Bay Banded Cuckoo.

TAXONOMY Sometimes placed in monotypic *Penthoceryx* (Cabanis & Heine 1863) because of adult's uniquely banded head and upperparts, type of banding on tail, longer, and from above, more blunt bill, no rufous morph, juvenile plumage very similar to adult (neoteny?), and order of primary replacement. However, vocalisations similar to other *Cacomantis*. MtDNA results show *sonneratii* well-embedded in *Cacomantis* clade, being sister to clade containing *merulinus*, *passerinus*, *sepulcralis* and *variolosus* (Sorenson & Payne 2005). Some revision of subspecific names and distribution may be required. Polytypic. Synonyms: *P. pravata* Horsfield 1821 replaced by *musicus*; *rufo-vittatus* Drapiez 1823 in *musicus*; *himalayanus* Vigors 1830–31, *Ololygon tenuirostris* Hume 1875 and *P. s. malayanus* Chasen & Kloss 1931, all in nominate; *venustus* Jerdon 1842 and *schlegeli* Junge 1948 in *fasciolatus*.

FIELD IDENTIFICATION 22–24cm. Smallish cuckoo, size of Common Myna but tail longer, more slender build with rather long curved bill; often perches in upright position. Sexes similar. **Adult** Barred red-brown and blackish above with broad whitish supercilium, dark eye-stripe and whitish cheeks, which with dark red-brown crown finely cross-barred blackish all over give striking head pattern; below whitish finely barred blackish, tail tipped white. **Juvenile** As adult but more white on head and upper mantle; underparts less barred. **Similar species** Hepatic morphs of many *Cuculus* species much larger and lack characteristic white and dark brown head pattern. Rufous morph of Plaintive about same size but head, throat and upper breast plain rufous and bill smaller.

VOICE Higher pitched and more shrill than Indian Cuckoo. Typical call 'smoke, your pepper' or *pi-pi-pew-pew*, last two notes descending. In Sri Lanka still higher pitched *whew-whew-whu-u-u*, *whu-u-u*, *whu-u-u* ending abruptly. In Malay Peninsula *pi-pi-piu-piu*, last two notes falling, and Thailand *pi-pi-pew-pew*. In south-east Asia generally *pi-hi-hi-hi-bee-bew-bew-bew*, call loud and fast, descending at end, or *pi-pi-pihihi-pi-pihihi-pi*, or *pi-pi-pi-pi-pi-pi-hu-hu-bi-pi-hu-hu-bi* higher and higher pitched. On Greater Sundas breeding call four rising, slow notes followed by 3–6 quicker notes of 2–3 syllables, gaining pitch with each repetition and ending abruptly; or *tay-ta-tee*. On Bali excited *trrr* "as if the bird was clearing its throat", shortly after followed by common four-note call *tee-tee-tee-tee* or rising *teee-teee-teetetee-tee-teetetee*. In south Thailand short down-turned staccato whistles *TEEN-\KER-TEEN-\ker*. Calls mostly at dawn and dusk, but whole day during overcast drizzly weather. Counter-calling, and when breeding rising call similar to Plaintive Cuckoo also given, peninsular Malaysia (Henry 1955, Ali & Ripley 1983, Lekagul & Round 1991, van Balen 1991, MacKinnon & Phillipps 1993, Grimmett *et al.* 1998, Jeyarajasingam & Pearson 1999, Robson 2000, Rasmussen & Anderton 2005, C. Khoo pers. comm.).

DESCRIPTION Nominate. **Adult** Crown and nape dark rufous cross-barred blackish-brown gradually becoming rufous on hindneck, diagnostic broad dark brown eye-stripe from lores to rear of rufous ear-coverts with narrow unbarred white line under eye, broad whitish and finely cross-barred supercilium, rest of head whitish finely cross-barred dark brown; rest of upperparts and wings rufous to bay broadly cross-barred dark brown, primaries blackish-brown finely tipped paler, underwing pale rufous, base whitish. Underparts whitish finely barred dark brown with buffish wash on flanks and belly, barring broader and wavy than on head. Tail rufous barred black and tipped white, broad black subterminal band, T1 barred blackish-brown near shaft; all rectrices narrower towards tip. **Juvenile** Like adult but crown and nape with some buff to white feather tips and whole head less contrasting, rufous on upperparts more pronounced, with many pale tips, particularly on wing-coverts; barring on underparts more irregular with buff tinge on breast. **Nestling** Striped rufous-red above and finely barred black below like woodcock. **Bare parts** Iris brown to yellow or red. Bill black, lower mandible greenish-grey tipped black. Feet grey, nestling and juvenile olive-green.

BIOMETRICS Nominate. Wing male 119–125mm, (n = 10) mean 121.1 ± 1.97, female 120–128mm, (n = 7) mean 123.7 ± 3.0. Bill male 19.0–21.4, mean 20.0 ± 0.9, female 18.6–21.1, mean 19.6 ± 0.9. Tail male 104–120mm, mean 113.6 ± 4.9, female 105–119mm, mean 111.9 ± 5.2. Tarsus male 15.7–18.4, mean 17.2 ± 0.9, female 15.2–19.3mm, mean 17.1 ± 1.5. Mass female 27.9, 34.3g; unsexed (n = 5) mean 34.9g (Price 1979, Becking 1981, Payne 2005).

MOULT Wings May–Oct. Female moulting heavily India, 3 Aug. Transilient primary moult, different to sequence in other *Cacomantis* (Abdulali 1943, Stresemann & Stresemann 1961, Wells 1999, 2007).

GEOGRAPHICAL VARIATION Four subspecies.
C. s. sonneratii (Latham, 1790). India, Nepal, Bhutan, Bangladesh, eastern and peninsular Burma, Thailand, Thai-Malay Peninsula, Indochina and south-west China. Described above.
C. s. waiti (E.C.S. Baker, 1919). Sri Lanka. Darker than nominate above and more brown than rufous; blacker, bolder barring below, large black subterminal spots on undertail; iris orange to brown, lower mandible often yellow. Wing male 119–126mm, (n = 6) mean 123 ± 2.7, female 122–125mm, (n = 3) mean 123.5 (Payne 2005).
C. s. fasciolatus (S. Müller, 1843). Sumatra and offshore islands (Bintang, Tanahmasa, Pini and Enggano); Borneo; Palawan. Mindoro, Lagdeaujuan and Luzon (MVZ, ROM). Smallest form; darker than nominate. Upperparts dark olive-brown glossed bronze. Iris light brown to orange, lower mandible yellow, feet greenish-yellow. Wing male 102–113mm, (n = 6) mean 107.7 ± 4.1, female 103–113mm, (n = 10) mean 106.6 ± 3.8 (Payne 2005); male 114mm worn (Ripley 1944).
C. s. musicus (Ljungh, 1803). Java; recent record Bali, probably this subspecies. Bright brown above, less barred below, iris grey to brown. Wing male 105–116mm, (n = 12) mean 109.6 ± 3.4, female 108–114mm, (n = 10) mean 110.8 ± 2.1 (Payne 2005).

DISTRIBUTION Oriental. Resident, but moves locally; northern population in Nepal and India perhaps migrates to south India. India from north Uttar Pradesh to Bhutan and Assam in north, west and north-east, but few records central India. Sri Lanka in lowlands in summer moving east to hills in winter. Local resident Bangladesh. Thailand, Burma, Indochina, peninsular Malaysia, including Singapore. South-west Sichuan and south Yunnan, China. Greater Sundas (Cheng 1987, Harvey 1990, MacKinnon & Phillipps

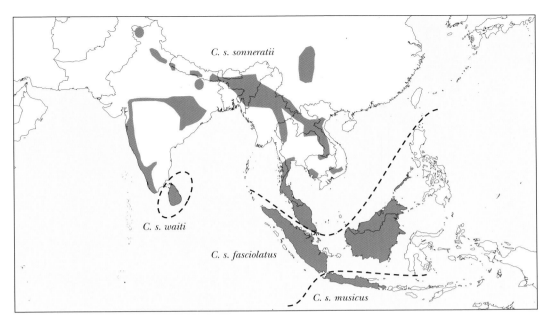

1993, 2000, Grimmett *et al.* 1998, Ollington *et al.* 1999 in Wells 2007). 13 Aug–Jan Palawan, Philippines; migratory patterns not known, perhaps winter visitor from Borneo (Dickinson *et al.* 1991, Kennedy *et al.* 2000); Silang, Luzon, and Lagdeaujuan and Mindoro (MVZ, ROM) presumably vagrants.

HABITAT Dense deciduous and evergreen forest, secondary growth, wooded country and farmland, India and Thailand. Prefers open forest in belt between wet and dry zones, and park-like habitats, Sri Lanka. Very often heard in heavily wooded areas around towns or old rubber plantations Thai-Malay Peninsula; once disturbed lower montane forest at 1,250m. Lowland and hill forest, secondary forest and scrub, farmlands, cocoa and *Albizia* plantations, gardens, casuarinas and old rubber plantations probably to 1,530m, Borneo. Nepal mainly below 250m but to 2,440m. India to 1,500m or higher. To 900m (rarely 1,500m) Greater Sundas generally (Phillips 1948, Smythies 1957, 1981, Ali & Ripley 1983, Lekagul & Round 1991, MacKinnon & Phillpps 1993, 2000, Grimmett *et al.* 1998, Jeyarajasingam & Pearson 1999, Wells 1999, 2007, Mann 2008).

BEHAVIOUR Mainly solitary. Calls mostly from exposed vantage points with tail depressed, raised rump feathers and drooping wings. Four counter-calling birds in single tree, 27 May. Sometimes catches flying insects (Ali & Ripley 1983, Grimmett *et al.* 1998, Wells 1999). Adults observed feeding fledgling cuckoos, peninsular Malaysia (C. Khoo pers. comm.).

BREEDING Parasitism Little trustworthy information of breeding biology exists due to confusion over egg identification (Becking 1981). Commonest host India Common Iora, also Scarlet and Small Minivets, *Pycnonotus* bulbuls and small babblers; Common Iora peninsular Malaysia and Singapore; Pied Fantail, Singapore; leafbirds and flycatchers preferred Java; Bar-winged Flycatcher-shrike and White-bellied Erpornis, south-east Asia (Batchelor 1958, Ali & Ripley 1969, MacKinnon & Phillpps 1993, Grimmett *et al.* 1998, Robson 2000, Lim 2008d, P.E. Lowther *in litt.*). **Season** India mainly Feb–Aug; oviduct egg, 3 Aug. Fledglings Thai-Malay Peninsula, Apr–late Jul, Dec. Egg Sumatra, 18 Apr. Breeding west Java from Oct, but otherwise Feb–Jun. Females in breeding condition Borneo, mid Mar; calls all months Brunei (Abdulali 1943, Voous 1951, Becking 1981, MacKinnon 1988, Smythies & Davison 1999, Wells 1999, Mann 2008, C. Khoo pers. comm.). **Eggs** Broadly oval; ground colour white to pink sparsely spotted red-brown or purplish-brown and underlying grey, most at broader end; *musicus* (n = 2), mean 17.6x13.5mm; *waiti* 17.4x13.5mm; *fasciolatus* 18.8x14.4mm (Becking 1981). **Chicks** Evict eggs or host chicks (Becking 1981). **Survival** No information.

FOOD Caterpillars, nymphs of Hemiptera, Red Silkcotton Bug, grasshoppers, locusts and crickets. Stomach contents: Orthoptera (grasshoppers, Tettigoniidae), Coleoptera, Hemiptera (Homoptera Fulgoridae, Flatidae; Heteroptera Coreidae, *Anoplecnemis* sp.), Lepidoptera (caterpillars of Pieridae *Terias hecabe*, Notodontidae *Tarsolepis javana*, hairy caterpillars of Arctiidae) (Ali & Ripley 1983, Sody 1989, Smythies & Davison 1999).

STATUS AND CONSERVATION Uncommon north India, but fairly common Deccan and south India, except in areas of heaviest rainfall (Ali & Ripley 1969); uncommon resident in Nameri NP, Assam (Barua & Sharma 1999); rare Bhutan; local Bangladesh; common Sri Lanka (Grimmett *et al.* 1998). Uncommon resident Thailand (Lekagul & Round 1991). Regular and more or less common Thai-Malay Peninsula (Wells 1999). Regularly heard but rarely seen China (MacKinnon & Phillpps 2000); rare (Cheng 1987, Lee *et al.* 2006). Widespread and moderately common, Sumatra (van Marle & Voous 1988), but only two records Gunung Leuser NP (Buij *et al.* 2006). Locally common to scarce Borneo (Mann 2008). Fairly common Java (MacKinnon 1988). Rare Philippines (Kennedy *et al.* 2000). Habitat preference forest, but is able to exploit disturbed forest, farmlands etc. suggesting no serious threat. Not globally threatened (BirdLife International 2011).

Banded Bay Cuckoo *Cacomantis sonneratii*. **Fig. 1.** Adult, *C. s. sonneratii*, with caterpillar, Singapore, May (*H. Y. Cheng*). **Fig. 2.** Juvenile, *C. s. sonneratii*, fed by female Common Iora, Tambun, Perak, Malaysia, July (*Khoo Siew Yoong*).

PLAINTIVE CUCKOO
Cacomantis merulinus   Plate 30

Cuculus merulinus Scopoli, 1786 (Panay, Philippines)

Alternative names: Brain-fever Bird (also used for *C. passerinus*, *C. pallidus*, *Hierococcyx varius* and *H. sparverioides*); Rufous-bellied Plaintive/Grey-headed Cuckoo; Grey-breasted Brush Cuckoo (*merulinus*).

TAXONOMY *Passerinus* recognised as species distinct from *merulinus* (Payne 2005), due chiefly to its grey belly and vocalisations; hosts similar. Forms superspecies with *passerinus*, and perhaps *variolosus* and *sepulcralis*. Polytypic. Synonyms: *Polyphasia tenuirostris* Jerdon 1862 in *querulus*; *Cacomantis m. subpallidus* Oberholser 1912 and *dysonymus* Heine 1863 in *threnodes*; *m. celebensis* Stresemann 1931 in *lanceolatus*.

FIELD IDENTIFICATION 20–23cm. Small, palish cuckoo. **Adult male** Pale grey head, rest of upperparts and wings dark grey slightly glossed, with some rufous on wing-coverts; rump and uppertail-coverts slaty-grey; chin to upper breast grey, rest of underparts pale rufous-buff; tail black tipped white, T5 barred white. **Adult female, grey morph** As male but abdomen barred whitish and rectrices more barred. **Adult female, rufous morph** Dark rufous heavily barred dark brown above; throat, breast and tail uniform dark rufous, rest of underparts rufous to whitish barred black. **Juvenile** Variable from dark grey only with very small white bars on abdomen, to rufous with dark brown barring on upperparts, rufous wider than dark brown bars, underparts white barred black; tail black barred white, some edged rufous. **Similar species** Banded Bay Cuckoo in all plumages has white supercilium, cross-barring on crown; whole underparts white finely banded brown. Rufous female like Banded Bay Cuckoo which has narrow tail tips, whereas Plaintive retains same width to rounded end. Adult Brush Cuckoo has entire underparts plain rusty. Rufous morph from Grey-bellied by duller rufous-brown upperparts with heavily barred dark, most of underparts washed rufous; T1 strongly barred or notched rufous, little barring, or dark line along shaft in Grey-bellied; juvenile's rufous crown and nape boldly streaked blackish.

VOICE Common call Greater Sundas and China *pee-pipee-pee… pipee-pee* ascending with higher pitches for every repetition, also rendered *tay-ta-tee*, or *tay-ta-tay*, *phi-phi-wi-wihi* or *phi-phi-phi-wi-hi* or quite short *wi-pihui*. On Sulawesi *ti-ter-wi* or *pee-to-pett*, second note lower and last higher than first note and higher pitched for every repetition. Another plaintive call, cadencing, trilling and descending *tee-tee-tee-tee-tita-tita-tita-tita-tee* or *peee-peee-peee-tcho-cho*. Call of female *piteer* or *tchree*. In Philippines *peet-to-peet… peet-to-peet* repeated four or more times in gradually ascending pitch. Very vocal, occasionally calling at night, Malay Peninsula, but silent in non-breeding season (duPont & Rabor 1973b, Ali & Ripley 1983, MacKinnon & Phillipps 1993, 2000, Payne 1997, Grimmett *et al.* 1998, Jeyarajasingam & Pearson 1999, Smythies & Davison 1999, Robson 2000, C.F. Mann).

DESCRIPTION Nominate. **Adult male** Head plain light grey, upperparts and wings dark grey slightly glossed green to bronze, slaty-grey rump and uppertail-coverts, greater wing-coverts edged buff-rufous, underside of wing with white area at base of secondaries and innermost primaries and white spot on inner web of outermost primary. Chin, throat and upper breast grey, lower breast to undertail-coverts pale rufous-buff. Graduated tail black narrowly tipped white except T1, T5 barred white. Rare morph sooty-blackish with tail tipped brown. **Adult female, grey morph** Like male but abdomen barred whitish and rectrices more barred. **Adult female, rufous morph** More common morph. Bright rufous on head with pale rufous-buff streak above and below eye in some, nape rufous; rest of upperparts and wings bright rufous barred dark brown; rump barred dark brown, sometimes plain; underparts whitish to buffish or pale rufous finely barred black, bars broader on abdomen and flanks; tail rufous with black subterminal band. **Juvenile** Varies from grey to pale rufous in ground colour with intermediates; grey variant very like adult but face darker and lacking gloss on upperparts; underparts grey with flanks and belly diffusely barred buffish-grey; tail dark grey-brown to blackish, inner rectrices notched rufous, outer rectrices notched white; rufous birds rufous on head and neck streaked darker brown, rest of upperparts barred dark and rufous, underparts like rufous female, but spots forming streaks from chin to upper breast. **Nestling** Undescribed. **Bare parts** Iris brown to red, juvenile grey becoming brown in centre; eye-ring grey. Bill black, sometimes with yellowish or orange lower mandible, gape orange. Feet pale yellow to yellow-olive.

BIOMETRICS *C. m. querulus*. Wing male 98–114mm (n = 34), mean 110.1 ± 5.1, female 106–117mm (n = 11), mean 110.1 ± 5.2. Tail male 94–122mm, mean 108.6 ± 7.4, female 95–114mm, mean 105.3 ± 6.7. Bill male 16.5–19.2mm, mean 17.1 ± 0.8, female 16.2–17.9mm, mean 17.1 ± 0.6. Tarsus male 14.2–17.9mm, mean 15.7 ± 1.2, female 12.9–18.6mm, mean 15.4 ± 1.5 (Payne 2005). Mass *threnodes* (Thai-Malay Peninsula) unsexed (n = 4) 21.3–26.6g (Wells 1999); (Brunei) 2 juveniles 19.5, 25.2g (C.F. Mann).

MOULT Most moult wings Jul–Sep, some starting earlier; primaries replaced at up to three loci. Post-juvenile wing moult Oct–Feb, often interrupted (Wells 1999).

GEOGRAPHICAL VARIATION Four subspecies.
C. m. merulinus (Scopoli, 1786). Philippines and Sulawesi. Wing male 97–118mm (n = 18), mean 103.9 ± 4.1, female 102–112mm (n = 11), mean 105.2 ± 3.6 (Payne 2005). Described above.
C. m. querulus Heine, 1863. Nepal (rare) to Arunachal Pradesh, Assam (Hume 1888), Bhutan, Bangladesh, south China (south of 27°N, including Hainan), Burma, Thailand, north Thai-Malay Peninsula and Indochina; some winter south India, Sri Lanka and south-east Asia. Colours stronger, forehead and crown brownish-grey, upperparts brown, rump streaked black tipped white, inner rectrices blackish notched buff, outer web of T5 completely barred white from below; chin to upper breast sometimes grey washed rufous, belly, undertail-coverts and underwing-coverts rufous, bend of wing white; some females as male (missexed?), but most rufous morph, similar to nominate but whitish below barred dark brown; juvenile streaked on head.
C. m. threnodes Cabanis & Heine, 1863. South Thai-Malay Peninsula, Sumatra, Nias, Siberut, Enggano and Borneo. Head more contrasting with rest of upperparts, paler buff-rufous below than *querulus*. Wing male 96–104 (n = 6), mean 99.8 ± 2.9, female 95–99mm (n = 7), mean 97.2 ± 2.4 (Payne 2005).
C. m. lanceolatus (S. Müller, 1843). Java. Bali? Underparts greyish-white becoming rufous-buff on belly and vent, some birds paler brown on upperparts. Wing male 96.0–107.5mm (n = 6), mean 103.9 ± 4.0, female 98–109mm (n = 7), mean 101.6 ± 4.4 (Payne 2005). Sometimes placed in nominate.

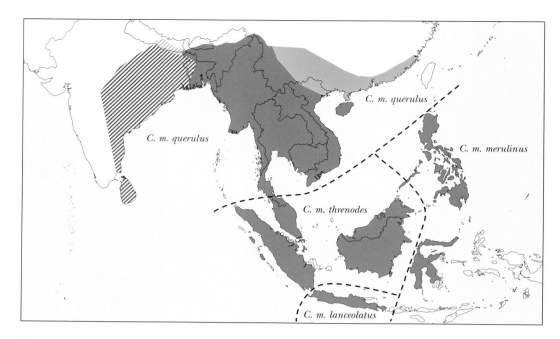

DISTRIBUTION Oriental. Occasional visitor Nepal, late Apr–Aug. Presumed partly resident north India with some local movements and apparently partly wintering over much of Indian subcontinent, particularly north-east Deccan, and Sri Lanka, where some over-summer; migrant and resident Manipur. Throughout whole Thai-Malay Peninsula, except Phang Nga, including following islands: Langgun, Langkawi, Lanta, Penang Tarutao, Yai, Yao, Singapore, Tekong and Ubin. All provinces Cambodia. In China breeding visitor Tibet (Zayu), south-west Sichuan, west Yunnan, south Guizhou, and east to Fujian; resident Hainan. Greater Sundas, including satellite islands of Sumatra (Nias, Siberut, Enggano), and Riau and Lingga archipelagos. Philippines on Balabac, Basilan, Biliran, Bohol, Cebu, Leyte, Luzon, Malamaui, Masbate, Mindanao, Mindoro, Palawan, Panay, Rasa, Sulawesi (Vaughan & Jones 1913, Inskipp & Inskipp 1985, Cheng 1987, MacKinnon & Phillipps 1993, Wells 1999, Kennedy *et al.* 2000, Thomas & Poole 2003, Mann 2008, Choudhury 2009).

HABITAT Open woodland, secondary growth, scrub, savanna grassland, groves, farmland, plantations, gardens and cities. Below 1,000m Assam. 400–2,000, occasionally 200m, in forest clearings and scrub near cultivation, Bhutan. Occasionally mangroves or heavily logged forests by rivers, or in poorly-managed teak and rubber plantations, and tea plantations, to 1,500m, Thai-Malay Peninsula. Frequently lowland rainforest, freshwater and peatswamp forests, and common in heavily disturbed forest and agricultural land, Siberut, west Sumatra. Occasionally primary forest but more often logged secondary forest, *Albizia* plantations and kerangas forest, scrub, grasslands and gardens to 1,220m, Borneo. To 1,000+m Philippines, often in isolated trees. Lowlands to 1,850m south-east Asia generally. To 3,000m, Likiang range, China. Lowlands and hills to 1,000m, Sumatra (Hume 1888, Becking 1981, Meyer de Schauensee 1984, van Marle & Voous 1988, Lekagul & Round 1991, Noske 1995, Wells 1999, Kemp 2000, Kennedy *et al.* 2000, Robson 2000, Spierenburg 2005, Mann 2008).

BEHAVIOUR Solitary; males call from favourite posts over many days suggesting territoriality. Male courtship-feeds female with caterpillars. Mainly in canopy of tall trees where difficult to find in dense foliage, but has rapid, hawk-like flight, usually from one tree to another, first in upwards direction ending in steep dive to foliage of another tree. Mobbed by small birds such as ioras and sunbirds (Jeyarajasingam & Pearson 1999, Smythies & Davison 1999, Wells 1999). Adults observed feeding fledgling cuckoos, peninsular Malaysia (C. Khoo pers. comm.).

BREEDING Parasitism At least 11 known host species, most with domed nests with narrow entrances. Zitting Cisticola; Hill, Striated, Grey-breasted and Tawny-flanked Prinias; Common Tailorbird, Assam. Yellow-bellied Prinia and Dark-necked Tailorbird, Thai-Malay Peninsula. On Borneo Dark-necked Tailorbird (and other tailorbirds?), Grey-breasted Spiderhunter, possibly Common Iora; Green Iora feeding juvenile. Grey-backed Tailorbird Philippines. Common Tailorbird China. Ashy, Olive-backed and Common Tailorbirds, and Eastern Crimson Sunbird, Java (Vaughan & Jones 1913, Smythies 1957, 1981, Schönwetter 1967, Ali & Ripley 1969, Becking 1981, Wilkinson *et al.* 1991, Hornskov 1996, Wells 1999). **Season** India Mar–Sep, peak Jul. Calling Assam Apr–Aug. Eggs Burma, May–Jul. Fledglings Jun–late Aug, Dec–Jan, Thai-Malay Peninsula. Male and female in breeding condition Feb–Mar and Jun, Borneo; fledglings May–Jul, Sep. Recent fledglings May–Jun, Philippines (Osmaston 1916, Ali & Ripley 1969, Becking 1981, Nash & Nash 1988, Wilkinson *et al.* 1991, Grimmett *et al.* 1998, Smythies & Davison 1999, Wells 1999, Kennedy *et al.* 2000, C. Khoo pers. comm.). **Eggs** Long oval, glossed, white to blue, blotched and spotted red-brown varying with host. Three *gentes* (Becking 1981): (1) plain chestnut, occasionally with some faint darker chestnut dots around thicker end, host Ashy Tailorbird; (2) pale blue sprinkled with spots and twisted lines of red, dark brown or purple, and greyish underlying markings, host Tawny-flanked Prinia; (3) pinkish or bluish ground colour with red-brown blotches and dots often forming ring around blunt end, hosts Common and Olive-backed Tailorbirds, and Zitting Cisticola; mean (*querulus*) 19.8x13.8mm (Baker 1908, 1934); mean India 19.1x13.3mm (n = 14) (Becking 1981). No further information.

FOOD Soft-bodied insects, mainly hairy, but also hairless, caterpillars, beetles, bugs, termite soldiers; fruit (Payne 1997, Jeyarajasingam & Pearson 1999, Smythies & Davison 1999).

STATUS AND CONSERVATION Common over large areas of range. Commoner than Grey-bellied Cuckoo, Bhutan (Spierenburg 2005). Common India (Rasmussen & Anderton 2005). Common and widespread Bangladesh (Harvey 1990). Regular and more or less common resident Thai-Malay Peninsula (Wells 1999), but residents in Singapore virtually disappeared by 2000 (R.F. Ollingon, SINGAV-14, both in Wells 2007). Common all provinces Cambodia (Thomas & Poole 2003). Very common resident Thailand (Lekagul and Round 1991). Uncommon China (Cheng 1987); common south-east China summer (Vaughn & Jones 1913). Very common in open and agricultural land central Sulawesi (Watling 1983). Commonest cuckoo in most areas of Borneo to 1,220m (Mann 2008). Common Java and Bali (MacKinnon & Phillipps 1993). Tolerant of habitat degradation. Not globally threatened (BirdLife International 2011).

Cacomantis

Plaintive Cuckoo *Cacomantis merulinus*. **Fig. 1.** Adult male, *C. m. threnodes*, Singapore, April (*H. Y. Cheng*). **Fig. 2.** Adult female rufous morph, *C. m. threnodes*, Juru, Penang, Malaysia, March (*Choy Wai Mun*). **Fig. 3.** Juvenile, *C. m. threnodes*, Singapore, August (*H. Y. Cheng*). **Fig. 4.** Adult female rufous morph, *C. m. querulus*, Hong Kong, January (*Martin Hale*). **Fig. 5.** Juvenile moulting to adult plumage, *C. m. querulus*, Dibrugarh, Assam, India, March (*Raj Kamal Phukan*). **Fig. 6.** Juvenile male moulting to adult plumage, *C. m. lanceolatus*, Carita FR, north-west Java, Indonesia, September (*Fabio Olmos*).

GREY-BELLIED CUCKOO
Cacomantis passerinus   Plate 30

Cuculus passerinus Vahl, 1797 (Tranquebar, India).

Alternative name: Indian Plaintive Cuckoo.

TAXONOMY Previously included in *merulinus* with which it is mostly allopatric (but sympatric in Bhutan and north-east India), and utilises similar hosts, but plumage and vocalisations differ (Payne 2005). Forms superspecies with *merulinus*. Monotypic. Synonyms: *Cuculus tenuirostris* Gray 1834, *niger var pyrommatus* J.E. & G.R. Gray 1846.

FIELD IDENTIFICATION 22cm. Small cuckoo. **Adult male** Head plain light grey, upperparts and wings dark grey slightly glossed green to bronze, slaty-grey rump and uppertail-coverts; underparts from grey all over except white vent and undertail-coverts, to grey on chin, throat and breast with belly and undertail-coverts whitish; tail black tipped white, T5 barred white. **Adult female, grey morph** Like male but abdomen barred whitish and rectrices more barred. **Adult female, rufous morph** More common. Bright rufous head with white streak above and below eye in some individuals, nape rufous; rest of upperparts and wings bright rufous barred dark grey-brown, rump plain rufous or finely barred; chin, throat, breast and tail rufous, finely barred on lower breast, tail with black subterminal band, rest of underparts whitish to buffish or pale rufous finely barred black. **Adult, blackish morph** Entirely sooty-blackish and dark brown. **Juvenile** Varies from grey to pale rufous in ground colour with intermediates. **Grey morph** Very like adult but face darker, upperparts lack gloss, underparts grey with diffusely barred buffish-grey on flanks and belly; tail dark, notched white and rufous. **Rufous morph** Rufous head and neck streaked darker brown, upperparts barred dark and rufous, underparts like rufous female. **Similar species** From Banded-bay Cuckoo by lack of white supra-loral streak. Rufous morph Grey-bellied has brighter rufous-brown upperparts less dark barring, rufous wash over most of underparts; T1 have little rufous notching or barring, but may have dark line along shaft; underparts much less rufous; similarly juvenile much less marked than Plaintive.

VOICE Plaintive whistle *ka-teer* 1/sec, or *peter-peter*, and other variants. Calls mostly at dusk and dawn and during first part of night, Pakistan. In India single plaintive whistle *piteer* or *kiveer* frequently repeated at short intervals; ascending whistled *pee-pipee-pe*, *pipee-pee* ending interrogatively, each repetition slightly higher (Ali 1962, 1996, Ali & Ripley 1969, Fleming *et al.* 1979, Roberts 1991, Martens & Eck 1995).

DESCRIPTION Adult male Head and back dark grey with light green gloss, wing-coverts dark grey, remiges dark greyish-brown to dark brown; greater wing-coverts edged pale buff-rufous. Below dark grey, undertail-coverts and sometimes belly whitish; bend of wing white, underside of wing with white at base of secondaries and innermost primaries and white spot on inner web of outermost primary, underwing-coverts barred white. Graduated tail blackish narrowly tipped white, except T1, T5 barred white. **Adult female, grey morph** As male (wrongly sexed?) but lightly barred whitish on belly, and tail barring not confined to T5, all rectrices edged white. **Adult female, rufous morph** Variable, with crown rufous, sometimes streaked dark brown, nape rufous, back bright rufous barred dark brown, dark bars equal to or narrower than rufous; rufous rump completely or nearly unbarred; face, throat and upper breast rufous, indistinctly and variably barred blackish; belly and undertail-coverts white barred blackish; bend of wing white, underwing-coverts barred white. Tail rufous, T5 with subterminal blackish bar. **Adult, blackish morph** (rare; sex?). Head to mantle and uppertail-coverts blackish-grey; back and remiges dark brown; below dark brownish-black. Tail blackish tipped brown without white, but paler fringes to rectrices. **Juvenile, grey morph** Variable; crown and back dark grey indistinctly barred rufous; rump and uppertail-coverts unbarred or feathers tipped rufous; face dark variably blotched grey or rufous; throat and upper breast whitish or rufous barred grey, belly narrowly barred whitish. Rectrices dark grey-brown to blackish, notched rufous on inner and white on outer webs. **Juvenile, rufous morph** (most or all females) Rufous, crown light rusty-brown, back and wings barred rusty and dark brown (bars equal width), rump rufous barred dark; face rufous spotted blackish; underparts whitish variably washed rusty and barred dark grey, some mostly whitish on abdomen and undertail-coverts; bend of wing white, underwing-coverts barred white and brownish. Tail brown with or without rufous barring and edges. **Nestling** Not described. **Bare parts** Eye-ring grey. Iris red with yellow outer ring to reddish-brown; juvenile blackish or grey becoming brown in centre. Bill brown to black, sometimes lower mandible yellowish or orange, gape orange; juvenile dark brown. Feet pale yellow to yellow-olive, juvenile light yellow.

BIOMETRICS Wing male 105–120mm (n = 18), mean 114.6 ± 4.5, female 110–115mm (n = 5), mean 113.0 ± 2.7. Bill male 15.3–17.7mm, mean 16.6 ± 0.8, female 15.9–18.3mm, mean 18.1. Tail male 98–115mm, mean 104.2 ± 5.1, female 87–115mm, mean 100.0 ± 5.2. Tarsus male 13.3–17.7mm, mean 15.4 ± 1.5, female 15.6–18.0mm, mean 17.1 ± 2.2 (Payne 2005).

MOULT Primary moult Sikkim, Jul, and Sri Lanka, Feb, Apr, Nov and Dec; interrupted primary moult Punjab, Jul (BMNH).

DISTRIBUTION Indian sub-continent. North-east Pakistan, north India in Himachal Pradesh east to Bhutan and West Bengal; some migrate in winter to south India, Sri Lanka and Maldives. Assam Dec. North Burma? In Pakistan limited to foothills of Murree Hills and subtropical pine zone where earliest arrival 20 May, and migrates end Sep. Nepal and Bangladesh summer. In Sri Lanka, where common in dry zone, rare in hills, and Maldives (rare) arrives with north-east winds Oct, departs Apr; Lakshadweep Is. Resident south India, but migratory in north as far south as Orissa (Henry 1955, Ali & Ripley 1983, Roberts 1991, Robson 2007). Kerman Province, Iran, Jun 1048 (AMNH). Three records Oman; 7 Nov 1988, and 1–2 18–27 Oct 2006, Masirah I., 3 Nov 2002 Dhofar (Eriksen *et al.* 2003, Eriksen 2010).

HABITAT Open woodland, secondary forest, scrubby hillsides, bush, cultivation including shade trees in tea plantations, gardens, human habitation, grassy plains, swamps. To 1,400–2,100m Nepal. Gardens and dry areas in lowlands more than hills, Sri Lanka. Forest clearings and open areas with scrub around cultivation, 400–2,300m, Bhutan. Prefers open scrub on hills more than dense pine forests in same area, Pakistan; to 1,800m (Ali & Ripley 1969, Becking 1981, Roberts 1991, Zacharias & Gaston 1993, Martens & Eck 1995, Payne 2005, Spierenburg 2005).

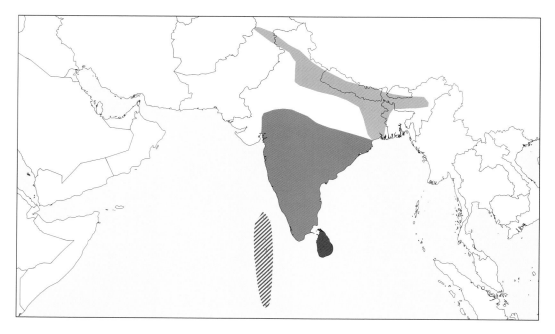

BEHAVIOUR Active, usually in canopy or tops of bushes. When calling tail depressed, rump arched with feathers fluffed out and wings drooping loosely. Feeds on ground or sallies for flying insects. Male territorial, calling from regular concealed sites in foliage (Ali & Ripley 1983, Payne 2005).

BREEDING Parasitism Parasitises warblers, chiefly Zitting Cisticola, Ashy and Plain Prinias, and tailorbirds, including Common, and Dark-fronted Babbler in India, and Common and Olive-backed Tailorbirds in Java, that build closed nests with small entrances; once Purple, and Purple-rumped Sunbirds, nestling in latter's nest also fed by Common Tailorbird. Presumably eggs introduced by bill. Suggested hosts: Yellow-eyed Babbler, Long-tailed Shrike, Red-vented Bulbul and Small Minivet, but no firm evidence (Baker 1906b, 1908a, 1934, 1942, Suter 1945, Becking 1981, Bharucha 1982, Payne 2005). **Season** Mar–Sep, synchronising with hosts, Pakistan. Calls Eastern Ghats Apr–Jun, Nilgiris Aug–Oct. Feathered nestling in prinia nest Kotagiri, Dec (Baker 1934, Ali & Ripley 1969, Price 1979, Roberts 1991, Payne 2005). **Eggs** Very similar to Plaintive Cuckoo. Long narrow ovals, much smaller at one end. Coloration mimics hosts' eggs: uniform brick red in Ashy Prinia, reddish or bluish-white spotted brownish red in Common Tailorbird; 19.9x14.0mm. Three *gentes* – (1) uniform chestnut- or mahogany-red, sometimes darker, faint spots at broad end, in Ashy Prinia's nest; (2) pale blue ground colour with sharply-defined blotches and twisted lines in blackish-brown, deep red or purple with greyer underlying marks, similar to but larger than that of host, Plain Prinia; (3) light pinkish or bluish-white ground colour with reddish-brown blotches or specks often forming ring around broad end, hosts Common and Olive-backed Tailorbirds' nests (S. Baker in Ali & Ripley 1969, Becking 1981, Payne 2005). **Incubation** Not recorded. **Chicks** In Purple Sunbirds' nest fed by hosts and tailorbirds. Eject host eggs and young (Suter 1945, Henry 1955). **Survival** Tailorbirds desert 20% of parasitised nests (Legge 1880).

FOOD Mostly caterpillars, particularly hairy (including moth *Dysdercus cingulatus*); other soft-bodied insects. Feeds on ground or sallies for flying insects. Once shaking hairy caterpillar intensely with bill before eating (Legge 1880, Ali & Ripley 1969, Roberts 1991).

STATUS AND CONSERVATION Common India, except drier montane regions (Baker 1934, Price 1979); fairly common summer visitor Himalayas (Grimmett *et al.* 1998). Frequent in summer Nepal below 1,400m (but to 2,135m) (Grimmett *et al.* 1998). Uncommon Bhutan (Spierenburg 2005). Local resident Bangladesh (Harvey 1990). Common in winter Sri Lanka in dry zone, occasionally oversummers (Kotagama & Fernando 1994, Grimmett *et al.* 1998). Vagrant Oman (Eriksen *et al.* 2003, Eriksen 2010). Not globally threatened (BirdLife International 2011).

Grey-bellied Cuckoo *Cacomantis passerinus*. **Fig. 1.** Adult male, Belgaum, Karnataka, India, March (*Niranjan Sant*). **Fig. 2.** Adult male. Talangama, Sri Lanka, March (*Gehan de Silva Wijeyeratne*). **Fig. 3.** Adult female rufous morph, Karkala, Karnataka, India, January (*ShivaShankar*). **Fig. 4.** Juvenile, Belgaum, Karnataka, India, March (*Niranjan Sant*).

BRUSH CUCKOO
Cacomantis variolosus   Plate 31

Cuculus variolosus Vigors & Horsfield, 1826 (Paramatta, New South Wales)

Alternative names: Square-tailed/Rufous-breasted Brush Cuckoo; Moluccan (Brush)/Heinrich's (Brush) Cuckoo ('*heinrichi*' = *infaustus*).

TAXONOMY *Sepulcralis,* together with subspecies *virescens, everetti* and *aeruginosus*, here considered species distinct from *variolosus* due to morphological and vocal differences, and possible sympatry. However Payne (2005), supporting conspecifity, states that calls similar, size and plumages intergrade, and no consistent difference in eye-ring colour. North Australian *dumetorum* (smaller, greyer upperparts and paler underparts) not recognised as it intergrades with nominate through Burdekin lowlands (Mason *et al.* 1984). '*Heinrichi*' Stresemann 1931 from Halmahera and Bacan was given species rank due to its rufous plumage and smaller size compared to more grey, possibly sympatric, *infaustus*. However, no consistent differences in plumage or measurements, or in vocalisations between those two forms, and some west NG *variolosus* similar in size and plumage to '*heinrichi*', and that genetic distance between latter and *variolosus* is less than between other forms of *variolosus*, we follow Payne (2005) in considering '*heinrichi*' synonym of *infaustus* (but see Tebb *et al.* 2008). Forms superspecies with *sepulcralis*, although perhaps some overlap in distribution. Polytypic. Other synonyms: *Cuculus tymbonomus* S. Müller 1843, *C. dumetorum* Gould 1845, *C. westwoodia* Mathews 1913, *lineatus* Dodd 1913, *pyrrhophanus* (sic) *vidgeni* (Mathews 1918), and *Cuculus variolosus whitei* Bruce 1986, all in nominate; *assimilis* Wallace 1863 in *virescens*; *assimilis* var *major* Salvadori 1880, *assimilis fortior* Rothschild & Hartert 1914, *v. oblitus* Hartert 1925, *v. obscuratus* Stresemann & Paludan 1932, *v. chivae* Mayr & Meyer de Schauensee 1939 and *v. obiensis* Jany 1955, all in *infaustus*; *Cuculus brisbanensis* Diggles 1872 and *v. tabarensis* Amadon 1942 in *macrocercus*.

FIELD IDENTIFICATION 20–23cm. Small cuckoo with upright posture; great intraspecific variation. Flight fast and slightly undulating. **Adult male** Nominate in Australia generally dull grey; head, neck and breast grey, upperparts and wings grey-brown; plain dull buff belly and undertail-coverts; tail blackish-brown tipped white and edged with fine white patches on outer webs, underside of tail dark boldly tipped white. **Adult female, unbarred morph** As male but less buff below, breast finely barred grey. **Adult female, barred morph** (rare) Differs in having upperparts more or less barred and streaked pale buff, chin to undertail-coverts whitish barred dark. **Juvenile** Above dark brown spotted and barred golden-rufous; whitish below streaked dark brown and washed buff. **Similar species** Sympatric Fan-tailed Cuckoo larger, has rufous breast, duller in female, and tail longer and rounded, with more white bars below; generally darker Fan-tailed juvenile less striated, and spotted brown on head, upperparts grey-brown, feathers edged rufous, below not barred but with arrow-like spots. Smaller Chestnut-breasted Cuckoo of Cape York area, Torres Strait and NG, has whole underparts deep chestnut, paler in female, juvenile has plain warm red-brown upperparts and off-white underparts with brown breast.

VOICE Various calls: 3–16 mournful notes, ventriloquial and high-pitched *peeyu-peeyu-peeyu-peeyu* … falling towards end, or *fear-fear-fear*, downslurred three-note call; shrill rising and hysterical (*too*)*-too-to-TEEI*, or 'where's the TEA', each note lasting *c.* 1sec; far-carrying descending call repeated many times only by male; female call abrupt *churrrt*; long and sharp shrill trilling call; low hissing call with open bill when mobbed. On Bougainville *addendus* call series of rising and falling notes (3–7), more trilled at end and quite different from other subspecies; this subspecies duetting on New Georgia, Solomons, produced great variety of calls. Very noisy in breeding season but otherwise quiet and unobtrusive; calls day and night, in some places in NG through much of year; immature birds also call (Campbell 1920, Sibley 1951, Diamond 1975, Marchant & Höhn 1980, Coates 1985, Coates & Bishop 1997, Higgins 1999). Far lengthier song scale than normal Brush and Rusty-breasted Cuckoos divided into repeated introductory scale and lengthy series of repeated notes believed to be '*heinrichi*' (Tebb *et al.* 2008).

DESCRIPTION Nominate. **Adult male** Crown to hindneck grey, sides of head grey, supercilium often paler; above grey-brown with hint of greenish gloss, in worn plumage longer shoulder feathers more brown, outer uppertail-coverts have varying white edges; wings plain grey-brown with bronze sheen, tertials more brown, very rarely fine off-white to buffish fringes on outer secondaries and outer greater secondary-coverts, remiges below dark grey with conspicuously white base, except 2–3 outer primaries, forming broad white band; forepart of underwing pink-buff and between that and white band, narrower grey band palest proximally; grey below, palest on chin and throat, gradually becoming buffish to rufous-brown on belly, rear flanks and undertail-coverts, few birds showing grey vermiculations on underparts. Graduated tail blackish-brown, with broad blackish subterminal band, tipped white and usually with small white notches on outer webs, some with narrow dark brown lower fringes, inner webs have small white half-bars not meeting shaft and largest on T5; undertail glossy grey broadly tipped white with white bars to inner webs. **Adult female, unbarred morph** As adult male but generally paler. Crown, nape and hindneck paler and browner, sides of head also paler; rest of upperparts little paler and browner when plumage worn, wings paler and more greyish-brown, varying number of secondary-coverts edged light rufous-brown and secondaries either edged or spotted rufous-brown; underparts off-white with light rufous-brown wash, often faintly vermiculated grey on breast, belly and flanks; tail spots often washed rufous-brown. **Adult female, barred morph** (often considered subadult but is rare female morph apparently homologous to hepatic/rufous morph in female *Cuculus* and other *Cacomantis*.) Top of head, nape, hindneck, upperparts and wings either uniform grey-brown or grey-brown barred and spotted buffish to pale buff, with spots and fringes on secondaries and wing-coverts; underparts white heavily cross-barred dark grey-brown, breast often washed rufous. Intermediates occur between these two morphs. **Juvenile** Like female barred morph, but more golden-buff to rufous-brown on head, large golden-rufous spots on upperparts and wings, barring below denser and darker, on breast more like black-brown spots; tail dark brown with broad light rufous-brown bars and tipped white. **Nestling** Naked; black on hatching, inside of mouth orange to red (Payne 2005). **Bare parts** Iris dark red-brown to black-brown, outer ring light; eye-ring cream, pale yellow, blue-grey or grey; juvenile olive-brown with dark spots, later dark brown; eye-ring grey. Bill grey-black, or black to dark brown, base of lower mandible buff to dull yellow; gape orange to vermilion; juvenile dark grey, basally horn-orange or creamy-buff base; mouth more yellow than adult. Feet pinkish orange, rufous-brown, greenish-yellow to olive, scutes on front of tarsus and toes brown, claws dark grey to black; juvenile feet greenish-grey (mostly Higgins 1999).

BIOMETRICS Nominate. Wing male 127–148mm (n = 17), mean 135.5 ± 6.05, female 128, 131, 132mm. Bill male 20.9–24.0mm, mean 22.4 ± 1.06, female (n = 4) 20.7–22.5mm, mean 21.9 ± 0.83. Tail male (n = 18) 105–128mm, mean 115.3 ± 6.07, female 102–113mm, mean 106.5 ± 5.07. Tarsus male 16.7–18.8mm, mean 17.8 ± 0.75, female 17.6–18.0mm, mean 17.8 ± 0.16. Mass male (southern population) 39–49g (n = 8) mean 41.8 ± 3.4, male (northern population) 28.0–37.5g (n = 22), mean 33.7 ± 2.4, female (northern population) 21.8–38.0g (n = 7), mean 32.1 ± 5.3 (Higgins 1999).

MOULT Australia, Jan–Aug; tail moult Oct. Two males Nov, New Georgia, Solomons. P9 and 7 simultaneously, P5–8–6, P4 and 1 simultaneously, P2–3, few individuals follow this sequence exactly. Post-juvenile moult complete or more often nearly so, retaining some secondaries and greater-coverts (Sibley 1951, Stresemann & Stresemann 1961, Higgins 1999, R. Clarke *in litt.*).

GEOGRAPHICAL VARIATION Seven subspecies.

C. v. variolosus (Vigors & Horsfield, 1826). North and east Australia, and Torres Strait; southern population winters south NG, Misool, and north to Aru Is, Taliabu and Kai (Moluccas), Timor and Kisar (Lesser Sundas); northern Australian population sedentary. Largest subspecies. Those of north Australia south to Burdekin R. '*dumetorum*' said to be smaller, wing male (n = 35) 118–134mm, mean 128.3 ± 4.19, female (n = 6) 119–132mm, mean 123.7 ± 4.55, and slightly paler; and intergrade with typical nominate at Burdekin R. (Higgins 1999). Described above.

C. v. infaustus Cabanis & Heine, 1863. Moluccas on Morotai, Tidore, Ternate, Halmahera, Bacan, Obi, Seram Laut (Gorong, Manawoka), Watubela, Kai, Tanimbar; WP islands (Gebe, Waigeu, Salawati, Kofiau, Misool, Goram), north and north-west NG east to Sepik R. and islands north-east of NG (Goodenough and Fergusson Is, D'Entrecasteaux Archipelago and perhaps Rook, Dampier and Vulcan). Larger; crown grey, upperparts darker brownish-olive with dull sheen, throat grey, breast and belly cinnamon to buff, rufous-grey or wholly grey; eye-ring grey or yellow; bill slender. Wing male 115–130mm (n = 24), mean 124.0 ± 4.3, female 114–130mm (n = 12), mean 122.1 ± 4.0 (Payne 2005). Those on Tanimbar may belong to *C. sepulcralis* (Bishop & Brickle 1999). Adult '*heinrichi*' has head dark grey, entire upperparts and wings dark olive-brown; chin and throat grey, rest of underparts dark rufous, undertail-coverts reddish. Tail dark olive-brown tipped white; fine white spots along outer edges, T5 obliquely barred white; iris carmine red with paler outer ring. Juvenile has head, upperparts and wings cross-barred dark brown and buffish, brown bars wider; underparts white cross-barred black; tail dark brown barred rufous. Wing male 112, 118, 122mm, female 113, 114mm; tail male 119, 121,123mm, female 116, 118mm (Stresemann 1931).

C. v. oreophilus Hartert, 1925. East and south NG highlands, west to Astrolabe Bay in north and Snow Mts. in south; Kakar, Manam and Umboi islands. Wing shorter, bill broader and thicker than *infaustus*, below grey-brown tinged rufous, eye-ring dark. Wing male 110–128mm (n = 23), mean 119.3 ± 4.1, female 107–120mm (n = 20), mean 116.7 ± 4.7 (Payne 2005).

C. v. websteri Hartert, 1898. Bismarck Archipelago (New Hanover). Underparts grey, breast occasionally washed rufous, undertail-coverts barred rufous or plain rufous; tail shorter. Wing male 115–122mm (n = 8), mean 117.9 ± 2.1, female 115mm (Payne 2005).

C. v. blandus Rothschild & Hartert, 1914. Bismarck Archipelago (Admiralty Is.). Smaller; back grey, uppertail-coverts blue-grey. Chin and throat grey, rest of underparts cinnamon to rufous. Wing male 102–117mm (n = 10),

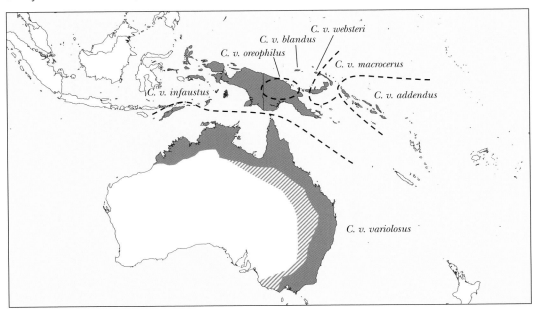

mean 112.1 ± 4.4, female 110–113mm (n = 7), mean 111.4 ± 1.3 (Payne 2005).

C. v. macrocercus Stresemann, 1921. Bismarck Archipelago (except New Hanover and Admiralty Is.); Tabar Is. Three morphs: (1) whole underparts grey, (2) belly rufous, (3) intermediate. Tail longer with white half-bars on inner edges. Wing male 112–139mm (n = 17), mean 121.7 ± 5.8, female 121, 125,127mm (Payne 2005).

C. v. addendus Rothschild & Hartert, 1901. Solomons (Bougainville, Kolombangara, New Georgia, Malaita, Guadacanal, Makira, Ulawa, and Santa Isabel). Upperparts uniform blue-black, chin and throat grey, rest of underparts variably dark rufous or rufous-buff; eye-ring yellow. Wing male 115–124mm (n = 12), mean 120.3 ± 3.4, female 117–126mm (n = 6), mean 120.8 ± 4.1 (Payne 2005).

DISTRIBUTION Wallacea and Australasia. Lombok, Sumbawa, Flores, Sumba, Sulawesi, Butung, Tukangbesi Is, Banggai Is, Sula Is, Buru, Ambon, Seram and Timor; austral migrant Kisar, resident Roti; NG; Bismarcks on New Ireland, Tabar, Tanga, New Hanover, Lihir and New Britain; Manus, Admiralty Is; Solomons; north and east Australia. Resident throughout range except nominate breeding in south-east Australia partly migratory to north Australia, NG and Wallacea in Feb–Apr returning Sep–Oct. 137 birds ringed between 1953–96 in Australia and NG, of which six recoveries, all within 10km of ringing locality (Peters 1940, Sibley 1951, Cain & Galbraith 1956, Higgins 1999, Fletcher 2000, Trainor 2003, 2005a, Dutson 2007). Population on Tanimbar, Lesser Sundas, may belong to this species or *C. sepulcralis*; specimens required (Bishop & Brickle 1999). Once Lord Howe I., Nov (McKean & Hindwood 1965); once Palau (Pratt *et al.* 1987).

HABITAT In Australia edges of dense monsoon rainforests, moist sclerophyll forests with *Eucalyptus*, gallery forests, mangroves, scrubs with vines, wattles (*Acacia*) or paperbarks (*Melaleuca*), but also in drier areas in open woodlands and forests and even gardens, with some changes in preferred habitat depending on season in north Australia: in dry season mangroves and monsoon forests and in wet season more dry open forests. Primary forests and at edges, secondary forests, logged forests, scrub, *Casuarina* groves, savannas, mangroves and gardens, to 1,800m, New Guinea; Bougainville 1,200m. To 1,200m Solomons. Generally woodlands, forest outskirts and farmlands, occasionally wooded urban areas, Wallacea generally. To 200m Halmahera and 300+m Timor (Crawford 1972, Coates 1985, White & Bruce 1986, Coates & Bishop 1997, Higgins 1999, Kennedy *et al.* 2000). '*Heinrichi*' in mountain forest at 1,000–1,500m Halmahera and 800–1,200m, but occasionally to 100–450m, Bacan; apparently confined to primary forest, whereas Brush Cuckoo normally in more open habitats (Coates & Bishop 1997, Tebb *et al.* 2008, BirdLife International 2011).

BEHAVIOUR Poorly known. Outside breeding mainly solitary but also in small groups. Often motionless for long periods; upright stance, slowly moving tail back and forth. Mostly in canopy where very difficult to identify; also forages on ground. Male seen feeding female. Some observations, e.g. caught at lighthouse at night, suggest nominate is nocturnal migrant (Bravery 1970, Hall 1974, Marchant & Höhn 1980, Hadden 1981, Noske 1981, Coates 1985, Higgins 1999).

BREEDING Parasitism Hosts in Australia mainly honey-eaters (35%) including Brown-backed, Bar-breasted, Yellow-faced, Yellow-throated, and Red-throated; fantails (22%) including New Zealand and Northern Fantail, and Willie-wagtail; monarchs (15%) including Leaden, Restless and Satin Flycatchers; Australasian robins (11%) including Scarlet, Red-capped, Rose and Flame, and Jacky Winter; fairy-wrens (7%) including Purple-crowned and Emperor. 38 host species recorded only from Australia of which only most common mentioned here as some of single host records from east Australia south of 20°S may be doubtful because sympatric Fan-tailed Cuckoo has similar eggs. White-shouldered Fairy-wren, Sooty Honeyeater, Brown-backed Honeyeater, Lemon-bellied Flycatcher, possibly Willie-wagtail, NG. Pied Bushchat, Flores. Black Sunbird Sula Is, Moluccas. Juvenile fed by Olive-backed Sunbird, Manus (Beruldsen 1978, Coates 1985, Brooker & Brooker 1989, Coates & Bishop 1997, Higgins 1999, Fletcher 2000, B. van Balen *in litt.*). **Season** Jan–Feb Northern WA; eggs May–Oct, Queensland, few Dec and Feb, fledglings Jan and Apr; eggs Aug and Oct–Jan, NSW, Nov and Jan Victoria. Five females laying Nov–Dec, south NG (Rand 1942a, b, Higgins 1999). Breeding condition ('*heinrichi*') early Oct (White & Bruce 1986). **Eggs** Oval close-grained, glossy or dull, ground colour white or creamy-white with brown or purplish-brown blotches and spots around thicker end with underlying grey to violet-grey spots; 16.0–21.6x11.4–16.0mm (n = 55), mean 18.0x13.7 (Higgins 1999). **Incubation** 12 days in Grey Fantail's nest; 12 ± 5.5 days (Marchant 1980, Higgins 1999). **Chicks** Hatch naked and blind; quills emerge 7–8th day, fully feathered 9–10 days later. Within 30 hours of hatching eject other eggs or young. Fledge after 17 ± 4.5 days and one fledgling fed by hosts for at least one month (Marchant 1980, Higgins 1999). **Survival** Of nine eggs, one was deserted, two failed to hatch and six fledged (Higgins 1999).

FOOD Hairy caterpillars, other insects including dragon-flies, grasshoppers, bugs, wasps, beetles; spiders; snails; fruit and grass seeds. (Hall 1974, Coates 1985, Payne 1997, Higgins 1999).

STATUS AND CONSERVATION In *Eucalyptus* forest, Victoria, 26 territories/km^2. Densities from different parts of Australia 0.02–0.1 birds/ha (Higgins 1999). Common Darwin area 1967–1972 (Crawford 1972). Moderately common Wallacea (Coates & Bishop 1997). Common Santa Isabel (Webb 1992), but considered rare by Kratter *et al.* (2001b). Common resident Timor-Leste (Trainor *et al.* 2008b). Fairly common resident throughout Solomons (Doughty *et al.* 1999). Very uncommon middle Sepik, NG (Gilliard & LeCroy 1966). Not globally threatened (BirdLife International 2011).

'*Heinrichi*' only known from five specimens collected 1931, and rarely recorded since (Coates & Bishop 1997, Tebb *et al.* 2008). Monitoring programme to decide status seems to be of highest priority. Owing to loss of forest habitat considered near threatened, population estimate 10,000–19,999 (BirdLife International 2011).

Brush Cuckoo *Cacomantis variolosus*. **Fig. 1.** Adult male, *C. v. variolosus*, Ashmore, Queensland, Australia, October (*Rohan Clarke*). **Fig. 2.** Adult female barred morph, *C. v. variolosus*, Ashmore, Queensland, Australia, November (*Rohan Clarke*). **Fig. 3.** Juvenile, *C. v. variolosus*, Broome, Western Australia, November (*Adrian Boyle*).**Fig. 4.** Adult male, '*heinrichi*', Foli, Halmahera, Moluccas, Indonesia, September (*Pete Morris*). **Fig. 5.** Adult male, probably '*heinrichi*', Obi Island, north Moluccas, Indonesia, March (*Marc Thibault*).

RUSTY-BREASTED CUCKOO
Cacomantis sepulcralis   Plate 31

Cuculus sepulcralis S. Müller, 1843 (Java and Sumatra)

TAXONOMY Sometimes considered conspecific with *variolosus* (Hartert 1925b, Payne 2005), but *sepulcralis* much smaller, has different vocalisations, and appears to be more closely related to *merulinus* sympatric from Malay Peninsula to Philippines (White & Bruce 1986). Payne (2005), supporting conspecifity, states that calls are similar, size and plumages intergrade, and no consistent difference in eye-ring colour. '*Heinrichi*' may belong here, perhaps in *aeruginosus* (Tebb *et al.* 2008). More acoustic and genetic work required. Possibly forms superspecies with *variolosus*, although perhaps partially sympatric with it. Polytypic. Synonyms: *variolosus stresemanni* Hartert 1925 in *aeruginosus*, *v. everetti* Hartert 1925 in nominate, *v. fistulator* Stresemann 1940 in *virescens*.

FIELD IDENTIFICATION 22cm. Small cuckoo. **Adult male** Grey head, grey-brown upperparts, wings more bronze, rump grey; in flight underwing-coverts pink-rufous with white area; below clear, uniform pink-rufous with grey tinge on chin and throat gradually becoming pink-rufous on breast. Tail blackish-brown tipped white and edged with white spots, except T1; T4–5 spots larger becoming incomplete bars; underside paler strongly tipped white. **Adult female, grey morph** As male but underparts finely barred grey. **Adult female, rufous morph** Dark brown above heavily barred and spotted rufous; below whitish heavily barred dark brown, spacing between bars much greater on undertail-coverts. **Juvenile** Head and upperparts buffish densely barred black, wings dark brown edged buffish, underparts white barred black, throat and breast white washed buffish; tail dark brown notched dull rufous. **Similar species** Sympatric Plaintive Cuckoo (*querulus*) very similar but lacks yellow eye-ring; sympatric *threnodes* has upper breast and throat pale grey instead of rufous; *lanceolatus* (Java) has only wash of rufous on belly; *merulinus* (Philippines) has pale buffish underparts with grey tinge on breast. Juvenile Plaintive has dull rufous tail patches on both webs of T1, where present species only has patches on outer webs which do not reach shaft. Rufous morph similar to Plaintive but darker, larger, tail appears longer, and blackish bars on upperparts, throat and breast broader; rufous notches on rectrices, not complete bars. Banded Bay has dark eye-stripe and banded head in all plumages.

VOICE Two calls recorded for nominate: 10–15 melancholy and ventriloquial *few-few-few*, each three notes lasting 2sec and descending; second call 3–4 notes *ti-ter-wee*, also interpreted as *ti-teter-wee*, higher pitched at end, also transcribed as *wheet to wheet*, repeated 4–5 times ascending each time. Trilling call on Sumba. On Java mellow descending note *hiet* reiterated 10–25 times, repeated *peewieiet* and call like that of Plaintive but faster *ti-ter-wi* or *pee-to-peet*. *Virescens*: *tee-too* first note rising, second falling; *fiew*, occasionally single but most often in series of 2–7 higher pitched rising notes, five notes/3sec; ascending and shrill *pee-peei-peeit-peeit-pee-towee-it* lasting 8.7sec, often without first two notes. *Aeruginosus*: c. 25 ascending piping whistles, repeated, staccato, little slower at end, lasting 4.5sec; 8–80 monotonous whistles, rather high pitched c. two notes/sec, occasionally shorter intervals. Sometimes calls throughout night. Vocalisations on Buru and Seram distinctive; series of several dozen notes partially level, partially rising, contrasting with those of other forms of this species and with Brush Cuckoo (Jones *et al.* 1994, Coates & Bishop 1997, Wells 1999, Kennedy *et al.* 2000, Rheindt & Hutchinson 2007).

DESCRIPTION Nominate. **Adult male** Top and sides of head to nape plain grey, upperparts and wings dark brown with bronzy sheen, rump and uppertail-coverts blackish-grey, underwing-coverts pink-rufous with prominent white area at base of inner webs of remiges; chin and throat grey tinged pink-rufous, rest of underparts plain rufous, some tinged pinkish, without sharp demarcation. Great individual variation, especially of colour of underparts, but perhaps due to fading of live birds or post-mortem colour changes in old skins. Tail blackish-brown to black, tipped white with row of white patches on outer webs of all except central rectrices, undertail brown with whitish patches or bars on inner webs. **Adult female, grey morph** As male but fine grey barring on underparts. **Adult female, rufous morph** Dark brown above heavily spotted on head and nape, and barred on mantle, back and upper tail-coverts buff-rufous; wing-coverts edged pale rufous; secondaries tipped rufous, spotted on edges; primaries narrowly tipped and edged pale rufous. Underparts whitish, heavily barred dark brown, bars further apart on undertail-coverts, with some rufous suffusion on sides of head to breast; large whitish area on underside of inner webs of remiges, becoming more buff on outer primaries, pale rufous notches on inner webs of secondaries. Tail dark brown tipped buff and white; T1 dark brown notched rufous, T2–5 heavily barred and notched pale to dark rufous. **Juvenile** As rufous morph but ground colour more buffish with broader black barring on upperparts and buffish tinge on throat and breast, underparts white finely barred dark brown. **Nestling** Undescribed. **Bare parts** Iris hazel-brown, juvenile grey-brown, eye-ring bright yellow, juvenile greenish-yellow. Bill black, sometimes pinkish yellow lower mandible. Feet yellow, juvenile pale yellow.

BIOMETRICS Nominate. Wing male 110–134mm (n = 24), mean 115.9 ± 2.7, female 104–120mm (n = 18), mean 114.7 ± 4.6. Bill male 15.6–18.4mm, mean 16.6 ± 1.0, female 14.4–19.1mm, mean 17.0 ± 1.3. Tail male 112–127mm, mean 116.8 ± 4.3, female 108–118mm, mean 116.2 ± 6.8. Tarsus male 15.8–19.2, mean 17.2 ± 1.0, female 14.8–18.4mm, mean 16.7 ± 1.2 (Payne 2005). Mass adult (n = 10) 28.2–38.0g (Wells 1999).

MOULT Juvenile to adult moult, Simeulue, Sumatra, Dec. Wing moult Thai-Malay Peninsula, late Apr–Sep; post-juvenile wing moult Sep and Jan (Ripley 1944, Wells 1999).

GEOGRAPHICAL VARIATION Three subspecies.
C. s. sepulcralis (S. Müller, 1843). Thai-Malay Peninsula (south of 11°55'N Tenasserim: Htin Hla *et al.* in prep. in Wells 2007, and south of 12°45'N Thailand: C. Robson pers. comm.); Sumatra and offshore islands (Simeulue, Enggano, Belitung); Borneo, Java, Bali, Lombok, Sumbawa, Flores, Alor, Sumba and Philippines. Described above.
C. s. virescens (Brüggemann, 1876). Sulawesi, Butung, Banggai, Tukangbesi (Tomea, Binongko); Peleng (BMNH). Darker, above more bluish-grey, below uniform dark chestnut, throat usually darker, and bill smaller. Wing male 102–118mm (n = 12), mean 111.6 ± 5.0, female 104–111mm (n = 4), mean 108.5 (Payne 2005).
C. s. aeruginosus Salvadori, 1878. Sula Is. (Taliabu, Seho, Mangole), Buru, Ambon, Seram. Slightly darker and less grey on throat and breast than nominate. Wing male 113–

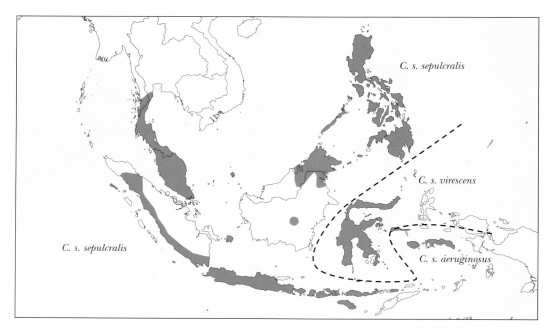

121mm (n = 9), mean 116.1 ± 3.4, female 117, 118, 119mm (Payne 2005). Vocalisations on Buru and Seram differ from nominate on Philippines (Rheindt & Hutchinson 2007). Vocalisations from Taliabu, and biogeographical considerations, suggest that birds from Sula Is. are closer to *virescens* (Rheindt 2010).

DISTRIBUTION Oriental. South Burma, Thai-Malay Peninsula, Greater Sundas, Philippines and Wallacea. Nominate possibly partial migrant as daily visitor to Alor Star, Langkawi Is. 7–9 Aug 1939, and three specimens supposedly of nominate subspecies Cocos Is, 1913. Population on Tanimbar, Lesser Sundas, may belong to this species or *variolosus* (specimens required). In Philippines on Bohol, Busuanga, Camiguin Sur, Catanduanes, Cebu, Leyte, Luzon, Marinduque, Masbate, Mindanao, Mindoro, Negros, Palawan, Panay, Samar, Siargao, Sibuyan, Siquijor, Tablas and Ticao. Resident throughout most of range. Two ringed birds at Dalton Pass, Luzon, recovered respectively 160 km south 16 months, and 40 km south 23 months later (Junge 1936, Bromley 1949, McClure 1998, Bishop & Brickle 1999, Kennedy *et al.* 2000, Payne 2005, Allen 2006).

HABITAT Lowland forests, both semi-evergreen and evergreen, logged or undisturbed, forest-edges including montane, mangroves and old plantations; prefers more forested and strongly wooded habitats than Plaintive Cuckoo. Found in both forest and cultivation north Sulawesi. Mangroves and mossy forest to 2,000m, Philippines. Wooded lowlands, beach scrub, peatswamp, lowland and hill dipterocarp, alluvial and logged forests, mangroves and old rubber plantations, most often in hilly lowlands to 1,700m, Borneo. To 600m Thai-Malay Peninsula. To 2,500m Sulawesi, 1,000+m Buru and Sumbawa, 1,500m Seram, 1,200m Lombok, 1,400m Flores and 950+m Sumba. Disturbed forest to 1,000m Taliabu (Coomans de Ruiter 1951, Coates & Bishop 1997, Jeyarajasingam & Pearson 1999, Smythies & Davison 1999, Wells 1999, 2007, Kennedy *et al.* 2000, Mann 2008, Rheindt 2010).

BEHAVIOUR Males calling from favourite posts suggests male territoriality. Easily overlooked; most often seen alone and when not vocal very difficult to find. Mistnetted near ground suggesting it forages there (Coates & Bishop 1997, Wells 1999).

BREEDING Parasitism Hosts, mostly south-east Asia: Long-tailed Shrike, Sooty-headed Bulbul, various fantail species, Chestnut-naped Forktail, Northern/Southern White-crowned Forktail, Pied Bushchat, flycatchers, Striated Grassbird, Mountain and other tailorbirds, and Olive-backed Sunbird. Pied Fantail, Singapore; Tickell's Blue Flycatcher, peninsular Malaysia; main host Flores Pied Bushchat; Long-tailed Shrike, Pied Bushchat and Snowy-browed Flycatcher, Java; Black Sunbird, Sula; Buru White-eye, Buru. Hosts unknown Borneo (perhaps Chestnut-naped Forktail), and Philippines (Hellebrekers & Hoogerwerf 1967, Schönwetter 1967, Ottow & Verheijen 1969, Cranbrook & Wells 1981, Rozendaal & Dekker 1989, Coates & Bishop 1997, Beisenherz 1998, Ollington & Loh 1999, Kennedy *et al.* 2000, Poh 2000 in Wells 2007). **Season** Fledglings Aug–Sep, Malay Peninsula, 17 Jul north Sulawesi, Mar–May Lesser Sundas. Laying female 12 Jan, Sulawesi. Flores, Jul–Oct. Newly fledged birds Mar and Apr, Mindanao (Ottow & Verheijen 1969, White & Bruce 1986, Beisenherz 1998, Ollington & Low 1999, Kennedy *et al.* 2000, Poh 2000 in Wells 2007). **Eggs** Flores 20.0x14.1mm; mean 19.5x14.8mm (Ottow & Verheijen 1969, Robson 2000). No further information.

FOOD Eats fruit, but presumably mainly insectivorous (Wells 1999, Walker 2007).

STATUS AND CONSERVATION Regular and locally common Thai-Malay Peninsula, but not in well-kept plantations (Wells 1999). Uncommon resident south-east Asia generally (Robson 2000). Largely overlooked Borneo, where uncommon and local to 1,700m, but widespread resident Sabah (Thompson 1966, Mann 2008). Common in wooded areas central Sulawesi (Watling 1983). Common Flores

(Ottow & Verheijen 1969). Fairly common Lesser Sundas (White & Bruce 1986). Widespread and moderately common resident Wallacea generally (Coates & Bishop 1997). Not globally threatened (BirdLife International 2011).

Rusty-breasted Cuckoo *Cacomantis sepulcralis*. **Fig. 1.** Adult male, *C. s. virescens*, Gunung Mahawu, Minehasa Peninsula, north Sulawesi, Indonesia, March (*Marc Thibault*). **Fig. 2.** Adult male, *C. s. sepulcralis*, Singapore, January (*H. Y. Cheng*). **Fig. 3.** Adult male, *C. s. sepulcralis*, Singapore, January (*H. Y. Cheng*). **Fig. 4.** Juvenile, *C. s. sepulcralis*, Quezon City, Luzon, Philippines, October (*Tina Sarmiento Mallari*).

Genus *Caliechthrus*

Cabanis & Heine, 1863 *Mus. Hein.* Th. 4, Heft 1: 31, note. New name for *Simotes* Blyth 1846.
Type, by monotypy, *Cuculus leucolophus* S. Müller 1840.
One species.

New Guinea; presumably brood-parasitic; sexes similar. Sole species, *leucolophus*, has been placed in *Eudynamis* (sic), which has slit nostrils, but this species has round nostrils as *Cacomantis*, in which genus it is placed by Payne (2005), but great difference in appearance to other members of that genus suggests that monotypic *Caliechthrus* is valid.

WHITE-CROWNED CUCKOO
Caliechthrus leucolophus Plate 30

Cuculus leucolophus S. Müller, 1840 (Lobo Bay, New Guinea)

Alternative name: White-crowned Koel.

TAXONOMY On mtDNA evidence Sorenson & Payne (2005) place it in *Cacomantis* closest to *pallidus*. However, appearance very different to others in this genus and we retain it in monotypic *Caliechthrus* pending further work. Monotypic.

FIELD IDENTIFICATION 30–35cm. Medium-sized black cuckoo with rather slender bill and undulating flight. **Adult** Dull blackish plumage with bluish gloss on upperparts and tail and conspicuously white stripe on top of head from bill to nape, tail tipped white. Female browner. **Subadult** Similar to adult but white stripe less prominent, more patchy; washed brown on mantle; barred whitish below. **Juvenile** Mostly white, with black face and throat to upper breast and grey remiges and rectrices. **Similar species** Common Koel male larger and totally black, and bill pale green to horn, not black.

VOICE Call short, loud and rapid *ka-ha-ha* or *ka-ha-ha-ha* resembling human laughter, and mournful falling whistled notes *too-too-too* or *too-too-too-too*; may switch calls. Calling occasionally ends with excited *week-week-week-week*; also single downslurred *whurr* (Coates 1985, Beehler *et al.* 1986).

DESCRIPTION Adult Forehead, crown and nape black with broad white stripe down centre, rest of head, neck and upperparts black glossed blue, wings black, underwing-coverts frequently with narrow white bars; underparts generally dull greyish-black, variable amount of white tipping to undertail-coverts forming bars; graduated tail black tinged blue and tipped white. Female browner on remiges and rectrices, and below. **Subadult** As adult, but whitish tips to feathers on breast, belly and flanks forming indistinct bars; undertail-coverts strongly barred white. **Juvenile** Forehead and lores dark brown; black streak from eye to bend of wing; white streak below eye. Crown, back and rump white streaked black; scapulars black streaked white; upperwing-coverts white streaked black; remiges grey notched buffy, edged white. Throat to upper breast black edged whitish; lower breast and belly white streaked dark grey. Rectrices blackish notched white. **Nestling** Down on underparts dark sooty-brown lacking white markings. **Bare parts** Eye-ring yellow, paler in juvenile. Iris brown, grey in

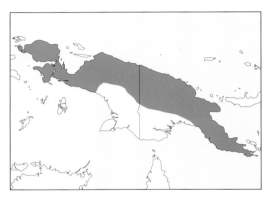

juvenile. Bill black. Feet pale grey-brown to dark grey.

BIOMETRICS Wing male 167–182mm (n = 8), mean 172.3 ± 5.2, female 165–178mm (n = 6), mean 168.5 ± 4.8. Bill male 24–29mm, mean 26.5 ± 2.9, female 22.6–27.8mm, mean 25.5 ± 2.2. Tail male 162–180mm, mean 169.6 ± 7.0, female 159–170mm, mean 163.8 ± 3.6. Tarsus male 23.2–25.0mm, mean 24.0 ± 0.7, female 20.6–24.5mm, mean 22.5 ± 1.3. Mass male 106.5, 112.9, 125.0g, female 99.0g (Payne 2005).

MOULT Post-juvenile moult Jan; two in wing moult, Nov. Primaries show disorderly transilient moult with no firm rule (Mayr 1931, Stresemann & Stresemann 1961, BMNH).

DISTRIBUTION NG, including Salawati I., WP. Widely distributed resident (Beehler *et al.* 1986).

HABITAT Forests, especially in hills; forest edges, clearings with high trees and secondary forest to 1,800m (Diamond 1972b).

BEHAVIOUR Solitary or in pairs, sometimes in small groups. Mostly only heard; sluggish frequenter of high canopy or midstorey, but also low in scrub (Coates 1985).

BREEDING Presumably parasitic. **Season** Two males 17 Jul and 28 Aug had small gonads, and female in breeding condition Oct (Diamond 1972b, Payne 2005). No further information.

FOOD Caterpillars and other insects; fruit (Coates 1985, Beehler *et al.* 1986).

STATUS AND CONSERVATION Widely distributed PNG and locally common (Coates 1985). Not globally threatened (BirdLife International 2011).

White-crowned Cuckoo *Caliechthrus leucolophus*. **Fig. 1.** Adult, Tabubil, PNG, June (*Pete Morris*). **Fig. 2.** Adult with traces of subadult plumage remaining, Boyston Road, near Kiunga, PNG, June (*Nik Borrow*).

Genus *Cercococcyx*

Cabanis 1882 *J. Orn.* 30: 230. Type, by original designation and monotypy, *Cercococcyx mechowi* Cabanis 1882. Three species.

Afrotropical; brood-parasitic. Small-bodied, with large rounded wings and very long tail; sexes similar. MtDNA analysis results in *Cercococcyx* and *Surniculus* forming two monophyletic clusters which are sister groups, these being sister to *Cuculus* and *Hierococcyx*; *Cacomantis* is sister group to these (Sorenson & Payne 2005).

DUSKY LONG-TAILED CUCKOO
Cercococcyx mechowi   Plate 32

Cercococcyx Mechowi Cabanis, 1882 (north Angola)

TAXONOMY Monotypic. Synonym: *m. wellsi* Bannerman 1919.

FIELD IDENTIFICATION 33cm. Small-bodied cuckoo with long, full tail. **Adult** Upperparts uniform dark sooty grey-brown, with slight purple-blue iridescence; below white boldly barred dark, breast and flanks variably washed buff in some; vent and undertail-coverts unmarked bright buff. Wings dark brown with small buff-brown spots on outer webs of remiges, large white spots on inner webs forming large white bar on underside. White-tipped rectrices variably spotted and barred white and tawny-white, more obvious on underside, T1 almost unmarked. **Juvenile** Similar to adult, but feathers of upperparts tipped rufous, giving barred appearance; rufous spots on both webs of remiges. Blackish throat; rest of underparts barred white and blackish-brown. Rectrices dark brown spotted and barred rufous. **Similar species** Long, full tail and relatively small body separate *Cercococcyx* cuckoos from *Cuculus* species, some of which occur in similar habitats. Best separated by voice from extremely similar Olive Long-tailed Cuckoo, but on average adult appears shorter-winged (wing 75% length of tail), darker sooty-grey above and broader bars below. Ground colour of breast pure white with less or no buffy wash. Vent usually unbarred plain buff, colour usually extending to flanks and deeper than in Olive Long-tailed. Most easily told from Mountain Long-tailed Cuckoo by voice, but adult Dusky has only remnant buff tips to greater and median coverts, and indistinctly barred primaries. Juvenile Dusky barred above and very similar to adult Mountain, but lacks buff ground colour on breast, has blackish throat and breast; narrow chestnut fringes to head, neck, throat and breast, and broader spaced bars below. Juvenile from Mountain by clean barring on underparts and white rather than buffy ground colour to underparts giving darker appearance, and dark barring below closer, but as birds moult to adult plumage they become equally pale. Juvenile Dusky from juvenile Olive by blackish-brown, not dark rufous, crown; rufous spots on remiges only on outer webs in Olive; rufous bars on tail across both webs in Olive, and broader, barring becomes whiter distally in Mountain, which also has white tips to all but central rectrices. Dusky much darker below and lacks tawny wash of Olive and Mountain. Undertail-coverts blackish-brown narrowly tipped white in Dusky, buffy-white with dark brown bars in Olive and Mountain. Adult Olive slightly more brownish above, paler undertail-coverts and different calls; juvenile has streaked underparts and never has blackish throat.

VOICE Call repeated three-note whistling *fwi-fwi-fwi*, or *tu-tu-to*, higher and sharper than Red-chested and Olive Long-tailed Cuckoos and all notes on same pitch or very slightly rising, 1sec pause between phrases which may be repeated for long periods at 25 phrases/min. Whistling descending series of *c*. 25–30 notes *teu, tew, du du du du du...* first rising then falling gently and slowing, lasting 10sec. Often calls at night. Two calls in west Africa, each with distinct western and eastern versions. In Upper Guinea (eastwards to west Cameroon) three rising notes, *he hee wheeu*, and less frequent whinnying series of plaintive notes, first accelerating, then slowing and descending *tiutiutiutiutiutiui-tiu-tui...* reminiscent of *Halcyon* kingfisher. In lower Guinea, first type faster, with 3 similar, less melodious notes, *wheet-wheet-wheet*; second, fast descending *wheewheewheewheewhee...* almost twice as fast as equivalent in Upper Guinea (Irwin 1988, Borrow & Demey 2001, C.F. Mann).

DESCRIPTION Adult Sexes generally alike, although some females may be less barred. Head, ear-coverts, nape and entire upperparts plain dark brown washed dark sooty-grey with slight purplish-blue iridescence, rump and uppertail-coverts blackish, outer webs of feathers flecked white. Wings dark brown, primaries with small buff-brown spots on outer webs and larger white spots on inner webs. Secondaries, greater coverts and scapulars dark brown spotted buff-brown on outer webs. Underwing-coverts white narrowly barred dark brown; large white spots on inner webs of primaries, broad white band on underside of remiges more extensive than in Mountain Long-tailed. Underparts from chin to belly white with little or no buff suffusion, strongly barred blackish-brown, vent and undertail-coverts plain tawny-buff. Long, full tail, strongly graduated, plain dark brown above with variable buff flecking on outer webs becoming whitish from T1 outwards, longer feathers very narrowly tipped white; T5 barred, spotted and tipped white. **Juvenile** Upperparts blackish-brown, crown feathers tipped rufous or chestnut; head and neck narrowly edged chestnut; rest of upperparts blackish-brown with rufous edging becoming pale rufous on back and rump, uppertail-coverts broadly tipped rufous; wings and tail dark brown barred rufous. Remiges spotted rufous on both webs, white proximally on inner webs. Throat blackish barred white with some rufous tipping; rest of underparts barred white and blackish-brown, heavier than in adult, undertail-coverts blackish-brown narrowly tipped white. Underwing-coverts white; large white spots on underside of remiges forming white bar on secondaries; distally spots become buff, more numerous on outer webs. Tail dark brown barred rufous, most bars incomplete; underside dark brown with large buffy-white notches on both webs mostly at edges, becoming paler on proximal third; rectrices pointed. **Nestling** Hatchling has black skin, yellow rump, pale feet (Brosset & Érard 1986). **Bare parts** Iris dark brown; eyelids yellow. Bill greenish-

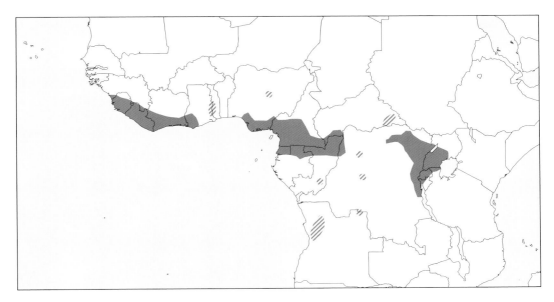

black, lower mandible dull green; mouth yellow. Feet chrome yellow.

BIOMETRICS Wing male (n = 32) 129–147mm, mean 135.7 ± 4.1; (n = 11) 132–140mm, mean 136.0 ± 4.3; unsexed adults (n =3) 133–134mm; juveniles (n = 9) 123–138mm. Bill male (n = 18) 17.8–21.9mm, mean 19.5 ± 1.0, female (n = 8) 16.6–22.2, mean 19.1 ± 1.6, unsexed (n = 14) 23–25mm, mean 24.0. Tail male (n = 30) 170–197mm, mean 185.1 ± 7.0; female 172–195mm (n = 11), mean 182.2 ± 7.9; unsexed adults 173–181mm; juveniles 142–195mm. Tarsus male (n = 18) 16.0–18.8mm, mean 17.6 ± 0.9, female (n = 8) 17.3–21.0, mean 18.3 ± 1.3, unsexed (n = 4) 17.0–20.4mm, mean 18.9. Mass male (n = 7) 52–60g, mean 57; female (n = 8) 50–61g, mean 56 (Irwin 1988, Louette & Herroelen 1994, Payne 2005); unsexed 47.0, 51.6, 59g (Baker & Baker 1994); female 43.4g (BMNH).

MOULT Adult plumage acquired after single post-juvenile moult. Wing moult Mt. Nimba, Liberia, Nov (Louette & Herroelen 1994, BMNH).

DISTRIBUTION Afrotropical; discontinuous distribution. Resident, but may wander. West Africa from southern Guinea and SL to south IC and south-west Ghana, then east from Omo Forest and Niger delta in Nigeria to northern RC, southern CAR, extreme north-west DRC and north Gabon. In eastern Africa, from north-east DRC to west and south Uganda, extreme north-west Tanzania, Rwanda and Burundi. Scattered records outside this range from south Guinea, Togo, Ghana, north and west Nigeria, CAR, central DRC and north Angola (including Kumbira Forest) (Chapin 1939, Irwin 1988, Baker & Baker 1994, Halleux 1994, Payne 1997, Baker 2001, Demey & Rainey 2004, Demey 2007).

HABITAT Forest specialist, avoiding small patches. Prefers forest with dense undergrowth and liana growth. Will tolerate tall secondary growth (including *Musanga*-dominated) and dense riparian undergrowth. Gallery forests in Cameroon and DRC, and up to edges of rivers in Gabon. To 1,830m (Chapin 1939, Marchant 1953, Irwin 1988).

BEHAVIOUR Often low in undergrowth, where it may keep to dark, dense tangled growth, but also frequently in mid-canopy where easier to detect. Often in pairs, calling throughout day from exposed perch in treetops, although can be silent in dry season. Joins mixed species foraging flocks of insectivorous birds (Ripley & Heinrich 1966a, Irwin 1988, Stevenson & Fanshawe 2002).

BREEDING Parasitism Adult mantling nest of Blue-headed Paradise-flycatcher, and fledgling fed by Brown Illadopsis, DRC. Nestling cuckoo in nest of Forest Robin ejected host young and had partly developed plumage of Dusky Long-tailed Cuckoo, but identity not confirmed; two eggs believed to be of this cuckoo in Forest Robin's nest, north-east Gabon (Chapin 1939, Brosset & Érard 1986). **Season** Calls most frequently in wet season, and breeding seemingly associated with rains. Calling throughout year Yapo Forest, IC; calling Apr–Dec, Kagoro, Nigeria; breeding condition Dec–Feb, Cameroon; apparently laying Nov and Jan, north-east Gabon; Oct–Nov, Angola; Jan–Apr (Itombwe), Apr–Jul (Ituri), Sep (Lutunguru), DRC, and fledgling there Jul; enlarged testes Cameroon, Jan and Feb, south Nigeria, Jan, Liberia, Aug; eggs forming Cameroon and SL, Feb; breeding condition Feb, Uganda and oviduct egg (Budongo), May; juvenile Cameroon, Dec (Chapin 1939, Brosset & Érard 1986, Dyer *et al.* 1986, Irwin 1988, Demey & Fishpool 1994, BMNH). No further information.

FOOD Mainly insects, particularly hairy caterpillars. Caterpillars in all 21 stomachs examined, several containing only caterpillars; also ants, beetles, spiders, snails, seeds (Chapin 1939, Irwin 1988, Payne 1997).

STATUS AND CONSERVATION Although elusive, not rare across most of range. Tolerates well-developed secondary growth, but appears to avoid very small forest patches, so could be affected by forest fragmentation. Uncommon Nigeria, and only once recorded south-east, at Umuagwu (Marchant 1953); not uncommon SL (Payne 1997); uncommon Yapo Forest, IC (Dewey & Fishpool 1994). Three records Togo (Cheke & Walsh 1996). Not globally threatened (BirdLife International 2011).

Dusky Long-tailed Cuckoo *Cercococcyx mechowi*. **Figs. 1–2.** Juvenile, Bwindi NP, Uganda, February (*Jonas Rosquist*).

OLIVE LONG-TAILED CUCKOO
Cercococcyx olivinus Plate 32

Cercococcyx olivinus Sassi, 1912 (mountain forest south of east edge of Rutshuru Plain, Congo)

TAXONOMY Forms superspecies with *montanus*. Monotypic.

FIELD IDENTIFICATION 33cm. Small-bodied cuckoo with long, full tail. **Adult** Upperparts plain olive-brown, sometimes glossed bronze. Below whitish with hint of buff wash on breast, dark bars below reaching to undertail-coverts; tail spotted and barred grey and white. **Juvenile** Dark rufous crown, nape to uppertail-coverts blackish-brown tipped rufous. Remiges dark brown spotted rufous. Underparts white washed tawny-buff with well-spaced dark brown bars. Underwing-coverts and large patch on secondaries white. Tail dark brown with complete, broad rufous bars; underside light and dark bars of equal width. **Similar species** Extremely similar to Dusky Long-tailed Cuckoo, and best separated on voice; adult Olive tends to appear longer-winged (wing >80% length of tail), less sooty and more olive-brown above, with hint of buffish wash on breast, and less distinct ventral barring. Below, buff restricted to vent, inner webs of primaries less distinctly barred than Dusky. From Mountain Long-tailed Cuckoo by lack of barring above, only indistinct barring on primaries and broad white band on underside of remiges more extensive. Juvenile from Dusky and Mountain by dark rufous crown; from Dusky by broader and complete rufous bars on tail, and dark bars on underparts much further apart giving generally paler appearance; from Mountain by lack of white tips to rectrices above and below. Juvenile has paler, more rufous throat than Dusky.

VOICE Call of male lazy repeated three-note *fi-fio-fiau*, descending, unlike Dusky Long-tailed, and very similar to Red-chested Cuckoo, but slightly more mellow and less clipped; series sometimes terminated with tremulous bubbling not unlike Eurasian Curlew; first note sometimes weak, and only second two notes audible, and constantly repeated; protracted series of *how* notes increasing in volume lasting 10–15sec. More frequent call long series of persistently repeated *feee-uu feee-uu feee-uu*, varying little in pitch or rising in quarter tones, at c. 45/min; emphatic, falling in pitch and repeated at ten calls/25sec. Loud and melodious call *do-you?*, first note higher and louder, second sliding down from first and may develop into series of rising bubbling notes. Generally calls in rainy season, often at night. In south-east Nigeria call at close range *cuk-kiew* becoming *whi-whièw* at distance; *whiew... whiew...* increasing in vigour (Chapin 1939, Marchant 1953, Irwin 1988, Payne 1997, Stevenson & Fanshawe 2002).

DESCRIPTION Adult Head, ear-coverts, nape and rest of upperparts dark olive-brown weakly glossed purple, rump and uppertail-coverts dark olive-brown, flecked white and buff. Remiges plain dark brown (small rusty spots on outer webs of primaries probably indicate younger birds), large white spots on inner webs on proximal half of primaries; secondaries and wing-coverts with very narrow white fringes distally. Chin to upper breast greyish-buff-white with dark brown feather margins giving scaling or barring effect; rest of underparts white, variably washed buff with well-spaced dark brown bars; undertail-coverts plain tawny-buff, longer feathers spotted dark brown. Underside of remiges have white spots, very large proximally, forming large white bar on open wing; underwing-coverts white barred brown. Tail long and full, strongly graduated, dark brown above with extensive grey and white spotting and barring, particularly on T5, T1 spotted rusty on margins; all rectrices tipped white. **Juvenile** Crown dark rufous; rest of upperparts blackish-brown tipped rufous; remiges dark brown with rufous spots on outer webs, becoming paler rufous-white on inner webs proximally; rectrices dark brown above with broad complete rufous bars. Underparts white washed tawny-buff with widely spaced dark brown bars; remiges with large white spots on inner webs forming large white patch on secondaries; spots become buff distally, with more on outer web; underwing-coverts white, innermost barred dark brown. Undersides of rectrices have dark and light bars (proximal third off white, distal two thirds dull buff) of equal width. No white tips to tail. **Nestling** Tawny overlay to white head, shoulders and mantle. Fledgling paler, wing-coverts and rectrices edged tawny (Louette & Herroelen 1994). **Bare parts** Iris dark brown; eye-ring and skin around eyes greenish-yellow. Bill slate-grey or blackish, lower mandible pale green basally; mouth yellow. Feet yellow, claws sepia (Irwin 1988).

BIOMETRICS Wing male (n = 44) 138–165mm, mean 146; female (n = 12) 126–154mm, mean 143.5. Bill unsexed (n = 18) 20–25mm, mean 23.7. Tail male (n = 45) 139–205mm, mean 167.7; female (n = 12) 136–183mm, mean 165.5. Tarsus male 17, 18, 19mm. Mass male 64, 66g (Irwin 1988, Louette & Herroelen 1994).

MOULT Post-juvenile moult Cameroon, Feb and Jul. Female almost completed moult to adult south-east Nigeria, Aug (G.L. Bates in Friedmann 1949, Serle 1957, BMNH).

DISTRIBUTION Afrotropical. Scattered localities in west Africa from SL (Gola Forest), Liberia (Mt. Nimba) and Guinea (Pic de Fon FR) through IC (Taï Forest, Yapo Forest), to Ghana, then east from Omo Forest and Niger delta in Nigeria through Cameroon, CAR, north Gabon, north RC, north, central and south DRC, west Uganda, north-west Angola, and north-west Zambia.

HABITAT Forest specialist, although will use small degraded fragments and gallery forest in southern part of range; secondary growth (including *Musanga*-dominated); fairly thick bush; to 1,800m (Marchant 1953, Irwin 1988, Payne 1997).

BEHAVIOUR Generally found in higher forest strata than

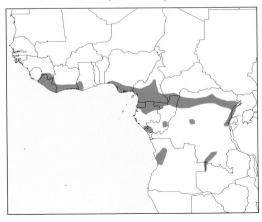

Dusky Long-tailed Cuckoo, usually mid-canopy or dense tops of forest trees. Elusive and difficult to observe, becoming silent if approached. Calls mainly early morning and evenings but also in night during wet season, particularly when humid and after thunderstorms. Joins mixed-species foraging flocks of insectivorous birds (Chapin 1939, Ripley & Heinrich 1966a, Irwin 1988).

BREEDING Parasitism Finsch's Flycatcher-thrush suggested as host since mimicry of its call reported Nigeria (Elgood 1984). Probable egg in nest of Pale-breasted Illadopsis (Brosset & Érard 1977). **Season** Breeding apparently associated with rains, although occurs during short dry seasons in wet areas. Calls mostly Aug–Mar, rarely Apr–May, Yapo Forest, IC; Jan–Apr, Jun, Oct, Nov, Liberia; throughout year Mt. Kupé, Cameroon; Apr south Guinea; Sep–Dec Mayombé, RC; Aug–Jan Zambia. Juvenile Cameroon, Feb. Possibly north-east Gabon, Dec–Feb. Breeding condition Angola, Sep–Nov; oviduct egg Uele, DRC, Sep during rains; enlarged testes Mt. Nimba, Liberia, Sep and Oct; enlarged gonads Angola, Sep and Nov. Young in post-juvenile moult Cameroon, Feb (Chapin 1928, 1939, G.L. Bates in Friedmann 1948b, Benson & Irwin 1965, Ripley & Heinrich 1966a, Brosset & Érard 1977, Irwin 1988, Dowsett-Lemaire et al. 1993, Demey & Fishpool 1994, Halleux 1994, Gatter 1997, Payne 1997, Aspinwall & Beel 1998, Dean 2000, Bowden 2001, BMNH). **Eggs** One, probably this species, deep blue with brown-violet spots (Brosset & Érard 1977). Oviduct egg white, unspotted, 23x16.4mm (Chapin 1928, 1939). No further information.

FOOD Insects, mainly hairy caterpillars (Chapin 1939).

STATUS AND CONSERVATION Elusive, and probably rarest *Cercococcyx*. Uncommon Nigeria (Elgood *et al.* 1994), but difficult to detect and little information available on status. Very common north-east Gabon (Brosset & Érard 1977) and Yapo Forest, IC (Demey & Fishpool 1994); rare south Guinea (Halleux 1994); rare south-east Nigeria (Marchant 1953); common Mayombé, RC, Sep–Dec, more local on coast (Dowsett-Lemaire *et al.* 1993). Not globally threatened (BirdLife International 2011).

Olive Long-tailed Cuckoo *Cercococcyx olivinus*. **Fig. 1.** Adult. Bwindi NP, Uganda, August (*Oleg Chernyshov*).

MOUNTAIN LONG-TAILED CUCKOO
Cercococcyx montanus Plate 32

Cercococcyx montanus Chapin, 1928 (Kalongi, *c.* 2,100m, Butahu Valley, Ruwenzori Range, Congo)

Alternative name: Barred (Long-tailed) Cuckoo.

TAXONOMY Forms superspecies with *olivinus*. Polytypic.

FIELD IDENTIFICATION 33cm. Small-bodied cuckoo with long, full tail. **Adult** Upperparts dark olive-brown with greenish sheen (paler without greenish sheen in southern birds), distinct buff-brown feather tips giving barred appearance above. Primaries strongly barred buff. Below whitish with variable buff wash on upper breast and vent, black-brown bars from throat to undertail-coverts, barring heavier on breast than throat or belly. **Juvenile** Medium to dark brown above, flecked brownish-white on crown, otherwise barred dull rufous; remiges dark brown spotted rufous; tail dark brown barred rufous, barring whiter distally; underparts white washed tawny-buff with widely spaced dark brown bars. T2–5 tipped white. **Similar species** Adults of congeners unbarred above, but beware confusion with immature Olive Long-tailed and Dusky Long-tailed Cuckoos. Adult Mountain glossed greenish on mantle (Dusky dark sooty-grey; Olive unglossed olive-brown), variable buffish wash to throat and stronger barring below than Olive. Juvenile from juvenile Olive by darker crown, from Olive and Dusky by more extensive barring on remiges, whitish bars on tail distally and rectrices tipped white; from Dusky by much paler underparts.

VOICE Individual call elements similar to Olive Long-tailed and Red-chested Cuckoos, but diagnostic in combination. Call of male long series of notes, firstly series of *c.* 10 *feee-uu* notes/10sec (similar to long call of Olive), then series of *c.* 10 clipped *pi-pi-tuu* notes at 1.5sec intervals (very similar to Olive), and often ending with fast *wit-wit-wu, wit-wit-wu*, occasionally also *pi-pi*; entire series lasts *c.* 40sec. Series sometimes repeated, becoming more rapid and emphatic. Occasionally triple note similar to Dusky Long-tailed; also leisurely, repeated *you-too*. Generally calls before sunrise, but during day if foggy and humid, also at night. Four (rarely three, occasionally five) note whistle 're-iterated without pause'; *whow, whow* doubling to *three-cow, three-cow*, lasting 10–15sec (Chapin 1939, Irwin 1988, Payne 1997, Stevenson & Fanshawe 2002).

DESCRIPTION Nominate. **Adult** Head, ear-coverts, nape and rest of upperparts dark olive-brown, with slight greenish sheen, barred rusty-tawny. Remiges barred rusty-tawny on outer webs, and wing coverts edged with same colour imparting barred appearance to wings; large buff notches on inner webs of primaries becoming white proximally. Chin to belly white barred dark brown, giving scaled effect, particularly on throat; usually buff wash on throat or upper breast, ground colour of flanks buff-white to white; some broken dark barring extends onto pale buff undertail-coverts. Undersides of remiges spotted tawny on outer webs, larger spots on inner webs, becoming white proximally, and forming large white area on open wing; underwing-coverts white with few dark brown bars. Tail long and full, strongly graduated, olive-brown above; rectrices have rusty buff spots on edges becoming larger and whiter to form complete bars from T1 to T5, and tips white; rectrices paler below with more obvious barring. **Juvenile** Crown medium brown flecked brownish-white; nape to uppertail-coverts medium to dark brown barred dull rufous with few white bars on mantle; remiges dark brown with rufous spots on outer webs extending down towards tips, becoming rufous-white proximally on inner webs. Tail dark brown above with broad rufous bars across both webs becoming whiter distally, with white tips to rectrices except T1. Underparts washed tawny to buff with well-spaced dark brown bars; large white spots on underside of remiges forming white bar on secondaries, spots becoming buff distally with more on outer web, coverts white. Undertail-coverts buffy-white with widely spaced dark brown bars; rectrices dark brown barred buffy-white becoming paler distally, and broadly tipped white. **Nestling** Undescribed. **Bare parts** Iris brown, eye-ring yellow. Bill black, greenish basally. Feet yellow.

BIOMETRICS Nominate. Wing male (n = 18) 134–148mm, mean 140.3 ± 3.6; female (n = 11) 133–145mm, mean 139.6 ± 3.8; juveniles unsexed (n = 6) 127–136mm. Tail male (n = 13) 180–190mm, mean 186.3 ± 3.4; female (n = 10) 182–187mm, mean 184.3 ± 2.7; juveniles unsexed 162–179mm. Tarsus unsexed 25–27mm. Mass unsexed 47.5–51.0g, mean 49.5. Mass male (n = 8) 56.5–60.5g, mean 58.7, female (n = 3) 52–64g, mean 57.8, unsexed 56.0g (Ripley & Heinrich 1966b, Britton 1977, Irwin 1988, Blake *et al.* 1990, Louette & Herroelen 1994, Hanmer 1995, Payne 2005).

MOULT Moulting adult Mozambique, Jun (Clancey 1968).

GEOGRAPHICAL VARIATION Two subspecies.
C. m. montanus Chapin, 1928. East-central Africa from south-west Uganda along Albertine Rift Mountains through east DRC, Rwanda, Burundi and north-west Tanzania to Lake Tanganyika. Described above.
C. m. patulus Friedmann, 1928. Central and south-east Kenya, east and south Tanzania, south Zambia, Malawi, Zimbabwe and Mozambique; two specimens from south DRC (Louette & Herroelen 1994). Larger, paler, more heavily barred above and with less sheen, crown feathers often with narrow buff shaft streaks, barring on dark underparts more widely spaced; crown, neck and mantle feathers edged cinnamon in juvenile, buffish-white underparts variable in shade. Wing male (n = 4) 142–152mm, mean 147; female (n = 4) 146–148mm, mean 147; also male (n = 9) 143–154mm,

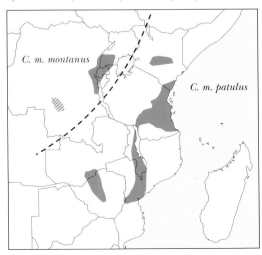

mean 144.8 ± 3.8; female 144, 154mm (Chapin 1939, Dowsett-Lemaire 1990, Louette & Herroelen 1994).

DISTRIBUTION Afrotropical. Highlands of small area of west-central DRC, west Uganda, Rwanda, Burundi and west Tanzania to Lake Tanganyika; Mt. Kenya, and from near coast in south Kenya through eastern half of Tanzania, Malawi, central Mozambique, Zimbabwe and along Gwembe Valley/middle Zambezi in Zambia. Some altitudinal movement, and appears to be breeding visitor to Zambia Dec–Feb during rains, and DRC, Kenya, Malawi and Zimbabwe, suggesting middle-distance migration (Britton 1977, 1980, Stjernstedt 1984, Louette & Herroelen 1994). One came to light at night in Tanzania, Nov (Moreau & Moreau 1939).

HABITAT Nominate in montane forest mostly 900+m; to 2,800m in Rwenzori (Uganda), 2,600m Rwanda, and 2,100 m Burundi. *Patulus* sea level to 1,700–2,100m east Kenya and Tanzania, in riparian forest, forest-savanna mosaics and coastal thicket; also breeds miombo woodland Tanzania. Dense riparian forest along Zambezi. Female shot in coffee plantation, Tanzania (Chapin 1939, Irwin 1988, Dowsett-Lemaire 1990, Payne 2005).

BEHAVIOUR Extremely elusive; low rapid flight making it very difficult to detect. Calls from perch high up in tall tree, particularly at night and in two hours before sunrise. May feed on ground on forest tracks although usually in foliage of trees (Clancey 1968, Irwin 1988, Stevenson & Fanshawe 2002).

BREEDING Parasitism Sharpe's Akalat possible host, as cuckoo seen to approach unattended nest in excited, vigilant state with tail fully spread, clumsily mantled nest, then removed the two half-incubated eggs in successive visits, Usambaras, Tanzania. Probably Archer's Robin-chat and Equatorial Akalat, Rwanda. Perhaps East Coast Akalat and African Broadbill Mozambique; eggs in nests of African Broadbill Tanzania (Usambaras) and probably Zimbabwe (Moreau & Moreau 1939, T.A. Bladock in Friedmann 1948b, Dean 1974, Dowsett-Lemaire 1990). **Season** Generally during rains. Rwanda, Oct; north Tanzania, Sep. Oviduct eggs Burundi, Aug; Kenya, Feb, Mar; laying East Usambaras, Tanzania, Nov, Dec–Feb; south Tanzania from early Sep. Breeding condition east Usambaras, Tanzania, Sep–Dec; enlarged testes north Zambia, Oct. Short-tailed juvenile Ulugurus, Tanzania, late Feb. Calls east DRC Dec–Mar; east Tanzania Nov–Feb, Sep–Dec; Malawi Dec–Apr; Mozambique from Oct; south Zambia Dec–Feb; Zimbabwe Nov, Jan (Sclater & Moreau 1932, Chapin 1939, Moreau & Moreau 1939, Friedmann 1948b, Mackworth-Praed & Grant 1957, Keith 1968, Schouteden 1968, Jensen & Jensen 1969, Benson & Benson 1977, Irwin 1988, Dowsett-Lemaire 1990, Payne 1997, 2005, BMNH). **Eggs** In captivity white with weak reddish band at blunt end, paler but similar to eggs of akalats; egg, later nestling, in African Broadbill nest white with faint band of brownish spots at broad end, 23x17mm, with white host's egg; well-developed oviduct egg pure white as African Broadbill, c. 21x15mm; oviduct egg Kibera NP, Burundi 24x13mm (Moreau & Moreau 1939, Dean et al. 1974, Payne 2005). No further information.

FOOD Insects, mainly hairy caterpillars; small snails in one stomach (Irwin 1988).

STATUS AND CONSERVATION Although elusive, common in parts of range, particularly in forested highlands of east DRC (Schouteden 1968). Common in forest 1,700–2,100m on east slopes of Mount Kenya and southern slopes of Aberdares (Kenya), and 900–1,600m Usambaras (Tanzania) (Zimmerman *et al.* 1996). Many museum specimens from Albertine Rift, suggesting previously common (Irwin 1988, Louette & Herroelen 1994, Payne 1997). Relatively tolerant of degraded forest and uses non-forest habitats. Not globally threatened (BirdLife International 2011).

Mountain Long-tailed Cuckoo *Cercococcyx montanus*. **Fig. 1.** Adult, *C. m. montanus*, Ruhija, Bwindi NP, Uganda, February (*Jonas Rosquist*).

Genus *Surniculus*

Lesson 1830 *Traité d'Orn.*, livre 2: 151. Type, by subsequent designation, *Cuculus lugubris* Horsfield 1821 (G.R. Gray 1855, *Cat. Gen. Subgen. Birds*: 97).
Four species.

Oriental; brood-parasitic. Typified by black adult plumage, rounded nostrils, and forked or square tail with outer feathers short; sexes alike. MtDNA analysis results in *Cuculus* and *Hierococcyx* forming monophyletic cluster which is sister to clade containing *Cercococcyx* and *Surniculus* (Sorenson & Payne 2005). Long regarded monotypic; due to different calls now seen as containing four species; in need of revision (Payne 2005, Collar & Pilgrim 2007). Includes *Pseudornis* (Hodgson 1839).

FORK-TAILED DRONGO-CUCKOO
Surniculus dicruroides   Plate 23

Pseudornis Dicruroides Hodgson, 1839 (Nepal).

Alternative name: Indian Drongo-cuckoo.

TAXONOMY Previously considered subspecies of *lugubris*. Payne (2005) and Rasmussen and Anderton (2005) gave *dicruroides* species status due to its deeply forked, flared tail, and vocalisations; said to be sympatric with *lugubris* in north-east India (Assam; Arunachal Pradesh) in breeding season, with no intermediates (Payne 2005). Among the large number of specimens at BMNH there are none that substantiate this claim, and many are indeterminate. Payne (2005) considers the species monotypic, placing *stewarti* in *lugubris*. Polytypic. Synonym: *massorhinus* Oberholser 1924 in nominate.

FIELD IDENTIFICATION 25cm. Smallish cuckoo, with forked tail. **Adult** Plumage black glossed steel-blue and long tail obviously forked. **Juvenile** Black spotted white. **Similar species** Slightly larger than Square-tailed Drongo-cuckoo, and easily separated when tail fully grown (T4 4–12mm or more longer than T1; 0–6mm in *lugubris*); outer web T3 narrower, becoming more tapered and broader distally than *lugubris*. Easily confused with drongos but small head and cuckoo-like bill distinguish it. On Indian subcontinent Black Drongo is all black, has deeply forked tail and bill is stronger, juvenile plain dark brown and immature has some white on belly and white-fringed uppertail-coverts. Crow-billed Drongo is black with stout bill, juvenile has white spots on breast and abdomen. Greater Racquet-tailed Drongo larger, glossy black with conspicuous crest and long, twisted tail-racquets, juvenile has less well developed crest and no racquets, and is duller black with white fringes on belly and vent.

VOICE Series of 5–6 loud piping whistles rising upscale, second note lower, or not higher, than first. Also long, bubbling crescendo with spread wings lifted above back; a shrill accelerating call rising then falling away; harsh buzzing series of note rising then falling; plaintive *whee wheep*, second note higher (Smythies 1940, Ali & Ripley 1969, Payne 2005).

DESCRIPTION Nominate. **Adult** Black glossed steel-blue, head and upper back less so, nape often has concealed white spot (retained juvenile plumage?); and long tail deeply forked, rectrices tipped black, T1–2 shorter than T3; T3 has outer web tip broad, T4 markedly curved outwards, outer web very narrow broadening at tip (in *lugubris* broad at tip and usually straight), and may have tiny white spots on shaft; T5 short, obliquely and narrowly barred white. Below black; long tarsal tufts have outer webs white, inner black or white; undertail-coverts narrowly barred whitish; bend of wing and underwing-coverts black; white on base of inner webs of P2–7 and S2–4 below forming bar, P9 with white spot near base of inner web. **Juvenile** Dull black with white spots on head, wings and breast; white nape patch and tips to rectrices; bend of wing barred and spotted white. Tail less deeply forked than adult, and with small white tips to rectrices. **Nestling** Mouth lining orange (Hume 1888). No further information. **Bare parts** Orbital skin black; iris brown; bill black; feet dark grey.

BIOMETRICS Nominate. Wing male (n = 23) 125–146mm, mean 137.2 ± 6.1, female (n = 7) 128–145mm, mean 135.9 ± 6.4. Tail male 124–137mm, mean 133.9 ± 7.4, female 122–143mm, mean 130.3 ± 8.3. Bill male 18.3–22.0mm, mean 19.8 ± 1.0, female 15.7–20.8mm, mean 18.7 ± 1.8. Tarsus male 14.0–16.5mm, mean 14.6 ± 0.9, female 13.9–15.9mm, mean 15.0 ± 0.6. Mass female 50.5g (Payne 2005). Unsexed autumn migrants Malay Peninsula (*lugubris*?) (n = 41) 32.5–43.3g (Wells 1999).

MOULT Adults moulting Sikkim, Aug–Sep (BMNH).

GEOGRAPHICAL VARIATION Two subspecies.
S. d. dicruroides (Hodgson, 1839). Indian subcontinent, Thailand, Indochina, south China. In winter south to Thai-Malay Peninsula, Sri Lanka?; rare Sumatra, west Java; once North Natuna Is. Described above.
S. d. stewarti E.C.S. Baker, 1919. Sri Lanka. South India? Similar to nominate, but shorter, less forked and curved tail. Wing male (n = 18) 122–135mm, mean 127.3 ± 4.5, female (n = 14) 119–141, mean 128.5 ± 6.5; tail male 118–147mm, mean 130.4 ± 7.9, female 104–140mm, mean 120.6 ± 9.0 (Payne 2005).

DISTRIBUTION Confusion with Square-tailed Drongo results in many from north Pakistan, north India, Nepal and Bhutan being considered dubious, though specimens are known from Himalayan foothills (identified by song in Nameri NP, Assam and west Arunachal Pradesh: Payne 2005). Indian subcontinent, including Eastern and Western Ghats, Punjab, north Uttar Pradesh (including Garhwal), north Madhya Pradesh east to Raipur, Andhra Pradesh, Himalayas east to Nepal (Hetora; Kathmandu Valley), Bhutan, Sikkim, Assam, Manipur, Arunachal Pradesh, Sri Lanka (perhaps also nominate in winter), Burma (Chin Hills; south to Tenasserim in winter), north Thailand, Indochina; winters south to Deccan, Thai-Malay Peninsula (including passage at Fraser's Hill from 25 Sep), Sumatra (mainland; Aruah Is.; Batu I.; once Nias), Lingga and Riau

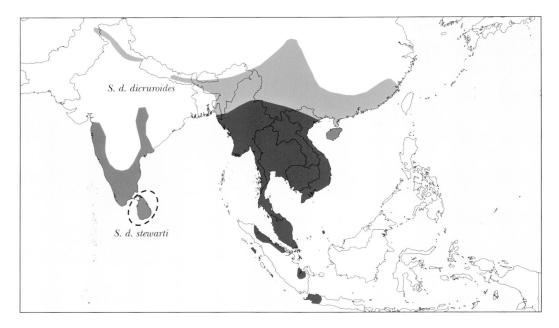

Archipelagos, Borneo (once North Natuna Is.) and Java. Both resident and migrant Manipur (Rattray 1905, Ripley 1944, Payne 2005, Choudhury 2009, BMNH). Maharastra, Kerala and Tamil Nadu, India, presumably migrants (Ripley 1961). In China breeding visitor Sichuan, Yunnan, Guizhou, Guangdong, Fujian and Guanxi Zhuang; resident Hainan (Cheng 1987). Andamans? (Tikader 1984). Twice Japan (Brazil 1991, Mita 2010); once South Korea, May (Birds Korea 2009). Occurs with monsoon rains in northern Indian subcontinent, but many from this area may be Square-tailed.

HABITAT Broadleaved and mixed forests, 1,000–2,600m, Bhutan. Forest, scrub jungle, bamboo, mainly in hills but also on plains, 200–1,800m, India. Dry forests, Sri Lanka. Lowlands to 2,100m, Yunnan; lowlands to 1,200m, Burma (Inglis *et al.* 1920, Rothschild 1926, Ali & Ripley 1969, Smythies 1986, Tymstra *et al.* 1997, Grimmett *et al.* 1998).

BEHAVIOUR Perches on stumps and saplings in burned clearings; smooth, direct flight. Gleans from branches and leaves in canopy and hawks insects drongo-like (Ripley 1950, Ali 1962, Ali & Ripley 1969).

BREEDING Parasitism Hosts in India shrikes, Sooty-headed Bulbul, Dark-fronted Babbler, Nepal Fulvetta, forktails, and Striated Grassbird; perhaps Zitting Cisticola and Black Drongo. Dark-fronted Babbler and perhaps Common Iora, Sri Lanka. Black and Hair-crested Drongos, China (Baker 1906c, 1934, La Touche 1931, Ali & Ripley 1969, Becking 1981, Grimmett *et al.* 1998). **Season** Eggs Nepal, Jun; north India Mar–Oct during monsoon; Sikkim Apr–Jun; oviduct eggs Bangladesh, Apr and Jun. Probably Jan–Mar Kerala and Dec–May Sri Lanka (nestling Mar). Calls north Burma, Apr (Baker 1934, Biswas 1960, Ali 1962, 1996, Ali & Ripley 1969, Payne 2005). **Eggs** Oviduct egg white with fine purple speckles, Nepal; in India 23.5x17.5mm (Baker 1934, Biswas 1960). **Incubation** No information. **Chicks** Evict host eggs and nestlings (Becking 1981). **Survival** No information.

FOOD Mostly insects, especially caterpillars; fruits such as *Ficus* figs (Ali & Ripley 1969).

STATUS AND CONSERVATION Fairly common much of peninsular India (Ali & Ripley 1969, Rasmussen & Anderton 2005). Common Bhutan, but not differentiated from Square-tailed (Spierenburg 2005). Uncommon China (Cheng 1987).

Fork-tailed Drongo-cuckoo *Surniculus dicruroides*. **Fig. 1.** Adult, *S. d. dicruroides*, Karkala, Karnataka, India, December (*ShivaShankar*).

PHILIPPINE DRONGO-CUCKOO
Surniculus velutinus Plate 23

Surniculus velutinus Sharpe, 1877 (Malamaui, Philippines).

TAXONOMY Previously considered subspecies of *lugubris*. Dickinson (2003) and Payne (2005) recognise *velutinus* as species with *chalybaeus* as subspecies, and latter also recognises *suluensis*. Marked difference between juvenile of nominate and other forms could suggest separation at species level. Polytypic. Synonym: *mindorensis* Ripley & Rabor 1958 in *chalybaeus*.

FIELD IDENTIFICATION 24cm. Smallish cuckoo. **Adult** Black, with white or black-and-white tarsal tufts. **Juvenile** Probably brown to rufous-brown in all subspecies. **Similar species** Drongos have stouter bills, and lack white tarsal tufts. Male Common Koel much larger with pale bill.

VOICE 8–10 whistled notes each rising in pitch, series rising slightly upscale, but not as much as in *lugubris* or *dicruroides*; also burry crescendo similar to that of congeners (Payne 2005). On Luzon 8–9 shrill *wi-wi-wi* notes becoming more and more high-pitched lasting 2.8sec, repeated constantly (Kennedy *et al.* 2000).

DESCRIPTION Nominate. **Adult** Head, upper back and throat to breast dull velvety black or purplish-black, not glossy; concealed white patch on nape in some (retained juvenile feature?); wings and tail black glossed steel-blue or purplish-blue or green. Outer web of T3–4 broad tipped, bending feather outwards; tail square or slightly emarginated and not forked; T5 shorter than other rectrices and narrowly and obliquely barred white, sometimes with very small white spots on vane, T2–3 may have narrow white edging to outer vanes. Below black; white outer webs of tarsal tufts, inner webs white or black; undertail and underwing-coverts black; inner webs of P2–7 and S2–4 show white bar below; P9 has white spot at base of inner web. **Juvenile** Brown to rufous-brown. **Nestling** Undescribed. **Bare parts** Bare orbital skin dark bluish-grey; iris brown; bill black; feet and toes dark grey.

BIOMETRICS Nominate. Wing male (n = 36) 112–130mm, mean 117.2 ± 4.0, female (n = 25) 115–121mm, mean 117.1 ± 2.5. Tail male 86–118mm, mean 101.6 ± 6.0, female 92–118mm, mean 105.0 ± 7.0. Bill male 17.0–21.9mm, mean 19.1 ± 1.3, female 17.0–20.6mm, mean 18.7 ± 1.2. Tarsus male 14.0–18.0mm, mean 15.1 ± 1.0; female 13–18.5mm, mean 16.1 ± 1.2 (Payne 2005).

MOULT Juvenile to adult moult 10 May, Bohol, and Samar 30 Jun (BMNH).

GEOGRAPHICAL VARIATION Three subspecies.
S. v. velutinus Sharpe, 1877. South Philippines on Malamaui, Basilan, Biliran, Bohol, Leyte, Mindanao and Samar; Samal? Described above.
S. v. suluensis Salomonsen, 1953. Jolo, Bongao and Tawi Tawi, Sulu Archipelago. Like *velutinus* but larger. Juvenile from Sulu Archipelago reported brown tipped buff (Payne 2005) is mistake (R.B. Payne pers. comm.). Wing male (n = 3) 125–130mm, mean 128.5, female (n = 6) 120–129mm, mean 123.8 ± 3.0. Tail male 105–126mm, mean 113.0 (Payne 2005).
S. v. chalybaeus Salomonsen, 1953. Philippines on Gigantes, Luzon, Mindoro and Negros. Presumably this subspecies Panay (Curio *et al.* 2001). Velvety blue-black, slightly glossed bluish-black on nape, back, wings, tail and underparts. Wing male (n = 15) 122–131mm, mean 126.8 ± 2.8, female (n = 6) 124–131mm, mean 126.8 ± 2.4. Tail male 98–121mm, mean 109.9 ± 5.9 (Payne 2005).

DISTRIBUTION Philippines on Basilan, Biliran, Bohol, Bongao, Gigantes, Jolo, Leyte, Luzon, Malamaui, Mindanao, Mindoro, Negros, Panay, Samal, Samar and Tawi Tawi (Kennedy *et al.* 2000).

HABITAT Canopy or midstorey of primary dipterocarp or secondary forest; forest edge; wooded country; mixed

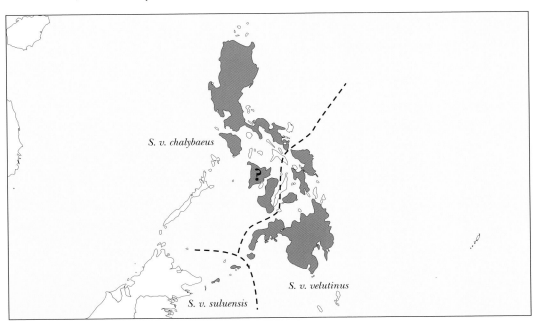

bamboo woodland; to 1,000m (Curio *et al.* 2001, Payne 2005).

BEHAVIOUR Secretive, but conspicuous by call. Solitary (Kennedy *et al.* 2000).

BREEDING Parasitism Hosts unknown. **Season** Oviduct eggs Mindoro, May and Bohol Apr–May (Ripley & Rabor 1958, Rand & Rabor 1960). No further information.

FOOD Insects (Payne 2005).

STATUS AND CONSERVATION Fairly common (Dickinson *et al.* 1991).

Philippine Drongo-cuckoo *Surniculus velutinus*. **Fig. 1.** Adult with traces of juvenile plumage, *S. v. velutinus*, Matina, Davao City, Mindanao, Philippines, May (*Martin Alvendia*). **Fig. 2.** Adult, *S. v. chalybaeus*, Subic, Zambales, Luzon, Philippines, August (*Tina Sarmiento Mallari*).

SQUARE-TAILED DRONGO-CUCKOO
Surniculus lugubris Plate 23

Cuculus lugubris Horsfield, 1821 (Java)

TAXONOMY Once placed in *Pseudornis* (Hodgson 1839). Said to differ in voice from *dicruroides* and to be sympatric with it in north-east India in breeding season with no intermediates known (Payne 2005), and therefore *dicruroides* separated at species level. However examination of numerous specimens at BMNH revealed many indeterminate. Some information from areas were both may occur cannot safely be ascribed to one or other. Payne (2005) places *S. d. stewarti* in this species. Residents of Sundaland (other than Java and Bali), with wings not above 129mm have been separated (but here synonymised with *barussarum*) as subspecies *brachyurus* Stresemann 1913 (synonym *minimus* Baker 1919). Larger individuals of *barussarum*, wing 128–136mm, breed further north, and some migrate into Sundaland at least as far as Sumatra (Mees 1986); includes type from Sumatra. Polytypic.

FIELD IDENTIFICATION 24–25cm. Smallish cuckoo with direct flight, silent and soft compared with energetic style of drongo; tail more or less square. Sexes similar, but female slightly duller. **Adult** Slender slightly decurved bill, some white on undertail-coverts and slightly forked tail, more square in *barussarum*. When perching white thighs usually visible and excellent field character, but found in all drongo-cuckoos. **Juvenile** Duller than adult, and spotted white. **Similar species** See Fork-tailed Drongo. Easily confused with drongos but small head and cuckoo-like bill distinguish it. On Indian subcontinent Black Drongo all black, has deeply forked tail and stronger bill, juvenile plain dark brown and immature has some white on belly and white-fringed uppertail-coverts. Crow-billed Drongo black with stout bill, juvenile has white spots on breast and abdomen. Greater Racquet-tailed Drongo larger, glossy black with conspicuous crest and long, twisted tail-rackets, juvenile has smaller crest and no racquets, and is duller black with white fringes on belly and vent. Male Common Koel larger, has red eyes, longer tail and whole plumage glossy black. More difficult to separate from moulting Fork-tailed Drongo-cuckoo which may show more or less square tail; many individuals indistinguishable unless calling.

VOICE Territorial call fast ascending *pi-pi-pi* or *pee-pee-pee* repeated quickly five-seven times; shrill version of 'brainfever' call of hawk-cuckoos, shrill *phew-phew-phewphewphewphewphew-phew-phew*, first rising, then falling. Five-six mellow whistles rising up scale, not piping as *dicruroides*; much higher-pitched, shriller and faster than *dicruroides*. In Malay Peninsula call *kree-kree-kree...* first ascending but falling on last two-three notes. On Palawan cadence of five-eight short ascending notes *wu-wu-wu-waa-waa-wee* lasting 1.6sec and continued every 4–5sec. Female call unknown. Calling Jan–Aug, occasionally beginning Nov, Thai-Malay Peninsula (Grimmett *et al.* 1998, Jeyarajasingam & Pearson 1999, Wells 1999, Kennedy *et al.* 2000, Robson 2000, Payne 2005, Rasmussen & Anderton 2005).

DESCRIPTION Nominate. **Adult** Sexes similar, but female duller. Glossy blue-black, greener on wing-coverts and inner secondaries and tertials, often concealed white patch on nape (retained juvenile character?). Underside of wing has variable white area on inner webs of P2–7 and S2–4 forming bar; P9 has white spot near base of inner web. Tarsal tufts with outer web white, inner black or white; narrow whitish barring on long undertail-coverts. Underparts black, slightly glossed on breast, belly and vent with brownish tinge. Tail square or slightly forked (up to 5mm deep), extreme tip white; T1–2 equal length, or slightly shorter than next two rectrices, T3 straight with outer web broad at tip, T4 straight (or very slightly curved outwards) with no, or only slight taper to outer web, which is imperfectly barred white on inner webs, T5 shorter, narrowly and obliquely barred white, with white spots on shaft. **Juvenile** As adult but duller blackish-brown with fine white spotting on head, wings and breast; tips of uppertail-coverts white; rectrices rounded and tipped white, often with white spotting on shafts and broader white bars. **Nestling** Black with triangular, subterminal white spots to many feathers. **Bare parts** Iris brown to dark grey, female yellow; skin around eye dark bluish-grey or black. Bill black, mouth pale yellow, gape of nestling bright vermilion. Feet dark grey, nestling vinaceous, claws black.

BIOMETRICS Nominate. Wing male (n = 17) 125–133mm, mean 127.4 ± 2.3, female (n = 5) 127–129mm, mean 128.0 ± 1.0. Bill male 19–22mm, mean 20.7 ± 0.9, female 19.0–21.5mm, mean 20.7 ± 1.1. Tail male 120–136mm, mean 128.8 ± 4.6, female 118–128mm, mean 123.2 ± 4.6. Tarsus male 14.2–17.6mm, mean 15.7 ± 1.1, female 14.4–16.0mm, mean 15.7 ± 1.4 (Payne 2005). Mass *barussarum* (n = 6) 26.0–35.8g (Wells 1999); adult (n = 28) 30.0–43.6g, mean 36.2 (Becking 1981). Also wing male 134.5mm (ZMUA).

MOULT Primaries moulted in up to three loci simultaneously, Jul–Sep after breeding, including migrants; usually P9 and P7, then P8, P6 and P5 in irregular succession, no pattern for P10, and inner primaries sequence most often P4, P3, P2, P1 (Stresemann & Stresemann 1961, Wells 1999).

GEOGRAPHICAL VARIATION Two subspecies.
S. l. lugubris (Horsfield, 1821). Java, Bali. Described above.
S. l. barussarum Oberholser, 1912. Nepal?; east Bangladesh; Assam; Garo, Lushai and Khasi Hills; Burma (Arakan, Pegu, Shan States, Tenasserim); Indochina; west and peninsular Thailand; Malay Peninsula, Sumatra, Bangka, Borneo and Philippines (Palawan, Balabac, Calauit). Smaller and shorter-tailed; underparts less brownish. Wing male (n = 16) 96–127mm, mean 121.6 ± 3.7, female (n = 8) 113–126mm, mean 120.9 ± 4.4. Tail male 98–120mm, mean 113.8 ± 5.2 (Payne 2005); type (female) 136mm (Oberholser 1912).

DISTRIBUTION Oriental. There is some confusion between this and *dicruroides*. Some northern populations migratory. Summer visitor north-east India, Nepal? and east Bangladesh; Burma. Present north Thailand Mar–Sep. In all provinces of Thai-Malay Peninsula except Ranong, and also on following islands: Jarak, Langkawi, Lanta, Libong, Penang, Perak, Pisang, Koh Samui, Sembilan group, Similans, Singapore, St John's, Tarutao, Ubin and Yao Yai. From Singapore north to at least Phuket, Nakhon Si Thammarat and Koh Samui. Philippines on Palawan, Balabac, Calauit. Uncommon passage migrant east Tonkin, Vietnam. Laos; Cambodia. Night migration Fraser's Hill, peninsular Malaysia, Aug–Sep (may include some *dicruroides*). Resident Sumatra throughout mainland and Bangka. Common resident throughout lowlands of Borneo. Java and Bali. Identified by song Assam (Kaziranga and Nameri NPs) and Thailand (Huai Kha Khaeng, Thaleban NP, Khao Phraa Bang Kran). Andamans? (Deignan 1945, Tikader 1984, van Marle & Voous 1988, Dickinson *et al.* 1991,

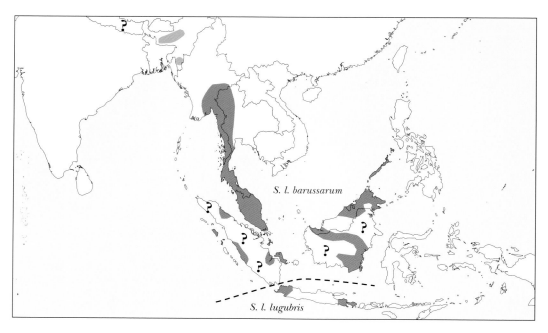

MacKinnon & Phillipps 1993, Duckworth *et al.* 1998, Wells 1999, Evans *et al.* 2000, Hill 2000, Robson 2000, Eames *et al.* 2002, Thomas & Poole 2003, Payne 2005, Rasmussen & Anderton 2005, Fuchs *et al.* 2007, Mann 2008).

HABITAT Forest edge, clearings, disturbed forest, scrub, small forest patches, orchards and high trees in tea plantations, India; semi-evergreen, swamp and riparian forests in Assam, or deciduous forest. Temperate broad-leaved forest, Bhutan. Cashew-nut and jackfruit orchards, and coconut and pepper plantations. Nepal; 250–2,250m, Bhutan. In Thai-Malay Peninsula lowland semi-evergreen and evergreen forest, pines, tree plantations and nearby wooded gardens to 800m; those to 1,200m probably migrants. In Thailand also mangroves and gardens but these are perhaps migrating birds; to 1,200m. Burma and Philippines also in bamboo thickets. On Sumatra common in peatswamp forest, and tall forest and forest edge to 900m. On Borneo lowland dipterocarp and peat-swamp forests, recently logged primary forest, forest edge, *Albizia* plantations, kerangas forest, ladang and fire padang, to 1,300m. Lowland primary forest Philippines. To 1,200m Java (Stanford & Ticehurst 1939, Deignan 1945, Hoogerwerf 1948, Nash & Nash 1985, van Marle & Voous 1988, Dickinson *et al.* 1991, Lekagul & Round 1991, Zacharias & Gaston 1993, Grimmett *et al.* 1998, Barua & Sharma 1999, Inskipp *et al.* 1999, Jeyarajasingam & Pearson 1999, Wells 1999, Kennedy *et al.* 2000, Payne 2005, Mann 2008).

BEHAVIOUR Usually in canopy or mid-level. Calls regularly from dead branches, and when calling often chased by Greater Racquet-tailed Drongos. Tame, but when not calling fairly secretive and solitary. Counter-calling from exposed tree perches suggesting territorial behaviour; sallies into termite swarms and hawks insects like drongo. Perches on saplings and dead stumps in burned clearings; flight smooth and direct. One mobbed by Striped Tit-babblers in Laos as it perched with head hunched forward, bill pointing downwards and wings held slightly forward. Joins mixed foraging flocks (McClure 1967, Duckworth 1997, Wells 1999, Kennedy *et al.* 2000, Strange 2001, Payne 2005).

BREEDING Parasitism Appearance suggests that drongos are preferred hosts, and Black Drongo claimed but no reliable record. Olive-winged Bulbul, Striped Tit-babbler, Rufous-fronted, Sooty-capped and Chestnut-winged Babblers feeding fledglings, Thai-Malay Peninsula, where perhaps also Horsfield's Babbler host. Possibly Chestnut-winged Babbler, Borneo, also perhaps Sumatra. Crescent-chested Babbler (6), Horsfield's Babbler (16), Grey-cheeked Tit-babbler (2 in one nest), Javan Fulvetta, Java. Also claimed Red-vented Bulbul, Brown Bush Warbler, Common Tailorbird, Spot-throated Babbler, Rufous-fronted Babbler and Brown-cheeked Fulvetta. Breeding not known in Philippines (Hellebrekers & Hoogerwerf 1967, Ali & Ripley 1969, Becking 1981, Cranbrook & Wells 1981, Jeyarajasingam & Pearson 1999, Wells 1999, Kennedy *et al.* 2000, SINGAV-14 in Wells 2007, Mulyawati *et al.* 2009, Lim 2011, P.E. Lowther *in litt.*, R. Noske *in litt.*, BMNH, ZMUA). **Season** In India (perhaps *dicruroides*) Dec–Oct but varying locally; mid Apr–Jul peaking May–Jun, Assam. Egg 14 Jul, and fledglings May–Jul, Thai-Malay Peninsula. Eggs Java Mar–May, Jul. Oviduct eggs Apr/May, Borneo. Eggs Java, Jan–May, Jul, Sep–Dec (Smythies 1957, Hellebrekers & Hoogerwerf 1967, Ali & Ripley 1969, Becking 1981, Kotagama & Fernando 1994, Grimmett *et al.* 1998, Wells 1999, Sheldon *et al.* 2001, BMNH, ZMUA). **Eggs** Mean India (perhaps *dicruroides*) 19.8x15.5mm; broad elliptical, ground colour pale pink spotted and scratched black and dark brown; 22.5x17.0mm; Java 18.6–23.0x14.4–16.2mm, (n = 25) mean 20.7x15.4; on Borneo oviduct egg white finely marked purple. On Java three egg colour *gentes*: rose-white ground colour spotted bluish-grey, reddish and chestnut, host Horsfield's Babbler; white with brown markings, host Grey-cheeked Tit-babbler; third pure white, host Crescent-chested Babbler (Smythies 1957, Hellebrekers

& Hoogerwerf 1967, Schönwetter 1967, Becking 1981, Wells 1999). **Incubation** No information. **Chicks** No nests found with cuckoo and eggs or chicks of host, strongly suggesting it evicts foster young (Becking 1981). **Survival** No information.

FOOD Foliage-gleans; aerial sallies for swarming termites. Hairy and hairless caterpillars (Pieridae, Lasiocampidae, Limacodidae), other soft-bodied insects, beetles, swarming termites; spiders, including very large one; more rarely banyan figs, other figs and fruits. One stomach contained 50 caterpillars (Thompson 1966, T. Harrisson in Smythies 1981, Sody 1989, Smythies & Davison 1999, Wells 1999).

STATUS AND CONSERVATION Local Nepal summer; in India locally common Himalayas, but not elsewhere; common and widespread Bangladesh (Grimmett *et al.* 1998). Common Bhutan (Spierenburg 2005). Uncommon Thailand (Lekagul & Round 1991); uncommon Burma (Smythies 1986). More or less common and not threatened Thai-Malay Peninsula (Wells 1999). Common at many sites, Laos (Duckworth *et al.* 1998). Widespread and often common resident Borneo (Mann 2008). Still fairly common in lowland forest in Philippines (Kennedy *et al.* 2000), but catastrophic forest destruction may detrimentally alter its status. Large range and wide habitat tolerance, which includes disturbed forest, give little cause for concern. Not globally threatened (BirdLife International 2011).

Square-tailed Drongo-cuckoo *Surniculus lugubris.* **Fig. 1.** Adult, *S. l. barussarum*, Kledang-Sayong FR, Ipoh, Perak, Malaysia, July (*Amar-Singh HSS*). **Fig. 2.** Juvenile, *S. l. barussarum*, Kledang-Sayong FR, Ipoh, Perak, Malaysia, September (*Amar-Singh HSS*).

MOLUCCAN DRONGO-CUCKOO
Surniculus musschenbroeki Plate 23

Surniculus musschenbroeki A.B. Meyer, 1878 (Batjan = Bacan, Moluccas or Celebes = Sulawesi: van Bemmel & Voous 1951)

Alternative name: Sulawesi Drongo-cuckoo.

TAXONOMY Previously considered subspecies of *lugubris*. Monotypic.

FIELD IDENTIFICATION 23cm. Smallish cuckoo, with slightly forked tail. **Adult** Slender slightly decurved bill, some white on undertail-coverts. When perching white thighs may be visible as in all drongo-cuckoos. **Juvenile** Duller than adult, and spotted white. **Similar species** Could be confused with drongos but small head and cuckoo-like bill, barring on tail and underwing bar distinguish it, as does direct flight, rather than swooping and swerving. If bird silent, male Common Koel can be separated by larger size, red eyes, longer tail and whole plumage glossy black.

VOICE *Ki-ki-ki...*; on Sulawesi 5–12, usually 8, rising, whistled notes each dropping in pitch, series repeated many times; on Halmahera 6–8 notes, occasionally 12. May call at night (Riley 1924, Lambert & Yong 1989, Coates & Bishop 1997, Payne 2005).

DESCRIPTION Adult Head, upper back and breast velvety purplish-black, small concealed white patch on nape, wings and square, or slightly forked, tail black glossed purplish, blue or green, latter tipped black; T3–4 broaden at tip, often curled upwards and outwards; T5 shorter with narrow, oblique white bars and spots near shaft at base. Below, including underwing and undertail-coverts, black; outer webs of tarsal tufts white, inner black or white. Inner webs of P2–7 and S2–4 white forming bar below; P9 with white spot near base of inner web. When perching white thighs usually visible and excellent field character, but found in all drongo-cuckoos. **Juvenile** Dull black with fine white spotting on crown, wing coverts, throat, breast and undertail coverts (Payne 2005). **Nestling** Undescribed (except for gape colour). **Bare parts** Iris brown, female yellow, bare orbital skin dark bluish-grey. Bill black, mouth pale yellow, gape of nestling bright vermilion. Feet dark grey, nestling vinaceous, claws black.

BIOMETRICS Wing male (n = 7) 125–136mm, mean 131.1 ± 4.2, female (n = 5) 122–135mm, mean 128.6. Tail male 122–139mm, mean 130.5 ± 7.8. Tail male 122–139mm, mean 130.5 ± 7.8, female 116–137mm, mean 121.5. Bill male 18.7, 19.2mm, female 18.3–21.4mm, mean 20.2 ± 1.1.

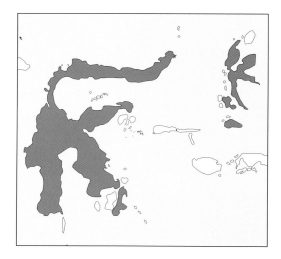

Tarsus male 15.5, 16.4, female 15.4–17.5mm, mean 16.3 ± 0.8 (Payne 2005).

MOULT Primaries moulted in up to three loci simultaneously usually beginning P9 and P7 followed by P8, P6 and P5 in irregular succession, no pattern for P10, and inner primaries sequence most often P4, P3, P2, P1 (Stresemann & Stresemann 1961).

DISTRIBUTION Wallacea. Sulawesi, Butung, Halmahera, Bacan and Obi (White & Bruce 1986, Lambert 1994).

HABITAT On Sulawesi wooded savannas and forest on ultra-basic rock; primarily hill forest; disturbed forest. 300–1,200m, Sulawesi; 250m, Halmahera; 360m, Obi; 100–200m, Bacan (Stresemann 1940–41, Lambert & Yong 1989, Lambert 1994, Coates & Bishop 1997, Wardill *et al.* 1999).

BEHAVIOUR Unrecorded.

BREEDING Parasitism Hosts unknown. **Season** Sings north Sulawesi Mar, Apr, Sep; displaying Oct (Rozendaal & Dekker 1989, Wardill *et al.* 1999). No further information.

FOOD No information. Presumably chiefly insects.

STATUS AND CONSERVATION Very uncommon throughout Sulawesi (White & Bruce 1986) but widespread (Watling 1983); however, Andrew and Holmes (1988) stated it was more common than earlier authors had recorded. Rare Halmahera, Bacan and Obi (Coates & Bishop 1997). At present large range and wide habitat tolerance, which includes disturbed forest, give little cause for concern.

Genus *Hierococcyx*

S. Müller 1845 *Verh. Nat. Gesch. Nederl., Land-en Volkenk.*, part 8: 233, note.
Type by original designation *Cuculus fugax* Horsfield. Cabanis & Heine (1863) substituted the non-admissable but etymologically-correct *Hiracococcyx*. Has been subsumed in *Cuculus* (e.g. Peters 1940).

Eight species.

Old World; brood-parasitic. Medium to large; broad or pointed wings, secondaries nearly as long as primaries; broadly barred tail; sexes similar. MtDNA analysis shows *Cuculus* and *Hierococcyx* forming two monophyletic clusters which are sister groups, this group is a sister group to clade containing *Cercococcyx* and *Surniculus*; *Cacomantis* forms sister group to these; these results place *bocki* as sister to clade consisting of *sparverioides* and *varius* (Sorenson & Payne 2005).

MOUSTACHED HAWK-CUCKOO
Hierococcyx vagans   Plate 22

Cuculus vagans S. Müller, 1845 (Java).

Alternative names: Lesser/Small/Dwarf Hawk-cuckoo.

TAXONOMY MtDNA evidence suggests that *vagans* forms sister clade to all congeners (Payne & Sorenson 2005). Everett (1889) suggests that some old records of *fugax* may refer to *vagans* as the two were previously confused. Monotypic. Synonym: *nanus* Hume 1877.

FIELD IDENTIFICATION 26cm. Rather small hawk-cuckoo, slimly built, with long tail and distinctive facial pattern. **Adult** Grey crown and nape, diagnostic white ear-coverts and blackish down-curved moustache; upperparts grey-brown, wings barred paler brown; white inner webs of remiges and coverts of underwing form pale patch in flight; underparts whitish with blackish streaks, tail grey with broad black bars and tipped white. **Subadult** From adult by grey and rufous barring on dark brown upperparts; rump and uppertail-coverts edged grey; pale remiges tipped rufous. **Juvenile** Crown, back and wings blackish; back faintly barred rufous; remiges barred and tipped pale rufous; rectrices grey barred black and tipped white; face, chin and throat blackish, with broad black line in front of eye continuous with throat. Breast thickly streaked black; undertail-coverts unmarked white. **Similar species** Javan, Rufous and Whistling Hawk-cuckoos larger and lack moustache, have dark chin, cheek and ear patch. Moustached has black moustache and line from behind eye backwards and downwards, and rectrices in all stages tipped whitish, rarely grey-buff, unlike other hawk-cuckoos which are usually rufous-buff, rarely with narrow grey tip.

VOICE Loud, monotonous *chu-chu* or *kang-koh* or *kan-ko*, melancholy, far carrying repeated monotonously every 2sec. Rarer mellow whistle *peu peu*, given as single notes, then hurried couplets accelerating to screaming climax *hee hee hee hee... hi-hi hi-hi hi-hi **hi-hi*** then coming to abrupt stop. Female voice unknown. Always calls from upper canopy as opposed to mid-level by Javan Hawk-cuckoo (Medway & Wells 1976, Thomas & Thomas 1994, Scharringa 1999, Smythies & Davison 1999, Wells 1999, Robson 2000).

DESCRIPTION Adult Forehead, crown and upper nape slaty-grey, sides of head white with contrasting blackish-brown moustache and dark crescent to rear of ear-coverts, hind nape and sides of neck streaked rusty, upperparts and wings dark brown with lighter rufous-brown barring on wings, white notches (some with rufous-brown suffusion) on inner webs of remiges, outer webs with rufous-brown notches; pale rufous fringes to back and uppertail-coverts; chin and throat white, some with deep buff tinge; rest of underparts creamy-white with generally buff suffusion, streaked black-brown on lower throat, breast, flanks, thighs and upper belly, undertail-coverts plain creamy-white; underwing-coverts buff, with few dark brown spots near edge of wing, underside of dark brown remiges buff proximally, with large buff notches on inner webs, small ones on outer webs; brownish-grey rectrices with four broad dark brown bars, terminal bar narrowly tipped greyish or pale rufous-white, bars gradually becoming wider towards end, underside of rectrices similar but paler. **Subadult** Differs from adult in dark brown upperparts barred grey and rufous, rump and uppertail-coverts edged grey, remiges tipped rufous. **Juvenile** Blackish brown crown and back, latter barred rufous, wings blackish, remiges barred and tipped pale rufous; tail as adult; face blackish with black streak in front of eye continuous with blackish throat and chin; rest of underparts white streaked black on breast. Fledgling has moustache and black-and-white tipped tail. **Nestling** Undescribed. **Bare parts** Iris dark brown to grey; eye-ring bright yellow. Short down-curved bill black, greenish at base, and lower mandible greenish-yellow. Feet orange-yellow.

BIOMETRICS Wing male (n = 13) 135–156mm, mean 145.0 ± 5.2, female (n = 10) 136–150mm, mean 143.1 ± 4.9. Bill male 19.3–21.7, mean 19.8 ± 0.8, female 19.6–22.6mm, mean 21.5 ± 1.0. Tail male 120–144mm, mean 131.4 ± 7.5, female 113–145mm, mean 125.2 ± 10.5. Tarsus male 17.2–23.1mm, mean 19.2 ± 0.8, female 17.2–20.2mm, mean 18.6 ± 1.2 (Payne 2005). Mass male 58.2–63.4g (Wells 1999); female 62g (Payne 2005); unsexed 53g (Wilkinson *et al.* 1991).

MOULT Primaries moult at up to four loci; active wing moult late Apr–Jun and early Sep, Thai-Malay Peninsula. Tail moult May, Burma (Wells 1999, BMNH).

DISTRIBUTION Sundaland. Resident Thai-Malay Peninsula in Tenasserim south from 14°N, including Mergui Archipelago, and Thailand south from 13°N to Singapore. South Laos (once); Bintan I. in Riau Archipelago, Borneo and west Java (Smythies 1940, Duckworth 1996, Rajathurai 1996, Wells 1999, Mann 2008, BMNH). Once Katembe, Aceh, Sumatra, possibly more widespread and resident (van Marle & Voous 1988), and once Bengkulu (MCZ).

HABITAT Lowlands to 915m, Thai-Malay Peninsula;

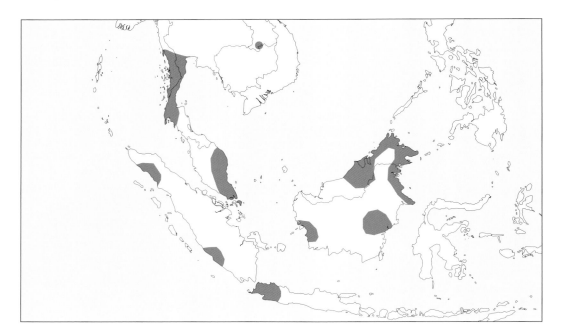

broadleaved semi-evergreen and evergreen forest, regenerating forest, forest edge and second growth; understratum to mid-canopy; bamboo forest. Commonest on lower hill slopes and lowlands. On Borneo lowland dipterocarp, alluvial, peat-swamp and secondary forests; usually lower storey, occasionally canopy; to 550m (Smythies 1981, Smythies & Davison 1999, Wells 1999, BirdLife International 2000, Robson 2000, Payne 2005, Mann 2008).

BEHAVIOUR Males have fixed song-posts and supposedly occupy territories, as one male captured three times in 13 months at same site. Frequents canopy, foraging and calling among thick foliage, but frequently also in lower and midstorey on Borneo. Joins mixed feeding flocks of woodpeckers, drongos, babblers and warblers in midstorey, Thailand (Jeyarajasingam & Pearson 1999, Wells 1999, Mann 2008, Pobprasert et al. 2008).

BREEDING Parasitism Juvenile attended by Abbott's Babbler, 24 Aug Krabi, Thailand; chick fed by female Rufous-winged Philentoma, Danum Valley, Sabah (Smythies & Davison 1999, P.D. Round in Wells 1999). **Season** Jun–Aug mainland south-east Asia. Calling central Malay Peninsula late Oct–May peaking Feb. Juvenile Thailand Aug. Calling Brunei Jan, Mar, May, Sep and Oct, Sabah Dec, Kalimantan, Mar. Calling Aug north Sumatra (van Marle & Voous 1988, Holmes 1997, Wells 1999, Robson 2000, Mann 2008). No further information.

FOOD Insects, including caterpillars and katydids; small fruits (MacKinnon 1988, Pobprasert et al. 2008).

STATUS AND CONSERVATION Nowhere common. Uncommon resident Tenasserim, south and south-east Thailand and south Laos (Robson 2000). Locally common to uncommon resident north Thai-Malay Peninsula, sparse south of Kedah (Wells 1999); occurs Hala-Bala WS (U. Treesucon in Wells 2007). First sighting Laos 1992 may be range extension but country poorly known (Duckworth 1996). Strays have reached west Java, but no recent records (MacKinnon & Phillipps 1993). Possibly overlooked Sumatra (van Marle & Voous 1988). Widespread but uncommon Borneo (Mann 2008). Most common in lowlands which therefore could be core habitat (Wells 1999). With deforestation over its entire range considered near threatened (BirdLife International 2011).

Moustached Hawk-cuckoo *Hierococcyx vagans*. **Fig. 1.** Adult, Borneo Rainforest Lodge, Danum, Sabah, Borneo, May (*Ron Hoff*).

DARK HAWK-CUCKOO
Hierococcyx bocki   **Plate 22**

Hierococcyx bocki R.G.W. Ramsay, 1886 (mountains of west Sumatra)

TAXONOMY Following Payne (2005) accepted as species separate from *sparverioides*. MtDNA analysis places *bocki* as sister to clade with *sparverioides* and *varius* (Sorenson & Payne 2005). Monotypic.

FIELD IDENTIFICATION 30–32cm. Medium-sized cuckoo. **Adult** Face dark; upperparts blackish slate-grey, wings brown, chin and throat grey and breast orange-rufous, rest of underparts white barred black; tail dark brown barred broadly buffy-grey to greyish-white, tipped white. **Juvenile** Crown slaty-brown; few white feathers on nape; back, wing-coverts and rump brownish-black barred rufous; chin to upper breast slaty-brown, latter with white feather bases; lower breast to undertail-coverts white barred dark; wings dark brown with rufous notches on outer edges of primaries and tipped buff; secondaries grey-brown with white notches, bend of wing white. Tail as adult. **Similar species** Considerably smaller than Large Hawk-cuckoo and lacks white on lores, and generally darker and subadult barred not streaked below. Pale tail bands narrower than in Rufous, Javan and Whistling Hawk-cuckoos. Juvenile uniformly brown above, with brown chin to breast, distinguishing it from other hawk-cuckoos.

VOICE Call *pit-piwit*; *CHUPCHUPchee*. Male gives 6–8 deliberately spaced, powerful and increasingly emphatic disyllables (once trisyllabic) rising steadily to screaming climax *pee-ha pee-ha pee-ha … … … pee-ha pee-hee* **pee-he;** also faster more hurried series of couplets *pi-hi* rising in semi-tones to shrill climax with brief tail-off of usually two simple notes; males counter-call (Wells 1999, Robson 2000, T. Harrisson in Payne 2005).

DESCRIPTION Adult Upperparts blackish slaty-grey; wings dark brown, inner webs of remiges with broad whitish bars becoming triangular wedges distally; chin and throat grey; breast dark rufous; belly and flanks white barred black; undertail-coverts white; bend of wing white, underwing-coverts white barred dark grey, marginal coverts rufous; tail dark brown barred narrowly black and broadly grey, with orange-buff on distal side of black bars, broad black subterminal bar, and tipped whitish. **Subadult** Chocolate-brown or grey-bronze crown, sides of face dark grey; back dark brown barred rufous; wings brown with rufous notches on outer webs of remiges, and extreme tips may be white, inner webs notched white to rufous-white; chin grey or brownish-grey, rest of underparts white to buff-white, broadly barred and spotted dark brown (or streaked blackish becoming spots on belly and chevron bars on rear flanks: Wells 1999), belly and undertail-coverts white. Tail as adult but with some rufous in grey. Some non-moulting individuals in this plumage but with underparts more as adult may indicate some subadult features retained by adult females (Wells 1999). **Juvenile** Crown slaty-brown; few white feathers on nape; back and rump brownish-black; chin to upper breast slaty-brown, latter with white feather bases; lower breast to undertail-coverts white barred dark; wings dark brown with rufous notches on outer edges of primaries and tipped buff; secondaries grey-brown, bend of wing white. Tail as adult. **Nestling** Unknown. **Bare parts** Eye-ring yellow; iris orange to buff; bill blackish above, greenish below; feet yellow.

BIOMETRICS Wing male (n = 16) 184–196mm, mean 186.3 ± 5.9, female (n = 7) 180–200mm, mean 189.7 ± 5.1; tail male 132–175mm, mean 151.0 ± 13.2, female 136–175mm, mean 155.0 ± 3.2; bill male 18.4–21.6mm, mean 21.1, female 19.5–23.2mm, mean 21.1; tarsus male 18.6–23.2mm, mean 20.3 ± 2.0, female 19.2–21.4mm, mean 19.9 (Payne 2005).

MOULT Females may acquire full adult plumage over more than one moult (Wells 1999).

DISTRIBUTION Sundaland resident. Peninsular Malaysia on Gunung Tahun, Gunung Benom and Larut Range, and Genting Highlands north to Cameron Highlands in Main Range. Sumatra at Bandarbaru, Berestagi and Mt. Sibayak, Utara; Mt. Sago, Barat; Rejang, Bengkulu; Gunung Leuser NP. Borneo at Gunung Murud, Kelabit uplands and Usun Apau Plateau, Sarawak; Gunung Kinabalu, Crocker Range, Gunung Trus Madi, Gunung Selidang, Kaingaran, Kiau and Poring, Sabah; upper Sungai Bahau, Kayan Mentarang NR,

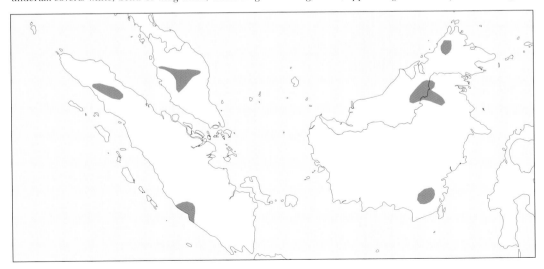

and Gunung Lunjut, East Kalimantan; Muratus Mts., South Kalimantan (van Marle & Voous 1988, Wells 1999, Buij *et al.* 2006, Mann 2008).

HABITAT Lower and tall upper montane forest in edge and interior canopy at 1,100–1,800m peninsular Malaysia. 900–1,600m Sumatra; also pine plantations. 800–2,000m Borneo (van Marle & Voous 1988, Wells 1999, Mann 2008).

BEHAVIOUR Males territorial, calling from tree tops (Wells 1999).

BREEDING Parasitism Host peninsular Malaysia Chestnut-capped Laughing-thrush (A. Chia *in litt.*); in Borneo Mountain Leaf Warbler (Phillipps 1970) but Moyle *et al.* (2001) suggest that it could have been hosting Himalayan (= Sunda Lesser) Cuckoo. **Season** Fledgling 19 Jun, peninsular Malaysia (A. Chia *in litt.*); 15 Aug, Borneo (Phillipps 1970), but see above; nestling, thought to be this species, Jun, south Barisan Range, Sumatra (van Marle & Voous 1988). No further information.

FOOD Caterpillars, Orthoptera, beetles, bugs, cockroaches, and ants; spiders; birds' eggs; berries and other fruit (Smythies 1981, Sody 1989).

STATUS AND CONSERVATION More or less common in range, peninsular Malaysia (Wells 1999). Local montane resident, Borneo (Mann 2008).

Dark Hawk-cuckoo *Hierococcyx bocki.* **Fig. 1.** Adult, Kuala Gula, Malaysia, November (*Choy Wai Mun*).

LARGE HAWK-CUCKOO
Hierococcyx sparverioides   Plate 22

Cuculus sparverioides Vigors, 1832 (Himalayas).

Alternative name: Brain-fever Bird (also for *Cacomantis merulinus*, *C. passerinus* and *Cuculus varius*).

TAXONOMY Previously included *bocki*, but mtDNA indicates that *sparverioides* is closer to *varius*, and *bocki* is sister to this clade (Sorenson & Payne 2005). Monotypic. Synonym: *strenuus* Gould 1856.

FIELD IDENTIFICATION 38–40cm. Largest hawk-cuckoo. Hawk-like; rather broad, rounded wings and tail; skulking. Flight low and buoyant, as sparrowhawk, with some fast wing beats followed by glide at the end to rise abruptly and land in tree. **Adult, dark morph** Head grey with yellow eye-ring and entire upperparts and tail grey-brown, tail strongly barred black, tipped white, and flight feathers strongly barred. Underparts white with grey streaks on throat, rusty with blackish stripes on upper breast and broad black barring on flanks and belly. **Adult, pale morph** Paler grey above than dark morph, with some even paler grey and rufous edging. Colours paler below than dark morph. Rufous and whitish throat; upper breast barred rufous and white; lower breast to belly and flanks creamy-white with rufous and grey bars and chevrons; vent and undertail-coverts white. **Juvenile** Brown upperparts, each feather edged paler and whitish underparts variable tinged rufous with drop-like brown streaks on breast, abdomen and flanks barred brown. **Similar species** Common Hawk-cuckoo smaller, more brownish-grey above, lacks blackish chin, plain rufous on breast and barring on belly, and flanks pale rufous; in subadult underparts more boldly streaked and barred; juvenile has flanks more spotted than barred. Pale morph superficially similar to large Common Hawk-cuckoo. Dark Hawk-cuckoo is smaller and darker. Moustached even smaller, with broad black moustache and lacks any rufous on breast. Juvenile Javan and Whistling Hawk-cuckoos have eye-catching yellow base to bill. Juvenile from other hawk-cuckoos by brown upperparts barred rufous, underparts streaked black on throat and breast, barred on belly and flanks. *Cuculus* spp. usually have obvious white bars on T5.

VOICE Shrill, loud *pee-pee-ah*, *pee-pee-ah*, or *brain fe-ver*, rising in speed and pitch and ending in 'hysterical crescendo', or *brain FE-ver* or *pi-PEE-ha* ending *bee-frever*. Rendering of sonogram series of loud, piercing whistles *PEEE-FW\uu* and calling of female series of short burry *dRUu-dRUu*. *Pwi pwee-wru*; when breeding calls at dawn and sunset and often through whole night especially in overcast weather (Fleming *et al.* 1979, Ali & Ripley 1983, Johnsgard 1997, Grimmett *et al.* 1998, Robson 2000, Rasmussen & Anderton 2005).

DESCRIPTION Adult, dark morph Head, nape and hindneck ashy-grey with some white on lores, upperparts plain grey-brown, remiges barred dark brown and white on inner webs, white on carpal joint, underwing-coverts white barred dark brown; middle of chin dark grey, rest of underparts white with grey streaks on throat becoming prominent blackish streaks on rusty breast and sides of breast, rusty extent variable, flanks and belly broadly barred dark brown, undertail-coverts white, longest with two concealed black bars; brown tail with 3–4 dark, broad bands, often with lighter buffish, and tipped white or buffy on both upper and underside. **Adult, pale morph** (Rare Burma and Thailand; perhaps only male). Whitish in front of eye and moustachial stripe; crown to uppertail-coverts medium grey (darker than *varius*) with some pale rufous and grey barring and fringes on mantle; tail same grey with broad blackish bar fringed pale rufous proximally, narrower band similar, even narrower band, narrow area of grey suffused pale rufous then broad blackish band fringed pale rufous distally and off-white tip. Wing-coverts similar grey with narrow whitish fringes; outer webs of secondaries grey-brown, inner webs brown with whitish notches becoming pale rufous towards shaft; primaries brown with fine whitish tips, large white notches on inner webs becoming rufous towards shaft. Rufous and whitish throat; upper breast mostly rufous; lower breast to belly and flanks creamy-white with rufous and grey bars and chevrons; vent and undertail coverts white; underside of rectrices silvery-grey barred blackish-brown and rufous, broad blackish bar near tip followed by narrow pale rufous bar and tip pale grey. **Subadult** Top of head grey-bronze, upperparts and wings dull dark brown barred rufous-brown; white to pale buffish underparts boldly streaked black with chevron-like bars on flanks. **Juvenile** Like adult but brown crown spotted rufous, upperparts barred rufous, dark streaked nape paler rufous and underparts white variable tinged buffish with strong drop-like brown markings or streaks on breast and belly, little or no rusty on breast; tail rufous barred black. **Nestling** Undescribed. **Bare parts** Iris tawny-yellow to orange, juvenile greyish-brown, yellow orbital ring. Downcurved bill black, below yellow-green, juvenile brownish-black with culmen tinged olive, yellow gape. Feet pale yellow to orange-yellow, soles dirty lemon-yellow.

BIOMETRICS Wing male (n = 11) 227–246mm, mean 236 ± 5.5, female (n = 9) 210–243mm, mean 226.7 ± 11.1. Bill male 24.0–26.7mm, mean 25.1 ± 0.9, female 23.3–26.7mm, mean 24.3 ± 1.1. Tail male 210–243mm, mean 223.4 ± 11.3, female 195–228mm, mean 213 ± 10.6. Tarsus male 24.0–29.4mm, mean 26.4 ± 1.5, female 22.2–24.8mm, mean 23.4 ± 0.9 (Payne 2005). Mass 116–131g (Ali & Ripley 1983); 140, 152.1, 160g, mean 150.7 (Payne 2005). Migrants Thai-Malay Peninsula 130, 131g (Wells 1999). Daytime mass (with visible subcutaneous fat) Nov/Dec Singapore, 159, 163g, 25% greater than autumn migrants at Fraser's Hill, peninsular Malaysia (P.R. Kennerley in Wells 2007).

MOULT Adult and subadult moult completed before migration into Thai-Malay Peninsula (Wells 1999).

DISTRIBUTION Oriental and marginally Palearctic. From Himalayas to south-east Asia; mostly summer visitor in north. North Pakistan (no recent records), Himalayas from Himachal Pradesh, east to Arunachal Pradesh, Nepal, Bhutan, through south-east Tibet, south-central China (stragglers further north), north and central Burma, north and west Thailand, Cambodia, Laos, west and east Tonkin. Resident south Yunnan and Hainan, China; breeding visitor south of Changjiang R., north to south Shaanxi, south-east Gansu, Henan, rarely Hebei, west to Sichuan, north-west Yunnan and Tibet. Resident and breeding migrant Manipur. Part of population winters in India south to Tamil Nadu, Bangladesh, central and south-east Thailand, Thai-Malay Peninsula, Cochinchina and Annam. Jan–May Great Nicobar I. In Thai-Malay Peninsula winters Surat Thani, Phang Nga, Phuket, Krabi and Trang, Pahang, Selangor, Johor and Singapore including St. John's I. and Langkawi Is, Nov–Feb; migratory autumn movements in Selangor and Singapore Oct–Nov. Rare west Java and perhaps Bali, winter. In Philippines uncommon Sep–May, mainly in north

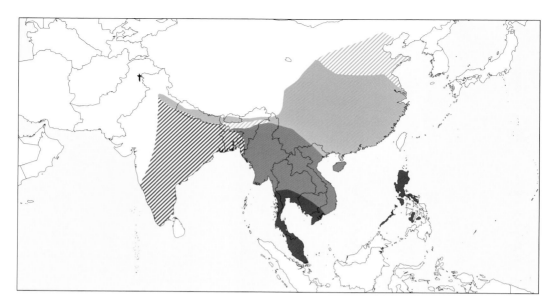

(Cheng 1987, MacKinnon & Phillipps 1993, Grimmett *et al.* 1998, Wells 1999, 2007, Kennedy *et al.* 2000, Sivakumar & Sankaran 2002, Choudhury 2009). Three Sumatra (Feb, Berestagi; undated Lampung; Dec, Lake Toba) (van Marle & Voous 1988, C.F. Mann). At least three Borneo (Nov, Feb, Kuching; Nov, Labuan; Sabah?) (Mann 2008). Less than ten records South Korea, Apr–May (Birds Korea 2009, 2010). Vagrant Taiwan (Cheng 1987). Vagrant north Sulawesi, Sep 1894 (Meyer & Wiglesworth 1896) and probable Tangkoko-Batuangus Reserve, north Sulawesi, Feb 1979 (Watling 1983). Vagrant Christmas I. (McAllan & James 2008).

HABITAT Broadleaved woodlands, deciduous, especially oak, evergreen and thickets. Mixed coniferous and broadleaved forests and groves, India and Bhutan. Oak-rhododendron forest, Nepal. Winter visitors in plantations, wooded gardens, tall secondary growth, mangroves, orchards and open country scrub, Thai-Malay Peninsula. Foothills to 2,500m, Thailand. Prefers open forests to 1,000m, China. Various habitats and all altitudes in winter, Philippines. Vagrants in lowlands, Borneo. Breeding 900–3,500m Himalayas, sometimes down to 200m Bhutan (Fleming *et al.* 1979, Lekagul & Round 1991, Tymstra *et al.* 1997, Grimmett *et al.* 1998, Jeyarajasingam & Pearson 1999, Wells 1999, Kennedy *et al.* 2000, MacKinnon & Phillipps 2000, Robson 2000, 2002, Spierenburg 2005, Mann 2008).

BEHAVIOUR Secretive tree-top and midstorey cuckoo; occasionally near ground on Philippines; migrants may forage on ground. May sit stationary for hours. Migrants solitary. Often mobbed by passerines (Stanford & Ticehurst 1939, Wells 1984–85, 1999, Coates & Bishop 1997, Jeyarajasingam & Pearson 1999, Kennedy *et al.* 2000).

BREEDING Parasitism Hosts: in Himalayas Lesser Shortwing, Streaked Spiderhunter and Little Spiderhunter; 68% of eggs collected were in Streaked Spiderhunters' nests, 2% in Little Spiderhunters' despite the latter being commoner there, and 17% in Lesser Shortwings' nests (Baker 1942). Becking (1981), using scanning electron-microscopy and protein electrophoresis data, suggested all other hosts in north-east India/north Burma are erroneous and the eggs belonged to other cuckoos or even non-cuckoos. These include Lesser Necklaced, Greater Necklaced and Moustached Laughing-thrushes, first with one to four and second with one Chestnut-winged Cuckoo eggs also per nest, and Rusty-cheeked Scimitar Babbler (Osmaston 1916, MacKenzie 1918); Baker's 'brown type' eggs in nests of Red-vented Bulbul, Long-tailed Shrike, Blue-throated Blue Flycatcher, Pygmy Blue Flycatcher, Small Niltava, Brownish-flanked Bush Warbler, Buff-chested Babbler, Grey-throated Babbler, Rufous-vented Yuhina, Whiskered Yuhina, Spot-throated Babbler, Puff-throated Babbler and Nepal Fulvetta; and Baker's 'blue type' eggs in nests of Blue Whistling Thrush, Orange-headed Thrush, Scaly Thrush, Grey-sided, Chestnut-crowned and Rufous-chinned Laughing-thrushes, Red-faced Liocichla, Rusty-cheeked Scimitar Babbler and Hoary-throated Barwing. White-browed Laughing-thrush, Sichuan, China (Jiang *et al.* 2007). Chinese Babax, China (Yang *et al.* 2012). **Season** Himalayas Apr–Jul, which is breeding season for Streaked Spiderhunter, Apr–Jun Assam. Calling mostly end Feb–Jun, eggs Apr–May, Burma. Fledgling Heinsin, north Chin Hills, May (Osmaston 1916, MacKenzie 1918, Smythies 1953, Ali & Ripley 1983, Grimmett *et al.* 1998, BMNH). **Eggs** Either light brownish-olive, rarely speckled with darker spots, like spiderhunters and shortwings, but larger with blunt oval shape; 24.5–29.0x17.3–20.6mm, (n = 70) mean 26.6x18.6mm (Becking 1981), or blue like laughing-thrushes, but Becking (1981) believes blue eggs from Baker's collection all belong to Common Cuckoo, subspecies *bakeri*. However, female has been collected with blue oviduct egg (Rattray 1905). White, laid in nest of Chinese Babax whose eggs are blue (Yang *et al.* 2012). No other information.

FOOD Mainly insects such as caterpillars, including hairy species, beetles, Hemiptera, ants, grasshoppers, crickets, cockroaches; spiders; birds' eggs; berries (Sody 1989, Roberts 1991).

STATUS AND CONSERVATION Fairly common summer India and Nepal, but only scattered winter records Nepal,

India and Bangladesh (Grimmett *et al.* 1998), perhaps because silent outside breeding season. Abundant summer, Bhutan (Spierenburg 2005). No records in recent decades from Pakistan but around 1905 common near Dunga Gali; last two specimens Murree, 2–3 June 1947 (Roberts 1991). Local and sparse in all winter months, Thai-Malay Peninsula (Wells 1999). Uncommon passage migrant and winter visitor, 27 Sep–May, Philippines (Dickinson *et al.* 1991, Kennedy *et al.* 2000). Uncommon summer breeder, China; accidental Taiwan and Hebei (Cheng 1987, MacKinnon & Phillipps 2000); 16–25 sightings in 41 days, May–Sep 2005, Hainan (Lee *et al.* 2005); uncommon Shiwandashan, Guangxi Zhuang, China (Robbins *et al.* 2006). Rare Sumatra, Java, ?Bali, Borneo, Sulawesi (Meyer & Wiglesworth 1896, Watling 1983, van Marle & Voous 1988, Mackinnon & Phillipps 1993, Mann 2008, C.F. Mann). Vagrant South Korea (Birds Korea 2009). Not globally threatened (BirdLife International 2011).

Large Hawk-cuckoo *Hierococcyx sparverioides*. **Fig. 1.** Subadult, Tinsukia, Assam, India, January (*Raj Kamal Phukan*). **Fig. 2.** Adult, Khao Yai NP, Thailand, February (*Peter Ericsson*). **Fig. 3.** Adult, Khao Yai NP, Thailand, February (*Peter Ericsson*).

COMMON HAWK-CUCKOO
Hierococcyx varius   **Plate 22**

Cuculus varius Vahl, 1797 (Tranquebar, India)

Alternative name: Brain-fever Bird (also used for *Cacomantis merulinus*, *C. passerinus*, *C. pallidus* and *Hierococcyx sparverioides*).

TAXONOMY MtDNA analysis shows *varius* to be closest to *sparverioides* (Payne 2005). Polytypic. Synonyms: *tenuirostris* Lesson 1830, *lathami* Gray 1832, *ejulans* Sundevall 1837, *nisoides* Blyth 1866, all in nominate.

FIELD IDENTIFICATION 34cm. Size as Common Cuckoo; hawk-like appearance and flight, but sometimes perches rather horizontally, drooping wings and slightly raising tail, very different from compact appearance of a hawk. **Adult** Above grey-brown with ashy-grey head, below white with plain, pale red-brown breast and pale brown barring on belly and flanks; wings uniform grey-brown, tail grey-brown with diagnostic 3–5 white and black bars, and pale rufous tip. Tail less graduated than in other *Hierococcyx* species. **Subadult** Rufous tips to wing-coverts and remiges barred rufous. Less rufous on throat and breast. **Juvenile** Brown above barred rufous, underparts white variously tinged buffy and streaked or spotted blackish-brown; tail lacks white on bright rufous and black bars. **Similar species** Large Hawk-cuckoo bigger, has white chin spotted black, white and rufous streaks on breast and broader tail bars; juvenile Large often has heavier and darker streaking on underparts but great variation, tail duller. Common Hawk-cuckoo occurs at lower altitude in Nepal than Large. Whistling Hawk-cuckoo is smaller with dark grey upperparts, black spot on white chin, much more rufous on breast and belly, and broader tail bands. From *Cuculus* spp. by white spots on shaft of T1. From Shikra by longer, slender bill, shorter legs and diagnostic barred tail.

VOICE Loud, shrill *pee-pee-ah.. pee-pee-ah* or *wee-piwhit* with accent on second syllable repeated 4–6 times in crescendo and breaking off abruptly, and repeated with monotonous persistency every minute or two, like Large Hawk-cuckoo but more shrill; sonogram interpreted as *bur-/FEE\Wer*. Calling may be almost non-stop on overcast days or moonlit nights, hence 'Brain-fever Bird' (also for some other cuckoos). Trilling note; female calls much less than male, with harsh grating call; strident trilling scream; *FYIEw-/FYIEw*. Fledglings *ke-ke* like chicks of babblers, or harsh, single screech like Ring-necked Parakeet. Calls Mar–Jul India peaking May, silent from Aug/Sep–Feb. Often calls at night in hot weather (Basil-Edwardes 1926, Ali & Whistler 1936, Henry 1955, Ali & Ripley 1983, Grimmett *et al* 1998, Robson 2000, Payne 2005, Rasmussen & Anderton 2005).

DESCRIPTION Nominate. **Adult** Head grey, slightly paler than upperparts, crown and nape same shade as mantle and wing-coverts, rest of upperparts and wings uniform grey-brown; inner webs of dark brown remiges have whitish notches with some brown suffusion; underwing-coverts faintly barred and tinged rufous, underside of dark brown remiges broadly notched white, with little brown suffusion on some; chin and throat white variably streaked or mottled dark, large pale rufous area on breast, flanks and belly faintly barred same colour; tail grey with 3–5 blackish bars bordered white, terminal bar broadest and tipped buffish; underside of rectrices silvery-grey barred dark brown and greyish-white, most distal brown bar widest, tip greyish-white. **Subadult** Above as adult with rufous tips to wing-coverts; remiges blackish barred rufous. Below less rufous on throat and breast than adult. **Juvenile** Forehead, forecrown and ear-coverts ashy-grey, whitish nape mottled brown, hind crown, sides of neck and upperparts brown faintly barred rufous, underparts buffy-white variably streaked with blackish-brown 'tear-drops', flanks usually less heavily streaked, tail with rufous and black banding (no white) and tipped rich rufous, dark banding on underside of tail. **Nestling** Undescribed. **Bare parts** Iris yellow to buffish, juvenile greyish-brown, eye-ring pale lemon-yellow, juvenile greenish-yellow. Bill yellowish, culmen and tip black, lower greenish, juvenile pale brown-yellow, mouth pink and yellow, juvenile yellow. Feet and claws yellow, juvenile pale yellow.

BIOMETRICS Nominate. Wing male (n = 12) 185–213mm, mean 200.5 ± 7.4, female (n = 7) 187–201mm, mean 191.4 ± 5.7. Bill male 21.6–24.8mm, mean 23.3 ± 1.2, female 22.2–24.9mm, mean 23.4 ±1.1. Tail male 160–182mm, mean 170.0 ± 7.0, female 146–178mm, mean 156.9 ± 10.7. Tarsus male 22.5–24.7mm, mean 23.8 ± 0.8, female 20.9–24.3mm, mean 22.6 ± 1.2. Mass male 93, 109.6g, female 105g (Payne 2005); fat female 104g (Ali & Ripley 1969).

MOULT Neither adult nor juvenile undergoes complete moult every year. In India primary moult, Jul–Sep; remiges moult, Mar; juvenile primary moult Jun; juvenile to adult secondary moult, Feb (Ali & Ripley 1983, BMNH).

GEOGRAPHICAL VARIATION Two subspecies.
H. v. varius (Vahl, 1797). Pakistan, Nepal, Bhutan, India, Bangladesh, and perhaps Burma. Described above.
H. v. ciceliae Phillips, 1949. Sri Lanka. Darker grey on upperparts, darker rufous on breast, more streaks on throat and barring below darker; dark tail-banding broader than in nominate.

DISTRIBUTION Indian subcontinent. Breeding visitor Bangladesh; resident and partial migrant north Pakistan, Nepal, Bhutan and India (including Sikkim); rains visitor in dry areas of Rajasthan and Gujarat. Unconfirmed Burma.

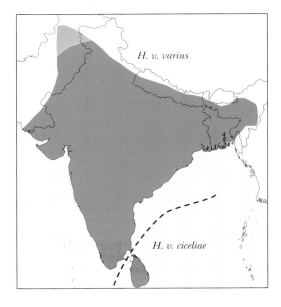

Resident and winter visitor Sri Lanka. Perhaps altitudinal and arid area migrant (Phillips 1948, 1949, Lushington 1949, Grimmett *et al.* 1998, Payne 2005). Twice Masirah I., Oman, juvenile Nov 1988 (BMNH), and juvenile 14 Jan 2010 (Balmer & Murdoch 2010). One peninsular Thailand, Apr 1914 (Robinson & Kloss 1923) apparently a mis-identification, but skin not traced (D.R. Wells pers. comm., C.F. Mann).

HABITAT Open deciduous and semi-evergreen woodland, gardens, orchards and cultivation; also mangroves, plantations, olive scrub and riverine forest in Pakistan. Below 1,000m Nepal (once 1,500m); 200–2,200m, mainly below 600m Bhutan; up to 900m Pakistan; to 1,200m India. Wetter hilly areas above 650m, including tea plantations, preferred Sri Lanka; migrants in lowlands (Henry 1955, Ali & Ripley 1983, Roberts 1991, Grimmett *et al.* 1998, Spierenburg 2005).

BEHAVIOUR Shy, but more often seen than Large Hawk-cuckoo when calling as it prefers less dense woodlands. Forages in foliage of trees and scrub; also takes food from ground. Recorded on telephone wire watching surrounding ground, and frequently flying low and swift in hawk-like manner down to grass to catch insects. Solitary most of year (Ali & Ripley 1983, Roberts 1991, Grimmett *et al.* 1998).

BREEDING Parasitism Following hosts claimed: Ashy-headed Laughing-thrush, Burma; Common Babbler favoured Sri Lanka; Yellow-billed, Jungle, Large Grey and Common Babblers, Lesser Necklaced, Ashy-headed, Rufous-necked, Grey-sided, Chestnut-crowned, Moustached and Rufous-vented Laughing-thrushes, Red-faced Liocichla, Rusty-fronted Barwing, Rufous-bellied Niltava, Asian Paradise-flycatcher and Asian Fairy-bluebird, India (Hopwood & Mackenzie 1917, Baker 1942, Henry 1955, Ali & Ripley 1969, 1983, Gaston & Zacharias 2000). However, some doubt about all host species, except Jungle Babbler, as Becking (1981) from egg-white analysis and shell morphology claimed that many Common Hawk-cuckoo eggs in Baker collection are wrongly identified Common Cuckoo and *Clamator* spp. eggs. Large Grey Babbler party feeding young of this species, India (Anon. 2009). **Season** Normally Jan–Jun India, but later months also recorded, varying locally (Grimmett *et al.* 1998), e.g. mostly Sep in Kerala which is also peak for Jungle Babbler, but also Mar. Sri Lanka Jan–Apr (Legge 1880, Phillips 1948, Robinson & Jackson 1992, Gaston & Zacharias 2000). **Eggs** Variously stated turquoise-blue, variable, or blue; 30.5x20.1mm; India 25x19mm (Baker 1906b, Henry 1955, Ali & Ripley 1969). However, the only two oviduct eggs in Baker's collection could not be located, and no other irrefutable eggs are known, therefore Becking (1981) preferred to declare the egg unknown. **Incubation** No information. **Chicks** Young seen evicting three babblers' eggs, Delhi (Gaston & Zacharias 2000). Fledgling in flock of Jungle Babblers fed by fosters as well as other members of party; if cuckoo flies, flock at once becomes nervous, some giving anti-hawk alarm call (Ali & Ripley 1983). Fledgling dependent for more than one month (Gaston & Zacharias 2000). **Survival** No information.

FOOD Mainly caterpillars (including cutworms), locusts, grasshoppers, crickets, ants, termite alates and other insects; spiders; lizards; fruit such as berries and wild banyan figs. (Ali 1946, Ali & Ripley 1983, Roberts 1991).

STATUS AND CONSERVATION Common and widespread through much of range. Summer visitor Pakistan, common only in Punjab (Grimmett *et al.* 1998). Common resident Nepal (usually below 1,000m) and Bangladesh, although rare Sundarbans (Khan 2005); widespread and locally common India (Grimmett *et al.* 1998), but rare Rajasthan and semi-desert of Gujarat (Ali & Ripley 1983); call among best known in India (Payne 1997). Uncommon Bhutan summer (Spierenburg 2005). Rare Sri Lanka (Kotagama & Fernando 1994). Not globally threatened (BirdLife International 2011).

Common Hawk-cuckoo *Hierococcyx varius*. **Fig. 1.** Adult, *H. v. varius*, Belgaum, Karnataka, India, March (*Niranjan Sant*). **Fig. 2.** Adult, *H. v. varius*, Karnataka, India, March (*Subharghya Das*). **Fig. 3.** Adult, *H. v. varius*, Belgaum, Karnataka, India, March (*Niranjan Sant*). **Fig. 4.** Adult, *H. v. varius*, Bandhavgah, India, March (*Martin Gottschling*).**Fig. 5.** Juvenile, *H. v. varius*, Howrah, West Bengal, India, May (*A. Das*). **Fig. 6.** Juvenile moulting to subadult, *H. v. varius,* Dharmaj, central Gujarat, India, June (*Arpit Deomurari*).

RUFOUS HAWK-CUCKOO
Hierococcyx hyperythrus   Plate 33

Cuculus hyperythrus Gould, 1856 (Shanghai, China)

Alternative name: Northern Hawk-cuckoo.

TAXONOMY Previously considered subspecies of *fugax*. Payne (1997) separated *pectoralis* from *fugax* at the species level, and King (2002) split *hyperythrus* and *nisicolor* from *fugax*, resulting in four species created from one, based on different vocalisations, plumage, wing and tail lengths, wing shape, and migratory pattern, and this is followed by Payne (2005). Forms superspecies with *pectoralis*, *fugax* and *nisicolor*, supported by mtDNA analysis (Sorenson & Payne 2005). Four-species arrangement followed here, although more studies and DNA analyses needed. Monotypic.

FIELD IDENTIFICATION 28–30cm. Smallish hawk-cuckoo; shy. **Adult** Slaty-grey with variable amount of white and rusty pink on underparts, undertail-coverts white and rather long banded tail. In flight pale area on underside of inner wings. Cheeks and chin slaty-grey, and narrow whitish vertical stripe from base of bill (in some extending in front of eyes) to chin. Some have large white patch on nape or white hind collar perhaps retained from subadult plumage. **Subadult** Grey head tipped buff, dark grey-brown barred rufous on upperparts, and underparts white with dark brown chevrons, spots or streaks; some have white on nape. **Juvenile** Dark grey-brown above with buff edgings; blackish throat; rest of underparts white streaked and spotted blackish from throat to belly. **Similar species** Moustached Hawk-cuckoo has conspicuous black moustache and tail proportionately longer. Large Hawk-cuckoo larger with dark brown upperparts, much larger black area on throat bordered with white, and dark barring on belly. Philippine Hawk-cuckoo has paler grey chin, is darker pink, not extending so far posteriorly, on breast with fine streaks and no sharp demarcation of breast and white belly; voices also differ. Adult from adult Javan by narrow penultimate dark band on tail, and from adult Javan and Whistling Hawk-cuckoos by lack of streaking below and larger size. Common Hawk-cuckoo lacks white vertical stripe before eyes, upperparts more brown than grey, underparts barred, not striped. For nestling see Oriental Cuckoo.

VOICE *Ju-ichi, ju ichi*, or *weeweepeeit* or *weeweepeeweit* or buzzing whistle *wee wee-pit*. Long call, *weeteetiditdidididittititititititi*, loud, shrill and higher and higher pitched until climax, and ending more calmly, or crescendo with higher long screeches with short *pee* notes; far-carrying and painful to the ears. Calls particularly frequent in overcast weather, during drizzle, or at twilight (Vorob'ev 1948 in Dement'ev and Gladkov 1966, Brazil 1991, MacKinnon & Phillipps 2000, King 2002, Payne 2005).

DESCRIPTION Adult Dark slaty-grey head and upperparts, rump feathers occasionally brownish edged, uppertail-coverts grey tipped black, shorter feathers narrowly edged grey, wings brownish-grey sometimes faintly edged rufous on outer webs of blackish rectrices, inner secondaries plain pale grey on inner webs and usually one of shorter tertials pale grey to white, bend of wing white, underside of remiges broadly barred and narrowly tipped white; whitish vertical line in front of eyes (or from base of bill) running into pale lower throat; chin blackish-grey, rest of throat white, lower cheeks and rest of underparts rusty-pink, in some feathers have obvious grey bases giving a slight scaled effect on breast and flanks, undertail-coverts white; tail grey with 3–4 black bars with some rufous, subterminal band fairly broad, terminal bar broad and tipped rufous and white. Some have white spot on nape, or narrow white collar (complete or incomplete). **Subadult** Head, neck, chin and sides of throat slate-grey washed brown, white band separates slate-black chin and throat sides, rest of throat white with dark speckling; sometimes also white hind collar; upperparts and wings blackish-brown narrowly barred dark rufous, often a white tertial; underparts buffy-white with dark brown chevrons, streaks or spots, feather shafts on belly black, undertail-coverts buffish-white, tail brownish-grey, occasionally tinged rufous, tipped rufous with broad black subterminal band and another 2–3 narrow black bands proximal to broad rufous bands. **Juvenile** Dull sooty-brown upperparts barred buffish, feathers of underparts white at base, tipped sooty with pale buffish margin, tail more pointed. **Nestling** Hatches naked, skin flesh-coloured; after few days skin above becomes blackish and below yellow including tarsus and wings. Plumage below whitish finely streaked; gape-coloured patches on wings. **Bare parts** Iris brown to orange-brown or dirty white, and eye-ring yellow. Bill black, inner half greenish-yellow. Feet yellow, claws whitish. Gape of nestling yellow, distal palate yellow, inner mouth pink with dark lines (Payne 2005).

BIOMETRICS Wing male (n = 13) 187–206mm, mean 199.0 ± 5.8; female (n = 5) 193–200mm, mean 196 ± 2.9. Tail male 127–144mm, mean 131.9 ± 5.8; female 126–143mm, mean 137.0 ± 8.1. Bill male 19.3–24.2mm. mean 21.1 ± 1.5; female 19.2–20.9mm, mean 20.3 ± 0.7. Tarsus male 18.2–24.0mm, mean 20.7 ± 1.6; female 18.2–22.5mm, mean 20.6 ± 2.0 (Payne 2005). Mass female Japan 115g; unsexed 99.0–147.8g (Enomoto 1941); male Palau, Feb, 92.5g (Ripley 1951).

MOULT Sep (Vorob'ev 1948 in Dement'ev & Gladkov 1966).

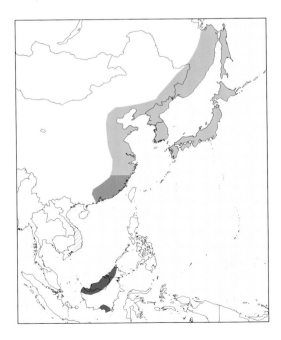

DISTRIBUTION Palearctic and Oriental. Summer visitor North-east Provinces and Hebei, China; passage Shandong, Jiangsu and Anhui; winter visitor and resident Fujian and Guangdong. Summer visitor in parts of North (probably breeds) and South Korea (uncommon breeder). Summer visitor Japan; Okinawa (passage only?). Summer visitor Ussuriland and Sakhalin (once). Passage Vietnam. About 12 records Sarawak, Sabah and South Kalimantan, Borneo (Delacour & Jabouille 1931, La Touche 1931, Dement'ev & Gladkov 1966, Neufeldt 1972, Flint *et al.* 1984, Fiebig 1993, Cheng 1987, 1991, Brazil 1991, 2009, Payne 2005, Duckworth & Moores 2008, Mann 2008, Birds Korea 2009). Vagrants: north Sulawesi 1870, 1885 and Dec 2007 (White & Bruce 1986, R. Jordan pers. comm.); Buru 1922 (White & Bruce 1986); Palau, west Micronesia, Feb (Ripley 1951); Hong Kong 1971 (Viney *et al.* 1994); Cocos Is, Jan 2011 (R. Clarke pers. comm.). Catanduanes, Philippines unsubstantiated (Dickinson *et al.* 1991).

HABITAT Lowlands to 2,800m in broadleaved evergreen, semi-evergreen and deciduous forest, secondary forest, bamboo jungle, moist ravines and plantations. In Russia in Siberian pine, silver fir, cedars, oaks, maple, ash, hornbeam and mixed taiga incised by deep ravines and rivers. Prefers spruce, fir, larch and cedar forests in Japanese Alps, Hokkaido; to 2,300m, Fujiyama (Dement'ev & Gladkov 1966, Flint *et al.* 1984, Brazil 1991, CCJB 2000).

BEHAVIOUR Perches in crowns of tall trees (Payne 2005).

BREEDING Parasitism In Japan Eurasian Skylark (rarely), Japanese Thrush, Brown-headed Thrush, Japanese Robin, Siberian Blue Robin, Orange-flanked Bush Robin, Siberian Stonechat, Asian Brown Flycatcher, Narcissus Flycatcher, Blue-and-white Flycatcher and Olive-backed Pipit (Kobayashi & Ishizawa 1932–1940, Brazil 1991). Also Japanese Paradise-flycatcher, (P.E. Lowther *in litt.*). **Season** Breeds Japan, May–Aug (Brazil 1991, Higuchi 1998, Yoshino 1999). Probably late May–Jul, South Korea (Birds Korea 2009). **Eggs** In Japan pale blue to greenish light blue (Kobayashi & Ishizawa 1932–1940), but Becking (1981) doubts these were definitely separated from Common Cuckoo (*telephonus*); pale blue, 28x20mm, 5.6g (Higuchi 1998, Yoshino 1999). **Incubation** No information. **Chicks** Display yellow, gape-coloured, patch on wing to host parents as they deliver food, simulating gaping display of more than one nestling and in this way getting more food (Tanaka & Ueda 2005); nestling period 19–20 days (Higuchi 1998, Yoshino 1999). **Survival** No information.

FOOD Mostly caterpillars, including silkworms and those of hawk-moths; larvae of sawflies (Tentheredinidae), beetles, cicadas, other insects; fruit. In Japan ants most important item after caterpillars (Dement'ev & Gladkov 1966, Ishizawa & Chiba 1966).

STATUS AND CONSERVATION Rare or uncommon over most of range. Uncommon Japan in summer (Brazil 1991), but rather common breeder Honshu and many smaller islands (CCJB 2000). Common in summer in parts of North and South Korea (Duckworth & Moores 2008, Birds Korea 2009); rare Socheong I., South Korea (Moores 2007). Uncommon China (Cheng 1987). Few records Borneo (Mann 2008). Rare Sulawesi (Riley *et al.* 2003).

Rufous Hawk-cuckoo *Hierococcyx hyperythrus*. **Figs. 1–2.** Adult, Azankei, Hokkaido, Japan, June (*Pete Morris*).

PHILIPPINE HAWK-CUCKOO
Hierococcyx pectoralis **Plate 33**

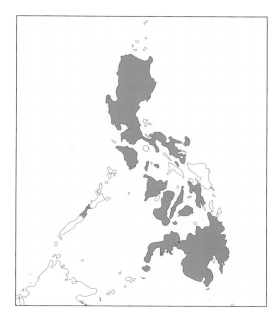

Hiracococcyx pectoralis Cabanis and Heine, 1863 (Philippines).

TAXONOMY Previously regarded as conspecific with *fugax* (see *H. hyperythrus*). Forms superspecies with *fugax*, *nisicolor* and *hyperythrus*, supported by mtDNA analysis (Sorenson & Payne 2005). Monotypic.

FIELD IDENTIFICATION 28cm. Smallish hawk-cuckoo; shy and inconspicuous, more often heard. **Adult** Head and upperparts slate-grey, paler on head, wings slate-grey with grey patch in flight; lores, throat and sides of neck white, breast and upper belly pale rufous, rest of underparts white, tail grey with black bands, tip rufous. **Juvenile** Black head barred rufous, upperparts grey barred rufous, underparts buffish heavily streaked dark brown with variable rufous tinge, tail as adult or washed rufous. **Similar species** Large Hawk-cuckoo is larger, with heavy dark brown barring below. Wing more rounded than Whistling Hawk-cuckoo. Pale grey chin separates from Javan, Rufous and Whistling Hawk-cuckoos. Lacks white spots on shaft of T1 found in *Cuculus*; juvenile is streaked not barred below, and tail has black subterminal bar tipped cinnamon. Small *Accipiter* hawks have different bill and long legs.

VOICE Repeated *wheet-wheet-wheet-wheet-tu*, each call with 5–7 notes lasting 1.5sec and given 9–10 times, each time louder, and ending in shrill crescendo. Whole series may last 25sec and then begins again, or *wee-wee-wee-tee-too* or *wee-wee-wee-tee-tee-too* or *wee-wee-wee-tee-too-too*. Also long, shrill call, on rising scale reaching crescendo and ending on descending scale *weetaweetaweetaweetaweetaweetaweetaweetatoo tootootootootoo-too*. Six short unmodulated whistles *wheet wheet...* first three rising, last three falling from note to note (Kennedy *et al.* 2000, King 2002, Payne 2005).

DESCRIPTION Adult Head slate-grey with horizontal white streak on lores continuing to cheeks and sides of neck; upperparts dark grey, wings slate-grey, remiges brownish-grey with inner webs broadly barred and notched buffy-white forming large pale area on wing below; bend of wing white; chin grey, throat white or pale grey and breast plain rufous or with fine grey streaks or brownish lanceolate centres to feathers of lower breast and belly, rest of underparts white, some with fine dark brown streaking; underwing-coverts buffy-white. Tail grey with 3–4 black bands with some cinnamon-white distally, subterminal band broadest, tipped cinnamon. **Subadult** Head, neck, chin and sides of throat slate-grey washed brown, white band separates grey chin and throat sides, sometimes also white hindneck band; upperparts and wings blackish-brown narrowly barred dark rufous; throat whitish finely streaked grey, rest of underparts buffy-white with dark brown chevrons, streaks or spots all edged rufous, undertail-coverts buffish-white with slight dark edging, tail brownish-grey occasionally tinged rufous, tipped rufous with broad black subterminal band, and 2–3 narrow black bands. **Juvenile** Head blackish, each feather tipped buffish, irregular white spot on nape; chin whitish streaked pale grey, whitish breast streaked blackish and washed rufous. **Nestling** Like juvenile but back brown and spots on breast rather large. **Bare parts** Iris yellow or red-brown to brown, juvenile pale brown, eye-ring yellow. Bill black with olive base and tip of upper mandible, lower olive tipped black, gape olive-yellow. Feet yellow to orange-yellow.

BIOMETRICS Wing male (n = 43) 168–189mm, mean 174.8 ± 4.7, female (n = 17) 166–182mm, mean 174.2 ± 4.5. Bill male 18.5–23.1mm, mean 20.6 ± 1.1, female 19.5–22.8mm, mean 20.8 ± 1.0. Tail male 112–153mm, mean 136.0 ± 12.3, female 116–154mm, mean 136.7 ± 11.7. Tarsus male 17.2–23.3mm, mean 20.2 ± 1.1, female 17.6–20.7mm, mean 20.1 ± 1.4. Mass male (n = 14) 70–80g, mean 76.2, female (n = 9) 67.3–89.2g, mean 78.6 (Payne 2005).

MOULT Secondary moult Jan (BMNH).

DISTRIBUTION Resident Philippines on Banton, Basilan, Bohol, Cagayancillo, Catanduanes, Cebu, Jolo, Leyte, Luzon, Mindanao, Mindoro, Negros, Palawan, Panay, Sibuyan, Siquijor, Tablas and Ticao (Kennedy *et al.* 2000). Samar?

HABITAT Primary and secondary forest; lowlands to 2,300m; virgin mossy dipterocarp forest; often near water (Ripley & Rabor 1958, Gonzales 1983, Payne 1997, Kennedy *et al.* 2000).

BEHAVIOUR Shy; difficult to observe even when calling. Forages in canopy and near ground level, singly or in pairs. Small numbers, presumably dispersing, caught at night autumn 1967–1969, Dalton Pass, Luzon (Hachisuka 1934, Rand & Rabor 1960, McClure 1969, Kennedy *et al.* 2000).

BREEDING Parasitism Presumably brood parasite. **Season** Enlarged ovaries and oviduct eggs Apr–May, Luzon and Negros; male with enlarged gonads 6 Apr, Panay. Juvenile female on Mindoro with enlarged ovary 18 May, and female from Negros in mixed subadult and adult plumage had large oviduct egg (Ripley & Rabor 1958, Dickinson *et al.* 1991, Miranda *et al.* 2000, Payne 2005). No further information.

FOOD Caterpillars, other insects; berries (Gonzales 1983).

STATUS AND CONSERVATION Uncommon to rare (Delacour & Mayr 1946, Dickinson *et al.* 1991), but may be overlooked. First recorded Panay 1992 (Miranda *et al.* 2000). Rapid deforestation indicates necessity of further research.

Philippine Hawk-cuckoo *Hierococcyx pectoralis*. **Fig. 1.** Adult, Quezon City, Luzon, Philippines, October (*Tina Sarmiento Mallari*). **Fig. 2.** Subadult, Quezon City, Luzon, Philippines, October (*Tina Sarmiento Mallari*).

JAVAN HAWK-CUCKOO
Hierococcyx fugax   **Plate 33**

Cuculus fugax Horsfield, 1821 (Java).

Alternative names: Hodgson's/Horsfield's/Malay Hawk-cuckoo.

TAXONOMY Forms superspecies with *pectoralis*, *hyperythrus* and *nisicolor*, supported by mtDNA analysis (Sorenson & Payne 2005); these forms were considered conspecific until recently (see *H. hyperythrus*). Everett (1889) suggests that some old records of *fugax* may refer to *vagans* as these two species previously confused. Monotypic.

FIELD IDENTIFICATION 28–30cm. Smallish hawk-cuckoo; shy. **Adult** Dark grey-brown above, white and pinkish-rusty on underparts with dark streaking, sometimes heavy, and rather long banded tail. In flight pale area on inner wings. Black spot on chin. **Subadult** Black crown tipped buff, dark grey barred rufous on upperparts, and underparts white with dark brown spots or streaks edged rufous. **Juvenile** Grey above with buff edgings. **Similar species** Moustached Hawk-cuckoo has conspicuous black moustache and tail proportionately longer. Adult Large Hawk-cuckoo larger with brown upperparts with greyish-black crown, large blackish area on throat bordered white at sides, and dark barring on belly and flanks. Common Hawk-cuckoo is larger, and has much pale rufous and grey from chin to breast, and breast to belly barred rufous and grey-brown. From Whistling and Rufous Hawk-cuckoos by brown, not slate-grey upperparts and lack of rufous below, and has more rounded wings and longer bill than former. Penultimate black band on tail much broader than in Whistling and Rufous Hawk-cuckoos. From *Cuculus* spp. by lack of white spots on T1, and juvenile by streaking below, and rufous subterminal tail bar.

VOICE Call series of very high pitched notes *pi-pwik*, or *pi-whit*, or *gee-whiizz*, or *fee-weet* or *wee-weet*, first call higher-pitched than second and repeated up to 20 times. Another rapid and shrill call often follows first call *ti-tu-tu* ending in climax and followed by slower and more even *tu-tu-tu-tu*. Long call, *wadawadawadawadawadaquedeequedeequedeequ edeetotototo-to-to-to*. Long, rapid, series of buzzing whistles *pee*, rising up scale in crescendo, then drop; also low *pik-kwik*. Voice of female unknown. In Thai-Malay Peninsula calls Feb–Aug, in Apr virtually the whole night when moonlit. Calls from midstorey, in contrast to Moustached Hawk-cuckoo which always calls from upper canopy (Payne 1997, 2005, Smythies & Davison 1999, Wells 1999, Robson 2000, King 2002).

DESCRIPTION Adult Dark slaty-grey head with whitish lores; some with incomplete white collar; mantle, wing-coverts and back bronze-brown, rump brownish edged rufous, uppertail-coverts brown with darker brown subterminal band and tipped rufous, tertials and secondaries brown edged greyish-white and notched rufous and white on inner webs, primaries darker brown finely tipped rufous-grey with white notches on inner webs; bend of wing white, underside of remiges brown broadly notched white or grey-white, underwing-coverts white. Dark grey chin, throat creamy-white meeting grey of cheeks and ear-coverts; rest of underparts white narrowly streaked brown and edged rufous; similar but broader arrow-shaped streaks on breast and chevrons on flanks; variable pinkish-rusty suffusion on breast to belly and flanks; vent and undertail-coverts plain white. Tail has four blackish-brown bands, subterminal being much broader, tip orange-rufous fringed white. **Subadult** Head, neck, chin and sides of throat slate-grey washed brown, white band separates grey chin and throat sides, sometimes also white hindneck band; upperparts and wings blackish-brown narrowly spotted and barred dark rufous, often one white tertial; underparts buffy-white with dark brown chevrons, streaks or spots, feather shafts on belly black, undertail-coverts buffish-white, tail brownish-grey, occasionally tinged rufous, tipped rufous with broad black subterminal band and 2–3 narrower blackish-brown bands behind broad rufous bands. **Juvenile** Dull sooty-brown upperparts barred and fringed rufous-buffish, feathers of underparts white at base, tipped sooty with pale buffish margin, tail more pointed. **Nestling** Hatches naked, skin flesh-coloured, after few days becomes blackish above and yellow below, including tarsus and wing. **Bare parts** Iris brown to orange-brown or dirty white, and eye-ring yellow. Bill black, inner half greenish-yellow. Feet yellow, claws whitish.

BIOMETRICS Wing male (n = 10) 168–180mm, mean 174.3 ± 7.2, female (n = 5) 165–183mm, mean 173.4. Bill male 20.0–25.2mm, mean 22.7 ± 1.8, female 20.8–24.8mm, mean 22.8. Tail male 114–148mm, mean 135.2 ± 10.1, female 122–146mm, mean 131.2. Tarsus male 17.5–23.3mm, mean 20.0 ± 2.4, female 17.9–22.3mm, mean 20.3 (Payne 2005). Mass unsexed 74.8–76.5g (n = 12) (Wells 1999); juvenile (Brunei) 62g (C.F. Mann).

MOULT Adults and subadults in Thai-Malay Peninsula moult primaries at up to 4 loci simultaneously; active wing moult Jul–Aug (Wells 1999).

DISTRIBUTION Sundaland. Resident (and migratory?). Thai-Malay Peninsula, from Isthmus of Kra southwards, including south Tenasserim. Throughout Sumatra, including Siberut, Batu Is, Rhio Archipelago, Bangka. Local throughout Borneo. West Java (Deignan 1963, van Marle & Voous 1988, Wells 1999, Robson 2000, Mann 2008). Northern Malay Peninsula populations should occur as winter visitors south to Singapore (Jeyarajasingam & Pearson 1999, Payne 2005), but questioned by Wells (1999).

HABITAT Lowland and hill dipterocarp forests, kerangas forest, peatswamp forest, secondary forest and cocoa plantations to 1,620m, Borneo. Breeds mainly at plains level to *c*. 250m, but also on slopes, in primary and secondary evergreen and semi-evergreen dry lowland forests, Malay Peninsula. Forest, riverine woods, bushes, bamboo and rubber plantations Sumatra; to 1,700m; dry hill forest, *c*. 450m, Gunung Leuser NP. Primary and secondary forest, Siberut, Mentawai Is. (van Marle & Voous 1988, Smythies & Davison 1999, Wells 1999, Kemp 2000, Buij *et al.* 2006, Mann 2008).

BEHAVIOUR Shy and secretive. Presumably territorial as males call from regular spots; one aggressively defended territory against Red-bearded Bee-eater. Foliage-gleans (Smythies & Davison 1999, Wells 1999, Buij *et al.* 2006).

BREEDING Parasitism White-rumped Shama feeding fledgling, Thailand, and feeding fledgling presumed to be this species Penang, peninsular Malaysia. On Borneo Grey-headed Canary-flycatcher. Juvenile fed by White-rumped Shama, Sumatra (Whitehead 1893, Gibson-Hill 1952, K. Kumar in Wells 1999, BCSTB-16 in Wells 2007, B. van Balen *in litt.*). Also on Borneo Black-throated Babbler (Cranbrook

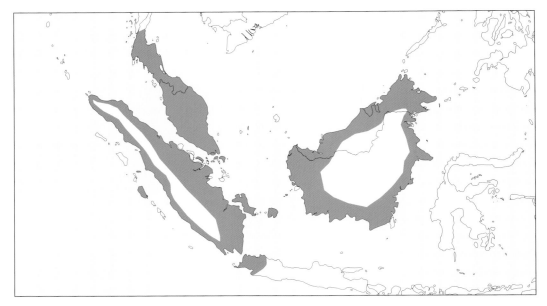

& Wells 1981), but cuckoo species queried by Payne (2005). **Season** Fledgling 27 Aug, peninsular Malaysia, presumed this species laying in nest of Grey-headed Canary-flycatcher Sabah, Borneo, 29 Apr (Gibson-Hill 1952, K. Kumar in Wells 1999). **Eggs** On Borneo creamy-white, encircled with brown and grey spots at broad end; 22x16mm (Sharpe & Whitehead 1890, Schönwetter 1967). No further information.

FOOD Most frequently caterpillars, also beetles, cicadas, other insects; fruit. On Borneo berries; small butterflies, locusts and beetles (Sody 1989, T. Harrisson in Smythies & Davison 1999).

STATUS AND CONSERVATION Rare or uncommon over most of range. Uncommon in south and east Burma and Tenasserim, Thailand, Malay Peninsula, north and central Laos, west Tonkin and central Annam (Robson 2000). Uncommon to more or less common Thai-Malay Peninsula (Wells 1999). Frequent Siberut, Sumatra (Kemp 2000). Sparingly distributed and local Borneo (Smythies and Davison 1999, Mann 2008).

Javan Hawk-cuckoo *Hierococcyx fugax*. **Fig. 1.** Adult, Bukit Larut, Perak, Malaysia, January (*Choy Wai Mun*).

WHISTLING HAWK-CUCKOO
Hierococcyx nisicolor   **Plate 33**

Cuculus nisicolor Blyth, 1843 (Nepal)

TAXONOMY Previously placed in *fugax* (see *H. hyperythrus*). Two long-winged migrants peninsular Malaysia may represent an undescribed taxon (Wells 1987). Monotypic.

FIELD IDENTIFICATION 28–30cm. Smallish hawk-cuckoo. **Adult** Blackish and dark grey above. Dark grey chin, face and sides of neck; undertail-coverts white; rest of underparts white streaked dark grey or brown with rufous suffusion. Underwing-coverts buffish with some brown barring in some; remiges broadly barred and notched white forming large white area on wing when seen from below. Tail grey banded black and tipped rufous. **Subadult** Back and wings barred rufous; some rufous on tail; rufous throat spotted white and streaked brown; rest of underparts white, streaked brown and rufous on breast to belly; buffy notching and barring on remiges. **Juvenile** Mostly brown above with whitish tipping, more blackish on crown and nape with buff edging; white spot(s) on nape or sides of neck in some; remiges dark brown barred rufous and white; tail as adult with broader rufous tip, extreme tip buffy-white. Throat to upper breast dark brown with buff-white fringes, often with some white spotting; lower breast to belly creamy-white spotted brown, streaked on lower flanks; undertail-coverts white. **Similar species** Pinkish below separates it from Javan Hawk-cuckoo, but less pinkish than Rufous Hawk-cuckoo; greyer above than former; shorter bill and more pointed wings than latter. More boldly marked than Common Hawk-cuckoo, which has some barring below and more pinkish-rufous. Penultimate black band on tail very narrow compared to Javan Hawk-cuckoo. Juveniles separated from *Cuculus* spp. by streaking below, not barring, and lack of white spots on shaft of T1. Lack of barring below and smaller size separates it from Large Hawk-cuckoo.

VOICE Call shrill thin piping whistle *gee-whiz*; 'long, rapid sequence of shrill buzzy whistles *pee* rising up scale in crescendo then dropping at end, *trrrrr-titititititirrrtrrr*'; churred call may be alarm (Robson 2000, various references in Payne 2005).

DESCRIPTION Adult Crown to nape grey-black; mantle brownish-black; back and rump dark grey; uppertail-coverts blackish fringed grey; rectrices grey with four black bands, penultimate narrow, distal broad, and tipped rufous, similar pattern below but paler; wing-coverts blackish, primaries brownish-black, secondaries blackish-grey. Dark grey chin, face and sides of neck; throat to upper breast white streaked dark grey; breast, flanks and belly white with varying amounts of dark grey or brown streaking and pinkish rufous suffusion, undertail-coverts white. Underwing-coverts buffish with some brown barring in some; remiges broadly barred and notched white forming large white area on wings when seen from below. **Subadult** As adult but back and wings barred rufous; tail has some rufous in grey areas; throat pale rufous with white spots and thin brown streaks; breast to belly white with pointed brown streaks edged rufous becoming chevrons on flanks; vent and undertail-coverts white; notching and barring on remiges tinged buffy. **Juvenile** Crown and nape blackish-brown, feathers edged pale buff; may have white spot(s) on nape or sides of neck; back to uppertail-coverts, and wing-coverts brown with dark brown subterminal bars and tipped buffy-white; remiges dark brown barred and notched rufous and white and fringed whitish; tail as adult with broader rufous tip, extreme tip buffy-white. Throat dark brown fringed buffy-white, often with some white spots; upper breast as chin with variable whitish areas; breast to belly creamy-white with large dark brown spots, becoming lanceolate on lower flanks; undertail-coverts white. **Nestling** Undescribed. **Bare parts** Eye-ring yellow; iris brown; bill black, base, lower mandible and tip green; feet yellow (Payne 2005).

BIOMETRICS Wing male (n = 6) 174–185mm, mean 178.3 ± 4.6, female (n = 5) 176–184mm, mean 181.0 ± 5.0. Tail male 122–143mm, mean 135.3 ± 9.6, female 128–143mm, mean 134 ± 7,4. Bill male 20.4–20.8mm, mean 20.6 ± 0.2, female 20.0–22.3mm, mean 20.9 ± 1.0. Tarsus male 18.9–20.1mm, mean 19.0, female 18.2–22.2mm, mean 19.4 (Payne 2005). Mass unsexed migrants Thai-Malay Peninsula 69.2–93.0g, most <80g (Wells 1999). Wing (two migrants peninsular Malaysia) 188, 193mm (Wells 1987).

MOULT Subadult to adult moult Assam, Aug; primary moult peninsular Malaysia, Feb, and Sikkim, Oct; tail moult Sikkim, Jul (fear moulting?) (BMNH).

DISTRIBUTION Oriental. Sikkim (Apr–Oct), Bhutan, north-east India to south-east Burma (perhaps only passage in south), north and east Thailand, north-central Laos, west Tonkin, central Annam. In Indian subcontinent, particularly Assam, summer visitor or sedentary with some altitudinal and migratory movements. Passage and winter south Thailand, peninsular Malaysia, Singapore. In China in summer south of Changjiang R. from Jiangsu west to Sichuan and south to Yunnan, Guangxi Zhuang, Guangdong and Hainan; Taiwan (once); some may winter in south; perhaps north-east provinces and Hebei in summer; Chinese population winters or on passage Thai-Malay Peninsula (Aug–Dec, Mar–May), Sumatra (Langkat, Padang Highlands, Palembang, Aruah Is, Bangka, Belitung), west Java and Borneo (rare Nov–Apr, Sabah and Sarawak) (Mayr 1938, Gibson-Hill 1956, Deignan 1963, Cheng 1987, van Marle & Voous 1988, Grimmett *et al.* 1998, Wells 1999,

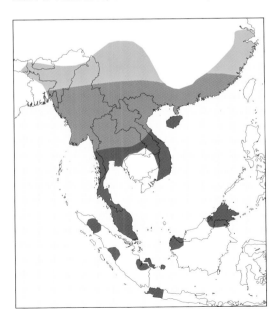

Robson 2000, Payne 2005, Yen 1933 and Li 1991 in Payne 2005, Rasmussen & Anderton 2005, Mann 2008). Three records Nepal (1824, 1829 and 1988) (Inskipp & Inskipp 1985). Once Hong Kong, Oct 1971 (Viney *et al.* 1994).

HABITAT Deciduous semi-evergreen, evergreen and secondary forests, pines in damp ravines, bamboo thickets and plantations. Various types of forest and plantations in winter quarters. Breeds mainly above 500m. Migrants in wooded suburban gardens, to 1,200m, Thai-Malay Peninsula. 600–2,300m Assam and 1,200–1,800m Arunachal Pradesh, India. Foothills 300–2,880m, usually 1,000–2,200m, Bhutan. 1,730–2,320m. 200–1,800m Hainan. Up to 1,000m Burma. Plains to 1,100m, Thailand (Baker 1927, Smythies 1940, Lekagul & Round 1991, Tymstra *et al.* 1997, Jeyarajasingam & Pearson 1999, Wells 1999, Inskipp *et al.* 2000, Robson 2000, Choudhury 2003, Lee *et al.* 2005, Spierenburg 2005).

BEHAVIOUR Shy and secretive. In India most often in low trees and bushes but higher when calling (Grimmett *et al.* 1998).

BREEDING Parasitism In India Lesser Shortwing, Small Niltava and other flycatchers, Rufous-fronted Babbler and Nepal Fulvetta; particularly Buff-breasted Babbler. Blue-throated Blue Flycatcher, Burma (Hopwood & Mackenzie 1917). Also White-browed Fantail, Asian Verditer Flycatcher, Spotted Forktail, Plumbeous Water Redstart, Eyebrowed Wren-babbler, Buff-chested Babbler, Yellow-throated Fulvetta, Grey-headed Parrotbill, Little and Streaked Spiderhunters (Baker 1927, 1934, 1942, P.E. Lowther *in litt.*). **Season** Calls India May–Aug, egg 14 Sep. Oviduct egg Sikkim, 5 Jun; also breeds Jul and once Sep; juveniles north Thailand, Jun and Jul; fledgling mid Aug (Hume & Oates 1890, Baker 1906b, 1934, 1937, Deignan 1945, BCSTB-16 in Wells 2007). **Eggs** Oviduct egg from Sikkim plain olive-brown, darker shade at larger end, 22.6x16.3mm. All other eggs from India (uniform olive-brown to green, darker at broad end; 22.5x15.4mm (Baker 1934), 24x16mm (Ali & Ripley 1969) not absolutely proven to belong to this species as eggs of sympatric Common Hawk-cuckoo unknown. No further information.

FOOD Most frequently caterpillars; cutworms, beetles, cicadas, crickets, grasshoppers, locusts, termite alates, ants; lizards; wild banyan figs, berries (Baker 1927, Ali 1946 in Payne 2005).

STATUS AND CONSERVATION Rare or uncommon over most of range. No recent records India; vagrant Nepal and Bangladesh (Grimmett *et al.* 1998). Uncommon winter and passage south Thailand, peninsular Malaysia and Singapore (Wells 1999, Robson 2000). Fairly common Bhutan, May (Tymstra *et al.* 1997). Rare Nov–Apr, Sarawak and Sabah, Borneo (Mann 2008). Uncommon China (Cheng 1987). Vagrant Hong Kong (Viney *et al.* 1994).

Whistling Hawk-cuckoo *Hierococcyx nisicolor*. **Figs. 1–2.** Subadult, Singapore, April (*Lee Tiah Khee*).

Genus *Cuculus*

Linnaeus 1758. *Syst. Nat.* (ed. 10) 1: 110. Type, by tautonymy, *Cuculus canorus* Linnaeus 1758.
Ten species.

Old World; brood-parasitic. Most pointed-winged; grey above, barred below; unstreaked juvenile plumage. MtDNA analysis results in *Cuculus* and *Hierococcyx* forming two monophyletic clusters which are sister groups, this group is sister to clade containing *Cercococcyx* and *Surniculus*; *Cacomantis* is sister group to these (Sorenson & Payne 2005). Includes *Nicoclarius* (Bonaparte 1854), and *Notococcyx* and *Surniculoides* (last two Roberts 1922).

BLACK CUCKOO
Cuculus clamosus   Plate 34

Cuculus clamosus Latham, 1801 (Cape of Good Hope; restricted to Cradock)

TAXONOMY Previously known as *C. cafer*, but name officially suppressed due to confusion with Levaillant's Cuckoo *Clamator levaillantii* (ICZN 1956). Polytypic. Synonyms: *aurivillii* Sjöstedt 1892, *jacksoni* Sharpe 1902 (intermediate between nominate and *gabonensis*), *chalybeus* Friedmann 1930, and *mabirae* van Someren 1915, all in *gabonensis*.

FIELD IDENTIFICATION 31cm. Medium-sized, fast-flying cuckoo; variable plumage on underparts. **Adult** Mainly black with white-tipped tail that may have white spotting or barring; remiges barred dull white; underparts variable, may show white or buff or reddish barring, with variable amount of chestnut from chin to breast. **Juvenile** Black, and lacks white tips to tail. **Similar species** From melanistic Jacobin and Levaillant's Cuckoos by lack of crest and white wing-bar. Heavier, darker and more thick-set than Red-chested Cuckoo; juvenile of latter buff-white from breast to undertail-coverts heavily barred blackish-brown. Melanistic Gabar Goshawk is more thick-set, has orange feet, barred tail, and inner webs of primaries mainly white.

VOICE Male's call mournful and hesitant 3–note whistle *oo-aa, aaa*, or *whoo whoo whee*, first note short, second note followed by pause, and third louder, higher-pitched and upslurred; third note may be repeated *oo-aa, aaa, aaa*. Call lasts 1–2sec, repeated every 4–5sec for 30+sec. Also bubbling trill (sex not known) of 20–30 notes lasting 5sec, rising to crescendo and then dying away *hohohohohoa hoa hoa HOara HOara HOarara HOarara HOarara huhuhuhuhuhuhu hu* or *gagagaGAGAGA*. Female's call *kwik-kwik-kwik-kwik*. Three slow whistles, rising in pitch *kwa, kwa, kwa*; hoarse, whistled paired syllables repeated 10 times, at first increasing and then dying away. Often calls at night. Only calls in savanna when breeding, but in forest throughout year (Chapin 1939, Irwin 1988, Payne 1997).

DESCRIPTION Very different in plumage from other cuckoos, and is perhaps mimetic (Louette & Herroelen 1993). Nominate. **Adult male** Black, glossed greenish or blue above. Remiges unglossed dark brown, with dull white barring on inner webs, and small amount of whitish barring on outer webs of primaries in some; bend of wing white, underwing-coverts creamy-white barred dark brown. Below dull black, darker and glossier on throat, sides of neck and breast, faintly glossed bluish; often partly or completely barred dull white or buff, and sometimes with reddish barring on breast; undertail-coverts uniform or strongly barred buff and white. Tail graduated with narrow white tips and often with partial white barring or spotting. **Adult female** Usually differs from male in having more prominent ventral barring. **Juvenile** Mostly black, slightly glossed above, without white tail tips; remiges dark brown with white patches and vermiculations on inner webs, not barring; rectrices more pointed. **Nestling** Naked on hatching with pale brownish-pink skin soon becoming purplish-black (Skead 1951). Inside of mouth pink. **Bare parts** Iris dark brown, sometimes grey in female; eyelids dusky. Bill black. Feet pinkish-brown; claws blackish.

BIOMETRICS Nominate. Wing male (n = 34) 166–187mm, mean 173; female (n = 13) 167–183mm, mean 174. Bill male (n = 30) 22–24mm, mean 23; female (n = 12) 22–24mm, mean 22.8. Tail male (n = 29) 138–156mm, mean 145; female (n = 13) 140–154mm, mean 146. Tarsus unsexed (n = 22) 16–18mm, mean 16.3. Mass male (n = 9) 78–94g, mean 85.0; female (n = 6) 79–92g, mean 87.4 (Irwin 1988); (RSA, Botswana, Zambia) male 78.0–103.5g, mean 90.2; female 86, 88.3g (after laying egg); female 102.7g (with oviduct egg) (Payne 2005); (DRC) unsexed (n = 3) 75.0–94.0g, mean 87.3 (Louette & Herroelen 1993); '*jacksoni*' male 81.5, 89g, female 78, 87g, fledgling 62g (Irwin 1988). Also wing male (n = 10) 163–180mm, mean 174.3 ± 4.7; female (n = 6) 173–186mm, mean 177.5 ± 4.4 (Payne 2005).

MOULT All secondaries, scapulars and rectrices moulted before primaries. Moulting north DRC Jan, Sep–Nov, in south Feb–Apr; Cameroon, Oct. Secondaries moulting Ethiopia, Jul, and Uganda, Apr (Louette & Herroelen 1993, BMNH).

GEOGRAPHICAL VARIATION Two subspecies.
C. c. clamosus Latham, 1801. Highlands of Ethiopia, Eritrea, Somalia (twice), west and central Kenya, Tanzania, south DRC and Angola south to north and central Namibia, north Botswana and east RSA, migrating to equatorial west Africa west to Senegambia. Described above.
C. c. gabonensis Lafresnaye, 1853. Probably non-migratory. Liberia, IC, Togo, Benin, Ghana, Nigeria, Gabon and north DRC to south Sudan, south-west Ethiopia, Uganda and south-west Kenya south to 9°S in Angola. Male differs from nominate in having chin, throat, sides of neck and breast rufous or chestnut (breast may be unbarred in some, and lower breast to belly pale tawny narrowly barred dark brown, undertail-coverts unbarred rich tawny, or whole underparts blackish-brown with few irregular pale bars); more obvious barring on remiges; underwing-coverts barred black and white. Female differs from nominate female in being duller, with underparts heavily barred black on dull white or buff, and broadly barred undertail-coverts; obvious white tail tips. Wing male (n = 10) 163–180mm, mean 174.3 ± 4.7; female (n = 6) 173–186mm, mean 177.5 ± 4.4 (Payne 2005).

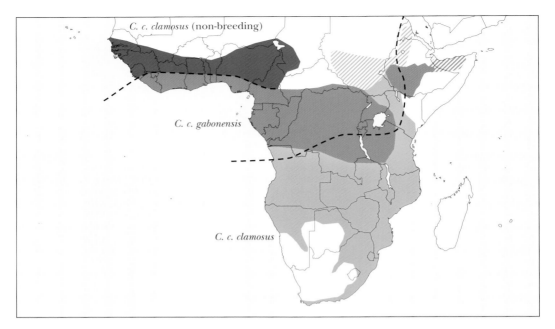

DISTRIBUTION Afrotropical; non-migratory, or intra-tropical migrant. Occurs over much of sub-Saharan Africa. Migrant to rainforests west of Benin; Senegambia; SL, Mar–Nov; Guinea, frequent Apr–Jun; Liberia; IC; Burkina Faso; Ghana; resident parts of Togo and Cameroon; resident south-east Nigeria, wet season visitor to north, but some Apr–Oct; Chad; RC, Aug, Nov; south-west Ethiopia and highlands, perhaps resident; south Sudan occasionally as far north as Nyala in South Darfur, Apr–Sep; Somalia (once north-west, once south); Uganda and Kenya (all months) to Tanzania (east of 35°E; all months in north, Oct–Apr in south-east) and south of equator from Western Rift Valley to DRC (in north Jan, Apr, Jun, Jul, Sep, Nov, in south Oct, Feb–May); Angola, Aug–Mar in north; Mozambique; Malawi, Sep–May, occasionally Jun–Jul; Zambia, late Aug–mid Mar, occasionally to mid May; Zimbabwe; north and central Namibia, Nov–Apr (arrival dependent on rains); Botswana; RSA (Transvaal, Natal and coastal Cape Province), Sep–Apr (Marchant 1953, Cave & MacDonald 1955, Serle 1965, Benson *et al.* 1971, Benson & Benson 1977, Irwin 1988, Dowsett & Dowsett-Lemaire 1989, 1993, Medland 1992, 1993, 1995, Louette & Herroelen 1993, Elgood *et al.* 1994, Halleux 1994, Cheke & Walsh 1996, Payne 1997, Ash & Miskell 1998, Salewski 2000).

HABITAT Lowland forest, gallery forest, miombo, woodland in drier savannas, *Acacia* thornveld, thickets, old riparian growth, and suburban gardens in RSA. In less luxuriant habitat in Malawi than Red-chested Cuckoo. Usually in canopy or mid stratum. Below 1,525m DRC and Malawi; below 2,000m Rwanda, and east Africa. Mostly 500+m Kenya (Chapin 1939, Benson *et al.* 1971, Benson & Benson 1977, Britton 1980, Irwin 1988, Lewis & Pomeroy 1989, Dowsett-Lemaire 1990, Payne 1997).

BEHAVIOUR Normally solitary but several may occur together in breeding season. Fast twisting and turning flight. Less shy than other *Cuculus* (Irwin 1988, C.F. Mann).

BREEDING Parasitism Preferred hosts of nominate differ with habitat. Localities and instances given where recorded: main hosts in bush country boubous and gonoleks *Laniarius*; White Helmet-shrike, Zambia; Black-backed Puffback, Lubumbashi, DRC; Southern Boubou, southern Africa (c. 30) in *Acacia* savanna; Tropical Boubou, east Africa (1), Zimbabwe; Sooty Boubou, Cameroon; Crimson-breasted Gonolek, RSA (22) in *Acacia* savanna, Zimbabwe and Namibia; Lühder's Bush Shrike; Olive Bush Shrike; African Golden Oriole, Zimbabwe and Zambia northwards, in miombo; Common Bulbul, removing host egg, Naivasha, Kenya; Sombre Greenbul, RSA; Red-capped Robin-chat, RSA, Zimbabwe; White-browed Robin-chat, Zimbabwe; Cape Robin-chat, RSA; Boulder Chat, Zimbabwe; Southern Black Flycatcher, Zambia (Ivy 1901, Bates 1911, Lees 1938, Hoesch 1940, Skead 1946, 1951, Vincent 1946, Friedmann 1948b, C.R.S. Pitman in Friedmann 1948b, Stocks 1948, Maclean 1957, Friedmann 1967, Jensen & Jensen 1969, Benson *et al.* 1971, Penry 1976, Colebrook-Robjent 1977, Irwin 1988, Pryce 1989, C.F. Mann). Possibly Tawny-flanked Prinia (Friedmann 1948b); possibly Karoo Prinia, RSA (Sharpe 1875). 2% boubou nests parasitised RSA, some every time; in Zimbabwe *c.* 2.5% Crimson-breasted Gonolek parasitised. Host of *gabonensis* unknown; fledgling attended by Sooty Boubou, Cameroon (Irwin 1988). Fledgling with 3+ Black-throated Malimbes and fed by one male, north RC, Nov; this species unlikely foster as nest has entrance tunnel 0.5–1.0m long, and suggested that due to colour resemblance between cuckoo fledgling and malimbe it could have been adopted after leaving nest of true foster parents (Dowsett-Lemaire 1996). **Season** Calling Yapo Forest, IC, Oct–Jan, rarely Jun, never Apr–May; calling Togo, Apr, Jun, Jul; calls throughout year, less Apr–Aug, south-east Nigeria; Cameroon, Sep; well-developed follicle Eritrea, Jul; oviduct egg north-west Ethiopia, May; male near breeding south Ethiopia, Mar; large ovarian eggs Eritrea, Jul and Ethiopia, May; Uganda, Apr–Oct; oviduct egg Eldama Ravine, Mar, nest-robbing Naivasha, Kenya, Apr; oviduct eggs Tanzania (except north-east), Mar–May, young Jul, and breeding,

May; Angola, Oct–Nov; fledgling north RC, Nov; oviduct egg north-east, and nestlings and fledglings south DRC, Mar, May, Jul, Aug, Oct, nestling Lubumbashi, DRC, Jan; Zambia, Sep–Dec; Malawi, Oct and post-breeding male Jan, and oviduct egg May; Zimbabwe, Nov–Jan; RSA, Sep–Apr; Namibia, Nov–Mar, mainly Feb; nestling and fledgling Botswana, Feb (arid areas dependent on onset of rains) (Bates 1911, 1924, Vincent 1934, Chapin 1939, Benson 1945, Friedmann 1948b, 1967, Marchant 1953, Douaud 1956b, Mackworth-Praed & Grant 1957, Maclean 1957, Smith 1957, Serle 1965, Jensen & Jensen 1969, Benson *et al.* 1971, Brown & Britton 1980, Irwin 1988, Louette & Herroelen 1993, Demey & Fishpool 1994, Cheke & Walsh 1996, Dowsett-Lemaire 1996, Payne 1997, Ash & Atkins 2009, C.F. Mann, BMNH). **Eggs** Host egg removed at time of laying. Up to 22 laid over 10 weeks. Elliptical; glossy; white or tinged green with variable amounts of reddish-brown and lilac spots and freckles. Oviduct egg of *gabonensis* ovate, cream profusely blotched and spotted reddish-brown and claret-brown forming ring at large end, resembling eggs of some forest bulbuls and boubous; others creamy-white spotted and speckled dull purplish or reddish-brown and lilac-grey. In south Africa sometimes exactly resembles hosts' eggs, but usually larger, longer with blunter ends, more finely and uniformly speckled and less blotched. May be good match for eggs of Crimson-breasted Gonolek, Tropical Boubou and African Golden Oriole; 23.6–27.3x17.5–19.1mm (n = 7), mean 25.1x18.6; also 27–28x21.5–22mm *clamosus*; 23.5x17mm *gabonensis* (Chapin 1939, Serle 1965, Irwin 1988). **Incubation** 13–14 days (3 days less than boubous). **Chicks** Evicts all host young or eggs within 16–30 hours (sometimes up to 3 days) of hatching. Eyes open 8th day; well-feathered 16th day. Leaves nest 20–21 days, but remains with foster parents for further 19–26 days (Skead 1951, Irwin 1988). **Survival** No information.

FOOD Mainly hairy caterpillars; other insects such as termites, including alates caught in air, ant alates, beetles, many Orthoptera; other larvae; eggs, presumably those of host, and nestlings (Chapin 1939, Serle 1957, Irwin 1988, Payne 1997).

STATUS AND CONSERVATION Generally uncommon to rare, but may be locally frequent or common, as in forests of lower Guinea, equatorial central and east Africa, and RSA (Chapin 1939, Irwin 1988, Payne 1997, Borrow & Demey 2001); not uncommon Yapo Forest, IC (Demey & Fishpool 1994); rare resident (*gabonensis*), and migrant (*clamosus*) Togo (Cheke & Walsh 1996); uncommon to rare south Sudan (Cave & MacDonald 1955, Nikolaus 1987); uncommon and local Uganda and Kenya (C.F. Mann); common Malawi below 1,525m (Benson & Benson 1977), although considered scarce by Stead (1979). Widely distributed southern Ethiopia, once Eritrea (Ash & Atkins 2009). Not globally threatened (BirdLife International 2011).

Black Cuckoo *Cuculus clamosus*. **Fig. 1.** Adult female, *C. c. gabonensis*, Maasai Mara NP, Kenya, January (*Adam Scott Kennedy*). **Fig. 2.** Adult female, *C. c. gabonensis*, Maasai Mara NP, Kenya, January (*Adam Scott Kennedy*). **Fig. 3.** Adult, *C. c. gabonensis*, Maasai Mara NP, Kenya, January (*Adam Scott Kennedy*). **Fig. 4.** Adult, *C. c. gabonensis*, Maasai Mara NP, Kenya, January (*Adam Scott Kennedy*).

RED-CHESTED CUCKOO
Cuculus solitarius Plate 34

Cuculus solitarius Stephens, 1815 (eastern Cape Province)

TAXONOMY Once placed in *Notococcyx* (Roberts 1922). Monotypic. Synonyms: *magnirostris* Amadon, 1953 and *heuglini* Emin, 1919.

FIELD IDENTIFICATION 31cm. Medium-sized cuckoo. Call loud and obvious. **Adult male** Dark grey above, sides of head and throat pale grey; white barring and spotting on tail. Breast cinnamon, barred or unbarred; rest of underparts pale buff to creamy-white with narrow blackish-brown bars. **Adult female** Similar to male but with less cinnamon on breast, and greater extent of barring. **Juvenile** Blackish above with white on hindcrown; throat to upper breast black, flecked or lightly barred white; rest of underparts barred black and buffish-white. **Similar species** Black Cuckoo much darker above and below and lacks white spots and bars on tail, and black-and-white barring on belly and undertail-coverts. Juvenile from same stage of Common and African Cuckoos by more slaty upperparts finely edged whitish.

VOICE Persistent, day and night, emphatic *ee-eye-ow*, or *pee-eye-fou* rendered 'it will rain' or 'piet mijn vrouw' (Afrikaans), last note sometimes slightly disyllabic, at 1kHz falling slightly in pitch, lasting 1.5sec with gaps of *c.* 0.5sec, *c.* 34 times in 72sec, but faster if more than one male calling. High pitched *quick quick quick quick!* Rising bubbling call. Female's call excited *hahe-hahe-hahe-hahe* with accent on *ha*, sometimes before copulation; abrupt repeated *kwik-kwik-kwik*, higher pitched than male. Two sexes sometimes duet. Loud, raucous *sschreep* from fledgling, becoming faster when foster appears; likened to sound made by stroking teeth of fine, stiff comb. Highly vocal when breeding; also calls at night (Chapin 1939, Beattie 1981, Irwin 1988, Payne 1997, Borrow & Demey 2001, Stevenson & Fanshawe 2002, C.F. Mann).

DESCRIPTION Adult male Dark bluish slate-grey above, lores and ear-coverts pale grey. Inner webs of primaries barred white except near tips; secondaries plain. Axillaries and underwing-coverts barred buff and blackish-brown. Throat pale grey; breast cinnamon or rufous, often barred; rest of underparts pale buff to creamy-white with narrow blackish-brown barring, except for unbarred buff or whitish undertail-coverts. Tail dark slate-grey tipped white, blackish along shafts, T1 irregularly spotted along shafts, T5 spotted and barred white. **Adult female** Differs from male in having less cinnamon (or paler rufous) on breast and being more barred (other *Cuculus* females browner or more rufous/cinnamon than males). Sometimes completely barred with sides of neck washed tawny. Throat buffy or pale grey unbarred. No distinct subadult plumage. **Juvenile** Blackish above, narrowly edged white; white hindcrown. Wings slaty black with faint cinnamon barring on inner webs of primaries; secondary-coverts flecked white on margins. Throat to upper breast black flecked white, occasionally slightly barred; lower breast to undertail-coverts more heavily barred black and buffish-white. Tail black, tipped and spotted white, T4 incompletely and T5 completely barred white. **Nestling** Hatches naked with bluish-black or purplish-brown skin; inside of mouth deep yellow rapidly becoming rich orange. **Bare parts** Iris brown to dark brown, juvenile black; olive-green eyelids surrounded by yellow skin. Bill black to blackish-horn tinged greenish, with yellow base and orange gape. Feet yellow or orange-yellow, orange in juvenile, flesh or yellow in nestling.

BIOMETRICS Wing male (n = 25) 167–196mm, mean 177; female (n = 13) 166–190mm, mean 176. Bill unsexed (n = 28) 20–23mm, mean 21.3. Tail male (n = 24) 137–160mm, mean 148; female (n = 12) 138–158mm, mean 148. Tarsus unsexed (n = 27) 15–17mm, mean 16.3. Mass male (n = 15) 68–90g, mean 75.3; female (n = 5) 67–74g, mean 71.6; fledgling 55+g (Irwin 1988), hatchling 4.9g (Liversidge 1955); male (n = 8) 66, 74–81g (Verheyen 1953). Also wing male (n = 20) 164–180mm, mean 172.0 ± 5.6; female (n = 11) 156–187mm, mean 169.0 ± 10.0 (Payne 2005).

MOULT In southern Africa juvenile commences moult to adult plumage Feb, probably completing moult in northern winter quarters. Juvenile to adult moult south-west Ethiopia, 1 Sep; adult secondary moult Tanzania, 2 Jan and 28 Dec, Uganda, May and Nov; moulting remiges Uganda, Aug (Irwin 1988, BMNH).

DISTRIBUTION Intra-African migrant throughout most of sub-Saharan Africa in suitable habitat. Bioko; Senegambia to Niger (once Sep); Togo, Nigeria, Gabon, CAR, RC, DRC; Sudan south of 13°N, mostly Mar–Apr, Jul–Nov; Ethiopia, east Africa, south to Angola, Malawi, Mozambique, Zambia, Zimbabwe, Namibia (Caprivi Strip and Brakwater), north Botswana, north Transvaal, Natal, and east and south Cape Province, RSA. Visitor to north savannas of west Africa Mar–Sep. Not in arid areas of south-west Africa and Horn of Africa, although once south Somalia, May. Arrives in southern Africa late Aug–Oct; Malawi, Sep–Apr; probably late Aug–Apr, Zambia; few may overwinter RSA. Mostly seasonal visitor east Africa, Oct–Apr coastal Tanzania and Kenya, and interior and south-east Tanzania, Oct–Mar; may occur throughout year as in parts of north and south-west Tanzania; absent from much of north and east Kenya, and north Uganda; mostly Nov–Mar south Uganda. Resident Ethiopia, but concentrations in early Mar suggest partially migratory. Throughout year DRC, but some perhaps migrants from south. Resident in or near rainforest zone west Africa, and visits northern savannas Mar–Sep (late dry season through rains). Calling intensively Mt. Nimba, Liberia, Dec–Apr; visits Senegambia in wet season (until Dec); in Nigeria moves from coastal areas to northern savannas during rains; Cameroon, at least Mar–Oct. Probably mainly passage late Sep–early Oct, Rwanda, when calls. In east Africa wanders widely in rains (Chapin 1939, Serle 1965, Benson *et al.* 1971, Benson & Benson 1977, Britton 1980, Nikolaus 1987, Irwin 1988, Dowsett-Lemaire 1990, Germain & Cornet 1994, Hines 1996, Ash & Miskell 1998, Borrow & Demey 2001, Stevenson & Fanshawe 2002, Christensen *et al.* 2005).

HABITAT Moist, well-wooded savanna, gallery forest, riparian vegetation, evergreen forest, semi-arid *Acacia*, edges of equatorial forest, rich *Brachystegia*, *Marquesia* forest, woodlands and gardens. Avoids extremely arid areas. To 3,000m; to 2,240m DRC; mainly 1,000–3,000m east Africa, but does occur down to sea level, in areas of 500+mm annual rainfall; breeds to 2,000m, Malawi; to 2,135m Zambia (Chapin 1939, Benson *et al.* 1971, Britton 1980, Johnstone-Stewart 1984, Lewis & Pomeroy 1989).

BEHAVIOUR Solitary. Probably territorial. Flies swiftly in twisting hawk-like manner. Male utilises regular calling perch in breeding season. When calling male droops wings and holds tail above horizontal, inflates throat and opens bill widely, and turns head from side to side. Several males

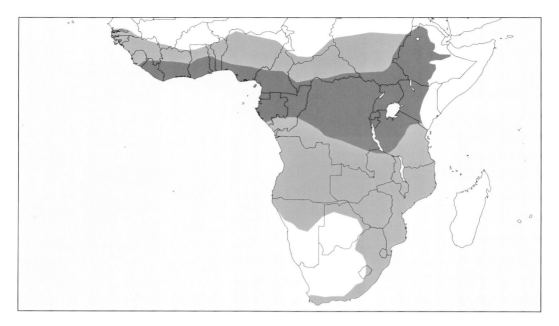

may call in one tree. During courtship female visits calling male and may duet. Male drops wings, fans tail and hops along branch; sometimes courtship feeding occurs. Pair may make short flight, particularly to host nest, which female may then inspect. Usually silent in southern Africa Jan–Feb. Searches canopy or other foliage, occasionally near ground, for insects, and occasionally hawks flying insects on wing. May fly from perch to catch insects in grass. Gut of caterpillars usually squeezed out and not swallowed (Irwin 1988, Payne 1997, 2005).

BREEDING Parasitism Female usually removes host egg when laying (Haydock 1950), or may lay before host (Reed 1969). Egg of this species and one of African Emerald Cuckoo together in nest of White-starred Robin, RSA (Oatley 1980). Nest of Cape Robin-chat with 2 cuckoo eggs, believed to have been laid by different females, and one of host, RSA (Tait 1952). Nest in Uganda of African Pied Wagtail with 3 hosts' eggs, one Dideric Cuckoo's, and one this species (Jackson 1938). Female recorded feeding young cuckoo, presumably hers, north Tanzania, Mar (B. Mwakamela *in litt.*). Known hosts (with instances): *Tchagra sp.* Baragoi, Kenya. African Paradise-flycatcher RSA (2), Zimbabwe. Red-bellied Paradise-flycatcher, Little Greenbul, White-tailed Ant Thrush and Rufous Flycatcher-thrush, Bioko. Yellow-bellied Greenbul, Kenya. Yellow-throated Greenbul, Uganda. Pair of *Lamprotornis sp.*, believed to be Miombo Blue-eared Starling, feeding young cuckoo Dec, Zambia. Kurrichane Thrush, RSA (5); Zimbabwe; Zambia. Olive Thrush, RSA (2). Fire-crested Alethe, north RC. White-starred Robin, RSA (5); Zimbabwe. Swynnerton's Robin, Zimbabwe (3). Forest Robin, Gabon. Archer's Robin-chat, DRC. Cape Robin-chat, RSA (commonest host; 90+); Zimbabwe; Malawi; Kenya; east Africa (18). White-throated Robin-chat, RSA (7); Zimbabwe. Blue-shouldered Robin-chat, Uganda. Rüppell's Robin-chat, east Africa (13). White-browed Robin-chat, RSA (c. 12); Zimbabwe; Zambia (3); Malawi; east Africa. Red-capped Robin-chat, RSA (2); Zimbabwe. Chorister Robin-chat, RSA. Eastern Bearded Scrub Robin, RSA (5); Zimbabwe; Mozambique. White-browed Scrub Robin, RSA (6); Zimbabwe; Zambia; east Africa; probably 'White-winged' Scrub Robin, Kenya. Perhaps Karoo Scrub Robin, RSA. African Stonechat, RSA (5); Zimbabwe; favourite host Uganda; Kenya. Mountain Wheatear, RSA. Familiar Chat, Zimbabwe. Mocking Cliffchat, Zimbabwe; probably RSA. Boulder Chat, Zambia; Zimbabwe (13). Cape Rock Thrush, RSA (5+). Southern Black Flycatcher, Zambia. Ashy Flycatcher, RSA. African Dusky Flycatcher, RSA (4). Yellow-tufted Malachite Sunbird, Natal, RSA. Cape Sparrow, RSA. Black-headed Weaver, RSA. Cape Wagtail, RSA (13); Kenya. Mountain Wagtail, east Africa; RSA. African Pied Wagtail, Uganda; Kenya; RSA. African Pipit, Kenya. Plain-backed Pipit, Uganda; Kenya (Levaillant 1806, Ivy 1901, Nehrkorn 1910, Seth-Smith 1913, van Someren 1916, 1956, Bates 1930, Jeffery 1931, H.F. Stoneham in Jeffrey 1931, Stoneham 1931, Vincent 1934, Sparrow 1936, Jackson 1938, Roberts 1939, Bergh 1942, Vincent 1946, Friedmann 1948b, 1949, 1956, 1967, Wiley 1948, Belcher 1949, Haydock 1950, Benson & Pitman 1956, Jessop 1960, Pitman 1961, 1964, Basilio 1963, R.I.G. Atwell, R.K. Brooke and W. Stanford in Friedmann 1967, Tree 1967, Jensen & Jensen 1969, Benson *et al.* 1971, Brosset 1976, Benson & Benson 1977, Fuggle 1980, Irwin 1988, Maclean & Maclean 1993, Dowsett-Lemaire 1996, Cherry & Bennett 2001, Kuiper & Cherry 2002). Unconfirmed hosts: White Helmet-shrike, Mozambique (Vincent 1934). White-browed Scrub-robin or African Stonechat, Baragoi, Kenya (Nehrkorn 1910). Yellow-crowned Canary, RSA (F. Barber in Sharpe 1875–1884). Cape Siskin, RSA (Steyn & Myburgh 1980). Snowy-crowned Robin-chat, Mabira Forest, Uganda (V.G.L. van Someren in Friedmann 1967), but Black Cuckoo also occurs there. Cape Penduline Tit, RSA (F. Levaillant in Friedmann 1948b) is almost certainly erroneous. Common Bulbul, Uganda (3), and Kenya (van Someren 1916), and RSA, not well-substantiated. Possibly Honeyguide Greenbul, Cameroon (Bates 1930). Vieillot's, Slender-billed and Village Weavers possible hosts south Uganda (C.R.S. Pitman in Friedmann 1948b). White-eyed

Slaty Flycatcher feeding young out of nest, but next day young fed by Collared Sunbird, east Africa (Harding 1948). In RSA 4.5% Cape Robin-chat nests parasitised, highest where host predominates (17% Natal, 22% Transvaal); elsewhere in RSA 11% available Cape Robin-chat nests in one area; 50% of small sample of Boulder Chat nests. In RSA rate of parasitism varied from 0.14 (Olive Thrush) to 12.5% (Eastern Bearded Scrub-robin) (nest record card data only), with % from all data as follows: Kurrichane Thrush 8, Olive Thrush 1, Cape Rock Thrush 6, African Stonechat 2, Chorister Robin-chat 2, White-browed Robin-chat 5, Red-headed Robin-chat 4, White-throated Robin-chat 6, White-starred Robin 10, Swynnerton's Robin 3, Boulder Chat 14, Eastern Bearded Scrub Robin 7, White-browed Scrub Robin 3, African Dusky Flycatcher 6, Southern Black Flycatcher 1, African Pied Wagtail 2, Cape Wagtail 15, Black-headed Weaver 1; bias of museum collections and nest record data should be considered (Kuiper & Cherry 2002). **Season** Breeds in early rains. Calls south Senegal and Gambia Jun–Aug; calls Yapo Forest, IC irregularly throughout year; Nigeria, Aug; Cameroon, Mar, Aug; fledgling north RC, early Feb; lays in rains north-east DRC; Sudan, probably Mar–Jun; Ethiopia, probably Apr–Jul; central and south Uganda, and south-west Kenya, Feb–Jul, Sep; nestlings Uganda, Jun (2); Tanzania (except north-east), Jan, May, Jun; Kenya (except west and coast) and north-east Tanzania, Jan, Mar–Jul, Dec; Nairobi, Apr–May, Dec; egg Baragoi, Kenya, 20 Dec; Zambia, Oct–Dec; Malawi, Jan, Sep, enlarged ovaries, Nov; Zimbabwe, Oct–Mar; RSA, Sep–Jan, peak Dec; Dusky Flycatcher with fledgling cuckoo Knysa, Cape Province, RSA, 26 Dec (Ivy 1901, van Someren 1916, Chapin 1939, Vincent 1946, Friedmann 1948b, Wiley 1948, Haydock 1950, Serle 1950, 1965, Bannerman 1951, Tait 1952, Jessop 1960, Pitman 1964, R.I.G. Attwell, H.H. Miles and W. Stanford in Friedmann 1967, Tree 1967, Steyn 1968, Jensen & Jensen 1969, Reed 1969, Brown & Britton 1980, Beattie 1981, Nikolaus 1987, Demey & Fishpool 1994, Payne 1997, B. Kamela *in litt.*, B. Mwakamela *in litt.*, BMNH). **Eggs** Broad oval and glossed. Polymorphic; 2 to 8 *gentes*, varying individually and geographically (Friedmann 1967, Grobler & Steyn 1980, J.F.R. Colebrook-Robjent in Irwin 1988, Cherry & Bennett 2001, Kuiper & Cherry 2002). Commonest uniform chocolate or olive-brown, highly glossed, distinct from common host Cape Robin-chat (Vincent 1946), but close to those of White-browed Robin-chat and to some of those of Red-capped Robin-chat. Those laid in Chorister Robin-chat's nest are exact match of brown eggs of host (T.B. Oatley in Irwin 1988). One in White-browed Robin-chat's nest ovate, deep cream with rough texture, and unlike host's (Haydock 1950). Some eggs unmarked pale greenish, olive-green or bluish. Those in Boulder Chat's nest closely match those of host being greenish-white speckled red-brown particularly at broad end and heavily capped; those laid in nests of Eastern Bearded Scrub Robin and thrushes vary considerably being pale green or olive-green to pale blue freckled and blotched pinkish-brown, generally matching those of host. Kuiper & Cherry (2002) accept 3 *gentes* in RSA with high degree of host-specificity: *gens* 1 (chocolate/coffee) found in nests of Kurrichane Thrush, Cape Rock Thrush, African Stonechat, Chorister Robin-chat, White-browed Robin-chat, Red-headed Robin-chat, Cape Robin-chat, White-starred Robin, Swynnerton's Robin, Southern Black Flycatcher; *gens* 2 (green/blue spotted red) in Kurrichane Thrush, Cape Robin-chat, White-throated Robin-chat, Boulder Chat, Eastern Bearded Scrub Robin; *gens* 3 (coffee/brown freckled) in Cape Robin-chat and Black-headed Weaver. Perfect or good matching eggs found in Chorister Robin-chat, White-browed Robin-chat, Red-headed Robin-chat; poor or very poor matches in Cape Robin-chat, African Stonechat, Cape Rock Thrush, Southern Black Flycatcher. 21.8–25.5x18.2–19.0mm (n = 13), 25.7x18.2mm, 24.5x18.0, 23.2x17.1mm (Vincent 1946, Haydock 1950, Irwin 1988). In Cape, White-browed, White-throated, Red-capped and Chorister Robin-chats, African Stonechat, Cape Rock Thrush, Kurrichane Thrush, Southern Black Flycatcher and Black-headed Weaver, host and cuckoo eggs are very highly matched for ultra-violet versus greenness component not detected by humans. Also, cuckoo eggs and host eggs are dissimilar in achromatic brightness not detectable by humans (Cherry & Bennett 2001). **Incubation** 11.5–16 days (R. Liversidge in Friedmann 1967, Reed 1969). **Chicks** Usually evict host eggs or young by 2nd day, well-feathered by 11th; show threat-response by 6th day, snapping, gaping, later spreading wings and tail, and lunging forward, snake-like, with bill closed giving impression of coiled tree-viper *Atheris sp.* with similar orange mouth colour to snake. Fledges after 17–22 days, leaving nest before able to fly, and dependent on fosters for further 20–25 days; feeds itself from 13th day (Liversidge 1955, Friedmann 1967, Reed 1969, Archer 1985). **Survival** No information.

FOOD Chiefly hairy caterpillars, but also beetles, winged ants, grasshoppers, spiders, centipedes, millipedes, slugs, snails, small frogs, lizards and berries. Nestlings fed by hosts on variety of insects, spiders, slugs, centipedes, millipedes, amphipods, small frogs, lizards and berries. Cape Robin-chat fed nestling on hairy caterpillar that would probably not form part of its own diet (Chapin 1939, Irwin 1988, Payne 1997).

STATUS AND CONSERVATION Not uncommon Togo (Cheke & Walsh 1996). "Almost ubiquitous" DRC (Chapin 1939). Fairly common resident south Sudan (Cave & MacDonald 1955). Common in western half of Ethiopia (Ash & Atkins 2009). Common at all altitudes, Malawi (Benson & Benson 1977). Common in most of Zambia (Benson *et al.* 1971). Not uncommon to uncommon from Senegambia to Liberia east to CAR and DRC, also Bioko (Borrow & Demey 2001). Common and widespread, east Africa (Stevenson & Fanshawe 2002). Common breeding visitor Mozambique, Zimbabwe, Swaziland, Lesotho and eastern RSA (Irwin 1988). Not globally threatened. (BirdLife International 2011).

Red-chested Cuckoo *Cuculus solitarius*. **Fig. 1.** Adult, Tarangiri NP, Tanzania, February (*Eyal Bartov*). **Fig. 2.** Adult male, Tarangiri NP, Tanzania, February (*Eyal Bartov*). **Fig. 3.** Adult female, Notwane, Botswana, October (*Ian White*). **Fig. 4.** Adult female, Maasai Mara NP, Kenya, June (*Adam Scott Kennedy*). **Fig. 5.** Juvenile, fed by White-browed Robin-chat, Maasai Mara NP, Kenya, January (*Adam Scott Kennedy*). **Fig. 6.** Juvenile, Maasai Mara NP, Kenya, January (*Adam Scott Kennedy*).

LESSER CUCKOO
Cuculus poliocephalus **Plate 35**

Cuculus poliocephalus Latham, 1790 (India)

Alternative names: Little/Small/Asian Lesser Cuckoo.

TAXONOMY Once considered conspecific with *lepidus* and *rochii* (Peters 1940). Monotypic. Synonyms: *himalayanus* Jerdon 1840, *Bartlettii* Layard 1854, *tamsuicus* Swinhoe 1865, *assamicus* Koelz 1952.

FIELD IDENTIFICATION 25cm. Small, slim cuckoo with long narrow wings, often with upright stance. Flight direct and fast with flat wing beats interrupted by short glides. **Adult male** Generally slaty blue-grey, paler on head, neck, throat and upper breast, becoming more blackish on uppertail-coverts; lower breast and rest of underparts white with bold black barring, undertail-coverts pink-buff barred black; graduated tail uniformly dark with white tips and white spots on all rectrices. **Adult female, grey morph** As male but washed rufous between grey and white of breast. **Adult female, hepatic morph** Rufous above with blackish barring all over except rump, and little on nape; underparts white barred black. Tail barred black and rufous, spotted and tipped white. **Juvenile, grey morph** Head blackish barred white; upperparts slaty-grey variably edged white to tawny; wings and coverts spotted and notched rufous and white; underparts barred black and white; tail blackish barred, notched and spotted white, more obvious on underside. **Juvenile, hepatic morph** Barred tawny and black above, and brown and buff below. **Similar species** Great similarity to Common, Oriental and Madagascar Cuckoos makes field identification very difficult if size comparison not available, the last being only marginally larger. Usually more rufous suffusion on upper breast than Madagascar. Black bands on underparts broader and fewer than Common, but similar to Oriental; broad blackish barring on buff-white undertail-coverts, usually little or no barring in Oriental, but some are barred; this barring narrower on Common, and absent or sparse on Madagascar. Juvenile darker on upperparts than Oriental, crown, cheeks and mantle more plain with finer white tipping, lower breast and belly with slightly narrower whitish bars, and darker on throat and breast than barred juvenile Oriental; hepatic female unbarred, or with few dark spots, on rump as most hepatic Common, whereas Oriental strongly barred. Bill weaker than other two. Indian larger and upperparts conspicuously browner, with broad black subterminal tail band. In winter range in Africa from African Cuckoo by latter's finer black barring below, paler lower abdomen, and absence of hepatic morph. Madagascar Cuckoo has longer wing, usually unbarred undertail-coverts and apparently lacks full hepatic morph. Iris in Asian and Madagascar Cuckoos brown in adults, whereas in African, Common and Oriental Cuckoos yellow, brown or orange. Best identified by call. For nestling see Oriental Cuckoo.

VOICE 5–6 note loud, rough, yelping *ho-to-ti-gi-su-ku* lasting 2-1.5sec giving Japanese name 'Hototogisu', first note rising, and after momentary pause remaining notes descending, repeated several times in quick succession, each time slightly lower than preceding; or rather even *wit-witi-witi-wit wit-witi-witi-wit*, or first rising and then falling *wit-it-iti-witu-wit-it-iti-witu*, *wit-it-iti-witu*, or *WHIK-/CHIK-/CHIK-\CHUK-\CHUK-\chuk*; rendered as 'eat your chóky pepper', in Korea as 'pot-pot-chip-chip-to-you', in Himalayas 'pretty-peel-lay-ka-beet' and Japan 'teppen-kake-taka', or 'kyoh-kyo-kyo-kyo-kyo'. Female's call resembles female Common Cuckoo but 'bubbling' slower in tempo. In Pakistan calls much more at night than Common, often among first birds in morning chorus; commonly calls noisily in flight. First days after arrival in Ussuriland some males call almost without pause, even all night, but otherwise peak of vocalisation late Jun–early Jul. Normally silent in winter but vocal before spring migration in Africa, but fat, silent specimens collected 31 Mar–12 Apr, Tanga, Tanzania; flushed cuckoos in winter range give load staccato rattle (Moreau and Moreau 1937, Panov 1973, Fleming *et al.* 1979, Ali & Ripley 1983, Irwin 1988, Brazil 1991, Roberts 1991, Grimmett *et al.* 1998, Lee *et al.* 2000, Mackinnon & Phillipps 2000, Robson 2000, Stevenson & Fanshawe 2002, Rasmussen & Anderton 2005).

DESCRIPTION Adult male Head pale slaty-grey becoming paler on chin, throat, sides of neck and upper breast; rest of upperparts uniform dark slaty blue-grey, sometimes blackish-grey on uppertail-coverts; wings dark greyish, remiges barred white on inner webs, bend of wing white, underwing-coverts white barred grey. Lower breast to belly white, often washed buff, with bold black transverse bars, undertail-coverts pink-buff with little or no barring. Graduated tail dark grey tipped white with white spots near shaft, T5 often barred white. **Adult female, grey morph** As male but breast washed rufous and faintly barred blackish. **Adult female, hepatic morph** Deep rufous heavily barred black above, but often with few or no bars on crown, nape, rump and uppertail-coverts; underparts white barred black; intermediate females frequent. **Juvenile, grey morph** Crown blackish-brown barred white, sides of head black barred white, indistinct white spot on nape; upperparts slaty-grey with variable white to tawny feather edging; wing-coverts and remiges tipped white spotted buff, latter with pale rufous spots on outer and large white notches on inner webs; throat and upper breast black finely barred white, and rest of underparts barred black and white; rectrices normally barred blackish-brown and white, and tipped white, or blackish-brown with white spots, notches and tips, barring more obvious on underside. **Juvenile, hepatic morph** Barred tawny and black above, and brown and buff below. **Nestling** Hatches naked, skin grey-rose, after three days becoming grey. Above slaty black to dark blackish-brown faintly fringed white; below similar but irregularly speckled white, or white broadly barred blackish. **Bare parts** Iris dark brown; eye-ring yellow. Bill black, yellow basally; at few days old bill pale rose, culmen yellowish and egg-tooth white; mouth and tongue orange, tip of tongue and gape pale yellow. Feet pale to tan-yellow.

BIOMETRICS Wing male (n = 31) 139–159mm, mean 149.8 ± 4.4, female (n = 11) 137–157mm, mean 145.8 ± 5.1. Bill male 15.7–19.2mm, mean 17.6 ± 0.8, female 16.9–18.7mm, mean 17.4 ± 0.7. Tail male 108–137mm, mean 121.7 ± 7.5, female 116–131mm, mean 119.4 ± 4.7. Tarsus male 15.9–20.4mm, mean 18.9 ± 1.4, female 16.4–18.4mm, mean 17.9 ± 0.7 (Payne 2005). Mass male (n = 3) 48–59g, mean 54.2, female 40g (Irwin 1988), unsexed (n = 10) 32–44g, mean 40.1 (Becking 1988).

MOULT No information.

DISTRIBUTION Palearctic and Oriental, wintering in Oriental and Afrotropical Regions. Summer breeding visitor north-east Afghanistan, north Pakistan, Kashmir, Arunachal Pradesh, Nepal, Sikkim, Bhutan, Assam, south

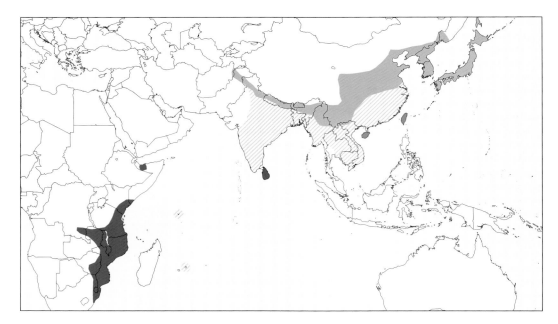

Ussuriland north to Iman R., Sakhalin, north Burma, north Laos, west Tonkin. Breeds China in North-east Provinces, Hebei, Shandong, south Shanxi, south Henan, Shaanxi, Gansu, Tibet, Sichuan, Yunnan, Guanxi Zhuang, Guizhou, Zhejiang; passage Fujian and Guangdong; five records Hong Kong. Resident Hainan and Taiwan where numbers augmented by winter visitors. Korea and Japan from Izu Is and Kyushu north to Hokkaido, late Apr–Sep. Arrives end May Ussuriland (departing Sep–early Oct) and Pakistan, later than other cuckoos; arrives montane India late Apr. Migrates through peninsular India south to Sri Lanka where few winter Sep–end Apr, and Andaman Is, where rarely winters, to eastern Africa. Birds presumably from China, Korea and Japan on migration or winter visitors in south China, Cochinchina, central Annam, and north, central and south Burma. In Africa Dec–mid Apr; Tanzania, Kenya, Zambia, Zimbabwe (rare), Malawi and Mozambique, presumably crossing Indian Ocean as recorded from Seychelles, Nov, Jan, Feb and Apr, and Mauritius, but not Middle East. Once Sakhalin (Osmaston 1916, La Touche 1931, Smith 1943, Benson 1951, Hachisuka & Udagawa 1953, Dement'ev & Gladkov 1966, Wells & Becking 1975, Netschajew 1977, Guangdong et al. 1983, Cheng 1987, 1991, Irwin 1988, Brazil 1991, Roberts 1991, Nechaev 1991, Kotagama & Fernando 1994, MacKinnon & Phillipps 2000, Rasmussen & Anderton 2005, Robson 2007, 2008, Demey 2010). Vagrants north Thailand May (3), Bangladesh (3) (Thompson et al. 1993); Uzbekistan and Turkmenistan (Kreuzberg-Mukhina & Kreuzberg 2005); Sakhalin (1) (Nechaev 1991); Bonin Is, Ryukyu Is. (Peters 1940); once Durban, RSA (Rowan 1983); twice Somalia (Ash & Miskell 1998); once DRC (Nouvelles Approches 2003). Recorded erroneously Greater Sundas and peninsular Malaysia.

HABITAT Light broadleaved forests of birch, oak, maple and hornbeam with much undergrowth, at forest edge and more often in riverine valleys, Ussuriland. Broadleaved montane forests, Japan, also second growth woodlands and open scrub. In Africa prefers denser habitat than Common Cuckoo in savanna, at forest edge, or in tangled ravines; only cuckoo in Africa found in pine forest plantations. Coniferous and deciduous forest, 1,500–3,660m, Himalayas generally; broadleaved and mixed forests, 1,000–2,800 m, Bhutan; 1,500–3,000 m, China, lower in north of range. To 1,200m (occasionally to 2,300m), Japan. At lower altitudes on migration (Dement'ev & Gladkov 1966, Panov 1973, Benson & Benson 1977, Netschajew 1977, Irwin 1988, Brazil 1991, Roberts 1991, Johnsgard 1997, Tymstra et al. 1997, MacKinnon & Phillipps 2000, Rasmussen & Anderton 2005).

BEHAVIOUR Secretive, but on first arrival in Pakistan often in gardens and less shy than Oriental Cuckoo, which is infrequently found near human settlement. On migration associates with Common Cuckoo Tanzania, Mar–Apr. In Kashmir forages on ground (Magrath 1923, Moreau & Moreau 1937, Irwin 1988, Roberts 1991).

BREEDING Parasitism Hosts: small passerines such as leaf warblers and wren-babblers, but Becking (1981) believes it prefers *Cettia* warblers, all breeding near ground. In Russia and Japan Japanese Bush Warbler commonest host, in Sikkim Pale-footed Bush Warbler. In India Chestnut-headed Tesia, *Cettia*, *Phylloscopus*, *Parus*, *Saxicola*, *Pnoepyga*, perhaps not all reliable. In Japan also Middendorff's, Eastern Crowned, and Ijima's Leaf Warblers, Winter Wren, Grey Thrush, Long-tailed Rosefinch and Black-faced Bunting (Dement'ev & Gladkov 1966, Becking 1981, Higuchi & Payne 1986, Brazil 1991). **Season** May–Jul (perhaps Apr–Aug) India; probably end May–early Aug, South Korea; Jun–Jul, Japan, later than other cuckoos (Ali & Ripley 1983, Brazil 1991, Birds Korea 2009). **Eggs** Plain brown to deep chocolate-brown or red-brown, one end slightly mottled with darker tones. Oviduct egg Darjeeling, India, terracotta-red. Brightest chestnut eggs in Japan (Osmaton 1904, Baker 1927). White eggs (Dement'ev & Gladkov 1966) may be misidentification as no *Cettia* has white eggs, but *Phylloscopus* warblers do, and are heavily parasitised by Oriental Cuckoo whose eggs are same size and could be mistaken. Becking

(1981) found no difference in ultra-structure of white and terracotta eggshells, but differ from those of Oriental, which makes it plausible that Lesser also lays white eggs. Eggs broad oval; 19.5–23.0x13.5–15.5 (n = 30), mean 21.0x14.2mm; mean 21.2x15.6mm, India; unmarked brown, 17.5x14.2, 19.0x15.3, 19.1x14.5, 21.7x15.7mm, Siberia, Ussuriland and Japan (Baker 1927, Neufeldt 1971, Higuchi 1989, 1998, Robson 2000). **Incubation** No data. **Chicks** Hatch naked. At 2–3 days ejects all host eggs and nestlings; eyes open 9th day; fledged 15th day (Netschajew 1977). **Survival** No data.

FOOD Mainly caterpillars and beetles; mantids, Hymenoptera. Caterpillars of Red Slug, pest on tea-bushes, Duars, India. Stomach contents of 59 cuckoos contained 90% caterpillars, Japan. Stomach contents Ussuriland contained caterpillars of Geometridae and Noctuidae. Chicks in nests of Japanese Bush Warblers fed grasshoppers, cicadas, butterflies, caterpillars, horseflies and other Diptera; spiders; snails (Inglis 1909, Dement'ev & Gladkov 1966, Ishizawa & Chiba 1966, Netschajew 1977).

STATUS AND CONSERVATION Frequent but local summer visitor Pakistan; fairly common Nepal; locally common India and Bhutan; rare, irregular winter visitor Sri Lanka (Grimmett *et al.* 1998). Not numerous summer visitor Russia generally (Flint *et al.* 1984), but common breeder, Ussuriland (Panov 1973). Scarce and local Burma (Smythies 1986). Uncommon China (Cheng 1987, MacKinnon & Phillipps 2000); rare resident Hainan (Cheng 1987); rare Hong Kong (Robson 2007, 2008). Not uncommon summer visitor Japan and Izu Is; common Okinawa (Brazil 1991). Winters widely in Africa, and often numerous (Irwin 1988). Not globally threatened (BirdLife International 2011).

Lesser Cuckoo *Cuculus poliocephalus*. **Fig. 1.** Adult male, Hegura Island, Ishikawa, Japan, June (*Tadao Shimba*). **Fig. 2.** Adult female hepatic morph, Hegura Island, Ishikawa, Japan, June (*Tadao Shimba*).

SULAWESI CUCKOO
Cuculus crassirostris Plate 35

Hierococcyx crassirostris Walden, 1872 (Northern Sulawesi).

Alternative name: Sulawesi Hawk-cuckoo.

TAXONOMY Retained in *Hierococcyx* due to wing structure and subadult plumage (e.g. Stresemann & Stresemann 1961), but mtDNA analysis indicates *Cuculus* (Sorenson & Payne 2005), as does call. Monotypic. *C. crassirostris* Stephens 1816 is *Eudynamys s. scolopacea*.

FIELD IDENTIFICATION 34–38cm. Medium-sized cuckoo. Considerable variation in plumage. **Adult** Entire head ashy-grey with yellow eye-ring, upperparts, wings and tail contrasting dark rufous-brown. Underparts white broadly barred black. Tail barred black and white. **Subadult** Head black-and-white; upperparts and wings rufous spotted dark brown, some remiges tipped white; below pale buff with sparse dark brown barring; nearly unbroken band across breast; underwing as adult but some rufous tips to primary-coverts; undertail-coverts buffy-white. Tail black barred rufous and white, tipped white; underside brownish-black barred and tipped white, T1–2 has buff-brown in white bars. **Juvenile** Head white with some black; upperparts bright rufous; underparts buffy-white. **Similar species** Large Hawk-cuckoo has rufous upper breast streaked dusky and white; T5 lacks large white bars. Rufous Hawk-cuckoo has dark chin, white streak from base of bill to throat; fine streaking and no barring on rufous underparts; juvenile heavily streaked below.

VOICE Loud, far-carrying *ho-oo*, second note lower than first; 3–4 note *chO-kO-ku-(ku)*, first two notes similar, third lower and fourth softer and shorter; duration 1sec; *to-ko-kuu* or *ka-ka-ku* with first syllable occasionally omitted and second higher than third; or similar to Common Cuckoo with second syllable lower, and first often doubled; *dong dong*. Calls at twilight and often persistently by night and on overcast mornings (Watling 1983, van den Berg & Bosman 1986, Rozendaal & Dekker 1989, Coates & Bishop 1997, Johnsgard 1997, Payne 2005).

DESCRIPTION Adult Great variation in plumage. Head and hindneck ashy-grey, some feathers with large white subterminal spots on back, scapulars, rump; wings uniform rufous-brown to grey-brown, primaries unbarred; bend of wing, underwing-coverts and base of remiges white. Underparts white with black spotting on throat, broad black band on upper breast (not obvious on some), and heavily barred black on rest of breast and belly, undertail-coverts white. Tail banded black, white and rufous, T1–2 dull rufous with black subterminal band and small white tip, T3–4 dull rufous with small white spots on shaft and indistinct black bands, T5 barred black and white with white notches on outer web; below T1 mostly buff-brown broadly tipped blackish-brown, and broadly tipped white; below black broadly barred white. **Subadult** Head black interspersed with white, back bright rufous, wings rufous with large dark brown spots and subterminal patches to secondaries and inner primaries, outer secondaries and inner primaries narrowly tipped white. Below pale buff with widely spaced dark brown bars; nearly unbroken band across breast; underwing as adult but some rufous tips to primary-coverts; undertail-coverts buffy-white. Tail more rufous than adult; underside brownish-black broadly barred and tipped white,

T1–2 with much buff-brown in white bars. **Juvenile** Head white with few black feathers on nape, upperparts and wings rufous-brown; unbarred buffy-white below, throat washed buffy; tail like adult but more white spots along shafts of T1. **Nestling** Undescribed. **Bare parts** Iris dark brown; eye-ring yellow. Bill blackish-brown with greenish base to lower mandible. Feet yellow.

BIOMETRICS Wing male (n = 10) 200–212mm, mean 206.1 ± 2.8, female 198mm. Bill male 25.0–28.2mm, mean 27.3 ± 1.6, female 26.0mm. Tail male 162–173mm, mean 168.2 ± 2.4, female 160mm. Tarsus male 22.8–26.7mm, mean 24.5 ± 1.8, female 27.7mm (Payne 2005).

MOULT No information.

DISTRIBUTION Sulawesi, on mountains in north, south-west and centre; Buton I.

HABITAT Primary, less often tall secondary, forests, mountain forest and forest edge. Generally 500–1,400m. Down to 200m Buton I. and 225m north Sulawesi. Eleven birds, all near fast flowing river or stream (Stresemann 1940–41, Rozendaal & Dekker 1989, Catterall 1997, Coates & Bishop 1997, Tebb *et al.* 2008).

BEHAVIOUR Shy and difficult to observe. Perches with upright posture among foliage, often in midstorey of large trees (Coates & Bishop 1997, Tebb *et al.* 2008).

BREEDING Parasitism Possibly parasitises Hair-crested Drongo (Stresemann 1940–41). **Season** Calls May–Aug (White & Bruce 1986). No further information.

FOOD Termites and other insects (Stresemann 1940–41).

STATUS AND CONSERVATION Rather rare, but because of skulking behaviour, difficult to monitor. Mostly only heard when vocal for short time each year (Coates & Bishop 1997). Stresemann (1940–41) had many villagers come to see collected bird that they only knew from call. Often recorded at different sites in Dumoga Bone NP (Payne 1997); frequently heard Lore Lindu NP, May and Aug (White & Bruce 1986, Jepson 1997). Not globally threatened (BirdLife International 2011).

INDIAN CUCKOO
Cuculus micropterus   Plate 35

Cuculus micropterus Gould, 1838 (Himalayas; type locality restricted to Simla-Almora Districts: Baker 1927).

Alternative name: Short-winged Cuckoo.

TAXONOMY Possibly undescribed subspecies (or species) resident Philippines (Kennedy *et al.* 2000). Polytypic. Synonyms: *michiaenus* Swinhoe 1870, *ognevi* Vorob'ev 1951 and *fatidicus* Koelz 1954, all in nominate.

FIELD IDENTIFICATION 32–33cm. Medium-sized cuckoo with hawk-like jizz. **Adult male** Head grey contrasting with grey-brown of rest of upperparts and wings; throat to breast grey, rest of underparts white with broad, widely spaced black barring. Tail grey with broad subterminal black band tipped white, white barring on T5 and shafts of T2–4. **Adult female** Browner; breast pale rufous. **Juvenile** Dark brown above; head to nape heavily spotted white to tawny-white; feathers of rest of upperparts broadly fringed white to rufous; blackish mask below eyes; below creamy-white variably barred brown. **Similar species** Common Cuckoo more grey on upperparts, finer barring on underparts, less white in uniform grey to blackish tail, longer and more pointed wings; no female hepatic in Indian; Oriental has head and upperparts ashy-grey like Common, but belly and undertail-coverts pale buffish; voice best recognition feature. Lesser smaller with uniform blackish tail tipped and sides spotted white. Juvenile from all other Asian *Cuculus* by extensive pale spotting on crown to nape.

VOICE Male call descending, loud flute-like four-note *kwer-kwah-kwah-kurh* lasting *c.* 1sec, often repeated 20–25 times/min; or *ko-ko-to-ko*; or *wa-ha-ha-hu* repeated 2–2/min; once five-note calls *wa-hu-ha-hu-ho* outside forest on Borneo, possibly by migrant; *whi-whi-whi-wu* or *wa-wa-wa-wu*, where last note lower, or rarely omitted. Sonogram rendered *KWIK-\KUH/KWIK(-)uh*. In Philippines three note *woop-woop-wuu* or *who-who-wu* lasting *c.* 1sec, repeated every 3–4sec. Wide range of local renderings, e.g. 'crossword puzzle', 'bola sepak', 'one more bottle' etc. Apparently male has no other call. Female call very much like Common Cuckoo's chuckle or 'water bubbling' but higher pitched, and often in flight or when copulating. Calls through night in breeding season, almost continuously in moonlight. Begging call of chicks similar to host's nestlings, shrill *tu-fju* (Hoffmann 1950, Ali 1962, Neufeldt 1966, 1967, Ali & Ripley 1983, Lekagul & Round 1991, Johnsgard 1997, Grimmett *et al.* 1998, Wells 1999, Kennedy *et al.* 2000, Robson 2000, Rasmussen & Anderton 2005, C.F. Mann).

DESCRIPTION Nominate. **Adult male** Head and neck uniform slaty-grey, upperparts and wings dark grey strongly washed brownish, bend of wing white, underwing-coverts white barred black; chin, throat and upper breast slaty-grey, darkest at sharp demarcation to rest of underparts, which are white broadly barred black, except plain white centre of belly and undertail-coverts. Tail dark grey tinged brown with broad diffuse subterminal black band tipped white, T5 broadly barred white; 4–5 large black-and-white shaft-spots on T2–4. **Adult female** As male but throat paler grey and some brownish to rufous on breast barred blackish, lower breast white barred black-brown, barring on belly narrower; no hepatic morph. **Subadult** (transitional?) Head and nape slaty-grey, sides of neck and breast rufous barred dark, back dark brown edged blackish, rump and uppertail-coverts grey-brown edged blackish and rufous; rest of underparts white barred black; tail as juvenile. **Juvenile** Head and nape blackish broadly spotted and barred white and rufous; broad black band speckled white below eyes; rest upperparts brown, scapulars and wing-coverts broadly edged dirty cream, isabelline or pale rufous; remiges dark brown tipped rufous or rufous and white. Underparts whitish or buff-white; throat and upper breast unevenly barred brown; rest of underparts barred brown. Tail brown with blackish subterminal bar and tipped white; spotted and notched rufous or rufous and white, becoming oblique bars on T2–5. **Nestling** Skin yellowish pink at 4–5 days becoming dark grey with violet tinge above, below remaining yellowish. **Bare parts** Iris brown to red-brown; eye-ring yellow-grey or grey; juvenile iris brown. Bill black, lower mandible greenish basally yellow; juvenile gape lemon-orange. Feet yellow.

BIOMETRICS Nominate. Wing male (n = 18) 188–205mm, mean 195.3 ± 5.7, female (n = 7) 189–203mm, mean 193.6 ± 6.8 (Indian subcontinent); male (n = 20) 189–215mm, mean 204.6 ± 6.3, female (n = 7) 198–214mm, mean 203.8 ± 6.3 (south-east Russia, Korea, north China); male (n = 10) 187–206mm, mean 194.9 ± 8.5, female (n = 5) 185–200mm, mean 193.2 ± 6.1 (Philippines). Bill male 23.5–26.4mm, mean 25.8 ± 0.9, female 24.2–27.0mm, mean 25.3 ± 1.2. Tail male 140–167mm, mean 158.6 ± 7.2, female 152–161mm, mean 157.0 ± 5.9. Tarsus male 19.2–24.6mm, mean 21.2 ± 1.9, female 19.2–22.0mm (Payne 2005). Mass male (n = 6) 112–129g, mean 119, female 119g (Neufeldt 1966); nominate unsexed autumn (n = 17) 77.5–121.8g (Wells 1999); male (n = 6) 85–168g, mean 115.8, female (n = 5) 110–138g, mean 120.8 (Shaw 1938a). Wing (migrants, Thai-Malay Peninsula) up to 221mm (Wells 1999).

MOULT Irregular tail moult (breeding female) 29 Apr. Young migrating birds arrive Thai-Malay Peninsula Jan either unmoulted or wing moult suspended with P1, P2 and P3 adult feathers; adults arrive in freshly moulted plumage. Some retain few juvenile feathers into first breeding season. Head moult Singapore, Nov (Neufeldt 1966, Wells 1999, H.Y. Cheng *in litt.*).

GEOGRAPHICAL VARIATION Two subspecies, differing only in size.
C. m. micropterus Gould, 1838. Breeds Pakistan, Nepal, north and east India, Bhutan, Bangladesh, Burma, Thailand, Siberia, China, Korea; northern populations winter south India, Sri Lanka, peninsular Malaysia, Andaman and Nicobar Is, Greater Sundas and Philippines. Described above.
C. m. concretus S. Müller, 1845. Resident south Thailand, peninsular Malaysia, Laos, Vietnam, Sumatra, west Java and Borneo. Smaller and slightly darker, wing <185mm. Wing male (n = 10) 152–168mm, mean 162.9 ± 5.9, female (n = 6) 152–170mm, mean 160.8 ± 9.0 (Payne 2005).

DISTRIBUTION Palearctic and Oriental. From Pakistan, Himalayas, India to Siberia, Indochina, China, Korea, Greater Sundas and Philippines. In Pakistan summer visitor in north-west Punjab, rarely from higher ranges of Murree Hills. In India breeding from Punjab east to Arunachal Pradesh and south through eastern part of subcontinent; rare Maldives. North-east Thailand and peninsular provinces. In Thai-Malay Peninsula *concretus* occurs north to Phatthalung, but two late Oct at Fraser's Hill with migrant nominates suggest possibly more extensive northern range. Migrants on small islands of Malacca Straits;

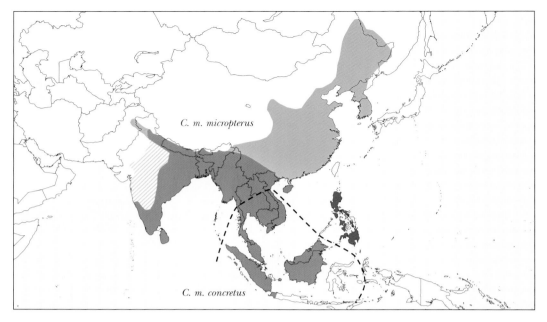

Singapore 14 Sep–17 May, peaking there and Fraser's Hill, Malaysia, Oct–Nov; Penang Apr and Nov. South-east Russia from upper reaches of Zeya R. to Komsomolsk on Amur R., migrating south second half Aug. Summer visitor China in Nei Mongol, North-east Provinces, Hebei, Henan, Shanxi, Sha'anxi, Gansu, Sichuan south to Yunnan, Guizou, Guangxi Zhuang, Guangdong and Fujian; resident Hainan; arrives Nanking end Apr, in Beijing about 14 days later, departs Aug–Sep. Passage May–Jun, Korea; some breed. On Sumatra *concretus* widespread, including Batu Is.; nominate winter visitor to mainland and Aruah Is, Lingga Archipelago, Bangka, Belitung and Nias. On Borneo *concretus* common resident to 1,220m; nominate rare migrant, once Sarawak, once Sabah late Oct, possibly Brunei late Jan; several heard Negara R. basin, South Kalimantan. In Philippines recorded Dinagat, Leyte, Luzon, Mindanao, Mindoro, Negros and Sanga Sanga, 26 Oct–29 May; Panay; at Dalton Pass only two taken per thousand birds caught, compared with 6–8 at Fraser's Hill, Malay Peninsula (Hoffmann 1950, Neufeldt 1967, Gore & Won 1971, Flint *et al.* 1984, Meyer de Schauensee 1984, Cheng 1987, van Marle & Voous 1988, Lekagul & Round 1991, Mann 1991a, 2008, Roberts 1991, van Balen & Prentice 1997, Grimmett *et al.* 1998, McClure 1998, Wells 1999, Kennedy *et al.* 2000, MacKinnon & Phillipps 2000, Curio *et al.* 2001, Choi Wai Mun in Wells 2007, Birds Korea 2009). Once Ternate, Moluccas, 6 Nov 1861 (Schlegel 1864). Three records Honshu, one Kyushu, Japan (CCJB 2000). Once Cocos Is, Feb 2011 (R. Clarke *in litt.*).

HABITAT Deciduous, evergreen or mixed forest, including Sal *Shorea robusta*, mixed *Shorea* and *Anogeissus* and pineswamp forest, but dense forests avoided; woodlands and groves with oaks; riparian woodland, disturbed areas (including forests and cultivation), plantations, scrubland; mixed forests in hills, grassy banks, scrubland, dried river beds and plantations; to 3,800m. Tropical semi-evergreen and evergreen forest, mixed secondary forest, subtropical and temperate broadleaved forest, pine forest and plantations to 200–2,600m, Arunachal Pradesh. 1,500–2,500m north India, occasionally higher in Himalayas. To 2,100m Nepal, rarely higher. 250–3,400m, Bhutan. Dry lowlands Sri Lanka, ubiquitous winter. Dry lowland evergreen and semi-evergreen forest, or regenerated after logging, to 600m, Thai-Malay Peninsula; silent birds, presumed migrants, also in gardens, parklands or low scrub. Stunted oak and larch forest influenced by logging and fire, larch taiga, oak groves on small hills or old pines with undergrowth of dwarf oak, Russia. 130–1,800m Vietnam. Lowland and hill forest, parks and gardens to 1,000m, China. Lowland and hill dipterocarp, alluvial, peatswamp, *Melaleuca* and occasionally secondary and logged forests, but most common riverine forest, to 1,220m, Borneo (Hoffmann 1950, Smythies 1957, 1981, Neufeldt 1966, 1967, Holmes & Burton 1987, Inskipp & Inskipp 1991, Robson *et al.* 1993, Pandey *et al.* 1995, Singh 1995, Tymstra *et al.* 1997, van Balen & Prentice 1997, Grimmett *et al.* 1998, Kalsi 1998, Barua & Sharma 1999, Wells 1999, MacKinnon & Phillipps 2000, Mann 2008).

BEHAVIOUR Generally high in canopy where more often heard than seen; females preferring middle and thickest parts of trees so more difficult to see. Males counter-call, suggesting territorial behaviour. Male droops wings with tail often spread and erected, simultaneously moving it slightly from side to side, and calls excitedly to attract female. Often in trios, mostly two males and one female; females so engaged very vocal. Confined to definite parts of a forest and even to individual trees year after year. Flight hawk-like. Night-migrant. Migrant observed taking caterpillars from ground. Catches termite alates in air (Neufeldt 1967, Becking 1981, Ali and Ripley 1983, MacKinnon 1990, Grimmett *et al.* 1998, McClure 1998, Wells 1999, Rasmussen & Anderton 2005, SUARENG 2000 in Wells 2007).

BREEDING Parasitism Drongos, e.g. Black and Ashy, primary hosts, India; fledgling fed by Streaked Spiderhunter, Sikkim; also Streaked Laughing-thrush, Tickell's Thrush, Siberian Stonechat, White-bellied Redstart and Indian Blue Robin. Fledgling fed by Greater Racket-tailed Drongo, and egg in nest of Ashy Drongo, Java. Brown Shrike, Siberia. Azure-winged Magpie preferred China, also Forest Wagtail, Black Drongo and probably Brown Shrike (Gammie 1877, Vaughn & Jones 1913, La Touche 1931, O'Donel 1936, Shaw 1938b, Jones 1941, Storrs 1944, Hoffmann 1950, Dement'ev & Gladkov 1966, Neufeldt 1966, Hellebrekers & Hoogerwerf 1967, Becking 1981, Johnsgard 1997, P.E. Lowther *in litt.*). Also *Calliope, Luscinia, Bombycilla* and *Acrocephalus* (Hoffmann 1950). Fledgling, perhaps this species, fed by Black-and-yellow Broadbills, Selangor, Malaysia (Wells 1999). Asian Paradise-flycatcher, Pakistan (Baker 1934) queried by Becking (1981) because egg was blue. **Season** Fledgling Sikkim, Sep; calling India, mid Mar–early Aug, egg laying Mar–Jun; most vocal May–Jul, main season of drongo hosts; five eggs May, Simla Region, Himachal Pradesh; Himalayas mid May–Jun, Indian plains Apr–Aug. Most calling Mar–Apr, Sri Lanka. Calling most intense Mar–Jun, Thai-Malay Peninsula; egg Johor 29 Apr; dependent fledgling Gombak 16 Sep. Young fledgling Luzon, 17 May 1961, only breeding record in Philippines. Eggs east Siberia and Beijing, Jun; north China, May–Jul. Egg Mt. Massigit, west Java, 10 Dec. South Korea probably mid May–Jul (Gammie 1877, La Touche 1931, Shaw 1938b, Hoffmann 1950, Dement'ev & Gladkov 1966, Neufeldt 1966, Becking 1981, Ali & Ripley 1983, Noramly & Jeyerajasingam 1985, Wells 1999, Payne 2005, Birds Korea 2009). **Eggs** Broad oval; ground colour white washed pinkish, spotted violet to carmine with underlying grey spots, matching both drongo and Black-and-yellow Broadbill eggs; 22.8–26.0x17.0–20.0mm, mean 24.0x18.2mm; in India mean 26.2x17.9mm, and 22.8–26.0x17.0–20.0 (n = 25), mean 23.7x18.1mm; Lower Amur, Siberia 25.2x19.5mm; *concretus* mean 23.6x17.7mm (Baker 1927, Schönwetter 1967, Becking 1981, Johnsgard 1997). Blue variant reported (Baker 1927, 1934), but scanning electron microscopy and biochemistry show these eggs are from Common Cuckoo (Becking 1981). **Incubation** 12 days, two days shorter than Brown Shrike (Johnsgard 1997). **Chicks** Hatch naked, skin yellowish pink, gape flange yellow and mouth orange-red, tongue orange-red tipped yellow, gape flanges, nostrils and tongue tip become black 8th day. Eyes open 7th day; 9–10th day feathers emerge from sheaths; body mass at hatching 4.7–4.9g. Ejection behaviour from 2nd day, but first seen at 3–4th days; Brown Shrike nest with six eggs had none evicted. Chick on edge of nest at 18th day, by 21st day fledged but incapable of flying until 30–40th; juvenile plumage completed by 45th day (Neufeldt 1966, Becking 1981). **Survival** No information.

FOOD Mostly hairy caterpillars; termite alates, grasshoppers, ants and butterflies; fruit. Mole Cricket and beetle (Melolonthidae) larva in stomach. Forages high in canopy and on forest floor (Mason & Maxwell-Lefroy 1912, Payne 1997, 2005, Wells 1999).

STATUS AND CONSERVATION Rare rains visitor Pakistan (Roberts 1991); common summer Nepal; common in breeding season, Himalayas, India, e.g. Arunachal Pradesh (Singh 1995), and locally common Manipur (Choudhury 2009), but uncommon rest of India, and status uncertain; common resident Sri Lanka; rare Maldives (Grimmett *et al.* 1998); common summer migrant Nameri NP, Assam (Barua & Sharma (2005); common at many sites Bhutan in summer (Spierenburg 2005); locally common Bangladesh (Harvey 1990). Fairly common resident and winter visitor Thailand (Lekagul & Round 1991). Heard commonly Sugihan WR, south Sumatra, Aug 1984–Jun 1985 (Nash & Nash 1985). Uncommon to common Thai-Malay Peninsula (Wells 1999). *Concretus* widespread and common Borneo (Mann 2008). Resident subspecies and winter visitors rare, Java (MacKinnon & Phillipps 1993). Uncommon Philippines (Kennedy *et al.* 2000). Not rare parts of Amur R. area, Russia (Dement'ev & Gladkov 1966). Common east and south China (Cheng 1987). Uncommon passage, perhaps rare summer resident, Korea (Gore & Won 1971). Rare Japan (CCJB 2000). Use of logged forest demonstrates flexibility of habitat choice. Not globally threatened (BirdLife International 2011).

1

Indian Cuckoo *Cuculus micropterus*. **Fig. 1.** Adult female with traces of subadult plumage, *C. m. micropterus*, Cocos Islands, Australia, February (*Rohan Clarke*). **Fig. 2.** Adult male, *C. m. concretus*, Kek Look Tong, Ipoh, Perak, Malaysia, December (*Khoo Siew Yoong*). **Fig. 3.** Probable adult male, with some subadult plumage obvious in tail, *C. m. micropterus*, Tanamalwila, Sri Lanka, March (*Amila Salgado*). **Fig. 4.** Subadult, *C. m. concretus*, Singapore, November (*H. Y. Cheng*). **Fig. 5.** Juvenile, *C. m. concretus*, Lake Kenyir, Terengganu, Malaysia, July (*Khoo Siew Yoong*); this individual has much more white in plumage than normal.

MADAGASCAR CUCKOO
Cuculus rochii   Plate 35

Cuculus rochii Hartlaub, 1863 (Madagascar)

Alternative name: Madagascar Lesser Cuckoo.

TAXONOMY Possibly forms superspecies with *poliocephalus*, and once considered conspecific (Peters 1940), but mtDNA analysis places it closest to *gularis* (Sorenson & Payne 2005). Monotypic. Synonym: *stormsi* Dubois 1887.

FIELD IDENTIFICATION 26–28cm. Smaller than other *Cuculus* in range except Lesser. **Adult male** Dark grey above. Throat and upper breast paler grey; undertail coverts pink-buff; rest of underparts white, barred heavily and broadly black. Tail dark grey, spotted and barred white. **Adult female** As male, but with some tawny on breast and sides of neck; undertail-coverts with broken blackish bars. Occurrence of hepatic morph female unresolved. **Juvenile** From adult by tawny nape and sides of neck, edges to crown feathers and barring on wings, breast, throat and tail; remiges and coverts tipped white; undertail-coverts white barred black. **Similar species** Smaller than Common and African Cuckoos. From Lesser Cuckoo by usually less rufous suffusion on breast, longer wing, usually unbarred undertail-coverts, voice, and female lacking full hepatic morph. Iris in Asian and Madagascar Cuckoos brown in adults, whereas in African, Common and Oriental Cuckoos it can be yellow, brown or orange. Best identified by call.

VOICE Four (or three) evenly spaced notes, first three (or two) on same pitch, fourth about third of octave lower *ho-ho-ho-hu* lasting *c.* 1sec, rather mellow and hollow; also described as *cuck-cuck-ooo*, with accent on 2nd syllable. May sing constantly up to 20min, throughout day, and sometimes at night (van Someren 1947, Irwin 1988, Langrand 1990).

DESCRIPTION Adult male Uniform dark slate-grey above, paler on head; uppertail-coverts blackish. Throat to breast light or mid grey; undertail-coverts pink-buff usually unbarred; belly white with bold, heavy black barring. Tail black, tipped and spotted white, T5 more spotted and barred. **Adult female** Differs from male by tawny suffusion on breast and sides of neck. Usually stated there is no hepatic morph, but partial hepatic plumage claimed (Becking 1988 in Clancey 1991). **Juvenile** From adult by tawny nape and edges to crown feathers with some white speckling; remiges, coverts and scapulars usually barred tawny and tipped whitish. Breast and throat white barred blackish-brown and washed tawny, sides of neck washed tawny, undertail-coverts barred black and white. Tail usually barred chestnut-brown. **Nestling** Feathered nestling has yellow gape flange, red mouth and tongue (Hawkins & Goodman 2003). **Bare parts** Iris brown, eyering yellow. Bill blackish or dark brown, base of lower mandible yellow. Feet yellow.

BIOMETRICS Wing male (n = 20) 155–175mm, mean 166.1 ± 4.4; female (n = 7) 153–168mm, mean 159.7 ± 5.6. Bill male 16.3–19.4mm, mean 18.7 ± 2.2; female 16–19mm, mean 17.5 ± 0.9. Tail male 124–152mm, mean 140.8 ± 8.6; female 137–151mm, mean 139.8 ± 2.5. Tarsus male 15.3–19.0mm, mean 16.6 ± 0.8; female 15.5–18.0mm, mean 16.7 ± 0.8 (Payne 2005). Mass male 64, 65g (Benson *et al.* 1976–77).

MOULT Active moult Jul (2), interrupted moult Jun–Aug (5), DRC; tail moult in female ready to lay Andapa, Aug;

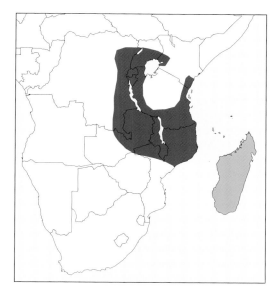

male Befandriana, Nov (Louette & Herroelen 1993, BMNH).

DISTRIBUTION Afrotropical. Migrant; breeds Madagascar where some remain throughout year; generally occurs continental Africa Apr–Sep. North Mozambique, chiefly Apr–Sep; Malawi; north and east Zambia, Nov, south-east Zambia, Apr; Tanzania in west, south (Sep) and coast; west and south coastal Kenya, May–Aug; south and west Uganda; once Feb and once Nov, east Africa; Rwanda, Jun–Aug; east DRC, Jun–Nov; Burundi; Natal RSA, once Feb (Durban) and once Nov; once Transvaal, calling Dec. Migrates in rains from forested east to arid west Madagascar. Occasional Mauritius (Chapin 1939, Benson *et al.* 1971, Aspinwall 1973, Backhurst *et al.* 1973, Benson & Benson 1977, EABR 1978, 1983, Britton 1980, Irwin 1988, T.J. Barnley in Lewis & Pomeroy 1989, Clancey 1991, Hockey 1992, Louette & Herroelen 1993, Payne 1997, Demey 2007).

HABITAT Forest, including gallery forest, forest edge, dense cover in savanna, scrub woodland, spiny subdesert woodland, marshes; *Eucalyptus* plantations and other human-influenced habitats; to 1,800m (Irwin 1988, Langrand 1990, Payne 1997).

BEHAVIOUR Solitary, secretive and shy; sings from perch in dense foliage. Prey taken from leaves and branches, occasionally from ground. Sometimes joins mixed feeding parties outside breeding season (Payne 2005).

BREEDING Parasitism Hosts: Madagascar Paradise-flycatcher, one in 8 clutches; Madagascar Cisticola, 10 in *c.* 210 clutches; Common Jery, one in *c.* 50 clutches; Grey-crowned Tetraka; Madagascar Sunbird (Milon 1946, P.E. Lowther *in litt.*); fledgling fed by Madagascar Magpie-robins (Anon. 2011a); fledgling fed by White-headed Vanga (Wilmé 1994). Unconfirmed hosts: African Stonechat, Forest Rock Thrush and Madagascar Bushlark (Cowan 1883, Sibree 1891, Milon 1959). **Season** Aug–Apr; ready to lay Andapa, Aug; enlarged testes Perinet, Nov; juvenile Belsileo, Apr (Rand 1936, Milon 1959, Langrand 1990, BMNH). **Eggs** Whitish, often tinged pink, spotted reddish-brown, not closely matching those of host; *c.* 18x13–14mm (Benson *et al.* 1972, Irwin 1988, Payne 1997, 2005). No further information.

FOOD Hairy caterpillars; other insects (Langrand 1990).

STATUS AND CONSERVATION Common throughout Madagascar except south where occasional (Langrand 1990). Regular locally continental Africa, but sparse and uncommon; probably overlooked (Irwin 1988, Payne 1997). Not globally threatened (BirdLife International 2011).

Madagascar Cuckoo *Cuculus rochii.* **Fig. 1.** Adult female, Salary, Toliara, Madagascar, December (*Louise Jasper*). **Fig. 2.** Adult female, Ranobe, Toliara, Madagascar, March (*Louise Jasper*). **Fig. 3.** Adult male, Anja, Ambalavao, Madagascar, October (*Louise Jasper*). **Fig. 4.** Juvenile, Antananarivo, Madagascar, March (*Bruno Boedts*).

AFRICAN CUCKOO
Cuculus gularis   Plate 34

Cuculus gularis Stephens, 1815 (Camdeboo, Cape Province)

Alternative name: Yellow-billed Cuckoo.

TAXONOMY Considered to form superspecies (once considered conspecific) with *canorus*, but mtDNA analysis places it as sister to *rochii*, these forming sister clade to *saturatus* and *canorus* (Sorenson & Payne 2005). Monotypic.

FIELD IDENTIFICATION 32cm. Long-tailed cuckoo with appearance of small raptor. Dashing, hawk-like flight. **Adult** Upperparts, including wings dark grey, paler grey on throat and breast, rest of underparts white barred dark. Yellow bill tipped dark, orange at base. **Juvenile, grey morph** From adult by white barring and tipping on upperparts, including wings, and patch on hind crown; pale tawny suffusion on head to upper back; uppertail-coverts brown broadly barred white; underparts creamy-white barred grey-brown, much more closely on chin to breast. Broad white spots and notches on tail, but no distinct barring except T5. **Juvenile, hepatic morph** Brownish instead of grey, tawny or buff instead of white. **Similar species** From Common Cuckoo, which is generally silent in winter quarters, by different call, very obvious yellow base to bill, and barred outer rectrices (spotted in Common). No adult female hepatic morph. Juvenile greyer above than that of Common, usually paler throat with narrower barring, broader white tips to crown when fresh, and noticeably broader white barring on upper surface, larger white spotting and barring on tail; hepatic juvenile brown, tawny and buff, much less rich than Common. Iris yellow or light brown; brown in adult Asian and Madagascar Cuckoos.

VOICE Call of male *oo-OO* with emphasis on slightly higher pitched second syllable, repeated indefinitely at rate of 16–25/10sec; similar to Common Hoopoe, more monotonous and flatter than Common Cuckoo and lacks falling cadence and initial *c*, although call sometimes rendered *coo-coó* or *coo-cook*. 'Tearing' call; insistent begging *chi-chi-chi...* or *ghiri-ghiri-ghiriri*; muted, excited *wick-wick-wick-wick*, higher and quicker in female; also from female ringing *kip-ip-p-p-p*, and bubbling *kwik-kwik-kwik!*; hard, harsh *krèk*. Calls at night (Lynes 1925, Chapin 1939, Rowan 1983, J.F.R. Colebrooke-Robjent in Irwin 1988, Payne 1997, Borrow & Demey 2001).

DESCRIPTION Adult male Slate-grey, or dark ash-grey above. Broad white barring on inner webs of primaries extending to secondaries. Throat and breast paler grey; rest of underparts white barred blackish-grey, belly and undertail-coverts less distinctly. White-tipped tail dark grey with blackish barring, T5 variably barred and spotted white. **Adult female** Less grey on chest than male and faintly barred; breast sometimes washed buff or tawny. **Juvenile, grey morph** Grey above, darker on crown and mantle; crown feathers barred and tipped white; white patch on hind crown; pale brownish-grey or grey spots or bars on remiges and wing-coverts. Tail above and below blackish-brown tipped white, T1 spotted white, T2–4 spotted and notched white, T5 barred white; broad white spots on tail, but no distinct barring except T5. **Juvenile, hepatic morph** Grey replaced by brownish and white by tawny to buff, although some show grey on dorsal surface (transitional?) (Friedmann 1948b, BMNH). **Nestling** Naked on hatching, with dark purplish or blackish skin and bright orange gape; grey above narrowly edged white, more extensively on head, similarly below but more narrowly on throat and chest; wings and tail as adult (Jensen & Jensen 1969). **Bare parts** Iris yellow (male), light brown (female); eyelids and eye-ring yellow. Bill yellow with blackish to horn tip, base orange; juvenile black, paler at base; brown in nestling. Interior of mouth reddish-orange. Feet yellow to orange-yellow, pale yellow in nestling.

BIOMETRICS Wing male (n = 15) 205–223mm, mean 213; female (n = 10) 202–218mm, mean 210. Bill unsexed 22–26mm (n = 20), mean 23.4. Tail male 152–168mm, mean 160; female 146–166mm, mean 155. Tarsus unsexed 19–21mm (n = 15), mean 20.0. Mass male (n = 6) 95–113g, mean 104; female 96, 99g (Irwin 1988); 19 day nestling 63g (Tarboton 1975). Wing male (n = 9) 204–226mm, mean 215.3 ± 8.4; female (n = 9) 197–209mm, mean 205.3 ± 5.1 (Payne 2005).

MOULT Juvenile finishing moult to adult, mid May, Darfur, Sudan. Juvenile, probably from northerly population, moulting Zimbabwe Sep; moulting juveniles Zambia, 1 and 17 Dec; Cameroon, 10 May; IC, 27 Mar (Lynes 1925, Irwin 1988, Payne 1977).

DISTRIBUTION Intra-African migrant, appearing with rains to breed. From Senegambia to Somalia to RSA (Cape). Breeding visitor south from DRC, where present in north-east savannas end Dec–early May, perhaps Oct–Dec, south-east Tanzania (Songea) and south Angola; less migratory north Angola where also present Jul–Sep; Malawi Sep–Apr; Zambia late Aug–late May. Interior Tanzania Nov–Apr, north-west May, Aug, south-east Oct–Dec; Uganda Mar–Nov; Kenya mainly Oct–May; Somalia (rare) May–Sep. Wet season visitor to north tropical savannas, Apr–Aug, where probably breeds throughout, north to *c*. 16°N in Gambia (once 2 Dec); north Nigeria to *c*. 12°N, not at one locality throughout year; migrant Sudan south of 15°N, Apr–Aug; Ethiopia and Eritrea Mar–Jun. Gambia at all seasons but not at one locality; Ghana and Togo, Jan–Sep, rarely Oct–Nov; north Burkina Faso, in rains end Jun–end Sep; once Guinea, Macenta Prefecture, Mar 1989; Dosso, Niger, May–Jun. North Nigeria late Feb–late Oct, but mostly south of 9°N Feb–Apr, very few records Nov–Jan; north-west Mar–Jul, mostly Mar and Jun. Abundant Cameroon (Bamenda Highlands); Yaoundé, Nov. RC Sep. Arrives late Jul–early Aug Ouadi Rime Reserve, Chad. West African birds probably spend early dry season (Nov–Dec) far to south-east. Many from savannas of northern and southern Africa may winter east Africa, with southern birds moving much further than northern ones (Chapin 1939, Benson *et al*. 1971, Urban & Brown 1971, Benson & Benson 1977, Britton 1980, Nikolaus 1987, Giraudoux *et al*. 1988, Irwin 1988, Medland 1992, 1993, Elgood *et al*. 1994, Halleux 1994, Cheke & Walsh 1996, Balança & de Visscher 1997, Payne 1997, Ash & Miskell 1998, Quantrill & Quantrill 1998, Dean 2000, Crewe 2001, de Bont 2002, Ash & Atkins 2009). Once Prince Edward I., 23 Nov 1982 (Newton *et al*. 1983 in Higgins 1999).

HABITAT Woodland and savanna, but not lowland forest; all savanna woodlands including *Acacia*; to 3,000m. Mostly areas of 250+mm rainfall, Kenya, in drier areas only in wet season. Light *Brachystegia*, but avoiding richer *Brachystegia*, to 1,370m, Malawi (Chapin 1939, Benson & Benson 1977, Britton 1980, Irwin 1988, Lewis & Pomeroy 1989).

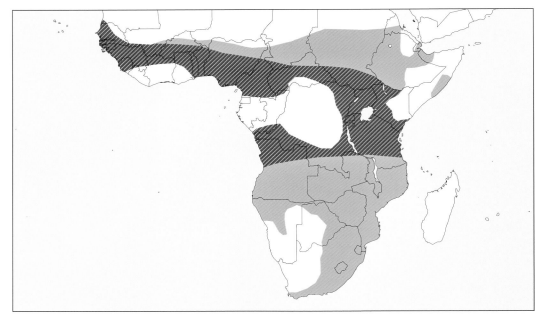

BEHAVIOUR Feeds in foliage and on ground. Perches almost vertically. Several call excitedly together at start of breeding season when calling most frequent. Monogamous; territorial. Territories probably >60 ha/pair Transvaal, RSA (Tarboton 1975, Irwin 1988).

BREEDING Parasitism Female removes host's egg at time of laying. Hosts in RSA: Southern Boubou, Bokmakierie, Common Fiscal, Fork-tailed Drongo, Common Bulbul, Cape Robin-chat, White-browed Scrub Robin, Kalahari Scrub Robin. Hosts in other areas: Yellow-billed Shrike Togo, Ghana, Nigeria, perhaps 'egg-dumping' as call of young cuckoo very different to young shrike; Fork-tailed Drongo, Senegal, Zambia, Ghana, Togo, Nigeria (>25), Namibia, Zimbabwe, Zambia; Common Bulbul Zimbabwe; unidentified babbler Zimbabwe; Arrow-marked Babbler; Kurrichane Thrush; White-browed Scrub Robin Zimbabwe; Southern Grey-headed Sparrow Zimbabwe (Mouritz 1915, Roberts 1928, Newby-Varty 1946, 1950, Friedmann 1948b, 1956, 1967, Plowes 1948, F. Levaillant and H.M. Miles in Friedmann 1956, Pitman 1957, G. Duve in Friedmann 1967, Jensen & Jensen 1969, Grimes 1979, Irwin 1988, J.F.R. Colebrook-Robjent in Irwin 1988, Cheke & Walsh 1996). Perhaps starling *Lamprotornis sp.* Zimbabwe (Newby-Varty 1946). Probably Eastern Black-headed Oriole, RSA (Tarboton 1975). 8% of Fork-tailed Drongo nests parasitised Choma, Zambia; for southern Africa generally 1.3% may be underestimate (Payne & Payne 1967, Jensen & Jensen 1969, Grimes 1979, Irwin 1988). **Season** Calling Gambia, Apr–May, nest searching Jun–Jul. Nestling Senegal, Jul. Fledglings Togo, Jul. Coastal Ghana, Jan–Jun, enlarged testes Ashanti, Jan; nestling May. North Nigeria, Feb–Apr, recently ovulated south-east, Feb, fledglings late Kishi, Apr–early May. Breeding female Cameroon, Feb. Yolked follicles Darfur, Sudan, May, calling May–Jun. Enlarged ovaries south Ethiopia, Mar. Rainy season breeder DRC, but breeds savannas of north-east in dry season; enlarged ovary Feb, Uelle District. Newly-fledged young Aug, Sep, Nov, ?Dec, Kenya. Breeding condition Angola, Oct–Dec. Perhaps Oct–Dec south-east Tanzania. Enlarged testes north Malawi Nov, Jan–Feb, juvenile Oct. Sep–Oct Zambia, nestlings 6 Sep, 30–31 Oct. Zimbabwe, Sep–Dec, egg 27 Oct. RSA, Sep–Jan, nestling Orange Free State, Dec. Namibia, Dec, fledgling Jan. Botswana, Dec–Jan (Mouritz 1915, Hopkinson 1922, Loveridge 1922, Lynes 1925, Roberts 1928, Bannerman 1933, Priest 1934, Lowe 1937, Chapin 1939, Benson 1945, Newby-Varty 1946, Friedmann 1948b, H.M. Miles in Friedmann 1956, Serle 1957, Heinrich 1958, G. Duve in Friedmann 1967, Jensen & Jensen 1969, Tarboton 1975, Benson & Benson 1977, Colebrook-Robjent 1977, Grimes 1979, Britton 1980, Pinto 1983, Irwin 1988, Elgood *et al.* 1994, Cheke & Walsh 1996, Zimmerman *et al.* 1996, Barlow *et al.* 1997, Payne 1997, 2005, P. Renaud in Payne 2005). **Eggs** Laid over nine weeks. Probably several *gentes*. Blunt oval; white, cream or pinkish buff spotted pale mauve or brown, sometimes bold rufous blotches overlying spots, or overlying irregular pale mauve and grey marks, or ring of dark reddish speckling at blunt end; also washed-out greeny blue, spotted pale mauve or brown, Zimbabwe, similar to Red-chested Cuckoo in DRC and Maasailand. Usually resembling and sometimes indistinguishable from highly variable eggs of Fork-tailed Drongo, and not dissimilar to those of Yellow-billed Shrike; 24.9x17.6mm (n = 6) (Chapin (1939, Newby-Varty 1946, A. Nehrkorn in Friedmann 1948b, Pitman 1957, Tarboton 1975, Irwin 1988). **Incubation** 12–17 days, probably *c.* 12 days whereas host 16 days (Tarboton 1975). **Chicks** Eject host young and eggs; all 5 eggs in Yellow-billed Shrikes' nest ejected (Grimes 1979). Eyes open 8th day; fledge 20–23rd day; remain dependent on foster parents for considerable time after that (Tarboton 1975, Irwin 1988, Payne 1997). **Survival** Young in Yellow-billed Shrikes' nest apparently abandoned by fosterers, and the group of shrikes started new nest (Grimes 1979). Breeding success RSA from egg to fledging 8–38% (Tarboton 1986).

FOOD Hairy and hairless caterpillars; beetles; termite alates (Chapin 1939, Irwin 1988).

STATUS AND CONSERVATION Common migrant Togo, Jan–Sep (Cheke & Walsh 1996). In west Africa generally uncommon to common intra-African migrant throughout, except lowland forest and arid north, moving north with rains; probably breeds throughout (Borrow & Demey 2001). Abundant Bamenda Highlands, Cameroon, Jan–Sep (Irwin 1988). Common in savannas of north-east DRC, late Dec–early May (Chapin 1939). Uncommon migrant Sudan (Nikolaus 1987). Fairly common non-breeding migrant in parts of Eritrea and Ethiopia (Ash & Atkins 2009). Scarce Somalia (Ash & Miskell 1998). Uncommon east Africa (Britton 1980). Locally to fairly common RSA (Payne 1997, 2005). Not globally threatened (BirdLife International 2011).

African Cuckoo *Cuculus gularis*. **Fig. 1.** Adult female, Etosha, Namibia, December (*Ralph Martin*). **Fig. 2.** Adult male, Etosha, Namibia, November (*Wil Leurs*).

ORIENTAL CUCKOO
Cuculus saturatus   Plate 36

Cuculus saturatus Blyth, 1843 (Nepal)

Alternative names: Blyth's/ Himalayan Cuckoo (all *saturatus*); Horsfield's Cuckoo (*optatus*).

TAXONOMY Junge (1937b) claimed that Gould's type of *optatus* synonymous with *saturatus* on measurements, and that it should be replaced by *horsfieldi* Moore 1857. Payne (2005) shows this is incorrect as wing measured on Gould's type had lost two longest primaries, and other wing was longer and in range of *optatus*. *Optatus* given species status by (King 2005, Payne 2005) due to larger size and different vocalisations, but this subspecies has northernmost distribution (Bergmann's Rule), and variants of its calls are within repertoire of single species (Wells & Becking 1975); plumage of *optatus* and *saturatus* extremely similar, including hepatic morph of female, and winter ranges overlap. Lindholm & Lindén (2007) found only minor differences in vocalisations between them, and these with some overlap, and reached no firm conclusions. *Lepidus* (with *insulindae*), previously considered subspecies of *saturatus*, treated by King (2005) and Payne (2005) as separate species due to different vocalisations and size, but Wells & Becking (1975) and Wells (1982) dispute this. This work tentatively accepts two-species arrangement, placing *saturatus* and *lepidus* in superspecies. Polytypic. Synonyms: *intermedius* Vahl 1797, *kelungensis* Swinhoe 1863 and *monosyllabicus* Swinhoe 1865 in nominate; *striatus* Drapiez 1823, *tenuirostris* and *barbatus* both Boie 1828, *canoroides* S. Müller 1845, *fucatus* Peale 1848, *brachyrhynchos* C.L. Brehm 1856, *horsfieldi* Moore 1857, *cantor* Cabanis & Heine 1863, *assimilis* C.L. Brehm 1865, *bubu* Dybowski & Parrex 1868, *peninsulae* Stejneger 1885, *belli* Mathews 1916 and *waigoui* Mathews 1918, all in *optatus*.

FIELD IDENTIFICATION 32–33cm. Medium-sized cuckoo with hawk-like jizz; rapid undulating flight more graceful than Common Cuckoo. **Adult male** Head, neck and upper breast plain ashy-grey; rest of upperparts and wings dark grey, white or white and barred dark grey edge of wing-shoulder; two prominent white bands on underwing; underparts from hindbreast to undertail-coverts white with bold and regular black barring. Graduated tail dark grey tipped white. **Adult female, grey morph** Like male but neck and upper breast washed rufous. **Adult female, hepatic morph** Rufous with dark barring on head, upperparts, wings and tail; off-white with fine black barring on sides of head, chin, throat and upper breast; rest of underparts white with black barring; tail tipped white. **Juvenile, grey morph** Grey-brown above, most feathers finely edged white, underparts white to buffish-white barred black, barring finer on sides of head and neck. **Juvenile female, hepatic morph** Like adult but with white feather fringes on upperparts and wings. **Similar species** Easily confused with Common Cuckoo when not calling, but broader tail (up to 1/3 wider) with fewer black bars, lack of barring on white leading edge of wings, much darker underwings with fewer white bars and spots, bill stouter, deeper and tip sharply curved on adult males, broader barring on underparts, including undertail-coverts (little or no barring in Common); grey morph female has coarser spots on more buffish neck sides and upper breast; hepatic females barred on rump and uppertail-coverts unlike most Common; juvenile in both morphs has bolder black barring above and on tail, and less white on nape than juvenile Common, and hepatic lacks narrow whitish tips to rump feathers and uppertail-coverts found in Common; perhaps no single character fully reliable. Oriental mostly found at lower altitudes than Common when sympatric. Sunda Cuckoo, usually at higher altitudes than Oriental in winter, smaller with shorter wings, darker overall plumage, and barring below broader. Lesser Cuckoo much smaller with contrasting blackish tail; hepatic morph has unbarred rump and coverts like Common. Indian Cuckoo more brownish on upperparts and wings, has broad black subterminal tail band, and female lacks hepatic morph; juvenile dark brown above feathers broadly tipped creamy or tawny white, particularly noticeable on head to mantle. Pallid Cuckoo very similar in size and silhouette but lacks broad black barring on underparts. Vocalisations more reliable for identification. In Japan black gape patches, and much darker plumage, in nestling distinguish it from Common and Lesser Cuckoos, and Rufous Hawk-cuckoo (Tojo *et al.* 2002).

VOICE Male's call similar to Common Hoopoe, but lower pitched and more muffled, and four notes not two or three; nominate two or three notes, higher in pitch than *optatus*, *(hoop), hoop-hoop*; four notes *hoop, hoop-hoop-hoop*, preceded by whistle. Individual variation as well as geographical. Hoarse croaks; chuckles; harsh *gaak-gaak-gak-ak-ak-ak*. Female's call bubbling *quick-quick-quick*. Call of *optatus* mellow whistled *hoop-hoop* usually not preceded by high (or whistled) grace note, lower in pitch than nominate (Dement'ev & Gladkov 1966, Roberts 1991, Grimmett *et al.* 1998, Payne 2005, Lindholm and Lindén 2007).

DESCRIPTION Nominate. **Adult male** Top of head and nape plain ashy-grey slightly washed blue, becoming slightly paler on sides of head, chin, throat and upper breast; mantle, scapulars, back, rump and uppertail-coverts darker grey, outer edge of tail-coverts tipped white. Wings dark grey, remiges faintly washed brown, often with narrow white stripe on leading edge below carpal joint; lesser primary coverts white or white barred dark grey, rest dark grey, underwing grey with 2–8 broad white spots on each primary; underwing-coverts buffish or creamy-white on leading edges and on base of secondaries forming two broad diagnostic white bands, innermost coverts narrowly barred dark brown and white; lower breast, flanks and belly white to pale buff gradually becoming more buff on vent and undertail-coverts with broad, regular black barring except on shorter undertail-coverts which have little or no barring, in some individuals buffish suffusion extends onto hind flanks and middle of belly. Graduated tail blackish-grey with variable small white spots down centre of T1; T2–5 have white on tips of outer webs, more visible from paler undertail. **Adult female, grey morph** Like male, slightly smaller but sizes overlap; variable rufous wash on breast which on some birds extends round sides of neck to hindneck, such individuals may also have rufous-brown spots on outer webs of remiges, tertials, longest scapulars and greater wing coverts. **Adult female, hepatic morph** Rufous-brown with dark barring on top of head, neck, upperparts and wings; chin, throat and upper breast off-white narrowly barred dark brown, rest of underparts white broadly barred dark brown, ground colour gradually becoming buffish on undertail-coverts; underwing as grey morph but greater primary-coverts finely barred pale rufous-brown. Tail rufous-brown with V-shaped dark bars, subterminal bar broadest, all rectrices tipped white and finely spotted white on shafts. **Juvenile, grey morph** Ground colour grey-brown with white edges to feathers on head, upperparts, wing-coverts and tertials, and fine rufous-brown spots on greater secondary-coverts, few white feathers on forehead and nape in fresh plumage but soon lost. Upper breast white with black-brown barring more narrow and less even than adult and often broken in midline of breast and abdomen. Tail dark brown notched black-brown and buff along edges. **Juvenile female, hepatic morph** Like adult female hepatic but duller and in fresh plumage has fine buff-white feather fringes, particularly to dark brown forehead and crown. Remiges fringed whitish. **Nestling** Naked initially. Light- and dark-coloured chicks occur in nominate; apparently only dark in *optatus* (Becking 1981, Higuchi & Payne 1986). Blackish brown above and below with narrow white barring; crown feathers erectile; some greater wing-coverts edged different shades of rufous; tail blackish-brown with fine barring and white tips; in dark coloured nestlings white barring much less conspicuous. Vermilion to orange mouth; gape yellow with four black triangular patches (latter appearing before 3rd day) (Tojo *et al.* 2002). **Bare parts** Iris yellow, rarely brownish orange (male); female similar or red to red-brown or rarely wholly brown, eye-ring yellow; juvenile iris creamy grey to black-brown. Bill blackish with orange-yellow to greenish-yellow base, palate orange, gape yellowish; juvenile like adult but duller base and gape. Feet yellow to orange, claws browner.

BIOMETRICS Nominate. Wing male (n = 35) 174–194mm, mean 183.4 ± 5.8, female (n = 15) 163–190mm, mean 175.4 ± 8.2. Bill male 17.9–22.1mm, mean 20.5 ± 1.1, female 18.5–23.0mm, mean 20.0 ± 1.4. Tail male 113–152mm, mean 136.4 ± 11.7, female 113–152mm, mean 136.6 ± 10.6. Tarsus male 14.2–20.2mm, mean 17.2 ± 2.3, female 14.2–20.4mm, mean 17.2 (Payne 2005). Mass *optatus* May–Jul male (n = 23) 92–139g, mean 116.2, female (n = 14) 74–156g, mean 93.8; Feb–May male (n = 5) 73–114g, mean 91.8, female (n = 14) 78–131g, mean 97.7 (Higgins 1999); Philippines Apr unsexed 79g (Gomez *et al.* 2009). Bill longer and upper mandible more hooked in adult male suggesting continuous growth until adulthood (Becking 1975).

MOULT Late Aug–early Apr, often with long suspension. Nominate wintering Indonesia finish moulting 1–2 months earlier than *optatus*. *Optatus* moulting remiges and rectrices Bali, 5 Mar. Sequence complex: P9 or P7 or both simultaneously, thereafter P5, P8, P6, P10, or P10 before P6; at same time moult starts at inner primaries with P4 or P1 or both simultaneously, thereafter P2 and P3. Great variance in sequence of tail moulting (Meise 1941, Stresemann & Stresemann 1961, 1966, Kuroda 1965, Cramp 1985).

GEOGRAPHICAL VARIATION Two subspecies.
C. s. saturatus Blyth, 1843. Southern Himalayas in north Pakistan, Kashmir, Nepal, Bhutan, Arunachal Pradesh, Assam, north and east Burma, Thailand to China south of 32°N, including Taiwan and Hainan; occasional Hong Kong; once South Korea. Winters north Deccan, and northern plains of India; Andamans and Nicobars; southeast Asia; Greater Sundas; Philippines; few west NG (Waigeu, Vogelkop Peninsula); possibly Australia (Ali & Ripley 1983, C.S. Roselaar in Cramp 1985, Kennedy *et al.* 2000, Robson 2008, Bird Korea 2009, Gomez *et al.* 2009). Twice Bangladesh (Harvey 1990). Described above.
C. s. optatus Gould, 1845. Breeds from European Russia east of 45°E, north to Arctic Circle in Siberia, and to Pacific coast, including Kamchatka, Kuriles and Sakhalin, south to Kazakhstan and Mongolia through Altai to north of 32°N in China (including Xinjiang, Nei Mongol, Northeast Provinces and Shaanxi south to Changjiang R.); Korea; Japan south to Honshu; passage/winter Hainan. Occasionally west to St. Petersburg; records from south Finland, including possible hybridisation with Common Cuckoo (Lindholm & Lindén 2003) not accepted (V. Rauste pers. comm.). Winters Indochina, Borneo, Java, Philippines, Wallacea, Bismarck Archipelago, Solomons, NG (including Waigeu;) Micronesia (Palau and Yap), north

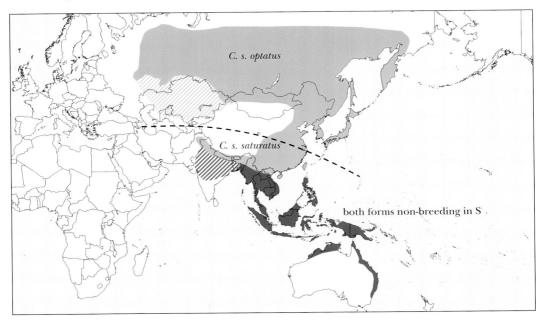

and east Australia. 17 records NZ, most beach-casts; eight Lord Howe; one each Norfolk, Snares and Christmas Is. (Higgins 1999). Three collected New Britain 1925 (Hartert 1926). Twice Israel (Shirihai 1996, Balmer & Betton 2008); once Crimea (Snow & Perrins 1998); once Zambia (Mann in prep.). Casual west Aleutian Is, Pribilof Is, St. Lawrence I., once Cape Prince of Wales, west Alaska (AOU 1983, 1998). Passage Ryu Kyu Is. and south-east Asia; possibly Taiwan (King 2005). Adult usually has slightly broader black barring on chest to belly which extends to vent; juvenile has white barring to black feathers of crown, nape, throat and breast, whereas nominate has buff-white edging. Wing male (n = 26) 214–223mm, mean 218.4 ± 1.9, female (n = 16) 213–222mm, mean 216.9 ± 2.3 (Higgins 1999); wing male (n = 66) 190–227mm, female (n = 52) 184–211mm (Payne 2005).

DISTRIBUTION Breeding (*optatus*) European Russia (perhaps occasionally west to south Finland, but see Geographical Variation) north to Arctic Circle in Siberia, to Pacific coast including Kuriles and Sakhalin, south to Kazakhstan and Mongolia through Altai to north of 32°N in China; Korea; Japan south to Honshu. Nominate Indian subcontinent in lower Himalayas from Gilgit, Kashmir, Ladakh, Himachal Pradesh, Garhwal, Kumaon, Nepal, Sikkim, Bhutan to Arunachal Pradesh and Assam Hills south of Brahmaputra R., end Mar–Sep. Arrives Murree Hills, Pakistan early May. North and east Burma, Thailand to east and south China, including Hong Kong, Hainan and Taiwan; nominate once apparently holding territory Hong Kong. Two forms have broadly overlapping winter ranges Indonesia, but proportion of *optatus* to nominate varies from 17% in Sumatra to 100% in south Moluccas and Timor. Winters south India and northern plains; Andamans and Nicobars; south-east Asia; Greater Sundas; Philippines (nominate: Alabat, Calicoan, Mangsi, Mindanao, Mindoro, Negros, rare Luzon; *optatus*: Biliran, Basilan, Batan, Bohol, Calayan, Calicoan, Ivojos, Leyte, Luzon, Marinduque, Mindanao, Mindoro, Palawan, Panay, Samar, Siquijor and Tawi-Tawi). 7 Oct–13 Nov and 14 Feb–23 Mar Thai-Malay Peninsula, but uncommon and probably overflies. On Sumatra lowland Deli, Padang Highlands, Aruah Is, Simuelue and Nias, Nov, Jan, Feb (all nominate), Nias (Feb; *optatus*). About equal numbers of nominate and *optatus* on passage, Borneo; most records Sarawak, Brunei and Sabah, fewer Kalimantan, Mar, Apr, Sep–Dec. Java; Wallacea; Bismarck Archipelago; Solomon Is.; Micronesia (Palau and Yap). In Australia, mostly north and east, Nov–May, few Jun–Aug, Kimberley mostly Oct–Apr; rare SA; throughout wet season Torres Strait; Cocos Is. NG Sep–Dec (mostly *optatus*; nominate Waigeu and Vogelkop), most probably on passage. Passage Ryu Kyu Is. and possibly Taiwan. Much vagrancy (Büttikofer 1896, Junge 1937a, b, Crawford 1972, Storr 1980, Ali & Ripley 1983, Blakers *et al.* 1984, C.S. Roselaar in Cramp 1985, Inskipp & Inskipp 1985, Smythies 1986, Cheng 1987, van Marle & Voous 1988, Harvey 1990, Johnstone & Burbridge 1991, Roberts 1991, Read 1995, Higgins 1999, Wells 1999, Kennedy *et al.* 2000, Robson 2000, 2008, King 2005, Payne 2005, Mann 2008, Gomez *et al.* 2009, R. Clarke pers. comm.).

HABITAT Breeds in mixed coniferous-deciduous, larch taiga, and broadleaved forests, thickets, woodlands, subtropical woodlands near streams, northern and eastern Asia; lower than Common Cuckoo and rarely above 1,000m, Altai. Wooded hill country and orchards, primarily montane (breeding *c.* 1,500–3,300m; birch above tree-line Kashmir), wintering in lowlands, India. 1,800+m Pakistan, but not alpine region occupied by Common Cuckoo, or subtropical pine where Indian and Grey-bellied Cuckoos occur. Conifers, 1,500–3,300m, Nepal. Broadleaf and mixed forests, 1,000–2,800m, Bhutan. Mixed pine and oak, Chin Hills, Burma. 1,300–2,700m (beech, 1,400–1,600m, Guanxi; to 4,500m, Sichuan), China. On Java all winter visitors except one (n = 87) collected in tea plantations mixed with *Albizia falcata* trees at lower elevation. Lower montane forest at 1,200–2,000m, Sumbawa. Primary and secondary forest, forest edge or partly cleared areas, savanna, gardens and teak plantations, to 1,600m, NG. Monsoon rainforest, wine-thickets, woodlands, open *Casuarina, Acacia, Eucalyptus* or wet sclerophyll forest, occasionally swamps or mangroves, often gardens and plantations, Australia. All levels in forest canopy and secondary growth up to higher elevations, Philippines. Mainly low elevations Malay Peninsula, Sumatra and Borneo (Baker 1927, Dement'ev & Gladkov 1966, Becking 1981, Ali & Ripley 1983, Coates 1985, Cramp 1985, Inskipp & Inskipp 1985, Beehler *et al.* 1986, van Marle & Voous 1988, Price & Jamdar 1990, Roberts 1991, Vuilleumier 1993, Tymstra *et al.* 1997, Robson *et al.* 1998, Higgins 1999, Jeyarajasingam & Pearson 1999, Kennedy *et al.* 2000, MacKinnon & Phillipps 2000, Trainor 2002, Mann 2008).

BEHAVIOUR Very similar to Common Cuckoo, e.g. courtship. Shy when breeding, especially female more silent. Outside breeding season less shy and approachable to within few metres, when less conspicuous back turned to intruders. Feeding birds keep distance of *c.*20m. Feeds in foliage of trees, scrub, more rarely on forest floor, in grassy areas or lawns. Forages alone or in small parties, often flapping wings among foliage to maintain balance while hunting prey; forms larger groups when food abundant. Perches sideways on large branches to catch caterpillars under loose bark. Where major food supply is found stays long periods, not leaving before food exhausted. Teak plantation (100ha) in Brown River area, NG, heavily infested by Teak Moth *Hyblaea puera* Nov–Dec 1976, attracted at least 300 cuckoos. Feeds on ground and in sally-flights. Gut contents of caterpillar removed by biting it from end to end before swallowing. Preening lasts up to 30min, with stretching of legs and wings, or wings stretched over head. Male calls most frequently at dawn and dusk from high vantage-point in tree-top, or in flight with wide-spread tail. Posture of advertising-calling male different from Common Cuckoo: bill closed, head lowered with throat 'ballooning out' at each vocalisation, back hunched, tail depressed simultaneously with call and lowered wings, sometimes interrupted by slow flight to another branch, often calling in flight. Male attempting to chase another male from territory perched moving tail slowly up and down before gliding flight towards intruder, tail fanned, attempting to drive him away. Male mobbed by warbler raised wing, fanned tail and gaped widely. When mobbed in Australia moves wings rapidly up and down in threatening position. Selects hosts with well-concealed nests. Feeds on branches and in foliage; flies from perch to catch insects on branches and trunks, or on ground. Fight between male Oriental and Common Cuckoo reported near Moscow; presumably normally separated by habitat and host choice. Migrates in loose groups of ten or more, often with Common Cuckoo, Kazakhstan (Stevens 1926, Baker 1927, Zillman 1965, Bravery 1966, Dement'ev & Gladkov 1966, Bell 1979, Becking 1981, Coates 1985,

Cramp 1985, Zykova & Ivanov 1967, Dolgushin *et al.* 1970, Panov 1973, last three in Cramp 1985, Beehler *et al.* 1986). Possibly hybridised with Common Cuckoo, south Finland (Lindholm & Lindén 2003).

BREEDING Parasitism Nominate: *Seicercus* and *Phylloscopus* warblers including Western Crowned and Blyth's Leaf Warblers. *Optatus*: in Russia Common Chiffchaff, Eastern Crowned Warbler, Pale-legged Leaf Warbler, other *Phylloscopus*, Lanceolated Warbler, Tree Pipit, Black-throated Accentor; in Japan, Japanese Bush Warbler (Hokkaido), Eastern Crowned Warbler, Asian Stubtail and Narcissus Flycatcher (Honshu) (Baker 1906a, 1927, 1934, Bates & Lowther 1952, Irisov 1967, Ali & Ripley 1969, Becking 1981, Cramp 1985, Knystautas and Sibnev 1987, Artyurkhin 1997, Payne 1997, Higuchi 1998). **Season** Nominate: India mainly May–Jun; Nepal Mar–Aug; Fujian, May; singing Chin Hills, Burma, Apr; apparently holding territory Hong Kong, 29 Apr–7 May or longer. *Optatus*: Jun–Jul, central Russia; early May–late Jun, Japan; probably late Apr–Jun, South Korea (La Touche 1931, Baker 1934, Dement'ev & Gladkov 1966, Cramp 1985, Inskipp & Inskipp 1985, Robson *et al.* 1998, Yoshino 1999, Robson 2008, Birds Korea 2009). **Eggs** Slightly glossy, long, elliptical, rather small for size of cuckoo, reflecting its small warbler hosts, and often formerly confused with sympatric Lesser Cuckoo. Nominate: pale blue or white stippled finely black and brown or reddish; 21.1x15.6mm; 20.0–25.4x12.0–16.2mm, (n = 45) mean 21.2x14.0. *Optatus*: two *gentes* in Japan – reddish-brown Hokkaido, as Japanese Bush Warbler; buff (in nest of Eastern Crowned Warbler), white or bluish-white marked brown elsewhere in Japan; 19.1–22.1x13.7–15.8mm (Baker 1906a, 1927, 1934, Schönwetter 1967, Makatsch 1971, Kiyosu 1978, Higuchi & Sato 1984, Higuchi 1998, Yoshino 1999, Tojo *et al.* 2002). **Incubation** No information. **Chicks** Young eject eggs or young of host (Dement'ev & Gladkov 1966, Roberts 1991). **Survival** No information.

FOOD Insects, mainly caterpillars (including Teak Moth), both hairy and hairless; also large beetles, grasshoppers, wasps, ants, crickets, cicadas, mantids, stick insects, flies; spiders; fruit; eggs removed from nests; pine shoots and needles (Stevens 1926, Baker 1927, Zillman 1965, Bravery 1966, Dement'ev & Gladkov 1966, Bell 1979, Coates 1985, Cramp 1985, Beehler *et al.*1986, Roberts 1991, Rasmussen & Anderton 2005).

STATUS AND CONSERVATION Common and widespread Himalayas (Inskipp & Inskipp 1985, Martens & Eck 1995). Fairly common in most parts of large breeding range, but more difficult to monitor in winter range due to skulking habits. Frequent but local Pakistan; fairly common Nepal; common Bhutan; locally common India (Grimmett *et al.* 1998). Uncommon to fairly common breeding visitor west and north Burma, north and central Laos, west Tonkin; scarce to uncommon winter and passage central and south Burma (common around Sinlum Kaba, Mar–Apr: Smythies 1986), Tenasserim, south Thailand, Cambodia, east Tonkin (Robson 2000). Infrequently recorded Thai-Malay Peninsula; many records from Java suggests that Peninsula is bypassed or overflown, e.g. netting on Fraser's Hill showed Oriental Cuckoo ten times rarer than Indian Cuckoo (Wells 1999). Seven records Sumatra (van Marle & Voous 1988). More than 50 definite records (both subspecies) Borneo (Mann 2008). Fairly common summer visitor, Japan and China (Brazil 1991, MacKinnon & Phillipps 2000), but not as frequent as Common in latter (Cheng 1987). In pine and spruce west Russia 0.4–0.5 males/km^2; in spruce-aspen Vyuchegda region 0.1 males/km^2; European population *c.* 5–10,000 breeding pairs (Hagemeijer & Blair 1997). Uncommon passage migrant and winter visitor Philippines (Dickinson *et al.* 1991). Common Sumbawa (Trainor 2002); locally fairly common, NG (Coates 1985). Rare Solomons (Doughty *et al.* 1999). Uncommon but possibly overlooked winter visitor, west New Britain lowlands (Bishop & Jones 2001). Much vagrancy. Not globally threatened (BirdLife International 2011).

Oriental Cuckoo *Cuculus saturatus*. **Fig. 1.** Adult female grey morph, probably *C. s. optatus*, Ashmore, Queensland, Australia, April (*Rohan Clarke*). **Fig. 2.** Adult male, *C. s. saturatus/optatus*, Kek Look Tong, Ipoh, Perak, Malaysia, December (*Khoo Siew Yoong*). **Fig. 3.** Juvenile hepatic morph, probably *C. s. optatus*, Ashmore, Queensland, Australia, November (*Rohan Clarke*). **Fig. 4.** Juvenile hepatic morph, *C. s. saturatus/optatus*, Singapore, November (*Lee Tiah Khee*). **Fig. 5.** Juvenile hepatic morph, probably *C. s. optatus*, Ashmore, Queensland, Australia, November (*Rohan Clarke*).

SUNDA CUCKOO
Cuculus lepidus   **Plate 36**

C. lepidus S. Müller, 1845 (W. Timor).

Alternative name: Sunda Lesser Cuckoo.

TAXONOMY Previously regarded as subspecies of *saturatus* (e.g. van Marle & Voous 1988, Wells 1999), but King (2005) and Payne (2005) gave it species status due to morphological and vocalisation differences; earlier as subspecies of *poliocephalus* (Peters 1940). Forms superspecies with *saturatus*. Polytypic.

FIELD IDENTIFICATION 29–30cm. **Adult male** Mostly dark grey above, tail barred white. Grey chin to breast, rest of underparts white barred blackish more buff on undertail-coverts. **Adult female, grey morph** Similar to male but ochraceous on breast, and narrower barring. **Adult female, rufous morph** (Sumatra and Java only: Wells 1999) Mostly barred chestnut and blackish, broadly barred black on rump. **Juvenile** Dark grey to blackish above; tail barred white; white nape patch; below barred black and white, more buffy on undertail-coverts. **Similar species** Much smaller and shorter-winged than Oriental Cuckoo with overall darker plumage; little contrast between sides of neck and dark upperwing-coverts; dark bars on lower breast and abdomen broader, about same width as pale bands; abdomen and undertail-coverts darker, richer, more rusty buff, than buffy; call higher, of 2–3, not 3–4, notes.

VOICE *Hoop, hoop-hoop*, higher pitch than Oriental Cuckoo (Payne 2005).

DESCRIPTION Nominate. **Adult male** Head and back dark grey, wings dark ashy-grey. Chin to upper breast ash grey, lower breast to belly white to buff-white barred black (bars 2–4mm); undertail-coverts buff-white, unbarred or finely barred black. Bend of wing unbarred white, underwing-coverts whitish finely barred black; greater coverts forming grey band, broad white band across secondaries and primaries except outer three or four. Tail blackish, T1 black with small white spots near shaft basally, tipped white; T2–4 narrow white spots on shaft, inner webs notched white; T5 incompletely barred with white inner notch continuous with large white spot on shaft. **Adult female, grey morph** As male, but tinged ochre on upper breast and black barring narrower. **Adult female, rufous morph** Barred chestnut and blackish-brown above; rump and uppertail-coverts rufous broadly barred black; wings barred dark brown and rufous. Below barred pale chestnut and blackish; lower breast tinged rufous; bend of wing white. **Juvenile** Some variation. Crown and back slate-grey narrowly edged white; white patch on nape; wing-coverts slate-grey narrowly edged white; remiges sooty-black, notched rufous on outer webs and tipped white. Chin to breast black with or without fine white bars; belly barred black and white; undertail-coverts buffy-white incompletely barred black. Bend of wing unbarred white; underwing-coverts almost or completely white. Tail blackish, rectrices tipped white and with several complete or incomplete white bars; T1 spotted white on shaft. **Nestling** Hatches naked; skin dark brown; mouth lining orange, inner edge of gape bright red. **Bare parts** Iris yellow to brown, brown in juvenile eye-ring yellow. Bill black, greenish to yellow basally. Feet yellow.

BIOMETRICS Nominate. Wing male (n = 16) 141–166mm, mean 154.1 ± 6.7, female (n = 20) 137–155mm, mean 144.8 ± 5.2. Tail male 114–134mm, mean 122.2 ± 7.8, female 103–117mm, mean 113.2 ± 10.3 (Payne 2005). Bill female 22.4mm; tarsus male 16.6mm (Wells 1999).

MOULT Active wing moult peninsular Malaysia, late Jul and early Sep (Wells 1999).

GEOGRAPHICAL VARIATION Two subspecies.
C. l. lepidus S. Müller, 1845. Peninsular Malaysia; Sumatra and satellite islands; Java, Bali; Lombok, Sumbawa, Flores, Pantar, Sumba, Wetar, Timor; Seram. Described above.
C. l. insulindae Hartert, 1912. Borneo. Buff underparts darker than nominate; buff to rufous on lower belly and undertail-coverts. Tail longer, male (n = 6) 146–152mm, mean 147.2 ± 2.6, female (n = 4) 128–164mm, mean 141.3 (Payne 2005).

DISTRIBUTION Sundaland and Wallacea. Resident. Malay Peninsula on Main Range from 5° 20'N south to Genting Highlands (Wells 1999); Sumatra on Barisan and Batak ranges from Aceh to Lampung; Borneo in Sarawak,

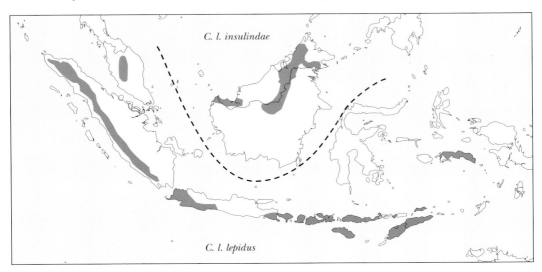

Sabah, East Kalimantan, perhaps South Kalimantan; Java; Bali; Lesser Sundas on Lombok, Sumbawa, Flores, Pantar, Sumba, Wetar and Timor; Moluccas on Seram (Vorderman 1901, White & Bruce 1986, van Marle & Voous 1988, Wells 1999, Payne 2005, Mann 2008, Trainor *et al.* 2008a).

HABITAT Montane forest 750–2,750+m Borneo. Primary and secondary forests, 700–1,200m Sumatra. 1,000+m Java. Malay Peninsula, 950–1,700m. Rainforest and casuarinas, 500–1,500m, Lesser Sundas (Rensch 1931, Smythies 1957, Becking 1975, van Marle & Voous 1988, Wells 1999, Prawiradilaga *et al.* 2003, Mann 2008, Trainor *et al.* 2008a, b).

BEHAVIOUR Resident males call from restricted areas suggesting territoriality (Wells 1999).

BREEDING Parasitism Hosts: Chestnut-crowned, Mountain Leaf and Yellow-breasted Warblers, Malay Peninsula; Mountain Leaf Warbler, and this species and Yellow-breasted Warbler together feeding fledgling, Borneo; Sunda Bush, Sunda and Mountain Leaf Warblers, Java; possibly Timor Leaf Warbler (Hoogerwerf 1949, Madoc 1956a, b, Hellebrekers & Hoogerwerf 1967, Becking 1975, 1981, Hitoshi *et al.* 1996, Smythies & Davison 1999, Wells 1999, Moyle *et al.* 2001, Robson 2007, ENGGANG-3 in Wells 2007, Schellekens 2009, Bakewell *et al.* 2011). **Season** Eggs Feb–Jun, fledglings late Feb–late Jul, Malay Peninsula; possibly Jun, Sumatra; eggs Apr–May and perhaps Jul, Borneo, fledglings May, late Jul–Aug; eggs Feb, Apr–Jul, Sep, Oct, Java; vocal peninsular Malaysia, May (Hoogerwerf 1949, Madoc 1956a, b, Hellebrekers & Hoogerwerf 1967, van Marle & Voous 1988, Hitoshi *et al.* 1996, Smythies & Davison 1999, Wells 1999, 2007, Moyle *et al.* 2001, Robson 2007). **Eggs** Creamy-white finely spotted brown; 21.2x13.7, 19x11.8mm (Hoogerwerf 1949, Madoc 1956a, b). No further information.

FOOD Insects (mainly hairy and hairless caterpillars), grasshoppers, crickets, cicadas, beetles, flies, ants; fruits (Coates 1985, Sody 1989).

STATUS AND CONSERVATION Local Sarawak, Sabah, East Kalimantan, perhaps South Kalimantan, Borneo (Mann 2008). More or less common peninsular Malaysia within restricted distribution, as yet not threatened by habitat destruction (Wells 1999). Common resident Timor-Leste (Trainor *et al.* 2008b). Not globally threatened (BirdLife International 2011).

Sunda Cuckoo *Cuculus lepidus*. **Fig. 1.** Adult female, *C. l. lepidus*, Danau Rana Mese, Flores, Lesser Sundas, Indonesia, July (*Erwin Collaerts*). **Fig. 2.** Adult male, *C. l. lepidus*, Danau Rana Mese, Flores, Lesser Sundas, Indonesia, July (*Erwin Collaerts*).

COMMON CUCKOO
Cuculus canorus   Plate 32

Cuculus canorus Linnaeus, 1758 (Sweden)

Alternative names: European/Eurasian Cuckoo; Khasi Hills Cuckoo (*bakeri*).

TAXONOMY Sometimes considered to form superspecies with *gularis*, but mtDNA evidence (Sorenson & Payne 2005) shows it to be closest to *saturatus*. Polytypic. Synonyms: *cinereus* Vieillot 1817, *borealis* Pallas 1827, *indicus* Blyth 1846, *minor* A.E. Brehm 1858, *telephonus* Heine 1863, *major*, *medius*, *alticeps*, *planiceps* last four A.E. Brehm 1866, *johanseni* Tschusi 1903, *rumenicus* Tschusi & Dombrowski 1904, *kleinschmidti* Schiebel 1910, *similis* Tschusi & Dombrowski 1912, *maximus* Neumann 1934, *sardus* Trischitta 1939, all in nominate; *fallax* Stresemann 1930 in *bakeri*; *kwenlunensis* Portenko 1931 in *subtelephonus*.

FIELD IDENTIFICATION 32–34cm. Medium-sized cuckoo. Common posture rather horizontal with drooping wings, but when calling body more upright. Fast flight with head, body and tail straight and bill pointing forward, with curiously feeble and constant wing-beats never above horizontal level (unlike Eurasian Sparrowhawk and Common Kestrel), round-shouldered and with pale area on centre of underwings. Flies mostly at low elevation between two areas, often with long glide before settling. Egg-laying female has gliding flight like sparrowhawk, and soaring also observed in courtship. Calling male partly spreads tail and lowers wings. **Adult male** Ash grey on head, upperparts, wings, throat and upper breast, rest of underparts white barred black, tail blackish tipped white. **Adult female, grey morph** As male but with some buffish wash on upper breast. **Adult female, hepatic morph** Red-brown barred black on whole upperparts and wings, barring on rump and uppertail-coverts slight or absent. **Juvenile, dark morph** Dark brown head, upperparts and wings, feathers edged white, wings spotted rufous, whole underparts white barred dark brown; tail dark brown tipped and spotted white mixed with rufous. **Juvenile, hepatic morph** Rufous-brown above fringed white and broadly barred black; rump slightly or extensively barred black with greyish-white tipping; uppertail-coverts rufous barred dark brown and tipped white; conspicuous white patch on nape, often some white feathers on forehead and crown; wings barred blackish-brown and red-brown; throat and breast barred blackish, flanks, belly and undertail-coverts buffish narrowly barred black; tail barred black and red-brown with elongated spots along shafts, and broadly tipped white. **Similar species** Adult African Cuckoo has yellow bill tipped black whereas yellow restricted to base of bill in Common, and bill is broader, outer rectrices with full white bars and black subterminal band broader; juvenile has no brown in plumage, crown entirely barred (not partially as Common), rump and uppertail-coverts barred, and white in tail broader; voice different; no hepatic morph in African. Oriental Cuckoo always has buffish undertail-coverts barred dark brown (rare in Common, and with little or no barring), and conspicuous broad white band across dark and nearly unspotted underside of secondaries; female hepatic Common has few or no dark bars on undertail-coverts which are broadly barred in Oriental, and dark bars on underparts much narrower; juvenile blackish with narrow white fringes, hepatic female morph lacks broad black bars on rump of Oriental; hepatic juvenile may have rump and uppertail-coverts barred blackish, but with fine white tipping which Oriental lacks. Indian Cuckoo has broader blackish-brown barring on underparts; juvenile blackish-brown above, most feathers tipped creamy to tawny white, particularly noticeable on crown to mantle. For nestling see Oriental Cuckoo.

VOICE Male responsible for most known calls; call far-carrying and ventriloquial making it difficult to locate, *cuck-oo* or *cu-coo*, first note louder and higher pitched and often in series of 10–20 or more with 1sec between series, but faster and first note often repeated when excited; another three-note call when excited: bubbling *gowk-gowk-gowk*. In flight *cu-coo-kuk...kuk.kuk-kuk-kuk-kuk-kuk-coo*. During mating soft *grorr-grorr-grorr*, or hissing call when chasing female. Female has quite different calls: bubbling *kwik-kwik-kwik*, reminiscent of woman's laughter; when mobbed *cak-cak-cak-cak-cak*. Prior to or just after egg-laying female gives bubbling call or soft mewing call only audible at close range. Nestling food call *seep*, louder with age and audible for about 50m, given so rapidly that it matches that of more than one host chick and so compensates for visual difference between its own gape area and that displayed by two or three nestlings and so stimulates foster parents to bring more food; later persistent *chizz-chizz-chizz* or *gi-gi-gi*, or single *chirp*. Most calling at dawn and dusk and by day from exposed post, especially in fine weather, but occasionally at night. Little variation in call within its huge range except faint variations on sonograms, and slight differences between cuckoos in Denmark and Ukraine. Juvenile imitates Meadow Pipit's call. Usually silent in winter quarters, except late Apr–late May when migrating birds heard in Somalia (Change 1940, O'Connor 1962, Wyllie 1981, Cramp 1985, Radford 1992, Thiede 1994, Ash & Miskell 1998, Kilner & Davies 1999, Lei *et al.* 2005, A.P. Møller *in litt.*).

DESCRIPTION Nominate. **Adult male** Head, nape, and neck plain dark ashy-grey, barely visible white spot on nape, upperparts and wings ashy-grey, carpal joint white barred grey, underwing with broad white bar on base of secondaries and inner primaries, broad pale grey band across greater wing-coverts and fine dark grey and white barring on rest of underwing-coverts; chin, throat and upper breast ashy-grey with sharp border to rest of underparts which are white barred black; graduated tail blackish-brown tipped and barred white except centres of plain T1–2, rest have white spots along shafts, underside barred white. **Adult female, grey morph** Like male but with pale rusty-buff wash on upper breast and sides, and often with finer black barring on underparts. **Adult female, hepatic morph** Rarer than grey morph. Bright rufous barred blackish on top of head, neck, upperparts and wings; usually little or no barring on rump (but more extensive on some), underwings medium brown with white and pale rufous barring becoming broader towards bases of feathers, white area at base of inner primaries, under wing-coverts and axillaries white with some black barring; chin to upper breast pale rufous barred black, rest of underparts white barred black except white on centre of belly and undertail-coverts, last often with faint buffish wash; tail rufous barred black with narrow white streaks along shafts and tipped white. **Juvenile, dark morph** Dark brown on head, upperparts and wings, each feather edged white or buffish, white area on nape, grey rump and uppertail-coverts, remiges spotted pale rufous on outer webs and broader pale rufous barring on inner webs of primaries; chin to upper breast off-white to pale buffish broadly barred black, rest of underparts white barred black, barring variable on undertail-coverts and rarely tinged buffish; tail dark brown variably spotted white along both sides of shaft, and diffuse pale rufous barring on inner webs, and tipped white. **Juvenile, hepatic morph** Not as rare as hepatic adult female. Rufous-brown above fringed white and broadly barred black except usually rump and uppertail-coverts, and if barring present there also greyish-white tipping to feathers; conspicuous white patch on nape, often some white feathers on forehead and crown; wings barred blackish-brown and red-brown; throat and breast barred blackish, flanks, belly and undertail-coverts buffish narrowly barred black; tail barred black and red-brown with elongated spots along shafts, and broadly tipped white. Hepatic morph increases southwards in Germany (Löhrl 1979). **Nestling** Hatches naked, flesh-coloured skin becoming blackish after three days, gape orange and mouth lining yellow to orange. Three colour morphs: grey-brown on upperparts, or brown with dark feather edges, or hepatic chestnut (link with juvenile morphs unclear). **Bare parts** Iris yellow, brownish-yellow or orange, hepatic female sometimes grey or hazel, juvenile brown, grey-brown, buffish or yellow, nestling grey, grey-brown or brown; adult eye-ring yellow to orange-yellow, paler in juvenile. Bill dark horn to blackish, base of upper mandible and ¾ of lower often yellow-green, juvenile bill brown; mouth of both adult and juvenile orange. Feet bright yellow to ochre-yellow, juvenile brown-yellow to orange-yellow. Albino recorded.

BIOMETRICS Nominate. Wing male 213–230mm, (n = 52) mean 221 ± 4.32, female 204–216mm, (n = 35) mean 210 ± 3.86. Bill male 25.5–31.2mm, mean 27.7 ± 1.51, female 25.2–28.6mm, mean 26.8 ± 1.01. Tail male 170–186mm, mean 177 ± 4.14, female 158–177mm, mean 167 ± 5.46. Mass male 114–133g (n = 121), female 106–112g (n = 76) (Cramp 1985); unsexed mean 117g ± 17 (n = 84) (Seel 1977a).

MOULT Some body feathers moulted Jun–early Jul, otherwise Sep–Apr with peak mid Oct–Mar, on wintering grounds, lasting *c.* 100 days (Seel 1984b).

GEOGRAPHICAL VARIATION Four subspecies.

C. c. canorus Linnaeus, 1758. Palearctic from Ireland to Kamchatka, China (except where *bakeri* occurs), Korea and Japan. Western populations winter in equatorial Africa, eastern populations in India and south-east Asia. Described above.

C. c. bangsi Oberholser, 1919. Iberia, Balearics, Morocco, Algeria and Tunisia. Probably winters Africa south of equator. Smaller; no hepatic morph female but often with rufous breast. Wing male (n = 7) 199–215mm, mean 206.7 ± 6.2, female (n = 5) 194–207mm, mean 200.6 ± 5.4 (Payne 2005).

C. c. subtelephonus Zarudny, 1914. Transcaspia to west Xinjiang and central Nei Mongol, China, and south to Iran and Afghanistan except north-east, and possibly Baluchistan (Ticehurst 1927); passage Iraq, Jordan, Israel, winters India? and Africa south of Sahara except south-west. Paler above; pure white underparts with narrower black bars (chiefly <1.5mm) than *canorus* (2.0–2.5mm), *bangsi* (1.2–2.6mm) and *bakeri* (1.9–2.3mm). Underwing-coverts and axillaries often less barred. Wing male (n = 30) 201.3–249mm, mean 221.9 ± 15.2, female (n = 18) 180–223mm, mean 200.5 ± 13.1 (Payne 2005).

C. c. bakeri Hartert, 1912. Northern India, Nepal, Bhutan, north Thailand, Burma, north Laos, northern Vietnam, south-east Tibet (and Qinghai), south and east China (except north-east). Winters India and south-east Asia. Darker above than *canorus*, and blackish bars below closer together. No hepatic morph. Wing male (n = 16) 198–232mm, mean 214.3 ± 9.9, female (n = 9) 195–215mm, mean 205.1 ± 6.8 (Payne 2005).

DISTRIBUTION Summer breeding visitor up to 66–70°N in Europe, Asian Russia including Siberia, China, Kurile Is, Korea and Japan, and south through whole of Europe to north Africa, Asia Minor, India and northern south-east Asia. Breeds throughout Europe, mostly in low densities (Norway one of lowest densities in Europe). Occasionally breeds Syria, Lebanon and Cyprus. In Pakistan mostly in north-west in summer. Common summer visitor Mar–Sep Nepal, India and Bhutan. Resident or local migrant India throughout peninsula to Kerala, but most in Himalayas

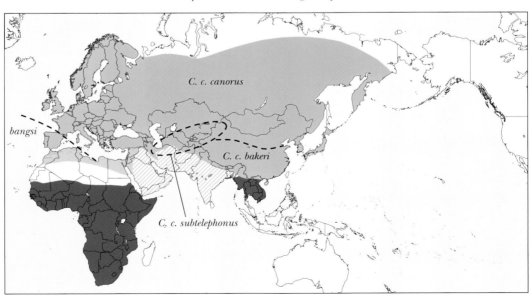

and north-east India. Widespread Kashmir, Apr–Sep. Both nominate and *bakeri* occur Kathmandu Valley, Nepal, Apr–Oct. Breeding migrant Manipur. 4–5 winter records Sri Lanka. Scarce passage migrant Bangladesh. Winter or passage visitor north-west Thailand; perhaps rare resident, or visitor north-west. Widely distributed Burma and probably breeds in north. Most parts of China, Xinjiang to central Nei Mongol (*subtelephonus*), Altai Mts. of north Xinjiang, north-east China, Sha'anxi, Hebei and north-east China (*canorus*), Qinghai and Sichuan to south Tibet and Yunnan (*bakeri*). Summer visitor Japan in Hokkaido and north alpine Honshu, rare elsewhere (Bates & Lowther 1952, Henry 1955, Deignan 1963, Dement'ev & Gladkov 1966, Ali & Ripley 1983, Cramp 1985, Inskipp & Inskipp 1985, Smythies 1986, Cheng 1987, Hollom *et al.* 1988, Harvey 1990, Brazil 1991, Lekagul & Round 1991, Roberts 1991, Fiebig 1993, Shirihai 1996, Grimmett *et al.* 1998, MacKinnon & Phillipps 2000, Choudhury 2009, A.P. Møller *in litt.*).

Usually appears on breeding grounds within 3–4 days of particular date every year (Wyllie 1981), but now cuckoos have advanced their arrival date in response to climate change, similar to rate for other long distance migrants (Saino *et al.* 2009). Arrives in Europe end Mar–end May, advancing northward *c.* 55km/day (De Smet 1970). Arrives Britain 17–23 Apr in south, 24–30 Apr in north (Wyllie 1981); males may depart as early as early Jun (Evans 2011), reaching Atlas Mts. by 13 Jul (I. Evans, *in litt.*) In Saxony, Germany, average arrival 26–27 Apr, departure 13 Sep (Bögershausen 1976). In Siberia its northern border coincident with timberline, *c.* 68–69°N, where arrives late May–early Jul, and from first half of Apr south Russia (Dement'ev & Gladkov 1966). In north and central Israel migrants widespread, most end Feb–mid May, and mainly mid Jul–mid Sep; breeding birds leave after egg-laying end Apr–May (Shirihai 1996).

Continental European birds move south-east, south or south-west in autumn, while British birds move south-east to Italy where they fatten before crossing Mediterranean and Sahara in single flight of *c.* 3,500km. In Britain most adults leave mid–end Jul while young birds migrate one month later, but also rarely still seen Oct–Nov. Most pass over Sahara, but also along west and east coasts of Africa and Nile. Scarce migrant Sinai. Most of Palearctic population thought to winter in Africa, but some more easterly populations in India and south-east Asia as well as Africa. Many passage records from Indian Ocean, UAE, Oman and south-west Arabia (including *subtelephonus* in first three) (Moreau 1972, Seel 1977a, 1984a, Gallagher & Woodcock 1980, Wyllie 1981, Hollom *et al.* 1988, Irwin 1988, Richardson & Chapman 1988, Shirihai 1996, Zimmerman *et al.* 1996, Wernham *et al.* 2002).

Migrates through Ethiopia mid Aug–end Sep, Chad 30 Aug–6 Sep, and Uganda early birds 29 Jul; most arrive equatorial Africa end Aug–mid Nov. Ringing recoveries in Africa few: UK bird recovered Tunisia, one Cameroon, Jan; Dutch bird Togo, Oct; Estonian bird Egypt, Sep; Swiss bird Mauritania, autumn. 1,376 museum skins reveal arrival in equatorial Africa Sep–Oct, and by Nov moves on to southern Africa, staying until Mar. Winters south of equator in Africa, except *bakeri* (and perhaps *bangsi*); *canorus* mostly in south-east Africa, *bangsi* in DRC and *subtelephonus* from east DRC southwards. Regular Maldives winter. Scarce passage migrant Seychelles and Aldabra (Wyllie 1981, Seel 1984a, Irwin 1988, Goodman *et al.* 1989, Ash & Shafeeg 1994, Hagemeijer & Blair 1997, Wernham *et al.* 2002, Rasmussen & Anderton 2005, Isenmann *et al.* 2010). Vagrants Prince Edward Is. (Irwin 1989). Straggler Palau Is, Micronesia (Baker 1951, Pratt *et al.* 1987). Few vagrants Andamans (Rasmussen & Anderton 2005). Once Borneo (Mann 2008). Vagrant Cape Verde Is. (Borrow & Demey 2001). Occasional passage migrant Canary Is. and Madeira (Bannerman 1963, Bannerman & Bannerman 1965). In North America casually to Commander Islands, Attu, Buldir, Kiska, Amchitka and Adak islands (west and central Aleutians), Pribilof Is, St. Lawrence I., St. Paul and Campbell Is, in west Alaska on Tutakoke River and Seward Peninsula; accidental Martha's Vineyard, Massachusetts, Barbados, Greenland (Jun 1968), Iceland (45 to 2009) and Faroe Is (21 to 2009) (AOU 1983, 1998, Boertmann 1994, Pétursson 2009, Olofson & Sørensen 2010). Twice Crozet Is, 24 Oct 1979 and 24 Nov 1981 (Stahl *et al.* 1984). Records from Sumatra and Lingga Archipelago unconfirmed (van Marle & Voous 1988). Records from Java, Taiwan and Philippines are *saturatus* (Mees 1979, Dickinson *et al.* 1991, Kennedy *et al.* 2000).

HABITAT Extremely diverse. In west Palearctic breeds in all climate zones except arctic tundra and desert. In Britain and Ireland favours most types of farmland but rare in intensive arable farming; moorland, parkland, light woodland and suburban fringes. Most 200–400m central Germany, Switzerland to 2,200m. Highest densities in Spain in oak woods. Farmland Russia, also dense broadleaved forests and taiga fringes, rose-willow thickets, plantations, orchards and forest edges. Breeds in bushy and rocky areas with low trees in semi-desert at 300–1,600m, Israel. Breeds only in mountains in Lebanon due to hunting pressure. Prefers altitudes above 1,800m with juniper forest-clad slopes in Baluchistan, Gilgit and Hazara Districts, Pakistan, to 3,650m in alpine meadow and dwarf juniper scrub. Lowlands to 3,800m Nepal. In Indian subcontinent generally, forest, secondary growth, well-wooded country, montane dwarf scrub and alpine meadows to 5,250m. Only *Cuculus* species found above timberline and mostly higher than Oriental. In Africa in mopane, miombo and *Acacia* woodlands, but avoids forests and arid and grassy areas. Forest edge, woodlands, grasslands and cultivated fields Japan; 280–1,800m Nagano Prefecture, central Japan (Makatsch 1955, Dement'ev & Gladkov 1966, Benson 1970, Rowan 1983, Cramp 1985, Inskipp & Inskipp 1985, Roberts 1991, Takasu *et al.* 1993, Díaz *et al.* 1996, Shirihai 1996, Grimmett *et al.* 1998, Richter 1999, CCJB 2000, Wernham *et al.* 2002, Rasmussen & Anderton 2005).

BEHAVIOUR In spring adult males often arrive before females and large gatherings reported although migrates solitarily, probably associated with locally abundant food. Ringing recoveries demonstrate that old birds generally return to last year's breeding area (males at same site for up to ten years) and young birds to birthplace (12 of 16 ringed as nestlings recovered within 40km of birthplace in subsequent years, mean dispersal 27km). Silent on feeding grounds, often some kilometres from egg-laying territories, where often in small congregations. When foraging on ground moves in hops and rather clumsily among branches. Drinking once observed (Harvie-Brown 1908, Seel 1977b, Cox *et al.* 1980, Wyllie 1981, Cramp 1985, Moffett 1985, Lindholm 1999).

Long thought to be promiscuous without pair-bond, but DNA fingerprinting reveals that females often monogamous for whole season (Jones *et al.* 1997). Copulation in morning as female perched on electric cable gave loud 'bubbling' call, and male flew directly towards her calling *cuc-koo* several times and made 1.5 circles *c*.5m diameter around her before landing directly on her back; after 15–20sec both birds flew off together; same behaviour thrice when mating, which took place in 3–4sec, and always in afternoon. Male twice used plant material in display, but courtship feeding never observed (Harrison 1974, King 1974, Wyllie 1981, Boddy & Boddy 1985).

Each female generally shows high level of host-specificity (but errors common) e.g. Reed Warbler *gens*, and molecular studies revealed existence of cuckoo *gentes* is restricted to female lineages. Moreover, females mate indiscriminately with males descended from females of any *gentes*, maintaining common gene pool (Chance 1922, Wyllie 1981, Moksnes & Røskaft 1995a, Davies & Brooke1989, Gibbs *et al*. 2000, Skjelseth *et al*. 2004). Female looking for new nest hides for hours in tree, lying along branch in horizontal position as nightjar. Nests nearest vantage points are therefore more often parasitised. Female cuckoo selects nests before host clutch is complete and lays at intervals of around 48hr, which gives more time to find suitable nest, and internal incubation of cuckoo embryos resulting in advanced hatching by 31hr gives advantage over host chicks. Eggs laid in afternoon or early evening when host away foraging, important because sight of cuckoo on host nest stimulates greater rejection of cuckoo eggs by both Reed Warblers and Meadow Pipits. Male does not assist female in egg-laying as in some other cuckoos, however two incidents where male distracted hosts, and after female had laid egg followed her away from mobbing hosts. Female lays directly into nests with narrow entrances (e.g. Winter Wren), and not by transferring egg first laid on ground in bill, as once thought. Eggs laid with great speed through extensible cloaca. On leaving she sometimes removes and eats one of host's eggs. Cuckoo trying to lay egg into Pied Bushchat nest in hole in ground was attacked by both hosts and laid her egg *c*. 7cm outside nest entrance with wings and tail spread, tail depressed onto ground, and then flew away without attempting to place egg in nest. Another laid egg between boards where opening was so narrow that female presumably placed egg with bill, and chick later would be unable to leave nest (Finckenstein 1936, Livesey 1938, Fr. Moeberts in Makatsch 1955, Lack 1968, Wyllie 1981, Davies & Brooke 1989, Davies 2000, Moksnes *et al*. 2000, Moskát & Honza 2000, Schulze-Hagen *et al*. 2009, Birkhead *et al*. 2010).

Egg-rejection (non-acceptance of 'foreign' egg) most effective and common defence against parasitism, but not always consistent; individual can be rejector one year and acceptor next (Soler *et al*. 2000a). 78% of Sedge Warblers, 72% of Marsh Warblers, 30% Great Reed Warblers and 29% Reed Warblers rejected Cuckoo eggs, Czech Republic (Kleven *et al*. 2004), compared to very rare rejection by Meadow Pipits in Scandinavia (Moksnes *et al*. 1993). In Hungary high similarity in rejection behaviour among three populations of Great Reed Warblers although in these populations rate of parasitism was 41–68%, 11% and >1%, suggesting high gene flow between populations (Moskát *et al*. 2008). Temperature and rainfall influence coloration of Reed Warblers' eggs: more brilliant, bluer and greener in springs with higher rainfall and lower temperature (Avilés *et al*. 2007). Host may also desert nest and build new one, or bury cuckoo egg in nest lining. Nest desertion commoner in smaller-billed species (Wyllie 1981, Davies & Brooke 1989). Cuckoos' egg shells more compact than other eggs of similar size to prevent host puncture and reduce chance of breakage if laid from a height. Egg mimicry is another important cuckoo adaptation against rejection behaviour. Natural selection has resulted in eggs that more and more closely mimic those of host species over time. As defence against this some host species (none of these hole-nesters) develop reduced intraclutch variation and more uniform spotting or 'signatures' for better recognition of own eggs, making it easier to identify cuckoos' eggs (Davies & Brooke 1989, Soler & Møller 1996). Reed Warblers learn to build their nests away from trees where female cuckoo can hide. Cuckoo spends only few seconds laying egg in host nest to avoid discovery (Davies 2000, Payne 2005). When parasitism rate becomes too high rejection rate becomes so great that this cuckoo *gens* dies out, e.g. recently in one Blackcap population in Czech Republic (Edvardsen *et al*. 2001) and other *gentes* in Israel and Japan (see Breeding). Common host, Dunnock, shows no egg discrimination, and despite its eggs being plain turquoise-blue, accepts model eggs of any colour, even spotted, and cuckoos specialising in Dunnocks lay normal spotted eggs in Dunnocks' nests. Common Redstarts' eggs are similarly blue and in central Europe and Finland are parasitised by cuckoo *gens* laying blue eggs (Davies 2000). Dunnocks may have been parasitised for >600 years (Davies & Brooke 1991), but have not developed counter adaptations, perhaps because parasitism rate is low (2%), and it may take thousands of years under this condition for rejecter genes to spread fully (Glue & Morgan 1972, Kelly 1987).

If egg-laying of host is too advanced female cuckoo may remove all eggs or nestlings, thus forcing host to start new clutch. Eggs normally laid in hosts with cup nest where female first takes host egg in bill and then lays egg directly in nest; in wren's domed nest she spreads her wings with tail against opening and ejects her egg directly in nest cavity. Fledglings burst out and escape from domed nests. Generally cuckoo egg hatches before host's, and in 8–36 hours young cuckoo evicts hosts' eggs or chicks by manoeuvring either onto slight hollow between scapulars, and moving up side of nest with wings raised upward and backward to prevent egg or young rolling off. If fosters present they watch without intervening, and make no attempt to retrieve young hanging outside nest. Ejection behaviour lasts 3–4 days. Huge orange gape and extreme begging intensity are super-stimulus to many passerines as fledged cuckoos often fed by species other than fosters, even by adult cuckoos. Begging call of cuckoo chick is not genetically fixed but develops through experience. Chick's exaggerated begging call mimics whole brood and results in increased investment from host parents, but must increase nest predation and reduce host survival (Milburn 1915, Burkill 1928, Heinroth & Heinroth 1966, Wyllie 1981, Redondo 1993, Kim 1996, Sealy & Lorenzana 1997, Madden & Davies 2006, Øistein *et al*. 2009). Polygynous Great Reed Warblers are more often successfully parasitized than nests of monogamous pairs (Trinka & Prokop 2011).

Coevolutionary 'arms race' has developed sophisticated defence mechanisms against cuckoo eggs, including ability to discriminate against sometimes accurately mimetic parasitic eggs, but enigmatically little chick discrimination although so different from host's own. Reed Warblers can discriminate against cuckoo chicks, not involving learning and/or recognition, but when duration of feeding to parasite are outside normal range in unparasitised nests (11 days versus 18 for cuckoo chicks) about 15% of fosters may stop feeding and desert nest, or may feed parasitic chick with low-quality food items. However, nearly all host species of Common Cuckoo have high rejection rate of parasitic eggs, and this, together with low parasitism rate, may explain rarity of nestling discrimination in these hosts (Dawkins & Krebs 1979, Brooke & Davies 1989, Redondo 1993, Grim *et al.* 2003, 2009, Grim 2006, Soler 2008). Lotem (1993) hypothesises that lack of nestling discrimination results from host, if parasitized in first year, imprints on cuckoo chicks, because own have been ejected and in future would not accept own young.

BREEDING Parasitism Eggs have been found in nests of about 280 species (at least 135 species in Europe alone), of which less than 10% are regular hosts; in each area only one to three are sometimes victimised. In *c.* 12,000 clutches (museum collections) containing cuckoo eggs most common hosts were: Eurasian Reed Warbler (2,655 with cuckoo eggs), Garden Warbler (1,266), White Wagtail (832), Meadow Pipit (708), Red-backed Shrike (708), Dunnock (547), Winter Wren (516), Greater Whitethroat (460), European Robin (459), Great Reed Warbler (415), Sedge Warbler (315), Common Redstart (299), Tree Pipit (213), Yellow Wagtail (197), Yellowhammer (195), Blackcap (180) and Marsh Warbler (141). Others record Meadow Pipit as most common European host. Rapid decline in parasitism rate in Red-backed Shrike in Czech Republic and Slovakia although it seems to have a stable population in former. Some variation in parasitism rates on hosts in different parts of Europe, e.g. Dunnock common host west Europe but rare east Europe; Common Redstart most common host Finland but rare elsewhere in Europe except Italy. In Britain, Netherlands, Belgium and north France main hosts Eurasian Reed Warbler, Meadow Pipit and Dunnock; Ireland Meadow Pipit; central Europe Garden Warbler; Hungary and central Asia Great Reed Warbler. Finland and Russia have Brambling and Chaffinch *gens*, both species egg-rejectors. Few successful rearings by European Greenfinch and Chaffinch hosts. Other hosts/areas: Bulgaria, Eurasian Skylark, Woodlark, European Robin, Common and Black Redstarts, Northern Wheatear, Common Nightingale, Common Stonechat, Eurasian Blackbird, Icterine, Eastern Olivaceous, Sedge, Eurasian Reed, Marsh and Barred Warblers, Greater and Lesser Whitethroats, Spotted Flycatcher, Dunnock, White and Grey Wagtails, Tree Pipit, European Greenfinch, European Goldfinch, Yellowhammer, Corn and Black-headed Buntings; Greece, Black-eared Wheatear, Common Stonechat, Orphean Warbler; west Russia, Common Redstart; Ukraine, White Wagtail, Great Reed Warbler, Greater Whitethroat, Red-backed Shrike, Spotted Flycatcher; north Urals/western Siberia, Yellow-breasted Bunting, Little Bunting; central former USSR, Isabelline Shrike, White Wagtail, Tree Pipit, European Robin, Common Stonechat, Great Reed Warbler; Turkmenistan, Streaked Scrub Warbler; south-west Siberia, Common Redstart, Siberian Rubythroat, Common Chiffchaff, Blyth's Reed Warbler, Grey Wagtail; Transbaikalia, Eurasian Skylark, Eye-browed Thrush, Chestnut Bunting, Chestnut-eared Bunting; maritime Amur, Brown Shrike, Thick-billed Warbler; South Korea, Vinous-throated Parrotbill; north Africa, Sardinian Warbler, Dartford Warbler, Woodchat Shrike; Morocco, Tristram's Warbler, Black-eared Wheatear; Algeria, Moussier's Redstart; Pakistan, Grey Bushchat and Paddyfield Pipit; India, Chestnut-crowned Laughing-thrush, Upland Pipit, Paddyfield Pipit, Crested Bunting; Burma, Paddyfield Pipit, Pied Bushchat, Striated Grassbird, Crested Bunting; China, Zitting Cisticola, Meadow Bunting, Vinous-throated Parrotbill. In Israel most common hosts Streaked Scrub Warbler and Long-billed Pipit, latter only since 1983 when Streaked Scrub Warbler became rarer. In Jordan Streaked Scrub Warbler. In Japan main species Bull-headed and Brown Shrikes, Oriental Reed Warbler, Chestnut-eared, Black-faced and Meadow Buntings, last establishing strong egg discriminating ability in past 60 years, and Azure-winged Magpie has instead been chosen as new host and cuckoo has expanded over almost host's entire breeding range, from parasitism rate of 0 to 30%; in Europe may exceed 50% (and Great Reed Warbler over 75% in some populations). However, such high level is normally only found when parasitism has recently started in population and will normally drop after few years. Some claimed hosts doubtful, including Little Grebe, Common Pheasant, 'snipes', Common Greenshank, Common Hoopoe, Stock Dove, Woodpigeon, European Turtle Dove, 'woodpeckers', Eurasian Jay, Common Magpie and Eurasian Golden Oriole. Two nestlings raised in Isabelline Shrikes' nest claimed: (Cock & Marshall 1873, Osmaston 1916, MacKenzie 1918, Chance 1922, La Touche 1931, Baker 1942, Dement'ev and Gladkov 1966, Bauer *et al.* 1969b, Glue & Morgan 1972, Seel & Davis 1981, Wyllie 1981, Jarvinen 1984, Khabaeva 1985, Sokolov & Sokolov 1986, Lopatin 1987, Balatskiy 1988, 1989, 1990, 2008, 2011, Davies & Brooke 1989, Moksnes *et al.* 1993, Braa *et al.* 1992, Schulze-Hagen 1992, Takasu *et al.* 1993, Moksnes & Røskaft 1995b, Sealy *et al.* 1996, Shirihai 1996, Takasu 1998, Thévenot *et al.* 2003, Nankinov 2007, Balmer & Betton 2008, Adamik *et al.* 2009, P.E. Lowther *in litt.*, A. P. Møller *in litt.*). A rapid change in host use due to climate change has been observed (Møller *et al.* 2010).
Season North-west Europe Apr–Jul, peak May–Jun; Algeria Apr–May; nestlings Jordan, 29 Apr; India Mar–Sep varying locally; Burma May–Jul; Japan late May–Jul with most Jun. Former USSR end Apr–mid Jul. Latest egg 22 Jul in Eurasian Reed Warbler nest, Netherlands, but state of incubation unknown. Generally in Britain begin laying three weeks after arrival, latest 14 Jul. Overall egg-laying period in Europe 12 weeks but individually does not exceed 6–7 weeks, mean 27.7 days. Eggs Siberia, Jun–Jul. Nestling China, May, eggs Jun. Eggs early May–early Jun, fledgling 15 Jun, Morocco (La Touche 1931, Paulussen 1957, Dement'ev & Gladkov 1966, Wyllie 1981, Cramp 1985, Irwin 1988, Brazil 1991, Grimmett *et al.* 1998, Thévenot *et al.* 2003, Balatskiy 2008, Balmer & Betton 2008). Breeds in Israel and Jordan when passing through on spring migration (D. Bercovic and M.C. Jennings pers. comm.). In China 100% of non-mimetic eggs were rejected by Great and Green-backed Tits (Liang *et al.* 2010). **Eggs** Much smaller (2.4% of body mass) compared to eggs of non-parasitic cuckoos. At least 15 *gentes* in Europe (including blue, whitish, pink, spotted brown, red-brown and plain) and every female parasitises one or group of species with rather similar eggs or nest sites. Degree of similarity differs (n = *c.* 12,000), 5.1% perfect mimics, 24.7% good mimics, others of more or less different ground colour or size and form of spots, but normally cuckoo eggs larger (except for those in Great Reed Warbler nests, and presumably Azure-winged Magpies), averaging 2mm greater in length and width, more rounded than hosts'. This is advantageous with regard to reserves for embryo and hosts prefer larger eggs to their own. Each female always lays identical eggs, probably evolved because female has behavioural response to

parasitise species which raised it. Most common egg morph in Europe like Garden Warbler egg constituting 31%, which is intermediate *Sylvia* egg type; 20.0–26.4x15.8–18.9mm, (n = 100) mean 23.0x17.2. Eggs of *bangsi* unspotted pale blue or white like Moussier's Redstart. Highest number of eggs laid 25 but mean 9.2 eggs for 46 cuckoo-years (surely estimate too low as difficult to find all nests parasitised). In South Korea Vinous-throated Parrotbill shows greater egg-rejection the greater the difference in egg colour regardless of size (Etchécopar & Hüe 1967, Schönwetter 1967, Lack 1968, Payne 1974, Alvarez *et al.* 1976, Wyllie 1981, Irwin 1988, Moksnes & Røskaft 1995a, b, Edvardsen *et al.* 2001, Lee 2008, Moksnes *et al.* 2008). **Incubation** 11.5–13.5 days (Wyllie 1981). **Chicks** Hatch naked, mass 3–4g, skin rose, changing soon to blackish violet, gape orange; mass 16g after four days, same mass as most foster parents; eyes open next day, after week wings in pin and until *c.* 12 days has porcupine-like appearance; *c.*17–21 days old fledges and juvenile plumage complete after six weeks, mass *c.* 90g. Cuckoos from Eurasian Reed Warbler nests become independent at 26–45 days, mean 33.1 days (Heinroth & Heinroth 1966, Wyllie 1981). **Survival** Mean adult survival in Europe perhaps *c.* 4–5 years. Among nestlings bad weather and predation most common mortality causes; rate 34% (n = 213). 74% hatched in Eurasian Reed Warbler nests, and of these 65% disappeared before independence, but 32 of 38 wing-tagged young reached independence (Owen 1933, Wyllie 1981).

FOOD Most common food hairy caterpillars, also hairless, often toxic and largely unexploited by other animals; large ones violently shaken to dispel gut contents; on arrival in Europe caterpillars often rare and small beetles taken on forest floor. 90–93% of food caterpillars, mostly of one species, but varying from time to time in China. Young female had only adult ladybirds in stomach; earthworms; dead Willow Warbler. Young cuckoo fed on Cinnabar Moths (*Hypocrita jacobaeae*), catching them in flight. In Tanzania most important food army worms (*Spodoptera exempta* caterpillars). Other caterpillars, grasshoppers, crickets and termites commonly in winter. Cuckoo fledgling fed with 25cm long Striped Keelback Snake (*Amphiesma stolata*), swallowing it after three attempts, India. Indigestible matter regurgitated as pellets (Bolam 1914, Bub 1943, 1982, Crawshaw 1963, Ishizawa & Chiba 1966, Wyllie 1981, Prins 1986, Mohan 1994, J. Erritzøe).

STATUS AND CONSERVATION Declining Britain, Ireland, Finland and Denmark (and possibly Netherlands and Germany), presumably due to intensive agriculture resulting in fewer insects and hosts (Glutz von Blotzheim & Bauer 1980, Gibbons *et al.* 1993, Grell 1998, Hewson & Noble 2009), but also climate change where important short-distance migrating host species have advanced their arrival more than cuckoos resulting in mismatch of nesting times (Saino *et al.* 2009). In Britain numbers decreased markedly especially since early 1950s (Cramp 1985) and numbers of females declined from *c.* 34,500 in 1939–1961 to 21,000 in 1972–1982 (Brooke & Davies 1987); significant reduction also 1991–96 (Wernham *et al.* 2002). Rare summer visitor Greece but regular on migration (Bauer *et al.* 1969b). Total European population *c.* 1.3–1.9 million, Russian 0.5–1.0 million and Turkish below 0.5 million (Hagemeijer & Blair 1997). Breeding north and central Israel with sharp annual fluctuations, but migrants widespread, most end Feb–mid May, and mainly mid Jul–mid Sep (Shirihai 1996). Common west, north and north-east Turkey, Armenia, Kashmir, Pakistan, Bhutan and China (Osmaston 1927, Kumerloeve 1961, Cheng 1987, Roberts 1991, Tymstra *et al.* 1997, MacKinnon & Phillipps 2000). Not globally threatened (BirdLife International 2011).

Common Cuckoo *Cuculus canorus*. **Fig. 1.** Adult male, *C. c. canorus*, Ouessant Island, France, May (*Aurélien Audevard*). **Fig. 2.** Probable adult female grey morph, with traces of juvenile plumage, *C. c. canorus*, Muradiye, Turkey, May (*Daniele Occhiato*). **Fig. 3.** Adult female hepatic morph, *C. c. canorus*, Akoursos, Paphos District, West Cyprus, May (*Alison McArthur*). **Fig. 4.** Fledgling, *C. c. canorus*, fed by Rock Pipit, Scilly Isles, U.K., August (*Martin Goodey*). **Fig. 5.** Juvenile hepatic morph, *C. c. canorus*, Ouessant Island, France, July (*Aurélien Audevard*). **Fig. 6.** Adult, *C. c. bakeri*, Taal Chappar, Rajasthan, India, September (*Niranjan Sant*).

APPENDIX

English and scientific names of birds other than cuckoos mentioned in the text, generally following Dickinson (2003)

Abbott's Babbler *Malacocincla abbotti*
African Blue Quail *Coturnix adansonii*
African Broadbill *Smithornis capensis*
African Dusky Flycatcher *Muscicapa adusta*
African Golden Oriole *Oriolus auratus*
African Golden Weaver *Ploceus subaureus*
African Grey Flycatcher *Melaenornis microrhynchus*
African Masked Weaver *Ploceus velatus*
African Openbill Stork *Anastomus lamelligerus*
African Paradise-flycatcher *Terpsiphone viridis*
African Pied Starling *Spreo bicolor*
African Pied Wagtail *Motacilla aguimp*
African Pipit *Anthus cinnamomeus*
African Reed Warbler *Acrocephalus baeticatus*
African Stonechat *Saxicola torquatus*
African Yellow White-eye *Zosterops senegalensis*
Alpine Chough *Pyrrhocorax graculus*
American Robin *Turdus migratorius*
American Woodcock *Scolopax minor*
Amethyst Starling *Cinnyricinclus leucogaster*
Amethyst Sunbird *Chalcomitra amethystina*
Annobón White-eye *Zosterops griseovirescens*
Aplomado Falcon *Falco femoralis*
Archer's Robin-chat *Cossypha archeri*
Arctic Warbler *Phylloscopus borealis*
Arrow-marked Babbler *Turdoides jardineii*
Ashy-bellied White-eye *Zosterops citrinella*
Ashy Drongo *Dicrurus leucophaeus*
Ashy Flycatcher *Muscicapa caerulescens*
Ashy Laughing-thrush *Garrulax cinereifrons*
Ashy Prinia *Prinia socialis*
Ashy Robin *Poecilodryas albispecularis*
Ashy Tailorbird *Orthotomus ruficeps*
Asian Brown Flycatcher *Muscicapa dauurica*
Asian Fairy-bluebird *Irena puella*
Asian Paradise-flycatcher *Terpsiphone paradisi*
Asian Stubtail *Urosphena squameiceps*
Asian Verditer Flycatcher *Eumyias thalassina*
Australasian Figbird *Sphecotheres vieilloti*
Australian Magpie *Gymnorhina tibicen*
Australian Raven *Corvus coronoides*
Australian Reed Warbler *Acrocephalus australis*
Azara's Spinetail *Synallaxis azarae*
Azure-winged Magpie *Cyanopica cyana*
Baglafecht Weaver *Ploceus baglafecht*
Balicassiao *Dicrurus balicassius*
Banded Pitta *Pitta guajana*

Banded Whiteface *Aphelocephala nigricincta*
Bar-breasted Honeyeater *Ramsayornis fasciatus*
Bare-cheeked Babbler *Turdoides gymnogenys*
Barratt's Scrub Warbler *Bradypterus barratti*
Barred Antshrike *Thamnophilus doliatus*
Barred Warbler *Sylvia nisoria*
Bar-throated Apalis *Apalis thoracica*
Bar-winged Flycatcher-shrike *Hemipus picatus*
Beautiful Firetail Finch *Stagonopleura bella*
Beautiful Sunbird *Cinnyris pulchellus*
Bell Miner *Manorina melanophrys*
Bicoloured Antbird *Gymnopithys leucaspis*
Black-and-white Triller *Lalage melanoleuca*
Black-and-yellow Broadbill *Eurylaimus ochromalus*
Black-backed Puffback *Dryoscopus cubla*
Blackbird *Turdus merula*
Black-breasted Thrush *Turdus dissimilis*
Black Butcherbird *Cracticus quoyi*
Blackcap *Sylvia atricapilla*
Blackcap Babbler *Turdoides reinwardtii*
Black-chested Prinia *Prinia flavicans*
Black-chinned Honeyeater *Melithreptus gularis*
Black-collared Hawk *Busarellus nigricollis*
Black-coloured Starling *Sturnus nigricollis*
Black Cuckoo-shrike *Campephaga flava*
Black Drongo *Dicrurus macrocercus*
Black-eared Miner *Manorina melanotis*
Black-eared Wheatear *Oenanthe hispanica*
Black-faced Babbler *Turdoides melanops*
Black-faced Bunting *Emberiza spodocephala*
Black-faced Cuckoo-shrike *Coracina novaehollandiae*
Black-faced Grassquit *Tiaris bicolor*
Black-faced Wood Swallow *Artamus cinereus*
Black-headed Bunting *Emberiza melanocephala*
Black-headed Duck *Heteronetta atricapilla*
Black-headed Honeyeater *Melithreptus affinis*
Black-headed Weaver *Ploceus melanocephalus*
Black Honeyeater *Certhionyx niger*
Blackish Cuckoo-shrike *Coracina coerulescens*
Black Magpie *Platysmurus leucopterus*
Black-necked Weaver *Ploceus nigricollis*
Black Redstart *Phoenicurus ochrurus*
Black-striped Sparrow *Arremonops conirostris*
Black Sunbird *Nectarinia sericea*
Black-throated Accentor *Prunella atrogularis*
Black-throated Babbler *Stachyris nigricollis*
Black-throated Butcherbird *Cracticus nigrogularis*

Appendix

Black-throated Finch *Poephila cincta*
Black-throated Malimbe *Malimbus cassini*
Black-throated Sparrow *Amphispiza bilineata*
Black-throated Wattle-eye *Platysteira peltata*
Black-winged Bishop *Euplectes hordeaceus*
Blue-and-white Flycatcher *Cyanoptila cyanomelaena*
Blue-breasted Fairy-wren *Malurus pulcherrimus*
Blue-faced Honeyeater *Entomyzon cyanotis*
Blue Crested Flycatcher *Elminia longicauda*
Blue-faced Honeyeater *Entomyzon cyanotis*
Blue-grey Robin *Peneothello cyanus*
Blue Ground Dove *Claravis pretiosa*
Blue Jay *Cyanocitta cristata*
Blue-headed Paradise-flycatcher *Trochocercus nitens*
Blue-mantled Paradise-flycatcher *Trochocercus cyanomelas*
Blue-shouldered Robin-chat *Cossypha cyanocampter*
Blue-throated Barbet *Megalaima asiatica*
Blue-throated Blue Flycatcher *Cyornis rubeculoides*
Blue-throated Roller *Eurystomus gularis*
Blue Whistling Thrush *Myophonus caeruleus*
Blue-winged Laughing-thrush *Garrulax squamatus*
Blyth's Leaf Warbler *Phylloscopus reguloides*
Blyth's Reed Warbler *Acrocephalus dumetorum*
Boat-tailed Grackle *Quiscalus major*
Bocage's Weaver *Ploceus temporalis*
Bokmakierie *Telophorus zeylonus*
Boulder Chat *Pinarornis plumosus*
Brambling *Fringilla montifringilla*
Broad-billed Flycatcher *Myiagra ruficollis*
Bronzy Sunbird *Nectarinia kilimensis*
Brown Babbler *Turdoides plebejus*
Brown-backed Honeyeater *Ramsayornis modestus*
Brown Bush-Warbler *Bradypterus luteoventris*
Brown-cheeked Fulvetta *Alcippe poioicephala*
Brown Cuckoo-dove *Macropygia amboinensis*
Brown Falcon *Falco berigora*
Brown Fish Owl *Bubo zeylonensis*
Brown Gerygone *Gerygone mouki*
Brown-headed Cowbird *Molothrus ater*
Brown-headed Honeyeater *Melithreptus brevirostris*
Brown-headed Thrush *Turdus chrysolaus*
Brown Honeyeater *Lichmera indistincta*
Brown-hooded Kingfisher *Halcyon albiventris*
Brown Illadopsis *Illadopsis fulvescens*
Brownish-flanked Bush Warbler *Cettia fortipes*
Brown-necked Raven *Corvus ruficollis*
Brown Shrike *Lanius cristatus*
Brown Thornbill *Acanthiza pusilla*
Brown-throated Sunbird *Anthreptes malacensis*
Brown-throated Wattle-eye *Platysteira cyanea*
Budgerigar *Melopsittacus undulatus*
Buff-bellied Warbler *Phyllolais pulchella*
Buff-breasted Babbler *Pellorneum tickelli*
Buff-chested Babbler *Stachyris ambigua*
Buff-fronted Foliage-gleaner *Philydor rufum*
Buff-rumped Thornbill *Acanthiza reguloides*
Bull-headed Shrike *Lanius bucephalus*
Burchell's Starling *Lamprotornis australis*
Buru White-eye *Zosterops buruensis*
Cameroon Sunbird *Cyanomitra oritis*
Cape Batis *Batis capensis*
Cape Bulbul *Pycnonotus capensis*
Cape Bunting *Emberiza capensis*
Cape Crow *Corvus capensis*
Cape Glossy Starling *Lamprotornis nitens*
Cape Penduline Tit *Anthoscopus minutus*
Cape Robin-chat *Cossypha caffra*
Cape Rock Thrush *Monticola rupestris*
Cape Siskin *Pseudochloroptila totta*
Cape Sparrow *Passer melanurus*
Cape Wagtail *Motacilla capensis*
Cape Weaver *Ploceus capensis*
Cape White-eye *Zosterops pallidus*
Capuchin Babbler *Phyllanthus atripennis*
Cardinal Woodpecker *Dendropicos fuscescens*
Carrion Crow *Corvus corone*
Cedar Waxwing *Bombycilla cedrorum*
Chabert Vanga *Leptopterus chabert*
Chaffinch *Fringilla coelebs*
Chalk-browed Mockingbird *Mimus saturninus*
Chatham Island Gerygone *Gerygone albofrontata*
Chattering Cisticola *Cisticola anonymus*
Checker-throated Antwren *Myrmotherula fulviventris*
Chestnut-backed Forktail *Enicurus ruficapillus*
Chestnut-bellied Fantail *Rhipidura hyperythra*
Chestnut-bellied Starling *Lamprotornis pulcher*
Chestnut Bunting *Emberiza rutila*
Chestnut-capped Laughing-thrush *Garrulax mitratus*
Chestnut-crowned Laughing-thrush *Garrulax erythrocephalus*
Chestnut-crowned Warbler *Seicercus castaniceps*
Chestnut-eared Bunting *Emberiza fucata*
Chestnut-fronted Helmet-shrike *Prionops scopifrons*
Chestnut-headed Tesia *Oligura castaneocoronata*
Chestnut-naped Forktail *Enicurus ruficapillus*
Chestnut-rumped Heathwren *Calamanthus pyrrhopygius*
Chestnut-rumped Thornbill *Acanthiza uropygialis*
Chestnut-winged Babbler *Stachyris erythroptera*
Chimango Caracara *Milvago chimango*
Chiming Wedgebill *Psophodes occidentalis*
Chinspot Batis *Batis molitor*
Chipping Sparrow *Spizella passerina*
Chorister Robin-chat *Cossypha dichroa*
Chotoy Spinetail *Schoeniophylax phryganophilus*
Cicadabird *Coracina tenuirostris*
Clamorous Reed Warbler *Acrocephalus stentoreus*

Cockatiel *Nymphicus hollandicus*
Collared Sparrowhawk *Accipiter cirrhocephalus*
Collared Sunbird *Hedydipna collaris*
Common Babbler *Turdoides caudata*
Common Bulbul *Pycnonotus barbatus*
Common Chiffchaff *Phylloscopus collybita*
Common Grackle *Quiscalus quiscula*
Common Greenshank *Tringa nebularia*
Common Fiscal *Lanius collaris*
Common Hoopoe *Upupa epops*
Common Iora *Aegithina tiphia*
Common Jery *Neomixis tenella*
Common Kestrel *Falco tinnunculus*
Common Magpie *Pica pica*
Common Myna *Acridotheres tristis*
Common Nightingale *Luscinia megarhynchos*
Common Pheasant *Phasianus colchicus*
Common Raven *Corvus corax*
Common Redstart *Phoenicurus phoenicurus*
Common Stonechat *Saxicola rubicola*
Common Tailorbird *Orthotomus sutorius*
Cooper's Hawk *Accipiter cooperii*
Copper Sunbird *Cinnyris cupreus*
Corn Bunting *Emberiza calandra*
Crescent-chested Babbler *Stachyris melanothorax*
Crescent Honeyeater *Phylidonyris pyrrhopterus*
Crested Bellbird *Oreoica gutturalis*
Crested Bunting *Melophus lathami*
Crested Drongo *Dicrurus forficatus*
Crested Malimbe *Malimbus malimbicus*
Crested Oropendola *Psarocolius decumanus*
Crimson-breasted Gonolek *Laniarius atrococcineus*
Crimson Chat *Epthianura tricolor*
Crimson Finch *Neochmia phaeton*
Crow-billed Drongo *Dicrurus annectans*
Cuckoo Finch *Anomalspiza imberbis*
Curl-crested Aracari *Pteroglossus beauharnaesii*
Dark-fronted Babbler *Rhopocichia atriceps*
Dark-necked Tailorbird *Orthotomus atrogularis*
Dartford Warbler *Sylvia undata*
Diamond Dove *Geopelia cuneata*
Diamond Firetail *Stagonopleura guttata*
Dickcissel *Spiza americana*
Dohrn's Thrush-babbler *Horizorhinus dohrni*
Double-barred Finch *Taeniopygia bichenovii*
Drab-breasted Pygmy Tyrant *Hemitriccus diops*
Dunnock *Prunella modularis*
Dusky Gerygone *Gerygone tenebrosa*
Dusky Robin *Melanodryas vittata*
Dusky Sunbird *Cinnyris fuscus*
Dusky Wood Swallow *Artamus cyanopterus*
Eared Pygmy Tyrant *Myiornis auricularis*
East Coast Akalat *Sheppardia gunningi*
Eastern Bearded Scrub Robin *Cercotrichas
quadrivirgata*
Eastern Black-headed Oriole *Oriolus larvatus*
Eastern Bristlebird *Dasyornis brachypterus*
Eastern Crimson Sunbird *Aethopyga siparaja*
Eastern Crowned Warbler *Phylloscopus coronatus*
Eastern Grey Plantain-eater *Crinifer zonurus*
Eastern Olivaceous Warbler *Hippolais pallida*
Eastern Olive Sunbird *Cyanomitra olivacea*
Eastern Shrike-tit *Falcunculus frontatus*
Eastern Spinebill *Acanthorhynchus tenuirostris*
Eastern Towhee *Pipilo erythrophthalmus*
Eastern Wood Pewee *Contopus virens*
Eastern Yellow Robin *Eopsaltria australis*
Eleonora's Falcon Falco *Falco eleonorae*
Emperor Fairy-wren *Malurus cyanocephalus*
Equatorial Akalat *Sheppardia aequatorialis*
Esmeraldas Antbird *Sipia rosenbergi*
Eurasian Blackbird *Turdus merula*
Eurasian Curlew *Numenius arquata*
Eurasian Golden Oriole *Oriolus oriolus*
Eurasian Jackdaw *Corvus monedula*
Eurasian Jay *Garrulus glandarius*
Eurasian Reed Warbler *Acrocephalus scirpaceus*
Eurasian Skylark *Alauda arvensis*
Eurasian Sparrowhawk *Accipiter nisus*
Eurasian Stone-curlew *Burhinus oedicnemus*
Eurasian Woodcock *Scolopax rustica*
European Goldfinch *Carduelis carduelis*
European Greenfinch *Carduelis chloris*
European Robin *Erithacus rubecula*
European Roller *Coracias garrulus*
European Turtle Dove *Streptopelia turtur*
Eyebrowed Thrush *Turdus obscurus*
Eyebrowed Wren-babbler *Napothera epilepidota*
Eye-ringed Flatbill *Rhynchocyclus brevirostris*
Fairy Gerygone *Gerygone palpebrosa*
Familiar Chat *Cercomela familiaris*
Fan-tailed Gerygone *Gerygone flavolateralis*
Fan-tailed Raven *Corvus rhipidurus*
Fiji Warbler *Cettia ruficapilla*
Finsch's Flycatcher-thrush *Stizorhina finschi*
Fire-crested Alethe *Alethe diademata*
Fiscal Shrike *Lanius collaris*
Flame Robin *Petroica phoenicea*
Flock Pigeon *Phaps histrionica*
Flores Crow *Corvus florensis*
Forest Raven *Corvus tasmanicus*
Forest Robin *Stiphrornis erythrothorx*
Forest Rock Thrush *Monticola sharpei*
Forest Wagtail *Dendronanthus indicus*
Fork-tailed Drongo *Dicrurus adsimilis*
Fork-tailed Flycatcher *Tyrannus savana*
Freckle-breasted Thornbird *Phacellodomus striaticollis*
Frilled Monarch *Arses telescopthalmus*

Fulvous Barbler *Turdoides fulva*
Fuscous Honeyeater *Lichenostomus fusca*
Gambel's Quail *Callipepla gambelii*
Garden Warbler *Sylvia borin*
Gilbert Whistler *Pachycephala inornata*
Goldcrest *Regulus regulus*
Golden-backed Weaver *Ploceus jacksoni*
Golden-bellied Gerygone *Gerygone sulphurea*
Golden-breasted Bunting *Emberiza flaviventris*
Golden-headed Cisticola *Cisticola exilis*
Golden-bellied Gerygone *Gerygone sulphurea*
Golden Monarch *Monarcha chrysomela*
Golden-naped Barbet *Megalaima pulcherrima*
Golden Palm Weaver *Ploceus bojeri*
Golden Whistler *Pachycephala pectoralis*
Gould's Sunbird *Aethopyga gouldiae*
Greater Blue-eared Starling *Lamprotornis chalybaeus*
Greater Double-collared Sunbird *Cinnyris afer*
Greater Flame-backed Woodpecker *Chrysocolaptes ucidus*
Greater Necklaced Laughing-thrush *Garrulax pectoralis*
Greater Racket-tailed Drongo *Dicrurus paradiseus*
Greater Swamp Warbler *Acrocephalus rufescens*
Greater Thornbird *Phacellodomus ruber*
Greater Whitethroat *Sylvia communis*
Great Kiskadee *Pitangus sulphuratus*
Great Reed Warbler *Acrocephalus arundinaceus*
Great Tit *Parus major*
Green-backed Gerygone *Gerygone chloronota*
Green-backed Tit *Parus monticolus*
Green Barbet *Stactolaema olivacea*
Green Crombec *Sylvietta virens*
Greenfinch *Carduelis chloris*
Green-headed Sunbird *Cyanomitra verticalis*
Green Iora *Aegithina viridissima*
Greenish Warbler *Phylloscopus trochiloides*
Green Wood Hoopoe *Phoeniculus purpureus*
Green Woodpecker *Picus viridis*
Grey-backed Camaroptera *Camaroptera brachyura*
Grey-backed Cisticola *Cisticola subruficapilla*
Grey-backed Tailorbird *Orthotomus derbianus*
Grey-bellied Tesia *Tesia cyaniventer*
Grey-breasted Prinia *Prinia hodgsonii*
Grey-breasted Spiderhunter *Arachnothera modesta*
Grey Bushchat *Saxicola ferreus*
Grey Butcherbird *Cracticus torquatus*
Grey Catbird *Dumetella carolinensis*
Grey-cheeked Tit-babbler *Macronous flavicolli*
Grey Crow *Corvus tristis*
Grey-crowned Tetraka *Xanthomixus cinereiceps*
Grey Fantail *Rhipidura albiscapa*
Grey-fronted Dove *Leptotila rufaxilla*
Grey-fronted Honeyeater *Lichenostomus plumulus*
Grey-headed Canary-flycatcher *Culicicapa ceylonensis*
Grey-headed Honeyeater *Meliphaga keartlandi*
Grey-headed Parrotbill *Paradoxornis gularis*
Grey-headed Social Weaver *Pseudonigrita arnaudi*
Grey-hooded Warbler *Seicercus xanthoschistos*
Grey Junglefowl *Gallus sonneratii*
Grey Shrike-thrush *Colluricincla harmonica*
Grey-sided Laughing-thrush *Garrulax caerulatus*
Grey-throated Babbler *Stachyris nigriceps*
Grey-throated Tit-flycatcher *Myioparus griseigularis*
Grey Thrush *Zoothera princei*
Grey Tit-flycatcher *Myioparus plumbeus*
Grey Wagtail *Motacilla cinerea*
Grey Warbler *Gerygone igata*
Grey Whistler *Pachycephala simplex*
Grosbeak Weaver *Amblyospiza albifrons*
Hair-crested Drongo *Dicrurus hottentottus*
Harris's Hawk *Parabuteo unicinctus*
Hartlaub's Babler *Turdoides hartlaubii*
Helmeted Friarbird *Philemon buceroides*
Helmeted Honeyeater *Lichenostomus melanops cassidix*
Heuglin's Masked Weaver *Ploceus heuglini*
Hildebrandt's Starling *Lamprotornis hildebrandti*
Hill Myna *Gracula religiosa*
Hill Prinia *Prinia atrogularis*
Hinde's Babbler *Turdoides hindei*
Hoary-throated Barwing *Actinodura nipalensis*
Holub's Golden Weaver *Ploceus xanthops*
Honeyguide Greenbul *Baeopogon indicator*
Hooded Crow *Corvus cornix*
Hooded Robin *Melanodryas cucullata*
Horsfield's Babbler *Malacocincla sepiaria*
Horsfield's Jungle Babbler *Malacocincla sepiaria*
House Crow *Corvus splendens*
House Sparrow *Passer domesticus*
Icterine Warbler *Hippolais icterina*
Ijima's Leaf Warbler *Phylloscopus ijimae*
Immaculate Antbird *Myrmeciza immaculata*
Indian Blue Robin *Luscinia brunnea*
Indigo Bunting *Passerina cyanea*
Inland Thornbill *Acanthiza apicalis*
Isabelline Shrike *Lanius isabellinus*
Jacky Winter *Microeca fascinans*
Japanese Bush Warbler *Cettia diphone*
Japanese Paradise-flycatcher *Terpsiphone atrocaudata*
Japanese Robin *Luscinia akahige*
Japanese Thrush *Turdus cardis*
Japanese White-eye *Zosterops japonicus*
Jay *Garrulus glandarius*
Jungle Babbler *Turdoides striata*
Jungle Crow *Corvus macrorhynchos*
Kalahari Scrub Robin *Cercotrichas paena*
Karoo Prinia *Prinia maculosa*
Karoo Scrub Robin *Cercotrichas coyphoeus*
Kenya Rufous Sparrow *Passer rufocinctus*

King of Saxony Bird of Paradise *Pteridophora alberti*
Kurrichane Thrush *Turdus libonyanus*
Lanceolated Warbler *Locustella lanceolata*
Large-billed Crow *Corvus macrorhynchos*
Large-billed Gerygone *Gerygone magnirostris*
Large-billed Scrubwren *Sericornis magnirostra*
Large Grey Babbler *Turdoides malcolmi*
Large Scrubwren *Sericornis nouhuysi*
Large Woodshrike *Tephrodornis virgatus*
Laughing Dove *Streptopelia senegalensis*
Leaden Flycatcher *Myiagra rubecula*
Lemon-bellied Flycatcher *Microeca flavigaster*
Lesser Masked Weaver *Ploceus intermedius*
Lesser Necklaced Laughing-thrush *Garrulax monileger*
Lesser Racket-tailed Drongo *Dicrurus remifer*
Lesser Shortwing *Brachypteryx leucophrys*
Lesser Swamp Warbler *Acrocephalus gracilirostris*
Lesser Whitethroat *Sylvia curruca*
Lesser Yellow-billed Kingfisher *Halcyon torororo*
Lewin's Honeyeater *Meliphaga lewinii*
Little Crow *Corvus bennetti*
Little Friarbird *Philemon citreogularis*
Little Grassbird *Megalurus gramineus*
Little Grebe *Tachybaptus ruficollis*
Little Greenbul *Andropadus virens*
Little Rush Warbler *Bradypterus baboecula*
Little Scrubwren *Sericornis beccarii*
Little Spiderhunter *Arachnothera longirostra*
Little Wattlebird *Anthochaera chrysoptera*
Little Wood Swallow *Artamus minor*
Lizard Buzzard *Kaupifalco monogrammicus*
Long-billed Crombec *Sylvietta rufescens*
Long-billed Pipit *Anthus similis*
Long-tailed Mockingbird *Mimus longicaudatus*
Long-tailed Rosefinch *Uragus sibiricus*
Long-tailed Shrike *Lanius schach*
Long-tailed Glossy Starling *Lamprotornis caudatus*
Long-tailed Starling *Aplonis magna*
Lovely Fairy-wren *Malurus amabilis*
Lühder's Bush Shrike *Laniarius luederi*
Macleay's Honeyeater *Xanthotis macleayanus*
Madagascar Bushlark *Mirafra hova*
Madagascar Cisticola *Cisticola cherina*
Madagascar Hoopoe *Upupa marginata*
Madagascar Magpie-robin *Copsychus albospecularis*
Madagascar Munia *Lonchura nana*
Madagascar Paradise-flycatcher *Terpsiphone mutata*
Madagascar Sunbird *Cinnyris sovimanga*
Magpie-lark *Grallina cyanoleuca*
Mangrove Gerygone *Gerygone levigaster*
Mangrove Golden Whistler *Pachycephala melanura*
Mangrove Honeyeater *Lichenostomus fasciogularis*
March Warbler *Acrocephalus palustris*
Mariqua Flycatcher *Melaenornis mariquensis*
Mariqua Sunbird *Cinnyris mariquensis*
Marsh Warbler *Acrocephalus palustris*
Masked Laughing-thrush *Garrulax perspicillatus*
Masked Wood Swallow *Artamus personatus*
Meadow Bunting *Emberiza cioides*
Meadow Pipit *Anthus pratensis*
Melodious Laughing-thrush *Garrulax canorus*
Middendorff's Warbler *Locustella ochotensis*
Miombo Blue-eared Starling *Lamprotornis elisabeth*
Mistletoebird *Dicaeum hirundinaceum*
Mocking Cliffchat *Thamnolaea cinnamomeiventris*
Mountain Leaf Warbler *Phylloscopus trivirgatus*
Mountain Tailorbird *Orthotomus cuculatus*
Mountain Wagtail *Motacilla clara*
Mountain Wheatear *Oenanthe monticola*
Mourning Dove *Zenaida macroura*
Mouse-coloured Sunbird *Cyanomitra veroxi*
Moussier's Redstart *Phoenicurus moussieri*
Moustached Laughing-thrush *Garrulus cineraceus*
Mrs Gould's Sunbird *Aethopyga gouldiae*
Narcissus Flycatcher *Ficedula narcissina*
Nepal Fulvetta *Alcippe nipalensis*
New Holland Honeyeater *Phylidonyris novaehollandiae*
Newton's Sunbird *Anabathmis newtoni*
New Zealand Bellbird *Anthornis melanura*
New Zealand Creeper *Finschia novaeseelandiae*
New Zealand Fantail *Rhipidura fuliginosa*
New Zealand Pipit *Anthus novaeseelandiae*
New Zealand Robin *Petroica australis*
New Zealand Tit *Petroica macrocephala*
Noisy Friarbird *Philemon corniculatus*
Noisy Miner *Manorina melanocephala*
Northern Cardinal *Cardinalis cardinalis*
Northern Fantail *Rhipidura rufiventris*
Northern Grey-headed Sparrow *Passer griseus*
Northern Masked Weaver *Ploceus taeniopterus*
Northern Mockingbird *Mimus polyglottus*
Northern Puffback *Dryoscopus gambensis*
Northern Wheatear *Oenanthe oenanthe*
Northern White-crowned Forktail *Enicurus sinensis*
Ocellated Antbird *Phaenostictus mcleannani*
Ochre-faced Tody-flycatcher *Poecilotriccus plumbeiceps*
Olive-backed Oriole *Oriolus sagittatus*
Olive-backed Pipit *Anthus hodgsoni*
Olive-backed Sunbird *Cinnyris jugularis*
Olive-backed Tailorbird *Orthotomus sepium*
Olive-bellied Sunbird *Cinnyris chloropygius*
Olive Bush Shrike *Chlorophoneus olivaceus*
Olive Thrush *Turdus olivaceus*
Olive Whistler *Pachycephala olivacea*
Olive-winged Bulbul *Pycnonotus plumosus*
Orange-billed Babbler *Turdoides rufescens*
Orange Chat *Epthianura aurifrons*
Orange-flanked Bush Robin *Luscinia cyanura*

Orange-headed Thrush *Zoothera citrina*
Oriental Magpie-robin *Copsychus saularis*
Oriental Reed Warbler *Acrocephalus orientalis*
Orphean Warbler *Sylvia hortensis*
Painted Honeyeater *Grantiella picta*
Paddyfield Pipit *Anthus rufulus*
Pale-breasted Illadopsis *Illadopsis rufipennis*
Pale-breasted Spinetail *Synallaxis albescens*
Pale Flycatcher *Melaenornis pallidus*
Pale-footed Bush Warbler *Cettia pallidipes*
Pale-legged Leaf Warbler *Phylloscopus tenellipes*
Pale-winged Starling *Onychognathus nabouroup*
Pale-yellow Robin *Tregellasia capito*
Pallas's Leaf Warbler *Phylloscopus proregulus*
Papuan Harpy Eagle *Harpyopsis novaeguineae*
Philippine Bulbul *Ixos philippinus*
Philippine Oriole *Oriolus steerii*
Pied Bushchat *Saxicola caprata*
Pied Butcherbird *Cracticus nigrogularis*
Pied Crow *Corvus albus*
Pied Currawong *Strepera graculina*
Pied Fantail *Rhipidura javanica*
Pied Honeyeater *Certhionyx variegates*
Pied Water Tyrant *Fluvicola pica*
Pilot Bird *Pycnoptilus floccosus*
Pink Robin *Petroica rodinogaster*
Piping Cisticola *Cisticola fulvicapilla*
Pipipi *Finschia novaeseelandiae*
Plain Antvireo *Dysithamnus mentalis*
Plain-backed Pipit *Anthus leucophrys*
Plain-brown Woodcreeper *Dendrocincla fuliginosa*
Plain-crowned Spinetail *Synallaxis gujanensis*
Plain Prinia *Prinia inornata*
Plain-throated Sunbird *Anthreptes malacensis*
Plain Wren *Thryothorus modestus*
Plumbeous Water Redstart *Rhyacornis fuliginosa*
Plum-headed Finch *Neochmia modesta*
Príncipe Speirops *Speirops leucophoeus*
Príncipe White-eye *Zosterops ficedulinus*
Pririt Batis *Batis pririt*
Puff-throated Babbler *Pellorneum ruficeps*
Purple-crowned Fairy-wren *Malurus coronatus*
Purple-gaped Honeyeater *Lichenostomus cratitius*
Purple-rumped Sunbird *Leptocoma zeylonica*
Purple Sunbird *Cinnyris asiaticus*
Purplish Jay *Cyanocorax cyanomelas*
Pygmy Blue Flycatcher *Muscicapella hodgsoni*
Pygmy Sunbird *Hedydipna platura*
Rainbow Bee-eater *Merops ornatus*
Rattling Cisticola *Cisticola chiniana*
Red-backed Fairy-wren *Malurus melanocephalus*
Red-backed Shrike *Lanius collurio*
Red-billed Blue Magpie *Urocissa erythrorhyncha*
Red-billed Chough *Pyrrhocorax pyrrhocorax*
Red-backed Fairy-wren *Malurus melanocephalus*
Red-backed Shrike *Lanius collurio*
Red-bearded Bee-eater *Nyctyornis amictus*
Red-bellied Paradise-flycatcher *Terpsiphone rufiventer*
Red-billed Helmet-shrike *Prionops caniceps*
Red-billed Pigeon *Columba flavirostris*
Red-billed Quelea *Quelea quelea*
Red-browed Finch *Neochmia temporalis*
Red-browed Treecreeper *Climacteris erythrops*
Red-capped Robin *Petroica goodenovii*
Red-capped Robin-chat *Cossypha natalensis*
Red-chested Sunbird *Cinnyris erythrocercus*
Red-collared Widowbird *Euplectes ardens*
Red-eyed Bulbul *Pycnonotus brunneus*
Red-eyed Thornbird *Phacellodomus erythrophthalmus*
Red-faced Cisticola *Cisticola erythrops*
Red-faced Liocichla *Liocichla phoenicea*
Red-faced Mousebird *Urocolius indicus*
Red Fody *Foudia madagascariensis*
Red-fronted Parakeet *Cyanoramphus novaezelandiae*
Red-headed Weaver *Anaplectes ruficeps*
Red-lored Whistler *Pachycephala rufogularis*
Red-pate Cisticola *Cisticola ruficeps*
Red-tailed Hawk *Buteo jamaicensis*
Redthroat *Pyrrholaemus brunneus*
Red-throated Honeyeater *Myzomela eques*
Red-vented Bulbul *Pycnonotus cafer*
Red Wattlebird *Anthochaera carunculata*
Red-whiskered Bulbul *Pycnonotus jocosus*
Red-winged Blackbird *Agelaius phoeniceus*
Red-winged Fairy-wren *Malurus elegans*
Red-winged Starling *Onychognathus morio*
Regent Honeyeater *Xanthomyza phrygia*
Restless Flycatcher *Myiagra inquieta*
Retz's Helmet-shrike *Prionops retzii*
Ring-necked Parakeet *Psitacula krameri*
Roadside Hawk *Buteo magnirostris*
Rockrunner *Chaetops pycnopygius*
Rock Warbler *Origma solitaria*
Rose-breasted Chat *Granatellus pelzelni*
Rose Robin *Petroica rosea*
Rufous-and-white Wren *Thryothorus rufalbus*
Rufous Bristlebird *Dasyornis broadbenti*
Rufous-bellied Niltava *Niltava sundara*
Rufous-breasted Spinetail *Synallaxis erythrothorax*
Rufous-capped Babbler *Stachyris ruficeps*
Rufous Chatterer *Turdoides rubiginosa*
Rufous-chinned Laughing-thrush *Garrulax rufogularis*
Rufous-crowned Emu-wren *Stipiturus ruficeps*
Rufous Fantail *Rhipidura rufifrons*
Rufous Field Wren *Calamanthus campestris*
Rufous Flycatcher-thrush *Stizorhina fraseri*
Rufous-fronted Babbler *Stachyris rufifrons*
Rufous-necked Laughing-thrush *Garrulax ruficollis*

Rufous-sided Gerygone *Gerygone dorsalis*
Rufous Songlark *Cinclorhamphus mathewsi*
Rufous-throated Honeyeater *Conopophila rufugularis*
Rufous-vented Laughing-thrush *Garrulax delesserti*
Rufous-vented Yuhina *Yuhina occipitalis*
Rufous Whistler *Pachycephala rufiventris*
Rufous-winged Philentoma *Philentoma pyrhoptera*
Rüppell's Glossy Starling *Lamprotornis purpuroptera*
Rüppell's Robin-chat *Cossypha semirufa*
Rüppell's Weaver *Ploceus galbula*
Russet-fronted Thornbird *Phacellodomus rufifrons*
Rusty-cheeked Scimitar Babbler *Pomatorhinus erythrogenys*
Rusty-fronted Barwing *Actinodura egertoni*
São Tomé Weaver *Ploceus sanctaethomae*
Sardinian Warbler *Sylvia melanocephala*
Satin Flycatcher *Myiagra cyanoleuca*
Scaled Dove *Columbina squammata*
Scaly Thrush *Zoothera dauma*
Scarlet-chested Sunbird *Chalcomitra senegalensis*
Scarlet Honeyeater *Myzomela sanguinolenta*
Scarlet Minivet *Pericrocotus flammeus*
Scarlet Robin *Petroica boodang*
Screaming Cowbird *Molothrus rufoaxillaris*
Scrub Tit *Acanthornis magna*
Sedge Warbler *Acrocephalus schoenobaenus*
Sharpe's Akalat *Sheppardia sharpei*
Shikra *Accipiter badius*
Shiny Cowbird *Molothrus bonariensis*
Short-beared Honeyeater *Melidectes nouhuysi*
Short-billed Canastero *Asthenes baeri*
Shy Heathwren *Calamanthus cautus*
Siberian Blue Robin *Luscinia cyane*
Siberian Rubythroat *Luscinia calliope*
Siberian Stonechat *Saxicola maurus*
Silver-crowned Friarbird *Philemon argenticeps*
Silver-eye *Zosterops lateralis*
Singing Cisticola *Cisticola cantans*
Singing Honeyeater *Lichenostomus virescens*
Slate-headed Tody-tyrant *Poecilotriccus Sylvia*
Slaty Spinetail *Synallaxis brachyura*
Slaty-backed Flycatcher *Ficedula hodgsonii*
Slaty-backed Forktail *Enicurus schistaceus*
Slaty-backed Thornbill *Acanthiza robustirostris*
Slender-billed Crow *Corvus enca*
Slender-billed Thornbill *Acanthiza iredalei*
Slender-billed Weaver *Ploceus pelzelni*
Small Minivet *Pericrocotus cinnamomeus*
Small Niltava *Niltava macgregoriae*
Snail Kite *Rostrhamus sociabilis*
Snowy-browed Flycatcher *Ficedula hyperythra*
Snowy-crowned Robin-chat *Cossypha niveicapilla*
Sombre Greenbul *Andropadus importunes*
Song Thrush *Turdus philomelos*

Sooty Boubou *Laniarius leucorhynchus*
Sooty-capped Babbler *Malacopteron affine*
Sooty-fronted Spinetail *Synallaxis frontalis*
Sooty-headed Bulbul *Pycnonotus aurigaster*
Sooty Honeyeater *Melidectes fuscus*
Sooty Woodpecker *Mulleripicus funebris*
Southern Black Flycatcher *Melaenornis pammelaina*
Southern Boubou *Laniarius ferrugineus*
Southern Brown-throated Weaver *Ploceus xanthopterus*
Southern Caracara *Caracara plancus*
Southern Double-collared Sunbird *Cinnyris chalybeus*
Southern Emu-wren *Stipiturus malachurus*
Southern Grey-headed Sparrow *Passer diffusus*
Southern Lapwing *Vanellus chilensis*
Southern Pied Babbler *Turdoides bicolor*
Southern Red Bishop *Euplectes orix*
Southern Rufous Sparrow *Passer motitensis*
Southern Scrub Robin *Drymodes brunneopygia*
Southern White-crowned Forktail *Enicurus leschenaulti*
Southern Whiteface *Aphelocephala leucopsis*
Spangled Drongo *Dicrurus bracteatus*
Speckled Mousebird *Colius striatus*
Speckled Warbler *Pyrrholaemus sagittatus*
Spectacled Monarch *Monarcha trivirgatus*
Spectacled Weaver *Ploceus ocularis*
Speke's Weaver *Ploceus spekei*
Spinifexbird *Eremiornis carteri*
Spix's Spinetail *Synallaxis spixi*
Spiny-checked Honeyeater *Acanthagenys rufogularis*
Splendid Fairy-wren *Malurus splendens*
Splendid Glossy Starling *Lamprotornis splendidus*
Spot-breasted Laughing-thrush *Garrulus merulinus*
Spotted Flycatcher *Muscicapa striata*
Spotted Forktail *Enicurus maculatus*
Spotted Pardalote *Pardalotus punctatus*
Spot-throated Babbler *Pellorneum albiventre*
Spangled Drongo *Dicrurus bracteatus*
Steel-blue Flycatcher *Myiagra ferrocyanea*
Stock Dove *Columba oenas*
Stout Cisticola *Cisticola robustus*
Streaked Laughing-thrush *Garrulax lineatus*
Streaked Scrub Warbler *Scotocerca inquieta*
Streaked Spiderhunter *Arachnothera magna*
Streaky-breasted Spiderhunter *Arachnothera affinis*
Striated Babbler *Turdoides earlei*
Striated Fieldwren *Calamanthus fuliginosus*
Striated Grassbird *Megalurus palustris*
Striated Grasswren *Amytornis striatus*
Striated Laughing-thrush *Garrulax striatus*
Striated Pardalote *Pardalotus striatus*
Striated Prinia *Prinia crinigera*
Striated Scrub Warbler *Scotocerca inquieta*
Striated Thornbill *Acanthiza lineata*

Stripe-breasted Shrike-thrush *Colluricincla boweri*
Stripe-breasted Spinetail *Synallaxis cinnamomea*
Striped Tit-babbler *Macronous gularis*
Striped Honeyeater *Plectrorhyncha lanceolata*
Stitchbird *Notiomystis cincta*
Strong-billed Honeyeater *Melithreptus validirostris*
Strong-footed Bush Warbler *Cettia fortipes*
Stygian Owl *Asio stygius*
Sulawesi Starling *Basilornis celebensis*
Sunda Bush Warbler *Cettia vulcania*
Sunda Warbler *Seicercus grammiceps*
Superb Fairy-wren *Malurus cyaneus*
Superb Starling *Lamprotornis superbus*
Swamp Flycatcher *Muscicapa aquatica*
Swynnerton's Robin *Swynnertonia swynnertoni*
Tasmanian Scrubwren *Sericornis frontalis humilis*
Tasmanian Thornbill *Acanthiza ewingii*
Tawny-breasted Honeyeater *Xanthotis flaviventer*
Tawny-crowned Honeyeater *Glyciphila melanops*
Tawny-flanked Prinia *Prinia subflava*
Tawny Grassbird *Megalurus timoriensis*
Terrestrial Greenbul *Phyllastrephus terrestris*
Tickell's Blue Flycatcher *Cyornis tickelliae*
Tickell's Thrush *Turdus unicolor*
Thick-billed Flowerpecker *Dicaeum agile*
Thick-billed Grasswren *Amytornis textilis*
Thick-billed Warbler *Phragmaticola aedon*
Timor Friarbird *Philemon inornatus*
Timor Leaf Warbler *Phylloscopus presbytes*
Tomtit *Petroica macrocephala*
Torresian Crow *Corvus orru*
Tree Martin *Petrochelidon nigricans*
Tree Pipit *Anthus trivialis*
Tristram's Warbler *Sylvia deserticola*
Tropical Boubou *Laniarius aethiopicus*
Tropical Kingbird *Tyrannus melancholicus*
Tufted Tit-spinetail *Leptasthenura platensis*
Tui *Prosthemadura novaeseelandiae*
Upland Pipit *Anthus sylvanus*
Variable Sunbird *Cinnyris venustus*
Varied Honeyeater *Lichenostomus versicolor*
Varied Sittella *Daphoenositta chrysoptera*
Variegated Fairy-wren *Malurus lamberti*
Veery *Catharus fuscescens*
Vermilion Flycatcher *Pyrocephalus rubinus*
Victoria's Riflebird *Ptiloris victoriae*
Vieillot's Weaver *Ploceus nigerrimus*
Village Weaver *Ploceus cucullatus*
Vinous-throated Parrotbill *Paradoxornis webbianus*
Wailing Cisticola *Cicticola lais*
Wallace's Fairy-wren *Sipodotus wallacii*
Weebill *Smicrornis brevirostris*
Welcome Swallow *Hirundo neoxena*
Western Crowned Warbler *Phylloscopus occipitalis*
Western Olive Sunbird *Cyanomitra obscura*
Western Spinebill *Acanthorhynchus superciliosus*
Western Thornbill *Acanthiza inornata*
Whiskered Yuhina *Yuhina flavicollis*
White-bellied Cuckoo-shrike *Coracina papuensis*
White-bellied Erpornis *Erpornis zantholeuca*
White-breasted Robin *Eopsaltria georgiana*
White-breasted Sunbird *Cinnyris talatala*
White-breasted Wood Swallow *Artamus leucorhynchus*
White-browed Babbler *Pomatostomus superciliosus*
White-browed Fantail *Rhipidura aureola*
White-browed Laughing-thrush *Garrulax sannio*
White-browed Robin *Poecilodryas superciliosa*
White-browed Robin-chat *Cossypha heuglini*
White-browed Scrub Robin *Cercotrichas leucophrys*
White-browed Scrubwren *Sericornis frontalis*
White-browed Sparrow-weaver *Plocepasser mahali*
White-browed Wood Swallow *Artamus superciliosus*
White-cheeked Honeyeater *Phylidonyris niger*
White-crested Laughing-thrush *Garrulax leucolophus*
White-crowned Forktail *Enicurus leschenaulti*
White-crowned Starling *Spreo albicapillus*
White-eared Honeyeater *Lichenostomus leucotis*
White-eyed Slaty Flycatcher *Melaenornis fischeri*
White-fronted Chat *Epthianura albifrons*
White-fronted Honeyeater *Phylidonyris albifrons*
White-gaped Honeyeater *Lichenostomus unicolor*
Whitehead *Mohoua albicilla*
White-headed Babbler *Turdoides leucocephala*
White-headed Marsh Tyrant *Arundinicola leucocephala*
White-headed Vanga *Artamella viridis*
White Helmet-shrike *Prionops plumatus*
White-naped Honeyeater *Melithreptus lunatus*
White Nile Rufous Sparrow *Passer shelleyi*
White-plumed Honeyeater *Lichenostomus penicillatus*
White-rumped Babbler *Turdoides leucopygia*
White-rumped Shama *Copsychus malabaricus*
White-shouldered Fairy-wren *Malurus alboscapulatus*
White-starred Robin *Pogonocichla stellata*
White-tailed Ant Thrush *Neocossyphus poensis*
White-tailed Cotinga *Xipholena lamellipennis*
White-tailed Gerygone *Gerygone fusca*
White-tailed Kite *Elanus leucurus*
White-throated Babbler *Turdoides gularis*
White-throated Gerygone *Gerygone olivacea*
White-throated Honeyeater *Melithreptus albogularis*
White-throated Robin-chat *Cossypha humeralis*
White-throated Treeceeper *Cormobates leucophaea*
White-tipped Plantcutter *Phytotoma rutila*
White Wagtail *Motacilla alba*
White-winged Chough *Corcorax melanorhamphos*
White-winged Fairy-wren *Malurus leucopterus*
White-winged Scrub Robin *Cercotrichas leucoptera*
White-winged Triller *Lalage tricolor*

White-winged Widowbird *Euplectes albonotatus*
Willie-wagtail *Rhipidura leucophrys*
Willow Warbler *Phylloscopus trochilus*
Winter Wren *Troglodytes troglodytes*
Woodlark *Lullula arborea*
Woodchat Shrike *Lanius senator*
Woodpigeon *Columba palumbus*
Wood Thrush *Hylocichla mustelina*
Yellow-bellied Eremomela *Eremomela icteropygialis*
Yellow-bellied Gerygone *Gerygone chrysogaster*
Yellow-bellied Greenbul *Chlorocichla flaviventris*
Yellow-bellied Prinia *Prinia flaviventris*
Yellow-billed Babbler *Turdoides affinis*
Yellow-billed Kingfisher *Syma torotoro*
Yellow-billed Shrike *Corvinella corvine*
Yellow Bishop *Euplectes capensis*
Yellow-breasted Bunting *Emberiza aureola*
Yellow-breasted Chat *Icteria virens*
Yellow-breasted Warbler *Seicercus montis*
Yellow-chinned Spinetail *Certhiaxis cinnamomeus*
Yellow-crowned Canary *Serinus flavivertex*
Yellow-eyed Babbler *Chrysomma sinense*
Yellow-faced Honeyeater *Lichenostomus chrysops*
Yellow-faced Myna *Mino dumontii*
Yellow-fronted Canary *Serinus mozambicus*
Yellow-fronted Honeyeater *Meliphaga plumula*
Yellowhammer *Emberiza citrinella*

Yellow-faced Honeyeater *Lichenostomus chrysops*
Yellowhead *Mohoura ochrocephala*
Yellow Honeyeater *Lichenostomus flavus*
Yellow-olive Flycatcher *Tolmomyias sulphurescens*
Yellow-plumed Honeyeater *Lichenostomus ornatus*
Yellow-rumped Thornbill *Acanthiza chrysorrhoa*
Yellow-spotted Petronia *Gymnorhis pyrgita*
Yellow-streaked Honeyeater *Meliphaga macleayana*
Yellow Thornbill *Acanthiza nana*
Yellow-throated Fulvetta *Alcippe cinerea*
Yellow-throated Greenbul *Phyllastrephus albigularis*
Yellow-throated Honeyeater *Lichenostomus flavicollis*
Yellow-throated Miner *Manorina flavigula*
Yellow-throated Petronia *Gymnorhis superciliaris*
Yellow-throated Scrubwren *Sericornis citreogularis*
Yellow-throated Woodland Warbler *Phylloscopus ruficapilla*
Yellow-tufted Honeyeater *Lichenostomus melanops*
Yellow-tufted Malachite Sunbird *Nectarinia famosa*
Yellow-vented Bulbul *Pycnonotus goiavier*
Yellow Wagtail *Motacilla flava*
Yellow Warbler *Dendroica petechia*
Yellow Wattlebird *Anthochaera paradoxa*
Yellow-whiskered Greenbul *Andropadus latirostris*
Zanzibar Red Bishop *Euplectes nigroventris*
Zebra Finch *Taeniopygia guttata*
Zitting Cisticola *Cisticola juncidis*

BIBLIOGRAPHY

Aagaard, C.J. 1930. *The common birds of Bangkok.* Chr. Backhausen, Copenhagen.

Abdulali, H. 1943. The eggs of the Indian Bay-banded Cuckoo (*Penthoceryx sonneratii*). *J. Bombay Nat. Hist. Soc.* 44: 122.

Abdulali, H. 1956. Some notes on the plumages of *Centropus sinensis* (Stephens). *J. Bombay Nat. Hist. Soc.* 54: 183-185.

Ackerman, M. 1841. Note sur le Coua. Famac-acora des Malgaches, Hache-escargot (traduction littérale) ou casseur d'escargots. *Rev. Zool.* 4: 209-210.

Adamik, P., Hušek, J. & Cepák, J. 2009. Rapid decline of Common Cuckoo *Cuculus canorus* parasitism in Red-backed Shrikes *Lanius collurio*. *Ardea* 97: 17-22.

Adams, I.C. 1933. Late nesting of the Yellow-billed Cuckoo in Missouri. *Auk* 50: 107.

Agro, D.J. 1994. Grasshoppers as food source for Black-billed Cuckoo. *Ontario Birds* 12: 28-29.

Aleixo, A. & Galetti, M. 1997. The conservation of the avifauna in a lowland Atlantic forest in south-east Brazil. *Bird Conserv. Int.* 7: 235-261.

Alexander, B. 1902. On the birds of the Gold Coast Colony and its hinterland; part 2. *Ibis* (8th series) 2: 355-377.

Alexander, B. 1908. On a new species of cuckoo. *Bull. Brit. Orn. Club* 21: 78.

Alexander, W.B. 1925. Visit to Coomooboolaroo. *Emu* 24: 236-237.

Ali, S.A. 1941. The birds of Bahawalpur (Punjab). *J. Bombay Nat. Hist. Soc.* 42: 704-747.

Ali, S.A. 1946. An ornithological pilgrimage to lake Manasarowar and mount Kailas. *J. Bombay Nat. Hist. Soc.* 46: 286-308.

Ali, S.A. 1962. *The birds of Sikkim.* OUP, London.

Ali, S. 1996. *The book of Indian birds.* 12th ed. Bombay Natural History Society, Bombay.

Ali, S. & Ripley, S.D. 1969. *Handbook of the birds of India and Pakistan.* Vol 3. OUP, Bombay.

Ali, S. & Ripley, S.D. 1983. *Handbook of the birds of India and Pakistan.* Compact edition. OUP, Delhi.

Ali, S.A. & Whistler, H. 1936. The ornithology of Travancore and Cochin. Part VI. *J. Bombay Nat. Hist. Soc.* 39: 3-35.

Allan, D.G. 2002. First record of Striped Cuckoo *Clamator levaillantii* parasitising Pied Babbler *Turdoides bicolor*. *Bird Numbers* 11(2): 30.

Allen, J.A. 1876. List of birds collected by Mr. Charles Linden, near Santarém, Brazil. *Bull. Essex Inst.* 8: 78-83.

Allen, J.A. 1893. On a collection of birds from Chapada, Matto Grosso, Brazil, made by Mr. H. H. Smith. Part III. Pipridae to Rheidae. *Bull. Amer. Mus. Nat. Hist.* 5: 107-158.

Allen, D., Española, C., Broad, G., Oliveros, C. & Gonzalez, J.C.T. 2006. New bird records for the Babuyan islands, Philippines, including two first records for the Philippines. *Forktail* 22: 57-70.

Alley, T.H. 1978. Observations on the diet of the fledgling Rufous-tailed Bronze-Cuckoo. *Corella* 2: 75-76.

Alonzo-Pasicolan, S. 1992: The bird-catchers at Dalton Pass. *Bull. OBC* 15: 33-36.

Allport, G.A. & Fanshawe, J.R. 1994. Is the Thick-billed Cuckoo *Pachycoccyx audeberti* a forest dependent species in West Africa? *Malimbus* 16: 52-53.

Al Sirhan, A. & Al Hajji, R. 2010. The first Asian Koel for Kuwait. *Phoenix* 26: 8.

Alvarez, F., de Reyna, L.A. & Segura, M. 1976. Experimental brood parasitism of the Magpie (*Pica pica*). *Anim. Behav.* 24: 907-916.

Alves, M.A.S., Pacheco, J.F., Gonzaga, L.P., Cavalcanti, R.B., Raposo, M.A., Yamashita, C., Maciel, N.C. & Castanheira, M. 2000. Aves. Pp. 113-124 in H.G. Bergallo, C.F.D. Rocha, M.A.S. Alves & M. van Sluys (Eds). *A fauna ameaçada de extinção do estado do Rio de Janeiro.* Ed. Universidade Estadual do Rio de Janeiro, Rio de Janeiro.

Amadon, D. 1942. Birds collected during the Whitney South Sea Expedition. 50. Notes on some non-passerine genera. 2. *Amer. Mus. Nov.* 1176: 1-21.

Amadon, D. 1953. Avian systematics and evolution in the Gulf of Guinea. *Bull. Amer. Mus. Nat. Hist.* 100: 399-451.

Ambrose, S.J. 1987. Adult Fan-tailed Cuckoo *Cuculus pyrrhophanus* feeds fledgling. *Emu* 87: 69.

Amos, E.J.R. 1991. *A guide to the birds of Bermuda.* Corncrake, Warwick, Bermuda.

Anciaux, M-R. 2000. Approche de la phénologie de la migration des migrateurs intra-africains de l'intérieur des terres du sud-Bénin (plateau d'Allada et sud de la depression de la Lama). Les non-Passériformes et les non-Coraciiformes. *Alauda* 68: 311-320.

Andersson, M. 1995. Evolution of reversed sex roles, sexual size dimorphism, and mating system in coucals (Centropodidae, Aves). *Biol. J. Linnean Soc.* 54 (2): 173–181.

Andrew, P. 1990. The status of the Sunda Coucal *Centropus nigrorufus*. *Kukila* 5 (1): 56-64.

Andrew, P. 1992. *The birds of Indonesia. A checklist (Peters' sequence).* Kukila Checklist No. 1. Indonesian Ornithological Society, Jakarta.

Andrew, P. & Holmes, D.A. 1988. Sulawesi bird report. *Kukila* 5: 4-26.

Andrews, C.W. 1900. *A monograph of Christmas Island.* British Museum (Natural History), London.

Andrews, I.J. 1995. *The Birds of the Hashemite Kingdom of Jordan.* Privately published, Midlothian.

Andrle, R.F. 1967. Birds of the Sierra de Tuxtla in Veracruz, Mexico. *Wilson Bull.* 79: 163-187.

Anon. 1937. Lista sistemática de las aves Argentinas. *Hornero* 6: 531-554.

Anon. 1958. Borneo bird notes, 1957. *Sarawak. Mus. J.* 8: 442-461.

Anon. 1980. Vier thailändische Laufkuckucke *Carpococcyx renauldi* neu im Zoo Berlin. *Gefiederte Welt* 104 (3): 59-60.

Anon. 1985. Nest record card data - Senegal Coucal, *Centropus senegalensis*. *Zambian Orn.. Soc. Newsl.* 15: 184-186.

Anon. 1995. European news. *Brit. Birds* 88: 273.

Anon. 2009. thebutterflydiaries.wordpress.com/2009/10/21/the-forcibly-adopted-child

Anon. 2010. avibase.bsc-eoc.org

Anon. 2011a. *Madagascar.* Pt. 2. Television program. British Broadcasting Corporation.

Anon. 2011b. www.birdguides.com.

Anthony, A.W. 1896. The Roadrunner as a rat-killer. *Auk* 13: 257-258.

Antunes, A.Z. 2005. Alterações na composição da comunidade de aves ao longo do tempo em um fragmento florestal do sudeste do Brasil. *Ararajuba* 13: 47-61.

AOU = American Ornithologists' Union 1983. *Check-list of North American birds,* 6th ed. AOU, Lawrence, Kansas.

AOU = American Ornithologists' Union 1998. *Check-list of North American birds,* 7th ed. AOU, Washington DC.

Appert, O. 1967. Die Rachenzeichnung beim Nestling des Braunkopf-Seidenkuckucks *Coua ruficeps olivaceiceps* Sharpe, von Madagaskar. *Orn. Beob.* 64: 52-56.

Appert, O. 1968. La répartition géographique des couas dans la région du Mongoky et la question de leur présence aux différentes époques de l'année. *Bull. Acad. Malgache (Nouvelle Sér.)* 44 (2): 29-39.

Appert, O. 1970. Zur Biologie einiger Kua-Arten Madagaskars (Aves Cuculi). *Zool. Jb. (Syst.)* 97: 424-453.

Appert, O. 1971. Über die Rachenzeichnung des jungen Seidenkuckucks. *Natur Mus. Frankf.* 101: 10-14.

Appert, O. 1980. Erste Farbaufnahmen der Rachenzeichnung junger Kuas von Madagaskar (Cuculi, Couinae). *Orn. Beob.* 77: 85-101.

Aragón, S., Møller, A.P., Soler, J.J. & Soler, M. 1999. Molecular phylogeny of cuckoos supports a polyphyletic origin of brood parasitism. *J. Evol. Biol.* 12: 495-506.

Aravena, R.O. 1928. Notas sobre la alimentación de las aves. *Hornero* 4: 153-166.

Archer, A.L. 1985. Protective behaviour in a nestling Red-chested Cuckoo *Cuculus solitarius*. *Scopus* 9: 52-53.

Arias de Reyna, L. & Hidalgo, S.J. 1982. Egg acceptance by Azure-winged Magpies and host-recognition by Great Spotted Cuckoo chicks. *Anim. Behav.* 30: 819-823.

Arias de Reyna, L., Recuerda, P., Trujillo, T., Corvillo, M. & Cruz, A. 1987. Territory in the Great Spotted Cuckoo (*Clamator glandarius*). *J. Orn.* 128: 231-239.

Arifin, H.W. 1997. Beberapa aspek ekologi burung bubut hitam (*Centropus nigrorufus* Cuvier) di Rph Cangkring, Bkph Indramayu, Kph Indramayu perum perhutani unit III Jawa Barat. Pp. 219–227 in N. Sugiri, D.M. Prawiradilaga, K.G. Wiryawan & A.S. Adhikerana (Eds). *Prosiding seminar nasional pelestarian burung dan ekosistemnya dalam pembangunan berkelanjutan di Indonesia* [*Proceedings of national seminar on the conservation of birds and their ecosystems for supporting sustainable development in Indonesia.*] LIPI-IPB, Bogor.

Arribas, M.A., Jammes, L. & Sagot, F. 1995. *Lista de las aves de Bolivia. A birdlist of Bolivia*, 4th ed. Asociación ARMONIA, Santa Cruz de la Sierra, Bolivia.

Artyukhin, Yu. B. 1997. [Lanceolated Warbler *Locustella lanceolata* - a new host of Oriental Cuckoo *Cuculus saturatus*.] *Russian Orn. J.* 6 (14): 19-20. [in Russian]

Ash, J.S. 1984. Bird observations on Bali. *Bull. Brit. Orn. Club* 104: 24-35.

Ash, J.S. 1990. Additions to the avifauna of Nigeria, with notes on distributional changes and breeding. *Malimbus* 11: 104-116.

Ash, J.S & Atkins, J. 2009. *Birds of Ethiopia and Eritrea: an atlas of distribution*. Christopher Helm, London.

Ash, J.S. & Miskell, J.E. 1983. *Birds of Somalia, their habitat, status and distribution*. *Scopus* Suppl. 1.

Ash, J.S. & Miskell, J.E. 1998. *Birds of Somalia*. Pica Press, Sussex.

Ash, J.S. & Shafeeg, A. 1994. Birds of the Maldive Islands, Indian Ocean. *Forktail* 10: 3-32.

Ashby, E. 1915. Notes on *Polophilus phasianinus melanurus*, Gld. (North-Western Coucal). *S. Aust. Orn.* 2: 72.

Ashton, C.B. 1985. The birds of the Aldinga-Sellicks Beach scrub. *S. Aust. Orn.* 29: 169-179.

Aspinwall, D.R. 1973. Bird notes from five provinces. *Bull. Zambian Ornithol. Soc.* 5 (2): 43-63.

Aspinwall, D.R. & Beel, C. 1998. *A field guide to the Zambian birds not found in southern Africa*. Zambian Ornithological Society, Lusaka.

Asterly Maberly, C.T. 1961. Observations on the Thick-billed Cuckoo *Pachycoccyx validus canescens*. *Ostrich* 32: 136-137.

Atkinson, R.W. 1982. Breeding the Renauld's or Coral-billed Ground Cuckoo *Carpococcyx renauldi* at the Metro Toronto Zoo. *Int. Zoo Yearb.* 22: 168-171.

Aubry, Y., Gosselin, M. & Yank, R. 1992. The autumn migration. August 1 - November 30, 1991. Quebec Region. *Am. Birds* 46: 57-60.

Aveledo-Hostos, R. & Ginés, R.H. 1950. Descripción de cuatro aves nuevas de Venezuela. *Mem. Soc. Cienc. Nat. La Salle* 10: 59-71.

Aveledo-Hostos, R. & Pérez-Chinchilla, L.A. 1994. Descripción de nueve subespecies nuevas y comentarios sobre dos especies de aves de Venezuela. *Bol. Soc. Venez. Cienc. Nat.* 148: 229-257.

Avilés, J.M. 2004. Egg rejection by Iberian Azure-winged Magpie *Cyanopica cyanus* in the absence of brood parasitism. *J. Avian Biol.* 35: 295-299.

Avilés, J.M., Soler, J.J., Perez-Contreras, T., Soler, M. & Møller, A.P. 2006. Ultraviolet reflectance of great spotted cuckoo eggs and egg discrimination by magpies. *Behav. Ecol.* 17: 310-314.

Avilés, J.M., Stokke, B.G., Moksnes, A., Røskaft, E. & Møller A.P. 2007. Environmental conditions influence egg color of reed warblers *Acrocephalus scirpaceus* and their parasite, the common cuckoo *Cuculus canorus*. *Behav. Ecol. Sociobiol.* 61: 475-485.

Ayres, J.M. 1985. On a new species of squirrel monkey, genus *Saimiri*, from Brazilian Amazonia (Primates, Cebidae). *Pap. Avuls. Zool. S. Paulo* 36: 147-164.

Azategui, A. 1975. Datos sobre aves de Santa Isabel (Córdoba). *Hornero* 11: 321.

Azevedo, M.A.G. & Ghizoni, I.R. Jnr. 2005. Novos registros de aves para o Estado de Santa Catarina, sul do Brasil. *Atual. Orn.* 126: 9, 12.

Azpiroz, A.B. 2001. *Aves del Uruguay. Lista e introducción a su biología y conservación*. Aves Uruguay - GUPECA. Graphis Editorial, Montevideo.

Azpiroz, A.B. & Menéndez, J.L. 2008. Three new species and novel distributional information for birds in Uruguay. *Bull. Brit. Orn. Club* 128: 38-56.

Babó, T. 1987. [Great Spotted Cuckoo (*Clamator glandarius*) – new species for the avifauna of Czechoslovakia.] *Tichodroma* 1: 6-11. [in Slovak]

Backhurst, G.C., Britton, P.L. & Mann, C.F. 1973. The less common Palaearctic migrant birds of Kenya and Tanzania. *J. E. Afr. Nat. Hist. Soc.* 140: 1-38.

Baepler, D.H. 1962. The avifauna of the Soloma region in Huehuetenango, Guatemala. *Condor* 64: 140-153.

Baerg, W.J. 1950. Occurrence of the Road-runner in Arkansas. *Condor* 52: 165.

Bahr, N., Neumann, F. & Schleussner, G. 1992. Über Handaufzuchten von Rennkuckucken im Vogelpark Walsrode. *Tropische Vögel* 13: 32-34.

Baker, E.C.S. 1906a. The oology of Indian parasitic cuckoos. Part I. *J. Bombay Nat. Hist. Soc.* 17: 72-83.

Baker, E.C.S. 1906b. The oology of Indian parasitic cuckoos. Part II. *J. Bombay Nat. Hist. Soc.* 17: 351-374.

Baker, E.C.S. 1906c. The oology of Indian parasitic cuckoos. Part III. *J. Bombay Nat. Hist. Soc.* 17: 678-696.

Baker, E.C.S. 1908. Additional cuckoo notes. *J. Bombay Nat. Hist. Soc.* 18: 275-279.

Baker, E.C.S. 1919. Some notes on the genus *Surniculus*. *Novit. Zool.* 26: 291-294.

Baker, E.C.S. 1919. On the races of *Penthoceryx sonnerati*. *Bull. Brit. Orn. Club* 39: 45-47.

Baker, E.C.S. 1926. Remarks on Oriental birds and descriptions of five new subspecies. *Bull. Brit.Orn. Club* 47: 41-45.

Baker, E.C.S. 1927. *The fauna of British India including Ceylon and Burma*. Birds 4. Taylor and Francis, London.

Baker, E.C.S. 1934. *The nidification of birds of the Indian Empire*. vol. 3. Taylor & Francis, London.

Baker, E.C.S. 1942. *Cuckoo problems*. Witherby, London.

Baker, N.E. & Baker, E.M. 1994. Dusky Long-tailed Cuckoo *Cercococcyx mechowi* and Papyrus Canary *Serinus koliensis*: two additions to the Tanzania list. *Scopus* 18: 122-123.

Baker, N.E. & Baker, E.M. (in prep). *The birds of Tanzania: an atlas of distribution and seasonality*.

Baker, R.H. 1951. The avifauna of Micronesia, its origin, evolution, and distribution. *Univ. Kansas Publ. Mus. Nat. Hist.* 3: 1-359.

Bakewell, D. 2010. New records for Malaysia. *BirdingASIA* 13: 22-29.

Bakewell, D.N., Lim, K.C. & Shahrul Anuar Mohd. Sahr. 2011. Brood-host record for Oriental (Sunda) Cuckoo *Cuculus (saturatus) lepidus* in the Cameron Highlands, Peninsular Malaysia. *BirdingASIA* 15: 48.

Balança, G. & Visscher, M-N. de. 1997. Composition et évolution saisonni re d'un peuplement d'oiseaux au nord du Burkina Faso (nord-Yatenga). *Malimbus* 19: 68-94.

Balatskiy, N.N. 1988. [*Cuculus canorus*. L. – the ecological races of *Acrocephalus dumetorum* Blyth and *Saxicola torquata* L. near Novosibirsk.] *Nauchn. Dokl. Vyssh. Shk. Biol. Nauki* 1988 (4): 57-61. [in Russian]

Balatskiy, N.N. 1989. [The ecological race of the Cuckoo – nest parasite of the Little Bunting in north Urals.] *Nauchn. Dokl. Vyssh. Shk. Biol. Nauki* 1989 (10): 46-49. [in Russian]

Balatskiy, N.N. 1990. [The ecological race of the Cuckoo – nest parasite of the Siberian Rubythroat in Siberia.] *Nauchn. Dokl. Vyssh. Shk. Biol. Nauki* 1990 (4): 37-44. [in Russian]

Balatskiy, N.N. 2008. [Cuckoo *Cuculus canorus* – nest parasite of

the Common Redstart near Novosibirsk.] *Russian Orn. J.* 17 (Express issue 455): 1835-1837. [in Russian]
Balatskiy, N.N. 2011. [*Cuculus canorus* in Russia.] http://www.balatskiy.ru/Cuc_Russii/Cuckoo.htm. [in Russian] accessed June 2011.
Baldwin, M. 1975. Birds of Inverell district, N.S.W. *Emu* 75: 113-120.
Balmer, D. & Betton, K. 2008. Around the region. *Sandgrouse* 30: 214-224.
Balmer, D. & Murdoch, D. 2009a. Around the Region. *Sandgrouse* 31: 95.
Balmer, D. & Murdoch, D. 2009b. Around the Region. *Sandgrouse* 31: 208-222.
Balmer, D. & Murdoch, D. 2010. Around the Region. *Sandgrouse* 32: 176-189.
Bangs, O. 1902. On a second collection of birds made in Chiriqui, by W.W. Brown, Jnr. *Proc. New England Zool. Club* 3: 15-70.
Bangs, O. 1907. A new race of the Mangrove Cuckoo, from Grenada and the Grenadines. *Proc. Biol. Soc. Wash.* 20: 53-54.
Bangs, O. 1918. Vertebrata from Madagascar. Aves. *Bull. Mus. Comp. Zool.* 61: 487-511.
Bangs, O. & Penard, T.E. 1918. Notes on a collection of Surinam birds. *Bull. Mus. Comp. Zool.* 62: 23-93.
Bangs, O. & Zappey, W.R. 1905. Birds of the Isle of Pines. *Am. Nat.* 39: 179-215.
Banks, R.C. 1988a. An old record of the Pearly-breasted Cuckoo in North America and a nomenclatural critique. *Bull. Brit. Orn. Club* 108: 87-91.
Banks, R.C. 1988b. Geographic variation in the Yellow-billed Cuckoo. *Condor* 90: 473-477.
Banks, R.C. 1990. Geographic variation in the Yellow-billed Cuckoo: corrections and comments. *Condor* 92: 538.
Banks, R.C., Cicero, C., Dunn, J.L., Kratter, A.W., Rasmussen, P.C., Remsen, J.V. Jnr, Rising, J.D. & Stotz, D.F. 2006. Forty-seventh supplement to the American Ornithologists' Union Check-List of North American Birds. *Auk* 123: 926-936.
Banks, R.C. & Hole, R., Jnr. 1991. Taxonomic review of the Mangrove Cuckoo, *Coccyzus minor* (Gmelin). *Caribb. J. Sci.* 27: 54-62.
Bannerman, D.A. 1922. On the Emerald and Golden Cuckoos of Africa. *Novit. Zool.* 29: 413–420.
Bannerman, D.A. 1933. *The birds of tropical West Africa. With special reference to those of the Gambia, Sierra Leone, the Gold Coast and Nigeria.* Vol. 3. Crown Agents, London.
Bannerman, D.A. 1951. *The birds of tropical West Africa.* Vol. 8. Crown Agents, London.
Bannerman, D.A. 1953. *The birds of West and Equatorial Africa.* Vol. 1. Oliver & Boyd, Edinburgh.
Bannerman, D.A. 1963. *Birds of the Atlantic Islands.*1. *A history of the birds of the Canary Islands and of the Salvages.* Oliver & Boyd, Edinburgh & London.
Bannerman, D.A. & Bannerman, W.M. 1958. *Birds of Cyprus.* Oliver & Boyd, Edinburgh.
Bannerman, D.A. & Bannerman, W.M. 1965. *Birds of the Atlantic Islands.* 2. *A history of the birds of Madeira, the Desertas, and the Porto Santo Islands.* Oliver & Boyd, Edinburgh & London.
Bannerman, D.A. & Bannerman, W.M. 1966. *Birds of the Atlantic Islands.* 3. *A history of the birds of the Azores.* Oliver & Boyd, Edinburgh.
Bannerman, D.A. & Bannerman, W.M. 1983. *The birds of the Balearics.* Croom Helm, London & Canberra.
Barbour, T. 1923. *The birds of Cuba.* (Memoirs; 6). Nuttall Ornithological Club, Cambridge, Mass.
Barbour, T. 1943. *Cuban ornithology.* (Memoirs; 9). Nuttall Ornithological Club, Cambridge, Mass.
Barclay, J.S. 1977. Roadrunner takes birds from mist net. *Bird-Banding* 48: 280.
Barlow, C.R. 2003. First conclusive evidence of breeding in Senegambia and parental behaviour of Black Coucal *Centropus grillii*. *Bull. ABC* 10: 53-54.
Barlow, C., Wacher, T. & Disley, T. 1997. *A field guide to birds of The Gambia and Senegal.* Pica Press, Sussex.

Barnard, H.G. 1926. Birds of the Cardwell district, Queensland. *Emu* 26: 1-13.
Barnett, J.M. &. Pearman, M. 2001. *Lista comentada de las aves argentinas. Annotated checklist of the birds of Argentina.* Lynx Edicions, Barcelona.
Barrett, G., Silcocks, A., Barry, S., Cunningham, R. & Poulter, R. 2003. *The new atlas of Australian birds.* Royal Australasian Ornithologists' Union, Hawthorn East.
Barros, R. & Schmitt, F. 2011. Resumen de avistamientos Septiembre 2010-Febrero 2011. *Chiricoca* 13: 30-50
Bartels, M. 1915–1931. Diaries and notebooks on Javan birds. Unpublished typescripts and manuscripts held at RMNH.
Barua, M. & Sharma, P. 1999. Birds of Kaziranga National Park, India. *Forktail* 15: 47-60.
Barua, M. & Sharma, P. 2005. The birds of Nameri National Park, Assam, India. *Forktail* 21: 15-26.
Basil-Edwardes, S. 1926. A contribution to the ornithology of Delhi. Part 2. *J. Bombay Nat. Hist. Soc.* 31: 567-578.
Basilio, A. 1963. *Aves de la isla de Fernando Poo.* Editorial Coculsa, Madrid.
Batchelor, D.M. 1958. A checklist of the birds seen at Asahan, Malacca: Part 1. *Malay Nat. J.* 12: 164-182.
Bates, J.M., Garvin, M.C., Schmitt, D.C. & Schmitt, C.G. 1989. Notes on bird distribution in northeastern Dpto. Santa Cruz, Bolivia, with 15 species new to Bolivia. *Bull. Brit. Orn. Club* 109: 236-244.
Bates, G.L. 1905. With notes by the collector, in Sharpe, R.B. On further collections of birds from the Efulen District of Camaroon, West Africa. Part 3. *Ibis* (8th series) 5: 461-476.
Bates, G.L. 1909. Field-notes on the birds of southern Kamerun, West Africa. *Ibis* (9th series) 3: 1-74.
Bates, G.L. 1911. Further notes on the birds of southern Cameroon. With descriptions of eggs by W.R. Ogilvie-Grant. *Ibis* (9th series) 5: 479-545, pls. 7-12.
Bates, G.L. 1924. On the birds collected in northwestern and northern Cameroon and parts of northern Nigeria. *Ibis* (11th series) 6: 1-45.
Bates, G.L. 1930. *Handbook of the birds of West Africa.* John Bale, Sons and Danielsson, London.
Bates, G.L. 1937. Two new Arabian birds. *Bull. Brit. Orn. Club* 57: 150-151.
Bates, J.M., Parker T.A., III, Capparella, A.P. & Davis, T.J. 1992. Observations on the *campo*, *cerrado* and forest avifaunas of eastern Dpto. Santa Cruz, Bolivia, including 21 species new to the country. *Bull. B.O.C.* 112: 86-98.
Bates, R.S.P. & Lowther, E.H.N. 1952. *Breeding birds of Kashmir.* Geoffrey Cumberlege & OUP, Oxford.
Bauer, W., von Helversen, O., Hodge, M. & Martens, J. 1969a. Bemerkenswerte Brutnachweise aus Griechenland. *J. Orn.* 79-83.
Bauer, W., von Helversen, O., Hodge, M. & Martens, J. 1969b. *Catalogus Fauna Graeciae.* Pars II Aves.
Beal, K.G. 1981. Winter foraging habits of the Roadrunner. *Bull. Oklahoma Orn. Soc.* 14: 13-15.
Beal, K.G. & Beal, R.E. 1978. Immature Cooper's Hawk attempts to capture Roadrunner. *Bull. Oklahoma Orn. Soc.* 11: 31.
Beal, K.G. & Gillam, L.D. 1979. On the function of prey beating by Roadrunners. *Condor* 81: 85-87.
Beaman, M., Madge, S. & Olsen, K.M. 1998. *Fuglene i Europa, Nordafrika og Mellemøsten.* Gads Forlag, Copenhagen.
Beattie, W.P. 1981. Parasitic behaviour of the Red-chested Cuckoo. *Bull. E. Afr. Nat. Hist. Soc.* (Jan.-Feb): 7.
Beavan, R.C. 1867. The avifauna of the Andaman Islands. *Ibis* (2nd series) 3: 314-334.
Becking, J.H. 1975. New evidence of the specific affinity of *Cuculus lepidus* Müller. *Ibis* 117: 275-284.
Becking, J.H. 1981. Notes on the breeding of Indian cuckoos. *J. Bombay Nat. Hist. Soc.* 78: 201-231.
Becking, J.H. 1988. The taxonomic status of the Madagascar Cuckoo *Cuculus poliocephalus rochii* and its occurrence on the African mainland, including southern Africa. *Bull. Brit. Orn. Club* 108: 195-206.
Beddard, F.E. 1901. On the anatomy of the Radiated Fruit-Cuckoo (*Carpococcyx radiatus*). *Ibis* (8th series) 1: 200-214.

Beehler, B.M. 1978. *Upland birds of northeastern New Guinea*. Wau Ecology Institute Handbook 4. Wau.

Beehler, B.M., Pratt, T.K. & Zimmerman, D.A. 1986. *Birds of New Guinea*. Princeton University Press, Princeton.

Begum, S., Moksnes, A., Røskaft, E. & Stokke, B.G. 2011. Factors influencing host nest use by the brood parasitic Asian Koel (*Eudynamys scolopacea*). *J. Orn.* 152: 793-800.

Behle, W.H. 1943. Birds of Pine Valley Mountain region, southwestern Utah. *Bull. Univ. Utah, Biol. Ser.* 34: 1-85.

Behle, W.H. & Selander, R.K. 1952. New and additional records of Utah birds. *Wilson Bull.* 64: 26-32.

Beisenherz, W. 1998. Mountain Tailorbird as host of Rusty-breasted Cuckoo. *Kukila* 10: 159.

Belcher, C.F. 1930. *The birds of Nyassaland*. Technical Press, London.

Belcher, C.F. 1949. Review [of Dr. Friedmann's parasitic cuckoos of Africa]. *Nature in E. Africa, Nairobi* 2: 14-20.

Belcher, C. & Smooker, G.D. 1936. Birds of the colony of Trinidad and Tobago. Part III. *Ibis* (13th series) 6: 1-35.

Bell, H.L. 1970. Field notes on the birds of Amazon Bay, Papua. *Emu* 70: 23-26.

Bell, H.L. 1979. The effects on rain-forest birds of plantings of Teak *Tectona grandis*, in Papua New Guinea. *Aust. Wildl. Res.* 6: 305-318.

Bell, H.L. 1982. A bird community of lowland rainforest in New Guinea. 1. Composition and density of avifauna. *Emu* 82: 24-41.

Bell, H.L. 1986. The participation by cuckoos in mixed-species flocks of insectivorous birds in south-eastern Australia. *Emu* 86: 249-253.

Belton, W. 1973. Some additional birds for the state of Rio Grande do Sul, Brazil. *Auk* 90: 94-99.

Belton, W. 1984. Birds of Rio Grande do Sul, Brazil. Part 1. Rheidae through Furnariidae. *Bull. Amer. Mus. Nat. Hist.* 178: 369-636.

Bencke, G.A. 2001. *Lista de referência das aves do Rio Grande do Sul*. Fundação Zoobotânica do Rio Grande do Sul, Porto Alegre.

Bender, R.O. 1961. Food competition among closely related sympatric species. *Wilson Bull.* 73: 214.

Bendire, C. 1895. *Life histories of North American birds, from the parrots to the grackles, with special reference to their breeding habits and eggs*. US Nat. Mus., Special Bull., 3. Gov. Printing Office, Washington DC.

Bennett, A.G. 1937. *Coccyzus melanocoryphus* in the Falkland Islands. *Ibis* (14th series) 1: 868.

Benson, C.W. 1945. Notes on the birds of southern Abyssinia. *Ibis* 87: 489-509.

Benson, C.W. 1946. Notes on the birds of southern Abyssinia. *Ibis* 88: 25-48, 180-205, 287-306, 444-461.

Benson, C.W. 1948. A new race of the coucal from Nyasaland. *Bull. Brit. Orn. Club* 68: 127-128.

Benson, C.W. 1951. The Lesser Cuckoo *Cuculus poliocephalus poliocephalus* Latham in Nyasaland. *Ibis* 93: 627-628.

Benson, C.W. 1953. *A check list of the birds of Nyasaland*. Nyasaland Society, Blantyre.

Benson, C.W. 1967. The birds of Aldabra and their status. *Atoll Res. Bull.* 118: 63-111.

Benson, C.W. & Benson, F.M. 1947. Some breeding records from southern Nyassaland. *Oolog. Rec.* 21 (4): 1-9.

Benson, C.W. & Benson, F.M. 1949. Notes on birds from northern Nyassaland and adjacent Tanganyika Territory. *Ann. Transvaal Mus.* 21: 155-177.

Benson, C.W. & Benson, F.M. 1977. *The birds of Malawi*. Montfort Press, Limbe, Malawi.

Benson, C.W., Brooke, R.K., Dowsett, R.J. & Irwin, M.P.S. 1971. *The birds of Zambia*. Collins, London.

Benson, C.W., Colebrook-Robjent, J.F.R. & Williams, A. 1976-1977. A contribution to the ornithology of Madagascar. *Oiseau RFO* 46: 103-134, 209-242, 368-386; 47: 41-64, 168-191.

Benson, C.W. & Irwin, M.P.S. 1965. The birds of *Cryptosepalum* forests, Zambia. *Arnoldia (Rhodesia)* 28: 1-12.

Benson, C.W. & Irwin, M.P.S. 1972. The Thick-billed Cuckoo *Pachycoccyx audeberti* (Schlegel) (Aves: Cuculidae). *Arnoldia (Rhodesia)* 5 (33): 1-24.

Benson, C.W. & Penny, M.J. 1971. The land birds of Aldabra. *Phil. Trans. Roy. Soc. London* (B) 260: 417-529.

Benson, C.W. & Pitman, C.R.S. 1956. Some breeding records from northern Rhodesia. *Oolog. Rec.* 30: 7-11, 21-27, 37-43.

Benson, C.W. & Pitman, C.R.S. 1964. Further breeding records from northern Rhodesia. Part 4. *Bull. Brit. Orn. Club* 84: 54-69.

Benson, C.W. & Schüz, E. 1971. A specimen of *Coua delalandei* (Temminck) (Cuculidae). *Bull. Brit. Orn. Club* 91: 159-160.

Benson, C.W. & Serventy, D.L. 1957. The feeding of the young Didric Cuckoo *Chrysococcyx caprius*. *Ibis* 99: 347-349.

Benson, C.W. & White, C.M.N. 1957. *Check-list of the birds of Northern Rhodesia*. Lusaka.

Benson, S.V. 1970. *Birds of Lebanon and the Jordan area*. ICBP and Warne, London & New York.

Bent, A.C. 1940. Life histories of North American cuckoos, goatsuckers, hummingbirds, and their allies. *Bull. US Nat. Mus.* 176: 1-506.

Berger, A.J. 1955. On the anatomy and relationships of glossy cuckoos of the genera *Chrysococcyx*, *Lampromorpha*, and *Chalcites*. *Proc. US Nat. Mus.* 103: 585-597.

Berger, A.J. 1960. Some anatomical characters of the Cuculidae and the Musophagidae. *Wilson Bull.* 72: 60-104.

Bergh, C.J. 1942. Some birds of Belfast, Transvaal. *Ostrich* 13: 98-99.

Berry, P.S.M. 1974. Bird records of note from the Luangwa Valley. *Bull. Zambian Orn. Soc.* 6: 33-34.

Bertoni, A.W. 1901. *Aves nuevas del Paraguay*. Talleres Nacionales de H. Kraus, Asunción.

Beruldsen, G.R. 1978. The egg of the Brush Cuckoo - *Cacomantis variolosus*. *Aust. Bird Watcher* 7: 281-282.

Beruldsen, G.R. 1980. *A Fieldguide to the Nests and Eggs of Australian Birds*. Rigby, Adelaide.

Beruldsen, G. 1990. Cape York in the wet. *Aust. Bird Watcher* 13: 209-217.

Berwick, E.J.H. 1952. Bird notes from Kelantan. *Bull. Raffles Mus.* 24: 183-198.

Best, L.B. & Stauffer, D.F. 1980. Factors affecting nesting success in riparian bird communities. *Condor* 82: 149-158.

Betts, M. 2002. A systematic list of the birds of Aldabra. *Bull. ABC* 9: 32-41.

Bharucha, E.K. 1982. Sunbirds fostering fledglings of the Plaintive Cuckoo. *J. Bombay Nat. Hist. Soc.* 79: 670-671.

Biddulph, C.H. 1956. Occurrence of the Redfaced Malkoha *Phoenicophaeus pyrrhocephalus* (Pennant) in Madura district, Madras Presidency. *J. Bombay Nat. Hist. Soc.* 53: 697-698.

Binford, L.C. 1971. Roadrunner captures Orchard Oriole in California. *Calif. Birds* 2: 139.

Binford, L.C. 1989. A distributional survey of the birds of the Mexican state of Oaxaca. *Orn. Monogr.* 43: 1-418.

BirdLife International. 2000. *Threatened birds of the world*. Lynx Edicions and BirdLife International, Barcelona and Cambridge, UK.

BirdLife International. 2001. *Threatened birds of Asia: the BirdLife International Red Data Book*. Part B. Pp. 1517-3038. BirdLife International, Cambridge, UK.

BirdLife International. 2003. BirdLife's online World Bird Database: the site for bird conservation. Version 2.0. BirdLife International, Cambridge, UK.

BirdLife International. 2011. Species factsheets. http://www.birdlife.org

Birds Korea. 2009. http://www.birdskorea.org

Birds Korea. 2010. http://www.birdskorea.org

Birkhead, T.R., Hemmings, N., Spottiswoode, C.N., Mikulica, O., Moskát, C., Bán, M. & Schulze-Hagen, K. 2010. Internal incubation and early hatching in brood parasitic birds. *Proc. R. Soc. B.* 278: 1019-1024.

Bishop, K.D. 1982. *Endemic birds of Biak Island. Report on the ICBP survey of Biak Island's endemic avifauna*. ICBP, Cambridge.

Bishop, D. & Brickle, N.W. 1999. An annotated checklist of the birds of the Tanimbar Islands. *Kukila* 10: 115-150.

Bishop, K.D. & Jones, D.N. 2001. The montane avifauna of

West New Britain, with special reference to the Nakanai Mountains. *Emu* 101: 205-220.
Biswas, B. 1960. The birds of Nepal. Part 2. *J. Bombay Nat. Hist. Soc.* 57: 516-546.
Blackburn, A. 1963. Selective feeding of Shining Cuckoo. *Notornis* 10: 189.
Blake, E.R. & Hanson, H.C. 1942. Notes on a collection of birds from Michoacán, Mexico. *Field Mus. Nat. Hist. Zool. Ser.* 22: 513-551.
Blake, J.G., Loiselle, B.A. & Vande Veghe, J.P. 1990.Weights and measurements of some Central African birds. *Gerfaut* 80: 3-11.
Blakers, M., Davies, S.J.J.F. & Reilly, P.N. 1984. *The atlas of Australian birds.* Melbourne University Press, Carlton.
Blanford, W.T. 1870. *Observations an the geology and zoology of Abyssinia, made during the progress of the British Expedition to that country in* 1867-68. London.
Blasius, W.A.H. & Nehrkorn, A. 1881. Beiträge zur Kenntnis der Vogelfauna von Borneo. *Jahresber. Ver. Naturwiss. Braunschw.* 1880-1881: 107-166.
Bleich, V.C. 1975. Roadrunner predation on ground squirrels in California. *Auk* 92: 147-149.
Bluntschli, H. 1938. Ein eigenartiges Gaumenorgan beim Nestling des madagassischen Buschkuckucks *Coua cristata* L. *Bio-Morphosis Basel* 1: 265-272.
Blyth, E. 1843. A Monograph of the Indian and Malayan species of Cuculidae, or Birds of the Cuckoo family. *J. Asiat. Soc. Bengal* 11: 897-928.
Blyth, E. 1843. Mr. Blyth's Report for December Meeting, 1842. *J. Asiat. Soc. Bengal* 12: 925-1011.
Blyth, E. 1845. Notices and descriptions of various New or Little Known Species of Birds. *J. Asiat. Soc. Bengal* 14: 546-602.
Blyth, E. 1846. Notices and Descriptions of various New or Little Known Species of Birds. *J. Asiat Soc. Bengal* 15: 1-54.
Blyth, E. 1849. A Supplemental Note to the Catalogue of the Birds in the Asiatic Society's Museum. *J. Asiat. Soc. Bengal* 18: 800-821.
Blyth, E. 1865. A few identifications and rectifications of synonymy. *Ibis* (2nd Series) 1: 27-50.
Blyth, E. 1866. The ornithology of India - a commentary on Dr. Jerdon's 'Birds of India'. *Ibis* (2nd series) 2: 225-258, 336-376.
Blyth, E. 1867. The ornithology of Ceylon. – a supplement to Dr. Jerdon's "Birds of India". *Ibis* (2nd series) 3: 294-314.
Boddaert, P. 1783. *Table des planches enluminéez d'histoire naturelle de M. D'Aubenton. Avec les denominations de M.M. de Buffon, Brisson, Edwards, Linnæus et Latham, precedé d'une notice des principaux ouvrages zoologiques enluminés.* Privately published, Utrecht.
Boddy, M. & Boddy, F.E. 1985. Cuckoos mating on overhead electricity cable. *Brit. Birds* 78: 239-240.
Bodrati, A. 2005. Nuevos aportes a la distribución de algunas aves argentinas. *Nuestras Aves* 50: 30-33.
Bodrati, A., Cockle, K.L., Matuchaka, V. & Maders, C. 2007. Reserva de la Biósfera Yabotí. Pp. 300-302 in A.S. Di Giacomo, M.V. de Francesco & E.G. Coconier (Eds). *Áreas importantes para la conservación de las aves en Argentina.* Aves Argentinas/Asociación Ornitológica del Plata, Buenos Aires.
Bodrati, A., Cockle, K.L., Segovia, J.M., Roesler, I., Areta, J.I. & Jordan, E. 2010. La avifauna del Parque Provincial Cruce Caballero, Provincia de Misiones, Argentina. *Cotinga* 32: 41-64.
Boertmann, D. 1994. An annotated checklist to the birds of Greenland. *Medd. Grønland, Biosci.* 38: 1-63.
Boesman, P. 1998. Some new information on the distribution of Venezuelan birds. *Cotinga* 9: 27-39.
Boesman, P. 1999. Birds of Venezuela/Aves de Venezuela: Photographs, sounds and distributions/fotografias, sonidos y distribuciones. CD-ROM.
Bögershausen, M. 1976. Der Kuckuck - *Cuculus canorus* L. 1758 - in Südniedersachsen. *Beitr. Naturk. Niedersachs.* 29: 67-71.
Bogert, C. 1937. Birds collected during the Whitney South Sea Expedition. 34. The distribution and the migration of the Long-tailed Cuckoo (*Urodynamis taitensis* Sparrman). *Amer. Mus. Nov.* 933: 1-12.
Boie, F. 1826. Generalübersicht der ornithologischen Ordnungen, Familien und Gattungen. *Isis* 19: col. 969-981.
Bolam, G. 1914. Foster-parents of the Cuckoo. *Brit. Birds* 7: 232-233.
Bolen, E.G. 1974. A note on the foraging behavior of Groove-billed Anis. *Bull. Texas Orn. Soc.* 7: 8.
Bonaparte, C.L. 1850. *Conspectus generum avium*, vol. 1. E. J. Brill, Academiae Typographum, Lugduni Batavorum.
Bonaparte, C.L. 1854. Conspectus Volucrum Zygodactylorum. *Ateneo Ital.* 2: 1-14.
Bond, J. 1928. The distribution and habits of the birds of the Republic of Haiti. *Proc. Acad. Nat. Sci. Philad.* 80: 483-521.
Bond, J. 1933. A new Lizard Cuckoo from the Dominican Republic with remarks on the Saona Palm Tanager. *Proc. Acad. Nat. Sci. Philad.* 85: 369.
Bond, J. 1939. Some birds from Montserrat, British West Indies. *Auk* 56: 193-195.
Bond, J. 1941. Nidification of the birds of Dominica, B. W. I. *Auk* 58: 364-375.
Bond, J. 1947. *Field guide to the birds of the West Indies: a guide to all the species known from the Greater Antilles, Lesser Antilles and Bahama Islands.* The Macmillan Company, New York.
Bond, J. 1950. Results of the Catherwood-Chaplin West Indies expedition, 1948. Part II. Birds of Cayo Largo (Cuba), San Andrés and Providencia. *Proc. Acad. Nat. Sci. Philad.* 102: 43-68.
Bond, J. 1956. *Check-list of the birds of the West Indies*, 4th ed. Acad. Nat. Sci.Philad., Philadelphia.
Bond, J. 1979. *Birds of the West Indies*. Collins, London.
Bond, J. 1984. *Twenty-fifth supplement to the Check-list of birds of the West Indies.* Acad. Nat. Sci. Philad., Philadelphia.
Bond, J. 1993. *A field guide to the birds of the West Indies*, 5th ed. Houghton Mifflin Company, Boston, New York.
Bond, J. & Meyer de Schauensee, R. 1940. On some birds from southern Colombia. *Proc. Acad. Nat. Sci. Philad.* 92: 153-169.
Bond, J. & Meyer de Schauensee, R. 1943. The birds of Bolivia. Part II. *Proc. Acad. Nat. Sci. Philad.* 95: 167-221.
Bonde, K. 1993. *Birds of Lesotho.* University of Natal Press, Pietermaritzburg.
Borges, S.H. 1994. Listagem e novos registros de aves para a região de Boa Vista, Roraima, Brasil. *Bol. Mus. Para. Emílio Goeldi, sér. Zool.* 10: 191-202.
Borges, S.H. 2004. Species poor but distinct: bird assemblages in white sand vegetation in Jaú National Park, Brazilian Amazon. *Ibis* 146: 114-124.
Borges, S.H., Cohn-Haft, M., Carvalhães, A.M.P., Henriques, L.M.P., Pacheco, J.F. & Whittaker, A. 2001. Birds of Jaú National Park, Brazilian Amazon: species check-list, biogeography and conservation. *Orn. Neotrop.* 12: 109-140.
Borghesio, L., Amakobe, B., Bakari, S., Balidy, H., Biasiol, D. & Menomussanga, M. 2009. A bird survey of the Ruvuma Delta, northern Mozambique. *Bull. ABC* 16: 197-203.
Bororing, R.F., Hunowu, I., Hunowu, Y., Maneasa, E., Mole, J., Nusalawo, M.H., Talangamin, F.S. & Wangko, M.F. 2000. Birds of the Manembonembo Nature Reserve, North Sulawesi, Indonesia. *Kukila* 11: 58-72.
Borrero H., J.I. 1946. Aves ocasionales en la Sabana de Bogotá. *Caldasia* 4: 169-173.
Borrow, N. & Demey, R. 2001. *Birds of Western Africa*. Christopher Helm, London.
Bosque, C. & Lentino R., M. 1987. The passage of North American migratory land birds through xerophytic habitats on the western coast of Venezuela. *Biotropica* 19: 267-273.
Bourns, F.S. & Worcester, D.C. 1894. Preliminary notes on the birds and mammals collected by the Menage scientific expedition to the Philippine Islands. *Occas. Papers Minnesota Acad. Nat. Sci.* 1: 1-64.
Bowden, C.G.R. 2001. The birds of Mount Kupe, southwest Cameroon. *Malimbus* 23: 13-44.
Bowdish, B.S. 1902. Birds of Porto Rico. *Auk* 19: 356-366.

Bowen, B.S. 2002. Groove-billed Ani *Crotophaga sulcirostris*. In A. Poole & F. Gill (Eds). *The birds of North America*. No. 612. 16 pp. The Birds of North America, Inc., Philadelphia.

Bowen, B.S., Koford, R.R. & Vehrencamp, S.L. 1989. Dispersal in the communally breeding Groove-billed Ani (*Crotophaga sulcirostris*). *Condor* 91: 52-64.

Bowen, B.S., Koford, R.R. & Vehrencamp, S.L. 1991. Seasonal pattern of reverse mounting in the Groove-billed Ani (*Crotophaga sulcirostris*). *Condor* 93: 159-163.

Bowen, W.W. 1932. Angolan birds collected during the second Gray African Expedition 1930. *Proc. Acad. Nat. Sci. Philad.* 84: 281-289.

Braa, A.T., Moksnes, A. & Røskaft, E. 1992. Adaptations of Bramblings and Chaffinches towards parasitism by Common Cuckoo. *Anim. Behav.* 43: 67-78.

Bradfield, R.D. 1931. The crested cuckoos of South Africa. *Ostrich* 2: 7-9.

Bradley, P.E. 1985. *Birds of the Cayman Islands*. P. E. Bradley, George Town, Grand Cayman.

Bradley, P.E. 2000. *The birds of the Cayman Islands*. British Ornithologists' Union Checklist 19. BOU, Tring.

Braun, M.J. & Wolf, D.E. 1987. Recent records of vagrant South American land birds in Panama. *Bull. Brit. Orn. Club* 107: 115-117.

Braun, R. 1931. Aufzucht von *Chrysococcyx klaasi* [sic] durch *Pycnonotus tricolor*. *Orn. Monatsb.* 39: 148-149.

Bravery, J.A. 1966. Field notes on the Oriental Cuckoo. *Emu* 66: 267-271.

Bravery, J.A. 1970. The birds of Atherton Shire, Queensland. *Emu* 70: 49-63.

Brazil, M. 2009. *Birds of East Asia*. A. & C. Black, London.

Brazil, M.A. 1991. *The birds of Japan*. Christopher Helm and A. & C. Black, London.

Bregulla, H.L. 1992. *Birds of Vanuatu*. Anthony Nelson, Oswestry.

Brendt, A.E. 1905. The Black-headed Honeyeater (*Melithreptus melanocephalus*). *Emu* 5: 12-15.

Brewster, C.A. 2000. Black Coucals *Centropus bengalensis* at Talana, eastern Botswana. *Babbler Gaborone* 37: 39-40.

Brichetti, P. & Fracasso, G. 2006. *Ornitologia italiana : identificazione, distribuzione, consistenza e movimenti degli uccelli italiani*. vol 3, *Stercorariidae – Picidae*. Alberto Perdisa, Bologna.

Brickle, N. 2007. Sumatran Ground Cuckoo *Carpococcyx viridis* seen in the wild for the first time. *Birding ASIA* 7: 52.

Briskie, J.V. 2003. Frequency of egg rejection by potential hosts of the New Zealand cuckoos. *Condor* 105: 719-727.

Britton, P.L. 1977. A Kenya record of *Cercococcyx montanus* at sea-level. *Scopus* 1: 23-24.

Britton, P.L. 1978. The Andersen collection from Tanzania. *Scopus* 2: 77-85.

Britton, P.L. (Ed.) 1980. *Birds of East Africa. Their habitat, status and distribution*. East Africa Natural History Society, Nairobi.

Brodkorb, P. 1937. Some birds of the Amazonian islands of Caviana and Marajó. *Occas. Pap. Mus. Zool. Univ. Michigan* 349: 1-7.

Brodkorb, P. 1940. New birds from southern Mexico. *Auk* 57: 542-549.

Bromley, E.H. 1949. Notes on the birds of some parts of Kedah. *Bull. Raffles Mus.* 19: 120-132.

Brooke, M. & Birkhead, T. (Eds.) 1991. *Ornithology*. Cambridge University Press, Cambridge.

Brooke, M.L. & Davies, N.B. 1987. Recent changes in host usage by cuckoos *Cuculus canorus* in Britain. *J. Anim. Ecol.* 56: 873-883.

Brooke, M.L. & Davies, N.B. 1989. Provisioning of nestling cuckoos *Cuculus canorus* by reed warbler *Acrocephalus scirpaceus* hosts. *Ibis* 131: 250-256.

Brooke, R.K. 1979. Two overlooked vagrants from the Tristan da Cunha group. *Bull. Brit. Orn. Club* 99: 81-82.

Brooker, L. & Brooker, M. 1999. www.users.bigpond.com/LesMikeBrooker/database.htm

Brooker, L.C., Brooker, M.G. & Brooker, A.M.H. 1990. An alternative populations genetics model for the evolution of egg mimesis and egg crypsis in cuckoos. *J. Theor. Biol.* 146: 344-358.

Brooker, M. & Brooker, L. 1986. Identification and development of the nestling cuckoos, *Chrysococcyx basalis* and *C. lucidus plagosus*, in Western Australia. *Aust. Wildl. Res.* 13: 197-202.

Brooker, M.G. & Brooker, L.C. 1989a. The comparative breeding behaviour of two sympatric cuckoos, Horsfield's Bronze-Cuckoo *Chrysococcyx basalis* and the Shining Bronze-Cuckoo *C. lucidus*, in Western Australia: a new model for the evolution of egg morphology and host specificity in avian brood parasites. *Ibis* 131: 528-547.

Brooker, M.G. & Brooker, L.C. 1989b. Cuckoo hosts in Australia. *Aust. Zool. Rev., R. Zool. Soc. New South Wales* 2: 1-67.

Brooker, M. & Brooker, L. 2003. Brood parasitism by Horsfield's Bronze-Cuckoo in a fragmented agricultural landscape in Western Australia. *Emu* 103: 357-361.

Brooker, M.G., Brooker, L.C. & Rowley, I. 1988. Egg deposition by the Bronze-Cuckoos *Chrysococcyx basalis* and *C. lucidus*. *Emu* 88: 107-109.

Brooker, M.G., Ridpath, M.G., Estbergs, A.J., Bywater, J., Hart, D.S. & Jones, M.S. 1979. Bird observations on the north-western Nullarbor Plain and neighbouring regions, 1967-1978. *Emu* 79: 176-187.

Brooks, D.J., Evans, M.I., Martins, R.P. & Porter, R.F. 1987. The status of birds in North Yemen, and the records of the OSME Expedition in autumn 1985. *Sandgrouse* 9: 4-66.

Brooks, D.M., Porzecanski, A.L., Weicker, J J., Honig, R.A., Saavedra Zankys, A.M. & Herrera, M. 2005. A preliminary assessment of avifauna of the Bolivian chiquitano and cerrado. *Orn. Neotrop.* 16: 85-99.

Brooks, T., Dutson, G., King, B. & Magsalay, P.M. 1995a. An annotated check-list of the forest birds of Rajah Sikatuna National Park, Bohol, Philippines. *Forktail* 11: 121-134.

Brooks, T., Magsalay, P., Dutson, G., King, B., & Allen, R. 1995b. Forest loss, extinctions and last hope for birds on Cebu. *Bull. OBC* 21: 24-27.

Brosset, A. 1976. Observations sur le parasitisme de la reproduction du coucou émeraude *Chrysococcyx cupreus* au Gabon. *Oiseau RFO* 46: 201-208.

Brosset, A. & Érard, C. 1986. *Les oiseaux des régions forestières du nord-est du Gabon. Écologie et comportement des especes*. Vol. 1. Société Nationale du Protection de la Nature, Paris.

Brotherton, J.F. 1965. Opportunistic feeding by Senegal Coucal. *Bull. Niger. Orn. Soc.* 2: 21.

Brown, L.H. & Britton, P.L. 1980. *The breeding seasons of East African birds*. East Africa Natural History Society, Nairobi.

Brown, L.N. 1963. Status of the Roadrunner in Missouri. *Condor* 65: 242-243.

Browning, M.R. 1978. An evaluation of the new species and subspecies proposed in Oberholser's *Bird Life of Texas*. *Proc. Biol. Soc. Wash.* 91: 85-122.

Browning, M.R. 1990. Taxa of North American birds described from 1957-1987. *Proc. Biol. Soc. Wash.* 103: 432-451.

Bruce, M.D. 1987. Additions to the birds of Wallacea. 1. Bird records from smaller islands in the Lesser Sundas. *Kukila* 3: 38-44.

Brudenell-Bruce, P.G.C. 1975. *The birds of New Providence and the Bahama Islands*. Collins, London.

Brüggemann, F. 1876. Beiträge zur Ornithologie von Celebes und Sangir. *Abh. Naturwiss. Ver. Bremen* 5: 35-102.

Bruggers, R.L. & Bortoli, L. 1979. Notes on breeding, parasitism and association with wasps of Heuglin's Weaver nesting on telephone wires in Mali. *Malimbus* 1: 135-144.

Brunel, J. & Thiollay, J-M. 1969. Liste préliminaire des oiseaux de Côte d'Ivoire. *Alauda* 37: 230-254, 315-337.

Bryant, H. 1864. Description of two new birds from the Bahama Islands, hitherto undescribed. *Proc. Boston Soc. Nat. Hist.* 9: 279-280.

Bryant, H.C. 1916. Habits and food of the Roadrunner in California. *Univ. Calif. Publ. Zool.* 17: 21-58.

Bub, H. 1943. Kuckuck frisst toten Fitislaubsänger. *Orn. Monatsb.* 51: 138.

Bub, H. 1982. Kuckuck (*Cuculus canorus*) isst Regenwürmer. *Orn. Mitt.* 34: 19.

Buck, H.A. 1988. Wallacea 1987. Unpublished report.

Buckley, P.A., Massiah, E.B., Hutt, M.B., Buckley, F.G. & Hutt, H.F. 2009. *The birds of Barbados*. British Ornithologists' Union/British Ornithologists' Club Checklist 24. BOU/BOC, Peterborough.

Bucknill, J.A.S. & Chasen, F.N. 1990. *Birds of Singapore and South-East Asia* (reprint). Tynron Press, Scotland.

Buden, D.W. 1987. *The birds of the southern Bahamas*. BOU Checklist 8. British Ornithologists' Union, London.

Buden, D.W. 1990. Report on the birds collected during the Armour-*Utowana* West Indian expeditions. *Bull. Brit. Orn. Club* 110: 14-20.

Buden, D.W. 1998. The birds of Kapingamarangi Atoll, including first record of the Shining Cuckoo (*Chrysococcyx lucidus*) from Micronesia. *Notornis* 45: 141-153.

Bugoni, L., Mohr, L.V., Scherer, A., Efe, M.A. & Scherer, S.B. 2002. Biometry, molt and brood patch parameters of birds in southern Brazil. *Ararajuba* 10: 85-94.

Buij, R., Fox, E.A. & Wich, S.A. 2006. Birds of Gunung Leuser National Park, Northern Sumatra. *Kukila* 13: 23-47.

Buitrón, G. & Freile, J.F. 2006. Registros inusuales de aves migratorias e de bosques subtropicales en Quito, Ecuador. *Cotinga* 26: 54-56.

Bundy, G. 1976. *The birds of Libya. An annotated check-list*. BOU Checklist 1. British Ornithologists' Union, London.

Burger, J. & Gochfeld, M. 2001. Smooth-billed Ani (*Crotophaga ani*) predation on butterflies in Mato Grosso, Brazil: risk decreases with increased group-size. *Behav. Ecol. Sociobiol.* 49: 482-492.

Burkill, H.J. 1928. Cuckoos taking eggs. *London Nat.*: 20-21.

Butler, A.L. 1908. A second contribution to the ornithology of the Egyptian Soudan. *Ibis* (9th series) 2: 205-263.

Butler, E.A. 1881. A tentative catalogue of the birds of the Deccan and south Mahratta Country. *Stray Feathers* 9: 367-442.

Büttikofer, J. 1896. On a collection of birds from Nias. *Notes Leyden Mus.* 18: 161-200.

Büttikofer, J. 1900. Zoological results of the Dutch scientific expedition to central Borneo. *Notes Leyden Mus.* 21: 145-276.

Buzzetti, D.R.C. 2000a. Avifauna do Parque. Pp. 69-96 in *Avaliação ecológica rápida do Parque Estadual do Cantão*. http://www.seplan.to.gov.br/dma/areas_protegidas/area_protegida_informacoes_uc.htm. Secretaria do Planejamento e Meio Ambiente, Governo do Estado do Tocantins, Palmas, Tocantins, Brazil.

Buzzetti, D.R.C. 2000b. Distribuição altitudinal de aves em Angra dos Reis e Parati, sul do estado do Rio de Janeiro, Brasil. Pp. 131-148 in M.A.S. Alves, J.M.C. da Silva, M. van Sluys, H.G. Bergallo & C.F.D. da Rocha (Eds.) *A ornitologia no Brasil: pesquisa atual e perspectivas*. Editora da Universidade do Estado do Rio de Janeiro, Rio de Janeiro.

Cabanis, J. 1848. Voegel. pp. 662-765 in Schomburgk, R. *Reisen in Britisch-Guiana in den Jahren* 1840-1844. Vol. III. Verlagsbuchhandlung von J. J. Weber, Leipzig.

Cabanis, J. 1862. Uebersicht der im Berliner Museum befindlichen Vögel von Costa Rica. *J. Orn.* 10: 161-176, 321-336.

Cabanis, J. 1873. *Coccygus euleri* n. sp. aus Brasilien. *J. Orn.* 21: 72-73.

Cabanis, J. 1874. Uebersicht der von Herrn Carl Euler im District Cantagallo, Provinz Rio de Janeiro, gesammelten Vögel. *J. Orn.* 22: 81-90, 225-231.

Cabanis, J. 1882. Bericht über die Februar-Sitzung... aus dem Sammlungen des Afrikareisenden Major v. Mechow... *J. Orn.* 30: 229-230.

Cabanis, J. & Heine, F., Jnr. 1863. *Museum Heineanum. Verzeichnis der ornithologischen Sammlung des Oberamtmann Ferdinand Heine. IV. Theil: Klettervögel, Heft I: Kuckuke und Faulvögel*. Frantz'schen Buchhandlung (G. Loose), Halberstadt.

Cain, A.J. & Galbraith, I.C.J. 1956. Field notes on the birds of the eastern Solomon Islands. *Ibis* 98: 100-134, 262-295.

Calder, D.R. 1951. A note on the Didric Cuckoo, *Chrysococcyx caprius* (Boddaert). *Ostrich* 22: 36.

Calder, W.A. 1967. Breeding behavior of the Roadrunner, *Geococcyx californianus*. *Auk* 84: 597-598.

Calder, W.A. 1968a. The diurnal activity of the Roadrunner, *Geococcyx californianus*. *Condor* 70: 84-85.

Calder, W.A. 1968b. Nest sanitation: a possible factor in the water economy of the Roadrunner. *Condor* 70: 279.

Cameron, E.D. 1983. Comments on the birds of Brunei. *Brunei Mus. J.* 5: 222-229.

Campbell, A.G. 1906. Observations on the rearing of a cuckoo. *Emu* 6: 120-126.

Campbell, A.J. 1900. *Nests and eggs of Australian birds*. A.J. Campbell, Melbourne.

Campbell, A.J. 1920. Notes on additions to the "H.L. White Collection". *Emu* 20: 49-66.

Campbell, B. & Lack, E. (Eds.) *A dictionary of birds*. T. & A.D. Poyser, Calton, U.K.

Canevari, M., Canevari, P., Carrizo, G.R., Harris, G., Mata, J.R. & Straneck, R.J. 1991. *Nueva guía de las aves argentinas*, vol. 1-2. Fundación Acindar, Buenos Aires.

Capper, D.R., Clay, R.P., Madroño Nieto, A., Barnett, J.M., Burfield, I.J., Esquivel, E.Z., Kennedy, C.P., Perrens, M. & Pople, R.G. 2001. First records, noteworthy observations and new distributional data for birds in Paraguay. *Bull. Brit. Orn. Club* 121: 23-37.

Capurro, H.A. & Bucher, E.H. 1988. Lista comentada de las aves del bosque chaqueño de Joaquín V. González, Salta, Argentina. *Hornero* 13: 39-46.

Cariello, M.O., Schwabl, H.G., Lee, R.W. & Macedo, R.H.F. 2002. Individual female clutch identification through yolk protein electrophoresis in the communally breeding guira cuckoo (*Guira guira*). *Mol. Ecol.* 11: 2417-2424.

Cariello, M.O., Lima, M.R., Schwabl, H.G. & Macedo, R.H.F. 2004. Egg characteristics are unreliable in determining maternity in communal clutches of guira cuckoos *Guira guira*. *J. Avian Biol.* 35: 117-124.

Cariello, M.O., Macedo, R.H.F. & Schwabl, H.G. 2006. Maternal androgens in eggs of communally breeding guira cuckoos (*Guira guira*). *Horm. Behav.* 49: 654-662.

Carriker, M.A., Jnr. 1910. An annotated list of the birds of Costa Rica including Cocos Island. *Ann. Carnegie Mus.* 6: 314-915.

Carriker, M.A., Jnr. 1935. Descriptions of new birds from Bolivia, with notes on other little known species. *Proc. Acad. Nat. Sci. Philad.* 87: 313-341.

Carriker, M.A., Jnr. 1955. Notes on the occurrence and distribution of certain species of Colombian birds. *Noved. Colomb.* 2: 48-64.

Carriker, M.A., Jnr. & Meyer de Schauensee, R. 1935. An annotated list of two collections of Guatemalan birds in the Academy of Natural Sciences of Philadelphia. *Proc. Acad. Nat. Sci. Philad.* 87: 411-455.

Carswell, M., Pomeroy, D., Reynolds, J. & Tushabe, H. 2005. *The bird atlas of Uganda*. British Ornithologists' Club/British Ornithologists' Union, Oxford.

Carter, C. 1973. A visit to Mwinilunga District. *Bull. Zambian Orn. Soc.* 5 (2): 74.

Carter, M. 1999. The north-western group of Torres Strait islands, with particular attention to Dauan. *Aust. Birding Mag.* 5 (2): 3-9.

Carter, M., Clarke, R., Pierce, F., Dooley, S., Swann, G. & Grant, M. 2010. Lesser Coucal *Centropus bengalensis* on Ashmore Reef: first record for Australia. *Australian Field Ornithology* 27: 109-118.

Carvalhães, A.M.P. 2001. Acréscimos à lista de aves da Chapada Diamantina, Bahia. P. 170 in F.C. Straube (Ed.) *Ornitologia sem fronteiras. Incluindo os Resumos do IX Congresso Brasileiro de Ornitologia*. Soc. Bras. Orn., Curitiba, Paraná.

Catterall, M. 1997. *Results of the 1996 bird survey of Buton Is., Sulawesi, Indonesia: Operation Wallacea*. Unpublished report.

Cavalcanti, R.B., Lemes, M.R. & Cintra, R. 1991. Egg losses in communal nests of the Guira Cuckoo. *J. Field Orn.* 62: 177-180.

Cave, F.O. & MacDonald, J.D. 1955. *Birds of the Sudan. Their identification and distribution*. Oliver & Boyd, Edinburgh & London.

Cawkell, E.M. 1965. Notes on Gambian birds. *Ibis* 107: 535-540.
Cawkell, E.M. & Moreau, R.E. 1963. Notes on birds in The Gambia. *Ibis* 105: 156-178.
CCJB = Committee for Check-List of Japanese Birds. 2000. *Check-list of Japanese birds*. 6th rev. ed. Ornithological Society of Japan.
Chaffer, N. 1973. Unusual behaviour of young Pallid Cuckoo. *Aust. Bird Watcher* 5: 10-12.
Chalton, D.O. 1976. Weight loss in Spectacled Weavers raising a cuckoo chick. *Ostrich* 47: 69.
Chalton, D.O. 1991. Development of a Diederik Cuckoo chick in a spectacled weaver nest. *Ostrich* 62: 84-85.
Chan, K.S. 2005. Recent Sightings. *Suara Enggang* 2005 (4): 28-39.
Chance, E.P. 1922. *The cuckoo´s secret*. Sedgwick & Jackson, London.
Chandi, M. 2007. On the sighting of the Lesser Coucal *Centropus bengalensis* in the Andaman and Nicobar Islands. *Jour. Bombay Nat. Hist. Soc.* 104: 213-214.
Chapin, J.P. 1928. The African cuckoos of the genus *Cercococcyx*. *Amer. Mus. Nov.* 313: 1-11.
Chapin, J.P. 1939. The birds of the Belgian Congo. Part 2. *Bull. Amer. Mus. Nat. Hist.* 75: i-vii, 1-632.
Chapin, J.P. 1953. The birds of the Belgian Congo. Part 3. *Bull. Amer. Mus. Nat. Hist.* 75A: 1-821.
Chapin, J.P. 1954. The birds of the Belgian Congo. Part 4. *Bull. Amer. Mus. Nat. Hist.* 75B: i-ix, 1-846.
Chapman, F.M. 1894. On the birds of the island of Trinidad. *Bull. Amer. Mus. Nat. Hist.* 6: 1-86.
Chapman, F.M. 1914. Descriptions of a new genus and species of birds from Venezuela. *Bull. Amer. Mus. Nat. Hist.* 33: 193-197.
Chapman, F.M. 1917. The distribution of bird life in Colombia: a contribution to a biological survey of South America. *Bull. Amer. Mus. Nat. Hist.* 36: 1-729.
Chapman, F.M. 1923. Descriptions of proposed new birds from Panama, Venezuela, Ecuador, Peru and Bolivia. *Amer. Mus. Nov.* 67: 1-12.
Chapman, F.M. 1926. The distribution of bird-life in Ecuador: a contribution to a study of the origin of Andean birdlife. *Bull. Amer. Mus. Nat. Hist.* 55: 1-784.
Chapman, F.M. 1928. Descriptions of new birds from eastern Ecuador and eastern Peru. *Amer. Mus. Nov.* 332: 1-12.
Chappuis, C. 1974. *Les oiseaux de l'ouest africain.* Disk 1: Columbidae et Cuculidae. *Alauda* sound suppl., 1. Paris [*Alauda* 42: 197-222, accompanying commentary.]
Chappuis, C. 2000. *Oiseaux d'Afrique/African bird sounds*. *Alauda*/Société Ornithologique de France, Paris (CDs and accompanying text.)
Charles-Dominique, P. 1976. Les gommes dans le régime alimentaire de *Coua cristata* à Madagascar. *Oiseau RFO* 46: 174-178.
Charter, M., Bouskila, A., Aviel, S. & Leshem, Y. 2005. First record of European Jackdaw (*Corvus monedula*) parasitism by the Great Spotted Cuckoo (*Clamator glandarius*) in Israel. *Wilson Bull.* 117: 201-204.
Chasen, F.N. 1934. Nine new races of Natuna birds. *Bull. Raffles Mus. Singapore* 9: 92-97.
Chasen, F.N. 1939. *The birds of the Malay Peninsula*, vol. 4. *The birds of the low-country jungle and scrub*. Witherby & Co. Ltd., London.
Chasen, F.N. & Kloss, C.B. 1929. Some new birds from North Borneo. *J. Orn. Festschr. Hartert*: 106-121.
Chasen, F.N. & Kloss, C.B. 1931. Five new Malaysian birds. *Bull. Raffles Mus. Singapore* 5: 82-86.
Chattopadhyay, S. 1986. New records of some birds from different parts of eastern India. *J. Bombay Nat. Hist. Soc.* 83: 668-669.
Chébez, J.C. 1992. Acerca de la presencia de algunas aves misioneras. *Hornero* 13: 257-258.
Chébez, J.C. 1993. Los manuscritos de William Henry Partridge. Aves Misioneras (VII). *Nuestras Aves* 10: 14-16.
Chébez, J.C. 1994. *Los que se van*. Albatros, Buenos Aires.
Chébez, J.C. 1996. *Fauna misionera, catálogo sistemático y zoogegráfico de los vertebrados de la provincia de Misiones, Argentina / Misiones fauna: a systematic and zoogeographical catalogue of the vertebrate fauna of the Province of Misiones, Argentina*. Literature of Latin America, Buenos Aires. (LOLA Monografia; 4).
Cheesman, R.F. & Sclater, W.L. 1935. On a collection of birds from north-western Abyssinia. Part 2. *Ibis* (13th series) 5: 297-329.
Cheke, R.A. & Walsh, J.F. 1996. *The birds of Togo*. British Ornithologists' Union Check-list 14, BOU, Tring.
Cheney, M. 1914. Notes from Wangaratta district. *Emu* 14: 98-99.
Cheng Tso-Hsin. 1987. *A synopsis of the avifauna of China*. Science Press, Beijing.
Cheng Tso-Hsin (Zheng Zuoxin). 1991. *Fauna Sinica, Aves*, vol. 6. *Columbiformes, Psittaciformes, Cuculiformes and Strigiformes*. Science Press, Beijing.
Cherrie, G.K. 1892. A preliminary list of the birds of San José, Costa Rica. *Auk* 9: 21-27, 247-251, 322-329.
Cherrie, G.K. 1896. Contribution to the ornithology of San Domingo. *Field Mus. Nat. Hist. Orn. Ser.* 1: 1-26.
Cherrie, G.K. 1916. New birds from the collection of the Collins-Day Expedition to South America. *Bull. Amer. Mus. Nat. Hist.* 35: 391-397.
Cherrie, G.K. 1916a. A contribution to the ornithology of the Orinoco region. *Mus. Brookl. Inst. Arts Sci., Sci. Bull.* 2: 133-374.
Cherry, M.I. & Bennett, A.T.D. 2001. Egg colour matching in an African cuckoo, as revealed by ultraviolet-visible reflectance spectrophotometry. *Proc. Royal Soc. London*, Series B 268: 565-571.
Chhangani, A.K. 2004. Frequency of avian road-kills in Kumbhalgarh Wildlife Sanctuary, Rajasthan, India. *Forktail* 20: 110-111.
Chisholm, A.H. 1962. The strange ways of parasitic birds. *Vict. Nat.* 78: 268-273.
Chisholm, A.H. 1973. Cuckoos are very resolute. *Aust. Bird Watcher* 5 (2): 49-54.
Choo, T.C. 2009. Pied Cuckoo *Clamator jacobinus* at Penanti, Pulau Pinang, 6 December 2009. *Suara Enggang* 17 (6): 22, 26.
Choudhury, A. 2009. Significant recent ornithological records from Manipur, north-east India, with an annotated checklist. *Forktail* 25: 71-89.
Chouteau, P. 1997. *Working Group Birds Madag. Region Newsl.* 7: 9-17.
Chouteau, P. & Pedrono, M. 2009. Breeding biology of Coquerel's Coua (*Coua coquereli*) in Western Madagascar. *J. Orn.* 150: 55-60.
Chow, C.P. & Lim, K.S. 2009. Bird Report June 2009. *Singapore Avifauna* 23 (6): 1-3.
Christensen, K.M., Tøttrup, A.P., Rahner, M.C. & Brouwer, J. 2005. First records for Niger of Red-chested Cuckoo *Cuculus solitarius*, Grassland Pipit *Anthus cinnamomeus*, Buff-bellied Warbler *Phyllolais pulchella* and Isabelline Shrike *Lanius isabellinus*. *Bull. ABC* 12: 162-164.
Christidis, L. & Boles, W.E. 2008. *Systematics and taxonomy of Australian birds*. Collingwood, Victoria.
Christy, C. 1897. Field-notes on the birds of the island of San Domingo. *Ibis* (7th series) 3: 317-343.
Christy, P. & Clarke, W. 1994. *Guide de oiseaux de la Réserve de la Lopé*. Ecofac Gabon, Libreville.
Chubb, C. 1910. On the birds of Paraguay. *Ibis* (9th series) 4: 53-78, 263-285, 517-534, 571-647.
Chubb, C. 1916. *The birds of British Guiana*, vol. 1. Bernard Quaritch, London.
Cintra, R. & Sanaiotti, T.M. 1990. *Osteocephalus taurinus* (treefrog). Predation. *Herpetological Review* 21: 59.
Cintra, R. & Yamashita, C. 1990. Habitats, abundância e ocorrência das espécies de aves do pantanal do Poconé, Mato Grosso, Brasil. *Pap. Avuls. Zool. S. Paulo* 37: 1-21.
Claessens, O., Brammer, F.P., Deville, T. & Renaudier, A. 2011. First documented records of Pearly-breasted Cuckoo *Coccyzus euleri* for French Guiana, and an overlooked specimen from Ecuador. *Bull. Brit. Orn. Club* 131: 128-133.

Clancey, P.A. 1951. Notes on birds of the South African subcontinent. *Ann. Natal Mus.* 12 (1): 137-152.

Clancey, P.A. 1962. Miscellaneous taxonomic notes on African birds. 19. *Durban Mus. Novit.* 6: 181-194.

Clancey, P.A. 1968. Interesting distributional records from Mozambique south of the Zambezi R. *Durban Mus. Novit.* 8 (13): 193-198.

Clancey, P.A. 1989. Four additional species of southern African endemic birds. *Durban Mus. Novit.* 14 (7): 140-152.

Clancey, P.A. 1990. Size-variation and post-breeding movement in the Didric Cuckoo *Chrysococcyx caprius* (Boddaert). *Bull. Brit. Orn. Club* 110: 130-137.

Clancey, P.A. 1991. Hepatic variation in *Cuculus* cuckoos occurring in the Afrotropics. *Albatross Durban* 309: 26-27.

Claramunt, S. & Cuello, J. P. 2004. Diversidad de la biota uruguaya. Aves. *An. Mus. Nac. Hist. Nat. Antrop.*, 2a Serie 10: 1-76.

Clark, A.H. 1905. Birds of the southern Lesser Antilles. *Proc. Boston Soc. Nat. Hist.* 32: 203-312.

Clark, J.E. & Clarke, J.A.C. 1985. Interactions between Arrow-marked Babblers and Striped Crested Cuckoos. *Nyala* 11: 28-29.

Clarke, M.M. 1982. Notes on a visit to Bensbach, 6th to 13th November 1982. *PNG Bird Soc. Newsletter* 195-196: 18-20.

Clarke, T. 2006. *Field guide to the birds of the Atlantic Islands.* Christopher Helm, London.

Clay, M. B. 1929. The Yellow-billed Cuckoo. *Bird-lore* 31: 189-190.

Clements, J.F. & Shany, N. 2001. *A field guide to the birds of Peru.* Lynx Edicions and Ibis Publishing Company, Temeluca, California.

Clunie, F. 1973. Fan-tailed Cuckoo parasitises Fiji Warbler. *Notornis* 20: 168.

Clunie, F. & Morse, P. 1984. *Birds of the Fiji bush.* Fiji Museum, Suva.

Coates, B.J. 1985. *The birds of Papua New Guinea.* Vol. 1. Dove Publications, Alderley.

Coates, B.J. 2001. *Birds of New Guinea and the Bismarck archipelago: a photographic guide.* Dove Publications, Alderley, Queensland.

Coates, B.J. & Bishop, K.D. 1997. *A guide to the birds of Wallacea. Sulawesi, the Moluccas and Lesser Sunda Islands, Indonesia.* Dove Publications, Alderley, Queensland.

Cochrane, J. & Cubitt, G. 2004. *The national parks and other wild places of Indonesia.* New Holland, London.

Cock, C. & Marshall, C.H.T. 1873. Notes on a collection of eggs made in and about Murree. *Stray Feathers* 1: 348-358.

Cockle, K.L., Leonard, M.L. & Bodrati, A. 2005. Presence and abundance of birds in an Atlantic forest reserve and adjacent plantation of shade-grown yerba mate, in Paraguay. *Biodiv. Conserv.* 14: 3265-3288.

Coelho, A.G.M. 1979. As aves da Estação Ecológica de Tapacurá, Pernambuco (com notas de campo). *Notulae Biol., N. S. Notula* 2: 1-18.

Coelho, A.G.M. 1987. Novas informações sobre a avifauna da Estação Ecológica do Tapacurá - Pernambuco. *Univ. Fed. Pernambuco Publ. Avulsa* 1: 1-16.

Cohn-Haft, M., Whittaker, A. & Stouffer, P.C. 1997. A new look at the "species-poor" central Amazon: The avifauna north of Manaus, Brazil. *Orn. Monogr.* 48: 205-235.

Cole, C.F. 1908. Observations on the finches as fosterparents to the cuckoos. *Emu* 8: 23-25.

Colebrook-Robjent, J.F.R. 1977. Cuckoos in Zambia. *Black Lechwe* 12 (3): 26-31.

Colebrook-Robjent, J.F.R. 1984. The breeding of the Didric Cuckoo *Chrysococcyx caprius* in Zambia. In J. Ledger (Ed.). *Proceedings of the Fifth Pan African Ornithological Congress.* Southern African Ornithological Society/Organizing Committee of the Fifth PAOC, Johannesburg.

Collar, N.J., Crosby, M.J. & Stattersfield, A.J. 1994. *Birds to watch 2: the world list of threatened birds.* BirdLife International, Cambridge, UK.

Collar, N.J., Gonzaga, L.P., Krabbe, N., Madroño Nieto, A., Naranjo, L.G., Parker, T.A. III, & Wege, D.C. 1992. *Threatened birds of the Americas: the ICBP/IUCN Red Data Book.* International Council for Bird Preservation, Cambridge, U.K.

Collar, N.J. & Long, A.J. 1996. Taxonomy and names of *Carpococcyx* cuckoos from the Greater Sundas. *Forktail* 11: 135-150.

Collar, N.J., Mallari, N.A.D. & Tabaranza, B.R. Jnr. 1999. *Threatened birds of the Philippines.* The Haribon Foundation/BirdLife International Red Data Book, Makati City, Philippines.

Collar, N.J. & Pilgrim, J.D. 2007. Species-level changes proposed for Asian birds, 2005-2006. *BirdingASIA* 8: 14-30.

Collar, N.J. & Stuart, S.N. 1985. *Threatened birds of Africa and related islands. The ICBP/IUCN Red Data Book.* Part 1. ICBP, Cambridge.

Collar, N.J. & Sykes, B.R. 2009. Lost, found and saved – OBC conservation: the next 25 years. *BirdingASIA* 11: 41-46.

Collias, N.E. & Collias, E.C. 1970. The behaviour of the West African Village Weaverbird. *Ibis* 112: 457-480.

Colombelli-Négrel, D., Robertson, J. & Kleindorfer, S. 2010. Mother to eggs communication. 25th IOC 22-28 August 2010. Abstracts, p. 210. www.i-o-c.org

Colston, P.R. & Curry-Lindahl, K. 1986. *The birds of Mount Nimba, Liberia.* British Museum (Natural History), London.

Colvin, W. 1935. Roadrunner nesting in Kansas. *Auk* 52: 88.

Condole, L. 1997. Trip report: Java and Bali (Indonesia), December 22, 1996-January 16, 1997. Unpublished.

Conradty, P. 1979. Ein weiterer Brutnachweis des Häkerkuckuck (*Clamator glandarius*) in Griechenland. *Anz. Orn. Gess. Bayern* 18: 192.

Contreras, A.O. 1993. Hallazgo del Yasy Yateré Guazú *Dromococcyx phasianellus phasianellus* (Spix, 1824) en la Isla Hú, Ñeembucú, sudeste del Paraguay oriental (Aves: Cuculidae, Neomorphinae). *Nótulas Faunísticas, Corrientes* 41: 1-2.

Contreras, J.R. 1979. Bird weights from northeastern Argentina. *Bull. Brit. Orn. Club* 99: 21-24.

Contreras, J.R., Berry, L.M., Contreras, A.O., Bertonatti, C.C. & Utges, E.E. 1990. *Atlas ornitogeográfico de la provincia del Chaco - República Argentina. No Passeriformes*, vol. 1. Fundación Vida Silvestre Argentina, Capítulo Corrientes, Corrientes. Cuadernos Técnicos Félix de Azara 1.

Coomans de Ruiter, L. 1951. Vogels van het dal van de Bodjo-rivier (Zuid-Celebes). *Ardea* 39: 261-318.

Cooper, R.P. 1958. Pallid Cuckoo feeding young. *Emu* 58: 67-68.

Cooper, R.P. 1967. Is the Helmeted Honeyeater doomed? *Aust. Bird Watcher* 3: 1–14.

Corlett, R. T. & Ping, Ice-Ko-Wai. 1995. Frugivory by Koels in Hong Kong. *Mem. Hong Kong Nat. Hist. Soc.* 20: 221-222.

Cornett, J. W. 1983. Early nesting of the Roadrunner, *Geococcyx californianus*, in California. *Am. Birds* 37: 236.

Cornwallis, L. & Porter, R.F. 1982. Spring observations on the birds of North Yemen. *Sandgrouse* 4: 1-36.

Cory, C.B. 1881. List of the birds of Haiti, taken in different parts of the island between January 1 and March 12, 1881. *Bull. Nuttall Orn. Club* 6: 151-155.

Cory, C.B. 1895. Description of two new species of birds from San Domingo. *Auk* 12: 278-279.

Cory, C.B. 1913. Descriptions of twenty-eight new species and subspecies of Neotropical birds. *Field Mus. Nat. Hist. Orn. Ser.* 1: 283-292.

Cory, C.B. 1915. Notes on South American birds, with descriptions of new subspecies. *Field Mus. Nat. Hist. Orn. Ser.* 1: 303-335.

Cory, C.B. 1919. Catalogue of birds of the Americas, Part II, no. 2. *Field Mus. Nat. Hist. Zool. Ser.* 13: 316-607.

Cowan, W.D. 1883. Notes on the Natural History of Madagascar. *Roy. Physical Soc.* 1883: 133-150.

Cox, J.R., Hurrell, H. G., Spencer, K.G., Headlam, C.G. & Rogers, M J. 1980. Large gathering of Cuckoos. *Brit. Birds* 73: 412-414.

Cracraft, J. 1964. Observations on sun-bathing in the Yellow-billed Cuckoo. *Wilson Bull.* 76: 187.

Cracraft, J., Barker, F.K., Braun, M.J., Harshman, J., Dyke, G.J., Feinstein, J., Stanley, S.E., Cibois, A., Schikler, P., Beresford, P., García-Moreno, J., Sorenson, M.D., Yuri, T. & Mindell, D.P. 2004. Phylogenetic relationships among modern birds (Neornithes): toward an avian tree of life. P. 468-489 in Cracraft, J. & Donoghue, M.J. (eds.) *Assembling the tree of life*. OUP, Oxford.

Cramp, S. (Ed.) 1985. *Handbook of the birds of Europe, the Middle East and North Africa. The birds of the Western Palaearctic. Vol. 4. Terns to woodpeckers*. OUP, Oxford & New York.

Cranbrook, Earl of, & Wells, D.R. 1981. Observations of fledgling cuckoos and their fosterers in Gunung Mulu National Park. *Sarawak Mus. J.* 29 No. 50: 147-149.

Crawford, D.N. 1972. Birds of the Darwin area, with some records from other parts of Northern Territory. *Emu* 72: 131-148.

Crawford, R.L. & Stevenson, H.M. 1984. Patterns of spring and fall migration in northwest Florida. *J. Field Orn.* 55: 196-203.

Crawshaw, K.R. 1963. Juvenile Cuckoo´s method of feeding upon Cinnabar Moth caterpillars. *Brit. Birds* 56: 28.

Crewe, M.D. 2001. Selected observations from the Gambia, 1997-1999, with comments on the identification of a number of species. *Bull. African Bird Club* 8: 113-116.

Crouther, M.M. 1985. Some breeding records of the Common Koel *Eudynamis [sic] scolopacea*. *Aust. Bird Watcher* 11: 89-92.

Cruz, A. 1973. Food and foraging ecology of the Chestnut-bellied Cuckoo. *Wilson Bull.* 85: 336-337.

Cruz, A. 1975. Ecology and behavior of the Jamaican Lizard Cuckoo. *Stud. Fauna Curaçao Caribb. Isl.* 46: 109-111.

Cuadros, T. 1991. Registro visual del cuco terrestre piquirrojo (*Neomorphus pucheranii*) en Colombia. *Bol. Soc. Antioq. Orn.* 2: 26-27.

Cuello, J. & Gerzenstein, E. 1962. Las aves del Uruguay: lista sistemática, distribución y notas. *Comun. Zool. Mus. Hist. Nat. Montevideo* 6: 1-191.

Cunningham, J.M. 1948. Rearing of Long-tailed Cuckoo. *New Zealand Bird Notes* 3: 176-178.

Curio, E. 1970. Eine Begattung des Eidechsenkuckucks (*Saurothera vetula*) auf Jamaika. *J. Orn.* 111: 240.

Curio, E., Hornbuckle, J., Soye, Y. de., Aston, P. & Lastimoza, L.L. 2001. New bird records for the island of Panay, Philippines, including the first record of the Asian Stubtail *Urosphena squameiceps* for the Philippines. *Bull. Brit. Orn. Club* 121: 183-197.

Curry, R.L. & Stoleson, S.H. 1988. New bird records from the Galápagos associated with the El Niño-Southern Oscillation. *Condor* 90: 505-507.

Cuvier, G.L.C.F.D. 1816. *Le Règne animal distribué d'après son organisation, pour servir de base à l'histoire naturelle des animaux et d'introduction à l'anatomie comparée* Vol. 1. Deterville, Paris.

Dabbene, R. 1910. Catálogo sistemático y descriptivo de las aves de la República Argentina. *An. Mus. Nac. Buenos Aires*, 3 11: i-xiv, 1-513.

da Fonseca, J.P. 1922. Notas biológicas sobre aves brasileiras. III. Novas notas biológicas sobre o Sacy *Tapera naevia* (Linn.). *Rev. Mus. Paulista* 13: 785-787.

Daguerre, J.B. 1922. Lista de aves coleccionadas y observadas en Rosas, F.C.S. *Hornero* 2: 259-271.

Daguerre, J.B. 1924. Apuntes sobre algunas aves de la Provincia de Buenos Aires. *Hornero* 3: 248-252.

Dahl, F. 1899. Das Leben der Vögel auf den Bismarckinseln. *Mitt. Zool. Mus. Berlin.* 1: 108-222.

D'Albertis, L.M. & Salvadori, T. 1879. Catalogo degli uccelli raccolti da L.M. D'Albertis durante la seconda e terza esplorazione del Fiume Fly negli anni 1876 e 1877. *Ann. Museo Civ. Storia Nat. Genova* 14: 21-147.

Danforth, S.T. 1928. Birds observed in Jamaica during the summer of 1926. *Auk* 45: 480-491.

Danforth, S.T. 1929. Notes on the birds of Hispaniola. *Auk* 46: 358-375.

Danielsen, F., Balete, D.S., Christensen, T.D., Heegaard, M., Jakobsen, O.F., Jensen, A., Lund, T. & Poulsen, M.K. 1994. *Conservation and biological diversity in the Sierra Madre Mountains of Isabella and southern Cagayan Province, the Philippines*. Department of Environment and Natural Resources, BirdLife International & Danish Ornithological Society, Manila & Copenhagen.

Darrieu, C.A. & Martínez, M.M. 1984. Estudios sobre la avifauna de Corrientes. I. Nuevos registros de aves (no passeres). *Rev. Mus. La Plata, Nueva Ser., Zool.* 13: 257-260.

Darwin, C. 1990 (1859). *Arternes oprindelse ved naturlig selektion eller ved de heldigst stillede formers sejr i kampen for tilværelsen*. Jørgen Paludans Forlag, Copenhagen. [Danish translation of 'On the origin of species….'.]

da Silva, J.M.C. & Oren, D.C. 1990. Introduced and invading birds in Belém, Brazil. *Wilson Bull.* 102: 309-313.

da Silva, J.M.C., Oren, D.C., Roma, J.C. & Henriques, L.M.P. 1997. Composition and distribution patterns of the avifauna of an Amazonian upland savanna, Amapá, Brazil. *Orn. Monogr.* 48: 743-762.

Davidar, P., Yoganand, T.R.K., Ganesh, T. & Joshi, N. 1997. An assessment of common and rare forest bird species of the Andaman Islands. *Forktail* 12: 99-105.

Davies, N.B. 1989. Co-evolution between the Cuckoo *Cuculus canorus* and its hosts. *J. Zool. London* 217: 521–522.

Davies, N.B. 2000. *Cuckoos, cowbirds and other cheats*. Poyser, London.

Davies, N.B. & Brooke, M. de L. 1989. An experimental study of co-evolution between the Cuckoo, *Cuculus canorus*, and its hosts. 1. Host egg discrimination. 2. Host egg markings, chick discrimination and general discussion. *J. Anim. Ecol.* 58: 207-236.

Davies, N.B. & Brooke, M. de L. 1991. Co-evolution of the Cuckoo and its hosts. *Sci. Am.* 264: 92-98.

Davis, D.E. 1940a. Social nesting habits of *Guira guira*. *Auk* 57: 472-484.

Davis, D.E. 1940b. A suggestion concerning territorialism in *Tapera naevia*. *Wilson Bull.* 52: 208.

Davis, D.E. 1940c. Social nesting habits of the Smooth-billed Ani. *Auk* 57: 179-218.

Davis, D.E. 1941. Social nesting habits of *Crotophaga major*. *Auk* 58: 179-183.

Davis, D.E. 1942. The phylogeny of social nesting habits in the Crotophaginae. *Quart. Rev. Biol.* 17: 115-134.

Davis, L.I. 1972. *A field guide to the birds of Mexico and Central America*. University of Texas Press, Austin & London.

Davis, S.E. 1993. Seasonal status, relative abundance, and behavior of the birds of Concepción, Departamento Santa Cruz, Bolivia. *Fieldiana Zool. N.S.* 71: 1-33.

Davis, W.B. 1944. Notes on summer birds of Guerrero. *Condor* 46: 9-14.

Davison, G.W.H. 1992. *Birds of Mount Kinabalu*. Natural History Publ. Sabah, & Koktas Sabah Berhad, Sabah.

Davison, G.W.H. 1999. Pulau Layang-Layang revisited. *Sabah Soc. J.* 16: 31-39.

Dawkins, R. & Krebs, J.R. 1979. Arms races between and within species. *Proc. R. Soc. London B* 205: 489-511

Dawn, W. 1955. Black-billed Cuckoo feeds on Monarch Butterfly. *Wilson Bull.* 67: 133.

Dawson, P., Dawson, D., Reynolds, I. & Reynolds, S. 1991. Notes on the birds of Logan Reserve, southeast Queensland, 1967-1990. *Sunbird* 21: 93-111.

de Albuquerque, E.P. 1996. Comunicação sobre a ocorrência do Peixe-frito-pavonino, *Dromococcyx pavoninus* Pelzeln, 1870, no Rio Grande do Sul, Brasil. *Acta Biol. Leopoldensia* 18: 165-166.

de Almeida, A., Couto, H.T.Z. do & Almeida, Á.F. de. 2003. Diversidade beta de aves em hábitats secundários da Pré-Amazônia maranhense e interação com modelos nulos. *Ararajuba* 11: 157-171.

Dean, W.R.J. 1971. Breeding data for the birds of Natal and Zululand. *Durban Mus. Novit.* 9(6): 59-61.

Dean, W.R.J. 1974. Breeding and some distributional notes on some Angolan birds. *Durban Mus. Novit.* 10: 109-125.

Dean, W.R.J. 2000. *The birds of Angola*. British Ornithologists' Union Checklist 18. BOU, Tring.

Dean, W.R.J., Macdonald, I.A.W. & Vernon, C.J. 1974. Possible breeding record of *Cercococcyx montanus*. *Ostrich* 45: 188.
de Bont, M. 2002. Avifauna of the Hwimo area, Nigeria. *Bull. ABC* 9: 101-106.
de Bont, M. 2009. Bird observations from south-east Sudan. *Bull. ABC* 16: 37-52.
Debout, G., Meister, P. & Ventelon, M. 2000. Notes complémentaires sur l'avifaune du Niger. *Malimbus* 22: 87-89.
Debus, S.J.S. 1989. A comment on the calls of Horsfield's Bronze-Cuckoo. *Aust. Bird Watcher* 13 (4): 135.
Dee, T.J. 1986. *The endemic birds of Madagascar*. International Council for Bird Preservation, Cambridge.
de Godoy, F. I. 2010. Registro documentado de *Micrococcyx cinereus* (Cuculiformes: Cuculidae), no município de Águas da Prata, nordeste do estado de São Paulo. *Atual. Orn.* 156: 4-5.
DeGraaf, R.M. & Rappole, J.H. 1995. *Neotropical migratory birds: natural history, distribution, and population change*. Cornell UP, Cornell.
Deignan, H.G. 1945. The birds of northern Thailand. *Bull. US Nat. Mus.* 186: 1-616.
Deignan, H.G. 1952. The correct name for the Malayo-Sumatran race of the chestnut-breasted malkoha (Cuculidae). *Bull. Raffles Mus.* 24: 219.
Deignan, H.G. 1955. Four new races of birds from East Asia. *Proc. Biol. Soc. Wash.* 68: 145-147.
Deignan, H.G. 1963. *Checklist of the birds of Thailand*. *Bull. US Nat. Mus.* 226: 1-263.
de Juana, E. & Comité Ibérico de Rarezas de la SEO 1996. Observaciones homologadas de aves raras en España y Portugal. Informe de 1994. *Ardeola* 43: 103-118.
Dekeyser, P.L. 1956. Le Parc National du Niokolo-Koba. Oiseaux. *Mem. Inst. Franç. Afr. Noire* 48: 79-141.
Delacour, J. 1931. Description de neuf oiseaux nouveaux de Madagascar. *Oiseau Rev. Fr. Orn.* 1: 473-486.
Delacour, J. 1930. Notes de Madagascar. *Oiseau RFO* 2: 65-77, 160-179.
Delacour, J. 1932a. Les oiseaux de la Mission Franco-Anglo-Américaine à Madagascar. *Oiseau RFO* 2: 1-96.
Delacour, J. 1932b. On the birds collected in Madagascar by the Franco-Anglo-American expedition, 1929-1931. *Ibis* (13th series) 2: 284-304.
Delacour, J. 1932c. La Mission Zoologique Franco-Anglo-Américaine à Madagascar. *Bull. Mus. National Hist. Nat.* 2: 212-219.
Delacour, J. 1947. *Birds of Malaysia*. Macmillan, New York.
Delacour, J. & Ezra, A. 1927. Renauld´s Ground Cuckoo (*Carpococcyx renauldi*). With notes on nesting in captivity. *Avicult. Mag.* 4: 229-232.
Delacour, J. & Jabouille, P. 1931. *Les oiseaux de l'Indochine Française*. Vol. 2. Exposition Coloniale Internationale, Paris.
Delacour, J. & Mayr, E. 1946. *Birds of the Philippines*. Macmillan, New York.
de la Peña, M.R. 1983. Nota sobre observaciones de nidificación de aves en la provincia de Santa Fe. *Hornero* 12: 129-131.
de la Peña, M.R. 1988. *Guía de aves argentinas. Columbiformes a Piciformes*, 1st ed., vol. 4. Facultad de Agronomía y Veterinaria, Universidad Nacional del Litoral, Santa Fé.
de la Peña, M.R. 1993. Parasitismo en el crespín, *Tapera naevia chochi* (Vieillot, 1817) en la provincia de Santa Fé, república Argentina. *Nótulas Faunísticas, Corrientes* 37: 1-4.
de la Peña, M.R. 1994. *Guía de aves argentinas. Columbiformes a Piciformes*, 2nd ed., vol. 3. Literature of Latin America, Buenos Aires.
de la Peña, M.R. 1997. *Lista y distribución de las aves de Santa Fe y Entre Ríos*. Monografía 15. Literature of Latin America, Buenos Aires.
de la Peña, M.R. 1999. *Aves argentinas. Lista y distribución*. Monografía 18. Literature of Latin America, Buenos Aires.
de la Peña, M.R. 2005. *Reproducción de las aves argentinas (con descripción de pichones)*. Literature of Latin America, Buenos Aires.
de la Peña, M. R. & M. Rumboll 1998. *Birds of southern South America and Antarctica*. HarperCollins, London.
de Magalhães, J.C.R. 1999. *As aves na fazenda Barreiro Rico*. Editora Plêiade, São Paulo.
de Mattos, G.T., Andrade, M.A. & Freitas, M.V. 1993. *Nova lista de aves do estado de Minas Gerais. Revisada, ampliada e ilustrada (check-list)*. Fundação Acangaú, Belo Horizonte.
de Melo, T.A., Jr. 1998. *Neomorphus geoffroyi dulcis* Snethlage, 1927. Pp. 286-287 in A.B.M. Machado, G.A.B. da Fonseca, R.B. Machado, L.M.S. Aguiar & L.V. Lins (Eds). *Livro vermelho das espécies ameaçadas de extinção da fauna de Minas Gerais*. Fundação Biodiversitas, Belo Horizonte.
Dement'ev, G.P. & Gladkov, N.A. (Eds). 1966. *Birds of the Soviet Union*. Vol. 1. Israel Program for Scientific Translations, Jerusalem.
Demey, R. 1995. Notes on the birds of the coastal and Kindia areas, Guinea. *Malimbus* 17: 85-99.
Demey, R. 2005. Recent Reports. *Bull ABC* 12: 191.
Demey, R. 2007. Recent Reports. *Bull ABC* 14: 92-105.
Demey, R. 2009. Recent Reports. *Bull ABC* 16: 227-241.
Demey, R. 2010. Recent Reports. *Bull ABC* 17: 239-252.
Demey, R. & Fishpool, L.D.C. 1994. The birds of Yapo Forest, Ivory Coast. *Malimbus* 16: 100-122.
Demey, R. & Rainey, H.J. 2004. The birds of Pic de Fon Forest Reserve, Guinea: a preliminary survey. *Bull. African Bird Club* 11: 126-138.
de Naurois, R. 1969. Peuplements et cycles de reproduction des oiseaux de la côte occidentale d'Afrique (du Cap Barbas, Sahara Espagnol, à la Frontiere de la République de Guinée). *Mem. Mus. National Hist. Nat.* Nouv. Série, Série A, *Zoologie*. 56.
de Naurois, R. 1979. The Emerald Cuckoo of São Tomé and Príncipe Islands (Gulf of Guinea). *Ostrich* 50: 88-93.
de Naurois, R. 1983. Les oiseaux réproducteurs des îles de São Tomé et Príncipe: liste systématique commentée et indications zoogéographiques. *Bonn. Zool. Beitr.* 34: 129-148.
Denny, T. & Dudman, D. 1979. Koel behaviour. *Sunbird* 10 (3-4): 78.
de Silva Wijeyeratne, G. 1999. Nesting Green-billed Coucal. *Bull. OBC* 30: 33.
de Silva Wijeyeratne, G., Perera, G.L., William, J., Warakagoda, D. & de Silva Wijeyeratne, N. 1997. A birdwatcher's guide to Sri Lanka. *Bull. OBC* Suppl.
de Silva Wijeyeratne, G., Warakagoda, D. & Zylva, T.S.U. de. 2000. *A photographic guide to birds of Sri Lanka*. New Holland, London.
De Smet, W.M.A. 1970. [Studies on the migration of the Cuckoo *Cuculus canorus* L.]. *Gerfaut* 60: 148-187.
de Vasconcelos, M.F. 1998. Urban environment utilization by the Squirrel Cuckoo, *Piaya cayana*: the importance of urban trees. *Ciência Cult.* 50: 462-464.
de Vasconcelos, M.F., d'Angelo Neto, S., Kirwan, G.M., Bornschein, M.R., Diniz, M.G. & da Silva, J.F. 2006. Important ornithological records from Minas Gerais state, Brazil. *Bull. Brit. Orn. Club* 126: 212-238.
Deville, É. 1851. Note sur quatre espèces nouvelles d'oiseaux provenant de l'expédition de M. Castelnau; le *Conurus Weddellii, C. jugularis, C. Luciani* et *Cultrides Pucheranii*. *Rev. Mag. Zool. Pure Appl.* (ser. 2) 3: 209-213.
de Visscher, M.N. & Moratorio, M. 1984. Notes on the feeding behaviour of the Smooth-billed Ani *Crotophaga ani*. *Gerfaut* 74: 71-74.
Dharmakumarsinji, R.S. 1954. *The birds of Saurashtra*. The author, Bombay.
Diamond, J.M. 1972a. Further examples of dual singing by southwest Pacific birds. *Auk* 89: 180-183.
Diamond, J.M. 1972b. *Avifauna of the Eastern Highlands of New Guinea*. Publ. Nuttall Orn. Club 12. Cambridge, Mass.
Diamond, J.M. 1975. Distributional ecology and habits of some Bougainville birds (Solomon Islands). *Condor* 77: 14-23.
Diamond, J.M. 2002. Dispersal, mimicry, and geographic variation in northern Melanesian birds. *Pacific Science* 56: 1-22.

Díaz, M., Asensio, B. & Tellería, J.L. 1996. *Aves Ibéricas. Vol. 1. No Passeriformes.* J.M. Reyero, Madrid.

Dickerman, R.W. 1994. Notes on birds from Africa with descriptions of three new subspecies. *Bull. Brit. Orn. Club* 114: 274-278.

Dickey, D.R. & van Rossem, A.J. 1938. The birds of El Salvador. *Field Mus. Nat. Hist. Zool. Ser.* 23: 1-609.

Dickinson, E.C. (Ed.). 2003. *The Howard and Moore complete checklist of the birds of the world*, 3rd ed. Princeton University Press, Princeton.

Dickinson, E.C., Kennedy, R.S. & Parkes, K.C. 1991. *The birds of the Philippines.* British Ornithologists' Union Check-list 12. BOU, London.

Diesselhorst, G. 1968. Beiträge zur Ökologie der Vögel Zentral- und Ost-Nepals. In W. Hellmich: *Khumbo Himal.* Universitätsverlag Wagner, Innsbruch-München.

Diggles, S. 1872. *Cuculus brisbanensis. Trans. Phil. Soc. Queensl.* 1872 (2): 12.

Diggles, S. 1876. *Lamprococcyx modesta. Trans. Phil. Soc. Queensl.* 1876 (2): 12.

Di Giacomo, A.G. 2005. Aves de la Reserva El Bagual. Pp. 201-465 in A.G. Di Giacomo & S.F. Krapovickas (Eds) *Historia natural y paisaje de la Reserva El Bagual, provincia de Formosa, Argentina. Inventario de los vertebrados y de la flora vascular de un área protegida del Chaco Húmedo.* Temas de Naturaleza y Conservación, 4. Aves Argentinas/Asociación Ornitológica del Plata, Buenos Aires.

Dinata, Y., Nugroho, A., Hairdir, I.A. & Linkie, M. 2008. Camera trapping rare and threatened avifauna in west-central Sumatra. *Bird Conserv. Int.* 18: 30-37.

Dodd, A.P. 1913. Foster-parents of cuckoos. *Emu* 12: 190-191.

Dolgushin, I.A., Korelov, M.N., Kuz'mina, M.A., Gavrilov, E.I., Gavrin, V.F., Kovshar', A.F., Borodikhin, I.F. & Rodionov, E.F. 1970. *Ptitsy Kazakhstana* 3. Alma-Ata.

do Rosário, L.A. 1996. *As aves em Santa Catarina: distribuição geográfica e meio ambiente.* FATMA (Fundação de Amparo à Tecnologia e ao Meio Ambiente), Florianópolis.

d'Orbigny, A.D. 1839: *Aves.* in R. de la Sagra (Ed.). *Histoire physique, politique et naturelle de l'ile de Cuba.* A. Bertrand, Paris.

Dorst, J. 1957. Contribution à l'étude ecologique des oiseaux du haut Marañon (Pérou septentrional). *Oiseau RFO* 27: 235-269.

Dorst, J. 1962. *The migrations of birds.* Houghton Mifflin, Chicago.

Doughty, C., Day, N. & Plant, A. 1999. *Birds of the Solomons, Vanuatu and New Caledonia.* Christopher Helm, London.

Douglas D.J.T., Newson, S.E., Leech, D.I., Noble, D.G. & Robinson, R.A. 2010. How important are climate-induced changes in host availability for population processes in an obligate brood parasite, the European cuckoo? *Oikos* 119: 1834-1840.

Dove, H.S. 1916. Cuckoo notes from Tasmania. *Emu* 16: 96-99.

Dowsett, R.J. 1969. Additions to the list of the birds of Kafue Flats. *Bull. Zambian Orn. Soc.* 1: 39-43.

Dowsett, R.J. (Ed.) 1989. *Enquête Faunistique dans la Forêt du Mayombé et Check-liste des Oiseaux et des Mammifères du Congo.* Tauraco Research Report 2. Tauraco Press, Liège.

Dowsett, R.J. & Dowsett-Lemaire, F. 1993. *A contribution to the distribution and taxonomy of Afrotropical and Malagasy birds.* Tauraco Research Report 5. Tauraco Press, Liège.

Dowsett, R.J. & Dowsett-Lemaire, F. 2005. Additions to the avifauna of Mali. *Bull. ABC* 12: 119-124.

Dowsett, R.J., Colebrook-Robjent, J.F.R. & Osborne, T.O. 1974. Further additions to the Nyika Plateau avifauna. *Bull. Zambian Orn. Soc.* 6: 40-43.

Dowsett, R.J. & Prigogine, A. 1974. The avifauna of the Marunga Highlands. *Exploration hydrobiologique du basin du lac Bangweolu et du Luapula: Cercle Hydrobiologique de Bruxelles* 19: 1-67.

Dowsett-Lemaire, F. 1990. Ecoethology, distribution and status of Nyungwe Forest birds (Ruanda). *Tauraco Res. Rep.* 3: 31-85.

Dowsett-Lemaire, F. 1996. Observations of two *Cuculus* species fed by forest hosts in the Congo. *Malimbus* 18: 153-154.

Dowsett-Lemaire F. 2008. Review: Bird sounds of Madagascar. An audio guide to the island's unique birds. By Frank Hawkins and Richard Ranft (2007). *Bull. ABC* 15: 280-281.

Dowsett-Lemaire F. & Dowsett, R.J. 2006. *The birds of Malawi.* Tauraco Press and Aves a.s.b.l., Liège.

Dowsett-Lemaire, F., Dowsett, R.J. & Bulene, P. 1993. Additions and corrections to the avifauna of Congo. *Malimbus* 15: 68-80.

Draffan, R.D.W., Garnett, S.T. & Malone, G.J. 1983. Birds of the Torres Strait; an annotated and biogeographical analysis. *Emu* 83: 207-234.

Drapiez, P.A.J. 1823. *Cuculus rufo-vittatus* n. sp. *Clas d'Hist. Nat.* IV: 568.

Dubs, B. 1992. *Birds of southwestern Brazil: catalogue and guide to the birds of the Pantanal of Mato Grosso and its border areas.* Betrona-Verlag, Küsnacht, Switzerland.

Duckett, J.E. 1990. Some notes on the Common (Greater) Coucal (*Centropus sinensis*). *Sarawak Mus. J.* 41 (62): 305-307.

Duckworth, J.W. 1997. Mobbing of a Drongo Cuckoo *Surniculus lugubris. Ibis* 139: 190-192.

Duckworth, J.W. & Kelsh, R. 1988. *A bird inventory of Similajau National Park.* ICBP Study Report No. 31. Cambridge.

Duckworth, J.W. & Moores, N. 2008. A re-evaluation of the pre-1948 Korean breeding avifauna: correcting a 'founder effect' in perceptions. *Forktail* 24: 25-47.

Duckworth, J.W., Tizard, R.J., Timmins, R.J., Thewlis, R.M., Robichaud, W.G. & Evans, T.D. 1998. Bird records from Laos, October 1994-August 1995. *Forktail* 13: 33-68.

Duckworth, J.W., Wilkinson, R.J., Tizard, R.J., Kelsh, R.N. & Evans, M.I. 1997. Bird records from Similajau National park, Sarawak, Malaysia. *Forktail* 12: 117-154.

Duckworth, W. 1996. Moustached Hawk-Cuckoo *Cuculus vagans* and Booted Eagle *Hieraaetus pennatus* in Laos: two species new for Indochina. *Forktail* 11: 159-160.

Dudley, S.P., Gee, M., Kehoe, C., Melling, T.M. & BOURC. 2006. The British List: A checklist of birds of Britain (7th ed.). *Ibis* 148: 526-563.

Duhart, F. & Descamps, M. 1963. Notes sur l'avifaune du Delta Central Nigérien et régions avoisinantes. *Oiseau RFO* 33 (No. spécial).

Dumont, C. 1818. Coucals. pp. 138-144 in Levrault, F.G. (Ed.) *Dictionnaire des sciences naturelles....*Vol. 11. Le Normant, Paris.

Dumont, C. 1823. Malkoha. pp. 449-451 in Levrault, F.G. (Ed.) *Dictionnaire des sciences naturelles....*Vol. 28. Le Normant, Paris.

Dunning, J.B., Jr. 1993. *CRC handbook of avian body masses.* CRC Press, Boca Raton, Florida.

Dunson, W.A., Dunson, M.K. & Ohmart, R.D. 1976. Evidence for the presence of nasal salt glands in the Roadrunner and the *Coturnix* quail. *J. Exp. Zool.* 198: 209-216.

duPont, J.E. & Rabor, D.S. 1973. Birds of Dinagat and Siargao, Philippines. *Nemouria* No. 10: 1-111.

Dutson, G. 2007. The birds of Djaul Island, New Ireland. *Muruk* 8: 133-139.

Dutson, G.C.L., Evans, T.D., Brooks, T.M., Asane, D.C., Timmins, R.J. & Toledo, A. 1992. Conservation status of birds of Mindoro, Philippines. *Bird Conserv. Int.* 2: 303-325.

Dybowski, B. & Parrex, A. 1868. Verzeichniss der während der Jahre 1866 und 1867 in Gebiete der Mineralwässer von Darasun in Darien beobachteten Vögel. *J. Orn.* 16: 330-339.

Dyer, M., Gartshore, M.E. & Sharland, R.E. 1986. The birds of Nindam Forest Reserve, Kagoro, Nigeria. *Malimbus* 8: 2-20.

Dymond, N. 1994. A survey of the birds of Nias Island, Sumatra. *Kukila* 7: 10-27.

EABR. 1978, 1980, 1982, 1983. *East African Bird Report.* East African Natural History Society, Nairobi.

Eames, J.C., Steinheimer, F.D. & Bansok, R. 2002. A collection of birds from the Cardamom Mountains, Cambodia, including a new subspecies of *Arborophila cambodiana. Forktail* 18: 67-86.

Easley, K. & Montoya, M. 2006. Observaciones ornitológicas en la Isla del Coco, Costa Rica (Mayo 2006). *Zeledonia* 10: 31-41.

Eastwood, C. 1996. A trip to Irian Jaya. *Muruk* 8: 12-23.

Eastwood, C. & Gregory, P. 1995. Interesting sightings during 1993 & 1994. *Muruk* 7: 128-142.

Eaton, M.A., Brown, A.F., Noble, D.G., Musgrove, A.J., Hearn, R., Aebischer, N.J., Gibbons, D.W., Evans, A. & Gregory, R.D. 2009. Birds of Conservation Concern 3: the population status of birds in the United Kingdom, Channel Islands and the Isle of Man. *British Birds* 102: 296-341.

Eaton, S.W. 1979. Notes on the reproductive behavior of the Yellow-billed Cuckoo. *Wilson Bull.* 91: 154-155.

Eccles, S.D. 1988. The birds of São Tomé - record of a visit, April 1987 with notes on the rediscovery of Bocage´s Longbill. *Malimbus* 10: 207-217.

Edgar, A.T. 1933. Notes on the nidification of some Perak birds. *Bull. Raffles Mus. Singapore* 8: 121-162.

Edvardsen, E., Moksnes, A., Røskaft, E., Øjen, I.J. & Honza, M. 2001. Egg mimicry in cuckoos parasitizing four sympatric species of *Acrocephalus* warblers. *Condor* 103: 829-837.

Edwards, C.O. 1925. A nest of the Pheasant Coucal. *Emu* 24: 304.

Edwards, E.P. 1957. Yellow-billed Cuckoo nesting in Yucatán. *Condor* 59: 69-70.

Edwards, E.P. & Lea, R.B. 1955. Birds of the Monserrate area, Chiapas, Mexico. *Condor* 57: 31-54.

Ehrhardt, J.P. 1971. Census of the birds of Clipperton Island, 1968. *Condor* 73: 476-480.

Eisentraut, M. 1935. Biologische Studien im bolivianischen Chaco. VI. Beitrag zur Biologie der Vogelfauna. *Mitt. Zool. Mus. Berlin* 20: 367-443.

Eisentraut, M. 1963. *Die Wirbeltiere des Kamerungebirges*. Verlag Paul Parey, Hamburg & Berlin.

Eisentraut, M. 1973. Die Wirbeltierfauna von Fernando Poo und Westkamerun. *Bonner zool. Monog.* 3.

Eisermann, K. & Avendaño, C. 2006. Diversidad de aves en Guatemala, con una lista bibliográfica. Pp. 525-623 in E.B. Cano (Ed.). *Biodiversidad de Guatemala*. Vol. 1. Universidad del Valle de Guatemala, Guatemala, Guatemala.

Eitniear, J.C. & Tapia, A.A. 2000. Red-billed Pigeon (*Columba flavirostris*) nest predated by Groove-billed Ani (*Crotophaga sulcirostris*). *Orn. Neotrop.* 11: 231-232.

Elgood, J.H. 1955. On the status of *Centropus epomidis*. *Ibis* 97: 586.

Elgood, J.H. 1973. Rufous phase Senegal Coucals in southwest Nigeria: an illustration of Gloger´s rule. *Bull. Brit. Orn. Club* 93: 173.

Elgood, J.H. 1984. *The birds of Nigeria*. British Ornithologists' Union Checklist Series. No. 4. BOU, London.

Elgood, J.H., Fry, C.H. & Dowsett, R.J. 1973. African migrants in Nigeria. *Ibis* 115: 1-45, 375-411.

Elgood, J.H., Heigham, J.B., Moore, A.M., Nason, A.M., Sharland, R.E. & Skinner, N.J. 1994. *The birds of Nigeria. An annotated checklist*. 2nd ed. British Ornithologists' Union Check-list No. 4. BOU, Tring.

Ellis, D.H., Kepler, C.B., Kepler, A.K. & Teebaki, K. 1990. Occurrence of the Long-tailed Cuckoo *Eudynamis taitensis* on Caroline Atoll, Kiribati. *Emu* 90: 202.

Emlen, J.T. 1974. An urban bird community in Tucson, Arizona: derivation, structure, regulation. *Condor* 76: 184-197.

Engbring, J. 1983. Avifauna of the Southwest Islands of Palau. *Atoll Res. Bull.* 267: 1-22.

Enomoto, K. 1941. *A guide to wild birds*, vol. 2. Wild Bird Society of Japan Chapter, Osaka. [in Japanese]

Erftemeijer, P., Allen, G., Zuwendra & Kosamah, S. 1991. Birds of the Bintuni Bay region, Irian Jaya. *Kukila* 5: 85-98.

Ericson, P.G.P., Anderson, C.L., Britton, T., Elzanowski, A., Johansson, U.S., Källersjö, M., Ohlson, J.I., Parsons, T.J., Zuccon, D. & Mayr, G. 2006. Diversification of Neoaves: integration of molecular sequence data and fossils. *Biol. Lett.* 2: 543-547.

Eriksen, J. 2003. *Oman Bird List* (3rd ed.). Centre for Environmental Studies and Research, Sultan Qaboos University, Muscat, Oman.

Eriksen, J. 2010. www.BirdsOman.com/ob16-update.html

Ervin, S. 1989. The nesting of the Dark-billed Cuckoo in the Galapagos. *Noticias Galáp.* 48: 8-10.

Etchécopar, R.D. & Hüe, F. 1967. *The birds of North Africa from the Canary Islands to the Red Sea*. Oliver & Boyd, Edinburgh and London.

Euler, C. 1867. Beiträge zur Naturgeschichte der Vögel Brasiliens. *J. Orn.* 15: 217-233.

Evans, I. 2011. [Satellite tracking of Common Cuckoos.] re-IG4-G4DG-39HIKH-BZ1GL@cmp.ctt-news.org.

Evans, T.D., Dutson, G.C.L. & Brooks, T.M. 1993. *Cambridge Philippines rainforest project* 1991. *Final report*. Study Report 54. BirdLife International, Cambridge.

Evans, T.D. & Timmins, R.J. 1998. Records of birds from Laos during January - July 1994. *Forktail* 13: 69-96.

Evans, T.D., Towell, H.C., Timmins, R.J., Thewlis, R.M., Stones, A., Robichaud, W.G. & Barzen, J. 2000. Ornithological records from the lowlands of southern Laos during December 1995 – September 1996, including areas on the Thai and Cambodian borders. *Forktail* 16: 29-52.

Everett, A.H. 1889. A list of the birds of the Bornean group of islands. *J. Straits Branch R. Asiatic Soc.* 20: 91-212.

Ewbank, D.A. 1985. Predation by Senegal Coucal. *Honeyguide* 31: 218.

Faaborg, J. 1980. Further observations on ecological release in Mona Island birds. *Auk* 97: 624-627.

Faaborg, J. 1985. Ecological constraints on West Indian bird distributions. *Orn. Monogr.* 36: 621-653.

Fain, M.G. & Houde, P. 2004. Parallel radiations in the primary clades of birds. *Evolution* 58: 2558-2573.

Falla, R.A., Sibson, R.B. & Turboff, E.G. 1981. *The new guide to the birds of New Zealand and outlying islands*. Collins, Auckland and London.

Farkus, T. 1962. Contribution to the bird fauna of Barberspan. *Ostrich* Suppl. 4: 1-39.

Farkus, T. 1966. The birds of Barberspan. III. Some structured changes in the avifauna of Barberspan Nature Reserve. *Ostrich* Suppl. 6: 463-491.

Farrell, L.L. 2006. Subspecies status of the western Yellow-billed Cuckoo (Cuculidae: *Coccyzus americanus occidentalis*): using cytochrome b to elucidate the enigma. MSc. Thesis. Lakehead University.

Favaloro, N.J. 1933. Fantail Cuckoo with Red-browed Finch. *Emu* 33: 141-142.

ffrench, R. 1980. *A guide to the birds of Trinidad and Tobago*, rev. ed. Harrowood Books, Newton Square, Pennsylvania.

ffrench, R. 1984. Further notes on the avifauna of Trinidad & Tobago. *Living World, J. Trinidad Tobago Field Nat. Club* 1983-1984: 32-34.

ffrench, R.P. 1991. *A guide to the birds of Trinidad and Tobago*, 2nd ed. Cornell University Press, Ithaca, New York

ffrench, R. 1996. *Checklist of the birds of Tobago*. Asa Wright Nature Centre.

ffrench, R. & Kenefick, M. 2003. Verification of rare bird records from Trinidad and Tobago. *Cotinga* 19: 75-79.

ffrench, R. & White, G. 1999. Verification of rare bird records from Trinidad & Tobago. *Cotinga* 12: 80-82.

Fiebig, J. 1993. Dreijährige ornithologische Studien in Nordkorea. 1. Allgemeiner Teil und Non-Passeriformes. *Mitt. Zool. Mus. Berlin* 69 (Suppl.). *Ann. Orn.* 17: 93-146.

Fiebrig, C. 1921. Algunos datos sobre aves del Paraguay. *Hornero* 2: 205-213.

Field, G.W. 1894. Notes on the birds of Port Henderson, Jamaica, West Indies. *Auk* 11: 117-127.

Finch, B. 2010. Kenya Mega 2nd to 27th April 2009 (26 days). www.travellingbirder.com/birdwatching/birding_Kenya.php

Finckenstein, H. 1936. Zur Fortpflanzungsbiologie des Kuckucks. *Orn. Monatsb.* 44: 1-3.

Finley, W.L. & Finley, I. 1915. With the Arizona Roadrunner. *Bird-lore* 17: 159-165.

Finsch, F.H.O. 1898. On the specific distinction of the ground cuckoos of Borneo and Sumatra (*Carpococcyx radiceus* and *C. viridis* = *C. radiceus viridis*). *Notes Leyden Mus.* 20: 97-100.

Finsch, F.H.O. 1900. Zur Catalogisierung der ornithologischen Abteilung. I. Cuculi. *Notes Leyden Mus.* 22: 75-125.

Fischer, D.H. 1979. Black-billed Cuckoo (*Coccyzus erythropthalmus*) breeding in south Texas. *Bull. Texas Orn. Soc.* 12: 25-26.

Fischer, G.A. 1879. Briefliche Reiseberichte aus Ost-Afrika. IV. *J. Orn.* 27: 275-304.

Fischer, G.A. 1880. Briefliche Berichte aus Ost-Afrika. *J. Orn.* 28: 187-193.

Fischer, G.A. & Reichenow, A. 1879. Uebersicht der von Dr. G. A. Fischer auf einer zweiten Reise durch das ostafricanische Küstenland von Mombassa bis Wito und am Tana-Fluss gesammelten Vögel. *J. Orn.* 27: 337-356.

Fisher, W.K. 1904. Road-runners eat young mockingbirds. *Condor* 6: 80.

Fitzgerald, M. 1960. Communal display and courtship feeding in the Shining Cuckoo. *Notornis* 9: 9-10.

Fjeldså, J. & Krabbe, N. 1990. *Birds of the High Andes*. Zoological Museum, University of Copenhagen, and Apollo Books, Svendborg, Denmark.

Fleischer, R.C. 2001. Taxonomic and evolutionarily significant unit (ESU) status of western Yellow-billed Cuckoos (*Coccyzus americanus*). Report to Sacramento Fish and Wildlife Service, Sacramento, California.

Fleischer, R.C., Murphy, M.T. & Hunt, L.E. 1985. Clutch size increase and intraspecific brood parasitism in the Yellow-billed Cuckoo. *Wilson Bull.* 97: 125-127.

Fleming, R.L. Jnr. 1977. *Comments on the endemic birds of Sri Lanka*. Ceylon Bird Club and Wildlife and Nature Protection Society of Sri Lanka, Colombo.

Fleming, R.L. & Traylor, M.A. 1961. Notes on Nepal birds. *Fieldiana Zool.* 35: 447-487.

Fleming, R.L., Fleming, R.L. Jnr. & Bangde, L.S. 1976 (reprint 1979). *Birds of Nepal with reference to Kashmir and Sikkim*. Avalok Publishers, Kathmandu.

Fletcher, B.S. 2000. Birds seen in the Bismarck and Admiralty Islands from 22 Oct–7 Nov 1996. *Muruk* 8: 45-60.

Flint, V.E., Boehme, R.L., Kostin, Y.V. & Kuznetsov, A.A. 1984. *A field guide to birds of the USSR*. Princeton University Press, Princeton.

Fogden, M.P.L. 1976. A census of the bird community in tropical rain forest in Sarawak. *Sarawak Mus. J.* 24: 251-267.

Folse, L.J., Jnr. & Arnold, K.A. 1978. Population ecology of Roadrunners (*Geococcyx californianus*) in south Texas. *Southw. Nat.* 23: 1-27.

Forbes, W.A. 1881. Eleven weeks in north-eastern Brazil. *Ibis* (4th series) 5: 312-362.

Forbush, E.H. 1927. *Birds of Massachussetts and other New England states. Part II. Land birds from bob-whites to grackles*. Norwood Press, Norwood.

Ford, J.R. 1981. Hybridization and migration in Australian populations of the Little and Rufous-breasted Bronze-Cuckoos. *Emu* 81: 209-222.

Ford, J.R. & Stone, P.S. 1957. Birds of the Kellerberrin/Kwolyin district, Western Australia. *Emu* 57: 9-21.

Forrester, B.C. 1993. *Birding Brazil: a check-list and site guide*. Privately published, Rankinston, Ayrshire.

Fossøy, F., Antonov, A., Moksnes, A., Røskaft, F., Vikan, J.R., Møller, A.P., Shykoff, J.A. & Stokke, B.G. 2011. Genetic differentiation among sympatric cuckoo host races – males matter. *Proc. R. Soc. Lond. B* 278:1639-1645.

Foster, M.S. 1975. The overlap of molting and breeding in some tropical birds. *Condor* 77: 304-314.

Franzreb, K.E. & Laymon, S.A. 1993. A reassessment of the taxonomic status of the Yellow-billed Cuckoo. *Western Birds* 24: 17-28.

Fraser, L. 1839. Two new species of birds from the island of Luzon. *Proc. Zool. Soc. London* 7: 112-113.

Fredriksson, G.M. & Nijman, V. 2004. Habitat use and conservation status of two elusive ground birds, *Carpococcyx radiatus* and *Polyplectron schleiermacheri* in the Sungai Wain Protection forest, East Kalimantan, Indonesian Borneo. *Oryx* 38: 297-303.

Freere, P.J. 1984. White-browed Coucal *Centropus superciliosus* robbing nest of Grosbeak Weaver *Amblyospiza albifrons* of eggs. *Scopus* 8: 48.

Freiberg, M.A. 1943. Enumeración sistemática de las aves de Entre Ríos y lista de los ejemplares que las representan en el Museu de Entre Ríos. *Mem. Mus. Entre Ríos Paraná* 21: 1-110.

Freile, J.F. & Chaves, J.A. 2004. Interesting distributional records and notes on the biology of bird species from a cloud forest reserve in north-west Ecuador. *Bull. Brit. Orn. Club* 124: 6-16.

Friedmann, H. 1927. Notes on some Argentina birds. *Bull. Mus. Comp. Zool.* 68: 139-236.

Friedmann, H. 1928. A new cuckoo from Tanganyika Territory. *Proc. New England Zoöl. Club* 10: 83-84.

Friedmann, H. 1930. Birds collected by the Childs Frick Expedition to Ethiopia and Kenya Colony, part 1. Non-passerines. *Bull. US National Mus.* 153: i-xiii, 1-516.

Friedmann, H. 1948a. Birds collected by the National Geographic Society's expeditions to northern Brazil and southern Venezuela. *Proc. US National Mus.* 97: 373-570.

Friedmann, H. 1948b. *The parasitic cuckoos of Africa*. Acad. Sci. Washington DC. Monogr. no. 1.

Friedmann, H. 1949. Additional data on African parasitic cuckoos. *Ibis* 91: 514-519.

Friedmann, H. 1956. Further data on African parasitic cuckoos. *Proc. US National Mus.* 106: 377-408.

Friedmann, H. 1967. Alloxenia in three sympatric African species of *Cuculus*. *Proc. US National Mus.* 124 (3633): 1-14.

Friedmann, H. 1968. The evolutionary history of the avian genus *Chrysococcyx*. *Bull. US National Mus.* 265: 1-137.

Friedmann, H. 1969. Additions to knowledge of the Yellow-throated Glossy Cuckoo, *Chrysococcyx flavigularis*. *J. Orn.* 110: 176-180.

Friedmann, H. 1971. Further information on the host relations of the parasitic cowbirds. *Auk* 88: 239-255.

Friedmann, H., Kiff, L.F. & Rothstein, S.I. 1977. A further contribution to knowledge of the host relations of the parasitic cowbirds. *Smithson. Contr. Zool.* 235: 1-75.

Frith, C.B. 1975. Field observations on *Centropus toulou insularis* on Aldabra Atoll. *Ostrich* 46: 251-257.

Frith, D.W. & Frith, C.B. 1995. *Cape York Peninsula. A natural history*. Reed Books Chatswood.

Frith, H.J. (Ed.). 1984. *Birds in the Australian High Country*. Reed, Sydney.

Frith, H.J. & Davies, S.J.J.F. 1961. Breeding season of birds in subcoastal Northern Territory. *Emu* 61: 97-111.

Fry, C.H. 1961. Notes on the birds of Annobon and other islands in the Gulf of Guinea. *Ibis* 103: 267-276.

Fuchs, J., Cibois, A., Duckworth, J.W., Eve, R., Robichaud, W.G., Tizard, T. & van Gansberghe, D. 2007. Birds of Phongsaly province and the Nam Ou river, Laos. *Forktail* 23: 22-86.

Fuggle, D. 1980. Parasitic behaviour of the Red-chested Cuckoo. *Bull. E. Afr. Nat. Hist. Soc.* (Sept.-Oct.): 82.

Fuller, R.A. & Erritzøe, J. 1997. Little-known Oriental bird: Red-faced Malkoha *Phaenicophaeus pyrrhocephalus*. *Bull. OBC* 26: 35-39.

Fulton, R. 1904. The Kohoperoa or Koekoea, Long-tailed Cuckoo (*Urodynamis taitensis*): an account of its habits, description of a nest containing its (supposed) egg, and a suggestion as to how the parasitic habit in birds has become estabished. *Trans. Proc. New Zealand Inst.* 36: 113-148.

Fulton, R. 1910. The Pipiwharauroa, or Bronze Cuckoo (*Chalcococcyx lucidus*), and an account of its habits. *Trans. New Zealand Inst.* 42: 392-408.

Gagliardi, R.L. 2005. Lista das Aves do Estado do Rio de Janeiro. http://ricardo-gagliardi.sites.uol.com.br/avesRJ.htm

Gaines, D. 1974. Review of the status of the Yellow-billed Cuckoo in California: Sacramento Valley populations. *Condor* 76: 204-209.

Gaines, D. & Laymon, S.A. 1984. Decline, status and preservation of the Yellow-billed Cuckoo in California. *Western Birds* 15: 49-80.

Galbraith, I.C.J. & Galbraith, E.H. 1962. Land birds of Guadalcanal and the San Cristoval group, eastern Solomon Islands. *Bull. Brit. Mus. Nat. Hist. Zool.* 9: 1-86.

Gallagher, M.D. 1986. Additional notes on the birds of Oman, eastern Arabia, 1980-1986. *Sandgrouse* 8: 93-101.

Gallagher, M. & Woodcock, M.W. 1980. *The birds of Oman.* Quartet Books, London.

Gallardo, J.M. 1984. Observaciones sobre el comportamiento del pirincho (*Guira guira*) Aves: Cuculiformes. *Rev. Mus. Arg. Cienc. Nat. Bernardino Rivadavia Inst. Nac. Invest. Cienc. Nat., Zool.* 13: 167-170.

Gambel, W. 1849. Remarks on the birds observed in upper California, with descriptions of new species. *J. Acad. Nat. Sci. Philad.* (2nd series) 1: 215-229.

Gammie, J.A. 1877. Occasional notes from Sikhim. *Stray Feathers* 5: 380-387.

Garbutt, N., Bradt, H. & Schuurman, D. 1996. *Madagascar wildlife: a visitor's guide.* Bradt Travel Guides, Chalfont St Peter, UK & The Globe Pequot Press Inc, Guilford, Conn.

Garland, B. 1989. Species profile: Greater Roadrunner - "King of the Road". *WildBird* 3: 14-19.

Garrido, O.H. 1971. Nueva raza del arriero, *Saurothera merlini* (Aves: Cuculidae) para Cuba. *Poeyana* 87: 1-12.

Garrido, O.H. & Kirkconnell, A. 2000. *Birds of Cuba.* Christopher Helm, London.

Garrigues, R.L. & Dean, R. 2007. *Field guide to the birds of Costa Rica.* Christopher Helm, London.

Gaston, A.J. 1976. Brood parasitism by the Pied Crested Cuckoo *Clamator jacobinus*. *J. Anim. Ecol.* 45: 331-348.

Gaston, A.J. & Zacharias, V.J. 2000. Hosts of the Common Hawk Cuckoo *Hierococcyx varius* in India. *Forktail* 16: 182.

Gatter, W. 1997. *Birds of Liberia.* Pica Press, Sussex.

Gaymer, R. 1967. Observations on the birds of Aldabra in 1964 and 1965. *Atoll Res. Bull.* 118: 113-125.

Gazari, R. 1967. Notas sobre algunas aves no señaladas o poco conocidas al sur del Río Colorado. *Hornero* 10: 451-454.

Geluso, K.N. 1969. Food and survival problems of Oklahoma Roadrunners in winter. *Bull. Oklahoma Orn. Soc.* 2: 5-6.

Geluso, K.N. 1970a. Additional notes on food and fat of Roadrunners in winter. *Bull. Oklahoma Orn. Soc.* 3: 6.

Geluso, K.N. 1970b. Feeding behavior of a Roadrunner in winter. *Bull. Oklahoma Orn. Soc.* 3: 32.

Germain, M. & Cornet, J-P. 1994. Oiseaux nouveaux pour la République Centrafricaine ou dont les notifications de ce pays sont peu nombreuses. *Malimbus* 16: 30-51.

Ghorpade, K.D. 1973. Preliminary notes on the ornithology of Sandur, Karnataka. *J. Bombay Nat. Hist. Soc.* 70: 499-531.

Giacomelli, E. 1923. Catálogo sistemático de las aves útiles y nocivas de la Provincia de La Rioja. *Hornero* 3: 66-84.

Giai, A.G. 1949. Sobre un ejemplar joven de *Dromococcyx pavoninus*, Pelzeln. *Hornero* 9: 84-87.

Giai, A.G. 1951. Notas sobre la avifauna de Salta y Misiones. *Hornero* 9: 247-276.

Gibbon, G. 1951 *Southern African bird sounds* (6 cassettes). Southern African Birding, Hillary.

Gibbons, D.W., Reid, J.B. & Chapman, R.A. (Eds) 1993. *The new atlas of breeding birds in Britain and Ireland: 1988-1991.* Poyser, London.

Gibbs, L.H., Sorenson, M.D., Marchetti, K., Brooke, M. De L., Davies, N.B. & Nakamura, H. 2000. Genetic evidence for female host-specific races of the Common Cuckoo. *Nature* 407: 183-186.

Gibson, D.D. & Kessel, B. 1992. Seventy-four new avian taxa documented in Alaska 1976-1991. *Condor* 94: 454-467.

Gibson, D.D. & Kessel, B. 1997. Inventory of the species and subspecies of Alaska birds. *Western Birds* 28: 45-95.

Gibson, E. 1885. Notes on the birds of Paisandú, Republic of Uruguay. *Ibis* (5th series) 3: 275-283.

Gibson-Hill, C.A. 1950. A collection of birds' eggs from North Borneo. *Bull. Raffles Mus.* 21: 106-121.

Gibson-Hill, C.A. 1952. The apparent breeding seasons of land birds in north Borneo and Malaya. *Bull. Raffles Mus.* 24: 270-294.

Gibson-Hill, C.A. 1956. Birds recorded from the Aroa Island, Malacca Strait. *Bull. Raffles Mus.* 27: 155-179.

Gifford, E.W. 1919. Field notes on the land birds of the Galapagos Islands and of Cocos Island, Costa Rica. *Proc. Calif. Acad. Sci.* (series 4) 2: 189-258.

Gilbert, P.A. 1934. The seasonal movements and migrations of birds in eastern New South Wales. *Emu* 34: 101-105, 200-209.

Gilbert, P.A. 1935. The seasonal movements and migrations of birds in eastern New South Wales. *Emu* 35: 17-27.

Gilbert, P.A. & Keane, H. 1913. The Black-eared Cuckoo (*Misocalius palliolatus*, Lath.). *Emu* 13: 80-82.

Gill, B.J. 1979. Breeding of the Shining Cuckoo at Kowhai Bush, Kaikoura. *New Zealand J. Ecol.* 2: 89-90.

Gill, B.J. 1980a. Foods of the Shining Cuckoo (*Chrysococcyx lucidus*, Aves: Cuculidae in New Zealand). *New Zealand J. Zool.* 3: 138-140.

Gill, B.J. 1980b. Foods of the Long-tailed Cuckoo. *Notornis* 27: 96.

Gill, B.J. 1982. Notes on the Shining Cuckoo (*Chrysococcyx lucidus*) in New Zealand. *Notornis* 29: 215-227.

Gill, B.J. 1983a. The Grey Warbler´s care of nestlings: a comparison between unparasitised broods and those comprising a Shining Bronze-Cuckoo. *Emu* 82: 177-181.

Gill, B.J. 1983b. Brood-parasitism by the Shining Cuckoo *Chrysococcyx lucidus* at Kaikoura, New Zealand. *Ibis* 125: 40-55.

Gill, B.J. 1983c. Morphology and migration of *Chrysococcyx lucidus*, an Australasian cuckoo. *New Zealand J. Zool.* 10: 371-381.

Gill, F.B. & Stokes, C.C. 1971. Predation on a netted bird by Smooth-billed Anis. *Wilson Bull.* 83: 101-102.

Gill, H.B. 1970. Birds of Innisfall and Hinterland. *Emu* 70: 105-116.

Gillard, L. 1987. *Southern African bird calls.* Gillard Bird Cassettes, Johannesburg.

Gilliard, E.T. 1950. Notes on a collection of birds from Bataan, Luzon, Philippine Islands. *Bull. Amer. Mus. Nat. Hist.* 94: 459-504.

Gilliard, E.T. & LeCroy, M. 1961. Birds of the Victor Emanuel and Hindenburg Mountains, New Guinea. Result of the American Museum of Natural History expedition to New Guinea in 1954. *Bull. Amer.Mus. Nat. Hist.* 123: 1-86.

Gilliard, E.T. & LeCroy, M. 1966. Birds of the Middle Sepik Region, New Guinea. Result of the American Museum of Natural History expedition to New Guinea in 1953-1954. *Bull. Amer. Mus. Nat. Hist.* 132: 245-275.

Gilliard, E.T. & LeCroy, M. 1967. Results of the 1958-59 Gilliard New Britain Expedition. 4. Annotated list of birds of the Whiteman Mountains, New Britain. *Bull. Amer. Mus. Nat. Hist.* 135: 173-216.

Giraudo, A.R. 1996. Adiciones a la avifauna de la provincia de Corrientes, Argentina y de zonas limítrofes del Paraguay. *Facena* 12: 49-53.

Giraudoux, P., Degauquier, R., Jones, P.J., Weigel, J. & Isenmann, P. 1988. Avifaune du Niger: état des connaissances en 1986. *Malimbus* 10: 1-140.

Gleffe, J.D., Collazo, J.A., Groom, M.J. & Miranda-Castro, L. 2006. Avian reproduction and the conservation value of shaded coffee plantations. *Orn. Neotrop.* 17: 271-282.

Gloger, C. 1827. Etwas über einige ornithologische Gattungsbenennungen. *Notiz. Geb. Nat. Heilk.* 16: column 275-279.

Gloger, C.W.L. 1854. Ein seltsamer Zug in der Fortpflanzung des amerikanischen Kuckuke. *J. Orn.* 2: 219-232.

Glue, D. & Morgan, R. 1972. Cuckoo hosts in British habitats. *Bird Study* 19: 187-192.

Glutz von Blotzheim, U.N. & Bauer, K.M. (Eds) 1980. *Handbuch der Vögel Mitteleuropas.* Vol. 9. Columbiformes-Piciformes. Akademische Verlagsgesellschaft, Wiesbaden.

Gmelin, J.F. 1788. *Caroli a Linné Systema Naturae per Regna tria Naturae, ed. 13, tomus 1, Pars 1.* Impensis Georg Emanuel Beer, Lipsiae.

Gochfeld, M., Keith, S. & Donahue, P. 1980. Records of rare or previously unrecorded birds from Colombia. *Bull. Brit. Orn. Club* 100: 196-201.

Goddard, M.T. & Marchant, S. 1983. The parasitic habits of the Channel-billed Cuckoo *Scythrops novaehollandiae* in Australia. *Aust. Birds* 17: 65-72.

Godfrey, R. 1939. The Black Crested Cuckoo. *Ostrich* 10: 21-27.

Godfrey, W.E. 1986. *Birds of Canada*. Rev. ed. University of Chicago Press, Chicago.

Goeldi, E.A. 1903. Ornithological results of an expedition up the Capim River, State of Pará, with critical remarks on the Cracidae of lower Amazonia. *Ibis* (8th series) 3: 472-500.

Goertz, J.W. & Mowbray, E.E. 1970. Nesting records for three species of Louisiana birds. *Southw. Nat.* 15: 265-266.

Gomez, J.P.S., Sison, R.V. & Lohman, D.J. 2009. New bird records for Alabat Island, Quezon Province, Philippines. *Forktail* 25: 147-150.

Gonzales, P.C. 1983. *Birds of Catanduanes*. Zoological Paper No. 2. National Museum, Manila, Philippines.

Gonzales, P.C. & Rees, C.P. 1988. *Birds of the Philippines*. Haribon Foundation for the Conservation of Natural Resources Inc., Manila.

Good, A.I. 1952. The birds of French Cameroon. Part 1. *Sciences Naturelles no. 2, Mémoires de l'Institut Français d'Afrique Noire (Centre du Cameroun)*.

Goodall, J.M. 1923. On American cuckoos and cowbirds. *Bull. Brit. Orn. Club* 43: 63-68.

Goodman, S.M. & Benstead, J.P. (Eds) 2003. *The natural history of Madagascar*. Univ. Chicago Press, Chicago.

Goodman, S.M. & Gonzales, P.C. 1990. The birds of Mt. Isarog National Park, southern Luzon, Philippines, with particular reference to altitudinal distribution. *Fieldiana Zool.* 60: i-iii, 1-39.

Goodman, S.M., Hawkins, A.F.A. & Razafimahaimodison, J-C. 2000. Birds of Parc National de Marojejy, Madagascar, with reference to altitudinal distribution. *Fieldiana Zool. N.S.* 97: 175-200.

Goodman, S.M., Meininger, P.L., Baha el Din, M., Hobbs, J.J. & Mulli, W.C. (Eds) 1989. *The birds of Egypt*. OUP, Oxford.

Goodman, S.M., Pidgeon, M., Hawkins, A.F.A. & Schulenberg, T.S. 1997. The birds of southeastern Madagascar. *Fieldiana Zool. N.S.* 87: 1-132.

Goodwin, W. 2001. Senegal Coucals and distasteful bush locusts. *Honeyguide* 47: 182.

Gore, M.E.J. 1968. A check-list of the birds of Sabah, Borneo. *Ibis* 110: 165-196.

Gore, M.E.J. 1990. *The birds of the Gambia. An annotated checklist.* British Ornithologists' Union Check-list No. 3 (2nd ed.). BOU, London.

Gore, M.E.J. & Gepp, A.R.M. 1978. *Las aves del Uruguay*. Mosca Hnos., Montevideo.

Gore, M.E.J. & Won, P-O. 1971. *The birds of Korea*. Royal Asiatic Society, Taewon Publishing Company, Charles E. Tuttle Co., Inc, Seoul, Rutland Vt., & Tokyo.

Gorman, M.L. 1975. Habitats of the land-birds of Viti Levu, Fiji Islands. *Ibis* 117: 152-161.

Gorsuch, D.M. 1932. The Roadrunner. *Ariz. Wild Life* 4: 1-2, 11.

Gosper, D. 1962. Breeding records of the Koel. *Aust. Bird Watcher* 1: 226-228.

Gosse, P.H. 1847. *The birds of Jamaica*. John van Voorst, London.

Gould, J. 1838. A valuable collection of birds from the Himalaya Mountains. *Proc. Zool. Soc. London* 5: 136-137.

Gould, J. 1843. July 25, 1843. Mr. Prince exhibited to the Meeting, on the part of Mr. Gould, nine new Birds, collected during the recent voyage of H. M. S. Sulphur. *Proc. Zool. Soc. London* 1843: 103-108.

Gould, J. 1845. A new species of *Trogon*, from South America, and seven new birds from Australia. *Proc. Zool. Soc. London* 13: 18-20.

Gould, J. 1847. *The birds of Australia*. Vol. 4. The author, London.

Gould, J. 1847. On eight new species of Australian birds; and on *Anthus minimus*, Vig. And Horsf., as the type of a new genus *Chthonicola*. *Proc. Zool. Soc. London* 1847: 31-35.

Gould, J. 1856. [Birds collected by Mr. John MacGillivray]. *Proc. Zool. Soc. London* 24: 135-138.

Gould, J. 1856. Descriptions of two new species of true cuckoos (genus *Cuculus* as restricted). *Proc. Zool. Soc. London* 24: 96.

Gould, J. 1859. Two new species of birds. *Proc. Zool. Soc. London* 1859: 128-129.

Gould, J. 1864. Description of a new species of *Chrysococcyx*. *Proc. Zool. Soc. London* 1864: 73-74.

Gould, J. 1867. On two new birds from eastern Australia. *Ann. & Mag. Nat. Hist.* (3) 20: 269-270.

Gould, J. 1868. On two new Australian birds. *Proc. Zool. Soc. London* 1868: 74-76.

Grandidier, A. 1867. Mammifères et oiseaux nouveaux découverts à Madagascar et décrits par M. Alfred Grandidier. *Rev. Mag. Zool. Pure Appl.*, ser. 2, 19: 84-88.

Grant, C.H.B. 1915. [Description of new subspecies of *Centropus* and *Melittophagus*]. *Bull. Brit. Orn. Club* 35: 54-55.

Grant, I. 1966. Egg removal by Horsfield's Bronze-Cuckoo. *Emu* 65: 124.

Grant, P.R. & de Vries, T. 1993. The unnatural colonization of Galapagos by Smooth-billed Anis (*Crotophaga ani*). *Noticias Galáp.* 52: 21-23.

Grantham, M.J. 2000. Birds of Alas Purwo National Park, East Java. *Kukila* 11: 97-121.

Granvik, S.H. 1923. Birds collected by the Swedish Mount Elgon Expedition 1920. *J. Orn.* 23 (Sonderheft): 1-280, 10 plates, 1 map.

Graves, G.R., Kratter, A.W. & Bates, J.M. 2002. First nesting record of Black-billed Cuckoo (*Coccyzus erythropthalmus*) in the lower Mississippi valley. *J. Louisiana Orn.* 5: 46-48.

Graves, G.R. & Zusi, R.L. 1990. Avian body weights from the lower Rio Xingu, Brazil. *Bull. Brit. Orn. Club* 110: 20-25.

Gray, G.R. 1832. *Cuculus lathami* in Hardwicke's *Illus. Ind. Zool.* vol. 2. pl. 34, fig. 2.

Gray, G.R. 1840. *A list of the genera of birds, with an indication of the typical species of each genus*. Richard and John E. Taylor, London.

Gray, G.R. 1846. *Oxylophus serratoides*, *Oxylophus coromandus* var. *rubramus*, *Cuculus niger* var. *pyrommatus*. Cat. Birds Nepal, pp. 85, 119, 220.

Gray, G.R. 1846. *The genera of birds: comprising their generic characters, a notice of the habits of each genus, and an extensive list of species referred to their several genera* Vol. 2. part 29, page 454, plate CXV. Longman, Brown, Green, and Longmans, London.

Gray, G.R. 1849. Description of a new species of the genus *Cultrides*. *Proc. Zool. Soc. London* 1849: 63.

Gray, G.R. 1858. A List of the Birds, with Descriptions of New Species obtained by Mr. Alfred R. Wallace in the Aru and Ké Islands. *Proc. Zool. Soc. London* 26: 169-198.

Gray, G.R. 1862. Remarks on, and descriptions of, new species of birds lately sent by Mr. A. R. Wallace from Waigiou, Mysol, and Gagie Islands. *Proc. Zool. Soc. London* 1861: 427-438.

Gray, G.R. 1870. *Hand-list of the genera and species of birds, distinguishing those contained in the British Museum*. Part II. British Museum, London.

Gray, J.E. 1831. *Illustrations of Indian Zoology; chiefly selected from the collection of Major-General Hardwicke, F. R. S.* Vol. I, pt. 6, pl. 28.

Green, A.A. & Carroll, R.W. 1991. The avifauna of Dzanga-Ndoki National Park and Dzanga-Sangha Rainforest Reserve, Central African Republic. *Malimbus* 13: 49-66.

Green, C.A. 1994. Roadrunner predation on Purple Martins. *Purple Martin Update* 5: 12.

Greenberg, D.A. 1975. Some breeding records for Zambia. *Bull. Zambian Orn. Soc.* 7: 7.

Greenway, J.C. Jnr. 1967. *Extinct and vanishing birds of the world*. 2nd ed. Dover Publications, New York.

Greenway, J.C. Jnr. 1978. Type specimens of birds in the American Museum of Natural History. Part 2. *Bull. Amer. Mus. Nat. Hist.* 161: 1-305.

Gregory, P. 1997. Range extensions and unusual sightings from Western Province, Papua New Guinea. *Bull. Brit. Orn. Club* 117: 304-311.

Gregory, P. 2007. Significant sightings from Tour Reports. *Muruk* 8: 99-128.

Greig-Smith, P.W. 1976. The composition and habitat preferences of the avifauna of the Mole National Park, Ghana. *Bull. Niger. Orn. Soc.* 12: 49-66.

Grell, M.B. 1998. *Fuglenes Danmark*. Gads Forlag & Dansk Ornitologisk Forening, Copenhagen.

Grim, T. 2006. Cuckoo growth performance in parasitized and unused hosts: not only host size matter. *Behav. Ecol. Sociobiol.* 60: 716-723.

Grim, T., Kleven, O. & Miculica, O. 2003. Nestling discrimination without recognition: a possible defence mechanism for hosts towards cuckoo parasitism? *Proc. R. Soc. B.* 270: S73-S75.

Grim, T., Rutila, J., Cassey, P. & Hauber M.E. 2009. Experimentally constrained virulence is costly for common cuckoo chicks. *J. Ethol.* 115: 14-22.

Grimes, L.G. 1979. The Yellow-billed Shrike *Corvinella corvina*: an abnormal host of the Yellow-billed Cuckoo *Cuculus gularis*. *Bull. Brit. Orn. Club* 99: 36-38.

Grimes, L.G. 1987. *The birds of Ghana. An annotated checklist*. British Ornithologists' Union Check-list No. 9. BOU, London.

Grimmett, R., Inskipp, C. & Inskipp, T. 1998. *Birds of the Indian Subcontinent*. Christopher Helm, London.

Grinnell, J. & Miller, A.H. 1944. The distribution of the birds of California. *Pacific Coast Avif.* 27: 1-608.

Griscom, L. 1930. Studies from the Dwight collection of Guatemala birds. II. *Amer. Mus. Nov.* 414: 1-8.

Griscom, L. 1932. The ornithology of the Caribbean coast of extreme eastern Panama. *Bull. Mus. Comp. Zool.* 72: 303-372.

Griscom, L. 1932a. The distribution of bird-life in Guatemala. *Bull. Amer. Mus. Nat. Hist.* 64: 1-439.

Griscom, L. & Greenway, J.C. Jnr. 1941. Birds of lower Amazonia. *Bull. Mus. Comp. Zool.* 88: 83-344.

Grobler, J.H. & Steyn, P. 1980. Breeding habits of the Boulder Chat and its parasitism by the Red-chested Cuckoo. *Ostrich* 51: 253-254.

Grote, H. 1924. Beitrag zur Kenntnis der Vogelfauna des Graslandes von Neukamerun. *J. Orn.* 72: 479-517, pl. 4.

Guichard, K.M. 1950. A summary of the birds of the Addis Ababa region, Ethiopia. *J. East Africa Nat. Hist. Soc.* 19: 154-178.

Guilherme, E. & Santos, M.P.D. 2009. Birds associated with bamboo forests in eastern Acre, Brazil. *Bull. Brit. Orn. Club* 129: 229-240.

Guimarães, A.C., Jnr. 1929. Ensaios sobre ornitologia (segunda contribuição). *Rev. Mus. Paulista* 16: 99-116.

Gunawardena, K., Warakagoda, D. & Ekanayake, U. (in prep.). Report of the range and distribution of the Green-billed Coucal *Centropus chlororhynchos* in Sri Lanka.

Gundlach, J. 1874. Neue Beiträge zur Ornithologie Cubas, nach eigenen 30jährigen Beobachtungen zusammengestellt. *J. Orn.* 22: 113-166.

Guthrie, T.H. 1973. Koels displaying. *Aust. Bird Watcher* 4: 168-169.

Guyra Paraguay 2005. *Atlas de las aves de Paraguay*. Guyra Paraguay, Asunción.

Gyldenstolpe, N. 1913. Birds collected by the Swedish Zoological Expedition to Siam 1911-1912. *Kungl. Svenska Vet.-Akad. Handl.* 50 (8): 1-76.

Gyldenstolpe, N. 1945a. The bird fauna of Rio Juruá in western Brazil. *Kungl. Svenska Vet.-Akad. Handl.* (Ser. 3) 22 (3): 1-338.

Gyldenstolpe, N. 1945b. A contribution to the ornithology of northern Bolivia. *Kungl. Svenska Vet.-Akad. Handl.* (Ser. 3) 23 (1): 1-300.

Gyldenstolpe, N. 1955. Birds collected by Dr. Sten Bergman during his expedition to Dutch New Guinea 1948-1949. *Ark. Zool.*, 3rd series, 8 (2): 183-397 + 6 plates.

Haagner, A.K. & Ivy, R.H. 1906. Notes on the nidification of the members of the genus *Chrysococcyx*. *J. South Afr. Orn. Union* 2: 35-39.

Hachisuka, M. 1930. Contributions to the birds of the Philippines. *Suppl. Publ. Orn. Soc. Japan* 2: 138-222.

Hachisuka, M. 1934. *The birds of the Philippine islands, with notes on the mammal fauna*. Vol. 2. Witherby, London.

Hachisuka, M. &. Udagawa, T. 1953. Contribution to the ornithology of the Ryukyu Islands. *Quart. J. Taiwan Mus* 6 (3-4): 141-279.

Hackett, S.J., Kimball, R.T., Reddy, S., Bowie, R.C.K., Braun, E.L., Braun, M.J., Chojnowski, J.L., Cox, W.A., Han, K-L., Harshman, J., Huddleston, C.J., Marks, B.D., Miglia, K.J., Moore, W.S., Sheldon, F.H., Steadman, D.W., Witt, C.C. & Yuri, T. 2008. A phylogenetic study of birds reveals their evolutionary history. *Science* 320: 1763-1768.

Hadden, D. 1981. *Birds of the North Solomons*. Wau Ecology Institute. Handbook No. 8, Wau.

Haffer, J. 1975. Avifauna of northwestern Colombia, South America. *Bonn. Zool. Monogr.* 7: 1-181.

Haffer, J. 1977. A systematic review of the Neotropical ground-cuckoos (Aves, *Neomorphus*). *Bonn. Zool. Beitr.* 28: 48-76.

Haffer, J. 1997. Contact zones between birds of southern Amazonia. *Orn. Monogr.* 48: 281-305.

Haga, H. & Nuka, T. 2009. The first record of Long-tailed Koel *Eudynamys taitensis* in Japan. *Japanese Jour. Orn.* 58: 206-207. [in Japanese]

Hagemeijer, W.J.M. & Blair, M.J. 1997. *The EBCC atlas of European breeding birds, their distribution and abundance*. European Bird Census Council and Poyser, London.

Hails, C. & Jarvis, F. 1987. *Birds of Singapore*. Times Editions, Singapore.

Hall, B.P. (Ed.). 1974. *Birds of the Harold Hall Australian Expedition 1962-70*. British Museum (Natural History), London.

Halleux, D. 1994. Annotated list of Macenta Prefecture, Guinea. *Malimbus* 16: 10-29.

Hallinan, T. 1924. Notes on some Panama Canal Zone birds with special reference to their food. *Auk* 41: 304-326.

Halterman, M.M. 2009. Sexual dimorphism, detection probability, home range, and parental care in the Yellow-billed Cuckoo. Ph.D. Dissertation. University of Nevada, Reno.

Hamel, P.B. & Kirkconnell, A. 2005. Composition of mixed-species flocks of migrant and resident birds in Cuba. *Cotinga* 24: 28-34.

Hamilton, T.H. 1962. The habitats of the avifauna of the mesquite plains of Texas. *Amer. Midland Nat.* 67: 85-105.

Hamilton, W.J. III & Hamilton, M.E. 1965. Breeding characteristics of Yellow-billed Cuckoos in Arizona. *Proc. Calif. Acad. Sci.* (4th Series) 32: 405-432.

Hamilton, W. J. & Orions, G.H. 1965. Evolution of brood parasitism in altricial birds. *Condor* 67: 361–382.

Hamling, H.H. 1937. Notes from Lomagundi District, Southern Rhodesia. *Ibis* (14th series) 1: 175-177.

Handrinos, G. & Akriotis, T. 1997. *The birds of Greece*. Christopher Helm, London.

Hanmer, D.B. 1982. Klaas Cuckoo parasitizing Whitebellied Sunbird. *Ostrich* 53: 58

Hanmer, D.B. 1995. Mensural and moult data on seven species of cuckoos and two species of coucals from Mozambique and Malawi. *Honeyguide* 41: 65-102.

Hanna, W.C. 1937. California Cuckoo in the San Bernardino Valley, California. *Condor* 39: 57-59.

Harding, A.K. 1948. Bird notes. *Nature in East Africa* 7: 2.

Hardy, J.W., Reynard, G.B. & Coffey, B.B. Jnr. 1990. *Voices of the New World cuckoos and trogons*. ARA Records, Gainesville, Florida.

Harris, A.H. & Crews, C.R. 1983. Conkling's Roadrunner - a subspecies of the California Roadrunner? *Southw. Nat.* 28: 407-412.

Harris, M.P. 1973. The Galápagos avifauna. *Condor* 75: 265-278.

Harris, M.P. 1982. *A field guide to the birds of Galápagos* (revised ed.). Collins, London.

Harrison, C. 1978. *A field guide to the nests, eggs and nestlings of North American birds*. Collins, Cleveland.

Harrison, C.J.O. 1970. The identification of the eggs of the smaller Indian cuckoos. *J. Bombay Nat. Hist. Soc.* 66: 478-488.

Harrison, C.J.O. 1971a. Notes on the identification of eggs, egg mimicry and distributional history and the status of the form *serratus* in the parasitic *Clamator* cuckoos. *Bull. Brit. Orn. Club* 91: 126-131.

Harrison, C.J.O. 1971b. Further notes on New Guinea birds. *Emu* 71: 85-86.

Harrison, C.J.O. 1971c. The sunbathing of the Roadrunner. *Avicult. Mag.* 77: 128.

Harrison, G.H. 2005. Comical cuckoo. *Birder's World* 19: 56-58.
Harrison, J.A., Allan, D.G., Underhill, L.G., Herremans, M., Tree, A.G., Parker, V. & Brown, C.J. (Eds) 1997. *The atlas of southern african birds including Botswana, Lesotho, Namibia, South Africa, Swaziland and Zimbabwe*. Vol. 1. Non-Passerines. BirdLife South Africa, Johannesburg.
Harrison, R. 1974. Male cuckoo using plant material in display. *Brit. Birds* 67: 514-515.
Hartert, E. 1896. On the ornithological collections made by Mr. Alfred Everett in Celebes and on the islands south of it. *Novit. Zool.* 3: 148-183.
Hartert, E. 1898. Appendix I. List of the Birds Collected on New Hanover. in H. Cayley-Webster. *Through New Guinea and the Cannibal Countries*. T. Fisher Unwin, London, UK. Pp. 369-375.
Hartert, E. 1898a: On a collection of birds from northwestern Ecuador, collected by W.F.H. Rosenberg. *Novit. Zool.* 5: 477-505.
Hartert, E. 1900. The birds of Buru, being a list of collections made on that island by Messrs. William Doherty and Dumas. *Novit. Zool.* 7: 226-242
Hartert, E. 1900. The birds of Dammer Island in the Banda Sea. *Novit. Zool.* 7: 12-24.
Hartert, E. 1903. On the birds collected on the Tukang-Besi Islands and Buton, south-east of Celebes, by Mr. Heinrich Kühn. *Novit. Zool.* 10: 18-38.
Hartert, E. 1910. The birds of Hainan. *Novit. Zool.* 17: 189-254.
Hartert, E. 1912. *Die Vögel der paläarktischen Fauna*. Band 2, Heft 1. Berlin, Friedländer.
Hartert, E. 1914. On the birds of Rook Island in the Bismarch Archipelago. *Novit. Zool.* 21: 207-218.
Hartert, E. 1915. List of a small collection of birds from Hausaland, northern Nigeria. *Novit. Zool.* 22: 244-266.
Hartert, E. 1925a. A collection of birds from New Ireland (Neu Mecklenburg). *Novit. Zool.* 32: 115-136.
Hartert, E. 1925b. Review of the genus *Cacomantis* Müll. *Novit. Zool.* 32: 164-174.
Hartert, E. 1926. On the birds of the district of Talasea in New Britain. *Novit. Zool.* 33: 122-145.
Hartert, E. & Ogilvie-Grant, W.R. 1905. On the birds of the Azores. *Novit. Zool.* 12: 80-128.
Hartert, E. & Stresemann, E. 1925. Über die Indo-australischen Glanzkuckucke (*Chalcites*). *Novit. Zool.* 32: 158-163.
Hartert, E. & Venturi, S. 1909. Notes sur les oiseaux de la République Argentine. *Novit. Zool.* 16: 159-267.
Hartlaub, G. 1844. Description de sept *Oiseaux* nouveaux de Guatemala. *Rev. Zool.* 7: 214-216.
Hartlaub, G. 1852. Ueber einige neue oder weniger bekannte Vögel Amerika's. *Naumannia* 2: 50-56.
Hartlaub, G. 1861. Ueber einige neue Vögel Westafrica's. *J. Orn.* 9: 11-13.
Hartlaub, G. 1863. On a new bird from the island of Madagascar. *Proc. Zool. Soc. London* 1862: 224.
Hartlaub, G. 1866. On five new species of birds from the Feejee Islands. *Ibis* (2nd series) 2: 171-173.
Hartlaub, K.J.G. 1877. *Die Vögel Madagascars und der benachbarten Inselgruppen*. H.W. Schmidt Verlag, Halle.
Harvey, W.G. 1990. *Birds of Bangladesh*. University Press Ltd., Dhaka.
Harvie-Brown, J.A. 1908. An invasion of Cuckoos. *Ann. Scott. Nat. Hist., Edinb.* 1908: 181-182.
Haverschmidt, F. 1948. Bird weights from Surinam. *Wilson Bull.* 60: 230-239.
Haverschmidt, F. 1955. *List of the birds of Surinam*. Publications of the Foundation for Scientific Research in Surinam and the Netherlands Antilles; 13, Utrecht.
Haverschmidt, F. 1961. Der Kuckuck *Tapera naevia* und seine Wirte in Surinam. *J. Orn.* 102: 353-359.
Haverschmidt, F. 1968. *Birds of Surinam*. Oliver & Boyd, Edinburgh & London.
Haverschmidt, F. 1970. Zum Brutparasitismus von *Tapera naevia*. *J. Orn.* 111: 105.
Haverschmidt, F. &. Mees, G.F. 1994. *The birds of Suriname*, 2nd ed. VACO N.V., Paramaribo, Suriname.

Hawkins, A.F.A. & Goodman, S.M. 2003. Introduction to the birds. Pp.1019-1044 in S.M. Goodman & J.P. Benstead (Eds) *The natural history of Madagascar*. Univerity of Chicago Press, Chicago.
Hawkins, A.F.A., Thiollay, J-M. & Goodman, S.M. 1998. The birds of the Réserve Spéciale d'Anjanaharibe-Sud Madagascar. Pp. 93-127 in A floral and fanunal inventory of the Réserve Spéciale d'Anjanaharibe-Sud, Madagascar: with reference to elevational variation. *Fieldiana Zool. N.S.* 90.
Haydock, E.L. 1950. Supplementary notes on African cuckoos. *Ibis* 92: 149-150.
Haydock, E.L. 1951. Extracts from a Northern Rhodesian diary. *Babbler Gaborone* 6: 1-3.
Haydock, E.L. 1952. Addition to Northern Rhodesian checklist. *Ibis* 94: 360.
Hayes, F.E. 1995. Status, distribution and biogeography of the birds of Paraguay. *Monogr. Field. Orn.* 1: 1-230. American Birding Association.
Hayes, F.E., Goodman, S.M., Fox, J.A., Granizo Tamayo, T. & López, N.E. 1990. North American bird migrants in Paraguay. *Condor* 92: 947-960.
Hayes, F.E., Scharf, P.A. & Ridgely, R.S. 1994. Austral bird migrants in Paraguay. *Condor* 96: 83-97.
Haynes-Sutton, A., Downer, A., Sutton, R. & Rey-Millet, Y-J. 2009. *A photographic guide to the birds of Jamaica*. Christopher Helm, London.
Hayward, C.L., Cottam, C., Woodbury, A.M. & Frost, H.H. 1976. *Birds of Utah*. Brigham Young University. Great Basin Nat. Mem. 1.
Heather, B.D. & Robertson, H.A. 1996. *The field guide to the birds of New Zealand*. OUP, Oxford.
Hector, D.P. 1985. The diet of the Aplomado Falcon (*Falco femoralis*) in eastern Mexico. *Condor* 87: 336-342.
Heim de Balsac, H. & Mayaud, N. 1962. *Les oiseaux du nord-ouest de l'Afrique*. Editions Paul Chevalier, Paris.
Heine, F. Jnr. 1863. Cuculidina. *J. Orn.* 11: 349-358.
Heinrich, G. 1958. Zur Verbreitung und Lebensweise der Vögel von Angola. Systematischer Teil I. *J. Orn.* 99: 322-362.
Heinroth, O. & Heinroth, M. (1926) 1966 (reprint). *Die Vögel Mitteleuropas in allen Lebens- und Entwicklungsstufen photographisch aufgenommen und in ihrem Seelenleben bei der Aufzucht von Ei ab beobachtet*. Band 1. Verlag für Kunst und Wissenschaft, Leipzig.
Helbig, A.J., Knox, A.G., Parkin, D.T., Sangster, G. & Collinson, M. 2002. Guidelines for assigning species rank. *Ibis* 144: 518-525.
Hellebrekers, W.P.J. 1942. Revision of the Penard oölogical collection from Surinam. *Zool. Med.* 24: 240-275.
Hellebrekers, W.P.J. & Hoogerwerf, A. 1967. A further contribution to our oological knowledge of the island of Java (Indonesia) *Zool. Verh. Leiden* 88: 1-164.
Hellmayr, C.E. 1906. On the birds of the island of Trinidad. *Novit. Zool.* 13: 1-60.
Hellmayr, C.E. 1913. Critical notes on the types of little-known species of Neotropical birds. Part II. *Novit. Zool.* 20: 227-256.
Hellmayr, C.E. 1929. A contribution to the ornithology of northeastern Brazil. *Field Mus. Nat. Hist. Zool. Ser.* 12: 233-501.
Hellmayr, C.E. 1932. The birds of Chile. *Field Mus. Nat. Hist. Zool. Ser.* 19: 1-472.
Hempel, A. 1949. Estudos da alimentação natural de aves silvestres do Brasil. *Arq. Inst. Biol., São Paulo* 19: 237-268.
Hemprich, F.W. & Ehrenberg, C.G. 1828. *Centropus superciliosus*, sp. n. *Symb. Phys., Aves*. Pl.11 & 12.
Hemprich, F.W. & Ehrenberg, C.G. 1833. *Cuculus pica*, sp. n. *Symb. Phys., Aves*, sig. r. note 2-3.
Hendricks, D.P. 1975. Copulatory behavior of a pair of Yellow-billed Cuckoos. *Auk* 92: 151.
Henriques, L.M.P. & Oren, D.C. 1997. The avifauna of Marajó, Caviana and Mexiana Islands, Amazon River estuary, Brazil. *Rev. Brasil. Biol.* 57: 357-382.
Henriques, L.M.P., Wunderle, J.M. Jnr. & Willig, M.R. 2003. Birds of the Tapajós National Forest, Brazilian Amazon: a preliminary assessment. *Orn. Neotrop.* 14: 307-338.

Henry, G.M. 1955. *A guide to the birds of Ceylon*. Geoffrey Cumberlege & OUP, Oxford.
Herklots, G.A.C. 1961. *The birds of Trinidad and Tobago*. Collins, London.
Hermes, N., Evans, O. & Evans, B. 1986. Norfolk Island birds: a review 1985. *Notornis* 33: 141-149.
Heron, M. 1977. Observations and bird notes (Red-necked Rail; Narrow-billed Bronze Cuckoo). *PNG Bird Soc. Newsletter* 134: 3.
Herreid, C.F. 1960. Roadrunner a predator of bats. *Condor* 62: 67.
Hewson, C.M. & Noble, D.G. 2009. Population trends of breeding birds in British woodlands over a 32-year period: relationships with food, habitat use and migratory behaviour. *Ibis* 151: 464-486.
Hermann, J. 1783. *Tabula affinitatum animalium olim academico specimine edita, nunc uberiore commentario illustrata cum annotationibus ad historiam naturalem animalium augendum facientibus*. J. G. Treuttel, Argentorati [Strasbourg].
Herrick, F.H. 1910. Life and behavior of the Cuckoo. *J. Exp. Zool.* 9: 169-233.
Hess, I.E. 1910. One hundred breeding birds of an Illinois ten-mile radius. *Auk* 27: 19-32.
Hicks, R.K. & Restall, R. 1992. Pheasant Coucal *Centropus phasianinus* attacking birds caught in a mist net. *Muruk* 5: 143.
Hidasi, J. 1973. *Catálogo das aves do Brasil especialmente do centro oeste e da Amazônia; lista dos exemplares existentes na coleção do Museu de Ornitologia*. Companhia de Desenvolvimento do Estado de Mato Grosso, Cuiabá.
Higgins, P.J. (Ed.) 1999. *Handbook of Australian, New Zealand and Antarctic birds. Vol. 4. Parrots to Dollarbird*. OUP, Melbourne.
Higuchi, H. 1989. Responses of the Bush Warbler *Cettia diphone* to artificial eggs of *Cuculus* cuckoos in Japan. *Ibis* 131: 94-98.
Higuchi, H. 1998. Host use and egg color of Japanese cuckoos. Pp. 80-93 in S. I. Rothstein & S. K. Robinson (Eds). *Parasitic birds and their hosts*. Oxford University Press, Oxford.
Higuchi, H. & Payne, R.B. 1986. Nestling and Fledgling plumages of *Cuculus saturatus horsfieldi* and *C. poliocephalus poliocephalus* in Japan. *Jap. J. Orn.* 35: 61-65.
Higuchi, H. & Sato, S. 1984. An example of character release in host selection and egg colour of cuckoos *Cuculus* spp. in Japan. *Ibis* 126: 398-404.
Hill, D. & Hill, C. 2010. Some observations of the Bornean Ground Cuckoo *Carpococcyx radiatus*. *BirdingASIA* 14: 44-45.
Hill, G.F. 1911. Field notes on the birds of the Kimberley, north-west Australia. *Emu* 10: 258-290.
Hill, M. 2000. Bird fauna of two protected forests in northern Vietnam. *Forktail* 16: 5-14.
Hilty, S.L. 2003. *Birds of Venezuela*. Christopher Helm, London.
Hilty, S.L. & Brown, W.L. 1986. *A guide to the birds of Colombia*. Princeton University Press, Princeton.
Hindwood, K.A. 1930. The habits of cuckoos. *Aust. Mus. Mag.* 4: 17-21.
Hindwood, K.A. 1935. Birds inhabiting mangroves in the neighbourhood of Sydney. *Emu* 34: 181-189.
Hindwood, K.A. 1940. The birds of Lord Howe Islands. *Emu* 40: 1-86.
Hindwood, K.A. 1953. Channel-billed Cuckoo in New Caledonia. *Emu* 53: 334-335.
Hindwood, K.A. 1957. The Coucal or Swamp-Pheasant. *Aust. Mus. Mag.* 12: 176-182.
Hindwood, K.A. & McGill, A.R. 1951. The "Derra Derra" 1950 camp-out of the RAOU. *Emu* 50: 217-238.
Hines, C. 1996. Namibia's Caprivi strip. *Bull. African Bird Club* 3: 113–128.
Hitoshi, N., Asama, S. & Biun, A. 1996. *A photographic guide to the birds of Mt. Kinabalu*. Bun-Ichi Sogoshuppan, Tokyo.
Hjartarson, G. 1995. Flækingsfuglar á Íslandi: Gaukar. *Náttúrufræðingurinn* 64: 255-263.
Hobbs, J.N. 1961. The birds of south-west New South Wales. *Emu* 61: 21-55.

Hockey, P.A.R. 1992. Rare birds in South Africa 1989-1990. Seventh report of the SAOS Rarities committee. *Birding in Southern Africa* 44: 38-44.
Hockey, P.A.R., Dean, W.R.J. & Ryan, P.G. (Eds). 2005. *Roberts – birds of Southern Africa, VII ed*. Trustees of John Voelcker Bird Book Fund, Cape Town.
Hodgson, B.H. 1839. On *Cuculus*. *J. Asiat. Soc. Bengal* 8: 136-137.
Hoesch, W. 1940. Ueber den Einfluss der Zivilisation auf das Brutverhalten der Vögel und über abweichende Brutgewohnheiten. *J. Orn.* 88: 576-586.
Hoffmann, A. 1950. Der Indische Kuckuck (*Cuculus micropterus* Gould). *Bonn. Zool. Beitr.* 1: 21-30.
Hoffmann, T.W. 1984. *National Red Data List of endangered and rare birds of Sri Lanka*. Ceylon Bird Club and Wild Life and Nature Protection Society of Sri Lanka.
Hoffmann, T.W. 1989. The Greenbilled or Ceylon Coucal *Centropus chlororhynchus* Blyth - Sri Lanka´s rarest endemic species. *J. Bombay Nat. Hist. Soc.* 86: 339-343.
Hoffmann, T.W. 1996. New bird records in Sri Lanka and some connected matters. *J. Bombay Nat. Hist. Soc.* 93: 382-387.
Hollom, P.A.D., Porter, R.F., Christensen, S. & Willis, I. 1988. *Birds of the Middle East and North Africa*. Poyser, Calton.
Holman, F.C. 1947. Birds of the Gold Coast. *Ibis* 89: 623-650.
Holmes, D.A. 1996. Sumatra bird report. *Kukila* 8: 9-56.
Holmes, D.A. 1997. Kalimantan bird report - 2. *Kukila* 9: 141-169.
Holmes, D.A. & Burton, K. 1987. Recent notes on the avifauna of Kalimantan. *Kukila* 3: 2-32.
Hoogerwerf, A. 1948. Korte mededelingen. Een exemplaar van *Centropus nigrorufus* (Cuv.) van Randoeblatoeng, gelegen in het centrum van Midden-Java. *Limosa* 21: 107.
Hoogerwerf, A. 1949. Bijdrage tot de oölogie van Java. *Limosa* 22: 1-279.
Hoogstraal, H. 1982. Ticks (*Acari:* Ixodoidea): a preliminary study. Pp. 537-44 in J.L. Gressitt (Ed.). *New Guinea biogeography*. W. Junk, The Hague.
Hopkins, N. 1948. Birds of Townsville and district. *Emu* 47: 331-347.
Hopkinson, E. 1922. Cuckoos in the Gambia. *Ibis* (11th series) 4: 745-746.
Hopwood, J.C. & Mackenzie, J.M.D. 1917. A list of birds from the North Chin Hills. *J. Bombay Nat. Hist. Soc.* 25: 72-91.
Hornbuckle, J. 1994. *Birdwatching in the Philippines*. Privately published.
Hornbuckle, J. 1997a. Two sightings of Banded Ground-Cuckoo *Neomorphus radiolosus* in Ecuador. *Cotinga* 8: 90.
Hornbuckle, J. 1997b. Eastern South Africa and Zimbabwe. 3rd Feb-5th March 1997. Unpublished report.
Hornskov, J. 1996. Recent observations of birds in the Philippine Archipelago. *Forktail* 11: 1-10.
Horsfield, T. 1821. Systematic arrangement and descriptions of birds from the island of Java. *Trans. Linn. Soc. London* 13: 133-200.
Howell, S.N.G. & Webb, S. 1992. Noteworthy bird observations from Baja California, Mexico. *Western Birds* 23: 153-163.
Howell, S.N.G. & Webb, S. 1995. *A guide to the birds of Mexico and northern Central America*. OUP, Oxford.
Howell, T.R. 1957. Birds of a second-growth rain forest area of Nicaragua. *Condor* 59: 73-111.
Höy, G. 1969. Addendas a la avifauna salteña. *Hornero* 11: 53-56.
Hubbard, J.P. & Crossin, R.S. 1974. Notes on northern Mexican birds. *Nemouria* 14: 1-41.
Huber, W. 1932. Birds collected in northeastern Nicaragua in 1922. *Proc. Acad. Nat. Sci. Philad.* 84: 205-249.
Hudson, W.H. 1874. Notes on the procreant instincts of the three species of *Molothrus* found in Buenos Ayres. *Proc. Zool. Soc. London* 1874: 153-174.
Hudson, W.H. 1920. *Birds of La Plata*, vol. 2. J. M. Dent & Sons, London & Toronto.
Hughes, J.M. 1996. Greater Roadrunner *Geococcyx californianus*. Pp. 1-24 in A. Poole & F. Gill (Eds). *The birds of North America*. Vol. 244. Acad. Nat. Sci. Philad., Philadelphia.

Hughes, J.M. 1997a. Vocal duetting by a mated pair of Coral-billed Ground-Cuckoos (*Carpococcyx renauldi*) at the Metro Toronto Zoo. *Zoo Biology* 16: 179-186.

Hughes, J. M. 1997b. Mangrove Cuckoo *Coccyzus minor* in A. Poole & F. Gill (Eds) *The birds of North America*. No. 299. Acad. Nat. Sci. Philad., Philadelphia.

Hughes, J. M. 1997c. Taxonomic significance of host-egg mimicry by facultative brood parasites of the avian genus *Coccyzus* (Cuculidae). *Can. J. Zool.* 75: 1380-1386.

Hughes, J.M. 1999a. Rare, local, little known, and declining North American breeders. A closer look: Mangrove Cuckoo. *Birding* 31: 22-27.

Hughes, J.M. 1999b. Yellow-billed Cuckoo *Coccyzus americanus*. in A. Poole & F. Gill (Eds) *The birds of North America*. No. 418. The Birds of North America, Inc., Philadelphia.

Hughes, J.M. 2000. Monophyly and phylogeny of cuckoos (Aves, Cuculidae) inferred from osteological characters. *Zool. Jour. Linn. Soc. London* 130: 263-307.

Hughes, J.M. 2001. Black-billed Cuckoo *Coccyzus erythropthalmus* in A. Poole & F. Gill (Eds) *The birds of North America*. No. 587. 24 pp. The Birds of North America, Inc., Philadelphia.

Hughes, J.M. 2003. Phylogeny of cooperatively-breeding cuckoos (Cuculidae, Crotophaginae) based on mitochondrial gene sequences. *Naturwiss.* 90: 231-233.

Hughes, J.M. 2006. Phylogeny of the cuckoo genus *Coccyzus* (Aves: Cuculidae): a test of monophyly. *Syst. Biodiv.* 4: 483-488.

Huguet, P. & Chappuis, C. 2003. *Bird Sounds of Madagascar, Mayotte, Comoros, Seychelles, Réunion and Mauritius*. 4 CDs, booklet. SEOF.

Hume, A.O. 1873. Acheen [India]. *Stray Feathers* 1 (6): 441-463.

Hume, A.O. & Davison, W. 1878. A revised list of the birds of Tenasserim. *Stray Feathers* 6: i-viii, 1-496, Appendices 1 & 2: 497-524.

Hunter, H.C. 1961. Parasitism of the Masked Weaver *Ploceus velatus arundinaceus*. *Ostrich* 32: 55-63.

Hustler, K. 1997. Notes on the breeding biology of the Coppery-tailed Coucal. *Honeyguide* 43 (4): 184-187.

Hustler, K., Best, G. & Voss, N. 1996. First breeding record of the Coppery-tailed Coucal in Zimbabwe. *Honeyguide* 42: 163-165.

Hutton, I. 1991. *Birds of Lord Howe Island: past and present*. Privately published, New South Wales.

ICZN = International Commission for Zoological Nomenclature. 1956. Opinion 407. *Opin. Decl. Int. Comm. Zool. Nomen.* 13: 131-146.

Illiger, J.K.W. 1811. *Prodromus systematis mammalium et avium*. C. Salfeld, Berlin.

Indrawan, M., Somadikarta, S., Supriatna, J., Bruce, M.D. & Djanubudiman, G. 2006. The birds of the Togian islands, Central Sulawesi, Indonesia. *Forktail* 22: 7-22.

Inglis, C.M. 1908. The oology of Indian parasitic cuckoos. *J. Bombay Nat. Hist. Soc.* 18: 681-682.

Inglis, C.M. 1909. Birds of the Duars. *J. Beng. Nat. Hist. Soc.* 35: 49-63.

Inglis, C.M., Travers, W.L., O'Donel, H.V. & Shebbeare, E.O. 1920. A tentative list of the vertebrates of the Jalpaiguri District, Bengal. Part II. *J. Bombay Nat. Hist. Soc.* 26: 151-162.

Ingram, C. 1912. The birds of Yunnan. *Novit. Zool.* 19: 269-310.

Insfran, F.R. 1931. Un caso raro de albinismo en la especie *Crotophaga ani* L. Anó moroti. *Rev. Soc. Cient. Paraguay* 3: 33.

Inskipp, C. & Inskipp, T. 1985 (reprinted 1991). *A guide to the birds of Nepal*. Christopher Helm, London.

Inskipp, C., Inskipp, T. & Grimmett, R. 1999. *Birds of Bhutan*. Christopher Helm, London.

Inskipp, C., Inskipp, T. & Sherub. 2000. The ornithological importance of Thrumshingla National Park, Bhutan. *Forktail* 16: 147-162.

Irisov, E.I. 1967. [The Himalayan Cuckoo in the south-eastern Altai.] *Ornitologiya* 8: 355-356. [in Russian]

Irwin, M.P.S. 1956. Field notes on a collection from Mozambique. *Ostrich* 27: 28-39.

Irwin, M.P.S. 1957. Some field notes on a collection of birds from Tanganyika Territory. *Ostrich* 28: 117.

Irwin, M.P.S. 1981. *The birds of Zimbabwe*. Quest Publishing, Harare.

Irwin, M.P.S. 1987. Geographical variation in plumage of female Klaas´s Cuckoo *Chrysococcyx klaas*. *Malimbus* 9: 43-46.

Irwin, M.P.S. 1988. Family Cuculidae in C.H. Fry, S. Keith, & E.K. Urban (Eds). *The birds of Africa* vol. III. Academic Press, London.

Irwin, M.P.S. 1989. Vagrant cuckoos in the southern Indian Ocean: a comment. *Cormorant* 17: 87-88.

Isenmann, P., Benmergui, M., Browne, P., Ba, A.D., Diagana, C.H., Diawara, Y. & El Abidine ould Sidaty, Z. 2010. *Oiseaux de Mauritanie/Birds of Mauritania*. Société d'Études Ornithologiques de France, Paris.

Isenmann, P., Gaultier, T., El Hili, A., Azafzaf, H. Dlensi, H. & Smart, M. 2005. *Oiseaux de Tunisie/Birds of Tunisia*. Société d'Études Ornithologiques de France, Paris.

Isenmann, P. & Moali, A. 2000. *Oiseaux d'Algérie/Birds of Algeria*. Société d'Études Ornithologiques de France, Paris.

Ishizawa, J. & Chiba, S. 1966. Food analyses of four species of cuckoos in Japan. *Misc. Rep. Yamashina Inst. Orn.* 4: 302-326.

Isler, M.L., Isler, P. R. & Brumfield, R.T. 2005. Clinal variation in vocalizations of an antbird (Thamnophilidae) and implications for defining species limits. *Auk* 122 (2): 433-444.

Iversen, E. & Hill, B. 1983. Diederick Cuckoo feeds fledgling of same species. *Bee-eater* 34: 47.

Ivy, R.H. 1901. Notes on the nesting and other habits of some South African birds. *Ibis* (8th series) 1: 18-28.

Jackson, F.T. 1938. *The birds of Kenya Colony and the Uganda Protectorate*. Gurney & Jackson, London.

Jackson, M.H. 1985. *Galápagos: a natural history guide*. The University of Calgary Press, Calgary, Alberta.

Jackson, W.M. 1992. Relative importance of parasitism by *Chrysococcyx* cuckoos versus conspecific nest parasitism in the Northern Masked Weaver *Ploceus taeniopterus*. *Ornis Scandinavia* 23: 203-206.

Jacobson, S. & Wallin, K. 1969. Skatgök *Clamator glandarius* för första gangen i Sverige. *Vår Fågelvärld* 28: 102-106.

Jaeger, E.C. 1947. Stone-turning habits of some desert birds. *Condor* 49: 171.

Jahn, A.E., Davis, S.E. & Saavedra Zankys, A.M. 2002. Patterns of austral bird migration in the Bolivian chaco. *J. Field Orn.* 73: 258-267.

Jaramillo, A. 2003. *A field guide to the birds of Chile including the Antarctic Peninsula, the Falkland Islands and South Georgia*. Princeton University Press, Princeton.

Jarvinen, A. 1984. Relationship between the Common Cuckoo *Cuculus canorus* and its host, the Redstart *Phoenicurus phoenicurus*. *Ornis Fennica* 61: 84-88.

Jefferey, G.W. 1931. The birds of Bahati Woods, Kenya Colony. Pt. III. *Bateleur* 3: 63-67.

Jennings, M.C. 1981. *The birds of Saudi Arabia: a checklist*. Privately published, Cambridge.

Jennings, M.C. 2010. *Fauna of Arabia. Atlas of the breeding birds of Arabia*. Senckenberg, Frankfurt and Riyadh.

Jenny, J. 1997. Unusual breeding behaviour of the Guira Cuckoo *Guira guira*. *Cotinga* 7: 18.

Jensen, R.A.C. & Clinning, C.F. 1974. Breeding biology of two cuckoos and their hosts in South West Africa. *Living Bird* 13: 5-50.

Jensen, R.A.C. & Jensen, M.K. 1969. On the breeding biology of Southern African cuckoos. *Ostrich* 40: 163-181.

Jepson, P. 1993. Recent ornithological observations from Buru. *Kukila* 6: 85-109.

Jepson, P. 1997. Birding Indonesia. *A bird watcher's guide to the world's largest archipelago*. Periplus Edition, Singapore.

Jerdon, T.C. 1840. Catalogue of the birds of the Peninsula of India, ... *Madras Jour. Lit. Sci.* 11: 223-239.

Jerome, C.H. 1943. Difficulties of amateur ornithologists in identifying the eggs of parasitic birds. *Ostrich* 14: 99-103.

Jessop, J.P. 1960. Red-chested Cuckoo egg in Cape Paradise Flycatcher nest. *Ostrich* 31: 179.

Jeyarajasingam, A. & Pearson, A. 1999. *A field guide to the birds of West Malaysia and Singapore.* OUP, Oxford.

Jiang, Ying-xin, Liang, Wei, Yang, Can-chao & Sun, Yue-hua. 2007. *Hierococcyx sparverioides* parasitized and reproduced in nest of Garrulax sannio. *Sichuan J. Zool.* 26: 509.

Johansson, U.S., Parsons, T.J., Irestedt, M. & Ericson, P.G.P. 2001. Clades within the "higher land birds", evaluated by nuclear DNA sequences. *J. Zool. Syst. Evol. Research* 39: 37-51.

Johns, A.D. 1986. Effects of selective logging on the ecological organization of a peninsular Malaysian rainforest avifauna. *Forktail* 1: 65-79.

Johnsgard, P.A. 1997. *The avian brood parasites. Deception at the nest.* OUP, New York, Oxford.

Johnsingh, A.J.T. & Paramanandham, K. 1982. Group care of White-headed Babblers *Turdoides affinis* for a Pied Crested Cuckoo *Clamator jacobinus* chick. *Ibis* 124: 179-183.

Johnson, A.W. 1967. *The birds of Chile and adjacent regions of Argentina, Bolivia, and Peru* Vol. 2. Platt Establecimientos Gráficos, Buenos Aires.

Johnson, A.W. 1972. *Supplement to the birds of Chile and adjacent regions of Argentina, Bolivia, and Peru.* Platt Establecimientos Gráficos, Buenos Aires.

Johnson, D.H., Bryant, M.D. & Miller, A.H. 1948. Vertebrate animals of the Providence Mountains area of California. *Univ. Calif. Publ. Zool.* 48: 219-375.

Johnson, J. & Tweit, B. 1989. The nesting season June 1-July 31, 1989. Northern Pacific coast region. *Am. Birds* 43: 1358-1361.

Johnson, K.P., Goodman, S.M. & Lanyon, S.M. 2000. A phylogenetic study of the Malagasy couas with insights into cuckoo relationships. *Mol. Phylog. Evol.* 14: 436-444.

Johnson, O.W. 1968a. Some morphological features of avian kidneys. *Auk* 85: 216-228.

Johnson, P.G. 1968b. Notes on nesting and diet of Burchell's Coucal. *Ostrich* 39: 194.

Johnston, R.F. 1964. The breeding birds of Kansas. *Univ. Kansas Publ. Mus. Nat. Hist.* 12: 575-655.

Johnstone, R.E. 1983. Birds of the Hammersley Range National Park. In B.G. Muir (Ed.). *A fauna survey of the Hammersley Range National Park, Western Australia.* National Parks Authority of Western Australia, Nedlands.

Johnstone, R.E. & Burbridge, A.H. 1991. The avifauna of Kimberley rainforests. Pp. 361-391 in N.L. McKenzie, R.B. Johnston & P.G. Kendrick (Eds) *Kimberley rainforests.* Surrey Beattu, Sydney.

Johnstone, R.E. & Storr, G.M. 2004. *Handbook of Western Australian birds.* Vol. II. Western Australian Museum, Perth.

Johnstone-Stewart, N.G.B. 1984. Evergreen forest birds in the southern third of Malawi. *Nyala* 10: 99-119.

Jones, A.E. 1941. Presumptive evidence of the nidification of the Indian Cuckoo (*Cuculus micropterus* Gould) *J. Bombay Nat. Hist. Soc.* 42: 931-933.

Jones, D.A., Gibbs, H.L., Matsuda, T., Brooke, M., Uchida, H. & Bayliss, M.J. 1997. The use of DNA fingerprinting to determine the possible mating system of an obligate brood parasitic bird, the Cuckoo, *Cuculus canorus. Ibis* 139: 560-562.

Jones, J.M.B. 1985. Striped Crested Cuckoo parasitizing Arrow-marked Babbler. *Honeyguide* 31: 170-171.

Jones, J.P.G., Ferry, C.D., Isherwood, C.E., Knight, C.G., Kumera, C.L. & Weerakoon, K. 1998. *A conservation review of three forest in south-west Sri Lanka.* CSB Conservation Reports, Cambridge, UK.

Jones, M., Juhaeni, D., Banjaransari, H., Banham, W., Lace, L., Linsley, M. & Marsden, S. 1994. *The status, ecology and conservation of the forest birds and butterflies of Sumba.* Manchester Metropolitan University, Department of Biological Sciences.

Joseph, L., Zeriga, T., Adcock, G.J. & Langmore, N.E. 2011. Phylogeography and taxonomy of the Little Bronze-Cuckoo (*Chalcites minutillus*) in Australia's monsoon tropics. *Emu* 111: 113-119.

Joubert, H.J. 1943. Some notes on birds. *Ostrich* 14: 1-7.

Jullien, M. & Thiollay, J-M. 1998. Multi-species territoriality and dynamics of Neotropical forest understorey bird flocks. *J. Anim. Ecol.* 67: 227-252.

Junge, G.C.A. 1936. Fauna Simalurensis - Aves. *Zool. Med. Temminckia & Novae Guinea* 1: 1-74.

Junge, G.C.A. 1937a. The birds of south New Guinea. Part I. Non-Passeres. *Zool. Med. Temminckia & Novae Guinea* 1: 125-184.

Junge, G.C.A. 1937b. Further notes on the birds of Simalur. *Temminckia* 2: 197–202.

Junge, G.C.A. 1937c. On Bonaparte's types of the cuckoos belonging to the genus *Piaya. Zool. Meded.* 19: 183-186.

Junge, G.C.A. 1938. Remarks on *Chalcites malayanus* (Raffles). *Zool. Meded. Leyden* 20: 237-239.

Junge, G.C.A. 1948. Notes on some Sumatran birds. *Zool. Meded. Leyden* 29: 311-326.

Junge, G.C.A. & Mees, G.F. 1958. The avifauna of Trinidad and Tobago. *Zool. Verhand.* 37: 1-172.

Kaempfer, E. 1924. Über das Vogelleben in Santo Domingo. *J. Orn.* 72: 178-184.

Kalsi, R.S. 1998. Birds of Kalesar Wildlife Sanctuary, Haryana, India. *Forktail* 13: 29-32.

Karubian, J. & Carrasco, L. 2008. Home range and habitat preferences of the Banded Ground-cuckoo (*Neomorphus radiolosus*). *Wilson J. Orn.* 120: 205-209.

Karubian, J., Carrasco, L., Cabrera, D., Cook, A. & Olivo, J. 2007. Nesting biology of the Banded Ground-Cuckoo (*Neomorphus radiolosus*). *Wilson J. Orn.* 119: 221-227.

Kasparek, M. 1992. *Die Vögel der Türkei.* Max Kasparek Verlag, Heidelberg.

Kaufman, K. 2000. *Birds of North America.* Houghton Mifflin, New York.

Kavanau, J.L. & Ramos, J. 1970. Roadrunners: activity of captive individuals. *Science* 169: 780-782.

Kaup, J.J. 1829. *Skizzirte Entwickelungs-Geschichte und natürliches System der Europäischen Thierwelt:* ….. Darmstadt & Leipzig, Germany.

Keith, A.R., Wiley, J.W., Latta, S.C. & Ottenwalder, J.A. 2003. *The birds of Hispaniola. Haiti and the Dominican Republic; an annotated checklist.* British Ornithologists' Union Checklist 21. BOU, Tring.

Keith, G.S. & Vernon, C.J. 1969 Bird notes from northern and eastern Zambia. *Puku* 5: 131-139.

Keith, S. 1968. Notes on some birds of East Africa, including additions to the avifauna. *Amer. Mus. Nov.* 2321: 1-15.

Kelly, C. 1987. A model to explore the rate of spread of mimicry and rejection in hypothetical populations of Cuckoos and their hosts. *J. Theor. Biol.* 125: 283-299.

Kelsall, H.J. 1914. Notes on a collection of birds from Sierra Leone. *Ibis* (10th series) 2: 192-228.

Kelsey, F.W. 1903. The home of the California Road-runner. *Condor* 5: 132-133.

Kemp, N. 2000. The birds of Siberut, Mentawi Islands, West Sumatra. *Kukila* 11: 73-96.

Kenefick, M. 2004. Verification of rare bird records from Trinidad & Tobago. *Cotinga* 22: 101-103.

Kenefick, M., Restall, R.L., Hayes, F.E. 2008. *Field guide to the birds of Trinidad and Tobago.* Yale University Press, New Haven.

Kenneally, K.F., Pirkopf, K.C. & Smith, J.A. 1979. Observations on the nest and young of the Pheasant Coucal. *W. Aust. Nat.* 14: 163-165.

Kennedy, R.S., Gonzales, P.C., Dickinson, E.C., Miranda, H.C. Jnr. & Fisher, T.H. 2000. *A guide to the birds of the Philippines.* OUP, Oxford.

Kent, T.H. 1988. Mapping vagrants. *Iowa Bird Life* 58: 101-105.

Kepler, C.B. & Kepler, A.K. 1978. Status and nesting of the Yellow-billed Cuckoo in Puerto Rico. *Auk* 95: 417-419.

Kerr, R. 1792. *Scythrops psittaceus* sp. nov. *Anim. Kingd.*: 619.

Keulemans, J.G. 1866. *Opmerkingen over de Vogels van de Kaap-Verdische Eilanden en van Prins-Eiland (Ilha do Principe) in de Bogt van Guinea gelegem.* Amsterdam.

Keulemans, J.G. 1907. The Golden Cuckoo. *Bird Notes* 5: 245–247.

Khabaeva, G.M 1985. [*Ecological populations of birds.*] Irkutsk. [in Russian]

Khajuria, H. 1984. The Crow-pheasant, *Centropus sinensis* (Stephens) (Aves: Cuculidae) of central and eastern Madhya Pradesh. *Records Zool. Survey India* 81: 89-93.

Khan, M.M.H. 2005. Species diversity, relative abundance and habitat use of the birds in the Sundarbans East Wildlife Sanctuary, Bangladesh. *Forktail* 21: 79-86.

Kiff, L.F. & Williams, A. 1978. Host records for the Striped Cuckoo from Costa Rica. *Wilson Bull.* 90: 138-139.

Kikkawa, J. 1966. Population distribution of land birds in temperate rainforest of southern New Zealand. *Trans. R. Soc. New Zealand (Zool.)* 7: 214-277.

Kikkawa, J. 1968. Courtship feeding in cuckoos. *Emu* 68: 213–214.

Kikkawa, J. & Dwyer, P.D. 1962. Who feeds the fledged Pallid Cuckoo? *Emu* 62: 169-171.

Kikkawa, J., Hore-Lacy, I. & Brereton, J.G. 1965. A preliminary report on the birds of the New England National Park. *Emu* 65: 139-143.

Kilner, R.M. & Davies, N. B. 1999. How selfish is a cuckoo chick? *Anim. Behav.* 58: 797-808.

Kim, Chang-Hoe. 1996. [Behavioral characteristics between the parasite and host: Crow Tits *Paradoxornis webbiana* and Common Cuckoos *Cuculus canorus*.] *Korean J. Orn.* 3: 51-57 [In Korean with English summary].

King, B. 1974. Male Cuckoo using plant material in display. *Brit. Birds* 67: 515.

King, B. 2005. The taxonomic status of the three subspecies of *Cuculus saturatus*. *Bull. Brit. Orn. Club* 125: 48-55.

King, B., Buck, H., Ferguson, R., Fisher, T., Goblet, C., Nickel, H. & Suter, W. 2001. Birds recorded during two expeditions to north Myanmar (Burma). *Forktail* 17: 29-40.

King, B.F. 2002. The *Hierococcyx fugax*, Hodgson's Hawk Cuckoo, complex. *Bull. Brit. Orn. Club* 122: 74-80.

King, B.F., Dickinson, E.C. & Woodcock, M.W. 1975. *A field guide to the birds of south-east Asia*. Collins, London.

Kirk, J. 1864. On the birds of the Zambesi region of eastern tropical Africa. *Ibis* (1st series) 6: 307-339.

Kirkconnell, A., Kirwan, G.M. & Garrido, O.H. in prep. *Annotated checklist of the birds of Cuba*.

Kirwan, G., Demirci, B., Welch, H., Boyla, K., Özen, M., Castell, P. & Marlow, T. 2008. *The birds of Turkey*. A. & C. Black, London.

Kiyosu, Y. 1966. [*The encyclopedia of wild birds*.] Tokyo-do shuppan, Tokyo. [In Japanese]

Kizungu, B. 2000. Observations d'un nid du Coucal noire *Centropus monachus* et attitude de l'homme face à ses oisillons à Irangi, République Démocratique du Congo. *Malimbus* 22: 88-90.

Klaas, E.E. 1968. Summer birds from the Yucatán Peninsula, Mexico. *Univ. Kansas Publ. Mus. Nat. Hist.* 17: 579-611.

Klapste, J. 1981. Observation of one Pallid Cuckoo feeding another. *Aust. Bird Watcher* 9: 27-28.

Klepfer, A. & Di Fraia, L. 2006. Nidificazione di cuculo dal cuiffo *Clamator glandarius* nel Lazio meridionale. *Alula* 13: 2000-2001.

Kleven, O., Moksnes, A., Røskaft, E., Rudolfsen, G., Stokke, B.G., Honza, S. & Honza, M. 2004. Breeding success of Common Cuckoo *Cuculus canorus* parasitising four sympatric species of *Acrocephalus* warblers. *J. Avian Biol.* 15: 394-398.

Klimaitis, J.F. & Moschione, F.N. 1987. *Aves de la Reserva Integral de Selva Marginal de Punta Lara y sus alrededores - reseña de sus relaciones con los principales ambientes y comunidades vegetales*. Dirección de Servicios Generales del Ministerio, Argentina.

Kloot, T. 1969. Courtship feeding in Horsfield Bronze Cuckoo. *Emu* 69: 116.

Kloss, C.B. 1930. An account of the Bornean birds in the Zoological Museum, Buitenzorg, with the description of a new race. *Treubia* 12: 395-424.

Knowlton, J.L. 2010. Breeding records of birds from the Tumbesian region of Ecuador. *Orn. Neotrop.* 21: 109-129.

Knystautas, A.J.V. & Sibnev, J.B. 1987. *Die Vogelwelt Ussuriens*. A. Ziemens Verlag, Lutherstadt.

Kobayashi, K. & Ishizawa, T. 1932-40. *Eggs of Japanese birds*. Privately published, Rokko, Kobe.

Koelz, W.N. 1954. Ornithological studies 1. New birds from Iran, Afghanistan and India. *Contrib. Inst. Regional Explor.* 32 pp. Privately published.

Koepcke, M. 1970. *The birds of the Department of Lima, Peru*, 2nd ed. Livingston Publishing Co., Wynnewood, Pennsylvania.

Komar, O. 1998. Avian diversity in El Salvador. *Wilson Bull.* 110: 511-533.

Komar, O. & Thurber, W.A. 2003. Predation on birds by a cuckoo (Cuculidae), mockingbird (Mimidae), and saltator (Cardinalidae). *Wilson Bull.* 115: 205-208.

Köster, F. 1971. Zum Nistverhalten des Ani, *Crotophaga ani*. *Bonn. Zool. Beitr.* 22: 4-27.

Kotagama, S.W. & Fernando, P. 1994. *A field guide to the birds of Sri Lanka*. Wildlife Heritage Trust of Sri Lanka, Colombo.

Kotagama, S.W. & Goodale, E. 2004. The composition and spatial organisation of mixed-species flocks in a Sri Lankan rainforest. *Forktail* 20: 63-70.

Kratter, A.W., Steadman, D.W., Smith, C.E. & Filardi, C.E. 2001a. Reproductive condition, moult, and body mass of birds from Isabel, Solomon Islands. *Bull. Brit. Orn. Club* 121: 128-144.

Kratter, A.W., Steadman, D.W., Smith, C.E., Filardi, C.E. & Webb, H.P. 2001b. Avifauna of a lowland forest on Isabel, Solomon Islands. *Auk* 118: 472-483.

Krauczuk, E.R. & Baldo, J.D. 2004. Contribuição para o conhecimento da avifauna de um fragmento de floresta com araucária em Misiones, Argentina. *Atual. Orn.* 119: 6.

Kreuzberg-Mukhina, E. & Kreuzberg, A. 2005. *Appendix 4 to the IBA criteria applied to the Central Asian Republics – bird species lists for Kazakhstan, Uzbekistan, Kyrgyzstan, Tadzhikistan and Turkmenistan*. (Word Table - online).

Kuiper, S.M. & Cherry, M.I. 2002. Brood parasitism and egg matching in the Red-chested Cuckoo *Cuculus solitarius* in southern Africa. *Ibis* 144: 632-639.

Kumerloeve, H. 1961. Zur Kenntnis der Avifauna Kleinasiens. *Bonner Zool. Beitr.* 12 Sonderheft. 1961:1-318.

Kuroda, N. 1933-1936. *Birds of the island of Java*. 2 vols. Privately published, Tokyo.

Kuroda, N. 1965. On the peculiar tail-molting of a Himalayan Cuckoo. *Annot. Zool. Jap.* 38: 151-153.

Lachenaud, O. 2003. On the plumages of the Senegal Coucal *Centropus senegalensis* and a putative observation of Black-throated Coucal *C. leucogaster* in Niger. *Malimbus* 25: 55-56.

Lack, D. 1959. Migration across the sea. *Ibis* 101: 374-399.

Lack, D. 1968. *Ecological adaptations for breeding in birds*. Methuen, London.

Lack, D. 1971. *Ecological isolation in birds*. Blackwell Scientific Publications, Oxford & Edinburgh.

Lack, D. 1976. *Island biology illustrated by the land birds of Jamaica*. Blackwell, Oxford.

Lack, P.C., Leuthold, W. & Smeenk, C. 1980. Checklist of the birds of Tsavo East National Park, Kenya. *J. E. Afr. Nat. Hist. Soc.* 170: 1-25.

Lack, P.C. & Quicke, D.L.J. 1978. Dietary notes on some Kenyan birds. *Scopus* 2: 86-91.

Lafresnaye, N.F.A.A. 1853. Quelques nouvelles espèces d'oiseaux. *Rev. Mag. Zool. Pure Appl.* (2° ser.) 5: 56-62.

Lahti, D.C. 2006. Persistence of egg recognition in the absence of cuckoo brood parasitism: pattern and mechanism. *Evolution* 60: 157-168.

Laman, T.G., Burnaford, J.L. & Knott, C.D. 1997. Bornean Ground-Cuckoo observations in Gunung Palung National Park, West Kalimantan. *Kukila* 9: 183-185.

Lamarche, B. 1980. Liste commentée des oiseaux du Mali. 1ère partie: Non-passereaux. *Malimbus* 2: 121-158.

Lamarche, B. 1988. *Liste commentée des oiseaux de Mauritanie*. Études des Naturalistes Sahariennes et Ouest-Africaines, Nouakchout.

Lamba, B.S. 1963. The nidification of some common Indian birds. *J. Bombay Nat. Hist. Soc.* 60: 121-133.

Lamba, B.S. 1967. The egg-laying of the Koel *Eudynamys scolopacea* (Linnaeus). *J. Bombay Nat. Hist. Soc.* 63: 750-751.

Lamba, B.S. 1969. The nidification of some common Indian birds. Part 12. *J. Bombay Nat. Hist. Soc.* 66: 72-80.

Lambert, F.R. 1994. Notes on the avifauna of Bacan, Kasiruta and Obi, north Moluccas. *Kukila* 7: 1-9.

Lambert, F.R. & Collar, N.J. 2002. The future for Sundaic lowland forest birds: long-term effects of commercial logging and fragmentation. *Forktail* 18: 127-146.

Lambert, F.R. & Yong, D. 1989. Some recent bird observations from Halmahera. *Kukila* 4: 30-33.

Lamm, D.W. 1948. Notes on the birds of the states of Pernambuco and Paraiba, Brazil. *Auk* 65: 261-283.

Lamm, D.W. 1955. Local migratory movements in southern Mozambique. *Ostrich* 26: 32-37.

Land, H.C. 1962. A collection of birds from the arid interior of eastern Guatemala. *Auk* 79: 1-11.

Land, H.C. 1970. *Birds of Guatemala*. Livingston Publ. Co., Wynnewood, Pennsylvania.

Langmore, N.E., Hunt, S. & Kilner, R.M. 2003. Escalation of a coevolutionary arms race through host rejection of brood parasitic young. *Nature* 422: 157-160.

Langmore, N.E., Maurer, G., Adcock, G.J. & Kilner, R.M. 2008. Socially acquired host-specific mimicry and the evolution of host races in Horsfield's bronce-cuckoo *Chalcites basalis*. *Evolution* 62: 1689-1699.

Langrand, O. 1990. *Guide to the birds of Madagascar*. Yale University Press, New Haven & London.

Langridge, H. 1990. Courtship feeding behaviour in the Mangrove Cuckoo (*Coccyzus minor*). *Florida Field Nat.* 18: 55-56.

Larkins, D. 1994. The Channel-billed Cuckoo: behaviour at nests of Pied Currawongs. *Aust. Birds* 28: 7-10.

Lasley, G.W. & Sexton, C. 1985. The nesting season June 1-July 31, 1985. South Texas region. *Am. Birds* 39: 933-936.

Latham, J. 1790. *Index Ornithologicus, sive systema ornithologiæ: complectens Avium divisionem in classes, ordines, genera, species, ipsarumque varietates*. London.

Latham, J. 1801. *Supplementum indicis ornithologici, sive systematis ornithologiae*. G. Leigh, J. & S. Sotheby, London.

La Touche, J.D.D. 1931. *A handbook of the birds of eastern China*. vol. 2, part 1. Taylor & Francis, London.

Latta, S.C. 2008. *Checklist for birds of Hispaniola - Listado de aves de la Hispaniola*. Committee of Records of the Sociedad Ornitológica de la Hispaniola (SOH). Version 1.4.

Latta, S.C. & Wunderle, J.M. Jnr. 1996. The composition and foraging ecology of mixed-species flocks in pine forest of Hispaniola. *Condor* 98: 595-607.

Latta, S.C., Rimmer, C.C., Keith, A.R., Wiley, J.W., Raffaele, H.A., McFarland, K.P. & Fernández, E. 2006. *Birds of the Dominican Republic and Haiti*. Christopher Helm, London.

Lau, P., Bosque, C. & Strahl, S.D. 1998. Nest predation in relation to nest placement in the Greater Ani (*Crotophaga major*). *Orn. Neotrop.* 9: 87-91.

Laubmann, A. 1930. *Wissenschaftliche Ergebnisse der Deutschen Gran Chaco-Expedition. Vögel*. Strecker & Schröder, Stuttgart.

Laubmann, A. 1939. *Wissenschaftliche Ergebnisse der Deutschen Gran Chaco-Expedition. Die Vögel von Paraguay*, vol. 1. Strecker & Schröder, Stuttgart.

Lavery, H.J. 1986. Breeding seasons of birds in north-eastern Australia. First supplement 1967-1974. *Emu* 86: 111-113.

Law, S.C. 1928. Little-noticed habits of some birds of the district of 24 - Parganas. *J. Asiat. Soc. Beng.* 22: 411-420.

Lawes, M.J. & Kirkman, S. 1996. Egg recognition and interspecific brood parasitism rates in Red Bishop (Aves: Ploceidae). *Anim. Behav.* 52: 553-563.

Lawrence, G.N. 1864. Descriptions of new species of birds of the families Tanagridae, Cuculidae, and Trochilidae, with a note on *Panterpe insignis*. *Ann. Lyc. Nat. Hist.* 8: 41-46.

Lawrence, G.N. 1867. Descriptions of five new species of Central American birds. *Proc. Acad. Nat. Sci. Philad.* 19: 232-234.

Lawson, W.J. 1962. Variation in the South African populations of the Coucal *Centropus superciliosus*. *Ostrich* 33: 45-47.

Layard, E.L. 1854. Notes on the ornithology of Ceylon, collected during an eight years' residence in the island. *Ann. Mag. Nat. Hist.* (2nd series) 13: 123-131, 212-218, 257-264, 446-453.

Layard, E.L. 1873. Notes on birds observed at Para. *Ibis* (3rd series) 3: 374-396.

Laymon, S.A. 2001. Cuckoos, roadrunners, and anis. Pp. 332-335 in C. Elphick, J.B. Dunning Jnr. & D. Sibley (Eds). *The Sibley guide to bird life and behaviour*. Christopher Helm, London.

Laymon, S.A. & Halterman, M.D. 1987. Can the western subspecies of the Yellow-billed Cuckoo be saved from extinction? *Western Birds* 18: 19-26.

Leach, W.E. 1814. *The zoological miscellany; being descriptions of new, or interesting animals*. Vol. 1, pp. 115-117. Nodder & Son, London.

LeCroy, M. & Peckover, W.S. 1983. Birds of the Kimbe area, west New Britain. *Condor* 85: 297-304.

Lee, K.S., Lok, C.B.P. & Li, S-N. 2005. Birds of Yinggeling, Hainan Island, China - with notes on new and important records. *BirdingASIA* 4: 68-79.

Lee, K.S., Lau, Wai-Neng, M., Fellowes, J.R. & Lok, C.B.P. 2006. Forest bird fauna of South China: notes on current distribution and status. *Forktail* 22: 23-38.

Lee, Woo-Shin, Ku, Tae-Hoe & Park, Jin-Young. 2000. A field guide to the birds of Korea. L.G. Evergreen Foundation, Seoul.

Lee, Y.Y. 2008. Egg discrimination by the Vinous-throated Parrotbill, a host of the Common Cuckoo that lays polychromatic eggs. MS Thesis, University of Manitoba.

Lees, A. 2003. Cristalino Jungle Lodge and Alta Floresta, Brazil birding trip report April-June 2003.

Lees, A.C., Davis, B.J.W., Oliveira, A.V.G. de & Peres, C.A. 2008. Avifauna of a structurally heterogeneous forest landscape in the Serra dos Caiabis, Mato Grosso, Brazil: a preliminary assessment. *Cotinga* 29: 149-159.

Lees, R.H. 1938. Further bird notes from northern Rhodesia. *Ostrich* 9: 13-19.

Legge, W.V. 1873. On *Phænicophaus pyrrhocephalus* Forster. *Stray Feathers* 1: 346-348.

Legge, W.V. 1880. *A history of the birds of Ceylon*. Published privately, London.

Lei, F-M., Zhao, H.F., Wang, A-Z., Yin, Z-H. & Payne, R.B. 2005. Vocalizations of the Common Cuckoo *Cuculus canorus* in China. *Acta Zoologica Sinica* 51: 31-37.

Lekagul, B. & Round, P.D. 1991. *A guide to the birds of Thailand*. Saha Karn Bhaet Co. Ltd, Bangkok.

Leonard, P.M. 1998. 1997 spring records. *Zambia Bird Report* 1997. Pp. 59-139.

Lesson, A. 1830. *Traité d'Orn.* livr. 2: 149.

Lesson, A. 1842. Note sur les oiseaux nouveaux ou peu connus rapportés de la mer du Sud. *Rev. Zool.* 5: 209-210.

Lesson, R.P. 1826. Espèce nouvelle du sous-genre Coucal (*Centropus* Illig.). *Bull. Univ. Sci. Industr.* (sect. 2) (*Bull. Sci. Nat. Geol.*) 8: 113.

Lesson, R.P. 1829. *Complément des ouevrées de Buffon, ou histoire naturelle des animaux rares découvertes par les naturalistes et les voyageurs depuis la mort de Buffon* Vol. 6. Baudouin Frères, Paris.

Lesson, R.P. 1830. *Traité d'ornithologie, ou description des oiseaux réunis dans les principales collections de France*. 2 vols. Levrault - P. Bertrand, Paris.

Lesson, R.P. & Garnot, P. 1828-1829. *Voyage autour du monde, Exécuté par Ordre du Roi, Sur la Corvette de Sa Majesté La Coquille, pendant les années 1822, 1823, 1824 et 1825*. Atlas (1828): pl. 33, Zool. 1, Part 2, Livr. 13 1829. Arthus Bertrand, Paris.

Levaillant, F. 1806. *Histoire naturelle des oiseaux d'Afrique*, vol. 5. Delachaussée. Paris.

Lévêque, R. 1964. Notes sur la reproduction des oiseaux aux Îles Galápagos. *Alauda* 32: 5-44, 81-96.

Lévêque, R. 1968. Über Verbreitung, Bestandsvermehrung und Zug des Häherkuckucks *Clamator glandarius* (L.) in Westeuropa. *Ornithol. Beob.* 65: 43-71.

Levy, C. 1984. Notes on the nesting of the Jamaican Lizard Cuckoo. *Gosse Bird Club Broadsheet* 42: 6-8.

Levy, C. 1996. Chestnut-bellied Cuckoo *Hyetornis pluvialis*. *Gosse Bird Club Broadsheet* 67: 12.

Lewington, I., Alström, P. & Colston, P. 1991. *A field guide to the rare birds of Britain and Europe*. Domino Books, Jersey.

Lewis, A.D. & Pomeroy, D.E. 1989. *A bird atlas of Kenya*. A.A. Balkema, Rotterdam.

Lewthwaite, R. 1995. Blue Magpie as a foster parent of Koel. *Hong Kong Bird Report* 1995: 222.

Liang, W., Yang, C., Antonov, A., Fossøy, F., Stokke, B.G., Moksnes, A., Røskaft, E., Shykoff, J., Møller, A.P. & Takasu, F. (in press). Sex roles in egg recognition and egg polymorphism in avian brood parasitism. *Behav. Ecol.*

Liang, W., Yang, C., Cai, Y. Stokke, B. & Shi, S. 2010. Cuckoos and their hosts in a multi-cuckoo species system in China. IOC 22-28 August 2010. www.i-o-c.org

Lichtenstein, M.H.C. 1823. *Verzeichniss der Doubletten des Zoologischen Museums der ... Universität ... nebst Beschreibung vieler bisher unbekannter Arten von Säugethieren, Vögeln, Amphibien und Fischen.* Berlin.

Lim, K.S. 2008a. Horsfield's Bronze Cuckoo *Chrysococcyx basalis* in Singapore. *BirdingASIA* 10: 30-35.

Lim, K.S. 2008b. Notes on the identification, status and distribution of Horsfield's Bronze Cuckoo *Chrysococcyx basalis* in Singapore *Singapore Avifauna* 22 (7): 15-25.

Lim, K.S. 2008c. Records Committee 2008 update – new species for Singapore. *Singapore Avifauna* 22 (10): 11.

Lim, K.S. 2008d. Report on the 9th Mid-year Bird Census. *Singapore Avifauna* 22 (7): 14.

Lim, K.S. 2011. Notes on a new brood-host of Drongo Cuckoo *Surniculus lugubris* in the Thai-Malay Peninsula. *BirdingASIA* 15: 49.

Lima, P.C. & Lima, R.C.F.R. 2004. Registro fotográfico de uma nova dieta alimentar para as espécies *Guira guira* e *Crotophaga ani. Atual. Orn.* 117: 3.

Lima, P.C., dos Santos, S.S. & Lima, R.C.F.R. 2003. Levantamento e anilhamento da ornitofauna na pátria da Arara-azul-de-Lear (*Anodorhynchus leari*, Bonaparte, 1856). *Atual. Orn.* 112: 11.

Limbrunner, A. 1979. Erstnachweis de Fortpflanzung des Häkerkuckuck (*Clamator glandarius*) in Griechenland. *Anz. Orn. Gess. Bayern* 18: 84-85.

Lindholm, A.K. 1999. Brood parasitism by the cuckoo on patchy Reed Warbler populations in Britain. *J. Anim. Ecol.* 68: 293-309.

Lindholm, A. & Lindén, A. 2003. Oriental Cuckoo in Finland. *Alula* 4:122-133.

Lindholm, A. & Lindén, A. 2007. Some notes on the distribution and songs of two Oriental Cuckoo taxa, *Cuculus (saturatus) saturatus* and *Cuculus (saturatus) optatus*. *Forktail* 23: 12-16.

Linnaeus, C. 1758. *Systema Naturæ per regna tria naturæ, secundum classes, ordines, genera, species, cum characteribus, differentiis, synonymis, locis*. 10th edn. Vol. 1. Salvius, Holmiae (Stockholm).

Linnaeus, C. 1766. *Systema Naturae*. 12th edn. Vol. 1. Salvius, Holmiae (Stockholm).

Linton, E.H. 1930. Some notes concerning two of the cuckoos (viz. *Lamprococcyx plagosus* Gould, and *Cacomantis flabelliformis* Sharpe). *Emu* 29: 304-307.

Littler, F.M. 1910. *A handbook of the birds of Tasmania and its dependencies*. F.M. Littler, Launceston.

Liu, Y., Zhang, Z-W., Li, J-Q., Zhang, Y., Lu, S. & Ruan, X-F. 2008. A survey of the birds of the Dabie Shan range, central China. *Forktail* 24: 80-91.

Livesey, T.R. 1938. Cuckoo problems. *J. Bombay Nat. Hist. Soc.* 40: 127, 329-330.

Liversidge, R. 1955. Observations on a Piet-my-vrou (*Cuculus solitarius*) and its host the Cape Robin (*Cossypha caffra*). *Ostrich* 27: 18-27.

Liversidge, R. 1961. Pre-incubation development of *Clamator jacobinus*. *Ibis* 103: 624.

Liversidge, R. 1969. The biology of the Jacobin Cuckoo *Clamator jacobinus*. *Ostrich* Suppl. 8: 117-137.

Livezey, B.C. & Zusi, B.C. 2007. Higher-order phylogeny of modern birds (Theropoda, Aves: Neornithes) based on comparative anatomy. II. Analysis and discussion. *Zool. Jour. Linn. Soc.* 149: 1-95.

Ljungh, S.I. (1803) 1804. *Cuculus musicus. Kong. Vet.-Akad. nya Handl.* 24: 309.

Lobb, M.G. 1983. Didric Cuckoo (*Chrysococcyx caprius*) - Akrotiri Cyprus - 27 June 1982. *RAF Orn. Soc. J.* 14: 154-156.

Loetscher, F.W., Jnr. 1952. Striped Cuckoo fed by Rufous-and-white Wren in Panama. *Condor* 54: 169.

Loflin, R.K. 1982. Ani male apparently killed by other anis while attempting to parasitize nest. *Auk* 99: 787-788.

Loflin, R.K. 1983. Communal behaviors of the Smooth-billed Ani (*Crotophaga ani*). Ph.D. Dissertation. Univ. of Miami, Coral Gables, Florida.

Logue, D.M. 2007. A Greater Ani (*Crotophaga major*) gives a nuptial gift while copulating. *Orn. Neotrop.* 18: 311-312.

Löhrl, H. 1979. Untersuchungen am Kuckuck *Cuculus canorus* (Biologie, Ethologie und Morphologie). *J. Orn.* 120: 139-173.

Long, A.J. & Collar, N.J. 2002. Distribution, status and natural history of the Bornean Ground Cuckoo *Carpococcyx radiatus*. *Forktail* 18: 111-119.

Lönnberg, E. 1903. Remarks on the type-specimens of certain birds named by the late Carl Peter Thunberg. *Ibis* (8th series) 3: 238-242.

Lönnberg, E. 1903a. On a collection of birds from north-western Argentina and the Bolivian chaco. *Ibis* (8th series) 3: 441-471.

Lönnberg, E. & Rendahl, H. 1922. A contribution to the ornithology of Ecuador. *Ark. Zool.* 14: 1-87.

Lopatin, V.V. 1987. [Turkestan Red-backed Shrikes feeding two cuckoo nestlings.] *Ornitologiya, Moscow Univ. Press* 22: 214. [in Russian]

Lopes, L.E., Goes, R., Souza, S. & Ferreira, R.M. 2004. Observations on a nest of the Stygian Owl (*Asio stygius*) in the central Brazilian cerrado. *Orn. Neotrop.* 15: 423-427.

Lopes, L.E., Fernandes, A.M. & Marini, M.Â. 2005a. Diet of some Atlantic forest birds. *Ararajuba* 13: 95-103.

Lopes, L.E., Leite, L.O., Pinho, J.B. & Goes, R. 2005b. New bird records to the Estação Ecológica de Águas Emendadas, Planaltina, Distrito Federal. *Ararajuba* 13: 107-108.

López-Lanús, B. 1999. Red-billed Ground-cuckoo *Neomorphus pucheranii*: hypothetical record in Ecuador. *Cotinga* 12: 73.

López-Lanús, B., Berg, K.S., Strewe, R. & Salaman, P.G.W. 1999. The ecology and vocalisations of Banded Ground-cuckoo *Neomorphus radiolosus*. *Cotinga* 11: 42-45.

Lorber, P. 1984. Unrecorded display of the Striped Cuckoo. *Honeyguide* 30: 76-77.

Lord, E.A.R. 1939. Notes on migratory birds - 1937-38. *Emu* 38: 372-376.

Lord, E.A.R. 1943. Migratory notes (1941-1942) from St. Murphy's Creek district, Queensland. *Emu* 43: 18-23.

Lord, E.A.R. 1956. The birds of the Murphy´s Creek district, southern Queensland. *Emu* 56: 100-128.

Lotem, A. 1993. Learning to recognize nestlings is maladaptive for Cuckoo *Cuculus canorus* hosts. *Nature* 362: 743-745.

Lotem, A., Nakamura, H., & Zahavi, A. 1995. Constraints on egg discrimination and cuckoo-host co-evolution. *Animal Behav.* 49 (5): 1185–1209.

Louette, M. 1981a. *The birds of Cameroon: an annotated checklist*. Paleis der Académiën, Brussels.

Louette, M. 1981b. Contribution to the Ornithology of Liberia. Part 5. *Revue Zool. Afr.* 95: 342-355.

Louette, M. 1986. Geographical contacts between the taxa of *Centropus* in Zaïre, with the description of a new race. *Bull. Brit. Orn. Club* 106: 126-133.

Louette, M. 1989. Additions and corrections to the avifauna of Zaïre (4). *Bull. Brit. Orn. Club* 109: 217-225.

Louette, M. & Herroelen, P. 1993. Status of migratory *Cuculus* cuckoos in Zaïre. *Bull. Brit. Orn. Club* 113: 147-152.

Louette, M. & Herroelen, P. 1994. A revised key for *Cercococcyx* cuckoos, taxonomic status of *C. montanus patulus* and its occurrence in Zaïre. *Bull. Brit. Orn. Club* 114: 144-149.

Loveridge, A. 1922. Notes on East African birds (chiefly nesting habits and stomach contents) collected 1915-1919. *Proc. Zool. Soc. London* 1922: 837-862.

Low, A. 2008a. Bird Report March 2008. *Singapore Avifauna* 22 (3): 4.

Low, A. 2008b. Bird Report June 2008. *Singapore Avifauna* 22 (6): 9.

Low, A. 2008c. Bird Report August 2008. *Singapore Avifauna* 22 (8): 8.

Low, A. 2008d. Bird Report September 2008. *Singapore Avifauna* 22 (9): 5.

Low, A. 2008e. Bird Report October 2008. *Singapore Avifauna* 22 (10): 13.

Lowe, W.P. 1937. Report on the Lowe-Waldron Expeditions to the Ashanti Forests and northern territories of the Gold Coast. Part 2. *Ibis* (14th series) 1: 635-662.

Lowen, J.C., Barnett, J.M., Pearman, M., Clay, R.P. & López-Lanús, B. 1997. New distributional information for 25 species in eastern Paraguay. *Ararajuba* 5: 240-243.

Lowery, G.H., Jr. 1974. *Louisiana birds*. 3rd ed. Louisiana State University Press, Baton Rouge.

Loye, J.E. & Zuk, M. (Eds) 1991. *Bird-parasite interactions. Ecology, evolution, and behaviour*. OUP, Oxford.

Loyn, R.H. 1980. Bird populations in a mixed eucalypt forest used for production of wood in Gippsland, Victoria. *Emu* 80: 145-156.

Lynes, H. 1925. On the birds of north and central Darfur, part IV. *Ibis* (12th series) 1: 344-416.

Lynes, H. 1938. Contribution to the ornithology of the southern Congo basin. Lynes-Vincent tour of 1933-34. *Rev. Zool. Bot. Afr.* 31: 1-128.

Lyons, J.A. & Pérez, V. 2000. New records and updates for Mindo, Ecuador. *Cotinga* 13: 67-69.

McAllan, I.A.W. & James, D.J. 2008. Large Hawk-cuckoo *Hierococcyx sparverioides* on Christmas I. *Australian Field Ornithology* 25: 31-35.

McClure, H.E. 1967. The composition of mixed-species flocks in lowland and submontane forests of Malaya. *Wilson Bull.* 79 131-154.

McClure, H.E. 1969. *Migratory animal pathological survey – annual progress report 1968*. Applied Scientific Research Corp. of Thailand, Bangkok.

McClure, H.E. 1998. *Migration and survival of the birds of Asia*. White Lotus, Bangkok.

MacDonagh, E.J. 1934. Materiales de *Piaya cayana* en el Museo de la Plata. *Notas Prelim. Mus. La Plata* 2: 51-64.

MacDonald, M.A. 1980. Observations on the Diederik Cuckoo in southern Ghana. *Ostrich* 51: 75-79.

Macedo, R.H. 1992. Reproductive patterns and social organization of the communal Guira Cuckoo (*Guira guira*) in central Brazil. *Auk* 109: 786-799.

Macedo, R.H. 1994. Inequities in parental effort and costs of communal breeding in the Guira Cuckoo. *Orn. Neotrop.* 5: 79-90.

Macedo, R.H.F. & Bianchi, C.A. 1997. When birds go bad: circumstantial evidence for infanticide in the communal South American Guira Cuckoo. *Ethol., Ecol. & Evol.* 9: 45-54.

Macedo, R.H.F. & Melo, C. 1999. Confirmation of infanticide in the communally breeding Guira Cuckoo. *Auk* 116: 847-851.

Macedo, R.H.F., Cariello, M.O. & Muniz, L.S.B. 2001. Context and frequency of infanticide in communally breeding Guira Cuckoos. *Condor* 103: 170-175.

Macedo, R.H.F., Cariello, M.O., Graves, J.A. & Schwabl, H.G. 2004a. Reproductive partitioning in communally breeding Guira Cuckoos, *Guira guira. Behav. Ecol. Sociobiol.* 55: 213-222.

Macedo, R.H.F., Cariello, M.O., Pacheco, A.M. & Schwabl, H.G. 2004b. Significance of social parameters on differential nutrient investment in Guira Cuckoo, *Guira guira*, eggs. *Anim. Behav.* 68: 485-494.

McEvey, A. 1965. The birds of the Rutherglen district. *Emu* 65: 1-56.

MacGillivray, W. 1914. Notes on some north Queensland birds. *Emu* 13: 132-186.

McGregor, R.C. 1904. The birds of Calayan and Fuga, Babuyan Group. *Bull. Philipp. Mus.* 4: 1-34.

Mack, G. 1930. Notes and descriptions of some rare eggs of Australian birds. *Emu* 29: 302-303.

MacKenzie, J.M.D. 1918. Some further notes on cuckoos in Maymyo. *J. Bombay Nat. Hist. Soc.* 25: 742-745.

MacKinnon, J. 1988 (reprint 1990). *Field guide to the birds of Java and Bali*. Gadjah Mada University Press, Yogyakarta.

MacKinnon, J. & Phillipps, K. 1993. *A field guide to the birds of Borneo, Sumatra, Java and Bali*. OUP, Oxford.

MacKinnon, J. & Phillipps, K. 2000. *A field guide to the birds of China*. OUP, Oxford.

Mackness, B. 1979. The ecology of the Pheasant Coucal *Centropus phasianinus* (Latham) in Australia. *Sunbird* 10: 1-8.

Mackworth-Praed, C.W. & Grant, C.H.B. 1957. *Birds of eastern and north eastern Africa*. Vol. 1. Longmans, Green, London.

Mackworth-Praed, C.W. & Grant, C.H.B. 1970. *Birds of west central and western Africa*. Vol. 1. Longmans, London.

Maclaren, P.I.R. 1950. *Centropus epomidus. Ibis* 92: 146-147.

Maclaren, P.I.R. 1952. Feeding by the Didric Cuckoo. *Ibis*. 94: 684.

McKean, J.L. & Hindwood, K.A. 1965. Additional notes on the birds of Lord Howe Island. *Emu* 64: 79-97.

Maclean, F. 1957. Black Cuckoo, *Cuculus cafer*, laying its egg in Heuglin´s Robin's nest and Sooty Babbler´s nest. *Ostrich* 28: 175.

Maclean, G.L. 1993. *Roberts' "Birds of Southern Africa"*. 6th ed. Trustees of the John Voelcker Bird Book Fund, Cape Town.

Maclean, G.L. & Maclean, C. 1993. A new host for the Redchested Cuckoo *Cuculus solitarius* in southern Africa. *Ostrich* 64: 136.

McLean, I.G. 1982. Whitehead breeding and parasitism by Long-tailed Cuckoos. *Notornis* 29: 156-158.

McLean, I.G. 1985. Breeding status of cuckoos on offshore islands of New Zealand: some corrections. *New Zealand J. Ecol.* 8: 123-124.

McLean, I.G. 1987. Response to a dangerous enemy: should a brood parasite be mobbed? *Ethology* 75: 235-245.

McLean, I.G. 1988. Breeding behaviour of the Long-tailed Cuckoo on Little Barrier Island. *Notornis* 35: 89-98.

McLean, I.G. & Gill, B.J. 1988. Breeding of an island-endemic bird: the New Zealand Whitehead *Mohoua albicilla*: Pachycephalinae. *Emu* 88: 177-182.

MacLeod, J.G.R. 1969. The birds of Hottentots Holland (part 3). *Ostrich* 40: 14.

MacLeod, J.G.R. & Hallack, M. 1956. Some notes on the breeding of Klaas´s Cuckoo. *Ostrich* 27: 2-5.

MacLeod, J.G.R. & Murray, C.C. 1952. Birds of Hottentots Holland, part 2. *Ostrich* 23: 16-25.

McNair, D.B. 1991. Copulation in the Mangrove Cuckoo (*Coccyzus minor*). *Florida Field Nat.* 19: 84-85.

McNair, D.B., Sibley, F.C., Massiah, E.B. & Frost, M.D. 2002. Ground-based Nearctic-Neotropic landbird migration during autumn in the eastern Caribbean. Pp. 86-103 in F. E. Hayes & S.A. Temple (Eds). *Studies in Trinidad & Tobago ornithology honouring Richard ffrench*. Univ. West Indies, St. Augustine, Trinidad.

Madarász, J. v. 1914. A contribution to the ornithology of the eastern Sudan. *Ann. Mus. Nat. Hung.* 12: 558.

Madden, J. & Davies, N. 2006. A host-race difference in begging calls of nestling cuckoos *Cuculus canorus* develops through experience and increases host provisioning. *Proc. Roy. Soc B*: 273.

Mader, W.J. 1979. Breeding behavior of a polyandrous trio of Harris's Hawks in southern Arizona. *Auk* 96: 776-788.

Madoc, G.C. 1956a. Observations on two Malayan cuckoos. *Malay. Nat. J.* 10: 97-103.

Madoc, G.C. 1956b. A further note on the Lesser Cuckoo. *Malay. Nat. J.* 10: 134-138.

Magrath, H.A.F. 1923. Kashmir bird notes. *J. Bombay Nat. Hist. Soc.* 28: 276-279.

Makatsch, W. 1955. *Der Brutparasitismus in der Vogelwelt*. Neumann Verlag, Radebeul & Berlin.

Makatsch, W. 1971. Einige Bemerkungen über die parasitären Kuckucke. *Zool. Abh. Staatl. Mus. Tierk. Dresden* 30: 247-283.

Maldonado-Coelho, M. & Marini, M.Â. 2003. Composição de bandos mistos de aves em fragmentos de mata atlântica no sudeste do Brasil. *Pap. Avuls. Zool. S. Paulo* 43: 31-54.

Maller, C.J. 1997. *An investigation of the vocalisations and mating system of a brood parasite: the Common Koel*. Honours thesis. Griffith University, Griffith, NSW.

Maller, C.J. & Jones, D.N. 2001. Vocal behaviour of the Common Koel *Eudynamys scolopacea*, and implications for mating systems. *Emu* 101: 105-112.

Mann, C.F. 1987. Notable bird observations from Brunei, Borneo. *Forktail* 3: 51-56.

Mann, C.F. 1991a. Bird report for Brunei Darussalam, July 1988 to June 1990. *Brunei Mus. J.* 7(3): 86-111.

Mann, C.F. 1991b. Sunda Frogmouth Batrachostomus cornutus carrying its young. *Forktail* 6: 77-78.

Mann, C.F. 2008. *The birds of Borneo*. British Ornithologists' Union/British Ornithologists' Club Checklist Series. No. 23. BOU, Peterborough.

Mann, C.F. in prep. A record of Oriental Cuckoo *Cuculus saturatus optatus* in Zambia; first occurrence of the species in Africa. [working title]

March, W.T. 1863. Notes on the birds of Jamaica. *Proc. Acad. Nat. Sci. Philad.* 15: 150-153, 283-304.

Marchant, S. 1949. Six west African migratory birds. *Nigerian Field* 14: 19-24.

Marchant, S. 1953. Notes on the birds of south-eastern Nigeria. *Ibis* 95: 38-69.

Marchant, S. 1958. The birds of the Santa Elena Peninsula, S.W. Ecuador. *Ibis* 100: 349-387.

Marchant, S. 1959. The breeding season in S.W. Ecuador. *Ibis* 101: 137-152.

Marchant, S. 1960. The breeding of some S.W. Ecuadorian birds. *Ibis* 102: 349-382, 584-599.

Marchant, S. 1980. Incubation and nestling periods of some Australian birds. *Corella* 4: 30-32.

Marchant, S. & Hohn, E.O. 1980. Field notes on the Fan-tailed Cuckoo. *Emu* 80: 77-80.

Marcodes, B. & Rinke, D. 2000. Der Hauben-Seidenkuckkuck. *Gefiederte Welt* 124: 189-191.

Markus, M.B. 1964. Egg polymorphism in the Diederik Cuckoo (*Chrysococcyx caprius*) (Boddaert) at Pretoria. *Ostrich* 35: 123.

Marler, P. & Hamilton, W.J., III. 1966. *Mechanisms of animal behavior*. John Wiley & Sons, New York.

Marshall, B.E. 1985. Notes on Kariba birds. *Honeyguide* 31: 111-113.

Martens, J. & Eck, S. 1995. Towards an ornithology of the Himalayas: systematics, ecology and vocalisations of Nepal birds. *Bonn. Zool. Monogr.* 38: 1-445.

Martin, L.D. & Mengel, R.M. 1984. A new cuckoo and a chachalaca from the early Miocene of Colorado. *Carneg. Mus. Nat. Hist. Spec. Publ.* 9: 171-177.

Martinez, J.G., Burke, T., Dawson, D., Soler, J.J., Soler, M. & Møller, A.P. 1998. Microsatellite typing reveals mating patterns in the brood parasitic great spotted cuckoo (*Clamator glandarius*). *Mol. Ecol.* 7: 289-297.

Martinez, J.G., Soler, M. & Soler, J.J. 1996. The effect of Magpie breeding density and synchrony on brood parasitism by Great Spotted Cuckoos. *Condor* 98: 272-278.

Martins, F.C. & Donatelli, R.J. 2001. Estratégia alimentar de *Guira guira* (Cuculidae, Crotophaginae) na região centro-oeste do Estado de São Paulo. *Ararajuba* 9: 89-94.

Mason, C.W. & Maxwell-Lefroy, H. 1912. *The food of birds in India*. Memoirs of the Department of Agriculture in India. Dept. Agric. India, Calcutta.

Mason, I.J. 1982. The identity of certain early Australian types referred to the Cuculidae. *Bull. Brit. Orn. Club* 102: 99-106.

Mason, I.J. 1996. [Two new subspecies of *Scythrops novaehollandiae*] in Mason & Forrester 1996.

Mason, I.J. & Forrester, R.I. 1996. Geographical differentiation in the Channel-billed Cuckoo *Scythrops novaehollandiae* Latham, with description of two new subspecies from Sulawesi and the Bismarck Archipelago. *Emu* 96: 217-233.

Mason, I.J., McKean, J.L. & Dudzinski, M.L. 1984. Geographical variation in the Pheasant Coucal *Centropus phasianinus* (Latham) and a description of a new subspecies from Timor. *Emu* 84: 1-15.

Mason, P. 1985. The nesting biology of some passerines of Buenos Aires, Argentina. *Orn. Monogr.* 36: 954-972.

Masterson, H.B. 1953. Klaas' Cuckoo, *Chrysococcyx klaas* parasiting Red-headed Weavers, *Anaplectes rubriceps*. *Ostrich* 24: 51.

Mata, J.R.R., Erize, F. & Rumboll, M. 2006. *A field guide to the birds of South America: Non-passerines*. HarperCollins, London.

Matheson, W.E. 1976. The birds of the Redcliff Point Development Area. *S. Aust. Orn.* 27: 125-130.

Mathews, G.M. 1912. A Reference-List to the birds of Australia. *Novit. Zool.* 18: 171-455.

Mathews, G.M. 1913. Additions and corrections to my reference list. *Aust. Av. Rec. London* 1: 187-194.

Mathews, G.M. 1914. Additions and corrections to my list of the birds of Australia. *Aust. Av. Rec. London* 2: 83-107.

Mathews, G.M. 1916. List of additions of new sub-species to, and changes in, my "List of the Birds of Australia". *Aust. Av. Rec. London* 3: 53-68.

Mathews, G.M. 1918. *The birds of Australia*. Part 7. Witherby, London.

Mathews, G.M. 1919. *The birds of Australia*. Vol. 7. Part 5. Witherby, London.

Mathews, G.M. 1922. Description of Australian birds. *Bull. Brit. Orn. Club* 43: 13-14.

Maumary, L. & Duperrex, H. 1991. Le Coucou-geai *Clamator glandarius* nicheur dans le sud marocain. *Alauda* 59: 53.

Maumary, L., Vallotton, L. & Knaus, P. 2007. *Die Vögel der Schweitz*. Station Ornithologique Suisse.

Maurício, G.N. & Dias, R.A. 2000. New distributional information for birds in southern Rio Grande do Sul, Brazil, and the first record of the Rufous Gnateater *Conopophaga lineata* for Uruguay. *Bull. Brit. Orn. Club* 120: 230-237.

Mayhew, W.W. 1971. Desert encounter: a Roadrunner and a sidewinder. *Fauna* 1: 17-19.

Mayr, E. 1931. Die Vögel des Saruwaged- und Herzoggebirges (NO-Neuguinea). *Mitt. Zool. Mus. Berlin* 17: 639-723.

Mayr, E. 1932. Birds collected during the Whitney South Sea Expedition. XIX. Notes on the Bronze Cuckoo *Chalcites lucidus* and its subspecies. *Amer. Mus. Nov.* 520: 1-9.

Mayr, E. 1937. Birds collected during the Whitney South Sea Expedition. XXXV. Notes on New Guinea birds. II. *Amer. Mus. Nov.* 939: 1-14.

Mayr, E. 1938. The birds of the Vernay-Hopwood Chindwin Expedition. *Ibis* (14th series) 2: 277-320.

Mayr, E. 1938. Notes on a collection of birds from South Borneo. *Bull. Raffles Mus. Singapore* 14: 5-46.

Mayr, E. 1944. Birds collected during the Whitney South Sea Expedition. 54. Noted on some genera from the southwest Pacific. *Amer. Mus. Nov.* 1269: 1-8.

Mayr, E. & Amadon, D. 1951. A classification of recent birds. *Amer. Mus. Novit.* 1496: 1-42.

Mayr, E. & Diamond, J.M. 2001. *The birds of northern Melanesia. Speciation, ecology and biogeography*. OUP, New York.

Mayr, G. & Ericson, P.G.P. 2004. Evidence for a sister group relationship between the Madagascan mesites (Mesitornithidae) and the cuckoos (Cuculidae). *Senckenb. Biol.* 84: 1-17.

Mayr, E. & Gilliard, E.T. 1954. Birds of Central New Guinea. Result of the American Museum of Natural History Expeditions to New Guinea in 1950 and 1952. *Bull. Amer. Mus. Nat. Hist.* 103: 315-374.

Mayr, E. & Meyer de Schauensee, R. 1939. Zoological results of the Denison-Crockett expedition to the South Pacific for the Academy of Natural Sciences of Philadelphia, 1937-1938. Part I. The birds of the island of Biak. *Proc. Acad. Nat. Sci. Philad.* 91: 1-37.

Mayr, E. & Rand, A.L. 1936. Results of the Archbold Expeditions. No. 10. Two new subspecies of birds from New Guinea. *Amer. Mus. Novit.* 868: 1-3

Mayr, E. & Rand, A.L. 1937. Results of the Archbold Expeditions. 14. Birds of the 1933-1934 Papuan Expedition. *Bull. Amer. Mus. Nat. Hist.* 73: 1-248.

Mayr, G., Manegold, A. & Johansson, U.S. 2003. Monophyletic groups within "higher land birds" – comparison of morphological and molecular data. *J. Zool. Syst. Evol. Research* 41: 233-248.

Mearns, E.A. 1907. Descriptions of a new genus and nine new

species of Philippine birds. *Philipp. J. Sci., Manila. A. General Sci.* 2 (5): 355-360.
Medland, R. D. (Ed.) 1992. Flora and fauna records. *Nyala* 16: 33-45.
Medland, R. D. (Ed.) 1993. Flora and fauna records. *Nyala* 17: 29-44.
Medland, R. D. (Ed.) 1995. Flora and fauna records. *Nyala* 18: 51-67.
Medway, L. 1972. The Gunong Benom Expedition 1967: 6. The distribution and altitudinal zonation of birds and mammals on Gunong Benom. *Bull. Brit. Mus. Nat. Hist. Zool.* 23: 103-154.
Medway, L. & Marshall, A.G. 1975. Terrestrial vertebrates of the New Hebrides. *Phil. Trans. Roy. Soc., London Series B* 272: 423-465.
Medway, L. & Wells, D.R. 1976. *The birds of the Malay Peninsula.* Vol. 5. H.F. & G. Witherby Ltd, London.
Mees, G.F. 1965. The avifauna of Misool. *Nova Guinea, Zool.* 31 (10): 139-205.
Mees, G.F. 1971. The Philippine subspecies of *Centropus bengalensis* (Gmelin) (Aves, Cuculidae). *Zool. Meded. Leiden* 45 (18): 189-191.
Mees, G.F. 1979. Die Nachweise von *Cuculus canorus* L. im Indo-Australischen Raum. *Mitt. Zool. Mus. Berlin* 55 (Suppl.): 127-134.
Mees, G.F. 1982. Birds from the lowlands of southern New Guinea (Merauke and Koembe). *Zool. Verhand. Leiden* 191: 1-188.
Mees, G.F. 1986. A list of the birds recorded from Bangka Island, Indonesia. *Zool. Verhand. Leiden* 232: 1-176.
Meinertzhagen, R. 1954. *Birds of Arabia.* Oliver & Boyd, Edinburgh & London.
Meinzer, W.P. 1993. *The Roadrunner.* Texas Tech University Press, Lubbock.
Meise, W. 1941. Über die Vogelwelt von Noesa Penida bei Bali nach einer Sammlung von Baron Viktor von Plessen. *J. Orn.* 89: 345-376.
Meise, W. 1950. Zur Deutung von A. Lichtensteins 1793 beschriebenen Vögeln aus Surinam und den übrigen Guianas. *Ardea* 38: 182-185.
Melo, C. & Macedo, R.H F. 1997. Mortalidade em ninhadas de *Guira guira* (Cuculidae): competição por recursos? *Ararajuba* 5: 49-56.
Ménégaux, A. 1911. Étude des oiseaux de l'Equateur, rapportés par le Dr. Rivet. Pp. 1-128 (chapter 2) *in: Mission du Service Géographique de l'Armée pour la mesure d'un arc méridien equatorial en Amérique du Sud sous le contrôle scientifique de l'Académie des Sciences*, 1899-1906. Vol. 9: Zoologie. Ministère de l'Instruction Publique, Paris.
Ménégaux, A. 1925. Étude d'une collection d'oiseaux faite par M.E. Wagner dans le chaco argentin. *Rev. Franç. d'Orn.* 17: 221-238, 279-297, 322-329.
Meneghetti, F. 1944. Further notes on the nestling of Paradise Flycatchers and on parasitism by Klaas' Cuckoo. *J. E. Afr. Nat. Hist. Soc.* 18: 96.
Menesse, N.H. 1939. The Punjab Sirkeer Cuckoo [*Taccocua leschenaultii sirkee* (Gray)] in Sind. *J. Bombay Nat. Hist. Soc.* 41: 172-173.
Mengel, R.M. 1965. The birds of Kentucky. *Orn. Monogr.* 3: 1-581.
Merritt, J.H. 1951. Little orphan ani. *Audubon* 53 (4): 224-231.
Meyer, A.B. 1874. Über neue und ungenügend bekannte Vögel von Neu-Guinea und den Inseln der Geelvinksbai. *Sitzungsb. Math.-Naturwiss. Cl. kaiserl. Akad. Wissensch.* 69: 81.
Meyer, A.B. 1878. Description of two species of birds from the Malay Archipelago. *Ornith. Misc.* 3: 163-166.
Meyer, A.B. 1879. Field-notes on the birds of Celebes. Part 1. Psittaci, Rapaces and Picaria. *Ibis* (4th series) 3: 43-70.
Meyer, A.B. & Wiglesworth, L.W. 1896. Bericht über die 5. bis 7. Vogelsammlung der Herren Dr P. und F. Sarasin aus Celebes. *Abh. Ber. K. Zool. Mus. Dresden* 6 (2): 1-16.
Meyer, O. 1931. Über Nest und Gelege von *Centropus violaceus. Orn. Monatsb.* 39: 148.
Meyer, O. 1936. *Die Vögel des Bismarckarchipel win hilfsbuch zur Erkennung der Vogelarten. Beobachtungen und Studien der Missionare vom Hlst. Herzen Jesu in der Südsee, Band* 2. Katholische Mission, Vunapope, Kokopo.
Meyer de Schauensee, R. 1949. The birds of the Republic of Colombia. Part 2. *Caldasia* 5: 381-644.
Meyer de Schauensee, R. 1958. The birds of the island of Bangka, Indonesia. *Proc. Acad. Nat. Sci. Philad.* 110: 279-299.
Meyer de Schauensee, R. 1966. *The species of birds of South America with their distribution.* Livingston Press, Narberth, Penn.
Meyer de Schauensee, R. 1982. *A guide to the birds of South America.* Livingston Publishing Company for Acad. Nat. Sci. Philad. (reprinted with addenda by ICBP Pan-American Section), Wynnewood, Penn.
Meyer de Schauensee, R. 1984. *The birds of China.* OUP, Oxford.
Meyer de Schauensee, R. & Phelps, W.H. Jnr. 1978. *A guide to the birds of Venezuela.* Princeton University Press, Princeton.
Michie, R.H. 1948. Habits of Shining Cuckoo. *New Zealand Bird Notes* 2: 196.
Milburn, C.E. 1915. Adult Cuckoo killing nestling Meadow Pipits. *Brit. Birds* 9: 95-96.
Miles, H.M. 1951. The cuckoos in Southern Rhodesia. *Babbler Gaborone* 6: 3-6.
Millar, H.M. 1943. The Emerald Cuckoo (Short Notes). *Ostrich* 14: 118.
Miller, A.H. 1932. Observations on some breeding birds of El Salvador, Central America. *Condor* 34: 8-17.
Miller, A.H. 1947. The tropical avifauna of the upper Magdalena Valley, Colombia. *Auk* 64: 351-381.
Miller, A.H. 1955. The avifauna of Sierra del Carmen of Coahuila, Mexico. *Condor* 57: 154-178.
Miller, G.O. 1988. Beep! Beep! The Roadrunner - notes on this comical and curious cuckoo. *Birder's World* 2: 16-18.
Miller, G.S., Jnr. 1894. The ground cuckoo of Andros Island. *Auk* 11: 164-165.
Miller, W. de W. 1924. Further notes on ptilosis. *Bull. Amer. Nat. Hist.* 50: 305-331.
Millet-Horsin, [H.] 1921. Souvenirs d'un naturaliste en Afrique occidentale Française. A. Au Togo. 6 Le Coq de pagode ou Coucal perons (*Centropus monachus* Rüpp.) *Oiseau* 2: 75-79.
Millington, S.J. & Price, T.D. 1982. Birds on Daphne Major 1979-1981. *Noticias Galáp.* 35: 25-27.
Mills, M.S.L. 2010. A record of brood parasitism of Barratt's Warbler by African Emerald Cuckoo. *Bull. ABC* 17: 212.
Mills, M.S.L., Caddick, J., Coetzer, I., Hoff, R., Myers, D., Reynolds, D., Walton, A. & Walton, R. 2007. First host species record for Yellow-throated Cuckoo *Chrysococcyx flavigularis. Bull. ABC* 14: 68-69.
Milne-Edwards, A. & Grandidier, A. 1879. *Histoire physique, naturelle et politique de Madagascar.* Vol. 12. *Histoire naturelle des oiseaux.* Tome 1. Société d'Éditions Géographiques, Maritimes et Coloniales, Paris.
Milon, P. 1946. Observations sur quelques oiseaux de Madagascar. *Oiseau RFO* 16: 82-86.
Milon, P. 1950. Description d'une sous-espèce nouvelle d'oiseau de Madagascar. *Bull. Mus. Hist. Nat. Paris* 22: 65-66.
Milon, P. 1952. Notes sur le genre Coua. *Oiseau RFO* 22: 75-90.
Milon, P. 1959. Sur la migration et la reproduction du Coucou de Madagascar *Cuculus poliocephalus rochii* Hartlaub. *Ostrich* Suppl. No. 3 (*Proc. Pan-Afr. Orn. Congr.*): 242-249.
Milon, P., Petter, J. & Randrianasolo, G. 1973. *Faune de Madagascar.* 35 *Oiseaux.* ORSTOM and CNRS, Tananarive and Paris.
Milstein, P. 1954. An egg problem. *Bokmakierie* 6: 4-5.
Miltschew, B.P. 1992. Häherkuckuck (*Clamator glandarius*) – Brutvogel in Bulgarien. *J. Orn.* 133: 86–88.
Minns, J., Buzzetti, D.R.C., Albano, C.G., Grosset, A., Whittaker, A. & Parrini, R. 2010. *Birds of Brazil: songs, calls and photos.* Vol. 1 - *Atlantic forest, cerrado, caatinga, pantanal, savannas and coast* (DVD-ROM). Avis Brasilis Editora, São Paulo.
Miranda, H.C. Jnr., Kennedy, R.S., Sison, R.V., Gonzales, P.C. & Ebreo, M.F. 2000. New records of birds from the island of Panay, Philippines. *Bull. Brit. Orn. Club* 120: 266-280.

Mita, N. 2010. Record of singing Asian Drongo-cuckoo at Tatsuta-Yama in Kumamoto City, Kyushu, Japan. *Bird Research* 6: S13-S16.

Mitsch, H. 1979. Der Fratzenkuckuck (*Scythrops novaehollandiae*), Stand-oder Zugvogel? *Orn. Mitt.* 31 (7): 155-158.

Mlodinow, S.G. 2006. Five new species of birds for Aruba, with notes on other significant sightings. *J. Caribb. Orn.* 19: 31-35.

Mlodinow, S.G. & Karlson, K.T. 1999. Anis in the United States and Canada. *North Amer. Birds* 53: 237-245.

Moffett, A.T. 1985. Cuckoo drinking. *Brit. Birds* 78: 351-353.

Mogensen, J. 1927. Nota sobre al parasitismo del "crespín" (*Tapera naevia*). *Hornero* 4: 68-70.

Mogensen, J. 1930. *Argentinas dyr*. Græbes Bogtrykkeri, København (Copenhagen).

Mohan, D. 1994. Rufous-backed Shrike (*Lanius schach* Linné) feeding a Striped Keelback (*Amphiesma stolata*) to Cuckoo (*Cuculus canorus* Linné) fledgling. *J. Bombay Nat. Hist. Soc.* 91: 143.

Moksnes, A. & Røskaft, E. 1995a. Evolution of interspecific brood parasitism in birds. *Fauna Norvegica Ser. C. Cinclus* 18: 1-18.

Moksnes, A. & Røskaft, E. 1995b. Egg-morphs and host preference in the Common Cuckoo (*Cuculus canorus*): an analysis of Cuckoo and host eggs from European museum collections. *J. Zool. London* 236: 625-648.

Moksnes, A., Røskaft, E., Bicik, V., Honza, M. & Øien, I.J. 1993. Cuckoo *Cuculus canorus* parasitism on *Acrocephalus* warblers in Southern Moravia in the Czech Republic. *J. Orn.* 134: 425-434.

Moksnes, A., Røskaft, E., Hagen, L.G., Honza, M., Mørk, C. & Olsen, P.H. 2000. Common cuckoo *Cuculus canorus* and host behaviour at reed warbler *Acrocephalus scirpaceus* nests. *Ibis* 142: 247-258.

Moksnes, A., Røskaft, E., Rudolfsen, G., Skjelseth, S., Stokke, B.G., Kleven, O., Gibbs, H.L., Honza, M., Taborsky, B., Teuschl, Y., Vogl, W. & Taborsky, M. 2008. Individual female common cuckoos *Cuculus canorus* lay constant egg types but egg appearance cannot be used to assign eggs to females. *J. Avian Biol.* 39: 238-241.

Møller, A. P. 2011. Climate Change and Birds, pp. 13-39 in del Hoyo, J., Elliott, A. & Christie, D. *Handbook of the Birds of the World*. Vol. 16. Lynx Edicions, Barcelona.

Møller, A.P., Saino, N., Adamík, P., Ambrosini, R., Antonov, A., Campobello, D., Stokke, B.G., Fossøy, F., Lehikoinen, E., Martin-Vivaldi, M., Moksnes, A., Moskat, C., Røskaft, E., Rubolini, D., Schulze-Hagen, K., Soler, M. & Shykoff, J.A. 2010. Rapid change in host use of the common cuckoo *Cuculus canorus* linked to climate change. *Proc. R. Soc. B.* 278: 733-738.

Monroe, B.L., Jnr. 1968. A distributional survey of the birds of Honduras. *Orn. Monogr.* 7: 1-458.

Monson, G. 1946. Road-runner preys on Poor-will. *Wilson Bull.* 58: 185.

Montoya, M. 2007. Aves de la Isla del Coco. Lista de especies. *Zeledonia* 7: 29-37.

Moojen, J., Carvalho, J.C. de & Lopes, H.S. 1941. Observações sobre o conteúdo gástrico das aves brasileiras. *Mem. Inst. Oswaldo Cruz* 36: 405-444.

Moore, A. 1984. Levaillant's Cuckoo *Clamator levaillanti* fed by Brown Babbler *Turdoides plebejus*. *Malimbus* 6: 94-95.

Moore, R.T. 1934. A review of the races of *Geococcyx velox*. *Trans. San Diego Soc. Nat. Hist.* 7: 455-464.

Moores, N. 2007. Selected records from Socheong Island, South Korea. *Forktail* 23: 102-124.

Moreau, R.E. 1949. Special review. Friedmann on African cuckoos. *Ibis* 91: 529-537.

Moreau, R.E. 1966. *The bird faunas of Africa and its islands*. Academic Press, New York and London.

Moreau, R.E. 1972. *The Palaearctic-African bird migration systems*. Academic Press, London.

Moreau, R.E. & Chapin, J.P. 1951. The African Emerald Cuckoo, *Chrysococcyx cupreus*. *Auk* 68: 174-189.

Moreau, R.E. & Moreau, W.M. 1937. Biological and other notes on some East African birds. Part 1. *Ibis* (14th series) 1: 152-174.

Moreau, R.E. & Moreau, W.M. 1939. Observations on some East African birds. *Ibis* (14th series) 3: 296-323.

Moreau, R.E. & Sclater, W.L. 1938. The avifauna of the mountains along the Rift Valley in north central Tanganyika Territory (Mbulu District). *Ibis* (14th series) 2: 1-32.

Morel, G.J. 1972. *Liste commentée des oiseaux du Sénégal et de la Gambie*. Off. Rech. Scient. Tech. Outre-mer, Dakar, Senegal.

Morel, G.J. & Morel, M.Y. 1990. *Les oiseaux de Sénégambie*. ORSTOM, Paris.

Morony, J.J. Jnr., Bock, W.J. & Farrand, J. Jnr. 1975. *Reference list of the birds of the world*. American Museum of Natural History, New York.

Morris, P. & Hawkins, F. 1998. *Birds of Madagascar. A photographic guide*. Pica Press, Sussex.

Morrison, C.M. 1947 Field notes on some Gold Coast birds. *Nigerian Field* 12: 59-64.

Morton, E.S. & Farabaugh, S.M. 1979. Infantici de and other adaptations of the nestling Striped Cuckoo *Tapera naevia*. *Ibis* 121: 212-213.

Moskát, C., Hansson, B., Barabás, L., Bártol, L. & Karcza, Z. 2008. Common cuckoo *Cuculus canorus* parasitism, antiparasite defence and gene flow in closely located populations of great reed warblers *Acrocephalus arundinaceus*. *J. Avian Biol.* 39: 663-671.

Moskát, C. & Honza, M. 2000. Effect of nest and nest site characteristics on the risk of cuckoo *Cuculus canorus* parasitism in the great reed warbler *Acrocephalus arundinaceus*. *Ecography* 23: 335-341.

Moskovits, D., Fitzpatrick, J.W. & Willard, D.E. 1985. Lista preliminar das aves da Estação Ecológica de Maracá, Território de Roraima, Brasil, e áreas adjacentes. *Pap. Avuls. Dep. Zool. São Paulo* 36: 51-68.

Mota, F. 2010. [Photos of *Coccyzus minor*, Piauí]. Wiki Aves - A Enciclopédia das Aves do Brasil. http://www.wikiaves.com/166780. Acessed 12 Jun 2011.

Mouritz, L.B. 1915. Notes on the ornithology of the Matapo District, Southern Rhodesia. *Ibis* (10th series) 3: 185-216.

Mousley, H. 1931. Reminiscences of the home life of the Black-billed Cuckoo. *Canad. Field Nat.* 45: 58-79.

Moyle, R. G., Biun, A., Butit, B. & Sumpongol, D. 2001. Brood hosts of Oriental Cuckoo *Cuculus saturatus* in Sabah, Malaysia. *Bull. Brit. Orn. Club* 121: 107-110.

Muck, C., Kempenaers, B., Kuhn, S. Valcu, M. & Goymann, W. 2009. Paternity in the classical polyandrous black coucal (*Centropus grillii*) – a cuckoo accepting cuckoldry? *Behavioural Ecology* 20: 1185-1193.

Mullarney, K., Svensson, L., Zetterström, D. & Grant, P.J. 2000. *Collins bird guide*. HarperCollins, London.

Muller, K.A. 1971. Physical and behavioral development of a Roadrunner raised at the National Zoological Park. *Wilson Bull.* 83: 186-193.

Müller, P.L.S. 1776. *Natursystems. Suppl. und Register Band. Des Ritters Carl von Linné...vollständiges Natursystem, nach der zwölften lateinischen Ausgabe und nach Anleitung des Holländischen Houttuynischen Werks, mit einer ausführlichen Erklärung ausgefertiget von P. L. S. Müller*. G. N. Raspe, Nürnberg.

Müller, S. 1835. Aanteekeningen, over de natuurlijke gesteldheid van een gedeelte der westkust en binnenlanden van Sumatra; met bijvoeging van eenige waarnemingen en beschrijvingen van verscheidene, op dit, en andere Sunda-eilanden voorkomende dieren *Tijdschr. Natuurl. Gesch. Phys.* 2: 315-355, 5 plates.

Müller, S. 1840. Bijdrage tot de kennis van Nieuw Guinea. *Verh. Nat. Gesch. Nederl., Land- en Volkenk.* 1: 1-80.

Müller, S. 1843. Bijdragen tot de kennis van Nieuw-Guinea. *Verh. Nat. Gesch. Nederl., Land- en Volkenk.* 6: 176-179.

Müller, S. 1845. Bijdrage tot de kennis van Timor en eenige andere naburige eilanden. *Verh. Nat. Gesch. Nederl., Land- en Volkenk.* 8: 129-320.

Mulyawati, D., Hua, F. & Zetra, B. 2009. First Indonesian record of probable brood parasitism of Chestnut-winged Babbler *Stachyris erythroptera* by Asian Drongo-cuckoo *Surniculus lugubris*. *Kukila* 14: 51-54.

Mundy, P.J. 1973. Vocal mimicry of their hosts by nestlings of the Great Spotted Cuckoo and Striped Crested Cuckoo. *Ibis* 115: 602-604.

Mundy, P.J. & Cook, A.W. 1977. Observations on the breeding of the Pied Crow and Great Spotted Cuckoo in northern Nigeria. *Ostrich* 48: 72-84.

Munn, C.A. 1985. Permanent canopy and understory flocks in Amazonia: Species composition and population density. *Orn. Monogr.* 36: 683-712.

Murdoch, D.A. & Betton, K.F. 2008. A checklist of the birds of Syria. *Sandgrouse* Suppl. 2. Orn. Soc. Middle East.

Nacinovic, J.B., Schloemp, I.M. & Luigi, G. 1990. Novas observações sobre a avifauna do Rio de Janeiro metropolitano. P. 177 in *Resumos XVII Congr. Brasil. Zool., Londrina.* Univ. Estadual de Londrina, Paraná, Brazil.

Naka, L.N. & Rodrigues, M. 2000. *As aves da Ilha de Santa Catarina.* Editora da Universidade Federal de Santa Catarina, Florianópolis.

Nanda, B.C. 1996. A Crake and a Malkoha. *Newsletter for Birdwatchers* 36 (5): 95.

Nankinov, D. 1999. Distribution of the Great Spotted Cuckoo, *Clamator glandarius*, in Bulgaria. *Zool. Middle East* 17: 9-14.

Nankinov, D.N. 2007. [The Common Cuckoo *Cuculus canorus* in Bulgaria]. *Russian Orn. J.* 16 (Express issue 367): 903-929. [In Russian].

Naranjo H., L.G. 1982. Notas adicionales a la historia natural de *Coccyzus pumilus* (Aves: Cuculidae). *Cespedesia* 11: 95-101.

Narosky, T. & Di Giacomo, A.G. 1993. *Las aves de la provincia de Buenos Aires: distribución y estatus.* Asociación Ornitológica del Plata & Vazquez Mazzini, L.O.L.A., Buenos Aires.

Narosky, T. & Yzurieta, D. 1987. *Guía para la identificación de las aves de la Argentina y Uruguay.* Vázquez Mazzini, Buenos Aires.

Nash, S.V. & Nash, A.D. 1985. A checklist of the forest and forest edge birds of the Padang - Sugihan Wildlife Reserve, South Sumatra. *Kukila* 2: 51-59.

Nash, S.V. & Nash, A.D. 1988. An annotated checklist of the birds of Tanjung Puting National Park, central Kalimantan. *Kukila* 3: 93-116.

Natarajan, V. 1993a. Food and feeding habits of the southern Crow-pheasant *Centropus sinensis parroti* Stresemann (Aves: Cuculidae) at Pt. Calimere, Tamil Nadu. *J. Bombay Nat. Hist. Soc.* 90: 11-16.

Natarajan, V. 1993b. Time budgeting by the southern Crow-pheasant *Centropus sinensis parroti* Stresemann (Aves: Cuculidae) at Pt. Calimere, Tamil Nadu. *J. Bombay Nat. Hist. Soc.* 90: 92-95.

Natarajan, V. 1997. Breeding biology of the Southern Crow-Pheasant *Centropus sinensis parroti* Stresemann (Aves: Cuculidae) at Point Calimere, Tamil Nadu. *J. Bombay Nat. Hist. Soc.* 94: 56-63.

Naumburg, E.M.B. 1930. The birds of Mato Grosso, Brazil. A report on the birds secured by the Roosevelt-Rondon Expedition. With field notes by George K. Cherrie. *Bull. Amer. Mus. Nat. Hist.* 60: 1-432.

Navarro-Sigüenza, A.G. & Peterson, A.T. 2004. An alternative species taxonomy of the birds of Mexico. *Biota Neotrop.* 4: 1-32.

Nechaev, V.A. 1991. *Ptitsy ostrova Sakhalin* [*Birds of Sakhalin Island*]. Institute of Biology and Pedology, Vladivostok. [in Russian]

Negret, A.J. 1991. Reportes recientes en el Parque Nacional Munchique de aves consideradas raras o amenazadas de extinción. *Noved. Colomb., Nueva Época* 3: 39-45.

Nehrkorn, A. 1881. Beschreibung yucatanischer Eier. *J. Orn.* 29: 65-69.

Nehrkorn, A. 1910. *Katalog der Eiersammlung nebst Beschreibungen der aussereuropäischen Eier*, 2nd ed. R. Friedländer & Sohn, Berlin.

Nellar, M. M. 1993. *Aves de la provincia de San Luis: lista y distribución.* Museo Privado de Ciencias Naturales e Investigaciones Ornitológicas 'Guillermo E. Hudson', San Luis, Argentina.

Nero, R.W. 1988. A birding moment: Courting cuckoos. *Birder's World* 2: 54-55.

Netschajew, V.A. 1977. Zur Biologie des Kleines Kuckucks im Ussurigebiet. *Falke* 24: 366-371.

Neufeldt, I. 1966. Der "viersilbige" Kuckuck. *Falke* 13: 197-201.

Neufeldt, I. 1967. Life history of the Indian Cuckoo, *Cuculus micropterus micropterus* Gould, in the Soviet Union. *J. Bombay Nat. Hist. Soc.* 63: 399-419.

Neufeldt, I.A. 1971. Der Kurzflügelsänger *Horeites diphone* (Kittlitz). *Falke* 18: 364-375.

Neufeldt, I.A. 1972. [Age and sexual dimorphism in the colour of *Hierococcyx fugax hyperythrus*]. *Ornitologiya* 10: 97-110. [In Russian].

Neumann, O. 1905. Vögel von Schoa und Süd-Äethiopien. *J. Orn.* 53: 335-360.

Neumann, O. 1908. New species and subspecies of African birds. *Bull. Brit. Orn. Club* 21: 76-78.

Neumann, O. 1934. Drei neue geographische Rassen aus dem paläarktischen Gebiet. *Anz. orn. Ges. Bayern* 2 (8): 331-334.

Neunteufel, A. 1951. Observaciones sobre el *Dromococcyx pavoninus* Pelzeln y el parasitismo de los cucúlidos. *Hornero* 9: 288-290.

Newby-Varty, B.V. 1946. Further notes on the birds on Umvukwe Ranch, Banket, S. Rhodesia. *Ostrich* 17: 343–347.

Newby-Varty, B.V. 1950. Some oological notes. *Ostrich* 21: 38.

Newman, K. 1980. *Birds of Southern Africa.* 1: *Kruger National Park.* MacMillan South Africa, Johannesburg.

Newton, A. & Newton, E. 1859. Observations on the birds of St. Croix, West Indies, made, between February 20th and August 6th 1857 by Alfred Newton, and, between March 4th and September 28th 1858 by Edward Newton. *Ibis* 1: 59-69, 138-150, 252-264, 365-379.

Nicéforo, H. & Olivares, A. 1966. Adiciones a la avifauna Colombiana - III (Columbidae-Caprimulgidae). *Bol. Soc. Venez. Cienc. Nat.* 26: 370-393.

Nichols, E.G. 1937. The Kodaikanal birds and how to name them. *J. Bombay Nat. Hist. Soc.* 39: 812-830.

Nicholson, F.G. 1897. Egg of South African Golden Cuckoo in nest of Cape Wagtail. *Zoologist, Ser.* 4, 1: 142-143.

Nickalls, E.M. 1983. Notes on Fiscal Shrike & Klaas' Cuckoo. *Bull. E. Afr. Nat. Hist. Soc.* (Sept.-Oct): 74.

Nickell, W.P. 1954. Red-wings hatch and raise a Yellow-billed Cuckoo. *Wilson Bull.* 66: 137-138.

Nicoll, M.J. 1906a. On some birds collected during the voyage of the "Valhalla" (1905-1906). *Bull. Brit. Orn. Club* 16: 102-106.

Nicoll, M.J. 1906b. On the birds collected and observed during the voyage of the "Valhalla", R.Y. S., from November 1905 to May 1906. *Ibis* (8th series) 6: 666-712.

Nielsen, L. 1996. *Birds of Queensland's wet tropics and Great Barrier Reef.* Gerard Industries, Bowden, SA.

Nikolaus, G. 1987. Distribution atlas of Sudan´s birds with notes on habitat and status. *Bonn. Zool. Monogr.* 25: 1-322.

Nikolov, S. Ch., Profirov, L., Gerdjikov, G. & Gradinarov, D. 2007. A review of the observations of Great Spotted Cuckoo *Clamator glandarius* in Bulgaria in 2006 and 2007. *Acrocephalus* 28 (133): 75-76.

Nolan, V., Jnr. 1963. Reproductive success of birds in a deciduous scrub habitat. *Ecology* 44: 305-313.

Nolan, V., Jnr. & Thompson, C.F. 1975. The occurrence and significance of anomalous reproductive activities in two North American non-parasitic cuckoos *Coccyzus* spp. *Ibis* 117: 496-503.

Noramly, G. & Jeyerajasingam, A. 1985. Bird notes 1984 from the Selangor Branch Bird Group. *Malayan Nat.* 39: 47-50.

Nores, M. & Yzurieta, D. 1979. Aves de costas marinas y de ambientes continentales, nuevas para la provincia de Córdoba. *Hornero* 12: 45-52.

Nores, M. & Yzurieta, D. 1981. Nuevas localidades para aves argentinas, parte 1. *Hist. Nat.* 2: 33-42.

Nores, M., Yzurieta, D. & Miatello, R. 1983. Lista y distribución de las aves de Córdoba, Argentina. *Bol. Acad. Nac. Cienc. Córdoba* 56: 1-114.

Nores, M., Yzurieta, D. & Salvador, S.A. 1991. Lista y distribución de las aves de Santiago del Estero, Argentina. *Bol. Acad. Nac. Cienc. Córdoba* 59: 157-196.

Norman, G.C. 1888. Notes on the Geographical Distribution of the Crested Cuckoos (*Coccystes*). *Ibis* (5th series) 6: 396-409.

Norris, D.J. & Elder, W.H. 1982. Decline of the Roadrunner in Missouri. *Wilson Bull.* 94: 354-355.

North, A.J. 1911. *Nests and eggs of birds found breeding in Australia and Tasmania*. Special Catalogue No. 1, Vol 3. Australian Museum, Sydney.

North, M.E.W. 1958. *Voices of African birds*. Laboratory of Ornithology, Cornell University, Ithaca, New York.

Northrop, J.I. 1891. The birds of Andros Island, Bahamas. *Auk* 8: 64-80.

Noske, R. 1981. Courtship feeding of cuckoos. *Aust. Bird Watcher* 9 (4): 136.

Noske, R. 1994. Shining Bronze Cuckoo and Channel-billed Cuckoo: first records for Timor. *Kukila* 7: 68-69.

Noske, R.A. 1995. The ecology of mangrove forest birds in Peninsular Malaysia. *Ibis* 137: 250-263.

Noske, R.A. 2001. The breeding biology of the Mangrove Gerygone, *Gerygone laevigaster*, in the Darwin region, with notes on brood parasitism by the Little Bronze-cuckoo, *Chrysococcyx minutillus*. *Emu* 101: 129-135.

Nouvelles Approches. 2003. *Checklists of National Parks of DRC*. www.nouvellesapproches.org.

Novaes, F.C. 1957. Contribuição à ornitologia do noroeste do Acre. *Bol. Mus. Par. Emílio Goeldi, Zool., nov. sér.* 9: 1-30.

Novaes, F.C. 1970. Distribuição ecológica e abundância das aves em um trecho da mata do baixo Rio Guamá (Estado do Pará). *Bol. Mus. Par. Emílio Goeldi, Zool., nov. sér.* 71: 1-54.

Novaes, F.C. 1974. Ornitologia do Território do Amapá I. *Publ. Avuls. Mus. Goeldi* 25: 1-122.

Novaes, F.C. & Lima, M.F.C. 1991. As aves do rio Peixoto de Azevedo, Mato Grosso, Brasil. *Rev. Bras. Zool.* 7: 351-381.

Novaes, F.C. & Lima, M.F.C. 1998. *Aves da Grande Belém: Municípios de Belém e Ananindeua, Pará*. Museu Paraense Emílio Goeldi, Belém, Pará, Brazil.

Nuttall, T. 1832. *A manual of the ornithology of the United States and Canada. Vol. I. The land birds*. Hilliard and Brown, Cambridge & Boston.

Oates, E.W. 1882. A List of the Birds of Pegu. *Stray Feathers* 10: 175-248.

Oatley, T. B. 1980. Eggs of two cuckoo genera in one nest, and a new host for Emerald Cuckoo. *Ostrich* 51: 126-127.

Oberholser, H.C. 1912. Description of one hundred and four new species and sub-species of birds from the Barussan Islands and Sumatra. *Smithson. Misc. Coll.* 60 (7): 1-22.

Oberholser, H.C. 1919. Mutanda Ornithologica V-VIII. *Proc. Biol. Soc. Wash.* 32: 7-8, 21-22, 127-128, 239-240.

Oberholser, H.C. 1924. Descriptions of new Treronidae and other non-Passerine birds from the East Indies. *J. Wash. Acad. Sci.* 14: 294-303.

Oberholser, H.C. 1932. The birds of the Natuna Islands. *Bull. US Nat. Mus.* 159: 1-137.

Oberholser, H.C. 1974. *Bird life of Texas*. Vol. 1. University of Texas Press, Austin.

Ochoa de Masramón, D. 1983. Lista de aves del nordeste de San Luis. *Hornero* Extraordinário: 77-87.

O'Connor, R.J. 1962. Juvenile Cuckoo apparently imitating Meadow Pipit's call. *Brit. Birds* 55: 481.

O'Donel, H.V. 1936. The Indian Cuckoo (*Cuculus m. micropterus* Gould). *J. Bombay Nat. Hist. Soc.* 39: 175.

Ogilvie-Grant, W.R. 1894. On the birds of the Philippine Islands. Part I. Mount Arajat, central Luzon. With field-notes by John Whitehead. *Ibis* (6th series) 6: 406–411.

Ogilvie-Grant, W.R. 1915. Report on the birds collected by the British Ornithologists' Union expedition and the Wollaston expedition in Dutch New Guinea. *Ibis Jub. Suppl.* (10th series) 2: i-xx, 1-336.

Ohmart, R.D. 1973. Observations on the breeding adaptations of the Roadrunner. *Condor* 75: 140-149.

Ohmart, R.D. 1989. A timid desert creature that appears to be half bird, half reptile. *Nat. Hist.* 98: 34-41.

Ohmart, R.D. & Lasiewski, R.C. 1971. Roadrunners: energy conservation by hypothermia and absorption of sunlight. *Science* 172: 67-69.

Øistein, H.H., Sætre, G-P., Slagsvold, T. & Stenseth, N.C. 2009. Parasites and supernormal manipulation. *Proc. R. Soc. London B* 268: 2551-2558.

Olivares, A. & Hernández Camacho, J.I. 1962. Aves de la Comisaría del Vaupés (Colombia). *Rev. Biol. Trop.* 10: 61-90.

Olivares, A. & Munves, J.A. 1973. Predatory behavior of Smooth-billed Ani. *Auk* 90: 891.

Oliver, W.R.B. 1955. *New Zealand birds*. A.H. & A.W. Reed. Wellington.

Ollington, R.F. & Loh, E. 1999. *The Birdline Singapore* 1995 *Singapore Bird Report*. Privately circulated, Singapore.

Ollington, R.F., Loh, E., & Subaraj Rajathurai. 1999. *The Birdline Singapore* 1998 *Singapore Bird Report*. Privately circulated, Singapore.

Olmos, F. 1993. Birds of Serra da Capivara National Park, in the "caatinga" of north-eastern Brazil. *Bird Conserv. Int.* 3: 21-36.

Olmos, F. 2003. Birds of Mata Estrela private reserve, Rio Grande do Norte, Brazil. *Cotinga* 20: 26-30.

Olofson, S. K. K. & Sørensen, S. 2010: *SU-listen for Færøerne*. Copenhagen: Dansk Ornitologisk Forening. www.dof.dk/sider/images/stories/gu/su/dokumenter/FaeroelistenSU_Jan2010.pdf

Olrog, C.C. 1979. Nueva lista de la avifauna argentina. *Opera Lilloana* 27: 1-324.

Olsen, K.M. 1992. *Danmarks fugle - en oversigt*. Dansk Ornitologisk Forening, Copenhagen.

Olson, S. L. 1975. Paleornithology of St Helena Island, South Atlantic Ocean. *Smithson. Contrib. Paleobiol.* 23: 1-49.

Olson, S.L. 1978. Greater Ani (*Crotophaga major*) in Mexico. *Auk* 95: 766-767.

Olson, S.L. 1993. Contributions to avian biogeography from the archipelago and lowlands of Bocas del Toro, Panamá. *Auk* 110: 100-108.

Olson, S.L. & Angle, J.P. 1977. Weights of some Puerto Rican birds. *Bull. Brit. Orn. Club* 97: 105-107.

Oniki, Y. 1972. Some temperatures of Panamanian birds. *Condor* 74: 209-215.

Oniki, Y. & Ricklefs, R.E. 1981. More growth rates of birds in the humid New World tropics. *Ibis* 123: 349-354.

Oniki, Y. & Willis, E.O. 1999. Nest building and early incubation in Squirrel Cuckoos (*Piaya cayana*). *Ararajuba* 7: 23-25.

Oren, D.C. 1991. As aves do estado do Maranhão, Brasil. *Goeldiana Zool.* 9: 1-55.

Orenstein, R.I. 1976. Birds of the Plesyumi area, central New Britain. *Condor* 78: 370-374.

Osborne, T.O. 1973. Additional notes on the birds of the Kafue Flats. *Puku* 7: 163-166.

Osmaston, B.B. 1916. Notes on cuckoos in Maymyo. *J. Bombay Nat. Hist. Soc.* 24: 359-363.

Osmaston, B.B. 1927. Notes on the birds of Kashmir. Part 2. *J. Bombay Nat. Hist. Soc.* 32: 134-153.

Ostende, L.W., Hoek, R.W. van den, Dekker, R.J. & Keijl, G.O. 1997. Type-specimens of birds in the National Museum of Natural History, Leiden, Part 1. Non-passerines. *National Museum of Nat. Hist., Leiden*. Technical Bull. 1.

Ottema, O.H., Ribot, J.H.J.M. & Spaans, A.L. 2009. *Annotated checklist of the birds of Suriname*. WWF Guianas, Paramaribo, Suriname.

Ottow, J. & Verheijen, J.A.J. 1969. Zur Lebensweise der Kuckucke von Flores. *J. Orn.* 110: 27-29.

Oustalet, J.-F.É. 1886. Notice sur quelque oiseaux du Congo. *Naturaliste* (1º série) 3: 299-300.

Oustalet, J.-F.É. 1896. Description de cinq espèces nouvelles d'oiseaux appartenant au Museum d'Histoire Naturelle et provenant de la Chine et de l'Indo-Chine. *Bull. Mus. Hist. Nat.* 2: 314-317.

Owen, J.H. 1933. The Cuckoo in the Felstead district. *Rep. Felstead Sch. Sci. Soc.*: 25-39.

Owen, R. 1861. On the nesting of some Guatemalan birds. *Ibis* (1st series) 3: 58-69.

Pacheco, J.F. 1993. As aves visitantes de inverno do médio Solimões, Amazonas. P. 29 in *Resumos III Congr. Bras. Orn.* Editora da Universidade Católica de Pelotas, Pelotas, Rio Grande do Sul, Brazil.

Pacheco, J.F. 2000. A ornitologia descobre o sertão: um balanço do conhecimento da avifauna da caatinga dos primórdios aos anos 1950. Pp. 11-70 in F.C. Straube, M.M. Argel-de-Oliveira & J.F. Cândido Jnr. (Eds). *Ornitologia brasileira no Século XX.* Univ. do Sul de Santa Catarina & Soc. Brasil. Ornitol., Curitiba.

Pacheco, J.F. & Olmos, F. 2005. Birds of a latitudinal transect in the Tapajós-Xingu interfluvium, eastern Brazilian Amazonia. *Ararajuba* 13: 29-46.

Pacheco, J.F. & Olmos, F. 2006. As aves do Tocantins 1: região sudeste. *Rev. Bras. Orn.* 14: 85-100.

Pacheco, M.A., Battistuzzi, F.U., Lentino, M., Aguilar, R.F., Kumar, S. & Escalante, A.A. 2011. Evolution of modern birds revealed by mitogenomics: timing the radiation and origin of major orders. *Mol. Biol. Evol.* 28: 1927–1942.

Pakenham, R.H.W. 1943. Field notes on the birds of Zanzibar and Pemba. *Ibis* 85: 165-189.

Pakenham, R.H.W. 1979. *The birds of Zanzibar and Pemba. An annotated checklist.* British Ornithologists' Union Checklist 2. BOU, London.

Pandey, S., Joshua, J., Rai, N.D., Mohan, D., Rawat, G.S., Sankar, K., Katti, M.V., Khati, D.V.S. & Johnsingh, A.J.T. 1995. Birds of Rajaji National Park, India. *Forktail* 10: 105-113.

Panov, E.N. 1973. [*The birds of South Ussuriland (Fauna, biology, behaviour)*]. Nauka Press, Siberian Branch, Novosibirsk. [in Russian]

Park, H.-S. & Kim, W.-B. 1995. First records of Chestnut-winged Cuckoo (*Clamator coromandus*), European Starling (*Sturnus vulgaris*) and Black Bittern (*Ixobrychus flavicollis*) in Korea. *Korean. Journ. Orn.* 2: 75-76.

Parker, R.E. & Campbell, E.G. 1985. Habitat use by wintering birds of prey in southeastern Arizona. *Western Birds* 15: 175-183.

Parker, S.A. 1981. Prolegomenon to further studies in the *Chrysococcyx "malayanus"* group (Aves, Cuculidae). *Zool. Verh. Leiden* 187: 1-56.

Parker, T.A. III. 1982. Observations of some unusual rainforest and marsh birds in southeastern Peru. *Wilson Bull.* 94: 477-493.

Parker, T.A. III & Goerck, J.M. 1997. The importance of national parks and biological reserves to bird conservation in the Atlantic forest region of Brazil. *Orn. Monogr.* 48: 527-541.

Parker, T.A., III & Hoke, P. 2002. Anexo/Appendix 6: Lista preliminar de especies de aves registradas durante la Expedición RAP a la Zona de Pando, Bolivia, 1992. Preliminary list of bird species recorded during the RAP Expedition to Pando, Bolivia, 1992. Pp. 114-125 *in*: Montambault, J.R. (ed.) *Informes de las evaluaciones biológicas de Pampas del Heath, Perú, Alto Madidi, Bolivia, y Pando, Bolivia.* RAP Bulletin of Biological Assessment, 24. Conservation International, Washington, DC, USA.

Parker, T.A. III & Remsen, J.V. Jnr. 1987. Fifty-two Amazonian bird species new to Bolivia. *Bull. Brit. Orn. Club* 107: 94-107.

Parker, T.A. III, Parker, S.A. & Plenge, M.A. 1982. *An annotated checklist of Peruvian birds.* Buteo Books, Vermillion, South Dakota.

Parker, T.A. III, Donahue, P.K. & Schulenberg, T.S. 1994. Appendix 3: Birds of the Tambopata Reserve (Explorer's Inn Reserve). Pp. 106-124 in R.B. Foster, T.A. Parker III, A.H. Gentry, L.H. Emmons, A. Chicchón, T.S. Schulenberg, L. Rodríguez, G. Lamas, H. Ortega, J. Icochea, W.H. Wust, M. Romo, J.A. Castillo, O. Phillips, C. Reynel, A.W. Kratter, P.K. Donahue & L.J. Barkley (Eds). *The Tambopata-Candamo reserved zone of southeastern Perú: a biological assessment.* Rapid Assessment Program Working Papers, 6. Conservation International, Washington DC.

Parker, V. 1999. *The atlas of the birds of Sul do Save, southern Mozambique.* Avian Demography Unit, Capetown.

Parkes, K. C. 1984. An apparent hybrid Black-billed x Yellow-billed Cuckoo. *Wilson Bull.* 96: 294-296.

Parkes, K.C. & Niles, D.M. 1988. Notes on Philippine birds, 12. An undescribed subspecies of *Centropus viridis. Bull. Brit. Orn. Club* 108: 193-194.

Parmley, D. 1982. Food items of Roadrunners from Palo Pinto County, north central Texas. *Texas J. Sci.* 34: 94-95.

Parrini, R., Raposo, M.A., Pacheco, J.F., Carvalhães, A.M.P., de Melo, T.A. Jnr. & Minns, J.C. 1999. Birds of the Chapada Diamantina, Bahia, Brazil. *Cotinga* 11: 86-95.

Parrot, C. 1907. Beiträge zur Ornithologie Sumatras und der Insel Banka mit besondere Zugrundelegung der von Dr. Hagen auf Banka gesammelten Vögel. *Abh. Königl. Bayer. Akad. Wiss. München. Kl.* 2 Band 24, Abt. 1: 186.

Parrott, S. & Andrew, P. 1996. An annotated checklist of the birds of Way Kambas National Park, Sumatra. *Kukila* 8: 57-85.

Partridge, W.H. 1961. Aves de Misiones nuevas para la Argentina. *Neotrópica* 7: 25-28.

Passburg, R.E. 1959. Bird notes from northern Iran. *Ibis* 101: 153-169.

Paulson, D.R., Orians, G.H. & Leck, C.F. 1969. Notes on birds of Isla San Andrés. *Auk* 86: 755-758.

Paulussen, W. 1957. Nieuwe gegevens over de eieren, de waarden, en de biologie, van de Koekoek, *Cuculus canorus* L. *Gerfaut* 47: 241-258.

Payne, R.B. 1968. The birds of Mopane Woodlands and other habitats of Hans Merensky Nature Reserve, Transvaal. *S. Afr. Avifauna Ser.* 56: 1-32.

Payne, R.B. 1973. Individual laying histories and the clutch size and numbers of eggs of parasitic cuckoos. *Condor* 75: 414-438.

Payne, R.B. 1974. The evolution of clutch size and reproductive rates in parasitic cuckoos. *Evolution* 28: 169-181.

Payne, R.B. 1977. Juvenile plumages of *Cuculus canorus* and *Cuculus gularis* in Africa. *Bull. Brit. Orn. Club* 97: 48-53.

Payne, R.B. 1997. Cuculidae. pp. 508-607 in J. del Hoyo, A. Elliott & J. Sargatal. *Handbook of the birds of the World.* Vol. 4. Lynx Edicions, Barcelona.

Payne, R.B. 2005. *The cuckoos.* OUP, Oxford.

Payne, R.B. & Payne, K. 1967. Cuckoo hosts in Southern Africa. *Ostrich* 38: 135-143.

Payne, R.B. & Payne, L.L. 1998. Nestling eviction and vocal begging behaviors in Australian glossy cuckoos *Chrysococcyx basalis* and *C. lucidus* in S. I. Rothstein & S. K. Robinson (Eds). *Parasitic birds and their hosts.* OUP, Oxford.

Paynter, R.A., Jnr. 1955. The ornithogeography of the Yucatán Peninsula. *Peabody Mus. Nat. Hist. Yale Univ. Bull.* 9: 1-347.

Paynter, R.A., Jnr. 1956. Birds of the Swan Islands. *Wilson Bull.* 68: 103-110.

Peale, T.R. 1848. *Narrative of the [Charles Wilkes's] United States Exploring Expedition [1838-1842] 8: Mammalia and ornithology.* P. 134, pl. xxxvii, fig. 1.

Pearman, M. 1993. The avifauna of the Río Machariapo dry forest, northern La Paz department, Bolivia: a preliminary investigation. *Bird Conserv. Int.* 3: 105-117.

Pearson, D.L. 1971. Vertical stratification of birds in a tropical dry forest. *Condor* 73: 46-55.

Peck, G.K. & James, R.D. 1983. *Breeding birds of Ontario: nidiology and distribution. Vol. 1: nonpasserines.* Royal Ontario Mus. Life Sci. Misc. Publ., Toronto.

Pehani, H. 1993. The co-evolution of the Cuckoo and its hosts. *Proteus* 55: 43–44.

Pemberton, J.R. 1925. Parasitism in the Road-runner. *Condor* 27: 35.

Peña, L.E. 1961. Explorations in the Antofagasta range, with observations on the fauna and flora. *Postilla* 49: 3-42.

Penard, F.P. & Penard, A.P. 1910. *De Vogels van Guyana (Suriname, Cayenne en Demerara),* vol. 2. Wed. F. P. Penard, Paramaribo.

Penhallurick, J. 2010. *World Bird Info.* www.worldbirdinfo.net

Pennant, T. 1769. *Indian Zoology.* 1st ed. Faulder, London.

Penry, E.H. 1976. Young Black Cuckoo soliciting African Golden Oriole. *Bull. Zambian Orn. Soc.* 8: 28-29.

Peres, C.A. 1992. Prey-capture benefits in a mixed-species group of Amazonian tamarins, *Saguinus fuscicollis* and *S. mystax. Behav. Ecol. Sociobiol.* 31: 339-347.

Peres, C.A. & Whittaker, A. 1991. Annotated checklist of the bird species of the upper Rio Urucu, Amazonas, Brazil. *Bull. Brit. Orn. Club* 111: 156-171.

Pereyra, J.A. 1933. Miscelanea ornitológica. *Hornero* 5: 215-219.

Pereyra, J.A. 1937. Contribución al estudio y observaciones ornitológicas de la zona norte de la Gobernación de La Pampa. *Mem. Jard. Zool. La Plata* 7: 197-326.

Pereyra, J.A. 1941. Miscelanea ornitológica. *Hornero* 8: 31-38.

Pérez del Val, J. 1996. *Las aves de Bioko, Guinea Equatorial: guia de campo*. Edilesa, Léon, Spain.

Pérez del Val, J., Castroviejo, J. & Purroy, F.J. 1997. Species rejected from and added to the avifauna of Bioko island (Equatorial Guinea). *Malimbus* 19: 19-31.

Pérez-Emán, J., Sharpe, C.J., Lentino, M., Prum, R.O. & Carreño, I.J. 2003. New records of birds from the summit of Cerro Guaiquinima, Estado Bolívar, Venezuela. *Bull. Brit. Orn. Club* 123: 78-90.

Perrins. C. (Ed.) 1998. *Cramp's The Complete Birds of the Western Palearctic on CD-ROM*. OUP, Oxford.

Perry, A., Kessler, M. & Helme, N. 1997. Birds of the central Río Tuichi Valley, with emphasis on dry forest, *Parque Nacional Madidi*, depto. La Paz, Bolivia. *Orn. Monogr.* 48: 557-576.

Peters, J.L. 1926. Two new birds from Argentina. *Occas. Pap. Boston Soc. Nat. Hist.* 5: 195-196.

Peters, J.L. 1927. Descriptions of new birds (*Aratinga, Coccyzus, Pentheres*). *Proc. New England Zool. Club* 9: 111-113.

Peters, J.L. 1929. Birds. In Vertebrates from the Corn Islands. *Bull. Mus. Comp. Zool.* 69: 130-138.

Peters, J.L. 1940. *Check-list of birds of the world*, vol. 4. Harvard University Press, Cambridge, Mass.

Peterson, R.T. 1960. *A field guide to the birds of Texas and adjacent states*. Houghton Mifflin, Boston.

Pétursson, G. 2009. *List of Icelandic bird species*. notendur.hi.is/yannk/1111.pdf

Petit, L. 1899. Ornithologie Congolaise. *Mem. Soc. Zool. France* 12: 59-106.

Pfeffer, P.L. 1960. Étude d'une collection d'oiseaux de Borneo. *Oiseau RFO* 30: 154-168, 191-218.

Phelps, W.H. & Phelps, W.H. Jnr. 1958. Lista de las aves de Venezuela con su distribución, tomo II, parte 1. No Passeriformes. *Bol. Soc. Venez. Cienc. Nat.* 19: 1-317.

Phillipps, Q. 1970. Some important nesting notes from Sabah. *Sabah Soc. J.* 8: 153-154.

Phillips, W.W.A. 1948. Cuckoo problems in Ceylon. *Spolia Zeylanica* 25 (2): 45-60.

Phillips, W.W.A. 1949. A new race of the Common Hawk Cuckoo from Ceylon. *Bull. Brit. Orn. Club* 69: 56-57.

Piacentini, V.Q. 2004. Aves raras ou falta de estudos? O exemplo da avifauna florestal de Santa Catarina. Pp. 56-57 in A.F. Testoni & S.L. Althoff (Eds). *Resumos do XII Congresso Brasileiro de Ornitologia*. FURB - Universidade Regional de Blumenau, Blumenau, Santa Catarina, Brazil.

Piacentini, V.Q., Ghizoni, I.R. Jnr., de Azevedo, M.A.G. & Kirwan, G.M. 2006. Sobre a distribuição de aves em Santa Catarina, Brasil, parte I: registros relevantes para o estado ou inéditos para a Ilha de Santa Catarina. *Cotinga* 26: 25-31.

Piaskowski, V.D., Teul, M., Williams, K.M. & Cal, R.N. 2006. Birds of the Sibun riverine forest, Belize. *Orn. Neotrop.* 17: 333-352.

Piechocki, R. 1971. Die Mauser der Kuckucke *Saurothera merlini* und *Crotophaga ani*. Ergebnisse der 1. Kubanisch-Deutschen "Alexander-von-Humbolt"-Expedition 1967/68. Nr. 15. *Zool. Jahrb. Abt. Syst. Ökol. Geogr. Tiere* 98: 1-10.

Pinto, A.A.R. 1959. Um esboço da avifauna sedentária da região da Gorongoza, Moçambique. Proc. 1st Pan-Afr. Orn. Congr. *Ostrich* Suppl. 3: 98-125.

Pinto, A.A.R. 1983. *Ornithologia de Angola*. Vol. 1. Non Passeres. Instituto de Investigação Científica Tropical, Lisboa.

Pinto, O.M.O. 1935. Aves da Bahia. Notas críticas e observações sobre uma collecção feita no Reconcavo e na parte meridional do estado. *Rev. Mus. Paulista* 19: 1-325.

Pinto, O.M.O. 1938. Catálogo das aves do Brasil e lista dos exemplares que as representam no Museu Paulista. *Rev. Mus. Paulista* 22: I-XVIII, 1-566.

Pinto, O.M.O. 1952. Súmula histórica e sistemática da ornitologia de Minas-Gerais. *Arq. Zool. São Paulo* 8: 1-51.

Pinto, O.M.O. 1953. Sobre a coleção Carlos Estevão de peles, ninhos e ovos das aves de Belém, Pará. *Pap. Avuls. Dep. Zool. São Paulo* 11: 113-224.

Pinto, O.M.O. 1962. Miscelânea ornitológica. VII. Notas sôbre a variação geográfica nas populações brasileiras de *Neomorphus geoffroyi*, com a descrição de uma subespécie nova. *Pap. Avuls. Dep. Zool. São Paulo* 15: 299-301.

Pinto, O.M.O. 1964. *Ornitologia Brasiliense. Catálogo descritivo e ilustrado das aves do Brasil. Parte introdutória e famílias Rheidae a Cuculidae*, vol. 1. Departamento de Zoologia da Secretaria da Agricultura do Estado de São Paulo, São Paulo, Brazil.

Pinto, O.M.O. 1966. Estudo critico e catálogo remissivo das aves do Território Federal de Roraima. *Cad. da Amaz.* 8: 1-176. Instituto Nacional de Pesquisas da Amazônia, Manaus, Brazil.

Pinto, O.M.O. 1978. *Novo catálogo das aves do Brasil*, vol. 1. Empresa Gráfica da Revista dos Tribunais, São Paulo, Brazil.

Pinto, O.M.O. & Camargo, E.A. 1955. Lista anotada de aves colecionadas nos limites occidentais do estado do Paraná. *Pap. Avuls. Dep. Zool. São Paulo* 12: 215-234.

Pinto, O.M.O. & Camargo, E.A. 1961. Resultados ornitológicos de quatro recentes expedições do Departamento de Zoologia ao Nordeste do Brasil, com a descrição de seis novas subespecies. *Arq. Zool. São Paulo* 11: 193-284.

Piratelli, A.J. & Pereira, M.R. 2002. Dieta de aves na região leste de Mato Grosso do Sul, Brasil. *Ararajuba* 10: 131-139.

Pistorius, A. 1985. Yellow-billed Cuckoo. Pp. 128-129 in S.B. Laughlin & D.P. Kibbe (Eds). *The atlas of breeding birds in Vermont*. Vermont Instit. Nat. Sci., Univ. Press of New England, Hanover, New Hampshire.

Pitman, C.R.S. 1928. The breeding of *Centropus superciliosus loandae* - the Central African White-browed Coucal in Uganda. *Oölog. Rec.* 8: 41-46.

Pitman, C.R.S. 1929. Various notes on E. African birds. *Bateleur Nairobi* 1: 98-101.

Pitman, C.R.S. 1957. On the egg of the African Cuckoo *Cuculus canorus gularis* Stephens. *Bull. Brit. Orn. Club* 77: 138-139.

Pitman, C.R.S. 1961. The Kurrichane Thrush *Turdus libonyanus tropicalis* Peters a host of the Red-chested Cuckoo *Cuculus solitarius* Stephens in Southern Rhodesia. *Bull. Brit. Orn. Club* 81: 48-49.

Pitman, C.R.S. 1964. A further note on the egg of the Red-chested Cuckoo *Cuculus solitarius* Stephens. *Bull. Brit. Orn. Club* 84: 140-141.

Pizzey, G. 1981. *A field guide to the birds of Australia*. Princeton University Press, Princeton.

Pizzey, G. & Knight, F. 1997. *Collins field guide birds of Australia*. HarperCollins, London.

Planque, B. & Vellinga, W-P. 2010. Xeno-canto: sharing bird songs from around the world. www.xeno-canto.org.

Plowes, D.C.H. 1948. Young African Cuckoo in Drongo nest. *Ostrich* 19: 99.

Pobprasert, K., Andrew, A.J. & Round, P.D. 2008. Four new bird records for Khao Yai National Park, Thailand. *BirdingASIA* 10: 98-99.

Pobprasert, K. & Pierce, A.J. 2010. Observations and predation of a Coral-billed Ground Cuckoo (*Carpococcyx renauldi*) nest in northeastern Thailand. *Wilson J. Orn.* 122: 173–177.

Portenko, L. 1931. Einige neue Unterarten paläarktischer Vögel. *Mitt. zool. Mus. Berlin* 17: 415-423.

Porter, R.F., Aspinall, S., Preddy, S. & Blair, M.J. 2010. *The Ornithological Society of the Middle East, the Caucasus and Central Asia List of bird taxa (OSME Region List)*. www.osme.org. [Updated biannually]

Porter, R.F., Christensen, S. & Schiermacker-Hansen, P. 1996. *Field guide to the birds of the Middle East*. Poyser, London.

Posso, S.R. & Donatelli, R.J. 2001. Cranial osteology and systematic implications in Crotophaginae (Aves, Cuculidae). *J. Zool. Syst. Evol. Research* 39: 247-256.

Potter, E.F. 1980. Notes on nesting Yellow-billed Cuckoos. *J. Field Orn.* 51: 17-29.

Potter, E.F. & Hauser, D.C. 1974. Relationship of anting and sunbathing to molting in wild birds. *Auk* 91: 537-563.

Poulsen, M.K. 1995. The threatened and near threatened birds of Luzon, Philippines, and the role of the Sierra Madre mountains in their conservation. *Bird Conserv. Int.* 5: 79-115.

Poulsen, M.K. & Lambert, F.R. 2000. Altitudinal distribution and habitat preferences of forest birds on Halmahera and Buru, Indonesia: implications for conservation of Moluccan avifaunas. *Ibis* 142: 566-586.

Pratt, H.D., Bruner, P.L. & Berrett, D.G. 1987. *The birds of Hawaii and the tropical Pacific*. Princeton University Press, Princeton.

Prawiradilaga, D.M, Marakarmah, A. & Wijamukti, S. 2003. *A photographic guide to the birds of Javan montane forest: Gunung Halimun National Park*. Biodiversity Conservation Project, LIPI, Bogor.

Preble, N.A. 1957. Nesting habits of the Yellow-billed Cuckoo. *Amer. Midland Nat.* 57: 474-482.

Price, J., Droege, S. & Price, A. 1995. *The summer atlas of North American birds*. Academic Press, London.

Price, T.D. 1979. The seasonality and occurrence of birds in eastern Ghats of Andhra Pradesh. *J. Bombay Nat. Hist. Soc.* 76: 379-422.

Price, T.D. & Jamdar, M. 1990. The breeding birds of Overa Wildlife Sanctuary, Kashmir. *J. Bombay Nat. Hist. Soc.* 87: 30-33.

Pridgeon, S. 1995. Life in the fast lane: the lifestyle of the Greater Roadrunner: up close and personal. *Birders' World* 9: 30-33.

Priest, C.D. 1934. *The birds of Southern Rhodesia*. Vol. 2. W. Clowes & Sons, London.

Prins, H.H.T. 1986. Spring migration of Cuckoo through the Rift Valley in northern Tanzania. *Ardea* 74: 215-217.

Prins, T.G., Reuter, J.H., Debrot, A.O., Wattel, J. & Nijman, V. 2009. Checklist of the birds of Aruba, Curaçao and Bonaire, South Caribbean. *Ardea* 97: 137-268.

Pruett, C.L., Gibson, D.D. & Winker, K. 2001. Molecular "cuckoo clock" suggests listing of western Yellow-billed Cuckoos may be warranted. *Wilson Bull.* 113: 228-231.

Pryce, E. 1989. A Black Cuckoo raised by Swamp Boubous. *Babbler Gaborone* 18: 38.

Prŷs-Jones, R.P. & Diamond, A.W. 1984. Ecology of the land birds on the granitic and coralline islands of the Seychelles, with particular reference to Cousin Island and Aldabra Atoll. In D.R. Stoddart (Ed.). *Biography and ecology of the Seychelles Islands. Mon. Biog.* Junk, The Hague.

Pucheran, J. 1845. Description de quelques espèces nouvelles d'oiseaux de Madagascar. *Rev. Zool.* 1848: 49-52.

Pyle, P. 1995. Incomplete flight feather molt and age in certain North American non-passerines. *N. Am. Bird Bander* 20: 15-26.

Pyle, P., Howell, S.N G., DeSante, D.F., Yunick, R.P. & Gustafson, M. 1997. *Identification guide to North American birds. Part I: Columbidae to Ploceidae*. Slate Creek Press, Bolinas, California.

Quantrill, B. & Quantrill, R. 1998. The birds of the Parcours Vita, Yaoundé, Cameroon. *Malimbus* 20: 1-14.

Quay, W.B. 1967. Comparative survey of the anal glands of birds. *Auk* 84: 379-389.

Quinn, J.S., Macedo, R.H.F. & White, B.N. 1994. Genetic relatedness in communally breeding Guira Cuckoos. *Animal Behaviour* 47: 515-529.

Quinn, J.S., Samuelsen, A., Barclay, M., Schmaltz, G. & Kahn, H. 2010. Circumstantial evidence for infanticide of chicks of the communal Smooth-billed Ani (*Crotophaga ani*). *Wilson J. Orn.* 122: 369-374.

Quinn, J.S. & Startek-Foote, J.M. 2000. Smooth-billed Ani *Crotophaga ani*. In A. Poole & F. Gill (Eds) *The birds of North America*. No. 539. 16 pp. The Birds of North America, Inc., Philadelphia.

Quiñonez, A.S. & Tello, A. 2011. Nuevos registros de *Coccyzus melacoryphus* en la costa del Perú. *Cotinga* 33: 130-132.

Quoy, J.R.C. & Gaimard, J.P. 1830. *Voyage de découvertes de l'Astrolabe, Zool.*, Vol. 1. J. Tastu, Paris.

Quý, Vo & Cu, Nguyên. 1999. *Checklist of the birds of Vietnam*. Hanoi.

Rabor, D.S. 1966. A report on the zoological expeditions in the Philippines for the period 1961-1966. *Silliman Journal* 13 (4): 605-616.

Rabor, D.S. 1977. *Philippine birds & mammals*. University of the Philippine Press, Quezon City.

Radford, P. 1992. Are young Cuckoos capable of acoustic mimicry? *Wildlife Sound* 6 (7): 27-29.

Raethel, H-S. 1992. Der Schuppenhelmkuckuck. *Gefiederte Welt* 116: 11-12.

Raffaele, H.A. 1989. *A guide to the birds of Puerto Rico and the Virgin Islands*, Revised ed. Princeton University Press, Princeton.

Raffaele, H.A., Wiley, J.W., Garrido, O.H., Keith, A.R. & Raffaele, J. 1998. *Birds of the West Indies*. Christopher Helm, London.

Raffles, T.S. 1822. Second Part of the Descriptive Catalogue of a Zoological Collection made in the Island of Sumatra and its vicinity. *Trans. Linn. Soc. London* 13: 277-340.

Rajasingh, S.G. & Rajasingh, I.V. 1970. Birds and mammals eating the fruits of yellow oleander (*Thevetia peruviana*). *J. Bombay Nat. Hist. Soc.* 67: 572-573.

Rajathurai, S. 1996. The birds of Batam and Bintan Islands, Riau Archipelago. *Kukila* 8: 86-113.

Ralph, C.P. 1975. Life style of *Coccyzus pumilus*, a tropical Cuckoo. *Condor* 77: 60-72.

Ralph, C.P. & Chaplin, S.J. 1973. Some birds of Isla Punta Arenas, Pacific coast, Colombia. *Condor* 75: 357-359.

Ramadan-Jaradi, G., Bara, T. & Ramadan-Jaradi, M. 2008. Revised checklist of the birds of Lebanon 1999-2007. *Sandgrouse* 30: 22-69.

Ramsay, R.G.W. 1886: Contributions to the ornithology of the Philippine Islands. No. 2. On additional collections of birds. *Ibis* (5th series) 4: 155-162.

Rand, A.L. 1936. The distribution and habits of Madagascar birds. *Bull. Amer. Mus. Nat. Hist.* 72: 143-499.

Rand, A.L. 1941. Courtship of the Roadrunner. *Auk* 58: 57-59.

Rand, A.L. 1942a. Results of the Archbold Expeditions. No. 42. Birds of the 1936-1937 New Guinea Expedition. *Bull. Amer. Mus. Nat. Hist.* 79: 289-366.

Rand, A.L. 1942b. Results of the Archbold Expeditions. No. 43. Birds of the 1938-1939 New Guinea Expedition. *Bull. Amer. Mus. Nat. Hist.* 179: 425-524.

Rand, A.L. 1953. Factors affecting feeding rates of anis. *Auk* 70: 26-30.

Rand, A.L., Friedmann, H. & Traylor, M.A. 1959. Birds from Gabon and Moyen Congo. *Fieldiana Zool.* 41: 1-191.

Rand, A.L. & Gilliard, E.T. 1967. *Handbook of New Guinea birds*. Weidenfeld & Nicolson, London.

Rand, A.L. & Rabor, D.S. 1960. Birds of the Philippine Islands. Siquijor, Mount Malindang, Bohol, and Samar. *Fieldiana Zool.* 35: 221-441.

Rand, A.L. & Rabor, D.S. 1967. New birds from Luzon, Philippine Islands. *Fieldiana Zool.* 51: 85-89.

Randrianary, V., Rifflet, S. & Roché, J.C. 1997. *Madagascar soundscapes* (CD). Sitelle, Le Verdier, France.

Raposo, M.A., Assis, C.P. & Firme, D.H. 2009. *Neomorphus squamiger* Todd, 1925 é uma espécie válida. P. 116 in J.E. Simon, M.A. Raposo, R.Stopiglia & J. Peres (Eds). *XVII Congresso Brasileiro de Ornitologia. Livro de resumos*. Tec Art Editora, São Paulo.

Rappole, J.H., Morton, E.S., Lovejoy, T.E. & Ruos, J.L. 1993. *Aves migratorias Neárticas en los Neotrópicos*. Conservation and Research Center, National Zoological Park, Smithsonian Institution, Front Royal, Virginia.

Rasmussen, P.C. & Anderton, J.C. 2005. *Birds of South Asia. The Ripley Guide. Vol. 1: Field guide. Vol. 2: Attributes and status*. Smithsonian Institution and Lynx Edicions, Washington DC and Barcelona.

Rattray, R.H. 1905. Birds nesting in the Murree Hills and Gullies. *J. Bombay Nat. Hist. Soc.* 16: 421-428, 657-663.

Read, J. 1995. First Australian record of the Oriental Cuckoo *Cuculus saturatus*. *S. Aust. Orn.* 32 (3): 62-63.

Redondo, T. 1993. Exploitation of host mechanisms for parental care by avian brood parasites. *Etologia* 3: 235-297.

Redondo, T. & Arias de Reyna, L. 1989. High breeding success in experimentally parasitized broods of Azure-winged Magpies (*Cyanopica cyana*). *Gerfaut* 79: 149-152.

Reed, R.A. 1953. Some observations on the Didric Cuckoo. *Ostrich* 24: 138-140.

Reed, R.A. 1968. Studies of the Diederik Cuckoo *Chrysococcyx caprius* in the Transvaal. *Ibis* 110: 321-331.

Reed, R.A. 1969. Notes on the Redchested Cuckoo in the Transvaal. *Ostrich* 40: 1-4.

Reed, S. 1980. Food of Long-tailed Cuckoo. *Notornis* 27: 96.

Reichenbach, H.G.L. 1849. *Avium systema naturale* (…). Taf. I-LI. Zool. Museum of Dresden, Leipzig, Germany.

Reichenow, A. 1879. [Eine grosse Anzahl neuer Arten…]. *Orn. Centralbl.* 4: 138-139.

Reichenow, A. 1887. Dr. Fischer's Ornithologische Sammlungen während der letzten Reise zum Victoria Njansa. Mit Benutzung der Tagebücher des Reisenden bearbeitet. *J. Orn.* 35: 38-78.

Reichenow, A. 1893. *Centropus flecki* n. sp. von Damaraland. *Orn. Monatsb.* 1: 84.

Reichenow, A. 1896. Bemerkungen über afrikanische Kukuke. *Orn. Monatsb.* 4: 53-54.

Reichenow, A. 1898. Neue Arten aus Afrika. *Orn. Monatsb.* 6: 22-23.

Reichenow, A. 1900. Neue Arten vom südöstlichen Neuguinea und Nordqueensland. *Orn. Monatsb.* 8: 185-188.

Reichenow, A. 1902. *Die Vögel Afrikas*. Zweiter Band. Neudamm.

Reichenow, A. 1915. Neue Arten. *J. Orn.* 63: 124-129.

Reichholf, J. 1974. Biotopwahl und Schwarmgrößen bei den neotropischen Kuckucksvögeln *Crotophaga ani* L. und *Guira guira* (Gmel.). *Bonn. Zool. Beitr.* 25: 118-122.

Reiser, O. 1926a. Ergebnisse der Zoolog. Expedition der Akad. der Wissenschaften nach Nordostbrasilien im Jahre 1903. Vögel. *Denkschr. Akad. Wiss. Wien, Math.-naturwiss. Kl.* 76: 107-252.

Reiser, O. 1926b. Liste der Vogelarten welche auf der von der Kaiserl. Akademie der Wissenschaften 1903 nach Nordostbrasilien entsendeten Expedition unter Leitung des Herrn Hofrates Dr. F. Steindachner gesammelt wurden. *Denkschr. Akad. Wiss. Wien, Math.-naturwiss. Kl.* 76: 55-100.

Remsen, J.V., Jnr. 2005. Pattern, process, and rigor meet classification. *Auk* 122: 403-413.

Remsen, J.V., Jnr & Ridgely, R.S. 1980. Additions to the avifauna of Bolivia. *Condor* 82: 69-75.

Remsen, J.V., Jnr & Traylor, M.A., Jnr. 1983. Additions to the avifauna of Bolivia, part 2. *Condor* 85: 95-98.

Remsen, J.V., Jnr & Traylor, M.A., Jnr. 1989. *An annotated list of the birds of Bolivia*. Buteo Books, Vermillion, South Dakota.

Remsen, J.V., Jnr., Traylor, M.A., Jnr. & Parkes, K.C. 1986. Range extension for some Bolivian birds, 2 (Columbidae to Rhinocryptidae). *Bull. Brit. Orn. Club* 106: 22-32.

Renaudier, A. & Comité d'Homologation de Guyane 2010. Rare birds in French Guiana in 2005-07. *Cotinga* 32: 75-83.

Rensch, B. 1931. Die Vogelwelt von Lombok, Sumbawa und Flores. *Mitt. Zool. Mus. Berlin* 17: 451-637.

Restall, R.L., Rodner, C. & Lentino, M. 2006. *Birds of northern South America: an identification guide.* 2 vols. Christopher Helm, London.

Reynaud, P.A. 1998. Changes in understory avifauna along the Sinnamary River (French Guyana, South America). *Orn. Neotrop.* 9: 51-69.

Rheindt, F.E. 2010. New biogeographic records for the avifauna of Taliabu (Sula Islands, Indonesia), with preliminary documentation of two previously undiscovered taxa. *Bull. Brit. Orn. Club* 130: 33-51.

Rheindt, F.E. & Hutchinson, R.O. 2007. A photoshot odyssey through the confused avian taxonomy of Seram and Buru (southern Moluccas). *BirdingASIA* 7: 18-38.

Richardson, C. & Chapman, J.A.D. 1988. Migration patterns through Dubai 1984-1988. *Sandgrouse* 10: 71-80.

Richardson, D. 1983. Didric Cuckoo: possible parasitising of Malachite Sunbird. *Promerops* 157: 12.

Richmond, C.W. & Swales, B.H. 1924. Descriptions of three new birds from Gonave Island, Haiti. *Proc. Biol. Soc. Wash.* 37: 105-107.

Richter, E. 1999. Umfrage zur Verbreitung des Kuckucks im Landkreis Waldeck-Frankenberg im Jahr 1998. *Vogelkdl. Hefte Edertal* 25: 66-69.

Ridgely, R.S. & Greenfield, P.J. 2001. *The birds of Ecuador*. Cornell University Press, Ithaca, NY.

Ridgely, R.S. & Gwynne, J.A. 1989. *A guide to the birds of Panama, with Costa Rica, Nicaragua and Honduras*, 2nd ed. Princeton University Press, Princeton.

Ridgely, R.S., Agro, D.J. & Joseph, L. 2005. Birds of Iwokrama Forest. *Proc. Acad. Nat. Sci. Philad.* 154: 109-121.

Ridgway, R. 1885. Description of an apparently new species of *Dromococcyx* from British Guiana. *Proc. US National Mus.* 8: 559.

Ridgway, R. 1887. *A manual of North American birds*. J.B. Lippincott, Philadelphia.

Ridgway, R. 1894. Descriptions of some new birds from Aldabra, Assumption, and Gloriosa Islands, collected by Dr. W. L. Abbott. *Proc. U. S. Natl. Mus.* 17: 371-373.

Ridgway, R. 1895. On birds collected by Doctor W.L. Abbott in the Seychelles, Amirantes, Gloriosa, Assumption, Aldabra, and adjacent islands, with notes on habits etc., by the collector. *Proc. US National Mus.* 18: 509-546.

Ridgway, R. 1912. Diagnoses of some new genera of American birds. *Proc. Biol. Soc. Wash.* 25: 97-101.

Ridgway, R. 1915. Descriptions of some new forms of American cuckoos, parrots and pigeons. *Proc. Biol. Soc. Wash.* 28: 105-108.

Ridgway, R. 1916. The birds of North and Middle America. Part 7. *Bull. US National Mus.* 50: 1-543.

Ridley, A.R. & Thompson, A.M. 2012. The effect of Jacobin Cuckoo *Clamator jacobinus* parasitism on the body mass and survival of young in a new host species. *Ibis* 154: 195-199.

Ridley, M.W., Percy, H.E. & Percy, R.C. 1953. The birds of Bwamba. Further additions. *Uganda Jour.* 17: 161-165.

Riehl, C. 2010. Egg ejection risk and hatching asynchrony predict egg mass in a communally breeding cuckoo, the Greater Ani (*Crotophaga major*). *Behav. Ecol.* 21: 676-683.

Riehl, C. & Jara, L. 2009. Natural history and reproductive biology of the communally breeding Greater Ani (*Crotophaga major*) at Gatún Lake, Panama. *Wilson J. Orn.* 121: 679-687.

Riker, C.B. 1891. A list of birds observed at Santarem, Brazil, with annotations by Frank M. Chapman. *Auk* 8: 24-31, 158-164.

Riley, J. & Mole, J. 2001. The birds of Gunung Ambang Nature Reserve, North Sulawesi, Indonesia. *Forktail* 17: 57-66.

Riley, J., Hunowu, Y., Mole, J. & Wangko, M.F. 2003. Noteworthy records of birds from the Panua Nature Reserve, North Sulawesi. *Kukila* 12: 17-26.

Riley, J.H. 1924. A collection of birds from north and north-central Celebes. *Proc. US National Mus.* 64 (16): 1-110.

Riley, J.H. 1938a. Birds from Siam and the Malay Peninsula in the United States National Museum collected by Drs. Hugh M. Smith & William L. Abbott. *Bull. US National Mus.* 172: i-iv, 1-581.

Riley, J.H. 1938b. Three new birds from Bangka and Borneo. *Proc. Biol. Soc. Wash.* 51: 95-96.

Ripley, S.D. 1942. Notes on Malaysian Cuckoos. *Auk* 59: 595-596.

Ripley, S.D. 1944. The bird fauna of the West Sumatran Islands. *Bull. Mus. Comp. Zool.* 94: 307-430.

Ripley, S.D. 1946. The Koels of the Bay of Bengal. *Auk* 63: 240-241.

Ripley, S.D. 1950. Birds from Nepal 1947-1949. *J. Bombay Nat. Hist. Soc.* 49: 354-417.

Ripley, S.D. 1951. Migrants and introduced species in the Palau archipelago. *Condor* 53: 299-300.

Ripley, S.D. 1964. A systematic and ecological study of birds of New Guinea. *Bull. Peabody Mus. Nat. Hist. Yale Univ.* 19: i-v, 1-87.

Ripley, S.D. 1982. *A synopsis of the birds of India and Pakistan*. (2nd ed.). Bombay Natural History Society, Bombay.

Ripley, S.D. & Beehler, B.M. 1989. Ornithogeographic affinities of the Andaman and Nicobar Islands. *J. Biogeogr.* 16: 323-332.

Ripley, S.D. & Bond, G.M. 1966. The birds of Socotra and Abd-el-Kuri. *Smithson. Misc. Coll.* 151: 1-37.
Ripley, S.D. & Heinrich, G.H. 1966a. Additions to the avifauna of northern Angola. 2. *Postilla* 95: 1-29.
Ripley, S.D. & Heinrich, G.H. 1966b. Comments on the avifauna of Tanzania. 1. *Postilla* 96: 1-45.
Ripley, S.D. & Rabor, D.S. 1958. Notes on a collection of birds from Mindoro Island, Philippines. *Bull. Peabody Mus. Nat. Hist.* 13: 1-83.
Rix, C.E. 1976. The birds of Sandy Creek Conservation Park, Part 1. *Aust. Bird Watcher* 6: 255-288.
Robbins, M.B., Dittmann, D.L., Dunn, J.L., Garrett, K.L., Heinl, S., Kratter, A.W., Lasley, G.W. & Mactavish, B. 2003. ABA Checklist Committee 2002 annual report. *Birding* 35: 138-144.
Robbins, M.B., Parker, T.A. III & Allen, S.E. 1985. The avifauna of Cerro Pirre, Darién, eastern Panamá. *Orn. Monogr.* 36: 198-232.
Robbins, M.B., Peterson, A.T., Nyari, A., Chen, G. & Davis, T.J. 2006. Ornithological surveys of two reserves in Guangxi province, China, 2004-2005. *Forktail* 22: 140-146.
Roberts, A. 1922. Review of the nomenclature of South African birds. *Ann. Transvaal Mus.* 8: 187-272.
Roberts, A. 1928. Birds and mammals from South-west Africa collected by Mr. R.D. Bradfield and presented to the Transvaal Museum, with taxonomic notes and descriptions of new forms. *Ann. Transvaal Mus.* 12: 289-329.
Roberts, A. 1939. Notes on the eggs of parasitic birds in South Africa. *Ostrich* 10: 1-20, 100-117.
Roberts, T.J. 1991. *The birds of Pakistan.* Vol. 1. Non-Passeriformes. OUP, Oxford.
Roberts, T.S. 1936. *The birds of Minnesota,* 2nd ed. Vol. 1. University of Minnesota Press, Minneapolis.
Robertson, C.J.R. (Ed.) 1985. *Reader´s Digest complete book of New Zealand Birds.* Reader´s Digest, Sydney.
Robertson, W.B., Jnr. 1978. Rare populations, Mangrove Cuckoo, *Coccyzus minor.* Pp. 57-58 in H.W. Kale II (Ed.). *Rare and endangered biota of Florida.* Vol. 2: Birds. University Presses of Florida, Gainesville.
Robiller, F., Michi, H. & Michi, M. 1992. Über den Renauldkuckuck (*Carpococcyx renauldi* Oustalet, 1896) und seine Zucht in der Forschungsstation Ornis Mallorca. *Tropische Vögel* 13 (3): 92-97.
Robinson, A. 1950. Observations on courtship feeding in some Australian birds. *W. Aust. Nat.* 2: 106-107.
Robinson, A.H. 1955. Nesting seasons of Western Australian birds - A further contribution. *W. Aust. Nat.* 4: 187-192.
Robinson, H.C. 1927. *The birds of the Malay Peninsula. Vol. 1: The commoner birds.* H.F. & G. Witherby, London.
Robinson, H.C. 1928. *The birds of the Malay Peninsula. Vol. 2: The birds of the Hill Stations.* H.F. & G. Witherby, London.
Robinson, H.C. & Kloss, C.B. 1910-1911. On birds from the northern portion of the Malay Peninsula, including the islands of Langkawi and Terutau; with notes on other rare Malayan species from the southern districts. *Ibis* (9th series) 4: 659-675; 5: 10-80.
Robinson, H.C. & Kloss, C.B. 1921. Nine new Oriental birds. *J. Fed. Malay States Mus.* 10: 203-206.
Robinson, H.C. & Kloss, C.B. 1923. The birds of south-west and Peninsular Siam. *J. Nat. Hist. Soc. Siam* 5: 89-218.
Robinson, H.C. & Kloss, C.B. 1924. On a large collection of birds chiefly from West Sumatra made by Mr. E. Jacobson. *J. Fed. Malay States Mus.* 11: 189-347.
Robinson, S.K. 1997. Birds of a Peruvian oxbow lake: populations, resources, predation, and social behavior. *Orn. Monogr.* 48: 613-639.
Robinson, S.K., Fitzpatrick, J.W. & Terborgh, J.W. 1995. Distribution and habitat use of Neotropical migrant landbirds in the Amazon basin and Andes. *Bird Conserv. Int.* 5: 305-323.
Robson, C. 1989. Recent Reports. *Bull. OBC* 9: 38-44.
Robson, C. 2000. *A guide to the birds of Southeast Asia.* Princeton University Press, Princeton.
Robson, C. 2007. From the field. *BirdingASIA* 7: 92-96.
Robson, C. 2008. From the field. *BirdingASIA* 10: 108-111.
Robson, C. 2010. From the field. *BirdingASIA* 13: 109-112.
Robson, C. 2011. From the field. *BirdingASIA* 15: 108-112.
Robson, C.R. 1990. Recent reports. Viet Nam. *Bull. OBC* 12: 43-44.
Robson, C.R., Buck, H., Farrow, D.S., Fisher, T. & King, B.F. 1998. A birdwatching visit to the Chin Hills, West Burma (Myanmar), with notes from nearby areas. *Forktail* 13: 109-120.
Robson, C.R., Eames, J.C., Cu, N. & Van La, T. 1993. Birds recorded during the third BirdLife/Forest Bird Working Group expedition in Viet Nam. *Forktail* 9: 89-119.
Rocamora, G. & Skerrett, A. 2001. Seychelles. In Fishpool, L.D.C. & Evans, M.I. (Eds) *Important Bird Areas in Africa and related islands: priority sites for conservation.* Cambridge, UK: BirdLife International.
Roché, J.C. 1971. *Birds of Madagascar* (1 disk). J.C. Roché, Aubenas-les-Alpes, France.
Rodner, C., Lentino R., M. & Restall, R.L. 2000. *Checklist of the birds of northern South America.* Yale University Press, New Haven & London.
Rodwell, S.P. 1996. Notes on the distribution and abundance of birds observed in Guinea-Bissau, 21 February to 3 April 1992. *Malimbus* 18: 25-43.
Rodwell, S.P., Sauvage, A., Rumsey, S.J.R. & Bräunlich, A. 1996. An annotated check-list of birds occurring at the Parc National des Oiseaux du Djoudj in Senegal, 1984-1994. *Malimbus* 18: 74-111.
Rojas-Soto, O.R., Sagahún-Sánchez, F.J. & Navarro-Sigüenza, A.G. 2001. Additional information on the avifauna of Querétaro, Mexico. *Cotinga* 15: 48-52.
Romè, A. & Tomei, P.E. 1977. Il cuculo dal ciuffo - *Clamator glandarius* (L.) in Toscana. *Atti Soc. Toscana Scienze Nat. Processi Verbali Memorie, Ser. B* 84: 151-160.
Root, T. 1988. *Atlas of wintering North American birds: an analysis of Christmas Bird Count data.* University of Chicago Press, Chicago, Illinois.
Rosenberg, D.K., Wilson, M.H. & Cruz, F. 1990. The distribution and abundance of the Smooth-billed Ani *Crotophaga ani* (L.) in the Galapagos Islands, Ecuador. *Biol. Conserv.* 51: 113-123.
Rosillo, M.A. 1943. Apuntes sistematico-ecologicos sobre Cuculiformes Argentinos. *Mem. Mus. Entre Ríos Paraná* 20: 1-16.
Rossouw, J. & Lindsell, J. 2001. Black-throated Coucal *Centropus leucogaster,* a first record for East Africa. *Scopus* 22: 63-65.
Rossouw, J. & Sacchi, M. 1998 *Where to watch birds in Uganda.* Uganda Tourist Board, Kampala.
Roth, P. 1981. A nest of the Rufous-vented Ground-Cuckoo (*Neomorphus geoffroyi*). *Condor* 83: 388.
Rothschild, W. 1904. [A new gigantic species from the Solomon Islands.] *Bull. Brit. Orn. Club* 14: 59.
Rothschild, W. 1926. On the avifauna of Yunnan, with critical notes. *Novit. Zool.* 33: 190-343, 395-400.
Rothschild, W. & Hartert, E. 1896. Contributions to the ornithology of the Papuan Islands. *Nov. Zool.* 3: 233-251.
Rothschild, W. & Hartert, E. 1901. List of a collection of birds from Kulambangra and Florida Islands, in the Solomons group. *Nov. Zool.* 8: 179-189.
Rothschild, W. & Hartert, E. 1902. List of a collection by A.S. Meek on Ysabel Island, Solomon Islands. *Nov. Zool.* 9: 581-594.
Rothschild, W. & Hartert, E. 1907. Notes on Papuan birds. *Nov. Zool.* 14: 433-446.
Rothschild, W. & Hartert, E. 1914. The birds of the Admiralty Islands, north of German New Guinea. *Nov. Zool.* 21: 281-298.
Rothschild, W. & Hartert, E. 1914a. On the birds of Rook Islands north of German New Guinea. *Nov. Zool.* 21: 207-218.
Rothschild, W. & Hartert, E. 1915. The birds of Vulcan Island. *Nov. Zool.* 22: 38-45.
Rothschild, W., Stresemann, E. & Paludan, K. 1932. Ornithologische Ergebnisse der Expedition Stein, 1931-1932. I. Die Vögel von Waigeu. II. Die Vögel von Numfor. III. Die Vögel von Japen. *Nov. Zool.* 38: 127-247.

Round, P. 2008. *The birds of the Bangkok area*. White Lotus, Bangkok.

Round, P.D. 1988. *Resident forest birds in Thailand: their status and conservation*. International Council for Bird Preservation. Monograph 2. Cambridge, UK.

Round, P.D. 1995. Recent reports. *Bird Conserv. Soc. Thailand Bull.* 12 (3): 14.

Rounsevell, D.E., Brothers, N.P., Blackhall, S.A., Riley, C. & Wilson, R.I. 1979. Some new or unusual records of birds for Tasmania during 1977. *Emu* 79: 139-140.

Rowan, M.K. 1983. *The doves, parrots, louries and cuckoos of southern Africa*. Croom Helm, London.

Rowley, I. & Russell, E. 1989. Splendid Fairy-wren. Pp. 233-252 in I. Newton (Ed.). *Lifetime reproduction in birds*. Academic Press, London.

Rowley, J.S. 1966. Breeding records of birds of the Sierra Madre del Sur, Oaxaca, Mexico. *Proc. West. Found. Vertebr. Zool.* 1: 107-204.

Rowley, J.S. 1984. Breeding records of land birds in Oaxaca, Mexico. *Proc. West. Found. Vertebr. Zool.* 2: 73-224.

Royston, A. 1981. Notes on two African species. *Avicult. Mag.* 87 (2): 111-112.

Rozendaal, F.G. & Dekker, R.W.R.J. 1989. Annotated checklist of the birds of the Dumoga-Bone National Park, North Sulawesi. *Kukila* 4: 85-109.

Ruggieri, L. & Festari, I. 2005. *A birdwatcher's guide to Italy*. Lynx Edicions, Barcelona.

Ruiz-Guerra, C., Johnston-Gonzáles, R., Cifuentes-Sarmiento, Y., Estala, F.A., Castillo, L.F. Hernández, C.E. & Naranjo, L.G. 2007. Noteworthy bird records from the southern Chocó of Colombia. *Bull. Brit. Orn. Club* 127: 283-293.

Rüppell, E.W.P.S. 1837. *Neue Wirbelthiere zu der Fauna von Abyssinien gehörig, entdeckt und beschrieben von Dr. Eduard Rüppell. Vögel*. Siegmund Schmerber, Frankfurt am Main. p. 37.

Ruschi, A. 1975. Aves do estado do Espírito Santo. Algumas observações sobre: *Neomorphus geoffroyi dulcis* Snethlage, 1927. *Bol. Mus. Biol. Prof. Mello Leitão, Sér. Zool.* 77: 1-3.

Ruschi, A. 1979. *Aves do Brasil*, vol. 1. Editora Rios, São Paulo.

Russell, S.M. 1964. A distributional study of the birds of British Honduras. *Orn. Monogr.* 1: 1-195.

Russell, S.M. 1980. Distribution and abundance of North American migrants in lowlands of northern Colombia. Pp. 249-252 in A. Keast & E.S. Morton (Eds). *Migrant birds in the neotropics: ecology, behavior, distribution and conservation*. Smithsonian Institution Press, Washington DC.

Russell, S.M., Barlow, J.C. & Lamm, D.W. 1979. Status of some birds on Isla San Andres and Isla Providencia, Colombia. *Condor* 81: 98-100.

Russell, S.M. & Monson, G. 1998. *The birds of Sonora*. University of Arizona Press, Tucson.

Ryan, J.V. 1980. A study of the Black-eared Cuckoo. *Bird Observer* 90.

Rylander, M.K. 1972. Winter dormitory of the Roadrunner, *Geococcyx californicus*, in west Texas. *Auk* 89: 896.

Ryves, V.W. 1959. An Emerald Cuckoo's foster parents. *J. E. Afr. Nat. Hist. Soc.* 23: 175.

Sagar, P.M. 1977. Birds of the 1976-77 Snares Islands Expedition. *Notornis* 24: 205-210.

Saino, N., Rubolini, D., Lehikoinen, E., Sokolov, L.V., Bonisoli-Alquati, A., Ambrosini, R., Boncoraglio, G. & Møller, A.P. 2009. Climate change effects on migration phenology may mismatch brood parasitic cuckoos and their hosts. *Biol. Letters* 5: 539-541.

Salaman, P.G.W., Cuadros, T., Jaramillo, J.G. & Weber, W.H. 200. *Checklist to the birds of Colombia*. Sociedad de Ornitología Antioqueña, Medellín.

Salewski, V. 2000. The birds of Comoé National Park, Ivory Coast. *Malimbus* 22: 55-76.

Salgado, A. 2006. Some observations on the diet of Red-faced Malkoha *Phaenicophaeus pyrrhocephalus* in Sri Lanka. *Forktail* 22: 122-123.

Salomonsen, F. 1953. Miscellaneous notes on Philippine birds. *Vidensk. Medd. Dansk Naturh. Foren.* 115: 205-281.

Salter, B.E. 1978. A note on the Channel-billed Cuckoo. *Bull. Fisheries Res. Devel. Agency Pusan* 559: 24.

Salvador, S.A. 1981. Desarrollo de una nidada comunal de Pirincho: *Guira guira* (Gmelin) (Aves: Cuculidae). *Hist. Nat.* 2: 29-31.

Salvador, S.A. 1982. Estudio de parasitismo del Crespín *Tapera naevia chochi* (Vieillot) (Aves: Cuculidae). *Hist. Nat.* 2: 65-70.

Salvadori, T. 1874. Catalogo sistematico degli uccelli di Borneo. *Annal. Mus. Civ. Storia Nat. Genova* 5: i-lii, 1-431.

Salvadori, T. 1874. Altre nuove specie di uccelli della Nuova Guinea e di Goram, raccolte dal Signor L. M. D'Albertis e descritte da Tommaso Salvadori. *Annal. Mus. Civ. Storia Nat. Genova* 6: 81-88.

Salvadori, T. 1878. Descrizione di trentuna specie nuove di ucceli della sottoregione papuana, e note intorno ad altre poco conosciute. *Annal. Mus. Civ. Storia Nat. Genova* 12: 317-347.

Salvadori, T. 1876. Descrizione di sei nuove specie di uccelli delle Molucche, delle Kei e delle Aru e del maschio della *Pachycephala lineolata*, Wall. *Annal. Mus. Civ. Storia Nat. Genova* 7: 983-990.

Salvadori, T. 1876. Descrizione di cinquantotto nuove specie di uccelli ed osservazioni intorno ad altre poco note, della Nuova Guinea e di altre Isole Papuane, raccolte dal Dr. Odoardo Beccari e dai cacciatori del Sig. A. A. Bruijn. *Annal. Mus. Civ. Storia Nat. Genova* 7: 896-976.

Salvadori, T. 1876. Catalogo di una seconda collezione di uccelli raccolti dal Sig. L. M. D'Albertis nell'Isola Yule e sulla vicina costa della Nuova Guinea e di una piccola collezione della regione bagnata dal Fiume Fly. *Annal. Mus. Civ. Storia Nat. Genova* 9: 7-49.

Salvadori, T. 1878. Prodromus Ornithologiae Papuasiae et Moluccarum. VI. Picariae. Fam. Cuculidae. Annal. Mus. Civ. Storia Nat. Genova 13: 456-463.

Salvadori, T. 1879. Catalogo di una collezione di uccelli fatta nella parte occidentale di Sumatra dal Prof. Odoardo Beccari e descritta da Tommaso Salvadori. *Annal. Mus. Civ. Storia Nat. Genova* 14: 169-253.

Salvadori, T. 1880. *Ornitologia della Papuasia e delle Molucche* Vol. 1. G. B. Paravia, Torino (Turin).

Salvadori, T. 1889. Aggiunte alla ornitologia della Papuasia e delle Molucche. Parte prima. Accipitres - Psittaci - Picariae. *Mem. R. Accad. Sci. Torino*, 2nd series, 40: 131-293.

Salvadori, T. 1897. Viaggio del Dott. Alfredo Borelli nel Chaco boliviano e nella Repubblica Argentina. VII. Uccelli. *Boll. Mus. Zool. Anat. Comp. R. Univ. Torino* 12: 1-36.

Salvin, O. & Godman, F.D. 1879-1904. *Biologia Centrali-Americana. Aves*. Taylor & Francis, London.

Sandoval, L., Biamonte, E. & Solano Ugalde, A. 2008. Previously unknown food items in the diet of six Neotropical bird species. *Wilson J. Orn.* 120: 214-216.

Santharam, V. 1996. Birds feeding on flowers in India. *Forktail* 12: 157-161.

Sassi, M. 1912. Eine neue Art des Genus *Cercococcyx. Cercococcyx olivinus* nov. spec. *Ann. K. K. Naturhist. Hofmus.* 26: 341-342.

Saunders, G.B. & Clark, E. 1962. Yellow-billed Cuckoo in stomach of Tiger Shark. *Auk* 79: 118.

Sazima, I. 2008. Validated cleaner: the cuculid bird *Crotophaga ani* picks ticks and pecks at sores of capybaras in southeastern Brazil. *Biota Neotrop.* 8: 213-216.

Schäfer, E. & Phelps, W.H. 1954. Las aves del Parque Nacional Henri Pittier (Rancho Grande) y sus funciones ecológicas. *Bol. Soc. Venez. Cienc. Nat.* 16: 3-167.

Schaldach, W.J., Jr. 1963. The avifauna of Colima and adjacent Jalisco, Mexico. *Proc. West. Found. Vertebr. Zool.* 1: 1-100.

Scharringa, J. 1999. *Birds of tropical Asia, sounds and sights* (CD-ROM). Bird Songs International, Westernieland, The Netherlands.

Schellekens, M. 2009. The Timor Leaf-Warbler *Phylloscopus presbytes* is a probable host of Sunda Cuckoo *Cuculus lepidus*. *Kukila* 14: 49-51.

Scherer-Neto, P., Carrano, E. & Ribas, C.F. 2001. Avifauna da

Estação Ecológica do Caiuá (Diamante do Norte, Paraná) e regiões adjacentes. Res. 192 in Straube, F.C. (Ed.). *Ornitologia sem Fronteiras. Incluindo os Resumos do IX Congresso Brasileiro de Ornitologia*. Soc. Bras. Orn., Curitiba, Paraná.

Scherer-Neto, P. & Straube, F.C. 1995. *Aves do Paraná: história, lista anotada e bibliografia*. Privately published, Curitiba.

Schiebel, G. 1910. Neue Vogelformen aus Corsica. *Orn. Jahrb.* 21: 102-103.

Schinz, H. R. 1821. *Das Thierreich eingetheilt nach dem Bau der Thiere als Grundlage ihrer Naturgeschichte und der vergleichenden Anatomie von dem Herrn Ritter von Cuvier. Aus dem Französischen frey übersetzt und mit vielen Zusätzen versehen. Vol. I. Säugethiere und Vögel.* J.G. Cotta, Stuttgart and Tübingen.

Schlegel, H. 1864. *Muséum d'Histoire naturelle des Pays-Bas. Revue méthodique et critique des collections déposées dans cet établissement*. Vol. 1. Leiden, E. J. Brill.

Schlegel, H. 1866. Observations zoologiques II. *Ned. Tijdschr. Dierk.* 3: 249-258.

Schlegel, H. 1879. On a new species of Cuckoo from Madagascar. *Notes R. Zool. Mus. Neth. Leyden*. 1: 99-100.

Schmaltz, G., Quinn, J.S. & Lentz, C. 2008. Competition and waste in the communally breeding Smooth-billed Ani: effects of group size on egg-laying behaviour. *Anim. Behav.* 76: 153-162.

Schmitt, C.G., Schmitt, D.C. & Remsen, J.V. Jnr. 1997. Birds of the Tambo area, an arid valley in the Bolivian Andes. *Orn. Monogr.* 48: 701-716.

Schneider, J.J. 1900. Nesting of the California Cuckoo in Los Angeles Co., Cal. *Condor* 2: 34.

Schodde, R., Fullagar, P. & Hermes, N. 1983. *A review of Norfolk Island birds: past and present*. Australian National Parks and Wildlife Service Special Publication 8.

Schodde, R. & Mason, I.J. 1997. *Zoological Catalogue of Australia: Aves. Columbidae to Coraciidae*. CSIRO, Collingwood.

Schodde, R. & Tidemann, S.C. (Eds) 1976. *Complete book of Australian birds*. Reader´s Digest, Sydney.

Schönwetter, M. 1928. Anmerkungen zum Nehrkorn-Katalog. *Beitr. Fortpflanz. Vögel Berücks. Ool.* 4: 49-53, 89-98, 129-136, 181-192, 200-204.

Schönwetter, M. 1967. *Handbuch der Oologie*. Band 1: Nonpasseres. Akademic-Verlag, Berlin.

Schouteden, H. 1968. La faune Ornithologique du Kivu. 1. Non Passereaux. *Doc. Zool. Mus. Roy. Afr. Centr.* 12: 1-168.

Schouteden, H. 1971. La faune ornithologique de la Province du Katanga. *Doc. Zool. Mus. Roy. Afr. Centr.* 17: 1-248.

Schubart, O., Aguirre, Á.C. & Sick, H. 1965. Contribuição para o conhecimento da alimentação das aves brasileiras. *Arq. Zool. São Paulo* 12: 95-249.

Schulenberg, T.S. & Parker, T.A. III. 1981. Status and distribution of some northwest Peruvian birds. *Condor* 83: 209-216.

Schulenberg, T.S., Stotz, D.F., Lane, D.F., O'Neill, J.P. & Parker, T.A. III. 2007. *Field guide to the birds of Peru*. Christopher Helm, London.

Schulze-Hagen, K. 1992. Parasitierung und Brutverluste durch den Kuckuck (*Cuculus canorus*) bei Teich- und Sumpfrohrsänger (*Acrocephalus scirpaceus, A. palustris*) in Mittel- und Westeuropa. *J. Orn.* 133: 237-249.

Schulze-Hagen, K., Stokke, B.G. & Birkhead, T.R. 2009. Reproductive biology of the European Cuckoo *Cuculus canorus*: early insights, persistent error and the acquisition of knowledge. *J. Orn.* 150: 1-16.

Schuster, L. 1926. Beiträge zu Verbreitung und Biologie der Vögel Deutsch-Ostafrikas. *J. Orn.* 133: 521-541.

Schwartz, A. & Klinikowski, R.F. 1965. Additional observations on West Indian birds. *Notulae Naturae* 376: 1-16.

Sclater, P.L. 1860. On a series of birds collected in the vicinity of Jalapa, in southern Mexico. *Proc. Zool. Soc. London* 27: 362-369.

Sclater, P.L. 1862. *Catalogue of a collection of American birds belonging to Philip Lutley Sclater*. Trübner, London.

Sclater, P.L. 1864. On the species of the American genus *Coccyzus*. *Proc. Zool. Soc. London* 1864: 119-122.

Sclater, P.L. 1866. On a new American cuckoo of the genus *Neomorphus*. *Proc. Zool. Soc. London* 1866: 59-60, pl. 5.

Sclater, P.L. & Hudson, W.H. 1889. *Argentine ornithology: a descriptive catalogue of the birds of the Argentine Republic*, vol. 2. R.H. Porter, London.

Sclater, P.L. & Salvin, O. 1878. Descriptions of three new species of birds from Ecuador. *Proc. Zool. Soc. London* 1878: 438-440.

Sclater, W.L. 1924. *Systema Avium Aethiopicarum*. Pt. 1. BOU, London.

Sclater, W.L. & Mackworth-Praed, C. 1919. A list of the birds of the Anglo-Egyptian Sudan, based on the collections of Mr. A L. Butler, Mr. A. Chapman and Capt. H. Lynes, R.N., and Major Cuthbert Christy, RAMC (T.F.). Part III. *Ibis* (11th series) 1: 628-707.

Sclater, W.L. & Moreau, R.E. 1932. Taxonomic and field notes on some birds of north-eastern Tanganyika Territory. Part 1. *Ibis* (13th series) 2: 487-522.

Scopoli, G.A. 1786. *Deliciae Florae et Fauna Insubricae*. Vol. 2. Ticini, Pavia.

Scott, D.A. & Brooke, M. de L. 1985. The endangered avifauna of southeastern Brazil: A report on the BOU/WWF expeditions of 1980/81 and 1981/82. Pp. 115-139 in A.W. Diamond & T.E. Lovejoy (Eds). *Conservation of tropical forest birds. Proceedings of ICBP Tropical Forest Birds Symposium, Cambridge, 1982*. ICBP Technical Publication, 4. ICBP, Cambridge, UK.

Scott, S.L., Swinson, L.M., Dickinson, M.B. & Howell, C.H. (Eds). 1983. *Field guide to the birds of North America*. 1st ed. National Geographic Society, Washington DC.

Scott, W.E. 1946. Birds observed on Espiritu Santo, New Hebrides. *Auk* 63: 362-368.

Scott, W.E.D. 1892. Observations on the birds of Jamaica, West Indies. II. A list of the birds recorded from the island, with annotations. *Auk* 9: 9-15, 120-129, 273-277, 369-375.

Sealy, S.G. 1978. Possible influence of food on egg-laying and clutch size in the Black-billed Cuckoo. *Condor* 80: 103-104.

Sealy, S.G. 1985. Erect posture of the young Black-billed Cuckoo: an adaptation for early mobility in a nomadic species. *Auk* 102: 889-892.

Sealy, S.G. 2003. Laying times and a case of conspecific nest parasitism in the Black-billed Cuckoo. *J. Field Orn.* 74: 257-260.

Sealy, S.G. & Lorenzana, J.C. 1997. Feeding of nestling and fledgling brood parasites by individuals other than the foster parents: a review. *Can. J. Zool.* 75: 1739-1752.

Sealy, S., O'Halloran, J. & Smiddy, P. 1996. Cuckoo hosts in Ireland. *Irish Birds* 5: 381-390.

Seaman, G.A. 1957. New bird records for Barbuda, British West Indies. *Wilson Bull.* 69: 109-110.

Seaton, C. 1962. The Yellow-breasted Sunbird as a host to the Rufous-breasted Bronze-Cuckoo. *Emu* 62: 174-176.

Sedgwick, E.H. 1951. Congregating of Bronze Cuckoos. *Notornis* 4: 48.

Sedgwick, E.H. 1955. Display by Bronze-Cuckoos. *Emu* 55: 254.

Seel, D.C. 1977a. Trapping season and body size in the Cuckoo. *Bird Study* 24: 114-118.

Seel, D.C. 1977b. Migration of the northwestern European population of the Cuckoo *Cuculus canorus*, as shown by ringing. *Ibis* 119: 309-322.

Seel, D.C. 1984a. Geographical distribution of the Cuckoo *Cuculus canorus* in the western Palaearctic and Afro-tropical regions. *Ann. Koninkl. Mus. Midden-Afrika Zool. Wetensch.* 239: 1-44.

Seel, D.C. 1984b. Moult and body weight in the Cuckoo *Cuculus canorus* in the western Palaearctic and Afro-tropical regions. *Ann. Koninkl. Mus. Midden-Afrika Zool. Wetensch.* 239: 49-91.

Seel, D.C. & Davis, P.R.K. 1988. Cuckoos reared by unusual host in Britain. *Bird Study* 28: 242-243.

Selander, R.K. & Giller, D.R. 1959. The avifauna of the Barranca de Oblatos, Jalisco, Mexico. *Condor* 61: 210-222.

Semple, J.B. 1937. Smooth-billed Ani in Florida. *Auk* 54: 391.

Seng, L.K. 2008. Horsfield's Bronze Cuckoo *Chrysococcyx basalis* in Singapore. *BirdingASIA* 10: 30-35.

Seow, G., Cheong, T.G. & Lim, K.S. 2010. Further notes on

the foraging behaviour of Horsfield's Bronze Cuckoo *Chrysococcyx basalis* in Singapore. *BirdingASIA* 13: 11.

Serié, P. 1923a. Un huevo de pirincho *Guira guira* en un nido de *Phytotoma rutila*. *Hornero* 3: 100.

Serié, P. 1923b. Huevos de pirincho en nido de chimango. *Hornero* 3: 189.

Serle, W. 1939. Field observations on some Northern Nigerian birds. *Ibis* (14th series) 3: 654-699.

Serle, W. 1943. Further field observations on Northern Nigerian birds. *Ibis* 85: 264-300, 413-437.

Serle, W. 1950. A contribution to the ornithology of the British Cameroons. Part 1. *Ibis* 92: 343-376.

Serle, W. 1954. A second contribution to the ornithology of the British Cameroons. *Ibis* 96: 47-80.

Serle, W. 1957. A contribution to the ornithology of the eastern region of Nigeria. *Ibis* 99: 371-418.

Serle, W. 1965. A third contribution to the ornithology of the British Cameroons. *Ibis* 107: 60-94.

Serle, W. 1977. The aberrant eggs of *Turdoides plebejus* in Nigeria and their relation to cooperative breeding and to victimisation by *Clamator* cuckoos. *Bull. Brit. Orn. Club* 97: 39-41.

Serra, G., Al Qaim, G., Abdallah, M.S., Kanani, A. & Assaed, A. K. 2005. A long-term bird survey in the central Syrian desert (2000-2004): Part 2 – a provisional annotated checklist. *Sandgrouse* 27: 104-125.

Servat, G. & Pearson, D.L. 1991. Natural history notes and records for seven poorly known bird species from Amazonian Peru. *Bull. Brit. Orn. Club* 111: 92-95.

Serventy, V.N. 1958. Bird notes from the Dumbleyung camp-out. *Emu* 58: 5-20.

Seth-Smith, L.M. 1913. Notes on the birds around Upuma, Uganda. *Ibis* (10th series) 1: 485-508.

Sferco, G.D. & Nores, M. 2003. Lista comentada de las aves de la Reserva Natural Chancaní, Córdoba, Argentina. *Hornero* 18: 21-29.

Sharland, M.S.R. 1943. Stray Feathers: Channel-billed Cuckoo in Tasmania. *Emu* 43: 75.

Sharp, C.S. 1907. The breeding birds of Escondido. *Condor* 9: 84-91.

Sharpe, R.B. 1873. On the Cuculidae of the Ethiopian Region. *Proc. Zool. Soc. Lond.* 1873: 578-624.

Sharpe, R.B. 1874. Descriptions of two new species of birds from Gaboon. *Proc. Zool. Soc. London* 1874: 204-205.

Sharpe, R.B. 1877. On the birds collected by Professor J.B. Steere in the Philippine Archipelago. *Trans. Linn. Soc. Zool.* 1: 307-355, 9 plates.

Sharpe, R.B. 1878. Contributions to the ornithology of Borneo. Part III. *Ibis* (4th series) 2: 414-419.

Sharpe, R.B. 1900. On a collection of birds made by Captain A. M. Farquhar, R. N., in the New Hebrides. *Ibis* (7th series) 6: 337-351.

Sharpe, R.B. 1904. On further collections of birds from the Efulen district of Cameroon, West Africa. *Ibis* (8th series) 4: 591-638.

Sharpe, R.B & Whitehead, J. 1890. On the ornithology of Northern Borneo. *Ibis* (6th series) 2: 1-24, 133-149, 273-292.

Shaw, G. 1792. *Museum Leverianum, containing select specimens from the museum of the late Sir Ashton Lever, Kt., with descriptions in Latin and English*. James Parkinson, London.

Shaw, G. 1810. *The naturalists' miscellany: or Coloured figures of natural objects; drawn and described immediately from nature* Vol. 21. Nodder, London, UK. '

Shaw, T.H. 1938a. An addition to the avifauna of Tsingtao. The body weight of the Indian Cuckoo. *China J.* 29: 208-209.

Shaw, T.H. 1938b. Nordchinesische Blauelstern als Pflegeeltern von *Cuculus micropterus*. *Orn. Monatsb.* 46: 154-155.

Sheehan, D. 1999. Winter calling by Klaas's Cuckoo. *Honeyguide* 45: 136-137.

Sheldon, F.H., Moyle, R.G. & Kennard, J. 2001. *Ornithology of Sabah: history, gazetteer, annotated checklist, and bibliography*. *Orn. Monogr.* 52: 1-285.

Sheldon, H. 1922. Top speed of the road-runner. *Condor* 24: 180.

Shelley, G.E. 1871. Contribution to the ornithology of Egypt. *Ibis* (3rd series) 1: 38-54, 131-147, 309-319.

Shelley, G.E. 1881. List of birds recently collected by Dr. Kirk in Eastern Africa. *Proc. Zool. Soc. Lond.* 1881: 561-602.

Shelley, G.E. 1891. Cuculidae. Pp. 209-434 in W.L. Slater & G.E. Shelley (Eds). *Catalogue of the birds in the British Museum. Catalogue of the Picariæ in the collection of the British Museum. Indicatoridæ, Capitonidæ, Cuculidæ, and Musophagidæ*. Vol. 19. British Museum (Natural History), London.

Shepherd, C.R. 2006. Special report. The bird trade in Medan, north Sumatra: an overview. *BirdingASIA* 5: 16-24.

Sherbrooke, W.C. 1990. Predatory behavior of captive Greater Roadrunners feeding on horned lizards. *Wilson Bull.* 102: 171-174.

Shirihai, H. 1996. *The birds of Israel*. Academic Press and Unipress, London and San Diego.

Short, L.L. 1972. Two avian species new to Paraguay. *Auk* 89: 895.

Short, L.L. 1974. Nesting of southern Sonoran birds during the summer rainy season. *Condor* 76: 21-32.

Short, L.L. 1975. A zoogeographical analysis of the South American Chaco avifauna. *Bull. Amer. Mus. Nat. Hist.* 154: 165-352.

Short, L.L. 1976. Notes on a collection of birds from the Paraguayan Chaco. *Amer. Mus. Nov.* 2597: 1-16.

Sibley, C.G. 1951. Notes on the birds of New Georgia, central Solomon Islands. *Condor* 53: 81-92.

Sibley, C.G. 1996. *Birds of the world 2.0* (computer diskette). Thayer, Cincinnati.

Sibley, C. G. & Ahlquist, J. E. 1990. *Phylogeny and classification of birds: A study in molecular evolution*. Yale University Press, New Haven & London

Sibley, C.G. & Monroe, B.L., Jnr. 1990. *Distribution and taxonomy of birds of the world*. Yale University Press, New Haven & London.

Sibley, D.A. 2000. *The Sibley guide to birds*. Alfred A. Knopf, New York.

Sibree, J., Jnr. 1891. On the birds of Madagascar, and their connection with native folk-lore, proverbs and superstitions. *Ibis* 33:194-228.

Sick, H. 1949. Beobachtungen an dem brasilianischen Bodenkuckuck *Neomorphus geoffroyi dulcis* Snethlage. Pp. 229-239 in E. Mayr & E. Schüz (Eds). *Ornithologie als biologische Wissenschaft. 28 Beiträge als Festschrift zum 60. Geburtstag von Erwin Stresemann*. Carl Winter Universitätsverlag, Heidelberg, Germany.

Sick, H. 1953a. Anotações sobre cucos brasileiros. *Rev. Brasil. Biol.* 13: 145-168.

Sick, H. 1953b. Zur Kenntnis der brasilianischen Lerchenkuckucke *Tapera* und *Dromococcyx*. *Bonn. Zool. Beitr.* 4: 305-326.

Sick, H. 1962. Escravismo em aves brasileiras. *Arq. Mus. Nac. Rio de Janeiro* 52: 185-192.

Sick, H. 1969. Aves brasileiras ameaçadas de extinção e noções gerais de conservação de aves do Brasil. *An. Acad. Brasil. Ci.* 41 (suplemento): 205-229.

Sick, H. 1993. *Birds in Brazil: a natural history*. Princeton University Press, Princeton.

Sick, H. 1997. *Ornitologia brasileira. Edição revista e ampliada*. Editora Nova Fronteira, Rio de Janeiro.

Sick, H. & Pabst, L.F. 1968. As aves do Rio de Janeiro (Guanabara), lista sistemática anotada. *Arq. Mus. Nac. Rio de Janeiro* 53: 99-160.

Sick, H. & Teixeira, D.M. 1979. Notas sobre aves brasileiras raras ou ameaçadas de extinção. *Publ. Avuls. Mus. Nac. Rio de Janeiro* 62: 1-39.

Siddle, C. 1992. The summer season June 1 - July 31, 1992. British Columbia/Yukon Region. *Am. Birds* 46: 1167-1170.

Siegel, C.E., Hamilton, J.M. & Castro, N.R. 1989. Observations of the Red-billed Ground-Cuckoo (*Neomorphus pucheranii*) in association with tamarins (*Saguinas*) in northeastern Amazonian Peru. *Condor* 91: 720-722.

Sieving, K.E. 1990. Pheasant Cuckoo foraging behavior, with notes on habits and possible social organization in Panama. *J. Field Orn.* 61: 41-46.

Silva, W.A.G. & Albano, C.G. 2002. *Lista remissiva da avifauna cearense*. Associação de Observadores de Aves de Pernambuco, Recife.

Silveira, L.F. 2008. *Neomorphus geoffroyi dulcis* Snethlage, 1927. Pp. 486-487 in A.B.M. Machado, G.M. Drummond & A.P. Paglia (Eds). *Livro vermelho da fauna brasileira ameaçada de extinção*. Ministério do Meio Ambiente & Fundação Biodiversitas, Brasília & Belo Horizonte.

Simon, J.E., Antas, P.T.Z., Pacheco, J.F., Efe, M.A., Ribon, R., Raposo, M.A., Laps, R.R., Musso, C., Passamani, J.A. & Paccagnella, S.G. 2007. As aves ameaçadas de extinção no estado do Espírito Santo. Pp. 47-64 in M. Passamani & S.L. Mendes (Eds). *Espécies da fauna ameaçadas de extinção no Estado do Espírito Santo*. Vitória, Ipema, Instituto de Pesquisas da Mata Atlântica.

Sinclair, I. & Langrand, O. 2003. *Birds of the Indian Ocean islands*. Struik, Cape Town.

Singh, P. 1995. Recent bird records from Arunachal Pradesh. *Forktail* 10: 65-104.

Singh-Dhindsa, M. & Toor, H.S. 1981. Some observations on a nest of a Common Crow-pheasant, *Centropus sinensis* (Stephens). *J. Bombay Nat. Hist. Soc.* 78: 600-602.

Sivakumar, K. & Sankaran, R. 2002. New records of birds from the Andaman and Nicobar Islands. *Forktail* 18: 149-150.

Skagen, S. K. 1988. Asynchronous hatching and food limitation: a test of Lack's hypothesis. *Auk* 105: 78-88.

Skead, G.J. 1946. Record of a young Black Cuckoo. *Ostrich* 17: 359.

Skead, G.J. 1951. Cuckoo studies on a South African farm (Part I). *Ostrich* 22: 163-175.

Skead, G.J. 1952. Cuckoo studies on a South African farm (Part II). *Ostrich* 23: 2-15.

Skead, G.J. 1971. A juvenile Klaas' Cuckoo *Chrysococcyx klaas* with its host in late June 1971. *Ostrich* 43: 134.

Skead, G.J. 1995. *Life-history notes on East Cape bird species (1940-1990)*. Vols. 1 & 2. Algoa Regional Services Council, Port Elizabeth.

Skerrett, A. & the Seychelles Bird Records Committee. 2001. The second report of the Seychelles Bird Records Committee. *Bull. ABC* 8: 23-29.

Skerrett, A., Betts, M., Bullock, I., Fisher, D., Gerlach, R., Lucking, R., Phillips J. & Scott, B. 2006. Third report of the Seychelles Bird Records Committee. *Bull. ABC* 13: 170-177.

Skinner, K.L. 1929. "Mr. K.L. Skinner exhibited". *Bull. Brit. Ool. Assoc.* 2 (20): 110-111.

Skinner, N.J. 1996. The breeding seasons of birds in Botswana. II. Non-passerine families (sandgrouse to woodpeckers). *Babbler Gaborone* 31: 6-16.

Skjelseth, S., Moksnes, A., Røskaft, E., Gibbs, H.L., Taborsky, M., Taborsky, B., Honza, M. & Kleven, O. 2004. Parentage and host preference in the Common Cuckoo *Cuculus canorus*. *J. Avian Biol.* 35: 21-24.

Skutch, A.F. 1959. Life history of the Groove-billed Ani. *Auk* 76: 281-317.

Skutch, A.F. 1966. Life history notes on three tropical American cuckoos. *Wilson Bull.* 78: 139-165.

Skutch, A.F. 1983. *Birds of tropical America*. University of Texas Press, Austin.

Skutch, A.F. 1999. *Trogons, laughing falcons, and other Neotropical birds*. Texas A&M University Press, College Station.

Slack, R. & Wallace, I. 2009. *Rare birds where and when: an analysis of status & distribution in Britain and Ireland. Volume 1: sandgrouse to New World orioles*. Rare Bird Books, York.

Slud, P. 1964. The birds of Costa Rica: distribution and ecology. *Bull. Amer. Mus. Nat. Hist.* 128: 1-430.

Slud, P. 1967. The birds of Cocos Island [Costa Rica]. *Bull. Amer. Mus. Nat. Hist.* 134: 261-295.

Slud, P. 1980. The birds of Hacienda Palo Verde, Guanacaste, Costa Rica. *son. Contr. Zool.* 292: 1-92.

Smedley, D.I. 1983. Mating of the Fan-tailed Cuckoo. *Corella* 7 (2): 43.

Smith, H.C. 1943. *Notes on birds of Burma*. Pivately published, Simla.

Smith, K.D. 1957. An annotated check list of the birds of Eritrea. *Ibis* 99: 307-337.

Smith, K.D. 1965. On the birds of Morocco. *Ibis* 107: 493-526.

Smith, M.H. 1981. Breeding the Roadrunner *Geococcyx californianus* at Columbia Zoo. *Int. Zoo Yb.* 21: 119-121.

Smith, S.M. 1971. The relationship of grazing cattle to foraging rates in anis. *Auk* 88: 876-880.

Smith, W.J. & Smith, A.M. 2000. Information about behavior is provided by songs of the Striped Cuckoo. *Wilson Bull.* 112: 491-497.

Smith, W.W. 1930. Feeding habits of the Shining Bronze-Cuckoo. *Emu* 30:217-218.

Smithe, F.B. & Paynter, R.A. Jnr. 1963. Birds of Tikal, Guatemala. *Bull. Mus. Comp. Zool.* 128: 245-324.

Smyth, C.H. 1928. Descripción de una colección de huevos de aves argentinas. *Hornero* 4: 125-152.

Smythies, B.E. 1940. *Birds of Burma*. American Baptist Mission Press, Rangoon.

Smythies, B.E. 1953. *Birds of Burma*. 2nd ed. Oliver & Boyd, Edinburgh.

Smythies, B.E. 1957. An annotated checklist of the birds of Borneo. *Sarawak Mus. J.* 7: 523-818.

Smythies, B.E. 1981. *Birds of Borneo* 3rd ed. Sabah Society & Malayan Nature Society, Kota Kinabalu, Malaysia.

Smythies, B E. 1986. *The birds of Burma*. 3rd ed. Nimrod Press Ltd & Silvio Mattacchione & Co., Liss, Hampshire & Pickering, Ontario.

Smythies, B.E. & Davison, G.W.H. 1999. *The birds of Borneo*. 4th ed. Natural History Publications, Kota Kinabalu.

Snethlage, E. 1908. Eine Vogelsammlung vom Rio Purús, Brasilien. *J. Orn.* 56: 7-24.

Snethlage, E. 1913. Über die Verbreitung der Vogelarten in Unteramazonien. *J. Orn.* 61: 469-539.

Snethlage, E. 1914. Catálogo das aves amazônicas. *Bol. Mus. Goeldi Hist. Nat. Ethnogr.* 8: 1-530.

Snethlage, E. 1927. Ein neuer Cuculide aus Südbrasilien. *Orn. Monatsber.* 35: 80-82.

Snethlage, E. 1935a. Beiträge zur Brutbiologie brasilianischer Vögel. *J. Orn.* 83: 1-24.

Snethlage, E. 1935b. Beiträge zur Fortpflanzungsbiologie brasilianischer Vögel. *J. Orn.* 83: 532-562.

Snethlage, H. 1927. Meine Reise durch Nordostbrasilien I. Reisebericht. *J. Orn.* 75: 453-484.

Snethlage, H. 1928. Meine Reise durch Nordostbrasilien III. Bausteine zur Biologie der angetroffenen Arten. *J. Orn.* 76: 668-738.

Snow, D.W. 1978. *An atlas of speciation in African Non-Passerine birds*. British Museum (Natural History), London.

Snow, D.W. & Perrins, C.M. 1998. *The birds of the western Palearctic*. Concise Edition. OUP, Oxford.

Snow, D.W. & Snow, B.K. 1964. Breeding seasons and annual cycles of Trinidad land-birds. *Zoologica* 49: 1-39.

Snyder, D.E. 1966. *The birds of Guyana*. Peabody Museum, Salem, Mass.

Sody, H.J.V. 1989. Diet of Javanese birds. Pp. 164-221 in J.H. Becking (Ed.). *Henri Jacob Victor Sody (1852-1959), his life and work*. E.J. Brill, Leiden.

Sokolov, E.P. & Sokolov, A.M. 1986. [New bird species hosts of Common Cuckoo.] *Trudy Zool. Inst. Akad. Nauk SSSR* 147: 145-146. [in Russian]

Soler, J.J., Avilés, J.M. Møller, A.P. & Moreno, J. (in press). Attractive blue egg coloration and cuckoo-host coevolution. *Biol. J. Linn. Soc. London*.

Soler, J.J., Martín-Gálvez, D., Martínez, J.G., Soler, M., Canestrari, D., Abad-Gómez, J.M. & Møller, A.P. 2011. Evolution of tolerance by magpies to brood parasitism by great spotted cuckoos. *Proc. R. Soc. B* 278: 2047-2052.

Soler, J.J., Martinez, J.G., Soler, M. & Møller, A.P. 1999. Genetic and geographic variation in rejection behavior of Cuckoo eggs by European Magpie populations: An experimental test of rejecter-gene flow. *Evolution* 53: 947-956.

Soler, J.J. & Møller, A.P. 1996. A comparative analysis of the evolution of variation in appearance of eggs of European passerines in relation to brood parasitism. *Behav. Ecol.* 7: 89-94.

Soler, M. 1990. Relationships between the Great Spotted

Cuckoo *Clamator glandarius* and its corvid hosts in a recently colonized area. *Orn. Scand.* 21: 212-223.

Soler, M. 2002. Breeding strategy and begging intensity: influences on food delivery by parents and host selection by parasitic cuckoos. pp. 413-428 in J. Wright & M. L. Leonard (Eds). *The evolution of begging: competition, co-operation, and communication*. Kluwer Academic Publishers, Dordrecht.

Soler, M. 2008. Do hosts of interspecific brood parasites feed parasitic chicks with lower quality prey? *Animal Behav.* 76: 1761-1763.

Soler, M. & Martinez, J.G. 2000. Is egg-damaging behavior by great spotted cuckoos an accident or an adaptation? *Behavioral Ecology* 11: 495-501.

Soler, M. & Møller, A.P. 1990. Duration of sympatry and the coevolution between the Great Spotted Cuckoo and its Magpie host. *Nature* 343 (6260): 748-750.

Soler, M., Palomino, J.J., Martinez, J.G. & Soler, J.J. 1994. Activity, survival, independence and migration of fledgling Great Spotted Cuckoos. *Condor* 96: 802-805.

Soler, M., Palomino, J.J., Martinez, J.G. & Soler, J.J. 1995a. Communal parental care by monogmous Magpie hosts of fledgling Great Spotted Cuckoos. *Condor* 97: 804-810.

Soler, M. & Soler, J.J. 1991. Growth and development of great spotted cuckoos and their magpie host. *Condor* 93: 49-54.

Soler, M., Soler, J.J., Martinez, J.G. & Møller, A.P. 1994. Micro-evolutionary change in host response to a brood parasite. *Behav. Ecol. Sociobiol.* 35: 295-301.

Soler, M., Soler, J.J., Martinez, J.G. & Møller, A.P. 1995b. Does the Great Spotted Cuckoo choose Magpie hosts according to their parenting ability? *Behav. Ecol. Sociobiol.* 36: 201-206.

Soler, M., Soler, J.J., Martinez, J.G. & Møller, A.P. 1995c. Chick recognition and acceptance: a weakness in Magpies exploited by the parasitic Great Spotted Cuckoo. *Behav. Ecol. Sociobiol.* 37: 243-248.

Soler, M., Soler, J.J., Martinez, J.G. & Møller, A.P. 1995d. Magpie host manipulation by Great Spotted Cuckoos: evidence for an avian mafia? *Evolution* 49: 770-775.

Soler, M., Palomino, J.J., Martin-Vivaldi, M. & Soler, J.J. 2000a. Lack of consistency in the response of Rufous-tailed Scrub Robins *Cerotrichas galactotes* towards parasitic Common Cuckoo eggs. *Ibis* 142: 151-154.

Soler, M., Soler, J.J. & Møller, A.P. 2000b. Effect of great spotted cuckoo presence on magpie rejection behaviour. *Behaviour* 137: 213-220.

Soler, M., Soler, J.J., Pérez-Contreras, T. & Martinez, J.G. 2002. Differential reproductive success of Great Spotted Cuckoos *Clamator glandarius* parasitising Magpies *Pica pica* and Carrion Crows *Corvus corone*: the importance of parasitism costs and host defences. *Avian Science* 1 (2): 25-32.

Sorenson, M.D., Oneal, E., García-Moreno, J. & Mindell, D.P. 2003. More taxa, more characters: the Hoatzin problem is still unresolved. *Mol. Biol. Evol.* 20: 1484-1498.

Sorenson, M.D. & Payne, R.B. 2005. A molecular genetic analysis of cuckoo phylogeny. Pp. 68-94 in R.B. Payne. *The cuckoos*. OUP, Oxford.

Sorrié, B. A. 1975. Observations on the birds of Vieques Island, Puerto Rico. *Caribb. J. Sci.* 15: 89-103.

Sparrman, A. 1786. *Museum Carlsonianum, in quo novas et selectas aves, coloribus ad vivum brevique descriptiones illustratas* Vol. 1. Ex Typographia Regia, Holmiae.

Sparrman, A. 1787. *Museum Carlsonianum, in quo novas et selectas aves, coloribus ad vivum brevique descriptiones illustratas* Vol. 2. Ex Typographia Regia, Holmiae.

Sparrow, R. 1936. Exhibition of a series of eggs of South African parasitic birds. *Bull. Brit. Ool. Assoc.* 5: 5-7.

Sparrow, S.C. 1984. Long-tailed Cuckoo flight speed. *Notornis* 31: 144.

Spencer, O.R. 1943. Nesting habits of the Black-billed Cuckoo. *Wilson Bull.* 55: 11-22.

Spennemann, A. 1928. Zur Brutbiologie von *Centropus javanicus* (Dumont). *Beitr. Fortpfl. Biol. Vögel Berlin* 4: 139-144.

Spierenburg, P. 2005. *Birds in Bhutan: status and distribution*. Oriental Bird Club, Bedford, UK.

Spittle, R.J. 1950. Nesting habits of some Singapore birds. *Bull. Raffles Mus.* 21: 184-204.

Spofford, S.H. 1976. Roadrunner catches hummingbird in flight. *Condor* 78: 142.

Spofford, S.H. 1978. Roadrunners as predators on birds in banding traps. *N. Am. Bird Bander* 3: 55.

Sprunt, A., Jnr. 1939. Smooth-billed Ani nesting in Florida. *Auk* 56: 335-336.

Stager, K.E. 1961. The Machris Brazilian Expedition. Ornithology: non-passerines. *Contrib. Sci. Los Angeles County Mus.* 41: 1-27.

Stager, K.E. 1964. The birds of Clipperton Island, eastern Pacific. *Condor* 66: 357-371.

Stagg, S.J. 1973. Parasitic birds of Kenton College, Nairobi. *Bull. E. Afr. Nat. Hist. Soc.* (April): 53-54.

Stahl, J.C., Weimerskirch, H., & Ridoux, V. 1984. Observations recentes d'oiseaux marins et terrestres visiteurs dans les îles Crozet, sud-ouest de l'Ocean Indien. *Gerfaut* 74: 39–46.

Stanford, J.K. & Ticehurst, C.B. 1939. On the birds of northern Burma. Parts V & VI. *Ibis* (14th series) 3: 1-45, 211-258.

Stannard, J. 1966. *Bird songs of Amanzi*. Percy Fitzpatrick Institute of African Ornithology, Cape Town.

Stark, A.C. & Sclater, W.L. 1903. *The birds of South Africa*. Vol. 3. R.H. Porter, London.

Stattersfield, A.J., Crosby, M.J., Long, A.J. & Wege, D.C. 1998. *Endemic bird areas of the world: priorities for biodiversity conservation*. BirdLife International, Cambridge, UK.

Stead, D.E. 1979. Liwonde National Park: part III - the birds. *Nyala* 5: 12-27.

Stead, E.F. 1936. The egg of the Long-tailed Cuckoo (*Eurodynamis taitensis*) [*sic*]. *Trans. Proc. Roy. Soc. N. Z.* 66: 182-184.

Steadman, W.D. 2008. Doves (Columbidae) and cuckoos (Cuculidae) from the early Miocene of Florida. *Bull. Florida Mus. Nat. Hist.* 48: 1-16.

Steere, J.B. 1890. *A list of the birds and mammals collected by the Steere Expedition to the Philippines, with localities, and with brief preliminary descriptions and supposed new species*. Courier Printers, Ann Arbor, Mich.

Stein, G.H.W. 1936. Ornithologische Ergebnisse der Expedition Stein 1931-1932. V. Beiträge zur Biologie papuanischer Vögel. *J. Orn.* 84: 21-57.

Stephens, J.F. 1815. *General Zoology or Systematic Natural History, commenced by the late George Shaw (...). Vol. IX. Part 1*. G. Wilkie, London.

Steullet, A. & Deautier, E.A. 1945. Catálogo sistemático de las aves de la República Argentina [pt. 4]. *Obra Cincuent. Mus. La Plata* 1: 733-932.

Stevens, H. 1926. Notes on the birds of the Sikkim Himalayas, Part 6. *J. Bombay Nat. Hist. Soc.* 30: 664-685.

Stevenson, H.M. & Anderson, B.H. 1994. *The birdlife of Florida*. University Press of Florida, Gainesville.

Stevenson, T. 1980. *The birds of Lake Baringo*. Sealpoint Publicity, Nairobi.

Stevenson, T. & Fanshawe, J. 2002. *Field guide to the birds of East Africa. Kenya, Tanzania, Uganda, Rwanda, Burundi*. Poyser, London.

Stevenson, T. & Fanshawe, J. 2004. *Birds of East Africa; Kenya, Tanzania, Uganda, Rwanda, Burundi*. Christopher Helm, London.

Steyn, P. 1968. Additional breeding data on the Red-chested Cuckoo, *Cuculus solitarius*. *Ostrich* 39: 267.

Steyn, P. 1970. Journey for birds. *Bokmakierie* 22: 86-92.

Steyn, P. 1972. The development of Senegal Coucal nestlings. *Ostrich* 43: 56-59.

Steyn, P. & Myburgh, N. 1980. Red-chested Cuckoo parasitizing Cape Siskin. *Ostrich* 51: 53.

Stiles, F.G. & Skutch, A.F. 1989. *A guide to the birds of Costa Rica*. Christopher Helm, London.

Stiles, F.G., Telleria, J.L. & Díaz, M. 1995. Observaciones sobre la composición, ecología y zoogeografía de la avifauna de la Sierra de Chiribiquete, Caquetá, Colombia. *Caldasia* 17: 481-500.

Stipcevic, M. 1991. Sixth record of Great Spotted Cuckoo *Clamator glandarius* in Croatia. *Acrocephalus* 12: 80-82.

Stipcevic, M. 1992. From the ornithological notebook: Great Spotted Cuckoo. *Acrocephalus* 13: 189.

Stjernstedt, R. 1984. First record of the Barred Long-tailed Cuckoo *Cercococcyx montanus* in Zambia. *Bull. Zambian Orn. Soc.* 16: 18-20.

Stjernstedt, R. 1993. *Birdsong in Zambia.* (3 cassettes). Privately published, Powys, UK.

Stocks, G.R. 1948. Host of the Black Cuckoo. *Ostrich* 19: 97.

Stockton de Dod, A. 1987. *Aves de la República Dominicana.* 2nd ed. Museo Nacional de Historia Natural, Santo Domingo.

Stoddart, H.L. & Norris, R.A. 1967. Bird casualties at a Leon County, Florida TV tower: An eleven-year study. *Tall Timbers Res. St. Bull.* 8: 1-104.

Stoddart, D.R., Benson, C.W. & Peake, J.F. 1970. Ecological change and effects of phosphate mining on Assumption Island. Pp. 121-145 in D.R. Stoddart (Ed.) Coral islands of the western Indian Ocean. *Atoll Research Bull.* No. 136.

Stokes, T., Merton, D., Hicks, J. & Tranter, J. 1987. Additional records of birds from Christmas Island, Indian Ocean. *Aust. Bird Watcher* 12 (1): 1-7.

Stone, W. 1899. A new species of *Coccyzus* from St. Andrews. *Proc. Acad. Nat. Sci. Philad.* 51: 301.

Stone, W. 1909. A review of the genus *Piaya* Lesson. *Proc. Acad. Nat. Sci. Philad.* 60: 492-501.

Stone, W. 1928. On a collection of birds from the Pará region, eastern Brazil (with field notes by James Bond and Rodolphe Meyer de Schauensee). *Proc. Acad. Nat. Sci. Philad.* 80: 149-176.

Stone, W. 1932. The birds of Honduras with special reference to a collection made in 1930 by John T. Emlen, Jr., and C. Brooke Worth. *Proc. Acad. Nat. Sci. Philad.* 84: 291-342.

Stoneham, H.F. 1931. Field-notes on a collection of birds from Uganda. *Ibis* (13th series) 1: 701-712.

Stones, A.J., Lucking, R.S., Davison, P.J. & Wahyu Raharjaningtrah. 1997. Checklist of the birds of the Sula Islands (1991), with particular reference to Taliabu Island. *Kukila* 9: 37-55.

Storer, R.W. 1966. Sexual dimorphism and food habits in three North American accipiters. *Auk* 83: 423-436.

Storer, R.W. 1989. Notes on Paraguayan birds. *Occas. Pap. Mus. Zool. Univ. Michigan* 719: 1-21.

Storr, G.M. 1953. Birds of Cooktown and Laura Districts, North Queensland. *Emu* 53: 225-248.

Storr, G.M. 1980. *Birds from the Kimberley Division, Western Australia. Rec. W. Aust. Mus.* Suppl. No. 11.

Storrs, H.T. 1944. Indian Cuckoo (*Cuculus m. micropterus*) laying in the nest of the Himalayan Black Drongo. *J. Bengal Nat. Hist. Soc.* 18: 116-120.

Stotz, D.F. & Bierregaard, R.O., Jnr. 1989. The birds of the fazendas Porto Alegre, Esteio and Dimona north of Manaus, Amazonas, Brazil. *Rev. Brasil. Biol.* 49: 861-872.

Stotz, D.F., Bierregaard, R.O., Jnr., Cohn-Haft, M., Petermann, P., Smith, J., Whittaker, A. & Wilson, S.V. 1992. The status of North American migrants in central Amazonian Brazil. *Condor* 94: 608-621.

Stotz, D.F., Fitzpatrick, J.W., Parker, T.A. III & Moskovits, D.K. 1996. *Neotropical birds: ecology and conservation.* Chicago University Press, Chicago.

Stotz, D.F., Lanyon, S.M., Schulenberg, T.S., Willard, D.E., Peterson, A.T. & Fitzpatrick, J.W. 1997. An avifaunal survey of two tropical forest localities on the middle Rio Jiparaná, Rondônia, Brazil. *Orn. Monogr.* 48: 763-781.

St. Paul, R. 1976. A bushman's seventeen years of noting birds. Part D. - Shining Cuckoo and Long-tailed Cuckoo. *Notornis* 23: 289-298.

Strahan, R. 1994. *Cuckoos, nightbirds and kingfishers of Australia.* Angus & Robertson Publication, Sydney.

Strange, M. 2000. *A photographic guide to the birds of southeast Asia.* Periplus Editions, Singapore.

Strange, M. 2001. *A phototographic guide to the birds of Indonesia.* Periplus Editions, Singapore.

Straube, F.C. & Bornschein, M.R. 1989. A contribuição de André Mayer à história natural no Paraná. I. Sobre uma coleção de aves do extremo noroeste do Paraná e sul do Mato Grosso do Sul. *Arq. Biol. Tecnol.* 32: 441-471.

Straube, F.C. & Bornschein, M.R. 1995. New or noteworthy records of birds from northwestern Paraná and adjacent areas (Brazil). *Bull. Brit. Orn. Club* 115: 219-225.

Straube, F.C., Krul, R. & Carrano, E. 2005. Coletânea da avifauna da região sul do estado do Paraná (Brasil). *Atual. Orn.* 125: 10.

Strauch, J.G., Jnr. 1977. Further bird weights from Panama. *Bull. Brit. Orn. Club* 97: 61-65.

Stresemann, E. 1912 Ornithologische Miszellen aus dem Indo-Australischen Gebiet. I-XVIII. *Nov. Zool.* 19: 311-351.

Stresemann, E. 1913. Ornithologische Miszellen aus dem Indo-Australischen Gebiet. II. Teil. XIX-XXVIII. *Nov. Zool.* 20: 289-324.

Stresemann, E. 1913a. Die Vögel von Bali. *Nov. Zool.* 20: 325-387.

Stresemann, E. 1920. The new names in J. Hermann's *Tabula Affinitatum Animalium*. *Nov. Zool.* 27: 327-332.

Stresemann, E. 1921. 11 neue Formen aus dem Stromgebiet des Sepik (nördl. Neuguinea). *Anz. Orn. Ges. Bayern* 5: 33-42.

Stresemann, E. 1927. *Centropus phasianus thierfelderi*, subsp. nova. *Orn. Monatsb.* 35: 111-112.

Stresemann, E. 1930. Neue Vogelrassen aus Kwangsi 2. *Orn. Monatsb.* 38: 47-49.

Stresemann, E. 1931. Vorläufiges über die ornithologischen Ergebnisse der Expedition Heinrich 1930-1931. II. Neue Vogelrassen aus dem Latimodjong-Gebirge. III. Zur Ornithologie des Matinang-Gebirges auf der Nordhalbinsel von Celebes. V. Zur Ornithologie von Halmahera und Batjan. *Orn. Monatsb.* 39: 44-47, 77-85, 167-171.

Stresemann, E. 1938. Vorläufiges über die ornithologischen Ergebnisse der Expedition Heinrich 1930-1931. IX. *Orn. Monatsb.* 45: 148.

Stresemann, E. 1940-41. Die Vögel von Celebes. Teil 3. Fortsetzung, Systematik und Biologie. Biologische Beiträge von Gerd Heinrich. *J. Orn.* 88: 389-487, 89: 1-102.

Stresemann, E. & Paludan, K. 1932. Vorläufiges über die ornithologischen Ergebnisse der Expedition Stein 1931-1932. I. Zur Ornithologie der Insel Waigeu. *Orn. Monatsb.* 40: 13-18.

Stresemann, E. & Stresemann, V. 1966. Die Mauser der Vögel. *J. Orn.* 107 (suppl.): 1-439.

Stresemann, E. & Stresemann, V. 1969. Die Mauser der Schopfkuckucke (*Clamator*). *J. Orn.* 110: 192-204.

Stresemann, V. & Stresemann, E. 1961. Die Handschwingen-Mauser der Kuckucke (Cuculidae). *J. Orn.* 102: 317-352.

Strewe, R. & Navarro, C. 2004. New and noteworthy records of birds from the Sierra Nevada de Santa Marta region, north-eastern Colombia. *Bull. Brit. Orn. Club* 124: 38-51.

Strickland, H.E. 1852. On a new species of *Coccyzus* from Trinidad. P. 28, pl. 82 in W. Jardine (Ed.). *Contributions to ornithology.* Vol. 2. Samuel Highley, London.

Strickland, H.E. 1847. Notes on certain species of birds from Malacca. *Proc. Zool. Soc. London* 14: 99-105.

Stronoch, N.R.H. 1981. Pallid Cuckoo *Cuculus pallidus*. A new species for Papua New Guinea. *Papua N. G. Bird Soc. Newsletter* No. 181-182 July-Aug: 23.

Struthers, P.H. 1923. Observations on the bird life of Porto Rico. *Auk* 40: 469-478.

Štumberger, B. 1987. [Young Great Spotted Cuckoo *Clamator glandarius*, recorded in the vicinity of Lake Skadar.] *Acrocephalus* 34: 53-55.

Supari, S. 2003. *Bird songs of Singapore* (CD). Nature Society, Singapore.

Suter, M.F. 1945. Plaintive Cuckoo (?) parasitising Purple Sunbird. *J. Bombay Nat. Hist. Soc.* 45: 235.

Sutton, G.M. 1913. A pet Road-runner. *Bird-lore* 15: 324-326.

Sutton, G.M. 1915. Suggestive methods of bird-study: pet Road-runners. *Bird-lore* 17: 57-61.

Sutton, G.M. 1922. Notes on the Road-runner at Fort Worth, Texas. *Wilson Bull.* 34: 2-20.

Sutton, G.M. 1940. *Geococcyx californianus* (Lesson), Roadrunner. pp. 36-51 in A.C. Bent (Ed.). *Life histories of North American cuckoos, goatsuckers, hummingbirds, and their allies.* US National Museum, Washington DC.

Sutton, G.M. 1951. Birds and an ant army in southern Tamaulipas. *Condor* 53: 16-18.
Sutton, G.M. 1967. *Oklahoma birds.* University of Oklahoma Press, Norman.
Sutton, G.M. 1972. Winter food of a central Oklahoma Roadrunner. *Bull. Oklahoma Orn. Soc.* 5: 30.
Swainson, W. 1827. A synopsis of the birds discovered in Mexico by W. Bullock, F.L.S. and H.S., and Mr. William Bullock, jun. *Philos. Mag., N. S.* 1: 364-369, 433-442.
Swainson, W. 1829. *Zoological Illustrations, or original figures and descriptions of new, rare, or interesting Animals, selected chiefly from the classes of Ornithology, Entomology, and Conchology, &c.* London.
Swainson, W. 1837. *On the natural history and classification of birds.* Vol. 2. London.
Swainson, W. 1837. *The natural history of the birds of Western Africa.* Edinburgh. 2 volumes. (*Ceuthmochares aereus flavirostris*)
Swainson, W. 1837. On the natural history and relations of the family of Cuculidæ or Cuckoos, with a view to determine the series of their variation. *Magazine of Zoology and Botany* 1: 213-225, 430-437.
Swainson, W. 1838. *Animals in menageries.* Longman, Orme, Brown, Green, Longmans and John Taylor, London. '
Swinhoe, R. 1861. Notes on the ornithology of Hongkong, Macao, and Canton, made during the latter end of February, March, April, and the beginning of May, 1860. *Ibis* 3: 23-57.
Swinhoe, R. 1870. On four new species of birds from China. *Ann. Mag. Nat. Hist.* (4th series) 6: 152-154.
Swynnerton, C.F.M. 1908. Further notes on the birds of Gazaland. *Ibis* (9th series) 2: 1-107, 392-443.
Swynnerton, C.F.M. 1911. On some nests and eggs from Mount Chirinda, southern Rhodesia. *J. South Afr. Orn. Union* 7: 15.
Sztolcman, J. 1926. Étude des collections ornithologiques de Paranà. *Ann. Zool. Mus. Polon. Hist. Nat.* 5: 107-196.
Tait, I.C. 1952. Some ornithological notes. *Ostrich* 23: 135.
Takasu, F., Kawasaki, K., Nakamura, H., Cohen, J.E. & Shigesada, N. 1993. Modeling the population dynamics of a Cuckoo-host association and the evolution of host defences. *Amer. Nat.* 142: 819-839.
Tanaka, K.D. & Ueda, K. 2005. Horsfield's hawk-cuckoo nestlings simulate multiple gapes for begging. *Science* 308: 653.
Taplin, A. & Beurteaux, Y. 1992. Aspects of the breeding biology of the Pheasant Coucal *Centropus phasianinus*. *Emu* 92: 141-146.
Tarboton, W. 1975. African Cuckoo parasitising Forktailed Drongo. *Ostrich* 46: 186-188.
Tarboton, W. 1986. African Cuckoo: the agony and ecstasy of being a parasite. *Bokmakierie* 38: 109-111.
Tarboton, W. 2001. *A guide to the nests and eggs of southern African birds.* Struik, Capetown.
Tarbuton, M.K. 1993. Is courtship a necessary prelude to mating? *Aust. Bird Watcher* 15 (2): 96-97.
Tashian, R.E. 1953. The birds of southeastern Guatemala. *Condor* 55: 198-210.
Taylor, E.C. 1864. Five months in the West Indies. Part II. Martinique, Dominica, and Porto Rico. *Ibis* 6: 157-173.
Taylor, L.E. 1906. The birds of Irene, near Pretoria, Transvaal. *J. South Afr. Orn. Union* 2: 55-83.
Tebb, G., Morris, P. & Los, P. 2008. Spotlight on Sulawesi: new and interesting bird records from Sulawesi and Halmahera, Indonesia. *BirdingASIA* 10: 67-76.
Telino, W.R. Jnr., Neves, R.M.L. & Nascimento, J.L.X. de 2005. Biologia e composição da avifauna em uma Reserva Particular de Patrimônio Natural da caatinga paraibana. *Ornithologia* 1: 49-57.
Temminck, C.J. 1820. *Nouveau recueil de planches coloriées d'oiseaux, pour servir de suite et de complément aux planches enluminées de Buffon. Imprimerie Royale,* 1778. Livraison 2, pl. 7. Gabr. Dufour, Paris.
Temminck, C.J. 1825. *Nouveau recueil de planches coloriées d'oiseaux, pour servier de suite et de complement aux planches enluminées de Buffon. Imprimerie Royale,* 1778. Gabr. Dufour, Paris. Livraison 59, pl. 349.

Temminck, C.J. 1827. *Nouveau recueil de planches coloriées d'oiseaux, pour servier de suite et de complement aux planches enluminées de Buffon. Imprimerie Royale,* 1778. Gabr. Dufour, Paris. Livraison 74, pl. 440.
Temminck, C.J. 1832. *Nouveau recueil de planches coloriées d'oiseaux, pour servier de suite et de complement aux planches enluminées de Buffon. Imprimerie Royale,* 1778. Gabr. Dufour, Paris. Livraison 91, pl. 538.
Temminck, C.J. 1838 *Nouveau recueil de planches coloriées d'oiseaux.* Table méthodique. G.F. Levrault, Paris.
Terborgh, J. 1983. *Five new world primates: a study in comparative ecology.* Princeton University Press, Princeton.
Terborgh, J.W., Fitzpatrick, J.W. & Emmons, L. 1984. Annotated checklist of bird and mammal species of Cocha Cashu Biological Station, Manu National Park, Peru. *Fieldiana Zool. N.S.* 21: 1-29.
Terborgh, J., Robinson, S. K., Parker, T. A. III, Munn, C. A. & Pierpont, N. 1990. Structure and organization of an Amazonian forest bird community. *Ecol. Monogr.* 60: 213-238.
Thayer, G.H. 1903. The mystery of the Black-billed Cuckoo. *Bird-lore* 5: 143-145.
Thévenot, M., Vernon, R. & Bergier, P. 2003. *The birds of Morocco.* British Ornithologists' Union/British Ornithologists' Club Checklist Series No. 20. BOU/BOC, Tring.
Thewlis, R.M., Duckworth, J.W., Anderson, G.Q.A., Dvorak, M., Evans, T.D., Nemeth, E., Timmins, R.J. & Wilkinson, R.J. 1996. Ornithological records from Laos, 1992-1993. *Forktail* 11: 47-100.
Thiede, W. 1994. Kuckucks-Fragen. Teil 7. - Der Dreisilbenkuckuck. *Orn. Mitt.* 46 (4): 102-104.
Thiollay, J-M. 1985. The birds of Ivory Coast: status and distribution. *Malimbus* 7: 1-59.
Thiollay, J-M. & Jullien, M. 1998. Flocking behaviour of foraging birds in a Neotropical rain forest and the antipredator defence hypothesis. *Ibis* 140: 382-394.
Thomas, B.T. 1978. The Dwarf Cuckoo in Venezuela. *Condor* 80: 105-106.
Thomas, B.T. 1979. The birds of a ranch in the Venezuelan llanos. Pp. 213-232 in J.F. Eisenberg (Ed.). *Vertebrate ecology in the northern Neotropics.* Smithsonian Institution Press, Washington DC.
Thomas, B.T. 1990. Additional weights of Venezuelan birds. *Bull. Brit. Orn. Club* 110: 48-51.
Thomas, B.T. 1995. Black-billed Cuckoo parasitizes the nest of a Yellow-breasted Chat. *Raven* 66: 3-5.
Thomas, R. & Thomas, S. 1994. *Birds of Sulawesi, Lesser Sundas and Sabah.* Cassette. Privately published, UK.
Thomas, W.W. & Poole, C.M. 2003. An annotated list of the birds of Cambodia from 1859 to 1970. *Forktail* 19: 103-127.
Thompson, H.A.F. 1982. The status of cuckoos Cuculidae in the Darwin area, Northern Territory, 1974-1980. *Northern Terr. Nat.* 5: 13-19.
Thompson, M.C. 1966. Birds from North Borneo. *Univ. Kansas Publ. Mus. Nat. Hist.* 17 (8): 377-433.
Thompson, P.M. & Johnson, D.L. 2003. Further notable bird records from Bangladesh. *Forktail* 19: 85-102.
Thonnérieux, Y., Walsh J.F. & Bortoli, L. 1989. L'avifaune de la ville de Ouagadougou et ses environs (Burkina Faso). *Malimbus* 11: 7-40.
Thunberg, C.P. 1819. *Götheb. Kongl. Vett. och Vitt. Samh. Nya Handl.* 3: 1.
Thurber, W.A., Serrano, J.F., Sermeño, A. & Benítez, M. 1987. Status of uncommon and previously unreported birds of El Salvador. *Proc. West. Found. Vertebr. Zool.* 3: 109-293.
Ticehurst, C.B. 1927. The birds of British Baluchistan. Part II. *Jour. Bombay Nat. Hist. Soc.* 31: 862-881.
Tikader, B.K. 1984. *Birds of Andaman & Nicobar Islands.* Zoological Survey of India, Calcutta.
Tirimanna, V. 1981. The Red-faced Malkoha. *Loris* 15: 310-312.
Tobish, T.G., Jnr. & Isleib, M.E. 1992. The autumn migration. August 1-November 30, 1991. Alaska Region. *Am. Birds* 46: 136-138.

Todd, W.E.C. 1912. Descriptions of seventeen new Neotropical birds. *Ann. Carnegie Mus.* 8: 198-214.
Todd, W.E.C. 1916. The birds of the Isle of Pines. *Ann. Carnegie Mus.* 10: 146-296.
Todd, W.E.C. 1925. Four new birds from Brazil. *Proc. Biol. Soc. Wash.* 38: 111-113.
Todd, W.E.C. 1947. The Venezuelan races of *Piaya cayana*. *Proc. Biol. Soc. Wash.* 60: 59-60.
Todd, W.E.C. & Carriker, M.A. Jnr. 1922. The birds of the Santa Marta region of Colombia: a study in altitudinal distribution. *Ann. Carnegie Mus.* 14: 1-611.
Tojo, H., Nakamura, S. & Higuchi, H. 2002. Gape patches in Oriental Cuckoo *Cuculus saturatus* nestlings. *Orn. Sci.* 1: 145-149.
Tokue, K. & Ueda, K. 2010. Mangrove Gerygones *Gerygone laevigaster* eject Little Bronze-cuckoo *Chalcites minutillus* from parasitised nests. *Ibis* 152: 835-839.
Tostain, O., Dujardin, J-L., Érard, C. & Thiollay, J-M. 1992. *Oiseaux de Guyane*. Société d'Études Ornithologiques, Brunoy, France.
Townsend, F.S. 1921. Eggs of *Coccystes cafer*. *S. Afr. J. Nat. Hist.* 3: 220-222.
Trainor, C. 2002. An expedition to Damar Island, south-west Maluku, Indonesia. *Bull. OBC* 36: 18-23.
Trainor, C.R. 2003. The birds of Lembata (Lomblen), Lesser Sundas. *Kukila* 12: 39-53.
Trainor, C.R. 2005a. Birds of Tapuafu peninsula, Roti island, Lesser Sundas, Indonesia. *Forktail* 21: 121-131.
Trainor, C.R. 2005b. Species richness, habitat use and conservation of birds of Alor Island, Lesser Sundas, Indonesia. *Emu* 105: 127-135.
Trainor, C.R. 2007. Birds of Damar Island, Banda Sea, Indonesia. *Bull. Br. Orn. Cl.* 127: 300-321.
Trainor, C.R., Imanuddin, Aldy, F. & Walker, J.S. 2008a. The status and conservation of the Endangered Wetar Ground-dove (*Gallicolumba hoedtii*) and other wildlife on Wetar Island, Indonesia, 2008. *Columbidae Conservation*. Final technical report.
Trainor, C.R., Santana, F., Pinto, P., Xavier, A.F., Safford, R. & Grimmett, R. 2008b. Birds, birding and conservation in Timor-Leste. *BirdingASIA* 9: 16-45.
Trautman, M.B. 1940. The birds of Buckeye Lake, Ohio. *Misc. Publ. Mus. Zool. Univ. Michigan* 44: 1-466.
Traylor, M.A. Jnr. 1958. Birds of northeastern Peru. *Fieldiana Zool.* 35: 87-141.
Traylor, M.A. Jnr. 1960. Notes on the birds of Angola, non-passeres. *Publ. Cult. Co. Diamantes Angola* 54: 129-186.
Traylor, M.A. Jnr. 1963. Check-list of Angolan birds. Co. Diamantes Angola, Museo do Dundo, Lisbon.
Traylor, M.A. Jnr. & Archer, A.L. 1982. Some results of the Field Museum 1977 expedition to south Sudan. *Scopus* 6: 5-12.
Trinka, A. & Prokop, P. 2011. Polygynous great reed warblers *Acrocephalus arundinaceus* suffer more cuckoo *Cuculus canorus* parasitism than monogamous pairs. *J. Avian Biol.* 42: 192-195.
Tubbs, P.K. 1992. *Coccyzus euleri* Cabanis, 1873 (Aves, Cuculiformes): specific name conserved. *Bull. Zool. Nomencl.* 49: 178-179.
Tuck, L.M. 1968. Recent Newfoundland bird records. *Auk* 85: 304-311.
Turbott, E.G. (Ed.) 1967. *Buller´s Birds of New Zealand*. (rev. ed.). East-West Centre Press, Honolulu.
Tweeddale, A.M. 1877. Descriptions of three New Species of Birds from the Indian Region. *Proc. Zool. Soc. London* 1877: 366-367.
Tweit, R.C. & Tweit, J.C. 1986. Urban development effects on the abundance of some common resident birds of the Tucson area of Arizona. *Am. Birds* 40: 431-436.
Tye, A. & Tye, H. 1991. Bird species on St. Andrew and Old Providence Islands, West Caribbean. *Wilson Bull.* 103: 493-497.
Tymstra, R., Connop, S. & Tshering, C. 1997. Some bird observations from central Bhutan, May 1994. *Forktail* 12: 49-60.

Urano, E., Yamagishi, S., Andrianarimisa, A. & Andriatsarafara, S. 1994. Different habitat use among three sympatric species of couas *Coua cristata*, *C. coquereli* and *C. ruficeps* in western Madagascar. *Ibis* 136: 485-487.
Urban, E.K. & Brown, L H. 1971. *A checklist of the birds of Ethiopia.* Adis Ababa University Press, Adis Ababa.
Vahl, M. 1797. Beskrivelse på trende ubekjendte fugle af gjøgeslægten. *Skriv. Naturhist.-Selsk., Kjøbenhavn* 4: 56-61.
van Balen, S. 1991. Faunistic notes from Bali, with some new records. *Kukila* 5: 125-132.
van Balen, S. 1999. *Birds on fragmented islands: persistence in the forests of Java and Bali*. Doctoral thesis. Wageningen University and Reserch Centre.
van Balen, B. & Prentice, C. 1997. Birds of the Negara River basin, South Kalimantan, Indonesia. *Kukila* 9: 81-107.
van Balen, S., Noske, R. & Supriatna, A.A. 2011. Around the Archipelago. *Kukila* 15: 126-143.
van Bemmel, A.C.V. 1948. A faunal list of the birds of the Moluccan Islands. *Treubia* 19: 323-402.
van Bemmel, A.C.V. & Voous, K.H. 1951. On the birds of the islands of Muna and Buton, S.E. Celebes. *Treubia* 21: 27–104.
van den Akker, M. 2003. Birds of Niaouli forest, southern Benin. *Bull. ABC* 10: 16-22.
van den Berg, A.B. 1994. WP reports. *Dutch Birding* 16: 77-83.
van den Berg, A.B. & Bosman, C.A.W. 1986. Supplementary notes on some birds of Lore Lindu Reserve, Central Sulawesi. *Forktail* 1: 7-13.
van den Berg, A.B. & Bosman, C.A.W. 1999. *Avifauna van Nederland 1: Zeldzame vogels van Nederland met vermelding van alle soorten.* GMB Uitgeverij & Stichting Uitgeverij van de KNNV, Haarlem & Utrecht.
van Marle, J.G. & Voous, K.H. 1988. *The birds of Sumatra*. British Ornithologists' Union Check-list 10. BOU, Tring.
van Oort, E.D. 1911. *Eudynamis minima*, an apparently new cuckoo from southwestern New Guinea. *Notes Leyden Mus.* 34: 54.
van Rossem, A.J. 1930. Some geographic variations in *Piaya cayana. Trans. San Diego Soc. Nat. Hist.* 6: 209-210.
van Rossem, A.J. 1934. Critical notes on Middle American birds. A. Notes on some species and subspecies of Guatemalan birds. *Bull. Mus. Comp. Zool.* 77: 387-490.
van Rossem, A.J. 1938. Descriptions of two new races of the Ground Cuckoo (*Morococcyx erythropygus* [Lesson]) from Mexico. *Proc. Biol. Soc. Wash.* 51: 169-171.
van Rossem, A.J. 1938a. The Groove-billed Ani of Lower California and northwestern Mexico. *Condor* 40: 91.
van Someren, V.D. 1947. Field notes on some Madagascar birds. *Ibis* 89: 235-267.
van Someren, V.G.L. 1915. [Three new birds from Uganda]. *Bull. Brit. Orn. Club* 35: 116.
van Someren, V.G.L. 1916. A list of birds collected in Uganda and British East Africa, with notes on their nesting and other habits. *Ibis* (10th series) 4: 193-252, 374-472.
van Someren, V.G.L. 1921. New birds from eastern Africa. *Bull. Brit. Orn. Club* 41: 121-126.
van Someren, V.G.L. 1922. Notes on the birds of East Africa. *Novit. Zool.* 29: 1-246.
van Someren, V.G.L. 1932. Birds of Kenya and Uganda, being addenda and corrigenda to my previous paper in "Novitates Zoologicae" XXIX, 1922. *Novit. Zool.* 37: 252-380.
van Someren, V.G.L. 1939. The birds of the Chyula Hills. *J. East Afr. Ug. Nat. Hist. Soc.* 14: 15-129.
van Someren, V.G.L. 1956. Days with birds. Studies of habits of some East African species. *Fieldiana Zool.* 38: 1-520.
van Someren, V.G.L. & van Someren, G.R.C. 1949. The birds of Bwamba. *Uganda Jour.* 13 (Special Supplement): 1-111.
van Tuinen, M., Sibley, C.G. & Hedges, S.B. 2000. The early history of modern birds inferred from DNA sequences of nuclear and mitochondrial ribosomal genes. *Mol. Biol. Evol.* 17: 451-457.
Van Tyne, J. & Sutton, G.M. 1937. The birds of Brewster County, Texas. *Misc. Publ. Mus. Zool. Univ. Michigan* 37: 1-119.
Varty, N. 1991. The status and conservation of Jamaica's

threatened and endemic forest avifauna and their habitats following Hurricane Gilbert. *Bird Conserv. Int.* 1: 135-151.

Vaughan, R. 1960. Notes on autumn migrants in Morocco. *Ibis* 102: 330-332.

Vaughan, R.E. & Jones, K.H. 1913. The birds of Hong Kong, Macao and the West River or Si Kiang in south-east China, with special reference to their nidification and seasonal movements. *Ibis* (10th Ser.) 1: 17-76, 163-201, 351-384.

Vaurie, C.H. 1957. Field notes on some Cuban birds. *Wilson Bull.* 69: 301-313.

Vehrencamp, S.L. 1977. Relative fecundity and parental effort in communally nesting anis, *Crotophaga sulcirostris*. *Science* 197: 403-405.

Vehrencamp, S.L. 1978. The adaptive significance of communal nesting in Groove-billed Anis (*Crotophaga sulcirostris*). *Behav. Ecol. Sociobiol.* 4: 1-33.

Vehrencamp, S.L. 1982a. Body temperatures of incubating versus non-incubating Roadrunners. *Condor* 84: 203-207.

Vehrencamp, S.L. 1982b. Testicular regression in relation to incubation effort in a tropical cuckoo. *Horm. Behav.* 16: 113-120.

Vehrencamp, S.L., Bowen, B.S. & Koford, R.R. 1986. Breeding roles and pairing patterns within communal groups of Groove-billed Anis. *Anim. Behav.* 34: 347-366.

Veiga, J.O., Filiberto, F C., Babarskas, M.P. & Savigny, C. 2005. *Aves de la provincia de Neuquén - Patagonia Argentina. Lista comentada y distribución*. R&C Editora, Buenos Aires.

Verea, C., Solórzano, A., Díaz, M., Parra, L., Araujo, M.A., Antón, F., Navas, O., Ruiz, O.J.L. & Fernández-Badillo, A. 2009. Registros de actividad reproductora y muda en algunas aves del norte de Venezuela. *Orn. Neotrop.* 20: 181-201.

Verheyen, R. 1953. *Oiseaux. Exploration du Parc National de l'Upemba*. Fasc. 19. Institut des Parcs Nationaux du Congo Belge, Bruxelles.

Verhoeye, J. & Holmes, D.A. 1999. The birds of the island of Flores – review. *Kukila* 10: 3–59.

Vernon, C.J. 1971a. Notes on the biology of the Black Coucal. *Ostrich* 42: 242-258.

Vernon, C.J. 1971b. Juvenile *Pachycoccyx audeberti* with *Prionops retzii*. *Ostrich* 42: 298.

Vernon, C.J. 1984. The breeding biology of the Thickbilled Cuckoo. Pp 825-840. in J. Ledger (Ed.). *Proc. Fifth Pan African Orn. Cong.* SAOS for the Organizing Committee, Johannesburg.

Vernon, C.J., Herremans, M., Underhill, L.G. & Berruti, A. 1997. Cuculidae: cuckoos and coucals. Pp. 544-573 in J.A. Harrison, D.G. Allan, L.G. Underhill, M. Herremans, A.J. Tree, V. Parker & C.J. Brown (Eds). *The atlas of southern African birds*. Vol. 1. Non-passerines. BirdLife South Africa, Johannesburg.

Verreaux, J. & Verreaux, É. 1855. Notice sur le genre *Phænicophæus*, Vieill., *Malcoha*, Cuv. *Rev. Mag. Zool. Pure Appl.*, 2nd ser., 7: 356-357.

Vesey-Fitzgerald, D. 1940. The birds of the Seychelles. I. The endemic birds. *Ibis* (14th series) 4: 480-489.

Vickery, P.D. 1977. The nesting season. 1 June-July 31, 1977. Northeastern maritime region. *Am. Birds* 31: 1110-1114.

Victoria, J.K. 1972. Clutch characteristics and egg discriminative ability of the African Village Weaverbird *Ploceus cucullatus*. *Ibis* 114: 367-376.

Vidoz, J.Q., Jahn, A.E. & Mamani, A.M. 2010. The avifauna of Estación Biológica Caparú, Bolivia. *Cotinga* 32: 5-22.

Vieillot, L.J.P. 1816. *Analyse d'une nouvelle ornithologie élémentaire*. Déterville, Paris.

Vieillot, L.J.P. 1817. *Nouveau dictionnaire d'histoire naturelle*, vol. 8. Déterville, Paris.

Vieillot, L.J.P. 1819. *Nouveau dictionnaire d'histoire naturelle*, vol. 34. Déterville, Paris.

Vigors, N.A. 1830. *Memoir of the life and public services of Sir T. S. Raffles (etc.)*. London.

Vigors, N.A. 1831. [Descriptions of new birds from a collection of birds from Manila]. *Proc. Committ. Sci. Corresp. Zool. Soc. London* 1: 96-98.

Vigors, N.A. 1832. Observations on a collection of birds from the Himalayan Mountains, with characters of new genera and species. *Proc. Committ. Sci. Corresp. Zool. Soc. London* 1: 170-176.

Vigors, N.A. & Horsfield, T. 1826. A description of the Australian birds in the collection of the Linnean Society; with an attempt at arranging them according to their natural affinities. *Trans. Linn. Soc. London* 15 (1): 170-331.

Vincent, A.W. 1946. On the breeding habits of some African Birds. *Ibis* 88: 48-67.

Vincent, A.W. 1947. On the breeding habits of some African Birds. *Ibis* 89: 163-204.

Vincent, J. 1934. The birds of northern Portuguese East Africa. Comprising a list of, and observations on, the collections made during the British Museum Expedition of 1931-32. Part v. *Ibis* (13th series) 4: 757-799.

Viney, C., Phillipps, K. & Lam Chiu Ying. 1994. *Birds of Hong Kong and South China*. 6th rev. ed. Privately published, Hong Kong.

Vogt, C.A. 2007. Range extensions and noteworthy records for mainland Ecuador. *Bull. Brit. Orn. Club* 127: 228-233.

von Berlepsch, H. 1895. Description of two new species of the genera *Phoenicophaes* and *Spilornis*, with a note on *Oriolus consobrinus* Rams. *Novit. Zool.* 2: 70-75.

von Berlepsch, H. & Hartert, E. 1902. On the birds of the Orinoco region. *Novit. Zool.* 9: 1-134.

von Berlepsch, H. & von Ihering, H. 1885. Die Vögel der Umgegend von Taquara do Mundo Novo, Prov. Rio Grande do Sul. *Z. Gesamte Orn.* 2: 97-184.

von Erlanger, C.F. 1905. Beiträge zur Vogelfauna Nordafrikas mit besonderer Berücksichtigung der Zoogeographie. *J. Orn.* 53: 433-499.

von Frisch, O. 1969. Die Entwicklung des Häherkuckucks (*Clamator glandarius*) im Nest der Wirtsvögel und seine Nachzucht in Gefangenschaft. *Z. Tierpsychol.* 26: 641-650.

von Haartman, L. 1981. Co-evolution of the Cuckoo *Cuculus canorus* and a regular Cuckoo host. *Ornis Fenn.* 58: 1–10.

von Ihering, H. 1902. Contribuições para o conhecimento da ornithologia de São Paulo. *Rev. Mus. Paulista* 5: 261-329.

von Ihering, H. 1904. O Rio Juruá. *Rev. Mus. Paulista* 6: 385-460.

von Ihering, H. 1914. Biologia e classificação das cuculidas brazileiras. *Rev. Mus. Paulista* 9: 371-390.

von Ihering, R. 1930. Contribution à la connaissance de l'ichthyophagie des oiseaux: L' "Anu-peixe". *Comptes Rend. Séances Soc. Biol. Filiales* 103: 1339-1340.

von Pelzeln, A. 1868-1871. *Zur Ornithologie Brasiliens. Resultate von Johann Natterers Reisen in den Jahren 1817 bis 1835*. Druck und Verlag von A. Pichler's Witwe und Sohn, Wien.

von Sneidern, K. 1954. Notas sobre algunas aves del Museo de História Natural de la Universidad del Cauca, Popayán, Colombia. *Noved. Colomb.* 1: 3-14.

von Spix, J.B. 1824. *Avium species novae, quas in itinere per Brasiliam annis 1817-1820 collegit et descripsit*, vol. 1. Hübschmanni, Munich.

Voous, K.H. 1955. *De vogels van de Nederlandse Antillen*. Natuurwetenschappelijke Werkgroep Nederlandse Antillen, Curaçao.

Voous, K.H. 1982. Straggling to islands - South American birds in the islands of Aruba, Curaçao, and Bonaire, south Caribbean. *J. Yamashina Inst. Orn.* 14: 171-178.

Voous, K.H. 1983. *Birds of the Netherlands Antilles*, 2nd ed. De Walburg Pers, Curaçao.

Vorderman, A.G. 1893. Bijdrage tot de kennis der vogels van den Kangean-Archipel. *Natuurk. Tijdschr. Nederl.-Indië* 52: 181-208.

Vorderman, A.G. 1898. Celebes-vogels. *Natuurk. Tijdschr. Nederl.-Indië* Part LVIII (10th Series, Part II): 26-121.

Vorderman, A.G. 1901. Systematisch overzicht der Vogels, welke op *Java* in inheemsch zijn/Naamlijst der *Java*-vogels. *Natuurk. Tijdschr. Nederl.-Indië* 60: 36-155.

Vorob'ev, K.A. 1951. [New Indian Cuckoo, *Cuculus micropterus ognevi*, subsp. n]. *Doklady Akad. Nauk SSSRNS.* 77 (3): 511. [In Russian]

Vowles, R.S. & Vowles, G.A. 1997. Some notes on the birds of Borneo. *Bull. Brit. Orn. Club* 105: 71-73.

Vuilleumier, F. 1993. Notes on birds observed in beech (*Fagus*) forests in the Maoershan Natural Reserve, Guangxi Autonomous Region, China. *Bull. Brit. Orn. Club* 113: 152-166.

Wagler, J.G. 1831. Einige Mittheilungen über Thiere Mexicos. *Isis* 24: col. 510-535.

Wagner, A.J. 1836. *Cuculus velox*, sp. nov. *Gel. Anz. Akad. Wiss. München* 3: col. 96.

Wait, W.E. 1931. Oceanic birds wanderers to Ceylon. *Spolia Zeylanica* 16:181-198.

Walden, A.V. 1872. Description of a supposed new species of cuckoo from Celebes. *Ann. Mag. Nat. Hist.* (4th series) 9: 305-306.

Walker, G.R. 1939. Notes on birds of Sierra Leone. *Ibis* (14th series) 3: 401-450.

Walker, J.S. 2007. Dietary specialization and fruit availability among frugivorous birds on Sulawesi. *Ibis* 149: 345-356.

Wallace, A.R. 1863. List of birds from the Sula Islands (east of Celebes), with descriptions of the new species. *Proc. Zool. Soc. London* 1862: 333-346.

Wallace, G.E., Alonso, H.G., McNicholl, M.K., Batista, D.R., Prieto, R.O., Llanes Sosa, A., Sanchez Oria, B. & Wallace, E.A.H. 1996. Winter surveys of forest-dwelling Neotropical migrant and resident birds in three regions of Cuba. *Condor* 98: 745-768.

Wallace, G.J. 1958. Notes on North American migrants in Colombia. *Auk* 75: 177-182.

Walter, H. 1979. *Eleonora's Falcon: Adaptations to Prey and Habitat in a Social Raptor.* Univ. of Chicago Press, Chicago.

Walters, M. 1995. *Birds' eggs: the visual guide to the eggs of over 500 species from around the world.* Dorling Kindersley, London, New York, Stuttgart.

Walters, M. 1996. The eggs of *Carpococcyx renauldi*: a correction. *Bull. Brit. Orn. Club* 116: 271.

Wardill, J.C., Fox, P.S., Hoare, D.J, Marthy, W. & Anggraini, K. 1999. Birds of the Rawa Aopa Watumohai National Park, south-east Sulawesi. *Kukila* 10: 91-114.

Warham, J. 1957. Cockatoo Island birds. *Emu* 57: 225-231.

Warner, D.W. 1951. A new race of the cuckoo, *Chalcites lucidus*, from the New Hebrides Islands. *Auk* 68: 106-107.

Warner, D.W. & Beer, J.R. 1957. Birds and mammals of the Mesa de San Diego, Puebla, Mexico. *Acta Zool. Mex.* 2: 1-17.

Watling, D. 1982. *Birds of Fiji, Tonga and Samoa.* Millwood Press, Wellington.

Watling, D. 1983. Ornithological notes from Sulawesi. *Emu* 83: 247-261.

Watson, J.S. & Bull, P.C. 1950. Communal display of the Shining Cuckoo. *New Zealand Bird Notes* 3: 226.

Webb, H.P. 1992. Field observations of the birds of Santa Isabel, Solomon Islands. *Emu* 92: 52-57.

Wedderburn, B. 2009. West Java Trip Report by Bruce Wedderburn – Aug 09. www.travellingbirder.com/tripreports/reports/indonesia_java_birding_trip_report_aug-09.pdf

Wehtje, W., Walter, H.S., Rodríguez-Estrella, R., Llinas, J. & Castellanos, A. 1993. An annotated checklist of the birds of Isla Socorro, Mexico. *Western Birds* 24: 1-16.

Wells, D.R. 1982. A confirmation of the specific relations of *Cuculus saturatus insulindae* Hartert. *Bull. Brit. Orn. Club* 102: 62-63.

Wells, D.R. 1984-85. Malayan Bird Report. *Malayan Nat. J.* 43: 148-171.

Wells, D.R. 1987. Malayan Bird Report: 1986 and 1987. *Malayan Nat. J.* 43: 172-210.

Wells, D.R. 1999. *The birds of the Thai-Malay Peninsula.* Vol 1. Non-passerines. London, Academic Press.

Wells, D.R. 2007. *The birds of the Thai-Malay Peninsula.* Vol 2. Passerines. London, Christopher Helm.

Wells, D.R. & Becking, J.H. 1975. Vocalizations and status of Little and Himalayan Cuckoos, *Cuculus poliocephalus* and *C. saturatus*, in southeast Asia. *Ibis* 117: 366-371.

Wendelken, P.W. & Martin, R.F. 1986. Recent data on the distribution of birds in Guatemala. *Bull. Brit. Orn. Club* 106: 16-21.

Wernham, C., Toms, M., Marchant, J., Clark, J., Siriwardena, G. & Baillie, S. 2002. *The Migration Atlas: movements of the birds of Britain and Ireland.* Poyser, London.

Wetmore, A. 1916. *Birds of Porto Rico.* Bulletin 326. US Department of Agriculture, Washington DC.

Wetmore, A. 1917. A new cuckoo from New Zealand. *Proc. Biol. Soc. Wash.* 30: 1-2.

Wetmore, A. 1926. Observations on the birds of Argentina, Paraguay, Uruguay, and Chile. *Bull. US National Mus.* 133: 1-448.

Wetmore, A. 1927. The birds of Porto Rico and the Virgin Islands. Psittaciformes to Passeriformes. Scientific Survey of Porto Rico and the Virgin Islands. *New York Acad. Sci., Scient. Surv. P. Rico Virgin Is.* 9: 409-571.

Wetmore, A. 1960. A classification for the birds of the world. *Smithson. Misc. Coll.* 139: 1-37.

Wetmore, A. 1968. Birds of the Republic of Panama. Part 2. *Smithson. Misc. Coll.* 150: 1-605.

Wetmore, A. & Swales, B.H. 1931. The birds of Haiti and the Dominican Republic. *Bull. US National Mus.* 155: i-iv, 1-483.

Whitaker, J.I.S. 1905. *Birds of Tunisia. II.* Porter, London.

White, C.M.N. & Bruce, M.D. 1986. *The birds of Wallacea. Sulawesi, the Moluccas & Lesser Sunda Islands, Indonesia.* BOU Check-list 7. British Ornithologists' Union, Tring.

White, G.L., Kenefick, M. & Murphy, W.L. 2007. The status and abundance of birds in Trinidad and Tobago. *Living World* 2007: 57-68.

Whitehead, J. 1893. *Exploration of Mt. Kina Balu, North Borneo.* Gurney & Jackson, London.

Whitney, B.M., Rowlett, J.L. & Rowlett, R.A. 1994. Distributional and other noteworthy records for some Bolivian birds. *Bull. Brit. Orn. Club* 114: 149-162.

Whitson, M.A. 1971. *Field and laboratory investigations of the ethology of courtship and copulation in the Greater Roadrunner (Geococcyx californianus - Aves, Cuculidae).* Ph.D. Dissertation. Univ. Oklahoma, Norman.

Whitson, M. 1975. Courtship behavior of the Greater Roadrunner. *Living Bird* 14: 215-255.

Whittaker, A. 1995. First report of *Coccyzus pumilus* for Brazil (Cuculiformes: Cuculidae). *Ararajuba* 3: 81.

Whittaker, A. 2004. Noteworthy ornithological records from Rondônia, Brazil, including a first country record, comments on austral migration, life history, taxonomy and distribution, with relevant data from neighbouring states, and a first record for Bolivia. *Bull. Brit. Orn. Club* 124: 239-271.

Whittaker, A. & Oren, D.C. 1999. Important ornithological records from the Rio Juruá, western Amazonia, including twelve additions to the Brazilian avifauna. *Bull. Brit. Orn. Club* 119: 235-260.

Widman, O. 1882. Cuckoos laying in the nests of other birds. *Bull. Nuttall Orn. Club* 7: 56-57.

Wied, M., Prinzen zu. 1832. *Beiträge zur Naturgeschichte von Brasilien* Vol. 4. Landes-Industrie-Comptoirs, Weimar.

Wiedenfeld, D.A. 2006. Aves, The Galapagos Islands, Ecuador. *Check List* 2: 1-27.

Wiens, J.A. 1965. Nest parasitism of the Dickcissel by the Yellow-billed Cuckoo in Marshall County, Oklahoma. *Southw. Nat.* 10: 142.

Wijesinghe, M. 1999. Nesting of Green-billed Coucals *Centropus chlororhynchos* in Sinharaja, Sri Lanka. *Forktail* 15: 43-45.

Wildash, P. 1968. *Birds of South Vietnam.* Charles E. Tuttle, Rutland, Vermont.

Wiley, A.J. 1948. Foster parents of the red-chested cuckoo. *Nature in East Africa* 7: 2.

Wilkinson, A.S. 1947. A cuckoo in the nest. *New Zealand Bird Notes* 2: 77-79.

Wilkinson, R., Dutson, G., Sheldon, B., Noor, D. & Noor, Y.R. 1991. The avifauna of the Barito Ulu Region, central Kalimantan. *Kukila* 5: 99-116.

Wilks, B.J. & Laughlin, H.E. 1961. Roadrunner preys on a bat. *J. Mamm.* 42: 98.

Willett, G. 1912. Birds of the Pacific slope of southern California. *Pacific Coast Avifauna* 7: 1-122.

Williams, N. 1946. *Alseonax minimus pumilus* (?) and *Lampromorpha klaasi. J. E. Afr. Nat. Hist. Soc.* 19: 138.

Willis, E.O. 1974. Populations and local extinctions of birds on Barro Colorado Island, Panama. *Ecol. Monogr.* 44: 153-169.

Willis, E.O. 1982. Ground-cuckoos (Aves, Cuculidae) as army ant followers. *Rev. Brasil. Biol.* 42: 753-756.

Willis, E.O. 1983a. Anis (Aves, Cuculidae) as army ant followers. *Rev. Brasil. Biol.* 43: 33-36.

Willis, E.O. 1983b. *Piaya* cuckoos (Aves, Cuculidae) as army ant followers. *Rev. Brasil. Biol.* 43: 29-32.

Willis, E.O. 1988. Behavioral notes, breeding records, and range extensions for Colombian birds. *Rev. Acad. Col. Cienc. Exact. Fis. Nat.* 16: 137-150.

Willis, E.O. & Eisenmann, E. 1979. A revised list of birds of Barro Colorado Island, Panama. *Smithson. Contr. Zool.* 291: 1-39.

Willis, E.O. & Oniki, Y. 1985. Bird specimens new for the state of São Paulo, Brazil. *Rev. Brasil. Biol.* 45: 105-108.

Willis, E.O. & Oniki, Y. 1990a. Case 2727. *Coccyzus euleri* Cabanis, 1873 (Aves, Cuculiformes): proposed conservation of the specific name. *Bull. Zool. Nomencl.* 47: 195-197.

Willis, E.O. & Oniki, Y. 1990b. Levantamento preliminar das aves de inverno em dez áreas do sudoeste de Mato Grosso, Brasil. *Ararajuba* 1: 19-38.

Willis, E.O. & Oniki, Y. 1991. Avifaunal transects across the open zones of northern Minas Gerais, Brazil. *Ararajuba* 2: 41-58.

Willis, E.O. & Oniki, Y. 2002. Birds of Santa Teresa, Espírito Santo, Brazil: do humans add or subtract species? *Pap. Avuls. Zool. S. Paulo* 42: 193-264.

Willis, E.O. & Oniki, Y. 2003. *Aves do estado de São Paulo*. Editora Divisa, Rio Claro, São Paulo, Brazil.

Wilmé, L. 1994. Note sur le parasitisme du Coucou de Madagascar *Cuculus rochii. Working group on birds in the Madagascar Region Newsletter.* 3(2): 4-5.

Wilson, A. 1811. *American ornithology or the natural history of the birds of the United States, illustrated with plates engraved and colored from original drawings taken from nature*. Vol. 4. Bradford and Inskeep, Philadelphia.

Wilson, J.K. 2000. Additional observations on precopulatory behavior of Yellow-billed Cuckoos. *Southw. Nat.* 45: 535-536.

Wilson, R.G. 1992. Parasitism of Yellow-olive Flycatcher by the Pheasant Cuckoo. *Euphonia* 1: 34-36.

Winterbottom, J.M. 1938. Further notes on some Northern Rhodesian birds. *Ibis* (14th series) 2: 269-277.

Winterbottom, J.M. 1939. Miscellaneous notes on some birds of Northern Rhodesia. *Ibis* (14th series) 3: 712-734.

Winterbottom, J.M. 1942. A contribution to the ornithology of Barotseland. *Ibis* (14th series) 6: 18-27, 337-389.

Winterbottom, J.M. 1951 *Common birds of the bush*. Longmans, Green, London.

Wolfe, D.H. 1994. Yellow-billed Cuckoo hatched in Mourning Dove nest. *Bull. Oklahoma Orn. Soc.* 27: 30-31.

Wolfe, L.R. 1938. Birds of central Luzon. *Auk* 55: 198-224.

Wolters, H.E. 1975-1982. *Die Vogelarten der Erde*. Paul Parey, Hamburg und Berlin.

Wood, J. 1911. Notes from East London. *J. South Afr. Orn. Union* 7: 51.

Woodell, R. 1976a. Notes on the Aldabran Coucal *Centropus toulou insularis. Ibis* 118: 263-268.

Woodell, R. 1976b. Variation in juvenile plumage of *Centropus toulou toulou* (Müller) and *Centropus toulou insularis* Ridgway. *Bull. Brit. Orn. Club* 96: 72-75.

Woods, R.S. 1960. Notes on the nesting of the Roadrunner. *Condor* 62: 483-484.

Worman, A.C. 1930. Male Emerald Cuckoo (*Chrysococcyx cupreus intermedius*) feeding young. *Oölog. Rec.* 10: 76-77.

Wright, R.E. 1973. Observations on the urban feeding habits of the Roadrunner (*Geococcyx californianus*). *Condor* 75: 246.

Wunderle, J.M., Jnr. 1981. Avian predaton upon *Anolis* lizards on Grenada, West Indies. *Herpetologia* 37: 104-108.

Wunderle, J.M., Jnr. & Latta, S.C. 1996. Avian abundance in sun and shade coffee plantations and remnant pine forest in the Cordillera Central, Dominican Republic. *Orn. Neotrop.* 7: 19-34.

Wunderle, J.M., Jnr. & Latta, S.C. 1998. Avian resource use in Dominican shade coffee plantations. *Wilson Bull.* 110: 271-281.

Wyllie, I. 1981. *The Cuckoo*. Batsford, London.

Yang, C., Antonov, A., Cai, Y., Stokke, B.G., Moksnes, A., Røskaft, E. & Liang, W. 2012. Large Hawk-Cuckoo *Hierococcyx sparverioides* parasitism on the Chinese Babax *Babax lanceolatus* may be an evolutionarily recent host–parasite system. *Ibis* 152: 200-204.

Yasukawa, K. 2010. Yellow-billed Cuckoo hatched and fed by a Red-winged Blackbird. *Wilson J. Orn.* 122: 402-405.

Yosef, R. 2002. Second breeding record of Great Spotted Cuckoo *Clamator glandarius* in Eilat. *Sandgrouse* 24: 142-144.

Yoshino, T. 1999. *Kakko – Nihon no takurancho* [*Cuckoos – broodparasitic birds in Japan*]. Bunichisogoshuppan, Tokyo. [in Japanese]

Young, C.G. 1925. Notes on the nests and eggs of some British Guiana birds. *Ibis* (12th series) 1: 465-475.

Young, C.G. 1929. A contribution to the ornithology of the coastland of British Guiana (Part II). *Ibis* (12th series) 5: 1-38.

Zacharias, V.J. & Gaston, A.J. 1993. The birds of Wynaad, southern India. *Forktail* 8: 11-23.

Zarudny, N.A. 1914. Notiz über die Kuckucke von Turkestan. *Orn. Mitt.* 5: 105-115 [in Russian, with German summary].

Zetra, B., Rafiastanto, A., Rombang, W.M. & Trainor, C.R. 2002. Rediscovery of the critically endangered Sumatran Ground Cuckoo *Carpococcyx viridis. Forktail* 18: 63-65.

Zietz, F.R. 1914. The avifauna of Melville Island, Northern Territory. *S. Aust. Orn.* 1 pt. 1: 11-18, pt. 2: 17.

Zillman, E.E. 1965. Observations of the Oriental Cuckoo at Gin Gin, Queensland. *Aust. Bird Watcher* 2: 148-151.

Zimmer, J.T. 1930. Birds of the Marshall Field Peruvian expedition, 1922-1923. *Field Mus. Nat. Hist. Zool. Ser.* 17: 231-493.

Zimmer, K.J. & Hilty, S.L. 1997. Avifauna of a locality in the upper Orinoco drainage of Amazonas, Venezuela. *Orn. Monogr.* 48: 865-885.

Zimmer, K.J., Parker T.A. III, Isler, M.L. & Isler, P.R. 1997. Survey of a southern Amazonian avifauna: the Alta Floresta region, Mato Grosso, Brazil. *Orn. Monogr.* 48: 887-918.

Zimmerman, D.A. 1970. Roadrunner predation on passerine birds. *Condor* 72: 475-476.

Zimmerman, D.A. 1972. The avifauna of the Kakamega Forest, western Kenya, including a bird population study. *Bull. Amer. Mus. Nat. Hist.* 140: 259-339.

Zimmerman, D.A. & Harry, G.B. 1951. Summer birds of Autlán, Jalisco. *Wilson Bull.* 63: 302-314.

Zimmerman, D.A., Turner, D.A. & Pearson, D.J. 1996. *Birds of Kenya and northern Tanzania*. Christopher Helm, London.

Zinner, D. 2001. Ornithological notes from a primate survey in Eritrea. *Bull. ABC* 8: 95-106.

Zotta, A.R. 1937. Notas ornitológicas. *Hornero* 6: 477-483.

INDEX

Numbers in **bold** refer to plate numbers, page numbers in *italic* refer to the caption text in the colour plate section. Other numbers refer to the first page only of the entry in the species accounts.

A

addendus, Cacomantis variolosus **31** *88*, 406
aegyptius, Centropus senegalensis **10** *46*, 165
aequatorialis, Neomorphus geoffroyi 138
aereus, Ceuthmochares **15** *56*, 235
aereus, Ceuthmochares aereus **15** *56*, 235
aeruginosus, Cacomantis sepulcralis 409
African Black Coucal **9** *44*, 189
African Cuckoo **34** *94*, 465
African Emerald Cuckoo **26** *78*, 354
African Striped Cuckoo 267
aheneus, Chrysococcyx poecilurus 374
alberti, Eudynamys scolopaceus 327
albidiventris, Centropus milo **4** *34*, 144
albifrons, Chrysococcyx minutillus **28** *82*, 370
American Striped Cuckoo **1** *28*, 115
americanus, Coccyzus **20** *66*, 290
amethystinus, Chrysococcyx xanthorhynchus 340
Andaman Coucal 180
andamanensis, Centropus sinensis **6** *38*, 180
ani, Crotophaga **1** *28*, 108
Annam Ground Cuckoo 210
anonymus, Centropus sinensis 180
anselli, Centropus **9** *44*, 162
arfakianus, Cacomantis castaneiventris 388
aruensis, Centropus menbeki 147
Ash-coloured Cuckoo **17** *60*, 285, 390
Asian Emerald Cuckoo **25** *76*, 337
Asian Koel 325
Asian Lesser Cuckoo 455
Asian Violet Cuckoo 340
ateralbus, Centropus **5** *36*, 145
audeberti, Pachycoccyx **22** *70*, 320
audeberti, Pachycoccyx audeberti 320
Australian Koel 325
australis, Ceuthmochares **15** *56*, 233
australis, Neomorphus geoffroyi 138
Azara's Cuckoo 287

B

bahamensis, Saurothera merlini 316
bakeri, Cuculus canorus 476
Banded Bay Cuckoo **30** *86*, 394
Banded Ground Cuckoo **3** *32*, 133

bangsi, Cuculus canorus 476
Bare-eyed Coucal 199
barnardi, Chrysococcyx minutillus 370
Barred Cuckoo 419
Barred Long-tailed Cuckoo 419
barussarum, Surniculus lugubris 425
basalis, Chrysococcyx **27** *80*, 357
Bay Banded Cuckoo 394
Bay-breasted Cuckoo 310
Bay Coucal **7** *40*, 160
bengalensis, Centropus **6** *38*, 195
bengalensis, Centropus bengalensis 195
bernsteini, Centropus **4** *34*, 200
Bernstein's Coucal 200
Biak Bronze Coucal 149
Biak Coucal **6** *38*, 149
Biak Island Coucal 149
Black-and-white Cuckoo 270
Black-and-white Roadrunner 131
Black-bellied Coucal 189
Black-bellied Cuckoo **13** *52*, 277
Black-bellied Malkoha **15** *56*, 247
Black-billed Cuckoo **20** *66*, 302
Black-billed Koel 325
Black-capped Cuckoo 323
Black-capped Koel 323
Black-chested Coucal 189
Black Coucal 189
Black Crested Cuckoo 270
Black Cuckoo **34** *94*, 447
Black-eared Cuckoo **27** *80*, 379
Black-faced Coucal **7** *40*, 154
Black-hooded Coucal 156
Black Jungle Coucal 147
Black Scrub Coucal 200
Black-throated Coucal **9** *44*, 163
blandus, Cacomantis variolosus 405
Blue Coua **7** *40*, 215
Blue Coucal 235
Blue-faced Malkoha **15** *56*, 246
Blue-headed Coucal **10** *46*, 170
Blue Malkoha 235
Blyth's Cuckoo 468
bocki, Hierococcyx **22** *70*, 431
Bornean Ground Cuckoo **11** *48*, 208

borneensis, Rhamphococcyx curvirostris **16** *58*, 251
borneensis, Rhopodytes diardi **15** *56*, 247
Brain-fever Bird 384, 397, 433, 436
brazzae, Pachycoccyx audeberti 320
Broad-billed Cuckoo 363
Brown Coucal 160, 180
Brush-cuckoo 388
Brush Cuckoo **31** *88*, 404
bubutus, Centropus sinensis 180
Buff-headed Coucal **4** *34*, 143
burchellii, Centropus superciliosus **10** *46*, 175
Burchell's Coucal 174

C

cabanisi, Piaya cayana 280
Cacomantis 384
caerulea, Coua **7** *40*, 215
cafer, see *levaillantii, Clamator*
cagayanensis, Dasylophus superciliosus **17** *60*, 256
Caliechthrus 412
californianus, Geococcyx **2** *30*, 126
California Roadrunner 126
calyorhynchus, Rhamphococcyx **16** *58*, 254
calyorhynchus, Rhamphococcyx calyorhynchus 254
canorus, Cuculus **32** *90*, 475
canorus, Cuculus canorus 475
caprius, Chrysococcyx **26** *78*, 343
carpenteri, Centropus viridis 192
Carpococcyx 206
castaneiventris, Cacomantis **29** *84*, 388
castaneiventris, Cacomantis castaneiventris **29** *84*, 388
cayana, Piaya **13** *52*, 279
cayana, Piaya cayana **13** *52*, 279
celebensis, Centropus **7** *40*, 160
celebensis, Centropus celebensis **7** *40*, 160
Celebes Coucal 160
Celebes Malkoha 254
CENTROPODINAE 143
Centropus 143
Cercococcyx 414
Ceuthmochares 233
Ceylon Coucal Pheasant 152
Chalcites, see *Chrysococcyx*
chalybaeus, Surniculus velutinus 423
chalybeus, Centropus **6** *38*, 149
Channel-billed Cuckoo **14** *54*, 334
Chapman's Ground Cuckoo 135
Chattering Yellowbill **15** *56*, 235
Chestnut-bellied Cuckoo **21** *68*, 308
Chestnut-bellied Malkoha **15** *56*, 244
Chestnut-breasted Brush-cuckoo 388

Chestnut-breasted Cuckoo **29** *84*, 388
Chestnut-breasted Malkoha **16** *58*, 251
Chestnut-winged Cuckoo **18** *62*, 260
chinensis, Eudynamys scolopaceus **24** *74*, 326
chlorophaea, Rhinortha **14** *54*, 230
chlororhynchos, Centropus **8** *42*, 152
Chrysococcyx 337
ciceliae, Hierococcyx varius 436
cinereus, Coccyzus **17** *60*, 285
circe, Piaya cayana **13** *52*, 280
Clamator 260
clamosus, Cuculus **34** *94*, 447
clamosus, Cuculus clamosus **34** *94*, 447
cleis, Chrysococcyx minutillus **28** *82*, 370
Coccycua, see *Piaya*
Coccyzus 283
Cocos Cuckoo **20** *66*, 301
Cocos Island Cuckoo 301
Common Crow-pheasant 180
Common Cuckoo **32** *90*, 475
Common Hawk-cuckoo **22** *70*, 436
Common Koel **24** *74*, 325
concretus, Cuculus micropterus 459
Congo Yellowbill 235
Coppery-tailed Coucal **9** *44*, 172
coquereli, Coua **12** *50*, 221
Coquerel's Coua **12** *50*, 221
Coquerel's Madagascar Coucal 221
Coral-billed Ground Cuckoo **11** *48*, 210
coromandus, Clamator **18** *62*, 260
corvinus, Eudynamys scolopaceus 326
Coua 212
COUINAE 206
crassirostris, Chrysococcyx **28** *82*, 377
crassirostris, Chrysococcyx crassirostris 377
crassirostris, Cuculus **35** *96*, 458
Crested Coua **11** *48*, 212
Crested Madagascar Coucal 212
Crimson-faced Malkoha 242
cristata, Coua **11** *48*, 212
cristata, Coua cristata **11** *48*, 212
Crotophaga 105
CROTOPHAGINAE 101
Cuban Lizard Cuckoo 316
CUCULIDAE 101
CUCULINAE 230
Cuculus 447
cumingi, Lepidogrammus **17** *60*, 258
cupreicaudus, Centropus **9** *44*, 172
cupreus, Chrysococcyx **26** *78*, 354
cursor, Coua **12** *50*, 223

curvirostris, Rhamphococcyx **16** *58*, 251
curvirostris, Rhamphococcyx curvirostris **16** *58*, 251
cyanocephalus, Eudynamys scolopaceus 327

D

Dark-backed Bronze Cuckoo 369
Dark-billed Cuckoo **21** *68*, 287
Dark Hawk-cuckoo **22** *70*, 431
Dasylophus 256
decolor, Saurothera merlini **19** *64*, 316
delalandei, Coua **13** *52*, 227
Delalande's Coua 227
deningeri, Rhamphococcyx curvirostris 251
diardi, Rhopodytes **15** *56*, 247
diardi, Rhopodytes diardi **15** *56*, 247
dicruroides, Surniculus **23** *72*, 421
dicruroides, Surniculus dicruroides 421
Dideric Cuckoo **26** *78*, 343
Didric Cuckoo 343
Diederic Cuckoo 343
Diederik Cuckoo 343
Diplopterus, see *Tapera*
Dromococcyx 118
dulcis, Neomorphus geoffroyi **3** *32*, 139
dumonti, Coua cristata 212
Dusky Long-tailed Cuckoo **32** *90*, 414
Dwarf Cuckoo **17** *60*, 283
Dwarf Hawk-cuckoo 429
Dwarf Koel **23** *72*, 323

E

efulenensis, Centropus leucogaster 163
elongatus, Rhopodytes tristis **15** *56*, 249
erythropthalmus, Coccyzus **20** *66*, 302
erythropygus, Morococcyx **2** *30*, 124
erythropygus, Morococcyx erythropygus 124
Eudynamys 325
euleri, Coccyzus **20** *66*, 295
Eurasian Cuckoo 475
European Cuckoo 475
excitus, Cacomantis flabelliformis **29** *84*, 391

F

Fan-tailed Brush Cuckoo 390
Fan-tailed Cuckoo **29** *84*, 390
fasciipygialis, Centropus superciliosus 175
fasciolatus, Cacomantis sonneratii 394
ferrugineus, Coccyzus **20** *66*, 301
Fiery-billed Malkoha 254
fischeri, Centropus monachus 170
flabelliformis, Cacomantis **29** *84*, 390

flabelliformis, Cacomantis flabelliformis **29** *84*, 390
flavigularis, Chrysococcyx **26** *78*, 352
flavirostris, Ceuthmochares aereus **15** *56*, 235
flecki, Centropus senegalensis 165
fordi, Scythrops novaehollandiae 335
Fork-tailed Drongo-cuckoo **23** *72*, 421
frater, Eudynamys scolopaceus 326
fugax, Hierococcyx **33** *92*, 443

G

Gabon Coucal **9** *44*, 162
gabonensis, Cuculus clamosus **34** *94*, 447
Geococcyx 126
geoffroyi, Neomorphus **3** *32*, 138
geoffroyi, Neomorphus geoffroyi **3** *32*, 138
Giant Coua **12** *50*, 225
Giant Coucal 184
Giant Cuckoo 334
Giant Forest Coucal 199
Giant Madagascar Coucal 225
gigas, Coua **12** *50*, 225
glandarius, Clamator **18** *62*, 263
Golden Bronze Cuckoo 363
Golden Cuckoo 354, 363
goliath, Centropus **4** *34*, 184
Goliath Coucal **4** *34*, 184
Gould's Bronze Cuckoo **28** *82*, 373
gracilis, Piaya minuta **21** *68*, 274
Great Ani 105
Great Coucal 163
Greater Ani **1** *28*, 105
Greater Black Coucal **4** *34*, 147
Greater Coucal **6** *38*, 180
Greater Green-billed Malkoha 249
Greater Roadrunner **2** *30*, 126
Great Lizard Cuckoo **19** *64*, 316
Great Spotted Cuckoo **18** *62*, 263
Green-billed Coucal **8** *42*, 152
Green-billed Ground Cuckoo 206, 208
Green-billed Malkoha 249
Green-cheeked Bronze Cuckoo 369
Green Coucal 192, 233
Green Cuckoo 354
Green Malkoha 233
Grey-bellied Cuckoo **30** *86*, 401
Grey-breasted Brush Cuckoo 397
Grey-capped Cuckoo **21** *68*, 306
Grey-headed Cuckoo 397
grillii, Centropus **9** *44*, 189
grisescens, Microdynamis parva 323
Groove-billed Ani **1** *28*, 112

Guira 101
Guira Cuckoo **1** *28*, 101
guira, Guira **1** *28*, 101
gularis, Cuculus **34** *94*, 465

H

hainanus, Rhopodytes tristis 249
Haitian Hyetornis 310
Haitian Lizard Cuckoo 318
harringtoni, Rhamphococcyx curvirostris **16** *58*, 252
harterti, Chrysococcyx lucidus 364
harterti, Eudynamys scolopaceus 326
'*heinrichi', Cacomantis variolosus* **31** *88*
Heinrich's Brush Cuckoo 404
Heinrich's Cuckoo 404
hellmayri, Piaya cayana 280
Hierococcyx 429
Himalayan Cuckoo 468
Hispaniolan Lizard Cuckoo **19** *64*, 318
Hodgson's Hawk-cuckoo 443
Horsfield's Bronze Cuckoo **27** *80*, 357
Horsfield's Cuckoo 468
Horsfield's Hawk-cuckoo 443
hybridus, Eudynamys scolopaceus 327
Hyetornis 308
hyperythrus, Hierococcyx **33** *92*, 439

I

Indian Cuckoo **35** *96*, 459
Indian Drongo-cuckoo 421
Indian Koel 325
Indian Plaintive Cuckoo 401
infaustus, Cacomantis variolosus **31** *88*, 405
infuscata, Taccocua leschenaultii 238
insulana, Piaya cayana 280
insularis, Centropus toulou 186
insulindae, Cuculus lepidus 473
intermedius, Centropus sinensis 180

J

Jacobin Cuckoo **18** *62*, 270
jacobinus, Clamator **18** *62*, 270
jacobinus, Clamator jacobinus **18** *62*, 270
Jamaican Hyetornis 308
Jamaican Lizard Cuckoo **19** *64*, 312
Javan Coucal **7** *40*, 178
javanensis, Centropus bengalensis 195
Javan Hawk-cuckoo **33** *92*, 443
javanicus, Zanclostomus **16** *58*, 240
javanicus, Zanclostomus javanicus 240
julieni, see *euleri*

jungei, Chrysococcyx poecilurus **28** *82*, 374

K

Kai Coucal 202
kangeanensis, Centropus sinensis **6** *38*, 180
kangeangensis, Rhopodytes tristis 249
Khasi Hills Cuckoo 475
klaas, Chrysococcyx **26** *78*, 348
Klaas's Cuckoo **26** *78*, 348

L

Lamprococcyx, see *Chrysococcyx*
lanceolatus, Cacomantis merulinus 397
lansbergi, Coccyzus **21** *68*, 306
Large Coucal 184
Large Green-billed Malkoha 249
Large Hawk-cuckoo **22** *70*, 433
Lark-heeled Cuckoo 174, 180, 195
layardi, Chrysococcyx lucidus **27** *80*, 364
Lepidogrammus 258
lepidophanes, Neomorphus pucheranii 137
lepidus, Cuculus **36** *98*, 473
lepidus, Cuculus lepidus 473
leschenaultii, Taccocua **16** *58*, 237
leschenaultii, Taccocua leschenaultii **16** *58*, 237
Lesser Black Coucal **4** *34*, 200
Lesser Coucal **6** *38*, 195
Lesser Cuckoo **35** *96*, 455
Lesser Green-billed Malkoha 246, 247
Lesser Ground Cuckoo **2** *30*, 124
Lesser Hawk-cuckoo 429
Lesser Roadrunner **2** *30*, 131
Lesser Sundan Bronze Cuckoo 369
Lesson Ground-Cuckoo 124
leucogaster, Centropus **9** *44*, 163
leucogaster, Centropus leucogaster 163
leucolophus, Caliechthrus **30** *86*, 412
levaillantii, Clamator **18** *62*, 267
Levaillant's Cuckoo **18** *62*, 267
lignator, Centropus bengalensis 195
Little Bronze Cuckoo **28** *82*, 369
Little Cuckoo **21** *68*, 274, 455
Little Koel 323, 382
Little Malkoha 230
loandae, Centropus superciliosus **10** *46*, 174
Long-billed Cuckoo **14** *54*, 382
longicaudatus, Rhopodytes tristis 249
longirostris, Saurothera **19** *64*, 318
longirostris, Saurothera longirostris **19** *64*, 318
Long-tailed Cuckoo 331
Long-tailed Koel **14** *54*, 331

Long-tailed Malkoha **15** *56*, 249
Long-tailed New Zealand Cuckoo 331
Long-tailed Pacific Cuckoo 331
lucidus, Chrysococcyx **27** *80*, 363
lucidus, Chrysococcyx lucidus **27** *80*, 363
lugubris, Surniculus **23** *72*, 425
lugubris, Surniculus lugubris 425

M

macrocercus, Cacomantis variolosus 406
macroura, Piaya cayana **13** *52*, 280
maculatus, Chrysococcyx **25** *76*, 337
Madagascar Coucal **8** *42*, 186
Madagascar Cuckoo **35** *96*, 463
Madagascar Lesser Cuckoo 463
major, Centropus viridis 192
major, Crotophaga **1** *28*, 105
malayanus, Eudynamys scolopaceus 326
Malay Cuckoo 369
Malay Green Cuckoo 369
Malay Ground Cuckoo 206, 208
Malay Hawk-cuckoo 443
malayanus, see *minutillus, Chrysococcyx*
Mangrove Cuckoo **20** *66*, 298
maxima, Coua cristata 212
maximiliani, Neomorphus geoffroyi **3** *32*, 139
mechowi, Cercococcyx **32** *90*, 414
medius, Centropus bengalensis 196
megarhynchus, Rhamphomantis **14** *54*, 382
mehleri, Piaya cayana 279
melacoryphus, Coccyzus **21** *68*, 287
melanogaster, Piaya **13** *52*, 277
melanops, Centropus **7** *40*, 154
melanorhynchus, Eudynamys scolopaceus **24** *74*, 326
melanurus, Centropus phasianinus 203
menbeki, Centropus **4** *34*, 147
menbeki, Centropus menbeki 147
Menbek's Coucal 147
meridionalis, Rhamphococcyx calyorhynchus 254
merlini, Saurothera **19** *64*, 316
merlini, Saurothera merlini **19** *64*, 316
merulinus, Cacomantis **30** *86*, 397
merulinus, Cacomantis merulinus **30** *86*, 397
mesura, Piaya cayana **13** *52*, 280
mexicana, Piaya cayana **13** *52*, 279
mexicanus, Morococcyx erythropygus 124
meyerii, Chrysococcyx **25** *76*, 367
Meyer's Bronze Cuckoo 367
Micrococcyx, see *Coccyzus*
Microdynamis 323
micropterus, Cuculus **35** *96*, 459

micropterus, Cuculus micropterus 459
milo, Centropus **4** *34*, 143
milo, Centropus milo **4** *34*, 143
mindanensis, Eudynamys scolopaceus **24** *74*, 326
mindorensis, Centropus viridis 192
minimus, Eudynamys scolopaceus 327
minor, Coccyzus **20** *66*, 298
minuta, Piaya **21** *68*, 274
minuta, Piaya minuta **21** *68*, 274
minutillus, Chrysococcyx **28** *82*, 369
minutillus, Chrysococcyx minutillus **28** *82*, 369
Misocalius 379
misoriensis, Chrysococcyx poecilurus 375
mogenseni, Piaya cayana 280
Moluccan Brush Cuckoo 404
Moluccan Cuckoo 404
Moluccan Drongo-cuckoo **23** *72*, 428
Moluccan Koel 325
monachus, Centropus **10** *46*, 170
monachus, Centropus monachus 170
montanus, Cercococcyx **32** *90*, 419
montanus, Cercococcyx montanus **32** *90*, 419
Morococcyx 124
Mountain Bronze Cuckoo 361, 367
Mountain Long-tailed Cuckoo **32** *90*, 419
Moustached Hawk-cuckoo **22** *70*, 429
mui, Centropus phasianinus **5** *36*, 203
musicus, Cacomantis sonneratii 394
musschenbroeki, Surniculus **23** *72*, 428

N

naevia, Tapera **1** *28*, 115
Narrow-billed Bronze Cuckoo 357
NEOMORPHINAE 115
Neomorphus 133
New Britain Coucal 145
neumanni, Centropus leucogaster 163
nigricans, Centropus phasianinus 203
nigricrissa, Piaya cayana 279
nigrorufus, Centropus **7** *40*, 178
nisicolor, Hierococcyx **33** *92*, 445
Northern Hawk-cuckoo 439
novaehollandiae, Scythrops **14** *54*, 334
novaehollandiae, Scythrops novaehollandiae 334

O

obscura, Piaya cayana **13** *52*, 280
occidentalis, Centropus monachus 170
oeneicaudus, Rhamphococcyx curvirostris **16** *58*, 251
Old Man Bird 308
Old Woman Bird 312

olivaceiceps, Coua ruficeps **12** *50*, 217
Olive Long-tailed Cuckoo **32** *90*, 417
olivinus, Cercococcyx **32** *90*, 417
optatus, Cuculus saturatus **36** *98*, 469
oreophilus, Cacomantis variolosus 405
Oriental Cuckoo **36** *98*, 468
Oriental Emerald Cuckoo 337
orientalis, Eudynamys scolopaceus **24** *74*, 326
osculans, Misocalius **27** *80*, 379
Oxylophus, see *Clamator*

P

Pachycoccyx 320
Pacific Koel 325
Palawan Malkoha 251
Pale Roadrunner 131
pallescens, Piaya cayana 280
Pallid Cuckoo **29** *84*, 384
pallidus, Cacomantis **29** *84*, 384
pallidus, Zanclostomus javanicus 240
parroti, Centropus sinensis **6** *38*, 180
parva, Microdynamis **23** *72*, 323
parva, Microdynamis parva 323
passerinus, Cacomantis **30** *86*, 401
patulus, Cercococcyx montanus **32** *90*, 419
Pavonine Cuckoo **2** *30*, 121
pavoninus, Dromococcyx **2** *30*, 121
Peacock Cuckoo 121
Pearly-breasted Cuckoo **20** *66*, 295
pectoralis, Hierococcyx **33** *92*, 441
peninsularis, Chrysococcyx minutillus 370
petersi, Saurothera longirostris **19** *64*, 318
Phaenicophaeus 242
phasianellus, Dromococcyx **2** *30*, 118
phasianinus, Centropus **5** *36*, 202
phasianinus, Centropus phasianinus **5** *36*, 202
Pheasant-coucal **5** *36*, 202
Pheasant-cuckoo **2** *30*, 118
Philippine Coucal **8** *42*, 192
Philippine Drongo-cuckoo **23** *72*, 423
Philippine Hawk-cuckoo **33** *92*, 441
philippinensis, Centropus bengalensis 195
Piaya 274
pica, Clamator jacobinus **18** *62*, 270
picatus, Eudynamys scolopaceus 326
Pied Bronze Cuckoo **28** *82*, 377
Pied Coucal **5** *36*, 145
Pied Crested Cuckoo 270
Pied Cuckoo 270
plagosus, Chrysococcyx lucidus **27** *80*, 364
Plaintive Cuckoo **30** *86*, 397

pluvialis, Hyetornis **21** *68*, 308
poecilurus, Chrysococcyx **28** *82*, 373
poecilurus, Chrysococcyx poecilurus 374
poliocephalus, Cuculus **35** *96*, 455
propinquus, Centropus phasianinus 203
pucheranii, Neomorphus **3** *32*, 137
pucheranii, Neomorphus pucheranii 137
Puerto Rican Lizard Cuckoo **19** *64*, 314
pumilus, Coccyzus **17** *60*, 283
pyropyga, Coua cristata **11** *48*, 212
pyrrhocephalus, Phaenicophaeus **17** *60*, 242
pyrrhophanus, Cacomantis flabelliformis 391

Q

querulus, Cacomantis merulinus **30** *86*, 397

R

radiatus, see *radiceus*
radiceus, Carpococcyx **11** *48*, 208
radiolosus, Neomorphus **3** *32*, 133
Raffles's Malkoha **14** *54*, 230
rectunguis, Centropus **6** *38*, 158
Red-billed Ground Cuckoo **3** *32*, 137, 210
Red-billed Malkoha **16** *58*, 240
Red-breasted Coua **12** *50*, 228
Red-capped Coua **12** *50*, 217
Red-capped Madagascar Coucal 217
Red-chested Cuckoo **34** *94*, 451
Red-crested Malkoha **17** *60*, 256
Reddish-throated Bronze Cuckoo 361
Red-eyed Bronze Cuckoo 369
Red-faced Malkoha **17** *60*, 242
Red-fronted Coua **7** *40*, 219
Red-fronted Madagascar Coucal 219
Red-winged Crested Cuckoo 260
Red-winged Cuckoo 260
renauldi, Carpococcyx **11** *48*, 210
Renauld's Ground Cuckoo 210
reynaudii, Coua **7** *40*, 219
Reynaud's Coua 219
Rhamphococcyx 251
Rhamphomantis 382
Rhinortha 230
Rhopodytes 244
Roadrunner 126
rochii, Cuculus **35** *96*, 463
Rough-crested Malkoha 256
rufescens, Centropus celebensis **7** *40*, 160
ruficeps, Coua **12** *50*, 217
ruficeps, Coua ruficeps **12** *50*, 217
ruficollis, Chrysococcyx **25** *76*, 361

rufigularis, Hyetornis **21** *68*, 310
rufiloris, Rhamphococcyx calyorhynchus 254
rufipennis, Neomorphus **3** *32*, 135
rufiventer, Eudynamys scolopaceus **24** *74*, 327
rufomerus, Chrysococcyx minutillus **28** *82*, 370
Rufous-bellied Coucal 165
Rufous-bellied Malkoha 244
Rufous-bellied Plaintive Cuckoo 397
Rufous-breasted Bronze Cuckoo 373
Rufous-breasted Brush Cuckoo 404
Rufous-breasted Cuckoo **21** *68*, 310
Rufous-breasted Madagascar Coucal 228
Rufous Bronze Cuckoo 373
Rufous Coucal **7** *40*, 150
Rufous Hawk-cuckoo **33** *92*, 439
Rufous-rumped Cuckoo 124
Rufous-rumped Ground-Cuckoo 124
Rufous-tailed Bronze Cuckoo 357
Rufous-throated Bronze Cuckoo **25** *76*, 361
Rufous-vented Cuckoo 138
Rufous-vented Ground Cuckoo **3** *32*, 138
Rufous-winged Ground Cuckoo **3** *32*, 135
Running Coua **12** *50*, 223
Running Coucal 223
russatus, Chrysococcyx poecilurus **28** *82*, 373
Rusty-breasted Cuckoo **31** *88*, 409
Rusty Roadrunner 131

S

saliens, Rhopodytes tristis 249
salvadorii, Chrysococcyx crassirostris 377
salvadorii, Eudynamys scolopaceus 327
salvini, Neomorphus geoffroyi **3** *32*, 138
santamariae, Saurothera merlini 316
sarasinorum, Centropus bengalensis 196
saturatus, Cuculus **36** *98*, 468
saturatus, Cuculus saturatus **36** *98*, 468
Saurothera 312
Scaled Ground Cuckoo **3** *32*, 142
Scale-feathered Cuckoo 258
Scale-feathered Malkoha **17** *60*, 258
Scaly Cuckoo 258
schistaceigularis, Cacomantis flabelliformis 391
schoddei, Scythrops novaehollandiae 335
scolopaceus, Eudynamys **24** *74*, 325
scolopaceus, Eudynamys scolopaceus **24** *74*, 326
Scrub Coucal 200
Scythrops 334
Senegal Coucal **10** *46*, 165
senegalensis, Centropus **10** *46*, 165
senegalensis, Centropus senegalensis **10** *46*, 165

Senegal Yellowbill 235
sepulcralis, Cacomantis **31** *88*, 409
sepulcralis, Cacomantis sepulcralis **31** *88*, 409
serratus, Clamator jacobinus **18** *62*, 271
serriana, Coua **12** *50*, 228
Shining Bronze Cuckoo **27** *80*, 363
Shining Cuckoo 363
Short-toed Coucal **6** *38*, 158
Short-winged Cuckoo 459
simalurensis, Eudynamys scolopaceus 326
simus, Cacomantis flabelliformis **29** *84*, 391
sinensis, Centropus **6** *38*, 180
sinensis, Centropus sinensis **6** *38*, 180
singularis, Rhamphococcyx curvirostris 251
Sirkeer 237
Sirkeer Cuckoo 237
Sirkeer Malkoha **16** *58*, 237
sirkee, Taccocua leschenaultii **16** *58*, 237
Small Coucal 195
Small Cuckoo 455
Small Green-billed Malkoha 246
Small Hawk-cuckoo 429
Smooth-billed Ani **1** *28*, 108
Snail-eating Coua **13** *52*, 227
sokotrae, Centropus superciliosus 175
solitarius, Cuculus **34** *94*, 451
sonneratii, Cacomantis **30** *86*, 394
sonneratii, Cacomantis sonneratii 394
South African Coucal 233
Southern Coucal 180
Southern Crested Coua 214
Southern Crested Madagascar Coucal 214
Southern Sirkeer 237
sparverioides, Hierococcyx **22** *70*, 433
spilopterus, Centropus phasianinus **5** *36*, 203
Spur-footed Cuckoo 202
squamiger, Neomorphus **3** *32*, 142
Square-tailed Cuckoo 404
Square-tailed Drongo-cuckoo **23** *72*, 425
Squirrel Cuckoo **13** *52*, 279
Sri Lanka Coucal 152
Steere's Coucal **8** *42*, 156
steerii, Centropus **8** *42*, 156
stewarti, Surniculus dicruroides 421
Striped Crested Cuckoo 267
Striped Cuckoo 267
subcyanocephalus, Eudynamys scolopaceus **24** *74* 327
subtelephonus, Cuculus canorus 476
Sulawesi Coucal 160
Sulawesi Cuckoo **35** *96*, 458
Sulawesi Drongo-cuckoo 428

Sulawesi Hawk-cuckoo 458
sulcirostris, Crotophaga **1** *28*, 112
suluensis, Surniculus velutinus 423
Sumatran Ground Cuckoo **11** *48*, 206
sumatranus, Rhopodytes **15** *56*, 244
Sunda Coucal 178
Sunda Cuckoo **36** *98*, 473
Sunda Ground Cuckoo 206, 208
Sunda Lesser Cuckoo 473
superciliosus, Centropus **10** *46*, 174
superciliosus, Centropus superciliosus **10** *46*, 175
superciliosus, Dasylophus **17** *60*, 256
superciliosus, Dasylophus superciliosus **17** *60*, 256
Surniculus 421
Swamp Pheasant 202

T

Taccocua 237
taitensis, Urodynamis **14** *54*, 331
Tapera 115
thermophila, Piaya cayana 279
Thick-billed Cuckoo **22** *70*, 320
thierfelderi, Centropus phasianinus 203
threnodes, Cacomantis merulinus 397
Timor Coucal 202
toulou, Centropus **8** *42*, 186
toulou, Centropus toulou 186
tristis, Rhopodytes **15** *56*, 249
tristis, Rhopodytes tristis **15** *56*, 249

U

Unadorned Cuckoo 384
unirufus, Centropus **7** *40*, 150
Urodynamis 331

V

vagans, Hierococcyx **22** *70*, 429
validus, Pachycoccyx audeberti 320
variolosus, Cacomantis **31** *88*, 404
variolosus, Cacomantis variolosus **31** *88*, 404
varius, Hierococcyx 70, 436
varius, Hierococcyx varius 436
velox, Geococcyx **2** *30*, 131
velutinus, Surniculus **23** *72*, 423
velutinus, Surniculus velutinus 423
verreauxi, Coua **11** *48*, 214
Verreaux's Coua **11** *48*, 214
vetula, Saurothera **19** *64*, 312
vieilloti, Saurothera **19** *64*, 314
Vieillot's Lizard Cuckoo 314
Violaceous Coucal **5** *36*, 199
violaceus, Centropus **5** *36*
Violet Coucal 199
Violet Cuckoo **25** *76*, 340
virescens, Cacomantis sepulcralis **31** *88*, 409
viridirostris, Rhopodytes **15** *56*, 246
viridis, Carpococcyx **11** *48*, 206
viridis, Centropus **8** *42*, 192
viridis, Centropus viridis 192

W

waiti, Cacomantis sonneratii 394
websteri, Cacomantis variolosus 405
weiskei, Cacomantis castaneiventris **29** *84*, 388
Western Yellowbill 235
Whistling Hawk-cuckoo **33** *92*, 445
Whistling Yellowbill **15** *56*, 233
White Ani 101
White-browed Coucal **10** *46*, 174
White-crowned Cuckoo **30** *86*, 412
White-crowned Koel 412
White-eared Bronze Cuckoo **25** *76*, 367
White-eared Cuckoo 367
White-necked Coucal 145

X

xanthorhynchus, Chrysococcyx **25** *76*, 340
xanthorhynchus, Chrysococcyx xanthorhynchus 340

Y

Yellow-billed Cuckoo **20** *66*, 290, 465
Yellow-billed Malkoha **16** *58*, 254
Yellow-throated Cuckoo **26** *78*, 352
Yellow-throated Glossy Cuckoo 352
Yellow-throated Green Cuckoo 352

Z

Zanclostomus 240